RAO'S COLLECTION

Chemistry Practice Book For Competitions

KINDLY NOTE:

The answers (A), (B), (C), (D) and (E) in the case of Assertion and Reasoning questions are based on:

(A) If both the assertion and reason are true, and the R is the correct explanation of A.

(B) If both the A and R are true, but the R is not a correct explanation of the A.

(C) If the A is true, but the R is false.

(D) If both the A and R are false.

(E) If the A is false, but the R is true.

RAO'S COLLECTION
Chemistry Practice Book For Competitions

T.S.S. Rao
M.Sc., Ph. D., Dr. Sc. (Germany)

T. Sampurna
M.Sc., Ph. D., Dr. rer. nat. (Germany)

CBS PUBLISHERS & DISTRIBUTORS
NEW DELHI • BANGALORE

ISBN : 81-239-1336-2

First Edition : 2006

Publishing Director : Vinod K. Jain

Published by :
Satish Kumar Jain for CBS Publishers & Distributors,
4596/1-A, 11 Darya Ganj, New Delhi - 110 002 (India)
E-mail: cbspubs@vsnl.com • Website: www.cbspd.com

Branch Office :
2975, 17th Cross, K.R. Road, Bansankari 2nd Stage, Bangalore-70
Fax : 080-26771680 • E-mail : cbsbng@vsnl.net

Printed at :
Asia Printograph, Delhi

PREFACE

The authors have been following the changing trends in competitive examinations like Science Quiz, NTSE, Pre-Medical, Pre-Engineering, PSC, UPSC, UGC and CSIR (NET/GATE) etc. since a long time. The students who are already grounded in the subject will pass over the sections of the book rapidly to encounter more challenging exams.

The authors tried to incorporate all basic aspects of every chapter. All types of numerical problems in Physical Chemistry appeared in the examinations from time to time have been solved. The basic conversions given will surely help the students to solve the multi-step organic chemistry porblems. The thought provoking questions with hints will induce the students to develop the aptitude level to anwer the basic/conceptual questions.

The authors are thankfull to all those students who have sent or personally handed over the questions of various examinations from time to time. The authors appreciate Shri Praveen Kumar Rathore, B. Sc., M.A., PGDCA for setting the manuscript in the present attractive form and Shri Vipin Jain, M.Com.for computing the Organic Chemistry part.

We are thankful to M/s CBS Publishers & Distributors, New Delhi for prompt Publication of the book.

C-99, University Campus
SAGAR (M.P.) 470 003
e-mail: *drtssrao@yahoo.co.in*

T. S. S. Rao
T. Sampurna

CONTENTS

Regarding the book

- ❋ The points to remember cover all the basic aspects and formulae of each chapter in brief.
- ❋ Basics required to answer the questions based on conversions, distinctions, processes, reaction mechanisms etc. are thoroughly prepared.
- ❋ All types of common and typical problems in physical and organic chemistry have been given as solved numerical problems.
- ❋ The thought provoking questions will surely help the students to answer the conceptual based typical and rare short answer type questions.

Points to Remember

Part I : Inorganic Chemistry

1. Periodic Table (general)

- The Periodic law of Mendeleev is based on atomic weight.
- Isotopes have the same atomic number e.g. C_6^{12}, C_6^{13}, C_6^{14}, Cl_{17}^{35}, Cl_{17}^{37}
- Isobars have the same atomic mass e.g. Ca_{20}^{40}, K_{19}^{40}, Ar_{18}^{40}; Xe_{54}^{130}, Ba_{56}^{130}
- Isotones have the same number of neutrons e.g. Si_{14}^{30}, P_{15}^{31}, S_{16}^{32}; K_{19}^{39}, Ca_{20}^{40}
- Nuclear isomers differ in radioactive property e.g. U_x has $t_{0.5}$, 1.4 min. while U_z has $t_{0.5}$, 6.7 hours.
- Bohrs classification is based on electronic configuration.
- Ionic radius : $A^+ < A < A^-$; $Fe^{+2} > Fe^{+3}$; $Mn^{2+} > Fe^{2+} > Ni^{2+} > Cu^{2+} > Zn^{2+}$.
- In Isoelectronic species, more the positive charge, lesser is the size and vice-versa e.g. $Na^+ > Mg^{+2} > Al^{+3} > Si^{+4}$; $O^{2-} > F^- > Ne > Na^+$; $N^{3-} > O^{2-} > F^- > Na^+ > Mg^{+2}$; $Ne > N^{3-} > O^{2-} > F^{-1}$
- Ionisation potential :
  IP_1: F>N>O>C; F>N>O>C>Be>B; Cl>P>S>Si;
  IP_2 : O>F>N>C
- Helium has the highest ionisation energy.
- The penetration effect of the electrons of the same principal Shell is s>p>d>f.
- Completely filled orbitals are more stable than half filled orbitals order of stability $p^6 > d^{10} > p^3 > d^5$
- Order of electron affinity : Cl>F>Br>I; S>Se>O; N>P>As. E_A of chlorine is the highest.
- Order of electronegativity : F>O>N>C; Cl>S>P>Si; F>O>N>Cl>C; F has the highest E.N. value. Hybrid orbitals with more s-character show more E.N. nature.
- Diagonal relationship is due to their similar polarizing power i.e. Ionic Charge/(Ionic radius)^2 and nearly the same electronegativity.
- In solids Osmium (Os) has the highest density while Hg has the highest among liquids.
- The element which exists as +ve ion and as –ve ion in acidic and alkaline medium respectively is called metalloid.
- Tungsten has the highest melting point among metals .
- Hg is the only metal which is liquid at room temperature.
- Carbon has the highest melting point among non metals Carbon exhibits different m.p. value due to different allotropic forms.
- Oxider : Basic : Na_2O, K_2O, Rb_2O, BaO, CaO etc.
- Neutral : CO, N_2O, NO, H_2O etc.
- Amphoteric : ZnO, Al_2O_3, BeO, Sb_2O_3, PbO, Cr_2O_3, Ar_2O_3, SnO, SnO_2 etc.
- Acidic : CO_2, B_2O_3, SiO_2, NO_2, N_2O_5, P_4O_{10}, SO_2, Cl_2O_7, I_2O_5, CrO_3 etc.
- Suboxides : C_3O_2, Pb_2O, N_2O (% of oxygen less than present in normal oxides)
- Superoxides : KO_2, CrO_2,(% of oxygen more than present in normal oxides); O_2^- ion is present in them.
- Peroxides : BaO_2, Na_2O_2, H_2O_2 etc.

2. 1st group elements

- Hydrogen is the only element which contains no neutrons.
- Hydrogen is the most abundant element in the universe.
- Very pure hydrogen for militiary purposes is prepared by Uyeno's method ($2Al+2KOH+H_2O \rightarrow 2KAlO_2+3H_2$)
- Bosch's process: $C+H_2O \xrightarrow{1000^\circ C} CO + H_2$ (Watergas)
- Ortho and para hydrogen are present as 75% +25% in ordinary H_2. In ortho hydrogen the proton spin is in the same direction while in para it is in the opposite direction.
- Heavy hydrogen or Deuterium is used as a tracer & projectile.
- D_2O is used as a moderator for slowing down the velocity of neutrons in nuclear reactors.
- Temporary hardness of water (due to bicarbonates of Ca & Mg) can be removed by boiling it alone or with slaked lime.
- Permanent hardness of water (due to chlorides & sulphates of Ca and Mg) can be removed by treating with washing soda, ion exchange resins, calgon, permutit etc.
- Peroxides e.g. Na_2O_2 or BaO_2 with H_2SO_4 produce H_2O_2. H_2O_2 is industrially prepared by electrolysis of cold 50% H_2SO_4. Marshall's acid is formed during the process. Autoxidation of 2-butyl anthraquinone or oxidation of 2-propanol also produces H_2O_2.
- H_2O_2 is an oxidising (in acid) reducing (in alkali) and bleaching agent.
- H_2O_2 oxidises acidified $K_2Cr_2O_7$ to blue peroxide of Chromium, CrO_5 which has butter fly structure.
- The aqueous solution of H_2O_2 is weakly acidic due to hydrolysis : $H_2O_2 + H_2O \rightleftharpoons H_3O^+ + HO_2^-$
- The strength of H_2O_2 is expressed in volumes.
- It is purified by vacuum distillation.
- Order of density : Na > K > Li (due to unusual increase in the size of K compared to mass).
- Order of hydreation of ions : $Li^+ > Na^+ > K^+ > Rb^+ > Cr^+$
- Order of conductivity in aqueous medium : $Cs^+ > Rb^+ > K^+ > Na^+ > Li^+$
- Li is the strongest reducing agent because of its greater heat of hydration and also Li has maximum E° value.
- Flame colour : Na-golden yellow; K-violet; Rb & Cs- violet red.

- Alkali metals give deep blue colour in ammonia due to solvated or ammoniated electrons.
- Basicity of hydroxides : CrOH>RbOH>KOH>NaOH>LiOH (Cr^+ is more soluble in water, therefore more basic). CsOH is the known strongest base.
- Na_2CO_3 is basic due to hydrolysis, $Na_2CO_3 + 2H_2O \rightarrow 2Na^+ + H_2CO_3$
- Li is the lightest metal.
- Li_3N and Mg_3N_2 are ionic nitrides.
- Li Cl and $MgCl_2$ undergo hydrolysis in hot water.
- Na_2O_2(oxone) is used for purification of air in submarines, aeroplanes, hospitals, theatres etc.
- Fusion mixture contains Na_2CO_3 and K_2CO_3.
- Cs, Rb, K etc. are used in photoelectric cells since the loosely held electrons are emitted under the influence of light easily.
- Metallic nature increases going down any group.
- Alkali metals are stored in kerosine. Due its low density Li is kept wrapped in paraffin wax.
- LiCl is soluble in organic solvents.
- Alkali metals are paramagnetic.
- All the alkali metals crystallize in bcc structure.
- Hypo $Na_2S_2O_3.5H_2O$ is used in photography as a fixing agent to dissolve the unreacted AgBr.

3. 2nd group elements

- Order of density : Ca < Mg
- Flame colour : Ca - brick red; Sr - crimson red; Ba-apple green, Be and Mg do not show flame colour as the electrons are tightly held and hence electron excitation becomes untikely in visible region.
- Chlorophyll is an important complex of Mg.
- BeH_2, MgH_2 are covalent hydrides while others are ionic.
- Basicity and solubility order : $Ba(OH)_2 > Sr(OH)_2 > Sr(OH)_2 > Ca(OH)_2 > Mg(OH)_2 > Be(OH)_2$
- Order of solubility in water : $BeSO_4 > MgSO_4 > CaSO_4 > SrSO_4 > BaSO_4$
- $BaSO_4$ is used in x-ray examination of intestine.
- Bicarbonates are stable only in solution. Decomposition of bicarbonates to carbonates is responsible for the formation of stalactites and stalagnites.
- Calcium oxalate mainly responsible for kidney stones is soluble in dilute acids (sufficientlystrong) but insoluble in alkaline solutions.
- Stablilty of carbonates : $BaCO_3 > SrCO_3 > CaCO_3 > MgCO_3 > BeCO_3$
- BeF_2 & $BeCl_2$ are covalent while others are electrovalent.
- Carbides of Be & Al produce methane on hydrolysis while CaC_2 gives ethyne (acetylene).
- $Ca(HSO_3)_2$ is used to manufacture paper pulpfrom wood.

4. 3rd group elements

- Tl is called platypus.
- Order of density : Al < B
- Al is a strong reducing agent due to high heat of energy.
- Al is the most abundant metal in the earth crust.
- Rubies and sapphires contain Al_2O_3 and aluminates.
- The blue coloured mineral Lapis - lazuli, a semi-precious stone is a sodium-alumino silicate- $Na_3Al_3Si_3S_3O_{12}$
- Order of Lewis acidity : $BI_3 > BBr_3 > BCl_3 > BF_3$ (Due to double bond character of B-F bond because of back donation of electrons).
- Boranes with B_nH_{n+4} formula are more stable than those with B_nH_{n+6} formula.
- B_2H_6 diborane is electron deficient compound and contains (banana bond) bridging three atoms is present.
- Υ-Al_2O_3 (activated alumina) is used as an adsorbent in column chromatography.
- BF_3, $AlCl_3$ are used as Friedel-Craft catalysts.
- In Borax bead test, the colours are due to metaborates. Cu(blue); Fe(green); Co(blue); Cr(green); Ni(brown) etc.
- Borazine or borazole is known as inorganic benzene.
- Unrefined borax is known as Tincal Suhaga (55% Borax).
- The aqueous solution of borax is alkaline due to hydrolysis.
- In Goldschmidt alumino thermic process Al acts as reducing agent.
- Boron, boron steel or Boron carbide rods are used in atomic reactors to control the nuclear reactions.
- B_4C(norbide) is the hardest artificial substance known.
- Al forms sodium meta aluminate and sodium aluminate with NaOH.
- Boric acid's slippery nature is due to hydrogen bonding which provides the acid a layer structure.
- BCl_3 & $AlCl_3$ are covalent in nature.
- Anhydrous $AlCl_3$ is covalent but hydrated $AlCl_3$ is ionic.
- Li AlH_4 and $NaBH_4$ are used as reducing agents.
- The halides undergo hydrolysis easily.
- Aluminium chloride exists as a dimer $(AlCl_3)_2$.

5. 14th group elements

- Pb^{2+} compounds are more stable than Pb^{4+}.
- In CaC_2, C_2^{2-}, ions are present.
- Stability of ions: $Si^{2+}<Ge^{2+}<Sn^{2+}<Pb^{2+}$; $Si^{4+}>Ge^{4+}>Sn^{4+}>Pb^{4+}$;
- In Ge and Sn, 4+ states are more stable.
- Inert pair effect is shown by heavy elements like Sn, Pb, Tl etc. and is responsible for variable valency of these elements.
- Carbon shows the highest catenation property.
- Diamond is the hardest and purest among the allotropes of carbon. Carbon atoms are in sp^3 hybridised state (tetra

hedral). It is a non conductor.

- Graphite is a good conductor of electricity. sp^2 hybridised, hexagonal.
- C_{60} is Buckminster fullerene. C_{70} is the recent one.
- Carbon can't form complexes due to the absence of d-orbitals.
- Thermal stability : $CH_4 > SiH_4 > GeH_4 > SnH_4 > PbH_4$
- Reducing nature : $CH_4 < SiH_4 < GeH_4 < SnH_4 < PbH_4$
- CO is neutral and sp hybridised.
- CCl_4 does not undergo hydrolysis.
- $PbBr_4$ and PbI_4 do not exit Pb^{4+} has a strong oxidising nature. Br^- and I^- being strong reducing agents will reduce Pb^{4+} to Pb^{2+}.
- Carbides of transition elements e.g. Ti, Zr, Hf, V, Nb, Mo, W etc. are interstital compounds.
- CO combines with Haemoglobin to form a stable carboxy haemoglobin leading to Asphyxia.
- Natural gas, Biogas and Gobar gas contain CH_4 mainly.
- Oil gas contains mainly $CH_4 + C_2H_4 + C_2H_2$.
- Coal gas contains H_2 mainly.
- Formation of $Pb(OH)_2$ form Pb in aerated water due to its dissolution is known as plumbo solvency effect.
- PbO_2 is used in lead storage battery.
- Crookes glass doesn't allow UV rays to pass through it. It contains CeO_2. Safety glass or shatter aroof glass contains a transparent plastic (vinyl acetate resin) between two layers of glass.
- Orthosilicates contain SiO_4^{4-} ion.
- Pyrosilicates contain $Si_2O_5^{6-}$ ion.
- Sheet silicates like talc, quartz, kaolin etc. contain SiO_4^{4-}
- Pb is used for making radiation shield.
- Graphite is more energetically stable than diamond.
- CO_2 is acidic in water, $CO_2 + H_2O \rightarrow HCO_3^- + H_3O^+$
- Liquified NH_3 bottle is cooled before opening as its vapour pressure is very high.

6. 15th group elements

- Red phosphorous has the least reactivity among allotropes of P.
- The reactivity of nitrogen is very low due to high decomposition energy of N-N triple bond (~946 k.J/mole).
- Hydrides :Basicity : $NH_3 > PH_3 > AsH_3 > SbH_3$
  Thermal stability: $NH_3 > PH_3 > AsH_3 > SbH_3$
  Bond angle : $NH_3 > PH_3 > AsH_3 > SbH_3$
  Boiling point : $SbH_3 > NH_3 > AsH_3 > PH_3$
- R_2O_3 type oxides form - ous acids in water.
- R_2O_5 type oxides form - ic acids in water.
- NF_3 is the only stable halide of nitrogen.
- Except NF_3 & PF_3, all others undergo hydrolysis.
- N can't form penta halides due to the absence of d-orbitals.
- Hyponitrous (HNO); nitrous(HNO_2); nitric(HNO_3) and peroxynitric acid(HNO_4) are the oxyacids of nitrogen.
- HNO_2 is a strong reducing agent. It is an oxidising and complex forming agent also.
- HNO_2 exhibits tautomerism.
- HNO_3 is known as aqua fortis and is prepared by Birkland and Eyde and Ostwald's process.
- More reactive metals like Fe, Zn, Sn etc. with cold or v. dil. HNO_3 give NH_4NO_3 or NH_2OH.
- Mg & Mn with v. dil. HNO_3 give hydrogen.
- Nascent chlorine ($HNO_3 + 3HCl \rightarrow NOCl + 2H_2O + 2Cl$) formed from aquaregia attacks the noble metals like Au, Ru, Rh, Pd or Ir, Pt etc. Ag is not attacked by aquaregia (difference from gold).
- Hypophosphorous acid (H_3PO_2, monobasic); phosphorous acid (H_3PO_3, dibasic); orthophosphoric or phosphoric acid(H_3PO_4, tribasic); metaphosphoric acid HPO_3 (monobasic); hypophosphoric acid ($H_2P_2O_6$, tetrabasic); pyrophosphoric acid ($H_2P_2O_7$, tetrabasic) and peroxomonophosphoric acid (H_3PO_5, tribasic) are the oxyacids of phosphorous.
- Nitrides of transition elements e.g. Fe, Mn, Mo, W etc. are interstitial compounds.
- N_2O (nitrous oxide) is known as laughing gas which is linear unsymmetrical and is neutral.
- NH_4NO_3 on heating gives N_2O.
- NO (nitric oxide) is paramagnetic and neutral.
- NO_2 is paramagnetic and brown gas.
- N_2O_5 is an anhydride of HNO_3. It is the strongest acidic oxide of nitrogen.
- White P is the most reactive form and is stored under water.
- White P shows phosphoresence with hot NaOH.
- White P is used as rat poison.
- Tartar Emetic is used in the treatment of kala - Zar.
- Persons working with phosphorous develop Phossy Jaw disease.
- P^{32} is employed in the treatment of leukemia.
- Sb_2S_3 is used to prepare safety matches.
- Red P is used as a dehydrating agent.
- AsH_3 (arsine) is used in Marsh's test for arsenic.
- PH_3 is used to prepare smoke screens in warfare. It is also used in Holme's signals for hlips to know the position of rocks in the sea.
- TiO_2 & $TiCl_4$ are used as smoke screens.

7. 16th group elements

- O, S, Se and Te are called chalcogens as they participate in ore formation.
- Oxygen is the most abundant element is the earth crust.
- Electron affinity : S > Se > O.

- Cesium compounds are used for scavenging oxygen and sulphur from other metals.
- The oxidation state of oxygen in H_2O_2, O_2F_2, OF_2 and super oxides is -1, +1, +2 and -1/2 respectively.
- Oxygen is paramagnetic.
- Sulphur forms a puckered ring of 8 atoms.
- Hydrides :acidity : $H_2O < H_2S < H_2Se < H_2Te$
  bond angle : $H_2O > H_2S > H_2Se > H_2Te$
  boiling point : $H_2O < H_2Te < H_2Se < H_2S$
  thermal stability : $H_2O > H_2S > H_2Se > H_2Te$
- RO_2 and RO_3 type of oxides form - ous and -ic acids in H_2O.
- Acidic strength : $H_2SO_3 > H_2SeO_3 > H_2TeO_3$
  $H_2SO_4 > H_2SeO_4 > H_2TeO_4$
- oxidising power : $H_2TeO_4 > H_2SeO_4 > H_2SO_4$
- H_2SO_4 is known as oil of vitriol.
- H_2SO_4 is used in lead accumulators and in storage batteries.
- H_2SO_4 is used as pickling agent.
- Caro's acid (H_2SO_5, peroxymono - sulphuric acid); leum ($H_2SO_4 + SO_3$, disulphuric acid or pyrosulphuric acid); Marshall's acid ($H_2S_2O_4$); Thiosulphuric acid ($H_2S_2O_3$) and Dithionic acid ($H_2S_2O_6$) are oxyacids of sulphur.
- H_2SO_4 is prepared by contact process and lead chamber process.
- SO_2 is an oxidising and a reducing agent.
- oxidation potential : $F > O > OF_2 > O_3$
- Tailing of mercury is due to the formation of Hg_2O ($2Hg+O_3 \rightarrow Hg_2O + O_2$).
- The proportion of dissolved oxygen in water is estimated by Winkler method.

8. 17th group elements

- I_2 can be obtained from sea weeds (Laminaria species).
- F can't show other than -1 oxidation state due to the absence of d-orbitals.
- Cl_2 can be obtained by the oxidation of HCl with MnO_2, $K_2Cr_2O_7$, $KMnO_4$, Pb_3O_4 etc.
- Cl_2 is manufactured by Weldon's and Deacon's process.
- Iodine is purified by sublimation.
- HCl is also known as muriatic acid.
- Bond dissociation energy : $Cl_2 > Br_2 > F_2 > I_2$
- Reducing power : $F^- < Cl^- < Br^- < I^-$
- Reactivity : $F_2 > Cl_2 > Br_2 > I_2$
- Acidic strength : L HI > HBr > HCl > HF
- Boiling point : H_2F_2 > HI > HBr > HCl
- Bond length : HF < HCl < HBr < HI
- Solubility in water at O°C : HF > HI > HBr >HCl
- Bond energy : HF > HCl. HBr > HI

- Oxyacids :

Thermal stability : $HClO_4 > HClO_3 > HClO_2 > HClO$
Acidic strength : $HClO_4 > HClO_3 > HClO_2 > HClO$
Oxidising power : $HOCl > HOClO > HOClO_2 > HOClO_3$

- F doesn't form oxyacids due to the absence of d-orbitals.
- HF can't be stored in glass bottles as it dissolves glass forming SiF_4 and H_2SiF_6. Therfore it is used for etching of glass.
- HF doesn't give ppt, with $AgNO_3$ and $Pb(CH_3COO)_2$.
- HF can be precipitated by adding $CaCl_2$ solution.
- $(CN)_2$; $(SCN)_2$; $(SeCN)_2$ and $(SCSN_3)_2$ etc. are pseudohalogens.
- Bleaching powder ($CaOCl_2$) can be prepared by Beckmann's process and Hasenclever method.
- Freon (CF_2Cl_2) is used in refrigerators and cold storage plant.
- I_2 is used in the treatment of goitre.
- UF_6 is used to separate the isotopes of Uranium.
- ICl is used for the estimation of iodine number of oils and fats.
- I_2O_5 is used to estimate CO present in the atmosphere.
- Each halogen will displace the halogen below it.
- I_2 dissolves in solutions containing iodide ions due to the formation of soluble triiodide I_3^- ion.
- Fluorine has the ability to bring out the highest oxidation state of an element e.g. OsF_8, SF_6 etc.
- $FeCl_3$, $AlCl_3$, $FeCl_2$ (covalent halides) etc. are easily hydrolysed.
- The chlorides of Be, B and Al are covalent.
- AgBr is used in making photographic films.
- Starch is used as an indicator in iodi & iodometric titrations with hypo.
- The solution of NaOCl in water is known as jewel water.

9. 18th group elements

- The existence of noble gases could not be predicted by Mendeleev.
- Ar is the priciple constituent of air among inert gases.
- Inert gases constitute around 1% of atmosphere.
- Ar, Kr and Xe mean lazy, hidden one and stranger respectively.
- He, Ne, Ar, Kr and Xe are known as aerogens.
- The radius for noble gases are only the van der Waal's radius.
- Noble gases are monoatomic, Cp/Cv =1.66
- The polarisability increases down the group.
- van der Waal's forces are proportional to the polarisability and inversely proportional to the ionisation potential.
- WHe_2, $HgHe_{10}$, PtHe, PdHe etc. are considered as the adsorption compounds of the inert gases on the dispersed

metal.

- Ar, Kr and Xe form clahurate compounds. He & Ne can't form clathrates as they are smaller in size.
- The noble gas hydrates are clathrate compounds.
- Neil Bartlett and Lohmann (1962) prepared the first chemical compound, $Xe^+ (PtF_6)^-$ of noble gases.
- XeF_6 can't be stored in quartz vessels.
- Xe fluorides are used as oxidising and fluorinating agents.
- Shapes : XeF_2 linear, XeF_4 square planar, XeF_6 distorted octahedral, XeO_3 pyramidal and XeO_4 tetrahedral.
- Helium is present in the Sun.
- Monozite, Pitch blende, Clevite etc. contain Helium.
- Dewar's method of separation of noble gases is based on the adsorption on activated coconut charcoal. At −100°C, Ar, Kr and Xe get adsorbed. At −180°C Ne is adsorbed.
- Noble gases are obtained in large scale by fractionation of liquid air (Rayleigh & Ramsay) method.
- In diving apparatus a mixture of He + O_2 (80+20) is used.
- Helium in liquid form is used as a cryogenic fluid.
- Sea divers suffer from bends disease.
- Airship tubes are filled with He.
- Ne glow lamps are used for advertising purpose in music beats, helipad, telephone tower etc. Orange red Ne tubes are used in televisions and plug testers.
- Ne lamps are used in green houses and botanical gardens as it stimulates the growth and also effective in the formation of chlorop last.
- Ar is filled in incandescent metal filament electric bulbs.
- Rn is used in the treatment of cancer (radiotherapy).
- Argon is used in geiger counter tubes.

10. Transition metals

- 3d: Sc^{21}–Zn^{30}; 4d: y^{39}–Cd^{48}; 5d: La^{57}, Hf^{72}–Hg^{80}; 6d: Ac^{89}, Rf^{104}, Db^{105}, Sg^{106}, Ns^{107}, Hs^{108}, Mt^{109}, Dt^{110}.
- The electronic configuration of transition elements is $(n-1)d^{1-10}$ n s 1 or 2..
- Exceptions of normal electronic configuration are : Cu^{29}: [Ar]$4s^1$,$3d^{10}$; Ag^{47} : [Kr] $5s^1$, $4d^{10}$; Au^{79}: [Xe]$4f^{14}$, $6s^1$, $5d^{10}$; Cr^{24}: [Ar] $4s^1$, $3d^5$; Mo^{42} : [Kr] $5s^1$, $3d^5$; W^{74} : [Xe] $4f^{14}$ $6s^2$ $5d^4$;Pt^{78} :[Xe] $4f^{14}$ $5d^{10}$ $6s^0$; Pd^{46} [Kr] $4d^{10}$ $5s^0$.
- Cu, Ag, Au, Zn, Cd and Hg have d^{10} configuration. They share a common feature of transition elements i.e. complex formation.
- Ag is the best conductor of heat and electricity.
- The m.p. & b.p.s. of transition elements are normally higer due to (i) interatomic attraction and (ii) covalent bonding.
- Alloy formation is due to the similar size of the d-block elements.
- Fe^{26}, Co^{27}, Ni^{28}, Ru^{44}, Rh^{45},Pd^{46}, Os^{76}, Ir^{77}, Pt^{78} do not form amalgam with Hg. These metals adsorb hydrogen gas.
- Hard metals have generally more unpaired electrons.
- Cu would not evolve H_2 gas with acid solutions (Cu has lower oxidation potential than that of the standard hydrogen electrode).
- Fe, Al, Cr become passive with conc. acids due to the formation of a thin impervious layer of an inert oxide.
- Fe is the most abundant transition metal.
- d-d electron transition is responsible for the colour of the complex ions.
- $KMnO_4$ and $K_2Cr_2O_7$ are coloured due to charge transfer spectrum (the charge on oxygen $MnO'_4{}^-$ or $Cr_2O_7{}^{2-}$ is transferred to the vacant d-orbitals).
- Colour of the ions : V^{3+}, Fe^{2+}, Ni^{2+} - green; Cu^{2+} -blue; Cr^{3+}, Mn^{3+} violet; Co^{2+}, Mn^{2+} - pink; $MnO'_4{}^-$ -purple.
- Presence of unpaired electrons in atoms, ions or molecules leads to paramagnetic property. Paramagnetic species can be detected by ESR (electron spin resonance) or EPR (electron paramagnetic resonance).
- Magnetic moment, $\mu_{eff} = \sqrt{n(n+2)}$ B.M (n=number of unpaired electrons).
- Fe, Co, Ni are ferromagnetic.
- Variable valency character is responsible for the catalytic properties of transition elements.
- Variable valencies and defects in the solid structure are responsible for the existence of non-stoichiometric compounds.
- H, B, C and N form interstitial compounds with transition elements.
- Van Arkel method is used for the purification of Ti.
- Kroll process is used for the extraction of Ti.
- Jantesh process is used for the preparation of V metal.
- Mond's or Orford's process is employed for pure Ni extraction.
- Beard and Crook's process is used to reduce V_2O_5 to V.
- Monzrite contains mainly the phosphates of rare earths.
- Electronic configuration of lanthanide is $4f^n5d^16s^2$ or $4f^{n+1}$,$6s^2$.
- +3 is the characteristic oxidation state of lanthanides. Ce^{4+} is a powerful oxidising agent.
- In lanthanides paramagnetic property is dependent on the electron spin and orbital motion.
- Basicity of lanthanones decreases as ion size decreases.
- Neodynium and Praseodynium are used for making coloured glasses for goggles.
- Misch metal contains rare earth element (94-95%), Iron (5%) and traces of S, C, Ca nd Al.

- Rare earth elements are used in the preparatio of tracer bullets, flints for lighters.
- Actinides are used in gas mantle industry.

11. Coordination Chemistry

- EDTA is a hexadentate ligand.
- Primary valencies are satisfied by negative ions. They are ionisable and tells about the oxidation state of the central metal atom/ion.
- Secondary valencies are satisfied either by neutral molecules or negative ions and tells about the coordination number of the central metal atom/ion.
- EAN rule was given by Sidgwick.
- $[Co(NH_3)_5SO_4]Cl$ and $[Co(NH_3)_5Cl]SO_4$ show are ionisation isomers (they give ppt with $AgNO_3$ and $BaCl_2$ solutions respectivelly).
- $[Co(NH_3)_5NO_2]Cl_2$ and $[Co(NH_3)_5ONO]Cl_2$ are linkage isomers.
- Ambidentate ligands e.g. NO_2^-, CN^-, CO, SCN^-, $S_2O_3^{2-}$ etc. are mainly responsible for linkage isomerism in complexes.
- $[Cr(H_2O)_6]Cl_3$ (violet), $[Cr(H_2O)_5Cl]Cl_2; H_2O$ (green) are hydrate isomers. H_2O molecules present outside the coordination sphere are water of crystallisation. on heating the water of crystallisation is lost.
- Square planar complexes with coordination number 4 of the type MA_2B_2 and MABCD show geometrical isomerism.
- Octahedral complexes of the type MA_4B_2 or MA_2B_4, MA_3B_3, $M(xy)A_3B$, $M(xy)A_2B_2$, $M(xy)_2A_2$ exhibit geometrical isomerism.
- Trans isomers have zero dipole moment.
- Square planar complexes do not exhibit optical activity as they possess plane of symmetry.
- Optical isomers of tetrahedral complexes having four different ligands can not be isolated due to the labile nature of the complexes.
- Tetrahedral complexes of Be, Cu(II), Zn(II) containing unsymmetrical bidentate ligands have been resolved into optically active isomers.
- In octahedral complexes, the trans form can not be resolved into optically active isomers (the mirror image is the same due to the symmetric structure). sis form can be resolved.
- $[Cu(NH_3)_4]SO_4$ is a square planar complex.
- Smaller the size and larger nuclear charge of the metal ion, more stable is the complex.
- Chelate formation increases the stability of the complexes.
- RMgX, ferrocene, TEL, Zeisse's salt etc. are organometallic compounds.
- $[Pt(NH_3)_2Cl]$ is considered to be useful as anticancer species.
- $KMnO_4$ is industrially prepared from pyrolusite (MnO_2).
- The equivalent wt. of $KMnO_4$ in acidic medium is equal to M/5 while in neutral or slightly alkaline medium is M/3.
- $KMnO_4$ crystal has Rhombohedral or trigonal shape.
- $K_2Cr_2O_7$ is known as "Surkh Kahi".

$K_2Cr_2O_7$ crystal is triclinic.

12. Metallurgy

- Mineral is a solid inorganic substance, occuring naturally in the Earth.
- Ore is a mineral from which the metal can be extracted profitably.
- The impurities associated with the ore is called gangue.
- Sulphide ores are generally concentrated by froth flotation method.
- Bauxite, native silver and gold ore concentrated by leaching.
- Roasting is generally an oxidation process. Sulphide ores are converted to oxides.
- Carbonates, oxide and hydrated ores are converted to oxides by calcination.
- Smelting is generally a reduction process.
- Highly electropositive metals e.g. alkali, alkali earth metals (Zn, Al etc.) are often extracted by electrolytic reduction.
- Zone refining method is used for the purification of semi conductors e.g. Si, Ge, Ga, In etc.
- Ti and Zr are purified by van Arkel method.
- $CaCl_2$, KCl, NaF are employed as flux. Flux is used either to reduce the m.p. of the ore or to combine with any infusible impurity that may be present.
- Down's or Castner process is employed in the metallurgy of Na.
- Na shows D_1 and D_2 lines at 5890Å and 5860Å in the yellow region of the spectrum.
- Aluminium is refined by Hoope's process (electrolytic refining).
- Bessemerisation in the case of Cu metallurgy is a self-reduction process ($CuSO_4$ or Cu_2O formed during roasting combines with the ore and the former is reduced to Cu).
- Blister copper contains 1–2% of Cu_2O.Refining of Cu is done by poling. It is a reduction process.
- Ruby glass and enamel contain Cu_2O.
- Copper salts show deep green colour with bunsen flame.
- The products of blast furnace are pig or cast iron.
- Stainless steel is an alloy of Fe, Cr and Ni-.

- Thomas slag $(Ca_3(PO_4)_2$. $CaSiO_3$ obtained in the basic bessemer process for the manufacture of steel is used as a fertilizer.
- Open-Hearth process has advantages over Bessemer process for the manufacture of steel.
- Wrought iron is the purest form of iron while pig iron is the most impure form.
- Removal of impurities e.g. Cu, Ag, Bi, Sb & Sr from leadores is known as softening of lead.
- Refining of Pb is done by Beltt's process.
- Silver is obtained from argentiferous lead by Parke's process or Pattison's process.
- AgBr is used in making photograpic films.
- $AgNO_3$ is used for making glass mirross in ammonical solutions.
- Colloidal solution of gold is called pruple of Cassius.
- Mac Arthur-Forrest cyanide process is employed in the metallurgy of gold.
- Gold forms $HAuCl_4$ (chlorauric acid) with aquaregia [con.HCl + con. HNO_3 (3:1)].
- $AuCl_3$ is used in photography for toning.
- Basic impurities are removed as nitrates by treating Hg with dil. HNO_3.
- Refining of Sn by poling is an oxidation process.
- Mg is obtained by electrolysis of $MgCl_2$ (obtained from sea water) by Dow's process.
- Cu, Ag and Au are called coinage metals.

13. Qualitative analysis

- In the test for CO_3^{2-} radical, lime water becomes milky due to the formation of $CaCO_3$. On passing excess of CO_2, the solution turns colourless due to the conversion of $CaCO_3$ to $Ca(HCO_3)_2$.
- In the test for SO_3^{2-} with $K_2Cr_2O_7$, the green colour of the solution is due to the formation of $Cr_2(SO_4)_3$.
- The brown colour of the solution in the test for NO_2^- or brown ring for NO_3^- redical is due to the formation of the complex, $[Fe(NO)]SO_4$ or $[Fe(NO)(H_2O)_5]SO_4$.
- AgCl ppt. dissolves in NH_4OH due to the formation of the soluble complex, $[Ag(NH_3)_2]Cl$.
- In chromyl chloride test for chloride red vapours formed are due to CrO_2Cl_2.
- The golden spongles, $PbCrO_4$ is formed when CrO_2Cl_2 vapours are passed through lead acetate solution.
- Chromyl chloride test fails to respond with chlorides of Pb, Ag, Sn and Hg. only ionic chlorides give this test.
- The canary yellow ppt. obtained for PO_4^{3-} is due to $(NH_4)_3 PO_4$. $12MoO_3$ (amononium phosphomolybdate)
- The green edge flame of borates with alcohol is due to the formation of $B(OC_2H_5)_3$.
- Basic radicals are grouped on the basis of their solubility product.
- Ist group : Ag^+, Pb^{2+}, Hg^+ (ous) are precipitated as chlorides. The group reagent is dil. HCl.
- 2nd group : 2-A: Hg^{2+}, Pb^{2+}, Cd^{2+}, Bi^{3+}, Cu^{2+}
  2-B : As^{3+} & As^{5+} : Sb^{3+} & Sb_{5+}; Sn^{2+} & Sn^{4+}
  are precipitated as sulphides in acidic medium. 2nd A & B radicals are separated by yellow ammonium sulphide. 2-A radicals are not soluble. The solubility product of 2nd group sulphides is less than the 4th group sulphides. To precipitate Cd^{2+}, the solution is diluted.
- 3rd group : Fe^{3+}, Al^{3+}, Cr^{3+} are precipitated as hydroxides $Al(OH)_3$ is obtained as silky or gelatinous white ppt. The group reagent is NH_4OH in the presence of NH_4Cl. The solubility product of hydroxides of 3rd group is lower than those of the 5th and 6th groups. NH_4Cl suppresses the ionisation of NH_4OH.
- 4th group : Ni^{2+}, Co^{2+}, Mn^{2+}, Zn^{2+} are precipitated as sulphides in alkaline medium.
- 5th group : Ba^{2+}, Sr^{2+}, Ca^{2+} are precipitated as carbonates.
- In 6th group Mg^{2+} is precipitated as phosphate.
- Zero group contains NH_4^+ ion. Alkaline solution of K_2HgI_4 is used for detecting NH_4^+ ions.
- Phenolphthalein (a weak acid) is used as an indicator between 8.00 - 9.8 pH range.
- Methylorange (a weak base) is used in the titrations of strong acid vs weak base.
- In the titration Na_2CO_3 vs strong acid, both phenolphthalein and methyl orange are used.
- In iodimetry, I_2 in solution is directly titrated against hypo.
- In iodometry first iodine present in the compund is liberated by oxidation and then titrated against hypo.

Some important compounds :

- Caustic Soda : NaOH (prepared by Gossage, Lowig, Nelson cell and Castner-kellner cell processes)
- Caustic potash : KOH
- Washing Soda : $Na_2CO_3.10H_2O$ (prepared by Leblanc and Solvay ammonia soda processes).
- Hypo : $Na_2S_2O_3.5H_2O$
- Baking Soda : $NaHCO_3$
- Microcosmic salt : Na (NH_4) $HPO_4.4H_2O$
- Glauber's salt : Na_2SO_4. $10H_2O$
- Chile saltpetre : $NaNO_3$
- Euchlorine : ClO_2
- Epsom salt : $MgSO_4$. $7H_2O$ (It is used as a purgative in medicine).
- Gypsum, alabaster or selenite : $CaSO_4.2H_2O$
- Plaster of paris : $CaSO_4.1/2$ H_2O or $2CaSO_4.H_2O$
- Lithopone : $BaSO_4$ + ZnS (white pigment)
- Sorel's cement : $MgCl_2.5MgO.xH_2O$

- Milk of magnesia : $Mg(OH)_2$
- Quick lime or burnt lime : CaO
- Slaked lime : $Ca(OH)_2$
- Magnesia alba : $3MgCO_3.Mg(OH)_2.3H_2O$
- Anhydrone : $(MgClO_4)_2$
- Potash alum : $K_2SO_4.Al_2(SO_4)_3.24H_2O$ (It is used as a mordant).
- Thallous alum : $Tl_2SO_4.\ Al_2(SO_4)_3.\ 24H_2O$
- Corrundum : α - Al_2O_3
- Permutit : $NaAlSiO_4.3H_2O$ (used in softening of water)
- Borax : $Na_2B_4O_7.10H_2O$; Jeweller's Borax : $Na_2B_4O_7.5H_2O$
- Drikold (dry ice) : Solid CO_2
- Water gas : $CO + H_2$ + a little of CO_2.
- Water glass : Na_2SiO_3. Producer gas : $CO + N_2$
- Pink salt : Ammonium chlorostannate, $(NH_4)_2\ SnCl_6$ (used as mordant in dyeing).
- Mosaic gold : SnS_2
- Type metal : Pb + Sb + Sn
- Carborendum : SiC
- Litharge or Massicoat : PbO
- Scheele's green HCu AsO_3 (Cupric arsenate)
- Swart's reagent : SbF_3
- Nitrolim : $CaCN_2$ + C (used as a fertilizer)
- Tartar Emetic : KOOC.CHOH.CHOH.COO(SbO)1/$2H_2O$
- Paris green : $(CH_3COO)_2Cu.3Cu(AsO_2)_2$ or $Cu_4(CH_3O)_2\ (ASO_3)_2$
- Amatol : NH_4NO_3 + TNT
- Ammonal : NH_4NO_3 + Al powder
- Tincture iodine : I_2 + KI + rectified spirit
- Blue vitriol : $CuSO_4.5H_2O$
- Green vitrio : $FeSO_4.7H_2O$
- White virtiol : $ZnSO_4.7H_2O$
- Guigrets green : $Cr_2O_3.2H_2O$
- Mohr's salt : $FeSO_4.(NH_4)_2SO_4.6H_2O$
- Prussian blue (Ferri ferrocyanide) : $Fe_4(Fe(CN)_6]_3$
- Turn bulls blue : $Fe_3[Fe(CN)_6]_2$
- Sodium nitroprusside : $Na_2[Fe(NO)(CN)_5].2H_2O$
- Fischer's salt : $K_3[Co(NO_2)_6]$
- Bordeau mixture : $CuSO_4$ + Lime
- Corrossive sublimate : $HgCl_2$
- Calomel : Hg_2Cl_2
- Philosophers wool : ZnO
- Rinman's green : $CoZnO_2$
- Brass : Cu + Zn
- Bronze : Cu + Sn (90 : 10)
- Bell metal : Cu + Sn (80 : 20)
- Aluminium bronze : Cu + Al(90 : 10)
- Gun metal : Cu + Sn + Zn
- German silver : Cu + Zn + Ni
- Monel metal : Cu + Ni + Fe + Mn
- Cryolite : Na_3AlF_6
- Drualumin : Al + Mg + Cu + Mn
- Magnalium : Al + Mg
- ϒ-alloy : Al + Cu + Mg + Ni
- Copper glance : Cu_2S
- Silver coins : Ag + Cu (92.5 + 7.5%)
- Haematite : Fe_2O_3
- Magnetite : Fe_3O_4
- Fool's gold : FeS_2
- Galena : Pbs
- Solder & pewter : Pb + Sn
- Sugar of lead : $Pb(CH_3COO)_2$
- Sindur (red lead) : Pb_3O_4. This red pigment is used to prepare antirust coatings.
- Argentite : Ag_2S
- Horn silver : AgCl
- Ruby silver (pyrargyrite) : $3Ag_2S.Sb_2S_3$
- Zinc blende : Zns
- Calamine : $ZnCO_3$ (used in ayurvedic medicines under the name ras sindoor and makaradhwaj)
- Sylvenite : a telluride of Au and Ag
- Cassiterite (Tinstone) : SnO_2
- Britannia metal : Sn + Sb + Cu (93 : 5 : 2)
- Babbit metal : Sn + Sb + Cu (90 : 7 : 3)
- α-stannic acid : $SnO(OH)_3$
- Dolamite : $MgCO_3.CaCO_3$
- Butter of Tin : $SnCl_4.5H_2O$
- Pearl white : BiOCl
- Lunar caustic : $AgNO_3$
- Massicot : PbO(yellow powder)
- Fischer salt : $K_3[Co(NO_3)_6]$
- Super phosphate of lime : $Ca(H_2PO_4)_2.H_2O$ $+CaSO_4.2H_2O$

Part II : Physical Chemistry

1. Atomic Structure

- Electron : Discovered by J. J. Thomson. Charge of an electron is -1.602×10^{-19}, coulomb. The charge was determined by Millikans oil drop method. Mass is 9.1×10^{-31} kg.e/m ratio is constant. The name 'electron' was given by J. S. Stony. Wave properties of electrons were first postulated by de Broglie. All electrons except those in the outer orbit are called Kernel electrons.
- Proton : Discovered by Goldstein. e/m ratio of positive ions is not constant. Mass of a proton is 1.00757 amu.
- Neutron : Discovered by Chadwick. Mass of a neutron is 1.00893 amu. Neutron decays as $n_0^1 \rightarrow H_1^1 + e^- + \upsilon$
- Other unstable particles : Positron - Anderson; Antiproton - Segre - Chamberlain; Neutrino and Anti-Neutrino - Fermi; Meson - Yukawa.
- Radius of the nucleus is 10^{-15} to 10^{-14} m.
- Radius of an atom is 10^{-10} m.
- Rutherford's atomic model could not explain the emission or absorption of electromagnetic radiation and therefore was not able to account for the existence of spectral lines.
- In vacuum all types of electromagnetic radiation regardless of their wavelength travel at the same speed i.e. 3×10^8m.
- The angular momentum $m v r = n.h / 2\pi$.
- Number of spectral lines when an electron drops from n^{th} level to ground level is given by, n(n-1)/2.
- Lyman series are observed in UV region while others are in Balmer - Visible; Paschen - NIR; Brackett, Pfound and Humphrey - FIR.
- Balmer series : n_2 to n_3 red; n_2 to n_4 green; n_2 to n_5 blue; n_2 to n_6 violet.
- The spectrum of white light ranges from violet to red. This spectrum is called continuous spectrum because violet merges into the blue, blue into green and so on.
- Line spectra show light only at specific wavelengths.
- Bohr's theory is extended only to hydrogen and solitary electron species e.g. He^+, Li^{2+}, Be^{3+} etc.
- Bohr radius, $r = \dfrac{n^2h^2}{4\pi^2ze^2m} = n^2 \times a^0$($a^0$=Bohr radius when n = 1)
- Bohr radius, $a^0 = 0.529$Å.
- Energy of an electron, $E = -\dfrac{2\pi^2z^2e^4m}{n^2h^2}$
- The energy of the electron in the nth orbit is, $En = -\dfrac{z^2}{n^2} \times 13.6$
- Frequency, $\nu = 3.289 \times 10^{15} (\dfrac{1}{n_1^2} - \dfrac{1}{n_2^2})$
- Wave number, $\overline{\nu} = 1.097 \times 10^7 (\dfrac{1}{n_1^2} - \dfrac{1}{n_2^2})$
- $E = h c \overline{\nu} = h \times c \times R_H z^2 (\dfrac{1}{n_1^2} - \dfrac{1}{n_2^2})$
- Photoelectric effect : It is a phenomenon in which surface of the metal emit electrons when electromagnetic radiation of high frequency fall on them.
- The escape energy which is the minimum energy required to remove an electron from the surface of a metal is called the workfunction of the metal.
- Bohr - Sommerfeld explained the fine structrue of the spectral lines in terms of elliptical orbits.
- Principal quantum number, n mainly determines the energy of the electron. As the value of n increases, the energy of the electron increases but its electronegativity decreases.
- The maximum number of electrons 'N' in any shell is given by $2n^2$.
- Azimuthal quantum number tells us the shape of the electron cloud.
- Azimuthal quantum number gives measure of energy due to orbital angular momentum of the electron, $m v r = \dfrac{h}{2\pi} \sqrt{l(l+1)}$
- Magnetic quantum number tells us the orientation of the electron in the space.
- The total number of orbitals in a given subshell can be obtained by the expression, $m = 2l + 1$
- Magnetic quantum number explains Zeeman effect.
- Splitting of spectral lines under the influence of strong magnetic field is called Zeeman effect.
- Increase in wavelength or decrease in energy of X-rays after scattering from the surface of an object is called Compton effect.
- Spin quantum number does not follow from the wave mechanical treatment.
- Pauli exclusion principle : No two electrons in any atom can have all the four quantum numbers the same.
- Aufbau principle : Electrons enter the various orbitals in the order of increasing energy.
- The sub-shell with lowest $n + l$ value is filled up first.
- Hunds rule of maximum multiplicity : Electron pairing is not possible until all the available orbitals of a given set contain one electron each.
- Orbit : Definite circular path at a definite distance from the nucleus in which electron is revolving round the nucleus.

- Orbitl : The maximum probability of the existence of an electron.
- de Broglie relationship, $k = \frac{h}{mv} = \frac{h}{p}$
- Heisenberg's uncertainity principle : It is impossible to measure simultaneously the exat position and velocity of a subatomic particle.
  $\Delta x \times m v \geq \frac{h}{4\pi}$ or $\Delta x \times \Delta p \geq \frac{h}{4\pi}$
- s orbital is spherical in shape
- p orbital has dumb-bell shape.
- p orbitals are oriented along the three mutually right angled x, y & z axes.
- d orbitals have double dumb-bell shape.
- The nuclei with 2, 8, 20, 50, 82, 126 protons or neutrons are observed to be stable. These numbers are called magic numbers.
- Packing fraction = $\frac{\text{(Isotopic mass)} - \text{(mass number)}}{\text{(mass number)}} \times 10^4$

2. Chemical Bonding

- Electrostatic attraction between the oppositely charged ions is called ionic bond.
- The theory of electrovalency was forwarded by Kossel.
- Ionic compounds are soluble in water when their hydration energy is greater than the lattice energy. AgCl, AgBr, AgI, CaF_2, $PbSO_4$, $BaSO_4$ etc. are ionic compounds but sparingly soluble in H_2O.
- The high bond energy of ionic compounds is due to high lattice energy.
- The ionic bond is non directional due to uniform distribution of electric field about the ion.
- Elements e.g. Pb, Sn, Tl etc. show different (valencies) oxidation states due to inert pair effect.
- Transition elements show variable electro valency due to nuclear charge.
- Ionic compounds undergo ionisation in water due to high dielectric constant of water.
- The dielectric constant is the property of a medium to separate the oppositely charged ions from the crystal lattice.
- % ionic character = $\frac{\text{Experimental value of dipole moment}}{\text{Theoretical value of dipole moment}} \times 100$
- Born-Haber cycle interrelates the various energy terms involved during the formation of an ionic bond.
- The high degree of covalency of an ionic bond is due to (i) small and high nuclear charge of the cation (ii) The large size of the anion.
- Polarisation : The distortion or deformation of the electron cloud of the oppositely charged ions leads to polarisaton.
- Molecules that posses strong van der Waal's forces are highly polarisable.
- The covalent bond formation was postulated by Lewis.
- The covalent bond is formed by the sharing of the electrons.
- The covalent bond between the atoms of the same element is non polar e.g. Cl_2, N_2 etc..
- Polar covalent bond is formed between atoms having small difference in their electronegativities.
- Polar molecules have unsymmetrical charge distribution, e.g. H_2O, HCl etc.
- CCl_4 is non polar eventhough it has polar C–Cl bonds. It is due to uniform distribution of charge.
- F, O and N having 1, 2 and 3 unpaired electrons in their outer shells show covalency of 1, 2 and 3 only as they do not contain d-orbitals to show variable covalency.
- He^{2+}, NO, NO^2 etc. contain three electron bonds.
- Co-ordinate covalent bond is formed when one of the atoms sharing the bond donates both the electrons. e.g. $H_3N \rightarrow BF_3$;NH_4^+, H_3O^+ etc.
- Octet theory fails to explain the formation and stability of $BeCl_2$, BCl_3, PCl_5, WF_6, OsF_8, many coordination compounds, metal carbonyls, nitrosyls etc.
- Metallic bond : The force that binds a metal atom to a number of electrons within the sphere of influence is known as metallic bond. It is mobile, non-directional, weaker and non localized.
- N, O, F are suitable hetero atoms for hydrogen bond formation.
- O-nitrophenol, salicylaldehyde show intramolecular hydrogen bonding. Stability of chloral (Cl_3CCHO)is due to intra molecular hydrogen bonding.
- Water, phenol, carboxylic acids etc. show intermolecular hydrogen bonding.
- Hydrogen bond stabilises proteins, DNA Nylon etc.
- Deusity of ice is less than water as the former contains more hydrogen bonds.
- The order of strength of hydrogen bond :
  H–F......H > H–O.......H > H – N ....H
- Each molecule of HF is involved in two hydrogen bonds while NH_3 and H_2O molecule involves in four hydrogen bond fromation each.
- The energy of hydrogen bond ranges from 8–40 k $Jmol^{-1}$.
- VSPER theory was proposed by Sidgwick & Powell and extended by Gillespie & Nyhom.
- VSPER theory helps to predict the shape or geometry

of a covalent molecule.

- Repulsion : Lone pair–Lone pair > Lone pair–Bond pair> Bond pair–Bond pair.
- CO, CO_2, BeF_2, C_2H_2 are sp hybridised and linear.
- $SnCl_2$ is sp^2 hybridized, plane triangle with < 120°.
- H_2O – sp^3 - Bent V shape, 104.5°.
- NH_3 – sp^3 - pyramidal, 107°.
- PCl_5 – sp^3d - trigonal bipyramidal, 120° and 90°.
- SF_6 – sp^3d^2 - octahedral, 90°.
- IF_7 – sp^3d^3 - pentagonal bipyramidal, 72° & 90°.
- The hybridisation, number of lone pairs and geometry can be deduced form, N = 1/2 (V + M – C + A) ; L = N – M – D (N = number of orbitals in the hybridised orbital; V = number of electrons in valence shell; M = number of monovalent atoms; C = charge on cation; A= charge on the anion; L = number of lone pairs; D = number of divalent atoms).
- VBT : Given by Heitler & London.
- MOT : Given by Hund and Mulliken.
- Molecular orbital theory explains the paramagnetic character of oxygen.
- O^2, O_2^{1-}, O_2^+ are paramagnetic.
- Bond order can be calculated from, B.O.=1/2(N_b-N_a).
- B.O. in benzene is 1.5.
- Bond energy increases with increase in Bond order.
- The order of energy of orbitals is s < sp < sp^2 < sp^3 < p
- Bond energy :$C \equiv C > C = C > C - H > C - C$;
  $Cl_2 > Br_2 > F_2 > I_2$;
  $N \equiv N > O = O > H - H$
- Order of stability : $H_2 > H_2^+ > H_2^-$; $O_2^+ > O_2 > O_2^- > O_2^{2-}$
- Isoelectronic species : BO_3^{3-}, BF_3, CO_3^{2-}, NO_3^-; CO, CN^-, N_2, NO^+
- NH_4Cl, $K_4[Fe(CN)_6]$ have ionic, covalent and coordinate bonds.
- NaOH, KCN have ionic and covalent bonds.
- H_2SO_4, $POCl_3$, H_3O^+ have covalent and coordinate bonds.
- BF_3 and BCl_3 have symmetrical structure and therfore the resultant dipole moment of any two bonds is zero.
- Trans alkenes have generally zero dipole moment.
- Unsymmetrical and non linear polyatomic molecules have dipole moment, e.g. H_2O, NH_3, SO_2, CH_3Cl etc.

3. Radioactivity

- The phenomenon of natural radioactivity was discovered by Henry Becquerel.
- The term radioactivity was coined by Madam Curie.
- Radioactive units : Curie (1ci = 3.7 x 10^{10} dps; Rutherford (1rd = 10^6 dps); Becquerel (1BQ = 1dps) is the modern SI unit.
- The Nuclei with N/P ratio greater than 1.5 become radioactive.
- If the N/P ratio of the nucleus is higher than required, the stability will be achieved through β – ray emission.
- If the N/P ratio is less than required, it increases by α–emission or a positron or by k electron capture. Positron emission or k electron capture generally takes place in artificial radioactive isotopes.
- Atoms with the same number of difference of neutron (s) and proton(s) are called isodiaphers.
- Nucleide and its product after α–emission are isodiaphers.
- Radioactivity is an exponential process.
- The activity of radio isotopes can be detected by electroscope, electrometer, Geiger-Muller counter, Wilson cloud chamber etc.
- γ – radiations are detected by Scintillation counter.
- γ – rays are neutral short electromagnetic radiations.
- Penetration power : $\alpha < \beta < \gamma$.
- Due to high velocity and small **wavelength, γ – rays** posses high penetrating power.
- Ionisation power : $\alpha > \beta > \gamma$.
- Ionisation power of γ – rays is extremely poor as they have no mass.
- α – rays effect photographic plate (ZnS plate) and produce luminosity, due to their high kinetic energy.
- Order of kinetic energy : $\alpha > \beta > \gamma$.
- β – rays or particles are produced by the nucleus due to the decay of neutron, $n_0^1 \rightarrow H_1^1 + \beta + \nu$.
- Mass defect = sum of the mass of all fundamental particles – the actual mass of the nucleus.
- Binding genergy = Mass defect x 931 (1amu = 931 MeV).
- Greater the binding energy per nucleon, more is its stability.
- Radioactive decay or group displacement law was given by Fajans, Fussel and Soddy.
- After α–decay, the element formed is present in the group less by two positions than the one from which it is formed.
- After β–decay, the element formed is present in the

group more by one position than the one from which it is formed.

- Radioactive process is purely a nuclear phenomenon and is independent of external conditions of temperature, pressure etc.
- Molecules having the same number of atoms and electrons are called isosters, e.g. CO_2, N_2O etc.
- Radioactive disintegration follows first order reaction.
- Decay constant, $\lambda = \frac{2.303}{t} \log \frac{N_0}{N}$
- Half life period, $t_{0.5} = \frac{0.693}{\lambda}$ or $\lambda = \frac{0.693}{t_{0.5}}$
- Average life, $t_{average} = \frac{1}{\lambda} = \frac{t_{0.5}}{0.693}$
- Amount of radioactive substance left after 'n' half life periods, $N = (\frac{1}{2})^n N_0$
- At steady state, $N_1\lambda_1 = N_2\lambda_2$ or $\frac{N_1}{N_2} = \frac{\lambda_1}{\lambda_2} = \frac{t_1 av}{t_2 av}$
- At steady state, $\frac{N_1}{N_2} = \frac{(t_{0.5})_1}{(t_{0.5})_2}$
- The first nuclear transmutation was observed by Rutherford.
- Cyclotron, Betaton, Bevatron, Synchrotron, (man made accelerators) are used to obtain high energy nuclear projectiles.
- Atom bomb is based on the principle of nuclear fission.
- The controlled chain reaction is the basis for nuclear reactor.
- Fissile materials : They produce a chin reaction leading to release of energy on bonbardment with slow neutrons directly. eg. U^{233}, U^{235}, Pu^{239} etc.
- Fertile materials : The materials which by themselves are non fissile but can be converted to fissile materials by reaction with neutrons are fertile materials, e.g. Th^{232}, U^{238} etc. The conversion is carried out in breeder reactors.
- H_2O, graphite, D_2O (in research reactors), He(most efficient) etc. are used as moderators to slow down neutrons.
- B, Cd rods are good neutron absorbers.
- Hydrogen bomb (thermonuclear reactions) is based on nuclear fusion.
- Fusion reactions take place in the Sun and Stars.
- Fusion reactions are responsible for generating new elements from hydrogen and helium.
- Quantitative determination of trace elements without destroying the sample is carried out by neutron activation analysis.
- C^{14} is used to determine the age of fossils.
- Uranium dating technique is used to determine the age of the earth.
- Age of rocks and minerals is determined by Uranium-Lead dating.
- CO^{60} isotope is used for the treatment of cancer.
- P^{32} isotope finds an application in blood cancer i.e. leukemia.
- I^{131} is used to determine the activity of thyroid gland.
- Radioactive series : 4n, 4n+1 and 4n+2 are natural radioactive series.

4n	Thorium series	$Th_{90}^{232} \longrightarrow Pb_{82}^{208}$
4n+1	Neptunium series (artificial series)	$Np_{93}^{237} \longrightarrow Bi_{83}^{209}$
4n+2	Uranium series	$U_{92}^{238} \longrightarrow Pb_{82}^{206}$
4n+3	Actinium series	$U_{92}^{235} \longrightarrow Pb_{82}^{207}$

4. Gaseous state

- Atmospheric pressure is measured by barometer.
- SI unit of pressure is Pascal (1atmosphere = 760 torr = 101.325kPa =101325 Nm^{-2}.
- The volume of any gas is zero at −273.15°c or zero kelvin.
- First direct evidence for molecular motion was furnished by Robert Brown.
- Boyles law : $V \propto \frac{1}{P}$; $PV = k$; $P_1V_1 = P_2V_2$; $\frac{P}{D} = k$; $P \propto D$
- Charles law : $V \propto T$; $\frac{V}{T} = k$; $\frac{V_1}{T_1} = \frac{V_2}{T_2}$; $D \propto \frac{1}{T}$
- Pressure temarature law : $P \propto T$; $\frac{P}{T} = k$; $\frac{P_1}{T_1} = \frac{P_2}{T_2}$
- General gas equation :$PV = nRT$ or $PV = \frac{W}{M} RT$
- The numerical value of R (gas constant) : In SI units: 8.314Joules $degree^{-1}$ mol^{-1}; In c.g.s units : 1.99 calories $degree^{-1}$ $mole^{-1}$.; when the pressure is in atmos. and volume lit. : 0.0821 lit.-atm. mol^{-1} degree k.
- The temperature of a gas is determined by the average kinetic energy of the molecules.
- The number of molecules present in 1C.C. of vapour at STP is known as Loschmidt number (its value is 2.617×10^{19}/C.C.).
- Graham's law of diffusion :$\frac{r_1}{r_2} = \frac{v_1/t_1}{v_2/t_2} = \sqrt{\frac{\rho_2}{\rho_1}} = \sqrt{\frac{M_2}{M_1}}$
- Isotopes of the same element from one another can be separated by the application of Graham's law of

diffusion.

- Partial pressure of a gas in a mixture of gases can be determined from the equation, $P_a = \left[\frac{n_a}{n_a + n_b}\right] P$

  (n_a & n_b are number of moles of gas a and b. P is the total pressure of the gaseous mixture).
- Kinetic gas equation : $PV = \frac{1}{3} mNu^2$
- Most probable velocity $\propto = \sqrt{\frac{2RT}{M}}$
- Average velocity $v = \sqrt{\frac{8RT}{\pi M}}$
- Root mean square velocity $u = \sqrt{\frac{3RT}{M}} = \sqrt{\frac{3PV}{M}} = \sqrt{\frac{3P}{D}}$
- $\propto : v : u = \sqrt{2} : \sqrt{8/\pi} : \sqrt{3} = 1 : 1.128 : 1.224$
- For a particular gas at a particular temperature $u > v > \propto$
- The temperature above which a gas can not be liquified under whatever pressure is called critical temperature, $Tc = \frac{8a}{27bR}$
- The temperature at which Joule - Thomson effect is zero is called the inversion temperature of the gas.
- Real gases tend to behave like ideal gases at low pressure and high temperature.
- Liquefiability : $SO_2 > NH_3 > H_2O > CO_2$
- Boyles temperature is 50°c & is given by $T_b = a/Rb$.
- Real gases obey van der Waals equation of state, $(P + \frac{an^2}{v^2})(v - nb) = nRT$ ('a' is the measure of intermolecular forces of attraction. 'b' is excluded volume. 'a' is higher for easily liquefiable gases).

  units : a = atm. L^2 $mole^{-2}$; SI unit : a = Nm^4 $mole^{-2}$

  b = L $mole^{-1}$ b = m^3 $mole^{-1}$
- Ratio of molar heat capacity : $\gamma = c_p/cv$: 1.66 for mono-atomic; 1.40 for diatomic and 1.33 for triatomic.

5. Solutions

- Molarity, M : number of moles of solute present in 1 lit. of the solution.
- One mole of solute present in one litre of the solution at 0°c is represented as one demal.
- Molality, m : number of moles of solute present in 1000g. of solvent.
- Normality : number of gram equivalents of the solute present in 1 lit. of the solution.
- mole fraction of the solute : $\frac{n}{n+N}$; for solvent it is $\frac{N}{n+N}$.
- Molality and mole fraction are independent of temperature (since they depend on weights and not volume).
- A liquid boils when its vapour pressure equals the atmospheric pressure.
- Raoults law : The relative lowering of vapour pressure is equal to the ratio of the molecules of the solute to the total number of molecules in the solution.

  $$\frac{p - p_s}{p} = \frac{n}{n+N} = \frac{w/m}{w/m + W/M}$$
- The electrolyte should be non volatile and the solution should be very dilute, $\frac{p - p_s}{p} = \frac{w}{m} X \frac{M}{W}$
- Non ideal solutions show deviation from Raoult's law.
- Wheather a mixture is ideal or shows deviations depends on the attractions and repulsions between different type of molecules present in it.
- Colligative properties depend only on the number and not the kind of particles.
- Colligative properties : Relative lowering of vapour pressure (determind by Ostwald & Walker); Elevation in boiling point (Landsberger or Cotterell's); Depression in the freezing point (Beckmann); Osmotic pressure (Berkeley & Hartley's or Morse & Frazer method) etc. are colligative properties.
- At the freezing point, the vapour pressure of a solid and its liquid are the same.
- Some relationships :

  $$m = \frac{1000\, K_b\, w}{\Delta T_b\, W}; \; K_b = \frac{RT^2}{1000\, L_v}; \; \Delta T_b = K_b \times m$$

  $$m = \frac{1000\, K_f\, w}{\Delta T_f\, W}; \; T_f = K_f.m$$

  K_b and K_f are ebullioscopic (moleal elevation constant) and cryoscopic (freezing point constant) constants respectively.
- Isotonic solutions have the same osmotic pressure $\left(\frac{w_1}{m_1 v_1} = \frac{w_2}{m_2 v_2}\right)$
- $\pi v = nRT$ or $\pi v = \frac{W_B}{m_B} RT$

  $\pi = \frac{n}{v} RT = CRT$
- Copper ferrocyanide acts as a good semipermeable membrane.
- When a plant or animal cell is placed in a hypertonic solution, it shrinks as the cell fluid flows out due to osmosis.
- Reverse osmosis is used for desalination of sea water.
- 0.91% solution of NaCl is isotonic with human RBC's.
- For ideal solutions $\Delta V_{mix} = 0$ and $\Delta H_{mix} = 0$.

- Ideal solutions obey Raoult's law at all concentrations and temperatures.
- Benzene + Toluene; $CH_2Cl_2 + CH_2Br_2$; n-hexane + n-pentane; $CCl_4 + SiCl_4$ etc. form ideal solution.
- Real solutions show deviation from Raoult's law.
- If the vapour pressures are greater than calculated, the solutions show positive deviation e.g. Benzene + C_2H_5OH; $C_2H_5OH + H_2O$; Benzene + CH_3COCH_3; $CHCl_3 + CCl_4$; $CHCl_3 + C_2H_5OH$ etc.
- If the vapour pressure are lesser than calculated, the solutions show negative deviation e.g. $H_2O + H_2SO_4$; $CHCl_3$ + Benzene; CH_3COOH + Pyridine etc.
- Abnormal melecular weight of the solute in solution is either due to association or dissociation of solute.
- Vant Hoff factor, $i = \dfrac{\text{normal molecular weight}}{\text{observed molecular weight}}$
- $\dfrac{\text{normal mol. wt.}}{\text{observed mol. wt.}} = \dfrac{\text{no. of solute particlesafter association or dissociation}}{\text{no. of normal particles}}$
- Azeotropic mixture boils at a constant temperature e.g. $C_2H_5OH + H_2O$; Benzene + C_2H_5OH; Pyridine + H_2O; H_2O + HCl; $CH_3COCH_3 + CHCl_3$ etc.
- Azeotropic mixture of alcohol with water is separated by azeotropic distillation by a adding some benzene to it.
- Deviatons from the partition law occur due to the dissociation or association of solute.
- If a gas doesn't react in solvent, the amount of gas dissolved in the given volume of the solvent at a constant temperature is proportional to the pressure applied, m=kxP (k =Henry's constant).

6. Solid State

- Crystalline substances have definite geometry, sharp melting point and are anisotropic. e.g. NaCl, Sugar, Sulphur etc.
- Amorphous substances do not have definite geometry and are isotropic.
- Liquids and gases are isotropic.
- Glass, rubber, plastics are infact super cooled liquids.
- Slower rate of crystallisation leads to bigger crystals.
- Higher rate of crystallisation leads to tiny but pure crystals.
- The crystals of one and the same substance may differ in shape and size but have the same interfacial angles.
- A cube has 23 elements of symmetry, 9 plane of symmetry (3 rectangular plane of symmetry + 6 diagonal plane of symmetry) + 13 axes of symmetry (3four fold symmetry + 4 three fold symmetry + 6 two fold symmetry) + 1 centre of symmetry).
- From geometrical considerations, there can be 32 elements of symmetry in total.
- The crystal may have any number of planes or axes but it has only one centre of symmetry.
- There are 7 crystal systems or habits.
- The total number of space lattices belonging to all the crystal systems is 14 (14 Bravais arrangements).
- There can be 230 different arrangements known as space groups (230 crystal forms are possible).
- Cube : In simple cube (e.g. Po) : There are points only at the corners. 48% emplty space is present.

  Number of atoms in a unit cell of simple cube : $8 \times \frac{1}{8} = 1$

  It has coordination number six.

  In bcc lattice : no. of atoms in a unit cell : $8 \times \frac{1}{8} + 1 = 2$

  It has 32% empty space (voids)

  $r = \frac{\sqrt{3}}{4} \times a$; $a = r \times \frac{4}{\sqrt{3}}$

  coordination number is eight

  In fcc lattice : no. of atoms in a unit cell :

  $8 \times \frac{1}{8} + 6 \times \frac{1}{2} = 4$

  It has 26% empty (voids).

  $r = \frac{\sqrt{2} \times a}{4} = \frac{a}{2\sqrt{2}}$; $a = \sqrt{8} \times r$

  coordination number is 12.
- Atoms at an edge count as 1/4.
- There are four types of crystals [Ionic (e.g. NaCl, CaF_2, KNO_3 etc.); covalent (e.g. quartz (SiO_2), graphite, diamond, BN, SiC etc.); Metallic (e.g. Na, Al, Cu, Fe and some alloys) and Molecular (polar, e.g. PCl_3, H_2O etc. and non polar, e.g. He, CO_2, ammonia solid, Ice etc.].
- Cholesteric liquid crystals are thread like and optically active. They are used for detecting tumours in the body by thermography.
- If an element crystallises in different forms, the process is known as allotropism.
- If one and the same substance can crystallie in more than one form, it is known as polymorphism.
- Allotropic forms that can undergo reversible change are said to be enantiotropic, e.g. $S_{rh} \underset{}{\overset{95.6°C}{\rightleftharpoons}} S_{monoclinic}$.
- The allotropic forms that can not change into one another reversibly are said to be monotropic, e.g. O_2 & O_3; Red P and Yellow P.
- The increase in the conductivity with increase in temperature of ionic solids is due to the defects in the

crystal structure.

- Metals are good conductors of electricity ~1 x 10^4 ohm^{-1} cm^{-1}. Conductivity of metals decrease with increase in temperature .
- Doping of B (with three elections) into the lattice of Si or Ge leadsto the p-type semiconductors while that of As or Sb (with five electrons) leads to the formation of n-type semiconductors.
- n – p junction can be formed when B and As are introduced as dopants from different sides. n – p juncton acts as a rectifier and can convert A.C. to D.C.
- Schottky and Frenkel defects are stoichiometric and exhibited by ionic solids.
- Schottky defect : It is due to the creation of paired cation & anion vacancies. Electric neutrality is mained although density decreases. e.g. NaCl, CsCl, KCl, KBr etc.
- Frenkel defect : It is due to shifting of an ion into interstitial position. Density remains the same. e.g. AgBr, AgI, ZnS, CaF_2 etc.
- F – centres : Electrons are trapped in the anoin vacancies. F – centres are responsible to impact the colour to the crystals. Excess Li in LiCl makes it pink while excess of K in KCl makes it violet. It is a non stoichiometric defect.
- Fools gold is an example for non-stoichiometric defect. Iron pyrites shine like gold due to free exchange of electrons between Fe^{2+} and Fe^{3+} which exist side by side.
- Hexagonal close packing is the most efficient way of packing of atoms in a crystal lattice.
- Hexagonal close packing has two versions : h c p or AB ..... AB. (e.g. Mg, Zn, Mo, Cd etc) and ccp or ABC.......ABC(e.g. Li, Na, K, Rb, Cs, Ba etc.).
- In NaCl each Na+ is surrounded by six Cl^- ions and vice-versa.
- Common interstitial sites in closely packed spheres :

r^+ / r^- ratio	coordination numbaer	Shape of the void	Example
0.155 – 0.225	3	Planar	B_2O_3
0.225 – 0.414	4	Tetrahedral	ZnS
0.414 – 0.732	6	Octahedral	NaCl
0.732 – 1.000	8	BCC	CsCl

- LiCl, KCl, NaI, MgO, Rb, AgF, AgCl, AgBr etc. are NaCl type crystals.
- CsBr, CsI, NH_4Cl etc. are CsCl type crystals.
- ZnS has zinc blende structure and wurzite structure.
- HgS, AgI etc. show zinc blende structure while wurzite structure is shown by ZnO, CdS etc.
- In CaF_2, each Ca^{2+} is in contact with $8F^-$ ions and each F^- ion is in contact with $4Ca^{2+}$ ions. BaF_2, SrF_2, PbF_2, CuF_2 etc. have fluorite structure.
- SiO_4^{4-} is the basic unit of all silicates.
- Quartz, Zeolites etc. are frame work silicates.
- Pyrosilicates contain $Si_2O_7^{6-}$ anion.
- Sheet silicates (e.g. Mica, Talc etc.) contain $(Si_2O_5)_n^{2-}$ ions.
- The phenomenon of offering no resistance to the flow of electric current is known as superconductivity.
- Superconductivity was discovered by Kammerlingh Onnes.
- Hg was the first metal to show superconductance at 4K.
- Ferroelectric : Dipoles are polarized even when no current is passed. e.g. Rnchelle salt, K_2HPO_4, B_2TiO_3 etc.
- Piezoelectric : Polar crystal which produces electricity on application of mechanical stress. Recod players work on this principle.
- Pyroelectric : on heating, some of the polar crystals produce feeble electric current, e.g. Tourmalin.
- Antiferroelectric : Inspite of having dipoles, they do not exhibit ferroelectric character as the polar polyhedra in them point up and down. The net dipole moment is zero, e.g. $PbZrO_3$ (lead zirconate).
- Bragg's equation is $n\lambda = 2d \sin\theta$
- Distance between parlle planes in a crystal,

$$d_{h, k, l} = \frac{a}{\sqrt{h^2 + k^2 + l^2}}$$

(a = length of the cube side; h, k, l are Miller indices of the plane)

- Multiples of the intercepts of unit plane are called Weiss indices.
- Reciprocal of Weiss indices are called Miller indices.
- Theoretical density of the crystal can be calculated from the equation, $\rho = \frac{nM}{NoV}$ or $\rho = \frac{zM}{N_A a^3}$

(n = no. of atoms in a unit cell; M = mol. wt.; V = volume of the unit cell; a = length of the cube side).

7. Chemical and Ionic Equilibria

- In irreversible reactions, the insolubles in the form of precipitate or volatile products formed serve as sink making the reaction non directional.

- Chemical equilibrium at a given temperature is characterised by constancy of concentration, pressure, colour or density.
- A catalyst does not alter the state of the equilibrium. It can hasten the approach of the equilibrium.
- Change in concentration, pressure and the use of catalyst have no effect on equilibrium constant. They can often change the rate at which equilibrium is achieved.
- At equilibrium the rate of forward and backward reactions is the same.
- At equilibrium, the change in free energy is zero ($\Delta G = 0$).
- Active mas of A, $[A] = \dfrac{\text{no. of grams of A/mol. wt. of A}}{\text{volume in litres}}$
- The active mass of solids is taken as unity as the molecules are very closely packed in them.
- Unit of active mass is gram molecule/litre.
- Equilibrium constant $K_{\text{forward reaction}} = \dfrac{1}{K_{\text{backward reaction}}}$
- $K_C = K_P \times \dfrac{1}{(RT)^{\Delta n}}$ or $K_P = K_C (RT)^{\Delta n}$

  Δn = number of moles of product - no. of moles of reactants

  when $\Delta n = 0$, $K_P = K_C (RT)^0 = K_C$

  when $\Delta n = +ve$, $K_P > K_C$; when $\Delta n = -ve$, $K_P < K_C$
- Units of K_C is $(\text{mole/lit})^{\Delta n}$; Units of K_P is $(\text{Atmos.})^{\Delta n}$
- Greater the value of K_C or K_P, greater is the yield of products.
- Variation of equilibrium constant with temperature as given by Vant Hoff is $\log \dfrac{k_2}{k_1} = \dfrac{\Delta H}{2.303R}\left[\dfrac{T_2 - T_1}{T_1 \times T_2}\right]$
- Standard free energy change and equilibrium constant are related by $\Delta H = -RT \ln k_{eq}$.
- The three main factors that alter the state of equilibrium are concentration, pressure and temperature.
- Le Chatelier's principle : If an equilibrium is subjected to a stress, the equilibrium shifts in such a way as to reduce the stress.
- Increasing the concentration of the reactants results in shifting the equilibrium in favour of the products and vice-versa.
- For an exothermic reaction, an increase in temperature favours reactants and thus the equilibrium constant lowers.
- In an endothermic reaction, increase in temperature favours the products, thus increaing the equlibrium constant value.
- Increase in pressure results in decrease of the volume of the system proportionately and therefore shifts the equilibrium in the direction where, the volume (products or reactants) is less.
- Pressure has no effect in equilibria where the volume of products and the reactants is the same, e.g. $H_2(g) + I_2(g) \rightleftharpoons 2HI(g)$; $N_2(g) + O_2(g) \rightleftharpoons 2NO(g)$ etc.
- Addition of intert gas will have no effect on equilibrium state as the equilibrium constant is independent of total volume and pressure.

  Effect of addition of inert gas on an equilibrium can be understood as

$\Delta V = 0$;	V = constant	$\Delta n = 0, +Ve, -Ve$	no effect
$\Delta V \neq 0$;	$V \neq$ constant	$\Delta n = 0$	no effect
$\Delta V \neq 0$;	$V \neq$ constant	$\Delta n > 0$	forward shift
$\Delta V \neq 0$;	$V \neq$ constant	$\Delta n < 0$	backward shift

- Increase of temperature on Liquid $\rightleftharpoons$ Vapour equilibrium shifts the equilibrium in the forward direction as the formation of vapour proceeds with absorption of heat. Increase in pressure results in the condensation of some vapour to liquid.
- Increase in pressure on Ice $\rightleftharpoons$ Water equilibrium results in the melting of ice to water as the volume of ice is more than water.
- Arrhenius concept is applicable only to aqueous solutions. According to this theory those substances which dissociate in water to give H^+ and OH^- ions are called acids and bases respectively.
- Lowry & Bronsted concept is applicable to both aqueous and non aqueous solutions. Acid is a proton donor while base is proton acceptor.
- The stronger an acid, the weaker must be its conjugated base and vice versa e.g. strength of acids : $HCl > CH_3COOH > H_2CO_3$; strength of the conjugated bases : $CO_3^{2-} > CH_3COO^- > Cl^-$
- Lewis concept : Acid is one which can accept a pair of electrons and base is one which can furnish a pair of electrons, e.g. $H_3N: \rightarrow BF_3$ etc.
- Precipitation of a substance occurs when the ionic product i.e. the product of the concentration of its ions present in a solution exceeds the value of solubility product of the substance.
- Basic radicals are grouped on the basis of their solubility products
- Solubility product of second group sulphides is lower than the fourth group sulphides.
- The solubility product of hydroxides of Fe, Al & Cr are much lower than the hydroxides of Mg & Zn.
- The solubility for binary salts (AB type e.g. AgCl, PbS etc.) can be calculated from $S = \sqrt{K_{sp}}$.

The solubility for ternary salts (A_2B type e.g. Ag_2CO_3 etc. or AB_2 type $PbBr_2$, $Mg(OH)_2$ etc.) is given by

$$S = \left[\frac{K_{sp}}{4}\right]^{1/3}$$

The solubility for AB_3 type compounds, e.g. $Al(OH)_3$ etc. is , $S = \left[\frac{K_{sp}}{27}\right]^{1/4}$.

- Common ion effect : The supression of the dissociation of a weak acid or a weak base on the addition of its own ions is called common ion effect.
- The phenomenon of the interaction of cations and anions of the salt with OH^- or H^+ ions furnished by water is known as salt hydrolysis.
- Salts of strong acid + strong base do not undergo hydrolysis. They are neutral e.g. KNO_3, NaCl etc.
- Salts of WA + SB undergo hydrolysis to give alkaline solutios. e.g. CH_3COONa, KCN etc.
- Salts of SA + WB undergo hydrolysis to give acidic solutions. e.g. $FeCl_3$, NH_4Cl etc.
- The nature of aqueous solution of the salts of WA + WB can be understood as : If $K_a = K_b$, neutral; $K_a > K_b$, acidic; $K_a < K_b$, basic.
- The degree of dissociation can also be calculated from density measurements, x = (D-d)/d (D = vapour density when there is no dissociation, d = observed vapour density at a particular temperature when degree of dissociation is x).

Type of salt	hydrolysis constant	degree of hydrolysis	pH
WA + SB	$K_h = \frac{K_w}{K_a}$	$x \text{ (or } \alpha) = \frac{\sqrt{K_h}}{C} = \frac{\sqrt{K_w}}{K_a x C}$	$pH = 7 + \frac{1}{2}pKa + \frac{1}{2}\log C$
SA + WB	$K_h = \frac{K_w}{K_b}$	$x \text{ (or } \alpha) = \frac{\sqrt{K_h}}{C} = \frac{\sqrt{K_w}}{K_b x C}$	$pH = 7 + \frac{1}{2}pKa - \frac{1}{2}\log C$
WA + WB	$K_h = \frac{K_w}{K_a x K_b}$	$x \text{ (or } \alpha) = \frac{\sqrt{K_w}}{K_a K_b}$	$pH = 7 + \frac{1}{2}pKa - \frac{1}{2}pK_b$

- pH scale was introduced by Sorensen.
- pH of a solution is the negative logarithm of the concentration of hydrogen ions which it contains, $pH = -\log [H^+]$.
- The sum of pH and pOH is equal to 14 in water or in any aqueous solution at 25°C.
- The concentration of H^+ and OH^- increse with increase in temperature (K_w also increases).
- pH for a neutral solution is 7. The solution with pH less than 7 is acidic while that with pH greater than 7 is alkaline.
- Sometimes pH of acid or base on calculation comes to the value more or less than 7 respectively which indicates that the solution is very dilute.
- $[H^+]$ for a mixture of weak and strong acids can be determined from the relationship, $C_2 + \frac{\sqrt{C_2^2 + 4k_aC_1}}{2}$ (C_1 and C_2 are concentration of weak and strong acid respectively).
- $[OH^-]$ of a mixture of two weak bases can be determined from the relationship, $\sqrt{k_1C_1 + k_2C_2}$.
- A buffer is one which can resist change in its pH value on the addition of a solution of an acid or a base.
- Acid buffers : e.g. $CH_3COOH + CH_3COONa$; Boric acid + Borax; Phthalic acid + Potassium phthalate.

  $[H^+] = K_a \left[\frac{\text{Acid}}{\text{Salt}}\right]$

  $pH = pK_a + \log\left[\frac{\text{Salt}}{\text{Acid}}\right]$ - Henderson equation
- Basic buffers : e.g. $NH_4OH + NH_4Cl$; glycine + glycine hydrochloride

  $[OH^-] = K_b \left[\frac{\text{Base}}{\text{Salt}}\right]$

  $pOH = pK_b + \log\left[\frac{\text{Salt}}{\text{Base}}\right]$; $pH = 14 - pOH$

8. Thermodynamics

- Thermodynamics is applicable only to macroscopic systems.
- Open system : A system which can exchange energy and matter with its surroundings.

  All the living beings are the examples of an open system.
- Closed system : when a system can exchange only energy and not the matter with its surroundings, it is called a closed system.
- Isolated system : A system which can neither exchange energy nor matter with its surroundings is called an isolated system.
- Pressure, temperature, volume, density, viscosity, surface tension, composition etc. are macroscopic properties.
- Macroscopic properties can change the original state of the system.
- Extensive property : The property which depends on the quantity or amount of matter present in the system is called extensive property. e.g. mass, number of moles, enthaply, entropy, Gibbs free energy etc.
- Intensive proerty : The property which is indeppendent of the amount of the substance present in the system.

e.g. pressure, temperature, density, viscosity, surface tension, concentration, refractive index, specific heat etc.

- For an isothermal process, $dT = 0$.
- For an adiabatic process $dQ = 0$.
- For an isobaric process $dP = 0$.
- For an isochoric process $dV = 0$.
- For a cyclic process, $dE = 0$ and $dH = 0$.
- The values of heat of reactions are given generally under standard conditions, i.e. 1 atmos. pressure and 25°C temperature.
- Change in the heat contents or in enthalpy of the system, $\Delta H = \Sigma H_P - \Sigma H_R$ or $\Delta H = H_P - H_R$.
- If $H_P > H_R$ or $\Delta H = +ve$, it is endothermic or endoergic.
- If $H_P < H_R$ or $\Delta H = -ve$, it is exothermic or exoergic.
- The heat of neutralisation of a strong acid against a strong base is always constant i.e. 13.7 k. cal or 57 kJ/mole and it is actually equal to the heat of formation of water.
- Heat of neutralisation of a weak acid or base against a strong base or acid is always less than 57 kJ/mole as a part of heat evolved is used up to ionize the weak acid or the base.
- Heat content and enthalpy are related as

  $\Delta H = \Delta E + \Delta nRT$ or $\Delta H = \Delta E + P\Delta V$ ($P\Delta V$ = work done = ΔW)

  $\Delta E = \Delta H - \Delta nRT$ (when $\Delta n = 0$, $\Delta E = \Delta H$; if $n > 0$, $\Delta H > \Delta E$; if $\Delta n < 0$, $\Delta H < \Delta E$).
- The heat of formation of an element is regarded as zero.
- Hess law : The amount of heat evolved or absorbed in a process, including the chemical change is the same whether the process takes place in one or several steps.
- The total mass and energy of an isolated system remains unchanged.
- Energy can neither be created nor destroyed.
- The first law of thermodynamics is also known in terms as law of conservation of energy.
- Internal energy of a system is made up of a number of components i.e. energy of translation + rotational + vibrational + coulombic energy of electron and nucleus in atoms + interaction energy.
- Internal energy, enthalpy, entropy, free energy etc. are state variables.
- Mathematical representation of first law of thermodynamics is, $\Delta E = Q + W$ (The net energy change of a closed system is equal to heat absorbed plus the work done on the system or $\Delta E = Q - W$ (the net energy change of a closed system is equal to heat absorbed minus the work done by the system.
- Work done in an isothermal expansion is given by,

  $W_{expansion} = -2.303\ nRT \log \frac{V_2}{V_1}$

  similarly $W = -2.303\ nRT \log \frac{P_1}{P_2}$
- Work done in an isothermal compression is giveb by

  $W_{compression} = 2.303\ nRT \log \frac{V_1}{V_2}$

  similarly $W = 2.303\ nRT \log \frac{P_2}{P_1}$
- Work done in an adiabatic expansion is given by, $P_1V_1^{\gamma} = P_2V_2^{\gamma}$ ($\gamma = c_p/c_v$, c_p and c_v are heat capacities of a gas at constant pressure and volume respectively).
- Kirchoff's equation : It explains the variation of heat of reaction with temperature, $\Delta C_p = C_p - C_p = \frac{\Delta H_2 - \Delta H_1}{T_2 - T_1}$

  If the reaction takes place at constant volume then the equation is, $\Delta C_v = C_v - C_v = \frac{\Delta E_2 - \Delta E_1}{T_2 - T_1}$
- The temperature at which the gas neither heats nor cools on exppansion is called the inversion temperature.
- Second law of thermodynamics : Work can be completely converted into heat, but it is not possible to convert heat completely into work.
- Carnot cycle : It determines the conversion of heat into work.

  Carnot equation is, $\frac{W}{Q_2} = \frac{Q_2 - Q_1}{Q_2} = \frac{T_2 - T_1}{T_2}$

  The total work done, W, in Carnot cycle is equal to the total heat absorbed i.e. $W = Q_2 - Q_1$.

  Efficiency of Carnot engine or heat engine, $\varepsilon = \frac{W}{Q_2}$
- Entropy : It is a measure of disorder. The entropy of the universe is increasing.
- The change in the entropy, $\Delta S = Q_{rev} / T$
- Units of entropy : Cal. degree^{-1} or J K^{-1}.
- Entropy change in isothermal expansion of an ideal gas is given by, $\Delta S = nR \times 2.303 \log \frac{V_2}{V_1}$
- Entropy change in a complete reversible cycle is given by, $\Delta S = 0$
- Entropy change in an irreversible process is given by, $\Delta S > 0$.
- Entropy change at constant pressure is, $\Delta S_P = C_P \times 2.303 \log \frac{T_2}{T_1}$
- Entropy change at constant volume is, $\Delta S_V = C_V \times 2.303 \log \frac{T_1}{T_2}$
- Entropy change for a process involving ideal gas

depends on the initial and final temperatures and volumes,

$$\Delta S = C_v \times 2.3030 \log \frac{T_2}{T_1} + R \times 2.303 \log \frac{V_2}{V_1}$$

- Entropy change accompanying change of solid to liquid phase is, $\Delta S_f = \frac{\Delta H_f}{T_f}$ (H_f = molar heat of fusion; liquid to gas phase, $\Delta S_v = \frac{\Delta H_v}{T_b}$; one form to another form, $\Delta S = \frac{\Delta H_t}{T_t}$ (T_t = transition temperature).
- Enthalpy and entropy are related by, $\Delta G = \Delta H - T\Delta S$ (ΔG = –ve, spontaneous: ΔG = +ve, non spontaneous; $\Delta G = 0$, the reaction is at equilibrium).
- Spontaneous change is one that occurs on its own without the external help.
- The free energy change ΔG and equilibrium constant K are related by the expression, $\Delta G = -2.303\ RT \log K$.
- The standard cell potential is related to the standard free energy by the relation, $\Delta G^o = -nFE^o$.
- Clausius - Clapeyron equation : It deals with equilibrium between two phases of one component system and is given by $\log = \frac{P_2}{P_1} = \frac{\Delta H}{2.303R}\left[\frac{T_2 - T_1}{T_2 \times T_1}\right]$
- Third law of thermodynamics : At absolute zero, the entropy of a perfect crystal is also zero, S = 0 at T = 0 K.

9. Kinetics

- For a reaction to take place, there must be effective collisions.
- Effective number of colisions, N =Pz (z = collision frequency, P = steric factor). z should be ~ 10^{25} to 10^{28}.
- Rate of a reaction depends upon the concentration of the reactants (exception : In zero order reactions, the rate is independent of concentration of reactants).
- The rate of a reaction generally increases with the increase in temperature. The rate doubles for every 10° (absolute) increase.
- The rate of those reactions which do not involve much bond rearrangements are generally higher.
- Catalyst normally enhances the rate of a reaction by lowering the activation energy.
- Radiation generally increases the rate of a reaction.
- Surface area of a reactant is of importance only in heterogeneous reactions. More the surface area of the reactant, greater is the rate of reaction.
- The order of rate of reaction based on physical state is, liquid > gases > solids. In liquid state, the molecules not only move relatively freely but also closer to one another.
- Rate constant is a constant at a particular temperature and is independent of the concentration of the reactant.
- Units of rate are, moles/lit. time^{-1}.
- Units of rate constant deppend on the order of the reaction.
- Order of a reaction can have +ve, zero, –ve and fractional values.
- Order of a reaction : It is the number of molecules of the reactants whose concentration alters appreciably as a result of chemical reactions. It is determined through experiment only.
- Molecularity : It is the number of molecules of the teactants involved in the reaction.
- The first order reactions whose molecularity is not one are called Pseudo-unimolecular reactions, e.g. Inversion of cane sugar, acid hydrolysis of esters etc.
- At unit concentration, the rate constant is called specific reaction rate.
- Zero order : The rate of a reaction becomes independent of concentration of a reactant when that reactant is taken in large excess. Rate law, $k = \frac{X}{T}$. Half time for zero order reaction is proportional to intial concentration, $t_{0.5} = a/2k$. Unit of k : mole / lit. time^{-1}. Photochemical and enzyme reactions are zero order reactions.
- First order : Rate law, $k = \frac{2.303}{t} \log \frac{a}{a-x}$; unit of k = time^{-1}; $t_{0.5}$ (even $t_{0.5}$ for any fraction=$\frac{0.693}{k}$) is independent of concentration of reactants; graph between t and log A gives a straight line. For acid hydrolysis of ester, $k = \frac{2.303}{t} \log \frac{T_\infty - T_0}{T_\infty - T_t}$; for optical rotation of sucrose etc. $k = \frac{2.303}{t} \log \frac{\theta_0 - \theta_\infty}{\theta_t - \theta_\infty}$ Radioactive decay, acid hydrolysis of esters, inversion of cane sugar, conversion of α – glucose to β – glucose etc. belong to first order reaction. First order reactions show exponentail change.
- Second order : Rate law, $k = \frac{1}{t} \times \frac{x}{a(a-x)}$; units of k = (mole/lit)$^{-1}$ time^{-1} or mole^{-1} lit.time^{-1}. Half life is inversely proportional to the initial concentration, $t_{0.5} \propto \frac{1}{a}$. Saponification of esters, Benzoin condensation etc. belong to second order reactions.
- For third order reactions, $t_{0.5} \propto \frac{1}{a^2}$
- Higher molecularity or higher order reactions greater than 3 are rare as the probability of simultaneous collisions between 3 or more particles is rare.

- In a multi-step reaction, the slow step is the rate determing step.
- Graphical method : log (a-x) against t gives a straight line for first order reaction while log $(a-x)^{-1}$ vs. t and log $(a-x)^2$ vs. t give straight line for second and third order respectively.
- $\frac{t_1}{t_2} = \left(\frac{a_2}{a_1}\right)^{n-1}$ (n = order reaction, a_1 & a_2 and t_1 & t_2 are the initial, concentrations and corresponding times in two different experiments) or $n = 1 + \frac{\log t_1 - \log t_2}{\log a_2 - \log a_1}$
- Activation energy is the minimum energy which a molecule must acquire before reaction can take place.
- Arrhenius equation $K = A.\, e^{-E_a/RT}$ or $\log k = \frac{-E_a}{2.303R} \times \frac{1}{T} \log A$.

  Graph between log k and $\frac{1}{T}$ gives a straight line whose slope is $\frac{-E_a}{2.303R}$.
- Integrated form of Arrhenius equation is
  $$\log\frac{k_2}{k_1} = \frac{E_a}{2.303R}\left[\frac{T_2 - T_1}{T_1 \times T_2}\right] = \frac{E_a}{2.303R}\left[\frac{1}{T_1} - \frac{1}{T_2}\right]$$
- Activation energy = Threshold energy - Energy possessed by the reactants, ($E_a = E_T - E_r$).
- The quantity $e^{-Ea/RT}$ in Arrhenius equation is known as Boltzmann factor.
- Rates of fast reactions can be determined by flash photolysis, relaxation etc. techniques.
- Quantum yield, $\phi = \frac{\text{number of molecules reacting in a given time}}{\text{number of quanta absorbed in the same time}}$

10. Electrochemistry and Redox reactions

- A substance which allows the flow of an electric current through it is called a conductor.
- Metallic conductors remain unchanged as current flows.
- Electrolytes are substances whose solutions conduct electric current and are decomposed by the passage of current.
- Conductance decreases on dilution as the number of ions per unit volume decreases.
- Conductance increses on heating as the viscosity lowers.
- Conductance increases on melting as ions are free to move.
- Carboxylic acids, H_2CO_3, C_6H_5COOH, HCN, NH_4OH etc. are weak electrolytes.
- Strong acids, bases and salts are strong electrolytes.
- Alcohols, ethers, esters, hydrocarbons, urea, sugar etc. are non electrolytes.
- Specific conductance, k (kappa) is reciprocal of resistivity, ρ, and is given, $k = \frac{1}{\rho} = \frac{1}{R} \times \frac{l}{A}$ ($\frac{1}{R}$ = observed conductance; $\frac{l}{A}$ = cell constant). Kappa has units, $ohm^{-1}\ cm^{-1}$ or mhos/cm; SI unit : Sm^{-1}.
- The conductance of a 1c.c. of solution is known as specific conductance, k, of that solution.
- Equivalent conductivity : It is the conductance due to all the ions produced by 1 gram equivalent of an electrolyte in a given volume of the solution. $\Lambda = k \times V$ (k = specific conductance, V = vol. of solution containing 1 g. equivalent of the electrolyte) or $\Lambda = \frac{1000 \times K}{N}$ (N = normality of the solution). Unit : $ohm^{-1}\ cm^2\ eq^{-1}$.
- Molar conductivity : It is the conductance of all the ions produced by one mole of an eectrolyte when dissolved in a certain volume.
- Conductivity, $\mu = K \times V$ or $\mu = \frac{1000 \times K}{M}$ (m = molarity); Unit : $cm^2\ mol^{-1}\ ohm^{-1}$.
- Kohlrausch's law : At infinite dilution (where the dissociation is complete) the equivalent conductance of any electrolyte is the contribution of cation and anion (these contributions are called ionic conductances at infinite dilution) $\Lambda^o = \lambda^o_c + \lambda^o_a$.
- At infinite dilution the percentage ionisation for both strong and weak electrolyles is 100%.
- Degree of dissociation of weak electrolyte, $\propto = \frac{\Lambda_c}{\Lambda_o}$ (Λ_c = equivalent conductance of the weak electrolyte at the given concentration; Λ_o = equivalent conductance at zero infinite dilution).
- Degree of dissociation of a weak electrolyte (like weak acids and weak bases can be determined from Ostwald's dilution law) is proportional to the square root of the dilution. e.g. $HA \rightleftharpoons H^+ + A^-$; $Ka = \frac{\propto^2}{(1-\propto)} = \frac{\propto^2}{(1-\propto)}$; $\propto = \sqrt{\frac{K}{C}}$
- For polybasic acids, e.g.

  $H_3PO_4 \rightleftharpoons H^+ + H_2PO_4^-$ K_1

  $H_2PO_4^- \rightleftharpoons H^+ + HPO_4^{2-}$ K_2

  $HPO_4^{2-} \rightleftharpoons H^+ + HPO_4^{3-}$ K_3
- The order of deprotonation is, $k_1 > k_2 > k_3$ as the proton comes from a neutral molecule in the first case while it is from the negative species (very difficult) in the second and third instances.
- Faraday's first law : The amount of any substance that is liberated or deposited at an electrode is directly proportional to the quantity of electricity passed through

the electrolyte.

$w \propto Q$	w = wt. of the ions liberated
w = Itz (since Q=It)	Q = quantity of electricity in coulombs
$z = \frac{M}{nF}$	I = strength of current in ampheres
$w = \frac{M}{nF} \times It$	z = proportionality constant or electrochemical equivalent
	M = molar mass

- Electrochemical equivalent of an ion is the weight of the ion deposited by one amphere current in one second.
- One Faraday of electricity (96,485 coulombs) will deposit one gram equivalent of the metal.
- Faraday's second law : When the same amount of electricity flows through different electrolytes, the amounts of different substances produced at the electrodes are directly proportional to their equivalent weights. $W \propto E$. $\frac{w_1}{E_1} = \frac{w_2}{E_2}$ or $\frac{w_1}{w_2} = \frac{E_1}{E_2}$
- The amount of charge carried by one gram equivalent of any ion is one Faraday i.e. 96,485 coulombs.
- In electrolytic cell, electrical energy causes chemical reactions to occur i.e. electrical energy is converted into chemical energy. In this cell cathode is assigned negative polarity as there is surplus of electrons at the cathode.
- In galvanic or voltaic cell, the chemical energy is converted into electrical energy. This cell is used in lightning of a torch light bulb, in hearing aids, car lights, transistor radios, camera flash lights etc. The cathode is assigned positive polarity due to the fact that the electrons enter the cell at this electrode.
- Salt bridge minimises the fluctuation of e.m.f. of the cell. It keeps the solutions in the two half cells electrically neutral.
- If the electrolyte solution contains Ag^+, Hg_2^{2+}, Ti ions, KCl can't be used in salt bridge (precipitaion of AgCl, Hg_2Cl_2 & TiCl take place.
- When an electrode is coupled with S H E, if the electrode reaction involves reduction i.e. positive ions take electrons from the electrode and deposit on it as atoms, electrode potential is given a posive sign. If oxidation occurs at the electrode, a negative sign is assigned.
- Standard electrode potential (E^o) is the potential obtained when the concentration (or the activity) of the ions in solution is unity and the temperature is 25°C.
- Potential of SHE is arbitrarily assigned as zero.
- Oxidation occurs at electrode with lower SRP (standard reduction potential)
- Reduction occurs at electrode with higher SRP.
- Metals with low reduction potentials are extracted by electrolytic method, e.g. Na, Al, Li, Mg etc.
- EMF is the potential difference between two electrodes or half cells.
- E^o_{cell} (EMF) = $E^o_R - E^o_L = E^0_{cathode} - E^o_{anode}$
- EMF is determined by potentiometers.
- Nernst equation : It gives the potential of any electrode or of a reversible cell, $E = E^o - \frac{0.0591}{n} \log_{10} K_C$ (At equilibrium, E = 0).
- Standard free energy change EMF are related by $\Delta G^o = -nFE^o_{cell}$.
- Maximum work obtainable from the cell, $w = -nFE^o_{cell}$.
- Primary cells (e.g. dry cell or leblanche cell) do not have indefinite life.
- Secondary cells (e.g. lead storage battery, Ni-Cd battery etc.) can be recharged.
- Compact source of power is called a fuel cell, e.g. hydrogen - oxygen fuel cell; pt electrode - conc. H_3PO_4; Hydrocarbons -air fuel etc.
- Rust is $Fe(OH)_3 + Fe_2O_3$
- Corrosion is accelerated by the presence of electrolytes.
- Process of coating iron with zinc is known as galvanisation.
- Prevention of corrosion and decorative property can be achieved simultaneously by metallic coating such as Sn, Ni, Cr etc.
- Galvanised iron containers can't be used for canning food as zinc will pass into the solution thereby poising it.
- O_3, H_2O_2, SO_2 etc.act both as oxidising and reducing agents.
- Equivalent weight of oxidising agent = $\frac{\text{mol. wt. of the oxidising agent}}{\text{number of electrons gained by 1 mole of it}}$
- Eq. wt. of reducing agent = $\frac{\text{mol. wt. of the reducing agent}}{\text{number of electrons lost by 1 mole of it}}$

11. Colloids

- Colloidal system is a heterogeneous system
- Blood, milk, food, rubber, wood etc. are colloidal in nature.
- Colloids have the particle size 10^{-6} to 10^{-9} m.
- Colloidal solution of graphite is called Aqua dag.
- Colloidal solution of gold is prepared by Bredig's arc method (a dispersion method).
- Colloidal solution of gold is purple of cassius.
- Dialysis is a method used to separate a crystalloid from

a colloid by diffusion through a parchment membrane.

- Colloidal solutions are purified by dialysis, ultrafiltration and ultracentrifuging.
- Colloidal solutions show colligative properties.
- Tyndall effect due to scattering of light from the surface of the colloidal particles.
- The projection beam of a cinema projector is visible due to Tyndall effect.
- Ultramicroscope is built on the principle of Tyndall effect.
- Tail of comet, twinkling of stars, blue colour of sky and sea water etc. are due to Tyndall effect.
- Brownian moment is due to the bombardment of colloidal particles by molecules of dispersion medium which are in constant motion.
- The migration of colloidal particles under the influence of an electric field is termed as cataphoresis or electrophoresis.
- The moment of dispersion medium under the influence of electric field when the dispersed particles are prevented from moving is known as electro-osmosis.
- The difference of potential between the fixed and diffused layers of colloidal sols is known as electrokinetic or zeta potential.
- Isoelectric point : At a particular H^+ ion concentration, the colloidal particles become neutral and exhibit no moment in an electric field. It is referred to as isoelectric point of the sol.
- The phenomenon of change of colloidal state to suspended state is known as coagulation or flocculation.
- Hardy-Schulze rule : greater the valency of the flocculating ion added, greater is its power to cause precipitation.
- Flocculation power : The order of coagulation of + vely charged sol.(e.g. $Fe(OH)_3$, $Al(OH)_3$, $Cr(OH)_3$, methylene blue sol etc.) is $[Fe(CN)_6]^{4-} > [PO_4]^{3-} > [SO_4]^{2} > Cl^-$. Coagulation of -vely charged sol (e.g. AS_2S_3, Sb_2S_3, Ag, Au, Cu etc.) is $Al^{3+} > Ba^{2+} > Na^+$.
- Flocculation value : The minimum concentration of an electrolyte required to cause coagulation or flocculation of a sol is called flocculation value. The order of flocculation value is inversely related to the flocculation power.
- Gold number : It was introduced by Zsigmondy. It is the weight in milligrams of a protective colloid which prevents the coagulation of a given gold sol on adding 1ml. of a 10% solution of NaCl. Gold number represents the protective power of colloids.
- Smaller, the gold number of a lyophilic colloid, greater is its protective power,
- Artificial rain can be caused by coagulating the mist in the air by spreading the electrified sand.
- Milk, vanishing cream etc. are emulsions of oil in water.
- Butter, cold cream etc. are emulsions of water in oil (or fat).
- Colloidal system containing a liquid dispersed in solid is called a gel. e.g. curd, table gelly, soaps, boot polish etc. have gel structure.
- On standing a gel may shrink and loose the liquid held by it. It is called Syneresis or weeping.
- The phenomenon of gels that liquify on shaking and reset when allowed to stand is known as thixotropy.
- Curd is a gel of water in casein.
- The phenomenon of higher concentration of any molecular species at the surface than in the bulk of a solid (or a liquid) is called adsorption.
- Adsorption of gases e.g. H_2 on the surface of finely divided metals e.g. Ni, Pt, Pd etc. is called occlusion.
- In physiosorption, nultimolecular layer forms while in chemisorption, a unimolecular layer is formed.
- In physiosorption, only van der Waal's forces are involved.
- A process in which a substance shows absorption and adsorption is called sorption, e.g. H_2/charcoal.
- The decrease of temperature and increase of pressure both tend to cause increase in the magnitude of adsorption of a gas on a solid.
- Rate of adsorption which is higher initially decreases till the equilibrium is obtained.
- Rate of absorption remains the same through out the process.
- Chemisorption is irreversible.
- In chemisorption a stable surface compound frequently reffered to as surface complex forms.
- Adsorption isobar represents the effect of temperature on the extent of adsorption.
- Adsorption isostere shows the relation of variation of pressure with temperature.
- Freundlich adsorption isotherm : It gives an empirical relationship between the quantity of gas adsorbed by a given amount of solid adsorbent surface and pressure at a particular temperature. It is expressed by the equation, $\frac{x}{m} = k.p^{1/n}$ or $\log\frac{x}{m} = \log k + \frac{1}{n}\log p$ or $\log\frac{x}{m} = \log k + \frac{1}{n}\log C$ (x is the weight of gas adsorbed by m grams of the adsorbent at a pressure p; k and n are constants at a

particular temperature and for a particular adsorbent and gas).

- The substances whose molecules aggregate spontaneously in a given solvent yielding stable particles of colloidal dimensions are called micelles.
- Micelles have both lyophilic and lyophobic groups.
- When a cloth with spots of dirt is soaked with soap solution, soap dissolves dirt by forming micelles.
- Sodium oleate, sodium stearate cetyltrimethyl ammonium bromide etc. form micelles.

12. Catalysis

- Substance which increases the rate of a chemical reaction without it permanently changed is called a catalyst.
- Negative catalysis decreases the velocity of a chemical reaction, e.g. addition of $(C_2H_5)_4$ Pb to petrol; antifreeze like glycol retards rusting of the machines; addition of H_3PO_4 retards the decomposition of H_2O_2. Addition of 1% C_2H_5OH to chloroform to prevent its oxidation to phosgene etc.
- The reaction, $KMnO_4 + C_2H_2O_4$ + dil. H_2SO_4 is slow initially but becomes brisk later because $MnSO_4$ (Mn^{2+} ions) formed accelerates the reaction.It is known as autocatalysis.
- Mo acts as a promoter of the activity of the catalyst (Fe of FeO) in Habers process for NH_3 synthesis.
- In homogeneous catalysis, the catalyst and the reactants are in the same phase.
- Only a small quantity of the catalyst is needed i.e. 10^{-6} to 10^{-8} g/l.
- The catalyst does not initiate the reaction.
- The catalyst can't alter the nature of the ractions.

  Exceptions :In the following reactions, the catalyst is specific and different products are formed.

e.g. (1) Toluene ($C_6H_5CH_3$) + Cl_2 $\xrightarrow{\text{catalyst}}$ o-chlorotoluene (CH_3, Cl on ring) and or p-chlorotoluene (CH_3, Cl on ring)

Toluene ($C_6H_5CH_3$) + Cl_2 $\xrightarrow{\text{sunlight}}$ $C_6H_5CH_2Cl$

(2)
$$CO + 3H_2 \xrightarrow{Ni} CH_4 + H_2O$$
$$CO + 2H_2 \xrightarrow{ZnO + Cr_2O_3} CH_3OH$$
$$CO + H_2 \xrightarrow{Cu} HCHO$$

- Co- enzyme acts as a promoter for catalyst.
- Enzyme is essentially a biological catalyst.
- Heterogeneous catalysis can be explained by adsorption theory.
- Wilkinson catalyst $[(Ph)_3P]_2$ RhCl selectively ctalyse non hindered double bonds.

Part III : Organic Chemistry

1. Classification, Nomenclature and purification of organic compounds

- Those compounds which follow Huckel's (4n +2) - π electron rule, are aromatic compounds.
- Furan, Pyrrole & Thophene are aromatic in nature as the lone pair of electrons of heteroatom in conjugation with the π- electrons take part in the formation of delocalized π-orbital of six electrons.
- Tropolone, Azulene are non-benzenoid aromatic compounds.
- pentadienyl anion ⊖ is aromatic in nature.
- If a molecule contains both C-double and triple bonds, the two are treated at par in seeking the lowest number of combination. If the sum of the numbers turns out to be the same, the lowest number is given to C=C bond.
  e.g. (a) $\underset{5}{H_3C} - \underset{4}{CH} = \underset{3}{CH} - \underset{2}{C} \equiv \underset{1}{CH}$ pent -3-en-1-yne
  (b) $\underset{5}{H_3C} - \underset{4}{C} \equiv \underset{3}{C} - \underset{2}{CH} = \underset{1}{CH_2}$ pent-1-en-3-yne
  (c) $\underset{5}{HC} \equiv \underset{4}{C} - \underset{3}{CH_2} - \underset{2}{CH} = \underset{1}{CH_2}$ pent-1-en-4-yne
- When more than one functional group is present, the rule of principal group is applied. Order of preference: Acids, acid derivatives, aldehydes, nitriles, ketones, alcohols, amines, ethers, double bond and triple bond.
- When the preference is not the highest, –OH is named as hydroxy, >C = O as oxo, –CHO as formyl.
- In alkoxyalkanes, smaller alkyl group is taken with oxygen as alkoxy, e.g. $H_3C–O–CH_2–CH_3$– methoxy ethane.
- Any two successive members of a homologus series differ by $–CH_2$.
- Soxhlet extraction is used to extract fats and oils from the seeds.
- Quick cooling gives tiny but purer crystals.
- Sublimation : Substances pass from solid to vapour state on heating and return to solid state on cooling without melting. The method is used to purify naphthalene, anthracene, benzoic acid, camphor, Iodine, NH_4Cl etc.
- Fractional distillation is used in petroleum refining.
- Purity of organic compounds can be determined by melting point.
- Steam distillation is used for purification of voltile liquids. e.g. Aniline, nitrobenzene, essential oils etc.
- Activated charcoal is employed to remove colouring matter from pure substances by adsorption.
- Mixed melting point can be used as criteria for determining the purity of organic solids.
- Distillation under reduced pressure is used to purify liquids which decompose below their boiling point, e.g. glycerol etc.
- Adsorption chromatography can be used to separate dyes, pigments, chlorophyll etc.

2. Empirical and molecular formulae of organic compounds

- Empirical formula : Simplest ratio of the constituent elements of a molecule
- Molecular formula : The actual number of the atoms of the constituent elements in a molecule.
- Molecular formula = empirical formula x n (n = molecular formula wt./emp. formula wt.)
- Lassaigne's test can be used for the detection of nitrogen, sulphur, halogens etc.
- Nitrogen containing compounds in Lassaine test give blue coloured solution or precipitate due to the formation of ferric ferrocyamide $Fe_4[Fe(CN)_6]_3$.
- In Lassignes test for sulphur containing compounds, the violet coloured solution or precipitate is due to the formation of sodium thionitroprusside, $Na_3[Fe(O = N–S–Na)(CN)_5]$
- If a compound contains both N and S, it gives blood red coloured solution or precipitate in Lassaignes test due to the formation of $[Fe(CNS)_3]$.
- Nitrogen present in diazo compounds can not be detected by Lassaigne's test.
- Beilstein test : It is used to detect halogens. But urea, thiourea, pyridine etc. respond positively to this test.
-

Element	Method of estimation
C and H	Liebigs method
N	Dumas or kjeldahl
Halogens	Carius or Stepanov's method
Sulphur	Messenger; Pirria & Schiff and Asboth's methods

- % estimation of different elements :

$$\% \text{ of H} = \frac{2}{18} \times \frac{\text{weight of } H_2O \text{ formed}}{\text{weight of organic compound}} \times 100$$

$$\% \text{ of C} = \frac{12}{44} \times \frac{\text{wt. of } CO_2 \text{ formed}}{\text{wt. of org. compound}} \times 100$$

$$\% \text{ of } N_2 \text{ (Dumas)} = \frac{28}{22400} \times \frac{\text{vol. of } N_2 \text{ at NTP}}{\text{wt. of org. compound}} \times 100$$

$$\% \text{ of N (kjeldahl)} = \frac{1.4 \times \text{vol. of acid} \times \text{normality of acid}}{\text{wt. of organic compound}}$$

$$\% \text{ of Cl} = \frac{35.5}{143.5} \times \frac{\text{wt. of AgCl formed}}{\text{wt. of org. compound}} \times 100$$

$$\% \text{ of Br} = \frac{80}{188} \times \frac{\text{wt. of AgBr formed}}{\text{wt. of org. compound}} \times 100$$

$$\% \text{ of I} = \frac{127}{235} \times \frac{\text{wt. of AgI formed}}{\text{wt. of org. compound}} \times 100$$

$$\% \text{ of S} = \frac{32}{233} \times \frac{\text{wt. of } BaSO_4 \text{ formed}}{\text{wt. of org. compound}} \times 100$$

- % of oxygen = 100 - Sum of the pecentages of all other elements
- Number of atoms of each element present in an org. compound:

$$H = \frac{\% \text{ of H}}{1};\ C = \frac{\% \text{ of C}}{12};\ N = \frac{\% \text{ of N}}{14};\ O = \frac{\% \text{ of O}}{16} \text{ etc.}$$

- Some relationships to determine the molecular weights

(i) Vapour density method :

molecular weight = 2 x vapour density

$$\text{Vapour density} = \frac{\text{wt. of x c. c. of vapour at NTP}}{\text{wt. of x c.c. of hydrogen at NTP}}$$

(one C.C. of hydrogen at NTP weighs 0.00009)

(ii) Silver salt method for acids :

$$\frac{\text{Equivalent weight of Silver salt}}{\text{Equivalent weight of Silver}} = \frac{\text{wt. of silver salt}}{\text{wt. of silver}}$$

$$\frac{E - 1 + 108}{108} = \frac{\text{wt. of silver salt}}{\text{wt. of silver}}$$

(iii) volumetric methods for acids :

(a) 1 lit of 1N solution of a substance contains 1g. equivalent of the substance.

(b) weight of 22,400 C.C. vapour of any substance at NTP is equal to its molecular weight.

3. Theoretical organic chemistry.

- In inductive effect only displacement of electrons takes place (polarisation of σ bonds). Permanent effect (shown even non reacting molecular).
- Electron withdrawing or attracting groups show –I effect. Order of –I effect: $-NO_2 > -CN > -F > -Cl > -I > -OH > -OCH_3 > -C_6H_5 > -H$ etc.
- Electron pushing or repelling groups show +I effect. Order of +I effect: $(CH_3)_3C- > (CH_3)_2CH- > CH_3 - CH_2- > -CH_3 > H-$
- Bond length decreases with increasing –I effect.
- More the inductive effect, more is the dipole moment.
- –I effect showing groups or atoms when attached will increases the acidic strength of carboxylic acids. e.g. $ClCH_2COOH > HCOOH > CH_3COOH > CH_3CH_2COOH$; $CH_3CH_2CHClCOOH > CH_3CHClCH_2COOH > ClCH_2CH_2CH_2COOH$
- +I effect groups or atoms increase the basicity of amines Basicity of amines : $2° > 1° > 3° > NH_3$ (This disturbance in the order is probably due to steric effect or solvation effects).
- +I effect, increases the reactivity of alkyl halides.
- Electromeric effect involves the complete transfer of π or lone pair of electrons. Temporary effect.
- +E effect is shown by electron attracting atoms or groups, e.g. F, Cl, Br, O, –OR etc.
- –E effect is shown by electron repelling atoms or groups, e.g. $> C = O$, $= NR$, –S etc.
- Mesomeric effect involves complete transfer of π or lone pair of electrons.
- It is a permanent effect. It is a conjugative effect which takes place in conjugated system via their π – orbitals.
- In compounds with cumulative double bonds, mesomeric effect is not possible, e.g. $H_2C = C = C = CH_2$
- Resonance includes conjugative effect and also all types of electronic displacements, e.g. $H - H \leftrightarrow H^+ - H^- \leftrightarrow H^- - H^+$ etc.
- When inductive and mesomeric effects oppose each other, mesomeric effect predominates (or overpowers).
- Resonance structures differ only in the distribution of π–electrons.
- Greater the resonance energy, higher is the stability of the molecule.
- Resonance increases the stability of the molecule/ion.
- $\text{Bond order} = \frac{\text{Total number of bonds between two atoms}}{\text{Total number of resonating structures}}$
- Conjugation of σ –electrons of a single bond with multiple bonds is observed in hyperconjugation (or Baker – Nathan effect).
- Hyperconjugation is called no bond resonance.
- Propene is more stable than ethene due to hyperconjugation.
- Free radicals are formed due to homolytic fission of a covalent bond.
- Free radical intermediates are formed in Wurtz reaction, anti Markownikoff's addition, substitution reactions of alkanes (nitration, halogenation, sulphonation) etc.
- Free radicals are neutral, odd electron species. They are detected by ESR (electron spin resonance) or EPR (electron peramagnetic resonance).
- Stability of alkyl free radicals is $3° > 2° > 1°$ (the stability can be explained on the basis of (a) +I effect of alkyl groups and (b) hyperconjugation).
- The odd electron containing carbon in free radicals is sp^2 hybridised.
- Carbocation (Carbonium ion) and Carbanion are

formed by heterolytic fission of the covalent bond.

- Order of stability of carbonium ions : (a) benzyl > allyl > 3° > 2° > 1° (b) $H_2C= CH -CH_2^+ > CH_3 - CH_2 - CH_2^+$ (c) $(Ph)_3C^+ > (Ph)_2 CH^+ > Ph -CH_2^+$
- Carbonium ions (carbocation) are formed in reactions like Friedel–Craft, Pinacol–Pinacolone, dehalogenation of haloalkanes, dehydration of akcohols, S_N1 of haloalkanes, diazonium salts etc.
- Benzyl carbocation is stabilized by tropylium rearrangement with the formation of tropylium carbonium ion.
- H_3C^+ (carbonium ion) is sp^2 hybridised while CH_3^- (carbanion) is sp^3 hybridised.
- +I effect increases the stability of carbonium ions while –I effect increases the stability of carbanions.
- Carbanion intermediates are formed in Aldol condensation, Cannizzaro, perkin, Knovenegel etc. reactions.
- The carbene carbon atom is sp hybridised, linear in the case of the triplet with electronic configuration , $1s^2, 2sp^1, 2sp^1, 2P_y^1, 2P_z^1$
- The carbene carbon in the case of the singlet is trigonal, sp^2 hybridised with $1s^2, 2(sp^2)^2, 2(sp^2)^1, 2(sp^2)^1\ 2Pz^o$.
- N-atom of nitrene, $\ddot{N}$: has one bond σ and two lone pairs of electrons. Nitrenes are neutral.
- Electrophiles can be neutral (e.g. BF_3, $AlCl_3$, $FeCl_3$, SO_3, $:CCl_2$) or cationic (e.g. R^+, RCO^+, $^+NO_2$, Cl^+).
- Nucleophiles can be neutral (e.g. H_2O, R–OH, R–O–R, NH_3, amines etc.) or anionic (e.g. Cl^-, OH^-, $^-OCH_3$, R^-)
- H_3O^+, R_4N^+ etc. can not act either as nucleophiles or electrophiles (as they have completed octet, they can't accept electron pair and as they are positively charged, they can't donate electron pair for the formation of a covalent bond).
- Saturated molecules without a heteroatom show substitution reactions.
- Unsaturated molecules and molecules with a heteroatom e.g. O, N, S etc. having lone pairs of electrons show addition reactions.
- Nucleophilic addition reactions shown by >C = O group are not shown by – COOH group and its derivatives due to resonance hybrid $R - C\begin{smallmatrix} O\ \delta- \\ O\ \delta+ \end{smallmatrix}$
- S_N1 reactions take place in two steps while S_N2 takes place in one step.
- S_N1 reactions lead to the formation of carbocation (or carbonium ion) while trnsition structure forms in S_N2 reactions.
- The order of basicity of nucleophiles is $CN^- > I^- > OH^- > Br^- > CH_3COO^- > Cl^-$.
- In case of S_N1 reactions, the rate of reacton is not normally effected by the concentration of the nucleophile as nucleophile doesnt take part in the rate determining first step.
- In the case of S_N2, more the concentration of the nucleophile, greater is the rate.
- In S_N1 reactions, optically active substrates form racemic mixture, while in S_N2 reactions, they form opposite enantiomorph due to inversion of the configuration.
- Polar conditions favour S_N1 reactions as they are favourable for the formation of Carbonium ions. Non polar conditions prefer S_N2 reactions.
- S_E1 takes place in two steps (Carbanion forms) while S_E2 takes place in one step.
- Nitration, halogenation, sulphonation, alkylation, acetylation, halogenation of aldehydes, ketones, carboxylic acids (Hell-Volhard-Zelinsky reaction) etc. are electrophilic substitution reactions.
- Organic compounds with carbon atoms ≥ 4 show chain isomerism. In alkynes, n ≥ 5.
- n-Butane and isobutane; n-pentane, isopentane and neopentane are chain isomers.
- C_6H_{14} exists as 5 chain isomers.
- $CH_3CH_2CH = CH_2$ and $CH_3 - CH = CH - CH_3$ are position isomers but $CH_3 - CH_2 - CH = CH_2$ and $CH_3 - C(CH_3) = CH_2$ are chain isomers.
- o, m and p- xylenes; cresols are position isomers.
- Propene & cyclopropane are chain -ring isomers (in reality functional isomers).
- For molecular formula of the type $C_nH_{2n+2}O$, alcohols and ethers are possible as isomers.
- For $C_nH_{2n}O$ aldehydes, ketones and unsaturated alcohols are the possible isomers.
- $C_nH_{2n}O_2$, acids and esters are possible isomers.
- Metamers have the same functional group but the arrangement of alkyl groups is different.
- For $C_4H_{10}O$, 7 isomers (2 primary alcohols one secondary and tertiary alcohol each and 3 ethers) are possible.
- Tautomerism is also known as cationotropy or desmotropism.
- Carbonyl compounds with α –H atoms show tautomerism.
- Nitrous acid, acetoacetic acid, HCN, acetylacetone, .

benzoylacetone etc. show tautomerism.

- At room temperature, acetoacetic ester is a mixture of 92.5% keto and 75% enol form.
- In tautomerism, a single substance behaves like two different compounds.
- Conformation is due to the free rotation of C-C single bond. Ethane, butane etc. show conformation.
- Stability : Staggered > Skew > Eclipsed (In eclipsed form, the atoms attached to carbon atoms are nearer to each other compared to other forms, thus leading to more repulsion).
- The structures that lie between eclipsed and staggered forms are known as skew forms.
- Conformers or rotational isomers or rotomers can't be separated as the energy barrier is very less i.e. 12.6 kJ/mole.
- Cyclohexane exhibits conformation. Chair form is more stable than boat form.
- Maleic acid - Fumaric acid, but-2-ene, cinnamic acid cyclopropane-dicarboxylic acid, benzaldoxime, azobenzene etc. show geometrical isomers.
- Benzaldoximes and ketoximes exhibit geometrical isomerism of syn and ant form.
- Trans form has zero dipole moment generally. Trans form is more stable and has higher m.p. Trans form has higher λ_{max} and E_{max} values.
- Geometrical isomers differ in physical and chemical properties.
- For a molecule to show optical activity, it shoul be chiral.
- Number of geometrical isomers, $N = 2^n$ (n = number of double bonds with different substituents at each end). When the ends of the alkene have the same substituents, the number of geometrical isomers, $N = 2^{n-1} + 2^{p-1}$, p is determined as : p = n/2 when the number of double bonds is even while p = n+1/2 when 'n' is odd.
- A chiral molecule does not posses plane of symmetry, an alternating axis of symmetry and centre of symmetry.
- All asymmetric molecules are dissymmetric but all dissymmetric molecules need not be asymmetric.
- Dissymmetric molecules are not superimposable on their mirror images (they may posses axis of symmetry).
- $KClO_3$, $KBrO_3$ and KIO_4 are optically acitve due to their crystalline structure.
- For molecules having 'n' nimber of chiral carbon atoms which can't be divided into two equal parts, number of active isomers, a= 2^n; no. of meso isomers, m = 0; no. of racemic mixures, r = a/2.

  If 'n' is odd and the molecule can be divided into two equal halves, $a = 2^{n-1} - 2^{0.5n-0.5}$; $m = 2^{0.5n-0.5}$.

  If 'n' is even and the molecule can bedivided into two equal parts $a = 2^{n-1}$; $m = 2^{0.5n-1}$; r = a/2.
- d and l are notations of rotation.
- Racemic mixture is optically inactive due to intermolecular or internal compensation (50% d and 50% l molecules are present in racemic mixture).
- Meso compounds are optically inactive due to intramolecular compensation or due to the presence of plane of symmetry.
- Chiral molecules which do not posses chiral carbon atoms i.e. allenes, spiranes, biphenyls etc.also exhibit optical isomerism.
- Stereo isomers which are not mirror reflections of each other are diastereomers e.g. d and meso or l and meso tartaric acids; geometrical isomers etc.
- In racemisation, 50% of the starting active enantiomer is converted into the other enantiomer thus leading to the formation of racemic mixture.
- Enantiomers differ in rotation and also in biological properties.
- In inversion e.g.Walden inversion which involves S_N2 mechanism, the active starting enantiomorph is converted to the other active enantiomorph.
- D and L are notations used to designate relative configuration while R and S are used to represent absolute configuration.

- **4. Hydrocarbons :**

- The general formulae of alkanes, alkenes and alkynes are respectively, $C_n H_{2n+2}$, $C_n H_{2n}$.
- Cycloalkanes are represented by general formula, C_nH_{2n}.
- $C_{20}H_{42}$ is called Eisocane while $C_{30}H_{62}$ is Triacontane.
- In Wurtz reaction, the reactive intermediates are free radicals or organometallic compounds.
- Wurtz reaction fails with tertiary alkyl halides.
- Methane can't be prepared by Wurtz reaction.
- Alkanes with odd number of carbon atoms can't be prepared in pure form by Wurtz reaction.
- In Frankland's reaction Zn is used instead of Na in Wurtz reaction.
- Disproportionation of $\dot{C}_2H_5$ radical leads to the formation of ethane and ethene.
- Alkanes are also prepared by Core-House reaction,

$R-X + R_2CuLi \rightarrow R - R + Cu - R + LiX$

- In Kolbes electrolytic reaction, sodium or potassium salt of fatty acids lead to the formation of alkanes. Free radicals and ions are reactive intermediates.
- The carbonyl group can be converted to $-CH_2$ group by clemmensen's or Wolff-Kishner reduction.
- The carbonyl compounds sensitive to bases (if they have α–hydrogen, they may undergo Aldol condensation) are converted to alkanes by Clemmesen's reduction while those sensitive to acids are converted by Wolff-Kishner reduction.
- Hydrogenation of alkenes or alkynes to alkanes in the presence Ni or Pt or Pd catalyst is called Sabatier and Senderens reaction.
- Grignard reagent with compounds having active hydrogen produces alkanes.
- Straight chain alkanes have higher boiling point than their isomeric ones as more space is available for van der Waal's forces to act in the former.
- Alkanes with even number of carbon atoms have higher m.p, than those containing odd number of carbon atoms.
- Alkanes don't show ionic reactions gererally as they are non polar.
- Alkanes show mainly substitution reactions. Free radicals are the reactive intermediates.
- The reaction of alkanes with I_2 (e.g. $CH_4 + I_2 \rightleftharpoons CH_3I + HI$) is a reversible reaction.
- Order of ease of substitution : 3°C–H>2°C–H>1°C–H
- n-hexane, n-heptane & n-octane are mainly converted to benzene toulene and xylene respectively in the presence of Al_2O_3 or Cr_2O_3 or MoO_3 as catalyst.
- Reed's reaction is employed for industrial preparation of alkyl sulphonates used as detergents.
- Paraffins containing 6 or more carbon atoms can be sulphonated with oleum $H_2S_2O_7$ ($SO_3 + H_2SO_4$) at 400°C.
- Carbon black for printing purpose is prepared from methane.
- The chief constituent of natural gas and gobar gas is methane.
- Marsh gas is mainly methane.
- Paraffin wax contains mainly C_{25} upward alkanes.
- Bee wax is an ester, mainly mericyl palmitate.
- Cycloalkanes are also known as polymethylenes.
- Cycloalkanes are prepared by Freund's reaction also known as intramolecular Wurtz reaction.
- The dehydration of alcohols to alkenes proceeds via carbonium ion formation.
- The dehydration is possible only if there is a β–hydrogen
- The rate ease of dehydration of alcohols is 3°>2°>1°
- Dehydrohalogenation of alkyl halides takes place in the presence of alcoholic KOH to produce alkenes.
- Saytzeff's rule : When two alkenes may be formed by elimination (dehydrohalogenation), the one which is more substituted predominates.
- Stability of alkenes :$R_2C = CR_2 > R_2C = CHR > R_2C = CH_2$, $RHC = CHR > RHC = CH_2 > H_2C = CH_2$.
- Less substituted alkene is produced in Hoffmann elimination reacton of Quaternary ammonium salt.
- Rate of dehydrohalogenation : 3° >2° >1°. Tertiary halides follow E_1 while primary halides follow E_2 mechanism. E_1 proceeds via carbocation while E_2 is a concerted reaction.
- Alk-1-enes do not show geometrical isomerism.
- Sodium or potassium salt of dicarboxylic acids on electrolysis (Kolbe's method) give alkenes.
- Controlled reduction of alkynes with $NaNH_2$/Na in NH_3 gives trans alkene.
- Electrophilic addition of halogen to alkenes preceeds via cyclic halonium ion intermediate.
- Addition of halogens on alkene is a trans addition. Halogenation of cis alkene leads to the formation of racemic mixture while trans gives meso compound.
- Addition of halogen acids to asymmetric alkenes follows the Markownikoff's addition meechanism with carbonium ion as reactive intermediates leading to the formation of 2-haloalkanes.
- Electrophilic addition to, $C = C$ bond is much faster than to $C \equiv C$ bond (as the sp hybridised C – atom in the later is more electronegative).
- $C \equiv C$ bond attracts nucleophiles more strongly than a $C = C$ bond.
- In the presence of peroxide, air etc. the HBr addition to unsymmetrical alkenes proceeds via kharasch effect (anti Morkownikoff's addition) with free radical inlermediates leading to the formation of 1-haloalkanes.
- Kharasch effect is not shown by HF, HCl and HI as the bond in the former two will not undergo homolytic fission to give free radicals due to high bond energy. HI breaks into free radicals but due to its bigger size, $I^\bullet$ is not reactive to add on a double bond, but gives $I^\bullet + I^\bullet \rightarrow I_2$.
- Hydroxylation of alkenes is a syn or cis addition

- Alkenes give epoxides in the presence of peracids in Prileschaiev reaction.
- Epoxides produce trans diols on acid hydrolysis.
- Baeyer's reagent and OsO_4 bring about the cis-hydroxylation of the hydrolysis of th double bond.
- In ozonolysis, H_2O_2 formed is decomposed by the addition of Zn. Otherwise H_2O_2 formed oxidizes aldehydes to acids.
- NBS (N-bromosuccinimide) is a specific reagent for allyl bromination which takes place via free radical mechanism.
- Ethene with sulphur monochloride forms mustard gas which was used as a poisonous gas in the world war.
- Conjugated dienes are more stable due to resonance.
- Conjugated diene (e.g. 1,3-butadiene) with HBr gives 1, 2 addition product at low temperature in non polar solvents while 1, 4-addition takes place in polar solvents at high temperature.
- Conjugated dienes with dienophiles give cyclic adducts in Diel's -Alder reaction.
- Benzyne intermediate participation in Diels - Alder can be seen in the reaction, 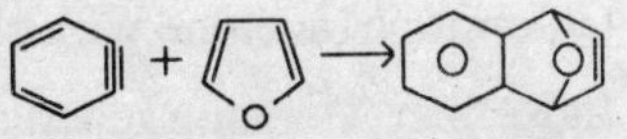
- Dehydrohalogenatin of gem or vicinal dihalides give alkynes
- For dehydrohalogenation of vinyl halide, a stronger base like $NaNH_2$ or KNO_3 than alcoholic KOH is required due to resonance stabilisation.
- Dehydrohalogenation of haloforms in the presence of Ag powder gives alkynes
- Chloroform on heating with Ag powder forms ethyne.
- Hydrolysis of CaC_2 gives ethyne.
- In terminal alkynes, the hydrogen atom attached to the carbon atom bound by triple bond is acidic in nature.
- 1-Alkynes on oxidative hydroboration yield aldehydes while ketones are produced by non terminal alkynes.
- Addition of borane to alkynes is a Syn-addition.
- Terminal alkynes have lower boiling points than their isomeric 2-alkynes etc.
- 1-alkynes can be distinguished from 2-alkynes with Tollen's reagent.
- Order of acidity : Ethyne > Ethene > Ethane
- Reduction of acetylenes in the presence of Lindlar's catalyst ($Pd/CaCO_3$) or liquid NH_3 + Na (or $NaNH_2$) gives trans or E alkenes as major products. Cis or Z alkenes are the main products of hydrogenation in presence of Pt or diborane.
- Ethyne gives acetaldehyde (formed by tautomerism from vinyl alcohol) in kucherov reaction with dil. H_2SO_4 in presence of 1% $HgSO_4$ (or Hg^{2+} ions) while all other acetylenes give ketones.
- Ethyne on oxidation with alcoholic $KMnO_4$ gives oxalic acid while with acidified $KMnO_4$, formic acid is formed. With weak oxidising agent like SeO_2, glyoxal (α–dicarbonyl compound) is formed.
- Ethyne on chlorination in presence of $FeCl_3$ gives . Westron ($Cl_2HC–CHCl_2$, a toxic solvent for cellulose acetate) which on reaction with $Ca(OH)_2$ gives Westrosol ($ClHC = CCl_2$, a non-toxic solvent for dry cleaning).
- Ethyne reacts with $AsCl_3$ to give Lewisite (a poisonous war gas).
- The terminal acetylenes like ethyne give copper acetylide (reddish ppt.) and silver acetylide (white ppt) with ammonical Cu_2Cl_2 and $AgNO_3$ respectively. These reactions are used to differentiate alkynes from alkenes in one step.
- Ethyne forms benzene on passing through a hot Fe or Si tube by cyclisation.
- Phenol when heated with Zn powder is reduced to benzene.
- Benzene (C_nH_n) although has high degree of unsaturation preferably gives electrophilic substitution reactions. Addition reactions are possible only under drastic reaction conditions as the π–electrons are delocalised and therefore are not available for addition reactions.
- –OH, –OR, $–NH_2$, –NHR, $–NR_2$, –Cl, –Br, –I, $–CH_3$, $–C_2H_5$, $–C_6H_5$ etc. are ortho and para directing groups in electrophilic substitution.
- $–NO_2$, –COOH, SO_3H, –CN, –CHO etc. groups are meta directing groups.
- Cl atom on benzene ring is o– and p– directing as +T i.e.(+M) + (+E) of Cl is more powerful than –I.
- Nitronium ion ($^+NO_2$) is the electrophile in nitration of benzene (conc. HNO_3 + conc.H_2SO_4 is nitration mixture).
- Carbonium ion is the electrophile in Friedel-Craft's alkylatino (electrophilic substitution) while RCO^+ (acylinium ion) is the electrophile in Friedel-Craft acylation.
- Dry $AlCl_3$ is used as catalyst in Friedel-Craft reaction.
- Benzene on chlorination in presence of light gives Benzene hexa chloride (BHC or 666) γ-isomer of BHC with chlorine atoms at 1, 2 and 3 positions as axial is known as gammexane (or Lindane) and is used as a

powerfur insecticide (pesticide also).

- In Birch reduction, benzene is partially reduced to 1, 4-hexadiene in the presence of Na + NH_3 in C_2H_5OH.
- Benzene on ozonolysis gives glyoxal.
- Benzene gives phenol and maleic anhydride at 300°C and 500°C respactively in presence of V_2O_5 as catalyst.
- $-NH_2$, $-OH$, $-OCH_3$, $-NHCOCH_3$, $-CH_3$ etc. are activating groups of benzene ring while $-NO_2$, $-COOH$, $-CHO$, $-SO_3H$ etc are deactivating groups in electrophilic substitution.
- The yield of gasoline or petrol is increased by cracking in the refineries.
- Quality of gasoline is expressed in terms of octane number [octane number is the percentage of iso-octane present in a mixture of n-heptane (octane number is zero) and iso-octane (octane number 100)].
- Order of knocking : Straight chain paraffins > branched chain paraffins > olefins > naphthenes > aromatics.
- TEL (tetraethyl lead, $(C_2H_5)_4$ Pb is used as antiknocking agent.
- Regular gasoline commonly used in motor cars, scooters etc. has octane number 74.
- Aviation gasoline has octane number 100.
- Cetane number tells us the quality of diesel. n-hexadecane (cetane) has 100 cetane number while 1-methyl naphthalene has zerocetane number. Straight chain hydrocarbons in diesel constitute a better fuel.
- The flash point (the minimum temperature at which an oil gives or produces sufficient vapours to form an explosive mixture with air) of kerosine oil in India is 44°C while in UK it is 22.8°C.

5. Halogen derivatives

- In Hunsdieker's method, silver salt of fatty acids on bromination gives bromo alkane via free radical mechanism.
- In Darzen's reaction alcohols react with thionyl chloride to give chloroalkane in pure form via nucleophilic substitution.
- In Rydon's method, alcohol reacts with halogen in the presence of triphenyl phosphate to give haloalkanes.
- Alkyl iodides are prepared by halogen exchange (Conant-Finkelstein) R–Cl or R–Br + KI $\xrightarrow{\text{acetone}}$ RI + KCl or KBr.
- The solutions of iodo alkanes develop brown colour on exposure to light due to decomposition.
- Trihalomethanes are called as haloforms.
- Haloalkanes with sodium alkoxide form ethers via S_N2 mechanism (3° halides don't give this reaction).
- Haloalkanes give alkylcyanides (or nitriles) with alcoholic KCN while isocyanides (or isonitriles) are formed with AgCN.
- Alkyl halides react with KSH to give mercaptans.
- Ethyl alcohol on boiling with bleaching powder produces chloroform, $CHCl_3$.
- Chloroform on exposure to light and air forms a poisonous phosgene gas ($COCl_2$).
- 1% C_2H_5OH is added to chloroform to prevent the oxidation of $CHCl_3$ to $COCl_2$.
- Chloroform is purified by treating it with salicylic anhydride.
- Carbylamine reaction is regarded as a test for primary anines. Aniline also responds to this test. Isocyanide or isonitrile (carbylamine) is the product. $:CCl_2$ (carbene) is the reactive intermediate.
- Chloropicrin (O_2NCCl_3) was used as an insecticide and also as a war gas.
- Chloroform on hydrolysis gives formic acid. Hydrolysis with NaOH leads to the formation of sodium formate.
- Chloretone (acetone with chloroform) is a hypnotic.
- In Reimer-Tiemann reaction, chloroform reacts with phenol in presence of a base to produce o-hydroxy benzaldehyde (salicylaldehyde). $:CCl_2$, carbene is the reactive intermediate.
- Chloroform is used as anaesthetic with 30% ether.
- Chloroform used for anaesthetic purposes is tested with $AgNO_3$ to know for the presence of $COCl_2$.
- Iodoform test is shown by compounds containing CH_3CO–group or by the compounds in which this group is produced very easily e.g. CH_3CHO, $CH_3CHOHCOOH$, CH_3COCH_3, CH_2CH_3OH etc.
- Iodoform test is used to distinguised between CH_3OH and C_2H_5OH; Ph–CO–CH_3 and Ph–CO–Ph etc.
- Phenol with CCl_4 in presence of a base produces salicylic acid.
- CCl_4 is used as fire extinguisher under the name pyrene.
- Allyl iodide undergoes nucleophilic substitution by S_N1 due to stabilisation of the resulting allyl carbonium ion.
- Freon is used as refrigerant and propellant in aerosol sprays.
- Chlorex β, β – (dichloroethyl ether) is used to remove aromatic impurities from lubricant oils.
- α, α – dichloroethyl ether was used as a lachrymator

(tear gas) in world war.

- Halogenation of ethers initially substitute α – hydrogen(s).
- Aldehydes and ketones produce gem dihalides with PCl_5.
- Glycols with PCl_5 and alkenes with halogens give vicinal dihalides
- Gem dihalides are less reactive than vicinal dihalides and alkyl halides.
- Vicinal dihalides produce pure olefins in the presence of Zn or Zn-Cu couple in alcoholic solution.
- Vicinal dihalides on hydrolysis give glycols. Terminal gem dihalides produce aldehydes while non terminal gem dihalides give ketones.
- Side chain aromatic halides are formed from alkylbenzenes in presence of light & heat by free radical mechanism. Their reactions are similar to haloalkanes.
- Aryl halides (nuclear substitution compounds) are formed at low temp. and in the presence of a catalyst e.g. $SbCl_5$, $AlCl_3$, Fe, Iodine etc. as halogen carrier.
- In Sandmeyer's reaction aryl halides are formed from benzene diazonium chloride in the presence of the catalyst, Cu_2Cl_2 / HCl or Cu_2 Br_2 / HBr with free radicals as reactive intermediates.
- Aryl fluorides are obtained by Schiemann reaction.
- In Fittig reaction, two aryl halide molecules react with Na in dry ether to give diphenyl.
- In Wurtz-Fittig reaction, aryl halide and an alkyl halide react with Na in dry ether to form alkylbenzenes.
- During oxidation, the halogenation atom in nuclear derivatives remains intact while the one in the side chain compounds is lost.
- Nuclear or aryl halides do not show Friedel-Craft reaction.
- S_N^1 or S_N^2 reactions characteristic of haloalkanes are not shown by aryl halides.
- Aryl halides are less reactive than alkyl halides as the resonance stabilises the former.
- Allyl halides are more reactive than alkyl halides and vinyl halides.
- Nitration of chlorobenzene is slower than the nitration of benzene.
- Chloral condenses with chlorobenzene in the presence of conc. H_2SO_4 to form DDT (dichloro diphenyl trichloroethane), a strong insecticide.
- Toulene on chlorination in presence of light gives benzyl, benzal and benzo trichlorides.
- Benzyl, benzal and benzotrichloride can be distinguished by ceric ammonium nitrate (red colour), Tolles's reagent (silver mirror) and $NaHCO_3$ (effervescence) respectively.
- Organometallic compounds are those compounds in which there should exist a direct bond between carbon atom and the metal, e.g. RMg X (Grignard reagent) ferrocene, tetra ethyl lead, Zeisse's salt etc.
- In RMg X, the bond between alkyl group and Mg is polar covalent bond while the one between Mg and halogen is essentially ionic bond.
- R Mg X reacts with compounds containing active hydrogen to give alkane. This principle is used to estimate – OH, – NH_2, – SH etc.
- Benzene on treatment with HCl and O_2 in the presence of $CuCl_2$ produces chlorobenzene in Rasching process.

6. Alkanols

- Primary amines react with HNO_2 to give alcohols.
- Methylamine with HNO_2 forms methyl nitrite generally while high concentration of methylamine leads to the formation of dimethyl ether.
- Bouveault - Blanc reduction (excess of Na and C_2H_5OH) of aldehydes or ketones or esters results in the formation of alcohol (acid derivatives also form alcohols).
- Reduction of unsaturated aldehydes, ketones etc with $LiAlH_4$ give unsaturated alcohols. The double bond i.e. the unsaturation is not affected.
- Glucose (carbohydrates) on fermentation with yeast and Zymase enzyme at 30°C gives ethyl alcohol.
- CH_3OH is obtained by heating CH_4 and air (9:1) in copper tubes at 200°C and 100 atmos. pressure.
- Ethyl alcohol is oftem known as grain alcohol as grains are employed as the source.
- CH_3OH is known as wood spirit and is obtained by destructive distillation of wood.
- The order of boiling points among isomeric alcohols is P >S > T.
- Branching of alcohols increase the solubility in H_2O. Among butanols, tertiary butanol is more soluble in H_2O.
- In reactons involving the O – H bond cleavage, the order of reactivity is 1° >2° >3°.
- In C-O bond cleavage, the order of reactivity is 1° >2° >3°.
- Addition of halogen acids to primary and secondary alcohols proceeds by S_N^2 while with tertiary it is by S_N^1.

- Rate of esterification with reference to alcohols is 1° >2° >3°.
- Esterification is a reversible reaction and therefore H_2SO_4 a dehydrating agent is used to avoid the hydrolysis of the ester formed.
- Tertiary alcohols (vapours) when passed over reduced copper at 300°C instead of dehydrogenation gives alkenes by dehydration.
- Rate of dehydration is more in alcohol that can produce conjugated dienes than those that produce non conjugated dienes.
- Primary alcohols on oxidation gives aldehydes and then carboxylic acids having the same number of C-atoms by oxidation.
- Secondary alcohols give ketones with the same number of C-atoms but on further oxidation carboxylic acids with less number of C-atoms are produced.
- Splitting of ketonic chain during oxidatin takes place according to Popoff's rule.
- Jone's reagent (aqueous solution of H_2CrO_4) oxidises primary & secondary alcohols to aldehydes and ketones respectively without effecting the double bond in the alcohols.
- Aluminium tertiary butoxide is a catalyst used in oppe naur oxidation to convert, 1° an 2° alcohols into aldehydes and ketones.
- Tertiary alcohols having no α – hydrogen do not undergo oxidation.
- Lucas reagent is conc. HCl / $ZnCl_2$.
- The order of reactivity of alcohols with Lucas reagent is T >S > P.
- In Victor Meyer's test primary and secondary alcohols give blood red coloured nitrolic acid and blue coloured pseudonitrol respectively.
- Excess of ethyl alcohol with conc. H_2SO_4 at 140°C gives diethylether while alcohol in the presence of excess of conc H_2SO_4 at 160°C produces ethene.
- Moisture present in alcohol is removed with metallic Mg.
- Pyroligneous acid contains acetic acid, CH_3OH and CH_3COCH_3.
- Rectified spirit is $C_2H_5OH + H_2O$ (94.5 + 5.5%).
- Rectified spirit is converted to absolute alcohol by treating it with CaO and distilling over CaO followed by drying with anhydrons $CuSO_4$ and consequently distilling it.
- Alcohol forms an azeotropic mixture of petrol + absolute alcohol (4:1) .
- Ethene with Baeyer's reagent produces ethylene glycol.
- The yield of glycols are better by OsO_4 hydroxylation than MnO_4^- hydroxylation.
- Glycol undergoes dehydration in presence of H_3PO_4 or H_2SO_4 to give dioxan.
- Glycol is used as an antifreeze under the name prestone.
- Glucose with yeast in presence of Na_2SO_3 gives glycerol.
- Glycerol with HI gives allyl iodide while in presence of excess of HI, 2-Iodopropane is formed.
- Glycerol with oxalic acid at 110°C and 260°C gives formic acid and allyl alcohol respectively.
- Glycerol on heating alone or with dehydrating reagents i.e. P_2O_5 or $KHSO_4$ produces acraldehyde (or acrolein)
- Glycerol on periodic acid (HIO_4) oxidation produces two molecules of formaldehyde and a molecule of formic acid.
- Glycerol is oxidised by Fenton's reagent (H_2O_2 + $FeSO_4$) or bromine water to glyceraldehyde and dihydroxy acetone.
- Glycerol with conc. HNO_3 forms glyceryl trinitrate (an ester) which is the chief constituent of dynamite.
- Nobels oil (glyceryl trinitrate) is used to treat heart ailment (angina pectoris) and asthama.
- Cellulose nitrate is known as gun cotton (smoleless powder).

7. Phenols

- If – OH group is directly attached to the benzene ring, the compounds are called Phenols.
- Phenols form coloured complexes with $FeCl_3$.
- Benzene diazonium chloride on heating with water gives phenol.
- Fenton's reagent (H_2O_2 + $FeSO_4$) oxidises benzene to phenol.
- Order of acidity : Carboxylic acids > phenols > alcohols > acetylenes; phenol > m-cresol > o-cresol > p-cresol; Nitrophenol > phenol > p-methoxy phenol > p-methyl phenol > p-amino phenol.
- Phenols can be separated from carboxylic acids by means of their insolubility in bicarbonate solutions (e.g. $NaHCO_3$)
- Carbolic acid is phenol containing 5% water which is liquid at room temperature.
- Phenols are soluble in water in appreciable amounts due to intermolecular hydrogen bonding.
- o-nitrophenol due to intramolecular hydrogen bonding has lower boiling point than its m and p-isomers.

- Phenol forms quinol in Elb's persulphate oxidatin ($K_2S_2O_8$)
- Phenol on exposure to air and light turns pink due to the formation of phenoquinone by oxidation.
- Bromine with phenol forms 2,4,6-tribromo phenol in presence of H_2O while o-bromo and p-bromo phenols are formed in presence of non polar solvent (e.g. CS_2 or CCl_4).
- Phenol on heating with conc. HNO_3 produces 2,4,6-trinitrophenol (picric acid) while with dil. HNO_3, o and p-nitrophenol are formed.
- Picric acid doesn't have acid (COOH) group.
- Sodium phenoxide reacts with CO_2 under pressure to give sodium salicylate which on hydrolysis produces salicylic acid in Schmidt-Kolbe reaction.
- Phenol with phthalic anhydride in presence of H_2SO_4 gives phenophthalein used as (indicator)
- Phenol and formaldehyde on heating in presence of a base form Backelite, a thermosetting resin.
- Lysol (50% soapy emulsion of cresol) is used as disinfectant.
- Resorcinol on nitration gives styphnic acid.
- Pyrogallol is used to absorb oxygen present as impurity from gaseous mixture.
- Pthalic anhydride reacts with resorcinol to produce fluorescein.
- The action of benzoyl chloride on hydroxyl compounds in alkaline medium is called Schotten-Baumann reaction.
- In the Liebermann's nitroso reaction sequential changes in the colour of phenol occurs as brown or red → green → red → deep blue.
- Phenol gives initially methylol phenol in the presence of acid or base with HCHO to produce Novolak (thermoplastic) or Backelite later respectively. The initial reaction is known as Lederer-Manasse reaction.
- Phenol-formaldehyde resins are used for demineralisation and softening of water.
- Gossypol a by-product of cotton oil is put in application in China in a major way as a male contraceptive.
- A mixture of ortho and para-nitro phenols (1:1) can be separated by distillation.
- Phenol and on heating with zinc powder gives benzene.

8. Ethers

- Ethers are Lewis bases.
- Williamson's synthesis proceeds by S_N^2 mechanism.
- Diethyl ether is called sulphuric ether as H_2SO_4 is employed in its preparation. Tertiarty halides give alkenes.
- Methyl ethers are prepared from alcohols by the action of diazo methane in presence of BF_3.
- In Drahowzal method, alcohols react with p-toulene alkyl sulphonates to give ethers.
- Ethers on treatment with cold (or dilute) HI give a molecule of alkyl Iodide and an alcohol . The iodine atom always goes with simpler alkyl group.
- Propyl isoproyl ether with cold HI gives

$$CH_3–CH_2–CH_2–O–CH \begin{matrix} CH_3 \\ CH_3 \end{matrix} + HI \xrightarrow{\text{Cold}} CH_3–CH_2–CH_2OH + CH_3–CHI–CH_3$$

- Ethers on treatment (or heating) with conc. HI give two molecules of Iodoalkane(s).
- Anisole ($C_6H_5 – O – CH_3$) with hot conc. HI gives phenol and CH_3I.
- Ethers when exposed to moist air (or O_3) and sunlight form peroxides.
- Presence of peroxide in ether sample cab be detected by the appearance of blue, colour with KI and starch.
- Diethyl ether on oxidation with acidic $K_2Cr_2O_7$ produces acetic acid.
- Ethers with acid halides produce a molecule of alkyl halide and an ester.
- PCl_5 cleaves ether as $C_2H_5OCH_3 + PCl_5 \rightarrow C_2H_5Cl + CH_3Cl + POCl_3$.
- Ethers with acid anhydrides give two molecules of ester.
- Ethers on hydrolysis (on heating) give alcohols.
- Ethers undergo dehydration in presence of dry Al_2O_3 at 380°C to preduce alkenes.
- Simple or symmetrical ethers can be obtained from excess of primary alcohols with conc. H_2SO_4 or H_3PO_4. 2° and 3° give predominantly alkenes.
- Ether mixed with C_2H_5OH and $CHCl_3$ is used as general anaesthetic .
- Alcohol and ether mixture under the trade name Natalite can be used as a substitute of petrol.
- Crown ethers are cyclic ethers containing several-4, 5, 6 or more oxygen atoms.
- Crown ethers are employed in ion chromatography, extractron-separation etc.
- Alkyl halide on heating with dry Ag_2O gives ether.

9. Akdehydes and ketones

- Friedel -Craft acylation of benzene in presence of dry

$AlCl_3$ leads to the formation of aromatic ketones. R^+CO (acylinium ion) is the reactive intermediate.

- In Reimer-Tiemann, reaction phenol reacts with chloroform in the presence of a strong base (NaOH or KOH) to give salicylaldehyde (o-hydroxy benzaldehyde). CCl_2 carbene is the reactive intermediate.
- In Rosenmund reaction, acid halides are reduced in the presence of Pd/$BaSO_4$ (it stops further reduction of RCHO to RCH_2OH) to aldehydes.
- In Etard reaction, toluene with chromyl chloride gives benzaldehyde.
- Alkyl nitriles (or cyanides) undergo reduction in the presence of $SnCl_2$/HCl followed by hydrolysis give aldehydes.
- Benzyl chloride in Sommelet reaction in the presence of hexamethylene tetramine produces benzaldehyde.
- Ethanal on industrial scale is made by Wacker process using $PdCl_2$ to convert a mixture of ethene and water.
- Calcium formate $(HCOO)_2Ca$ and calcium acetate $(RCOO)_2$ Ca on dry distillation produce formaldehyde and acetaldehyde respectively.
- Gattermann-Koch synthesis (ArH + CO + HCl $\rightarrow$ ArCHO + HCl) is used to prepare benzaldehyde. ^+CHO is the reactive intermediate.
- Nucleophilic addition is the typical reaction of carbonyl compounds.
- Order of reactivity : HCHO > CH_3CHO > CH_3COCH_3.
- Ketones are converted to secondary alcohols in the presence of $[(CH_3)_2CHO]Al$ in Meerwin Pondorf Verley reduction.
- Aluminium isopropoxide doesn't effect the unsaturation (C = C or C ≡ C or NO_2 present in the carbonyl compounds.
- Aldehydes and ketones having α – hydrogen exhibit tautomerism.
- Ketones on reduction in neutral or alkaline medium (Mg-Hg/H_2O) give prinacols.
- Aldehydes give esters in the presence of $(C_2H_5O)_3Al$ in Tischenko reaction.
- Aldehydes and ketones having α – hydrogen atom undergo aldol condensation. Carbanions are the reactive intermediates.
- HCHO, C_6H_5CHO, $(CH_3)_3CCHO$ etc. do not undergo aldol condensation.
- Reverse aldol condensation can be represented as $H_2C = CH - CHO + H_2O \xrightarrow{NaOH} HCHO + CH_3CHO$.
- Aldehydes having no α – hydrogen give Cannizzaro's reaction.
- Certain aldehydes, $(CH_3)_2CHCHO$ having α – hydrogen also undergo Cannizzaro's reaction when heated with NaOH at high temperature.
- Cannizzaro's reaction is a redox reaction.
- Intramolecular Cannizzaro's reaction can be represented as, $OHC.CHO \xrightarrow{NaOH} HOH_2C.COONa$ (sodium glycollate)
- Two different aldehydes having no α – hydrogen undergo crossed Cannizzaro's reaction with a base.
- Benzaldehyde with KCN in the presence of alcohol undergoes Benzoin condensation to give benzoin.
- The presence of carbonyl group can be detected by 2, 4-dinitrophenyl hydrazine.
- Aldehydes reduce Fehling solution to Cu_2O (red precipitate).
- Benzaldehyde can't reduce Fehling solution as it gives Cannizzaro's reaction in the presence of NaOH present in the reagent.
- Aldehydes reduce Tollen's reagent (ammonical silver nitrate) to Ag (silver mirror).
- Aldehydes when treated with Schiff's reagent (Rosanaline hydrochloride), the pink colour of the dye is restored.
- Aldehydes give Cu_2O red precipitate with Benedict's reagent ($CuSO_4$ + Sodium citrate + Na_2CO_3).
- Formaldehyde reacts with ammonia to give hexamethy lene tetramine known as urotropine which is used as urinary antiseptic medicine.
- In Reformatsky reaction, ketones with bromoesters in the presence of zinc produce β – hydroxy esters.Organometallic compound is formed as intermediate.
- In Wittig reaction aldehydes or ketones are converted to alkenes in the presence of methylene triphenyl phosphorane.
- Benzaldehyde with acetic anhydride in the presence of sodium acetate gives cinnamic acid (α, β – unsaturated acid) in Perkin reaction.Carbanion is the reactive intermediate.
- Aldehyde or ketone with active methylene group containing compound e.g. malonic ester in the presence of pyridine gives α, β – unsaturated acid - Knoevenagel reaction.
- Formaldehyde when allowed to stand or its aqueous solution is treated with 2% H_2SO_4, trioxane is formed.
- Acetaldehyde on treatment with conc. H_2SO_4 gives paraldehyde, which is used as hypnotic and sporific

(sleep producing).

- Acetone when left in the saturated HCl gas at 0°C forms Mesityl oxide.
- Three molecules of acetone condence to give phorone.
- Acetone on distillation with conc. H_2SO_4 (1:2 w/w) forms mesitylene.
- Acetaldehyde undergoes oxidation with SeO_2 (mild oxidising agent) to give glyoxal (an α – dicarbonyl compound).
- Formalin (40% HCHO + 4% methanol + 56% H_2O) is used to preserve biological specimen.
- Golalith (milk stone) is formaldehyde + casein
- Formamint, a mixture of formaldehyde and lactic acid is used as throat lozenges.

10. Carboxylic acids and their derivatives

- Acidic character of carboxylic acids is due to the presence of replaceable hydrogen.
- Ants contain formic acid; milk-lactic acid; vinegar-acetic acid; butter-butanoic acid; goat fat-capric caproic and caprylic acids.
- Fats and oils develop unpleasant smell on long exposure to moist air due to rancidity.
- Reichert-Meiss value (R/M value) represents the steam volatile fatty acids present in the oil or the fat.
- Hydrolysis of trihalogen derivatives leads to the formation of acids.
- Acid amides with nitrous acid produce acids.
- Grignard reagent with CO_2 followed by hydrolysis gives acids.
- Alkyl cyanides on acid hydrolysis give acids.
- Gem dicarboxylic acids on heating give monocarboxylic acids.
- RCOOH can be converted to RCH_2COOH by Arndt-Eistert synthesis.
- Order of acidity :o-nitrobenzoic acid > p-nitro benzoic acid > m-nitro benzoic acid > benzoic acid.
  o-hydroxy benzoic acid > m-hydroxy benzoic acid > benzoic acid > p-hydroxy benzoic acid.
- The reactivity of acid derivatives towards hydrolysis is RCOX > $(RCO)_2O$ >RCOOR > $RCONH_2$. The electrophilic nature of carbonyl group also decreases in the some order.
- Tolune on oxidation with $KMnO_4$ or $K_2Cr_2O_7$ gives benzoic acid.
- Naphthalene on oxidation in the presence of V_2O_5 catalyst forms phthalic acid.
- α – keto acids on heating with conc. H_2SO_4 give carboxylic acids by decarbonylation.
- Carboxylic acids exist in dimeric state due to hydrogen bonding.
- Lower volatility, higher molecular weights, higher solubility in H_2O, higher boiling point etc. of lower carboxylic acids is due to hydrogen bonding.
- Kolbe synthesised acetic acid, the first organic compound from the elements.
- HCOOH is the strongest acid among the fatty acids.
- HCOOH reduces Fehling reagent, Tollen's reagent and $HgCl_2$ respectively.
- Glacial acetic acid (means icy) solidifies at 16°Cand is 16N. 100% acetic acid shines like glacier.
- In Hell-Volhard -Zelinsky reaction, the α – hydrogen containing carboxylic acids are brominated at α – position.
- Carboxylic acids react with hydrazoic acid in the presence of conc. H_2SO_4 followed by hydrolysis give primary amines in Schmidt reaction.
- Carboxylic acids are reduced with $LiAlH_4$ to alcohols while alkanes are formed with H_2/Ni.
- $LiAlH_4$ being a stronger reducing agent than $NaBH_4$ reduces carboxylic acids and esters to 10 alcohols while $NaBH_4$ can not.
- order of acidity : RCOOH > H_2O > ROH > $HC \equiv CH$.
- More the pk_a value, lesser is the acidic strength of carboxylic acids.
- Sucrose on oxidation with V_2O_5/HNO_3 gives oxalic acid.
- Potassium tetra oxalate can be used for removing ink stains.
- Betel stains can be removed by oxalic acid solution.
- Blanc's rule : Gem dicarboxylic acids on heating produce monocarboxylic acids. 1, 4 and 1, 5 dicarboxylic acids form cyclic anhydrides. 1, 6 and 1, 7- dicarboxylic acids on heating with salicylic acid produce aspirin.
- Acid chlorides with Grignard reagent in 1:1 give ketones while tertiary alcohols are produced in 1:2 or more .
- Acid halides with NaN_3 (sodium azide) followed by hydrolysis produce primary amines.
- Acetyl chloride is used to determine $-NH_2$, >NH, –SH etc. groups.
- Two molecules of ethyl acetate condense to give acetoacetic ester in Claisen condensation method.

Carbanion is the reactive intermediate.

- Ketones on oxidation with peracids give esters in Baeyer-Villeger reaction.
- In the case of hydroxy acids where a 5 or 6 membered ring can be formed, intramolecular esterification tekes place and a lactone is formed.
- Esterification of alcohols with acid halides is an irreversible reaction.
- Alkaline hydrolysis of esters is known as saponification.
- Esters present in fruits/flowers : Apple-Isoamyl valerate; Banana-Isoamyl acetate; Pine apple - Methyl butyrate; Jasmine - Benzyl butyrate; Orange - octyl acetate; Winter green-Methyl salicylate; Rum odour-Ethyl formate; clove oil - Methyl benzoate.
- Acetoacetic ester, cyano-ethyl malonate and diethyl malonate have active methylene group.
- Isoamyl acetate is used as thinner in paints.
- Acetamide is obtained in laboratory by heating CH_3COONH_4.
- Amides are less basic than amines due to resonance.
- Acetamide shows keto-enol isomerism (tautomerism).
- Due to intermolecular hydrogen bonding, acetamide is more soluble in water.
- Acid amides on dehydration with P_2O_5 give alkylcyanides.
- Acid amides on Hoffmann degradation (or bromamide reaction) in the presence of Br_2 and KOH produce amines. R–N = C = O (isocyanate) is the stable intermediate. Nitrene is the reactive intermediate.
- Amides give ketones with Grignard reagent.
- Urea stibamine is used in the treatment of Kal -azar fever.
- Urea is the first organic compound synthesised in the laboratory of Wohler.
- Urea reacts with malonic ester and acetoacetic ester to give barbutyric acid and 4-methyl uracil (used in the synthesis of RNA) respectively.
- Esters with acid halide or ketene or acid (in presence of P_2O_5) give acid anhydrides.
- Esters show tautomerism among themselves.
- Methyl and ethyl esters have lower boiling points than the corresponding acids as they are non ionic.
- Acetic anhydride and acetyl chlorid are used as acetylating agents.
- Methylation of carboxylic acids can be better achieved with BF_3–CH_3OH than CH_2N_2.

11. Nitrogen containing compounds

- By Gabriel's phthalimide synthesis, aliphatic primary amines can be prepared.
- Acid azide with chloroform gives primary amines in Curtius reaction.
- Alkyl cyanides are reduced to primary amines while isocyanides are reduced to secondary amines in Mendius reaction in the presence of Na/C_2H_5OH.
- In Ritter reaction, tertiary alcohols are converted to primary amines in the presence of HCN & H_2SO_4.
- 1°, 2° and 3° amines can be separated by Hinsberg's method (benzene supphonyl chloride) or Hoffmann's method (ethyloxalate).
- Among isomeric amines, the order of boiling point is 1°>2°> 3°.
- Hydrogen bonding is possible in primary and secondary amines.
- The order of basicity is aliphatic amines > ammonia > aromatic amines.
- Liebermann's nitroso reaction is shown by secondary amines.
- Mustard oil reaction is given by primary amines. Alkyl isothiocyanate, R-N=C=S is formed.
- In Hoffmann elimination reaction, quaternary ammonium salts undergo degradation to give a tertiary amine, alkene and water.
- Both aliphatic and aromatic amines react with aldehydes to form Schiff's base.
- Primary amines of the type RCH_2NH_2 and R_2CHNH_2 on oxidation with acidic $KMnO_4$ give aldehydes and ketones respectively.
- Aniline on bromination in aqueous medium gives 2, 4, 6-tribromo aniline while in non polar solvents (e.g. CS_2, CCl_4 etc.) o-bromo and p-bromo aniline are formed.
- In halogenation and nitration of aniline, the amine group is protected by converting it to acetanilide.
- Aniline on diazotisation followed by the addition of β – naphthol, produces α – phenyl diazo – β – naphthol an orange red azodye in dye test.
- Due to high molecular weight sulphanilic acid is insoluble in H_2O.
- Aniline on sulphonation forms sulphanilic acid as major product and orthanilic and metennilic. acids as minor products.
- Sulphanilic acid exists as zwitter ion.
- Aniline with neutral $KMnO_4$ gives nitrobenzene while azobenzene is produced with alkaline $KMnO_4$. Aniline with acidic $Na_2Cr_2O_7$ oxidation gives benzoquinone.
- Aloxyl is used in the treatment of sleeping sickness.
- $C \equiv N$ bond is stronger than $C \equiv C$ bond.

- Alkyl halides produce alkyl cyanides with alcoholic KCN while alkyl isocyanides are produced with AgCN.
- In Stephen's reaction alkyl cyanides are converted to aldehydes by hydrolysis of aldimine intermediate in the presence of $SnCl_2/HCl$.
- Alkyl cyanides with Grignard reagent give ketones.
- Alkyl cyanides have higher boiling points than their isomeric isocyanides.
- Alkyl cyanides are more soluble than alkyl isocyanides.
- Alkyl halides (halo alkanes) with alcoholic KNO_2 give alkyl nitrites while nitro alkanes are formed with alcoholic $AgNO_2$.
- Alkyl nitrites being esters undergo hydrolysis to give alcohols.
- Nitroalkanes on reduction in acidic medium give alkyl primary amines, while N-alkyl hydroxylamines are produced in neutral medium.
- In Mannich reaction, secondary or tertiary nitroamines are produced from nitroalkanes on heating in the presence of HCHO and NH_4Cl.
- During oxidation, the side chain nitro group is lost while the nuclear substituted one remains intact.
- Nitrobenzene on nitration produces m-dinitrobenzene.
- $LiAlH_4$ reduces $-NO_2$ group to $-NH_2$ group while $NaBH_4$ can't.
- Spirit of nitre (4% alcoholic solution of ethyl nitrite) is used as a diuretic.
- Order of basicity : $\underset{\text{Amine}}{R'NH_2} > \underset{\text{Imine}}{R'CH = NR} > \underset{\text{Nitrite}}{R{-}C \equiv N}$

12. Polymers

- In polymers, the repeating units are held together by covalent bonds.
- Natural polymers are also known as biological polymers, e.g. proteins, starch, cellulose, nucleic acids, rubber, wool etc.
- Phenol-formaldehyde polymer is the first synthetic polymer.
- Monomers of different polymers : polythene-ethene; PVC (poly vinyl chloride) vinyl chloride (chloroethene); Teflon-tetrafluroethene; polystyrene-styrene; polymethyl methacrylate (also known as plexiglas, perspex or Leusite is used to make lenses, windowpanes of aeroplanes etc.) -methyl methacrylate; orlon or polyacrylon-acrylonitrile; Nylon 6, 6 -adipic acid and hexa-methylene diamine; Nylon, 6-caprolactam; Terylene (or Dacron or Vinyon) -polyester of terephthalic acid and ethylene glycol; Alkyd resins or glyptals-polyester of phthalic anhydride and glycerol.
- Nylon is also known as polyamide. Molecular chains in Nylon are held together by hydrogen bonds.
- Linear polymers are generally thermoplastic, e.g. polythene, nylon, orlon etc.
- Polymers which are changed irreversably into hard and rigid materials on heating are called thermosetting resins, e.g. Backelite, Araldite, Melamine etc.
- Rubber is in the form of colloidal solution in water, known as latex. Rubber is coagulated by adding a little amount of acetic acid.
- The monomer of rubber is isoprene (2-methyl-1, 3-butadiene).
- The monomer of Neoprene is 2-chloroisoprene.
- Ziegler-Natta polymerisation is useful to produce high density polymers.
- $TiCl_4$ and $(C_2H_5)_3$ Al is employed as catalyst in ZieglerNatta polymerisation.
- Arrangement of groups along the carbon backbone of the polymer chain can be : (a) Atactic : random pattern (b) Syndiotactic : regular alternating pattern of groups (c) Isotactic : groups on the same side of the backbone.
- Nomex is used for preparing heat resistant space suits.
- I_2, O_2, amines, phenols quinones etc. are used as initiators.
- Reagents employed to regulate the molecular weight of the polymer are called telomers or chain transfer reagents.
- Cellophane used for wrapping is cellulose acetate.
- The monomer units of polypeptides are amino acids.
- Herculon is the polymer of polypropylene.
- Perlon is Nylon.
- Propyne on polymerisation gives mesitylene.
- Treatment of rubber with sulphur is called vulcanisation. The linear rubber molecules get cross linked through sulphur atoms.
- Ebonite is an inorganic polymer obtained by vulcanisation of rubbers.

13. Biomolecules

- Carbohydrates respond positively to Molisch's test.
- Plants store energy by converting glucose into starch.
- Animals store energy converting glucose into glycogen.
- Glucose is known as dextrose as it is optically active and dextrorotatory.
- Glucose on reduction with Ni/H_2 or Na–Hg gives sorbitol while n-hexane is obtained with HI.

- Glucose on oxidation with Br_2/H_2O (mild oxidising agent) produces gluconic acid while glucaric acid (saccharic acid) is produced with dil. HNO_3 (strong oxidising agent).
- Osazone reaction is given by α – hydroxy carbonyl compounds.
- Fructose although contains a keto group is a reducing sugar as it undergoes Lobry de Bruyn van Ekenstein rearrangement in the presence of dilute aqueous alkalies (glucose/fructose/mannose undergo this rearrangement and all the three are formed).
- Yeast contains invertase, zymase and maltase.
- Glucose on fermentation in presence of zymase and yeast forms alcohol.
- α – D(+) glucose and β – D(+) glucose show mutarotation. All reducing sugars undergo mutarotation.
- Anomers are diastereomers that differ in configuration at C_1 position.
- Glucose and Mannose are epimers which differ in configuration at C_2 position.
- In periodic oxidation –CHOH and –CHO groups are converted to HCOOH while $-CH_2OH$ gives HCHO.
- Fructose is known as laevulose as it is laevorotatory.
- Fructose is the sweetest of the sugars.
- Non sugars like Monallein, Saccharin etc. are sweeter than sugars.
- Fructose is conmercially prepared from Inulin.
- Fructose can be oxidised very easily as it actses as a strong reducing agent.
- Glucose and fructose can be distinguished by Furfural test, Pinoff's test, Selivanoff's test etc.
- Sucrose is a non reducing sugar.
- Sucrose on hydrolysis gives glucose and fructose.
- Lactose is known as milk sugar .C_4 of glucose is joined with C_1 of β –D–Galactose.
- Maltose and lactose are reducing sugars.
- In maltose C_4 of one glucose unit is joined with C_1 of the other glucose unit.
- Keratin, Haemoglobin, Insulin, Enzymes (biological catalysts) etc. are proteinous in nature.
- Myoglobin is oxygen storing protein of muscle.
- Proteins are polyamides of α – amino acids.
- Proline has secondary amino group while all others have primary one.
- All amino acids except glycine H_2N-CH_2-COOH contain chiral (asymmetric) carbon atom.
- Amino acids are separated by electrophoresis.
- Arginine, Histidine, Isoleucine, Leucine, Lysine, Methionine, Phenylalanine, Threonine, Tryptophane and Valine are essential amino acids.
- The serum proteins in blood behave like a buffer solution.
- Lecithin is a phospholipid present in cell membrane.
- At isoelectric point, the Zwitter ion concentration is maximum.
- Amino acids show bluish violet colour with ninhydrin (proline gives yellow colour).
- Millons test is given by proteins containing phenolic – OH group.
- Proteins can also be detected by xanthoproteic test.
- Nitroprusside test is given by proteins containing –SH group.
- Hopkins-Cole test is given by Tryptophan.
- For identification of N-terminal residue analysis, Sanger's reagent (2,4-dinitrofluorobenzene-DNFB) is used. Edman's method can also be used (phenyl isothio-cyanate, Ph-N=C=S is used).
- For C-terminal residue analysis, carboxy-peptidase is used.
- Primary structure of proleins tells us the number and sequence of amino acids.
- Secondary structure of proteins i.e. the helical one is mainly studied by x-ray analysis.
- Starch is converted to maltose by amylose while maltose is changed to glucose by maltase.
- Sucrose is converted to glucose and fructose by invertase.
- Proteins are converted to amino acids by pepsin or trypsin
- α – glycosides are hydrolysed by maltase
- β – glycosides are hydrolysed by emulsin.
- Nucleotide unit is composed of a base, a sugar and an inorganic phosphate group.
- A nucleoside is made up of a base unit and a sugar moiety.
- DNA is responsible for heredity traits.
- The law of complimentarity of DNA was proposed by Watson and Crick.
- The bases found in DNA are Adenine, Gaunine, Thymine and Cytosine.
- Codon is a coding unit.
- Genetic code is the order of the codons on the strands of DNA.

14. Chemistry in Action

- The colour of a compound is due to chromophores that they contain.
- Auxochromes don't produce colour themselves but when present together with chromophore they alter the intensity of the colour.
- Azodyes have azo (–N=N–) group, e.g. methyl orange, methyl red, congo red etc.
- Anthraquinone dyes contain anthraquinone basic moiety, e.g. Alizarin.
- Indigo is an indigoid dye.
- Mordant dyes have no natural affinity for the fabric and therefore are applied with the help of certain mordants, e.g. $FeCl_3$, $Al_2(SO_4)_3$, $SnCl_4$ etc. The metal ions in these compounds act as mordants.
- Vat dyes are insoluble in water and can't be applied directly. These dyes on reduction with $NaHSO_3$ in a vat form a soluble compound which has greater affinity for cotton and other cellulose fibres e.g. Indigo.
- Chemicals used for treating diseases are called drugs or medicines.
- Treatment of diseases by chemical substances which do not destroy the tissues of the host while destroying the micro-organisms responsible is called chemotherapy.
- Analgesics are pain killers or pain relievers. Morphine, Codein, Heroin etc. are norcotic analgesics while Aspirin, Analgin etc. are non norcotic.
- Antipyretics reduce body temperature in fever but do not cause hypothermia in normothermic individuals e.g. Paracetamol, Phenacetin, Phenylbutazone, Ibuprofen, Diclofenac sodium etc.
- Agents used to inhibit or kill microbes on contact on living surfaces are called antiseptics. Agents used for inanimate objects are called disinfectants. Antiseptics, e.g. Dettol, Salol, Savlon, Mercurochrome, Iodoform Chloramine, Boric acid etc.Disinfectants e.g. DDT, Phenol, $KMnO_4$, Bleaching powder, H_2O_2, O_3, SO_2 etc.
- Drugs which upon topical application or local injection cause reversible loss of sensory perception, specially of pain in restricted area of the body are called local anaesthetics e.g.procaine, lignocaine, cocaine, benzocaine, xylocaine etc.
- Drugs which produce reversible loss of all sensation and conciousness are called general anaesthetics, e.g. N_2O, ether, halothane, Isoflurane, Sodium pentothal cyclopropane etc.
- Drugs which alter mood, behaviour, thought and manner are called psychedelics, e.g. LSD (Lysergic acid diethylamide), Mescaline, Cannabinoids: Bhang, Ganja charas (these are drugs of abuse).
- Drugs which reduce mental tension and produce calmness without inducing sleep or depressing mental faculties. Barbutyrates have been popular hypnotics and sedatives
- Chloroquine is active against plasmodium in blood (a single dose of 600mg is recommended).
- Primaquine is active against plasmodium in liver (15 mg base daily for 14 days).
- Mefloquine is used for chloroquine-resistant and multidrug-resistant malaria.
- Substances produced by microorganisms which supress the growth of or kill other microorganisms at very low concentration are known an antibiotics. Penicillin, the first antibiotic was discovered by Alexander Flemming. Ampicillin, Streptomycin, Tetracycline, Ciprofloxacin etc. are antibiotics.
- Sulphadiazine, Trimethoprim, Sulphamethoxazole are some important sulpha drugs.
- Pheromones are secreted by animals which affect the behaviour of other animals of the same species, e.g. Muscone (Tibetan musk deer), Civetone (civet cat) etc.
- Contraceptives are used for reversible supression of fertility.
- Female contraceptives : Combined pill contains an estrogen and a progestin.
- Gossypol, a by product of cotton seed oil is put in application in a major way in China as male contraceptive.
- Vitamin A is derivative of β – Carotene and its deficiency leads to night blindness, xerophthalmia, keratomalacia.
- Vitamin B_{12} contains Co metal in its structure. This vitamin is absent in plants as it is synthesized exclusively by microorganisms.
- Vitamin C deficiency leads to scurvy.
- Vitamin D is a steroid pro-harmone. It is derived from either 7-dehydrocholesterol or ergosterol by UV radiation. Its deficiency causes rickets in children and osteomalacia in adults.
- Vitamin E (tocopherol) deficiency leads to anaemia in newly born and also degeneration of reproduction power.
- Vitamin K (phylloquinone) is responsible for clotting of blood. Its deficiency leads to uncontrolled bleeding, hemorrhage.
- UDMH -unsymmetrical dimethyl hydrazine and MMH-monomethyl hydrazine are liquid propellants.
- Liquid propellants are more efficient than solid propellants.
- Parathion, Malathion etc. are organic insecticides.

Organic Reactions of Importance

Würtz reaction :

$$R\text{–}X + 2Na + X\text{–}R \xrightarrow{\text{ether}} R\text{–}R + 2NaX$$

Methane can't be prepared by this method. Hydrocarbons with odd number of carbon atoms can't be prepared in pure state. The reaction fails with tertiary alkyl halides.

Reactive intermediates: alkyl free radicals (as presented by A) or organometallic compounds (as represented by B)

A) (i) $H_3C\text{—}CH_2\text{—}X + Na^{\bullet} \longrightarrow NaX + H_3C\text{—}\dot{C}H_2$

(ii) $H_3C\text{–}CH_2 + H_2\dot{C}\text{—}CH_3 \rightarrow H_3C\text{–}CH_2\text{—}CH_2\text{—}CH_3$

B) (i) $H_3C\text{–}I + 2Na \rightarrow H_3\bar{C}\text{—}Na^+ + NaI$

(ii) $H_3C\text{–}I + H_3\bar{C}\text{—}Na^+ \rightarrow H_3C\text{—}CH_3 + NaI$

disproportionation of ethyl free radical:

$$H_5\dot{C}_2 + \dot{C}_2H_5 \rightarrow C_2H_6 + C_2H_4$$

2. Core–House reaction

$$R'\text{–}X + R_2CuLi \rightarrow R'\text{–}R + Cu\text{–}R + Li\text{–}X$$

3. Kolbe's electrolytic reaction: Electrolysis of K or Na salt of fatty acid

$$R\text{—}\overset{O}{\overset{\|}{C}}\text{—}ONa \rightleftharpoons R\text{—}\overset{O}{\overset{\|}{C}}\text{—}\bar{O} + \overset{+}{Na}$$

$$R\text{—}\overset{O}{\overset{\|}{C}}\text{—}\dot{O} + \bar{e} \xrightarrow{-CO_2} \dot{R}$$

4. By Freund's reaction / Intramolecular Würtz reaction: α or ω dihalides give this reaction

$$H_2C\begin{matrix}\diagup CH_2X \\ \diagdown CH_2X\end{matrix} + Na \text{ (or Zn dust)} \xrightarrow[\Delta]{\text{ether}} H_2C\begin{matrix}\diagup CH_2 \\ \diagdown CH_2\end{matrix}$$

5. Clemmensen's reduction for compounds sensitive to bases. Aldehydes and ketones having α-hydrogen in the presence of a base undergo aldol condesation :

$$>C{=}O \xrightarrow[\text{Conc. HCl}]{\text{Zn - Hg /}} >CH_2 + H_2O$$

6. Wolff - kishner reduction : for compounds sensitive to acids

$$R\text{–}\overset{H}{\overset{|}{C}}{=}O + H_2NNHCONH_2 \text{ or } NH_2\text{—}NH_2 \xrightarrow{-H_2O} RCH{=}NNHCONH_2$$

$$RCH{=}NNHCONH_2 \xrightarrow{C_2H_5ONa,\ -CO_2,\ -NH_3} RCH{=}NNH_2 \longrightarrow N_2 + RCH_3$$

7. Huang-Minlon reaction : It is a simple modification of Wolff Kishner reaction.

$$>C{=}O \xrightarrow{H_2N\text{—}NH_2} >C{=}N\text{–}NH_2 \xrightarrow[\text{KOH } 200°C]{\text{diethylene glycol}} >CH_2$$

in Wolff-Kishner, the second step proceeds with sodium ethoxide at about 100°C.

8. Derfin's reaction : Methane when heated at high temperature in the presence of nickel produces ethane

$$2CH_4 \xrightarrow[-H_2]{650°C} C_2H_6$$

9. Zeriwitinoff's method : Compounds having active hydrogen e.g. H_2O, ROH, –SH, NH_3, RNH_2, R_2NH etc. react with RMgX. For every active H– atom, one paraffin molecule is produced. This principle is used for the estimation of –OH, $-NH_2$, –SH etc. groups in Zeriwitinoff's method. e.g.

$$R\text{—}Mg\text{---}X + ROH \rightarrow RH + Mg\begin{matrix}\diagup X \\ \diagdown OR\end{matrix}$$

$$RMgX + R_2NH \rightarrow RH + Mg\begin{matrix}\diagup X \\ \diagdown NR_2\end{matrix}$$

10. Sabatier and Senderens reaction:

$$\text{Alkene or Alkyne} \xrightarrow{Ni;\ Pt;\ Pd\ /\ H_2} \text{alkanes}$$

11. Pschorr synthesis : Phenanthrene, a polynuclear hydrocarbon is prepared from ortho nitrobenzaldehyde as given below

o-nitrobenzaldehyde (CHO, NO_2) + $H_2C\text{—}COONa$ (phenyl) $\xrightarrow{(CH_3CO)_2O}$ $CH{=}C\text{—}COOH$ (with NO_2 ring and phenyl)

$\xrightarrow{\text{reduction}}$ $CH{=}C\text{—}COOH$ (with NH_2 ring) $\xrightarrow{NaNO_2/H_2SO_4}$ $CH{=}C\text{—}COOH$ (with $N_2^+HSO_4^-$ ring)

$\xrightarrow{Cu/H_2SO_4}$ $CH{=}C\text{—}COOH$ (biaryl-linked) $\xrightarrow[\Delta]{-CO_2}$ phenanathrene

12. Reed reaction: This reaction is very useful to prepare detergents industrially.

$$R\text{—}SO_2Cl + NaOH \xrightarrow{-NaCl} RSO_3H \xrightarrow{NaOH} RSO_3^- \overset{+}{Na} + H_2O$$

alkane sulphonates

$$R\text{—}H + SO_2 + Cl_2 \xrightarrow{h\nu} R\text{—}SO_2Cl + HCl$$

(where R ~ C_9 to C_{15})

13. Brown Hydroboration: Alkenes with diborane give an addition compound which on oxidation yield alcohols.

$$6H_2C{=}CH_2 + B_2H_6 \longrightarrow 2(CH_3CH_2)_3B \xrightarrow{3H_2O_2} 3H_3C\text{—}CH_2OH + H_3BO_3$$

Here, addition of water takes place by antimorkownikoff's rule.

14. Chugaev reaction : Methyl Xanthates on heating give alkenes

$$R_2\text{—}CH\text{—}CH_2\text{—}\underset{S}{\underset{\|}{C}}\text{—}CH_3 \xrightarrow[200°C]{\Delta} R_2\text{—}CH{=}CH_2 + O{=}C{=}S + H_3C\text{–}SH$$

15. Cope reaction : Pyrolysis of tri alkyl amine oxide

at 150°C gives an alkene along with N,N'–dialkylhydroxylamine

$$R-CH_2-CH_2-\overset{+}{N}(R')(R'')-O^- \xrightarrow{150°C} R-CH=CH_2 + N(R')(R'')-OH$$

16. Diels' Alder reaction: (4+2) cyclo addition Diene + Dienophile → Adduct. Diene can be $H_2C=CH-CH=CH_2$, enyne $H_2C=CH-C\equiv CH$ or diyne $HC\equiv C-C\equiv CH$. The two double bonds must be present in cisoid conformation. Transoid conjugated double bonds do not show Diel's–Alder reaction (e.g. [bicyclic diene]).

$$\text{diene } (HC(=CH_2)-HC=CH_2) + \text{dienophile } (CH_2=CH_2) \longrightarrow \text{cyclohexene}$$

17. Prileschaiev reaction : Alkenes react with peracids to give oxirane (epoxide).

$$H_3C-CH=CH_2 \xrightarrow{\text{(peracids)}} H_3C-\underset{\diagdown O \diagup}{HC-CH_2}$$

18. Kucherov reaction (Addition of water):

$$HC\equiv CH \xrightarrow[1\%\ HgSO_4,\ 60\%]{40\%\ H_2SO_4} \left[\begin{array}{c} CH_2 \\ \| \\ CHOH \end{array}\right] \rightleftharpoons \begin{array}{c} CH_3 \\ | \\ CHO \end{array}$$

Vinyl alcohol (unstable)

Other acetylenes give ketones

$$R-C\equiv CH \longrightarrow R-\underset{\underset{O}{\|}}{C}-CH_3$$

19. Gattermann Koch reaction:

$$C_6H_6 + CO + HCl \xrightarrow{AlCl_3} C_6H_5-CHO$$

$$CO + HCl \longrightarrow \underset{\text{formylchloride}}{HCOCl} \xrightarrow{+AlCl_3} AlCl_4^- + H\overset{+}{C}O$$

20. Gattermann aldehyde synthesis:

$$C_6H_6 + HCN + HCl \xrightarrow{AlCl_3^*} C_6H_5-CH=\overset{+}{N}H_2Cl^- \xrightarrow{H_2O} NH_4Cl + C_6H_5-CHO$$

(* dry BF_3, $SnCl_2$, $ZnCl_2$, $FeBr_3$ etc. can also be used)

$$H-C\equiv N + HCl \rightarrow ClHC=NH \xrightarrow{AlCl_3} AlCl_4^{\ominus} + HC\overset{\oplus}{\equiv}NH$$

21. Birch reduction:

$$C_6H_6 + H_2 \xrightarrow[\text{in } C_2H_5OH]{Na + liq.\ NH_3} \text{1,4-cyclohexadiene} \quad \text{(partial reduction)}$$

22. Groves reaction : Alcohols react with HCl in the presence of anhydrous zinc chloride to give chloroalkanes $R-OH + HCl \xrightarrow[-H_2O]{\text{anhydrous } ZnCl_2} R-Cl$

23. Hunsdiecker's method: Silver salt of fatty acids on bromination give halo derivatives.

$$RCOOAg + Br_2 \longrightarrow R-Br + CO_2 + AgBr$$

Mechanism: $RCOOAg \xrightarrow{Br-Br} RCO\dot{O} + Br^{\bullet}$

$$RCO\dot{O} \longrightarrow \dot{R} + CO_2\ ;\ \dot{R} + Br^{\bullet} \longrightarrow RBr$$

Free radical intermediates.

24. Kinkelstein reaction : Halogen exchange in haloalkanes takes place with sodium or potassium halides in acetone, $R-Cl + KI \xrightarrow{\text{acetone}} R-I + KCl$

25. Darzen's reaction : Halo alkanes are obtained in pure form by this method. Nucleophilic substitution reaction. (i) $ROH + SOCl_2 \longrightarrow RCl + SO_2 + HCl$

(ii) $ROH + HX \text{ (or } PX_3) \longrightarrow RX + H_2O$

(iii) $C_2H_5OH \xrightarrow{P + I_2} C_2H_5I$

Reactivity: T > S > P; HI > HBr > HCl

26. Rydon's method: mixture of alcohol and triphenyl phosphate is treated with halogen

$$ROH + X_2 + (C_6H_5O)_3P \longrightarrow R-X + (C_6H_5O)_2POX$$

27. Williamson's synthesis: it follows S_N2

$$C_2H_5Br + NaOR \xrightarrow{\Delta} C_2H_5OR + NaBr$$

Mechanism

$$RONa \rightleftharpoons RO^- + Na^+$$

$$RO^- + R-X \xrightarrow{slow} [R-\overset{\delta-}{O}\cdots R\cdots\overset{\delta-}{X}] \xrightarrow{fast} ROR + X^-\ ;\ Na^+ + X^- \longrightarrow NaX$$

with tertiary halides, the reaction fails.

$$H_3C-CH_2Br + NaO-C(CH_3)_3 \ \checkmark \longrightarrow H_3C-CH_2O-C(CH_3)_3$$

$$H_3C-C(CH_3)_2-Br + NaO-CH_2-CH_3 \ \times$$

28. Fridel–Craft reaction:

(i) Alkylation: $C_6H_6 + R-X \xrightarrow[-HX]{\text{dry } AlCl_3^*} C_6H_5-R$

(* BF_3 / $ZnCl_2$ / $SnCl_4$ / $FeCl_3$ etc. can also be

employed). ($R—X + AlCl_3 \rightarrow R^+ + AlCl_3X^-$)

(ii) Acylation:

$$C_6H_6 + R—\overset{O}{\overset{\|}{C}}—X \xrightarrow[-HX]{\text{dry } AlCl_3} C_6H_5—\overset{O}{\overset{\|}{C}}—R$$

$$(R—\overset{O}{\overset{\|}{C}}—X + AlCl_3 \rightarrow R—\overset{O}{\overset{\|}{C}}^{\oplus} + AlCl_3X^-)$$

29. Carbylamine reaction: it is regarded as a test for primary amines. Aniline also gives this test.

$$RNH_2 + CHCl_3 + 3KOH \xrightarrow{\Delta} RNC + 3KCl + 3H_2O$$

$$C_6H_5—NH_2 + CHCl_3 + 3KOH \xrightarrow{\text{warm}} C_6H_5—NC + 3KCl + 3H_2O$$

Mechanism:

$$H\ddot{O}^- + H—CCl_3 \xrightarrow{-H_2O} {}^-CCl_3 \xrightarrow{-Cl^-} :CCl_2$$ dichloro carbene

$$(R—\overset{O}{\overset{\|}{C}}—X + AlCl_3 \rightarrow R—\overset{O}{\overset{\|}{C}}^{\oplus} + AlCl_3X^-)$$

30. Haloform reaction:

$$H_3C—\overset{O}{\overset{\|}{C}}—CH_3 + 3NaOCl \rightarrow H_3C—\overset{O}{\overset{\|}{C}}—CCl_3 + 3NaOH$$

$$\downarrow$$

$$CHCl_3 + H_3C—\overset{O}{\overset{\|}{C}}—ONa$$

31. Swart reaction : Halogen exchange in halo alkanes takes place with silver halide in the presence of glycol. Halogen exchange is generally a reversible reaction.

$$R—I + AgF \xrightarrow{\text{ethylene glycol}} R—F + AgI$$

32. Reimer–Tiemann reaction:

$$C_6H_5—OH + CHCl_3 + KOH \xrightarrow{\text{warm}} o\text{-}OHC—C_6H_4—OH + OHC—C_6H_4—OH$$

Mechanism:

$KOH \rightleftharpoons K^+ + OH^-; CHCl_3 + OH^- \rightarrow {}^-CCl_3 + H_2O$;

$${}^-CCl_3 \xrightarrow{-Cl^-} :CCl_2$$

$$C_6H_5\ddot{O}—H \xrightarrow[-H^+]{NaOH} C_6H_5O^- \rightarrow \xrightarrow{:CCl_2} o\text{-}CHCl_2C_6H_4O^- \xrightarrow{NaOH} o\text{-}CHO\text{-}C_6H_4O^- \xrightarrow{H^+} o\text{-}HO\text{-}C_6H_4\text{-}CHO$$

33. Iodoform (CHI_3) Test: Laboratory method:
$C_2H_5OH + 4I_2 + 6NaOH \rightarrow CHI_3 + HCOONa + 5NaI + 5H_2O$

Industrial method:
$CH_3COCH_3 + 3I_2 + 2Na_2CO_3 \rightarrow CHI_3 + 3NaI + CH_3COONa + 5H_2O$

(KI on electrolysis liberates iodine which takes part in this reaction). Iodoform test is given by compounds containing CH_3CO– group or by those compounds in which this group is produced very easily. e.g. CH_3CHO, $CH_3CHOHCOOH$, CH_3COCH_3, CH_3CH_2OH etc. Iodoform is used to distinguish CH_3OH, C_2H_5OH ; $C_6H_5—\overset{O}{\overset{\|}{C}}—CH_3$ (acetophenone), $C_6H_5—\overset{O}{\overset{\|}{C}}—C_6H_5$ (benzophenone)

34. Sandmeyer's reaction (Laboratory method)

$$C_6H_5—NH_2 \xrightarrow[0-5°C]{NaNO_2 / HCl} C_6H_5—\overset{+}{N}\equiv NCl^- \xrightarrow{Cu_2Cl_2 / HCl} C_6H_5—Cl + N_2$$

benzene diazonium chloride → chlorobenzene

Mechanism:

(i) $C_6H_5N_2Cl + CuCl \rightarrow C_6H_5N_2[CuCl_2]^- \rightarrow C_6H_5^{\bullet} + N_2 + CuCl_2$

(ii) $C_6H_5^{\bullet} + CuCl_2 \rightarrow C_6H_5Cl + CuCl$

35. Gattermann reaction: Using metallic copper or freshly precipitated copper powder (Zn dust and copper sulphate solution) as catalyst

$C_6H_5—N_2Cl \xrightarrow{Cu\ ;\ HCl} C_6H_5—Cl + N_2$; $C_6H_5—N_2Cl \xrightarrow{Cu/KCN} C_6H_5—CN + N_2$

36. Blace reaction : Benzene reacts with formaldehyde and HCl in the presence of anhydrous zinc chloride to produce benzyl chloride

$$Ph—H + O{=}CH_2 + HCl \xrightarrow{-H_2O} Ph—CH_2—Cl$$

37. Schiemann reaction: For aryl fluorides.

$$C_6H_5—NH_2 \xrightarrow[NaNO_2 / H_3O^+ ;\ HBF_4]{\text{diazotisation}} Ph\overset{+}{N}\equiv N\ {}^-BF_4 \xrightarrow{\Delta} PhF$$

38. Würtz–fittig reaction:

$$C_6H_5—X + 2Na + X—R \xrightarrow{\text{ether}} C_6H_5—R$$

39.Fittig reaction:

$$C_6H_5—X + 2Na + X—C_6H_5 \xrightarrow{\text{ether}} C_6H_5—C_6H_5$$

40. Ullmann synthesis:

$$2C_6H_5I + 2Cu \rightarrow C_6H_5—C_6H_5$$

Aryl fluorides do not react in Ullmann reaction.

41. Guberts reaction : Two molecules of an alcohol condense to give a higher alcohol on heating in the presence of sodium alkoxide.

42. Bouveault–Blanc reduction:

$$\text{aldehydes; ketones; esters} \xrightarrow[Na\ \&\ C_2H_5OH]{\text{excess of}} \text{alcohol}$$

($C_2H_5OH + Na \rightarrow C_2H_5ONa + H$, nascent hydrogen responsible for reduction) Acid derivatives also can deliver alcohols through this route.

43. Lucas test: The Lucas reagent is conc. HCl / dry $ZnCl_2$ (concentrated HCl saturated with anhydrous $ZnCl_2$). Carbonium ion intermediates form and

therefore the reactivity order is T>S>P.

3° reacts immediately; 2° reacts in ~ 5 minutes; 1° do not react at room temperature.

44. Victor Meyer's test: (Red, Blue and White test)

$H_3C-CH_2-CH_2-CH_2OH \xrightarrow{HI \text{ or } P+I_2} H_3C-CH_2-CH_2-CH_2I \xrightarrow{AgNO_2} H_3C-CH_2-CH_2-CH_2NO_2 \xrightarrow{HONO} H_3C-CH_2-CH_2-C(=N-OH)NO_2$ (nitrolic acid) $\xrightarrow{NaOH} H_3C-CH_2-CH_2-C(=NO^-Na^+)NO_2$

Blood red colour

$(CH_3)(C_2H_5)CHOH \rightarrow (CH_3)(C_2H_5)CHI \rightarrow (CH_3)(C_2H_5)CHNO_2 \xrightarrow{HONO} H_3C-CH_2-C(CH_3)(N=O)-NO_2$

pseudonitrol Blue colour

$(CH_3)_3C-OH \rightarrow (CH_3)_3C-I \rightarrow (CH_3)_3C-NO_2 \xrightarrow{HNO_2}$ No Reaction

45. Hoesch reaction : Polyhydric phenols react with alkyl nitriles and HCl gas in the presence of $ZnCl_2$ in ether solution to give acyl derivative at o-position to phenolic –OH group. Phenol does not show this reaction.

3,5-dihydroxytoluene $\xrightarrow[ZnCl_2]{R-C\equiv N + HCl}$ ketimine hydrochloride ($C=NH.HCl$, R) $\xrightarrow[\Delta]{H_2O}$ acyl derivative (COR)

46. Elbs persulphate oxidation:

$C_6H_5-OH \xrightarrow[OH^-]{K_2S_2O_8} HO-C_6H_4-OH$ (quinol)

47. Kolbe – Schmidt reaction:

$C_6H_5ONa + CO_2 \xrightarrow[\text{pressure}]{393-413\ K}$ o-$HOC_6H_4COONa \xrightarrow{H_3O^+}$ o-HOC_6H_4COOH (salicylic acid)

Mechanism: $C_6H_5ONa \rightarrow C_6H_5O^- + Na^+$;

(phenoxide resonance structures; carbanion at ortho/para position attacks CO_2 to give COO^-, followed by tautomerisation)

48. Duff's reaction : Salicylaldehyde is prepared from phenol

$C_6H_5OH \xrightarrow[\text{boric acid}]{(CH_2)_6N_4}$ o-$HOC_6H_4CH=N-CH_3 \xrightarrow{H_3O^+}$ o-HOC_6H_4CHO

49. Schotten-Baumann reaction:

$C_6H_5OH + Cl-CO-CH_3 \xrightarrow[\text{or pyridine}]{K_2CO_3} C_6H_5O-CO-CH_3$ (phenyl acetate)

50. Fries rearrangement : The ester, phenylacetate undergoes rearrangement as:

$C_6H_5O-CO-CH_3 \xrightarrow[\text{dry } AlCl_3]{CS_2}$ o-$HOC_6H_4COCH_3$ + p-$HOC_6H_4COCH_3$

51. Liebermann's nitroso-reaction:

$C_6H_5-OH \xrightarrow{NaNO_2 / H_2SO_4} ON-C_6H_4-OH$ (p-nitrosophenol) $\rightleftharpoons HON=C_6H_4=O$ (quinone oxime)

$O=C_6H_4=NOH + H-C_6H_4-OH \xrightarrow[-H_2O]{\text{conc. } H_2SO_4} [H^+O=C_6H_4=N-C_6H_4-O]\ HSO_4^- \xrightarrow{H_2O} O=C_6H_4=N-C_6H_4-OH$ (indophenol (Red)) $\xrightarrow{NaOH} O=C_6H_4=N-C_6H_4-O^-Na^+$ (sodium salt (Blue))

52. Lederer–Manasse reaction : Phenol with HCHO reacts as

$C_6H_5OH + HCHO \xrightarrow{1:1;\ H^+}$ novolak, thermoplastic; linear

$C_6H_5OH + HCHO \xrightarrow{OH^-}$ Backelite; thermosetting; three dimentional (more HCHO ratio)

Formation of methylol phenol (o-$HOC_6H_4CH_2OH$) in the first step is known as Lederer–Manasse reaction.

53. Drahowzal method:

$ROH + RO_3S-C_6H_4-CH_3 \rightarrow HO_3S-C_6H_4-CH_3 + ROR$

54. Williamson's continuous etherification: $C_2H_5OC_2H_5$ is also called as sulphuric ether as H_2SO_4 is employed in this reaction.

$C_2H_5OH + H_2SO_4 \xrightarrow{\sim 100°C} C_2H_5HSO_4 \xrightarrow[140°C]{C_2H_5OH\ \text{excess}} C_2H_5OC_2H_5$

$C_2H_5HSO_4 \xrightarrow[\text{excess},\ 160°c]{\text{conc. } H_2SO_4} H_2C=CH_2$

55. Dieckmann's method : Cyclic ketones are obtained from esters of dicarboxilic acids by interamolecular condensation in the presence of sodium alkoxide.

56. Zeisel's method With HI:

(i) $CH_3OCH_3 + HI\ (Cold) \rightarrow CH_3OH + CH_3I$

(ii) $CH_3 \vdots OCH_2CH_3 + HI \xrightarrow{\text{cold}} CH_3I + CH_3CH_2OH$

(iii) $CH_3 - O - CH_3 + 2HI \xrightarrow{\Delta} 2CH_3I + H_2O$

(iv) $CH_3 - O - CH_2CH_3 + 2HI \xrightarrow{\Delta} CH_3I + H_2O + CH_3CH_2I$

In the case of mixed ethers, the halogen atom attaches to a smaller and less complex alkyl group.

$$R-O-R + H^+ \longrightarrow R-\overset{+}{\underset{|}{O}}(H)-R$$

S_N1 : (i) $R-\overset{+}{O}(H)-R \xrightarrow{\text{slow}} R^+ + ROH$

(ii) $R^+ + X^- \xrightarrow{\text{fast}} R-X$

S_N2 : $X^- + R-\overset{+}{O}(H)-R \xrightarrow{\text{slow}} \left[\overset{\delta^-}{X}\cdots R\cdots \overset{\delta^+}{O}(H)-R\right] \longrightarrow R-X + ROH$

Order or reactivity: HI > HBr > HCl. This reaction is the basis of Zeisel's method (Estimation of alkoxy group)

$$C_6H_5-O-CH_3 \text{ (anisole)} \xrightarrow{\text{hot conc.HI}} C_6H_5-OH + CH_3I$$

57. Rosenmund's reaction:

$$RCOCl + 2H \xrightarrow{Pd / BaSO_4} RCHO + HCl$$

This catalyst system is to stop further reduction of RCHO to RCH_2OH. Use of $LiAlH_4$ or $NaBH_4$ leads to the formation of alcohols.

58. Etard reaction:

$$C_6H_5-CH_3 \xrightarrow[\text{(Chromyl chloride)}]{CrO_2Cl_2;\ H_2O} C_6H_5-CHO$$

59. Vilsmeier formylation reaction : Aldehydes are produced when activated aromatic nucleus reacts with N-methyl formanilide or N–N–dimethluformamide in the presence of phosphorous oxychloride. It is an electrophilic substitution reaction.

$$(H_3C)_2N-C_6H_5 \text{ (N,Ndimethylaniline)} + OCHN(CH_3)_2 \text{ (N,N - dimethylformamide)} \longrightarrow (H_3C)_2N-C_6H_4-CH(OH)-N-(CH_3)_2 \text{ (Unstable)}$$

$$\longrightarrow (H_3C)_2N-C_6H_4-CHO + (CH_3)_2NH \text{ (p-dimethylaminobenzaldehyde)}$$

60. Stephen's reaction:

$$R-C\equiv N \xrightarrow[2[H]]{SnCl_2\ /HCl} RCH=NH.HCl \xrightarrow{H_2O} RCHO$$

61. Oxo reaction / Hydroformylation : Aldehydes can be obtained by passing the mixture of an alkene, carbon monoxide and hydrogen over cobalt or dicobalt octacarbonyl $[Co_2(CO)_8]$ catalyst at high temperature and pressure.

$$R-CH_2-CH=CH_2 + CO + H_2 \xrightarrow{[Co_2(CO)_8]} R-CH_2-CH_2-CHO$$

62. Sommelet reaction:

$$C_6H_5-CH_2Cl \xrightarrow{(CH_2)_6N_4} C_6H_5-CHO$$

63. Collman's reaction : Collman's reagent is Sodium tetra carboxyl ferrate (II), $Na_2[Fe(CO)_4]$. The reagent reacts with haloalkanes in the presence of triphenyl phosphine $(Ph)_3P$ to give aldehyde having one carbon atom more than the ones present in R–X.

$$R-C\equiv N \xrightarrow[2[H]]{SnCl_2\ /HCl} RCH=NH.HCl \xrightarrow{H_2O} RCHO$$

R–CHO is obtained from the product of the above reaction by quenching it with acetic acid.

64. Nef reaction : Depending upon the nature of nitro alkane, aldehyde or ketone is produced in the presence of a base followed by hydrolysis.

$$R-CH_2-NO_2 \xrightarrow{OH^-} R-CH=NO-OH \xrightarrow{H_3O^+} R-CHO$$

$$(R)(R_1)CH-NO_2 \xrightarrow{OH^-} (R)(R_1)C=NO-OH \xrightarrow{H_3O^+} (R)(R_1)C=O$$

65. Wislicenus method : Calcium or barium salts of dicarboxilic acids on dry distillation give cyclic ketones having one carbon atom less

$$\begin{matrix}CH_2-CH_2-COO\\ |\\ CH_2-CH_2-COO\end{matrix}\!\!>Ca \text{ or } Ba \xrightarrow[\Delta]{-CaCO_3} \begin{matrix}CH_2-CH_2\\ |\\ CH_2-CH_2\end{matrix}\!\!>CO$$

66. Meerwin Pondorf verley reduction:

$$R_2CO + [(CH_3)_2CHO)]_3Al \longrightarrow R_2CHOH$$

67. Tischenko reaction: aldehydes only give this reaction.

$$CH_3CHO \xrightarrow{Al(OC_2H_5)_3} CH_3COOC_2H_5$$

68. Aldol condensation: Aldehydes and ketones having α-hydrogen atom undergo this condensation. HCHO, H_6H_5CHO, $(CH_3)_3C.CHO$ do not give this reaction.

$$OH^- + H-CH_2-CHO \xrightarrow{-H_2O} \overset{-}{:}CH_2CHO;$$

$$CH_3-\overset{O}{\underset{H}{C}} + \bar{:}CH_2CHO \longrightarrow CH_3-\overset{O^-}{\underset{H}{C}}-CH_2CHO \overset{H^+}{\rightleftharpoons} \text{Aldol}$$

69. Claisen reaction : The condensation between an aromatic aldehyde or a ketone having no α-hydrogen and aliphatic aldehyde, ketone or ester having an active atom at α-position in the presence of dilute alkali for form α,β–unsaturated compound is known as Claisen or Claisen Schmidt condensation.

$$C_6H_5CHO + HCH_2CHO \overset{NaOH}{\rightleftharpoons} C_6H_5-\underset{OH}{CH}-CH_2-CHO$$

$$\xrightarrow{\Delta\ -H_2O} C_6H_5-CH=CH-CHO$$

70. Cannizzaro's reaction: Aldehydes having no α-hydrogen respond to this reaction

$$2\,HCHO + NaOH \longrightarrow HCOONa + CH_3OH$$

$$C_6H_5-CHO + NaOH \longrightarrow C_6H_5-COONa + C_6H_5-CH_2OH$$

$$CH_3CHO \xrightarrow{Al(OC_2H_5)_3} CH_3COOC_2H_5$$

$$H-\overset{OH}{\underset{H}{C}}-O^- + \overset{H}{\underset{H}{C}}=O \longrightarrow H-\overset{OH}{C}=O + H-\overset{H}{\underset{H}{C}}-O^-$$

$$H-\overset{OH}{C}=O + H-\overset{H}{\underset{H}{C}}-O^- \xrightarrow[\text{exchange}]{\text{Proton}} H-\overset{O^-}{C}=O + H-\overset{H}{\underset{H}{C}}-OH$$

$$\downarrow Na^+$$

$$H-\overset{ONa}{C}=O$$

Certain aldehydes having hydrogen at α-position also undergo connizzaro's reaction when heated with NaOH at high temperature.

$$2(CH_3)_2CHCHO + NaOH \xrightarrow{200°C} (CH_3)_2CHCH_2OH + (CH_3)_2CHCOONa$$

2-methylpropanal → 2-methylpropanol + sodium salt of 2-methylpropanoic acid

Crossed cannizzaro's reaction:

$$\text{Anisaldehyde } (p\text{-}CH_3OC_6H_4CHO) + HCHO \xrightarrow[NaOH]{conc.} p\text{-}CH_3OC_6H_4CH_2OH + HCOONa$$

$$\text{veratraldehyde} + HCHO \xrightarrow[\substack{NaOH \\ 65°C}]{50\%} (\text{CH}_2\text{OH product, ring with } CH_3, OCH_3) + HCOONa$$

Intramolecular Cannizzaro's reaction

$$CHO.CHO \xrightarrow{NaOH} CH_2OH.COONa$$

Sod. glycollate

71. Benzoin condensation:

$$2\,C_6H_5-CHO \xrightarrow[C_2H_5OH,\,H_2O]{KCN\,;} C_6H_5-\overset{O}{C}-\underset{H}{\overset{OH}{C}}-C_6H_5$$

$$\text{(i)}\ Ph-\underset{H}{\overset{O}{C}} \overset{CN^-}{\rightleftharpoons} Ph-\underset{H}{\overset{O^-}{C}}-CN \underset{-OH^-}{\overset{H_2O}{\rightleftharpoons}} Ph-\underset{H}{\overset{OH}{C}}-CN \underset{H_2O}{\overset{-OH^-}{\rightleftharpoons}} Ph-\underset{CN}{\overset{OH}{C}}:$$

Cynohydrin

$$\text{(ii)}\ Ph-\underset{CN}{\overset{OH}{C}}: + \underset{H}{\overset{O}{C}}-Ph \rightleftharpoons Ph-\underset{CN}{\overset{OH}{C}}-\underset{H}{\overset{O^-}{C}}-Ph \underset{-OH^-}{\overset{H_2O}{\rightleftharpoons}} Ph-\underset{CN}{\overset{OH}{C}}-\underset{H}{\overset{OH}{C}}-Ph$$

Intermediate — Cyanohydrin (of benzoin)

$$\text{(iii)}\ Ph-\underset{CN}{\overset{OH}{C}}-\underset{H}{\overset{OH}{C}}-Ph \underset{-H_2O}{\overset{OH^-}{\rightleftharpoons}} Ph-\underset{CN}{\overset{O^-}{C}}-\underset{H}{\overset{OH}{C}}-Ph \overset{CN^-}{\rightleftharpoons} Ph-\overset{O}{C}-\underset{H}{\overset{OH}{C}}-Ph$$

72. Reformatsky reaction:

$$RCOR' + Zn + BrCH_2COOC_2H_5 \longrightarrow R-\underset{OH}{\overset{R'}{C}}-CH_2COOC_2H_5$$

$$R-\underset{Br}{CH}-COOC_2H_5 + Zn \xrightarrow{\text{ehter}} Br-Zn-\overset{R}{CH}-COOC_2H_5$$

$$R-\underset{R}{\overset{O}{C}} + Br\,Zn-\overset{R}{CH}-COOC_2H_5 \longrightarrow R-\underset{R}{\overset{OZnBr}{C}}-\underset{R}{CH}-COOC_2H_5$$

organometallic compound

$$\downarrow H_3O^+$$

$$R-\underset{R}{\overset{OH}{C}}-\underset{R}{CH}-COOC_2H_5$$

β-hydroxy ester

73. Wittig reaction: Conversion of an aldehyde or ketone into alkene

$Ph_2C = O + Ph_3P = CH_2 \rightarrow Ph_2C = CH_2$

Methylene triphenyl phosphorane Nucleophilic addition at the carbonyl centre

$$Ph_3P{=}C(R')(R) \longleftrightarrow Ph_3P^{+}{-}\bar{C}(R')(R)$$

$$R_2C{=}O + Ph_3\overset{+}{P}{-}\bar{C}(R')(R) \rightarrow R_2C(O^-){-}C(R')(R){-}\overset{+}{P}Ph_3 \rightarrow R_2C{=}C(R')(R) + Ph_3P{\rightarrow}O$$

74. Perkin reaction: it is shown by aromatic aldehydes.

$$C_6H_5{-}CHO + (CH_3CO)_2O \xrightarrow{CH_3COONa} C_6H_5{-}CH{=}CH{-}COOH$$ (cinnamic acid)

$$H_3C{-}\overset{O}{\overset{\|}{C}}{-}O{-}\overset{O}{\overset{\|}{C}}{-}CH_3 + CH_3COO^- \rightleftharpoons \bar{C}H_2{-}\overset{O}{\overset{\|}{C}}{-}O{-}\overset{O}{\overset{\|}{C}}{-}CH_3 + CH_3COONa$$

$$Ph{-}\underset{H}{\overset{O}{\overset{\|}{C}}} + \bar{C}H_2{-}\overset{O}{\overset{\|}{C}}{-}O{-}\overset{O}{\overset{\|}{C}}{-}CH_3 \rightarrow Ph{-}\underset{H}{\overset{O^-}{C}}{-}CH_2{-}\overset{O}{\overset{\|}{C}}{-}O{-}\overset{O}{\overset{\|}{C}}{-}CH_3$$

$$\downarrow CH_3COOH$$

$$Ph{-}CH{=}CH{-}\overset{O}{\overset{\|}{C}}{-}O{-}\overset{O}{\overset{\|}{C}}{-}CH_3 \xleftarrow{-H_2O} Ph{-}\underset{H}{\overset{OH}{C}}{-}CH_2{-}\overset{O}{\overset{\|}{C}}{-}O{-}\overset{O}{\overset{\|}{C}}{-}CH_3$$

$$\downarrow H_2O$$

$$Ph{-}CH{=}CH{-}\overset{O}{\overset{\|}{C}}{-}OH + CH_3COOH$$

75. Knoevenagel reaction: Aldehyde or ketone with active methylene group containing compound gives α, β-unsaturated acid.

$$RCHO(\text{or ketone}) + \underset{\text{malonic ester}}{CH_2(COOC_2H_5)_2} \xrightarrow{Pyridine} RCH{=}CHCOOH$$

$$\ddot{B} + H{-}\underset{H}{C}{-}(COOC_2H_5)_2 \rightarrow \overset{\ominus}{C}H{-}(COOC_2H_5)_2 + BH$$

$$R{-}\underset{H}{\overset{O}{\overset{\|}{C}}} + \bar{C}H(COOC_2H_5)_2 \rightarrow R{-}\underset{H}{\overset{O^-}{C}}{-}CH(COOC_2H_5)_2$$

$$\downarrow BH$$

$$R{-}CH{=}C(COOC_2H_5)_2 \xleftarrow{-H_2O} R{-}\underset{H}{\overset{OH}{C}}{-}CH(COOC_2H_5)_2$$

$$\downarrow H_2O$$

$$R{-}CH{=}C(COOH)_2 \xrightarrow[-CO_2]{\Delta} R{-}CH{=}CH{-}COOH$$

76. Oppenauer oxidation : Secondary alcohols are oxidised to ketones by treating with aluminium tert – butoxide in excess of acetone or cyclohexanone

$$(R)(R_1)CH{-}OH + CH_3COCH_3 \xrightarrow{[(CH_3)_3CHO]_3Al} (R)(R_1)C{=}O + CH_3CHOH$$

77. Hell-Volhard Zelinsky reaction:
Here α-bromination of carboxylic acids takes place. It is a substitution reaction in alkyl or phenyl group.

Acids without α–hydrogen do not show this reaction.

$$R_2{-}\underset{H}{C}{-}COOH + Br_2 \xrightarrow{PCl_3} R_2{-}\underset{Br}{C}{-}COOH + HBr$$

78. Arndt-Eistert synthesis : An acid (aliphatic, alicyclic or aromatic) is converted to the next higher homologue. The ketene obtained in Wolff rearrangement is converted to acid on hydrolysis

$$R{-}COOH \xrightarrow{SOCl_2} R{-}COCl \xrightarrow{CH_2N_2} RCOCHN_2 \longrightarrow R{-}CH{=}C{=}O \xrightarrow{H_3O^+} RCH_2COOH$$

79. Schmidt reaction: Primary amines are formed

$$R{-}COOH + \underset{\text{hydrazoic acid}}{HN_3} \xrightarrow{\text{conc. } H_2SO_4} RCON_3 \xrightarrow{H_2O} R{-}NH_2 + CO_2 + N_2$$

80. Balnc's rule: on heating

(i) Two carboxylic acid groups on the same carbon atom leads to the formation of mono carboxylic acid. e.g.

$$\underset{\text{malonic acid}}{H_2C(COOH)_2} \xrightarrow[-CO_2]{\Delta} \underset{\text{acetic acid}}{H_3C{-}\overset{O}{\overset{\|}{C}}{-}OH}$$

(ii) I,4 and 1,5 dicarboxylic acids form cyclic anhydrides. e.g.

$$\underset{\text{succinic acid (1,4)}}{CH_2{\cdot}COOH / CH_2{\cdot}COOH} \xrightarrow[-H_2O]{\Delta} CH_2{\cdot}CO / CH_2{\cdot}CO \text{ (}{>}O\text{)};$$

$$\underset{\text{glutaric acid (1,5)}}{H_2C(CH_2{\cdot}COOH)_2} \xrightarrow[-H_2O]{\Delta} H_2C(CH_2{-}CO)_2O$$

(iii) 1,6 and 1,7– dicarboxylic acids on heating with acetic anhydride form cyclic ketons.

$$\underset{\text{adipic acid (1,6)}}{CH_2{\cdot}CH_2{\cdot}COOH / CH_2\,CH_2\,COOH} \xrightarrow[(CH_3CO)_2O]{\Delta} CH_2{-}CH_2 / CH_2{-}CH_2 \text{ (}{>}CO\text{)};$$

$$\underset{\text{pimelic acid (1,7)}}{H_2C(CH_2{\cdot}CH_2{\cdot}COOH)_2} \xrightarrow[(CH_3CO)_2O]{\Delta} H_2C(CH_2{-}CH_2)_2CO$$

81. Intramolecular esterification: A hydroxy acid is both alcohol and acid. In such cases, where a 5 or 6 membered ring can be formed, intramolecular esterification takes place and a lactone is formed. e.g.

$$R{-}\underset{OH}{CH}{-}CH_2{-}CH_2{-}COOH \rightarrow R{-}CH(OH)\,H_2C{-}CH_2\,(HOOC) \xrightarrow{-H_2O} \text{lactone: } R{-}CH{-}CH_2{-}CH_2{-}CO{-}O \text{ (ring)}$$

82. Trans esterification or ester interchange:

$$R-\overset{O}{\overset{\|}{C}}-OR' + R''-OH \underset{\text{or HCl gas}}{\overset{\text{conc. } H_2SO_4}{\rightleftharpoons}} R-\overset{O}{\overset{\|}{C}}-OR'' + R'-OH$$

high boiling ester, high boiling alcohol → higher boiling ester, lower boiling alcohol

83. Claisen condensation:

$$H_3C-\overset{O}{\overset{\|}{C}}-O-CH_2CH_3 + H\,CH_2-\overset{O}{\overset{\|}{C}}-O-CH_2CH_3 \xrightarrow{-C_2H_5OH} H_3C-\overset{O}{\overset{\|}{C}}-CH_2-\overset{O}{\overset{\|}{C}}-O-CH_2CH_3$$

acetoacetic ester

carbanion is the reactive intermediate in this reaction.

$$(B.. + H\,CH_2-\overset{O}{\overset{\|}{C}}-O-CH_2CH_3 \xrightarrow{-BH} H_2\overset{..}{\overset{-}{C}}-\overset{O}{\overset{\|}{C}}-O-CH_2CH_3)$$

84. Baeyer – Villeger reaction: Ketones on oxidation with peracids give esters.

$$R-\overset{O}{\overset{\|}{C}}-R' + R-\overset{O}{\overset{\|}{C}}-O-O-H \longrightarrow \underset{\text{major}}{R-\overset{O}{\overset{\|}{C}}-OR'} + \underset{\text{minor}}{R'-\overset{O}{\overset{\|}{C}}-OR}$$

85. Hoffmann degradation/ bromamide reaction:

$$H_3C-\overset{O}{\overset{\|}{C}}-\underset{H}{\underset{|}{\ddot{N}}}H + KOH + Br-Br \xrightarrow{\text{bromination}} H_3C-\overset{O}{\overset{\|}{C}}-\underset{H}{\underset{|}{\ddot{N}}}-Br \text{ (acetobromamide)}$$

$$\xrightarrow{\text{dehydro bromination}} \left[H_3C-\overset{O}{\overset{\|}{C}}-\ddot{N}:\right] \text{ (acetyl nitrene)} \xrightarrow{\text{rearrangement}} H_3C-\ddot{N}=C=O \text{ (methyl isocyanate)} \xrightarrow[\text{hydrolysis}]{KOH} \underset{\text{methyl amine}}{CH_3NH_2} + K_2CO_3$$

R–N=C=O (isocyanate) is the satble intermedicate. N– alkyl or N–aryl amides do not show this degradation.

86. Hoffmann's method:

$$R-X + NH_3 \rightarrow R-NH_2 + R_2-NH + R_3-N + R_4\overset{+}{N}X^-$$

depending upon the ratio of the reactants a mixture of primary, secondary, tertiary and quaternary ammonium salt is formed.

87. Bischler – Napieralski synthesis :
α-phenylethylamide undergoes cyclo dehydration to form 3,4-dihydroisoquinoline derivative

$$C_6H_5-CH_2-CH_2-NH-CO-R \xrightarrow[\text{or } PCl_5 / 140°C]{P_2O_5/\Delta \text{or } POCl_3 / 140°C} \text{1-R-3,4-dihydroisoquinoline} \xrightarrow[-H_2]{\text{Se or S}/\Delta} \text{1-R-isoquinoline}$$

88. Gabriel's pthalimide synthesis: only aliphatic primary amines can be prepared.

$$\underset{\text{phthalimide}}{C_6H_4(CO)_2N-H} + \text{alc. KOH} \xrightarrow{-H_2O} \underset{\text{pot. phthalimide}}{C_6H_4(CO)_2N-K} \xrightarrow[-KX]{RX} \underset{\text{N-alkyl phthalimide}}{C_6H_4(CO)_2N-R} \xrightarrow{20\%\ HCl} C_6H_4(COOH)_2 + RNH_2$$

89. Leuckart reaction : Carbonyl compounds on heating with ammonium formate or formamide produce primary amines.

$$R_2C=O \xrightarrow{NH_3} R_2C\begin{matrix}OH\\NH_2\end{matrix} \xrightarrow{-H_2O} R_2-C=NH$$

$$\xrightarrow[-CO_2]{HCOOH} R_2-CH-NH_2$$

the participation of HCOOH in the last step is due to the reaction,

$$HCOONH_4 \rightleftharpoons [HCOOH + NH_3] \xleftarrow{H_2O} HCONH_2$$

90. Curtius reaction:

$$\underset{\text{acid azide}}{R-CON_3} \xrightarrow[-N_2]{CHCl_3} \underset{\text{alkyl isocyanate}}{R-N=C=O} \xrightarrow{NaOH} R-NH_2 + Na_2CO_3 + H_2O$$

$$R-\overset{O}{\overset{\|}{C}}-X + NaN_3 \longrightarrow R-\overset{O}{\overset{\|}{C}}-\ddot{N}-\overset{+}{N}\equiv N: \xrightarrow{-N_2} \left[R-\overset{O}{\overset{\|}{C}}-\ddot{N}\right] \longrightarrow O=C=N-R$$

91. Erlenmeyer azlactone synthesis : Benzoyl glycine (hippuric acid) on heating with aromatic aldehyde in the presence of CH_3COONa or $(CH_3CO)_2O$ produced azlactones.

$$Ph-CHO + \underset{NHCOPh}{\overset{CH_2-CH_3}{|}} \xrightarrow[(CH_3CO)_2O]{CH_3COONa \text{ or}} Ph-CH=C-CO \text{ (ring: N, O, C–Ph)}$$

92. Mannich reaction or amino methylation

$$R-\underset{NO_2}{\overset{R}{C}}-H + HCHO + NH_4Cl \xrightarrow[-H_2O]{\Delta} R-\underset{NO_2}{\overset{R}{C}}-CH_2NH_2.HCl$$

depending upon the nature of nitroalkane employed in the reaction secondary or tertiary nitro amines are produced. Primary or secondary amines are condensed with HCHO and a compound having active hydrogen atom. The active hydrogen is substituted by amino methyl group

$$PhCOCH_3 + HCHO + R_2NH \xrightarrow{-H_2O} \underset{\text{mannich base}}{Ph-CO-CH_2-CH_2-NR_2}$$

Nitro compounds like 2-nitropropane containing active hydrogen can also be employed.

93. Mendius reaction:

$$H_3C-C\equiv N + 4[H] \xrightarrow{Na/C_2H_5OH} CH_3CH_2NH_2 \text{ primary}$$

$$H_3C-N\equiv C + 4[H] \xrightarrow{Na/C_2H_5OH} CH_3NHCH_3 \text{ secondary}$$

94. Sorensen's synthesis : α–amino acids are synthesized. Malonic ester and pthalimide are employed in the reaction.

$$CH_2(COOC_2H_5)_2 \xrightarrow{Br_2} BrHC(COOC_2H_5)_2$$

$$C_6H_4(CO)_2N-K + BrHC(COOC_2H_5)_2 \rightarrow C_6H_4(CO)_2N-CH(COOC_2H_5)_2$$

$$\xrightarrow{C_2H_5ONa} C_6H_4(CO)_2N-CNa(COOC_2H_5)_2 \xrightarrow{BrCH_2Ph} C_6H_4(CO)_2N-C(CH_2-Ph)(COOC_2H_5)_2$$

$$\xrightarrow{hydrolysis} C_6H_4(COOH)CO-NH-C(CH_2-Ph)(COOC_2H_5)_2 \xrightarrow{conc.\ HCl} C_6H_4(COOH)_2 + CO_2 + PhCH_2-CH(NH_2)-COOH$$

95. Ritter reaction: Tertiary alcohols are converted into primary amines

$$H_3C-C(CH_3)_2-OH + HCN + H_2SO_4 \rightarrow H_3C-C(CH_3)_2-NH_2$$

96. Lossen rearrangement : Hydroxamic acids by dehydrative rearrangement by heating in the presence of a very little amount of conc. HCl form isocyanates. Amines can be obtained from the isocyanates by hydrolysis with aqueous KOH.

97. Hinsberg's method: The reagent employed is benzene sulphonyl chloride

1° : $C_6H_5-SO_2Cl + H_2NR \xrightarrow{-HCl} C_6H_5-SO_2NHR \xrightarrow{KOH} C_6H_5-SO_2NKR$

2° : $C_6H_5-SO_2Cl + HNR_2 \xrightarrow{-HCl} C_6H_5-SO_2NR_2 \not\rightarrow$ no reaction

with 3° amines, no reaction is possible as active H atom is absent.

98. Dakins method : The amino acids obtained by the hydrolysis of proteins with dilute acids are extracted with water, saturated butanol and then separated either by fractional distillation or fractional crystallisation of esters. The basic amino acids are precipitated from the residue after butanol extraction by treating with phosphotunustic acid.

99. Hoffmann's method: The reagent employed is ethyl oxalate

with 1° $\begin{matrix} COOC_2H_5 \\ | \\ COOC_2H_5 \end{matrix} + \begin{matrix} H_2NR \\ H_2NR \end{matrix} \longrightarrow \begin{matrix} CONHR \\ | \\ CONHR \end{matrix} + 2C_2H_5OH$

dialkyl oxamide

with 2° $\begin{matrix} COOC_2H_5 \\ | \\ COOC_2H_5 \end{matrix} + HNR_2 \longrightarrow \begin{matrix} CONR_2 \\ | \\ COOC_2H_5 \end{matrix} + C_2H_5OH$

dialkyl oxamic ester

with 3° amines no reaction is possible

100. Chichibabin or Tschitschibabin reaction : This is an amination reaction where amino group is introduced.

$$\text{pyridine} + NaNH_2 \xrightarrow{110°C} \text{2 aminopyridine} \xrightarrow[\Delta]{NaNH_2} \text{2, 6 diaminopyridine}$$

101. Mustard oil reaction:

(i) $2R-NH_2 + S=C=S \longrightarrow S=C(NHR)(SH.NH_2R) \xrightarrow{HgCl_2} R-N=C=S + HgS + 2HCl$

N-alkyl thiocarbamic acid salt; alkyl isothiocynate (mustard oil)

(ii) $2R_2-NH + S=C=S \longrightarrow S=C(NR_2)(SH.NHR_2) \xrightarrow{HgCl_2}$ No reaction

dialkyl dithiocarbamic acid salt

Tertiary amines do not show reaction with CS_2. In fact this test can also be used to distinguish 1°, 2° and 3° amines.

102. Liebermann's nitroso reaction is shown by secondary amines

$$(CH_3)_2N\boxed{H + HO}NO \xrightarrow{-H_2O} (CH_3)_2N-N=O$$

dimethyl nitrosoamine

103. Demjamov Reaction : Demjamov ring expansion or ring contraction as shown below:

(a) cyclopropyl methyl amine → Ring expansion

$$2\ \text{cyclopropyl}-CH_2NH_2 + HNO_2 \longrightarrow \text{cyclopropyl}-CH_2OH + \text{cyclobutanol (OH)}$$

(b) cyclobutylamine → Ring contraction

$$2\ \text{cyclobutyl}-NH_2 + HNO_2 \longrightarrow \text{cyclobutyl}-OH + \text{cyclopropyl}-CH_2OH$$

104. Diazo Coupling reaction: Benzene diazonium chloride couples with phenols and amines to give azo compounds. These are used as dyes or dye intermediates.

(i) 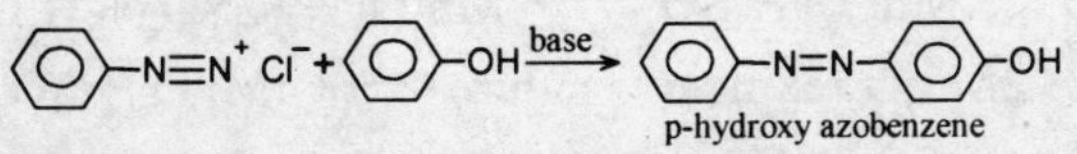 $C_6H_5-N\equiv N^+Cl^- + C_6H_5-OH \xrightarrow{base} C_6H_5-N=N-C_6H_4-OH$
p-hydroxy azobenzene

(ii) 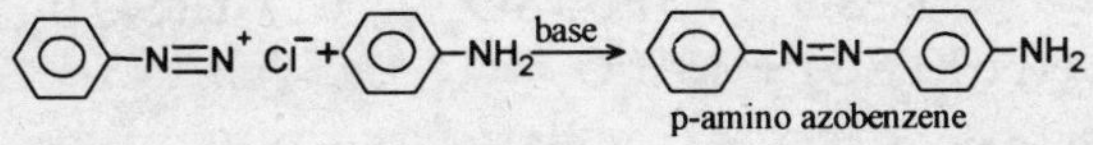 $C_6H_5-N\equiv N^+Cl^- + C_6H_5-NH_2 \xrightarrow{base} C_6H_5-N=N-C_6H_4-NH_2$
p-amino azobenzene

(iii) 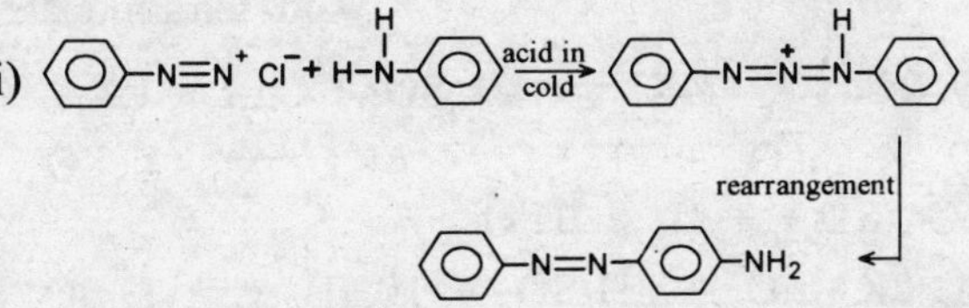 $C_6H_5-N\equiv N^+Cl^- + H-NH-C_6H_5 \xrightarrow{\text{acid in cold}} C_6H_5-N=\overset{+}{N}H-NH-C_6H_5 \xrightarrow{rearrangement} C_6H_5-N=N-C_6H_4-NH_2$

105. Hoffmann elimination: Degradation of quaternary ammonium salts

$R_4N^+OH^- \xrightarrow{\Delta}$ Tertiary amine + H_2O + alkene

$H_3C-CH_2-CH_2-CH_2N^+(CH_3)_3OH^- \xrightarrow{\Delta} H_3C-CH_2-CH=CH_2 + N(CH_3)_3 + H_2O$

106. Thorpe or Thorpe – Ziegler reaction : Two molecules of an alkyl nitrile condense in the presence of sodium and dry ether to give an adduct

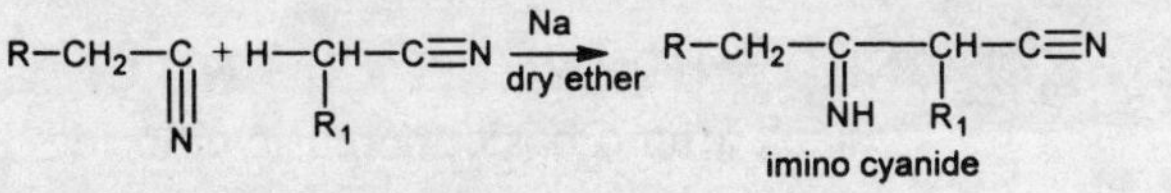

$R-CH_2-C\equiv N + H-CH(R_1)-C\equiv N \xrightarrow[\text{dry ether}]{Na} R-CH_2-C(=NH)-CH(R_1)-C\equiv N$
imino cyanide

107. Strecker synthesis:

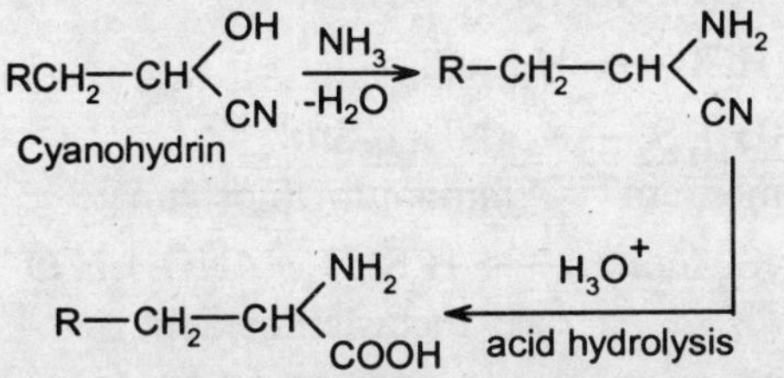

$RCH_2-CH(OH)CN \xrightarrow[-H_2O]{NH_3} R-CH_2-CH(NH_2)CN \xrightarrow[\text{acid hydrolysis}]{H_3O^+} R-CH_2-CH(NH_2)COOH$

Cyanohydrin

112. Baeyer – Fisher's synthesis : Urea is converted to ureic acid.

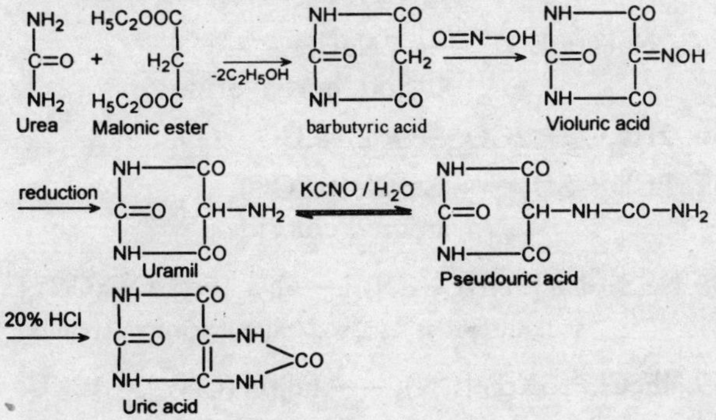

Urea + Malonic ester $\xrightarrow{-2C_2H_5OH}$ barbutyric acid $\xrightarrow{O=N-OH}$ Violuric acid $\xrightarrow{reduction}$ Uramil $\xrightarrow{KCNO / H_2O}$ Pseudouric acid $\xrightarrow{20\%\ HCl}$ Uric acid

Pseudo ureic acid was converted to uric acid by Fischer.

109. Lobry de Bruyn van Ekenstein rearrangement: In the presence of dilute aqueous alkalies glucose / fructose / mannose undergoes this rearrangement and all the three are formed.

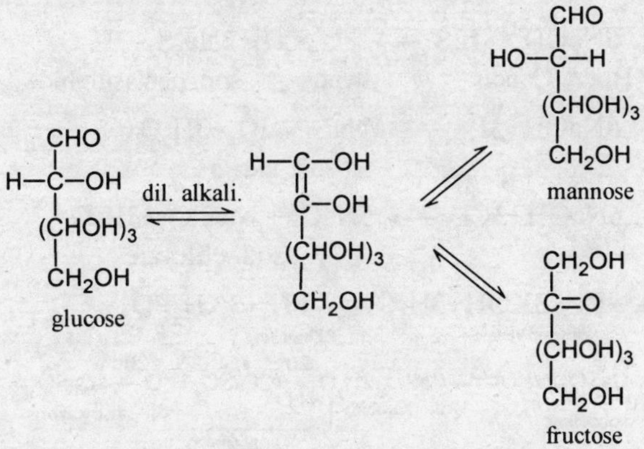

glucose (CHO–CHOH–(CHOH)$_3$–CH_2OH) $\underset{\text{dil. alkali.}}{\rightleftharpoons}$ enediol (H–C–OH=C–OH–(CHOH)$_3$–CH_2OH) $\rightleftharpoons$ mannose (CHO–HO–C–H–(CHOH)$_3$–CH_2OH) ; $\rightleftharpoons$ fructose (CH_2OH–C=O–(CHOH)$_3$–CH_2OH)

Fructose, although contains a keto group, it is a reducing sugar because of this rearrangement in which glucose and mannose which contain aldehyde group are formed.

110. Skraup synthesis : Quinoline and its derivatives are obtained by heating aniline with corresponding nitro compound (nitro benzene), glycerol and conc. H_2SO_4.

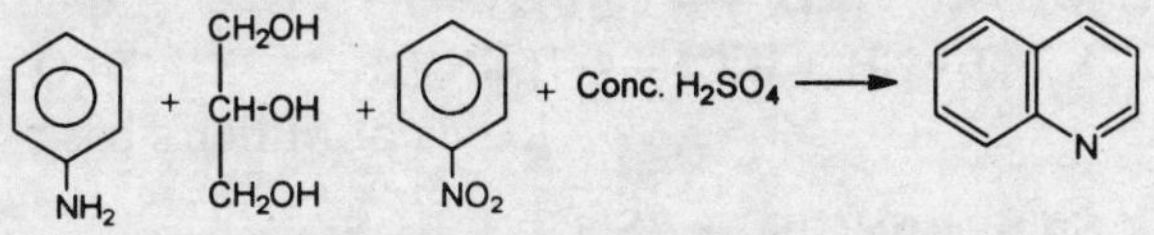

Aniline + $CH_2OH-CH\text{-}OH-CH_2OH$ + Nitrobenzene + Conc. H_2SO_4 ⟶ Quinoline

Some important Inorganic reactions :

1. $CO + 2H_2 \xrightarrow[Cro_3,\ 300°c]{ZnO\ /} CH_3OH$

2. Na_2O_2 (or BaO_2) + $H_2SO_4 \longrightarrow Na_2SO_4 + H_2O_2$

3. $K_2Cr_2O_7 + H_2SO_4 + 4H_2O_2 \longrightarrow 2CrO_5 + K_2SO_4 + 5H_2O$

4. $Na \xrightarrow{O_2} Na_2O \xrightarrow{H_2O} NaOH \xrightarrow{CO_2} Na_2CO_3$; $Na_2O \xrightarrow{O_2} Na_2O_2$

5. $2K_4[Fe(CN)_6] + H_2SO_4 \longrightarrow 2K_3[Fe(CN)_6] + K_2SO_4 + 2H_2O$

6. $2KI + H_2SO_4 + H_2O_2 \longrightarrow K_2SO_4 + 2H_2O + I_2$

7. $ZnCl_2$ or $ZnSO_4 + 2NaOH \longrightarrow Zn(OH)_2 + 2NaCl$

$Zn(OH)_2 + 2NaOH \longrightarrow Na_2ZnO_2 + 2H_2O$
Sodium Zincate

8. $Al_2(SO_4)_3 + 6NaOH \longrightarrow 2Al(OH)_3 + 3Na_2SO_4$

$Al(OH)_3 + NaOH \longrightarrow NaAlO_2 + 2H_2O$
Sodium meta aluminate

9. $2Al + 2NaOH + 2H_2O \longrightarrow 2NaAlO_2 + 3H_2$
Solution — Sod. meta aluminate

10. $2Al + 6NaOH \xrightarrow{\text{fused}} 2NaAlO_2 + 3H_2$

11. $2KI + HgCl_2 \longrightarrow HgI_2 + 2KCl$

$HgI_2 + 2KI \longrightarrow K_2HgI_4$

Nessler's reagent

12. $6NaOH + 12S \longrightarrow Na_2S_2O_3 + 2Na_2S_5 + 3H_2O$

Hot & Conc. hypo sod. pentasulphide

13. $6NaOH + 3I_2 \longrightarrow 5NaI + NaIO_3 + 3H_2O$

sod. iodate

14. $6NaOH + 3Cl_2 \longrightarrow 5NaCl + NaClO_3 + 3H_2O$

sod. chlorate

15. $4P + 3KOH + 3H_2O \longrightarrow PH_3 + 3KH_2PO_2$

White or Yellow Phosphine

16. $CaSO_4.2H_2O \xrightarrow{\Delta} CaSO_4.2H_2O \xrightarrow[H_2O]{120^\circ c} 2CaSO_4.H_2O \xrightarrow{200^\circ c} CaSO_4$

monoclinic Orthorhombic plaster of paris anhydrous

17. $B_2H_6 + 2NH_3 \xrightarrow{\Delta} B_2H_6.2NH_3 \xrightarrow{200^\circ C} B_3N_3H_6$

Diborane Borazine

18. $H_2N - NH_2 + HNO_2 \longrightarrow N_3H + 2H_2O$

hydrazine hydrazoic acid

19. $NaNH_2 + N_2O \longrightarrow NaN_3 + H_2O$

sodium azide

20. $FeSO_4 + NO + 5H_2O \longrightarrow [Fe(NO)(H_2O)_5]SO_4$

hydrated nitrosyl complex

21. $Mg_3N_2 + 6H_2O \longrightarrow 3Mg(OH)_2 + 2NH_3$

22. $2K_2HgI_4 + NH_3 + 3KOH \longrightarrow H_2NHgOHgI + 7KI + 2H_2O$

Iodide of Millon's base

23. $Sb_2S_3 + 6HCl \longrightarrow 2SbCl_3 + 3H_2S$

24. $2Hg + O_3 \longrightarrow Hg_2O + O_2$ (Tailing of mercury)

25. Moist $I_2 + H_2O + 5O_3 \longrightarrow 2HIO_3 + 5O_2$

iodic acid

$KI + 3O_3 \longrightarrow KIO_3$

pot. iodate

$KI + 5O_3 \longrightarrow KIO_4$

pot. periodate

26. $SiO_2 + 4HF \longrightarrow SiF_4 + 2H_2O$

$SiF_4 + 2HF \longrightarrow H_2SiF_6$ (hydrofluosilicic acid)

27. $Ca(OH)_2 + Cl_2 \longrightarrow CaOCl_2$

slaked lime Bleaching powder

28. $2FeSO_4 + CaOCl_2 + H_2SO_4 \longrightarrow Fe_2(SO_4)_3 + CaCl_2 + H_2O$

29. $3KBr$ (or KI) $+ H_3PO_4 \longrightarrow K_3PO_4 + 3HBr$

30. Springs reaction: $Na_2S + Na_2SO_3 + I_2 \rightarrow Na_2S_2O_3 + 2NaI$

$Na_2CO_3 + 4SO_2 + 2Na_2S \rightarrow 3Na_2S_2O_3 + CO_2$.

31. $3XeF_4 + 6H_2O \longrightarrow 2Xe + XeO_3 + 12HF + 1\frac{1}{2}O_2$

32. $XeF_6 + 3H_2O \longrightarrow XeO_3 + 6HF$

33. $XeF_6 + H_2O$(partial hydrolysis) $\longrightarrow XeOF_4 + 2HF$

34. $I_2 + 10HNO_3 \longrightarrow 2HIO_3 + 4H_2O + 10NO_2$

35. $2KMnO_4 \xrightarrow{240^\circ C} K_2MnO_4 + MnO_2 + O_2$

36. $3KMnO_4 + H_2SO_4 \longrightarrow Mn_2O_7 + K_2SO_4 + H_2O$

$2Mn_2O_7$(explosive nature) $\longrightarrow 4MnO_2 + 3O_2$

37. $6FeSO_4(NH_4)_2SO_4.6H_2O + K_2Cr_2O_7 + 7H_2SO_4 \longrightarrow K_2SO_4 + 3Fe_2(SO_4)_3 + Cr_2(SO_4)_3 + 6(NH_4)_2SO_4 + 13H_2O$

38. $2Na_2S_2O_3 + I_2 \longrightarrow 2NaI + Na_2S_4O_6$

hypo sod. tetrathionate

39. $2CuSO_4 + 4KI \longrightarrow 2k_2SO_4 + Cu_2I_2 + I_2$

40. $K_2Cr_2O_7 + 7H_2SO_4 + 6KI \longrightarrow 4K_2SO_4 + Cr_2(SO_4)_3 + 3I_2 + 7H_2O$

41. $2KMnO_4 + 3H_2SO_4 + 10HBr \longrightarrow K_2SO_4 + 2MnSO_4 + 8H_2O + 5Br_2$

42. $2KMnO_4 + 4H_2S \longrightarrow 2MnS + S + K_2SO_4 + 4H_2O$

43. $K_2CrO_4 + H_2SO_4 \longrightarrow K_2Cr_2O_7 + K_2SO_4 + H_2O$

44. $K_2Cr_2O_7 + 2KOH \longrightarrow K_2CrO_4 + H_2O$

45. $K_2Cr_2O_7 + 4NaCl + 6H_2SO_4 \longrightarrow 2KHSO_4 + 4NaHSO_4 + 2CrO_2Cl_2 + 8H_2O$

chromyl chloride

46. $AgCl\ (Ag_2S) + 4NaCN \longrightarrow Na[Ag(CN)_2] + NaCl$

47. $4Fe + 10HNO_3$(dil.) $\longrightarrow 4Fe(NO_3)_2 + NH_4NO_3 + 3H_2O$

48. $Hg_2Cl_2 + 2NH_3 \longrightarrow [Hg + Hg(NH_2)Cl] + NH_4Cl$

black amino basic mercuric chloride

49. $SnCl_4 + 4NH_4OH \longrightarrow 4NH_4Cl + Sn(OH)_4$

$Sn(OH)_4 \longrightarrow SnO(OH)_2 + H_2O$

50. $SnCl_4 + 2HCl \longrightarrow H_2SnCl_6$

51. $SnS + (NH_4)_2S_2 \longrightarrow (NH_4)_2SnS_3$

Yellow ammonium Ammonium thiostannate

52. $Sn + 4HNO_3$ (conc.) $\longrightarrow H_2SnO_3 + 4NO_2 + H_2O$

metastannic acid

53. $AgBr + 2Na_2S_2O_3 \longrightarrow Na_3[Ag(S_2O_3)_2] + NaBr$

hypo Sod.argentothiosulphate

54. $4LiH + AlCl_3 \longrightarrow LiAlH_4 + 3LiCl$

Lithium aluminium hydride

55. $2NaH + B_2H_6 \longrightarrow 2NaBH_4$

Sod. borohydride

56. $2Na_2O_2 + 2CO_2 \longrightarrow 2Na_2CO_3 + O_2$

57. $PCl_5 + SO_2 \longrightarrow SOCl_2 + POCl_3$

thionyl chloride

58. $Na_2S + Na_2[Fe(NO)(CN)_5] \longrightarrow Na_4[Fe(NOS)(CN)_5]$

sod. nitroprusside sod. thionitroprusside(violet)

59. $4FeCl_3 + 3K_4Fe[CN)_6 \longrightarrow Fe_4[Fe(CN)_6]_3 + 12KCl$

prussian blue

Some Processes in Chemistry

1. Lane's process : Manufacture of hydrogen.
2. Bosch process : Manufacture of hydrogen and nitrogen simultaneously.
3. Uyeno's process : Preparation of pure hydrogen.
4. Permutit process : Softening of water.
5. Calgon process : Softening of water.
6. Chloramine process : Sterilization of water.
7. Clark's method : For removal of temporary hardness of water.
8. Taylor, Eyring and Frost process : Preparation of D_2O.
9. Mercks process : Manufacture of H_2O_2.
10. Castner process : Manufacture of Na.
11. Down's process : Manufacture of Na.
12. Kellner-Solvay process : Manufacture of NaOH.
13. Causticizing process : Manufacture of NaOH.
14. Lowig process : Manufacture of NaOH.
15. Nelson cell process : Manufacture of NaOH.
16. Castner–Kellner cell : Manufacture of NaOH.
17. Gossage process : Manufacture of NaOH.
18. LeBlanc process : Manufacture of Na_2CO_3.
19. Ammonia Soda or Solvay process : Manufacture of Na_2CO_3.
20. Magnesia process or Precht's process : Manufacture of pearl ash (K_2CO_3).
21. Dow's process : Manufacture of Mg from sea water by electrolysis of $MgCl_2$.
22. Baeyer's process : Purification of Bauxite when FeO is present as impurity (wet process)
23. Hall's process : Purification of Bauxite when FeO is present as impurity (dry process).
24. Serpeck's process : Purification of Bauxite when SiO_2 is present as impurity.
25. Goldschmidt-Alumino-Thermic process : Used in thermite welding.
26. Hoope's process : Refining of aluminium.
27. Zairvogel process : Preparation of anhydrous $AlCl_3$.
28. Acheson Process : Manufacture of graphite.
29. Dutch process : Preparation of White lead.
30. Carter process : Manufacture of White lead.
31. Bett's process : Refining of lead.
32. Claude's process : Manufacture of N_2 gas.
33. Linde's process : Manufacture of N_2 gas.
34. Brikland – Eyde process : Manufacture of HNO_3.
35. Ostwald's process : Manufacture of HNO_3.
36. Haber's process : Manufacture of NH_3.
37. Raschig's method : Preparation of hydrazine.
38. Silician process : Manufacture of sulphur.
39. Froth–Flotation process : Purification of sulphide ores.
40. Lead chamber process : Manufacture of H_2SO_4.
41. Contact process : Manufacture of H_2SO_4.
42. Whytlaw Gray's method : Manufacture of fluorine gas.
43. Deacon's process : Manufacture of chlorine gas.
44. Weldon's process : Manufacture of chlorine gas.
45. Hasenclever method : Manufacture of bleaching powder.
46. Beckmann's process : Manufacture of bleaching powder.
47. Fischer and Ringes method : Large scale production of Noble gases.
48. Dewar's method : Separation of noble gases with activated animal charcoal.
49. Rayleigh and Ramsay method : Large scale preparation of noble gases.
50. Claude's process : Production of liquid O_2, N_2 and also noble gases.
51. Puddling process : Manufacture of wrought iron from cast iron.
52. LD process : Manufacture of steel.
53. Siemens – Martin's open process : Manufacture of steel.
54. Bessemer process : Manufacture of steel.
55. Parke's process : Extraction of Ag from argentiferous lead ores.
56. Pattisons process : Extraction of Ag from argentiferous lead ores.
57. Cupellation process : Purification of Ag, (Removal of lead from argentiferous lead).
58. Amalgamation process : Used in gold and silver metallurgy.
59. Moebius electrolytic process : Separation of Ag and Au.
60. Van Arkel de Boer method : Purification of Ti and Zr.
61. Kroll process : Manufacture of Ti and Zr.
62. Mond's process : Refining of Ni.
63. Jantesh Process : Preparation of Vanadium metal.
64. Beard and Crook's method : Reduction of V_2O_5 to V.
65. Mac Arthur – Forrest – Cyanide process : Used in the metallurgy of gold.
66. Belgian process : Reduction of ZnO to zinc metal.
67. Bessemerisation process : In copper metallurgy, it is a self reduction of Cu_2O and Cu_2S to cooper.
68. Zone refining : Purification of semi conductors, e.g. Si, Ge, Ga, In, etc.
69. Quenching : Cooling iron suddenly by plunging in oil or H_2O. The process is also known as hardening.
70. Tempering : Hardened steel is heated below redness and cooled slowly.
71. Nitriding : Heating steel in an atmosphere of NH_3. The surface is coated with hard iron nitride.
72. Annealing : Cooling slowly. This process is used in glass and iron industries.
73. Bergius process : Heavy oil petroleum fractions are mixed with powdered carbon and subjected

to hydrogenation at high pressures and temperatures in the presence of suitable catalysts to produce low boiling hydrocarbons like, petrol, kerosene, diesel, etc.

74. Bethelot method : Alkyl iodides are reduced by HI in the presence of red phosphorous to give hydrocarbons $R-I+HI \xrightarrow{\text{red P; -}I_2} RH$

75. Fischer–Tropsch process : Artificial petrol is obtained on passing a mixture of hydrogen and carbon monoxide (2:1) over ThO_2 + finely powered Fe, Co and Ni at high pressure and temperature.

76. Wig method : Iodine value of oils and fats is determined. ICl dissolved in acetic acid is used as the reagent.

77. Dam method : Iodine value of oils and fats is determined with a solution containing glacial acetic acid, bromine, H_2SO_4 and pyridine.

78. Hubl method : Iodine value of oils and fats is determined with oil/fat solution in CCl_4 and iodine in alcohol followed by the addition of a small amount of $HgCl_2$.

79. Ziesel's method : Estimation of alkoxy group with HI.

80. Quick Vinegar process : Wash containing ~15% C_2H_5OH obtained from alcoholic fermentation is oxidised to vinegar (contains ~8% CH_3COOH) by acetobacter aceti (bacteria) at 35°C.

81. Kharasch and Brown method : Alkanes are converted to carboxylic acids with oxalyl chloride in presence of light.

$$R-H+(COCl)_2 \xrightarrow{h\nu\ -CO_2;\ -HCl} RCOCl \xrightarrow{\text{hydrolysis}\ ;\ -HCl} RCOOH$$

82. Enflevvrage process : Essential oils are extracted from flowers by this method.

83. Hufener's method : Urea in urine or other physiological specimen is estimated. $H_2N-CO-NH_2+3NaOBr \rightarrow N_2+3NaBr+CO_2+2H_2O$

84. Rancidification : Unpleasant smell which fats and oils develop on long exposure to moist air is due to rancidity. By partial saponification, fatty acids are set free which have a strong smell.

85. Raschig process : Chlorobenzene is hydrolysed by superheated steam to form phenol.

86. Schroller process : Preparation of ethanol from saw dust.

$$\text{Saw dust (cellulose)} \xrightarrow{\text{dil. } H_2SO_4;\ \Delta;\ \text{high pressure}} \text{glucose} \xrightarrow{\text{Zymase (yeast)}} \text{ethanol}$$

87. Rayon process : Rayon is mainly prepared from cellulose via cellulose xanthate by this process.

Important Distinctions

1. Ethane – Ethene : Bromine water will be decolourised by ethene due to unsaturation.
2. Ethene – Ethyne : Ammonical Cu_2Cl_2 or ammonical $AgNO_3$ reacts with ethyne to form copper acetylide (reddish ppt.) and silver acetylide (white ppt.)
3. Benzene – Toluene : Toluene is oxidised to benzoic acid by dil. HNO_3 or alk. $KMnO_4$.
4. Ethyl benzene – o-xylene : Ethyl benzene on oxidation with dil. HNO_3 or alkaline $KMnO_4$ gives benzoic acid while phthalic acid is formed from o-xylene.
5. m-xylene – p-xylene : m-xylene dissolves in conc. H_2SO_4 to form m-xylene sulphonic acid while the later doesn't react. P-xylene is easily oxidised to terephthalic acid while isophthalic acid is formed with difficulty from m-xylene.
6. 1-Butene – 2-Butene : 1-Butene on oxidation with hot $KMnO_4$ solution gives propanoic acid $H_3C-CH_2-CH=CH_2 \xrightarrow{(O)} CH_3CH_2COOH$ while 2-Butene gives two molecules of acetic acid.
7. Prop-1-yne–Prop-2-yne : $H_3C-C\equiv C-H$ forms red ppt. with ammonical Cu_2Cl_2 and a white ppt. with Tollen's reagent. Prop-2-yne being a non-terminal acetylene does not give these reactions.
8. Chloroform – Carbon tetrachloride : Chloroform reacts with primary amines (aniline also) in the presence of alc. KOH to give a bad smelling carbylamine. Chloroform gives red colour solution on warming with alkaline solution.
9. p–Bromotoluene – Benzylbromide : p–Bromotoluene on oxidation is converted to p–bromobenzoic acid while benzyl bromide forms benzoic acid. Nuclear substituted bromine remains intact during oxidation.
10. Chlorobenzene – Benzylchloride : Benzylchloride on oxidation gives benzoic acid.
11. Chloroform – chlorobenzene : chlorobenzene burns with black sooty flame as it is aromatic. Chloroform forms a bad selling carbylamine with a primary amine and alc. KOH.
12. Benzyl chloride – p-chlorotoluene : On oxidation benzyl chloride gives benzoic acid while p-chlorobenzoic acid is formed from the later.
13. 1-chloropropane ($CH_2CH_2CH_2Cl$) – Propionyl chloride (CH_3–CH_2COCl) : Propionyl chloride on warning gives white ppt. of AgCl. Propionyl chloride with water gives HCl fumes along with propanoic acid.
14. Methanol – ethanol : Ethanol gives positive iodoform test while the former does not give. Methanol with salicylic acid and conc. H_2SO_4 gives oil of winter green (methyl salicylate).
15. Ethyl alcohol – Benzyl alcohol : Ethyl alcohol is soluble in water while benzyl alcohol is not. Ethyl alcohol gives iodoform with I_2 and NaOH.

16. Ethyl alcohol – Ethanal – Propane : Ethanal gives silver mirror and violet colour with Tollen's and Schiff's reagents respectively. Propanone forms a red coloured solution with sodium nitroprusside and NaOH. Ethyl alcohol gives red colour with ceric nitrate solution and a pleasant smelling ester, ethyl acetate with acetic acid and a few drops of H_2SO_4.
17. 1-propanol – 2-propanol : 1-propanol on oxidation gives propanoic acid having the same number of carbon atom as in the starting compound. 2-propanol gives acid with less number of carbon atoms (CH_3COOH + HCOOH).
18. 1-pentanol – 2-pentanol : 2-pentanol gives iodoform with I_2 and NaOH while other does not. 1-pentanol in Victor Meyer's test gives red colour while the other gives blue colour.
19. Resorcinol – pyrogallol : Resorcinol on warming with conc. H_2SO_4 and Phthalic anhydride followed by the addition of NaOH gives greenish yellow fluorescein.
20. Phenol – α – Naphthol : α – Naphthol reduces ammonical silver nitrate.
21. Resorcinol – α – Naphthol : Resorcinol gives fluorescein test. α – Naphthol gives blue colour on warming with NaOH, Cu powder in CCl_4.
22. o-Nitrophenol – p-Nitrophenol : p-Nitrophenol has higher melting point due to intermolecular hydrogen bonding. o-nitrophenol gives orange red colour in NaOH solution while p-isomer gives yellow colour.
23. p-Methoxy phenol – p-methoxy benzyl alcohol : p-methoxy phenol will give colour with $FeCl_3$ solution.
24. Methyl ethyl ketone ($CH_3COCH_2CH_3$) – Diethyl ketone ($CH_3CH_2COCH_2CH_3$) methyl ethyl ketone on heating with I_2 + NaOH solution gives iodoform. When these compounds are subjected to Clemmenson reduction give n-butane and n-pentane respectively.
25. Methoxy methane (CH_3OCH_3) – Ethoxy ethane ($C_2H_5OC_2H_5$) : On treatment with dil. HI :

$$\text{(i) } CH_3OCH_3 + \text{dil. HI} \xrightarrow{\text{cold}} CH_3I + CH_3OH$$

$$\text{(i) } C_2H_5OCH_3 + \text{dil. HI} \xrightarrow{\text{cold}} C_2H_5I + C_2H_5OH$$

Now the reaction mixture (ii) containing ethyl alcohol gives positive iodoform test.
26. Ethoxy ethane ($C_2H_5OC_2H_5$) – Methoxy ropane ($CH_3OC_3H_7$) – Mixture :
$CH_3OC_3H_7 + HI \rightarrow CH_3I + C_3H_7OH$, which doesn't respond positively to iodoform test while other one responds.
27. Benzyldehyde – Acetophenone : Acetophenone gives iodoform test with iodine and NaOH. Benzyldehyde with alk. $KMnO_4$ followed by acidification gives benzoic acid ppt.
28. Benzaldehyde – Saliclaldehyde : Benzaldehyde shows Cannizzaro's reaction. Saliclaldehyde on reaction with acetyl chloride gives aspirin.
29. Benzamide ($PhCONH_2$) – Acetaniliide ($PhNHCOCH_3$): Acetaniliide gives p-bromoacetanilide ppt. with Br_2 solution.
30. N-methyl benzamide ($PhCONHCH_3$) – N-phenylacetamide ($CH_3CONHPh$) : when heated with NaOH, N-methylbenzamide gives sodium benzoate and methylamine, which gives the ppt. of benzoic acid. N-phenylacetamide gives aniline and acetic acid with NaOH.
31. Ethylacetate ($CH_3COOC_2H_5$) – Methyl propionate ($C_2H_5COOCH_3$) : Ethyl acetate on distillation with NaOH followed by the addition of I_2 and NaOH gives iodoform while methyl propionate does not.
32. Malonic ester [$CH_2(COOC_2H_5)_2$] – Acetoacetic ester ($CH_3COCH_2COOC_2H_5$) : Malonic ester on heating with H_2SO_4 produces acetic acid while acetone is formed from acetoacetic ester.
33. Methanoic acid – Ethanoic acid : Methanoic acid reduces (i) ammonical silver nitrate to give silver mirror and (ii) $HgCl_2$ to white ppt. of Hg_2Cl_2.
34. Propanoic acid ($CH_3CHCOOH$) – Acrylic acid (CH_2=CH–COOH) : Acrylic acid decolourises bromine water and reacts with Baeyer's reagent to give 1,2–diol.
35. Stearic acid - Oleic acid : Oleic acid decolourises bromine and iodine solution as it contains unsaturation.
36. Maleic acid – Fumaric acid : Maleic acid on heating at 110°C produces maleic anhydride by dehydration as both acid groups are on the same side (cis).
37. α–chloropropionic acid – α–chloropropionic acid : The two can be distinguished by boiling with KOH.

$$CH_3CHClCOOH + KOH \rightarrow \underset{\alpha\text{–hydroxy propionic acid}}{CH_3CHOHCOOH} + KCl$$

$$CH_2ClCH_2COOH + KOH \rightarrow \underset{\text{acrylic acid}}{H_2C{=}CH{-}COOH} + KCl + H_2O$$

38. Vegetable oil – Mineral Oil : Vegetable oils are generally the triglycerides of higher fatty acids. Mineral oil is petroleum. Mineral oil burns readily. Vegetable oils form soap on heating with strong alkali solution.
39. Ethylamine – Diethylamine – Triethylamine : They can be distinguished by Hinsberg's and Hoffmann's method. Hinsberg's method: The reagent employed is benzene sulphonyl chloride

$$1° : C_6H_5{-}SO_2Cl + H_2NR \xrightarrow{-HCl} C_6H_5{-}SO_2NHR \xrightarrow{KOH} C_6H_5{-}SO_2NKR$$

$$2° : C_6H_5{-}SO_2Cl + HNR_2 \xrightarrow{-HCl} C_6H_5{-}SO_2NR_2 \nrightarrow \text{no reaction}$$

with 3° amines, no reaction is possible as active H atom is absent.

Hoffmann's method: The reagent employed is ethyl oxalate

$$\text{with } 1^\circ \quad \begin{array}{l} COOC_2H_5 \\ | \\ COOC_2H_5 \end{array} + \begin{array}{l} H_2NR \\ \\ H_2NR \end{array} \longrightarrow \begin{array}{l} CONHR \\ | \\ CONHR \end{array} + 2C_2H_5OH$$

dialkyl oxamide

$$\text{with } 2^\circ \quad \begin{array}{l} COOC_2H_5 \\ | \\ COOC_2H_5 \end{array} + HNR_2 \longrightarrow \begin{array}{l} CONR_2 \\ | \\ COOC_2H_5 \end{array} + C_2H_5OH$$

dialkyl oxamic ester

40. Ethyl amine ($CH_3CH_2NH_2$) – Acetamide (CH_3CONH_2) : Ethylamine gives a foul smelling carbylamine with chloroform and alkali. Acetamide gives methylamine by Hoffmann bromamide reaction with NaOBr.
41. Phenyl nitromethane ($Ph–CH_2–NO_2$) – Benzyl nitrite ($Ph–CH_2–O–N=O$) : Phenyl nitromethane forms nitrolic acid with HNO_2 which gives blood red colour with alkali. On reduction benzyl nitrite gives benzylalcohol while the other forms aniline.
42. Ethyl cyanide – Ethyl isocyanide : Ethyl cyanide gives propanaldehyde by Stephen's reduction reaction. At about 250°C isocyanide changes to cyanide while the reverse is not possible.
43. Glucose – Sucrose : Glucose reduces ammonical silver nitrate to silver mirror. Glucose gives red ppt. of Cu_2O with Fehling's solution. Sucrose doesn't respond to these tests as such.
44. Glucose – Fructose : Glucose forms soluble calcium glucosate with lime water while fructose forms insoluble calcium fructosate. Fructose gives red colour ppt. with resorcinol and conc. HCl (Selivanoff's test).
45. Glucose – Starch : Starch gives blue colour with iodine solution. Glucose gives red Cu_2O ppt. with Fehling's solution.
46. Amylose – Amylopectin : Amylose gives blue colour with iodine solution while amylopectin gives brown colour.
47. o–, m– and p– dibromo benzene : on nitration o–, m– and p– dibromo benzene gives two dibromo nitro derivatives respectively

Br–C₆H₄–Br (p) —nitration→ (Br, Br, NO_2)

Br–C₆H₄–Br (o) —nitration→ (Br, Br, O_2N) + O_2N (Br, Br)

Br–C₆H₄–Br (m) —nitration→ (Br, Br, NO_2) + O_2N (Br, Br) + O_2N (Br, Br)

Some Reagents in Organic Chemistry

1. Lassaignes test : It is used for the detection of nitrogen, sulphur and halogens present in the organic compounds. Nitrogen present in diazo compounds can't be detected as it undergoes decomposition even before fusion with sodium metal.
2. Beilstein test : It is used for the detection of halogens present in organic compounds. Urea, thiourea and pyridine give positive test although they don't have halogens.
3. Liebig's method : For the estimation of carbon and hydrogen.
4. Kjeldahl method : Estimation of nitrogen present in organic compounds by converting it to $(NH_4)_2SO_4$ and subjecting it to distillation with ~40% NaOH when NH_3 has is liberated.
5. Dumas method : N_2 gas formed from the compound at NTP is measured for the estimation of nitrogen.
6. Messenger's method : Estimation of sulphur.
7. Pirria & Schiff's method : Estimation of Sulphur (fusion method).
8. Asboths method : Estimation of sulphur.
9. Ter Meulen's method : Estimation of oxygen by converting it to CO, CO_2 and H_2O vapours.
10. NBS – (N-Bromosuccinimide) : Used for allylic and benzylic substitution by bromine.
11. Aluises method : Oxygen present in organic compounds in converted to CO and estimated on the basis of the given reaction,

$$CO + I_2O_5 \xrightarrow{175\ ^\circ C} 5CO_2 + I_2$$

12. Carius method : Estimation of halogens and sulphur.
13. Stepanov's method : Estimation of halogens.
14. Frankland reagent : Dialkyl Zinc in ether is used for the synthesis of alkanes (Zn in ether is present as dialkyl zinc).
15. Alcoholic KOH : used for dehydrohalogenation.
16. Aqueous KOH : Halogen derivates are hydrolysed to alkanol.
17. Ammonical Cu_2Cl_2 : Terminal alkynes form copper acetylides (reddish ppt.)
18. Ammonical $AgNO_3$: Terminal alkynes form silver acetylides (white ppt.)
19. Alcoholic KCN : Haloalkanes are converted to alkyl nitriles (KCN is an ionic compound).
20. Alcoholic AgCN : Haloalkanes are converted to alkyl isonitriles (AgCN is a covalent compound).
21. Legals test : This test is given by methyl ketones. The reagent used is sodium nitroprusside in dil. NaOH (Alkaline medium).
22. Iodoform test : The compounds having $–COCH_3$ group when heated with I_2+NaOH give CHI_3 (Iodoform) precipitate. Compounds like CH_3CH_2OH, $CH_3CHOHCOOH$ etc. also give

this test as these compounds are converted easily to CH_3CHO and $CH_3COCOOH$ respectively.

23. Tilden reagent : The reagent is NOCl (nitrosyl chlorides). Primary amines including CH_3NH_2 are converted to the corresponding haloalkenes.
24. PCl_5 : It is used to replace the –OH group present in organic compound, e.g. R-OH, R-COOH, etc.
25. $SOCl_2$: It is better than PCl_5 to replace the –OH group present in organic compound as the by-products are gases.
26. Isocynaide test : It is used to detect the primary amine group. Primary amines with $CHCl_3$ and alc. KOH give isocyanides (Carbylamine).
27. Inophenin test : Used to detect the presence of thiophene in benzene. Benzene is washed with cold conc. H_2SO_4 in the presence of isatin (formation of coloured solution indicates the positive test).
28. $NaNO_2$/(HCl) acid : Amines are converted to alcohols and amides to carboxylic acids. Used to diazotize amino group ($-NH_2$).
29. Lucas reagent : Anhydrous $ZnCl_2$ saturated with conc. HCl. Used to distinguish 1°, 2° and 3° alcohols.
30. Victor Mayer test : 1° and 2° alcohols give blood red colour and blue colour respectively. 3° alcohols give white colour.
31. HIO_4 : It is a strong oxidising agent. In general $-CH_2OH$ group is oxidised to HCHO while –CHOH is changed to HCOOH.
32. Baeyer's reagent : Alkaline $KMnO_4$. Alkenes are converted to 1,2 –diols.
33. Fenton's reagent : H_2O_2+$FeSO_4$. It is a mild oxidising agent.
34. Ceric ammounium nitrate : With alcohol, it shows red colour. Phenols, aliphatic and aromatic amines also give coloured solutions with this reagent.
35. Aqueous and alcoholic $FeCl_3$: Phenols form coloured complexes with it.
36. GRB test : Phenols in Liebermann's nitroso reaction gives green, red and blue coloured solutions in sequence.
37. Fehling solution : It is 1:1 mixture of Fehling solution A (copper sulphate solution) and Fehling solution B (Rochelle salt i.e. sodium potassium tartarate). Mainly used to distinguish aldehydes from ketones. HCOOH is reduced by it.
38. Tollen's reagent : Ammonical silver nitrate solution. Used to distinguish aldehydes from ketones.
39. 2,4–dinitro phenylhydrazine : Presence of carbonyl group can be detected (both aldehydes and ketones).
40. Schiffs reagent : It is Rosaniline hydrochloride. Aldehydes restore the pink colour of the dye.
41. Benedict's reagent : It is $CuSO_4$ + sodium citrate + Na_2CO_3. Aldehydes give Cu_2O red ppt.
42. Mulliken and Barker test : Nitrobenzene is detected by this test.
43. Br_2/NaOH or NaOBr : Hoffmann bromamide reagent is used for degradation of amides to primary amines.
44. Hinsberg reagent : Benzene sulphonyl chloride. It is used to separate 1°, 2° & 3° amines.
45. Hoffmann's reagent : Ethyl oxalate. It is used to separate 1°, 2° and 3° amines.
46. Girard reagent : The reagent is used to separate ketonic from non-ketonic materials. Mainly employed for separation of oestrone from concentrated urine samples.
47. Molisch test : All carbohydrates respond to this test.
48. Barfoed's reagent : It is cupric acetate in acetic acid.
49. Furfural test : Fructose gives violet colour with 1% alcoholic solution of α–naphthol and a few drops of HCl. Glucose gives no colour.
50. Pinoff's test : Fructose gives initially greenish blue colour which turns to bluish green with ammonium molybdate and a few drops of acetic acid. Glucose gives light blue colour.
51. Selivanoff's test : Fructose gives reddish brown precipitate with dilute solution of resorcinol in dil. HCl. Glucose gives no colour.
52. Ninhydrin test : All amino acids give bluish – violet colour except proline (a secondary amine) which gives yellow colour.
53. Biuret test : All compounds with –CO–NH– group respond positively giving violet colour. Urea on heating gives biuret
$$H_2NCONH_2 + H_2NCONH_2 \xrightarrow{-NH_3} H_2NCONHCONH_2$$
to which is added a few drops of alkaline copper sulphate solution and the resulting solution is used as the reagent.
54. Xanthoproteic test : Proteins gives yellow colour with conc. HNO_3 in this test.
55. Millions test : Proteins with phenolic –OH group give red ppt. on heating with Millon's reagent (mercurous and mercuric nitrate in nitric acid).
56. Hopkin's cole test : Proteins containing tryptophan give bluish violet colour on the addition of glyoxalic acid in the presence of H_2SO_4.
57. Nitroprusside test : Proteins containing thio group (–SH) give transient red colour.
58. Sanger's reagent : It is 2,4 – dinitro fluro benzene (DNFB) It is used for identification of N–terminal residue analysis of proteins.
59. Edman's reagent : It is phenyl isothiocyanate (Ph–N=C=S) used for N–terminal residue analysis of proteins.
60. Maltose : The enzyme is specific for hydrolysis of α–glycosides.
61. Emulsin : The enzyme in specific for hydrolysis of α–glycosides.
62. Urease : It is used to estimate urea.

Some Basic Conversions in Organic Chemistry

1. Methane → Ethane

$$CH_4 + Cl_2 \xrightarrow{h\nu} CH_3Cl\,;$$

$$CH_3Cl + 2Na + ClH_3C \xrightarrow{\text{dry ether}} H_3C - CH_3$$

2. Ethane → Methane

$$H_3C - CH_3 + Cl_2 \xrightarrow{h\nu} CH_3CH_2Cl$$

$$\xrightarrow{\text{aq. KOH}} CH_3CH_2OH \xrightarrow{K_2Cr_2O_7 / H_2SO_4}$$

$$CH_3COOH \xrightarrow{NaOH} CH_3COONa \xrightarrow{CaO + NaOH} CH_4$$

3. Methane → Ethane → Ethene

Convert methane to ethane as in the 1st conversion. Convert ethane to ethanol as in 2nd conversion and then

$$CH_3CH_2OH \xrightarrow{\text{conc. } H_2SO_4,\, 160°C} H_2C = CH_2$$

4. Methane → Ethene → Ethyne

Convert methane to ethene as in the 3rd conversion. Now

$$H_2C = CH_2 + Br_2 \longrightarrow BrH_2C - CH_2Br$$

$$\xrightarrow{\text{alc.KOH}, \Delta; -2HBr} HC \equiv CH$$

5. Ethene → Propene

$$H_2C = CH_2 + HBr \longrightarrow H_3C - CH_2Br$$

$$\xrightarrow{\text{alc.KCN}} H_3C - CH_2CN \xrightarrow{C_2H_5OH/Na}$$

$$H_3C - CH_2 - CH_2 - NH_2 \xrightarrow{HNO_2}$$

$$H_3C - CH_2 - CH_2 - OH \xrightarrow{\text{Conc. } H_2SO_4}$$

$$H_3C - CH = CH_2$$

6. Propene → Ethene

$$H_3C - CH = CH_2 \xrightarrow{\text{Ozonolysis}} H_3C - CHO \xrightarrow{LiAlH_4}$$

$$H_3C - CH_2 - OH \xrightarrow{\text{Conc. } H_2SO_4} H_2C = CH_2$$

7. Ethene → Propyne

Convert ethene → ethyne as in the 4th conversion and then,

$$HC \equiv CH \xrightarrow{NaNH_2 / NH_3} HC \equiv CNa \xrightarrow{CH_3Br}$$

$$HC \equiv C - CH_3$$

8. Propyne → Ethyne

$$CH_3 - C \equiv CH \xrightarrow{H_2 / \text{Lindlar's catalysts}} CH_3 - CH = CH_2$$

Now convert propene to ethene as in 6th conversion and then convert ethene to ethyne as in the 4th conversion

9. Primary alcohol → Secondary alcohol

$$H_3C\text{-}CH_2\text{-}CH_2OH \xrightarrow{PCl_5} H_3C\text{-}CH_2\text{-}CH_2Cl \xrightarrow[\text{KOH}]{\text{alc.}} H_3C\text{-}CH{=}CH_2$$

($H_3C\text{-}CH_2\text{-}CH_2OH \xrightarrow{\text{conc. } H_2SO_4;\ -H_2O} H_3C\text{-}CH{=}CH_2$)

$$H_3C\text{-}CH{=}CH_2 \xrightarrow{HI} H_3C\text{-}CHI\text{-}CH_3 \xrightarrow[\text{KOH}]{\text{aq.}} H_3C\text{-}CHOH\text{-}CH_3$$

10. Primary alcohol → Tertiary alcohol

$$H_3C\text{-}CH_2\text{-}CH_2OH \xrightarrow[-H_2O]{\text{conc. } H_2SO_4} H_3C\text{-}CH{=}CH_2 \xrightarrow{HI} H_3C\text{-}CHI\text{-}CH_3 \xrightarrow[\text{KOH}]{\text{aq.}} H_3C\text{-}CHOH\text{-}CH_3$$

$$H_3C\text{-}CHOH\text{-}CH_3 \xrightarrow{(O)} H_3C\text{-}CO\text{-}CH_3 \xrightarrow[H_3O^+]{RMgX} H_3C\text{-}C(R)(OH)\text{-}CH_3$$

11. Primary alcohol → Isomeric tertiary alcohol

$$H_3C\text{-}CH(CH_3)\text{-}CH_2OH \xrightarrow[-H_2O]{\text{conc. } H_2SO_4} (H_3C)_2\text{-}C{=}CH_2 \xrightarrow{HI} (H_3C)_2\text{-}CI\text{-}CH_3 \xrightarrow[\text{KOH}]{\text{aq.}} (H_3C)_2\text{-}C(OH)\text{-}CH_3$$

12. Secondary alcohol → Tertiary alcohol

$$(H_3C)_2\text{-}CH\text{-}OH \xrightarrow[(O)]{K_2Cr_2O_7} H_3C\text{-}CO\text{-}CH_3 \xrightarrow[H_3O^+]{RMgX} H_3C\text{-}C(OH)(CH_3)\text{-}R$$

13. Secondary alcohol → Isomeric tertiary alcohol

$$(H_3C)_2\text{-}CH\text{-}CHOH\text{-}CH_3 \xrightarrow[-H_2O]{\text{conc. } H_2SO_4} (H_3C)_2\text{-}C{=}CH\text{-}CH_3 \xrightarrow{HI} (H_3C)_2\text{-}CI\text{-}CH_2CH_3 \xrightarrow[\text{KOH}]{\text{aq.}} (H_3C)_2\text{-}C(OH)\text{-}CH_2CH_3$$

14. Primary alcohol → Higher primary alcohol

$$CH_3OH \xrightarrow{P + I_2} CH_3I \xrightarrow[\text{ether}]{Mg} CH_3MgI \xrightarrow[H_3O^+]{HCHO} CH_3CH_2OH$$

$$CH_3I \xrightarrow{KCN} CH_3CN \xrightarrow{Ni / [H]} CH_3CH_2NH_2 \xrightarrow{HNO_2} CH_3CH_2OH$$

15. Primary alcohol → Higher secondary alcohol

$$CH_3CH_2OH \xrightarrow{(O)} CH_3CHO \xrightarrow[H_3O^+]{RMgX} H_3C\text{-}CH(R)\text{-}OH$$

16. Secondary alcohol → Higher tertiary alcohol

$$H_3C\text{-}CHOH\text{-}CH_3 \xrightarrow{(O)} H_3C\text{-}CO\text{-}CH_3 \xrightarrow[H_3O^+]{RMgX} H_3C\text{-}C(R)(OH)\text{-}CH_3$$

17. Primary alcohol → Lower primary alcohol

$$CH_3CH_2OH \xrightarrow{(O)} CH_3CHO \xrightarrow{(O)} CH_3COOH \xrightarrow{NaOH} CH_3COONa \xrightarrow[-CO_2]{CaO + NaOH} CH_4$$

$$CH_4 \xrightarrow{Cl_2} CH_3Cl \xrightarrow[\text{or AgOH}]{\text{aq. KOH}} CH_3OH$$

Methanol can also be obtained from acetic acid as,

$CH_3COOH \xrightarrow{NH_3} CH_3COONH_4 \xrightarrow{\Delta, -H_2O} CH_3CONH_2 \xrightarrow{Br_2/KOH} CH_3NH_2 \xrightarrow{NOCl} CH_3Cl \xrightarrow{Ag_2O/H_2O} CH_3OH$

18. Higher secondary alcohol → Lower primary alcohol

$H_3C\text{-}CH_2\text{-}CHOH\text{-}CH_3 \xrightarrow[-H_2O]{conc.\ H_2SO_4} H_3C\text{-}CH{=}CH\text{-}CH_3 \xrightarrow{O_3;\ H_2O} 2CH_3CHO \xrightarrow{LiAlH_4} CH_3CH_2OH$

19. Tertiary alcohol → Primary and secondary alcohol

$(H_3C)_3\text{-}C\text{-}OH \xrightarrow[-H_2O]{} (H_3C)_2\text{-}C{=}CH_2 \xrightarrow{O_3;\ H_2O} (H_3C)_2\text{-}C{=}O + HCHO$

$(H_3C)_2\text{-}C{=}O \xrightarrow{LiAlH_4} (H_3C)_2\text{-}CH\text{-}OH$; $HCHO \xrightarrow{LiAlH_4} CH_3OH$

20. Methanal → Ethanal

(i) $HCHO \xrightarrow{Ni/H_2} CH_3OH \xrightarrow{PCl_5} CH_3Cl \xrightarrow{KCN} CH_3CN$

$\xrightarrow{C_2H_5OH/Na} CH_3CH_2NH_2 \xrightarrow{NaNO_2/HCl}$

$CH_3CH_2OH \xrightarrow{K_2Cr_2O_7/H_2SO_4} CH_3CHO$

(ii) $HCHO + CH_3MgX \longrightarrow CH_3CH_2OH \xrightarrow[Cu]{300°C,} CH_3CHO$

21. Ethanal → Methanal

$CH_3CHO \xrightarrow{K_2Cr_2O_7/H_2SO_4} CH_3COOH \xrightarrow{NaOH} CH_3COONa$

$\xrightarrow[(Sodalime)]{NaOH+CaO} CH_4 \xrightarrow{hv;\ Cl_2} CH_3Cl$

$\xrightarrow{AgOH} CH_3OH \xrightarrow[300°C]{Cu} HCHO$

22. Ethanal → Propanone

$H_3C-CHO \xrightarrow{K_2Cr_2O_7/H_2SO_4} H_3C-COOH$

$\xrightarrow{CaO} (H_3C-COO)_2Ca \xrightarrow{dry\ distillation} H_3C-CO-CH_3$

23. Propanone → Ethanal

$H_3C-CO-CH_3 \xrightarrow{LiAlH_4} H_3C-HCOH-CH_3$

$\xrightarrow{conc.\ H_2SO_4,\Delta} H_3C-CH=CH_2 \xrightarrow{Ozonolysis} H_3C-CHO$

24. Methanoic acid → Ethanoic acid

$HCOOH \xrightarrow{Ca(OH)_2} (HCOO)_2Ca \xrightarrow{dry\ distillation}$

$HCHO \xrightarrow{H_2/Ni} CH_3OH \xrightarrow{SOCl_2\ or\ PCl_5} CH_3Cl$

$\xrightarrow{alc.\ KCN} CH_3-CN \xrightarrow{H_3O^+} CH_3COOH$

25. Ethanoic acid → Methanoic acid

$CH_3COOH \xrightarrow{NH_3} CH_3COONH_4 \xrightarrow{\Delta, -H_2O}$

$CH_3CONH_2 \xrightarrow{Br_2/KOH} CH_3NH_2 \xrightarrow{NOCl}$

$CH_3CN \xrightarrow{aq.\ KOH} CH_3OH \xrightarrow{O} HCOOH$

26. Methanoic acid → Oxalic acid

$HCOOH \xrightarrow{NaOH} HCOONa \xrightarrow{400°C}$

$NaOOC-COONa \xrightarrow{Ca(OH)_2} (COO)_2Ca \xrightarrow{dil.\ H_2SO_4}$

$HOOC-COOH$

27. Oxalic acid → Methanoic acid

$HOOC-COOH.2H_2O \xrightarrow{100°C} HOOC-COOH$

$\xrightarrow{200°C,\ -CO_2} HCOOH$

28. Ethanoic acid → Malonic acid

$CH_3COOH \xrightarrow{Cl_2} ClCH_2COOH \xrightarrow{K_2CO_3}$

$ClCH_2COOK \xrightarrow{KCN} CNCH_2COOK \xrightarrow{H_3O^+} CH_2(COOH)_2$

29. Malonic acid → Ethanoic acid

$CH_2(COOH)_2 \xrightarrow{150°C} CH_3COOH$

30. Aminomethane → Aminoethane

$CH_3NH_2 \xrightarrow{NOCl} CH_3Cl \xrightarrow{alc.KCN}$

$CH_3CN \xrightarrow{C_2H_5OH/Na} CH_3CH_2NH_2$

31. Aminoethane → Aminomethane

$CH_3CH_2NH_2 \xrightarrow{NaNO_2/HCl} CH_3CH_2OH \xrightarrow{K_2Cr_2O_7/H_2SO_4}$

$CH_3COOH \xrightarrow{NH_3} CH_3COONH_4 \xrightarrow{\Delta,-H_2O}$

$CH_3CONH_2 \xrightarrow{Br_2/KOH} CH_3NH_2$

An alternative route to convert ethanoic acid to aminomethane can be represented as

$CH_3COOH \xrightarrow{NaOH} CH_3COONa \xrightarrow{CaO+NaOH}$

$CH_4 \xrightarrow{Cl_2,hv} CH_3Cl \xrightarrow{NH_3} CH_3NH_2$

32. Trans-2-butene → Cis-2-butene

$H_3C\text{-}CH{=}CH\text{-}CH_3$ (Trans) $\xrightarrow{Br_2} H_3C\text{-}CHBr\text{-}CHBr\text{-}CH_3 \xrightarrow[\Delta]{alc.\ KOH}$

$H_3C-C{\equiv}C-CH_3 \xrightarrow{Pt\text{-}Ni\ boride} H_3C\text{-}CH{=}CH\text{-}CH_3$ (cis)

33. Cis-2-butene → Trans-2-butene

$H_3C\text{-}CH{=}CH\text{-}CH_3$ (cis) $\xrightarrow{Br_2} H_3C-CHBr-CHBr-CH_3 \xrightarrow[\Delta]{alc.\ KOH}$

$H_3C-C{\equiv}C-H$ $\xrightarrow{C_2H_5OH/Na} H_3C\text{-}CH{=}CH\text{-}CH_3$ (Trans)

34. Cyclopentanone → Cyclohexanone

Cyclopentanone $\xrightarrow[\text{or } CH_2N_2 \text{ (diazomethane)}]{H_2C=C=O \text{ (Ketene)}}$ Cyclohexanone + N_2 / CO

(N_2 is formed when diazomethane is used and CO is formed when ketene is used).

35. Cyclohexanone → Cyclopentanone

Cyclohexanone $\xrightarrow[\text{(oxidation)}]{HNO_3}$ $CH_2\text{-}CH_2\text{-}COOH$ / $CH_2\text{-}CH_2COOH$ $\xrightarrow[\text{heat}]{BaCO_3}$

$CH_2\text{-}CH_2\text{-}COO$ / $CH_2\text{-}CH_2COO$ >Ba $\xrightarrow[\text{heat}]{-BaCO_3}$ Cyclopentanone

36. Propene → Allyl alcohol

$H_3C-CH=CH_2 \xrightarrow{\text{N-bromo-succinimide}} BrCH_2-CH=CH_2 \xrightarrow{\text{aq. KOH}}$

$HOH_2C-CH=CH_2$

37. R–COOH → R–CH_2COOH (Lower Carboxylic acid → Next higher Carboxylic acid)

This can be achieved by Arndt – Eistert synthesis

$RCOOH \xrightarrow[\text{or } PCl_5]{SOCl_2} RCOCl \xrightarrow[-HCl]{CH_2N_2} RCOCHN_2 \xrightarrow{\Delta}$

$R(H)C=C=O \xrightarrow[H_2O]{\Delta} RCH_2COOH$

38. n-Hexanoic acid → n-Pentanoic acid

$H_3C-CH_2-CH_2-CH_2-CH_2-COOH \xrightarrow{MeOH/HCl} H_3C-CH_2-CH_2-CH_2-CH_2-C(=O)-O-CH_3$

$\xrightarrow{PhMgBr} H_3C-CH_2-CH_2-CH_2-CH_2-C(OMgBr)(Ph)-O-CH_3 \xrightarrow[-BrMgOCH_3]{PhMgBr}$

$H_3C-CH_2-CH_2-CH_2-CH_2-C(OMgBr)(Ph)-Ph \xrightarrow{\text{hydrolysis}} H_3C-CH_2-CH_2-CH_2-CH_2-C(OH)(Ph)-Ph$

$\xrightarrow[-H_2O]{H_3O^+} H_3C-CH_2-CH_2-CH_2-CH=C(Ph)-Ph \xrightarrow{CrO_3/CH_3COOH}$

$H_3C-CH_2-CH_2-CH_2-COOH + OC(Ph)-Ph$

39. n-Butyl alcohol → n-Propyl amine

$H_3C-CH_2-CH_2-CH_2OH \xrightarrow{K_2Cr_2O_7/H_2SO_4} H_3C-CH_2-CH_2-COOH$

$\xrightarrow{SOCl_2} H_3C-CH_2-CH_2-COCl \xrightarrow{NH_3} H_3C-CH_2-CH_2-CONH_2 \xrightarrow{Br_2/KOH}$

$H_3C-CH_2-CH_2-NH_2$

40. Aniline → Benzylamine

$Ph-NH_2 \xrightarrow[\text{diazotization}]{NaNO_2/HCl, 0°C} Ph-N_2Cl \xrightarrow{CuCN/KCN}$

$Ph-C\equiv N \xrightarrow{Ni/H_2} Ph-CH_2-NH_2$

41. Glucose → Fructose (Aldose → Ketose)

HC=O / CHOH / $(CHOH)_3$ / CH_2OH $\xrightarrow{3PhNHNH_2}$ HC=NNHPh / C=NNHPh / $(CHOH)_3$ / CH_2OH $\xrightarrow{H_3O^+}$ HC=O / C=O / $(CHOH)_3$ / CH_2OH

$\xrightarrow{CH_3COOH/Zn}$ CH_2OH / C=O / $(CHOH)_3$ / CH_2OH

42. Fructose → Glucose (Ketose → Aldose)

CH_2OH / C=O / $(CHOH)_3$ / CH_2OH $\xrightarrow{Ni/H_2}$ CH_2OH / CH OH / $(CHOH)_3$ / CH_2OH $\xrightarrow{(O)}$ COOH / CH OH / $(CHOH)_3$ / CH_2OH $\xrightarrow{\Delta}$

CO– / CHOH / CHOH / CH– (lactone, –O–) / CHOH / CH_2OH $\xrightarrow[\text{[H]}]{Na/Hg}$ HC=O / CHOH / $(CHOH)_3$ / CH_2OH

43. Aldohexose → Aldopentose

(i) Wohl's method

HC=O / CHOH / $(CHOH)_3$ / CH_2OH $\xrightarrow{NH_2OH}$ HC=NOH / CHOH / $(CHOH)_3$ / CH_2OH $\xrightarrow[-H_2O]{(CH_3CO)_2O}$ C≡N / CHO.COCH$_3$ / $(CHOCOCH_3)_3$ / CH_2OH

$\xrightarrow[-AgCN]{AgOH}$ HC=O / $(CHOH)_3$ / CH_2OH

(ii) Ruff's method

HC=O / CHOH / $(CHOH)_3$ / CH_2OH $\xrightarrow{Br_2/H_2O}$ COOH / CHOH / $(CHOH)_3$ / CH_2OH $\xrightarrow{CaCO_3}$ COOCa/2 / CHOH / $(CHOH)_3$ / CH_2OH

$\xrightarrow[-CO_2]{\text{Fenton's reagent}}$ HC=O / $(CHOH)_3$ / CH_2OH

(iii) Weerman's method

HC=O / CHOH / $(CHOH)_3$ / CH_2OH $\xrightarrow{Br_2/H_2O}$ COOH / CHOH / $(CHOH)_3$ / CH_2OH $\xrightarrow[\Delta]{NH_3}$ CONH$_2$ / CHOH / $(CHOH)_3$ / CH_2OH

$\xrightarrow{NaOH/NaOCl}$ NCO / CHOH / $(CHOH)_3$ / CH_2OH $\xrightarrow[-NaNCO]{NaOH}$ HC=O / $(CHOH)_3$ / CH_2OH

44. **Aldopentose → Aldohexose**

$$HC{=}O-(CHOH)_3-CH_2OH \xrightarrow{HCN} CN-CHOH-(CHOH)_3-CH_2OH \xrightarrow{Ba(OH)_2;\ H_2SO_4} COOH-CHOH-(CHOH)_3-CH_2OH \xrightarrow[-H_2O]{\Delta}$$

$$\text{lactone (CO-CHOH-CHOH-CH(-O-)-CHOH-CH}_2\text{OH)} \xrightarrow{Na/Hg,\ H_2SO_4} HC{=}O-CHOH-(CHOH)_3-CH_2OH$$

45. **Benzene → Toluene**

$$C_6H_6 + CH_3{-}X \xrightarrow[-HX]{dry\ AlCl_3} C_6H_5{-}CH_3$$

46. **Benzene → Styrene**

(i) $C_6H_6 + C_2H_5X \xrightarrow{dry\ AlCl_3} Ph-CH_2-CH_3 \xrightarrow{ZnO; 600°C, -H_2} Ph-CH=CH_2$

(ii) $C_6H_6 + H_2C\overset{O}{—}CH_2 \xrightarrow{dry\ AlCl_3} Ph{-}CH_2{-}CH_2OH \xrightarrow{\Delta} Ph{-}CH{=}CH_2$

(iii) $C_6H_6 + Mg + Br_2 \xrightarrow{dry\ ether} Ph{-}MgBr \xrightarrow{CH_3CHO} Ph{-}C(OMgBr)(H){-}CH_3 \xrightarrow{hydrolysis}$

$Ph{-}CHOH{-}CH_3 \xrightarrow[-H_2O]{\Delta} Ph{-}CH{=}CH_2$

47. **Benzene → Phenyl acetylene**

Convert benzene to styrene as in 46th conversion and then

$Ph{-}CH{=}CH_2 + Br_2 \longrightarrow Ph{-}CHBr{-}CH_2Br \xrightarrow[\Delta]{alc.\ KOH} Ph{-}C{\equiv}CH$

48. **Benzene → Chlorbenzene**

(i) $C_6H_6 + Cl_2 \xrightarrow{Fe, -HCl} Ph-Cl$

(ii) $2C_6H_6 + 2HCl + O_2 \xrightarrow{CaCl_2} 2Ph-Cl + 2H_2O$

(Raschig method)

49. **Benzene → Benzyl chloride**

First convert benzene to toluene and then

$C_6H_5-CH_3 + Cl_2 \xrightarrow{Light} C_6H_5-CH_2Cl$

50. **Benzene → benzyl alcohol**

Benzene is converted to toluene, then to benzyl chloride and then

(i) $C_6H_5-CH_2Cl \xrightarrow{aq.\ NaOH} C_6H_5-CH_2OH$

(ii) $C_6H_5MgBr + HCHO \longrightarrow C_6H_5-CH_2MgBr \xrightarrow{H_3O^+} C_6H_5CH_2OH$

51. **Benzene → Phenol**

First convert benzene → chlorobenzene as in 48th conversion and then

(i) $C_6H_5Cl + H_2O \xrightarrow{700K,\ Steam} C_6H_5OH + HCl$

(ii) $C_6H_6 \xrightarrow{conc.\ H_2SO_4} C_6H_5{-}SO_3H \xrightarrow{NaOH} C_6H_5{-}SO_3Na$ (sodium benzene sulphonate) $\xrightarrow{-NaHSO_3} C_6H_5{-}ONa \xrightarrow{V.\ dil.\ HCl} C_6H_5{-}OH$

52. **Benzene → Anisole**

First convert benzene to phenol as in 51st conversion and then

(i) $C_6H_5OH \xrightarrow{NaOH} C_6H_5ONa \xrightarrow{CH_3I, \Delta} C_6H_5OCH_3$

(ii) $C_6H_5OH + CH_2NH_2 \xrightarrow{-N_2} C_6H_5OCH_3$

53. **Benzene → Benzaldehyde**

(i) $C_6H_6 + CO + HCl \xrightarrow{AlCl_3, -HCl} C_6H_5CHO$

(ii) $C_6H_6 + HCN + \xrightarrow{ZnCl_2} C_6H_5CH{=}^+NH_2Cl \xrightarrow{H_2O} C_6H_5CHO$

(iii) $C_6H_6 + CH_3X \xrightarrow{dry AlCl_3} C_6H_5CH_3 \xrightarrow{2Cl_2, \Delta} C_6H_5CHCl_2 \xrightarrow{Ca(OH)_2, \Delta} C_6H_5CHO$

(iv) $2C_6H_5CH_2Cl \xrightarrow{Cu(NO_3)_2} 2C_6H_5CHO$

54. **Benzene → Salicylaldehyde**

Benzene is converted to phenol and then

(i) $C_6H_5{-}OH + CHCl_3 + KOH \xrightarrow{warm} C_6H_4(CHO)(OH)$

(ii) $C_6H_5{-}OH + HCN + HCl \xrightarrow{dry\ AlCl_3} C_6H_4(CHO)(OH)$

55. **Benzene → Acetophenone**

(i) $C_6H_6 + R{-}\overset{O}{\overset{\|}{C}}{-}X \xrightarrow[-HX]{dry\ AlCl_3} C_6H_5{-}\overset{O}{\overset{\|}{C}}{-}R$

(ii) $C_6H_6 + ClH_5C_2 \xrightarrow{dry\ AlCl_3} C_6H_5CH_2CH_3 \xrightarrow{O_2, V_2O_5, 500°C} C_6H_5COCH_3$

56. **Benzene → Benzophenone**

$C_6H_6 + ClCOC_6H_5 \xrightarrow{dry\ AlCl_3} C_6H_5COC_6H_5$

57. Benzene → Benzoic acid

(i) $C_6H_6 + COCl_2 \rightarrow C_6H_5COCl \xrightarrow{H_2O} C_6H_5COOH$

(ii) $C_6H_6 + ClH_3C \xrightarrow{\text{dry } AlCl_3} C_6H_5CH_3 \xrightarrow{3Cl_2, \Delta}$

$C_6H_5CCl_3 \xrightarrow{CaOH_2, \text{ Fe catalyst}} C_6H_5COOH$

(iii) $C_6H_5CH_3 \xrightarrow[H_2SO_4]{K_2Cr_2O_7 \text{ or } KMnO_4 /} C_6H_5COOH$

58. Benzene → Benzoyl chloride
Conversion is carried out as per 57(i).

59. Benzene → Benazamide
Benzoyl chloride obtained is reacted with NH_3

$C_6H_5COCl + 2NH_3 \longrightarrow C_6H_5CONH_2 + NH_4Cl$

60. Benzene → Benzo nitrile
After obtaining benzamide, it is subjected to dehydration

$C_6H_5CONH_2 \xrightarrow{P_2O_5, \Delta, -H_2O} C_6H_5C \equiv N$

61. Benzene → Nitrobenzene

$C_6H_6 + HONO_2 \xrightarrow{\text{conc.}H_2SO_4, -H_2O} C_6H_5NO_2$

62. Benzene → Aniline
Nitrobenzene obtained is subjected to reduction

(i) $C_6H_5NO_2 \xrightarrow{Fe+HCl} C_6H_5NH_2$

(ii) $C_6H_5NO_2 \xrightarrow{Sn+HCl} C_6H_5NH_2$

63. Benzene → Benzylamine
Benzene is converted to toluene as in 45th conversion, then to benzyl chloride as in 49th conversion and then treat with NH_3

$C_6H_5CH_2Cl \xrightarrow{NH_3, -HCl} C_6H_5CH_2NH_2$

64. Benzene → Salicylic acid
Benzene is converted to sodium phenoxide as in 51st conversion and then

$C_6H_5ONa + CO_2 \xrightarrow[\text{pressure}]{393 - 413 \text{ K}} o\text{-}HOC_6H_4COONa \xrightarrow{H_3O^+} o\text{-}HOC_6H_4COOH$ (salicylic acid)

65. Benzene → Cinamic acid
Benzene is converted to Benzaldehyde as in 53th conversion and then subjected to Perkin reaction

$C_6H_5\text{—}CHO + (CH_3CO)_2O \xrightarrow{CH_3COONa} C_6H_5\text{—}CH{=}CH\text{—}COOH$ (cinnamic acid)

66. Benzene → p-Toluidine

$C_6H_6 + CH_3\text{-}X \xrightarrow[-HX]{\text{dry } AlCl_3} C_6H_5CH_3 \xrightarrow{\text{Nitration}} o\text{-}CH_3C_6H_4NO_2 + p\text{-}CH_3C_6H_4NO_2$

o-nitrophenol is separated by distillation under reduced pressure and then p-nitro toluene is reduced to p-toluidine

$p\text{-}CH_3C_6H_4NO_2 \xrightarrow{SnCl_2/HCl} p\text{-}CH_3C_6H_4NH_2$

67. Aniline → p-Bromoaniline

$C_6H_5NH_2 \xrightarrow[AlCl_3]{CH_3COCl} C_6H_5NHCOCH_3 \xrightarrow{\text{bromination}} o\text{-}BrC_6H_4NHCOCH_3 \text{ (o-bromo acetanilide)} + p\text{-}BrC_6H_4NHCOCH_3 \text{ (p-bromo acetanilide)}$

$\xrightarrow{\text{hydrolysis}} o\text{-}BrC_6H_4NH_2 \text{ (o-bromo aniline)} + p\text{-}BrC_6H_4NH_2 \text{ (p-bromo aniline)}$

In halogenation and nitration, the amine group is protected by converting it to acetanilide as they have the oxidising property (which may decompose the benzene ring, which is a activated one).

68. Aniline → p-Nitroaniline

$C_6H_5NH_2 \xrightarrow[AlCl_3]{CH_3COCl} C_6H_5NHCOCH_3 \xrightarrow{\text{nitration}} o\text{-}O_2NC_6H_4NHCOCH_3 \text{ (o-nitro acetanilide)} + p\text{-}O_2NC_6H_4NHCOCH_3 \text{ (p-nitro acetanilide)}$

$\xrightarrow{\text{hydrolysis}} o\text{-}O_2NC_6H_4NH_2 \text{ (o-nitro aniline)} + p\text{-}O_2NC_6H_4NH_2 \text{ (p-nitro aniline)}$

69. Toluene → Benzene

$C_6H_5 - CH_3 \xrightarrow{\text{dil. }HNO_3} C_6H_5COOH \xrightarrow{CaO/NaOH} C_6H_6$

70. Bromobenzene → Benzene

$C_6H_5 - Br + Mg \xrightarrow{\text{ether}} C_6H_5MgBr \xrightarrow{HOH} C_6H_6$

71. Nitrobenzene → Benzene

$C_6H_5NO_2 \xrightarrow{Zn/HCl} C_6H_5NH_2 \xrightarrow{NaNO_2/HCl}$

$C_6H_5N_2Cl \xrightarrow{C_2H_5OH, \text{boil}} C_6H_6$

72. Benzaldehyde → Benzene

$C_6H_5CHO \xrightarrow{O} C_6H_5COOH \xrightarrow{NaOH}$

$C_6H_5COONa \xrightarrow[-Na_2CO_3]{CaO+NaOH} C_6H_6$

73. Benzamide → Benzene

(i) $C_6H_5CONH_2 \xrightarrow{NaOH, \Delta} C_6H_5COONa \xrightarrow[-Na_2CO_3]{CaO+NaOH} C_6H_6$

(ii) $C_6H_5CONH_2 \xrightarrow{Br_2/KOH} C_6H_5NH_2 \xrightarrow{NaNO_2/HCl}$

$C_6H_5N_2Cl \xrightarrow{C_2H_5OH, \text{boil}} C_6H_6$

74. Phenol → Benzene

$$C_6H_5\text{-}OH + Zn \xrightarrow{\Delta} C_6H_6$$

75. Benzaldehyde → Benzylamine

$$C_6H_5CHO \xrightarrow{[H]} C_6H_5CH_2OH \xrightarrow{SOCl_2 \text{ or } PCl_5}$$

$$C_6H_5CH_2Cl \xrightarrow{NH_3} C_6H_5CH_2NH_2$$

76. Ethyl benzene → Acetophenone

$$C_6H_5CH_2CH_3 \xrightarrow{[O]} C_6H_5COOH \xrightarrow{Ca(OH)_2}$$

$$(C_6H_5COO)_2Ca \xrightarrow{Ca(CH_3COO)_2} C_6H_5COCH_3$$

77. Benzoic acid → Benzophenone

$$C_6H_5COOH \xrightarrow{CaCO_3} (C_6H_5COO)_2Ca \xrightarrow{\text{distillation}} C_6H_5COC_6H_5$$

78. p-Nitrotoluene → m-Nitrotoluene

$$H_3C\text{-}C_6H_4\text{-}NO_2 \xrightarrow{Sn + HCl} H_3C\text{-}C_6H_4\text{-}NH_2 \xrightarrow{(CH_3CO)_2O}$$

$$H_3C\text{-}C_6H_4\text{-}NHCOCH_3 \xrightarrow[H_2SO_4]{HNO_3 +} H_3C\text{-}C_6H_3(NO_2)\text{-}NHCOCH_3$$

$$\xrightarrow{H_3O^+} H_3C\text{-}C_6H_3(NO_2)\text{-}NH_2 \xrightarrow[0-5°]{NaNO_2/HCl} H_3C\text{-}C_6H_3(NO_2)\text{-}N_2Cl$$

$$\xrightarrow{H_3PO_2} H_3C\text{-}C_6H_4\text{-}NO_2 \text{ (meta)}$$

79. Benzene → m-Chlorotoluene

$$C_6H_6 \xrightarrow{\text{nitration}} C_6H_5NO_2 \xrightarrow[\text{dry } AlCl_3]{CH_3Cl} m\text{-}NO_2C_6H_4CH_3 \xrightarrow{Sn/HCl}$$

$$m\text{-}NH_2C_6H_4CH_3 \xrightarrow[0-5°C]{NaNO_2/HCl} m\text{-}N_2ClC_6H_4CH_3 \xrightarrow{CuCl/HCl} m\text{-}ClC_6H_4CH_3$$

80. Benzene → Nitrobenzaldehyde

$$C_6H_6 \xrightarrow[\text{dry } AlCl_3]{CH_3Cl} C_6H_5CH_3 \xrightarrow{\text{dil. } HNO_3} p\text{-}CH_3C_6H_4NO_2 \xrightarrow{(O)} p\text{-}CHOC_6H_4NO_2$$

81. Benzene → p-Bromobenzoic acid

$$C_6H_6 \xrightarrow[\text{dry } AlCl_3]{CH_3Cl} C_6H_5CH_3 \xrightarrow{Br_2/Fe} p\text{-}CH_3C_6H_4Br \xrightarrow[KMnO_4]{(O)} p\text{-}COOHC_6H_4Br$$

82. Benzene → m-Bromobenzoic acid

$$C_6H_6 \xrightarrow[\text{dry } AlCl_3]{CH_3Cl} C_6H_5CH_3 \xrightarrow[\Delta]{KMnO_4} C_6H_5COOH \xrightarrow{Br_2/Fe} m\text{-}COOHC_6H_4Br$$

83. Benzene → o-Bromobenzoic acid

$$C_6H_6 \xrightarrow[\text{dry } AlCl_3]{CH_3Cl} C_6H_5CH_3 \xrightarrow[SO_3]{H_2SO_4} p\text{-}CH_3C_6H_4SO_3H \xrightarrow{Br_2/Fe}$$

$$CH_3C_6H_3(Br)SO_3H \xrightarrow[\text{steam}]{H_2SO_4} o\text{-}CH_3C_6H_4Br \xrightarrow[\Delta]{KMnO_4} o\text{-}COOHC_6H_4Br$$

84. Naphthalene → α- Naphthol

$$C_{10}H_8 \xrightarrow[150°C]{H_2SO_4} C_{10}H_7SO_3H \xrightarrow[H_3O^+]{NaOH, \text{ fuse}} C_{10}H_7OH$$

Atomic Structure

(1) The wave length of the line of the highest wave length of a particular series in the atomic spectrum of hydrogen is at 656.3 n m. what is the series. ($R_H = 1.097 \times 10^7 m^{-1}$)

Sol. $\frac{1}{\lambda} = R_H \left(\frac{1}{n_1^2} - \frac{1}{n_2^2}\right)$

$\frac{1}{656.3 \times 10^{-9}} = 1.097 \times 10^7 \left(\frac{1}{2^2} - \frac{1}{x^2}\right)$

$\frac{1}{656.3 \times 10^{-9} \times 1.097 \times 10^7} = \left(\frac{1}{2^2} - \frac{1}{x^2}\right); x = 3$

(2) Calculate the wave number and energy of the spectral line (emitted by an excited hydrogen atom) corresponding to n_1=3 and n_2 = 5

Sol. $\bar{v} = 1.097 \times 10^7 \left(\frac{1}{3^2} - \frac{1}{5^2}\right)$

$= 1.097 \times 10^7 \times \frac{16}{225} = 7.8 \times 10^5 m^{-2}$

$E = hv = h\frac{c}{\lambda} = hc\bar{v} = 6.626 \times 10^{-34} \times 3 \times 10^8 \times 7.8 \times 10^5$

$= 1.55 \times 10^{-19} J$

(3) The wave length of the first member of Balmer series of hydrogen is 6563×10^{-10}m. Calculate the wave length of its second member.

Sol. For the first member $\bar{v}_1 = R_H\left(\frac{1}{2^2} - \frac{1}{3^2}\right) = \frac{5}{36} R_H$

For the second member $\bar{v}_2 = R_H\left(\frac{1}{2^2} - \frac{1}{4^2}\right) = \frac{3}{16} R_H$

$\frac{\bar{v}_1}{\bar{v}_2} = \frac{20}{27} \qquad \bar{v}_2 = \frac{27}{20} \times \bar{v}_1$

$\frac{1}{\lambda_2} = \frac{27}{20} \times \frac{1}{6563 \times 10^{-10}}$

$\lambda_2 = \frac{20}{27} \times 6563 \times 10^{-10} = 4861 \times 10^{-10} m.$

(4) What should be the kinetic energy and potential energy respectively of the electron present in the third orbit of hydrogen atom ?

Sol. We know, $E = -\frac{z^2}{n^2} \times 13.6$

Total energy of the 3rd orbit = $\frac{1}{9} \times 13.6 = -1.5$ eV

kinetic energy = - total energy = - (-1.5) = 1.5 eV

potential energy = 2 x Total energy = 2 x -1.5 = -3.0 eV

(5) What should be the ionisation energy of He^+ (The ionisation energy of hydrogen is 13.6 ev).

Sol. we know $E_n = E_1 \times \frac{z^2}{n^2}$

$-\frac{z^2}{n^2} \times 13.6 = -\frac{2^2}{1^2} \times 13.6 = -54.4ev = 54.4ev.$

(6) What should be the emitted energy when the electron present in hydrogen atom jumps from n = 4 to n = 2?

Sol. $E_2 = -\frac{z^2}{n^2} \times 21.79 \times 10^{-19}$ joules

$= -\frac{1^2 \times 21.79 \times 10^{-19}}{2^2} = -5.4 \times 10^{-19}$ J.

$E_4 = -\frac{1^2 \times 21.79 \times 10^{-19}}{4^2} = -1.36 \times 10^{-19}$

$E_4 - E_2 = -1.36 \times 10^{-19} - (-5.4 \times 10^{-19}) = 4.04 \times 10^{-19} J$

(7) Calculate the frequency of the line in the emission spectrum of hydrogen when the atoms of the gas contain electrons in the third energy level.

Sol. When it falls from 3rd → 2nd level

$v = 3.289 \times 10^{15} \left(\frac{1}{2^2} - \frac{1}{3^2}\right) = 3.289 \times 10^{15} \times \frac{5}{36}$

$= 4.568 \times 10^{14}$ Hz

when it falls from 3rd → 1st level

$v = 3.289 \times 10^{15} \left(\frac{1}{1^2} - \frac{1}{3^2}\right) = 2.924 \times 10^{15}$Hz

(8) The ionisation energy of hydrogen atom in the ground state is 2.17×10^{-11} erg./atom. Calculate the wavelength of the emitted radiation in Å unit, if electron returns from 2nd to 1st orbit.

Sol. I. E. = $E_\alpha - E_1 = 0 - (2.17 \times 10^{-11}) = -2.17 \times 10^{-11}$

$\Delta E = hv = h\frac{c}{\lambda}$

$\lambda = \frac{h \times c}{\Delta E} = 1.221 \times 10^2$ Å

(9) The energy of electron in the second orbit of H atom is - 3.41 eV.Calculate the energy of second orbit of He^+ ion

Sol. Energy of an electon in n^{th} orbit of hydrogen like species is given by $E_n = -\frac{13.6z^2}{n^2}$ eV/atom

For H-atom z = 1; n=2

$E_2 = -\frac{13.6 \times 1^2}{2^2} = -3.41eV$

for He^+, z = 2; n=2;

$E_{2'} = -\frac{13.6 \times 2^2}{2^2}$

$\frac{E_{2'}}{3.41} = \frac{-13.6 \times 2^2 \times 2^2}{2^2 \times 13.6 \times 1^2}$ or $\frac{E_{2'}}{3.41} = -4$

$E_{2'} = -4 \times 3.41 = -13.6$ eV

(10) Calculate and compare the energies of two radiations one with a wave length of 800nm and the other with 400nm.

Sol. $E_1 = hv_1 = h\frac{c}{\lambda_1}$; $E_2 = hv_2 = h\frac{c}{\lambda_2}$

$\frac{E_1}{E_2} = \frac{hc}{\lambda_1} \times \frac{\lambda_2}{hc} = \frac{\lambda_2}{\lambda_1} = \frac{800 \times 10^{-9}}{400 \times 10^{-9}} = \frac{2}{1}$

$E_1 = 2E_2$

(11) The threshold wavelength for the ejection of electron from metal M is 330 nm. Calculate the photoelectric emission.

Sol. Work function $\phi = hv_0 = \frac{hc}{\lambda_0} = \frac{6.6 \times 10^{-34} \times 3 \times 10^8}{330 \times 10^{-9}}$

$= 6.0 \times 10^{-19}$

(12) The work function of gold metal is 7.7×10^{-19}J. Calculate the kinetic energy and velocity of the electrons ejected when sodium metal is exposed to UVradiation of

wavelength 2.0 x 10^5 pm.

Sol. $\phi = h\nu_o = 7.7 \times 10^{-19}$ J

$$\nu = \frac{c}{\lambda} = \frac{3 \times 10^8}{2.0 \times 10^5 \times \frac{1}{10^{12}}} = 1.5 \times 10^{15} \text{ s}^{-1}$$

$KE = h\nu - h\nu_0$

$= 6.626 \times 10^{-34} \times 1.5 \times 10^{15} - 7.7 \times 10^{-19} = 2.239 \times 10^{-19}$J

$KE = \frac{1}{2} mv^2$

$$v^2 = \frac{2KE}{m};\ v = \sqrt{\frac{2KE}{m}} = \sqrt{\frac{2 \times 2.239 \times 10^{-19}}{9.1 \times 10^{-31}}}$$

$= \sqrt{49 \times 10^{10}} = 7 \times 10^5 \text{ ms}^{-1}$

(13) Calculate the energy of the spectral line of minimum energy in visible region obtained due to electronic transition in 1 gm atom of hydrogen.

Sol. The transition corresponds to Balmer series

$n_1 = 2;\ n_2 = 3$

$$\frac{1}{\lambda} = \bar{\nu} = 1.1 \times 10^7 \left(\frac{1}{2^2} - \frac{1}{3^2}\right) = 1.1 \times 10^7 \times \frac{5}{36} \text{m}^{-1}$$

$\lambda = 6.55 \times 10^{-7}$ m

$$E = h\nu = h \times \frac{c}{\lambda} = \frac{6.62 \times 10^{-34} \times 3 \times 10^8}{6.55 \times 10^{-7}} = 3.03 \times 10^{-19} \text{ J.}$$

For 1gm atom of hydrogen

$E_{total} = 3.03 \times 10^{-19} \times 6.023 \times 10^{23} = 182.5$ K.J.

(14) An electron is moving with a kinetic energy of 4.55 x 10^{-25} J. calculate the de Broglie wave length for it.

Sol. Kinetic energy $\frac{1}{2}$ $mv^2 = 4.55 \times 10^{-25}$ J

$$\frac{1}{2}mv^2 \div m = \frac{1}{2}v^2 = \frac{4.55 \times 10^{-25}}{9.1 \times 10^{-31}} = 0.5 \times 10^6$$

$v^2 = 2 \times 0.5 \times 10^6 = 1 \times 10^6$

$v = \sqrt{1 \times 10^6} = 1 \times 10^3 \text{ ms}^{-1}$

$$\lambda = \frac{h}{mv} = \frac{6.6 \times 10^{-34}}{9.1 \times 10^{-31} \times 1 \times 10^3}$$

$= 7.253 \times 10^{-7}$m or 725.3 nm

(15) Calculate the wave length of a body of mass 1 mg. moving with a velocity of 10 ms^{-1}.

$$\text{Sol. } \lambda = \frac{h}{mv} = \frac{6.6 \times 10^{-34}}{10^{-6} \times 10} = 6.6 \times 10^{-29}\text{m.}$$

(16) Calculate the uncertainity in the velocity of a cricket ball of mass 150g, if the uncertainity in its position is the order of 1Å.

$$\text{Sol. } \Delta v = \frac{h m \Delta x}{4\pi} = \frac{6.6 \times 10^{-34}}{4 \times 3.14 \times 1 \times 10^{-10} \times 150 \times 10^{-3}}$$

$= 3.5 \times 10^{-24}\text{ms}^{-1}$

(17) An electron has a velocity of 300m s^{-1} with an uncertainity of 0.001%. what is the uncertainity in locating its position?

Sol. Velocity of electron = 300ms^{-1}

uncertainity in veocity = 0.001% $\therefore \Delta v = \frac{0.001}{100} \times 300$

$= 0.003 \text{ ms}^{-1}$

$$\Delta x = \frac{h m \Delta v}{4\pi} = \frac{6.626 \times 10^{-34}}{4 \times 3.14 \times 9.1 \times 10^{-31} \times 3 \times 10^{-3}}$$

$= 1.932 \times 10^{-2}$m

(18) The ionisation energy of the He^+ is 19.6x10^{-18} J atom^{-1}. Calculate the ionisation energy of the stationary state of Li^{2+}.

Sol. IE_1 for $Li^{2+} = IE_1$ for H x Z^2

$= IE_1$ for H x 3^2

IE_1 for $He^+ = IE_1$ for H x Z^2

$= IE_1$ for H x 2^2

$$\frac{IE_1 \text{ for } Li^{2+}}{IE_1 \text{ for } He^+} = \frac{9}{4}$$

IE_1 for $Li^{2+} = \frac{9}{4} \times 19.6 \times 10^{-18} = 44.1 \times 10^{-18}$ Jatom^{-1}

(19) Calculate the energy required to excite one litre of hydrogen gas at 1 atm. and 298 K to the first excited state of atomic hydrogen. The energy for the dissociation of H-H bond is 436 kJ mol^{-1}.

Sol. Energy required to excite one H-atom from I^{st} to 2nd energy Level= $13.6 \left(\frac{1}{n_1^2} - \frac{1}{n_2^2}\right)$

$$= 13.6 \left(\frac{1}{1^2} - \frac{1}{2^2}\right) = 13.6 \times \frac{3}{4} = 10.2 \text{ eV}$$

$= 10.2 \times 1.6 \times 10^{-19} = 16.32 \times 10^{-19}$ J

No. of moles of hydrogen can be calculated from PV = nRT

$$n = \frac{PV}{RT} = \frac{1 \times 1}{0.0821 \times 298} = 0.0409 \text{ mole}$$

No. of hydrogen atoms that 0.0409 mole contain = 0.0818

Energy required for dissociation of 1 mole of H-H bonds = 436 kJ

Energy required for the dissociation of 0.0409 mole = 436x0.0409 = 17.83 kJ

Energy required to excite 0.0818 mole of H-atoms

$= 0.0818 \times 6.02 \times 10^{23} \times 16.32 \times 10^{-19}$

$= 8.0366 \times 10^4$ J = 80.37 kJ

Total energy required = Energy required to dissociate 0.0409 mole of H-H bonds + Energy required to excite 0.0818 mole of H-atoms= 17.83 + 80.37 = 98.20 kJ.

(20) The threshold frequency for a metal is 7.0X10^{14} s^{-1}. Calculate the kinetic energy of an electron emitted when radiation of frequency, $\nu = 1.0 \times 10^{15}\text{s}^{-1}$ hits the metal.

Sol. We know, kinetic energy = 1/2 $mv^2 = h(\nu - \nu_0)$

Substituting the values,

K.E. = $6.0626 \times 10^{-34}(1.0 \times 10^{15} - 7.0 \times 10^{14})$

$= 1.988 \times 10^{-19}$J

(21) The kinetic energy of an electron is 5.76X10^{-15} J. Calculate the wavelength associated with the electrons of the mass 9.1x10^{-31} Kg.

Sol. We know that K.E. = 1/2 mv^2

E. = 1/2m. m^2v^2

or $m^2v^2 = 2me$ or $mv = \sqrt{2me}$

we know de Broglie relation, $\lambda = h/mv$

therefore, $\lambda = h/\sqrt{2mE}$

Substituting the values,

$$\lambda = \frac{6.0626 \times 10^{-34}}{\sqrt{2 \times 9.1 \times 10^{-31} \times 5.76 \times 10^{-15}}} = 6.47 \times 10^{-12} \text{m}$$

Bonding

(1) Predict the hybridisation, shape and number of lone pairs present in $Cl F_3$.

Sol. We know that the required parameters can be deduced from, N=1/2 (V+M-C+A); L=N - M - D

(N = no.of orbitals in the hybridised orbital; V = no. of electrons in valence shell; M= no. of monovalent atoms; C = Charge on cation; A = Charge on anion ; L= no. of lone pairs; D = nol of divalent atoms)

for $Cl F_3$, N= 1/2 (7+3-0+0) = 1/2x10 = 5

L = 5-3 = 2

The no. of atomic orbitals taken part in hybridisation = 5

∴ It is sp^3d (dsp^3 is observed in transition elements). The shape is trigonal bipyramidal. Two lone pains are present.

(2) Predict the bond order in (i) H_2^+ (ii)He_2 and (iii) O_2 molecules / ions.

Sol. We know that, Bond order = 1/2 (N_b-N_a)

(N_b = no. of bonding electrons ; N_a = no. of antibonding electrons).

To know the values of N_b and N_a we should arrange the electrons in molecular orbitals as, $\sigma 1s$, $\sigma^* 1s$; $\sigma 2s$, $\sigma^* 2s$; $\sigma 2p$, $\pi_y 2p$, $\pi_z 2p$; $\pi_y^* 2p$, $\pi_z^* 2p$; $\sigma^* 2p$.........

(i) In H_2^+, only one electron is there and it enter $\sigma 1s$

B.O. = 1/2 (1-0) = 1/2x1=0.5

(ii) In He_2, four electrons are present and therefore they enter $\sigma 1s^2$ and $\sigma^* 1s^2$,

B.O. = 1/2 (2-2) = 1/2x0=0

It means that there exists no bond between two atoms of He.

(iii) O_2 has 16 electrons. They enter the orbitals as, $\sigma 1s^2$, $\sigma^* 1s^2$, $\sigma 2s^2$, $\sigma^* 2s^2$, $\sigma 2p^2$, $\pi_y 2p^2$, $\pi_z 2p^2$, $\pi_y^* 2p^1$, $\pi_z^* 2p^1$.

In π_y^* and π_z^* orbitals, one electron each enters according to Hund's rule of maximum multiplicity as these orbitals have the same energy.

B.O. = 1/2 (10-6) = 1/2x4=2

Note (a) O = O representation is for singlet Oxygen.

(b) $\dot{O}$ – $\dot{O}$ representation is for triplet Oxygen.

(3) Predict the bond order of 'C' atoms in benzene.

Sol. Here, one can use the formula for bond order

$$B.O. = \frac{\text{Total number of bonds between two atoms}}{\text{Number of stable resonating structures.}}$$

Benzene has two stable resonating structures of Kekule,

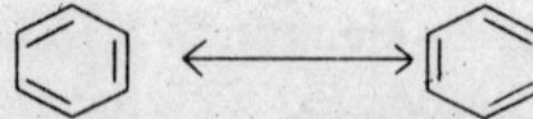

By substituting the values, B.O. = 3/2 = 1.5

21. Calculate the % ionic character of a bond having 1.355Å length and 1.05 D dipole moment.

Sol. we know that, % ionic character =

$$\frac{\text{observed dipole moment}}{\text{dipole moment for 100\% ionic bond}} \times 100$$

μ for 100% ionic bond = e × d

$= 4.8 \times 10^{-10}$ e.s.u. $\times 1.355 \times 10^{-8}$ cm

$= 6.50$ D (∴ 1D = 1×10^{-18} e.s.u. cm)

(or 10^{-29} coulomb metre)

$$\% \text{ ionic character} = \frac{1.05}{6.50} \times 100 = 16.15$$

22. The dipole moment of KCl is 3.336×10^{-29} coulomb metre which indicates that it is a highly polar molecule. The interatomic distance between K^+ and Cl^- in this molecule is 2.6×10^{-10}m. Calculate the dipole moment of KCl molecule if there were opposite charges of one fundamental unit located at each nucleus. Calculate the percentage ionic character of KCl.

Sol. we know that dipole moment, μ = e × d

For the separation of unit charge (charge of the electron) = 1.602×10^{-19}C

For KCl, μ = $1.602 \times 10^{-19} \times 2.6 \times 10^{-10}$

$= 4.1652 \times 10^{-29}$ coulombs metre

% ionic character of KCl =

$$\frac{\text{observed dipole moment}}{\text{dipole moment for 100\% ionic bond}} \times 100$$

$$= \frac{3.336 \times 10^{-29}}{4.1652 \times 10^{-29}} \times 100 = 80.09\%$$

Radioactivity

(1) How many α and β - particles will be emitted by Ra^{218}_{84} in changing to a stable isotope of lead Pb^{206}_{82}.

Sol. Let x and y be the number of α and β particles involved in bringing about the change. Then we can represent the change as

$$Ra^{218}_{84} \rightarrow Pb^{206}_{82} + x\,^{4}_{2} + y\,^{0}_{-1}$$

comparing the mass numbers : 218 = 206 + 4x + 0y

218 - 206 = 4x ; x = $\frac{12}{4}$ = 3

comparing the atomic numbers : 84 = 82 + 2x - y

substituting the value of x : 84 = 82 + 2 x 3 - y

84 - 88 = -y

- 4 = - y ; y = 4

3 α and 4 β - particles are emitted.

(2) Calculate the packing fraction, mass defect and energy released in the formation of argon atom, Ar^{40}_{18}. Mass of the isotope of argon is 39.962384 amu.

$$\text{Sol. Packing fraction} = \frac{\text{isotopic mass-mass number}}{\text{mass number}} \times 10^4$$

$$= \frac{39.962384 - 40}{40} \times 10^4 = -9.4025$$

mass defect (ΔM) = M' – M

M' = 18 x 1.007825 + 22 x 1.008665 = 40.33148

M = 39.962384

ΔM = 40.33148 - 39.962384 = 0.369096

Energy released = 0.369096 x 931 = 343.628 MeV.

(3) The fission fragments in the thermal neutron fission of U^{235} are found to be Mo^{98}_{42} and Xe^{136}_{54}. What are the elementary particles released in the reaction?If the masses of U^{235}, Xe^{136}, Mo^{90} and n^1_0 are 235.044, 135.907, 97.906 and 1.0086 amu respectively, calculate the binding energy.

Sol. The balanced nuclear reaction is

$U^{235}_{92} + n^1_0 \rightarrow Mo^{98}_{42} + Xe^{136}_{54} + 4e^0_1 + 2n^1_0$

Mass defect = mass of products – mass of reactants

= (97.906 + 135.9070 + 4 x 0.005486 + 2 x 1.0086) – (235.044 + 1.0086)

= 235.854 - 236.0526 = - 0.1986 (decrease in the mass)

Energy released due to a decrease in mass = 0.1986 x 931 = 184.90 MeV.

(4) The mass defect of Cl^{35}_{17} is found to be 0.320 amu. Calculate the binding energy per nucleon.

Sol. Mass defect = 0.320

Binding energy = 0.320 x 931 = 297.92 MeV.

Binding energy per nucleon = $\frac{297.92}{35}$=8.512 MeV.

(5) In the fission of U^{235}_{92} with a neutron, Mo^{95}, La^{139} and two neutrons are formed. The atomic masses of U, Mo, La and n^1_0 are 235.118, 94.936, 138.950 and 1.009 amu respectively. Calculate the energy released by the fission of 1g. of U^{235}.

Sol. We can write the balanced equation as

$U^{235}_{92} + n^1_0 \rightarrow Mo^{95} + La^{139} + 2n^1_0$

mass defect=(235.118+1.009) - (94.936+138.950+2x1.009)

= 236.127 - 235.904 = 0.223 amu.

Binding energy = 0.223 x 931 = 207.613 MeV

No. of atoms in 1g. of U^{235} = $\frac{1}{235}$x 6.623 x 10^{23}

Energy released by U^{235}_{92} atom = 207.613 MeV

∴ Energy released by the fission of $\frac{1}{235}$x 6.023 x 10^{23} is

$= \frac{207.613 \text{ x } 6.023 \text{ x } 10^{23}}{235} = 5.321 \text{ x } 10^{23}$ MeV

(6) If half life period of an element x is 20 years, calculate the time required to reduce its activity to 1/8 of its original value.

Sol. $\frac{1}{8} = (\frac{1}{2})^x = (\frac{1}{2})^3$

Number of $t_{0.5}$ = 3; time required = 3 x 20 = 60 years

(7) $t_{0.5}$ of a radioactive subtance is 10 years. What percentage of the substance will be left behind after 40 years ?

Sol. $t_{0.5}$ = 10 years or 10 years = $t_{0.5}$

$(x = \frac{\text{time passed}}{\text{half life period}})$

x = number of half lives

40 years = 4

$(\frac{1}{2})^x = (\frac{1}{2})^4 = \frac{1}{16}$ = 6.625%

(8) Half life of a radioactive isotope is 10 days. Its 125 g. is left behind. What was the original mass 40 days back ?

Sol. $t_{0.5}$ = 10 days

$x = (\frac{1}{2})^4 = \frac{x}{16} = 125$

x = 125 x 16 = 2000 g.

(9) The half life period of Po^{210}_{84} is 140 days. In how many days 1g. of this isotope is reduced to 0.25 g ?

Sol. Initial quantity of the isotope N_o= 1g.

Final quantity of isotope N = 0.25 g.

We know that N = $(\frac{1}{2})^n$ No

$\frac{1}{4} = (\frac{1}{2})^n \text{ x } 1$ or $(\frac{1}{2})^2 = (\frac{1}{2})^n$

n = 2

Time taken t = n x $t_{0.5}$ = 2 x 140 = 280 days.

(10) A radioactive element x is spread over a room. It has $t_{0.5}$ equal to 30 days. Its activity is 50 times the permissible value. After how many days will the room be safe ?

Sol. $\lambda = \frac{0.693}{30}$

$t = \frac{2.303}{0.693} \text{ x } 30 \log \frac{50 N}{N} = 169.38$ days

(11) If Co^{60} has a $t_{0.5}$ of 5.2 years, how much of a 10g. of a sample be radioactive after 31.2 years ?

Sol. Number of half - lives $=\frac{312}{5.2} = 6$

Sample remaining = 10 x $(\frac{1}{2})^6$ = 10 x$\frac{1}{64}$= 0.156 g.

(12) Radio-isotope P^{32}_{15} has a half life of 15 days. Calculate the time in which the radioactivity of its 1mg. quantity will fall to 10% of its initial value.

Sol. $\lambda = \frac{0.693}{t_{0.5}} = \frac{0.693}{15}$

$t = \frac{2.303 \text{ x } 15}{0.693} \log \frac{100}{10} = 49.85$ days

(13) 1g. of Mo decays by β - emission to 0.12 5g. in 200 hours. How much more time will elapse until only 0.01g of it is left ?

Sol. $\lambda = \frac{2.303}{t} \log \frac{No}{Nt} = \frac{2.303}{200} \log \frac{1}{0.125} = 0.0104 \text{ hr}^{-1}$

Now $t = \frac{2.303}{0.0104} \log \frac{0.125}{0.01} = 242.9$ hr.

(14) Calculate the ratio of U^{238} to Ra^{226} atoms in natural uranium. $t_{0.5}$ of U^{238} is 4.49 x 10^9 years and $t_{0.5}$ of Ra^{226} is 1622 years.

Sol. $\frac{N_1}{N_2} = \frac{(t_{0.5})_1}{(t_{0.5})_2} = \frac{4.49 \text{ x } 10^9}{1622} = 2.77 \text{ x } 10^6$

∴ U^{238} : Ra^{226} = 2.77 x 10^6 : 1

(15) A piece of wood was found to have C^{14}/C^{12} ratio 0.7 times that in a living plant. Calculate the period when the plant died. $t_{0.5}$ of C^{14} is 5760 years.

Sol. $\lambda = \frac{0.693}{t_{0.5}} = \frac{0.693}{5760}$

$t = \frac{2.303}{0.693} \text{ x } 5760 \text{ x } \log \frac{1}{0.7} = 2970$ years

(16) Calculate the half life of Na^{24} if a 2.4 x 10^{-5} g. sample disintegrates at the rate of 7.73 x 10^{12} atoms per second.

Sol. $\lambda = \frac{\text{No. of atoms disintegrating per second}}{\text{No. of atoms present}}$

$$= \frac{7.73 \times 10^{12}}{2.4 \times 10^{-5} \times \frac{1}{24} \times 6.02 \times 10^{23}} = 1.28 \times 10^{-5}\ sec^{-1}$$

$$t_{0.5} = \frac{0.693}{1.28 \times 10^{-5}} = 5.41 \times 10^4\ sec.$$

(17) Uranium ore is found to contain one radium atom for every 2.78 x 10^6 uranium atom. Radium is known to decay with a half life of 1620 years. Find the half life of uranium.

Sol. $\lambda_1 N_1 = \lambda_2 N_2$

$$\lambda_1 \times 2.78 \times 10^6 = \frac{0.693}{1620} \times 1$$

$$\lambda = \frac{0.693 \times 1}{1620 \times 2.78 \times 10^6}$$

$$t_{0.5} = \frac{0.693 \times 1620 \times 2.78 \times 10^6}{0.693} = 4.5 \times 10^9\ years$$

(18) Analysis of a sample of uranium ore showed that it contained 0.277 g. of Pb^{206} and 1.667 g. of U^{238}. Calculate the age of the ore assuming that all the Pb^{206} had accumulated due to decay of U^{238}. $t_{0.5}$ of U^{238} is 4.5 x 10^9 years.

Sol. As Pb^{206} had accumulated due to decay of U^{238}, we have to find out how much U^{238} was initially present.
If the wt. of U is 238 we obtain 206 Pb
from how much 0.277 g.Pb

$$\frac{0.277}{206} \times 238 = 0.320g.$$

Initial wt. of U^{238} = 1.667 + 0.320 = 1.987g.

Now we calculate $\lambda = \frac{0.693}{4.5 \times 10^9}$

$$t = \frac{2.303 \times 4.5 \times 10^9}{0.693} \log \frac{1.987}{1.667} = 1.129 \times 10^9\ years$$

(19) The amount of C^{14} isotope in a piece of wood is found to be 1/6 of its amount present in a fresh piece of wood. Calculate the age of the wood. ($t_{0.5}$ of ^{14}C is 5577 years)

Sol. $\lambda = \frac{0.693}{t_{0.5}} = \frac{0.693}{5577}$

$$t = \frac{2.303}{\lambda} \log \frac{No}{Nt} = \frac{2.303 \times 5577}{0.693} \log \frac{1}{6}$$

$$= \frac{2.303 \times 5577}{0.693} \times 0.7782$$

$$= 14422.\dot{9}\ years = \sim 14423\ years$$

(20) Po^{210}_{84} decays with α - emission to Pb^{206}_{82} and has $t_{0.5}$ of 138.4 days. If 1 g. of Po^{210} is placed in a sealed tube, how much Helium will accumulate in 69.2 days?

Sol. $\lambda = \frac{0.693}{138.4} = 5.007 \times 10^{-3}$ per day

Now applying the decay constant equation

$$\lambda = \frac{2.303}{t} \log \frac{No}{Nt},\ \text{we get}$$

$$5.007 \times 10^{-3} = \frac{2.303}{69.2} \log \frac{1}{Nt}$$

$$\log \frac{1}{Nt} = \frac{5.007 \times 10^{-3} \times 69.2}{2.303} = 0.1504$$

$$\frac{1}{N} = 1.414$$

$$N = \frac{1}{1.414} = 0.707\ g.$$

Amount of Po^{210} decayed = 1 - 0.707 = 0.293

from the equation $Po^{210}_{84} \rightarrow Pb^{206}_{82} + He^4_2$

We can say 210 g. of Po produce He i.e. equivalent to 22,400 c.c. at STP

0.293 g. Po will produce He equivalent to

$$= \frac{22400 \times 0.293}{210} = 31.25\ c.c.$$

(21) Predict the atomic number and atomic mass of the element X formed from Y_{90}^{232} after the emission of 6-α and 4-β particles. $Y_{90}^{232} \longrightarrow X + 6\alpha + 4\beta$

Emission of 6-α particles reduce the mass by 6X4 = 24 units (Since the mass of β– particle is negligible)

Emission of 6-α particles reduces the charge by, 6x2 = 12 units.

Emissions of 4–β particles increases the charge by, 4x1 = 4 units.

(Since the emission of a β– particle from the nucleus due to the decomposition of neutron as, $n_o \longrightarrow H_1^1 + \beta + \nu$ increases the charge by one unit). Therefore, the element formed can be represented as X_{82}^{208}

22. Cu^{64}_{29}(half life = 12.8 hr) decays by β^- emission (38%), β^+ emission (19%) and electron capture (43%). Write the decay products and calculate partial half-lives for each of the decay processes.

Sol. β^- emission reaction, $Cu^{64}_{29} \rightarrow Zn^{64}_{30} + \beta^o_{-1}$ .... (i)

β^+ emission reaction, $Cu^{64}_{29} \rightarrow Ni^{64}_{28} + \beta^o_{+1}$ .... (ii)

electron capture reaction, $Cu^{64}_{29} + e^o_{-1} \rightarrow Ni^{64}_{28}$ ....(iii)

Let the rections (i), (ii) and (iii) have k_1, k_2 and k_3 rate constants respectively. Then the overall rate of the reaction, $K = k_1 + k_2 + k_3$

Rate constant $K = \frac{0.693}{t_{0.5}} = \frac{0.693}{12.8} = 0.054$

$$\therefore k_1 = \frac{38}{100} \times 0.054 = 0.0205$$

Now $t_{0.5}$ for β^- emission $= \frac{0.693}{k_1} = \frac{0.693}{0.0205} = 33.80$ hr

$$k_2 = \frac{19}{100} \times 0.054 = 0.0103$$

$\therefore t_{0.5}$ for β^+ emission $= \frac{0.693}{k_2} = \frac{0.693}{0.0103} = 67.28$ hr

$$k_3 = \frac{43}{100} \times 0.054 = 0.0232$$

$\therefore t_{0.5}$ for electron capture $= \frac{0.693}{k_3} = \frac{0.693}{0.0232} = 29.87$ hr.

Gaseous State

(1) The volume of a gas with some marble measures 100ml at 760mm pressure. On raising the pressure to 1000mm, the total volume becomes 80ml. Calculate the volume occupied by the marble.

Sol. let V be the volume of the marble

Now apply $P_1V_1 = P_2V_2$

$$(100 - v) \times 760 = (80 - v) \times 1000$$

$$24v = 400 \; ; \; v = 16.6\text{ml.}$$

(2) A vessel filled with cold gas can withstand a pressure of 14.9 atmosphere. The pressure gauge attached to the vessel reads 12 atmosphere at 27°C. The building in which the vessel is kept suddenly catches fire and the temperature starts rising. Calculate the temperature at which the vessel will explode.

Sol. We know at constant volume, $\frac{P_1}{T_1} = \frac{P_2}{T_2}$

$$\frac{12}{300} = \frac{14.9}{T_2}$$

$$T_2 = \frac{300 \times 14.9}{12} = 372.5\text{k i.e. } 99.5^\circ\text{C.}$$

∴ The vessel will explode at any temperature beyond 99.5°C.

(3) Calculate the weight of 10 litres of nitrogen gas at 27°C and 74 cm pressure.

Sol. First we calculate the volume equivalent at NTP

we know $\frac{P_1V_1}{T_1} = \frac{P_2V_2}{T_2}$

$$V_2 = \frac{P_1V_1T_2}{T_1P_2} = \frac{74 \times 10 \times 273}{300 \times 76} = 8.86 \text{ litres}$$

22.4 lit. of volume of nitrogen at NTP weighs 28g.

8.86 lit. of volume of nitrogen at NTP weighs

$$= \frac{8.86 \times 28}{22.4} = 11.075\text{g.}$$

(4) 2.8g. of a gas at 27°C and 180 mm pressure occupies 3.8 lit. Calculate the molecular weight of the gas

Sol. $PV = nRT$

$$= \frac{w}{m}RT$$

$$\frac{180}{760} \times 3.8 = \frac{2.8}{m} \times 0.0821 \times 300 \; ; \; m = 76.44$$

(5) Calculate the relative rates of diffusion of $^{235}UF_6$ and $^{238}UF_6$ in the gaseous state

Sol. Molecular mass of $^{235}UF_6 = 235 + 6 \times 19 = 349$

Molecular mass of $^{238}UF_6 = 238 + 6 \times 19 = 352$

$$\frac{r_1}{r_2} = \sqrt{\frac{M_2}{M_1}} = \sqrt{\frac{352}{349}} = 1.0043$$

$r_1 : r_2 = 1.0043 : 1.0000$

(6) The rate of diffusion of a gas B with respect to another diatomic gas A, is 6 : 5. If the atomic weight of A is 36, calculate the molecular weight of B.

Sol. We know $\frac{r_B}{r_A} = \sqrt{\frac{M_A}{M_B}}$; mol. wt. of A = 2 x 36 = 72

$$\frac{6}{5} = \sqrt{\frac{72}{M_B}}$$

$$M_B = \frac{72 \times 5 \times 5}{6 \times 6} = 50$$

(7) 180ml. of a hydrocarbon diffuse in 15 minutes, whereas 120ml. of SO_2 diffuse in 20 minutes under the similar conditions. If the molecular weight of SO_2 is 64, calculate the molecular weight of the hydrocarbon.

Sol. $\frac{r_{hy}}{r_{SO_2}} = \frac{v_{hy}/t}{v_{SO_2}/t} = \sqrt{\frac{64}{M_{hy}}}$

$$\frac{180/15}{120/20} = \sqrt{\frac{64}{M_{hy}}}$$

$$\frac{12}{6} = \sqrt{\frac{64}{M_{hy}}}$$

$$M_{hy} = \frac{64 \times 6 \times 6}{12 \times 12} = 16$$

(8) At the top of a mountain, the thermometer reads 10°C and the barometer reads 700mm Hg. At the bottom of the mountain, the temp. is 27°C and the pressure is 760mm Hg. Compare the density of the air at the top with that at the bottom.

Sol. $PV = \frac{w}{m}RT$

$$P = \frac{w}{V} \times \frac{RT}{m}$$

$$= d\frac{RT}{m} \quad (\because \frac{w}{V} = d)$$

or $d = \frac{P}{T} \cdot \frac{m}{R}$

$d \propto \frac{P}{T}$ [m/R being constant]

or $\frac{d_{top}}{d_{bottom}} = \frac{P_{top}}{T_{top}} \times \frac{T_{bottom}}{P_{bottom}} = \frac{700 \times 300}{760 \times 283} = 0.986$

(9) A gaseous mixture of 0.336g. N_2 and 11.2g. litre O_2 is taken at NTP in a 10 lit. flask at 27°C. Calculate the partial pressure and total pressure of the gases.

Sol. Number of moles of $N_2 = \frac{\text{mass of } N_2}{\text{molecular mass}}$

$$= \frac{0.336}{28} = 0.012$$

Number of moles of $O_2 = \frac{11.2}{22.4} = 0.5$

P_{N_2} (partial pressure of N_2) $= \frac{nRT}{V}$

$$= \frac{0.012 \times 0.0821 \times 300}{10} = 0.0296 \text{ atm.}$$

$$P_{O_2} = \frac{0.5 \times 0.082 \times 300}{10} = 1.2315 \text{ atm.}$$

Total pressure = 0.0296 + 1.2315 = 1.2611 atm.

(10) A flask is filled with 250ml. of Kr at 500mm pressure. Another flask contains 500ml. He at 950mm. Both the gases are mixed at the same temperature. Calculate the combined volume and the total pressure exerted by the mixture.

Sol. Total volume = 250 + 500 = 750ml.

$P_1V_1 = P_{kr} \times V$

$P_{kr} = \frac{250 \times 500}{750} = 166.6mm$

$P_{He} = \frac{950 \times 500}{750} = 633.3mm$

Total pressure of the mixture $P_{kr} + P_{He} = 166.6 + 633.3 = 800$ mm

(11) At 28°C and 730 torr pressure, the mass of 525 cc of a gaseous compound was 0.90g. Calculate the molecular weight of the compound.

Sol. $PV = nRT$

$= \frac{w}{m}RT$

$m = \frac{wRT}{PV} = \frac{0.90 \times 82.1 \times 301}{(730/760) \times 525} = 44.1$

(12) Calculate the total and averge kinetic energy in joules, of the molecules in 8g. of methane at 27°C.

Sol. Average kinetic energy of one molecule $= \frac{3}{2}$ KT

$= \frac{3}{2} \times \frac{8.314 \times 300}{6.022 \times 10^{23}} = 6.213 \times 10^{-21}$J.

No. of moles of methane $= \frac{8}{16} = 0.5$

Total kinetic energy $= \frac{3}{2} \times 0.5 \times 8.314 \times 300 = 1870.65$ J.

(13) A gas bulb of 1 lit. capacity contains 2.0×10^{21} molecules of nitrogen exerting a pressure of 7.57×10^3 Nm^{-2}. Calculate the root mean square speed and temperature of the gas molecules. If the ratio of most probable speed to the RMS speed is 0.82, calculate the most probable speed of these molecules at this temperature.

Sol. $PV = nRT$ or $T = \frac{PV}{nR}$ $n = \frac{2.0 \times 10^{21}}{6.023 \times 10^{23}}$ mole

$T = \frac{7.57 \times 10^3 \times 6.023 \times 10^{23}}{2.0 \times 10^{21} \times 8.314} = 274.2K$

RMS speed $= \sqrt{\frac{3RT}{M}} = \sqrt{\frac{3 \times 8.314 \times 274.2}{28 \times 10^{-3}}} = 494.22\ ms^{-1}$

Most probable speed = RMS x 0.8

$= 494.22 \times 0.8 = 395.38\ ms^{-1}$

(14) 2 mole of NH_3 gas is kept in a closed container of 2 lit. capacity at 27°C. Calculate the pressure exerted by the gas when it behaves as (i) an ideal gas and (ii) a real gas.(a = 4.17 lit.² atm. mole²; b = 0.037 lit. mole⁻¹.)

Sol. As an ideal gas, $PV = nRT$

$P = \frac{nRT}{V} = \frac{2 \times 0.0821 \times 300}{2} = 24.63$ atm.

As a real gas, $(P + a\frac{n^2}{V^2}) - (V - nb) = nRT$

$P = \frac{nRT}{V\text{-}nb} - \frac{an^2}{V^2}$

$= \frac{2 \times 0.0821 \times 300}{2 - (2 \times 0.037)} - \frac{4.17 \times 2^2}{2^2} = 21.48$ atm.

(15) Calculate form the van der Waal's equation the temperature at which 3 moles of SO_2 would occupy a volume of 10 litres at 15 atmospheric pressure from the given data :

a = 6.7 atm lit²/mole² ; b = 0.0564 lit/mole ;
R = 0.082 litre atm/degree/mole.

Sol. We know $(P + \frac{an^2}{V^2})(V - nb) = nRT$

$T = (P + \frac{an^2}{V^2})(V - nb) \times 1/nR$

$= (15 + \frac{6.7 \times 3^2}{10^2})(10 - 3 \times 0.0564) \times \frac{1}{3 \times 0.082}$

$= 15.603 \times 9.8308 \times \frac{1}{0.246}$

= 623.5 k

T = 623.5 - 273 = 350.5°C

(16) The pressure exerted by 12g of an ideal gas at temperature t°c in a vessel of volume V litre is one atmosphere. When the temperature is increases by 10 degrees at the same volume, the pressure increases by 10%. Calculate the temperature T and volume V.

Sol. We know, PV=nRT or PV = w/m RT

For the gas at t°c, $1 \text{x} V = \frac{12}{120} \text{x}\ 0.0821 \text{x} (T+273)$-(i)

For the gas at (t°+10)°c, $1.1 \text{x} V = \frac{12}{120} \text{x} 0.0821 \text{x} (T+273)$--(ii)

Dividing (ii) by (i), we get $\frac{1.1}{1} = \frac{t+273}{t+273}$

1.1t+300.3=t+283

0.1t = t-t +283-300.3= -17.3°c

$t = \frac{17.3}{0.1} = -173$°c or 100 K

Now substituting the value in (i)

$V = \frac{12}{120} \times 0.0821 \times 100 = 0.821$ lit.

(17) An LPG cylinder weighs 14.8 Kg. when empty, when full it weighs 29.0 Kg and shows a pressure of 2.5 atm. In the course of use at 27°c, The weight of full cylinder is reduced to 23.2 Kg. Find out the volume of the gas in cubic metres used up at the normal usage conditions and the final pressure inside the cylinder. Assume LPG to be n-butane with normal boiling point at 0°c($R = 0.0821\ d\ m^3 atm.\ k^{-1}\ mol^{-1} = 8.314\ Jk^{-1} mol^{-1}$).

Sol. Weight of the cylinder reduced due to usage of gas = 29.0 - 23.2 = 5.8 Kg.

Number of moles of gas used

= Wt. of the gas used / mol. wt. of the gas x 1000

= 5.8 x 1000 /58 = 100 (mol. wt. of C_4H_{10}=58)

The volume of the gas used can be calculated from,

$PV = nRT;\ V = \frac{nRT}{P} = \frac{100 \times 0.0821 \times 300}{1}$

$= 2463\ d\ m^3 = 2.463m^3$

Wt. of gas left in the cylinder = 23.2 -14.8 = 8.4 Kg.

No. of moles of gas left = 8.4/58 x 1000 = 144.8

The volume of the gas left in the cylinder under the conditions is much higher than the volume of the cylinder.

Therefore the gas inside the cylinder is expected to be in liquid state. We know that at a constant temperature, the vapour pressure of the gas is constant Therefore, the pressure of the gas left in the cylinder must be 2.5 atmosphere.

(18) A gas expands from a volume of 3.0 dm^3 to 5.0 d m^3 against a constant external pressure of 3.0 atm. The work done during the expansion is used to heat 10.0 ml of water of 290.0 K. Calculate the final temperature of water. Specific heat of water = 4.184 $Jg^{-1}k^{-1}$

Sol. We know, $W = P\Delta V$

$= 3x2x10^3x76x13.6x981$

$= 6.08 \times 10^9$ ergs = 608 J(since1 Joule = 10^7ergs)

Wt. of 10 mole of H_2o = 10 x 18 = 180 g.

Rise in temperature of $H_2O = \frac{608}{180x4.184} = 0.81K$

Final temperature of water = 290.0 + 0.81 = 290.81 K or 17.81°c.

(19) Calculate the pressure exerted by one mole of CO_2 gas at 273 K if the van der Waals constant, a=3.592 d m^6 atm. mol^{-2}. Assume that the volume occupied by CO_2 molecule is negligible.

Sol. We know van der Waals equation $(P + \frac{a}{V^2})(V-b) = RT$

The value of b=0 as the volume occupied by CO_2 molecules is negligible $P = \frac{RT}{V} - \frac{a}{V^2} = \frac{RTV-a}{V^2}$

or $PV^2 = RTV - a = PV^2 - RTV + a = 0$

In terms of V, the equation is quadratic.

$(V = \frac{+RT \pm \sqrt{R^2T^2-4ap}}{2a})$

Under the given conditions of temperature and pressure, V can have only one value. Therefore the numerical value of discriminant is treated as Zero.

Therefore, we can write the equation, $R^2T^2 = 4ap$

or $P = \frac{R^2T^2}{4a} = \frac{(0.0821)^2 \times (273)^2}{4 \times 3.592} = 34.96$ atm.

(20) The compression factor (compressibility factor) for 1 mole of a van der Waals gas at 0°c and 100 atmosphere pressure is found to be 0.5 . Assuming that the volume of gas molecule is negligible, calculate the van der Waal's constant a.

Sol. We know that compressibility factor $Z = \frac{PV}{nRT}$

From the equation we calculate, $V = \frac{ZnRT}{P}$

$= \frac{0.5 \times 1 \times 0.082 \times 273}{100} = 0.1121$ lit.

From van der Waals equation, $(P + \frac{an^2}{V^2})(V - nb) = nRT$

We calculate a As 'b' is negligible,

$$\frac{100+a}{(0.1121)^2} = \frac{0.0821 \times 273}{0.1121}$$

$$\frac{a}{(0.1121)^2} = 199.94 - 100$$

$a = 99.94 \times (0.1121)^2 = 1.256$ lit^2 mol^2 atm.

21. A spherical balloon of 21 cm diameter is to be filled with hydrogen at N.T.P. from a cylinder containing the gas at 20 atmosphere at 27°C. If the cylinder can hold 2.82 litres of water, calculate the number of balloons that can be filled up.

Sol. The volume of the balloon, $V = 4/3\ \pi r^3$

$\therefore V = \frac{4}{3} \times 3.14 \times (\frac{21}{2})^3 = 4847$ ml.

volume of hydrogen gas in the cylinder at NTP can be calculated from the relation,

$$\frac{P_1V_1}{T_1} = \frac{P_2V_2}{T_2}$$

$$\frac{1 \times V_1}{273} = \frac{20 \times 2.82}{300}$$

$$V_1 = \frac{20 \times 2.82 \times 273}{300} = 51324 \text{ ml}$$

Actual volume of hydrogen gas = 51324 - 2820 = 48504ml

$\therefore$ number of balloons that can be filled up =

$$\frac{\text{volume of gas available}}{\text{volume of a balloon}} = \frac{48504}{4847} = 10$$

22. Oxygen is present in a litre flask at a pressure of 7.6×10^{-10} mm of Hg. Calculate the number of oxygen molecules in the flask at 0°C.

Sol. we know, PV = nRT or n = PV/RT

$$= \frac{7.6 \times 10^{-10} \times 1}{760 \times 0.0821 \times 273} = 4.46 \times 10^{-14} \text{ moles.}$$

1 mole contains 6.023×10^{23} molecules

$\therefore 4.46 \times 10^{-14}$ moles contain $= 6.023 \times 10^{23} \times 4.46 \times 10^{-14}$

$= 2.86 \times 10^{10}$ molecules.

23. One mole of nitrogen gas at 0.8 atm. takes 38 seconds to diffuses through a pin hole whereas one mole of an unknown compound of xenon with fluorine at 1.6 atm. takes 57 seconds to diffuse through the same hole. Calculate the molecular formula of the compound.

Sol. we know that for diffusion of a gas at pressur, P, the rate of diffusion is given as, $r \propto P/\sqrt{M}$ or $r = KP/\sqrt{M}$

$r_{N_2} = 1/38$ mol s^{-1}; $r_x = 1/57$ mol s^{-1}

Now substituting in the above equation for N_2,

$$\frac{1}{38} = \frac{K \times 0.8}{\sqrt{28}} \quad \text{........ (i)}$$

For the unknown gas, $\frac{1}{57} = \frac{K \times 1.6}{\sqrt{M_x}}$ ....(ii)

Dividing (i) by (ii), $\frac{1}{38} \times \frac{57}{1} = \frac{K}{K} \times \frac{0.8}{1.6} \times \sqrt{\frac{M_x}{28}}$

$1.5 = 0.5\sqrt{Mx/28}$

$$Mx = \frac{1.5 \times 1.5 \times 28}{0.5 \times 0.5} = 252 \text{ g/mole}$$

The compound is fluoride of xenon

$\therefore 131 + (x \times 19) = 252$

$$x = \frac{252 - 131}{19} = \frac{121}{19} = 6.37 \sim 6$$

$\therefore$ The formula of the xenon fluoride is XeF_6

Solutions

1. Calculate the molality of a 20% (by weight) aqueous solution of KOH.

Sol. A 20% solution of KOH contains 20g. of KOH dissolved in 80 of H_2O.

No. of moles of KOH $= \frac{20}{56}$

$\frac{20}{56}$ moles of KOH are dissolved in 80 g. of H_2O

No. of moles of KOH in 1000g. of $H_2O = \frac{20}{56} \times \frac{1000}{80} = 4.464$

Molality $= \frac{\text{no. of moles}}{\text{no. of kg}} = \frac{4.464}{1} = 4.464$ m

2. What should be the molarity of 1 lit. solution of 93% w/w H_2SO_4, if the density is 1.84 ?

sol. Molarity $= \frac{\text{wt. in gram x density x 1000}}{\text{Molecular wt. x 100}}$

$= \frac{93 \times 1.84 \times 1000}{98 \times 100} = 17.46$

3. A solution contains 410.3 g. of H_2SO_4 per litre of solution at 20°C. If the density is 1.243 g/ml, calculate the molarity and molality of the solution.

Sol. no. of moles of $H_2SO_4 = \frac{410.3}{98} = 4.187$ M

weight of 1000 ml of H_2SO_4 solution = 1000 x 1.243 = 1243 g.

Weight of water = 1243 – 410.3 = 832.7 g.

Molality of the solution $= \frac{\text{no. of moles of } H_2SO_4}{\text{wt. of water}}$

$= \frac{4.187 \times 1000}{832.7} = 5.028$ m

4. Calculate the molality of a 1M solution of $NaNO_3$ whose density is 1.25 g cm^{-3}.

Sol. Molality $= \frac{\text{no. of moles of } NaNO_3}{\text{Mass of water in kg}}$

wt. of 1lit. of solution = 1000 x 1.25 = 1250 g.

Mass of $NaNO_3$ in the solution = 85 g. (since it is 1M solution)

Mass of water = 1250 – 85 = 1165 g.

Molality $= \frac{1}{1.165} = 0.858$

5. What volume of 95% H_2SO_4 (density = 1.85 g/cm^3) and what mass of water must be taken to prepare 100cc of 15% solution of H_2SO_4 (density = 1.10 g/cm^3).

Sol. 95% H_2SO_4 means 95 g. of H_2SO_4 is present in 100g. of solution.

Volume of the solution $= \frac{100}{1.85} = 54.05$ ($\because \frac{M}{D} = V$)

Molarity $= \frac{95}{98} \times \frac{1000}{54.05} = 17.93$

15% H_2SO_4 means 15g. of H_2SO_4 is present in 100g. of solutionvolume of this solution $= \frac{100}{1.10} = 90.91$

Molarity $= \frac{15}{98} \times \frac{1000}{90.91} = 1.68$

$M_1V_1 = M_2V_2$

$V_1 = \frac{1.68 \times 100}{17.93} = 9.37$ cc.

Volume of 95% H_2SO_4 required = 9.37 c.c.

Mass of 100 c.c.of 15% H_2SO_4 to be prepared = 100 x 1.10=110g.

Mass of 9.37 c.c. of 95% H_2SO_4 = 9.37 x 1.85 = 17.3 g.

Mass of H_2O to be taken = 110 - 17.3 = 92.7 g.

6. Upon mixing 45ml. of 0.25 molar $Pb(NO_3)_2$ solution with 25ml. of 0.1 molar $Cr_2(SO_4)_3$ solution precipitation of $PbSO_4$ takes place. How many moles of $PbSO_4$ are formed. Calculate the molar concentrations of the species left behind in the final solution (assume that $PbSO_4$ is completely insoluble)

Sol. On balancing the reaction we get

$$3Pb(NO_3)_2 + Cr_2(SO_4)_3 \rightarrow 3PbSO_4 + 2Cr(NO_3)_3.$$

3 mole 1 mole 3 mole 2 mole

Initial concentration

$0.25 \times \frac{45}{1000}$ $0.1 \times \frac{25}{1000}$ 0 0

$= 11.25 \times 10^{-3}$moles $= 2.5 \times 10^{-3}$moles

After reaction, number of moles left unused

$= 11.25 \times 10^{-3}$ $-3 \times 2.5 \times 10^{-3} = 3.75 \times 10^{-3}$.

consumed, $3 \times 2.5 \times 10^{-3} = 7.5 \times 10^{-3}$ $2 \times 2.5 \times 10^{-3} = 5.0 \times 10^{-3}$.

Total volume of the solution = 45 + 25 = 70ml.

molarity of $Pb(NO_3)_2$ left unused $= \frac{3.75 \times 10^{-3}}{70 \times 10^{-3}} = 0.0535$ M

molarity of $Cr(NO_3)_3$ formed $= \frac{5.0 \times 10^{-3}}{70 \times 10^{-3}} = 0.0714$ M

7. 1000 ml. of a water sample contain 1mg. of $CaCl_2$ and 1mg. of $MgCl_2$. Find the hardness of water in terms of parts of $CaCO_3$.

Sol. 1mg. of $CaCl_2 = \frac{100}{111}$ x1 of $CaCO_3$ ($\because \underset{111}{CaCl_2} \rightarrow \underset{100}{CaCO_3}$)

1mg. of $CaCl_2 = \frac{100}{95}$x1 of $CaCO_3$ ($\underset{95}{CaCl_2} \rightarrow \underset{100}{CaCO_3}$)

Total weight of $CaCO_3 = \frac{100}{111} + \frac{100}{95} = 1.9535$ mg.

Thus 1.9535 mg. of $CaCO_3$ is present in 10^6mg of water. Hardness is therefore 1.9535 ppm.

8. 10g. of a solute was dissolved in 80g. of acetone at 30°C. The vapour pressure of the solution was found to be 271 mm of mercury. Calculate the molecular weight of the solute. Vapour pressure of acetone at 30°C is 283mm. Assume that the solution is not very dilute.

Sol.

$$\frac{p - p_s}{p} = \frac{n}{n + N} = \frac{\frac{w}{m}}{\frac{w}{m} + \frac{W}{M}}$$

$$\frac{283-271}{283} = \frac{\frac{10}{m}}{\frac{10}{m} + \frac{80}{58}} \qquad m = 163.5$$

9. Air was drawn through a solution containing 38.08g. of solute in 100g. of water and through water. The loss in weight of water was 0.0551g. and total weight of water absorbed in H_2SO_4 tube was 2.2117g. Calculate the molecular weight of the dissolved substance.

Sol. $\frac{p - p_s}{p} = \frac{w}{m} \times \frac{W}{M}$

$$\frac{0.0551}{2.2117} = \frac{38.08 \times 18}{m \times 100}$$

$$m = \frac{38.08 \times 18 \times 2.2117}{0.0551 \times 100} = 267.7$$

10. Dry air was passed successively through a solution of 5g. of a solute in 80g. of water and then through pure water. The loss in weight of solution was 2.5g. and that of the pure solvent 0.04g. Calculate the molecular weight of the solute.

Sol.: $\frac{p - p_s}{p} = \frac{w}{m} \times \frac{W}{M}$

$$\frac{0.04}{2.54} = \frac{5 \times 18}{m \times 80}$$

$$m = \frac{5 \times 18 \times 2.54}{0.04 \times 80} = 71.44$$

11. The molar heat of vapourisation of water at 100°C is 40.585 kJ. At what temperature will a solution containing 5.60g. of glucose per 1000g. of water boil ?

($R=8.314 JK^{-1}mol^{-1}$)

Sol. Molar heat of vapourisation of water at 100°C

$L_v = 40.585$ kJ = 40585 J

$\therefore$ Heat of vapourisation per gram, $L_v = \frac{40585}{18} = 2254.7J$

First we find out k_b from $k_b = \frac{RT^2}{L_v \times 1000}$

$$= \frac{8.314 \times 373 \times 373}{2254.7 \times 1000} = 0.513$$

Now we calculate ΔT from, $\Delta T = \frac{1000\ k_b\ w}{W.m}$

$$= \frac{1000 \times 0.513 \times 5.6}{1000 \times 180} = 0.016$$

Boiling point of the solution = 100 + 0.016 = 100.016°C

12. A solution of urea in H_2O has a b.p. of 100.128°C. Calculate the freezing point of the solution. K_f and k_b for water are 1.86 km^{-1} and 0.52 km^{-1} respectively.

Sol. $\Delta T_b = k_b m$

$$m = \frac{\Delta T_b}{k_b} = \frac{100.128° - 100°}{0.52} = 0.246 \text{ m}$$

$\Delta T_f = k_f m$

$T_f° - T_f = 1.86 \times 0.246 = 0.458°$

($T_f°$ = freezing point of the pure solvent, T_f = freezing point of the solution)

$T_f = T_f° - 0.458°$

$= 0 - 0.458 = -0.458°$

13. The latent heat of fusion of benzene is 126.6 Jg^{-1}. 0.0802g. of a substance when dissolved in 13.03g. of benzene lowered the freezing point by 0.490K. If pure benzene freezes at 278.4K, calculate the molecular weight of the substance.

Sol. First we calculate k_f from, $k_f = \frac{RT^2}{1000 \times L_f}$

$$= \frac{8.314 \times (278.4)^2}{1000 \times 126.6} = 5.088k \text{ kg. mol}^{-1}$$

Now $m = \frac{1000\ k_f\ w}{\Delta T_f . W} = \frac{1000 \times 5.088 \times 0.0802 \times 10^{-3}}{0.49 \times 13.03 \times 10^{-3}} = 63.92$

14. How much urea must be dissolved in 10^{-3} m^3 water to yield a solution with an osmotic pressure of 7.65×10^5 Mm^{-2} at 310 k ?

Sol. $PV = \frac{w}{m} RT$

$$= \frac{PVm}{RT} = \frac{7.65 \times 10^5 \times 10^{-3} \times 60}{8.314 \times 310} = 17.81 \text{ g.}$$

15. A solution containg 8.6 g per litre of urea (mol. wt.60) was found to be isotonic with a 5% solution of an organic non-volatile solution. Calculate the molecular weight of the other.

Sol. The isotonic solutions have the same molar concentrations, $\frac{50}{M} = \frac{8.6}{60}$

$$M = \frac{60 \times 50}{8.6} = 348.9$$

16. 0.85% aqueous solution of $NaNO_3$ is apparently 90% dissociated. Calculate the osmotic pressure of the solution.

Sol. Molality $= \frac{0.85 \times 1000}{100 \times 85} = 0.1 \text{ mol lit}^{-1}$.

$$\begin{matrix} NaNO_3 & \rightarrow & Na^+ & + & NO_3^- \\ 1-x & & x & & x \end{matrix}$$

$$i = \frac{1-x+x+x}{1} = 1+x = 1+0.90 = 1.90$$

$$\pi = iCRT = 1.90 \times 0.1 \times 0.082 \times 300 = 4.674 \text{ atm.}$$

17. Acetic acid associates in benzene to form double molecules. 1.65 g. of acetic acid when dissolved in 100g. of benzene raised the b.p. by 0.36°C. Calculate the degree of association of acetic acid if the molal elevation constant of benzene is 2.57°C.

Sol. $m = \frac{1000\ k_b\ w}{\Delta T \times W} = \frac{2.57 \times 1.65 \times 1000}{100 \times 0.36} = 118$

$$\begin{matrix} 1 & & 0 \\ CH_3COOH & \rightleftharpoons & (CH_3COOH)_2 \\ 1-x & & x/2 \end{matrix}$$

No. of molecules after association $= 1 - x + \frac{x}{2}$

$$\frac{\text{normal mol. wt.}}{\text{observed mol. wt.}} = \frac{1 - x + \frac{x}{2}}{1}$$

$$\therefore \frac{60}{118} = \frac{1 - x + \frac{x}{2}}{1}$$

x = 0.983

degree of association is 98.3%

18. The freezing point depression of a M/200 solution of Na_2SO_4 in H_2O was found to be 0.0265°C. Calculate the degree of dissociation of the salt at this concentration. k_f for water is 1.86°.

Sol. concentration of the solution $= \frac{M}{200} = \frac{142}{200} = 0.71$g./l

observed molecular weight $m = \frac{1000\ k_f\ w}{\Delta TW}$

$$= \frac{1000 \times 1.86 \times 0.71}{0.0265 \times 1000}$$

= 49.84

Let x be the dissociation of Na_2SO_4

$$Na_2SO_4 \rightleftharpoons 2Na^+ + SO_4^{2-}$$

At eq. 1-x 2x x

Number of molecules after dissociation = 1-x+2x+x = 1+2x

$$\frac{\text{normal molecular weight}}{\text{observed molecular weight}} = \frac{\text{no. of particles after dissociation}}{\text{no. of normal particles}}$$

$$\frac{142}{49.84} = 1 + 2x$$

x = 0.925 ∴ Degree of dissociation = 92.5%

19. 0.5 percent aqueous solution of KCl was found to freeze at –0.24°C. Calculate the vant Hoff factor and degree of dissociation of the solute at this concentration (k_f for H_2O is 1.86°)

Sol. First we calculate the observed molecular weight from

$$m = \frac{1000\ k_f\ w}{\Delta T_f \times W}$$

$$= \frac{1000 \times 1.86 \times 0.50}{100 \times 0.24} = 38.75$$

$$\text{Vant Hoff factor, } i = \frac{\text{normal mol. wt.}}{\text{observed mol. wt.}}$$

$$= \frac{74.5}{38.75} = 1.92$$

Let x be the degree of dissociation of KCl

$$KCl \rightleftharpoons K^+ + Cl^-$$

concentration at equilibrium 1-x x x

The total no. of molecules after dissociation = 1-x+x+x = 1+x

$$\frac{\text{normal mol. wt.}}{\text{observed mol. wt.}} = \frac{\text{no. of particles after dissociation}}{\text{no. of normal particles}}$$

$$\frac{74.5}{38.75} = 1 + x$$

x = 0.923 ∴ Degree of dissociation = 92.3%

20. A solution containing 110 g. of $Ba(NO_3)_2$ per dm^3 of H_2O has a b.p. of 373.466 K at one atmosphere pressure. Calculate the degree of ionisation of the salt in the solution. The molal elevation constant for H_2O is 0.52 K kg mol^{-1}.

$$\text{Sol. } \Delta T_b \text{ calculated} = \frac{1000\ k_b\ w}{mW}$$

$$= \frac{1000 \times 0.52 \times 110}{261.34 \times 1000} = 0.219$$

$$\text{we know } \frac{\Delta T_b \text{ observed}}{\Delta T_b \text{ calculated}} = \frac{\text{no. of particles after dissociation}}{\text{no. of normal particles}}$$

$$Ba(NO_3)_2 \rightleftharpoons Ba^{+2} + 2NO_3^-$$

1-x x 2x

Let x be the degree of dissociation

$$\frac{0.466}{0.219} = \frac{1+2x}{1}$$

x = 0.5635 or 56.35%

21. The density of a solution containing 13% by mass of H_2SO_4 is 0.109 g/ml. Calculate the molarity and normality of the solution.

$$\text{Sol.: Molarity} = \frac{\%\text{ strength of solution} \times \text{density of solution} \times 10}{\text{molecular mass}}$$

$$= \frac{13 \times 0.109 \times 10}{98} = 1.445\ M$$

Normality = molarity x n = 1.445 x 2 = 2.89 N

(22) (i) What is the weight of sodium bromate and molarity of solution necessary to prepare 85.5 ml. of 0.672 N solution when the half cell reaction is

$BrO_3^- + 6H^+ + 6^- \longrightarrow Br^- + 3H_2O$.

(ii) What would be the weight and molarity if the half cell reaction is, $2BrO_3^- + 2H^+ + 10e^- \longrightarrow Br_2 + 6H_2O$

Sol. (i) Wt. of Na BrO_3 required,

$$W = \frac{ENV}{1000};\ E = \frac{\text{mol. wt.}}{6} = \frac{151}{6}$$

$$= \frac{151 \times 0.672 \times 85.5}{6 \times 1000} = 1.446\ g.$$

$$\text{Molarity} = \frac{N \times \text{eq.wt}}{\text{mol. wt.}} = \frac{0.672 \times 151}{151 \times 6} = 0.112$$

$$\text{(ii) wt. of } NaBrO_3,\ w = \frac{151 \times 0.672 \times 85.5}{5 \times 1000} = 1.735\ g.$$

$$\text{molarity} = \frac{0.672 \times 151}{5 \times 151} = 0.1344$$

(23) How many ml. of 0.5 M H_2So_4 are needed to dissolve 0.5 g. of copper (ii) carbonate ?

Sol. We can write the reactionas, $CuCO_3 + H_2SO_4 \rightarrow CuSO_4 + CO_2 + H_2O$

$$\text{No. of moles of } CuCO_3 = \frac{0.5}{123.5}$$

$$\text{No. of moles of } H_2SO_4 \text{ required would be} = \frac{0.5}{123.5}$$

$$\text{No. of moles of } H_2SO_4 = \frac{0.5 \times V}{1000}$$

$$\frac{0.5}{123.5} = \frac{0.5 \times V}{1000}$$

$$V = \frac{0.5}{123.5} \times \frac{1000}{0.5} = 8.1\ ml.$$

(24) 50 ml of an aqueous solution of hydrogen peroxide was added into excess of a solution of KI in dil H_2SO_4. The liberated iodine required 20 ml. of 0.1 N $Na_2S_2O_3$ solution for complete reaction. Find the concentration of H_2O_2 in gram per litre.

Sol. No. of milli equivalents of $Na_2S_2O_3$ = N x V = 0.1 x 20 = 2

No. of milli equivalents of I_2 liberated = 2

No. of milli equivalents of H_2O_2 = 2

No. of equivalents of H_2O_2 = 2 x 10^{-3} - 0.002

Mass of H_2O_2 = No. of equivalents x Eq. wt. of H_2O_2 = 0.002 x 17 = 0.034 g.

$$\text{Concentration of } H_2O_2 = 0.034 \times \frac{1000}{0.5} = 0.068\ g.$$

(25) What volume of 0.05 M $KMnO_4$ solution will be required for complete oxidation of an aqueous and

acidified solution of 2.0 grams of $FeSO_4$?

Sol. The oxidation reaction is written as,

$10FeSO_4 + 2KMnO_4 + 8H_2O \longrightarrow K_2SO_4 + 2MnSO_4 + 5Fe_2(SO_4)_3 + 8H_2O$

As per the balance equation, 10 moles of $FeSO_4$ reacts with 2 moles of $KMnO_4$ i.e. 10 x 151.8 g. of $FeSO_4$ reacts with 2 x 158 g. of $KMnO_4$

$$2 \text{ g of } FeSO_4 \text{ reacts with} = \frac{2 \times 158 \times 2}{10 \times 15.18}$$

$$= 0.4163 \text{ g. of } KMnO_4$$

$$\text{Molarity} = \frac{\text{mass of the solute / mol. wt.}}{V/1000}$$

$$V = \frac{0.4163/158}{0.05/1000} = \frac{0.00263481}{0.00005} = 52.69$$

V = 52.7 ml.

(26) The vapour pressure of pure water at 25°c is 23.62 mm. What will be the vapour pressure of a solution of 1.5 g of urea in 50g. water ?

Sol. We know that according to Raoult's Law,

$$\frac{P-P_s}{P} = \frac{n}{n+N}$$

Number of moles of urea = wt. of urea / molecular wt.

= 1.5/60 = 0.025

Number of moles of water = 50/18 = 2.778

$$\text{Substituting the values, } \frac{23.62 - P_s}{23.62} = \frac{0.025}{0.025 + 2.778}$$

P_s = 23.41 mm.

(27) Calculate normality and molarity for a 20 volume H_2O_2 solution.

Sol. 20 volumes means 1 ml. 20 volume H_2O_2 at NTP gives 20 ml. O_2

We Know that, 22,400 ml. of O_2 is obtained from 68 g. H_2O_2 20 ml of O_2 can be obtained from

$$\frac{20 \times 68}{22,400} = 0.0607 \text{ g. of } H_2O_2$$

Normality = strength per lit./equivalent weight = 60.7/17 = 3.57 N

Molarity = Normality / basicity = 3.57 / 2 = 1.785 M

(28) Calculate the normality of 10 volume H_2O_2 solution

Sol. We know that 10 Volume H_2O_2 means

1 ml. of 10 volume H_2O_2 at NTP gives 10 ml O_2

1000 ml. of 10 volume H_2O_2 at NTP gives = $\frac{1000 \times 10}{1}$ = 10 lit.

We know 22400 ml. of O_2 is obtained from 68g. H_2O_2

$$\therefore 10 \text{ lit. of } O_2 \text{ is obtained from} = \frac{10 \times 68}{22.4} = 30.36\text{g. } H_2O_2$$

No. of moles of H_2O_2 = wt of H_2O_2 in 1 lit / 34

= 30.36/34 = 0.89 M

Normality of the vol. H_2O_2 solution = 0.89 x2 = 1.78 N

(29) Calculate the volume strength of 2.5 N H_2O_2 solution.

Sol. 2.5 N H_2O_2 contains 2.5 x 17 = 42.5 g H_2O_2 (since1 NH_2O_2 contains 17 g Per lit.)

We know that, 68 g. of H_2O_2 evolves 22.4 lit of O_2 at NTP

42.5 g. of H_2O_2 evolves ? lit. of O_2

$$= \frac{42.5}{68} \times 22.4 = 14.0 \text{ litre of } O_2 \text{ at NTP}$$

We know that 14.0 lit. of O_2 at NTP is evolved from 42.5 g. of H_2O_2 present in one litre

therefore, the strength of H_2O_2 = 14 volume.

30. The degree of dissociation of calcium nitrate in dilute aqueous solution containing 7.0g. of the salt per 100g. of water at 100°C is 70%. If the vapour pressure of water at 100°C is 760mm. Calculate the vapour pressure of the solution.

$$\text{Sol. we know that, } \frac{p - p_s}{p} = \frac{n}{n+N} \quad \text{.............(i)}$$

No. of moles of water i.e. solvent = 100/18 = 5.56

degree of dissociation of $Ca(NO_3)_2$ = 70/100 = 0.7

No. of moles of the solute i.e. $Ca(NO_3)_2$ is calculated as follows

dissociation of $Ca(NO_3)_2$ be represented as,

$$Ca(NO_3)_2 \rightleftharpoons Ca^{2+} + 2NO_3^-$$

initial conc.	1	0	0
equilibrium conc.	1–0.7	0.7	2×0.7

No. of moles at equi. = 1– 0.7 + 0.7 + (2 × 0.7) = 2.4

$$\text{No. of moles of solute, } n = \frac{\text{wt. of the solute}}{\text{mol. wt.}} \times \text{total no. of moles at eq.}$$

$$= \frac{7.0}{164} \times 2.4 = 0.102$$

$$\text{Substituting in (i), } \frac{760 - p_s}{760} = \frac{0.102}{0.102 + 5.56}$$

$$\frac{760 - p_s}{760} = 0.018$$

p_s = 760 - (760 x 0.018) = 746.32 mm of Hg.

31. How many millilitres of 0.5 M H_2SO_4 are needed to dissolve 0.5 g of copper (II) Carbonate ?

Sol. Numbe of milliequivalents of H_2SO_4 = no. of m eq. of $CuCO_3$

meq of $CuCO_3$ = (wt/ Eq. wt)x1000 = (0.5/61.75) x 1000

meq of H_2SO_4 = Normality × volume

= 2 × 0.5 × v (since Molarity = Normality/ basicity)

∴ 2 × 0.5 × v = (0.5/61.75) × 1000

$$v = \frac{0.5 \times 1000}{61.75 \times 2 \times 0.5} = 8.1 \text{ ml.}$$

Solid State

(1) The three planes in a crystalline solid intersect the crystal axes at (i) 2a , b, c (ii) a, -b, c (iii) -a, b, ∝. Write the Miller indices.

Sol.

Intercepts	2, 1, 1	1, -1, 1	-1, 1, ∝
Weiss indices	2, 1, 1	1, -1, 1	-1, 1, ∝
Reciprocal of WI	1/2, 1/1, 1/1	1/1 -1/1, 1/1	-1/1,1/1,1/∝
Clear fractions	1, 2, 2	1, -1, 1	-1, 1, 0
Miller indices	(1 2 2)	$(1\,\bar{1}\,1)$	$(\bar{1}\,1\,0)$

(2) How do the spacings of three planes (100), (110) and (111) of cubic lattice vary ?

Sol. The distances between the parallel planes in a crystal are designated by $d_{h,k,l}$. It is given by

$$d_{h,k,l} = \frac{a}{\sqrt{h^2+k^2+l^2}}$$ where a = length of the cube side h, k, l are the Miller indices of the plane.

$$d_{(100)} = \frac{a}{\sqrt{1^2+0+0}} = a; \quad d_{(110)} = \frac{a}{\sqrt{1^2+1^2+0}} = \frac{a}{\sqrt{2}}$$

$$d_{(111)} = \frac{a}{\sqrt{1^2+1^2+1^2}} = \frac{a}{\sqrt{3}}$$

Thus $d_{(100)} : d_{(110)} : d_{(111)} = 1 : \frac{1}{\sqrt{2}} : \frac{1}{\sqrt{3}} = 1 : 0.707 : 0.577$

(3) Perdict the type of crystal structure of (i) NaCl and (ii) Li F. What is the coordination number of each ion in the crystal ? Ionic radii of the ions are :

Li^+ = 60 pm; Na^+= 95 pm; Cl^- = 181 pm and F^-= 136 pm

Sol. In the case of NaCl = r^+/r^- = 95/181 = 0.525

This value is less than 0.732 and greater than 0.414. As such the cation will occrpy the octahedral void with coordination no. 6. Hence, it should crystallise in ccp. In case of LiF r^+/r^- = 60/136 = 0.441. In this case also the larger F^- ions will form cubic close packing in which smaller Li^+ ions fit into the octahedral holes. Coordination number is 6.

(4) Sodium crystallises in a bcc cell with 'a' is 0.424 n m at 298 K. Calculate the theoretical density of sodium.

Sol. $\rho = \frac{n\,M}{N_o V}$ or $\rho = \frac{zM}{N_A a^3}$ (n = z = no. of atoms in a unit cell)

number of atoms in a unit cell of bcc = $8 \times \frac{1}{8} + 1 = 2$

$$\rho = \frac{2 \times (23 \times 10^{-3})}{(6.022 \times 10^{23})\,(0.424 \times 10^{-9})^3} = 1 \times 10^3 \text{ Kg m}^{-3}$$

(5) Sodium chloride (mol. mass : 58.45 g mol^{-1})crystallises in fcc lattice arrangement. If the density of NaCl is 2.163 g cm^{-3}. What is the unit cell length ?

Sol. $\rho = \frac{z\,M}{N_A a^3}$ $\quad a^3 = \frac{4 \times 58.45}{2.163 \times 6.022 \times 10^{23}}$

$$a = \left(\frac{4 \times 58.45}{2.163 \times 6.022 \times 10^{23}}\right)^{\frac{1}{3}} = 5.64 \times 10^{-8} \text{ cm.}$$

(6) An element occurs in the bcc structure with a cell edge length of 228 pm. The density of the element is 7.2g cm^{-3}. How many atoms of the element does 208 g of the element contain ?

$$\text{Sol. } N = \frac{M}{\rho V} = \frac{208 \times 10^{-3}}{7.2 \times 10^3 \times (228 \times 10^{-12})^3} = 24.2 \times 10^{23} \text{ atoms}$$

(7) Calculate the Avogadro number from the following data : The fcc of silver (at. wt. 107.9 amu) has side of length 4.08 x 10^{-10}m. Its density is 10.53 x 10^3 kg. m^{-3}.

Sol. $\rho = \frac{z\,M}{N_A a^3}$

$$N_A = \frac{z\,M}{\rho a^3} = \frac{4 \times 107.9 \times 10^{-3}}{10.53 \times 10^3 \times (4.08 \times 10^{-10})^3} = 6.052 \times 10^{23} \text{ mol}^{-1}$$

(8) Al crystallises as fcc lattice and it has a density of 2.7x10^3 kg m^{-3}. Calculate the edge length of the unit cell.

Sol.

$$\rho = \frac{zM}{N_A a^3}; \quad a^3 = \frac{zM}{\rho N_A} = \frac{4 \times 27 \times 10^{-3}}{2.7 \times 10^3 \times 6.02 \times 10^{23}} = 0.6645 \times 10^{-30}$$

a = $(0.6645)^{1/3}$ x 10^{10} = 0.87 x 10^{-10} = 0.087 nm.

(9) Calculate the density and atomic radii of elementary Ag which crystallizes in a fcc with unit cell length = 4.086 x 10^{-10} m.

Sol. In fcc lattice, 2 halves of the atom lie at each corner and one atom lies in the centre of diagonal, so that diagonal = 4r = $\sqrt{2}a$;

$$r = \frac{\sqrt{2}a}{4} = \frac{\sqrt{2} \times 4.086 \times 10^{-10}}{4} = 1.445 \times 10^{-10} \text{ m.}$$

(10) NH_4Cl crystallises in a bcc lattice with a unit distance of 387 pm. Calculate (a) the distance between oppositely charged ion in the lattice and (b) the radius of the NH_4^+ ion, if the radius of the Cl^- ion is 181 pm.

Sol. (a) In bcc lattice oppositely charged ions touch each other along the cross-diagonal of the cube

So $2r_c + 2r_a = \sqrt{3}a$

$$r_c + r_a = \frac{\sqrt{3}}{2}a = \frac{\sqrt{3}}{2} \times 387 = 335.15 \text{ pm.}$$

(b) given r_a = 181 pm, r_c = 335.15 - 181.0 = 154.15 pm

(11) The unit cell cube length for LiCl(NaCl structure) is 5.14Å. Assuming anion-anion contact, calculate the ionic radius of Cl^- ion.

Sol. In a fcc lattice anions touch each other along the fcc diagonal of the cube; $4r_{Cl} = \sqrt{2}a$

$$r_{Cl} = \frac{\sqrt{2}}{4}a = \frac{\sqrt{2}}{4} \times 5.14 = 1.82 \text{ Å}$$

Cl^-

2.57

Li^+ 2.57 Cl^-

(12) An element crystallises in fcc structure having an unit cell of an edge 200 pm. Calculate its density, if 200g. of this element contains 24x10^{23} atoms.

Sol.

Length of the edge of the unit cell = 200pm = 2 x 10^{-8} cm.

Volume of the unit cell = $(2 \times 10^{-8})^3$ = 8 x 10^{-24} cm^3

$$\text{Number of unit cells} = \frac{\text{Total number of atoms}}{\text{number of atoms per unit cell}}$$

$$= \frac{24 \times 10^{23}}{4} = 6 \times 10^{23}$$

Volume of the substance = no. of unit cells x volume of a unit cell

$$= 6 \times 10^{23} \times 8 \times 10^{-24} = 4.8 \text{ cm}^3$$

$$\text{Density } \rho = \frac{\text{Mass}}{\text{Volume}} = \frac{200}{4.8} = 41.7 \text{ g / cm}^3$$

(13) x-rays of wavelength 1.54 Å strike a crystal and are observed to be deflected at an angle of 22.5°. Assuming that n=1, calculate the spacing between the planes of atoms that are responsible for the reflection.

Sol.
$$d = \frac{1.54}{2 \text{ Sin } 22.5} = \frac{1.54}{2 \times 0.383} = 2.01 \text{ Å}$$

(14) The first order reflection of a beam of x-rays of wavelength 1.54 Å from the (100) face of a crystal of the simple cubic type occurs at an angle of 11.29°. Calculate the length of the unit cell

Sol.
$$d = \frac{1.54 \times 10^{-8}}{2 \times \text{Sin } 11.29} = \frac{1.54 \times 10^{-8}}{2 \times 0.1991} = 3.68 \times 10^{-8} \text{ cm.}$$

$$d_{hkl} = \frac{a}{\sqrt{h^2 + k^2 + l^2}} = a; \quad a = 3.68 \times 10^{-8} \text{ cm.}$$

(15) The first order reflection maximum from crystals of NaCl is obtained at an angle 5.9°. If $d_{h,k,l}$ is 0.282×10^{-9} m, what is the wave length of the x-rays used.

Sol. $2 d \sin\theta = n \lambda$

$2 \times 0.282 \times 10^{-9} \sin\theta = 1 \times \lambda$

$\lambda = 0.058$ nm

(16) At what angle would a first order reflection be observed in the x-ray diffraction of a set of crystal planes for which d = 0.303 nm., if the x-rays used have a wavelength of 0.071 nm.

Sol.
$$\sin\theta = \frac{n\lambda}{2d} = \frac{1 \times 7.1 \times 10^{-11}}{2 \times 3.03 \times 10^{-10}} = 0.1172 \quad \therefore \theta = 6.73°$$

(17) An element occurs in bcc structure with a cell edge of 288 pm. The density of metal is 7.2 g cm^{-3}. How many atoms does 208 g of the element contain ?

Sol. Volume of unit cell = $(288 \times 10^{-10})^3 = 23.9 \times 10^{-24}$ cm^3

Volume of 208 g of the element $= \frac{208}{7.2} = 28.88$ cm^3

No. of unit cells in 28.88cm^3 $= \frac{28.88}{23.9 \times 10^{-24}}$

$= 12.08 \times 10^{23}$ unit cells

each bcc structure contains 2 atoms

∴ Total no. of atoms in 12.08×10^{23} unit cells

$= 2 \times 12.08 \times 10^{23} = 24.16 \times 10^{23}$

(18) Li forms bcc crystals. Calculate the atomic radius of Li, if the length of the side of a unit cel of Li is 351 pm.

Sol. In body centred cubic crystals, $r = \frac{\sqrt{3}}{4} a$

$$= \frac{\sqrt{3}}{4} \times 351 = 151.98 \text{ m}$$

(19) Find the lattice spacing in a cubic crystal of NaCl having a density of 2.16×10^3 kg.m^{-3}. Mol. wt. of NaCl is 58.

Sol. If M is the mol. wt. of NaCl and N, the Avagadro's number, since there are 2N ions or diffracting centres i.e. N ions of Na^+ and N ions of Cl^- in one g. molecule, the mass associatied with each ion is M/2 N.

As there is only one ion for each elementary cube in the system either sodium or chlorine

$$\frac{M}{2N} = \rho d^3; \ d = \left(\frac{M}{2N\rho}\right)^{1/3} = \left(\frac{58 \times 10^{-3}}{2 \times 6.023 \times 10^{23} \times 2.16 \times 10^3}\right)^{1/3}$$

$$= 2.814 \times 10^{-10} \text{ m.}$$

(20) A compound formed by elements A and B crystallizes in cubic structure where A atoms are at the corners of the cube and B atoms are at face centre. What is the formula of the compound ?

Sol. no. of A atoms in the unit cell $= 8 \times \frac{1}{8} = 1$

no.of B atoms in the unit cell $= \frac{1}{2} \times 6 = 3$

∴ The formula is AB_3.

(21) The crystal AB (rock salt structure) has molecular weight 6.023 Y a m u where Y is an arbitrary number. If the minimum distance between cation and anion is $Y^{1/3}$ nm and the observed density is 20 kg/m^3. Find the (a) density in kg/m^3 and (b) type of defect.

Sol. The crystal AB has fcc unit cell (as it has rock salt structure)

The crystal has 4 molecules for unit cell.

Edge length of the unit cell = $2 (r^+ + r^-)$

$= 2 \times Y^{1/3}$ nm

Volume of the unit cell = $a^3 = (2 \times Y^{1/3})^3$

$= 8 \times Y$ nm^3

$= 8 \times Y \times (10^{-9})^3$ m^3

Mass of one molcule of AB

$$= \frac{6.023 \times Y}{6.023 \times 10^{23}} \times 1 \times 10^{-3} = Y \times 10^{-26}$$

Mass of one unit cell = $4 \times Y \times 10^{-26}$ (1 unit cell contains 4 molecules)

$$\text{Density} = \frac{\text{mass of one unit cell}}{\text{volume of the unit cell}} = \frac{4 \times Y \times 10^{-26}}{8 \times Y \times (10^{-9})^3}$$

$= 5$ kg m^{-3}.

The observed density i.e. 20 kgm^{-3} is much higher than the theoretical (expected) one. Therefore the crystal can't contains Schottky or Frenkel defects as in the former the density is lowered while in the later it remains constant. It has excess of metal.

22. The nucleus of a platinum atom has a radius of 5.0×10^{-14}. Platinum has an atomic weight of 195.2 and a density, of 21.4g per c.c. If platinum atom is spherical in shape, how many times larger is the radius of the platinum

atom the radius of the nucleus ?

Sol. Total volume occupied by Pt = $\frac{\text{atomic wt.}}{\text{density}}$

= 195.2/21.4 = 9.122 c.c

volume occupied by one Pt atom = $\frac{\text{total volume occupied}}{\text{Avogadro number}}$

$= \frac{9.122}{6.02 \times 10^{23}} = 1.52 \times 10^{-23}$ c.c.

we know that for a sphere, $V = 4/3\ \pi r^3$

$\therefore r^3 = 3v/4\pi = \frac{3 \times 1.52 \times 10^{-23}}{4 \times 3.14}$

$r = \sqrt{\frac{3 \times 1.52 \times 10^{-23}}{4 \times 3.14}} = 1.53 \times 10^{-8}$ cm.

Ratio of the atom to the radius of the nucleus =

$\frac{1.53 \times 10^{-8}}{5.0 \times 10^{-14}} = 3.06 \times 10^5$

23. If the radius of an Argon atom is 1.54×10^{-8} cm, what percent of the total volume occupied by one mole of argon gas at STP is empty space ?

Sol. we know, $V = 4/3\ \pi r^3$

$\therefore$ volume of an argon atom $= \frac{4}{3} \times 3.14 \times (1.54 \times 10^{-8})^3$

$= 1.53 \times 10^{-23}$

Total volume of 1 mole of argon gas

$= 1.53 \times 10^{-23} \times 6.02 \times 10^{23} = 9.21$ cc

% of empty space = $\frac{22400 - 9.21}{22400} \times 100 = 99.96$

Chemical & Ionic Equilibrium

(1) Given : $Ag(NH_3)_2 \rightleftharpoons Ag^+ + 2NH_3$, $K_c = 6.2 \times 10^{-8}$ and K_{sp} of AgCl = 1.8×10^{-10} at 298 K. Calculate the concentation of the complex in 1.0 M aqueous ammonia.

Sol. Let the concentration of the complex in 1.0 M aqueous ammnia be x mol per litre

Therefore, The concentration of Cl^- = x

k_c for the dissociation of $[Ag(NH_3)_2]^+$ is very small and at the same time, the concentration of NH_3 is higher

therefore, $[Ag(NH_3)_2]^+ \approx x$

$$k_c = \frac{[Ag]^+[NH_3]^2}{[Ag(NH_3)_2]^+}$$

$$6.2 \times 10^{-8} = \frac{[Ag^+]^2 \times 1^2}{x} \quad \text{(i)}$$

$k_{sp}(AgCl) = [Ag^+][Cl^-]$

$1.8 \times 10^{-10} = [Ag^+] \times x$ (ii)

Dividing (ii) by (i) we obtain, $\frac{1.8 \times 10^{-10}}{6.2 \times 10^{-8}} = \frac{[Ag^+] \times x \times x}{[Ag^+] \times 1^2}$

$x^2 = \frac{1.8 \times 10^{-10}}{6.2 \times 10^{-8}} = 0.2903 \times 10^{-2}$

$x = \sqrt{0.2903 \times 10^{-2}} = 0.0538 = 5.38 \times 10^{-2}$

(2) Calculate the p^H value of 2 N NaOH solution. Assume complete ionisation of NaOH.

Sol. The solution is 2 N and the ionisation is assumed to be complete

therefore, $[OH^-] = 2$

We know, $[H^+][OH^-] = 10^{-14}$

$[H^+] = \frac{10^{-14}}{2}$

$pH = -\log\frac{[10^{-14}]}{2} = -\log 10^{-14} + \log 2$

$= 14 + 0.3010 = 14.3010$

The pH of a strongly basic solution can be more than 14. This condition arises when the solutions are 2 N, 3 N and so on.

(3) Calculate the pH for 10^{-8} MHcl solution

Sol. If we calculate the pH from the formula,

$pH = -\log[H^-]$

We obtain pH=8, which is not possible for acidic solution In such cases (highly diluted solutions) H^+ contribution from the decomposition of water should also be considered.

Thus total $[H^+] = (1.0 \times 10^{-7} + 1.0 \times 10^{-8})$

$= 1.0 \times 10^{-7} + 0.1 \times 10^{-7}$

$= 1.1 \times 10^{-7}$

$pH = -\log[H^+] = -\log[1.1 \times 10^{-7}] = 6.96$

(4) Calculate the pH of 10^{-8} M solution of NaOH.

Sol. From the formula, on calculation, we get pH = 6 which is not possible for a basic solution. Therefore and as it is a very dilute solution, we should consider OH^- contribution from water.

Thus total pOH = $1.0 \times 10^{-7} \times 1.0 \times 10^{-8}$

$= 1.0 \times 10^{-7} \times 0.1 \times 10^{-7}$

$= 6.9586$

pH = 14 - pOH = 14-6.9586 = 7.0414 = 7.04

(5) What is the hydrogen ion concentration of 0.1 N CH_3COOH solution ? The ionisation constant of CH_3COOH is 1.8×10^{-5}, what is the pH of the solution ?

Sol. We know $K_a = \alpha^2 x C$

$1.8 \times 10^{-5} = \alpha^2 x 0.1$

$\alpha = \sqrt{\frac{1.8 \times 10^{-5}}{0.1}} = 1.34 \times 10^{-2}$

$[H^+] = \alpha \times C = 1.34 \times 10^{-2} \times 0.1$

$= 1.34 \times 10^{-3}$

$pH = -\log[H^+]$

$= -\log 1.34 \times 10^{-3} = 2.8729 = 2.87$

(6) Calculate the pH at which $Mg(OH)_2$ begins to percipitate from a solution containing 0.10 M Mg^{2+} ions. K_{sp} for $Mg(OH)_2 = 1.0 \times 10^{-11}$

Sol. $K_{sp} = [Mg^{2+}][OH^-]^2$

$[OH^-]^2 = \frac{Ksp}{[Mg^{2+}]} = \frac{1.0 \times 10^{-11}}{0.10} = 1.0 \times 10^{-10}$

$[OH^-] = \sqrt{1.0 \times 10^{-10}} = 1.0 \times 10^{-5}$

$H^+ = \frac{10^{-14}}{10^{-5}} = 10^{-9}$

$pH = -\log[H^+] = -\log[10^{-9}] = 9$

(7) For the reactions $N_2 + 3H_2 \rightleftharpoons 2NH_3$
$\frac{1}{2}N_2 + \frac{3}{2}H_2 \rightleftharpoons NH_3$
write down the expression for equilibrium constants k_c and k_c'. How is k_c related to k_c'?

Sol. $N_2 + 3H_2 \rightleftharpoons 2NH_3$; for $\frac{1}{2}N_2 + \frac{3}{2}H_2 \rightleftharpoons NH_3$

$$k_c = \frac{[NH_3]^2}{[N_2][H_2]^3} \qquad k_c' = \frac{[NH_3]}{[N_2]^{1/2}[H_2]^{3/2}}$$

$$\text{squaring; } (k_c')^2 = \frac{[NH_3]^2}{[N_2][H_2]^3}$$

$$\text{Thus } k_c = (k_c')^2, \quad \sqrt{k_c} = k_c'$$

(8) At a certain temp. equilibrium constant k_c is 16 for the reaction $SO_2(g) + NO_2(g) \rightleftharpoons SO_3(g) + NO(g)$; if we take one mole each of the four gases in one lit. container, what would be the equilibrium concentration of NO and NO_2?

Sol.	SO_2 +	NO_2 $\rightleftharpoons$	SO_3 +	NO
Initial conc.	1	1	1	1
Eq. conc.	1-x	1-x	1+x	1+x

$$k_c = \frac{[SO_3][NO]}{[SO_2][NO_2]} = \frac{(1+x)(1+x)}{(1-x)(1-x)} = 16$$

$$\frac{1+x}{1-x} = 4 \text{ or } 1+x = 4-4x$$

$$5x = 3;\ x = \frac{3}{5} = 0.6$$

conc. of NO_2 at equilibrium = 1-0.6 = 0.4 moles
conc. of NO at equilibrium = 1+0.6 = 1.6 moles

(9) What is the percent decomposition of H_2S if 1.0 mole of H_2S is introduced into 1.1 lit. vessel at 1000 K? k_c for the reaction $2H_2S(g) \rightleftharpoons 2H_2(g) + S_2(g)$ is 1×10^{-6}.

Sol.	$2H_2S$ $\rightleftharpoons$	$2H_2$ +	S_2
Initail conc.	1	0	0
Eq. conc.	1-x	x	x / 2

$$k_c = \frac{[H_2]^2[S_2]}{[H_2S]^2} = \frac{\left(\frac{x}{1.1}\right)^2 \times \frac{x}{2.2}}{\left(\frac{1-x}{1.1}\right)^2}$$

$$1.0 \times 10^{-6} = \frac{x^3}{2.2 \times (1-x)^2} = \frac{x^3}{2.2} \text{ (as x is very small)}$$

$x^3 = 2.2 \times 10^{-6}$
$x = 1.3 \times 10^{-2}$
% decomposition of H_2S is 1.3

(10) At 250°C and 1 atmospheric pressure, the vapour density of PCl_5 is 57.9. Calculate k_p for the reaction
$PCl_5 \rightleftharpoons PCl_3 + PCl_2$.

Sol. degree of dissociation $\alpha = \frac{D-d}{d} = \frac{104.25 - 57.9}{57.9} = 0.80$

At equilibrium $PCl_5 \rightleftharpoons PCl_3 + PCl_2$
$1-\alpha \quad \alpha \quad \alpha$
$(1-0.80) \quad 0.80 \quad 0.80$

Total no. of moles = $(1+\alpha) = (1+0.80) = 1.80$

$$p\,PCl_5 = \frac{0.2}{1.80} \times 1 = \frac{1}{9};\ pPCl_3 = \frac{0.80}{1.80} \times 1 = \frac{4}{9};$$

$$pPCl_2 = \frac{0.80}{1.80} \times 1 = \frac{4}{9}$$

$$k_p = \frac{pPCl_3 \times pCl_2}{pPCl_5} = \frac{4/9 \times 4/9}{1/9} = \frac{16}{9} = 1.78$$

(11) One mole of nitrogen and three moles of hydrogen are mixed in a 4 lit. container. If 0.25% of nitrogen is converted to ammonia by the following reaction;

$$N_2 + 3H_2 \rightleftharpoons 2NH_3$$

calculate the equilibrium constant (k_C). What will be the value of k for the equilibrium $\frac{1}{2}N_2 + \frac{3}{2}H_2 \rightleftharpoons NH_3$

Sol.	N_2 +	$3H_2$ $\rightleftharpoons$	$2NH_3$
At equilibrium	(1-x)	(3-3x)	2x (x = 0.0025)
Active masses	$\frac{(1-0.0025)}{4}$	$\frac{(3-0.0075)}{4}$	$\frac{(0.0050)}{4}$

$$k_c = \frac{[NH_3]^2}{[N_2][H_2]^3} = \frac{\left(\frac{0.0050}{4}\right)^2}{\left(\frac{0.9975}{4}\right)\left(\frac{2.9925}{4}\right)^3} = 1.49 \times 10^{-5} \text{ lit.}^2\text{mol}^{-2}$$

k for the reaction $\frac{1}{2}N_2 + \frac{3}{2}H_2 \rightleftharpoons NH_3$ is,

$$k = \sqrt{k_c} = \sqrt{1.49 \times 10^{-5}} = 3.86 \times 10^{-3} \text{ lit mol}^{-1},$$

(12) A mixture of 7.9 ml. H_2 and 33.1 ml I_2 vapour was heated at 444°C until the equilibrium was reached. If the equilibrium constant for the reaction $H_2 + I_2 \rightleftharpoons 2HI$ is 36.68, calculate the number of ml. of H_2 present when equilibrium is obtained.

Sol. Let x ml. of H_2 and I_2 be changed to HI
volume of H_2 at equilibrium = (7.9 – x) ml.
volume of I_2 at equilibrium = (33.1 – x) ml.
volume of HI formed at equilibrium = 2x

$$k = \frac{[HI]^2}{[H_2][I_2]} \qquad 36.68 = \frac{(2x)^2}{(7.9-x)(33.01-x)}$$

$4x^2 = 36.68\,(7.9-x)(33.1-x)$
$\frac{4}{36.68}x^2 = (7.9-x)(33.1-x)$
$0.109\,x^2 = 261.49 - 41x + x^2$
$x^2 - 0.109x^2 - 41x + 261.49 = 0$
x = 7.64 or 8.37. x can't be more than 7.9.

(The roots of a quadratic equation are given by the relation $x = -b \pm \sqrt{\frac{b^2 - 4ac}{2a}}$)

(13) When sulphur in the form of S_8 is heated at 900 K, the initial pressure of 1 atm. falls by 29% at equilibrium. This is because of conversion of some S_8 to S_2. Find the equilibrium constant.

Sol.	S_8 $\rightleftharpoons$	$4S_2$
initial conc.	1	0
equilibrium conc.	1-x	4x
	(1- 0.29)	(4x0.29)
	= 0.71	= 1.16

$$k_p = \frac{(pS_2)^4}{(pS_8)} = \frac{(1.16)^4}{0.71} = 2.5582 \text{ atm}^3$$

(14) The value of k_C for the reaction, $N_2(g) + 3H_2(g) \rightleftharpoons 2NH_3(g)$ is 0.50 at 400°C. Find the value of k_p at 400°C when, concentrations are expressed in mol lit^{-1} and pressure in atomsphere.

Sol. $k_p = k_C (RT)^{\Delta n}$
$= 0.05\,(0.082 \times 673)^{-2} = 1.64 \times 10^{-4}$

(15) 1 mole of nitrogen and 3.0 moles of PCl_5 are placed in 100 lit. vessel and heated to 227°C. The equilibrium pressure is 2.05 atmos. Assuming ideal gas behaviour,

calculate the degree of dissociation for PCl_5 and k_p for the reaction $PCl_5 \rightleftharpoons PCl_3 + Cl_2$

Sol. $PCl_5 \rightleftharpoons PCl_3 + Cl_2$

Initial con. 3 0 0

At equilibrium 3(1-x) 3x 3x

1 mole of nitrogen is present ; hence actual no. of moles at equlibrium = 3 (1+x) + 1

we know PV = nRT

$$n = \frac{2.05 \times 100}{0.082 \times 500} = 5$$

$3(1+x) + 1 = 5$

$3x = 1$ or $x = 0.333$

Thus $3(1+x) + 1 = 3 + 3 \times 0.333 + 1 = 4 + 0.999$

At equilibrium $pPCl_5 = \frac{3(1-x)}{(3x+4)} \times 2.05$; $pPCl_3 = \frac{3x}{3x+4} \times 2.05$

$$k_p = \frac{\frac{3x}{(3x-4)} \times 2.05 \times \frac{3x}{(3x+4)} \times 2.05}{\frac{3(1-x)}{(3x+4)} \times 2.05} = \frac{9x^2 \times 2.05}{3(3x+4)(1-x)}$$

$$= \frac{3x(0.333)^2 \times 2.05}{(4+0.999)(0.667)} = 0.204 \text{ atm}$$

(16) 2 gram molecules of PCl_5 were heated in a closed two litre vessel. When equilibrium is obtained the PCl_5 is 40% dissociated into PCl_3 and Cl_2.Calculate the equilibrium constant.

Sol. $PCl_5 \rightleftharpoons PCl_3 + Cl_2$

amount of PCl_5 at equilibrium $= 2 - \frac{2 \times 40}{100} = 2 - 0.8 = 1.2$ mole

amount of PCl_3 and Cl_2 at equilibrium respectively is 0.8 mole each

$[PCl_5] = \frac{1.2}{2} = 0.6$; $[PCl_3] = \frac{0.8}{2} = 0.4$; $[Cl_2] = \frac{0.8}{2} = 0.4$

$$k = \frac{[PCl_3][Cl_2]}{[PCl_5]} = \frac{[0.4][0.4]}{[0.6]} = 0.267$$

(17) 60 ml. of H_2 and 42 ml. of I_2 are heated in a closed vessel. At equilibrium the vessel contains 28 ml. HI. Calculate the degree of dissociation of HI.

Sol. $H_2 + I_2 \rightleftharpoons 2HI$

volume at equilibrium (60-x) (42-x) 2x

2x = 28, x = 28/2 =14

$$k_c = \frac{28 \times 28}{46 \times 28} = \frac{28}{46} = 0.61$$

(18) The vapour density of N_2O_4 at a certain temp. is 30 Calculate the percentage dissociation of N_2O_4 at this temperature.

Sol. $N_2O_4 \rightleftharpoons 2NO_2$

mol. wt. of $N_2O_4 = 92$ ∴ Vapour density $D = \frac{92}{2} = 46$

$$x = \frac{D-d}{d} = \frac{46-30}{30} = \frac{16}{30} = 0.533 \text{ i.e. } 53.3\%$$

(19) A 0.01M solution of acetic acid is 5% ionised at 25°C. Calculate its dissociation constant.

Sol. $k_a = \frac{\alpha^2}{(1-\alpha)V} = \frac{0.05 \times 0.05}{(1-0.05)100} = 2.63 \times 10^{-5}$

(20) What will be the dissociation constant of 0.1N aqueous ammonia solution in terms of degree of dissociation 'α' ? What will be the value if the concentration is 0.01N ?

Sol. $NH_4OH \rightleftharpoons NH_4^+ + OH^-$

$1-\alpha$ α α

$$k_b = \frac{\alpha^2}{(1-\alpha)10}$$

for 0.01 N solution, k_b remains same at the same temp. but degree of dissociation bemomes different.

(21) The equilibrium constant, k_p for the reaction $PCl_5 \rightleftharpoons PCl_3 + Cl_2$ at 523K is 0.82. Calculate the degree of dissociation at the same temperature and at a pressure of 5 atmosphere.

Sol. $PCl_5 \rightleftharpoons PCl_3 + Cl_2$

1-x x x x is the degree of dissociation Total number of molecules at equilibrium = 1-x + x+x=1+x Total pressure is 5 atmosphere

$pPCl_5 = \frac{1-x}{1+x} \times 5$; $pPCl_3 = \frac{x \times 5}{1+x}$; $pCl_2 = \frac{x \times 5}{1+x}$

$$k_p = \frac{[pPCl_3][pCl_2]}{[pPCl_5]} \quad 0.82 = \frac{\frac{5x}{1+x} \times \frac{5x}{1+x}}{\frac{1-x}{1+x} \times 5} \quad 0.82 = \frac{5x^2}{1-x^2}$$

$0.82 - 0.82x^2 = 5x^2$; $5.82x^2 = 0.82$, $x = \sqrt{\frac{0.82}{5.82}} = 0.375 = 37.5\%$

(22) Calculate the standard equilibrium constant at 298 K. Given that the value of ΔH^o and ΔS^o of the reaction at 298k are 77.2 kJ $mol^{-1}$and 122 J K^{-1} mol^{-1}.

Sol. $\Delta G^o = \Delta H^o - T\Delta S^o = 77200 - 298 \times 122 = 40844$ J mol^{-1}

Let the equilibrium constant k_c^o.

We know $\Delta G^o = -2.303RT \log k_c^o$

$$\log k_c^o = -\frac{\Delta G^o}{2.303 \times 8.314 \times 298} = \frac{40844}{2.303 \times 8.314 \times 298} = -7.158$$

$k_c^o = 6.95 \times 10^{-8}$

(23) A 0.0128 N solution of acetic acid has Λ=14 mhos $equiv^{-1}$ and Λ_∞ = 391 mhos $equiv^{-1}$ at 25°C. Calculate the dissociation constant of acetic acid.

Sol. $\alpha = \frac{\Lambda}{\Lambda_\alpha} = \frac{14}{391} = 3.58 \times 10^{-2}$

$$k_a = \frac{\alpha^2}{(1-\alpha)v} = 3.58 \times 10^{-2} \times 3.58 \times 10^{-2} \times \frac{1}{0.128} = 1.64 \times 10^{-5}$$

(24) ΔG^o for $\frac{1}{2}N_2(g) + \frac{3}{2}H_2(g) \rightleftharpoons NH_3(g)$ is -16.5 $kJmol^{-1}$ Find out k_p for the reaction at 25°C. Also report k_p and ΔG^o for $N_2(g) + 3H_2(g) \rightleftharpoons 2NH_3(g)$ at 25°C.

Sol. $\log k_p = \frac{-\Delta G^o}{2.303RT} = \frac{-(-16.5 \times 10^3)}{2.303 \times 8.314 \times 298} = 2.8917$

$k_p = 779.41$

k_p for the reaction

$N_2(g) + 3H_2(g) \rightleftharpoons 2NH_3(g)$, $= (779.41)^2 = 6.07 \times 10^5$

$\Delta G^o = -2.303 \times 8.314 \times 298 \log 6.07 \times 10^5 = -32.99$ $kJmol^{-1}$

(25) The equilibrium constant k_p for the reaction $A+B \rightleftharpoons C+D$ is 1×10^{-12} at 600 K and 1×10^{-7} at 800 K. Calculate the heat of reaction.

Sol. we know $\log \frac{kp_2}{kp_1} = \frac{\Delta H}{2.303R}\left[\frac{T_2 - T_1}{T_1 x T_2}\right]$

$$\log(1\times10^{-7}) - \log(1\times10^{-12}) = \frac{\Delta H}{2.303 \times 8.314}\left[\frac{800-600}{600 \times 800}\right]$$

$$-7+12 = \frac{\Delta H}{2.303 \times 8.314}\left[\frac{200}{480000}\right]$$

$$\Delta H = \frac{5 \times 2.303 \times 8.314 \times 480000}{200} = 229.77 \text{ kJ mol}^{-1}$$

(26) 500 ml. of a solution contains 15 g. acetic acid and 8 g. methanol. What will be their active mass ?

Sol. $[X] = \frac{\text{mass of x (g.) / molar mass of x (g mol}^{-1})}{\text{volume (l)}}$

$[CH_3COOH] = \frac{15/60}{0.5} = 0.5 \text{ mol l}^{-1}$

$[CH_3OH] = \frac{8/32}{0.5} = 0.5 \text{ mol l}^{-1}$

(27) How many moles of calcium hydroxide must be dissolved to produce 250 ml. of an aqueous solution of p^H 10.65 ? Assume complete ionisation.

Sol. pOH = 14-10.65 = 3.35

$[OH^-] = 10^{-3.35} = 10^{-4} \times 10^{0.65}$

$= 4.47 \times 10^{-4}$ M

one mole of $Ca(OH)_2$ furnishes 2 OH^- ions

Hence, concentration of Ca $(OH)_2 = \frac{4.47\times10^{-4}}{2}$

$= 2.235 \times 10^{-4}$ M

No. of moles in 250 ml $= \frac{2.235\times10^{-4}}{4} = 5.58 \times 10^{-5}$

(28) The pH of 0.1M hydrocyanic acid solution. is 5.2. What is the value of k_a for hydrocyanic acid ?

Sol. pH = – log $[H^+]$

log $[H^+]$ = – pH = –5.2

$[H^+] = 10^{-5.2} = 10^{-6} \times 10^{-0.8} = 6.3 \times 10^{-6}$ M

'α', degree of dissociation $= \frac{[H^+]}{C} = \frac{6.3\times10^{-6}}{0.1} = 6.3\times10^{-5}$

According to Ostwald's formula for weak electrolyte,

$k_a = \alpha^2 c = 6.3 \times 10^{-5} \times 6.3 \times 10^{-5} \times 0.1 = 3.69 \times 10^{-10}$

(29) Calculate for 0.01N solution of sodium acetate (a) hydrolysis constant (b) degree of hydrolysis and (c) pH $k_a = 1.9 \times 10^{-5}$.

Sol. $k_h = \frac{k_w}{k_a} = \frac{10^{-14}}{1.9 \times 10^{-5}} = 5.26 \times 10^{-10}$

$x = \sqrt{\frac{k_h}{c}} = \sqrt{\frac{5.26\times10^{-10}}{0.01}} = 2.29 \times 10^{-4}$

$CH_3COO^- + H_2O \rightleftharpoons CH_3COOH + OH^-$

$[OH^-] = c \times h = 0.01 \times 2.29 \times 10^{-4} = 2.29 \times 10^{-6}$ M

$[H^+] = \frac{kw}{[OH^-]} = \frac{10^{-14}}{2.29 \times 10^{-6}} = 4.37 \times 10^{-9}$ M

pH = - log $[H^+]$ = -log (4.37×10^{-9}) = 8.36

(30) If 250 c.c. of 0.050 M $Ba(NO_3)_2$ are mixed with 250 c.c. of 0.020M NaF, will any BaF_2 precipitate ? k_{sp} of BaF_2 is 1.7×10^{-6} at 298 k.

Sol. As equal volumes of two solutions are mixed, the concentration will be halved,

$[Ba^{2+}] = \frac{0.05}{2} = 0.025$ M, $[F^-] = \frac{0.02}{2} = 0.01$ M

$BaF_2 \rightleftharpoons [Ba^{2+}] + 2F^-$

Ionic product = $0.025 \times (0.01)^2 = 2.5 \times 10^{-6}$

Ionic product is greater than the solubility product. Hence precipitation will take place.

(31) The solubility product of magnesium hydroxide at 25°C is 1.4×10^{-11}. What is the solubility of $Mg(OH)_2$?

Sol. $Mg(OH)_2 \rightleftharpoons Mg^{2+} + 2OH^-$

Let the solubility be x moles per litre

$\therefore$ concentration of Mg^{2+} = x moles per litre

$x \times (2x)^2 = 1.4 \times 10^{-11}$

$4x^3 = 1.4 \times 10^{-11}$

$x = 1.5 \times 10^{-4}$ mole / lit.

$\therefore$ Solubility of $Mg(OH)_2 = 1.5 \times 10^{-4}$ mole / lit.

$= 58 \times 1.5 \times 10^{-4} = 0.0087$ g / lit.

(32) The solubility product of $PbBr_2$ is 8×10^{-5}. If the salt is 80% dissociaed in saturated solution, find out the solubility of the salt.

Sol. $PbBr_2 \rightleftharpoons Pb^{2+} + 2Br^-$

Let x be the number of gram moles ionised

$k_{sp} = [x] \times [2x]^2 = 4x^3 = 8 \times 10^{-5}$

$x = \sqrt[3]{\frac{8\times10^{-5}}{4}} = 2.714 \times 10^{-2}$

$\therefore$ we have 2.714×10^{-2} moles ionized per litre

The total number of moles per lit. $= 2.714 \times 10^{-2} \times \frac{100}{80}$

$= 3.392 \times 10^{-2}$

Molecular weight of $PbBr_2$ = 207 + 160 = 367

Solubility = $3.392 \times 10^{-2} \times 367$ = 12.45 g / lit.

(33) Calculate the pH value of a solution obtained by mixing 50 ml of 0.2 N HCl with 50 ml. of 0.1N NaOH.

Sol.: The normality of the resulting solution

$= \frac{N_1V_1 - N_2V_2}{\text{Total Volume}}$

$= \frac{0.2 \times 50 - 0.1 \times 50}{100} = 0.05N$

As the acid is monobasic the H^+ Concentration

= 0.05 mole/lit.

pH of the solution = - log $[H^+]$

= - log [0.05] = 1.30

(34) The ionisation constant of aniline, acetic acid and water at 25°C are respectively 3.83×10^{-10}, 1.75×10^{-5} and 1.008×10^{-14}. Calculate the percentage hydrolysis of aniline acetate in decinormal solution.

Sol. $x = \sqrt{\frac{kw}{ka \times kb}} = \sqrt{\frac{1.008 \times 10^{-14}}{1.75 \times 10^{-5} \times 3.83 \times 10^{-10}}} = 1.219$

x can't be more than 1. It means that the degree of hydrolysis in this case is very large.

$\therefore$ We have to use $\frac{x}{1-x} = 1.219$

x = 1.219 - 1.219 x ; x + 1.219 x = 1.219

2.219x = 1.219

$x = \frac{1.219}{2.219} = 0.5495$ i.e. 54.95%

(35) Calculate H^+ ion concentration of a solution which has a pH of 4.5.

Sol. $pH = -\log[H^+]$

$4.5 = -\log[H^+]$

$\log[H^+] = -4.5$

$[H^+] = \text{Antilog}[-4.5] = 0.000031622 = 3.1622\times10^{-5}$

(36) 7.3×10^{-3} g. of HCl is dissolved in 200 ml. of water. Calculate the pH of the solution.

Sol. Molarity $= 7.3 \times 10^{-3} \times \frac{1000}{200} \times \frac{1}{36.5} = 10^{-3}$ M

$pH = -\log[H^+] = -\log 10^{-3} = 3$

(37) What will be the pH value of a solution obtained by mixing 5gm of acetic acid and 7.5 gm of sodium acetate and making the volume equal to 500 ml. Dissociation constant of acetic acid at 25°C is 1.8×10^{-5}.

Sol. We know for acid buffer $pH = pk_a + \log\frac{[Salt]}{[Acid]}$

$pk_a = -\log k_a = -\log(1.8\times10^{-5}) = 4.7447$

$[Salt] = \frac{1000}{500} \times \frac{7.5}{82} = 0.1829$ mole/lit.

$[Acid] = \frac{1000}{500} \times \frac{5}{60} = 0.1666$ mole/lit.

$pH = 4.7447 + \log\frac{0.1829}{0.1666} = 4.7447 + 0.0435 = 4.7882$

$pH = 4.79$

(38) A buffer solution contains 0.20 mole of NH_4OH and 0.25 mole of NH_4Cl per litre. Calculte the pH of the solution. Dissociation constant of NH_4OH at the room temperature is 1.80×10^{-5} and $k_w = 10^{-14}$.

Sol. We know $pOH = pk_b + \log\frac{[Salt]}{[Base]}$

$pk_b = -\log k_b = -\log(1.80\times10^{-5}) = 4.7447$

$pOH = 4.7447 + \log\frac{0.25}{0.20} = 4.8419$

$pH + pOH = 14$

$pH = 14 - 4.8419 = 9.1581$

Thus the pH of the solution is 9.16.

(39) Calculate approximately pH of a 0.100 M aqueous solution. K_1 and K_2 for H_2S are 1.00×10^{-7} and 1.30×10^{-13} respectively at 25°c.

Sol. The ionisation of H_2S is represented as,

1st step, $H_2S \rightleftharpoons H^+ + HS^-$

$$K_1 = \frac{[H^+][HS^-]}{[H_2S]} \quad (i)$$

2nd step, $HS^- \rightleftharpoons H^+ + S^{2-}$ (ii)

$$K_2 = \frac{[H^+][S^{2-}]}{[HS]}$$

From K_1 and K_2 values, we can conclude that the $[H^+]$ is mainly due to 1 st step reaction since it is 10^6 times more than that in the 2 nd step.

H_2S is a weak acid and therefore will undergo feeble ionisation. The equilibrium concentration of H_2S is about the same as the initial one.

Substituting the values in equation (i)

$$1.00 \times 10^{-7} = \frac{[H^+][H^+]}{0.100} \quad \{\text{since } [HS^-]=[H^+]\}$$

$[H^+]^2 = 1.00\times10^{-7}\times0.100 = 1.0\times10^{-8}$

$[H^+] = \sqrt{1.0 \times 10^{-8}} = 1.0 \times 10^{-4}$

$pH = -\log[H^+] = -\log[1.0\times10^{-4}] = 4$

(40) Calculate the pH at the equivalence point when a solution of 0.10 M acetic acid is titrated with a solution of 0.10 M sodium hydroxide. K_a for acetic acid $=1.9\times10^{-5}$

Sol. The neutralisation reaction is,

$CH_3COOH + NaOH \rightarrow CH_3COONa + H_2O$

At equivalence $[CH_3COONa] = 1/2 \times 0.10 = 0.05$ M

Let x mole/lit. of CH_3COONa is hydrolysed

	$CH_3COO^- + Na^+ + H_2O \rightleftharpoons$	$CH_3COOH + Na^+ +$	OH^-
Initial con.	0.05	0	0
con. at equi.	0.05-x	x	x

$$K_h = \frac{[CH_3COOH][OH^-]}{[CH_3COOH]}$$

$$K_h = \frac{x \times x}{0.05 - x}$$

We know $K_h = \frac{K_w}{K_a}$

$$\frac{x \times x}{0.05 - x} = \frac{1 \times 10^{-14}}{1.9 \times 10^{-5}}$$

$$\frac{x^2}{0.05} = \frac{1\times10^{-14}}{1.9 \times 10^{-5}}$$ (since K_a is very small; x is also very small and therefore neglected in the denominator is)

$$x = \sqrt{\frac{0.05 \times 1 \times 10^{-14}}{1.9 \times 10^{-5}}} = 5.12 \times 10^{-6}$$

Since $[OH^-] = 5.12 \times 10^{-6}$

$$[H^+] = \frac{1 \times 10^{-14}}{5.12 \times 10^{-6}} = 1.95 \times 10^{-9}$$

$pH = -\log[H^+] = -\log[1.95 \times 10^{-9}]$

$= 8.71$

(41) An acid indicator, HIn differs in colour from its conjugate base (In^{-1}). The human eye is sensitive to colour differences only when the ratio $[In^{-1}] / [HIn]$ is greater than 10 or smaller than 0.1. What should be the minimum change in the pH of the solution to observe a complete colour change ($K_a = 1.0 \times 10^{-5}$) ?

Sol. The indicator dissocition equilibrium can be represented as, $HIn \rightleftharpoons H^+ + In^-$ being an acid

$$pH = pKa + \log\frac{[In^-]}{[HIn]} \quad (i)$$

The conditions when the colour of the indicator will be visible can be derived from the equation (i)

1 st condition : $pH = 5 + \log 10 = 6$

2 nd condition : $pH = 5 + \log 0.1 = 4$

The minimum change in pH = 2

(42) Calculate the pH of an aqueous solution of 1.0 M ammonuum formate assuming complete dissociation. (pK_a of formic acid = 3.8 and pK_b of ammonia = 4.8).

Sol. K_a formic acid $= 10^{-pKa} = 10^{-3.8} = 1.58 \times 10^{-4}$

K_b ammonia $= 10^{-pka} = 10^{-4.8} = 1.58 \times 10^{-5}$

$HCOONH_4$ is salt of weak acid and a weak base

$$K_h = \frac{K_w}{Ka \times Kb} = \frac{1 \times 10^{-14}}{1.58 \times 10^{-4} \times 1.58 \times 10^{-5}}$$
$$= 4.0 \times 10^{-6}$$

The hydrolysis of $HCOONH_4$ can be represented as

$$HCOONH_4 + H_2O \rightleftharpoons HCOOH + NH_4OH$$

$$K_h = \frac{[HCOOH][NH_4OH]}{[HCOONH_4]}$$

Let the concentration of NH_4OH be x.

The concentration of HCOOH will also be x

$\therefore$ [NH4+] = [HCOO$^-$] = [HCOO NH_4] = 1.0

Substituting the values in equation (i),

$$4.0 \times 10^{-6} = \frac{x \times x}{1.0 \times 1.0}$$

$$x^2 = 4.0 \times 10^{-6} \times 1.0 \times 1.0 = 4.0 \times 10^{-6}$$

$$x = \sqrt{4.0 \times 10^{-6}} = 2.0 \times 10^{-3}$$

Ionisation of formicacid can be represented as,

$$HCOOH \rightleftharpoons H^+ + HCOO^-$$

$$Ka = \frac{[HCOO^-][H^+]}{[HCOOH]}$$

Substituting the values, $1.58 \times 10^{-4} = \frac{1.0 \times [H^+]}{2.0 \times 10^{-3}}$

$$[H+] = \frac{1.58 \times 10^{-4} \times 2.0 \times 10^{-3}}{1.0} = 3.16 \times 10^{-7}$$

pH = -log[H$^+$]

= - log [3.16×10^{-7}] = 6.50

43. Calculate the amount of NH_3 and NH_4Cl required to prepare a buffer solution of pH 9.0, when total concentration of buffering reagent is 0.6 mol l^{-1}. pk_b for NH_3 = 4.7, log 2 = 0.30.

Sol. Let the concentration of NH_4Cl be = x mol l^{-1}

Then, the concentration of NH_3 = 0.6 – x

we know that for basic buffer,

$$p^{OH} = pk_b + \log \frac{[Salt]}{[Base]}$$

$$14 - 9.0 = 4.7 = \log \frac{x}{0.6 - x}$$

$$5 = 4.7 + \log \frac{x}{0.6 - x}$$

$$5 - 4.7 = \log \frac{x}{1.6 - x}$$

$$\log 2 = \log \frac{x}{0.6 - x}$$

$$2 = \frac{x}{0.6 - x}$$

$$x = 1.2 - 2x = 0.4$$

$\therefore$ [NH_4Cl] = 0.4 mol l^{-1}

[NH_3] = 0.2 mol l^{-1}

44. A sample of hard water contain 0.005 moles of $CaCl_2$ per litre. What is the minimum concentration of Na_2SO_4 which must be added for removing Ca^{2+} ions from the water sample. k_{sp} for Ca SO_4 is 2.4×10^{-5} at 25°C.

Sol. ksp = [Ca^{2+}] [SO_4^{2-}]

2.4×10^{-5} = [0.005] [SO_4^{2-}]

$$[SO_4^{2-}] = \frac{2.4 \times 10^{-5}}{0.005} = 4.8 \times 10^{-3} \text{ M.}$$

45. The solulbility of $Pb(OH)_2$ in water is 6.7×10^{-6} M. Calculate the solubility of $Pb(OH)_2$ in a buffer solution of pOH = 8.

Sol. $Pb(OH)_2 \rightleftharpoons Pb^{2+} + 2OH^-$

$$k_{sp} = [Pb^{2+}][OH^-]^2$$
$$= 6.7 \times 10^{-6} \times (2 \times 6.7 \times 10^{-6})^2$$
$$= (6.7 \times 10^{-6})^3 \times 4 = 1.2 \times 10^{-15}$$

pH of the buffer solution = 14 – pH = 14 – 8 = 6

$\therefore$ [OH^-] = 10^{-6}

$$[Pb^{2+}] = \frac{1.2 \times 10^{-15}}{(10^{-6})^2} = 1.2 \times 10^{-3} \text{ M.}$$

46. The average concentration of SO_2 in the atmosphere over a city on a certain day is 10 ppm, when the average temperature is 298 K. Given that the solubility of SO_2 in water at 298K is 1.3653 moles litre^{-1} and the pk_a of H_2SO_3 is 1.92, estimate the pH of rain water on that day.

Sol. Concentration of SO_2 in atmosphere = 10 ppm

$\therefore$ Amount of SO_2 in atmosphere = $10/10^6$

Molar concentration of SO_2 = concentration of SO_2 × solubility of SO_2 in H_2O

$$= 10 \times 10^{-6} \times 1.3653 = 1.3653 \times 10^{-5} M$$

The chemical reaction can be represented as,

	H_2SO_3	$\rightarrow$	H^+ +	HSO_3^-
conc. at eq.	$1.3653 \times 10^{-5} - x$		x	x

Dissociation constant of H_2SO_3, $k_a = \frac{[H^+][HSO_3^-]}{[H_2SO_3]}$

$$k_a = \frac{x^2}{1.3653 \times 10^{-5} - x}$$

$$1.2 \times 10^{-2} = \frac{x^2}{(1.3653 \times 10^{-5} - x)} \quad (\therefore -\log k_a = pk_a)$$

$$x^2 = 1.2 \times 10^{-2} \times (1.3653 \times 10^{-15} - x)$$

$$x = 1.36 \times 10^{-5}$$

$$pH = -\log[1.36 \times 10^{-5}] = 4.9$$

Thermodynamics

(1) The heat of combustion of CH_4 (g) at constant volume is measured in a bomb calorimeter at 298.2K and is found to be -885389 J/mole. Find the value of enthalpy change.

Sol. on combustion : $CH_4(g) + 2O_2(g) \rightarrow CO_2(g) + 2H_2O(l)$

$$\Delta H = \Delta E + \Delta nRT \text{ or } \Delta H = \Delta E + P\Delta V$$
$$= -885389 + (-2)(8.314)x298$$
$$= -885389 \text{ Jmol}^{-1} - 4959 \text{ Jmol}^{-1}$$
$$= -890348 \text{ J mol}^{-1} \text{ or } -890.348 \text{ k. J. /mole.}$$
$$\text{or } \Delta H = \Delta E \text{ since } \Delta V = 0$$

(2) Calculate standard enthalpy change and standard internal energy change for the reaction

$OF_2(g) + H_2O(g) \rightarrow O_2(g) + 2HF(g)$

ΔH_f OF_2(g) 23; H_2O(g) = -241.8 & HF(g) = -268.6 kJmol^{-1}

Sol. $\Delta H_f = \Sigma\Delta H_p$ (products) - $\Sigma\Delta H_f$ (reactants)

$$= 2(-268.6) - \{23.0 - 241.8\}$$
$$= -537.2 + 218.8 = -318.4$$

(The enthalpy of formation of O_2 is zero)

$$\Delta E = \Delta H - \Delta nRT$$
$$= -318.4 - (1) \times 8.31 \times 298.2$$

$= 320.9$ k. J. mol^{-1}

(3) Calculate the enthalpy of formation of C_2H_5OH at 25°C from the following data

(i) $C_2H_5OH(l) + 3O_2(g) \rightarrow 2CO_2(g) + 3H_2O(l)$; $\Delta H = -1368.2$ k J

(ii) $C(s) + O_2(g) \rightarrow CO_2(g)$; $\Delta H = -393.5$ kJ

(iii) $H_2(g) + \frac{1}{2} O_2(g) \rightarrow H_2O(l)$; $\Delta H = -286.2$ kJ

Sol. We aim for the equation

$2C(s) + 3H_2(g) + \frac{1}{2} O_2(g) \rightarrow C_2H_5OH(l)$; $\Delta H = ?$

To obtain this equation, multiply equation (ii) by 2 and (iii) by 3. (iv) $2C(s) + 2O_2(g) \rightarrow 2CO_2(g)$; $\Delta H = -787.0$ kJ

(v) $3H_2(g) + 1\frac{1}{2} O_2(g) \rightarrow 3H_2O(l)$; $\Delta H = -858.5$ kJ

Add (iv) and (v) and subtract (i)

on adding $2C(s) + 3H_2(g) + 3\frac{1}{2} O_2(g) \rightarrow 2CO_2(g) + 3H_2O(l)$; $\Delta H = -1645.5$ kJ

$C_2H_5OH(l) + 3O_2(g) \rightarrow 2CO_2(g) + 3H_2O(l)$; $\Delta H = -1368.2$kJ

$-C_2H_5OH(l) + 2C(s) + 3H_2(g) + \frac{1}{2} O_2(g) \rightarrow -277.3$ kJ

finally we can write $2C(s) + 3H_2(g) + \frac{1}{2}O_2(g) \rightarrow C_2H_5OH(l)$; $\Delta H = -277.3$ kJ

(4) Calculate the heat of neutralisation of aqueous NH_3 with aqueous HCl from the following data

(i) $NH_3(g) + aq \rightarrow NH_3(aq) + 8,400$ cal.

(ii) $HCl(g) + aq \rightarrow HCl(aq) + 17,300$ cal.

(iii) $NH_3(g) + HCl(g) + aq \rightarrow NH_4Cl(aq) + 38,200$ cal.

Sol. We aim for the reaction

$NH_3(aq) + HCl(aq) \rightarrow NH_4Cl(aq) + \Delta H$

To obtain this equation, rearrange the equation (i) and (ii) and add the equation (iii) to their sum

$NH_3(aq) \rightarrow NH_3(g) + aq - 8,400$ cal.

$HCl(aq) \rightarrow HCl(g) + aq - 17,300$ cal.

$NH_3(aq) + HCl(aq) \rightarrow NH_3(g) + HCl(g) + aq - 25700$ cal add $NH_3(g) + HCl(g) + aq \rightarrow NH_4Cl(aq) + 38200$ cal.

$NH_3(aq) + HCl(aq) \rightarrow NH_4Cl(aq) + 12500$ cal

Thus, heat of neutralisation = 12500 cal

(5) If 500 calories of heat energy are added to a system and the system does 350 calories of work, on the surroundings, what is the energy change of the system ?

Sol. $\Delta E = Q + W = 500 + (-350) = 150$ calories

(6) Calculate the work done when 1.0 mole of water at 373K vaporizes against an atmospheric pressure of 1.0 atmosphere. Assume ideal gas behaviour.

Sol. $V = \frac{nRT}{P} = \frac{1.0 \times 0.0821 \times 373}{1.0} = 31.0$ lit.

$W = -P_{ext}. \Delta V = -1 \times 31.0 = -31.0$ lit-atmo.

$= -31 \times 101.3 = -3140.3$ J

(7) Calculate the work done in the reversible compression of 10.0g. of helium gas from an initial pressure of 1.00 atmos. to a final pressure of 5.00 atmos. at a constant temp of 27°C.

Sol. $W = nRT \times 2.303 \log \frac{P_2}{P_1}$

$= \frac{10}{4} \times 1.987 \times 300 \times 2.303 \log 5 = 2400$ cal.

(8) 5 moles of oxygen are heated at constant volume from 10°C to 20°C. What will be the change in the internal energy of the gas ? The molar heat of oxygen at constant pressure, $C_P = 7.03$ cal. mol^{-} deg^{-1}and R = 8.31 J mol^{-1} deg.$^{-1}$.

Sol. We know $C_P - C_V = R$

$C_V - C_P - R = 7.03 - 1.99 = 5.04$ cal. mol^{-1} deg.$^{-1}$

Heat absorbed by 5 moles of oxygen in heating from 10°C to 20°C $= 5 \times C_V \times \Delta T = 5 \times 5.04 \times 10 = 252$ cal.

since gas is heated at constant volume, no external work is done, $W = 0$ $\therefore \Delta E = Q + W = 252 + 0 = 252$ cal.

(9) Two moles of an ideal gas initially at 27°C and one atm. are compresed reversibly to half the initial volume. Calculate Q, W, ΔE and ΔH when the process is carried on isothermally.

Sol. $W = 2.303\, nRT \log \frac{V_2}{V_1}$

$= 2.303 \times 2 \times 1.98 \times 300 \log \frac{1}{2} = -1914$ cal.

Since the process is being carried on isothermally i.e. no change of temp., $\therefore \Delta H = 0$ and $\Delta E = 0$

According to 1st law of thermodynamics

$Q = \Delta E + P\Delta V$, or $Q = \Delta E + W$

$\therefore W = -1914 + 0 = -1914$ cal.

(10) A gas contained in a cylinder fitted with a frictionless piston expands against a constant pressure 1 atmo. from a volume of 4 lit. to 14 litres. In doing so, it absorbs 800J thermal energy from surroundings. Determine ΔE for the process.

Sol. $\Delta E = Q + W$

$W = -P \times \Delta V = -1 \times 10 = -10$ lit.- atmos.

But 0.082 lit- atmos = 1.987 cal.

$W = \frac{10 \times 1.987}{0.082} = -242.3$ cals.$= 242.3 \times 4.184 = -1013.7$J

$\Delta E = 800 - 1013.7$ J $= -213.7$ J.

(11) 0.5 g. of benzoic acid was subjected to combustion in a bomb calorimeter. The temp. of the calorimeter system including water was found to rise by 0.5°C. Calculate the heat of combustion of methane at (i) constant volume and (ii) constant pressure. The thermal capacity of calorimeter system is 17.7 kJ K^{-1}.

Sol. Heat of combustin at constant volume,

$\Delta E = \frac{\text{Heat capacity of}}{\text{calorimeter system}} \times \text{rise in temp.} \times \frac{\text{mol. mass of a comp.}}{\text{mass of a compd.}}$

$= 17.7 \times 0.5 \times \frac{16}{0.16} = 885$

$\Delta E = -885$ k.J.mol^{-1}

$CH_4(g) + 2O_2(g) \rightarrow CO_2(g) + 2H_2O(l)$

$\Delta H = \Delta E + \Delta nRT$

$= -885 + (-2) \times 8.314 \times 10^{-3} \times 300$

$= -885 - 4.988 = -889.988$ k J mol^{-1}.

(12) Heat supplied to a Carnot engine is 453.6 k. cal. How much useful work can be done by the engine which works between 10°C and 100°C ?

Sol.

$\frac{W}{Q_2} = \frac{T_2 - T_1}{T_2}$ $\therefore W = Q_2 \times \frac{T_2 - T_1}{T_2}$

$= 1897.86 \times \frac{373 - 283}{373}$

$= \frac{1897.86 \times 90}{373} = 457.92$ k J.

(13) The standard molar enthalpy of formation of ammonia is - 46.2 k J mol^{-1}. Given H_{H-H} = 435 k J mol^{-1} and H_{N-H} = 390 k J mol^{-1}. Calculate the bond energy in $N_2(g)$

Sol. $N_2(g) + 3H_2(g) \rightarrow 2NH_3(g)$

sum of bond energies of reactants = $H_{N\equiv N} + 3 \times H_{H-H}$

$= H_{N\equiv N} + 3 \times 435 = H_{N\equiv N} + 1305$

sum of bond energies of products = $2 \times 3H_{N-H}$

$= 6\ H_{N-H} = 6 \times 390 = 2340$ kJ

ΔH = sum of bond energies of reactants - sum of bond energies of products

$-92.4 = H_{N\equiv N} + 1305 - 2340$

$H_{N\equiv N} = -92.4 - 1305 + 2340 = 942.6$ k J mol^{-1}.

(14) Calculate the temperature at which liquid water will be in equilibrium with water vapour.

$\Delta H_{vap} = 40.73$ and $\Delta S_{vap} = 0.109$ k J/ mol/ k

Sol. $\Delta G = \Delta H - T\Delta S$

At equilibrium $\Delta G = 0$

$0 = 40.73 - T \times 0.109$

$T = \frac{40.73}{0.109} = 373.6$ k

(15) Calculate ΔG for the isothermal expansion of 2.00 moles of an ideal gas at 25°C from a pressure P_1 = 1 atm. to P_2 = 0.100 atmos.

Sol. $\Delta G = nRT \ln \frac{P_2}{P_1}$

$= 2.303\ nRT \log \frac{P_2}{P_1}$

$= 2.303 \times 2 \times 1.987 \log \frac{0.100}{1.00} = -2720$ cals.

(16) One mole of liquid water at 100°C is placed in a thermostat maintained at 25°C. Calculate the entropy change in the process. The molar heat capacity of liquid water is 18 cal. degree^{-1}.

Sol. Entropy change of water $\Delta S_1 = 2.303\ C_p \log \frac{T_2}{T_1}$

$= 2.303 \times 18 \log \frac{298}{373}$

$= -4.042$ cal. degree^{-1}

Entropy change in the thermostat $\Delta S_2 = \frac{Q}{T_1}$

$= 18 \times 1 \times (T_2 - T_1)$

$= 18 \times 1 \times 75 = 1350$ cals.

$\Delta S_2 = \frac{1350}{298} = 4.53$ cal / deg.

Net entropy change = 4.53 - 4.042 = 0.488 cal. deg^{-1}.

(17) The enthalpy and entropy changes for the reaction C(s) diamond + $O_2 \rightarrow CO_2(g)$ at 25°C and 1 atm. are -393.4 k J and 0.006 k J/K respectively. Is the conversion at room temp. a spontaneous process ?

Sol. $\Delta G = \Delta H - T\Delta S$

$= 393.4 - (298)(0.006) = -395.2$ kJ.

Since ΔG = -ve, the reaction is spontaneous

(18) The enthalpy change for the transition of liquid water to steam is 40.8 kJ mol^{-1} at 373K. Calculate $\Delta S°$ for the process.

Sol. $\Delta S_{vap} = \frac{\Delta H_{vap}}{T} = \frac{40.8 \times 1000}{373} = 109$ J K^{-1} mol^{-1}.

(19) Calculate the standard entropy change and the free energy change when 1 mole of water is formed from its elements at 25°C. The standard entropy of H_2, O_2 and H_2O are 31.21, 49.0 and 16.73 cal.deg^{-1}mole^{-1} respectively.

Sol. $H_2(g) + \frac{1}{2}O_2(g) \rightarrow H_2O(l)$

$\Delta S°_f = S°_{H_2O} - S°_{H_2} - \frac{1}{2}S°_{O_2}$

$= 16.73 - 31.21 - 24.50 = -38.98$

$\Delta G = \Delta H - T\Delta S = -68,320 - 298(-38.98) = -56,700$ cal mol^{-1}.

(20) ΔH and ΔS for the system $H_2O\ (l) \rightleftharpoons H_2O\ (g)$ at 1 atmos pressure are 40.63 kJ mol^{-1} and 108.8 J k^{-1} mol^{-1} respectively. Calculate the temp. at which the rates of forward and backward reactions will be the same. Predict the sign of free energy for this transformation above this temperature.

Sol. $\Delta G = \Delta H - T\Delta S$

$O = 40.63 - T \times 0.1088$

$T = \frac{40.63}{0.1088} = 373.4$ k

$\Delta G = \Delta H - T\Delta S$

$= 40.63 - 374 \times 0.1088 = 40.63 - 40.69 = -0.06$ kJ.

ΔG is negative, hence, the reaction will be spontaneous.

(21) Calculate $\Delta G°$ of the following reaction

$CO(g) + \frac{1}{2}O_2(g) \rightarrow CO_2(g)\ \Delta H° = -282.84$ kJ

given $S°_{CO_2(g)} = 213.8$ Jk^{-1}mol^{-1}, $S°_{CO(g)} = 197.9$ Jk^{-1}mol^{-1}, $S°_{O_2(g)} = 205.0$ Jk^{-1}mol^{-1}.

Sol. $\Delta S° = \Sigma S°_{products} - \Sigma S°_{reactants}$

$= 213.8 - 197.9 + \frac{1}{2} \times 205.0 = -86.6$ JK^{-1} = -0.0866 kJK^{-1}.

$\Delta G = \Delta H - T\Delta S$ (T=25°C = 298 K as the values of $\Delta H°$ & $\Delta S°$ are at standard conditions = -282.84 - 298 x -0.0866

= -282.84 - 25.8 = -308.64 kJ

(22) The standard voltage of the cell

$2H_2(g) + O_2(g) \rightleftharpoons 2H_2O(l)$ is 1.23 V at 25°C. The standard enthalpy of the reaction is -571.6 kJmol^{-1}.

(i) Calculate the $\Delta G°$, $\Delta S°$ and k_{eq} values of the reaction.

Sol. $2H^+(aq) + 2e^- \rightarrow H_2(g)$ ---- anode reaction

$2H_2O(l) \rightarrow O_2(g) + 4H^+(aq) + 4e^-$ ----- cathode reaction

for balancing purpose, we have to multiply the anode reaction by 2

(i) $\Delta G° = -nFE° = 4 \times 96485 \times 1.23 = 474706$ J (i i)

$\Delta G° = \Delta H° - T\Delta S°$

$\Delta S° = \frac{\Delta H° - \Delta G°}{T} = \frac{-571.6 \times 10^3 - (-474706)}{298} = 325.1$Jk^{-1}.

(iii) $\Delta G° = -2.303\ RT \log k_{eq}$

$\log k_{eq} = \frac{-\Delta G°}{2.303RT} = \frac{-(-474706)}{2.303 \times 8.314 \times 298} = 83.2$

k = Antilog of 83.2 = 1.28×10^{83}.

(23) Calculate the free energy change accompanying the compression of 1 mole of a gas at 57°C from 25 to 200 atmospheres. The fugacities of the gas at 57°C may be taken as 23 and 91 atmospheres, respectively, at pressures of 25 and 200 atmospheres.

Sol.: we know $\Delta G = nRT \times 2.303 \log \frac{P_2}{P_1}$

$= 1 \times 8.314 \times 330 \times 2.303 \log \frac{200}{25}$

$= 5702.8$ J

for more accurate value; we should use equation involving fugacities i.e. $\Delta G = nRT \times 2.303 \log \frac{f_2}{f_1}$

$= 1 \times 8.314 \times 330 \times 2.303 \log \frac{91}{23}$

$= 3774.0$ J

(24) Equilibrium constant for the reaction

$PCl_5(g) \rightleftharpoons PCl_3(g) + Cl_2(g)$ at 298 k is 1.8×10^{-7}.

Calculate the value for ΔG^o.

Sol. we know $\Delta G^o = -2.303 \text{ RT} \log k_{eq}$

$= -2.303 \times 8.31 \times 298 \times \log(1.8 \times 10^{-7})$

$= -2.303 \times 8.31 \times 298 \times -6.9664$

$= 39.73$ k J mol^{-1}.

(25) Calculate the equilibrium constant for the reaction $2NOCl(g) \rightleftharpoons 2NO(g) + Cl_2(g)$ at 400k if $\Delta H^o = 77.2 kJmol^{-1}$ and $\Delta S^o = 122$ J K^-.

Sol. $\Delta G^o = \Delta H - T\Delta S$

$= 77200 - 400 \times 122 = 28400$ J mol^{-1}

we know $\Delta G = 2.303$ RT $\log k_{eq}$.

$\log k_{eq} = \frac{-\Delta G}{2.303RT} = \frac{-28400}{2.303 \times 8.31 \times 400}$

$\log k_{eq} = -3.70$

k_{eq} = Antilog (-3.70) = 1.995×10^{-4}

(26) The activity of 2.5 moles of a substance changes from 0.05 to 0.35. What would be the change in its free energy at 27°C ?

Sol.: we know that $\Delta G = nRT \times 2.303 \log \frac{a_2}{a_1}$

$= 2.5 \times 8.314 \times 300 \times \log \frac{0.35}{0.05} = 12133.6$ J

(27). Calculate the resonance energy of N_2O from the follwing data. ΔH^o_f of N_2O = 82 kJ/mole. Bond energies of N≡N, N=N, O=O and N=O bonds are 946, 418, 498 and 607 kJ/mole respectively.

sol. Formation of N_2O is ; $N_2(g) + \frac{1}{2} O_2(g) \rightarrow N_2O(g)$

ΔH = 82 kJ/mol. Structure of N_2O is, $\bar{N}{\equiv}N{\rightarrow}O \leftrightarrow N{=}N{=}O$

ΔH^o_f of $NO_2 = \Delta H_{N\equiv N} + \frac{1}{2} \times \Delta H_{O=O} - \Delta H_{N=N} - \Delta H_{N=O}$

Resonance energy = heat of formation from expermantal - bond energy

$= 946 + \frac{1}{2} \times 498 - 418 - 607 = 170$ kJ/mole

Resonance energy = 82 - 170 = -88 kJ $mole^{-1}$.

(28). The heat liberated on complete combustion of 7.8 g. benzene in 327 kJ. This heat has been measured at constant volume and at 27°c. Calculate the heat of combustion of benzene at constant pressure.

sol. Combustion of benzene can be represented as,

$C_6H_6(l) + 7\frac{1}{2}O_2(g) \rightarrow 6CO_2(g) + 3H_2O(l)$

ΔH at constant pressure = $\Delta E + \Delta nRT$ ------------ (i)

7.8 g. of benzene on complete combustion liberates 327kJ.

1 mole of benzene i.e. 78g. liberates $= \frac{78}{7.8} \times 327 = 3270$ kJ.

n = no. of moles of products - no. of moles of reactants in gaseous state.

$n = 6 - \frac{15}{2} = -\frac{3}{2} = -1.5$

Substituting the values in equation (i), we get,

$\Delta H = -3270 + (-1.5) \times 8.314 \times 10^{-3} \times 300$

$= -3273.741$ kJ

(29). 10 g. of argon gas is compressed isothermally and reversibly at a temperature, of 27°c from 10 L to 5L. Calculate q, w, ΔE and ΔH. R = 2.0 cal $k^{-1} mol^{-1}$, log 2 = 0.30, atomic weight of Argon = 40.

sol. ΔE = O (since the process is isothermal)

$\Delta H = \Delta E + P \Delta V$

= O + O = O (since at constant temp. PV is constant)

workdone in isothermal system, $w = -2.303 \times nRT \times \log \frac{V_2}{V_1}$

$= -2.303 \times \frac{10}{40} \times 2 \times 300 \times \log \frac{5}{10}$

$= -103.64$ cal

(30). An athlete is given 100 g. of glucose of energy equivalent to 1560 kJ. He utilizes 50 percent of this gained energy in the event. In order to avoid storage of energy in the body, calculate the weight of water he would need to perspire, The enthalpy of evaporation of water is 44 kJ/mole.

sol. 100 g. of glucose = 1560 kJ energy

Energy utilized $= \frac{50}{100} \times 1560 = 780$ kJ

44 kJ of energy is required to evaporate 18 g of H_2O

780 kJ of energy will be spent to evaporate,

$\frac{780}{44} \times 18 = 319.09$ g. H_2O

(31). The standard molar enthalpies of formation of cyclohexane (l) and benzene (l) at 25°C are –156 an +49 kJ mol^{-1} respectively. The standard enthalpy of hydrogenation of cyclohexene (l) at 25°C is – 119 kJ $mole^{-1}$. Use these data to estimate the magnitude of the resonance energy of benzene.

Sol. For the reaction (cyclohexene) $\xrightarrow{\text{hydrogenation}}$ (cyclohexane) i.e. for hydrogenation of one double bond required –119kJ mol^{-1}.

Enthalpy of hydrogenation for (benzene) $\xrightarrow{\text{hydrogenation}}$ (cyclohexane) is

$3 \times -119 = -357$ kJ/mol

$\Delta H = \Delta Hp - \Delta Hr = -156 - (49 + 0) = -205$ kJ/mol

Resonance energy = $\Delta H_{observed} - \Delta H_{calculated}$

$= -357 - (-205) = -152$ kJ mole

Kinetics

1. For the reaction P + Q + R → products, the rate equation is : Rate = k [P] [Q] [R]^2. What would be the rate of reaction when the concentration of P and Q are doubled while that of R is halved ?

Sol. Rate = k [P] [Q] [R]^2.

Substituting the new concentrations in the equation

$$Rate = [2]\,[2]\,[\tfrac{1}{2}]^2 = 1$$

∴ It remains the same.

2. The half life of a first order chemical reaction A → B is 10 min. What percentage of A remains after 1 hour ?

Sol. $t_{0.5} = \frac{0.693}{k}$, $k = \frac{0.693}{10} = 0.0693$

using first order reaction, $k = \frac{2.303}{t} \log \frac{a}{a-x}$

$$0.0693 = \frac{2.303}{60} \log \frac{100}{C}$$

$$C = 1.565\%$$

3. For the reaction : A + B → C + D, the initial rates for different initial concentrations are found as follows :

Initial conc. [A]	[B]	Initial rate mol l^{-1} s^{-1}
1.0	1.0	2.0 x 10^{-3}
2.0	1.0	4.0 x 10^{-3}
4.0	1.0	8.0 x 10^{-3}
1.0	2.0	2.0 x 10^{-3}
1.0	4.0	2.0 x 10^{-3}

What is the order with respect to A and B ? Overall order? Write the rate law equation. Calculate the rate constant. Suggest a mechanism.

Sol. By doubling [A] and keeping [B] constant, the initial rate also doubles. Hence with respect to 'A' first order. Doubling [B] and keeping [A] constant, rate constant doesn't change, hence with respect to 'B' is zero order. Overall order 1+0 = 1 i.e. first order .

Rate law expression, r = k [A]

Rate constant, $k = \frac{r}{[A]} = \frac{4 \times 10^{-3}\ \text{mol lit}^{-1}\ \text{sec}^{-1}}{2.0\ \text{mol lit}^{-1}} = 2 \times 10^{-3}\ \text{sec}^{-1}$

Mechanism : A → x (slow)

x + B → C + D (fast)

4. The rate constant of a reaction with respect to reactant A is 6 min^{-1}. If we start with [A] = 0.8 mol L^{-1}, when would [A] reach the value of 0.08 mol L^{-1}.

Sol. $t = \frac{2.303}{k} \log \frac{[A]_0}{[A]_t}$

$$= \frac{2.303}{6} \log \frac{0.8}{0.08} = \frac{2.303}{6} \log 10 = 0.38\ \text{min}$$

5. Show that in case of a first order reaction, the time required for 99.9% of the reaction to take place is 10 times that required for half of the reaction.

Sol. $t_{50\%} = \frac{2.303}{k} \log \frac{1}{1-0.5} = \frac{2.303}{k} \log 2$ (a)

$t_{99.9\%} = \frac{2.303}{k} \log \frac{1}{1-0.999}$

$= \frac{2.303}{k} \log \frac{1}{0.001} = \frac{2.303}{k} \log 10^3$ (b)

Now dividing (b) by (a), we get

$$\frac{t_{99.9\%}}{t_{50\%}} = \frac{2.303}{k} \times \frac{k}{2.303} \times \frac{\log 10^3}{\log 2} = \frac{3.000}{0.3010} = 9.967 \approx 10$$

6. For a reaction A + B + C → products, the rate of the reaction is given by rate = k [A]^a [B]^b [C]^c.

From the following data determine (i) the order of reaction with respect to A, B and C (ii) the overall order (iii) the rate constant

	Initial conc. (mol. d m^{-3}) [A]	[B]	[C]	Initial rate (mol.dm^{-3} s^{-1})
(i)	0.010	0.005	0.010	5.0 x 10^{-5}
(ii)	0.010	0.005	0.015	5.0 x 10^{-5}
(iii)	0.010	0.010	0.010	2.5 x 10^{-5}
(iv)	0.020	0.005	0.010	15 x 10^{-5}

Sol. comparing (i) and (iv) the conc. of [B] and [C] are constant. When the [A] is doubled the rate changes to 3 fold. Thus $a = \frac{3}{2}$ i.e. the order is $\frac{3}{2}$ with respect to A.

Comparing (i) and (iii) where [A] and [C] are constant. By doubling the [B], the rate changes to half. Thus, b= -1 i.e. the order is -1 with respect to B.

Comparing (i) and (ii) where [A] and [B] are constant. When [C] is changed $\frac{3}{2}$ times, the rate remains unaltered. The order with respect to C is zero.

(i) overall order = $\frac{3}{2} - 1 + 0 = \frac{1}{2}$

(ii) rate = k $[A]^{3/2}$ $[B]^{-1}$ $[C]^0$ = k $[A]^{3/2}$ $[B]^{-1}$

$$k = \frac{\text{rate}}{[A]^{3/2}[B]^{-1}} = \frac{15 \times 10^{-5}\ \text{mol d m}^{-3}\ \text{s}^{-1}}{(0.020)^{3/2}\,(0.005)^{-1}}$$

$$= 2.65 \times 10^{-4}\ (\text{mol. d m}^{-1})^{1/2}\ \text{s}^{-1}.$$

7. Nitrogen pentoxide when allowed to decompose in an inert solvent gave the following volumes of oxygen

t(s)	0	300	600	900	∞
vol. (cc)	0	3.42	6.30	8.95	34.75

Show that it is of first order reaction

$$k = \frac{2.303}{t} \log \frac{V\infty}{V\infty - Vt}$$

Sol. $K_{300} = \frac{2.303}{300} \log \frac{34.75}{31.33} = 3.454 \times 10^{-4}\ s^{-1}$

$K_{600} = \frac{2.303}{600} \log \frac{34.75}{28.45} = 3.334 \times 10^{-4}\ s^{-1}$

$K_{900} = \frac{2.303}{900} \log \frac{34.75}{25.80} = 3.310 \times 10^{-4}\ s^{-1}$

8. 5 ml. of methyl acetate was added to the reaction vessel containing 100 ml. of 1.89 N HCl maintained at a room temp. of 298 k, 5 ml. of reaction mixture was withdrawn and titrated with a standard alkali and the following results were obtained

Time (s)	0	900	1800	3000	∞
ml. of alkali used	11.03	11.70	12.15	12.75	14.10

from the above data, find out the order of the reaction

Sol. $k = \frac{2.303}{t} \log \frac{V\infty - Vo}{V\infty - Vt}$

$K_{900} = \frac{2.303}{900} \log \frac{14.10 - 11.03}{14.10 - 11.70} = 0.2735 \times 10^{-3}\ s^{-1}$

$K_{1800} = \frac{2.303}{1800} \log \frac{14.10 - 11.03}{14.10 - 12.15} = 0.255 \times 10^{-3}\ s^{-1}$

$K_{3000} = \frac{2.303}{3000} \log \frac{14.10 - 11.03}{14.10 - 12.75} = 0.2738 \times 10^{-3}\ s^{-1}$

As the values of k are almost constant, the order is one.

(9) The optical rotation of sucrose in 0.5 N HCl at 308k and at various time intervals are given below. Find out the order of the reaction.

Time (min)	0	10	20	30	40	∞
Rotation (degree)	+32.4	+28.8	+25.5	+22.4	+19.6	−11.1

Sol. $k = \frac{2.303}{t} \log \frac{\theta_o - \theta_\infty}{\theta_t - \theta_\infty}$; $\theta_o - \theta_\infty = +32.4 - (-11.1) = 43.5$

$$K_{10} = \frac{2.303}{10} \log \frac{43.5}{28.8-(-11.1)} = 8.636 \times 10^{-3}\ min^{-1}$$

$$K_{20} = \frac{2.303}{20} \log \frac{43.5}{25.5-(-11.1)} = 8.638 \times 10^{-3}\ min^{-1}$$

$$K_{30} = \frac{2.303}{30} \log \frac{43.5}{22.4-(-11.1)} = 8.706 \times 10^{-3}\ min^{-1}$$

$$K_{40} = \frac{2.303}{40} \log \frac{43.5}{19.6-(-11.1)} = 8.658 \times 10^{-3}\ min^{-1}$$

Since the value of k is fairly constant, it is first order reaction.

10. The decomposition of H_2O_2 was studied by titrating it at different intervals with $KMnO_4$. Calculate the velocity constant for it from the following data assuming it to be of the first order.

Time (min)	0	10	20	50
Vol. of $KMnO_4$(ml)	23.05	14.90	6.15	2.6

Sol. $K_{10} = \frac{2.303}{10} \log \frac{23.05}{14.90} = 0.04362\ min^{-1}$

$$K_{30} = \frac{2.303}{30} \log \frac{23.05}{6.15} = 0.04404\ min^{-1}$$

$$K_{50} = \frac{2.303}{50} \log \frac{23.05}{2.6} = 0.04364\ min^{-1}$$

constant values of k prove that it is of first order.

11. The half change time of a second order reaction is 20 min., When the initial concentration of the single reactant is 12×10^{-2} moles. Calculate the value of k.

Sol. For second order, $t_{0.5} \propto \frac{1}{a}$

or $t_{0.5} = \frac{1}{ka}$

$$k = \frac{1}{t_{1/2} .a} = \frac{1}{20 \times 12 \times 10^{-2}} = 0.416\ d\ m^3\ mol^{-1}\ min^{-1}.$$

12. In the saponification of ethylacetate by NaOH at 283k, y ml. of 0.043 N HCl were required to neutralize 100 ml. of the reaction mixture, t minutes after the commencement of the reaction.

t (min)	0	4.89	10.37	28.18	∝
y (ml)	61.95	50.59	42.40	29.35	14.92

find out the order of reaction.

Sol. $k = \frac{1}{t} \times \frac{x}{a(a-x)}$

t	a-x	x
0	61.95(=a)	0

$$K_{4.89} = \frac{1}{61.95 \times 4.89} \times \frac{11.36}{50.59} = 7.412 \times 10^{-4}\ ml^{-1}\ min^{-1}$$

$$K_{10.37} = \frac{1}{61.95 \times 10.37} \times \frac{19.55}{42.40} \Rightarrow 7.177 \times 10^{-4}\ ml^{-1}\ min^{-1}$$

$$K_{28.18} = \frac{1}{61.95 \times 28.18} \times \frac{32.60}{29.35} = 6.362 \times 10^{-4}\ ml^{-1}\ min^{-1}$$

As the k values are constant, it is second order.

13. In Arrhenius equation for a certain reaction, the value of A and E_a are $4 \times 10^{13}\ sec^{-1}$ and 98.6 kJ. mol^{-1} respectively. At what temp. the reaction will have specific rate constant $1.1 \times 10^{-3}\ sec^{-1}$?

Sol. $2.303 \log k = 2.303 \log A - \frac{E_a}{RT}$,

$$2.303 \log (1.1 \times 10^{-3}) = 2.303 \log (4 \times 10^{13}) \times - \frac{98.6 \times 10^3}{RT}$$

$$T = \frac{98.6 \times 10^3}{8.314 \times 2.303 \times 16.56} = 310.96\ K$$

14. The slope of the curve found in a graph between $\frac{1}{T}$ vs. $\log_{10} k$ for the decomposition of acetaldehyde was - 9920. Calculate the energy of activation for this reaction.

Sol. we know that $\frac{E_a}{2.303R} = slope$

$$E_a = slope \times 2.303 \times R$$

$$= -9920 \times 2.303 \times 8.314 = 189940\ J/mole$$

15. A first order reaction is 50% completed in 30 minutes at 27°C and in 10 min. at 47°C. Calculate the reaction rate constant at 27°C and the energy of activation of the reaction.

Sol. At 27°C, $k_{27°C} = \frac{0.693}{30} = 0.0231\ min^{-1}$

At 47°C, $k_{47°C} = \frac{0.693}{10} = 0.0693\ min^{-1}$

$$\log \frac{k_1}{k_2} = \frac{-Ea}{2.303R} [\frac{T_2 - T_1}{T_1 x T_2}]$$

$$\log \frac{0.0231}{0.0693} = \frac{-Ea}{2.303 x 8.314} \times [\frac{320-300}{320 \times 300}]$$

$$E_a = -\frac{19.1471 \times 96000}{20} \times \log 0.3333$$

$$= -91906 \times (-0.4772) = 43857\ J\ mol^{-1}$$

$$= 43.86\ k\ J/\ mole$$

16. The following values for the first order rate constant were obtained for a certain reaction :

temp. (C)	$k \times 10^{-5}\ sec^{-1}$
25	3.46
35	13.50

Calculate the Arrhenius frequency factor.

Sol. $\log \frac{k_2}{k_1} = \frac{E_a}{2.303R} [\frac{T_2 - T_1}{T_1 x T_2}]$

$$E_a = 2.303 \times \frac{1.987 \times 298 \times 308}{308 - 298} \log \frac{13.5 \times 10^{-5}}{3.46 \times 10^{-5}}$$

$$= 24830\ calories$$

$$\because \log k_1 = -\frac{E_A}{2.303\ RT} + \log A$$

$$\log 3.46 \times 10^{-5} = \frac{24830}{2.303 \times 1.987 \times 298} + \log A$$

$$\log A = 13.7491$$

$$A = 5.611 \times 10^{13}$$

17. For the bimolecular decomposition of NO_2 into NO and O_2, the velocity constant at 600k is 83.9 and at 645k is 407. Calculate the energy of activation for 2 moles of NO_2 (given log 407 =2.6096 and log 83.9 = 1.9298)

Sol. $1.9298 - 2.6096 = \frac{E_a}{2.303 \times 8.314} \times (\frac{45}{645 \times 600})$

$$E_a = \frac{0.6798 \times 2.303 \times 8.314 \times 645 \times 600}{45} = 111.94\ kJmol^{-1}.$$

18. The time required for 10% completion of a first order reaction at 298 K is equal to that required for its 25% completion at 308 K. If the pre-exponential factor for the reaction is 3.56×10^9 sec^{-1}, calculate its rate at 318 K and also the energy of activation.

Sol. $t = \frac{2.303}{k_{298}} \log \frac{100}{90} = \frac{2.303}{k_{308}} \log \frac{100}{75}$

$\frac{k_{308}}{k_{298}} = \frac{E_a}{2.303 x R} [\frac{T_2 - T_1}{T_1 x T_2}]$

$\frac{\log 100/75}{\log 100/90} = \frac{E_a}{2.303 \times 8.314} \times \frac{10}{298 \times 308}$

$E_a = 76.66$

Now apply $k = A e^{-E_a/RT}$

$k_{318} = 3.56 \times 10^9 e^{-76.66 \times 10^3/(8.314 \times 318)}$
$= 9.1 \times 10^{-4}$ sec^{-1}.

19. Raising concentration of a reactant 4 fold increases the reaction rate 64 times. What is the kinetic order of this reaction.

Sol. $R \propto [A]^a$ --------- (i)

$64 \propto [4A]^a$ --------- (ii)

Dividing (ii) by (i)

we obtain $\frac{64 R}{R} = \frac{[4A]^a}{[A]^a}$

$[4]^3 = [4]^a$

$a = 3$ $\therefore$ third order reaction

20. At a certain temp. the $t_{0.5}$ of the decomposition of ammonia in contact with Tungsten were as follows:

pressure (mm of Hg)	50	100	200
$t_{0.5}$ (relative)	3.52	1.82	0.93

find the order of the reaction.

Sol. First method :

1st order $t_{0.5}$ independent of initial concentration, a

2nd order $t_{0.5} \propto \frac{1}{a}$ or $t \times a$ = constant

3rd order $t_{0.5} \propto \frac{1}{a^2}$ or $t \times a^2$ = constant

t x a	t x a²
3.52 x 50 = 176	3.52 x (50)² = 8800
100 x 1.82 = 182	1.82 x (100)² = 18200
200 x 0.93 = 186	0.93 x (200)² = 37200

$\because$ t x a is almost constant, the reaction is second order

Second method :

$\frac{t_1}{t_2} = (\frac{a_2}{a_1})^{n-1}$

$n = 1 + \frac{\log t_1 - \log t_2}{\log a_2 - \log a_1}$

I. $n = 1 + \frac{\log 3.52 - \log 1.82}{\log 100 - \log 50} = 1.95$

II. $n = 1 + \frac{\log 1.82 - \log 0.93}{\log 200 - \log 100} = 1.97$

Thus order of reaction is 2.

21. What will be the initial rate of a reaction if its rate constant is 10^{-3} min^{-1} and the concentration of the reactant is 0.2 mol dm^{-3} ? How much of the reactant will be converted into the products in 200 minutes ?

Sol. It is of a first order reaction (due to the units of k)

we know, Rate $= k \times$ [cocentration]

$= 10^{-3} \times 0.2 = 2.0 \times 10^{-4}$ mol dm^{-3} min^{-1}

The equation for first order, $k = \frac{2.303}{t} \log \frac{a}{(a-x)}$

$10^{-3} = \frac{2.303}{200} \log \frac{0.2}{(0.2-x)}$

solving, we get x = 0.036 mol dm^{-3}.

22. The rate constant for an isomerisation reaction $A \rightarrow B$ is 4.5×10^{-3} min^{-1}. If the initial concentration of A is 1M, calcualte the rate of reaction after 1 hour.

Sol. Rate = 4.5×10^{-3} min^{-1}

From the unit of the rate, the reaction is of first order

$\therefore k = \frac{2.303}{t} \log \frac{1}{1-x}$

Substituting the values in the equation,

$4.5 \times 10^{-3} = \frac{2.303}{60} \log \frac{1}{1-x}$

x = 0.237

concentration of B after 1 hour = 0.237 M

concentration of A = 1 – 0.237 = 0.763 M

we know, Rate = k [A]

$= 4.5 \times 10^{-3} \times [0.763]$

$= 3.43 \times 10^{-3}$ mole/ minute

Electrochemistry

1. 0.5N solution of a salt surrounding two platinum electrodes 0.02m apart and 4×10^{-4} m^2 in area was found to offer a resistance of 214 ohm. Calculate (i) conductance (ii) cell constant (iii) electrolytic conductivity and (iv) the equivalent conductivity of the solution.

Sol. R = 25 ohm

conductance $= \frac{1}{R} = \frac{1}{25} = 0.04$ S

cell constant $= \frac{1}{A} = \frac{0.02}{4 \times 10^{-4}} = 50$ m^{-1}

Electrolytic conductivity, k = cell constant x conductance

= 50 x 0.04 = 2.0 Sm^{-1} = 0.02 Scm^{-1}

Equivalent conductivity = $\frac{1000\ k}{N} = \frac{1000 \times 0.02}{0.5}$

= 40 S cm^2 eq^{-1}.

2. The resistance of a 0.01N solution of an electrolyte was found to be 210 ohm at 25°C. Calculate the equivalent conductance of the solution at 25°C. Cell constant is 0.88

Sol. we know $\Lambda_{eq} = \frac{1000 \times k}{N}$ or $\Lambda_{eq} = k \times v$

(v = volume of the solution containing 1gram equivalent of the substance)

Specific conductance $k = \frac{\text{cell constant}}{\text{resistance}} = \frac{0.88}{210}$ S cm^{-1}

$\Lambda_{eq} = \frac{1000 \times 0.88}{210 \times 0.01} = 419.05$ S cm^2 equiv.$^{-1}$(ohm^{-1}cm^2 eq^{-1})

3. The cell constant of a particular cell is 1.52 and it gives a resistance of 72.18 ohms for a decinormal solution of

$ZnSO_4$. Find the equivalent and molecular conductance of $ZnSO_4$ solution.

Sol. Specific conductance,

k = observed conductance $(\frac{1}{R})$ x cell constant

$= \frac{1}{72.18} \times 1.52 = 0.021058$ mhos

$\Lambda_{eq} = \frac{1000 \times k}{N} = \frac{1000 \times 0.021058}{0.10}$

$= 210.58$ mhos or s cm^2 $equiv^{-1}$.

$\Lambda_m = \frac{1000 \times k}{M} = \frac{1000 \times 0.021058}{0.05}$

$= 421.16$ mhos or s cm^2 mol^{-1}.

4. If the molar ionic conductances of Al^{3+} is 189 s cm^2 mol^{-1} and SO_4^{2-} is 160 s cm^2 mol^{-1}, calculate the molar conductance of $Al_2(SO_4)_3$.

Sol. $Al_2(SO_4)_3 \quad 2Al^{3+} + 3SO_4^{2-}$

$\Lambda^{\infty}_m = \lambda^{\infty} \nu + \lambda^{\infty} \nu$

$= (189 \times 2) + (160 \times 3) = 858$ s cm^2 mol^{-1}

5. The equivalent conductances of sodium acetate, hydrochloric acid and sodium chloride at infinite dilution are 91.0, 426.0 and 126.0 ohm^{-1} respectively at 25°C. Calculate the equivalent conductance at infinite dilution for acetic acid.

Sol. $\Lambda^{\infty}_m CH_3COONa = \lambda^{\infty}_{Na^+} + \lambda^{\infty}_{CH_3COO^-} = 91.0\ ohm^{-1}$

$\Lambda^{\infty}_m HCl = \lambda^{\infty}_{H^+} + \lambda^{\infty}_{Cl^-} = 426.0\ ohm^{-1}$

$\Lambda^{\infty}_m NaCl = \lambda^{\infty}_{Na^+} + \lambda^{\infty}_{Cl^-} = 126.0\ ohm^{-1}$

from the data we can write the equation as

$CH_3COO Na + HCl \rightarrow CH_3COOH + NaCl$

in other words, $CH_3COOH \rightarrow CH_3COONa + HCl - NaCl$

$\Lambda^{\infty}_m CH_3 COONa = 91.0 + 426.0 - 126.0 = 391 ohm^{-1}$.

6. The equivalent conductivity of a normal solution at 18°C is 98.2 and for infinite dilution at the same temperature was 131. Calculate the degree of ionisation of KCl at this dilution.

Sol. we know that $\alpha = \frac{\lambda_c}{\lambda_o} = \frac{98.2}{131} = 0.75$ or 75%

7. A current of 0.8 amphere was passed through a voltameter containing copper sulphate solution for 40 minutes. Calculate the amount of copper that would be deposited.

Sol.: we know, $W \propto Q$

$Q = It = 0.8 \times 2400 = 1920$ coulombs

$CuSO_4 \rightleftharpoons Cu^{2+} + SO_4^{2-}$

From the reaction at cathode $Cu^{2+} + 2e^- \rightarrow Cu$ (deposits) one can draw the conclusion that 2 electrons are involved per atom of copper ∴ Eq. weight $= \frac{\text{Atomic wt}}{2} = \frac{63.6}{2} = 31.8$

96,500 coulombs of electricity deposits 31.8 gm of Cu

1920 coulombs of electricity deposits $\frac{1920}{96,500} \times 31.8$

$= 0.6327$ g.

8. A current was passed through a solution of a salt of a metal 'A' and a solution of zinc sulphate connected in series. After a certain time 1.264 g. of zinc and 0.348 g. of 'A' were deposited. Calculate the equivalent weight of 'A' (Eq. wt. of Zn is 32.7)

Sol. we know that $\frac{W_1}{W_2} = \frac{E_1}{E_2}$

$\frac{\text{Eq. wt. of 'A'}}{\text{Eq. wt. of Zn}} = \frac{\text{wt. of x}}{\text{wt. of Zn}}$

substituting the values, $\frac{\text{Eq. wt. of 'A'}}{32.7} = \frac{0.348}{1.264}$

Eq. wt. of 'A' $= \frac{32.7 \times 0.348}{1.264} = 9.0$

9. Consider the reaction $2Ag^+ + Cd \rightarrow 2Ag + Cd^{2+}$ The standard electrode potentials for $Ag^+ \rightarrow Ag$ and $Cd^{2+} \rightarrow Cd$ couples are 0.80 and -0.40 volt, respectively. (i) what is the standard potential E° for this reaction ? (ii) for the electrochemical cell in which this reaction takes place, which electrode is -ve electrode ?

Sol. $2Ag^+ + 2e^- \rightarrow 2Ag; \quad E^o_{Ag^+/Ag} = 0.80$ volt

Reduction

$Cd \rightarrow Cd^{2+} + 2e^-; \quad E^o_{Cd^{2+}/Cd} = -0.40$ volt

Oxidation $\quad E^o_{Cd/Cd^{2+}} = +0.40$ volt

$E^o = E^o_{Cd/Cd^{2+}} + E^o_{Ag^+/Ag} = 0.40 + 0.80 = 1.20$ volts

(ii) the negative electrode is always the electrode whose reduction potential has smaller value or the electrode where oxidaton occurs. Thus Cd electrode is the negative electrode.

10. Calculate the e.m.f. of the cell, Mg(s) /Mg^{2+} (0.2M)/ / $Ag^+(1 \times 10^{-3}M)$/Ag. $E^o_{Ag^+/Ag} = 0.8$ volt; $E^o_{Mg^{2+}/Mg} = -2.37$ volt What will be the effect on e.m.f. if concentration of Mg^{2+} ion is decreased to 0.1M ?

Sol. $E^o_{cell} = E^o_{cathode} - E^o_{anode}$

$= 0.80 - (-2.37) = 3.17$ volts

cell reaction $\quad Mg + 2Ag^+ \rightarrow 2Ag + Mg^{2+}$

$E_{cell} = E^o_{cell} - \frac{0.0591}{n} \log \frac{[Mg^{2+}]}{[Ag^+]}$

$= 3.17 - \frac{0.0591}{2} \log \frac{0.2}{[1 \times 10^{-3}]^2}$

$= 3.17 - 0.1566 = 0.0134$ volts

$Mg^{2+} = 0.1$ M

$E_{cell} = E^o_{cell} = \frac{0.0591}{2} \log \frac{0.1}{(1 \times 10^{-3})^2}$

$= (3.17 - 0.1477)$ volt

$= 3.0223$ volt

11. Calculate the electrode potential at a copper electrode dipped in a 0.1M solution of copper sulphate at 25°C. The standard electrode potential of Cu^{2+}/Cu system is 0.34 volt at 298 K.

Sol. we know that $E_{red} = E^o_{red} + \frac{0.0591}{n} \log_{10} [\text{ion}]$

$= 0.34 + \frac{0.0591}{2} \log_{10} [0.1]$

$= 0.34 + 0.02955 \times (-1) = 0.34 - 0.02955 = 0.31045$ volt.

12. What is the single electrode potential of a half cell for Zinc electrode dipping in 0.01M $ZnSO_4$ solution at 25°C ? The standard electrode potential of Zn / Zn^{2+} system is 0.763 volt at 25°C.

Sol. we know that

$$E_{oxi} = E^o_{oxi} - \frac{0.0591}{n}\log_{10}[\text{ion}]$$

$$= 0.763 - \frac{0.0591}{2}\log_{10}[0.01]$$

$= 0.763 - 0.02955 \times (-2) = 0.763 + 0.0591 = 0.8221$ volt.

13. For the cell reaction $Mg / Mg^{2+}(aq) // Ag^+(aq) / Ag$, calculate the equilibrium constant at 25°C and maximum work that can be obtained by operating the cell.

$E^o_{Mg^{2+}/Mg} = -2.37$ volts and $E^o_{Ag^+/Ag} = +0.80$ volt

Sol. $E^o_{cell} = 0.80 + 2.37 = 3.17$

$$\log k_c = \frac{nE^o_{cell}}{0.0591} = \frac{2 \times 3.17}{0.0591} = 107.2758$$

$$k_c = 1.89 \times 10^{107}$$

$-\Delta G$ = maximum work $= nFE^o = 2 \times 96500 \times 3.17$

$= 611810$ J.

14. Calculate the equilibrium constant for the electrochemical reaction in the $Zn/Zn^{2+}(IM)//Ag^+(IM)/Ag$ at 25°C from the given data; $E^o(Zn^{2+}/Zn) = -0.765$ V and $E^o(Ag^+/Ag) = 0.80$ V.

Sol. we know $E = E^o \frac{2.303}{n F} R T \log_{10} k_{eq}$.

At equilibrium, $E = 0$

$$\therefore E^o = \frac{2.303}{n F} RT \log_{10} k_{eq} = \frac{0.0591}{n} RT \log_{10} k_{eq}.$$

$E^o_{cell} = E^o_R - E^o_L = 0.80 - (-0.765) = 1.565$V

Substituting the values

$$1.565 = \frac{0.0591}{2} RT \log_{10} k_{eq}$$

$$1.565 = 0.0295 \log_{10} k_{eq}.$$

$$\log_{10} k_{eq} = \frac{1.565}{0.0295} = 53.0508$$

k_{eq} = Antilog of $53.0508 = 1.124 \times 10^{53}$.

15. Calculate the free energy change taking place in hydrogen/oxgen fuel cell in which the following reactions take place :

(i) $O_2 + 4H^+ + 4e^- \rightarrow 2H_2O$ at the cathode; $E^o = 1.229$ V

(ii) $2H_2 \rightarrow 4H^+ + 4e^-$ at the anode

Sol. $E^o_{cell} = E^o_R - E^o_L = 1.229 - 0.000 = 1.229$ V

we know $\Delta G^o = - n F E^o$

Substituting the values $\Delta G^o = - 4 \times 96{,}500 \times 1.229$

$= - 474.4$ k J

16. Electrolysis of a solution of $MnSO_4$ in aqueous sulphuric acid is a method for the preparation of MnO_2 as per the reaction, $Mn^{2+}(aq) + H_2O \rightarrow MnO_2(s) + 2H^+(aq) + H_2(g)$. passing a current of 27A for 24 hours gives one Kg. of MnO_2. What is the value of current efficiency ? Write the reactions taking place at the cathode and at the anode.

sol. From the given reaction, we can conclude, Mn^{2+} is oxidised to MnO_2

$\therefore$ Anode reaction is, $Mn^{2+} + 2H_2O \rightarrow MnO_2 + 4H^+ + 22e^-$

cathode reaction is, $2H_2O + 2e^- \rightarrow H_2 + 2OH^-$

or we can also represent it as $2H^+ + 2e^- \rightarrow H_2$

The equivalent wt. of $MnO_2 = \frac{\text{mol. wt}}{2}$ (since 2 moles of electrons are involved)

$= 86.9/2 = 43.45$

Electrochemical equivalent, $z = \frac{\text{eq. wt}}{96{,}485} = \frac{43.45}{96{,}485} = 4.5 \times 10^{-4}$

we know, $w = I t z$

Amount of MnO_2 produced $= 27 \times 24 \times 60 \times 60 \times 4.5 \times 10^{-4}$

$= 1050$ g $= 1.059$kg.

current efficiency $= \frac{\text{Experimental yield}}{\text{Theoretical yield}} \times 100 = \frac{1.0}{1.05} \times 100$

$= 95.24\%$

17. A solution containing 4.5 m M of Cr_2O_7 and 15m M of Cr^{3+} shows a pH of 2.0. Calculate potential of half-reaction. Standard potential of the reaction, $Cr_2O_7^{2-} \rightarrow Cr^{3+}$ is 1.33 V.

sol. The half cell reaction is represented as,

$Cr_2O_7^{2-} + 14H^+ + 6e^- \rightleftharpoons 2Cr^{3+} + 7H_2O$

Nernst equation, $E = E^o - \frac{0.059}{6} \log \frac{[Cr^{3+}]^2}{[Cr_2O_7^-][H^+]^{14}}$

Substituting the values, $E = 1.33 - \frac{0.059}{6} \log \frac{(15 \times 10^{-3})^2}{(4.5 \times 10^{-3})(10^{-2})^{14}}$

$= 1.07$ V.

18. For the $Pb \rightarrow Pb^{+2}$ half – cell reaction, what would the oxidation potential be at 25°C if the Pb^{2+}ion concentration is 0.010 M ? $E^o_{Pb/Pb^{2+}} = 0.126$ V.

Sol. we know that, $E = Eo - \frac{0.059}{n} \log \frac{[Pb^{+2}]}{[Pb]}$

$= 0.126 - \frac{0.059}{2} \log \frac{[0.010]}{[1]}$

$= 0.126 - (0.0295)(-2.0)$

$= 0.126 + 0.059 = 0.185 = 0.19$V

19. Represent schematically the cell made up of the following half cell ractions. $Mg \rightarrow Mg^{2+}$ (0.10M) + $2e^-$, $E^o = 2.34$ volts. Sn^{2+} (0.10M) + $2e^-$ Sn, $E^o = -0.136$ volts. Calculate the potentials for each half cell and the total e.m.f. at 25°C for the cell.

Sol. $Mg/Mg^{2+}(0.010M)// Sn^{2+} (0.10)/Sn$

$EMF = E^o_{ox} + E^o_{red}$

$E_{ox} = E^o_{Mg/Mg^{2+}} - \frac{0.059}{2} \log \frac{[Mg^{+2}]}{[Mg]}$

$= 2.34 - \frac{0.059}{2} \log [0.010]$

$= 2.34 - 0.0295 \times -2.0 = 2.34 + 0.59 = 2.399 = 2.4$V

$Ered = - 0.136 \frac{0.059}{2} \log \frac{1}{[0.010]}$

$= - 0.136 - 0.0295(1) = 0.1655 = - 0.166$V

$E = 2.40 - 0.166 = 2.234 = 2.23$V

20. How many grams of silver could be plated out on a serving tray by electrolysis of a solution containg silver in +1 oxidation state for a period of 8.0 hours at a current of 8.46 amperes ? What is the area of the tray if the thickness of the silver plating is 0.00254 cm? Density of silver is 10.5 g/cm³.

Sol. Substituting the values in the equation, $w = M/nF \times It$,

we can know the weight of silver deposited,

$w = \frac{107.87}{1 \times 96500} \times 8.46 \times 8 \times 60 \times 60 = 272.356$ g.

volume of Ag = $\frac{272.356}{10.5} = 25.94$ $(\because V = M/D)$

The area of the tray = $\frac{\text{Volume of Ag}}{\text{Thickness of the layer}} = \frac{25.94}{0.00254}$

$= 1.02 \times 10^4$ cm^2.

21. The standard reduction potential for Cu^{2+}/Cu is +0.34V. Calculate the reduction potential at pH = 14 for the above couple; k_{sp} of $Cu(OH)_2$ is 1.0×10^{-19}.

Sol. $[H^+] = 10^{-14}$ $(\because pH = 14)$

$[OH^-] = 10^{-0} = 1$

$k_{sp} = [Cu^{2+}][OH^-]^2$

$1.0 \times 10^{-9} = [Cu^{2+}][1]^2$

$[Cu^{2+}] = 1 \times 10^{-9}$

cell reaction, $Cu^{2+} + 2e^- \rightarrow Cu$ at pH 14

substituting the values in, $E = E^\circ - \frac{0.059}{n} \log \frac{[\text{Products}]}{[\text{Reactants}]}$

$= 0.34 - \frac{0.059}{2} \log \frac{1}{1 \times 10^{-19}} = 0.22V$

22. A cell contains two hydrogen electrodes. The negative electrode is in contact with a solution of 10^{-6} M hydrogen ions. The EMF of the cell is 0.118 V at 25°C. Calculate the concentration of hydrogen ions at the positive electrode.

Sol. As both electrodes are made of same element, hydrogen it is a concentration cell.

substituting the values in the equation,

$E_{cell} = \frac{0.059}{n} \log \frac{C_1}{C_2}$

$0.118 = \frac{0.059}{1} \log \frac{C_{H^+}}{10^{-6}}$

$\log \frac{C_{H^+}}{10^{-6}} = \frac{0.118}{0.059} = 2$

Concentration of H^+ at positive electrode = 10^{-4} M.

23. The Edison storage cell is represented as

$Fe(s)/FeO(s)/KOH(aq)/Ni_2O_3(s)/Ni(s)$

The half cell reactions are

$Ni_2O_3(s) + H_2O(l) = 2e^- \rightleftharpoons 2NiO(s) + 2H_2O^-; E^\circ = +0.40V$

$FeO(s) + H_2O(l) + 2e^- \rightleftharpoons Fe(s) + 2OH^-; E^\circ = -0.87V$

(i) What is the cell reaction ?

(ii) What is the cell e.m.f. ? How does it depend on the concentration of KOH ?

(iii) What is the maximum amount of electrical energy that can be obtained from one mole of Ni_2O_3 ?

Sol. (i) Given data,

$E^\circ_{Ni_2O_3/NiO} = +0.40$ V; $E^\circ_{FeO/Fe} = -0.87V$

$E^\circ_{NiO/Ni_2O_3} = -0.40$ V; $E^\circ_{Fe/FeO} = +0.87V$

The oxidation potential of Fe/FeO is greater than the oxidation potential of for NiO/Ni_2O_3.

$\therefore$ We can write the cathode and anode reactions as,

At anode : $Fe(s) + 2OH^- \rightarrow FeO(s) + H_2O\,(l) + 2e^-$

At cathode : $Ni_2O_3(s) + H_2O\,(l) + 2e^- \rightarrow 2NiO(s) + 2OH^-$

cell reation : $Fe\,(s) + Ni_2O_3\,(s) \rightarrow FeO\,(s) + 2NiO\,(s)$

(ii) $E = E_R - E_L = 0.87 + 0.40 = 1.27$ V

It is independent of the concentration of KOH.

(iii) Electrical energy obtainable $= nFE^\circ_{cell}$

$= 2 \times 96500 \times 1.27 = 2.45 \times 10^5$ J.

24. A 100 watt. 110 volt incandescent lamp is connected in series with an electrolyte cell containing cadmium sulphate solution. What weight of cadmium will be deposited by the current flowing for 10 hours ?

Sol. we know, watt = volt x current

100 = 110 x current

$\therefore$ current = 100/110 = 0.91 ampere

substituting the values in the equation, $W = (M/nF) \times It$

amount of Cd deposited,

$W = \frac{112.41}{2 \times 96500} \times 0.91 \times 10 \times 60 \times 60 = 19.08$ g.

Colloids

(1) The coagulation of 100ml of a colloidal solution of gold is completely prevented by adding 0.25 g. of starch to it before adding 10 ml. of 10% NaCl solution. Calculate the gold number of starch.

Sol. : we know that gold number is the number of milligrams of a lyophilic colloid that is required to just prevent the precipitaion of 10 ml. of a gold sol on the addition of 1 ml. of 10% NaCl solution.

10 ml of 10% NaCl solution is added to 100 ml. gold sol. one can say 1 ml. of 10% NaCl solution is added to 10 ml. of gold sol.

100 ml. gold sol required 0.25g. starch = 0.25×10^3 mg.

10 ml. gold sol require = $\frac{10}{100} \times 0.25 \times 10^3 = 25$ mg. starch

$\therefore$ gold number of starch is 25.

(2) 5 ml. of 1M NaCl is required for the coagulation of 100 ml. of As_2S_3 sol. Calculate the coagulation power of NaCl.

Sol.: we know that the minimum concentration of an electrolyte in milli-moles per litre required to cause precipititaion is to be calculated.

Resultant volume after the addition of 5 ml. of 1M NaCl solution = 105

concentration in millimoles / litre $= \frac{5}{105} \times 1000 = 47.6$

(3) The coagulation of 10ml. of gold solution is completely prevented by addition of 0.025 g. of starch to it before adding 1ml. of 10% NaCl solution. Find out the gold number of starch.

Sol. : 10 ml of gold sol required 0.025 g. of starch = 0.025×10^3 mg.

$\therefore$ gold number of starch is 25.

Empirical & Molecular formulae-

The empirical formula of an organic compound is calculated as follows:

step1: The percentage of each element if not given is calculated from the following formulae-

1. $\% \text{ of H} = \frac{2}{18} \times \frac{\text{wt. of } H_2O \text{ formed}}{\text{wt. of organic compound}} \times 100$

2. $\% \text{ of C} = \frac{12}{44} \times \frac{\text{wt. of } CO_2 \text{ formed}}{\text{wt. of organic compound}} \times 100$

3. % of N_2 (Duma's method)

$$= \frac{28}{22{,}400} \times \frac{\text{Vol. of } N_2 \text{ at NTP}}{\text{wt. of org. compound}} \times 100$$

4. $\% \text{ of N} = \frac{1.4 \times \text{vol. of acid required} \times \text{normbility of acid}}{\text{wt. of organic compund}}$ (Kjeldahl's Method)

5. $\% \text{ of Cl} = \frac{35.5}{143.5} \times \frac{\text{Wt. of AgCl formed}}{\text{wt. of organic compound}} \times 100$

6. $\% \text{ of Br} = \frac{80}{188} \times \frac{\text{wt. of AgBr formed}}{\text{wt. of organic compound}} \times 100$

7. $\% \text{ of I} = \frac{127}{235} \times \frac{\text{wt. of AgI formed}}{\text{wt. of organic compound}} \times 100$

8. $\% \text{ of S} = \frac{32}{233} \times \frac{\text{Wt. of Ba } SO_4 \text{ formed}}{\text{wt. of organic compund}} \times 100$

9. % of O = 100 - sum of the percentages of all the elements present in the organic compound.

Step:2 : The relative number of atoms of each element present in the organic compound is found by dividing the percentage of the element by atomic mass.

Step:3 : The simplest ratio of the relative number of atoms of the elemetns is calculated.

Step:4 : If the simplest ratio of the atoms of the elements present in organic compounds obtained is in fraction (s), it is converted to get the whole number. This gives empirical formula.

Empirical Formula: Simplest ratio of the constituent elements of a molecule, e.g. $C_2H_4O_2$ and $C_6H_{12}O_6$ have the same empirical formula, CH_2O.

Some relationships for determining the molecular weights:

(i). Molecular formula = empirical formula x n

n = molecular formula wt./empirical formula wt.

(ii). Vapour density method :

molecular wt. = 2 x vapour density.

$$\text{Vapour density} = \frac{\text{wt. of 'x' c.c. of vapour at N.T.P.}}{\text{wt. of 'x' c.c.c hydrogen at NTP}}$$

(1CC of hydrogen at NTP weighs 0.00009)

(iii). Volumetric method for acids

a. 1 litre of IN solution of a substance contains one gram equivalent of the substance.

b. Weight of 22,400 cc vapour of any substance at NTP is equal to the molecular mass of the substance.

(iv). Silver salt method for acids :

$$\frac{\text{Equivalent wt. of silver salt}}{\text{Equivalent wt. of sivler}} = \frac{\text{wt. of silver salt}}{\text{wt. of silver}}$$

$$\frac{E\text{-}1+108}{108} = \frac{\text{wt. of silver slat}}{\text{wt. of silver}}$$

(v) Platinchloride method for bases :

The organic base with chloroplatinic acid (H_2PtCl_6) to form insoluble double salts is known as chloroplatinates (or platinchlorides). Chloroplatinates on ignition leave platinum as residue. The equivalent weight of the organic base can be calculated if the amounts of chloroplatinate and the residual platinum are known, the molecular weight of the base is then calculated by multiplying the equivalent weight with acidity of the base.

$$\text{mol. wt. of the base} = \frac{n}{2} \times \left(\frac{\text{wt. of platinchloride}}{\text{wt. of platinum}} \times 195\right) - 410$$

(n = acidity of the base)

(vi) Eudiometry :

The method is used to determine the molecular weights of gaseous hydrocarbons. The gaseous hydrocarbon is subjected to explosion (combustion) with excess of known volume of oxygen in the eudiometer. The reaction is represented as

$$C_xH_y(g) + (x + \frac{y}{4})\, O_2(g) \rightarrow xCo_2(g) + \frac{y}{2}\, H_2O(g)$$

The volume of CO_2 can be measured by absorbing it in KOH solution (which reduces the volume in eudiometer). The gas left behind is unused oxygen and its volume can be calculated by absorbing it in alkaline pyrogallol solution. Knowing the volume of the gaseous hydrocarbon taken, carbon dioxide formed and the oxygen left unused, one can calculate the molecular formula of the gaseous hydrocarbon.

(vii) The molecular weights of non volatile organic compounds can be calculated from their colligative properties as follows :

(a) Relative lowering of vapour pressure (Raoult's law)

$$\frac{p - p_s}{p} = \frac{n}{n + N} = \frac{\frac{w}{m}}{\frac{w}{m} + \frac{W}{M}}$$

Since w/m for a dilute solution is negligible, it is neglected in the denominator. For sufficiently dilute solutions,

$$\frac{p - p_s}{p} = \frac{w\,M}{m\,W}$$

(w and W are the weights of solute and solvent while m and M are the molecular weights of solute and solvent respectively)

(b) Elevation in the boiling point (Landsberger method)

$$m = \frac{1000\, k_b\, w}{\Delta T\, W}$$ (k_b = molal elevation constant or boiling point constant)

The value of k_b can be calculated from, $k_b = \frac{RT^2}{1000\, L_v}$

(L_v = heat of vapourisation)

When 1 mole of a solute that does not associate or dissociate is dissolved in 1kg. of the solvent, the boiling point elevation is called molal elevation constant.

$\Delta T_b = k_b \times m$

$\Delta T_b = k_b$ (when molality is one)

(c) By depression of freezing point method : (Beckmann method)

$$m = \frac{1000\ k_f\ w}{\Delta T_f\ W} \quad (k_f = \text{freezing point constant})$$

$$\Delta T_f = k_f .\ m$$

(d) By osmotic pressure method : (Berkeley & Hartley's method)

By osmotic pressure, one can calculate the molecular weight from the following relations.

$$PV = nRT \text{ or } PV = \frac{w_B}{m_B} RT \quad (w_B = \text{wt. of the solute},\ m_B = \text{molar mass})$$

$$P = \frac{n}{V} RT \quad = CRT$$

Isotonic solutions have the same osmotic pressure since they have the same molar concentration.

$$\frac{w_1}{m_1 v_1} = \frac{w_2}{m_2 v_2}$$

1. An organic compund contains 4% sulphur. Determine its minimum molecular weight.

Sol. for every 100 units of molecular wt. 4 units of S is presnt. Even if only one 'S' atom is present in the organic compound, its mass is 32.

∴ 32 units of mass of one S is present in 32/4 x 100=800

The minimum mol. wt. should be 800.

2. The empirical formula of a compound is CH_2. One molecule of this compound has a mass of 42g. Find out its molecular formula.

Sol. we know,

molecular formula=empirical formula x n

$$n = \frac{\text{molecular formula wt.}}{\text{empirical formulawt.}} = \frac{42}{14} = 3$$

molecular formula = CH_2 x 3 = C_3H_6

3. A compund (60g) on analysis gave, C= 24g; H = 4g; O = 32g. what is its empirical formula ?

Sol. Relative number of atoms of C = 24/12 = 2

Relative number of atoms of H = 4/1 = 4

Relative number of atoms of O = 32/16 = 2

Simplest ratio is 2 : 4 : 2 = 1 : 2 : 1

Empirical formula = CH_2O

4. A volatile organic compound contained 10%C, 0.84%, H and 89.12% Cl. In Victor Meyer's method, 0.6g. of the substance displaced 112c.c. of air at NTP. Find out the empirical and molecular formula of the compound.

Sol. Relative no. of C atoms = 10/12 = 0.83

Relative no. of H atoms = 0.84/1 = 0.84

Relative no. of Cl atoms = 89.12/35.5=2.51

Simplest ratio = 0.83/0.83; 0.84/0.83; 2.51/0.83

= 1 : 1 : 3

∴ The empirical formula = $CHCl_3$

112C.C. of air is displaced at NTP by 0.6g.

22,400 C.C. of air is displaced at NTP by $= \frac{22400}{112}$ x 0.6

= 120

∴ The molecular wt. of the compound = 120

n = molecular formula wt./empirical formula wt.

= 120/119.5=1

∴ molecular formula = empirical formula x 1= $CHCl_3$.

5. An organic monobasic acid has been analysed as follows:

(a) 0.1935g. gave 0.132g. of CO_2 and 0.0270g. of H_2O.

(b) It contains 55.04% of chlorine.

(c) 1.6125g of acid dissolved in H_2O neutralised 25 cc N/2-NaOH. Calculate the empirical and molecular formulae of the acid.

Sol.

$$\% \text{ of C} = \frac{12}{44} \times \frac{0.132}{0.1935} \times 100 = 18.60$$

$$\% \text{ of H} = \frac{2}{18} \times \frac{0.0270}{0.1935} \times 100 = 1.56$$

% of Oxygen=100-(18.60+1.56+ 55.04)=24.8

$$\text{Relative no. of atoms of C} = \frac{18.60}{12} = 1.55$$

$$\text{Realtive no. of atoms of H} = \frac{1.56}{1} = 1.56$$

$$\text{Realtive no. of atoms of Cl} = \frac{55.04}{35.5} = 1.55$$

$$\text{Realtive no. of atoms of O} = \frac{24.8}{16} = 1.55$$

Simplest ratio = 1 : 1 : 1 : 1

Therefore, the empirical formula = CHClO

25c.c. of N/2- NaOH ≡ 1.6125g. of the substance

$$1000 \text{ c.c.of N-NaOH} \equiv \frac{1000}{25} \times \frac{1}{0.5} \times 1.6125 = 129$$

Molecular wt. of the acid=129

$$n = \frac{\text{molecular wt.}}{\text{empirical formula wt.}} = \frac{129}{64.5} = 2$$

molecular formula= empirical formula x 2

= CHClO x 2 = $C_2H_2O_2Cl_2$

6. 0.225g. of an organic nitrogenous compound on being digested with conc. H_2SO_4 followed by distillation with NaOH evolved ammonia which was absorbed in 50c.c. of N/10-H_2SO_4. After absorption, the excess of acid required 10cc of N/5 NaOH. Calculate the percentage of N in the organic compound.

sol. we know that the estimation of N is by Kjeldahl method.

$$\% \text{ of N} = \frac{1.4 \times \text{vol. of acid} \times \text{normality of acid}}{\text{weight of organic compound}}$$

we have to find out the vol. of H_2SO_4 neutralised by ammonia evolved.

50c.c. of N/10-H_2SO_4 was taken

10c.c. of N/5-H_2SO_4 was left (∵10c.c. of N/5-NaOH≡ 10c.c. of N/5-H_2SO_4). Here the two solutions have different normality and therefore they should be equated with the same normality. Let us convert 50cc of N/10-H_2SO_4 to N/5-H_2SO_4

$N_1V_1 = N_2V_2 = 50 \times 0.1 = 0.2 \times V_2$

$V_2 = 50 \times 0.1/0.2 = 25cc$

Therefore 15c.c of N/5-H_2SO_4 is neutralised by ammonia evolved.

$$\% \text{ of N} = \frac{1.4 \times 15 \times 0.2}{0.225} = 18.67$$

7. 0.76g. of a silver salt of a diabsic acid on ignition gave 0.54g. of silver.Calculate the molecular weight of the acid

sol. First we calculate the equivalent wt. of the acid from,

$$\frac{\text{Eq. wt. of silver salt}}{\text{Eq. wt. of silver}} = \frac{\text{wt. of silver salt}}{\text{wt. of silver}}$$

$$\frac{E - 1 + 108}{108} = \frac{0.76}{0.54}$$

$$\frac{E + 107}{108} = \frac{0.76}{0.54}$$

$$E = \left[\frac{0.76 \times 108}{0.54}\right] - 107 = 45$$

Molecular wt. of the acid = Eq. wt. x basicity

= 45 x 2 = 90

8. In a Carius determination, 0.2595g. of an organic compound gave 0.35g. $BaSO_4$.Calculate the percentage of sulphur.

$$\text{Sol. \% of S} = \frac{\text{wt. of sulphur}}{\text{wt. of } BaSO_4} \times \frac{\text{wt. of } BaSO_4 \text{ formed}}{\text{wt. of org. compound}} \times 100$$

$$= \frac{32}{233} \times \frac{0.35}{0.2595} \times 100 = 18.52$$

9. An organic compound of vapour density 23 contains 52% C and 13.04% %H. On reaction with conc. H_2SO_4 at 140°C, it gives a compound, B. B when reacted with cold HI gives back A and another compound C which also gives A on hydrolysis. Write the structures of A, B and C.

Sol. Relative no. of C atoms = 52/12=4.33

Relative no. of H atome = 13.04/1=13.04

% of oxygen=100 - (52 + 13.04) = 34.96

Relative no. of O atoms = 34.96/16=2.18

$$\text{Simplest ratio} = \frac{4.33}{2.18} : \frac{13.04}{2.18} : \frac{2.18}{2.18} = 2:6:1$$

Empirical formula = C_2H_6O

molecular wt. = Vapour density x 2

= 23 x 2 = 46

∴ Molecular formula = C_2H_6O (since, mol.wt.is 46)

The compound is CH_3CH_2OH

$2CH_3CH_2OH + \text{conc.}H_2SO_4 \rightarrow CH_3CH_2OCH_2CH_3 + H_2O$

$H_3CCH_2OCH_2CH_3 + HI \rightarrow CH_3CH_2OH + CH_3CH_2I$

$CH_3CH_2I + H_2O \rightarrow CH_3CH_2OH + HI$

A = C_2H_5OH; B = $C_2H_5OC_2H_5$and C = C_2H_5I

10. Two compounds A and B have the same empirical formula CH. A is a gaseous, aliphatic compound with molecular weight 26, while B is a liquid aromatic comound with molecular wegiht 78. Both of them burn with a sooty flame. (i) Give the names and structures (ii) what type of reactions are generally shown by them.

Sol. For compound A,

$$n = \frac{\text{Molecular wt.}}{\text{empirical formula wt.}} = \frac{26}{13} = 2$$

Molecular fromula of A is = CH x 2 = C_2H_2

Compund A is ethyne.

$$\text{For compound B, } n = \frac{\text{Molecular wt.}}{\text{empirical formulawt}} = \frac{78}{13} = 6$$

Molecular formula of compound is B = C_6H_6

B is aromatic and therefore it is benzene. Structure of 'A' is H-C≡C-H and 'B' is ⌬

Compound 'A' shows mainly addition reactions

Compound 'B' shows mainly substitution reactions as the π electrons are delocalised and therefore are not available for addition reactions under normal conditions.

11. An organic compound 'A' ($C_4H_{10}O$) is optically active. On mild oxidation it gives a compound 'B' (C_4H_8O), but on vigorous oxidation, it gives another compound C($C_2H_4O_2$). Write the structure of A,B and C.

Sol. 'A' should be an alcohol and it can be $H_3C\text{-}CH_2\text{-}CH_2\text{-}CH_2OH$ or $H_3C\text{-}CH_2\text{-}CHOH\text{-}CH_3$.

'A' is 2-butanol as it is optically active (1-butanol is not optically active).

$$CH_3CH_2CHOHCH_3 \xrightarrow[\text{oxidation}]{\text{mild}} CH_3CH_2COCH_3(C_4H_8O)$$

'B' 2-butanone

$$CH_3CH_2CHOHCH_3 \xrightarrow[\text{oxidation}]{\text{vigorous}} 2CH_3COOH$$

'C' aceti acid

12. 0.40g. of the platinum chloride of an organic monoacid base gas on ignition 0.125g. of platinum. Calculate the molecular weight of the base.

Sol. Let B be the equivalent weight of the base

Then the mol. wt.of platinum chloride = 2B + 410

(since the molecular formula of chloroplatinate is $B_2H_2PtCl_6$ as the acid H_2PtCl_6 is dibasic).

0.125g. of platinum is obtained from 0.40g. of platinum chloride.

$$\text{195g. of platinum is obtained from} = \frac{0.40}{0.125} \times 195 = \mathbf{624g.}$$

2B + 410 = 624

2B = 624 - 410 = 214

B = 214/2 = 107

Mol. wt. of the base = Eq.wt. x acidity =107 x 1=107

13. 30 ml. of a gaseous hydrocarbon were exploded with 210 ml. of oxygen. After cooling, the residual gas occupied 150 ml. On treatment with strong solution of caustic potash, the volume was further reduced to 60ml. Determine the molecular formula of the hydrocarbon. All values were recorded under the same conditions of temeprature and pressure.

sol. Let the molecular formula of the hydrocarbon be C_xH_x.

we know, $C_xH_y + (x+y/4)\ O_2 \rightarrow xCO_2 + y/2H_2O$
30 ml 309(x+y/4) 30 ml
volume of the hydrocarbon = 30ml.
volume of CO_2 produced= 150 - 60 = 90ml. (since CO_2 is absorbed by strong KOH solution)
volume of oxygen used = 210 - 60 = 150ml.
volume of CO_2 produced = 30 x ml.($\because$ the vol. of hydrocarbon = 30ml)
From the experiment, the vol. of CO_2 produced = 90ml.
30 x = 90; x = 90/30 = 3
The amount of oxygen consumed from the equation should be, 30 (x+y/4) ml
volume of oxygen as per experimental data = 150ml.
30 (x+y/4) = 150
30 (3+y/4) = 150
30y/4 = 150 - 90 or 36y = 60 x 4 = 240
y = 240/30 =8
The molecular formula is C_3H_8.

14. 15ml of a gaseous hydrocarbon was exploded with excess of oxygen. On cooling, the reduction in volume was 45ml. If the vapour density of the hydrocarbon be 22, find the molecular formula of it.

Sol. We know that the combustion of the hydrocarbon is represented as
$C_xH_y + (x+y/4)O_2 \rightarrow xCO_2+y/2H_2O$
15 ml 15 (x+y/4)ml 15x ml
The anount of oxygen consumed from the equation should be
= 15 + 15 (x + y/4) - 15x = 15 (1 +y/4)
From experimental data, the reduction in volume = 45ml.
$\therefore$ 15 (1 +y/4) = 45
y = 8
The molecular formula of the hydrocarbon = C_xH_8
molecular wt = vapourdensity x 2
= 22 x 2 = 44
Mol. wt. accoding to $C_xH_8 = C_x+8$
xc + 8 = 44 = 44 - 8 = 36
x = 36/12 = 3
$\therefore$ The molecular formula of the hydrocarbon = C_3H_8.

15. 15ml. of a gaseous hydrocarbon was exploded with oxygen. After explosion, there was a contraction of 30 ml.in volume. On shaking the residual gaseous mixture with KOH, there was a further contraction of 30ml. in volume. Calculate the molecular formula of the hydrocarbon. All volumes were recorded under the same condtions of temperature and pressure.

Sol. Combustion of the hydrocarbon is represented as,
$C_xH_y + (x+y/4)O_2 \rightarrow xCO_2 + y/2H_2O$
15ml. 15 (x+y/4ml.) 15xml.
Volume of hydrocarbon = 15ml.
Volume of CO_2 produced = 30ml.
The first reduction in volume after first explosion
= 15 + 15 (x+y/4) -15x = 15 + 15y/4ml.
From the data of the problem, it is 30ml.
15 + 15y/4 = 30
15y/4 = 30 - 15 = 15
15y = 4x15 = 60
y = 60/15 = 4
similarly, 15ml of CO_2 produced from the equation is given as 30ml in the problem
$\therefore$ 15x = 30 or x = 30/15 = 2
molecular formula = C_2H_4.

16. A compound 'A' gave the following results upon combustion,
(i) 0.118g. of 'A'yielded 0.264g. CO_2 and O. 162g. H_2O
(ii) 0.1475g. of 'A' produced 28cc of nitrogen at N T P (or STP) by Dumas method.
(iii) 'A' is a monoacidic base whose platinchloride contains 36.5% pt.
Findout the empirical and molecular of the compound 'A'.

Sol. $\% \text{ of C} = \frac{12}{44} \times \frac{0.264}{0.118} \times 100 = 61.02$

$\% \text{ of H} = \frac{2}{18} \times \frac{0.162}{0.118} \times 100 = 15.25$

$\% \text{ of } N_2 = \frac{28}{22400} \times \frac{28}{0.1475} \times 100 = 23.72$

Relative no. of atoms of C = 61.02/12 = 5.08

Relative no. of atoms of H = 15.25/1=15.25

Relative no. of N atoms = 23.72/14=1.69

SImplest ratio = 5.08/1.69 : 15.25/1.69 : 1.69/1.69 = 3 : 9 : 1

The empirical formula = C_3H_9N

molecular weight of the base can be calculated from

$= \frac{n}{2}\left(\frac{\text{wt. of platin chloride}}{\text{wt. of platinum}} \times 195\right) - 410$

$= \frac{1}{2}\left[\left(\frac{100}{36.5} \times 195\right) - 410\right] = 62.12$

n = mol.wt./eq.wt. = 62.12/59 = 1.05
Molecular formula = Empirical formula x n
= C_3H_9N x 1 = C_3H_9N

17. When 0.0088g. of compound (A) was dissolved in 0.50g. of camphor, the melting point of camphor was lowered by 8^{0}c. Analysis of (A)gave 68.18% carbon and 13.61% hydrogen. Compound (A) showed the following reactions:
(i) It reacted with acetyl chloride and evolved hydrogen with sodium.
(ii) when reacted with HCl and $ZnCl_2$, a dense oily layer separated out immediately. Compound (A) was passed over Al_2O_3 at 350^{0}c to give compound (B) which on ozonolysis followed by hydrolysis gave two neutral compounds (C) and (D) which gave positive test with carbonly reagents but only (C) gave a positive test with Fehling's solution and resinous substance with NaOH.

Identity (A), (B), (C) and (D) with proper reasoning (K_f for camphor = 40K mole^{-1}kg.)

Sol. Relative no. of C atoms = 68.18/12 = 5.68

Relative no. of H atoms = 13.61/1 = 13.61

% of xygen = 100 - (68.18+13.61) = 18.21

Relative no. of C-atoms = 18.21/16 = 1.14

Ratio of C:H:O = 5.68/1.14 : 13.61/1.14 : 1.14/1.14

= 5 : 12 : 1

The empirical formula of (A) = $C_5H_{12}O$

Now, we can calculate the molecular wt. of the compound (A) from the equation,

$$m = 1000K_f w / \Delta T_f w$$

$$= \frac{1000 \times 40 \times 0.0088}{8 \times 0.50} = 88$$

$$n = \frac{\text{mol. wt.}}{\text{eq. wt.}} = \frac{88}{88} = 1$$

$\therefore$ molecular formula of (A) = $C_5H_{12}O$

Compound (A) having one oxygen atom reacts with Na to evolve hydrogen

$\therefore$ (A) should be an alcohol

As it reacts with Lucas reagent immediately, it is a tertiary alcohol

$$(A) \xrightarrow{Al_2O_3;\ 350^0c} (B) \xrightarrow{\text{ozonolysis}} \underset{\text{Carbonyl Compounds}}{(C) + (D)}$$

(C) Gave positive test with Fehling's solution, therefore, it is an aldehyde and the aldehyde should not be formaldehyde.

(D) Should be a Ketone

(A) is converted to (B) through dehydration

From the molecular formula, (A) can have the structure

$$\underset{(A)}{H_3C\text{-}CH_2\text{-}\underset{OH}{\overset{CH_3}{C}}\text{-}CH_3} \xrightarrow{Al_2O_3;\ 350^0C} \underset{(B)}{H_3C - CH{=}\overset{CH_3}{C}\text{-}CH_3}$$

$\xrightarrow{\text{ozonolysis}}$ CH_3CHO (C) + CH_3COCH_3(D)

Compound (A) reacts with acetyl chloride and evolves hydrogen with sodium as follows:

$$H_3C\text{-}CH_2\text{-}\underset{OCOCH_3}{\overset{CH_3}{C}}\text{-}CH_3 \xleftarrow{CH_3COCl} H_3C - CH_2\text{-}\underset{OH}{\overset{CH_3}{C}}\text{-}CH_3 \xrightarrow{Na} H_3C\text{-}CH_2\text{-}\underset{O^-Na^+}{\overset{CH_3}{C}}\text{-}CH_3$$

18. 0.45g. of an aromatic organic compound (A) on ignition gives 0.905g. CO_2 and 0.185g. H_2O. 0.305g. of (A) on boiling with HNO_3 and adding $AgNO_3$ solution gives 0.50g. of AgCl. The vapour density of (A) is 87.5. (A) On hydrolysis with $Ca(OH)_2$ yields (B) which on mild reduction gives an optically active compound (C). On heating (C) with I_2 and NaOH, iodoform is produced along with (D). With HCl (D) gives a solid which is markedly more soluble in hot water than in cold water.

Sol. $\% \text{ of C} = \frac{12}{44} \times \frac{0.905}{0.45} \times 100 = 54.85$

$$\% \text{ of H} = \frac{2}{18} \times \frac{0.185}{0.45} \times 100 = 4.57$$

$$\% \text{ of Cl} = \frac{35.5}{143.5} \times \frac{0.50}{0.305} \times 100 = 40.56$$

mol. wt. of (A) = V.P. x 2 = 87.5 x 2 = 175

$$\text{No. of C atoms} = \frac{\% \text{ of C}}{100} \times \frac{\text{Mol. wt.}}{\text{atomic. wt.}}$$

$$= \frac{54.85}{100} \times \frac{175}{12} = 8$$

$$\text{No. of H atoms} = \frac{4.57}{100} \times \frac{175}{1} = 8$$

$$\text{No. of Cl atoms} = \frac{40.56}{100} \times \frac{175}{35.5} = 2$$

Molecular formula of (A) is $C_8H_8Cl_2$

The reaction scheme as per the problem can be shown as,

$$(A) \xrightarrow{Ca(OH)_2} (B) \xrightarrow{\text{mild reduction}} (C) \xrightarrow{I_2} (D) + CHI_3 \xrightarrow{HCl} \text{Solid}$$

(A) is an organic compound and therefore it can be represented as Ph-$C_2H_3Cl_2$ now the nature of the side chain has to be decided.

(A) on hydrolysis gives (B) which gives iodoform test. Therefore, there shoulbe be a -$COCH_3$ group in the side chain. This is possible only when (A) is a gem dihalide. The reactions can be represented as,

$$\underset{(A)}{Ph\text{-}\underset{Cl}{\overset{Cl}{C}}\text{-}CH_3} \xrightarrow[\text{hydrolysis}]{Ca\ (OH)_2} Ph\text{-}\underset{OH}{\overset{OH}{C}}\text{-}CH_3 \xrightarrow{-H_2O} \underset{(B)}{Ph\text{-}\overset{O}{\overset{\|}{C}}\text{-}CH_3}$$

$$Ph\text{-}COCH_3 \xrightarrow[-CHI_3]{I_2+NaOH} \underset{(D)}{Ph\text{-}COO^-Na^+} \xrightarrow{HCl} \underset{(Solid)}{Ph\text{-}COCH_3 + NaCl}$$

The solubility of benzoic acid in hot water is more than in cold.

(B) On mild reduction gives optically active compound.

$$Ph\text{-}COCH_3 \xrightarrow[\text{reduction}]{\text{mild}} Ph\text{-}\underset{OH}{\overset{H}{C}}\text{-}CH_3 \text{(optically active)}$$

19. An organic compound (A) $C_4H_{10}O_2$ weighing 0.90g. reacts with sodium to give 224ml. hydrogen at NTP. The compound (A) can be separated into two parts (B) and (C) by crystallisation. Only (B) can be resloved into optical isomers (D) and (E). Write the structural formulae of (C), (D) and (E) giving proper reasoning.

Sol. Molecular wt. of (A) is calculated as follows

224ml. of H_2 is obtained from 0.90g of compound (A)

22400ml. of H_2 is obtaine from =22400/224x0.90=90

Compound (A) gives hydrogen with Na, therefore it must be an alchohol. Two oxygen atoms account for two OH groups. The two OH groups should be attached to different C-atoms, otherwise it will not be stable. Secondly B is optically active. Thereore, the two structures possible for (A) are

```
H3C-CH-CH-CH3      H3C-CH2-CH-CH2OH
    |   |                  |
    OH OH                  OH
     (B)                  (C)
```

The two isomers that can be obtained from(B) on resolution are

```
     CH3              CH3
      |                |
  H - C - OH      OH - C - H
      |                |
  HO-C - H         H-C - OH
      |                |
     CH3              CH3
     (D)              (E)
```

Both are optically active

20. Compound (x) containing chlorine, on treatment with strong ammonia gives a solid (y) which is free from chlorine. 'Y' analysed as C=49.31%, H=9.59% and N=19.18% and reacts with Br_2 and caustic soda to give a basic compound Z. 'Z' reacts with HNO_2 to give ethanol. Suggest structures for (X), (Y) and (Z).

Sol. Relative no. of C - atoms $= \frac{49.31}{12} = 4.11$

Relative no. of H - atoms $= \frac{9.59}{1} = 9.59$

Relative no. of N - atoms $= \frac{19.18}{14} = 1.37$

% of oxygen = 100 - (49.31 + 9.59 + 19.18) = 21.92

Relative no. of O - atoms $= \frac{21.92}{16} = 1.37$

The simplest ratio of , C : H : N : O

$$= \frac{4.11}{1.37} : \frac{9.59}{1.37} : \frac{1.37}{1.37} : \frac{1.37}{1.37}$$

The empirical formula of (Y) is C_3H_7NO

The reaction schem of the problem can be represented as,

$$(X) \xrightarrow{NH_3} (Y) \xrightarrow{NaOH/Br_2} (Z) \xrightarrow{HNO_2} C_2H_5OH$$

(Z) must be ethylamine; (Y) must be propanamide as ethylamine is formed by Hoffmann degradation reaction. (X) must be propanoyl chloride as it contains chlorine and is converted to propanamide on treatment with strong NH_3.

$$\underset{(X)}{H_3C-CH_2-COCl} + H_3N \xrightarrow{-HCl} \underset{(Y)}{H_3C-CH_2-CONH_2} \xrightarrow[Br_2]{NaOH}$$

$$H_3C-CH_2NH_2 \xrightarrow[-(N_2+H_2O)]{HNO_2} H_3C-CH_2OH$$

21. A basic volatile, nitrogen compound gave a foul smelling gas when treated with chloroform and alcoholic potash. A 0. 295 g. sample of the substance dissolved in aq. HCl and treated with $NaNO_2$ solution at 0°C, liberated a colourless, odourless gas whose volume corresponded to 112 ml at STP. After the evolution of the gas was complete, the aqueous solution was distilled to give an organic liquid which did not contain nitrogen and which on warming with alkali and iodine gave a yellow precipitate. Identify the original substance, assuming that it contains only one nitrogen atom.

Sol. From the problem, we understand that it is a carbylamine reaction,

$$RNH_2 + CHCl_3 + \text{alc. } KOH \rightarrow \underset{\text{foul smelling}}{RNC} + KCl + 2KCl$$

The compound under scrutiny is a primary amine.

It's reaction with HNO_2 can be represented as,

$$RNH_2 \xrightarrow{NaNO_2/HCl} ROH + N_2 + H_2O$$

For each mole of the amine, one mole of N_2 is given we can calculate the number of moles of amine from,

$$PV = nRT$$

$$n = \frac{PV}{RT} = \frac{1 \times 0.112}{0.0821 \times 273} = 0.00499$$

$$\text{mol. wt. of the amine} = \frac{\text{weight}}{\text{no. of moles}} = \frac{0.295}{0.00499} = 59$$

since the amine is a primary one and has mol. wt. 59, the following structures are probale.

$H_3C - CH_2 - CH_2 NH_2$ or $H_3C - CH(CH_3) NH_2$.

As the alcohol obtained from the amine on action with HNO_2 shows positive iodoform test, it should be, $H_3C - CH(CH_3) NH_2$.

Now, the reactions can be represented as,

$H_3C - CH(CH_3)NH_2 + CHCl_3 + \text{alc. } KHO \rightarrow H_3C - CH(CH_3)NC$

$$H_3C - CH(CH_3)NH_2 \xrightarrow{NaNO_2/HCl} \underset{\text{propan-2-ol}}{H_3C - CH(CH_3)OH}$$

Propan-2-ol gives iodoform test as it can be easily oxidised to $H_3C - CO - CH_3$ which contains $-COCH_3$ group.

22. An acid (A) contains carbon 40.7%, hydrogen 5.1% and its silver salt contains 65.1% silver. The ethyl ester of (A) has vapour density of 87(H_2 = 1). What structure (A) may have ? How would you distinguish between isomers ?

Sol. Relative no. of C - atoms $= \frac{40.7}{12} = 3.39$

Relative no. of H - atoms $= \frac{5.1}{1} = 5.1$

% of oxygen = 100 - (40.7 + 5.1) = 54.2

Relative no. of O -atoms $= \frac{54.2}{16} = 3.39$

Ratio of C : H : O $= \frac{3.39}{3.39} : \frac{5.1}{3.39} : \frac{3.39}{3.39}$

= 1 : 1.50 : 1

simplest ratio in whole numbers = 2 : 3 : 2

∴ The empirical formula is $C_2H_3O_2$

Equivalent wt. of the acid can be calculated from silver

salt method,

$$\frac{E+107}{108} = \frac{100}{65.1}$$

$$E = \frac{100}{65.1} \times 108 - 107 = 58.90$$

molecular wt. of ethyl ester of the acid = 2 x 87 = 174

molecular formula of the acid = $(C_2H_3O_2)_n$

∴ The expected formula of the acid = $(C_2H_3O_2) = C_4H_6O_4$

The formula of the ethyl acetate = $C_4H_4O_4\ (C_2H_5)_2$

the structure of acdi (A) can be,

$$\begin{array}{l} H_2C - COOH \\ \quad | \\ H_2C - COOH \end{array} \text{ or } H_3C - CH\begin{array}{l} COOH \\ COOH \end{array}$$

These isomers can be distinguished by heating

$$\begin{array}{l} H_2C - COOH \\ \quad | \\ H_2C - COOH \end{array} \xrightarrow[-H_2O]{\Delta} \begin{array}{l} H_2C - CO \\ \qquad\quad\ \ > O \\ H_2C - CO \end{array}$$

succinic acid → succinic anhydride

$$H_3C - CH\begin{array}{l} COOH \\ COOH \end{array} \xrightarrow[-H_2O]{\Delta} H_3C - CH_2 - COOH$$

Propanoic acid

23. An organic compound containing C, H and O exists in two isomeric forms (A) and (B). An amount of 0.108 g. of the isomers gives on combustion 0.308 g. CO_2 and 0.072 g. H_2O. (A) is insoluble in NaOH and $NaHCO_3$ while (B) is soluble in NaOH. (A) reacts with conc. HI to give (C) and (D). (C) can be separated from (D) by ethanolic $AgNO_3$ solution and (D) is soluble in NaOH. (B) reacts readily with bromine water to give compound (E) of molecular formula. $C_7H_5OBr_3$. Identify A, B, C, D and E with justification and give their structures.

Sol. $\% \text{ of C} = \frac{12}{44} \times \frac{0.308}{0.108} \times 100 = 77.78$

$\% \text{ of H} = \frac{2}{18} \times \frac{0.072}{0.108} \times 100 = 7.41$

% of oxygen = 100 - (77.75 + 7.41) = 14.81

Relative no. of C - atoms = $\frac{77.78}{12}$ = 6.48

Relative no. of H - atoms = $\frac{7.41}{1}$ = 7.41

Relative no. of O -atoms = $\frac{14.81}{16}$ = 0.93

simplest Ratio of C : H : O = $\frac{6.48}{0.93} : \frac{7.41}{0.93} : \frac{0.93}{0.93}$

= 7: 8 : 1

∴ The empirical formula is C_7H_8O

compound (A) is insoluble in NaOH and Na HCO_3. This means that there is no free - OH group in it. Therfore we can say a - OCH_3 group is present and the structure can be $C_6H_5OCH_3$ (A). (B) is soluble is NaOH. It means it contains a free -OH gruoup and (B) is an isomer of (A). Therfore the structure of (B) is $CH_3C_6H_4OH$. The -CH_3 group should be present at meta position as the compound (B) gives a tribromoderivative of molecular formula, $C_7H_5OBr_3$.

$$\underset{(B)}{m\text{-}CH_3C_6H_4OH} + 3Br_2 \longrightarrow \text{2,4,6-tribromo-3-methylphenol}$$

$$\underset{(A)}{C_6H_5OCH_3} + HI \longrightarrow \underset{(C)}{C_6H_5OH} + \underset{(D)}{CH_3I}$$

(C) and (D) can be separated by ethanolic $AgNO_3$ solution.

24. An aromatic compound contains 69.4% C and 5.8% H. A sample of 0.303 g of this compound was analysed for nitrogen by kjeldahl's method. The ammonia evolved was absorbed in 50 ml. of 0.05M sulphuric acid. The excess of acid required 25ml of 0.1M sodium hydroxide for neutralisation. Determine the molecular formula of the compound if its molecular weight of 121. Draw two possible structures for this compound.

Sol. %N from the kjeldahl method can be calculated from,

$$\% N = \frac{1.4 \times \text{vol. of acid required} \times \text{normality of acid}}{\text{wt. of organic compound}}$$

50 ml of 0.1N H_2SO_4 was taken ($N = \frac{\text{molarity x mol. wt}}{\text{eq. wt.}}$)

25 ml of 0.1 N H_2SO_4 was left (∵25 ml of 0.1N NaOH = 25 ml of 0.1 N H_2SO_4)

25 ml of 0.1 N H_2SO_4 was used to neutralise NH_3 formed.

$$\% N = \frac{1.4 \times 25 \times 1.00}{0.303} = 11.55$$

Relative no. of C - atoms = $\frac{69.4}{12}$ = 5.78

Relative no. of H - atoms = $\frac{5.8}{1}$ = 5.8

Relative no. of N -atoms = $\frac{11.5}{14}$ = 0.83

% of oxygen = 100 - (69.4 +5.8 + 11.55) = 13.25

Relative no. of O -atoms = $\frac{13.25}{16}$ = 0.83

simplest Ratio of C : H : O = $\frac{5.78}{0.83} : \frac{5.16}{0.83} : \frac{0.83}{0.83} : \frac{0.83}{0.83}$

= 7: 7 : 1 : 1

∴ The empirical formula is C_7H_7NO

$$n = \frac{\text{molecular formula weight}}{\text{empirical formula weight}} = \frac{121}{121} = 1$$

∴ Molecular formula of the compound is , C_7H_7NO.

As the compound is aromatic, it is, Ph-CH = NOH (benzaldoxime). The two geometrical isomers of the compound are

$$\begin{array}{c} Ph - C - H \\ \| \\ N - OH \end{array} \qquad \begin{array}{c} Ph - C - H \\ \| \\ HO - N \end{array}$$

syn-benzaldoxime anti - benzaldoxime

Volumetry

Acid-Base titrations:

Typical acid-base titration uses the neutralisation reaction between oxonium ions and hydroxide ions

$H_3O^+ (aq.) + OH^- (aq.) \rightleftharpoons 2H_2O (l)$

Phenolphthalein: It is a colourless weak acid.

HPh $\rightleftharpoons$ $H^+ + Ph^-$. The indicator is colourless in acid and pink in alkali solution. It can be used between 8.00 - 9.8 pH range.

Methyl orange:It is a weak base: MeOH $\rightleftharpoons$ $Me^+ + OH^-$

It remains yellow in alkaline medium while in acid solution, it gives red colour

Titration	Indicator used
SA Vs S B	Phenolphthalein, methyl red, methyl orange
W A Vs S B	Phenolphthalein,
S A Vs W B	Methyl red, methyl orange
Na_2CO_3 Vs S A	Phenolphthalein&methyl orange

Redox titrations

(i) Titration of iorn (II) with $K_2Cr_2O_7$ (VI)

$6Fe^{2+}$ (aq) + $Cr_2O_7^{2-}$ (aq)+ $14H^+$ (aq)$\rightarrow$ $6Fe^{3+}$ (aq) + $2Cr^{3+}$+ (aq) + $7H_2O$ (l)

(ii) Titration of oxalate ions with potassium permanganate (VII)

$5C_2O_4^{2-}$ (aq) + $2MnO_4^-$ (aq) + $8H^+$ (aq) $\rightarrow$ $10CO_2$ (g) + $2Mn^{2+}$ (aq) + $4H_2O$ (l)

Titrations involving iodine:

(i) Iodine ions are easily oxidised

$2I^-$ (aq) - $2e^- \rightarrow I_2$ (s)

e.g. by iodate (V) ions

IO_3^- (aq) + $5I^-$ (aq) + $6H^+$ (aq) $\rightarrow$ $3I_2$ (s) + $3H_2O$ (l)

(ii) Iodine can be titrated with thiosulphate solution using starch as an indicator

$2S_2O_3^{2-}$ (aq) + I_2 (s) $\rightarrow$ $S_4O_6^{2-}$ (aq) + $2I^-$ (aq)

Silver nitrate titrations

Silver ions can be titrated with chloride ions using a solution of chromate (VI) ions, CrO_4^{2-} as indicator

Ag^+ (aq) + Cl^- (aq)$\rightarrow$ AgCl (s)

Solid silver chromate (Ag_2CrO_4) is red and is precipitated once the majority of chloride ions are removed.

1. 15 ml. 12 N- H_2SO_4, 30 ml. 10N- HCl, 10 ml. 2N- HNO_3 and 100 ml. 3N- NaOH were mixed and the volume was made up to 1000 ml. Find the normality of the mixture.

Sol. $\text{Normality} = \dfrac{(N_1V_1+N_2V_2------) - (N_3V_3+ N_4V_4---)}{\text{Total Volume}}$

$\dfrac{(15 \times 12 + 30 \times 10 + 10 \times 2)- (3 \times 100)}{1000} = \dfrac{200}{1000}$

$= 0.2$ or N /5

2. 1.325g., of anhydrous Na_2CO_3 was dissolved in 250 ml. of distilled water. 25 ml, of this solution required 20 ml. H_2SO_4 for complete neutralisation . Find out the normality of the acid. How much water is to be added to 450 ml. of this acid solution to make it N/12 solution?

Sol. Normality of Na_2CO_3 solution

$N = \dfrac{W \times 1000}{E \times V} = \dfrac{1.325 \times 1000}{53 \times 250} = \dfrac{1}{10}$

Normatity of the acid, $N_1V_1 = N_2V_2$

$= \dfrac{1}{10} \times \dfrac{25}{20} = \dfrac{1}{8} = 0.125$ N

Amount of water to be added, $N_2\ V_2 = N_3\ V_3$

$12 \times 0.125 \times 450 = V_3 = 675$ ml.

225 ml. of water is to be added to make it $\frac{N}{12}$ solution

3. A sample of Na_2CO_3, H_2O weighing o.62 g is added to 100 ml. of 0.1N- H_2SO_4 solution. Will the resulting solution be acidic, basic or neutral?

Sol. Equivalent wt. of

$Na_2CO_3.H_2O = \dfrac{\text{mol.wt.}}{2} = \dfrac{124}{2} = 62$

0.62 g. of Na_2CO_3. H_2O is equal to $\dfrac{0.62}{62} = \dfrac{1}{100}$ of eq.wt.

The amount of H_2SO_4 in 1000 ml of 0.1N H_2SO_4 is

$W = \dfrac{ENV}{1000} = \dfrac{49 \text{ X } 0.1 \text{ X } 100}{1000} = 0.49$ g.

0.49 g of $H_2SO_4 = \dfrac{0.49}{49} = \dfrac{1}{100}$ th of eq.wt.

Since the equivalent weight of Na_2CO_3. H_2O and H_2SO_4 is 1/100 th of their equivalent weights and because they are a base and an acid, the resulting solution is neutral.

4. 25C.C. of a solution of iron (II) ammonium sulphate needed excatly 20 C.C. of a solution of $K_2Cr_2O_7$ (VI) of concentration 0.05 mol dm^{-3}. What was the concentration of the iron (II) solution?

Sol: First, we work out the number of moles of dicromate (VI) ions used in the titration

1000 C.C. of soultion contains 0.05 mol.

So, 20 C.C. of solution contains, $0.05 \times \dfrac{20}{1000} = 0.001$ mol.

From the equation, we know that 1 mol $Cr_2O_7^{2-}$ reacts with 6 mol Fe^{2+}.

$\therefore$ the number of moles of iron (II) should be 6 times i.e. 0.006 mol

Concentration of iron (II) $= 0.006 \times \dfrac{1000}{25} = 0.24$ mol $=$ .24 M

5. What volume of the manganate (VII) solution of concentration 0.038 mol dm^{-3} would be needed to give the end point with 25 CC of sodium oxalate of 0.1 mol dm^{-3} concentration.

Sol: No. of moles of oxalate ions in 25 CC

$= 0.1 \times \dfrac{25}{1000} = 0.0025$ mol

From the eq., we know that 5 mol $C_2O_4^{2-} \equiv 2$ mol MnO_4^-

or 1 mol $C_2O_4^{2-} \equiv 0.4$ mol MnO_4^-
$\therefore$ 0.0025 mol 4 $0.0025 \times 0.4 = 0.001$ mol MnO_4^-
0.038 mol are contained in 1000 CC of solution

0.001 mol are contained in $= \frac{1000 \times 0.01}{0.038} = 26.32$

The volume of manganate (VII) solution required is 26.32 CC

6. An impure sample of iron of mass 2.55 g was dissolved in dilute H_2SO_4 and the solution made upto 250 CC. The solution contained Fe(II) ions together with the impurities. 25 CC of this solution was titrated with potassium manganate (VII) solution of 0.02 mol dm^{-3} concentration. The end point was reached at 28.50 CC titre reading. (i) How many moles of manganate (VII) solution were used in the titration? (ii) How many moles of Fe(II) ions will react with 1 mol of manganate (VII) ions? (iii) How many moles of Fe(II) must have been present in 25 CC of solution? (iv) How many moles of Fe(II) ions were there in the 250 CC stock solution? (v) what is the mass of this number of moles of Fe(II)? and (vi) What was the percentage purity of the sample of iron?

Sol: (i) No. of moles of $MnO_4^- = 0.02 \times \frac{28.50}{1000} = 5.7 \times 10^{-4}$ mol

(ii) Manganate (VII) ions take 5 electrons each Fe(II) ion can lose only one electron
$\therefore$ 1 moles of $MnO_4^- \equiv 5$ mol Fe^{2+}

(iii) There were $5 \times 5.7 \times 10^{-4} = 2.85 \times 10^{-3}$ mol
(iv) No. of mol present in 250 CC stock solution =

$2.85 \times 10^{-3} \times 10 = 2.85 \times 10^{-2}$ mol

(v) Mass = $2.85 \times 10^{-2} \times 56 = 1.596$g

(vi) Percentage purity = $\frac{1.596}{2.55} \times 100 = 62.6\%$

7. 25 CC of a solution of potassium iodate (V), KIO_3, were placed in a titration flask. 20 CC of a solution of KI and 10 CC of dil H_2SO_4 were added. The iodine liberated required 24 CC of 0.2 mol dm^{-3} solution of sodium thio sulphate to reach the end point. What was the concentration of the original iodate solution?
Sol: From the equations, we know
1 mol $IO_3^- \equiv 3$ mol I_2
and 1 mol $I_2 \equiv 2$ mol $S_2O_3^{2-}$
Therefore, 3 mol $I_2 \equiv 6$ mol $S_2O_3^{2-}$
and overall 1 mol $IO^{3-} \equiv 6$ mol $S_2O_3^{2-}$

No. of moles of $S_2O_3^{2-}$ ions used $= 0.2 \times \frac{24}{1000} = 0.0048$mol

No. of moles of IO_3^- present = $0.0048 \times \frac{1}{6} = 0.0008$ mol

concentration of KIO_3 = $0.0008 \times \frac{1000}{25} = 0.032$ mol

8. 25 CC of KCl solution required 22.5 CC of 0.02 mol dm^{-3} of $AgNO_3$ to reach the end point using yellow potassium chromate (VI) as indicator. Find out the concentration of the KCl solution?

Sol: No. of moles of Ag^+ added $= 0.02 \times \frac{22.5}{1000} = 0.00045$ mol
from the equation, we know that 1 mol $Ag^+ \equiv 1$ mol Cl^-
$\therefore$ No. of moles of ions = 0.00045

concentration of KCl solution $= 0.00045 \times \frac{1000}{25} = 0.018$ mol

9. Calculate the amount of H_2O_2 in g. in 10 ml. 10 volume of H_2O_2.
Sol. 1ml. of 10 vol. H_2O_2 will give 10 ml. O_2 at NTP
10 ml. of 10 vol. H_2O_2 gives 100 ml. O_2 at NTP
$2\ H_2O_2 \longrightarrow 2H_2O + O_2$
2 (2+32) $\longrightarrow$ 22.4 lit of O_2
22,400 ml.of O_2 is obtained from 68g.of H_2O_2
100 ml " " " $\frac{68 \times 100}{22{,}400} = 0.3035$g. H_2O_2
Sometimes the concentration of H_2O_2 is expressed as percentage of H_2O_2 in soln i.e. g / 100 ml.
Here 3.035 % or 30.35 g / lit

Normality $= \frac{\text{conc. or strength per lit}}{\text{equivalent weight}} = \frac{30.35}{17} = 1.785$N

10. A 5.0 cm³ soln. of H_2O_2 liberates 0.508 g. of iodine from acidified KI solution. Calculate the strength of H_2O_2 solution in terms of volume strength at STP.
Sol.: $H_2O_2 + H_2SO_4 + 2KI \longrightarrow K_2SO_4 + I_2 + 2H_2O$ 34 g
2 x 127 = 254g
254g. of I_2 is obtained from 34 g of H_2O_2
0.508 g of I_2 " " H_2O_2 = (34/254) x 0.508
= 0.068 g.

5 ml of soln contains 0.068 g of H_2O_2
1 ml of soln contains = 0.068/5=0.0136 g of H_2O_2
$2H_2O_2 \longrightarrow 2H_2O + O_2$
68 g of H_2O_2 gives at STP, 22400 ml oxygen
0.0136 g of " " = (22400/68) x 0.0136
= 4.48 ml.

i.e. 1 ml of H_2O_2 soln. gives 4.48 ml of oxygen.
so, strength of H_2O_2 soln. = 4.48 volume.

11. 50 ml. of an aqueous solutions of H_2O_2 was treated with an excess of KI solution and dil H_2SO_4. The liberated iodine required 20 ml. of 0.1 N $Na_2S_2O_3$ solution for complete interaction. Calculate the concentration of H_2O_2 in g/lit.
Sol. The involved reactions are
(i) $H_2O_2 + 2KI + H_2SO_4 \rightarrow K_2SO_4 + I_2 + 2H_2O$
(ii) $I_2 + 2Na_2S_2O_3 \rightarrow Na_2S_4O_6 + 2NaI$
1 mole of $H_2O_2 \equiv 1$ mole of $I_2 \equiv 2$ moles of $Na_2S_2O_3$
no. of moles of $Na_2S_2O_3$ = weight / mol. wt.

$W = \frac{ENV}{1000} = \frac{158 \times 0.1 \times 20}{1000} = 0.316$ g.

no. of moles of $Na_2S_2O_3 = 0.316 / 158 = 2.0 \times 10^{-3}$

Moles of $H_2O_2 = \frac{1}{2}$ of the no. of moles of $Na_2S_2O_3$

$\therefore$ no. of moles of $H_2O_2 = \frac{1}{2} \times 2.0 \times 10^{-3} = 1.0 \times 10^{-3}$

wt. of H_2O_2 = no. of moles $\times$ not.wt.

$= 1.0 \times 10^{-3} \times 34 = 0.034$ g.

50 ml of H_2O_2 contains 0.024 g

1000 ml contains $= \frac{1000}{50} \times 0.034 = 0.68$ g.

12. 50 ml.. of a water sample required 10 ml. of N/50 HCl for complete neutralisation. Calculate the temparary hardness of water. 200 ml. of the above solution was then boiled with 100 ml. of N/50 soda reagent. After boiling, the volume was again made to 200 ml. and the solution filtered. 25ml. of the filtrate required 8.2 ml. of N/50 HCl for neutralisation. Calculate the permanent hardness.

Sol. 10 ml. of N/50 HCl was required for neutralisation

Amout of HCl required, $w = \frac{1000}{50} = \frac{36.5 \times 1 \times 10}{50 \times 1000} = 7.3 \times 10^{-3}$g.

The reaction can be represented as,

$CaCO_3 + 2HCl \rightarrow CaCl_2 + H_2O + CO_2$

From the equation, we know that,

73 g. of HCl reacts with 100g, of $CaCO_3$

$\therefore$ 7.3×10^{-3} g. of HCl reacts $= \frac{7.3 \times 10^{-3}}{73} \times 100 = 0.01$ g. of $CaCO_3$

50 ml. of water contains 0.01 g.$CaCO_3$

10^6 ml. of water contains $= \frac{10^6 \times 0.01}{50} = 200$ ppm

$\therefore$ Temparary hardness of the water sample = 200 ppm

25ml of filtrate required 8.2 ml of N/50 HCl to neutralise the excess of Na_2CO_3 left in the water

25 ml.of filtrate $\equiv$ 8.2 ml N/50 HCl.

200ml. of filtrate $\equiv \frac{200}{25} \times 8.2$ of N/50 HCl

$\equiv$ 65.6 ml. of N/50 HCl.

Vol. of Na_2CO_3 equivalent to 65.6 ml. of N/50 HCl can be known from,

$N_1V_1(HCl) = N_2V_2(Na_2CO_3)$

$$V_2 = \frac{1/50 \times 65.6}{1/50} = 65.6 \text{ ml. } Na_2CO_3$$

volume of Na_2CO_3 consumed = (100-65.6) ml. of N/50 Na_2CO_3

= 34.4 ml. N/50 Na_2CO_3

34.4 ml. of N/50 Na_2CO_3 = 34.4 ml. of N/50 $CaCO_3$

wt. of $CaCO_3$, $w = \frac{ENV}{1000} = \frac{50 \times 1 \times 34.4}{50 \times 1000} = 0.0344$g

200 ml. water contains 0.0344g. $CaCO_3$

10^6 ml water contains $= \frac{10^6}{200} \times 0.0344 = 172$ ppm

13. A mixture of HCOOH and $(COOH)_2$ is heated with concentrated H_2SO_4. The gas produced is collected and on its treatment with KOH solution, the volume of the gas is decreased by 1/6 th. Calculate the molar ratio of two acids in original mixture.

Sol. let a and b moles of HCOOH and $(COOH)_2$ respectively are present

(i) $HCOOH \xrightarrow{H_2SO_4} CO + H_2O$

a moles a moles

(ii) $HOOC - COOH \xrightarrow{H_2SO_4} CO_2 + CO + H_2O$

b moles b mole b mole

total moies of CO and CO_2 = a + 2b

mole fraction of CO_2 = b / (a+2b)

b / (a+2b) = 1 / 6 (since only CO_2 is absorbed in KOH)

6b = a + 2b or 4b = a

$\therefore$ a : b = 4 : 1 ; HCOOH : $(COOH)_2$ = 4 : 1

14. A solution contains Na_2CO_3 and $NaHCO_3$ 10 ml of the solution requires 2.5 ml. of 0.1 M H_2SO_4 for neutralisation using phenolphthalein as indicator. Methyl orange is then added when a further 2.5 ml. of 0.2 M H_2SO_4 was required. Calculate the amount of Na_2CO_3 and $NaHCO_3$ in one litre of the solution.

Sol. 2.5 ml. of 0.1 M H_2SO_4 was used for half neutralisation of Na_2CO_3

Normality of this solution $= \frac{\text{mol. wt}}{\text{eq. wt}} \times$ molarity

$= \frac{98}{49} \times 0.1 = 0.2$

since 5 ml. of 0.2 N H_2SO_4 was used for complete neutralisation of Na_2CO_3

Total volume of H_2SO_4 used = 2.5 ml of 0.2 N H_2SO_4 + 2.5 ml of 0.2 M H_2SO_4

= 2.5 ml of 0.2 N H_2SO_4 + 5 ml of 0.2 N H_2SO_4

= 7.5 ml of 0.2 N H_2SO_4

volume of 0.2 N required for the nertralisation of $NaHCO_3$

= 7.5 - 5.0 = 2.5 ml.

Normality of Na_2CO_3 can be calculated as,

$N_1V_1 = N_2V_2$

(for 0.2 N H_2SO_4) (for Na_2CO_3)

$N_2 = \frac{5 \times 0.2}{10} = 0.1$ N

strength of N_2CO_3 in 1 lit. = normality $\times$ eq. wt

$= 0.1 \times 53 = 5.3$ g / l

Normality of $NaHCO_3 = \frac{2.5 \times 0.2}{10} = 0.05$ N

strength of $NaHCO_3$ = N$\times$eq. wt $= 0.05 \times 84 = 4.2$ g/l

15. 25 c.c. of a mixture of Na_2CO_3 and $NaHCO_3$ required 10c.c. of N/20 HCl when titrated in presence of phenolphthalein, but 25 c.c. of the same when titrated separately in presence of methyl orange required 25 c.c. of N/20 HCl. Calculate the strength of the both.

Sol. The neutralisation reactions can be written as,

$Na_2CO_3 + 2HCl \rightarrow 2NaCl + H_2O + CO_2$

$NaHCO_3 + HCl \rightarrow NaCl + H_2O + CO_2$

10 C.C. of N/20 HCl was required for half titration of Na_2CO_3

20 c.c. of N/20 HCl should be required for complete neutralisation of Na_2CO_3

25c.c. of N/20 HCl was required for complete neutralisation of Na_2CO_3 and $NaHCO_3$

$\therefore$ 5 c.c. of N/20 HCl was required to neutralise $NaHCO_3$.

Normality of the Na_2CO_3 can be calculated as

$$N_1V_1 = N_2V_2 \quad ; \quad N_2 = 20 \times \frac{1}{20} \times \frac{1}{25} = 0.04$$

(HCl) $\quad (Na_2CO_3)$

strength of Na_2CO_3 = N × eq.wt. = 0.04 × 53 = 2.12 g/l

similarly, the normality of $NaHCO_3$ is,

$$N_1V_1 = N_2V_2 \quad ; N_2 = \frac{1}{20} \times 5 \times \frac{1}{25} = 0.01$$

(HCl) $\quad (NaHCO_3)$

strength of $NaHCO_3$ = N×eq. wt.= 0.01 × 84 = 0.84 g/l

16. 2.0 g. of a mixture of carbonate. bicarbonate and chloride of sodium on heating, produced 56 ml of N-HCl solution for neutralisation. Calculate the percentage of Na_2CO_3, $NaHCO_3$ and NaCl in the mixture.

Sol. $NaHCO_3$ alone present in the mixture decomposes on heating as

$2NaHCO_3 \rightarrow Na_2CO_3 + H_2O + CO_2$

equivalent of $NaHCO_3$= eq. of CO_2= 56/11200 (1 eq. of CO_2 occupies 11200 ml at NTP)

$\therefore$ wt. of $NaHCO_3 = \frac{56}{11200} \times 84 = 0.42$ g.

% of $NaHCO_3 = \frac{100}{2} \times 0.42 = 21$

Let x be the wt. of NaCl in 1.6 g. of the mixture

wt. of $NaHCO_3$ in the mixture = $1.6 \times \frac{21}{100} = 0.336$ g.

wt. of Na_2CO_3 in the mixture = 1.6 – 0.336 – xg =1.264– xg.

neutralisation of Na_2CO_3 & $NaHCO_3$ with HCl takes place as,

$Na_2CO_3 + 2HCl \rightarrow 2NaCl + H_2O + CO_2$

$NaHCO_3 + HCl \rightarrow NaCl + H_2O + CO_2$

equivalent of Na_2CO_3 × 1000 + eq. of $NaHCO_3$ X 1000 = milli equivalent of HCl

$$\frac{1.264 - x}{53} \times 1000 + \frac{0.336}{84} \times 1000 = 25 \times 1$$

x = 0.151 g. ; $\therefore$ % of NaCl = $\frac{0.151}{1.6} \times 100 = 9.44$

% of Na_2CO_3 = 100 - (21+9.44)= 69.56

17. A one gram sample of H_2O_2 solution containing x% of H_2O_2 by weight requires x ml. of $KMnO_4$ solution for complete oxidation under acidic conditions. Calculate the normality of $KMnO_4$ solution.

Sol. Reaction involved is,

$2KMnO_4+5H_2O_2+3H_2SO_4 \rightarrow K_2SO_4+2MnSO_4+8H_2O+5O_2$

$\therefore$ 2 mole $KMnO_4$ = 5 mole H_2O_2

wt. of H_2O_2 = x/100 g ; No. of moles of H_2O_2= x/100 × 34

No of moles of $KMnO_4 = \frac{\text{weight}}{\text{mol. wt.}}$

$$= \frac{31.6 \times N \times x}{1000 \times 158} \quad \therefore w = \frac{ENV}{1000} = \frac{Nx}{5000}$$

No. of moles of H_2O_2 = 2.5 × no. of moles of $KMnO_4$

$$\frac{x}{100 \times 34} = 2.5 \times \frac{Nx}{5000}$$

$$N = \frac{x \times 5000}{100 \times 34 \times 2.5x} = 0.588$$

18. 25 g. of a sample of ferrous sulphate was dissolved in water containing dil. H_2SO_4 and the volume made upto 1 lit. 25 c.c. of this solution required 20 c.c. of N/10 $KMnO_4$ solution for complete oxidation. Calculate the percentage of $FeSO_4.7H_2O$ in the sample.

Sol. The reaction is represented as,

$2KMnO_4+3H_2SO_4 \rightarrow K_2SO_4+ 2MnSO_4+ 3H_2O+5(O)$

$5 \times [2FeSO_4.7H_2O + H_2SO_4+ O \rightarrow Fe_2(SO_4)_3 +15H_2O]$

1 mole of $KMnO_4$ = 5 moles of $Fe_2SO_4.7H_2O$

no. of moles of $KMnO_4$ consumed = Weight / molecular wt.

$$= \frac{31.6 \times 0.1 \times 20 \times 40}{1000 \times 158} = 0.016$$

(The volume of $KMnO_4$ consumed = 20 × 40=800ml.)

$\therefore$ no. of moles of $FeSO_4.7H_2O$ = 0.016 × 5 = 0.08

weight of $FeSO4.7H_2O$ = no. of moles × mol. wt.

= 0.08 × 278 = 22.24 g.

22.24 g. of $FeSO_4.7H_2O$ is present in 25 g. of the sample

% of purity of $FeSO_4.7H_2O = \frac{22.4}{25} \times 100 = 88.96$

19. An aqueous solution containing 0.1 g KIO_3 (mol. wt. 214) treated with excess KI solution. The solution was acidified with the liberated iodine consumed 45 ml. of thiosulphate solution to decolourise the blue starch iodine solution. Calculate molarity of the sodium thiosulphate solution.

Sol. The reactions are represented as,

$KIO_3 + 5KI \rightarrow 3H_2O + 3I_2$

$2Na_2S_2O_3 + I_2 \rightarrow 2 NaI + Na_2S_4O_6$

1 mole of I_2 reacts with 2 moles of $Na_2S_2O_3$

no. of moles of KIO_3 = weight / mol. wt. = 0.1/214

no. of moles of $I_2 = 3 \times \frac{0.1}{214}$

no. of moles of $Na_2S_2O_3 = 2 \times 3 \times \frac{0.1}{214} = 2.804 \times 10^{-3}$

45 ml. of hypo contain = 2.804×10^{-3} moles

1000 ml. of hypo contains = $2.804 \times 10^{-3} \times \frac{1000}{45} = 0.0623$

Molarity of hypo = 0.0623

20. Calculate the weight of CaO required to remove the hardness of 10,00,000 litre of water containing 1.62g. of $Ca(HCO_3)_2$ per litre.

Sol. Total amount of $Ca(HCO_3)_2$ present in lthe water = 1.62 × 10,00,000 = 1.62×10^6 g = 1620 Kg.

The reaction of the removal of hardness can be written as,
$Ca(HCO_3)_2 + CaO \rightarrow 2CaCO_3 + H_2O$
From the equation, we can say that
162 kg, of $Ca(HCO_3)_2$ requires 56 kg. of CaO
$\therefore$ 1620 kg of $Ca(HCO_3)_2$ requires 560 Kg. of CaO

21. A solid mixture 5.0 g consisting of lead nitrate and sodium nitrate was heated below 600°c until the weight of residue was constant. if the loss of the weight is 28% find the amount of lead nitrate and sodium nitrate in the mixture.

Sol. let x g. of lead nitrate is present in the mixture
$\therefore$ weight of sodium nitrate in the mixture = 5.0 – x g.
The reactions involved are represented as,

$$\underset{331}{2Pb(NO_3)_2} \rightarrow \underset{223}{2\,PbO} + 4NO_2 + O_2$$

$$\underset{85}{2\,NaNO_3} \rightarrow \underset{69}{2NaNO_2} + O_2$$

331 g. of $Pb(NO_3)_2$ gives 223 g. of PbO
$\therefore$ x g. of $Pb(NO_3)_2$ gives $= \frac{223}{331} \times x$ g. of PbO
85 g. of $NaNO_3$ gives 69 g.of $NaNO_2$
(5 – x) g. of $NaNO_3$ gives $= \frac{69}{85} \times$ (5-x) g. of $NaNO_2$
The residue contains Pbo + $NaNO_2$
$\therefore$ Total wt. of the residue $= \frac{223}{331} \times x + \frac{69}{85} \times (5-x)$

weight of the residue $= \frac{72}{100} \times 5 = 3.6$ g.

Now we can write, $\frac{223}{331} \times x + \frac{69}{85} \times (5-x) = 3.6g$

$0.6737\,x + 0.8118\,(5 - x) = 3.6$
$0.6737\,x + 4.059 - 0.8118\,x = 3.6$
$-0.1381\,x = -0.459$
$x = 0.459 / 0.1381 = 3.324$ g.
mass of $Pb(NO_3)_2$ = 3.324 g.
mass of $NaNO_3$ = 5-3.324 = 1.676g.

Thought Provoking Questions with Hints

1. Atomic volume is not a direct measure of the size of the atom, why ?

Hint: Atoms can pack together in different crystal structures. This means that, for example, atoms that pack in very open structure will have a large atomic volume; slightly larger atoms that pack in a tighter arrangement will have a smaller atomic volume. This is the reverse of the pattern in their sizes.

2. Why is it not strictly correct to consider electro negativity values as physical properties of elements?

Hint: We can't measure electro negativities because they have been invented by chemists particularly Linus Pauling. They are not properties of atoms. However, the values have been conjured from measurements we can make. e.g. ionisation potentials.

3. Give a short explanation of diagonal relationship in the periodic table in terms of changes in nuclear charge and shielding of electrons.

Hint: Moving one place to the right across a period the increased nuclear charge holds the electrons more tightly to the atom. Moving down one place in the group, the extra shell of electron lessens the attraction of the nucleus for the outer electrons. A diagonal move means that these effects tend to compensate for one another.

4. What is the equation for the nuclear reaction that takes place when tritium undergoes beta decay? What is produced in the reaction?

Hint: $H_1^3 \rightarrow He_2^3 + e_{-1}^0$. The product is an isotope of helium.

5. The mass spectrum of a sample of hydrogen gas corresponding to a mass to charge ratio (m/e) value of 2. However there are other lines at higher m/e value. What would their m/e values be?

Hint: A sample of hydrogen gas will contain H_2, HD and D_2 molecules which have m/e values of 2, 3, and 4 respectively. As you would expect, the later two peaks are extremely small in intensities.

6. Explain why AlH_3 would be expected to bond with a hydride ion?

Hint: Like $AlCl_3$, there is an empty orbital on the aluminium which can accept a lone pair of electrons. The hydride ion has the necessary pair of electrons.

7. Write down the formula of the molecule made when $LiAlH_4$ reacts with $H_2C{=}CHCOOH$.

Hint: $H_2C{=}CHCH_2OH$. The double bond is not affected during the reduction. The acid group is converted to an alcohol.

8. Suggest a way of making PH_3, starting with PCl_3.

Hint: One possibility is to react phosphorous trichloride with $LiAlH_4$

$3LiAlH_4 + 4PCl_3 \rightarrow 3LiAl + 3AlCl_3 + 4PH_3$

9. What does molecular orbital theory have to say about the presence of lone pairs in a water molecule?

Hint: The lone pairs are not to be found in molecular orbital theory. This goes to show that we should not mistake the results of mathematical theories as a direct picture of realities.

10. Why is $K^{39.098}$ placed after $Ar^{39.948}$ in the periodic table?

Hint: The elements are arranged in the order of increasing atomic number in the modern classifications. Therefore Ar_{18} is placed before K_{19}.

11. Water shows an amphoteric oxide behaviour when it reacts with HCl and NH_3. Explain.

Hint: $\underset{\text{base}}{H_2O\,(l)} + \underset{\text{acid}}{HCl\,(g)} \rightarrow \underset{\text{acid}}{H_3O^+\,(aq.)} + \underset{\text{base}}{Cl^-\,(aq.)}$

$\underset{\text{acid}}{H_2O\,(l)} + \underset{\text{base}}{NH_3\,(aq.)} \rightleftharpoons \underset{\text{acid}}{NH_4^+\,(aq.)} + \underset{\text{base}}{OH^-\,(aq.)}$

In the first case, H_2O is a proton acceptor and in the second a proton donor according to Bronsted-Lowry theory of acids and bases.

12. Calculate the electro negativity of fluorine from the following data. $E_{H-H} = 104.2$ k.cal.mol^{-1}, $E_{F-F} = 36.6$ k.cal.mol^{-1}, $E_{H-F} = 134.6$ k.cal.mol^{-1}; $X_H = 2.1$.

Hint: Substituting the values in Pauling's equation,

$X_F - X_H = 0.208\,[E_{H-F} - (E_{F-F} \times E_{H-H})^{½}]^{½}$

we get, $X_F - 2.1 = 0.208\,[134.6 - (104.2 \times 36.6)^{½}]^{½}$

$X_F = 3.4$

13. Calculate the electronegativity of fluorine if its IE and EA are respectively 17.5 and 3.5 eV.

Hint: Substituting the values in Mulliken equation,

X = IE + EA/ 5.6

we get, for $X_F = 17.5 + 3.5 / 5.6 = 3.75$

14. Calculate the electro negativity of an element with atomic no. 14 by Allred–Rochow method if its covalent radius is 1.18 Å.

Hint: First we calculate the Z_{eff} from the Allred–Rochow equation,

$$X = 0.359 \times \frac{Z_{eff}}{r^2} + 0.744$$

$$Z_{eff} = 14 - (0.35{\times}4 + 0.85{\times}8 + 2{\times}1) = 3.80$$

Now substituting the value we get,

$$X = 0.359\frac{3.80}{(1.18)^2} + 0.744 = 1.72$$

15. What would you expect to happen if you heat washing soda crystals?

Hint: The crystals dissolve in their own water of crystallisation and the steam is given off.

16. Why are washing soda crystals used in bath salts together with a little of colouring and perfume?

Hint: The carbonate ions present in these crystals give a deposit of $CaCO_3$ with Ca^{2+} ions often found in hard water.

17. Why is it that lithium salts have a grater degree of covalent characters than the other halides of the group?

Hint: The small size of Li^+ gives it a huge polarising power. This leads to covalency.

18. Suggest a reason why LiF has the lowest solubility of the group 1 metal halides?
Hint: The small size of both the Li^+ and F^- ions leads to a very large lattice energy, which means the crystal is very hard to break.
19. How many electrons does a hydride ion possess? What does this tell about the statement that all noble gas structures are very stable?
Hint: Two; $1s^2$. It is wrong.
20. A student suggested that calcium should be made if CaO is reacted with Al powder. Was the student correct?
ΔG_f^0 (CaO) = –604.2 kJ mol^{-1};
ΔG_f^0 (Al_2O_3) = –1582.4 kJ mol^{-1}.
Hint: The equation is,
$3CaO (s) + 2Al (s) \rightarrow 3Ca (s) + Al_2O_3 (s)$
$\Delta G^0 = \Delta G_f^0(Al_2O_3) - 3\,\Delta G_f^0(CaO) = +230.2$ kJ mol^{-1}
The positive sign tells us that the reaction cannot occur under standard conditions. But at high temperatures, the free energy change is negative and the reaction takes place. This method is used to extract calcium in industry.
21. How are Zinti compounds formed?
Hint: The first group alkali elements form a number of compounds with 13–16 group members in which they are present as complex polynuclear anions.
22. Why is MgO used for the lining of the steel making furnace?
Hint: MgO is a basic flux. It is helpful in the removal of acidic impurities such as SO_2, P_2O_5, SiO_2 etc from the steel if present forming $MgSO_3$, $Mg_3(PO_4)_2$ and $MgSiO_3$ as slag in each case.
23. Why is the white ash obtained on burning Mg in air gives odour of ammonia on treatment with water?
Hint: Mg_3N_2 (magnesium nitride) formed is present in small quantity in the ash. It undergoes hydrolysis to give ammonia.
$Mg_3N_2 + 6H_2O \rightarrow Mg(OH)_2 + 2NH_3$
24. Why do crystalline salts of alkaline earth metals (2nd group) contain more water of crystallisation than the corresponding salts of alkali metals (1st group)?
Hint: As the ionic radii of second group metals are smaller compared to the first group metal ions, their degree of hydration is higher.
25. Apart from lattice energy, which other energy change is important in accounting for solubilities?
Hint: The solvation energies of ions.
26. Beryllium makes complex ion BeF_4^{2-}. Predict the shape of this ion?
Hint: The electronic configuration of Be^{2+} is $1s^2$. The ion can make four bonds by making use of its empty 2s and 2p orbital. As there are no lone pairs and only four bond pairs to the four fluorine ions, the electron repulsion predicts that the BeF_4^{2-} ion should be tetrahedral in shape and it is.
27. At a temperature of 15°C the energy absorbed when water changes from liquid to gas is about 2500 kJ Kg^{-1}. Estimate the energy used in the evaporation of water over land and sea each year. How much energy is released over land each year if around 125×10^{12} tonnes of precipitation takes place? (Total fall of water to the earth's surface is ~500×10^{12} tonnes and about one quarter of this precipitation for water falling is rain, snow etc.)
Hint: $500\times10^{12}\times10^3$ Kg × 2500 kJ Kg^{-1} = 1.25×10^{21} kJ. This energy is absorbed and leads to a lowering of temperature.
$125\times10^{12}\times10^3$ Kg × 2500 kJ Kg^{-1} = 0.31×10^{21} kJ. This energy is released and thereby tends to increase the temperature of the atmosphere.
28. Why I_2 dissolves more in KI solution than in water?
Hint: Being a covalent molecule, I_2 is sparingly soluble in water. In KI solution it forms a polyhalide KI_3, an ionic compound which is more soluble.
29. Why is neon used in warning signal illuminations?
Hint: The neon illuminated signals are visible from long distances. They are visible under the mist, fog etc. conditions also.
30. Why Tl has a greater tendency to give Tl^+ ions in solution rather than Tl^{3+} ions?
Hint: $E^\circ_{Tl^+/Tl}$ is more negative than $E^\circ_{Tl^{3+}/Tl}$ which suggest that Tl has a greater tendency to turn into Tl^+ in solution.
31. How can the thermit reaction not be stopped by water like an ordinary fire?
Hint: In an ordinary fire the reactions are kept going by the material reacting with oxygen in the air. Water not only cools the fire, but also keeps oxygen out. In the thermit reaction, the oxygen involved is in the metal oxide form, so keeping atmospheric reaction out has no effect. Also, at the temperatures of the thermit reactions ~1000°C water will give off hydrogen with aluminium.
32. What type of bonding holds the layers together in graphite? Why will graphite conduct electricity well in a direction parallel to the planes of the hexagon?
Hint: van der Waal's bonding holds the layers together. The delocalised orbitals in graphite lie parallel to the planes. Electrons can move along the planes making use of these orbitals. Conduction is good parallel to the planes. There is no overlap of orbitals perpendicular to the planes, so conduction is poor in this direction.
33. Which theory other than VBT could be used to explain the bonding in SiF_6^{2-} and predict the shape?
Hint: Molecular orbital theory could be used. Octahedral shape.
34. Copper articles corrode to produce green patches on the surface. What is the chemistry of this corrosion?
Hint: The copper present in the articles on exposure to moist air (oxygen) and carbon dioxide present in atmosphere is converted into basic carbonate which is seen on the surface of layers as green patches.

$2Cu + CO_2 + H_2O + O_2 \rightarrow [Cu(OH)_2CuCO_3]$

35. Why is it reasonable to think that the release of chloride ions from $SiCl_4$ is a favourable process?

Hint: Chloride ions, Cl^- are kinetically and energetically stable in water and they can be solvated by water molecules, so they are good leaving groups in the reaction.

36. Why is the shape of PCl_5 trigonal bipyramidal?

Hint: It minimises the repulsion.

37. Triple superphosphate $Ca(H_2PO_4)_2$ commands a higher price than ordinary super phosphate. Why?

Hint: It contains no anhydride $CaSO_4$, so each kilogram is richer in fertilizer.

38. Even reactive metals like Na do not react with liquid oxygen. Why?

Hint: At the temperature of liquid oxygen (less than 90K) the sodium and oxygen have insufficient energy to get over the energy barrier.

39. How do CO and CN^- ions stop haemoglobin to take up oxygen?

Hint: CO and CN^- are able to bond more strongly to the iron atom in haemoglobin than can oxygen. Haemoglobin would become saturated with these species blocking it to take up oxygen leading to death.

40. Why is it necessary to remove sulphur compounds from oil and natural gas?

Hint: Sulphur compounds easily poison the catalyst used in many industrial processes.

41. Why is $SOCl_2$ preferred to PCl_5 for replacement of OH group in organic reactions?

Hint: The side products from $SOCl_2$ are gases and are easily removed. PCl_5 gives non-volatile $POCl_3$.

42. SO_2 and H_2S react together according to

$2H_2S\ (g) + SO_2\ (g) \rightarrow 3S\ (s) + 2H_2O\ (l)$. Both have reducing properties. Which is more powerful reducing agent?

Hint: The oxidation number of sulphur in SO_2 is +4, and in H_2S is –2. In both cases the sulphur is left with an oxidation number of 0. The SO_2 gains electrons, the H_2S losses them; therefore H_2S reducing agent.

43. Explain why the displacement reactions of the halogens are best regarded as redox reactions.

Hint: When, for example chlorine converts iodide ions into iodine, the iodide ions lose electrons and the chlorine atoms gain electrons. i.e. chlorine oxidises the iodide.

44. Is it true that iodine will not displace any other halogen?

Hint: No. It would displace Astatine from solution.

45. Why does ClF_3 have a distorted T–shape?

Hint: Each fluorine will provide one of the electrons in the three bonding pairs. Chlorine has 7 electrons in its valence shells. This makes ten electrons in total. As six of them are used in bonding, four are left. These make the two lone pairs. Lone pair – bond pair repulsion causes distortion of the molecule.

46. Why should you expect chromium (VI) compounds to show oxidising properties?

Hint: Ions in high oxidation states tend to take electrons from another chemical and return to a lower oxidation state. As it loses electrons the other chemical must have been oxidised.

47. Explain how to know the concentration of a solution of chloride in a silver nitrate solution.

Hint: In titrations $AgNO_3$ is mostly used to determine the concentrations of solutions that contain chloride ions. When all the chloride ions are used up there will be no more precipitate made. To know this point, one can use K_2CrO_4 (potassium chromate) solution as indicator. CrO_4^{2-} ions give a red precipitate of silver chromate. (Ag_2CrO_4) with silver ions which takes place after the precipitation of AgCl is completed.

48. Which is the stronger oxidising agent, manganate(VII) ions in acid or alkaline conditions?

Hint: In alkaline solution. It has the more positive redox potential.

49. Solutions of iron (III) salts are often acidic. Why is it so?

Hint: One may assume that $[Fe(H_2O)_6]^{3+}$ ions are present in the solution. The complex ion can act as a proton donor,

$[Fe(H_2O)_6]^{3+} + H_2O\ (l)$®$[Fe(H_2O)_5OH]^{3+}(aq) + H_3O^+(l)$

50. For dichlorobis (ethane–1,2–diamine) cobalt (III) $[Co(NH_2CH_2CH_2NH_2)_2]Cl_2$ three isomers are possible two are optically active while one is not. Explain.

Hint:

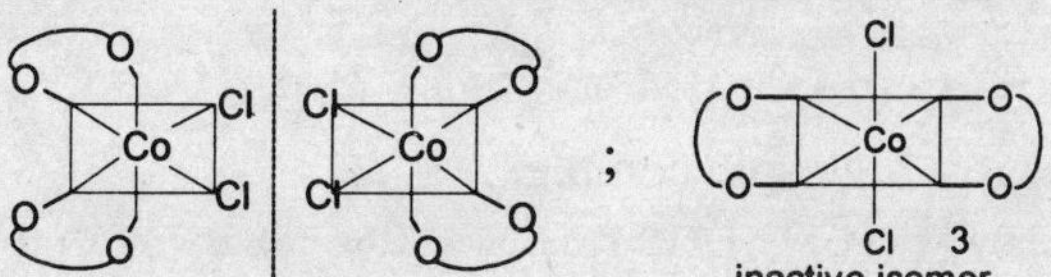

1 optically active 2 ; 3 inactive isomer

The structures 1 and 2 are enantiomers and are optically active. The third one is inactive isomer as it posses both a plane of symmetry and centre of symmetry.

51. What appears to be the oxidation state of iron atoms in $FeFe(CN)^-_6$?

Hint: Each cyanide carries a charge of -1. The two iron atoms take up five negative charges so their average oxidation number is +2.5. In practice there is a mixture of Fe^{2+} and Fe^{3+} ions.

52. Zinc(II) chloride is extremely deliquescent. What might be the reason for the insolubility of zinc(II) fluoride?

Hint: ZnF_2 has very high lattice energy compared to $ZnCl_2$.

53. Dimercury(I) iodide Hg_2I_2, is a greenish colour and is precipitated if iodide ions are added to a solution of dimercury(I) sulphate. Likewise the red mercury(II) iodide HgI_2, is precipitated from a solution of mercury(II) sulphate. However both precipitates dissolve in excess iodide solution. What might be the reason for it?

Hint: A complex ion HgI_4^{2-} is produced. This is a mercury (II) complex. Dimercury(I) complexes do not occur. If a complex can be made in a reaction, it is common for dimercury(I) to change its oxidation state, giving the mercury(II) complex. Infact disproportionation reactions occur.

e.g. Hg_2^{2+} (aq) +4I^- (aq) $\rightarrow$ HgI_4^{2-} (aq) + Hg(I)

54. Why is mercury(II) nitrate an exception to the normal pattern of decomposition by heat?

Hint: Because HgO is easily decomposed by heat, the product of reaction is liquid mercury rather than the oxide.

55. How do you separate ZnS from CdS?

Hint: The solubility product of ZnS is greater than that of CdS. If the acid is present the concentration of free sulphide ions is reduced and becomes too low for ZnS to be precipitated. However CdS will be precipitated. Then add a little acid to the solution before adding the sulphide solution to the mixture. Filter the precipitate of the CdS. Then add little alkali to filtrate to neutralise the acid. Now ZnS will precipitate.

56. How do you account for low spin and high spin complexes with a suitable example?

Hint: d^4, d^5, d^6, d^7 configuration in octahedral complexes exhibits low and high spin.

e.g. (i) $[Fe(H_2O)_6]^{3+}$ in this complex iron is in +3 state i.e. the ion possess 23 electrons. (d^5 configuration)

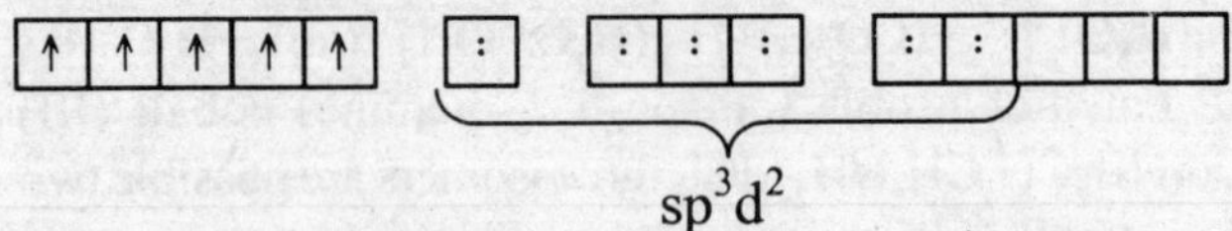

There are 5 unpaired electrons.

$\mu_{eff} = \sqrt{n(n+2)}$ BM $= \sqrt{5(5+2)} = \sqrt{35}$ =5.92 BM.

Thus it is high spin complex.

e.g. (ii) $[Fe(CN)_6]^{3-}$ in this case the electrons in iron +3 state can be represented as

↑↓ ↑↓ ↑ : : | : | : : :

d^2sp^3

$\mu_{eff} = \sqrt{n(n+2)}$ BM $= \sqrt{1(1+2)} = \sqrt{2} = 1.73$ BM

Thus it is low spin complex.

57. Why have the energy levels negative sign?

Hint: Because the zero of potential energy is the world beyond the atom, electrons inside the atoms are lower in potential energy, so they have negative values. We have to put energy in to get the electrons out.

58. What happens to gaps between the energy levels as 'n' increases?

Hint: The gap becomes increasingly small; eventually zero.

59. Why could the mass of a neutron not determined in a mass spectrometer?

Hint: The neutron has no charge. Only charged particles can be deflected and detected in a mass spectrometer.

60. The relative atomic mass of chlorine is quoted as 35.5. A sample of chlorine made in the laboratory is a mixture of Cl_{17}^{35} and Cl_{17}^{37}. What are the percentages of the two isotopes in chlorine?

Hint: We must have $\frac{X}{100} \times 37 + \left[\frac{100-X}{100}\right] \times 35 = 35.5$

X = 25%; therefore, chlorine contains 75% Cl_{17}^{35} and 25% Cl_{17}^{37}.

61. Power stations are about 30% efficient i.e. they convert about 30% of heat energy to electrical energy suppose a nuclear power station uses fuel that generates electrical power at the rate of 1MW g^{-1}, what would be the thermal power generated? How many grams of fuel would have to undergo fission each day? Assuming that 1 tonne of coal can release 25 GJ, how many tonnes of coal would have to be burnt in order to generate 1000 MW of electrical power?

Hint: If 25 GJ i.e. 25000 MJ were released in one second the power would be 25000 MW. So to produce 3 MW would need 0.12 tonnes of coal.

If the power is needed at this level for the whole of one day, we need = 0.12×24×60×60 = 10368 tonnes.

62. 236 g U^{236} isotope could provide about 1.6×10^{13} J of heat energy. Because 1 J s^{-1} is the same as one watt, if the 236 g suffered fission in 1 second the power output would be 1.6×10^{13} W. What would be the power output if 1 g of the isotope were used up in one day?

Hint: The power generated is

$$\frac{1.6\text{x}10^{13}}{236 \text{ x } 24 \text{ x } 60 \text{ x } 60} = 7.8\times10^5 \text{ W which rounds up to 1 MW.}$$

63. Suppose a carbon –12 atom is formed in a star, which fusion reaction would convert it into an atom of nitrogen–13?

Hint: ${}_6C^{12} + {}_1H^1 \rightarrow {}_7N^{13}$

64. It has been estimated that the C^{14} in the atmosphere is responsible for producing 60 atoms of N^{14} and 60 electrons every hour for each gram of carbon. We can quote this disintegration rate as 60 counts hr^{-1} g^{-1}. A sample of a sea shell found near a sea shore was found to have a count of 4 counts $hr^{-1}g^{-1}$. Estimate the age of the shell. ($t_{0.5}$ of is 5730 years).

Hint: Every 5730 years the rate would have halved.

Count rate / counts $hr^{-1}g^{-1}$	60	30	15	7.5	3.75
No. of $t_{0.5}$	0	1	2	3	4

If the count has reduced to 3.75 counts $hr^{-1}g^{-1}$, the lapse of time would be 4×5370 = 21480. The actual count rate is a little higher than 3.75 counts $hr^{-1}g^{-1}$. So the shell must be less than the calculated age i.e. ~21000 years.

65. What is the maximum distance away from the nucleus that the electron can get? What has happened to the atom then?

Hint: Infinity. It has been ionised.

66. Why could we not discover the ionisation energy of the hydrogen atom by looking at the line in the Balmer series?

Hint: The ionisation energy is the energy difference between the energy levels with n = 1 and n = ∞. The Balmer series only show transition to the level n = 2. We have to look in Lyman series as they contain transitions to the lowest energy level, n = 1.

67. Methanoic acid $H-C(=O)-O-H$ has one carbon–oxygen bond of length 123 pm, and another of 136 pm. But the both carbon oxygen bond in $HCOO^-$ ion have the same length of 127 pm. What do you infer from this?

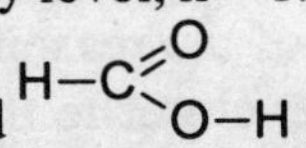

Hint: In $H-C(=O)-O-H$ the C=O bond is shorter than the other. In $HCOO^-$, the two bonds have the same length due to resonating structures.

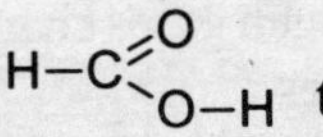

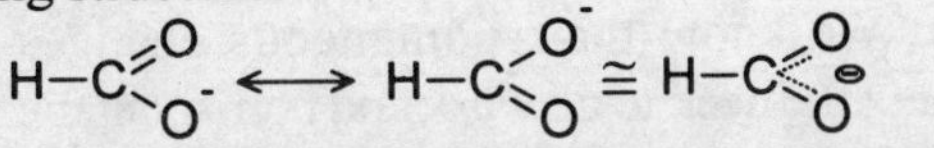

$$H-C(=O)-O^- \longleftrightarrow H-C(-O^-)=O \equiv H-C(\cdots O)(\cdots O)^{\ominus}$$

68. What is an important difference between the ions as they appear in the equations for ionisation energies and electron affinities, and the ions in a real ionic compound?

Hint: The key difference is that the equations for the ionisation energies are for atoms and ions in the gaseous state. In practice the ions are found in (solid) crystals. It is the attractions in the crystal that make all the difference in deciding weather an ionic compound exist.

69. Which of the halogens have the highest polarisibility and why?

Hint: It is iodine. It has the largest number of electrons and therefore the most shielding and the greatest polarisability.

70. Whether intermolecular forces exist in NaCl?

Hint: NaCl is an ionic substance. As such it is held together by electrostatic attractions between the ions and not by intermolecular forces. There are no molecules in ionic substances.

71. Which type of intermolecular force would hold H_2, benzene and pentan–3–one together?

Hint: H_2 and benzene are non-polar. These molecules are held together in liquid or solid by dispersion forces. Pentan–3–one has a permanent dipole moment, so the forces are between permanent dipoles.

72. Why does smoke from a fire or a cigarette often have a blue tinge to its colour?

Hint: The smoke is colloidal, so when it is viewed with an angle to the source light, it appears blue due to Tyndall effect.

73. What type of forces hold the layer of carbon atoms in graphite together?

Hint: van der Waal's forces.

74. Why is ZnS not completely ionic in nature?

Hint: The sulphide ion is large and squashy. It is easily polarised by the much smaller zinc ion, and covalence results.

75. Explain what happens to the pressure of an ideal gas if its volume is halved at the same temperature.

Hint: The pressure increases. Halving the volume will double number of collisions of the particles with walls of the container as particles are travelling across the container have a shorter distance to travel before they reach a wall.

76. 100 cc of a solution of HCl were neutralised by 0.12grams of Mg ribbon. What was the concentration of the acid (mol. Wt. of Mg is 24)?

Hint: $Mg\ (s) + 2HCl\ (aq) \rightarrow MgCl_2\ (aq) + H_2O\ (g)$

From this equation we understand that 1 mole of Mg will react with 2 moles of HCl.

No. of moles of Mg used = 0.12 / 24 = 0.005 mole.

No. of moles of HCl used = 0.01 mole, since 1 mol of Mg reacts with two moles of HCl.

Concentration of the acid = 0.1 mole.

77. Why is it important to wash a pipette and especially a burette with distilled water after a titration using alkali?

Hint: If alkali is left in the pipette, it will become encrusted with solid Na_2CO_3 and the jet may become blocked. This can be removed using dil. HNO_3. Alkali left in the burette may block the jet. However in the case of a burette it can led to a deposit of solid sodium carbonate in the socket of the tap. The alkali can also attack the grease that may be used to lubricate the socket.

78. Which of the following reactions represent redox reaction?

(a) $BaCl_2\ (aq) + H\text{-}_2SO_4(aq) \rightarrow BaSO_4(s) + 2HCl\ (aq)$

(b) $ClO_3^-(aq)+3Zn(s)+6H^+(aq) \rightarrow Cl^-(aq) + 3Zn^{2+}(aq) + 3H_2O\ (l)$

(c) $2Cu^{2+}(aq) + 4I^-(aq) \rightarrow 2CuI\ (s) + I_2\ (aq)$

Hint: (a) This is a precipitation reaction not a redox reaction.

(b) Zn atoms are converted into Zn ions. The oxidation number changes from 0 to +2. Zn atoms have lost electrons. They have been oxidised by chlorate (V) ClO_3^- ions. Chlorine in ClO_3^- has +5 oxidation number. As zinc atoms have given up electrons, they are the reducing agents. This is a redox reaction.

(c) The copper(II) ions are converted to copper(I) ions. Some, but not all of the iodide ions which start with an oxidation number of -1 are converted to molecules, which have an oxidation number of 0. Iodide ions have been oxidised. Therefore, in this reaction Cu^{2+} ions are oxidising agents while I^- are reducing agents.

79. Which is likely to be stronger oxidising agent: PbO or PbO_2?

Hint: PbO_2 is a better oxidising agent as Pb is in (IV) oxidation state i.e. higher oxidation state.In PbO, it is (II).

80. In an exothermic reaction how do the strengths of the bonds in the reactants compare with the strengths of the bonds in the products?

Hint: The bonds in the products are (overall) stronger than in the reactants. The stronger the bonds that are made, the more energy is released.

81. 1,3-butadiene has structure that is often shown as

Its heat of hydration is -239 kJmol^{-1}.

The heat of hydrogentation of ethene is -132 kJ mol^{-1}. Could there be delocalised pi bonds in 1,3–butadiene?

Hint: If there were two entirely separate double bonds in 1,3–butadiene we would expect its heat of hydrogenation to be twice that of ethene i.e. -264 kJ mol^{-1}. Infact the value is -239 kJ mol^{-1}. The value is 25 kJ mol^{-1} less than expected from the values, it appears that the pi bond in this molecule is delocalised.

82. Hydrazine N_2H_4 is an endothermic compound. Its heat of formation is 50.6 kJ mol^{-1}. Calculate the enthalpy change for the reaction of burning hydrazine in oxygen: $N_2H_4(l)+O_2 \rightarrow N_2(g) + 2H_2O\ (l)$. Why do you think hydrazine has had some success as a rocket fuel? $\Delta H_f(H_2O) = -285.9$ kJ mol^{-1} : $\Delta H_f(N_2H_4) = 50.6$ kJ mol^{-1}.

Hint: $\Delta H_{(reaction)} = -2\ \Delta H_{f}\ (H_2O) - \Delta H_f\ (N_2H_4)$

$= 2\times(-285.9) - (50.6) = -622.4$ kJ mol^{-1}.

The reaction is strongly exothermic. It is an explosively fast reaction. Both the qualities are prime for a rocket fuel.

83. Ammonium nitrate NH_4NO_3 is widely used as a fertilizer. Unfortunately it is also found use as an explosive by terrorists. $\Delta H_f\ (NH_4NO_3) = -364.6$ kJ mol^{-1}. How will you account for the statement?

Hint: We know

$2NH_4NO_3\ (s) \rightarrow 2N_2(g) + 4H_2O(g) + O_2(g)$.

The enthalpy change for the reaction is

$\Delta H_f\ (H_2O) - 2\Delta H_f(NH_4NO_3)$

$= 4\times(-285.9) - 2\times(-364.4) = -414.4$ kJ mol^{-1}.

A good explosive must decompose rapidly and produce large changes in volume by releasing gases. NH_4NO_3 fulfils these requirements. The 7 moles of the gas produced have a volume of over 150000 cc. It is the rapid expansion that forces shock waves through the air which causes the damage.

84. Calculate the lattice energy for KF from the following data

(i) $K(s) + \frac{1}{2} F_2 \rightarrow KF(s)$	-562.6 kJ mol^{-1}
(ii) $K(s) \rightarrow K\ (g)$	89.6 kJ mol^{-1}
(iii) $K\ (g) \rightarrow K^+\ (g)$	419.0 kJ mol^{-1}
(iv) $\frac{1}{2} F_2\ (g) \rightarrow F\ (g)$	79.1 kJ mol^{-1}
(v) $F\ (g) \rightarrow F^-\ (g)$	-332.6 kJ mol^{-1}

Hint: We know

– (lattice energy) = – (heat of formation) + heats of atomization + ionisation energies + electron affinities.

On substitution we get

= -(-562.6)+89.6+79.1+419-332.6

= 562.6+89.6+79.1+419-332.6 = 817.7

Lattice energy = -817.7 kJ mol^{-1}

85. What relation can be found between ionic radii and the heats of hydration?

Hint: The smaller the ionic radius, the greater is the hydration energy. One of the reason for this is that the water molecules can get closer to the centre of charge, because of which the attractions are increased.

86. Which of the following are spontaneous changes:

(a) a ball rolling downhill

(b) A mixture of nitrogen and oxygen gases changing into separate samples of the two gasses

(c) Instant coffee granules dissolving in water

(d) Petrol burning?

Hint: We know that spontaneous changes take place without the need to do work and non-spontaneous changes take place when work is done.

(a) (c) and (d) do not require us to do work, so they are spontaneous.

87. Why are many metal oxides insoluble in water?

Hint: The oxide ion is small and has a negative charge. Both factors lead to large lattice energies of metal oxides and therefore to insolubility.

88. Why NaI is more soluble in water than NaCl?

Hint: Lattice energy of NaCl is greater than that of NaI by about 82 kJ mol^{-1}. We know that high solubility is favoured by a small lattice energy and a high hydration energy.

89. Would you expect a mixture of (a) Hexane and pentane; (b) propan–1–ol and water will show deviations from Raoult's law?

Hint: (a) Both are non-polar hydrocarbons and therefore no deviation.

(b) propan–1–ol is polar and it can take part in hydrogen bonding. Therefore this mixture shows deviation.

90. Why is benzene added to ethanol and water mixture before distillation?

Hint: Ethanol and water give an azeotropic mixture containing about 95.6 percent ethanol. If benzene is added to this mixture then azeotropic mixture does not form. This enables one to separate ethanol from the mixture completely pure.

91. A solution contains equal concentrations of zinc ions and nickel(II) ions. What would you expect to happen if H_2S is passed through the solution?

Hint: The solubility product of nickel(II) sulphide is 100 times smaller than that of zinc sulphide. Therefore nickel(II) sulphide will be precipitated first. Soon after zinc sulphide would also be precipitated as the solubility products are not very different.

92. Explain why blocks of magnesium can be attached to the hulls of the ships or attached to iron pipes with the aim of preventing rusting.

Hint: Because Mg has a more negative electrode potential than iron. Magnesium acts as an anode. Mg is worn away before the iron.

93. Briefly explain why equal number of electrons must be lost from the cathode as gained at the anode during electrolysis.

Hint: When electricity passes between two points charges are on the move. However the charges are not used up. Therefore the electrons that leave the cathode must enter into anode.

94. When copper(II) sulphate solution is electrolysed using platinum electrodes, the blue colour of the solution gradually fades. Why? After some time when the solution is almost colourless bubbles appear at the cathode. Why do they appear, what are they?

Hint: The blue colour is due to the presence of Cu^{2+} (aq). During the electrolysis, these ions are removed as they change into Cu(s). Hence the colour fades. When the concentration of Cu(II) ions becomes very small, water molecules begin to be discharged. The bubbles are hydrogen.

95. When does electrolysis of Cu(II) sulphate using copper electrodes start?

Hint: When the anode has been worn away so much that it no longer dips properly into the electrolyte.

96. Li^+ has radius of 68 pm and Na^+ has 98 pm radius. Why does the Li^+ travel less easily than Na^+ ion?

Hint: Li^+ ion has a very dense centre of charge and water molecules are attached to it extremely strongly. The ion carries several layers of water molecules around it which greatly increase the effective size of the ion. Hence Li^+ ion does not conduct so well as Na^+ ion.

97. A bottle containing an acid was marked with pH 5. What can we understand from it?

Hint: A weak acid will have a pH in this range. However it is possible that the bottle contains a very dilute solution of a strong acid like H_2SO_4. The pH alone does not tell us whether a solution contains a strong or a weak acid. pH depends upon concentration.

98. Imagine a hydrogen atom and a chlorine atom coming together with a total energy of 10×10^{-19} J. Why is it that the molecule is likely to fall apart almost as soon as it is made?

Hint: The energy available is greater than the energy needed to bring the bond between them. The outcome is that the molecule breaks up and the atoms fly apart as soon as it is made. If the collision is to be successful there has to be a way for the hydrogen chloride molecule to lose energy before it has time to break apart. This can be achieved in the presence of, say nitrogen, where the molecule can give up some of its energy to the nitrogen molecule. It can also happen that gaseous reactions are influenced by collisions of the molecules with the walls of the container.

99. Suggest the best method for measuring the rate of the reaction for $5C_2O_4^{2-}(aq) + 2MnO_4^-(aq) + 16H^+(aq) \rightarrow 10CO_2(g)+2Mn^{2+}(aq)+ 8H_2O$.

Hint: Manganate(VII) ions (MnO_4^-) are purple in solution while Mn^{2+} ions are pale pink. Colorimetric method is the best. It could also be done by measuring the volume of CO_2 given off. This approach has the disadvantage that CO_2 is somewhat soluble in water, so initial measurements of gas volume at the start of the experiment will be erroneous.

100. Two bond energies are sometimes quoted for the bond in the chlorine molecule 242 kJ mol^{-1} and 1150kJmol^{-1}. Which do you think refers to homolysis and to heterolysis?

Hint: The smaller value is for the homolytic fission. Heterolytic fission requires a process very much like ionisation to occur. This requires a great deal of energy.

101. The reaction between hydrogen and bromine is $H_2(g)+Br_2(g) \rightarrow 2HBr\ (g)$, predict the rate law.

Hint: Rate = $k_1[H_2][Br_2]^{3/2} / k_2[Br_2]+[HBr]$.

The mechanism involves free radicals. Therefore it is wrong to represent it as rate = $k[H_2(g)][Br_2(g)]$.

102. How does the addition of an inert gas at constant volume and constant pressure effect the equilibrium constant?

Hint: The addition of an inert gas at constant volume causes a increase in the pressure of the system. Since K_p is independent of pressure, the degree of advancement of the reaction is not affected. The addition of inert gas at constant pressure causes an increase in the volume of the system. Thus the reaction moves towards larger number of gaseous reactants or products.

103. How is the energy of activation different from the energy of intermediate complex?

Hint: By definition the energy of activation is the difference in the average of reactants and the energy of intermediate complex which is formed while going from reactants to products. The intermediate complex generally lies above the average energy of products.

104. What will happen to the pressure of a gas containing fixed number of molecules at constant volume if the temperature is raised?

Hint: The pressure of the gas is due to the collisions of the molecules with the side of the vessel. An increase in temperature causes increase in average molecular speeds. As the volume is kept constant the molecules will collide more frequently with the sides of the vessel and also impart more change in momentum. Therefore the pressure increases.

105. How many years it would take to spend one Avogadro's number of Rupees at a rate of 10 lakh of rupees in one second?

Hint: Rupees spent in one second = 1000000 i.e. 10^6.
Taking 365 days equivalent to a year
Rs. that can be spent = $10^6 \times 365 \times 24 \times 60 \times 60$
Avogadro's number = 6.02×10^{23}
$\therefore$ Number of years = $(6.02 \times 10^{23}) / (10^6 \times 365 \times 24 \times 60 \times 60)$
= 1.91×10^{10} years.

106. The composition of a sample of Wüstite is $Fe_{0.93}O_{1.00}$. What percentage of the iron is present in the form of Fe(III)?
Hint: The compound is non-stoichiometric where electrical neutrality is achieved by converting appropriate number of Fe^{2+} to Fe^{3+} ions. 7 Fe^{2+} ions are missing out of the expected 100 Fe^{2+} ions. The missing 2×7 positive charge is compensated by the presence of Fe^{3+} ions. Replacement of one Fe^{2+} ion by a Fe^{3+} ion causes the positive charge to increase by 1 unit. Thus the 14 positive charge is compensated by 14 Fe^{3+} ions out of a total of 93 iron ions.

107. What could be inferred from the following graph?

$P_A°$ $P_B°$
$X_A = 1$ $X_A = 0$
$X_B = 0$ $X_B = 1$

Hint: This shows that there is a positive deviation from Raoult's law.

108. What causes the explosions in houses built on land that were used for rubbish dumps?
Hint: Methane is released when the rubbish decomposes under the ground. It can happen that methane escapes into houses built over the site. The explosion is the rapid burning of methane if, for example a spark starts the reaction.

109. Explain the hybridisation of the carbon atoms in 1,2–butadiene.
Hint: The formula for 1,2–butadiene is

$H_2C=C=CH-CH_3$ (carbons numbered 1 2 3 4)

It is an example of cumulative dienes. The hybridisation of carbon atoms is $sp^2(C_1)$ $sp(C_2)$ $sp^2(C_3)$ and $sp^3(C_4)$. The cumulative double bonds cannot participate in resonance and thereby the delocalisation or displacement of electrons is not possible as is seen in conjugated dienes.

110. Draw the structures of the molecules having $C_2H_2Cl_2$ formula and note which of them show dipole moment?
Hint:

I: H(Cl)C=C(Cl)H (trans) II: H(Cl)C=C(H)Cl (cis) III: $H_2C=CCl_2$

The Structure I is trans 1,2–dichloroethene it has no dipole moment. The structure II which is cis 1,2–dichloroethene and the III, 1,1–dichloroethene have dipole moment.

111. If bromine water is used to test for an alkene, the water contains HOBr. What will be made if this molecule adds across the double in ethene?
Hint: 2-bromoethan-1-ol(ethylene bromohydrin) is formed. The solution is decolourised.

H–C(Br)(H)–C(OH)(H)–H

112. After an ozonolysis experiment the only product obtained is CH_3CHO. What could be the starting compound?
Hint: The Starting alkene could be any of the following as both will deliver acetaldehyde.

$H_3C(H)C=C(H)CH_3$ — cis-but-2-ene; $H_3C(H)C=C(H)CH_3$ — trans-but-2-ene

113. Why is ethyne used in oxy-acetylene welding?
Hint: The exothermic nature of the reaction allows the very high temperature to melt metals to be reached.

114. How do you account for the stability of acylinium ion?
Hint: among the resonance structuresin the acylinium ion (II),every atom is complete with octet configuration while in the structure I the Carbon atom has only six electrons.

$R-C^+=\ddot{O}:$ (I) $\longleftrightarrow$ $R-C\equiv O:^+$ (II)

115. How do you account for the oil slick formation on the surface of the ocean (in case of oil tanker accidents)?
Hint: Petroleum which mainly contains paraffins along with other fractions has lower density than water therefore it forms oil slicks on the surface.

116. Predict the structure of alcohols that will be made by the addition of H_2SO_4 followed with reaction by water to but–1–ene and but–2–ene.
Hint:

$H_3C-CH_2-CH=CH_2 + H_2SO_4 \rightarrow H_3C-CH_2-CH(OSO_3H)-CH_3 \xrightarrow{H_2O} H_2SO_4 + H_3C-CH_2-CH(OH)-CH_3$ (butan-2-ol)

With but–2–ene also the product formed is butan–2–ol just in the case of but–1–ene.

117. Why is that if benzene is burned in air, it produces a very smoky flame and large amounts of soot are sent into the air?
Hint: Going to the very large proportion of carbon in benzene, a great deal of oxygen is needed to convert all of it to carbon dioxide.

$C_6H_6(l) + 15O_2(g) \rightarrow 12CO_2(g) + 6H_2O(g)$.

In air there is not enough oxygen available close to the flame and some of the carbon remains as soot.

118. Suggest a reason why benzene diazonium ion attacks the para rather than the ortho position on phenol.

Hint: There is steric hindrance at work. The large benzene diazonium ion cannot easily occupy the ortho positions to the phenolic OH group.

119. PCl_5 and $SOCl_2$ can be used to convert ethanol to chloroethane. In which method chloroethane can be separated easily and why?

Hint: $C_2H_5OH + SOCl_2 \rightarrow C_2H_5Cl + SO_2 + HCl$

Sulphur dioxide and hydrogen chloride are gasses and therefore bubble out of the reaction mixture, leaving a reasonably pure product. While with PCl_5, the products

$C_2H_5OH + PCl_5 \rightarrow C_2H_5Cl + POCl_3 + HCl$

formed are not volatile as in the former case.

120. What is alcoholometry?

Hint: The method or process by which the percentage of alcohol present in a sample is determined is known as alcoholometry. In this method, the specific gravity of the given sample is measured with the help of a hydrometer. The alcohol percentage is noted from the standard tables coinciding the measured specific gravity (percentage of alcohol in some beverages : Beer 3 to 6, Wine 8 to 10, Whisky 35 to 40, Brandy 40 to 45, Gin 40 to 45, Rum 45 to 55, etc.).

121. How is Pyrene helpful in extinguishing fire?

Hint: CCl_4 is known as pyrene. It is used as fire extinguisher. Its vapour density is very high. The vapours of CCl_4 form a layer around the burning object thus cutting off oxygen supply from the air needed to continue the combustion.

122.Mention the self addition compounds of acetylene.

Hint:(a) $3HC{\equiv}CH \longrightarrow$ (benzene)

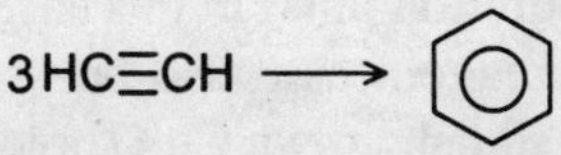

(b) $2HC{\equiv}CH \longrightarrow H_2C{=}CH{-}C{\equiv}CH$
mono vinyl acetylene

$H_2C{=}CH{-}C{\equiv}C{-}CH{=}CH_2$
divinyl acetylene

(acetylene)

(c) 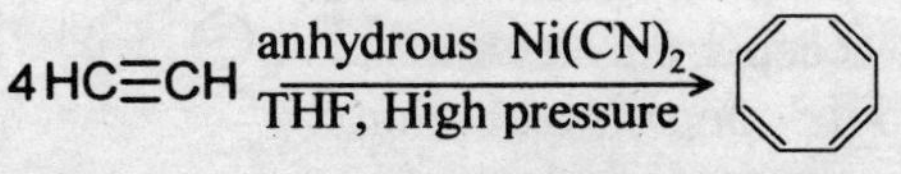 $4HC{\equiv}CH \xrightarrow[\text{THF, High pressure}]{\text{anhydrous } Ni(CN)_2}$ (cyclo-octatetraene)

cyclo-octatetraene

123. How do you account for the reaction of HI with $C_6H_5{-}O{-}CH_3$?

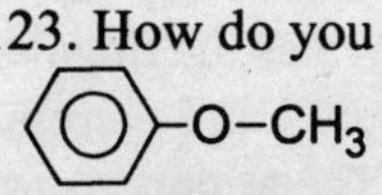

Hint: The possibilities are

(i) 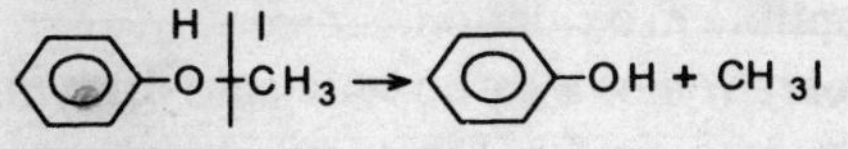 $C_6H_5{-}O{+}CH_3 \rightarrow C_6H_5{-}OH + CH_3I$

(ii) 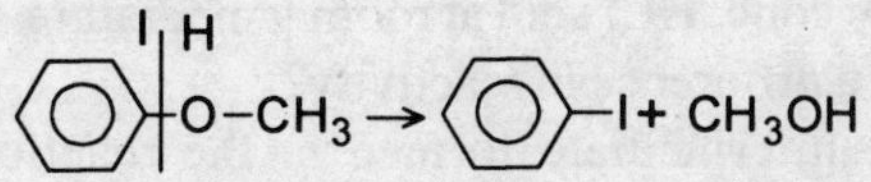 $C_6H_5{+}O{-}CH_3 \rightarrow C_6H_5{-}I + CH_3OH$

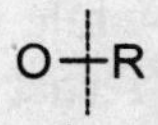

O–R bond cleavage leads to a more stable R^+ (carboaction) which in turn reacts with I^- to give RI.

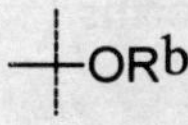

–OR bond cleavage leads to a less stable $C_6H_5^+$ species. Nucleophilic substitution on the benzene ring ($^SN^2$) is not possible while the high energetic species $C_6H_5^+$ can't form by ($^SN^1$) mechanism.

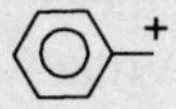

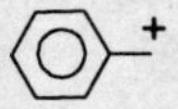

124. Why benzoic acid is stronger acid than phenol?

Hint: Phenoxide ion shows more resonating structures than the benzoate ion.

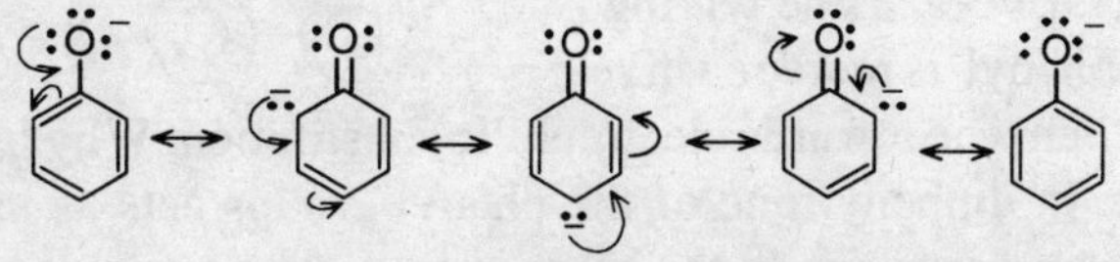

(resonance structures of phenoxide ion)

The benzoate ion is more stabilised because the electronegative charge is carried by the oxygen atom.

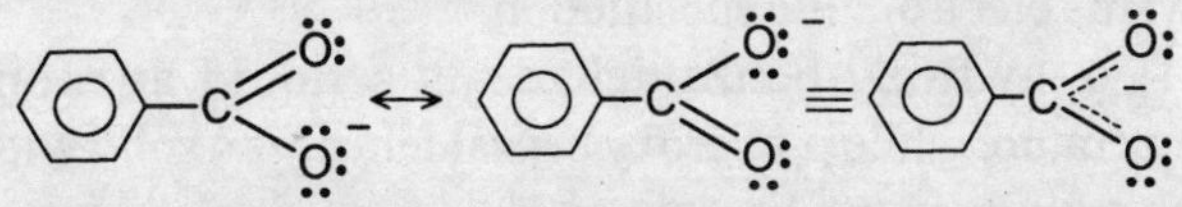

(resonance structures of benzoate ion)

In the resonating structures of phenoxide ions, the negative charge is present on the less electronegative carbon atoms of the ring.

125. Why is it easier to carry out experiments with iodoethane rather than chloroethane?

Hint: C_2H_5Cl is a gas while C_2H_5I is a liquid at room temperature and atmospheric pressure. It is much easier to handle liquids rather than gases in a reaction.

126. Why do alcohols with a long carbon backbone become immiscible with water?

Hint: The miscibility of alcohols with water depends on (i) the attractions between the molecules owing to the hydrogen bonds between them and (ii) the entropy of mixing. If the carbon chain becomes too long. Its presence in water disrupts many hydrogen bonds between the water molecules. This is energetically unfavourable. Eventually the enthalpy change for mixing becomes positive and overcomes the favourable entropy of mixing.

127. What will be the result of using pent–2–ene, (i) by absorbing it in conc. H_2SO_4 followed by dilution with H_2O (ii) by reacting it with diborane followed by H_2O_2?

Hint: (i) The addition is Markownikoff addition. The product formed is $CH_3{-}CH_2{-}CH_2{-}CHOH{-}CH_3$ (pentan–2–ol).

(ii) It is anti-Markownikoff addition. The product formed is $CH_3{-}CH_2{-}CH_2{-}CH_2{-}CH_2OH$ (pentan–1–ol).

128. To convert butan–1–ol to 1–chlorobutane using HCl requires heat and a $ZnCl_2$ catalyst. The isomeric 2–methyl–propan–2–ol is converted to 2–chloro–2–methyl

propane by conc. HCl acid at room temperature. Why is there such a difference in reactivity?

Hint: The intermediate formed in the reaction with 2-methyl propan-2-ol is a tertiary carbocation while a primary carbocation is obtained from butan–1–ol. Tertiary carbocation formation is energetically favoured and also more the stability of the carbocation, earlier it forms.

129. What type of alcohol is 1–phenyl ethanol (Ph–CHOH–CH_3). What is the product formed on oxidation with $K_2Cr_2O_7$ solution. What product is expected when this alcohol is subjected to dehydration with conc. H_2SO_4?

Hint: It is a secondary alcohol. Acetophenone, a ketone is formed. On dehydration it forms styrene (phenyl ethene).

130. Phenyl group is known to exert negative inductive effect. But each phenyl ring in diphenyl is more reactive than benzene towards electrophilic substitution. Why?

Hint: In diphenyl one of the phenyl groups acts as an electron donor while the other is an electron acceptor. This makes phenyl ring more reactive than benzene towards electrophilic substitution.

131. o–hydroxy benzaldehyde is a liquid at room temperature while p-hydroxy benzaldehyde is a solid with high melting point. Explain.

Hint: In o–hydroxy benzaldehyde intramolecular hydrogen bonding is present while intermolecular hydrogen bonding in the case of p-hydroxy benzaldehyde allows the association of the molecules leading to the solid nature.

o-hydroxy benzaldehyde p-hydroxy benzaldehyde

132. Why is propanone a liquid at room temperature and pressure while propane is a gas?

Hint: The attractions between dipole moments on different molecules are sufficiently strong to keep the molecules together as a liquid.

133. From which alkene, butanone and pentan–2–one can be obtained?

$H_3CH_2CH_2C(H_3C)C{=}C(CH_2CH_3)(CH_3)$ cis $\quad$ $H_3CH_2CH_2C(H_3C)C{=}C(CH_3)(CH_2CH_3)$ trans

3,4–dimethylhept–3–ene

Hint: both will deliver the same products on ozonolysis.

134. What are crown ethers?

Hint: They are heterocyclic polyethers generally having 4 or more oxygen atoms. They have crown like appearance: The complex of 18–Crown–6 and $K^+ [SCN]^-$ is shown here. There are 18 atoms present in total in the ring. The ring contains 6 oxygen atoms.

135. The labels of different reagent bottles containing CH_3OH, C_2H_5OH, CH_3COOH and CH_3COCH_3 were destroyed. How would you identify them?

Hint: The following tests are performed in each case.

(i) The one which forms a white precipitate with $NaHSO_3$ is CH_3COOCH_3

(ii) The one which gives brisk effervescence (due to release of CO_2 gas) is CH_3COOH.

Now methanol and ethanol are left. They can be distinguished by iodoform test. C_2H_5OH responds to the test with I_2 and NaOH forming a yellow precipitate.

136. Why does a solution of 2–hydroxypropanoic acid (lactic acid made by the method) $CH_3CHO \rightarrow CH_3CH(OH)CN \rightarrow CH_3CH(OH)COOH$ is optically inactive?

Hint: The cyanide ion can attack the aldehyde from above or below the plane of molecule. The products are mirror images of each other. The resulting solution is optically inactive as it contains equal amount of d and l varieties.

137. What happens when iodine solution is added to propanol?

Hint: Iodine is decolourised. The reaction involves keto–enol tautomerism.

$$H_3C-\overset{O}{\overset{\|}{C}}-CH_3 \rightleftharpoons H_3C-\overset{OH}{\overset{|}{C}}=CH_2$$

138. If the labels of different reagent bottles containing CH_3COCH_3, CH_3CHICH_3, CH_3CH_2CHO and heptane were destroyed, how would you test the samples and identify them?

Hint: (i) CH_3COCH_3 would response iodoform test giving yellow precipitate

(ii) CH_3CH_2CHO would give positive test with schiff's reagent, turning the solution pink. In this case Fehling's solution or Tollen's reagent can also be used for the identification where reddish brown Cu_2O precipitate and silver mirror is produced respectively.

(iii) CH_3CHICH_3 can be identified by adding $AgNO_3$ solution which gives yellow precipitate.

139. How do you account for the stability of α–amino naphthalene and α–nitro naphthalene towards oxidation process?

NH_2 $\quad$ NO_2

Hint: Oxidising agent is one which accepts the electrons given out during the oxidation process. Therefore the oxidising agent is an electron acceptor one. In 1–nitro naphthalene, the nitro group is an electron withdrawing and thus deactivates the ring i.e. the electron density of the ring is lowered and thereby making it resistant towards oxidation. The amino group being electron repelling group activates the ring and thus making it susceptible to oxidation.

140. In the preparation of a carboxylic acid from nitrile using alkaline hydrolysis. (i) How would you know that

the reaction is complete? (ii) Why very little ammonia is given off when acid is used for hydrolysis?
Hint: (i) No more ammonia is given off.
(ii) ammonia formed in the reaction is converted to NH_4^+ ions by taking H^+ ions from the acid.
141. Why is $CaCl_2$ guard tube used in the preparation of ethanoyl chloride?
Hint: They are used to prevent water vapour entering from the atmosphere and reacting with ethanoyl chloride.
142. How do you account for the b.p. of carboxylic acids and esters?
Hint: Carboxylic acids have higher b.p. due to intermolecular hydrogen bonding (association of molecules takes place). In the case of esters hydrogen bonding formation is not possible.
143. How do you account for the basicity of $CH_3CH_2CH_2NH_2$, $CH_2=CHCH_2NH_2$ and $HC\equiv CCH_2NH_2$?
Hint: In all the cases the nitrogen atom is having a lone pair of electrons showing the basic nature. It can be observed that at β–position with amino group the carbon atom is attached to a single, double and triple bond respectively by sp^3, sp^2 and sp hybradisation. More the s-character in the hybradised orbitals forming the bonds, more the electronegativity and thus increasing the electron – withdrawing capability. Therefore the order is $CH_3CH_2CH_2NH_2 > CH_2=CHCH_2NH_2 > HC\equiv CCH_2NH_2$.
144. How do you account for the acid hydrolysis and alkaline hydrolysis of ester?
Hint: Esters undergo hydrolysis very slowly even with boiling water. In the presence of acids or alkalies, the rate of hydrolysis is higher. As the acid hydrolysis is reversible, it is slower than alkaline hydrolysis which is irreversible.

Acid hydrolysis $RCOOR' + H_2O \overset{H^+}{\rightleftharpoons} RCOOH + R'OH$

$$R{-}C(=O){-}OR' \overset{H^+}{\rightleftharpoons} R{-}\overset{+}{C}(OH){-}OR' \overset{H_2O}{\rightleftharpoons} R{-}C(OH)(\overset{+}{O}H_2){-}OR'$$

$$\rightleftharpoons R{-}C(OH)_2{-}\overset{+}{O}(H)R' \overset{-R'OH}{\rightleftharpoons} R{-}\overset{+}{C}(OH)_2 \overset{-H^+}{\rightleftharpoons} R{-}C(=O){-}OH$$

Alkaline hydrolysis

$$HO^- + R{-}C(=O){-}OR' \longrightarrow \left[HO^{\delta-}\cdots C(=O)(R)\cdots OR'^{\delta-}\right] \longrightarrow HO{-}C(=O){-}R + R'O^- \longrightarrow {}^-O{-}C(=O){-}R + R'OH$$

transition state

145. How would you distinguish CH_3COONa from $CH_3COOC_2H_5$?
Hint: Sodium acetate is a salt formed by the neutralisation reaction,
$CH_3COOH + NaOH \rightarrow CH_3COONa + H_2O$. It is an ionic solid compound which is soluble in water and undergoes ionisation. Salts of W.A. + S.B. or W.B. + S.A. undergo hydrolysis. Ethyl acetate is an ester formed by esterification,
$CH_3COOH + HOH_5C_2 \rightarrow CH_3COOC_2H_5$.
It is a covalent compound and is a liquid. It undergoes hydrolysis in the presence of acid, base or even water to give acetic acid and ethyl alcohol.
146. Why cyanides are soluble in water while isocynaides are sparingly soluble?
Hint: Alkyl cyanides or nitriles are polar molecules. They form intermolecular hydrogen bonding with water. Thus accounting for higher solubilites. In isonitriles the hydrogen bonding with water molecule is not possible and therefore are insoluble.
147. Why is ethoxyethane more soluble in acids like HCl
Hint: Just as the oxygen atom in water has one of its lone pairs promoted to make the oxonium ion, so a lone pair on the oxygen atom too in an ether can be promoted.Hydrogen bonding can occur between oxygen atom on an ether and hydrogen atoms on surrounding water molecules. Hydrogen bonding cannot occur between oxygen atoms in water molecules and hydrogen atoms bonded to carbon in ether. These bonds are not sufficiently polar to give the hydrogen atoms an appreciable positive charge.

$$H_5C_2{-}\ddot{O}{-}C_2H_5 \xrightarrow{H^+} H_5C_2{-}\overset{+}{\ddot{O}}(H){-}C_2H_5 \quad ; \quad H_5C_2{-}\ddot{O}(\cdots H{-}\ddot{O}{-}H){-}C_2H_5$$

Protonation Hydrogen bonding

148. What would happen if you were to add silver nitrate solution to quaternary ammonium salts?
Hint: The quaternary salt is ionic and when it dissolves, the component ions are free to move through water. The chloride and silver ions give a white precipitate of AgCl.
149. Why are b–keto acids unstable?
Hint: It has been observed that some b–keto acids start loosing CO_2 even at normal temperatures. The same will be brisk at higher temperatures ~100°C.

$$R{-}C(=O){-}CH_2{-}C(=O){-}O{-}H \rightleftharpoons R{-}C(=\overset{+}{O}H){-}CH_2{-}C(=O){-}O^- \overset{-CO_2}{\rightleftharpoons}$$

$$\underset{\text{keto form}}{R{-}C(=O){-}CH_3} \rightleftharpoons \underset{\text{enol form}}{R{-}C(OH){=}CH_2}$$

This is due to the possibility of a cyclic transition structure.

150. Why aniline doesn't undergo Friedel-Craft's reaction?

Hint: The lone pair of electrons on nitrogen atom of NH_2 group form a coordinate covalent bond with the Friedel-Craft catalyst i.e. dry $AlCl_3$ and the complex formed is an electron withdrawing thus deactivating the ring towards electrophilic substitution

151. How do you account for the basicity of primary amine, alkyl nitrile and imine?

Hint. Consider, $H_3C—NH_2$, $H_3C—C{\equiv}N$ and and as examples The hybridisation of carbon attached to the nitrogen in the above structures are sp^3, sp and sp^2 respectively. As 's' character in sp hybridised orbital is more than the other two, the electronegativity of the carbon atom attached to nitrogen atom in alkyl nitrile is more and therefore the lone pair of electrons on nitrogen atom can't be easily available for donation. Therefore the order of basicity is $H_3C—NH_2 > H_3C—C{\equiv}N > H_3C—CH{=}NR$.

152. How is reverse-hyperconjugation different from hyperconjugation?

Hint. In hyperconjugation, sigma electrons of a single bond are in conjugation with multiple bonds is observed, e.g.

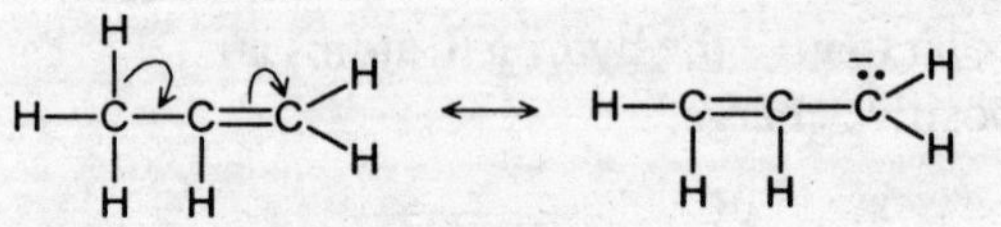

$$H_3C{-}CH{=}CH_2 \longleftrightarrow H{-}C(H){=}C(H){-}\ddot{C}^{-}H_2\ (H^+)$$

In reverse hyperconjugation, p-electrons are attracted from the alkene system, e.g.

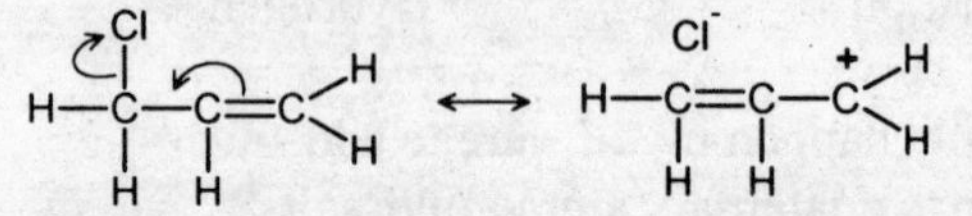

$$ClCH_2{-}CH{=}CH_2 \longleftrightarrow Cl^-\ \ H{-}C(H){=}C(H){-}\overset{+}{C}H_2$$

153. Write the major contributing structure of methylacetate

Hint. One can write three resonating structures

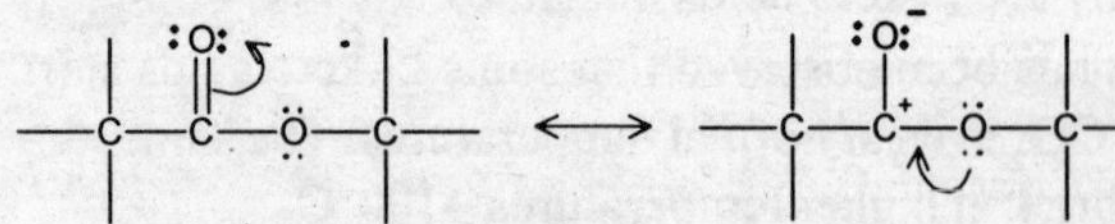

$$\underset{\text{I}}{—C—C({=}\ddot{O}:)—\ddot{O}—C—} \longleftrightarrow \underset{\text{II}}{—C—\overset{+}{C}(—\ddot{O}:^-)—\ddot{O}—C—}$$

$$\longleftrightarrow \underset{\text{III}}{—C—C(—\ddot{O}:^-){=}\overset{+}{O}—C—}$$

II and III structures contain charge separation. Structure II has an electron deficient carbon atom (trivalent) structure of methyl acetate as it has no charge and also all the carbon and oxygen atom have octet stable electronic structure.

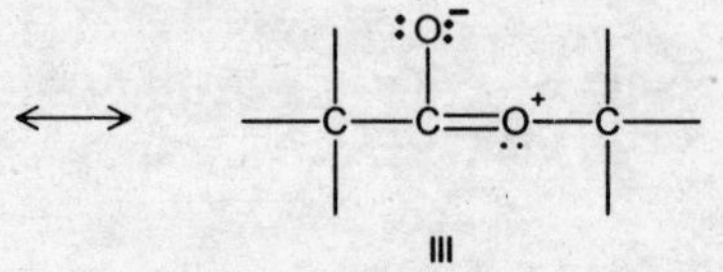

154. Why is the chlorinated product of a-1-chloro-2-methyl butane optically inactive?

Hint. The reaction can be represented as follows

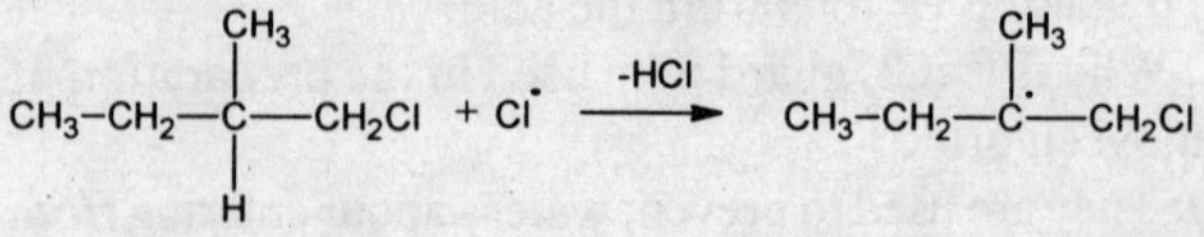

$$CH_3{-}CH_2{-}C(CH_3)(H){-}CH_2Cl + Cl^{\bullet} \xrightarrow{-HCl} CH_3{-}CH_2{-}\dot{C}(CH_3){-}CH_2Cl$$

The tertiary free radical formed here is flat which combines with chlorine to give both d and l compounds leading to the formation of dl (racemic) mixture which is optically inactive.

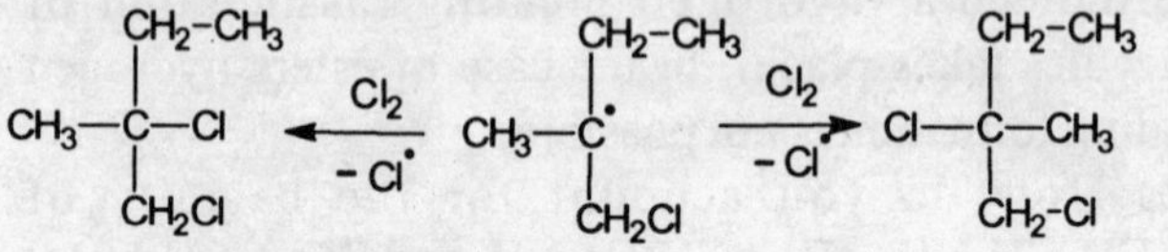

$$CH_3{-}C(CH_2{-}CH_3)(Cl)(CH_2Cl) \xleftarrow[-Cl^{\bullet}]{Cl_2} CH_3{-}\dot{C}(CH_2{-}CH_3)(CH_2Cl) \xrightarrow[-Cl]{Cl_2} Cl{-}C(CH_2{-}CH_3)(CH_3)(CH_2{-}Cl)$$

155. Explain the difference in $^SN^2$ reaction of alkyl halides and aromatic halides.

Hint. $^SN^2$ proceeds in a one step mechanism in the case of alkyl halides while it occurs in a two step process in aromatic halides. In aliphatic halides the nuclephilic substitution occurs via the unstable transition structure containing a pentavalent carbon, $\overset{\delta-}{HO}\text{----}C\text{----}\overset{\delta-}{Cl}$ while a tetrahedral carbon atom with a negative charge distribution on the ring (H, Cl on ring carbon, ring bearing −) is the reactive intermediate in aromatic nucleophilic substitution.

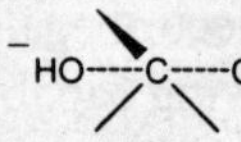

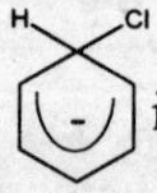

156. What do you understand of cine substitution?

Hint. It can be understood taking into account the conversion of chlorobenzene to aniline,

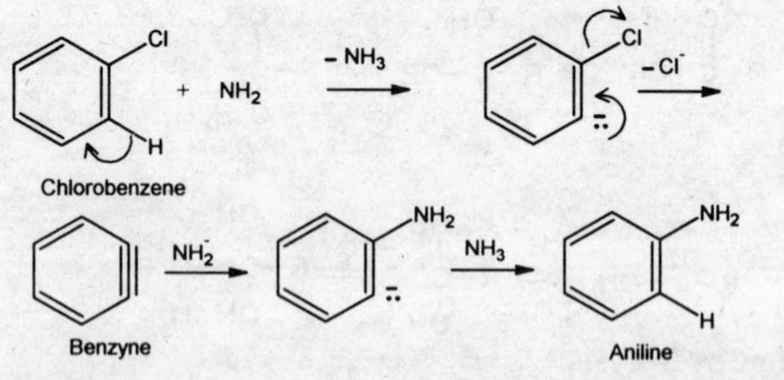

Chlorobenzene $+ NH_2 \xrightarrow{-NH_3}$ (carbanion) $\xrightarrow{-Cl^-}$ Benzyne $\xrightarrow{NH_2}$ (aryl anion) $\xrightarrow{NH_3}$ Aniline

The conversion via benzene intermediate is called as cine substitution.

157. How do you account for the formation racemic mixture from an optically active molecule in $^SN^1$ reactions?

Hint. We know that $^SN^1$ takes place in two steps, first step is the formation of a carbocation which can be represented as

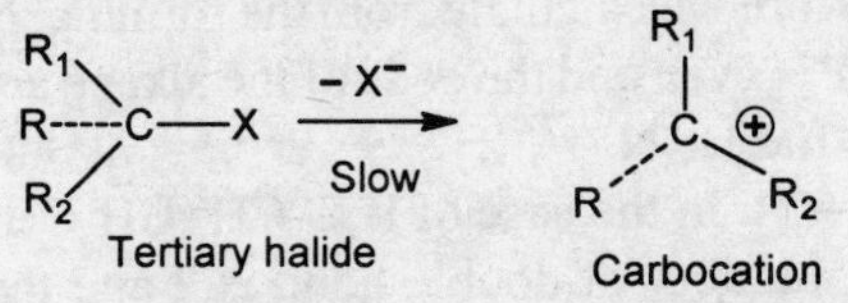

$$R_1R_2RC{-}X \xrightarrow[\text{Slow}]{-X^-} R_1R_2RC^{\oplus}$$

Tertiary halide — Carbocation

In this step, the optically active halide (sp^3 hybridised carbon atom with three different alkyl groups and a halogen atom is a chiral molecule) is converted to a carbocation (or carbonium ion) with sp^2 hybridised trivalent carbon atom which is achiral with trigonal planer stracture. In the second step the nucleophile can attack either from front side or back side leading to the formation of a product with retained or inverted configuration resulting in racemic mixture.

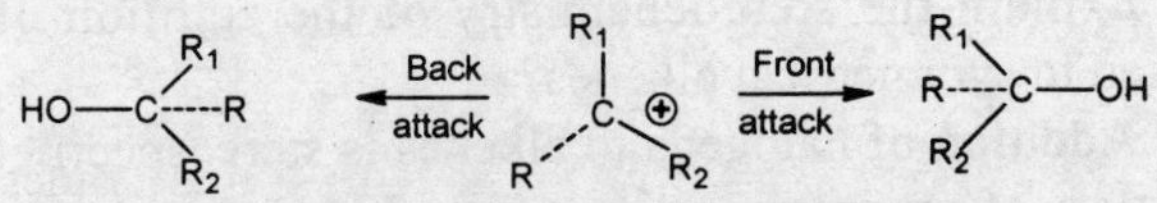

$$HO{-}CR_1R_2R \xleftarrow{\text{Back attack}} R_1R_2RC^{\oplus} \xrightarrow{\text{Front attack}} R_1R_2RC{-}OH$$

158. Write the details of nucleophilic aromatic substitution mechanisms.

Hint. Aromatic nucleophilic substitution can take place by $^SN^1$, $^SN^2$ or benzene mechanism which can be represented as

(a) $^SN^1$:

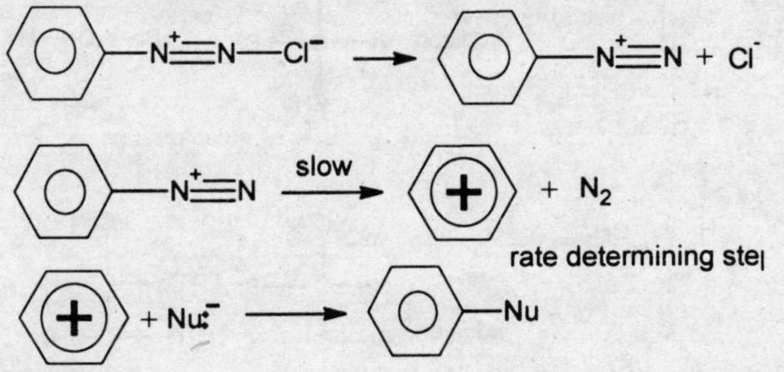

$$C_6H_5{-}\overset{+}{N}{\equiv}N{-}Cl^- \longrightarrow C_6H_5{-}\overset{+}{N}{\equiv}N + Cl^-$$

$$C_6H_5{-}\overset{+}{N}{\equiv}N \xrightarrow{\text{slow}} C_6H_5^{+} + N_2$$

rate determining step

$$C_6H_5^{+} + Nu^- \longrightarrow C_6H_5{-}Nu$$

(b) $^SN^2$:

$$C_6H_5X + Nu^- \xrightarrow{\text{slow}} [C_6H_5(X)(Nu)]^- \longleftrightarrow \longleftrightarrow \equiv$$

$$[C_6H_5(X)(Nu)]^- \xrightarrow[-X]{\text{fast}} C_6H_5Nu$$

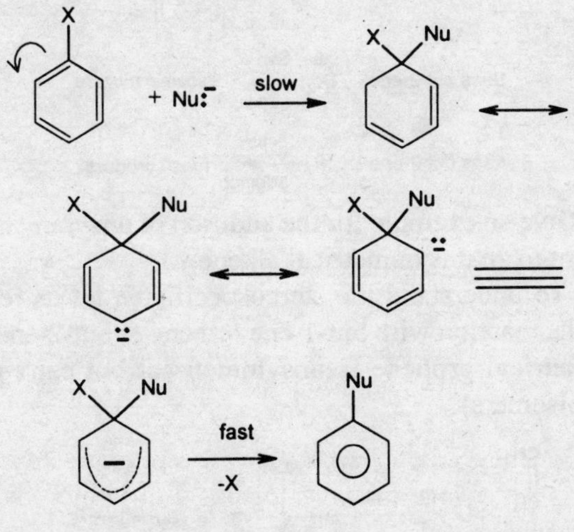

(c) Benzene mechanism : Aryl halides with strong bases for e.g. with NH_2 produce aromatic amines which can be represented as

$$C_6H_5Cl \xrightarrow{-NH_2} \text{R-C}_6H_4\text{-}NH_2\ (meta) + \text{R-C}_6H_4\text{-}NH_2\ (ortho)$$

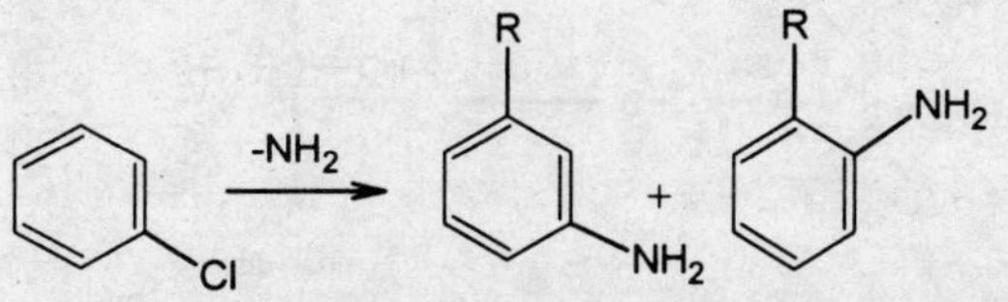

The entering nucleophile doesn't always take the position of the leaving the nucleophile and therefore a mixture of compounds is formed. The benzene mechanism takes place in two steps i.e. elimination followed by an addition step.

(i) Elimination step :

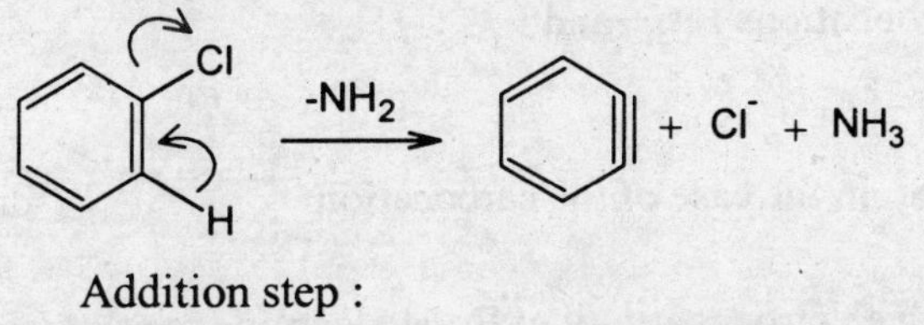

$$C_6H_5Cl \xrightarrow{-NH_2} C_6H_4\ (\text{benzyne}) + Cl^- + NH_3$$

(ii) Addition step :

$$C_6H_4 + NH_3\ (\text{liquid}) \longrightarrow C_6H_5NH_2 \text{ or } C_6H_5NH_2$$

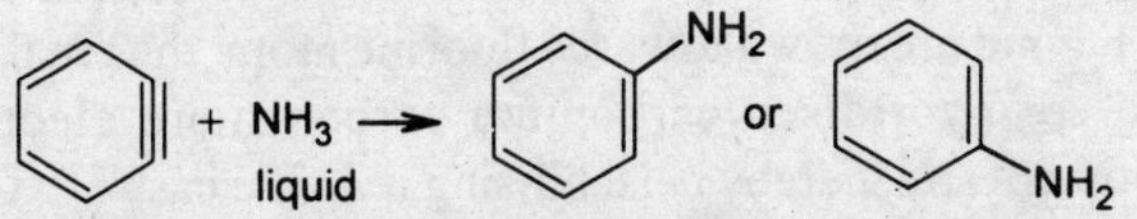

This reaction is not shown by the compounds having no α-hydrogen on the adjacent carbon atom to the carbon bearing the halogen atom, e.g.

159. Why is the dehydration of alcohols to alkenes not possible via OH^- elimination?

Hint. The dehydration proceeds via the formation carbonium ion. The carbocation formation can be represented as,

(a) $R{-}CH_2{-}CH_2{-}OH \longrightarrow R{-}CH_2{-}\overset{+}{C}H_2 + OH^-$

(b) $R{-}CH_2{-}CH_2{-}\ddot{O}{-}H \xrightarrow{H^+} R{-}CH_2{-}CH_2{-}\overset{+}{O}H{-}H$

$\longrightarrow R{-}CH_2{-}\overset{+}{C}H_2 + H_2O$

The path (a) is very difficult as it requires very high amount of energy. H_2SO_4 which is a dehydrating agent also acts as a catalyst to weaker the C-O bond thus leading to the form of the carbonium ion.

160. Why do both cis and trans but-2-ene yield optically inactive butane-2,3-diol on oxidation with $KMnO_4$ solution?

Hint. The reaction involves cis-hydroxylation of the double bond. The cis alkene gives a meso diol while the trans one leads to the formation the optically inactive racemic mixtures as

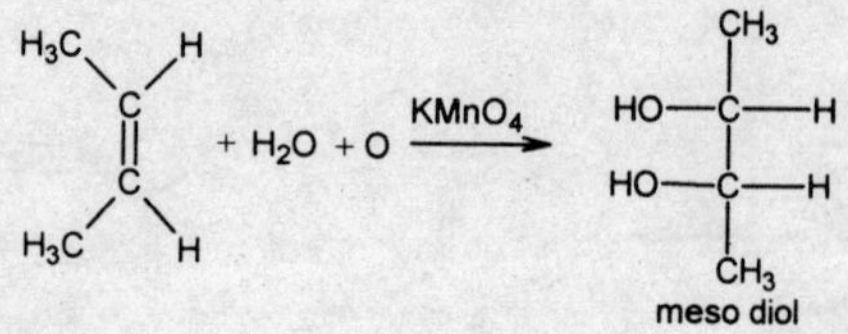

$H_3C-CH=CH-CH_3$ (cis) + H_2O + O $\xrightarrow{KMnO_4}$ meso diol

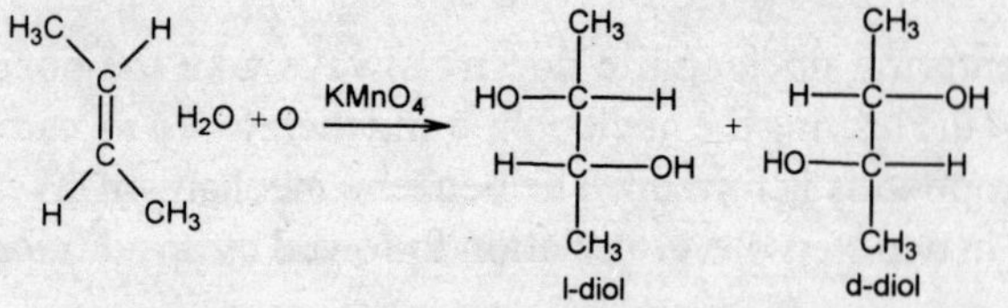

$H_3C-CH=CH-CH_3$ (trans) + H_2O + O $\xrightarrow{KMnO_4}$ l-diol + d-diol

161. How do you account for the stability of the carbocations F_3C+ and F_3C-C+H_2

Hint. In the case of the carbocation $F_3C-CH_2^+$ due to high electronegativity of F, the electrons present in each C-F bond are drawn towards fluorine atom, thus making the sp^2 hybridised carbonium carbon more electron deficient and thereby destabilising it. In the case of $+CF_3$, the fluorine atoms with completely filled p-orbitals try to overlap with the vacant orbital of the sp^2 hybridised carbon of the carbonium ion thus giving the double bond character to the C-F bonds thereby stabilising the carbonium ion by charge distribution

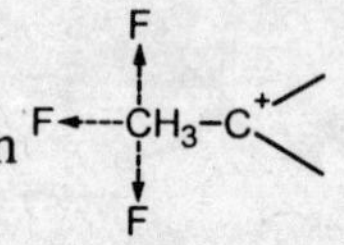

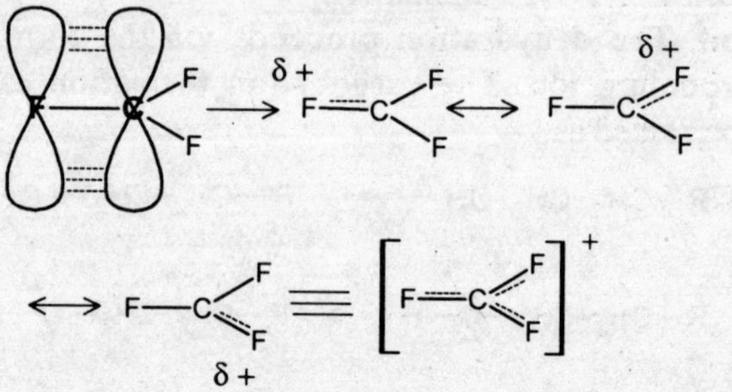

162. How many optically active, meso and racemic varieties are possible for $HOOC-(CHOH)_2-COOH$?

Hint. In the compound, the number of chiral carbon atoms is two i.e. even and the molecule can be divided into two identical halves. Therefore we calculate using

$a = 2^{n-1}$; $m = 2^{0.5n-1}$; $r = a/2$

$a = 2^{2-1} = 2^1 = 2$; $m = 2^{0.5\times 2-1} = 2^0 = 1$; $r = 2/2 = 1$

Therefore, 2 optically active, 1 meso and 1 racemic forms are possible.

163. Calculate the number of geometrical isomers possible for $H_3C-CH=CH-CH=CH-CH_3$ and $H_3C-CH=CH-CH-CH=CH-CH_2-CH_3$?

Hint. In the case of $H_3C-CH=CH-CH=CH-CH_3$, there are two double bonds and the ends of the alkene (diene) are the same; therefore we calculate from the formula, $N = 2^{n-1}+2^{p-1}$ (when 'n' is even and the ends of the alkene are the same $p = n/2$); $N = 2^{2-1} + 2^{1-1} = 2^1+2^0 = 3$. (here $p = 2/2 = 1$). In the case of $H_3C-CH=CH-CH-CH=CH-CH_2-CH_3$ number of double bond is 2 and the ends are different, therefore we calculate the no. of geometrical isomers from the formula, $N = 2^n = 2^2 = 4$. Let us also try to understand the calculation for $H_3C-CH=CH-CH-CH=CH-CH_2-CH_3$. Here n = 3 and the ends are different. We calculate the no. of geometrical isomers from equation, $N = 2^{n-1}+2^{p-1}$ (and the ends of the alkene are different when n is odd $p = n+1/2$) $N = 2^{3-1}+2^{2-1} = 2^2+2^1 = 4+2 = 6$.

164. Explain the stereochemistry of the addition of halogen to symmetrical alkenes?

Hint. Addition of halogens to alkenes is stereospecific. Addition of symmetrical reagents like halogens of symmetrical alkenes can take place either by syn or anti addition which can be exemplified with but-2-ene as follows:

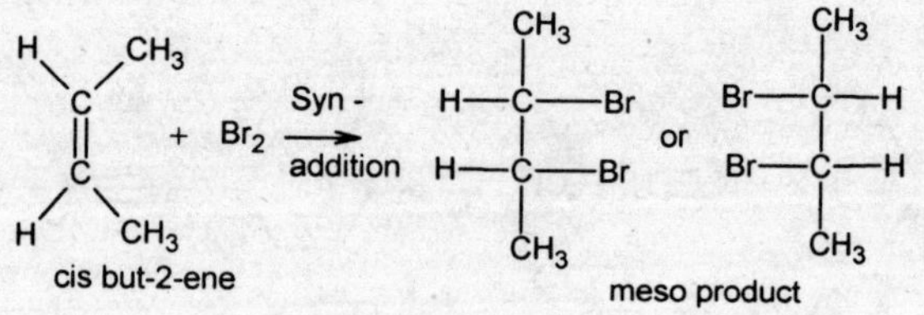

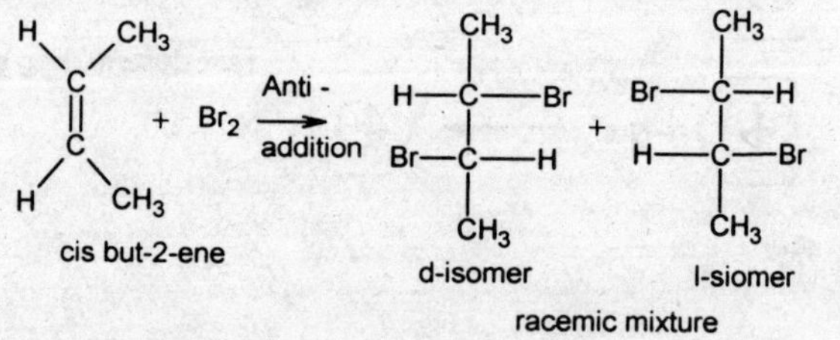

cis but-2-ene + Br_2 $\xrightarrow{\text{Syn - addition}}$ meso product

cis but-2-ene + Br_2 $\xrightarrow{\text{Anti - addition}}$ d-isomer + l-siomer (racemic mixture)

trans but-2-ene + Br_2 $\xrightarrow{\text{Syn - addition}}$ racemic mixture

trans but-2-ene + Br_2 $\xrightarrow{\text{Anti - addition}}$ meso product

165. Give an example for the addition of unsymmetrical reagent to an unsymmetrical alkene.

Hint. To understand the stereospecificity, let us represent the reaction with but-1-ene (ethene & but-2-ene are symmetrical propene is unsymmetrical but can't give stereoisomers)

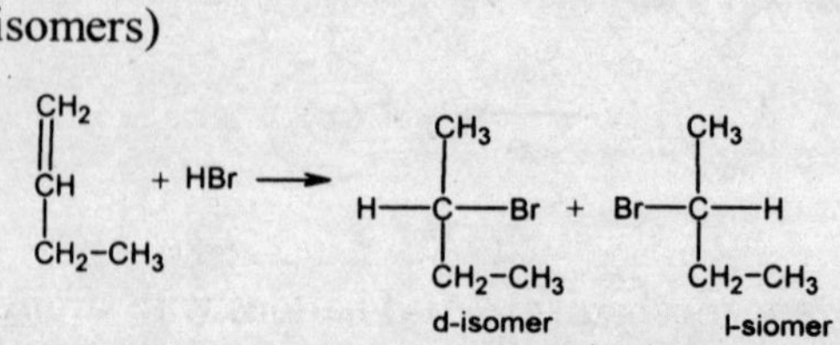

$CH_2=CH-CH_2-CH_3$ + HBr → d-isomer + l-siomer (racemic mixture)

166. Suggest a better method for preparation of an alkyl benzene?

Hint. The formation of an alkyl benzene can be brought about by Würtz-Fittig and Friedel-Craft reactions. The former is better when one has to introduce long n-alkyl chain, e.g.

$$C_6H_5\text{–}Br + 2Na + BrCH_2(CH_2)_3CH_3 \xrightarrow[-2NaBr]{\text{dry ether}} C_6H_5\text{–}(CH_2)_4CH_3$$

n-pentyl benzene

167. Write a note on the esterification of alcohols by carboxylic acids or their derivatives.

Hint. Alcohols react with carboxylic acids to form esters in the presence of dil. acids H_2SO_4 (~10%) or HCl.

$$R\text{–}\overset{O}{\overset{\|}{C}}\text{–}OH + HO\text{–}R' \overset{H^+}{\rightleftharpoons} R\text{–}\overset{O}{\overset{\|}{C}}\text{–}O\text{–}R' + H_2O$$

This reaction is a reversible reaction. The forward reaction is esterification while the backward reaction is a hydrolysis reaction. Acid catalyses both the reactions. If acid derivatives, acid halides, amides and anahydrides) are employed instead of carboxylic acids, esterification will be an irreversible reaction as water is not formed in these reactions to hydrolyse the ester formed e.g.

$$R\text{–}\overset{O}{\overset{\|}{C}}\text{–}X + HO\text{–}R' \longrightarrow R\text{–}\overset{O}{\overset{\|}{C}}\text{–}O\text{–}R + HX$$

168. Why C_2H_5OH reacts with HI but not with HCN?

Hint. Ethanol is a weak base which reacts with HI a strong acid. HCN being a weak acid can't enter into reaction with the alcohol, a weak base.

169. What way would you suggest to prepare Phenetole?

Hint. Phenetole is Ph-O-CH_2-CH_3. This ether can be prepared by Williamson's synthesis via. $^SN^2$ mechanism. The possible representations are

(a)Ph-Br + $NaOCH_2CH_3 \rightarrow$ no reaction

(b)Ph-ONa + $BrCH_2CH_3 \rightarrow$ Ph-O-CH_2-CH_3

The pathway (a) cannot deliver the required product as the aryl bromide is more stable due to resonance and therefore is less reactive. The second pathway takes place as the bromoalkane is more reactive. The reaction (a) can be made possible by selecting the aryl bromide having -I effect atom or group like NO_2 etc. at ortho or para position to the bromine atom.

170. Comment on the nucleophilic substitution of alkyl and benzyl halides.

Hint. Allyl and benzyl halides can undergo both $^SN^1$ and $^SN^2$ reactions. In $^SN^1$, carbocation (carbonium ion) intermediate is formed while transition state (with a pentavalent carbon character) forms in $^SN^2$. In the case of allyl and benzyl halides both the carbonium ion and the transition state are stabilised. The presence of polar protic solvents prefers the $^SN^1$ pathway with any type of nuclophiles while the presence of polar aprotic solvents with concentrated nucleophiles leads the $^NS^2$ pathway.

171. Explain why vinyl and aryl halides do not undergo $^SN^1$ reaction?

Hint. Vinyl halide,H_2C=CHX and aryl halide don't undergo $^SN^1$ due to (a) $^SN^1$ proceeds in two steps, the first step, i.e. the formation of carbonium ion (or carbocation) and is the rate determining step.

$$H_2C{=}CH\text{–}X \longrightarrow H_2C{=}\overset{+}{C}H + X^-$$

$$C_6H_5\text{–}X \longrightarrow C_6H_5^+$$

(b)the carbonium ions formed from these halides are highly unstable compared to the one obtained from primary haloalkanes. In both the cases the halogen atom is attached to the carbon atom which is sp^2 hybridised and therefore a large amount of energy is required for the C-X bond dissociation. The hybridisation of the carbon of the carbocation is in sp hybridisation which makes it further unstable.

172. How is ozone destroyed by chlorofluro hydrocarbons?

Hint. Chlorofluro hydrocarbons are highly stable and non-bio degradable. The released CFCS when reach the upper atmospheric layer are subjected to continuous UV-radiation exposure, thus undergoing a series of reactions by free radical mechanism

$$CF_2Cl_2 \xrightarrow{\text{UV rays}} {}^{\bullet}CF_2Cl + Cl^{\bullet} \text{ or } {}^{\bullet}CF_2 + 2Cl^{\bullet}$$

Ozone is attacked as,

$$O_3 + Cl^{\bullet} \longrightarrow ClO + O_2\ ,\ ClO + O \longrightarrow Cl^{\bullet} + O_2$$

Chlorine free radical produced in the last step may initiate the chain reaction in both cases of CFC's and ozone dissociation.

173. Explain why alcohols can't undergo nucleophilic substitution with a halide ion while haloalkanes undergo this reaction with OH^- ion?

Hint. We know that the order of basicity of nucleophiles is $CN^- > I^- > OH^- > Br^- >$ CH3COO$^-$ > Cl^-. Generally the weak leaking nucleophile will be displaced by a stronger nucleophile. Therefore X^- in haloalkanes being weaker is substituted by the OH^- nucleophile. Replacement of OH^- by X^- in alcohols can be achieved in presence of acid catalyst as shown,

$$R-\ddot{O}-H + H^+ \longrightarrow R-\overset{+H}{O}-H \xrightarrow{-H_2O} R^+$$

The $^+OH_2$ is a good leaving nucleaophile which can give nucleophilic substitution with alkali halides, e.g. NaCl etc.

174. Why bromine with phenol gives different products in polar and non-polar solvents?

Hint. Bromine undergoes ionisation in polar solvents like H_2O to a greater extent leading to high Br^+ concentration. It therefore forms 2,4,6–tribromophenol. In non polar solvents like CCl_4, CS_2, etc. less concentration of Br+ ions leads to the formation of ortho and p–bromophenol.

175. Write the importance of Thorpe reaction?

Hint. In the presence of a strong base C_2H_5OH/Na two molecules of alkyl nitriles of the same or different nature condense to give b–iminonitrile

$$H_3C-C\equiv N \longrightarrow H_3C-C^+=N^- \quad ;$$

$$H-CH_2-C\equiv N \xrightarrow{H^+} \bar{C}H_2-C\equiv N$$

$$H_3C-C^+(=N^-) + \bar{C}H_2-C\equiv N \longrightarrow H_3C-C(=N^-)-CH_2-C\equiv N$$

$$\xrightarrow{H^+} H_3C-C(=NH)-CH_2-C\equiv N$$

176. Give the details of the reaction of nitrous acid with amines?

Hint. Primary amines except methyl amine generally form alkanols

$$R-NH_2 + HNO_2 \longrightarrow R-OH + N_2 + H_2O$$

methylamine+ HNO_2 — 1:1 → dimethyl ether; 1:2 → methyl nitrite

Secondary amines show Lieberman's nitroso reaction

$$(CH_3)_2NH + OHNO \xrightarrow{-H_2O} (CH_3)_2N-N=O$$

dimethylnitrosoamine

Tertiary amines give amine salt which on decomposition gives nitrosoamine

$$R_3-N + HNO_2 \longrightarrow R_3N\text{-}HNO_2$$

$$\xrightarrow{\Delta} R_2NNO + ROH$$

Cyclic amines depending upon their structure undergo Demjamov ring expansion or ring contraction as shown below:

(a) cyclopropyl methyl amine Ring expansion

2 (cyclopropyl)$-CH_2NH_2$ + HNO_2 ⟶ (cyclopropyl)$-CH_2OH$ + (cyclobutanol, OH)

(b) cyclobutylamine Ring contraction

2 (cyclobutyl)$-NH_2$ + HNO_2 ⟶ (cyclobutyl)$-OH$ + (cyclopropyl)$-CH_2OH$

177. Explain why aniline gives m-nitroaniline on direct nitration?

Hint. In nitration, halogenation etc. the $-NH_2$ group is protected by converting it generally to acetanilide as they have the oxidising property. If aniline is subjected to nitration directly,

$$C_6H_5-NH_2 \xrightarrow{HNO_2/H_2SO_4} C_6H_5-\overset{+}{N}H_3$$

$$\xrightarrow{HNO_2/H_2SO_4} (NO_2SO_4^-)\ C_6H_5-\overset{+}{N}H_3$$

The anilium ion formed deactivates the ring. Although $-NH_3$ group is a ring activator through its lone pair electrons, anilium ion can't activate the ring due to (i) lone pair on N atom is involved in bond formation and (ii) being a positively charged group, it withdraws electrons from the ring thus directing the electrophile to the more electron density position i.e. meta.

178. Write the importance of Collman's reagent?

Hint. Collman's reagent is Sodium tetra carboxyl ferrate (II), $Na_2[Fe(CO)_4]$. The reagent reacts with haloalkanes in the presence of triphenyl phosphine $(Ph)_3P$ to give aldehyde having one carbon atom more than the ones present in R–X.

$$R-X + Na_2[Fe(CO_4)] \xrightarrow{(ph)_3P} RCOFe(CO)_3P(ph)_3$$

R–CHO is obtained from the product of the above reaction by quenching it with acetic acid.

179. How is glacial acetic acid prepared?

Hint. Absolutely pure acetic acid as it shines like a glacier. It is prepared from its aqueous solution by the following steps.

(a) First step: The aqueous solution is neutralised with Na2CO3,

$$2H_3C\text{–}COOH + Na_2CO_3 ® 2H_3C\text{–}COONa + H_2O + CO_2$$

(b) Second step : The reaction mixture obtained in the first step is subjected to distillation in the presence of conc. H_2SO_4

$$H_3C\text{–}COONa + conc.\ H_2SO_4 ® H_3C\text{–}COOH + NaHSO_4$$

180. Comment on the advantages and disadvantages of reduction by $LiAlH_4$ and $NaBH_4$.

Hint. $LiAlH_4$ can't be used in protic solvent as it reacts with them. Being a stronger reducing agent than $NaBH_4$ it can reduce acid derivatives like acid halides, esters etc. to primary alcohols and haloalkanes to alkanes while $NaBH_4$ can't bring out these changes. $NaBH_4$ can be used in protic solvents like H_2O, alcohol etc. as it is less reactive. Reduction of sugars in aqueous medium can be carried out with $NaBH_4$.

181. Why is the –OH group in o–hydroxy benzaldehyde is less reactive than that in meta or para isomers and how does it effect other properties?

Hint. o-hydroxy benzyldehyde is Salicylaldehyde. The –OH group in o-isomer is involved in intra-molecular hydrogen bonding and therefore is not readily available for reactions. Due to intermolecular hydrogen bonding; is has lower b.p., i.e. high volatility and less solubility in water than m and p-isomers in which intermolecular hydrogen bonding is present. The later two isomers can also form intermolecular hydrogen bonding with water molecules thus making them more soluble in water.

182. Mention the reactions which distinguish aniline from nitrobenzene.

Hint. Aniline with HCl can form a soluble anilinium chloride while nitrobenzene is not soluble in HCl.

$$ph—\ddot{N}H_2 + HCl \rightarrow ph\overset{+}{N}H_3\overset{-}{Cl}$$

Nitrobenzene on reduction in the presence of Z in and aqueous ammonium chloride is converted to phenyl hydroxylamine which reduces the Tollen's reagent to silver (or silver mirroring). Reduction of the reagent by aniline is not possible.

$$ph—NO_2 \xrightarrow[NH_4Cl]{Zn\ \&} ph—NHOH$$

$$\xrightarrow[Reagent]{Tollen's} ph—NO_2 + H_2O + Ag$$

183. How is aniline distinguished from benzylamide?

Hint. Aniline and benzylamine can be distinguished by diazotisation reaction. Benzylamine with HNO_2 produces benzyl alcohol just as the primary alkylamines (CH_3NH_2 exception) Aniline forms diazonium compound with HNO_2 which in turn reacts with alkaline α-naphthol solution to give an orange-red dye.

$$\underset{\text{benzlyamine}}{ph—CH_2—NH_2} + HNO_2 \longrightarrow ph—CH_2OH + N_2 + H_2O$$

$$ph—NH_2 + NaNO_2/HCl\ (\text{or } HNO_2) \xrightarrow{0\text{ - }4°C} ph—\overset{+}{N_2}\overset{-}{Cl}$$

$$\xrightarrow[\beta\ \text{naphthol}]{\text{alkaline}}$$ (1-phenylazo-2-naphthol: naphthalene ring bearing N=N—ph and OH)

184.Explain why benzaldehyde reduces Tollen's reagent but not the Fehling & Benedict's solution.

Hint. Fehling solution in a 1:1 mixture of $CuSO_4$ solution + Rochelle salt. Benedict's reagent is $CuSO_4$ + Sodium citrate + Na_2CO_3. Tollen's reagent is ammonical silver nitrate solution. Tollen's reagent is comparatively a stronger oxidising agent than the other two reagents and therefore oxidises benzaldehyde. Benzyldehyde on the other hand undergoes Cannizzaro's reaction in presence of base present in Fehling's and Benedict's reagent solutions.

185. How does a line spectrum differ from continuous spectrum?

Hint. A ray of white light has a series of seven colours in the visible region. In this case, we obtain a continuous spectrum as violet merges into blue, blue into green and so on. Line spectrum does not show a continuous spread of wavelengths but emit light at specific wavelengths. The emission spectra produced are generally the line spectra.

186. Explain why the energy levels have negative signs.

Hint. Because the zero of potential energy is the world beyond the atom, electrons inside are lower in potential energy. Energy of an electron at infinite distance from the nucleus is zero. When an electron approaches the nucleus, the energy of the electron decreases and it becomes negative. The negative sign of the electron also indicates that there is attraction between +vely charged nucleus and the negatively charged electrons.

187. Can you know the ionisation energy of the hydrogen atom from Balmer series?

Hint. No, the ionisation energy is the energy difference when the electron is in the energy levels with n = 1 and n = a. The Balmer series can give the energy transition with n = 2. In the case of Lyman series ni corresponds to 1 and therefore can be helpful to calculate the ionisation energy.

188. The $I.E_1$ and $I.E_2$ of Li atom are 520 and 7298 K J mol-1. Can we say that the 1s orbital requires energy of the order of 7298 K J mol-1 for an electron to be removed?

Hint. Li has $1s^2$, $2s^1$ electronic configuration while Li+ has $1s^2$. It is correct to say that the 2s orbital requires 520 KJ mol-1 but it is wrong to state that an amount of 7298 K J mol-1 is required to remove an electron from 1s orbital. Here Li+ has $1s^2$ configuration which is different from an atom having $1s^2$ configuration. The nuclear charge in Li+ is higher. Therefore it is more attracted towards the nucleus and thus requires an extra energy than required to remove from an electron from 1s orbital of neutral atom.

189. Can there be intermolecular forces in NaCl?

Hint. No, NaCl is an ionic solid. The Na^+ and Cl^- ions are held by electrostatic lines of forces. As there are no molecules in ionic substances, there can be no intermolecular forces.

190. Why do metals conduct heat and electricity?

Hint. The atoms in a metal lattice are embedded in a sea of electrons. The electrons exist in energy bonds that spread throughout the lattice. These electrons are free to move, and are responsible metals being good conductors of heat and electricity.

191. How are insulators different from semiconductors?

Hint. In insulators, the energy gap between the conduction and valence bands is much greater which doesn't permit the excitation of valence electrons. In intrinsic semiconductors e.g. Si, Ge etc. the valence electrons can be excited into the conduction band at room temperature while in the case of extrinsic semiconductors, the conduction is mainly caused by the movement of positively charged holes (p–type) or by the movement of electrons (n–type).

192. Explain why graphite can mark paper?

Hint. The layers of the carbon atoms in graphite are held together by van der Waal's forces. As these forces are very week, the layers can slide over one another. At room temperature and pressure, the sliding of the layers is assisted by small molecules e.g. from air that are trapped between layers.

193. Why does the pressure of a fixed volume of gas increase with increasing temperature?

Hint. The average speed of the gas molecules increases with increase in temperature. The momentum of the particles also increases as a consequence of increase in the speed. Therefore, the force, with which the particles collide on the walls of the container also increases, thus an increase in the pressure is observed.

44. Comment on the relation between the ionic radii and the heats of hydration?

Hint. Smaller the size of ion, greater is the hydration energy. The water molecules get closer to the centre of the charge and as a result of this, the attractions are increased and thus the ion carries several layers of water molecules around it. In other words, we can say that in smaller ions, the degree of hydration is high.

195. Comment on the spontaneity of the reaction :

$\Delta G = 4\ \Delta G_f\,(Al_2O_3)\ (s) - 3\ \Delta G_f\,(Fe_3O_4)\ (s)$

{Given that $4\Delta G_f\ (Al_2O_3)\ (s) = -1582.4$ K J mol-1; $3\ \Delta G_f\,(Fe_3O_4)\ (s) = -1014.2$ K J mol-1}.

Hint. $\Delta G = 4\Delta G_f\ (Al_2O_3)\ (s) - 3\ \Delta G_f\ (Fe_3O_4)\ (s)$ substituting the values, $\Delta G = 4 \times (-1582.4) - 3 \times (-1014.2) = -3287$ KJ.

From the large negative value of ΔG_f one concludes that it is a spontaneous reaction. This reaction is a violent one but when it is ignited at high temperature. Infact this reaction doesn't take place at room temperature (or ordinary temperature). The free energy value doesn't tell us all information. Spontaneity has to be found out by doing experiment carefully.

196. What would you predict about the quantities of reactants and products of a reaction whose standard free energy change is approximately -250 K J mol-1?

Hint. At equilibrium, the amount of products will certainly be more than the reactants. The value of the free energy change will not tell us, how fast the reaction takes place. Equilibrium may or may not be achieved in any reasonable time skate. This is so because thermodynamics do not allow us to predict anything about the reaction rates.

197. What would happen if a very little amount of $Na_2S_2O_3$ crystals are added to a solution containing KI and I_2?

Hint. I_2 reacts with $Na_2S_2O_3$, $I_2(s)+2S_2O_3^{2-}\ (aq)$ ® $2I^-\ (aq) + S_4O_6^{2-}\ (aq)$.

The thiosulphate ions remove I_2 molecules by converting them into iodide ions. Therefore, the equilibrium, $I_2\ (s) + I^-\ (aq)$ ® $I^{3-}\ (aq)$ moves to the left and the colour of reaction gradually changes to colourless.

198. What special care has to be taken during the study of ester equilibrium?

Hint. Let us represent an equilibrium for an ester.

$CH_3COOH\ (l) + CH_3OH\ (l) \rightleftharpoons CH_3COOCH_3\ (l) + H_2O$

The sample taken out at any time interval for study should be run into iced water to slow down the reaction. The reaction flask should be stoppered, otherwise the components i.e. acetic acid, methal and methyl acetate which are volatile would escape. As a consequence of this, equilibrium can't be achieved.

199. Ionic solids are expected to some extent in ethanol. Explain why it is not a good solvent as water?

Hint. Ethanol $H_3C{-}C{-}\overset{\delta-}{O}{-}\overset{\delta+}{H}$ is a polar molecule.

Ethanol molecules with partial negative and positive

charges on oxygen and hydrogen respective can gather around positive and negative ions like water molecules. Ethanol molecules can solvate the ions and form salvation spheres. But ions present in ethanol can easily combine together to make crystals than they do in H_2O (as solvent) due to relative lower permittivity of ethanol.

200. Explain the function of air bleed during vacuum distillation.

Hint. If the pressure is reduced quickly, the liquid boils violently leading to possiblc cxplosion. The rate at which the vacuum pressure is reduced is created can be checked or controlled by air bubbles entering the apparatus. The air bubbles can also enhance the boiling.

201. Why should a catalyst be preferred in an industrial process which works at much higher temperatures than the room temperatures?

Hint. For an industrial process say working at ~200°C, a lot of heat has to be generated to attain this temperature. Not only this but also we have to use methods of cooling products. This can be a very costly affair. Therefore it is always better to find a suitable catalyst which can work at room temperature for the process.

202. Explain the oxidation state or iron atoms in $[Fe\,Fe(CN)_6]^-$

Hint. Each cyanide ion carries a charge of –1. the two iron atoms account for five positive charges. We should expect that there is a mixture of Fe^{2+} and Fe^{3+} ions. In practice also we see the existence of these two ions side by side e.g. in fool's gold, etc.

ratio of total mass for a given element are identical.

203. The law of multiple proportions is stated in trans of mass. Can we restate this law in trans of ratio of atoms? Why?

Hint. Yes, since each atom has a definite mass, twice the number of atoms would have twice the mass. Therefore, the ratio of atoms and the ratio of total mass for a given element are identical.

204. Give the reason for mass defect.

Hint. Since the nucleus is made up of neutrons and protons (nucleons), its mass should be equal to the sum of the constituent protons and neutrons. However, when a stable nucleus is formed, the actual mass of the nucleus, M, is always less than the sum of the masses of the nucleons, M'. The difference is known as mass defect $\Delta M = M' - M$. The mass defect is has been attributed to the packing of nucleons into tiny elements of volume.

205. A graph of binding energy per nucleon, B.E./A, vs. the mass number, A is given below. Approximately what mass number has the most stable nuclei?

Hint. Binding energy is equal to the energy lost in forming the nucleus. As more stable nuclei have higher binding energy per nucleon, the mass number of the most stable nucleus corresponds to the maximum point of the graph.

206. How is the N/P ratio responsible for the stability of the nucleus.

Hint. Both attractive and repulsive forces exist in the nucleus. The attractive forces are responsible for the stabilisation of the nucleus exists between neutrons and protons. These forces arise due to energy exchange between the nucleons. The repulsive forces decrease the stability of the nucleus. These forces exist between protons as they posses positive charge and therefore repel each other. For stability of the nucleus, there should be a balance between attractive and repulsive forces. As protons and neutrons take part in energy exchange process, the attractive forces will dominate if their number is the same.

207. Mention the ways of readjustment of an unfavourable ratio to a more favourable ratio.

Hint. N/P ratio greater than one can be decreased by α–emission i.e. transforming neutron into proton. The ratio can be increased by α–emission. The N/P ration can also be lowered by neutron emission. But such emissions do not occur naturally. K – electron capture also reduces the ratio (e.g. $Au_{79}^{194} + e_{-1}^{0} \rightarrow Pt_{78}^{194}$) but it is possible only in artificial radioactive isotopes. If necessary, the N/P ratio may be increased by positron emission. A positron is an electron with a positive charge. In effect, a proton emits its positive charge, thus becoming a neutron. $\rho_1^1 + n_0^1 \rightarrow \beta^+$.

208. How could one differentiate between absorption and emission spectra?

Hint. As absorption spectrum arises from the interaction of certain wave lengths of light with an atom while emission spectrum is due to the emission of certain wavelengths by an excited atom corresponding to the emission of energy from the atom.

209. How would you account for the probability of finding an electron in the shells?

Hint. The probability of finding an electron which is nearer to the nucleus is greater while it decreases with increasing distance of the shell from the nucleus. The redial distribution function gives the probability of finding an electron at a certain distance from the nucleus. The r.d.f. is a product of (i) the variation in the probability of finding an electron in a cell at various distances from the nucleus and (ii) the number of cells in spherical shells at various distances from nucleus. Charge cloud boundaries are used to enclose all the regions in which the chance of finding a given electron is greater than 1 percent.

210. Is there any difference between the energy of an electron in orbitals represented by the Quantum numbers n = 3, l = 2, m = +2 and n = 3, l = 2 , m = -1?
Hint. There is no difference between the energy of an electron in orbitals with the same principal and angular momentum quantum numbers have same energy.
211. Can you see a spectral line for the transition, $2p_x \rightarrow 2\ p_y$?
We can't see any line. All the three 2p orbitals have the same energy. Therefore in such transitions, no energy is needed in going from one 2p – orbital to another 2P – orbital in the same atom.
212. What does the term degenerate orbitals mean?
Hint. Orbitals which have the same energy for e.g. the three p-orbitals of a particular shell. We can better understand by taking an example. Say in oxygen having $1s^2$, $2s^2$, $2p^4$ electronic configuration, the three occupied 2p orbitals are degenerate orbitals. In helium with $1s^2$ configuration, no degenerate orbitals are present as only one occupied s-orbital is present.
213. Explain the statement – "atomic radius decreases with increasing atomic number within each period".
Hint. With the increase in the atomic number, the nuclear charge increases within a period. The outermost electrons are in the same energy shell and therefore a greater attraction between the positive nucleus and negative electrons leads to the concentration of the electron cloud giving rise to a smaller atom.
214. Why are the radii of Zr and Hf are nearly equal?
Hint. Zr^{40} has the electronic configuration [Kr] $4d^2$, $5s^2$. Hf^{72} has the electronic configuration, [Xe] $4f^{14}$, $5d^2$, $6s^2$. The addition of electrons to the Hf level adds no new energy levels to the atom. The addition of protons to the nucleus causes a large attraction between the nucleus and the electrons. This produces a contraction in the size of electron shell. The contraction in the size of the Hf atom because of the 14 protons just offsets the increase in size due to additional electronic energy level.
215. Explain – why the isolation energy decreases down the same group although the atomic number increases.
Hint. As the atomic number increases down the group, additional energy levels are added successively. Therefore the valence electrons become farther from the nucleus resulting in lesser attraction from the nucleus. The inner orbital electrons will also cause partial shielding effecting thereby further decrease in the attraction between the nucleus and the outer shell valence electrons. Therefore ionisation energy required decreases down the same group.
216. Can you distinguish the coordinate covalent bond in NH^+_4 ion from the other three bonds?
Hint. It is not possible to distinguish it from the other. All electrons are identical. Therefore it is not possible to know whether two electrons forming the bond come from nitrogen atom alone or one each from nitrogen and oxygen atoms.
217. How does pressure effect condensation of ages?
Hint. At high pressures, the intermolecular attraction increases leading to the condensation of the gas. Every gas has a critical temperature above which it cannot be condense into liquid anymore regardless of high pressure.
218. What type of forces can be accounted for between molecules in liquid state?
Hint. Van der Waal's attractions, dipole interactions, hydrogen bonding etc.
219. Why does a block of ice feel cold to your touch?
Hint. The block of ice is at a much lower temperature of the finger. Therefore heat is transferred from the finger to the ice block. The loss of heat from the finger makes the block of ice feel cold.
220. Two ideal gases were taken in different gas bottles initially with a closed concentration between them. The concentration was opened that the ideal gases mix. Comment on the disorder of the process.
Hint. Initially, the two gases were present in separate bottles and therefore there is more order. In the final state sue to mixing of the gases the disorder increases.
221. Explain why, "At absolute zero, the entropy of all pure elements and compounds is zero."
Hint. The kinetic energy is zero at absolute zero. The system is also rigid under the conditions. Due to these factors the system is a completely ordered one. Lack of any disorder results in zero entropy.
222. Whether the ratio of rate constants of an equilibrium reaction remain constant with temperature changes or not?
Hint. The ratio doesn't remain constant. The rate constants of the forward and the backward reactions change with the change in temperature. The change will not always be to the same extent.
223. What does a negative emf for a cell indicate?
Hint. Electrons flow through the external circuit in the reverse direction. Therefore, the opposite half cell reactions take place spontaneously.
224. Write the energy terms and their significance that are included in the oxidation potential?
Hint. Sublimation energy, ionisation energy of the gaseous atom and the hydration energy of two atoms are the same, then the atom with higher sublimation energy will have lower oxidation potential as more energy is required to sublime the atom. If the sublimation and hydration energies of two atoms are the same then the atom with higher ionisation energy will have lower

oxidation potential as more energy is required to remove an electron from the gaseous atom. If the sublimation and ionisation of two atoms are same, then the atom with higher hydration energy will have higher oxidation potential as more energy is resealed when the ion with higher hydration energy is dissolved in water.

225. Would gaseous HCl be considered as Arrhenius acid?

Hint. According to Arrhenius, any substance which produces H^+ ion in water is an acid. The theory is applicable only to aqueous solutions and therefore cannot be applied to gases or solids.

226. Comment critically on the given equilibrium, H^+,

Hint. In water molecule the oxygen atom has two lone pairs of unshared electrons along with a partial negative charge. The proton would be strongly attracted towards the oxygen atom thereby forming hydrated proton referred to as hydronium ion. The large equilibrium constant indicates that the ratio of $[H_3O^+]/[H^+]$ is very high thus leading to the conclusion that H^+ ions do not practically exist in aqueous solution.

227. Comment on the following equilibria

(i) $CH_3\text{-}COOH + H\,SO_4^- \rightarrow H_2SO_4 + CH_3COO^-$

(ii) $HClO_4 + H\,SO_4^- \rightarrow H_2SO_4 + ClO_4^-$

Hint. We know that in a reaction involving the transfer of a proton, the weaker base and the weaker acid tend to form. Sulphuric acid is a stronger acid than acetic acid. Therefore, the equilibrium (i) shifts more towards the left, it can be better represented as, $CH_3\text{-}COOH + H\,SO_4^- \rightarrow H_2SO_4 + CH_3COO^-$ while (ii) should be $HClO_4 + H\,SO_4^- \rightarrow H_2SO_4 + ClO_4^-$. As H_2SO_4 is a weaker acid than $HClO_4$, the equilibrium favours to shift more towards the right.

228. How do you account for the stability of ions?

Hint. The base strength of negative ions will be inversely proportional to the stability. The relative stability of ions can be predicted from the electro negativity of the central atom. As the electro negativity increases, the stability of the ion increases. The order of basicity $F^- < OH^- < NH_2^- < CH_3^-$ can be explained as : F^- ion is the most stable ion as the electro negativity value of F^- is the highest and thus making it the weakest base. The distribution of the charge in the ion is another factor that affects the stability. Distribution of the charge in a larger volume tends to stabilise the ion.

229. Comment on the basicity of ethylamine and acetamide?

Hint. $H_3C—\ddot{N}H_2$ is a base $H_3C—C(=O)\ddot{N}H_2$ is not a base as acyl group ($-COCH_3$) increases the electro negativity of nitrogen atom. Due to this affect, nitrogen will not be in a position to donate electrons to act as a base.

230. How would you account for the nature of oxides and hydroxides in general?

Hint. H, Be, Cr^{+3}, Zn, Al, Ga, In, Ge, Sn, Pb and Sb from amphoteric oxides or hydroxides. In general moving down the group, the trend in case of oxides and hydroxides change from acidic to an amphoteric to basic properties. If an element forms an amphoteric hydroxide in M^{+n} oxidation state, the hydroxide formed in higher oxidation state i.e. M^{n+1}, will be more acidic while those in lower oxidation state i.e. M^{n-1} will be more basic. But we also see exceptions, e.g. both SnO and SnO_2 are amphoteric. Basic oxides and hydroxides are more soluble in acidic solutions than in water. Amphoteric oxides however are more soluble in either basic or acidic solution than in H_2O while acidic oxides are more soluble in basic solutions than in water.

231. What factors are mainly helpful to separate and identify ions?

Hint. Ions may be separated and identified by reactions which form complexes. Amphoteric properties are also highly useful. Using NaOH and NH_4OH, a large number of ions may be separated and identified.

232. What is the significance of Dulong and Petit law?

Hint. The law states that the atomic weight of a solid element is approximately equal to 6.3 calories per mole gram divided by the specific heat of the element in categories per gram per degree. If the specific heat of an element say is 0.122 calories per gram degree, than the approximate atomic weight is 6.3 calories per mole per degree / 0.122 calories per gram degree which comes to 51.64 g / mole. Therefore we can say that atomic weight is 6.3 / specific heat.

233. An atom emits energy by a $4 \rightarrow 2$ transition. How may transitions are possible to reach the ground state?

Hint. We know that the total number of spectral lines can be determined from the relation, $n(n-1)/2$. Number of spectral lines $= 4\,(4-1)/2 = 16 - 4/2 = 6$. We can know the transitions responsible for these 6 spectral from the given graphical representation. The possible transitions are:

$n_4 \rightarrow n_3$, $n_4 \rightarrow n_2$; $n_4 \rightarrow n_1$,

$n_3 \rightarrow n_2$;$n_3 \rightarrow n_1$, $n_2 \rightarrow n_1$.

234. What happens when atom absorbs energy in terms of charge cloud?

Hint. The energy absorbed by an atom is mainly transferred to the electrons present in it. The electron after absorption of energy moves from the shell of lower value to a shell of higher value. The electron with higher value of n would have a charge cloud which extends further from the nucleus of the atom. The electronegativity of electron in higher 'n' value is lower and therefore the attraction between the nucleus and it also decreases. In the process the electron is transferred to a more loosely bound location in the atom.

235. How would you account for new lines of an atom when it placed in a magnetic field?

Hint. We generally say that the orbitals having the same energy are degenerate orbitals, e.g. the electrons present in $2p_x$, $2p_y$ and $2p_z$ orbitals. When an atom is placed under the influence of a magnetic field, the same kind or set of orbitals will no more have the same energy. This is because that some of the orbitals would be parallel to the applied magnetic field while others are not. The extra lines are due to the excitement of the electrons from one of these orbitals to another. Except under these conditions, the orbitals having the same 'n' and values will have same energy.

236. Explain the paramagnetic and diamagnetic characteristics of atoms?

Hint. A spinning charge gives rise to a magnetic filed. Electrons are charged particles / waves and are revolving in stationary orbitals around the nucleus. The electrons give rise to magnetic fields oriented in opposite directions as they have spin quantum numbers of $+\frac{1}{2}$ and $-\frac{1}{2}$. An atom containing orbitals with unpaired electrons will be attracted to a magnetic field. Such atoms are said to be paramagnetic. On the other hand if a given atom contains orbaitals with paired electrons, the magnetic field caused by the electrons will be repelled by the magnetic fields thus making them diamagnetic.

237. What observations can be made from the given graph. Which represents the heating at a constant rate of substances?

Hint. Line ab represents the increase in temperature of the solid with time, t. bc is the line where heat is added to the equilibrium containing solid and liquid at the melting temperature h. Although heat is added, there is no increase in temperature along bc as the heat added is used to do work against the attractive forces in the crystal when the crystal melts. The kinetic energy of solid at 'a' and that of liquid at 'b' is same since according to kinetic molecular theory the kinetic energy is the same if the temperature is the same. The potential energy at b and c is not the same as work is done against the attractive forces in the crystal. The line cd represents the increase in temperature of the liquid. There will be no inceasein temperature along the line de as the addition of heat cannot increase the kinetic energy at boiling point. The line ef represents the increase in temperature of the gas with time, t.

238. Write the expected observations from the following cooling curve for the crystallisation of a liquid.

Hint. ab curve represents the cooling of the liquid. Line bc represents no change in temperature as it remains constant till all the liquid solidify. Curve cd represents the cooling of the solid. At time 'c' there exists an equilibrium between the liquid and solid. Point f is the freezing point of the liquid or melting point of the solid.

239. An ideal gas is kept in a flask which is connected to another flask with vacuum with a closed stop cork in between. What happens to the gas when the stop cork is opened?

Hint. Firstly, the gas flows from the flask containing it to the other flask till the pressure in both the flasks is the same. There will be no change in enthalpy as the process involves molecules moving further apart. As the gas is an ideal one, there are no attractive forces between the molecules. Therefore, no energy will be absorbed or released by the molecules. The disorder in the system before opening the stop cork was lesser compared to the one after opening it. In the final state, the gas molecules have more volume to move around.

240. Discuss the order of decomposition reactions on metal surfaces.

Hint. Decomposition reactions on metal surfaces are the most common examples of zero order reactions

e.g. $2N_2O(g) \xrightarrow{Pt} 2N_2(g) + O_2(g)$ The reaction takes place on a hot platinum surface. When the Pt surface is covered by N_2O molecules, the increase in the concentration of N_2O does not have any influence on the rate as only those N_2O molecules which are in direct contact of the surface would react.

241. How does nucleophilicity differ from basicity?

Hint. The ability of the nucleophile to donate a pair of electrons to the electrophile i.e. the carbocation (carbonium ion) is nuceolhilicity. The ability is determined from the rate constant of the reaction

$H_3\overset{+}{C} + \overset{-}{:}Nu \longrightarrow H_3C-Nu$ Basicity is mainly determined by the capability of donating an electron pair to a proton. $\overset{-}{B}: + H:A \longrightarrow B:H + \overset{-}{A}:$

242. Explain the statement "2–pentanol or 3–pentanol reacts with HCl to give a mixture of 2–chloropentane and 3–chloropentane?

Hint. The carbonium ions formed by dehydration form the alcohols are interconverted due to hydride shift which then react with Cl^- (nucleophile) to give the 2 and 3–Chloropentanes

$$H_3C-CH_2-CH_2-\underset{OH}{\underset{|}{CH}}-CH_3 \xrightarrow{H^+} H_3C-CH_2-CH_2-\underset{OH_2^+}{\underset{|}{CH}}-CH_3 \xrightarrow{-H_2O} H_3C-CH_2-CH_2-\overset{+}{CH}-CH_3$$

$$\updownarrow \; H^- \text{ shift}$$

$$H_3C-CH_2-CH_2-\underset{OH}{\underset{|}{CH}}-CH_3 \xrightarrow{H^+} H_3C-CH_2-CH_2-\underset{OH_2^+}{\underset{|}{CH}}-CH_3 \xrightarrow{-H_2O} H_3C-CH_2-CH_2-\overset{+}{CH}-CH_3$$

NOBEL LAUREATES IN CHEMISTRY

Year	Name	Discovery/Contribution
1901	Jacobus H. van 't Hoff	The laws of chemical dynamics and osmotic pressure in solutions.
1902	Emil Fischer	Sugar and purine syntheses.
1903	Svante Arrhenius	Electrolytic theory of dissociation.
1904	Sir William Ramsay	Inert gaseous elements in air, and determination of their place in the periodic system.
1905	Adolf von Baeyer	Organic dyes and hydro-aromatic compounds.
1906	Henri Moissan	Isolation of the element fluorine, and the adoption in the service of science of the electric furnace called after him.
1907	Eduard Buchner	Biochemical researches and discovery of cellfree fermentation.
1908	Ernest Rutherford	Disintegration of the elements, and the chemistry of radioactive substances.
1909	Wilhelm Ostwald	Catalysis and the fundamental principles governing chemical equilibria and rates of reaction.
1910	Otto Wallach	Pioneer work in the field of alicyclic compounds.
1911	Marie Curie	Discovery of the elements radium and polonium.
1912	Victor Grignard	Grignard reagent, which in recent years has greatly advanced the progress of organic chemistry.
	Paul Sabatier	Hydrogenation of organic compounds in presence of finely disintegrated metals.
1913	Alfred Werner	The linkage of atoms in molecules especially in inorganic chemistry.
1914	Theodore W. Richards	Accurate determinations of the atomic weight of a large number of chemical elements.
1915	Richard Willstätter	Plant pigments, especially chlorophyll.
1918	Fritz Haber	Synthesis of ammonia from its elements.
1920	Walther Nernst	Thermochemistry.
1921	Frederick Soddy	Chemistry of radioactive substances, investigations into the origin and nature of isotopes.
1922	Francis W. Aston	Mass Spetrograph of isotopes, in a large number of non-radioactive elements and enunciation of the whole-number rule.
1923	Fritz Pregl	Micro-analysis of organic substances.
1925	Richard Zsigmondy	Heterogenous nature of colloid solutions.
1926	The Svedberg	Disperse systems.
1927	Heinrich Wieland	Constitution of the bile acids and related substances.
1928	Adolf Windaus	Constitution of the sterols and their connection with the vitamins.

NOBEL LAUREATES IN CHEMISTRY

Year	Name	Discovery/Contribution
1929	Arthur Harden and Hans von Euler-Chelpin	Fermentation of sugar and fermentative enzymes.
1930	Hans Fischer	Constitution of haemin and chlorophyll and synthesis of haemin.
1931	Carl Bosch and Friedrich Bergius	Chemical high pressure methods.
1932	Irving Langmuir	Surface chemistry.
1934	Harold C. Urey	Heavy hydrogen.
1935	Frédéric Joliot and Irène Joliot-Curie	Synthesis of new radioactive elements.
1936	Peter Debye	Dipole moments and diffraction of X-rays and electrons in gases.
1937	Norman Haworth and Paul Karrer	Investigations on carbohydrates and vitaminC. Investigations on carotenoids, flavins and vitamins A and B2.
1938	Richard Kuhn	Carotenoids and vitamins.
1939	Adolf Butenandt and Leopold Ruzicka	Sex hormones. Polymethylenes and higher terpenes.
1943	George de Hevesy	Use of isotopes as tracers in the study of chemical processes.
1944	Otto Hahn	Fission of heavy nuclei.
1945	Artturi Virtanen	Agricultural and nutrition chemistry, especially for fodder preservation method.
1946	James B. Sumner and John H. Northrop, Wendell M. Stanley	Crystallisation of enzymes. Preparation of enzymes and virus proteins in a pure form.
1947	Sir Robert Robinson	Plant products of biological importance, the alkaloids.
1948	Arne Tiselius	Electrophoresis and adsorption analysis, especially the complex nature of the serum proteins.
1949	William F. Giauque	Chemical thermodynamics, particularly concerning behaviour of substances at very low temperatures.
1950	Otto Diels and Kurt Alder	Diene synthesis.
1951	Edwin M. McMillan and Glenn T. Seaborg	Chemistry of the transuranium elements.
1952	Archer J.P. Martin and Richard L.M. Synge	Partition chromatography.
1953	Hermann Staudinger	Macromolecular chemistry.
1954	Linus Pauling	Nature of the chemical bond and its application to

NOBEL LAUREATES IN CHEMISTRY

Year	Name	Discovery/Contribution
		the elucidaion of the structure of complex substances.
1955	Vincent du Vigneaud	Biochemically important sulphur compounds, especially for the elucidation of the structure of complex substances.
1956	Sir Cyril Hinshelwood and Nikolay Semenov	Mechanism of chemical reactions.
1957	Lord Todd	Nucleotides and nucleotidic co-enzymes.
1958	Frederick Sanger	Structure of proteins, especially that of Insulin.
1959	Jaroslav Heyrovsky	Polarographic methods of analysis.
1960	Willard F. Libby	Use of Carbon-14 for age determination in archaeology, geology, geophysics and other branches of science.
1961	Melvin Calvin	Carbon dioxide assimilation in plants.
1962	Max F. Perutz and John C. Kendrew	Structure of globular proteins
1963	Karl Ziegler and Giulio Natta	Chemistry and technology of high polymers.
1964	Dorothy Crowfoot Hodgkin	Determination by X-ray techniques of the structures of important biochemical substances.
1965	Robert B. Woodward	For the art of organic synthesis.
1966	Robert S. Mulliken	Chemical bonds and the electronic structure of molecules by the molecular orbital method.
1967	Manfred Eigen, Ronald G.W. Norrish, George Porter	Extremely fast chemical reactions, affected by disturbing the equilibrium by means of very short pulses of energy.
1968	Lars Onsager	The reciprocal relations bearing his name, which are fundamental for the thermodynamics of irreversible processes.
1969	Derek Barton, Odd Hassel	Development of the concept of conformation and its application in chemistry.
1970	Luis Leloir	Sugar nucleotides and their role in the biosynthesis of carbohydrates.
1971	Gerhard Herzberg	Electronic structure and geometry of molecules, particularly free radicals.
1972	Christian Anfinsen,	Ribonuclease, especially concerning the connection between the amino acid sequence and the biologically active confirmation.
	Stanford Moore, William H. Stein	Understanding of he connection between chemical structure and catalytic activity of the active centre of the ribonuclease molecule.

NOBEL LAUREATES IN CHEMISTRY

Year	Name	Discovery/Contribution
1973	Ernst Otto Fischer, Geoffrey Wilkinson	Chemistry of the organometallic, so called sandwich compounds.
1974	Paul J. Flory	Both theoretical and experimental, in the physical chemistry of the macromolecules.
1975	John Cornforth, Vladimir Prelog	Stereochemistry of enzyme-catalyzed reactions . Stereochemistry of organic molecules and reactions.
1976	William Lipscomb	Structure of boranes, illuminating problems of chemical bonding.
1977	Ilya Prigogine	Non-equilibrium thermodynamics, particularly the theory of dissipative structures.
1978	Peter Mitchell	Understanding of biological energy transfer through the formulation of the chemiosmotic theory.
1979	Herbert C. Brown, Georg Wittig	Use of boron-and phosohorus-containing compounds, respectively, into the reagents in organic synthesis.
1980	Paul Berg,	Biochemistry of nucleic acids, with particular regard to recombinant-DNA.
	Walter Gilbert, Frederick Sanger	Determination of base sequences in nucleic acids.
1981	Kenichi Fukui, Roald Hoffmann	Theories, developed independently, concerning the course of chemical reactions.
1982	Aaron Klug	Crystallographic electron microscopy and structural elucidation of biologically important nucleic acid-pro tein complexes.
1983	Henry Taube	Mechanisms of electron transfer reactions, especially in metal complexes.
1984	Bruce Merrifield	Methodology of chemical synthesis on a solid matrix.
1985	Herbert A. Hauptman, Jerome Karle	Direct methods for the determination of crystal structures.
1986	Dudley R. Herschbach, Yuan T. Lee, John C. Polanyi	Dynamics of chemical elementary processes.
1987	Donald J. Cram, Jean-Marie Lehn, Charles J. Pedersen	Development and use of molecules with structure-specific interactions of high selectivity.
1988	Johann Deisenhofer, Robert Huber, Hartmut Michel	Determination of the three-dimensional structure of a photosynthetic reaction centre.
1989	Sidney Altman, Thomas R. Cech	Catalytic properties of R.N.A.

NOBEL LAUREATES IN CHEMISTRY

Year	Name	Discovery/Contribution
1990	Elias James Corey	Development of the theory and methodology of organic synthesis.
1991	Richard R. Ernst	Methodology of high resolution nuclear magnetic resonance (NMR) spectroscopy.
1992	Rudolph A. Marcus	Theory of electron transfer reactions in chemical systems.
1993	Kary B. Mullis, Michael Smith	Polymerase chain reaction (PCR). Oligonucleotide based site-directed mutagenesis and its development for protein studies
1994	George A. Olah	Carbocation chemistry.
1995	Paul J. Crutzen, Mario J. Molina, F. Sherwood Rowland	Atmospheric chemistry particulary concerning the formation and decomposition of ozone.
1996	Robert F. Curl Jr., Sir Harold Kroto, Richard E. Smalley	Fullerenes.
1997	Paul D. Boyer, John E. Walker, Jens C. Skou	For their elucidation of the enzymatic mechanism underlying the synthesis of adenosine triphosphate (A.T.P). Ion-transporting enzyme, $Na^{+}K^{+}$-ATPase.
1998	Walter Kohn, John Pople	Density-functional theory. Computational methods in quantum chemistry.
1999	Ahmed Zewail	Studies of transition states and chemical reactions using femtosecond spectroscopy.
2000	Alan Heeger, Alan G. MacDiarmid, Hideki Shirakawa	Conductive polymers.
2001	William S. Knowles, Ryoji Noyori, K. Barry Sharpless	Catalytic asymmetric synthesis and development of new drugs and material.
2002	John B. Fenn, Koichi Tanaka, Kurt Wüthrich	For developing electro-spray ionisation technique for mass spectrometry. Soft LASER desorption technique for mass spectroscopy 3-D NMR for the identification of biological molecules.
2003	Peter Agre, Roderick MacKinnon	Fundamental chemical knowledge on how cell functions.
2004	Avram Hershko, Aaron Ciechanover, Irwin Rose	Cell regulation of unwanted proteins

AFMC - 1991

1. 2, 4 D is a
(A) Fungicide (B) Rodenticide
(C) Insecticide (D) Herbicide

2. pH of 0.002 NH_2SO_4 is
(A) 0.002 (B) 2.699 (C) 2 (D) 1

3. To become a carbohydrate a compound must contain at least
(A) Six carbons (B) Three carbons
(C) Four carbons (D) Two carbons

4. When the specific heat of a metallic element is 0.214 cal/gm, the atomic weight will be close to which of the following ?
(A) 30 (B) 68 (C) 6.6 (D) 12

5. The force present in the crystals of Naphthalene is
(A) van der Waal (B) Electrostatic
(C) Hydrogen bonding (D) None of these

6. Efficiency of machine A is 80% and that of B is 70%. Which will be having more frictional resistance ?
(A) B (B) A (C) Both A&B (D) None of these

7. Solubility of salt M_2X_3 is x gm mole/l. The solubility product of the salt will be
(A) x^5 (B) $126x^2$ (C) $96x^5$ (D) $108x^5$

8. All fats are
(A) Hydrocarbons (B) Carbohydrates
(C) Aldehydes (D) Esters

9. C-C bond length is minimum in
(A) Benzene (B) Ethyne (C) Ethane (D) Ethylene

10. Nucleophile substitution reaction is more easier in
(A) Nitrobenzene (B) Aniline (C) Phenol (D) Benzene

11. The acid present in tomatoes is
(A) Citric acid (B) Tartaric acid
(C) Lactic acid (D) Oxalic acid

12. The compound which doe not contain OH group is
(A) Anisol (B) Alcohol (C) Methanoic acid (D) Methanol

13. The free radical formed is due to
(A) Heterolytic fission (B) Homolytic fission
(C) Electrophilic attack (D) Nucleophilic attack

14. Which of the following is not a Colloid ?
(A) Smoke (B) Rubber latex
(C) Saline (D) Blood

15. The recent discovery of cytology is that of
(A) Enzymes (B) Co-enzymes
(C) Genetic code (D) Respiration

16. Rutile is an important source for
(A) Ruthenium (B) Radium (C) Titanium (D) Iron

17. The Osmotic pressure in atomosphere of 10% solution of cane-sugar at 69°C is
(A) 8.21 (B) 7.21 (C) 724 (D) 824

18. The conjugate base of a weak acid is
(A) A strong base (B) A weak base
(C) A neutral species (D) May be a weak or strong base

19. The % of H_2 in H_2O_2 is (H = 1, O = 16)
(A) 6.25 (B) 55.92 (C) 55.88 (D) 5.8

20. Electron is an ore of
(A) Mg (B) Ca (C) Sr (D) K

21. The catalyst used in Etard's reaction is
(A) Chromyl chloride (B) Chromic oxide
(C) CCl_4 (D) CCl_3

22. The compound that is not Lewis acid is
(A) BF_3 (B) C_2H_5OH (C) $AlCl_3$ (D) BCl_3

23. The halide that is not hydrolysed is
(A) SiF_4 (B) $PbCl_4$ (C) $SiCl_4$ (D) CCl_4

24. Addition of HCN to Ethyne in presence of $Ba(CN)_2$ as catalyst gives
(A) 1, 1 dicyano ethane (B) Ethyl cyanide
(C) Vinyl cyanide (D) Divinyl cyanide

25. Benzene vapour mixed with air when passed over V_2O_5 catalyst at 775 K gives
(A) Glyoxal (B) Oxalic acid
(C) Maleic anhydride (D) Fumaric acid

26. The size of the colloidal particles is
(A) More than 10^{-7} m (B) Between 10^{-10} and 10^{-7} m
(C) Less than 10^{-10} m (D) More than 10^{-10} m

27. When acetic acid is dissolved in benzene, its molecular mass
(A) Decreases (B) Increases
(C) Either increases or decreases (D) Suffers no change

28. Sulphur molecule is
(A) Diatomic (B) Triatomic
(C) Tetratomic (D) Octatomic

29. Euchlorine is a mixture of
(A) O_2 and Cl_2 (B) O_2 and ClO_2
(C) Cl_2 and ClO_2 (D) ClO_2 and Cl_2O

30. The oxidation number of chromium in $K_2Cr_2O_7$ is
(A) 4 (B) 6 (C) 5 (D) 3

31. The purest form of iron is
(A) Cast iron (B) Wrought iron
(C) Mild steel (D) Stainless steel

32. Which of the following undergoes Cannizaro's reaction ?
(A) Propanal (B) Benzaldehyde
(C) Butanal (D) None of these

33. Esters of amino acids are
(A) Acidic (B) Amphoteric (C) Neutral (D) Basic

34. Neoprene is a polymer of
(A) Butadiene (B) Isoprene
(C) Styrene (D) Chloroprene

35. Inorganic compounds are mostly soluble in
(A) Acids (B) Bases
(C) Non-polar solvents (D) Polar solvents

36. When propyne is treated with dil. H_2SO_4 in presence of $HgSO_4$, the major product is
(A) Propanal (B) Propane
(C) Acetone (D) Propyl H_2SO_4

37. The number of π-electrons present in naphthalene is
(A) 6 (B) 4 (C) 10 (D) 14

38. Unit of specific conductivity is
(A) mho cm (B) ohm cm^{-1}
(C) mho (D) mho cm^{-1}

39. Mixing of two gases by diffusion is
(A) Reversible (B) Irreversible
(C) Exothermic (D) Endothermic

40. Heat of combustion is always
(A) Neutral (B) Positive
(C) Negative (D) All of the above

AFMC - 1992

1. The oxidation of benzene releases 718 k.cal. If the mass of benzene formed is 39 gm, then the heat of formation of benzene is
(A) 309 kcal (B) 1436 kcal (C) 718 kcal (D) 1359 kcal

2. When copper nitrate is strongly heated it gets converted into
(A) Copper metal (B) Cupric oxide
(C) Cuprous oxide (D) Copper nitrite

3. Which of the following is used as refrigerant ?
(A) Benzene (B) Ethyl alcohol
(C) CCl_4 (D) Acetylene

4. Which of the following is a hydrocarbon ?
(A) Urea (B) Benzene
(C) Ammonium cyanate (D) Phenol

5. The compound having blue colour is
(A) $CuSO_4$ (B) $CuSO_4.5H_2O$
(C) $PbSO_4$ (D) $HgSO_4$

6. In the electrolysis of NaCl
(A) Cl^- is reduced at anode
(B) Cl^- is oxidised at cathode
(C) Cl^- is oxidised at anode
(D) Cl^- is neither reduced nor oxidised

7. When Cl_2 is dissolved in KOH, the product formed is
(A) KClO (B) $KClO_2$ (C) $KClO_3$ (D) $KClO_4$

8. Aspirin is
(A) Acetyl salicylic acid (B) Phenyl salicylic acid
(C) Salicylic acid (D) Benzoic acid

9. Which of the following phosphorus is most stable ?
(A) Black (B) White (C) Red (D) All stable

10. Which has the highest electro-negativity ?
(A) Cl (B) Br (C) F (D) I

11. An example of colligative property is
(A) Boiling point (B) Osmotic pressure
(C) Vapour pressure (D) Freezing point

12. As_2S_3 is a
(A) Positive colloid (B) Negative colloid
(C) Neutral colloid (D)None of the above

13. Gammexane is
(A) Benzene (B) Benzene hexachloride
(C) Nitrobenzene (D) Benzenaldehyde

14. Decomposition of H_2O_2 is favoured by
(A) Traces of acid (B) Alcohol
(C) Acetanilide (D) MnO_2

15. D.D.T. is formed by
(A) Bromo and chlorobenzene
(B) Chloral and benzene
(C) Chloral and bromobenzene
(D) Chloral and chlorobenzene

16. The raw material to form Nylon is
(A) Isoprene (B) Ethylene
(C) Adipic acid (D) Butadiene

17. Hydrogen bonding does not play any role in boiling of
(A) HI (B) C_2H_5OH (C) Ammonia (D) H_2O

18. Which is used as an antifreeze ?
(A) Glycol (B) Ethyl alcohol
(C) Water (D) Methanol

19. In the titration of Na_2CO_3 and HCl, the indicator used is
(A) Methyl orange (B) Methylene blue
(C) Litmus (D) Phenolphthalein

20. The molarity of a solution of Na_2CO_3 in 10.6 gm/500 ml is
(A) 0.2 M (B) 2 M (C) 20 M (D) 0.02 M

21. In the hydrogenation of oils, the catalyst used is
(A) Pd (B) Pt (C) Co (D) Ni

22. Lithopone is a combination of ZnS and
(A) $PbSO_4$ (B) $CaSO_4$ (C) $SrSO_4$ (D) $BaSO_4$

23. SO_3 is formed by heating
(A) Fe SO_4 (B) $CuSO_4$ (C) $PbSO_4$ (D) $NiSO_4$

24. Nitrogen dioxide is released by heating
(A) $PbNO_3$ (B) KNO_3 (C) $Fe(NO_3)_3$ (D) $AgNO_3$

25. XeF_4 has a shape of
(A) Spherical (B) Trigonal bipyramidal
(C) Square planar (D) Tetrahedral

26. Hess's law of heat summation includes
(A) Initial reactants and final products
(B) Initial reactants only
(C) Final products only
(D) Intermediates only

27. The valency of inter gases is
(A) Zero (B) One (C) Three (D) Two

28. Which one of the following is endothermic ?
(A) $CH_3Br + NaOH \rightarrow CH_3OH + HBr$
(B) $CH_4 + 2O_2 \rightarrow CO_2 + 2H_2O$
(C) $2NO + O_2 \rightarrow NO_2$
(D) $2CH_3OH + O_2 \rightarrow 2HCHO + 2H_2O$

29. Compounds having same number of electrons but different mass are
(A) Isotopes (B) Isobar (C) Isotone (D) Allotropes

30. Kerosine oils is a/an
(A) Aromatic compound (B) Plant oil
(C) Gasoline (D) Hydrocarbon

31. van der Waal's forces are applied to
(A) Inert gases (B) Real gases
(C) Mixture of gases (D) Elementary gases

32. The first element of rare earth metals is
(A) Cerium (B) Actinium
(C) Uranium (D) Lanthanide

33. Which of the following is both oxidizing and reducing agent ?
(A) H_2SO_4 (B) HCl (C) HNO_3 (D) HNO_2

34. When current is passed through an electrolyte containing ions,
(A) Ions do not move
(B) All ions moves to respective charged plates
(C) All ions move to oppositively charged plates
(D) All the above are false

35. Pure NaCl is prepared by saturating a cold staturated solution of common salt in water with HCl gas. The principle used is
(A) Le Chatelier principle (B) Displacement law
(C) The common ion effect (D)Fractional distillation

36. Non-formation of meniscus in mercury is due to the formation of
(A) Mercuric oxide (B) Mercurous oxide
(C) Mercuric chloride (D) Mercurous chloride

37. A gas of volume 100 cc is kept in a vessel at pressure 104P maintained at temperature 24°C. If now the pressure is increased to 105P keeping the temperature constant, then the volume of gas will become
(A) 10 cc (B) 100 cc (C) 1.0 cc (D) 1000 cc

38. Which is not a characteristic of a catalyst ?
(A) It changes the equilibrium constant
(B) It initiates the reaction
(C) It increases the rate of reaction
(D) It increases the average K.E. of the molecules.

39. In the preparation of sodium carbonate (Na_2CO_3) which of the following is used ?
(A) Slaked lime (B) Quick lime
(C) Limestone (D) Sodium hydroxide

40. Which of the following is not characteristic of alcohols ?
(A) Higher alcohols are stronger and have bitter taste
(B) Lower alcohols are stronger and have bitter taste
(C) The boiling points of alcohols increase with increasing molecular mass
(D) The lower alcohols are soluble in water

AFMC - 1993

1. According to Lewis concept, an acid is
(A) Electron pair donor (B) electron pair acceptor
(C) Proton donor (D) Proton acceptor

2. Ostwald's dilution law is applicable to
(A) Both strong and weak electrolytes
(B) Strong electrolytes only (C) Weak electrolytes only
(D) Non - electrolytes only

3. A base, as defined by Bronsted theory, is a substance which
(A) Can accept protons (B) Can lose a pair of electrons
(C) Can donate protons (D) Can gain a pair of electrons

4. In absorption, the solute
(A) Stays in contact with the solvent
(B) Remains in the surface of the solvent
(C) Moves into the body of the solvent
(D) None of the above

5. Osmotic pressure of a solution increases by
(A) Decreasing the temperature
(B) Increasing the volume
(C) Increasing the number of molecules of the solute
(D) None of the above

6. Crystalloids differ from colloids mainly in the aspect of
(A) Electrical behaviour (B) Particle size
(C) Particle nature (D) Solublity

7. For an exothermic reaction to be spontaneous
(A) Temperature is zero
(B) Temperature should be low
(C) Temperature should be very high
(D) Temperature has no impact

8. The temperature at which a real gas behaves as an ideal gas for a wide range of pressure is
(A) Inversion temperature (B) Boyle temperature
(C) Critical temperature (D)Reduction temperature

9. The amount of electricity that can deposit 108 gm of silver from $AgNO_3$ solution is
(A) 1 Ampere (B) 1 Coulomb
(C) 1 Faraday (D) None of these

10. Diethyl ester is not associated with which one of these isomers
(A) Butanoic acid (B) Methyl propanoate
(C) Stereoisomerism (D) Tautomerism

11. Reaction of NaOH with fats is called as
(A) Hydrolysis (B) Saponification

(C) Esterification (D) None of these

12. Benzaldehyde can be prepared by the dry distillation of
(A) Calcium benzoate and benzoic acid
(B) Calcium benzoate and acetic acid
(C) Calcium benzoate and formic acid
(D) Calcium benzoate and calcium formate

13. Iron is tougher than sodium because
(A) Iron atoms are more closely packed
(B) Metallic bonds are stronger in Iron
(C) Iron atom is smaller
(D) None of the above

14. Glass is a/an
(A) Polymer (B) Amorphous solid
(C) Crystalline solid (D) Special type of crystal

15. Which one of these is the weakest ?
(A) Ionic bond (B) Covalent bond
(C) Metallic (D) van der Waal's forces

16. Copper nitrate $Cu(NO_3)_2$ on heating gives
(A) Copper metal (B) Cupric oxide
(C) Cuprous oxide (D) Cuprous nitrite

17. The catalyst used in Haber's process is
(A) Finely divided platinum
(B) Finely divided iron (C) Finely divided nickel
(D) Finely divided molybdenium

18. Brass is an alloy of copper and
(A) Al (B) Al and Ni (C) Zn (D) Sn

19. Pure nitrogen can be prepared from
(A) NH_4NO_2 (B) BaO (C) NH_4OH (D) Ca_3N_2

20. The oxide which reacts with both NaOH and HCl is
(A) CaO (B) ZnO (C) N_2O_5 (D) CO_2

21. The gas used in electric discharges for decoration is
(A) Helium (B) Neon (C) Radon (D) Krypton

22. Metallic bonds do not play a role in
(A) Germanium (B) Zinc (C) Brass (D) Copper

23. Blasting of TNT is done by mixing it with
(A) NH_4Cl (B) NH_4NO_3 (C) NH_4NO_2 (D) $(NH_4)_2SO_4$

24. Oxidation state of chromium in $K_2Cr_2O_7$ is
(A) 4 (B) 6 (C) 7 (D) 8

25. Hypo is used in photography because it is a
(A) Strong oxidizing agent (B) Strong reducing agent
(C) Strong complexing agent (D) None of the above

26. When Iodine is dissolved in CCl_4, the colour that results is
(A) Colourless (B) Bluish green (C) Brown (D) Violet

27. The reciprocal of the radioactive decay constant is called
(A) Half-life period (B) Whole life-period
(C) Mean life period (D) None of these

28. The van der Waal's equation is for
(A) Ideal gases (B) Real gases
(C) Mixture of gases (D) Inert gases

29. A gas is initially at 1 atm. pressure. To compress it to 1/4th of its initial volume, pressure to be applied is
(A) 1 atm. (B) 2 atm. (C) 4 atm. (D) 1/4 atm.

30. Which of the following is optically active ?
(A) Ethanol (B) Succinic acid
(C) Butanoic acid (D) Lactic acid

31. Formaldehyde and formic acid can be distinguished using
(A) Tollen's reagent (B) Fehling's solution
(C) Ferric chloride (D) Sodium carbonate

32. When diethyl ether reacts with excess of HI, the product is
(A) Ethyl iodide (B) Hydrogen (C) Ethanol (D) Water

33. Aspirin is
(A) Phenyl salicylic acid (B) Acetyl salicylic acid
(C) Methyl salicylate (D) Salicyladehyde

34. Electrolysis of sodium salt of acetic acid liberates at the anode
(A) Hydrogen (B) Ethane
(C) Ethylene (D) Acetylene

35. The hydrogen atom present in which one of the following can be replaced by a metal
(A) Ethylene (B) Ethyne
(C) Dimethyl acetylene (D) Benzene

36. Gammexane is
(A) Chlorobenzene (B) Benzene hexachloride
(C) Bromobenzne (D) Benzyl chloride

37. Which of the statements is not true for alcohols ?
(A) Lower alcohols are water insoluble and their solubility increases with molecular weight
(B) Lower alcohols have fiery taste and strong smell
(C) As molecular mass increases, boiling point increases
(D) Lower alcohols are water soluble and their solubility decreases with molecular weight.

38. Kerosene is a mixture of
(A) Alkenes (B) Alkanes (C) Alkynes (D) Arenes

39. Sour taste of lemon is due to
(A) Ascorbic acid (B) Lactic acid
(C) Citric acid (D) Tartaric acid

40. Chromatography is used to separate
(A) Small samples of mixtures (B) Plant pigments
(C) Dye stuffs (D) All the above

41. Nylon is made using
(A) Succinic acid (B) Adipic acid
(C) Benzaldehyde (D) Phenol

42. Terelyne is a condensation polymer of ethylene glycol and
(A) Benzoic acid (B) Phthallic acid

(C) Terephthallic acid (D) Salicylic acid

43. Purity of organic compounds can be determined by knowing
(A) Melting point (B) Mixed melting point
(C) Solubility (D) None of the above

44. Which of the following is a characteristic of a reversible reaction ?
(A) It can be influenced by a catalyst
(B) It never proceeds to completion
(C) Number of moles of reactants and products are equal
(D) None of the above

45. Which of the following is not a semi-conductor ?
(A) As (B) B (C) Si (D) Ge

46. According to molecular orbital theory, which is non-existant
(A) H_2^- (B) O_2^- (C) O_2^+ (D) He_2

47. In the hydrogenation of oils, the catalyst used is
(A) Co (B) Ni (C) Pd (D) Pt

48. The temperature at which the second virial coefficient of a real gas is zero is called
(A) Crticial temperature (B) Eutectic point
(C) Boiling point (D) Boyle's temperature

49. Formalin is
(A) 40% acetaldehyde (B) 7% formic acid
(C) 40% formaldehyde (D) 7% acetic acid

50. Which of the following is used in fire extinguishers ?
(A) CH_4 (B) $CHCl_3$ (C) CH_2Cl_2 (D) CCl_4

AFMC - 1994

1. The reducing agent is
(A) HCl (B) HBr (C) HF (D) HI

2. Which of the following has unchanged valency ?
(A) Fe (B) Oxygen (C) H (D) Na

3. Which of the following gives brick red colour to the flame test ?
(A) Calcium (B) Barium
(C) Strontium (D) None of these

4. Soaps can be classified as
(A) Carbohydrates (B) Ethers
(C) Organic compounds (D) Salts of fatty acid

5. When ethyl chloride and alcoholic KOH are heated, the compound obtained is
(A) $C_2H_5OC_2H_5$ (B) C_2H_6 (C) C_2H_4 (D) C_2H_5OH

6. Which of the following elements always contains -1 oxidation state ?
(A) Carbon (B) Oxygen (C) Fluorine (D) Nitrogen

7. Isotonic solutions have the same
(A) Density (B) Molar concentration
(C) Normality (D) None of the above

8. Living in the atmosphere of CO is dangerous because it
(A) Dries up the blood
(B) Combines with O_2 present inside to form CO_2
(C) Reduces organic matter of tissues
(D) Combines with hemoglobin and makes it incapable to absorb oxygen

9. The polar compound among the following is
(A) C_2H_6 (B) HCl (C) CH_4 (D) CCl_4

10. At equilibrium ΔG is
(A) Positive (B) Negative (C) Zero (D) None of the above

11. The units of a first order rate constant are
(A) Mol $dm^{-3}s^{-1}$ (B) dm^3 mol^{-1} s^{-1}
(C) s^{-1} (D) None of the above

12. The solubility product of a sparingly soluble salt AB at room temperature is 1.21 x 10^{-6} M^2. Its molar solubility is
(A) 1.21 x 10^{-6} M (B) 1.1 x 10^{-4} M
(C) 1.1 x 10^{-3} M (D) None of the above

13. The turbidity of a polymer solution measures
(A) The light transmitted by the solution
(B) The light scattered by the solution
(C) The light absorbed by the solution
(D) None of the above

14. Isomers which can be interconverted through rotation around a single bond are
(A) Conformers (B) Diastereomers
(C) Enantiomers (D) Positional isomers

15. Which of the following has highest pH in water ?
(A) $NaHCO_3$ (B) KCl (C) NaCl (D) Na_2CO_3

16. If the hydrogen ion concentration of an acid is decreased ten times, its pH
(A) Increases by one (B) Decreases by one
(C) Remains unchanged (D) Increases by 10

17. What is formed when formaldehyde reacts with KOH ?
(A) Methanol (B) Methane
(C) Ethyl formation (D) Acetylene

18. What is formed when oxalic acid is dehydrated by conc. H_2SO_4 ?
(A) CO_2 (B) CO + CO_2 (C) C +CO_2 (D) CO

19. Vinegar contains how much percentage of acetic acid?
(A) 10-12% (B) 4-5% (C) 90-100% (D) 90-98%

20. Beverages is made up of
(A) Ethyl alcohol (B) Acetic acid
(C) Formic acid (D) None of these

21. Aniline is purified by ?
(A) Fractional distillation (B) Steam distillation
(C) Distillation (D) None of these

22. Which of the following is electron deficient ?

(A) PCl_5 (B) NH_3 (C) BCl_3 (D) PCl_3

23. Complex ion formation is possible with
(A) Ag (B) Au (C) Cu (D) All of these

24. In the titration of strong acid and weak base, indicator used is
(A) Methyl orange (B) Phenolphthalein
(C) Thymol (D) None of these

25. Gammexane is the name of
(A) DDT (B) Benzene hexachloride
(C) Chlorobenzene (D) None of these

26. Chromium has most stable oxidation state in
(A) + 2 (B) + 4 (C) +5 (D) + 3

27. Ethanamide is
(A) Feebly acidic (B) Basic
(C) Neutral (D) None of these

28. Which of the following is Polyamide
(A) Nylon (B) Terylene (C) Orlon (D) Teflon

29. Petroleum is mainlya mixture of
(A) Alkanes (B) Alkenes
(C) Alkynes (D) Alkylhalides

30. The first fractional product of petroleum is
(A) Petroleum ether (B) Kerosene
(C) Diesel (D) None of these

31. Isoprene is used for the preparation of
(A) Teflon (B) Nylon (C) Rubber (D) None of these

32. Atomic number is equal to
(A) Number of protons in the nucleus
(B) The sum of number of neutrons and protons
(C) Number of neutrons in the nucleus
(D) Atomic weight of element

33. Arrange in the increasing order of ionic radii of the following elements O, C, F, Cl, Br
(A) F, O, C, Cl, Br (B) F, C, O, Cl, Br
(C) F, Cl, Br, O, C, (D) C, O, P, Cl, Br

34. Melting point is highest for
(A) B (B) AI (C) Ga (D) In

35. Melting point is lowest for
(A) Ca (B) Sr (C) Be (D) Mg

36. The Lightest metal is
(A) Li (B) Fe (C) Cu (D) Ag

37. Which of the following phenomenon will occur when two atom elements of same spin react ?
(A) Bonding will not occur
(B) Orbital overlap will not occur
(C) Both (A) and (B)
(D) All of these

38. Hydrogen can not be obtained by
(A) Zn and dil. H_2SO_4 (B) Zn + dil. HNO_3
(C) Mg + H_2SO_4 (D) Mg and H_2O

39. Pure nascent hydrogen is best obtained by
(A) Na and C_2H_5OH (B) Zn and dil. H_2SO_4
(C) A! and NaOH (D) All of these

40. Which of the following gases is formed when CaC_2 is reacted with H_2O ?
(A) Acetylene (B) Ethylene (C) Methane (D) Ethane

41. What is formed when in presence of alkali, chloroform is treated with acetone ?
(A) Chloretone (B) Chloral
(C) Carbon tetrachloride (D) None of these

42. CO_2 is obtained by heating
(A) Na_2CO_3 (B) K_2CO_3
(C) $NaHCO_3$ (D) None of the above

43. Which of the following is most polarized among Noble gases ?
(A) He (B) Xe (C) Kr (D) Rn

44. If the rate of diffusion of A is 5 times that of B, what will the density ratio of A and B ?
(A) $\frac{1}{25}$ (B) $\frac{1}{5}$ (C) 25 (D) 5

45. Hydrolysis of ester in alkaline medium is
(A) Second order reaction with molecularity two
(B) First order reaction with molecularity two
(C) First order reaction with molecularity one
(D) Second order reaction with molecularity one

46. Starch is converted into maltose by the enzyme
(A) Maltase (B) Invertase (C) Zymase (D) Diastase

47. A violet colour with sodium nitroprusside in the test of sulphur in organic compound is due to the formation of
(A) $Na_4[Fe(CN)_5NOS]$ (B) $Na_3[Fe(CN)_6]$
(C) $Na_2[Fe(CN)_5S]$ (D) None of these

48. Fischer-Topsch process is used in the manufacture of
(A) Synthetic petrol (B) Ethanol
(C) Benzene (D) Ethanoic acid

49. Formaldehyde when treated with KOH gives
(A) CH_3OH (B) C_2H_4
(C) CH_3CHO (D) C_2H_5

50. When acetylene is reacted with hydrochloric acid in the presence of mercuric chloride, the product obtained is
(A) Methyl chloride (B) Acetaldehyde
(C) Ethyl chloride (D) Formaldehyde

AFMC - 1995

1. The formula of freon is
(A) $CCIF_2$ (B) $CFCl_2$
(C) CCl_2F_2 (D) None of these

2. Nitrous Oxide is
(A) Soluble in cold water (B) Soluble in hot water
(C) Acidic in nature (D) None of these

3. $AlCl_3$ is
(A) Anhydrous and covalent (B) Anhydrous and ionic

(C) Covalent and basic (D) Co-ordinate and acidic

4. Beilstein test is used for
(A) Na (B) CO_2 (C) N_2 (D) Cl

5. Glycerol reacts with $KHSO_4$ to form
(A) Allyl alcohol (B) Acrolein
(C) Aldehyde (D) Acetone

6. Structure of benzene is
(A) Planar (B) Linear
(C) Trigonal bipyramidal (D) Tetrahedral

7. Cyanohydrin of which of the following forms lactic acid ?
(A) CH_3CHO (B) CH_3CH_2CHO
(C) HCHO (D) CH_3 CO CH_3

8. The isomerism exhibited by alkyl cyanide and alkyl isocyanide is
(A) Functional (B) Positional
(C) Tautomerism (D) Metamerism

9. In hydrolysis of aniline the reagent used is
(A) Dil. HCl (B) Acetal chloride
(C) CH_3OH (D) None of these

10. The number of electrons in the nucleus of $C1_2$ is
(A) 0 (B) 3 (C) 6 (D) 12

11. If pH of A, B, C and D are 9.5, 2.5, 3.5, and 5.5 respectively, then the strongest acid is
(A) D (B) B (C) A (D) C

12. Calcination is used in metallergy for removal of
(A) Water and sulphide (B) Water and CO_2
(C) CO_2 and H_2S (D) H_2O and H_2S

13. Catalyst only
(A) Increases activation energy
(B) Changes the equilibrium
(C) Decreases activation energy
(D) All of the above

14. Osmotic pressure can be measured by
(A) Berkley and Hartley Method
(B) Solvage Method
(C) Ostwald Method
(D) Haber's Method

15. XeF_4 on partial hydrolysis produces
(A) XeF_2 (B) $XeOF_2$ (C) $XeOF_4$ (D) XeO_3

16. Last molecule of H_2O is evolved from H_2O_2 by
(A) Crystallisation (B) Evaporation
(C) Distillation under reduced pressure (D) Electrolysis

17. Which of the following is not linear ?
(A) CO_2 (B) HF (C) H_2O (D) HCl

18. H_2S reacts with O_2 to form
(A) $H_2O + S$ (B) $H_2O + SO_2$
(C) $H_2O + SO_3$ (D) $H_2SO_4 + S$

19. Bromine water reacts with SO_2 to form
(A) H_2O and HBr (B) H_2SO_4 + HBr
(C) HBr and S (D) S and H_2O

20. Which of the following behaves both as oxidizing and reducing agent ?
(A) H_2S (B) HNO_3 (C) H_2SO_4 (D) SO_2

21. Which of the following conducts electricity ?
(A) Fused NaCl (B) CO_2 (C) Br_2 (D) Si

22. In which case NH_3 is not used ?
(A) Cold storage (B) Anaeshthetic
(C) Manufacture of rayon and plastic (D) None of these

23. In NH_3 and PH_3 common feature is
(A) Odour (B) Combustibility
(C) Basic nature (D) None of these

24. If α-particle is bombarded in gold foil, then
(A) Some α-particles are deflected
(B) Maximum α-particles are deflected
(C) Most α-particles travel in straight line
(D) All α-particles return to original path

25. The isotopes of elements was discovered by
(A) Thomson (B) Chadwick (C) Soddy (D) Curie

26. Which of the following does not give Borax bead text ?
(A) Chromium (B) Ferrous salt (C) Sodium (D) Cobalt

27. Which Oxide of Manganese is amphoteric ?
(A) Mn_2O_7 (B) MnO (C) MnO_2 (D) Mn_2O_3

28. The most common lanthanide is
(A) Lanthanum (B) Cerium (C) Samarium (D) Plutonium

29. The rate of reaction depends upon
(A) Temperature (B) Pressure (C) Volume
(D) Initial concentration of reactant

30. Which of the following undergoes sublimation ?
(A) $ZnCl_2$ (B) $CuCl_2$ (C) AgCl (D) NH_4Cl

31. Which of the following reacts with water with high rate?
(A) Li (B) K (C) Na (D) Rb

32. In-electrolysis of aqueous copper sulphate, the gas at anode and cathode is
(A) O_2 and H_2 (B) SO_2 and H_2
(C) H_2 and O_2 (D) SO_3 and O_2

33. The most electron affinity of the elements is
(A) F (B) Cl (C) Br (D) I

34. Arrange the following in increasing order of their atomic radius : Na, K, Mg, Rb
(A) Mg < K < Na < Rb (B) Mg < Na < K < Rb
(C) Mg < Na < K < Rb (D) Na < K < Rb < Mg

35. In which of the following reactions $K_P > K_C$?
(A) $N_2 + 3H_2 \rightleftharpoons 2NH_3$ (B) $H_2 + I_2 \rightleftharpoons 2HI$
(C) $PCl_3 + Cl_2 \rightleftharpoons PCl_5$ (D) $2SO_3 \rightleftharpoons O_2 + 2\,SO_2$

36. Splitting of spectral line under the influence of magnetic field is called
(A) Stark effect (B) Zeeman's effect
(C) Photoelectric effect (D) None of these

37. Isotonic solutions have same
(A) Molar concentration (B) Molality

(C) Normality (D) None of these

38. Fog is an example of
(A) Solid dispresed in gas (B) Liquid dispresed in gas
(C) Liquid dispresed in solid (D) Solid dispresed in liquid

39. In the formation of a molecule by an atom
(A) Repulsive forces operate
(B) Both attractive & repulsive forces operate
(C) Attractive forces operate
(D) None of them

40. The precipitation occurs if ionic concentration is
(A) More than solubility product
(B) Less than solubility product
(C) Equal to solubility product (D) None of them

41. Use of electrolysis is
(A) Electroplating (B) Electrorefining
(C) Both (A) and (B) (D) None of these

42. The hydride ions are iso-electronic with
(A) He (B) Be (C) H (D) He^+

43. In electro refining of copper some gold is deposited at
(A) Anode mud (B) Cathode mud
(C) Cathode (D) Electrolyte

44. Cl_2 reacts with CS_2 is presence of I_2 to form
(A) $CHCl_3$ (B) CCl_4 (C) C_2H_5Cl (D) C_2H_6

45. The solubility of I_2 increases in water in presence of
(A) $KMnO_4$ (B) H_2S (C) KI (D) H_2SO_4

46. The structure of PF_5 molecule is
(A) Tetrahedral (B) Trigonal bipyramidal
(C) Square planar (D) Pentagonal bipyramidal

47. Fusion mixture is
(A) $K_2CO_3 + Na_2CO_3$ (B) $KHSO_4 + NaHSO_4$
(C) $K_2CO_3 + NaHSO_4$ (D) $KHSO_4 + Na_2SO_3$

48. Which of the following phosphorus is most reactive ?
(A) Red phosphorus (B) White phosphorus
(C) Scarlet phosphorus (D) Violet phosphorus

49. Optical activity is measured by
(A) Polarimeter (B) Refractometer
(C) Spectrograph (D) Tracer technique

50. Pyruvic acid is obtained by
(A) Oxidation of acetone cyanohydrin
(B) Oxidation of formaldehyde cyanohydrin
(C) Oxidation of acetaldehyde cyanohydrin
(D) None of these

AFMC - 1996

1. Which of the following statements regarding hydrogen peroxide is false ?
(A) It behaves as a reducing agent towards acidified $KMnO_4$
(B) It is decomposed by MnO_2
(C) It is more stable in basic solution
(D) It is strong oxidizing agent

2. Milk is
(A) Fat dispersed in water (B) Water dispersed in fat
(C) Fat dispersed in milk (D) Water dispersed in oil

3. Which of the following is endothermic ?
(A) $N + \frac{1}{2} O_2 \rightarrow NO$ (B) $C + O_2 \rightarrow CO_2$
(C) $3H_2 + N_2 \rightarrow 2NH_3$ (D) $PCl_3 + Cl_2 \rightarrow PCl_5$

4. Formula of Plaster of Paris
(A) $CaSO_4.H_2O$ (B) $CaSO_4.\ 2H_2O$
(C) $CaSO_4\ .\frac{1}{2}H_2O$ (D) $CaSO_4.\ 1\frac{1}{2}\ H_2O$

5. Which of the following set belongs to same period ?
(A) Cu, Ni, Zn (B) F, Cl, Br (C) Li, Na, K (D) Li, Mg, Ca

6. Colloidal sol effective in treating eye disease
(A) Colloidal S. (B) Colloidal antimony
(C) Colloidal gold (D) Colloidal silver

7. T*l* shows variable valency due to
(A) Lone pair effect (B) Inert pair effect
(C) High M.P. (D) High B.P.

8. Dipole moment is zero for
(A) CH_2Cl_2 (B) CH_3Cl (C) CCl_4 (D) $CHCl_3$

9. Aspirin is acetyl salicylic acid, it
(A) ppt. in stomach but not in intestine
(B) Does not ppt. in stomach but ppt. in intestine
(C) Both of these (D) None of these

10. $NaHCO_3$ is prepared by
(A) Solvey process (B) Bosch process
(C) Down process (D) None of these

11. Which of the following structures is planar ?
(A) C_2H_2 (B) C_2H_6 (C) CH_3^- (D) C_2H_4

12. Constituent of light oil is
(A) Benzene (B) Phenol (C) Anthracene (D) Aniline

13. Vanaspati ghee is manufactured by
(A) Hydrogenation of oil (B) Oxidation of oil
(C) Reduction of oil (D) None of these

14. Dulong and Petit's Law used specific heat for determination of atomic weight for
(A) Liquid (B) Solid (C) Gas (D) All of these

15. In towns, a layer of air is condensed as smoke due to pollution, is called
(A) Smog (B) Mist (C) Smoke (D) Fog

16. Natural gas is mainly
(A) C_3H_8 (B) C_2H_2 (C) C_6H_6 (D) CH_4

17. Which of the following is correct sequence for ionic radius ?
(A) $Na^+ < Al^{+3} < Cl^- < Ar$ (B) $Cl^- > Ar > Na^+ > Al^{3+}$
(C) $Na^+ < Al^{+3} > Cl^- < Ar$ (D) $Cl^- > Ar > Al^{+3} > Na^+$

18. CCl_4 is used as fire extinguisher due to
(A) High melting point (B) Covalent bonds
(C) Low boiling point (D) Its incombustible vapour

19. NH_3 and H_2O form NH_4OH by
(A) Coordinate bond (B) Covalent bond
(C) Electrovalent bond (D) None of these

20. Diamond is used in glass cutting due to
(A) Its hardness (B) High R.I.
(C) High M.P. (D) High metallic bonding

21. Brownian moment is found in
(A) Colloidal solution (B) Suspension
(C) Saturated solution (D) Unsaturated solution

22. Unit of molarity is
(A) kg/lit (B) gm/lit (C) mole/lit (D) None of these

23. Electroplating is discovered by
(A) Newton (B) Dalton (C) Faraday (D) Arrhenious

24. Which of the following is not a mode of dissociation of radioactive substance ?
(A) Electron capture (B) α-particle emission
(C) β-emission (D) Fission

25. Indicator used in titration of oxalic acid and NaOH is
(A) Methyl orange (B) Methyl red
(C) Phenolphthalein (D) None of these

26. Glass is soluble in
(A) $HClO_4$ (B) Aqua regia (C) HF (D) H_2SO_4

27. Paraldehyde is formed by polymerization of
(A) CH_3CHO (B) HCHO (C) CH_3OH (D) CH_3CH_2OH

28. Which of the following is metalloid ?
(A) Se (B) At (C) Te (D) S

29. Which of the following is used as "Control rod" ?
(A) Cadmium rod (B) Graphite rod
(C) Steel (D) All of these

30. Which of the following is used as Moderator ?
(A) D_2O (B) Alum (C) H_2O (D) None of these

31. pH of a solution is 4, then OH^- concentration is
(A) 10^{-14} (B) 10^{-4} (C) 10×10^{-5} (D) 10^{-10}

32. Osmotic pressure can be increased by
(A) Increasing temperature (B) Decreasing temperature
(C) Increasing volume (D) None of these

33. Chloroform on warming with Ag power gives
(A) C_6H_6 (B) C_2H_4 (C) C_2H_2 (D) None of these

34. Potash Alum is used as
(A) Water softner (B) Disinfectant
(C) Mordant (D) Coolant

35. Which of the following is used as "Anaesthesia"
(A) N_2O (B) N_2 (C) CH_4 (D) CO_2

36. Stainless steel contains iron with
(A) Cr + Ni (B) Cr +Zn (C) Sn + Pb (D) Bi + Ni

37. In photography HYPO is used as
(A) Oxidising agent (B) Reducing agent
(C) Complexing agent (D) All of the above

38. A war gas Lewisite is formed by reaction of arsenic chloride with
(A) CH_4 (B) C_6H_6 (C) C_2H_2 (D) C_2H_4

39. Dynamite contains
(A) Nitroglycerine mixed with H_2SO_4
(B) Nitroglycerine mixed with saw dust
(C) Both (A) & (B) (D) None of these

40. Natural rubber is polymer of
(A) Trans-isoprene (B) Cis-isoprene
(C) Co-cis-trans isoprene (D) None of these

41. Sea-diver goes deep in sea with mixture of which of the following gases ?
(A) O_2 and He (B) O_2 and Ar
(C) O_2 and CO_2 (D) CO_2 and Ar

42. A mixture contains Benzene & Chlorobenzene is separated by
(A) Sublimation (B) Separating funnel
(C) Crystallization (D) Distillation

43. Kerosene oil has compounds having
(A) $C_5 - C_8$ (B) $C_8 - C_{12}$ (C) $C_{12} - C_{16}$ (D) $C_{16} - C_{18}$

44. Which of the following is used as refrigerant ?
(A) SiC (B) CF_2Cl_2 (C) CO_2 (D) $CHCl_3$

45. Higher percentage of carbon is found in
(A) Peat (B) Anthracene (C) Bituminous (D) All equal

46. Lead tetraethyl is used as
(A) Fire extinguisher (B) Pain killer
(C) Anti-knock agent (D) Mosquito repellent

47. A neutral atom (Atomic No. >1) contain
(A) Proton only (B) eutron + Proton
(C) Neutron + Electron (D) Neutron+ Proton + Electron

48. Atomic number means
(A) No. of electrons (B) No. of protons
(C) No. of neutrons (D) No. of nucleons

49. It is impossible to know simultaneously the position and momentum of a moving particle with absolute exactness at any instant. This is known as
(A) Aufbau principle (B) Hund's rule
(C) Heisenberg principle (D) Pauli's principle

50. Ethane or Methane is prepared in one step from
(A) CH_3Br (B) C_2H_5Br (C) C_2H_5OH (D) C_2H_5CHO

AFMC - 1997

1. Which statement is not correct about alcohol ?
(A) Alcohol of less number of carbon atoms is less soluble in water than alcohol of higher number of carbon atoms
(B) Alcohol is lighter than water
(C) Alcohol evaporates quickly
(D) All are correct

2. Which of the following is paramagnetic ?
(A) $[Ni(CO)_4]^{2+}$ (B) $Co[(NH_3)_6]^{2+}$
(C) $[Ni(CN)_4]^{2-}$ (D) $[NiCl_4]^{2-}$

3. Which of the following has square planar structure ?
(A) $[Ni(Cl)_4]^{2+}$ (B) $[Ni(CO)_4]^{2+}$
(C) $[Ni(CN)_4]^{2-}$ (D) None of these

4. Which is the most stable atom ?
(A) U (B) Pb (C) Bi (D) Al

5. Which of the following statements is correct ?
(A) H-has same electronegativity as halogens
(B) H has oxidation number -1 and +1

(C) Hydrogen has same ionization potential as alkali metals
(D) It will not be liberated at anode

6. The electronic configuration, $1s^2 2s^2 2p_x^1 2p_y^1 2p_z^1$ is of
(A) Oxygen (B) Nitrogen (C) Hydrogen (D) Fluorine

7. The test of O_3 can be done by
(A) Au (B) Ag (C) Hg (D) None of these

8. Heavy water is
(A) $H_2\,^{18}O$ (B) $H_2\,^{16}O$ (C) H_2O_3 (D) D_2O

9. Cyanide and isocyanide are isomers of type
(A) Tautomer (B) Positional
(C) Functional (D) All of these

10. How much quick lime can be obtained from 25 gm of $CaCO_3$?
(A) 28 gm (B) 14 gm (C) 56 gm (D) None of these

11. Which of the following is not efflorescent ?
(A) $CuSO_4$ (B) Hydrated $CuSO_4$
(C) NaOH (D) All of these

12. Which of the following can be used for dating a fossil ?
(A) $^{14}_{6}C$ (B) $^{12}_{6}C$ (C) $^{15}_{7}N$ (D) $^{40}_{20}Ca$

13. K_{sp} for $HgSO_4$ is 6.4×10^{-5}, then solubility of the salt is
(A) 8×10^{-3} (B) 8×10^{-6} (C) 6.4×10^{-5} (D) None of these

14. $C_{12}O_{22}O_{11} + H_2O \xrightarrow{dil.\ H_2SO_4} C_6H_{12}O_6 + C_6H_{12}O_6$. In this reaction dil. H_2SO_4 is
(A) Homogeneous catalysis (B) Homogeneous catalyst
(C) Heterogeneous catalysis (D) Heterogeneous catalyst

15. Acetaldoxime reacts with P_2O_5 (Phosphorus pentaoxide) to give
(A) Methyl cyanide (B) Methyl cyanate
(C) Ethyl cyanide (D) None of these

16. Volume occupied by 4.4g of CO_2 is
(A) 2.24 litres (B) 22.4 litres
(C) 224 litres (D) None of these

17. Which has the highest weight ?
(A) $1m^3$ water (B) 10 litre of Hg
(C) A normal adult man (D) all have same weight

18. What is the value of decay constant of a compound having half-life $t_{1/2} = 2.95$ days ?
(A) $2.9 \times 10^{-5}\ s^{-1}$ (B) $2.9 \times 10^{6}\ s^{-1}$
(C) $2.9 \times 10^{-6}\ s^{-1}$ (D) $3 \times 10^{5}\ s^{-1}$

19. Positron is
(A) $_1H^1$ (B) $_{-1}e^o$ (C) $_{+1}e^o$ (D) None of these

20. Number of protons, neutrons and electrons in the element Y^{231}_{89} is
(A) 89, 231, 89 (B) 89, 89, 242
(C) 89, 142, 89 (D) 89, 71, 89

21. Which cannot be formed ?
(A) He (B) He_2 (C) He^{2+} (D) He^+

22. Arrange the elements, Na, Rb, K, Mg in increasing order of atomic radius
(A) Na, K, Mg, Rb (B) K, Na, Mg, Rb
(C) Mg, Na, K, Rb (D) Rb, K, Mg,. Na

23. If electron falls form n = 3 to n =2, then emitted energy is
(A) 1.9 eV (B) 0.64 eV (C) 10.2 eV (D) 12.09 eV

24. IUPAC name of the compound given below

$CH_3 - CH(CH_2CH_3) - CH_2 - CHOH - CH_3$

(A) 4-Methyl-3-hexanol (B) Heptanol
(C) 4-Methyl-2-hexanol (D) None of these

25. Which of the following atomic number of Uranium is the most radio-active ?
(A) 226 (B) 248 (C) 238 (D) 235

26. Arrange the following in order of increasing acidity H_3PO_4, H_2CO_3, HCl, HI
(A) H_3PO_4, HCl, H_2CO_3, HI
(B) H_3PO_4, H_2CO_3, HCl, HI
(C) H_2CO_3, H_3PO_4, HCl, HI (D)None of these

27. Arrange increasing order of basicity of CH_3NH_2, $(CH_3)_2NH$, $C_6H_5NH_2$, $(CH_3)_3N$
(A) $C_6H_5NH_2 < (CH_3)N_3 < CH_3NH_2 < (CH_3)_2NH$
(B) $(CH_3)_3N < (CH_3)_2NH < CH_3NH_2 < C_6H_5NH_2$
(C) Reverse of (a)
(D) Reverse of (c)

28. Which of the following is simple ether ?
(A) $C_2H_5OCH_3$ (B) CH_3OCH_3
(C) Phenyl-O-CH_3 (D) All are simple ether

29. Aromatic hydrocarbon shows mostly
(A) Electrophilic substitution
(B) Electrophilic addition (C) Nucleophilic addition
(D) Nucleophilic substitution

30. $C_3H_8 + Cl_2 \xrightarrow{Light} C_3H_7Cl + HCl$ is an example of
(A) Substitution (B) Elimination
(C) Addition (D) Rearrangement reaction

31. Which of the following shows geometrical isomerism ?
(A) 1-Butene (B) 2-Butene
(C) Propene (D) 1,1-dichlorobutane

32. Which of the following is ferroelectric compounds ?
(A) $BaTiO_3$ (B) $K_4[Fe(CN)_6]$
(C) Pb_2O_3 (D) None of these

33. If concentration of reactants is increased by 'X', then the K becomes
(A) K + x (B) K (C) In K/x (D) K/x

34. In which case K, is less then K ?
(A) $PCl_5 \rightleftharpoons PCl_3 + Cl_2$ (B) $H_2 + Cl_2 \rightleftharpoons 2HCl$
(C) $2SO_2 + O_2 \rightleftharpoons 2SO_3$ (D) All of these

35. Azeotropic mixture of HCl and H_2O has
(A) 48% HCl (B) 22.2% HCl
(C) 36% HCl (D) 20.2% HCl

36. If a neutral solution has pKw = 13.36 at 50° C, then pH of the solution is
(A) 7.63 (B) 7 (C) 6.63 (D) None of these

37. BF_3 is acid according to
(A) Arrhenius (B) Lewis
(C) Bronsted & Lowry (D) All of these

38. Iodoethane reacts with sodium in presence of ether. What is the product ?
(A) Butene (B) Butane (C) Pentane (D) Propane

39. Which is a mixed ketone ?
(A) Pentanone (B) Benzophenone
(C) Benzenone (D) All of these

40. In Friedal-Craft's reaction, besides $AlCl_3$ what are the reactants ?
(A) $C_6H_6 + CH_3Cl$ (B) $C_6H_6 + CH_4$
(C) $C_6H_6 + NH_2$ (D) None of these

41. Number of isomers of C_4H_{10} is
(A) 4 (B) 3 (C) 2 (D) None of these

42. Which of the following is correct relationship
(A) $\Delta G = \Delta H + T \Delta S$ (B) $\Delta H = \Delta G - T \Delta S$
(C) $\Delta G = \Delta H - T \Delta S$ (D) $\Delta G + \Delta H = T \Delta S$

43. First law of thermodynamics is represented by the equation
(A) $\Delta E = Q + W$ (B) $\Delta E = Q - W$
(C) $W = Q + \Delta E$ (D) None of these

44. Which of the following reacts with chloroform and base to form phenylisocyanide ?
(A) Aniline (B) Phenol (C) Benzene (D) Nitrobenzene

45. Electronic formula of chromium is
(A) $3d^6 4s^1$ (B) $3d^5 4s^1$ (C) $3d^4 4s^1$ (D) $3d^4 4s^2$

46. Which lanthanide is most commonly used ?
(A) Lanthanum (B) Nobilium
(C) Thorium (D) Cerium

47. Which of the following is true for diamond ?
(A) It is a good conductor of electricity
(B) It is soft
(C) It is a bad conductor of heat
(D) It is made up of C H and O

48. Marsh gas contains
(A) C_2H_2 (B) CH_4 (C) H_2S (D) CO

49. Orlon is a polymer of
(A) Styrene (B) Tetrafluroethylene
(C) Vinyl chloride (D) Acrylonitrile

50. Which of the following is correct for Hydrogen ?
(A) It is always collected at anode
(B) It has very high ionization potential
(C) It can form bonds in +1 as well as -1 oxidation state
(D) None of these

AFMC - 1998

1. The freezing point of a solution prepared from 1.25 gm of a non-electrolyte and 20 gm of water, is 271.9 K. If molar depression constant is 1.86 K $mole^{-1}$, then molar mass of the solute will be
(A) 115.3 (B) 93.9 (C) 105.7 (D) 106.7

2. Ozone is prepared by passing silent electric discharge through oxygen. In this reaction
(A) Oxygen is loaded with energy
(B) Energy is given out (C) Energy is absorbed
(D) Oxygen is dissociated into atoms

3. In the equation $4M + 8CN^- + 2H_2O + O_2 \rightarrow 4[M(CN_2)]^- + 4OH^-$, identify the metal M.
(A) Gold (B) Zinc (C) Copper (D) Iron

4. The reaction of aromatic acyl chloride and phenol in the presence of a base NaOH or pyridine is called
(A) Sandmayer's reaction
(B) Kolbe's reaction (C) Perkin's reaction
(D) Schotten Baumann reaction

5. Galactose is converted into glucose in
(A) Liver (B) An intestine (C) Mouth (D) Stomach

6. Identify Z in the reaction : $I \rightarrow X \rightarrow Y \rightarrow Z$
(A) $CH_3CH(CN)_2$ (B) $CNCH_2CH_2CN$
(C) $BrCH_2CH_2CN$ (D) $BrCH = CHCN$

7. Which of the following is obtained when N_2 reacts with calcium carbide ?
(A) Calcium cyanate (B) Calcium acetate
(C) Calcium cyanamide (D) Calcium carbonate

8. In the Liebermann's nitroso reaction sequential changes in the colour of phenol occurs as
(A) red → deep blue → green
(B) Brown or red → green → red → deep blue
(C) red → green → white (D) white → red → green

9. Aspirin is obtained by reaction of salicylic acid with
(A) Acetone (B) Acetaldehyde
(C) Acetyl chloride (D) Acetic anhydride

10. Which of the following is the strongest acid ?
(A) CF_3COOH (B) CBr_3COOH
(C) CH_3COOH (D) CCl_3COOH

11. The polymer, which has amide linkage, is
(A) Teflon (B) Bakelite (C) Nylon-6,6 (D) Terylene

12. If acetylchlofide is reduced in the presence of $BaSO_4$ and Pd, then .... is formed.
(A) CH_3CHO (B) CH_3CH_2OH
(C) CH_3COOH (D) CH_3COCH_3

13. 1-Chlorobutane, on reaction with alcohlic potash (KOH), gives
(A) 2-butene (B) 1- butanal (C) 1-butene (D) 1-butanol

14. Both methane and ethane can be prepared in one step by the reaction of
(A) CH_3Br (B) CH_3CH_2OH (C) C_2H_4 (D) CH_3OH

15. If the volume of 2 moles of an ideal gas at 540 K is 44.8 litre, then its pressure will be
(A) 1 atmosphere (B) 2 atmosphere
(C) 3 atmosphere (D) 4 atmosphere

16. Which of the following will not displace hydrogen ?
(A) Hg (B) Sn (C) Ba (D) Pb

17. Night-blindness is caused due to the deficiency of
(A) Vitamin B_{12} (B) Vitamin A
(C) Vitamin C (D) Vitamin E

18. Ag can be obtained from $AgNO_3$ by reacting it with
(A) Na_2CO_3 (B) NH_3 (C) PH_3 (D) AsH_3

19. The number of electrons shared by each outermost shell of N_2 is
(A) 2 (B) 3 (C) 4 (B) 5

20. The pH value of 10^{-7} M HCl is
(A) Equal to 7 (B) Less than 7
(C) More than 7 (D) Either (A) or (C)

21. Which of the following has zero dipole moment ?
(A) PH_4 (B) CH_2Cl_2 (C) PH_3 (D) CH_4

22. Benzaldehyde reacts with ammonia to form
(A) Aniline (B) Benzamide
(C) Hydro benzamide (D) Phenyl Cyanide

23. Which of the following metals is present in brass, bronze and German silver ?
(A) Cu (B) Al (C) Zn (D) Mg

24. Nitrogen can be obtained from air by removing
(A) Oxygen (B) Hydrogen
(C) Carbon dioxide (D) Both (A) and (C)

25. When enthalpy and entropy change for a chemical reaction are - 2.5 x 10^3 cals and 7.4 cals deg^{-1} respectively, the reaction at 298 K would be
(A) Spontaneous (B) Reversible
(C) Irreversible (D) Non-spontaneous

26. The reaction N_2O_5(in CCl_4 solution)$\rightarrow$2 NO_2 (solution) + 1/2 O_2 (g) is of first order with rate constant 6.2 x 10^{-1} s^{-1}. What is the value of rate of reaction when $[N_2O_5]$ = 1.25 mol L^{-1} ?
(A) 7.75 x 10^{-4} mole L^{-1} s^{-1} (B) 6.35 x 10^{-3} mole L^{-1} s^{-1}
(C) 5.15 x 10^{-2} mole L^{-1} s^{-1} (D) 3.85 x 10^{-1} mole L^{-1} s^{-1}

27. If 5.85 grams of NaCl (molecular weight = 58.5) is dissolved in 90 grams of water, the mole fraction of NaCl will be
(A) 0.0196 (B) 0.2 (C) 0.01 (D) 0.1

28. The correct order of the increasing ionic character is
(A) $BeCl_2 < MgCl_2 < BaCl_2 < CaCl_2$
(B) $BeCl_2 < BaCl_2 < MgCl_2 < CaCl_2$
(C) $BeCl_2 < MgCl_2 < CaCl_2 < BaCl_2$
(D) $BaCl_2 < CaCl_2 < MgCl_2 < BeCl_2$

29. A compound is treated with $NaNH_2$ to give sodium salt. Identify the compound :
(A) C_2H_6 (B) C_2H_4 (C) C_2H_2 (D) C_6H_6

30. Which of the following compounds does not belong to lipids ?
(A) Fats (B) Amino acids
(C) Phospho-lipid (D) Carbohydrates

31. Oxalic acid, when heated with concentrated H_2SO_4 gives
(A) H_2O_2 and CO_2 (B) CO and CO_2
(C) H_2O_2 and CO_2 (D) CO_2 and H_2S

32. Chlorine is liberated, when we heat
(A) $Pb_2(NO_3)_4$ + MnO_2 (B) $K_2Cr_2O_7$ + HCl
(C) $KMnO_4$ + NaCl (D) $K_2Cr_2O_7$ + MnO_2

33. A sudden large jump between the values of second and third ionization energies of an element would be associated with which the following electronic configuration ?
(A) $1s^22s^2p^63s^1$ (B) $1s^22s^2p^63s^23p^1$
(C) $1s^22s^2p^63s^23p^2$ (D) $1s^22s^2p^63s^2$

34. One mole of CO_2 contains
(A) 18.11 x 10^{23} molecules of CO_2
(B) 6.023 x 10^{23} atoms of oxygen
(C) 3 grams atoms of CO_2
(D) 6.023 x 10^{23} atoms of carbon

35. When cold potassium permanganate ($KMnO_4$) is added to ethylene, it gives
(A) Glycerol (B) Ethanol
(C) Methanol (D) Ethylene glycol

36. The lustre of a metal is duet to
(A) Its high density (B) Its high polishing nature
(C) Its chemical inertness (D) Presence of free electrons

37. The hybridization of carbon in diamond, graphite and acetylene is in the order of
(A) sp^3 sp sp^2 (B) sp^3 sp^2 sp
(C) sp sp^2 sp^3 (D) sp^2 sp^3 sp^2

38. Which of the following is a calomel ?
(A) Hg_2Cl_2 (B) $HgSO_4$ (C) $HgCl_2$ (D) $HgBr_2$

39. The correct order of acidic strength is
(A) HF < HCl < HBr < HI (B) HCl > HBr < HF > HI
(C) HBr < HCl < HI < HF (D) HI < HBr < HCl < HF

40. Lyophilic colloids are stable due to
(A) Small size of the particle
(B) Charge on the particle (C) Large size of article
(D) Layer of dispersion medium on the particles.

41. Anhydrous $AlCl_3$ is prepared from
(A) Aluminium and Cl_2
(B) Dry HCl gas + heated aluminium metal
(C) Conc. HCl and aluminium metal
(D) Dilute HCl and aluminium metal

42. Chloroform on treatment with concentrated HNO_3 gives
(A) $CHCl_2NO_2$ (B) $CHCl_2HNO_3$
(C) CCl_3NO_2 (D) $CHCl_2NO_3$

43. If $_{90}Th^{228}$ disintegrates to $_{83}Bi^{212}$, then the number of α and β particles emitted are respectively
(A) 4α only (B) 7β only (C) 4α and 7β (D) 4α and 1β

44. A biological catalyst is essentially
(A) An amino acid (B) An enzyme
(C) A carbohydrate (D) A nitrogen molecule

45. The uncertainty in the momentum of an electron is 1×10^{-4} kg - ms^{-1}. The uncertainty in its position will be (h = 6.62×10^{-34} kg-m^{-2} s^{-1})
(A) 1.05×10^{-28} m (B) 1.05×10^{-28} m
(C) 5.27×10^{-30} m (D) 5.25×10^{-28} m

46. An organic compound A reacts with sodium metal and forms B. On heating with conc. H_2SO_4, A gives diethyl ether. Then A and B are
(A) C_2H_5OH and C_2H_5ONa
(B) C_3H_7OH and CH_3ONa
(C) CH_3OH and CH_3ONa (D) C_4H_9OH and C_4H_9ONa

47. Nitroso amines (R_2N-N=O) are soluble in water. On heating them with conc. H_2SO_4, they give secondary amines. This reaction is called
(A) Perkin reaction (B) Fries reaction
(C) Liebermann nitroso reaction (D) Etard's reaction

48. Example of unit cell with crystallographic dimensions $a \neq b \neq c$, $\alpha = \gamma = 90^o$, $\beta \neq 90^o$ is
(A) Calcite (B) Graphite
(C) Rhombic sulphur (D) Monoclinic sulphur

49. The oxidation of toluene to benzaldehyde by chromyl chloride is called
(A) Wurtz reaction (B) Etard reaction
(C) Fittig reaction (D) Rosenmund reaction

50. If 0.2 gm of an organic compound containing carbon, hydrogen and oxygen on combustion yields 0.147 gram of carbon dioxide and 0.12 gram of water, what will be the content of oxygen in compound ?
(A) 83.23% (B) 89.50 (C) 73.29% (D) 78.45%

AFMC - 1999

1. The acid formed, when iodine reacts with concentrated nitric acid, is
(A) Iodic acid (B) Nitrous acid
(C) Per-iodic acid (D) Hydroiodic acid

2. Carnalite has the composition
(A) Na_2AlF_5 (B) KCl. $MgCl_2$ $6H_2O$
(C) Fe_3O_4 (D) Na_3AlF_6

3. Calcium formate on dry heating yields
(A) Acetone (B) Formaldehyde
(C) Acetic acid (D) Acetaldehyde

4. The molecular formula of cryolite is
(A) Na_3AlF_6 (B) Fe_3O_4 (C) $Na_2Al_2O_3$ (D) All of these

5. In the manufacture of NH_3, the most favourable condition is
(A) High temperature and high pressure
(B) Low temperature and high pressure
(C) Low temperature and low pressure
(D) High temperature and low pressure

6. Ratio of the energy of a photon of wavelengths 3000 Å and 6000 Å is
(A) 1 : 2 (B) 1 : 3 (C) 3 : 1 (D) 2 : 1

7. The correct order of reactivity of halogen is
(A) F < Cl < Br > I (B) F > Cl < Br < I
(C) F > Cl > Br > I (D) F < Cl < Br < I

8. The compound containing co-ordinate bond
(A) SO_3 (B) H_2SO_4 (C) O_3 (D) All of these

9. The oxidation state of chromium in potassium dichromate $K_2Cr_2O_7$ is
(A) +2 (B) - 2 (C) -5 (D) +6

10. Boiling point of chloroform was raised by 0.323 K when 0.5143 g of anthracene was dissolved in its 35 g. Molecular mass of anthracene is
K_b for $CHCl_3$ = 3.9 k - kg mol^{-1}
(A) 177.42 g/mol (B) 79.42 g/mol
(C) 132.32 g/mol (D) 242.32 g/mol

11. How many grams of a dibasic acid (Mo. wt. = 200) should be present in 100 ml of the aqueous solution to give N/10 strength ?
(A) 2g (B) 1g (C) 5g (D) 10g

12. The de-Broglie wavelength of a particle with mass 1 g and velocity 100 m/s is
(A) 6.63×10^{-35}m (B) 6.65×10^{-35}m
(C) 6.63×10^{-33} m (D) 6.63×10^{-34} m

13. The half-life of $_6C^{14}$, if its wavelength (λ) is 2.31×10^{-4}, is
(A) 2×10^2 year (B) 4×10^3 year
(C) 3.5×10^4 year (D) 3×10^3 year

14. pH value of a solution, whose hydronium ion concentration is 6.2×10^{-9} mol/1, is
(A) 7.75 (B) 7.21 (C) 8.21 (D) 6.21

15. Solutions with same osmotic pressure are called
(A) Hypertonic (B) Normal
(C) Isotonic (D) None of these

16. The compounds CH_3 - O - C_3H_7 and C_2H_5 - O - C_2H_5 exhibit
(A) Optical isomerism (B) Cis-trans isomerism
(C) Metamerism (D) Chain isomerism

17. Empirical formula of a compound is CH_2O. If its molecular weight is 180, then the molecular formula of the compound is
(A) $C_3H_6O_3$ (B) $C_4H_8O_4$ (C) $C_6H_{12}O_6$ (D) $C_5H_{10}O_5$

18. In the preparation of sulphuric acid, V_2O_5 is used in the reaction. It is
(A) $SO_2 + H_2O \rightarrow H_2SO_4$ (B) $S + O_2 \rightarrow SO_2$
(C) $N_2 + 3H_2 \rightarrow 2NH_3$ (D) $2SO_2 + O_2 \rightarrow 2SO_3$

19. A base, as defined by Bronsted theory, is a substance which can
(A) Gain a pair of electrons (B) Accept protons
(C) Loose a pair of electrons (D) Donate protons

20. Tyndall effect can be observed in a
(A) Colloidal solution (B) Solution
(C) Solvent (D) Precipitate

21. In which of the following hydrocarbons, hydrogen is most acidic ?
(A) $CH \equiv CH$ (B) $CH_3 - CH_3$
(C) $CH_2 = CH_2$ (D) None of these

22. Which of the following substances is used, in laboratory, for fast drying of neutral gases ?
(A) Phosphorus pentoxide
(B) Active charcoal (C) Sodium phosphate
(D) Anhydrous calcium chloride

23. Aufbau principle is obeyed in which of the following electronic configuration ?
(A) $1s^2 3s^2 3p^6$ (B) $1s^2 2s^2 3s^2$ (C) $1s^2 2s^2 2p^6$ (D) $1s^2 3s^2 3p^2$

24. The number of electrons in $[_{19}K^{40}]^{-1}$ is
(A) 18 (B) 19 (C) 20 (D) 40

25. The volume of a gas measured at 27°C and 1 atm pressure is 10 litres. What final temperature would be required to reduce the volume to 5 litres at 1 atm pressure ?
(A) 450 K (B) 100°C (C) 150 K (D) 327 K

26. According to Bohr's theory of hydrogen atom, which of the following is quantised for an electron ?
(A) Angular momentum (B) Angular acceleration
(C) Acceleration (D) Velocity

27. Magnetic quantum number specifies
(A) Nuclear stability (B) Orientation of orbitals
(C) size of orbitals (D) Shape of irbital

28. Which of the following is the most electrovalent ?
(A) Co (B) Cu (C) Ni (D) Fe

29. Which of the following has the highest boiling point ?
(A) n-octane (B) Butane (C) Pentane (D) Ethane

30. Nessler's reagent is represented by
(A) $K_2 [HgI_4]$ (B) $HgOHgI_2$ (C) $K_2 [HgI_2]$ (D) HgI_2

31. Half-life of radium is 1580 years. Its average life will be
(A) 8.25×10^2 years (B) 1.8×10^3 years
(C) 2.5×10^3 years (D) 2.275×10^3 years

32. Nylon-6,6 is made by using
(A) Phenol (B) Benzaldehyde
(C) Adipic acid (D) Succinic acid

33. Heating a mixture of sodium benzoate and soad lime gives
(A) Benzene (B) Calcium benzoate
(C) Methane (D) Sodium benzoate

34. Which of the following halogen is solid at room temperature ?
(A) Iodine (B) Bromine (C) Fluorine (D) Chlorine

35. The oxidation of toluene to benzaldehyde by chromyl chloride is called
(A) Rosenmund reaction (B) Wurtz reaction
(C) Etard reaction (D) Fittig reaction

36. In the reaction , $HNO_3 + P_4O_{10} \rightarrow 4HPO_3 + X$, the product X is
(A) N_2O_5 (B) H_2O (C) NO_2 (D) N_2O_3

37. The Beilstein test for organic compounds is used to detect
(A) Nitrogen (B) Sulphur (C) Carbon (D) Halognes

38. When primary alcohol is oxidised with chlorine, it gives
(A) CCl_3CHO (B) C_3H_7CHO
(C) HCHO (D) CH_3CHO

39. Which of the following will not displace hydrogen ?
(A) Sn (B) Hg (C) Ba (D) Pb

40. Chloroform on treatment with concentrated HNO_3 gives
(A) $CHCl_2NO_2$ (B) CCl_3NO_2
(C) $CHCl_2NO_3$ (D) $CHCl_2HNO_3$

41. Which of the following statements is true for fuel cells ?
(A) They run till the reactants are active
(B) They are more efficient
(C) They are free from pollution (D) All of these

42. The structure of DNA is
(A) Single helix (B) Dobule helix
(C) Triple helix (D) Linear

43. Which of the following is the monomer of natural rubber ?
(A) Glycol (B) Ethylene (C) Isoprene (D) Adipic acid

44. Emery consists of
(A) Impure carborundum (B) Impure graphite
(C) Purest form of iron (D) Impure corundum

45. When acetylene reacts with arsenic trichloride in the presence of anhydrous aluminum chloride, it produces
(A) β-chlorovinyl dichloroarsine (B) Lewisite
(C) Nitrobenzene (D) Both (A) and (B)

46. Which of the following sets has the strongest tendency to form anions ?
(A) N, O and F (B) V, Cr and Mn
(C) Ga, In and T (D) Na, Mg and Al

47. The reaction $CH_2 = CHCH_3 + HBr \rightarrow CH_3CHBrCH_3$ is
(A) Electrophilic addition reaction
(B) Nucleophilic addition reaction
(C) Free radical addition reaction
(D) Electrophonic substitution reaction

48. Which of the following is the most stable oxidation state of aluminium ?
(A) +2 (B) +1 (C) -1 (D) + 3

49. Oxidation of acetaldehyde with selenium dioxide produces
(A) Ethanoic acid (B) Methanoic acid
(C) Glyoxal (D) Oxalic acid

50. Baeyer's reagent is used in the laboratory for the
(A) Detection of double bond (B) Reduction process
(C) Oxidation process (D) Detection of glucose

AFMC - 2000

1. Arrange the following in order of their decreasing thermal conductivity
(A) Ag, Cu, Al (B) Al, Cu, Ag
(C) Al, Ag, Cu (D) Cu, Ag, Al

2. If heat of formation of $C + O_2 \rightarrow CO_2 = -94$ K cal and $2C + O_2 \rightarrow 2CO = -110.5$ K cal, then heat of combustion of carbon is
(A) 94 K cal (B) 110.5 K cal
(C) 16.5 K cal (D) 15 K cal

3. Which of the following reactions is endothermic ?
(A) $H_2SO_4 + 2NaOH \rightarrow Na_2SO_4 + 2H_2O$
(B) $N_2 + O_2 \rightarrow 2NO$ (C) $H_2 + Cl_2 \rightarrow 2HCl$
(D) None of these

4. What happens to the yield of this reaction when temperature is increased ? $N_2 + O_2 \leftrightarrow 2NO$; 43.2 Kcal.
(A) Remains same (B) Increases
(C) Decreases (D) None of these

5. Which of the following lose charge at cathode
(A) Anions (B) Ions (C) Cations (D) None of these

6. The number of essential amino acids in man is
(A) 18 (B) 20 (C) 8 (D) 10

7. An invert sugar is
(A) Isorotatory (B) Dextrorotatory
(C) Laevorotatory (D) Optically inactive

8. Physical adsorption is inversely proportional to
(A) Temperature (B) Concentration
(C) Volume (D) All of these

9. The movement of colloidal particles towards the oppositely charged electrodes on passing electricity is known as
(A) Cataphoresis (B) Tyndall effect
(C) Brownian (D) None of these

10. Acetaldehyde and acetone can be identified by
(A) Iodoform test (B) Bromoform test
(C) Schiff's test (D) Molisch test

11. When a slow neutron goes sufficiently close to U^{235} nucleus, the process is
(A) Fusion of U^{235} (B) Fission of U^{235}
(C) Fusion of neutron (D) First (A) then (B)

12. Which of the following does not give Borax bead test ?
(A) Sodium (B) Cobalt
(C) Chromium (D) Ferrous salt

13. Acidified solution of potassium dichromate on treatment with hydrogen peroxide yields
(A) $CrO_3 + H_2O$ (B) $CrO_5 + H_2O$
(C) $H_2Cr_2O_7 + H_2O$ (D) $Cr_2O_3 + H_2O + O_2$

14. Ester and acetamide are distinguished by
(A) Hydrolysis with strong acids or alkali
(B) Derivatives of fatty acids
(C) Grignard reagent (D) Both (A) and (B)

15. After two hours, the radioactive substance becomes 1/16th of original amount. Then half-life is
(A) 60 minutes (B) 120 minutes
(C) 30 minutes (D) 15 minutes

16. The variable valency is the characteristic of
(A) Halogens (B) Transition elements
(C) Alkali metals (D) Noble gases

17. Compounds are isobar if .... emission takes place
(A) 1 α and 1 β emission (B) 2 α and 1 β emission
(C) 1 α emission (D) 1 β emission

18. Gas evolved in Bhopal gas tragedy was
(A) Methane (B) CO_2
(C) Methyl isocyanate (D) Butane

19. Colloidal sol is
(A) True solution (B) Suspension
(C) Heterogeneous sol (D) Homogeneous sol

20. The oxidation number of oxygen in O_2PtF_6 is
(A) + 1 (B) - 1/2 (C) Zero (D) + 1/2

21. Chloroform when reacts with nitric acid gives
(A) $C_2H_5C(NO)_5SH$ (B) CCl_3CHOH
(C) $CCl_3\ NO_2$ (D) CCl_3NO_3

22. PV/T = constant. A real gas will behave as ideal gas when
(A) Both pressure and temperature are low
(B) Temperature is high and pressure is low
(C) Both pressure and temperature are high
(D) Pressure is high and temperature is low

23. Neutral divalent carbon species released as reaction intermediate in carbylamine reaction is
(A) Carbocation (B) Carbene
(C) Carbanion (D) Free radical

24. The empirical formula of an acid is CH_2O_2, the probable molecular formula of acid may be
(A) $C_2H_4O_2$ (B) $C_3H_6O_4$ (C) CH_2O (D) CH_2O_2

25. Which of the following has lowest pH value ?
(A) 1 M HCl (B) 1 M NaOH
(C) 1 M H_2SO_4 (D) $C_3H_6O_4$

26. Neuron was discovered by
(A) Yukank (B) Dalton (C) Chadwick (D) Rutherford

27. Meso tartaric acid is inactive due to
(A) Internal compensation (B) External compensation
(C) Molecular dissymmetry (D) None of these

28. Rate constant depends on
(A) Initial concentration (B) Time
(C) Temperature (D) None of these

29. Soap removes grease by
(A) Adsorption (B) Emulsification
(C) Coagulation (D) None of these

30. A compound has C, H, N in the following percentage C = 40%, H = 13.33%, N = 46.67% What is its empirical formula ?

(A) CH_3N (B) CH_4N (C) CH_2N (D) C_2H_5N

31. The movement of colloidal particles towards the charged electrodes on passing electricity is
(A) Brownian movement (B) Cataphoresis
(C) Tyndall effect (D) None of these

32. Which of the following has least electron affinity ?
(A) Nitrogen (B) Boron (C) Oxygen (D) Argon

33. Which of the following set has the same number of electrons in the outermost orbit ?
(A) As, Bi (B) P, Ca (C) P, Sb (D) N, O

34. $C_6H_6Cl_6 \xrightarrow[\text{KOH (alc.)}]{\text{Heat or}}$?
(A) Benzene (B) Phenol
(C) $C_6H_5(OH)_6$ (D) Chlorobenzene

35. The active constituent of bleaching power is
(A) $Ca(OCl)_2$ (B) $Ca(OCl)Cl$
(C) $Ca(ClO_2)_2$ (D) $Ca(ClO_2)Cl$

36. Which of the following will not show resonance ?
(A) Ozone (B) Oxygen (C) H_2O_2 (D) Allene

37. During electrolysis of water, hydrogen and oxygen are liberated in the ratio of
(A) 1: 1 (B) 2 :1 (C) 1 :2 (D) 1 : 18

38. Rectified spirit has ethanol
(A) 100% (B) 80% (C) 5% (D) 95%

39. The vapour density of a gas is 22. What is its molecular weight ?
(A) 22 (B) 11 (C) 44 (D) 66

40. Which of the following impurities are present in maximum concentration in pig iron ?
(A) Phosphorus (B) Sand (C) Sulphur (D) Iron

41. $K_4[Fe(CN)_6]$ is a/an
(A) Complex salt (B) Double salt
(C) Alum (D) None of these

42. The reactivity of a catalyst depends on
(A) Particle size (B) Solubility
(C) Molecular weight (D) None of these

43. Isotopes have same number of
(A) Nucleons (B) Positions (C) Protons (D) Neutrons

44. Which of the following pairs is isoelectronic ?
(A) Ar and Cl (B) Na^+ and Ne
(C) Na^+ and Mg (D) Mg and Ne

45. The boiling point of heavy water is
(A) 99°C (B) 110°C (C) 100°C (D) 101.4°C

46. Which one is most viscous ?
(A) CH_3OH (B) C_2H_5OH
(C) HOH_2C-CH_2OH (D) None of these

47. Condensation of two molecules of acetaldehyde is
(A) Perkin reaction (B) Cannizaro reaction
(C) Coupling reaction (D) Aldol condensation

48. Which of the following does not form phenol or phenoxide ?
(A) C_6H_5Cl (B) C_6H_5COOH
(C) $C_6H_5N_2Cl$ (D) $C_6H_5SO_3Na$

49. The reaction of benzene with ethyl chloride, in the presence of aluminium chloride, is called
(A) Wurtz reaction (B) Carbylamine reaction
(C) Perkin's reaction (D) Friedel-Craft's reaction

50. When aniline in hydrochloric acid is treated with ice cold solution of sodium nitrite, it gives
(A) Chlorobenzene (B) Chloroaniline
(C) Nitroaniline (D) Benzene diazonium chloride

AFMC - 2001

1. Which of the following is the most basic in nature ?
(A) $(CH_3)_2NH$ (B) $C_6H_5N(CH_3)_2$
(C) NH_3 (D) CH_3NH_2

2. The H-O-H angle in water molecule is about
(A) 102° (B) 105° (C) 90° (D) 180°

3. The number of α and β particles emitted in nuclear reaction ${}_{90}Th^{238} \rightarrow {}_{83}Bi^{212}$ are respectively
(A) 8, 1 (B) 4, 7 (C) 4, 1 (D) 3, 7

4. At 25°C the pH value of a solution is 6, the nature of the solution is
(A) Neutral (B) Alkaline (C) Acidic (D) None

5. The temperature below which a gas does not exist is called its
(A) Inversion temperature (B) Critical temperature
(C) Neutral temperature (D) Curie point

6. The boiling point of alcohol is higher than that of ether because of
(A) The presence of hydroxyl group
(B) Higher molecular weight (C) Association of molecules
(D) Hydrogen bonding between the molecules

7. The mass of $CaCO_3$ produced when carbon dioxide is bubbled through 500 ml of 0.5 M $Ca(OH)_2$ will be
(A) 50 g (B) 25g (C) 10 g (D) 20 g

8. The element with highest electron affinity among halogens is
(A) F (B) Cl (C) Br (D) I

9. When two ice cubes are pressed over each other, they unite to form one cube. Which of the following forces is responsible to hold them together ?
(A) Hydrogen bond formation
(B) van der Waal's Forces
(C) Covalent attraction (D) Ionic interaction

10. A buffer solution changes its pH
(A) Very rapidly (B) Very slowly
(C) Does not change (D) None of these

11. $FeCl_3$ solution is
(A) Neutral (B) Amphoteric (C) Acidic (D) Basic

12. For reaction $A+B \rightleftharpoons C+D$, doubling the concentration of both the reactants increases the reaction rate by 8

times and doubling initial concentration of only B simply doubles the reaction rate. The rate law is given as
(A) $r = K [A] [B]$ (B) $r = K [A]^2 [B]$
(C) $r = K [A] [B]^2$ (D) $r = K [A]^{1/2} [B]^{1/2}$

13. Effect of temperature on reaction rate is given by
(A) Arrhenius equation
(B) Claisen - Clapeyron equation
(C) Gibbs-Helmholtz equation (D) Kirchhoffs equation

14. Which of the following is used as "Moderator" in nuclear reactor ?
(A) Alum (B) H_2O (C) D_2O (D) Any of these

15. Which of the following laws directly explains the law of "conservation of mass" ?
(A) Dalton's law (B) Avogadro's law
(C) Berzelius' hypothesis (D) Hund's rule

16. Which of the following is not true about vitamins ?
(A) They are vital for life
(B) Their deficiency causes diseases
(C) They were named by "Funk"
(D) They help in digestion

17. Solutions having same osmotic pressure are called
(A) Hypotonic (B) Hypertonic (C) Isotonic (D) Normal

18. Organic catalysts differ from inorganic catalysts in the fact that
(A) They are proteinous in nature
(B) They are not much effective
(C) They are consumed
(D) They have no difference

19. Compounds having same molecular formula but differ in their structure are called
(A) Homomers (B) Mesomers (C) Isotopes (D) Isomers

20. Cellulose is a
(A) Disaccharide (B) Structural Polysaccharide
(C) Structural Oligosaceharide
(D) Structural Monosaccharide

21. Perlon is
(A) Terylene (B) Oxlon (C) Rubber (D) Nylon

22. Water dissolved in HCl
(A) Shows a positive deviation from Raoult's law.
(B) Shows a negative deviation from Raoult's law.
(C) Obey Raoult's law. (D) Is an ideal solution

23. Alcohol which is used as a beverage is
(A) Propanol (B) Butanol (C) Ethanol (D) Methanol

24. Bonding present in DNA double helix is
(A) Covalent (B) Ionic
(C) van der Waall (D) Hydrogen

25. Rutherford's atomic model syggests the existence of
(A) α-particle (B) Mesons (C) Atom (D) Nucleus

26. An "antibiotic" with a broad spectrum is one
(A) That acts on a specific antigen
(B) That acts on both the antigens and antibodies
(C) That acts on different antigens
(D) That kills the antibodies

27. Light scattering takes place in
(A) Solutions of electrolyte (B) Colloidal solution
(C) Electrolysis (D) Electroplating

28. Blue colour of water in sea is due to
(A) Refraction of blue light by impurities in sea water
(B) Refraction of blue sky by water
(C) Scattering of light by water (D) None of these

29. The reaction , ${}_7N^{14} + {}_0n^1 \rightarrow {}_6C^{14} + x$ requires emission of
(A) ${}^4_2He$ (B) ${}^1_1H$ (C) ${}^1_0n$ (D) ${}^0_{-1}e$

30. Sidewise overlap of p-p orbitals forms
(A) Covalent bond (B) H bond
(C) Sigma bond (D) Pi bond

31. Antiknocking character of gasoline can be improved by addition of
(A) Lead sulphate (B) Tetraethyl lead
(C) Lead bromide (D) Diethyl lead

32. Gammexane is
(A) D.D.T. (B) Benzene hexachloride
(C) Chloral (D) Hexachloroethane

33. Marsh gas contains mainly
(A) C_2H_6 (B) N_2 (C) CH_4 (D) CO_2

34. Reaction between alkyl halides and sodium metal in the presence of dry ether is called
(A) Wurtz reaction (B) Kolbe's reaction
(C) Clemmenson's reaction (D) None of these

35. Terelene is a polymer of
(A) Terephthalic acid and ethylene glycol
(B) Adipic acid & hexamethylene diamine
(C) Phenol and formaldehyde (D) Vinyl cyanide

36. Hydrolysis of sucrose is called
(A) Hydration (B) Saponification
(C) Esterification (D) Inversion

37. Which of the following is not correct for transition meatls ?
(A) Their compounds are generally coloured
(B) They can form ionic or covalent compounds
(C) Their melting and boiling points are high
(D) They do not exhibit variable valency

38. $K_3 [Al (C_2O_4)_3]$ is called
(A) Potassium alumino (III) oxalate
(B) Potassium trioxalatoaluminate
(C) Potassium alumino oxalate
(D) Potassium trioxalatoaluminate (III)

39. Metal often found in free state is
(A) Gold (B) Silver (C) Copper (D) Sodium

40. Which of the noble gases has highest polarisability ?
(A) Kr (B) Xe (C) He (D) Ar

41. Shape of ClO_3 is

(A) Triangular pyramidal (B) Tetrahedral
(C) Triangular planar (D) Triangular bipyramidal

42. Hypo is used in photography because of its
(A) Oxidising behavriour (B) Reducing behaviour
(C) Complex forming behaviour
(D) Reaction with light

43. Which of the following sets of atomic numbers belong to that of alkali metals ?
(A) 1, 12, 30, 4, 62 (B) 37, 19, 3, 55
(C) 9, 17, 35, 53 (D) 12, 20, 56, 88

44. When conc. H_2SO_4 is brought in contact with sugar, it becomes black due to
(A) Hydrolysis (B) Hydration
(C) Decolourisation (D) Dehydration

45. Which of the following is diamagnetic ?
(A) O_2^+ (B) O_2 (C) O_2^- (D) O_2^{2-}

46. In Kjeldahl's method, $CuSO_4$ acts as
(A) Oxidising agent (B) Reducing agent
(C) Hydrolysing agent (D) Catalytic agent

47. Positron was discovered by
(A) Rutherford (B) Fermi
(C) Vulrava (D) Anderson

48. Toluene reacts with chromyl chloride to form
(A) Chlorotoluene (B) Benzaldehyde
(C) Benzyl chloride (D) Benzoic acid

49. Laws of electrolysis were given by
(A) Lamarck (B) Ostwald
(C) Faraday (D) Arrhenius

50. Modern Periodic Law states that properties of elements are periodic functions of
(A) Atomic mass (B) Atomic volume
(C) Ionisation potential (D) Atomic number

AFMC - 2002

1. pH of a solution can be expressed as
(A) $-\log_e [H^+]$ (B) $\log_{10} [H^+]$
(C) $-\log [H^+]$ (D) $-\log_{10} [H^+]$

2. The element with atomic number 51 exists in block
(A) p (B) f (C) s (D) d

3. Which of the following has changeable valency ?
(A) He (B) Br (C) B (D) None of these

4. Which of the following is asbestos ?
(A) Sodium - calcium silicate
(B) Calcium - magnesium silicate
(C) Sodium - magnesium silicate (D) None of these

5. Which of the following is not true in case of reaction with heated copper at 300° C ?
(A) Primary alcohol → Aldehyde
(B) Secondary alcohol → Ketone
(C) Phenol → Benzyl alcohol
(D) Tertiary alcohol → alkene

6. Which of the following carbon atoms is likely to possess geometry ?

$$H_2\overset{1}{C} = \overset{2}{C}H - \overset{3}{C}H_2 - \overset{4}{C}OOH$$

(A) 1 (B) 2 (C) 3 (D) 4

7. Which if the following will not give iodofom test ?
(A) Ethanol (B) Ethanal
(C) Isopropyl alcohol (D) Bezyl alcohol

8. Propyne on polymerization gives
(A) Mesitylene (B) Benzene
(C) Ethyl benzene (D) Propyl benzene

9. In amines, the hybridization state of N is
(A) sp^3 (B) sp^3d (C) sp (D) sp^2

10. Which of the following is most electronegative ?
(A) C (B) Si (C) Pb (D) Sn

11. Which of the order for ionization energy is correct ?
(A) Be > b > C > N > O (B) B < Be < C < O < N
(C) B < Be < C < N < N (D) B < Be < N < C < O

12. Which of the following is a aldohexose ?
(A) Raffinose (B) Cellulose
(C) Glucose (D) Sucrose

13. A metal ion N^{3+} loses 3 electrons its oxidation number will be
(A) 0 (B) - 3 (C) 3+ (D) +6

14. Orthophosporic acid is
(A) Monobasic (B) Dibasic
(C) Tribasic (D) Tetrabasic

15. Small liquid droplets dispersed in another liquid is called
(A) Suspension (B) Emulsion
(C) Gel (D) True solution

16. Internal energy and pressure of a gas per unit volume are related as
(A) $P = \frac{2}{3}E$ (B) $P = \frac{3}{2}E$ (C) $P = \frac{1}{2}E$ (D) $P = 2E$

17. Propene and propyne may be differentiated by
(A) $Br_2 + H_2O$ (B) Alkaline $KMnO_4$
(C) Acidic $K_2Cr_2O_7$ (D) Ammonical Cu_2Cl_2

18. Oxidation number of iron in $K_4[Fe(CN)_6]$ is
(A) + 10 (B) + 1 (C) +3 (D) +2

19. Which of the following is most soluble ?
(A) Bi_2S_3 [$K_{sp} = 1 \times 10^{-70}$] (B) MnS [$K_{sp} = 7 \times 10^{-16}$]
(C) CuS [$K_{sp} = 8 \times 10^{-37}$] (D) Ag_2S [$K_{sp} = 6 \times 10^{-51}$]

20. A molecule in which sp^2 hybrid orbitals are used by the central atom in forming covalent bond is
(A) He_2 (B) SO_2 (C) PCl_5 (D) N_2

21. Consider the following reaction occurring in an automobile : $2C_8H_{18}(g)+25O_2(g) \rightarrow 16CO_2(g) + 18H_2O(g)$ The sign of ΔH, ΔS and ΔG would be
(A) +, −, + (B) −, +, − (C) −, +, + (D) +, +, −

22. Grignard reagent is prepared by the reaction between
(A) Magnesium and alkyl halide
(B) Zinc and alkyl halide (C) Magnesium and alkane

(D) Magnesium and aromatic hydrocarbon

23. In the given the numbers, 161 cm, 0.161 cm, 0.0161cm, the number of significant figures is
(A) 3, 4 and 5 (B) 3, 3 and 3
(C) 3, 3 and 4 (D) 3, 4 and 4

24. The edge length of face centred unit cubic cell is 508 pm. If the radius of the cation is 110 pm, the radius of the anion is
(A) 618 pm (B) 398 pm (C) 110 pm (D) 249 pm

25. For the cell reaction $Cu(s)/Cu^{2+}(aq) // Ag+(aq) /Ag(s)$
$E^{o}_{cu^{2+}/cu} = + 0.34$ V, $E^{o}_{Ag+/Ag} = + 0.80$ V
Standard EMF of the cell will be
(A) – 0.46 V (B) + 0.46 V (C) – 1.14 V (D) + 1.14 V

26. Reaction of $CH_2 - CH_2$ (bridged by O, epoxide) with R-MgX leads to the formation of
(A) R - CHOH - R (B) $R - CHOHCH_3$
(C) $R_2 > CH - CH_2 - OH$
(D) $R - CH_2 - CH_2 - OH$

27. Aspirin is an acetylation product of
(A) p-dihydroxybenzene (B) o-hydroxybenzoic acid
(C) o-dihydroxybenzene (D) p-hydroxybenzoic acid

28. Which of the following will give colour to an aqueous solution ?
(A) Cu^{+} (B) Zn^{+2} (C) Cr^{+3} (D) Ti^{+4}

29. Strongest hydrogen bonding is shown by
(A) Water (B) Ammonia
(C) Hydrogen fluoride (D) Hydrogen sulphide

30. Electronic configuration of calcium atom may be written as
(A) [Ne] $4s^2$ (B) [Kr] $4p^2$ (C) [Ne] $4p^2$ (D) [Ar] $4s^2$

31. Which of the following acid is present in citrus fruits ?
(A) Citric acid (B) Tartaric acid
(C) Malic acid (D) All of these

32. Nitrolim is
(A) $CaC_2 + N_2$ (B) $Ca(CN)_2 + Ca(NO_3)_2$
(C) $CaCN_2 + C$ (D) None of these

33. Which of the following subshell is not possible ?
(A) 6p (B) 3s (C) 2d (D) 4f

34. How many isomers of $C_5H_{11}OH$ will be primary alcohols ?
(A) 5 (B) 4 (C) 2 (D) 3

35. Baeyer's reagent is used in the laboratory for the
(A) Detection of double bonds (B) Detection of glucose
(C) Reduction (D) Oxidation

36. For the carbylamine reaction we need hot alcoholic KOH and
(A) Any amine and chloroform
(B) Chloroform and silver power
(C) A primary amine and an alkyl halide
(D) A monoalkylamine and trichloromethane

37. Bakelite is a product of reaction between
(A) Formaldehyde and NaOH (B) Phenol and methanal
(C) Aniline and urea (D) Phenol and methanol

38. N/10 NaOH 100 ml and N/5 HCl 50 ml are mixed together then pH of the resulting solution is
(A) 0 (B) 7 (C) Less than 7 (D) Greater than 7

39. Based on the following thermochemical equations
$H_2O(g) + C(s) = CO(g) + H_2(g)$ $\Delta H = 131$ kJ/mole
$CO(g) + ½O_2(g) = CO_2(g)$ $\Delta H = -282$ kJ/mole
$H_2(g) + ½O_2(g) = H_2O(g)$ $\Delta H = -242$ kJ/mole
$C(s) + O_2(g) = CO_2(g)$ $\Delta H = x$ kJ/mole
the value of x will be
(A) - 393 kJ (B) - 655 kJ (C) 393 kJ (D) 655 kJ

40. Increasing order of acid strength is
(A) p-Nitrophenol, p-Methoxyphenol, p-Methylphenol
(B) p-Methylphenol, p-Methoxyphenol, p-Nitrophenol
(C) p-Nitrophenol, p-Methylphenol, p-Methoxyphenol
(D) p-Methoxyphenol, p-Methylphenol, p-Nitrophenol

41. Conjugate base of NH_4^+ is
(A) KOH (B) NH_4OH (C) NH_3 (D) OH^-

42. Number of atoms in one unit cell of a FCC lattice is
(A) 2 (B) 6 (C) 4 (D) 1

43. Maximum electrons in a d-orbital are
(A) 2 (B) 10 (C) 6 (D) 14

44. Amino acids are the building blocks of
(A) Carbohydrates (B) Vitamins
(C) Fats (D) Proteins

45. Which quantum number will determine the shape of the subshell ?
(A) n (B) l (C) m (D) s

46. Value of universal gas constant depends upon
(A) Temperature of the gas (B) Volume of the gas
(C) Number of moles of the gas (D) None of these

47. The reaction $2SO_2 + 2O_2 \rightleftharpoons SO_3$ will be favoured by
(A) High temperature and low pressure
(B) Low temperature and high pressure
(C) low temperature and low pressure
(D) High temperature and high pressure

48. The alcohol obtained by the hydrolysis of oils and fats is
(A) Pentanol (B) Propanol
(C) Glycerol (D) Glycol

49. The base present in DNA but not in RNA is
(A) Adenine (B) Thymine
(C) Uracil (D) Guanine

50. $KMnO_4$ will oxidise acetylene to
(A) Ethylene glycol (B) Ethyl alcohol
(C) Oxalic acid (D) Acetic acid

AFMC-2003

1. The structure of H_2O_2 is
(A) Planar (B) Linear
(C) Non-planar (D) Three dimensional

2. Hydrogen burns in air with a
(A) Yellow flame (B) Green flame
(C) Light bluish flame (D) None of these

3. CO is isoelectronic with
(A) N_2 (B) O_2 (C) NH_3 (D) NO_2

4. Which of the following does not exist
(A)PbF_4 (B)$PbCl_4$ (C)$PbBr_4$ (D) PbI_4

5. Elements having same number of nucleons and different number of protons are known as
(A)Isotopes (B) Isobars
(C) Isotones (D) None of these

6.Which of the following is a non-metal
(A) Iridium (B) Boron (C) Gallium (D) Aluminium

7. The element with the highest electron affinity is
(A) F (B) Cl (C) Br (D) I

8. Diamond and emerald are formed of
(A) Carbon-Carbon (B) Silica-Silica
(C) Carbon-Silica (D) Silica-Carbon

9. Proton is likely to have high
(A) Hydration energy (B) Electron affinity
(C) Atomic size (D) Atomic mass

10. Number of elements present in fifth period is
(A) 8 (B) 18 (C) 24 (D)32

11. To study the relative arrangement of atoms in a molecule, we study
(A) Molecular formula (B) Structural formula
(C) Emperical formula (D) None of these

12. Gobar gas contains
(A) Butane (B) Ethane
(C) Methane (D) Propane

13. Number of orbitals present in third shell is
(A) 1 (B) 3 (C) 9 (D) 18

14. Bohr's theory was modified to give the concept of elliptical orbits by
(A) Rutherford (B) Hund
(C) Sommerfeld (D)Davisson

15. Isotopes have the same
(A) Mass number (B) Atomic number
(C) Nucleons (D) None of these

16. An element has half life period of 1600 years. The mass left after 6400 years will be
(A) 1/4 (B) 1/12 (C) 1/16 (D)1/32

17. α–particles are
(A) Protons (B) Electrons
(C) Positrons (D) Charged helium atoms

18. Which of the following can be used for carrying out nuclear reactions?
(A) Uranium-238 (B) Plutonium-239
(C) Neptunium-239 (D) Thorium-232

19. Process of converting vapour into liquid is known as
(A) Condensation (B) Vapourisation
(C) Freezing (D) Melting

20. Diffusion of solvent through a semipermeable membrane is called
(A) Osmosis (B) Diffusion
(C) Absorption (D) Plasmolysis

21. Surface water contains
(A) Salt (B) Suspended impurities
(C) Organic compounds
(D) Salt and organic compounds

22. Which of the following has the highest % of nitrogen?
(A) Ammonium sulphate (B) Nitrolim
(C) Calcium ammonium nitrate (D) Urea

23. Which of the following forms of phosphorus is most reactive?
(A) Red (B) White (C) Violet (D) Scarlet

24. SO_2 reacts with Cl in sunlight to form
(A) Sulphur dioxide (B) Sulphuryl chloride
(C) Sulphanyl chloride (D) None of these

25. Argon gas was discovered by
(A) Rayleigh (B) Ramsay
(C) Lockerey (D) None of these

26. Aqua regia is a mixture of HCl and HNO_3 in the ratio of
(A) 1:3 (B) 3:1 (C) 3:2 (D) 2:3

27. Silvering of mirror is done by
(A) Ag_2O_3 (B) Fe_2O_3 (C) Al_2O_3 (D) $AgNO_3$

28. When CuSO4 is hydrated, it becomes
(A) Acidic (B) Basic (C) Neutral (D) Amphoteric

29. Formula of ferric sulphate is
(A) $FeSO_4$ (B) Fe_2SO_4 (C) $Fe_2(SO_4)_3$ (D)$Fe(SO_4)_2$

30. The product formed on heating ethyl alcohol with sulphuric acid at 140°C is
(A) Diethyl ether (B) Ethene
(C) Ethyl sulphate (D) Ethanoyl sulphate

31. pH value of water is
(A) Pressure dependent
(B) Temperature dependent
(C) Temperature independent (D) None of these

32. Which of the following is an intensive property
(A) Temperature (B) Entropy
(C) Volume (D) Mass

33. If in a container mass and no heat exchange occurs, then it is
(A) Closed system (B) Open system

(C) Imaginary system (D) Isolated system

34. Absolute temperature is the temperature at which
(A) Mass becomes zero (B) Volume ceases to zero
(C) All molecular motion ceases (D) None of these

35. Rate of a first order reaction depends upon
(A) Time (B) Temperature
(C) Concentration of the reactants
(D) All of these

36. Which of the following 1 M soltions conducts more electricity ?
(A) Boric acid (B) Nitric acid
(C) Phosphorous acid (D) Sulphuric acid

37.Galvanization includes plating of iron with
(A) Aluminium (B) Zinc
(C) Copper (D) Tin

38. Which of the following is used in electroplating ?
(A) $AgNO_3$ (B) AgCl
(C) Sodium argentocyanide (D) None of these

39. Which of the following is not a natural polymer ?
(A) Cellulose (B) PVC
(C) Protein (D) Nucleic acid

40. Nylon is a
(A) Polyester (B) Polyamide
(C) Polysaccharide (D) None of these

41. Treatment of rubber with sulphur is
(A) Annealing (B) Quenching
(C) Vulcanisation (D) None of these

42. Process of heating the ore in air to remove sulphur is
(A) Calcination (B) Roasting
(C) Smelting (D) None of these

43. In electrorefining of Copper, some gold is deposited as
(A) Anode mud (B) Cathode
(C) Cathode mud (D) None of these

44. Copper and Tin are costituents of
(A) Bronze (B) Brass
(C) German silver (D) None of these

45. Which of the following has the highest carbon content?
(A) Stainless steel (B) Cast iron
(C) Wrought iron (D) None of these

46. Electrolytic refining of aluminium is done by .... process.
(A) Serpeck's (B) Hall's
(C) Baeyer's (D) Hoop's

47. In NaCl, Na^+ is called
(A) Acid radical (B) Basic radical
(C) Free radical (D) Salt

48. Precipitation of Group 4 cations takes place when H_2S is
(A) Highly ionized (B) Less ionized
(C) Not ionized (D) None of these

49. Which of the following tests is used to identify primary, secondary and tertiary alcohols ?
(A) Perkin's test (B) Victor Meyer's test
(C) Hoffmann test (D) All of these

50. Which of the following compounds is resistant to nucleophilic attack by hydroxyl ion ?
(A) Acetamide (B) Acetonitrile
(C) Methylacetate (D) Diethylether

AFMC -2004

1. Which of the following has no unit ?
(A) electronegativity (B) electron affinity
(C) ionisation energy (D) excitation potential

2. Structure of H_2O_2 is
(A) planar (B) non-planar
(C) linear (D) three-dimensional

3. Which of the following metals has stable carbonates ?
(A) Na (B) Mg (C) Al (D) Si

4. Which of the following cuts ultraviolet rays ?
(A) soda glass (B) Crooke's glass
(C) pyrex (D) none of these.

5. Colemnite is
(A) $Ca[B_3O_4(OH)_2]2H_2O$ (B) $Ca_2B_6O_{11}5H_2O$
(C) $Ca(OH)_2$ (D) $Na_2B_4O_72H_2O$

6. Lapis lazuli is
(A) ferrous sulphate (B) copper sulphate
(C) sodium alumino silicate (D) zinc sulphate

7. Photoelectric effect is maximum in
(A) Cs (B) Na (C) K (D) Li

8. Which of the following can't be used in Friedel-Craft's reactions ?
(A) $FeCl_3$ (B) $FeBr_2$ (C) $AlCl_3$ (D) NaCl

9. Mg and Li are similar in their properties due to
(A) same e/m ratio (B) same electron affinity
(C) same group (D) same ionic potential

10. Write the IUPAC name of CH_3CH_2COOH.
(A) ethyl formic acid (B) ethyl carboxylic acid
(C) ethane methanoic acid (D) propanoic acid

11. Which of the following yields both alkane and alkene ?
(A) Kolbe's reaction (B) Williamson's synthesis
(C) Wurtz reaction (D) Sandmeyer reaction

12. "The addition of unsymmetrical reagents to unsymmetrical alkenes occurs in such a way that the negative part of the addendum goes to that carbon atom of the double bond which carries lesser number of hydrogen atoms" is called
(A) Saytzeff rule (B) Markownikoff's rule
(C) Kharasch effect (D) Anti-Saytzeff rule

13. Gasoline has composition
(A) C_8 - C_{12} (B) C_2 - C_5

(C) C_6 - C_{11} (D) none of these

14. Octane number can be changed by
(A) isomerisation (B) alkylation
(C) cyclisation (D) all of these

15. Acetone is mixed with bleaching powder to give
(A) chloroform (B) acetaldehyde
(C) ethanol (D) phosgene

16. Acetone reacts with Grignard reagent to form
(A) 3^0 alcohol (B) 2^0 alcohol
(C) ether (D) no reaction

17. Proton was discovered by
(A) Chadwick (B)Thomson
(C) Goldstein (D) Bohr

18. Oxygen molecule is
(A) diamagnetic (B) paramagnetic
(C) ferromagnetic (D) ferrimagnetic

19..The ratio of area covered by second orbital to the first orbital is
(A) 1 : 2 (B) 1 : 16 (C) 8 : 1 (D) 16 : 1

20. The element used for dating the ancient remains is
(A) Ni (B) C - 14 (C) C-12 (D) Rd

21.The molecule of CO_2 has angle 180°. It can be explained on the basis of
(A) sp^3 hyvbridisation (B) sp^2 hybridisation
(C) sp hybridisation (D) d^2sp^3 hybridisation

22. Fusion bomb involves
(A) combination of lighter nuclei into bigger nucleus
(B) destrucion of heavy nucleus into smaller nuclei
(C) combustion of oxygen (D) explosion of TNT

23. How will you separate mixture of two gases ?
(A) fractional distillation technique
(B) Grahams law of diffusion technique
(C) osmosis (D) chromatography

24. 4.4g of CO_2 contains how many litres of CO_2 at STP ?
(A) 2.4 litre (B) 2.24 litre
(C) 44 litre (D) 22.4 litre

25. Containers A and B have same gases. Pressure, volume and temperature of A are all twice that of B, then the ratio of number of molecules of A and B are
(A) 1 : 2 (B) 2 (C) 1 : 4 (D) 4

26. Triple point of water is
(A) 273 K (B) 373 K (C) 203 K (D) 193 K

27. "Relative lowering in vapour pressure of solution containing non-volatile solute is directly proportional to mole fraction of solute". Above statement is
(A) Henry law (B) Dulong and Petit law
(C) Raoult's law (D) Le-Chatelier's principle.

28. Naphthalene is a/an
(A) ionic solid (B) covalent solid
(C) ethanol (D) phosgene.

29. Which of the following is the buffer solution ?
(A) $CH_3COOH + CH_3COONa$
(B) $CH_3COOH + CH_3COONH_4$
(C) $CH_3COOH + NH_4Cl$
(D) NaOH + NaCl

30. Which of the following gas mixture is used by the divers inside the sea ?
(A) O_2 + He (B) O_2 + Xe
(C) O_2 + Ar (D) $O_2 + N_2$

31. Which of the following oxides of nitrogen is solid ?
(A) NO_2 (B) N_2O (C) N_2O_3 (D) N_2O_5

32. What is the % of acetic acid present in vinegar ?
(A) 6 - 10% (B) 70 - 80%
(C) 7 - 8% (D) 90 - 100%

33. Lanthanides and actinides resemble in
(A) electronic configuration (B) oxidation state
(C) ionization energy (D) formation of complexes

34. EDTA has coordination number ?
(A) 3 (B) 4 (C) 5 (D) 6

35. What is the net charge on ferrous ion ?
(A) +2 (B) +3 (C) +4 (D) +5

36. Stainless steel is an alloy of
(A) Copper (B) nickel and chromium
(C) manganese (D) zinc

37. Nitrous oxide is known as
(A) breathing gas (B) laughing gas
(C) exercising gas (D) laboratory gas

38. Which of these do not contain - COOH group ?
(A) aspirin (B) benzoic acid
(C) picric acid (D) salicylic acid

39. Wood spirit is known as
(A) methanol (B) ethanol
(C) acetone (D) benzene

40. Orlon has a unit
(A) vinyl cyanide (B) acrolein
(C) glycol (D) isoprene

41. Which of the following is a basic dye ?
(A) congo red (B) aniline yellow
(C) alizarin (D) indigo.

42. Aniline reacts with which of these to form Schiff base ?
(A) acetic acid (B) benzaldehyde
(C) acetone (D) NH_3

43. $\Delta G = \Delta H - T\Delta S$ was given by
(A) Faraday (B) Kirchoff
(C) Einstein (D) Gibbs-Helmholtz

44. Internal energy is
(A) partly potential and partly kinetic
(B) totally kinetic (C) totally potential
(D) none of these

45. Among them intensive property is

(A) mass (B) volume

(C) surface tension (D) enthalpy

46. Bredig arc method cannot be used to prepare colloidal solution of which of the following?

(A) Pt (B) Fe (C) Ag (D) Au

47. Percentage of silver in German silver is

(A) 0% (B) 1% (C) 5% (D) none of these

48. Monds process is used for

(A) Ni (B) Al (C) Fe (D) Cu

49. Oxidation number of carbon in CH_2Cl_2 is

(A) 0 (B) 2 (C) 3 (D) 5

50. Bessemer converter is used for

(A) steel (B) wrought iron

(C) pig iron (D) cast iron

AFMC - 1991

1. (D) 22. (B)
2. (B) 23. (D)
3. (A) 24. (C)
4. (A) 25. (C)
5. (A) 26. (B)
6. (A) 27. (B)
7. (D) 28. (D)
8. (D) 29. (C)
9. (B) 30. (B)
10. (B) 31. (B)
11. (D) 32. (B)
12. (A) 33. (D)
13. (B) 34. (D)
14. (C) 35. (D)
15. (C) 36. (C)
16. (C) 37. (C)
17. (D) 38. (D)
18. (A) 39. (B)
19. (D) 40. (C)
20. (A)
21. (A)

AFMC - 1992

1. (A) 22. (D)
2. (B) 23. (A)
3. (C) 24. (D)
4. (B) 25. (C)
5. (B) 26. (A)
6. (C) 27. (A)
7. (A) 28. (D)
8. (A) 29. (A)
9. (A) 30. (D)
10. (C) 31. (A)
11. (B) 32. (A)
12. (B) 33. (D)
13. (B) 34. (C)
14. (D) 35. (C)
15. (D) 36. (B)
16. (C) 37. (A)
17. (A) 38. (A)
18. (A) 39. (C)
19. (A) 40. (A)
20. (A)
21. (D)

AFMC - 1993

1. (B) 27. (C)
2. (C) 28. (B)
3. (A) 29. (C)
4. (C) 30. (D)
5. (C) 31. (D)
6. (B) 32. (A)
7. (B) 33. (B)
8. (B) 34. (B)
9. (C) 35. (B)
10. (C) 36. (B)
11. (B) 37. (A)
12. (D) 38. (A)
13. (B) 39. (C)
14. (B) 40. (D)
15. (D) 41. (B)
16. (B) 42. (C)
17. (B) 43. (A)
18. (C) 44. (B)
19. (A) 45. (A)
20. (B) 46. (D)
21. (B) 47. (B)
22. (A) 48. (D)
23. (B) 49. (C)
24. (B) 50. (D)
25. (C)
26. (D)

AFMC - 1994

1. (D) 27. (B)
2. (D) 28. (A)
3. (A) 29. (A)
4. (D) 30. (A)
5. (C) 31. (C)
6. (C) 32. (A)
7. (B) 33. (A)
8. (D) 34. (B)
9. (B) 35. (B)
10. (C) 36. (A)
11. (C) 37. (D)
12. (C) 38. (B)
13. (B) 39. (D)
14. (A) 40. (A)
15. (D) 41. (A)
16. (A) 42. (C)
17. (A) 43. (B)
18. (B) 44. (A)
19. (A) 45. (A)
20. (A) 46. (A)
21. (B) 47. (C)
22. (C) 48. (A)
23. (D) 49. (A)
24. (A) 50. (B)
25. (B)
26. (D)

AFMC - 1995

1. (C) 27. (C)
2. (A) 28. (B)
3. (A) 29. (D)
4. (D) 30. (D)
5. (B) 31. (D)
6. (A) 32. (A)
7. (A) 33. (B)
8. (A) 34. (C)
9. (A) 35. (D)
10. (A) 36. (B)
11. (B) 37. (A)
12. (B) 38. (B)
13. (C) 39. (B)
14. (A) 40. (A)
15. (D) 41. (C)
16. (A) 42. (A)
17. (C) 43. (A)
18. (A) 44. (B)
19. (B) 45. (C)
20. (D) 46. (B)
21. (A) 47. (A)
22. (B) 48. (B)
23. (C) 49. (A)
24. (C) 50. (C)
25. (A)
26. (C)

AFMC - 1996

1. (C) 27. (A)
2. (A) 28. (C)
3. (A) 29. (A)
4. (C) 30. (A)
5. (A) 31. (D)
6. (D) 32. (A)
7. (B) 33. (C)
8. (C) 34. (A)
9. (B) 35. (A)
10. (A) 36. (A)
11. (D) 37. (C)
12. (A) 38. (C)
13. (A) 39. (B)
14. (B) 40. (B)
15. (A) 41. (A)
16. (D) 42. (D)
17. (B) 43. (C)
18. (D) 44. (B)
19. (C) 45. (B)
20. (A) 46. (C)
21. (A) 47. (D)
22. (C) 48. (B)
23. (C) 49. (C)
24. (D) 50. (A)
25. (C)
26. (C)

AFMC- 1997

1. (A) 27. (A)
2. (D) 28. (B)
3. (C) 29. (A)
4. (B) 30. (A)
5. (B) 31. (B)
6. (B) 32. (A)
7. (C) 33. (B)
8. (D) 34. (C)
9. (C) 35. (D)
10. (B) 36. (C)
11. (C) 37. (B)
12. (A) 38. (B)
13. (A) 39. (B)
14. (B) 40. (A)
15. (A) 41. (C)
16. (A) 42. (C)
17. (A) 43. (B)
18. (C) 44. (A)
19. (C) 45. (B)
20. (C) 46. (D)
21. (B) 47. (C)
22. (C) 48. (B)
23. (A) 49. (D)
24. (C) 50. (C)
25. (D)
26. (C)

AFMC- 1998

1. (C) 27. (A)
2. (C) 28. (C)
3. (A) 29. (C)
4. (D) 30. (D)
5. (A) 31. (B)
6. (B) 32. (B)
7. (C) 33. (D)
8. (B) 34. (D)
9. (D) 35. (D)
10. (A) 36. (D)
11. (C) 37. (B)
12. (A) 38. (A)
13. (C) 39. (A)
14. (A) 40. (A)
15. (B) 41. (B)
16. (A) 42. (C)
17. (B) 43. (D)
18. (A) 44. (B)
19. (B) 45. (C)
20. (B) 46. (A)
21. (D) 47. (C)
22. (C) 48. (D)
23. (A) 49. (B)
24. (D) 50. (C)
25. (A)
26. (A)

AFMC- 1999

1. (A) 27. (B)
2. (B) 28. (B)
3. (B) 29. (A)
4. (A) 30. (A)
5. (B) 31. (D)
6. (D) 32. (C)
7. (C) 33. (A)
8. (D) 34. (A)
9. (D) 35. (C)
10. (A) 36. (A)
11. (B) 37. (D)
12. (C) 38. (A)
13. (D) 39. (B)
14. (C) 40. (B)
15. (C) 41. (D)
16. (C) 42. (B)
17. (C) 43. (C)
18. (D) 44. (D)
19. (B) 45. (D)
20. (A) 46. (A)
21. (A) 47. (A)
22. (D) 48. (D)
23. (C) 49. (C)
24. (C) 50. (A)
25. (C)
26. (A)

AFMC- 2000

1. (D) 27. (A)
2. (A) 28. (C)
3. (B) 29. (B)
4. (C) 30. (B)
5. (A) 31. (B)
6. (D) 32. (D)
7. (B) 33. (A)
8. (A) 34. (B)
9. (A) 35. (B)
10. (C) 36. (D)
11. (B) 37. (B)
12. (A) 38. (D)
13. (B) 39. (C)
14. (D) 40. (A)
15. (C) 41. (A)
16. (B) 42. (A)
17. (D) 43. (C)
18. (C) 44. (B)
19. (D) 45. (D)
20. (A) 46. (C)
21. (C) 47. (D)
22. (B) 48. (B)
23. (B) 49. (D)
24. (D) 50. (D)
25. (C)
26. (C)

AFMC - 2001

1. (A)	27. (B)
2. (B)	28. (C)
3. (C)	29. (B)
4. (C)	30. (D)
5. (B)	31. (B)
6. (D)	32. (B)
7. (B)	33. (C)
8. (B)	34. (A)
9. (A)	35. (A)
10. (C)	36. (D)
11. (C)	37. (D)
12. (B)	38. (D)
13. (A)	39. (A)
14. (C)	40. (B)
15. (A)	41. (B)
16. (D)	42. (C)
17. (C)	43. (B)
18. (A)	44. (D)
19. (D)	45. (D)
20. (B)	46. (D)
21. (D)	47. (D)
22. (B)	48. (B)
23. (C)	49. (C)
24. (D)	50. (D)
25. (D)	
26. (C)	

AFMC- 2002

1. (D)	27. (B)
2. (A)	28. (C)
3. (B)	29. (C)
4. (B)	30. (D)
5. (C)	31. (A)
6. (C)	32. (C)
7. (D)	33. (C)
8. (A)	34. (B)
9. (A)	35. (A)
10. (A)	36. (D)
11. (B)	37. (B)
12. (C)	38. (B)
13. (D)	39. (A)
14. (C)	40. (B)
15. (C)	41. (C)
16. (A)	42. (C)
17. (D)	43. (B)
18. (D)	44. (D)
19. (B)	45. (B)
20. (B)	46. (D)
21. (B)	47. (B)
22. (A)	48. (C)
23. (B)	49. (B)
24. (D)	50. (C)
25. (B)	
26. (D)	

AFMC- 2003

1. (C)	26. (B)
2. (C)	27. (D)
3. (A)	28. (A)
4. (D)	29. (C)
5. (B)	30. (B)
6. (B)	31. (C)
7. (B)	32. (A)
8. (A)	33. (D)
9. (A)	34. (C)
10. (B)	35. (C)
11. (B)	36. (D)
12. (C)	37. (B)
13. (C)	38. (A)
14. (C)	39. (B)
15. (B)	40. (B)
16. (C)	41. (C)
17. (D)	42. (B)
18. (B)	43. (C)
19. (A)	44. (A)
20. (A)	45. (B)
21. (B)	46. (D)
22. (D)	47. (A)
23. (B)	48. (A)
24. (B)	49. (B)
25. (B)	50. (D)

AFMC- 2004

1. (A)	26. (A)
2. (B)	27. (C)
3. (A)	28. (B)
4. (B)	29. (A)
5. (B)	30. (A)
6. (C)	31. (D)
7. (A)	32. (A)
8. (D)	33. (B)
9. (D)	34. (D)
10. (D)	35. (A)
11. (A)	36. (B)
12. (B)	37. (B)
13. (C)	38. (C)
14. (D)	39. (A)
15. (A)	40. (A)
16. (A)	41. (B)
17. (C)	42. (B)
18. (B)	43. (D)
19. (D)	44. (A)
20. (B)	45. (C)
21. (C)	46. (B)
22. (A)	47. (D)
23. (B)	48. (A)
24. (B)	49. (A)
25. (B)	50. (C)

1. For the reactions
$C + O_2 \rightarrow CO_2; \Delta H = -393$ J
$2Zn + O_2 \rightarrow 2ZnO; \Delta H = -412$ J
(A) Oxidation of carbon is not feasible
(B) Oxidation of Zn is not feasible
(C) Carbon can oxidize Zn
(D) Zn can oxidize carbon

2. Which of the following ions has the maximum magnetic moment ?
(A) Mn^{+2} (B) Fe^{+2} (C) Ti^{+2} (D) Gr^{+2}

3. Which of the following set is arranged in an increasing order of their bond strengths ?
(A) $O_2^- < O_2 < O_2^+ < O_2^{2-}$ (B) $O_2^{2-} < O_2^- < O_2 < O_2^+$
(C) $O_2^- < O_2^{2-} < O_2 < O_2^+$ (D) $O_2^+ < O_2 < O_2^- < O_2^{2-}$

4. If an endothermic reaction is non-spontaneous at freezing point of water and becomes feasible at its boiling point, then
(A) ΔH and ΔS both are +ve
(B) ΔH is –ve, ΔS is +ve
(C) ΔH and ΔS both are –ve (D) ΔH is +ve, ΔS is –ve

5. A heat engine absorbs heat Q_1 at temperature T_1 and heat Q_2 at temperature T_2. work done by the engine is J (Q_1+Q_2) This data
(A) Violates 1st law of thermodynamics
(B) Violates 1st law of thermodynamics if Q_1 is –ve
(C) Violates 1st law of thermodynamics if Q_2 is –ve
(D) Does not violate 1st law of thermodynamics

6. Most common oxidation states of Ce (cerium) are
(A) +2, +3 (B) +2, +4 (C) +3, +4 (D) +3,+5

7. Arrange Ce^{+3}, La^{+3}, Pm^{+3} and Yb^{+3} in increasing order of their ionic radii.
(A) $Ce^{+3} < Yb^{+3} < Pm^{+3} < La^{+3}$
(B) $Yb^{+3} < Pm^{+3} < La^{+3} < Ce^{+3}$
(C) $Yb^{+3} < Pm^{+3} < Ce^{+3} < La^{+3}$
(D) $Pm^{+3} < La^{+3} < Ce^{+3} < Yb^{+3}$

8. A metal M readily forms its sulphate MSO_4 which is water soluble. It forms its oxide MO which becomes inert on heating. It forms an insoluble hydroxide $M(OH)_2$ which is soluble in NaOH solution. Then M is
(A) Mg (B) Ba (C) Ca (D) Be

9. If ϕ denotes reduction potential, then which is true?
(A) $E^o_{cell} = \phi_{right} - \phi_{left}$ (B) $E^o_{cell} = \phi_{left} + \phi_{right}$
(C) $E^o_{cell} = \phi_{left} - \phi_{right}$ (D) $E^o_{cell} = -(\phi_{left} + \phi_{right})$

10. What is the product when acetylene reacts with hypochlorous acid ?
(A) Cl_2CHCHO (B) $ClCH_2COOH$
(C) CH_3COCl (D) $ClCH_2CHO$

11. On vigorous oxidation by permanganate solution, $(CH_3)_2C = CH - CH_2 - CHO$ gives

(A) $CH_3 - C(OH)(CH_3) - CH(OH) - CH_2CH_3$

(B) $(CH_3)_2 > COOH + CH_3CH_2COOH$

(C) $(CH_3)_2 > CH - OH + CH_3CH_2CH_2OH$

(D) $(CH_3)_2 > C = O + CH_3CH_2CHO$

12. The compound (benzene ring with OCOCH$_3$ and COOH at ortho positions) is used as
(A) Antiseptic (B) Antibiotic
(C) Analgesic (D) Pesticide

13. What will be the emf for the given cell
$Pt | H_2 (P_1) | H^+ (aq) || H_2 (P_2) | Pt$
(A) $\frac{RT}{f} \log \frac{P_1}{P_2}$ (B) $\frac{RT}{2f} \log \frac{P_1}{P_2}$
(C) $\frac{RT}{f} \log \frac{P_2}{P_1}$ (D) None of these

14. When primary amine reacts with chloroform in ethanoic KOH the product formed is
(A) An isocyanide (B) An aldehyde
(C) A cyanide (D) An alconol

15. Which of the following reaction is possible at anode ?
(A) $2Cr^{3+} + 7H_2O \rightarrow Cr_2O_7^{2-} + 14H^+$ (B) $F_2 \rightarrow 2F^-$
(C) $(1/2)O_2 + 2H^+ \rightarrow H_2O$ (D) None of these

16. The reaction
$(CH_3)_3C - Br \xrightarrow{H_2O} (CH_3)_3 - C - OH$
(A) Elimination reaction (B) Substitution reaction
(C) Free radical reaction (D) Displacement reaction

17. If half-life of a substance is 5 yrs, then the total amount of substance left after 15 years, when initial amount is 64 grams is
(A) 32 Grams (B) 8 Grams (C) 16 Grams (D) 2 Grams

18. Cyanide process is used for the extraction of
(A) Barium (B) Aluminum (C) Boron (D) Silver

19. Which is the correct order of ionic sizes ?
(A) Lu > Yb > Sn > Ce (B) Sn > Yb>Ce>Lu
(C) Ce > Sn > Yb > Lu (D) Sn > Ce >Lu >Yb
(At. Nos.: Ce=58, Sn=50, Yb = 70 and Lu 71)

20. In case of nitrogen, NCl_3 is possible but not NCl_5 while in case of phosphorous, PCl_3 as well as PCl_5 are possible. It is due to
(A) Lower electronegativity of P than N
(B) Lower tendency of H-bond formation in P than N
(C) Availability of vacant d orbitals in P but not in N

(D) Occurrence of P in solid while N in gaseous state a room temperature

21. For an ideal gas, number of moles per liter in terms of its pressure P, gas constant R and temperature T is
(A) P/RT (B) RT/P (C) PT/R (D) PRT

22. Na and Mg crystallize in BCC and FCC type crystals respectively. The number of atoms of Na and Mg present in the unit cell of their respective crystal is
(A) 4 and 2 (B) 9 and 14 (C) 14 and 9 (D) 2 and 4

23. Number of atoms in 558.5 gram Fe (at. wt. of Fe = 55.85 g mol^{-1}) is
(A) Twice that in 60 g carbon (B) 6.023×10^{22}
(C) Half that in 8g He (D) $558.5 \times 6.023 \times 10^{23}$

24. When $KMnO_4$ acts as an oxidizing agent and ultimately forms $[MnO_4]^{-1}$, MnO_2, Mn_2O_3, Mn^{+2},the number of electrons transferred in each case respectively is
(A) 4, 3, 1, 5 (B) 1, 5, 3, 7
(C) 1, 3, 4, 5 (D) 3, 5, 7, 1

25. Which of the following is a redox reaction ?
(A) $CaC_2O_4 + 2HCl \rightarrow CaCl_2 + H_2C_2O_4$
(B) $Mg(OH)_2 + 2NH_4Cl \rightarrow MgCl_2 + 2NH_4OH$
(C) $NaCl + KNO_3 \rightarrow NaNO_3 + KCl$
(D) $Zn + 2AgCN \rightarrow 2Ag + Zn(CN)_2$

26. For the reaction.
$CO(g) + (1/2)\ O_2(g) = CO_2(g)$, K_p / K_c is
(A) RT (B) $(RT)^{-1}$ (C) $(RT)^{-1/2}$ (D) $(RT)^{1/2}$

27. Conductivity (unit Siemen's S) is directly proportional to area of the vessel and the concentration of the solution in it and is inversely proportional to the length of the vessel then the unit of the constant of proportionality is
(A) $Sm\ mol^{-1}$ (B) $Sm^2\ mol^{-1}$
(C) $S^{-2}m^2\ mol$ (D) $S^2m^2\ mol^{-2}$

28. In a hydrogen atom, if energy of an electron in ground state is 13.6 eV, then that in the 2nd excited state is
(A) 6.04 eV (B) 13.6 eV (C) 1.51 eV (D) 3.4 eV

29. Which of the following statements is true?
(A) Chemical bond formation take place when forces of attraction overcome the forces of repulsion
(B) HF is less polar than HBr
(C) Absolutely pure water does not contain any ions
(D) In covalency transference of electron takes place.

30. Which of the following compounds has wrong IUPAC name ?
(A) $CH_3-CH_2-CH_2-COO-CH_2CH_3 \rightarrow$ ethyl butanoate
(B) $CH_3-CH(CH_3)-CH_2-CHO \rightarrow$ 3- methyl -butanal
(C) $CH_3-CH(OH)-CH(CH_3)-CH_3 \rightarrow$ 2-methy 1-3-butanol
(D) $CH_3-CH(CH_3)-C(=O)-CH_2-CH_3 \rightarrow$ 2 methyl-3pentanone

31. $CH_3CH_2COOH \xrightarrow[\text{red P}]{Cl_2} A \xrightarrow{\text{alc. KOH}} B$.
What is B?
(A) CH_3CH_2COCl (B) CH_3CH_2CHO
(C) $CH_2 = CHCOOH$ (D) $ClCH_2CH_2COOH$

32. When H_2S is passed through Hg_2S se get
(A) $Hg_2S + Hg$ (B) Hg_2S (C) HgS (D) $HgS + Hg_2S$

33. Alum helps in purifying water by
(A) Sulphate part which combines with the dirt removes it
(B) Coagulating the mud particles
(C) Forming Si complex with clay particles
(D) Making mud water soluble

34. A square planar complex is formed by hybridization of which atomic orbitals ?
(A) s, p_x, p_y, d_{yz} (B) s, p_x, p_y, $d_{x^2-y^2}$
(C) s, p_x, p_y, d_{z^2} (D) s, p_y, p_z, d_{xy}

35. Polymer formation from monomers starts by
(A) Coordinate reaction between monomers
(B) Conversion of monomer to monomer ions by protons
(C) Condensation reaction between monomers
(D) Hydrolysis of monomers

36. The type of isomerism present in nitropentaammine chromium (III) chloride is
(A) Optical (B) Linkage (C) Ionization (D) Polymerisation

37. Uncertainty in position of a minute particle of mass 25g in space is 10^{-5} m. What is the uncertainty in its velocity (in ms^{-1}) ?
(A) 2.1×10^{-34} (B) 0.5×10^{-34}
(C) 2.1×10^{-28} (D) 0.5×10^{-23}

38. Which of these will not react with acetylene ?
(A) NaOH (B) Ammonical $AgNO_3$ (C) Na (D) HCl

39. Change in volume of the system does not alter the number of moles in which of the following equilibria ?
(A) $N_2(g) + O_2(g) \rightleftharpoons 2NO(g)$
(B) $PCl_5(g) \rightleftharpoons PCl_3(g) + Cl_2(g)$
(C) $N_2(g) + 3H_2(g) \rightleftharpoons 2NH_3(g)$
(D) $SO_2Cl_2(g) \rightleftharpoons SO_2(g) + Cl_2(g)$

40. Aluminium is extracted by the electrolysis of
(A) Molten cryolite. (B) Bauxite (C) Alumina
(D) Alumina mixed with molten cryolite

41. Value of gas constant R is
(A) 0.082 litre atm (B) 0.987 cal $mol^{-1}K^{-1}$
(C) 8.3 J mol^{-1} K^{-1} (D) 83 erg mol^{-1} K^{-1}

42. The metal extracted by leaching with a cyanide is.
(A) Mg (B) Ag (C) Cu (D) Na

43. Freezing point of an aqueous solution is (–0.186)°C. Elevation of boiling point of the same solution is K_b=0.512°C, K_f= 1.86°C, find the increase in boiling point.

(A) 0.092°C (B) 0.2372°C (C) 0.186°C (D) 0.0512°C

44. EMF of a cell in terms of reduction potential of its left and right electrodes is

(A) $E = E_{left} - E_{right}$ (B) $E = E_{left} + E_{right}$
(C) $E = E_{right} - E_{left}$ (D) $E = -(E_{right} + E_{left})$

45. Arrangement of $(CH_3)_3$–C–, $(CH_3)_2$–CH–, CH_3–CH_2–when attached to benzyl or an unsaturated group in increasing order of inductive effect is

(A) $(CH_3)_3$ –C– < $(CH_3)_2$ –CH – < CH_3 – CH_2
(B) CH_3 – CH_2– < $(CH_3)_2$ – CH– < $(CH_3)_3$–C–
(C) $(CH_3)_2$ –CH– < $(CH_3)_3$ –C– < CH_3 –CH_2
(D) $(CH_3)_3$ – C– < CH_3 –CH_2– < $(CH_3)_2$ –CH–

46. CH_3–Mg – Br is an organo metallic compound due to

(A) Mg – Br bond (B) C – Mg bond
(C) C – Br bond (D) C – H bond

47. 1 M NaCl and 1 M HCl are present in an aqueous solution. The solution is

(A) Not a buffer solution with pH > 7
(B) A buffer solution with pH < 7
(C) Not a buffer solution with pH < 7
(D) A buffer solution with pH > 7

48. Species acting as both Bronsted acid and base is

(A) HSO_4^{-1} (B) Na_2CO_3 (C) NH_3 (D) OH^{-1}

49. Let the solubility of an aqueous solution of $Mg(OH)_2$ be x then its K_{sp} is

(A) $27x^4$ (B) 9x (C) $4x^3$ (D) $108x^5$

50. Units of rate constant of first and zero order reactions in terms of molarity M unit are respectively.

(A) sec^{-1} $Msec^{-1}$ (B) sec^{-1}, M
(C) $Msec^{-1}$, Sec^{-1} (D) M, sec^{-1}

51. In XeF_2, XeF_4, XeF_6 the number of lone pairs on Xe are respectively

(A) 2, 3, 1 (B) 1, 2, 3 (C) 4, 1, 2 (D) 3, 2, 1

52. In which of the following species the interatomic bond angle is 109° 28' ?

(A) NH_3, BF_4^{-1} (B) $(NH_4)^+$, BF_3
(C) NH_3, BF_4 (D) $(NH_2)^{-1}$, BF_3

53. For the reaction A + 2B → C, rate is given by R = [A] $[B]^2$ then the order of the reaction is

(A) 3 (B) 6 (C) 5 (D) 7

54. RNA is different from DNA because RNA contains

(A) Ribose sugar and uracil
(B) Deoxyribose sugar and thymine
(C) Ribose sugar and thymine
(D) Deoxyribose sugar and uracil

55. In which of the following species is the underlined carbon having sp^3 hybridisation ?

(A) $CH_3\underline{C}OOH$ (B) $CH_3\underline{C}H_2OH$
(C) $CH_3\underline{C}OCH_3$ (D) $CH_2 = \underline{C}H- CH_3$

56. Racemic mixture is formed by mixing two.

(A) Isomeric compounds (B) Chiral compounds
(C) Meso compounds (D) Optical isomers

57. The differntial rate law for the reaction

$H_2 + I_2 \rightarrow 2HI$ is

(A) $-\frac{d[H_2]}{dt} = \frac{d[I_2]}{dt} = -\frac{d[HI]}{dt}$

(B) $\frac{d[H_2]}{dt} = \frac{d[I_2]}{dt} = \frac{1}{2}\frac{d[HI]}{dt}$

(C) $\frac{1}{2}\frac{d[H_2]}{dt} = \frac{1}{2}\frac{d[I_2]}{dt} = -\frac{d[HI]}{dt}$

(D) $-2\frac{d[H_2]}{dt} = -2\frac{d[I_2]}{dt} = \frac{d[HI]}{dt}$

58. Number of sigma bonds in P_4O_{10} is

(A) 6 (B) 7 (C) 17 (D) 16

59. Kinetic theory of gases proves

(A) Only Boyle's law (B) Only Charle's law
(C) Only Avogadro's law (D) All of these

60. KO_2 (potassium super oxide) is used in oxygen cylinders in space and submarines because it

(A) Eliminates moisture
(B) Absorbs CO_2 and increases O_2 content
(C) Absorbs CO_2 (D) Produces ozone

61. A similarity between optical and geometrical isomerism is that

(A) If in a compound one is present then so is the other
(B) Both are included in stereoisomerism
(C) Each forms equal number of isomers for a given compound
(D) They have no similarity.

62. Which of the following does not show geometrical isomerism ?

(A) 1,2-dichloro-1-pentene (B) 1,3 dichloro-2-pentene
(C) 1,1 dichloro-2-pentene (D) 1,4 dichloro-2-pentene

63. With increase of temperature, the one that changes is

(A) Molality (B) Mole fraction
(C) Weight fraction of solute
(D) Fraction of solute present in water

64. The integrated rate equation is $Rt = \log C_o - \log C_t$ The straight line graph is obtained by plotting

(A) time vs $\log C_t$ (B) $\frac{1}{time}$ vs C_t
(C) time vs C_t (D) $\frac{1}{time}$ vs $\frac{1}{C}$

65. In which of the following reactions, increase in the volume at constant temperature does not affect the number of moles at equilibrium

(A) $C(g) + ½O_2(g) \rightarrow CO(g)$
(B) $2NH_3 \rightarrow N_2 + 3H_2$
(C) $H_2(g) + O_2(g) \rightarrow H_2O_2(g)$ (D) None of these

66. When the sample of copper with zinc impurity is to be purified by electrolysis, the appropriate electrodes are

Cathode	Anode
(A) Impure zinc	Impure sample
(B) Pure copper	Impure sample
(C) Pure zinc	Pure copper
(D) Impure sample	Pure copper

67. The most stable ion is.
(A) $[Fe(OH)_3]^{3-}$ (B) $[Fe(Cl)_6]^{3-}$
(C) $[Fe(CN)_6]^{3-}$ (D) $[Fe(H_2O)_6]^{3+}$

68. β-particle is emitted in radioactivity by.
(A) Conversion of proton to neutron
(B) From outermost orbit
(C) Conversion of neutron to proton
(D) β-Particle is not emitted.

69. A mixture of A and B components show –ve deviation as
(A) $\Delta V_{mix} > 0$ (B) $\Delta H_{mix} < 0$
(C) A – B interaction is weaker than A – A and B – B interaction
(D) A – B interaction is stronger than A – A and B – B interaction

70. In a compound C, H and N atoms are present in 9:1:35 by weight. Molecular weight of compound is 108. Molecular formula of compound is.
(A) $C_2H_6N_2$ (B) C_3H_4N (C) $C_6H_8N_2$ (D) $C_9H_{12}N_3$

71. The functional group, which is found in amino acid is.
(A) –COOH group (B) $-NH_2$ group
(C) $-CH_3$ group (D) Both (a) and (b)

72. The heat required to raise the temperature of body by 1 K is called.
(A) Specific heat (B) Thermal capacity
(C) Water equivalent (D) None of these

73. The formation of gas at the surface of tungsten due to adsorption is the reaction of order.
(A) 0 (B) 1 (C) 2 (D) Insufficient data

74. The solubility of $Mg(OH)_2$ is S moles/litre. The solubility product of it under the same condition is.
(A) $4s^3$ (B) $3s^4$ (C) $4s^2$ (D) s^3

75. How do we differentiate between Fe^{3+} and Cr^{3+} in group III ?
(A) By decreasing OH^- ion concentration
(B) By increasing NH_4^+ ion concentration
(C) By taking excess of NH_4OH solution
(D) Both (b) and (c)

AIEEE- 2003

1. For reaction system :
$2NO(g) + O_2(g) \rightarrow 2NO_2(g)$ volume is suddenly reduced to half its value by increasing the pressure on it. If the reaction is of first order with respect to O_2 and second order with respect to NO, the rate of reaction will
(A) Diminish to one-eight of its initial value
(B) Increase to eight times of its initial value
(C) Diminish to one-fourth of its initial value
(D) Increase to four times of its initial value

2. For a cell reaction involving a two-electron change, the standard emf of the cell is found to be 0.295 V at 25°C. The equilibrium constant of the reaction at 25°C will be
(A) 10 (B) 1×10^{10} (C) 1×10^{-10} (D) 29.5×10^{-2}

3. In an irreversible process taking place at constant T and P and in which only pressure- volume work is being done, the change in Gibbs free energy (ΔG) and change in entropy (ΔS), satisfy the criteria
(A) $(\Delta S)_{V,E} < 0, (\Delta G)_{T,P} < 0$ (B) $(\Delta S)_{V,E} > 0, (\Delta G)_{T,P} < 0$
(C) $(\Delta S)_{V,E} = 0, (\Delta G)_{T,P} = 0$ (D) $(\Delta S)_{V,E} = 0, (\Delta G)_{T,P} > 0$

4. Which one of the following characteristics is not correct for physical adsorption ?
(A) Adsorption increases with increase in temperature
(B) Adsorption is spontaneous
(C) Adsorption on solids is reversible
(D) Both enthalpy and entropy of adsorption are negative

5. In respect of the equation $k = Ae^{-Ea/RT}$ in chemical kinetics, which one of the following statements is correct ?
(A) k is equilibrium constant (B) A is adsorption factor
(C) E_a is energy of activation(D) R is Rydberg's constant

6. Standard reduction electrode potentials of three metals A, B and C are respectively + 0.5 V, –3.0 V and –1.2 V. The reducing powers of these metals are
(A) B > C > A (B) A > B > C
(C) C > B > A (D) A > C > B

7. Which one of the following substances has the highest proton affinity ?
(A) NH_3 (B) PH_3 (C) H_2O (D) H_2S

8. Which one of the following is an amphoteric oxide ?
(A) SO_2 (B) B_2O_3 (C) ZnO (D) Na_2O

9. A red solid is insoluble in water. However it becomes soluble if some KI is added to water. Heating the red solid in a test tube result in liberation of some violet coloured fumes and droplets of a metal appear on the cooler parts of the test tube. The red solid is
(A) $(NH_4)_2Cr_2O_7$ (B) HgI_2 (C) HgO (D) Pb_3O_4

10. Concentrated hydrochloric acid when kept in open air sometimes produces a cloud of white fumes. The explanation for it is that
(A) Strong affinity of HCl gas for moisture in air results in forming of droplets of liquid solution which appears like a cloudy smoke.
(B) Concentrated hydrochloric acid emits strongly smelling HCl gas all the time.
(C) Oxygen in air reacts with the emitted HCl gas to form a cloud of chlorine gas.

(D) Due to strong affinity for water, concentrated hydrochloric acid pulls moisture forms droplets of water and hence the cloud

11. What may be expected to happen when phosphine gas is mixed with chlorine gas ?
(A) The mixture only cools down
(B) PCl_3 and HCl are formed and the mixture cools down
(C) PCl_5 and HCl are formed and the mixture cools down
(D) $PH_3.Cl_2$ is formed with warming up

12. The number of d-electrons retained in Fe^{2+} (At. no. of Fe = 26) ion is
(A) 3 (B) 4 (C) 5 (D) 6

13. What would happen when a solution of potassium chromate is treated with an excess of dilute nitric acid ?
(A) CrO_7^{2-} is reduced to +3 states of Cr
(B) CrO_4^{3-} is oxidized to +7 state of Cr
(C) Cr^{3+} and $Cr_2O_7^{2-}$ are formed
(D) $Cr_2O_7^{2-}$ and H_2O are formed

14. In the coordination compound, $K_4[Ni(CN)_4]$, the oxidation state of nickel is
(A) –1 (B) 0 (C) +1 (D) +2

15. Ammonia forms the complex ion $[Cu(NH_3)_4]^{2+}$ with copper ions in alkaline solutions but not in acidic solutions. What is the reason for it ?
(A) In acidic solution protons coordinate with ammonia molecules forming NH_4^+ ions and NH_3 molecules are not available
(B) In alkaline solutions insoluble $Cu(OH)_2$ is precipitated which is soluble in excess of any alkali
(C) In acidic solution hydration protects copper ions
(D) Copper hydroxide is an amphtoeric substance

16. One mole of the complex compound $Co(NH_3)_5Cl_3$, give 3 moles of ions on dissolution in water. One mole of the same complex reacts with two moles of $AgNO_3$ solution to yield two moles of AgCl (s). The structure of the complex is
(A) $[Co(NH_3)_5Cl]\ Cl_2$ (B) $[Co\ (NH_3)_3Cl].2NH_3$
(C) $[Co(NH_3)_4Cl_2]Cl.NH_3$ (D) $[Co(NH_3)_4Cl]\ Cl_2.NH_3$

17. Which one of the following groupings represents a collection of isoelectronic species (At. nos.: Cs-55, Br-35)
(A) Na^+, Ca^{2+}, Mg^{2+} (B) N^{3-}, F^-, Na^+
(C) Be, Al^{3+}, Cl^- (D) Ca^{2+}, Cs^+, Br

18. In the anion $HCOO^-$ the two carbon-oxygen bonds are found to be of equal length. What is the reason for it ?
(A) The C = O bond is weaker than the C–O bond
(B) The anion $HCOO^-$ has two resonating structures
(C) Electronic orbitals of carbon atom are hybridized
(D) The anion is obtained by removal of a proton from the acid molecule

19. The pair of species having identical shapes for molecules of both species is
(A) CF_4, SF_4 (B) XeF_2, CO_2
(C) BF_3, PCl_3 (D) PF_5, IF_5

20. The atomic numbers of vanadium (V), chromium (Cr), manganese (Mn) and iron (Fe) are respectively 23, 24, 25 and 26. Which one of these may be expected to have the highest second ionization enthalpy ?
(A) Mn (B) Fe (C) V (D) Cr

21. Consider the reaction equilibrium ,

$$2SO_2(g) + O_2(g) \rightleftharpoons 2SO_3(g)\ ;\ \Delta H^\circ = -198\ kJ$$

On the basis of Le Chatelier's principle, the condition favourable for the forward reaction is
(A) Lowering of temperature as well as pressure
(B) Increasing temperature as well as pressure
(C) Lowering the temperature and increasing the pressure.
(D) Any value of temperature and pressure

22. What volume of hydrogen gas, at 273 K and 1 atm. pressure will be consumed in obtaining 21.6 g of elemental boron (atomic mass=10.8) from the reduction of boron trichloride by hydrogen ?
(A) 44.8 L (B) 22.4 L (C) 89.6 L (D) 67.2 L

23. For the reaction equilibrium $N_2O_4(g) \rightleftharpoons 2NO_2\ (g)$ the concentrations of N_2O_4 and NO_2 at equilibrium are 4.8×10^{-2} and 1.2×10^{-2} mol L^{-1} respectively. The value of K_C for the reaction is
(A) 3.3×10^2 mol L^{-1} (B) 3×10^{-1} mol L^{-1}
(C) 3×10^{-3} mol L^{-1} (D) 3×10^3 mol L^{-1}

24. The solubility in water of a sparingly soluble salt AB_2 is 1.0×10^{-5} mol L^{-1}. Its solubility product will be
(A) 1×10^{-15} (B) 1×10^{-10} (C) 4×10^{-15} (D) 4×10^{-10}

25. When during electrolysis of a solution of $AgNO_3$ 9650 coulombs of charge pass through the electroplating bath, the mass of silver deposited on the cathode will be
(A) 1.08 g. (B) 10.8 g. (C) 21.6 g. (D) 108 g.

26. For the redox reaction ,
Zn (s) + Cu^{2+} (0.1 M) → Zn^{2+}(1M) + Cu(S) taking place in a cell, E°_{cell} is 1.10 volt. E_{cell} for the cell will be

$$\left(2.303\frac{RT}{F} = 0.0591\right)$$

(A) 1.07 volt (B) 0.82 volt (C) 2.14 volt (D) 1.80 volt

27. In a 0.2 molal aqueous solution of a weak acid HX the degree of ionization is 0.3. Taking k_f for water as 1.85, the freezing point of the solution will be nearest to
(A) –0.480°C (B) –0.360°C
(C) –0.260°C (D) +0.480°C

28. The rate law for a reaction between the substances A and B is given by Rate = $k[A]^n[B]^m$
One doubling the concentration of A and halving the concentration of B, the ratio of the new rate to the earlier rate of the reaction will be as
(A) $\frac{1}{2^{(n-m)}}$ (B) (m + n) (C) (n – m) (D) $2^{(n-m)}$

29. 25ml of a solution of barium hydroxide on titration with a

0.1 molar solution of hydrochloric acid gave a titre value of 35ml. The molarity of barium hydroxide solution was
(A) 0.28 (B) 0.35 (C) 0.07 (D) 0.14

30. The correct relationship between free energy change in a reaction and the corresponding equilibrium constant K_c is
(A) $\Delta G = RT \ln K_c$ (B) $-\Delta G = RT \ln K_c$
(C) $\Delta G = RT \ln K_c$ (D) $-\Delta G^o = RT \ln K_c$

31. If at 298 K the bond energies of C – H, C – C, C = C and H – H bond are respectively 414, 347, 615 and 435 kJ mol^{-1}, the enthalpy change for the reaction $H_2C = CH_2 (g) + H_2 (g) \rightarrow H_3C - CH_3 (g)$ at 298 K will be
(A) +250 kJ (B) –250 kJ (C) +125 kJ (D) –125 kJ

32. The enthalpy change for a reaction does not depend upon
(A) Use of different reactants for the same product
(B) The nature of intermediate reaction steps
(C) The physical states of reactants and products
(D) The differences in initial or final temperatures of involved substances

33. A pressure cooker reduces cooking time for food because
(A) Boiling point of water involved in cooking is increased.
(B) The higher pressure inside the cooker crushes the food material
(C) Heat is more evenly distributed in the cooking space.
(D) Cooking involves chemical changes helped by a rise in temperature

34. If liquids A and B form an ideal solution
(A) The enthalpy of mixing is zero
(B) The entropy of mixing is zero
(C) The free energy of mixing is zero
(D) The free energy as well as the entropy of mixing are each zero

35. In Bohr series of lines of hydrogen spectrum, the third line from the red end corresponds to which one of the following inter-orbit jumps of the electron for Bohr orbits in an atom of hydrogen ?
(A) $4 \rightarrow 1$ (B) $2 \rightarrow 5$ (C) $3 \rightarrow 2$ (D) $5 \rightarrow 2$

36. The de Broglie wavelength of a tennis ball of mass 60 g moving with a velocity of 10 meters per second is approximately
(A) 10^{-33} meters (B) 10^{-31} meters
(C) 10^{-16} meters (D) 10^{-25} meters
Planck's constant, h = 6.63 x 10-34 Js

37. The orbital angular momentum for an electron revolving in an orbit is given by $\sqrt{l(l+1)}.\frac{h}{2\pi}$. This momentum for an s-electron will be given by
(A) $+\frac{1}{2}.\frac{h}{2\pi}$ (B) Zero (C) $\frac{h}{2\pi}$ (D) $\sqrt{2}.\frac{h}{2\pi}$

38. How many unit cells are present in a cube-shaped ideal crystal of NaCl of mass 1.00 ? [Atomic masses : Na = 23, Cl = 35.5]
(A) 2.57 x 10^{21} unit cell (B) 5.14 x 10^{21} unit cell
(C) 1.28 x 10^{21} unit cell (D) 1.71 x 10^{21} unit cell

39. Glass is a
(A) Gel (B) Polymeric mixture
(C) Micro-cystalline solid (D) Super-cooled liquid

40. Which one of the following statements is correct ?
(A) From a mixed precipitate of AgCl and AgI, ammonia solution dissolves only AgCl
(B) Ferric ions give a deep green precipitate on adding potassium ferrocyanide solution
(C) Manganese salts give a violet borax bead test in the reducing flame
(D) On boiling a solution having K^+, Ca^{2+} and HCO_3^- ions we get the precipitate of $K_2Ca(CO_3)_2$

41. According to the Periodic law of elements, the variation in properties of elements is related to their
(A) Atomic masses (B) Nuclear masses
(C) Atomic numbers
(D) Nuclear neutron- proton number ratios

42. Graphite is a soft lubricant extremely difficult to melt. The reason for this anomalous behaviour is that graphite
(A) is a non-crystalline substance
(B) is an allotropic form of diamond
(C) has molecules of variable molecular masses like polymers
(D) has carbon atoms arranged in large plates of rings of strongly bound carbon atoms with weak interplate bonds.

43. The IUPAC name of $CH_3COCH(CH_3)_2$ is
(A) Isopropylmethyl ketone (B) 2-methyl-3 butanone
(C) 4-methylisoproply ketone (D) 3-methyl-2-butanone

44. When $CH_2 = CH - COOH$ is reduced with $LiAlH_4$, the compound obtained will be
(A) CH_3-CH_2-COOH (B) $CH_2 = CH - CH_2OH$
(C) $CH_3 - CH_2 - CH_2OH$ (D) $CH_3 - CH_2 - CHO$

45. According to the kinetic theory of gases, in an ideal gas, between two successive collisions a gas molecule travels
(A) in a straight line path
(B) in a circular path (C) in a wavy path
(D) with an accelerated velocity

46. A reduction in atomic size with increase in atomic number is a characteristic of
(A) High atomic masses (B) d-block
(C) f-block (D) Radioactive series

47. The general formula $C_nH_{2n}O_2$ could be for open chain
(A) Diketones (B) Carboxylic acids
(C) Diols (D) Dialdehydes

48. An ether is more volatile than an alcohol having the same molecular formula. This is due to
(A) Inter-molecular hydrogen bonding in ethers
(B) Inter-molecular hydrogen bonding in alcohols
(C) Dipolar character of ethers
(D) Alcohols having resonance structures

49. Among the following four structures I to IV,

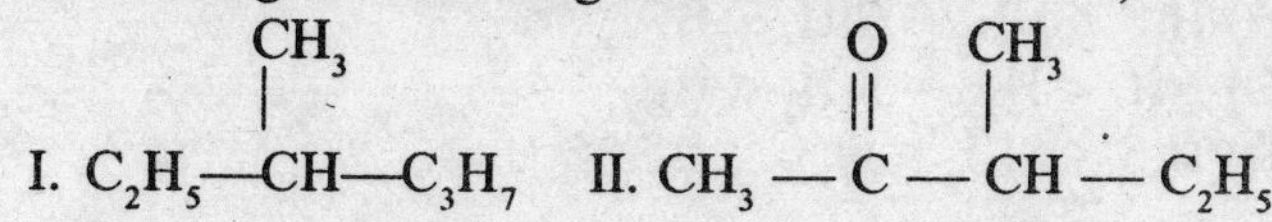

$$\text{III. } H-\overset{\displaystyle H}{\underset{\displaystyle H}{\overset{|}{\underset{|}{C^{\oplus}}}}} \qquad \text{IV. } C_2H_5-\overset{\displaystyle CH_3}{\overset{|}{C}H}-C_2H_5$$

It is true that

(A) All four are chiral compounds
(B) Only I and II are chiral compounds
(C) Only III is a chiral compounds
(D) Only II and IV are chiral compounds

50. Which one of the following processes will produce hard water ?

(A) Saturation of water with $MgCO_3$
(B) Saturation of water with $CaSO_4$
(C) Saturation of water with $CaCO_3$
(D) Addition of Na_2SO_4 to water

51. Which one of the following compounds has the smallest bond angle in its molecule ?

(A) SO_2 (B) OH_2 (C) H_2S (D) NH_3

52. Which one of the following pairs of molecules will have permanent dipole moments for both members ?

(A) SiF_4 and NO_2 (B) NO_2 and CO_2
(C) NO_2 and O_3 (D) SiF_4 and CO_2

53. The radius of La^{3+} (Atomic number of La = 57) is 1.06 Å. Which one of the following given value will be close to the radius of Lu^{3+}(Atomic number of Lu = 71) ?

(A) 1.60 Å (B) 1.40 Å (C) 1.06 Å (D) 0.85 Å

54. The radionucleide $^{234}_{90}Th$ undergoes two successive β-decays followed by one α-decay. The atomic number and the mass number respectively of the resulting radionucleide are

(A) 92 and 234 (B) 94 and 230
(C) 90 and 230 (D) 92 and 230

55. The half life of a radioactive isotope is three hours. If the initial mass of the isotope were 256 g, the mass of it remaining undecayed after 18 hours would be

(A) 12.0g (B) 16.0g (C) 4.0g (D) 8.0g

56. Several blocks of magnesium are fixed to the bottom of a ship to

(A) Prevent action of water and salt
(B) Keep away the sharks (C) Make the ship lighter
(D) Prevent puncturing by under-sea rocks

57. In curing cement plasters water is sprinkled from time to time. This helps in

(A) Keeping in cool (B) Converting sand into silicic acid
(C) Developing interlocking needlelike crystals of hydrated silicates
(D) Hydrating sand and gravels mixed with cement

58. Which one of the following statements is not true ?

(A) The conjugate base of $H_2PO_4^-$ is HPO_4^{2-}
(B) pH + pOH = 14 for all aqueous solutions
(C) The pH of 1×10^{-5} M HCl is 8
(D) 96,500 coulombs of electricity when passed through a $CuSO_4$ solution deposits 1 gram equivalent of copper at the cathode

59. The correct order of increasing basic nature for the bases NH_3, CH_3NH_2 and $(CH_3)_2NH$ is

(A) $(CH_3)_2NH < NH_3 < CH_3NH_2$
(B) $NH_3 < CH_3NH_2 < (CH_3)_2NH$
(C) $CH_3NH_2 < NH_3 < (CH_3)_2NH$
(D) $CH_3NH_2 < (CH_3)_2NH < NH_3$

60. Butene -1 may be converted to butane by reaction with

(A) Zn–HCl (B) Sn–HCl (C) Zn–Hg (D) Pd/H_2

61. The solubilities of carbonates decrease down the magnesium group due to a decrease in

(A) Hydration energies of cations
(B) Lattice energies of solids (C) Inter-ionic attraction
(D) Entropy of solution formation

62. During dehydration of alcohols to alkenes by heating with conc. H_2SO_4 the initiation step is

(A) Protonation of alcohol molecule
(B) Formation of carbocation
(C) Elimination of water (D) Formation of an ester

63. Which one of the following nitrates will leave behind a metal on strong heating ?

(A) Ferric nitrate (B) Copper nitrate
(C) Manganese nitrate (D) Silver nitrate

64. When rain is accompanied by a thunderstorm, the collected rain water will have a pH value.

(A) Slightly lower than that of rain water without thunderstorm
(B) Slightly higher than that of when the thunderstorm is not there.
(C) Uninfluenced by occurrence of thunderstorm
(D) Which depends on the amount of dust in air

65. Complete hydrolysis of cellulose gives

(A) D-fructose (B) D-ribose
(C) D-glucose (D) L-glucose

66. For making good quality mirrors, plates of float glass are used. These are obtained by floating molten glass over a liquid metal which does not solidify before glass. The metal used can be

(A) Sodium (B) Magnesium (C) Mercury (D) Tin

67. The substance not likely to contain $CaCO_3$ is

(A) A marble (B) Calcined gypsum
(C) Sea shells (D) Dolomite

68. The reason for double helical structure of DNA is due to operation of

(A) Van der Waals force (B) Dipole-dipole interaction
(C) Hydrogen bonding (D) Electrostatic attractions

69. Bottles containing C_6H_5I and $C_6H_5CH_2I$ lost their original labels. They were labeled A and B for testing. A and B were separately taken in test tubes and boiled with NaOH solution. The end solution in each tube was made acidic with dilute HNO_3 and then some $AgNO_3$ solution was added. Substance B gave a yellow precipitate. Which one of the following statements is true for this experiment ?

(A) B was C_6H_5I

(B) A was C_6H_5I (C) A was $C_6H_5CH_2I$

(D) Addition of HNO_3 was unnecessary

70. Nylon threads are made of

(A) Polyvinyl polymer (B) Polyester polymer

(C) Polyamide polymer (D) Polyethylene polymer

71. On mixing a certain alkane with chlorine and irradiating it with ultraviolet light, it forms only one monochloroalkane. This alkane could be.

(A) Propane (B) Pentane

(C) Isopentane (D) Neopentane

72. Which of the following could act as a propellant for rockets ?

(A) Liquid hydrogen + liquid nitrogen

(B) Liquid oxygen + liquid argon

(C) Liquid oxygen + liquid hydrogen

(D) Liquid nitrogen + liquid oxygen

73. Ethyl isocyanide on hydrolysis in acidic medium generates

(A) Ethylamine salt and methanoic acid

(B) Propanoic acid and ammonium salt

(C) Ethanoic acid and ammonium salt

(D) Methylamine salt and ethanoic acid

74. The internal energy change when a system goes from state A to B is 40 kJ/mole. If the system goes from A to B by a reversible path and returns to state A by an irreversible path what would be the net change in internal energy ?

(A) 40 kJ (B) > 40 kJ (C) < 40 kJ (D) Zero

75. The reaction of chloroform with alcoholic KOH and p-toluidine forms

(A) $H_3C-C_6H_4-CHO$

(B) $H_3C-C_6H_4-N_2Cl$

(C) $H_3C-C_6H_4-NHCHCl_2$

(D) $H_3C-C_6H_4-NC$

AIEEE -2004

1. As the temperature is raised from 20° C to 40° C, the average kinetic energy of neon atoms changes by a factor of which of the following ?

(A) 2 (B) 1/2 (C) 313/293 (D) $(\sqrt{313/293})$

2. The correct order of bond angles (smallest first) in H_2S, NH_3, BF_3 and SiH_4 is

(A) $H_2S < NH_3 < BF_3 < SiH_4$

(B) $H_2S < NH_3 < SiH_4 < BF_3$

(C) $NH_3 < H_2S < SiH_4 < BF_3$

(D) $H_2S < SiH_4 < NH_3 < BF_3$

3. Which one of the following aqueous solution will exhibit highest boiling point ?

(A) 0.015 M glucose (B) 0.01 M KNO_3

(C) 0.01 M Na_2SO_4 (D) 0.015 M urea

4. The rate equation for the reaction 2A + B → C is found to be : rate = k[A][B]. The correct statement in relation to this reaction, is that the

(A) $t_{1/2}$ isa constant (B) unit of k must be s^{-1}

(C) value of k is independent of the initial concentrations of A and B

(D) rate of formation of C is twice the rate of disappearance of A

5. The maximum number of 90° angles between bond pair-bond pair of electrons is observed in

(A) dsp^2 hybridisation (B) sp^3d^2 hybridisation

(C) sp^3d^2 hybridisation (D) dsp^3 hybridisation

6. Among Al_2O_3, SiO_2, P_2O_3 and SO_2, the correct order of acid strenght is

(A) $Al_2O_3 < SiO_2 < SO_2 < P_2O_3$

(B) $Al_2O_3 < SiO_2 < P_2O_3 < SO_2$

(C) $SO_2 < P_2O_3 < SiO_2 < Al_2O_3$

(D) $SiO_2 < SO_2 < Al_2O_3 < P_2O_3$

7. The formation of the oxide ion O^{2-} (g) requires first an exothermic and then an endothermic step as shown below

$O(g) + e^- = O^-(g)$ $\Delta H^o = -142$ kJmol^{-1}

$O^-(g) + e^- = O^{2-}(g)$ $\Delta H^o = 844$ kJmol^{-1}

This is because

(A) O^- ion has comparatively larger size than oxygen atom

(B) oxygen has high electron affinity

(C) O^- ion will tend to resist the addition of another electron

(D) oxygen is more electronegative

8. Consider the following E° values

$E^0_{Fe^{3+}/Fe^{2+}} = +0.77$ V

$E^0_{Sn^{2+}/Sn} = -0.14$ V

Under standard conditions, the potential for the reaction, $Sn(s) + 2Fe^{3+}(aq) \longrightarrow 2Fe^{2+}(aq) + Sn^{2+}(aq)$ is

(A) 1.40 V (B) 0.91 V (C) 1.68 V (D) 0.63 V

9. What is the equilibrium expression for the reaction $P_4(s) + 5O_2(g) \rightleftharpoons P_4O_{10}(s)$?

(A) $K_c = [P_4O_{10}]/5[P_4][O_2]$

(B) $K_c = [P_4O_{10}]/[P_4][O_2]^5$

(C) $K_c = 1/[O_2]$ (D) $Kc = [O_2]^5$

10. The bond order in NO is 2.5 while that in NO^+ is 3. Which of the following statements is true for these two

species ?

(A) Bond length in NO^+ is greater than in NO

(B) Bond length in unpredictable

(C) Bond length in NO^+ is equal to that in NO

(D) Bond length in NO is greater than in NO^+

11.For the reaction,

$CO\ (g) + Cl_2\ (g) \rightleftharpoons COCl_2\ (g)$

the K_p/K_c is equal to

(A) RT (B) 1/RT (C) 1.0 (D) $\sqrt{RT}$

12. The molar solubility (in mol^{-1}) of a sparingly soluble salt MX_4 is 's'. The corresponding solubility product is K_{sp}. s is given in terms of K_{sp} by the reaction

(A) $s = (256\ K_{sp})^{1/5}$ (B) $s = (K_{sp}/256)^{1/5}$

(C) $s = (128\ K_{sp})^{1/4}$ (D) $s = (K_{sp}/128)^{1/4}$

13. The states of hybridigation of boron and oxygen atoms in boric acid (H_3BO_3) are respectively

(A) sp^2 and sp^3 (B) sp^2 and sp^2

(C) sp^3 and sp^3 (D) sp^3 and sp^2

14. The equilibrium constant for the reaction $N_2\ (g) + O_2\ (g) \rightleftharpoons 2NO\ (g)$ at temperature T is 4×10^{-4}. The value of K_C for the reaction

$NO\ (g) \rightleftharpoons \frac{1}{2} N_2\ (g) + \frac{1}{2} O_2\ (g)$

at the same temperature is

(A) 4×10^{-4} (B) 2.5×10^2 (C) 0.02 (D) 50

15. Which one of the following sets of ions represents the collection of isoelectronic species ?

(A) $Na^+, Mg^{2+}, Al^{3+}, Cl^-$ (B) $K^+, Cl^-, Mg^{2+}, Sc^{3+}$

(C) $Na^+, Ca^{2+}, Sc^{3+}, F^-$ (D) $K^+, Ca^{2+}, Sc^{3+}, Cl^-$

16. Of the following, outer electronic configurations of atoms, the highest oxidation state is achived by which one of them ?

(A) $(n-1)d^5\ ns^2$ (B) $(n-1)d^5\ ns^1$

(C) $(n-1)d^3\ ns^2$ (D) $(n-1)d^8\ ns^2$

17. Which one of the following has the regular tetrahedral structure ?

(A) SF_4 (B) XeF_4 (C) $[Ni(CN)_4]^{2-}$ (D) BF_4^-

18. The ammonia evolved from the treatment of 0.30 g of an organic compound for the estimation of nitrogen was passed in 100 ml of 0.1 M sulphuric acid. The excess of acid required 20 ml of 0.5 M sodium hydroxide solution for complete neutralisation. The organic compound is

(A) urea (B) acetamide

(C) Thiourea (D) benzamide

19. Which one of the following complexes is an outer orbital complex ?

(A) $[Co(NH_3)_6]^{3+}$ (B) $[Mn(CN)_6]^{4-}$

(C) $[Fe(CN)_6]^{4-}$ (D) $[Ni(NH_3)_6]^{2+}$

20. For which of the following parameters the structural isomers C_2H_5OH and CH_3OCH_3 would be expected to have the same values ? (Assume ideal behaviour)

(A) Heat of vaporisation (B) Boilig points

(C) Vapour pressure at the same temperature

(D) Gaseous densities at the same temperature and pressure

21. Which among the following factors is the most important in making fluorine the strongest oxidising halogen ?

(A) Electron affinity (B) Hydration enthalpy

(C) Ionisation enthalpy (D) Bond dissociation energy

22. Which one of the following statements is false?

(A) Two sucrose solutions of same molality prepared in different solvents will have the same freezing point depression

(B) Raoult's law states that the vapour pressure of a component over a solution is proportional to its mole fraction

(C) The correct order of osmatic pressure for 0.01 M aqueous solution of each compound is $BaCl_2 > KCl > CH_3COOH >$ sucrose

(D) The osmatic pressure (π) of a solution is given by the equation $\pi = MRT$, where M is the molarity of the solution

23. 6.02×10^{20} molecules of urea are present in 100ml of its solution. The concentration of urea solution is

(A) 0.1 M (B) 0.01 M (C) 0.02 M (D) 0.001 M

24. What type of crystal defect is indicated in the diagram below ?

Na^+ Cl^- Na^+ Cl^- Na^+ Cl^-

Cl^- □ Cl^- Na^+ □ Na^+

Na^+ Cl^- □ Cl^- Na^+ Cl^-

Cl^- Na^+ Cl^- Na^+ □ Na^+

(A) Interstitial defect (B) Frenkel defect

(C) Schottky defect

(D) Frenkel and Schottky defect

25. The IUPAC name of the compound [structure: cyclohexane ring with HO and two methyl groups] is

HO

(A) 1, 1-dimethyl-3cyclohexane

(B) 3, 3-dimethyl-1cyclohexanol

(C) 3, 3-dimethyl-1-hydroxy cyclohexane

(D) 1, 1-dimethyl-3-hydroxy cyclohexane

26. In van der Waals equation of state of the gas law, the constant b is a measure of

(A) intermolecular repulsions

(B) intermolecular attraction

(C) intermolecular collisions per unit volume

(D) volume occupied by the molecules

27. Rate of the reaction,

$$R-C(=O)-Z + Nu^{\ominus} \longrightarrow R-C(=O)-Nu + Z^{\ominus}$$

is fastest when Z is

(A) Cl (B) NH_2 (C) OC_2H_5 (D) $OCOCH_3$

28. In a first order reaction, the concentration of the reactant, decreases from 0.8 M to 0.4 M in 15 minutes. The time taken for the concentration to change from 0.1 M to 0.025 M is

(A) 60 minutes (B) 15 minutes

(C) 30 minutes (D) 7.5 minutes

29. The conjugate base of $H_2PO_4^-$ is

(A) PO_4^{3-} (B) P_2O_5 (C) HPO_4^{2-} (D) H_3PO_4

30. An ideal gas expands in volume from 1×10^{-3} m³ to 1×10^{-2} m³ at 300K against a constant pressure of 1×10^5 Nm^{-2}. The work done is
(A) -900 KJ (B) -900 J (C) 900 KJ (D) 270 KJ

31. Which one of the following does not have sp^2 hybridised carbon ?
(A) Acetone (B) Acetic acid
(C) Acetamide (D) Acetonitrile

32. The compound formed in the positive test for nitrogen with the Lassaigne solution of an organic compound is
(A) $Fe(CN)_3$ (B) $Na_4[Fe(CN)_5NOS]$
(C) $Fe_4[Fe(CN)_6]_3$ (D) $Na_3[Fe(CN)_6]$

33. In a hydrogen-oxygen fuel cell, combustion of hydrogen occurs to
(A) generate heat (B) produce high purity water
(C) create potential difference between the two electrodes
(D) remove adsorbed oxygen from electrode surfaces

34. To neutralise completely 20 ml of 0.1 M aqueous solution of phosphorus acid (H_3PO_3), the volume of 0.1 M aqueous KOH solution required is
(A) 60 ml. (B) 10 ml. (C) 40 ml. (D) 20 ml.

35. The half-life of a radioisotope is four hours. If the initial mass of the isotope was 200 g, the mass remaining after 24 hours undecayed is
(A) 1.042 g (B) 3.125 g (C) 2.084 g (D) 4.167 g

36. Which of the following liquid pairs shows a positive deviation from Raoult's law ?
(A) Benzene - methanol (B) water - nitric acid
(C) Acetone - chloroform
(D) water - hydrochloric acid

37. Which one of the following has the minimum boiling point ?
(A) 1-Butyne (B) n-Butane
(C) 1-Butene (D) isobutane

38. Which of the following will have a meso-isomer also ?
(A) 2-Hydroxypropanoic acid (B) 2-Chlorobutane
(C) 2, 3- Dichloropentane (D) 2, 3-Dichlorobutane

39. Beryllium and aluminium exhibit many properties which are similar. But the two elements differ in
(A) exhibiting maximum covalency in compounds
(B) forming polymeric hydrides
(C) forming covalent halides
(D) exhibiting amphoteric nature in their oxides

40. Coordination compound have great importance in biological systems. In this context which of the following statments is incorrect ?
(A) Haemoglobin is the red pigment of blood and contains iron
(B) Chlorophylls are green pigments in plants and contain calcium
(C) Carboxypeptidase-A is an enzyme and contains zinc
(D) Cyanocobalamine is B_{12} and contains cobalt

41. Which one of the following ions has the highest value of ionic radius ?
(A) Li^+ (B) B^{3+} (C) O^{2-} (D) F^-

42. Which one of the following ores is best concentrated by froth-floatation method ?
(A) Galena (B) Malachite
(C) Magnetite (D) Cassiterite

43. Which one of the following has largest number of isomers ?
(A) $[Co(en)_2Cl_2]^+$ (B) $[Co(NH_3)_5Cl]^{2+}$
(C) $[Ir(PR_3)_2H(CO)]^{2+}$ (D) $[Ru(NH_3)_4Cl_2]^+$

44. one mole of magnesium nitride on the reaction with an excess of water gives
(A) one mole of ammonia (B) one mole of nitric acid
(C) two moles of ammonia (D) two moles of nitric acid

45. Consider the ground state of Cr atom (Z = 24). The number of electrons with the azimuthal quantum numbers, l = 1 and 2 are, respectively
(A) 12 and 5(B) 12 and 4 (C) 16 and 4 (D) 16 and 5

46. Cerium (Z = 58) is an important member of the lanthanoids. Which of the following statements about cerium is incorrect ?
(A) The +4 oxidation state of cerium is not known in solution
(B) Cerium (IV) acts as an oxidising agent
(C) The common oxidation states of cerium are +3 and +4
(D) The +3 oxidation state of cerium is more stable than the +4 oxidation state

47. Identify the correct statement regarding enzymes.
(A) Enzymes are normaly heterogeneous catalysts that are very specific in the action
(B) Enzymes are specific biological catalysts that can not be poisoned
(C) Enzymes are specific biological catalysts that can normally function at very high temperatures(T~1000K)
(D) Enzymes are specific biological catalyst that possess well-defined active sites

48. The wavelength of the radiation emitted, when in a hydrogen atom electron falls from infinity to stationary state 1, would be (Rydberg constant = 1.097×10^7 m^{-1})
(A) 9.1×10^{-8} nm (B) 192 nm (C) 91 nm (D) 406 nm

49. The correct order of magnetic moments (spin onlt values in B.M.) among is
(Atomic nos. : Mn = 25, Fe = 26, Co = 27)
(A) $[Fe(CN)_6]^{4-} > [CoCl_4]^{2-} > [MnCl_4]^{2-}$
(B) $[MnCl_4]^{2-} > [Fe(CN)_6]^{4-} > [Co(CN)]^{2-}$
(C) $[Fe(CN)_6]^{4-} > [MnCl_4]^{2-} > [CoCl_4]^{2-}$
(D) $[MnCl_4]^{2-} > [CoCl_4]^{2-} > [Fe(CN)]^{4-}$

50. Which one of the following statements regarding helium is incorrect ?
(A) It is used to fill gas balloons instead of hydrogen because it is lighter and non-inflammable

(B) It is used in gas-cooled nuclear reactor.

(C) It is used to produce and sustain powerful superconducting magnets

(D) It is used as a cryogenic agent for carrying out experiments at low temperatures

51. The standard e.m.f. of a cell involving one electron change is found to be 0.591 V at 25°C. The equilibrium contant of the reaction is (F = 96,500 C mol^{-1}; R = 8.314 JK^{-1} mol^{-1})

(A) 1.0×10^5 (B) 1.0×10^{10}

(A) 1.0×10^{30} (D) 1.0×10^1

52. Consider the following nuclear reactions

$$^{238}_{92}M \longrightarrow ^{X}_{Y}N + 2\ ^{4}_{2}He$$

$$^{X}_{Y}N \longrightarrow ^{A}_{B}L + 2\ \beta^+$$

The number of neutrons in the element L is

(A) 140 (B) 142 (C) 144 (D) 146

53. The enthalpies of combustion of carbon and carbon monoxide are -393.5 and -283 KJ mol-1 respectively. The enthalpy of formation of carbon monoxide per mole is

(A) 110.5 KJ (B) 676.5 KJ

(C) -110.5 KJ (D) -676.5 KJ

54. Which of the follwing sets of quantum numbers is correct for an electron in 4f orbital ?

(A) $n = 4, l = 3, m = +4, s = +\frac{1}{2}$

(B) $n = 4, l = 4, m = -4, s = -\frac{1}{2}$

(C) $n = 3, l = 2, m = -2, s = +\frac{1}{2}$

(D) $n = 4, l = 3, m = +1, s = +\frac{1}{2}$

55. The limiting molar conductivities Λ^o for NaCl, KBr and KCl are 126, 152 and 150 S cm^2 mol^{-1} respectively. The Λ^o for NaBr is

(A) 176 S cm^2 mol^{-1} (B) 302 S cm^2 mol^{-1}

(C) 128 S cm^2 mol^{-1} (D) 278 S cm^2 mol^{-1}

56. In a cell that utilises the reaction,

$Zn(s) + 2H^+(aq) \longrightarrow Zn^{2+}(aq) + H_2(g)$

addition of H_2SO_4 to cathode compartment, will

(A) lower the E and shift equilibrium to the left

(B) increase the E and shift equilibrium to the left

(C) lower the E and shift equilibrium to the right

(D) increase the E and shift equilibrium to the right

57. The coordination number of a central metal atom in a complex is determined by

(A) the number of only anion ic ligands bonded to the metal ion

(B) the number of ligands around a metal ion bonded by sigma and pi-bonds both

(C) the nuber of ligands around a metal ion bonded by sigma bonds

(D) the number of ligands around a metal ion bonded by pi-bonds

58. Which one of the following is reduced with zinc and hydrochloric acid to give the corresponding hydrocarbon ?

(A) Butan-2-one (B) Acetamide

(C) Ethyl acetate (D) Acetic acid

59. Aluminium chloride exists as dimer, Al_2Cl_6 in solid state as well as in solution of non-polar solvents such as benzene. When dissolved in water, it gives

(A) $Al_2O_3 + 6HCl$ (B) $[Al(H_2O)_6]^{3+} + 3Cl^-$

(C) $[Al(OH)_6]^{3-} + 3HCl$ (D) $Al^{3+} + 3Cl^-$

60. Which of the following compounds is not chiral ?

(A) 3- chloro-2-methyl pentane

(B) 1-chloro-2-methyl pentane

(C) 1-chloropentane

(D) 2-chloropentane

61. Acetyl bromide reacts with excess of CH_3Mg followed by treatment with a saturated solution NH_4Cl gives

(A) acetone (B) acetyl iodide

(C) acetamide (D) 2-methyl-2-propanol

62. The soldiers of Napolean army while at Alps during freezing winter suffered a serious problem as regards to the tin buttons of thier uniforms. White metallic tin buttons got converted to gray powder. This transformation is related to

(A) a change in the crystalline structure of tin

(B) an iteraction with water vapour contained in the humid air

(C) a change in the partial pressure of oxygen in the air

(D) an interaction with nitrogen of the air at very low temperatures

63. Which one of the following undergoes reaction with 50% sodium hydroxide solution to give the corresponding alcohol and acid ?

(A) Benzoic acid (B) Benzaldehyde

(C) Phenol (D) Butanal

64. The $E^0_{M^{3+}/M^{2+}}$ values for Cr, Mn, Fe and Co are 0.41, +1.57, +0.77 and +1.97 V respectively. For which one of these metals, the change in oxidation state from +2 to +3 is easiest ?

(A) Mn (B) Co (C) Fe (D) Cr

65. On mixing ethyl acetate with aqueous sodium chloride, the composition of the resultant solution is

(A) $CH_3COCl + C_2H_5OH + NaOH$

(B) $CH_3COONa + C_2H_5OH$

(C) $CH_3Cl + C_2H_5COONa$

(D) $CH_3COOC_2H_5 + NaCl$

66. Which base is present in RNA but not in DNA ?

(A) Uracil (B) Guanine (C) Thymine (D) Cytosine

67. Excess of KI reacts with $CuSO_4$ solution and then $Na_2S_2O_3$ solution is added to it. Which of the statements is incorrect for this reaction ?

(A) Evolved I_2 is reduced (B) Cu_2I_2 is formed

(C) $Na_2S_2O_3$ is oxidised (D) CuI_2 is formed

68. Insulin production and its action in human body are responsible for the level of diabetes. This compound belongs to which of the following categories ?

(A) A Co-enzyme (B) An enzyme
(C) A hormone (D) An antibiotic

69. The compound formed on heating chlorobenzene with chloral in the presence of concentrated sulphuric acid is
(A) freon (B) hexachloroethane
(C) gammexane (D) DDT

70. The smog is essentially caused by the presence of
(A) O_2 and N_2 (B) O_3 and N_2
(C) O_2 andO_3 (D) Oxides of sulphur and nitrogen

71. Among the properties (a) reducing, (b) oxidising, (c) complexing, the set of properties shown by CN^- ion towards metal species is
(A) a,b,c (B) b,c (C) a,b (D) c,a

72. Consider the acidity of the carboxylic acids :
I. PhCOOH
II. o-$NO_2C_4H_4COOH$
III. p-$NO_2C_6H_4COOH$
IV. m-$NO_2C_6H_4COOH$
which of the following order is correct ?
(A) II > III > IV > I (B) II > IV > I > III
(C) I > II > III > IV (D) II > IV > III > I

73. Among the following compounds, which can be dehydrated very easily is
(A) $CH_3CH_2CH(CH_3)CH_2CH_2OH$

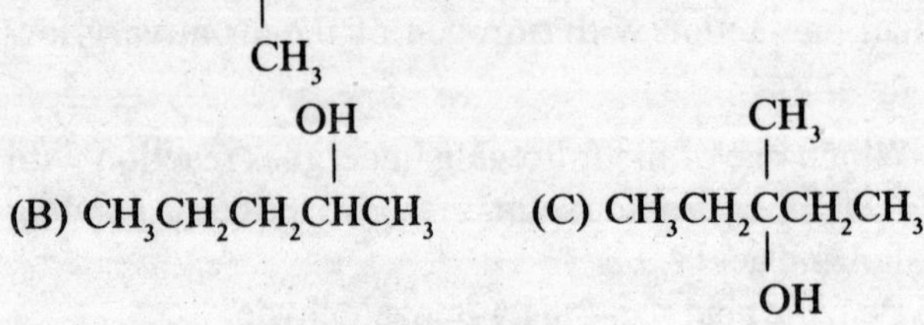

(B) $CH_3CH_2CH_2CH(OH)CH_3$ (C) $CH_3CH_2C(CH_3)(OH)CH_2CH_3$

(D) $CH_3CH_2CH_2CH_2CH_2OH$

74. Amongst the following compounds, the optically active alkane having lowest molecular mass is
(A) $CH_3—CH_2—CH_2—CH_3$
(B) $CH_3—C(H)(C_2H_5)—$◁ (C) $CH_3—CH_2—CH(CH_3)—CH_3$
(D) $CH_3—CH_2—C{\equiv}CH$

75. Which of the following is the strongest base ?
(A) C_6H_5–CH_2NH_2 (B) C_6H_5–NH_2
(C) C_6H_5–$NHCH_3$ (D) 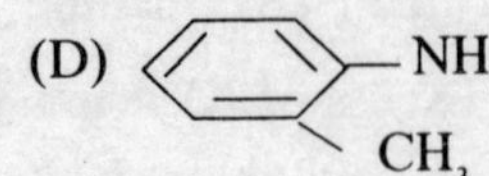 NH_2, CH_3

AIEEE - 2005

1. Which of the following oxides is amphoteric in character ?
(A) CaO (B) CO_2 (C) SiO_2 (D) SnO_2

2. Which one of the following species is diamagnetic in nature?
(A) He_2^+ (B) H_2 (C) H_2^+ (D) H_2^-

3. If α is the degree of dissociation of Na_2SO_4, the vant Hoffs factor (i) used for calculating the molecular mass is
(A) $1 + \alpha$ (B) $1 - \alpha$ (C) $1 + 2\alpha$ (D) $1 - 2\alpha$

4. The oxidation state of Cr in $[Cr(NH_3)_4Cl]^{2-}$ is
(A) + 3 (B) + 2 (C) + 1 (D) 0

5. Hydrogen bomb is based on the principle of
(A) nuclear fission (B) natural radioactivity
(C) nuclear fusion (D) artificial radioactivity

6. An ionic compound has a unit cell consisting of A ions at the corners of a cube and B ions on the centres of the faces of the cube. The empirical formula for this compound would be
(A) AB (B) A_2B (C) AB_3 (D) A_3B

7. For a spontaneous reaction the ΔG, equilibrium constant (K) and E^o_{Cell} will be respectively
(A) –ve, > 1, + ve (B) + ve, >1, –ve
(C) –ve, < 1, –ve (D) –ve, > 1, –ve

8. Which of the following is a polyamide ?
(A) Teflon (B) Nylon-66 (C) Terylene (D) Bakelite

9. Which one of the following types of drugs reduces fever ?
(A) Analgesic (B) Antipyretic
(C) Antibiotic (D) Tranquiliser

10. Due to the presence of an unpaired electron, free radicals are:
(A) Chemically reactive (B) Chemically inactive
(C) Anions (D) Cations

11. Lattice energy of an ionic compound depends upon
(A) Charge on the ion only
(B) Size of the ion only (C) Packing of ions only
(D) Charge on the ion and size of the ion

12. The highest electrical conductivity of the following aqueous solutions is of
(A) 0.1 M acetic acid (B) 0.1 M chloroacetic acid
(C) 0.1 M fluoroacetic acid
(D) 0.1 M difluoroacetic acid

13. Aluminium oxide may be electrolysed at 1000°C to furnish aluminium metal (At. Mass = 27 amu; 1 Faraday = 96,500 Coulombs). The cathod reaction is
$Al^{3+} + 3e^- \rightarrow Al^o$. To prepare 5.12 kg of aluminium metal by this method would require
(A) 5.49×10^7 C of electricity

(B) 1.83×10^7 C of electricity
(C) 5.49×10^4 C of electricity
(D) 5.49×10^1 C of electricity

14. Consider an endothermic reaction X →Y with the activation energies Eb and Ef for the backward and forward reactions, respectively. In general
(A) $E_b < E_f$ (B) $E_b > E_f$ (C) $E_b = E_f$
(D) there is no definite relation between E_b and E_f

15. Consider the reaction: $N_2 + 3H_2 \rightarrow 2NH_3$ carried out at constant temperature and pressure. If ΔH and ΔU are the enthalpy and internal energy changes for the reaction, which of the following expressions is true?
(A) $\Delta H = 0$ (B) $\Delta H = \Delta U$
(C) $\Delta H < \Delta U$ (D) $\Delta H > \Delta U$

16. Which one of the following statements is NOT true about the effect of an increase in temperature on the distribution of molecular speeds in a gas?
(A) The most probable speed increases
(B) The fraction of the molecules with the most probable speed increases
(C) The distribution becomes broader
(D) The area under the distribution curve remains the same as under the lower temperature

17. The volume of a colloidal particle, VC as compared to the volume of a solute particle in a true solution VS, could be

(A) $\frac{V_C}{V_S} \simeq 1$ (B) $\frac{V_C}{V_S} \simeq 10^{23}$ (C) $\frac{V_C}{V_S} \simeq 10^{-3}$ (D) $\frac{V_C}{V_S} \simeq 10^{3}$

18. The solubility product of a salt having general formula MX_2, in water is: 4×10^{-12}. The concentration of M^{2+} ions in the aqueous solution of the salt is
(A) 2.0×10^{-6} M (B) 1.0×10^{-4} M
(C) 1.6×10^{-4} M (D) 4.0×10^{-10} M

19. Benzene and toluene form nearly ideal solutions. At 20°C, the vapour pressure of benzene is 75 torr and that of toluene is 22 torr.. The partial vapour pressure of benzene at 20°C for a solution containing 78 g of benzene and 46 g of toluene in torr is
(A) 50 (B) 25 (C) 37.5 (D) 53.5

20. The exothermic formation of ClF_3 is represented by the equation: $Cl_2(g) + 3F_2(g) \rightleftharpoons 2ClF_3(g)$; $\Delta rH = -329$ kJ
Which of the following will increase the quantity of ClF_3 in an equilibrium mixture of Cl_2, F_2 and ClF_3?
(A) Increasing the temperature (B) Removing Cl_2
(C) Increasing the volume of the container
(D) Adding F_2

21. Two solutions of a substances (non electrolyte) are mixed in the following manner. 480 ml of 1.5 M first solution + 520 mL of 1.2 M second solution. What is the molarity of the final mixture ?
(A) 1.20 M (B) 1.50 M (C) 1.344 M (D) 2.70 M

22. For the reaction
$2NO_2(g) \rightleftharpoons 2NO(g) + O_2(g)$ (Kc = 1.8 . 10"6 at 184 °C) (R = 0.0831 kJ/(mol.K)) When Kp and Kc are compared at 184°C it is found that
(A) Kp is greater than Kc (B) Kp is less than Kc
(C) Kp = Kc
(D) Whether Kp is greater than, less than or equal to Kc depends upon the total gas pressure

23. Hydrogen ion concentration in mol/L in a solution of pH = 5.4 will be
(A) 3.98×10^8 (B) 3.88×10^6
(C) 3.68×10^{-6} (D) 3.98×10^{-6}

24. A reaction involving two different reactants can never be
(A) unimolecular reaction (B) first order reaction
(C) second order reaction (D) bimolecular reaction

25. If we consider that 1/6, in place of 1/12, mass of carbon atom is taken to be the relative atomic mass unit, the mass of one mole of a substance will
(A) decrease twice (B) increase tow fold
(C) remain unchanged
(D) be a function of the molecular mass of the substance

26. In a multi-electron atom, which of the following orbitals described by the three quantum members will have the same energy in the absence of magnetic and electric fields ?
(a) n = 1, l = 0, m = 0 (b) n = 2, l = 0, m = 0
(c) n = 2, l = 1, m = 1 (d) n = 3, l = 2, m = 1
(e) n = 3, l = 2 m = 0
(A) (a) and (b) (B) (b) and (c)
(C) (c) and (d) (D) (d) and (e)

27. During the process of electrolytic refining of copper, some metals present as impurity settle as anode mud. There are
(A) Sn and Ag (B) Pb and Zn
(C) Ag and Au (D) Fe and Ni

28.

Electrolyte:	KCl	KNO_3	HCl	NaOAc	NaCl
Λ^∞ (S cm² mol⁻¹):	149.9	145.0	426.2	91.0	126.5

Calculate H^∞_{OAc} using appropriate molar conductances of the electrolytes listed above at infinite dilution in H_2O at 25°C
(A) 517.2 (B) 552.7 (C) 390.7 (D) 217.5

29. A schematic plot of ln Keq versus inverse of temperature for a reaction is shown below

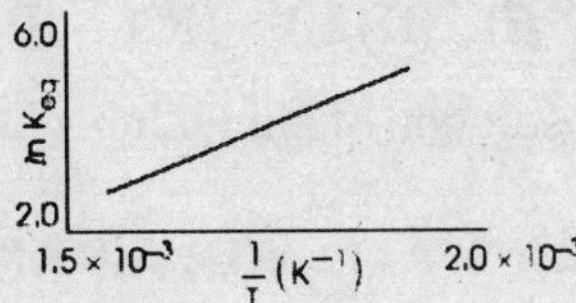

The reaction must be

(A) exothermic (B) endothermic
(C) one with negligible enthalpy change
(D) highly spontaneous at ordinary temperature

30. The disperse phase in colloidal iron (III) hydroxide and colloidal gold is positively and negatively charged, respectively. Which of the following statement is Not correct ?
(A) Magnesium chloride solution coagulates, the gold sol more readily than the iron (III) hydroxide sol.
(B) Sodium sulphate solution causes coagulation in both sols
(C) Mixing the sols has no effect
(D) Coagulation in both sols can be brought about by electrophoresis

31. Based on lattice energy and other considerations which one of the following alkali metal chlorides is expected to have the highest melting point ?
(A) LiCl (B) NaCl (C) KCl (D) Rb Cl

32. Heating mixture of Cu_2O and Cu_2S will give
(A) Cu + SO_2 (B) Cu + SO_3
(C) CuO + CuS (D) Cu_2 SO_3

33. The molecular shapes of SF_4, CF_4 and XeF_4 are
(A) the same with 2, 0 and 1 lone pairs of electrons on the central atom, respectively
(B) the same with 1, 1 and 1 lone pair of electrons on the central atoms, respectively
(C) different with 0, 1 and 2 lone pairs of electrons on the central atom, respectively
(D) different with 1, 0 and 2 lone pairs of electrons on the central atom, respectively

34. The number and type of bonds between two carbon atoms in calcium carbide are
(A) One sigma, one pi (B) One sigma, two pi
(C) Two sigma, one pi (D) Two sigma, two pi

35. The oxidation state of chromium in the final product formed by the reaction between KI and acidified potassium dichromate solution is
(A) + 4 (B) + 6 (C) + 2 (D) + 3

36. The number of hydrogen atom(s) attached to phosphorus atom in hypophosphorous acid is
(A) zero (B) two (C) one (D) three

37. What is the conjugate base of OH^-?
(A) O_2 (B) H_2O (C) O^- (D) O_2^-

38. The correct order of the thermal stability of hydrogen halides (H -X) is
(A) HI > HBr > HCl > HF (B) HF > HCl > HBr > HI
(C) HCl < HF > HBr < HI (D) HI > HCl < HF > HBr

39. Heating an aqueous solution of aluminium chloride to dryness will give
(A) $AlCl_3$ (B) Al_2Cl_6 (C) Al_2O_3 (D) $Al(OH)Cl_2$

40. Calomel (Hg_2Cl_2) on reaction with ammonium hydroxide gives
(A) $HgNH_2Cl$ (B) NH_2 - Hg - Hg - Cl
(C) Hg_2O (D) HgO

41. In which of the following arrangements, the order is NOT according to the property indicated against it?
(A) $Al^{3+} < Mg^{2+} < Na^+ < F^-$: Increasing ionic size
(B) B < C < N < O:Increasing first ionization enthalpy
(C) I < Br < F < Cl: Increasing electron gain enthalpy (with negative sign)
(D) Li < Na < K < Rb: Increasing metallic radius

42. In silicon dioxide
(A) each silicon atom is surrounded by four oxygen atoms and each oxygen atom is bonded to two silicon atoms
(B) each silicon atom is surrounded by two oxygen atoms and each oxygen atom is bonded to two silicon atoms
(C) silicon atom is bonded to two oxygen atoms
(D) there are double bonds between silicon and oxygen atoms

43. Of the following sets, which one does NOT contain isoelectronic species ?
(A) PO_4^{3-}, SO_4^{2-}, ClO_4^- (B) CN^-, N_2, C_2^{2-}
(C) SO_3^{2-}, CO_3^{2-}, NO_3^- (D) BO_3^{3-}, CO_3^{2-}, NO_3^-

44. The lanthanide contraction is responsible for the fact that
(A) Zr and Y have about the same radius
(B) Zr and Nb have similar oxidation state
(C) Zr and Hf have about the same radius
(D) Zr and Zn have the same oxidation state

45. The IUPAC name of the coordination compound $K_3[Fe(CN)_6]$ is
(A) Potassiumhexacyanoferrate (II)
(B) Potassiumhexacyanoferrate (III)
(C) Potassiumhexacyanoiron (II)
(D) Tripotassiumhexacyanoiron (II).

46. Which of the following compounds shows optical isomerism ?
(A) $[Cu(NH_3)_4]^{2+}$ (B) $[ZnCl_4]^{2-}$
(C) $[Cr(C_2O_4)_3]^{3-}$ (D) $[Co(CN)_6]^{3-}$

47. Which one of the following cyano complexes would exhibit the lowest value of paramagnetic behaviour?
(A) $[Cr(CN)_6]^{3-}$ (B) $[Mn(CN)_6]^{3-}$
(C) $[Fe(CN)_6]^{3-}$ (D) $[Co(CN)_6]^{3-}$

48. 2-Methylbutane on reacting with bromine in the presence of sunlight gives mainly
(A) 1-bromo-2-methylbutane
(B) 2-bromo-2-methylbutane
(C) 2-bromo-3-methylbutane
(D) 1-bromo-3-methylbutane

49. A photon of hard gamma radiation knocks a proton out of ${}^{12}_{24}Mg$ nucleus to form
(A) the isotope of parent nucleus
(B) the isobar of parent nucleus

(C) the nuclide $^{11}_{23}Na$ (D) the isobar of $^{11}_{23}Na$

50. The best reagent to convert pent-3-en-2-ol into pent-3-in-2-one is
(A) Acidic permanganate (B) Acidic dichromate
(C) Chromic anhydride in glacial acetic acid
(D) Pyridinium chloro-chromate

51. Tertiary alkyl halides are practically inert to subsitution by S_N2 mechanism because of
(A) insolubility (B) instability
(C) inductive effect (D) steric hindrance

52. In both DNA and RNA, heterocylic base and phosphate ester linkages are at
(A) C'_5 and C'_2 respectively of the sugar molecule
(B) C'_2 and C'_5 respectively of the sugar molecule
(C) C'_1 and C'_5 respectively of the sugar molecule
(D) C'_5 and C'_1 respectively of the sugar molecule

53. Reaction of one molecule of HBr with one molecule of 1, 3-butadiene at 40°C gives predominantly
(A) 3-bromobutene under kinetically controlled conditions
(B) 1-bromo-2-butene under thermodynamically controlled conditions
(C) 3-brom obutene under thermodynamically controlled conditions
(D) 1-bromo-2 butene under kinetically controlled conditions

54. Among the following acids which has the lowest pKa value ?
(A) CH_3COOH (B) HCOOH
(C) $(CH_3)_2$ CH - COOH (D) CH_3CH_2COOH

55. The decreasing order of nucleophilicity among the nuclephiles
(A) $CH_3-\underset{\underset{O}{\|}}{C}-O^-$ (B) CH^3O^-
(C) CN^- (D) $H_3C-C_6H_4-\underset{\underset{O}{\|}}{\overset{\overset{O}{\|}}{S}}-O^-$ is
(A) (a), (b), (c), (d) (B) (d), (c), (b), (a)
(C) (b), (c), (a), (d) (D) (c), (b), (a), (d)

56. Which one of the following methods is neither meant for the synthesis nor for separation of amines?
(A) Hinsberg method (B) Hofmann method
(C) Wurtz reaction (D) Curtius reaction

57. Which of the following is fully fluorinated polymer?
(A) Neoprene (B) Teflon (C) Thiokol (D) PVC

58. Of the five isomeric hexanes, the isomer which can give two monochlorinated compounds is
(A) n-hexane (B) 2, 3-dimethylbutane
(C) 2, 2-dimethylbutane (D) 2-methylpentane

59. Alkyl halides react with dialkyl copper reagents to give
(A) alkenes (B) alkyl copper halides
(C) alkanes (D) alkenyl halides

60. Acid catalyzed hydration of alkenes except ethene leads to the formation of
(A) primary alcohol (B) secondary or tertiary alcohol
(C) mixture of primary and secondary alcohols
(D) mixture of secondary and tertiary alcohols

61. Amongst the following the most basic compound is
(A) benzylamine (B) aniline
(C) acetanilide (D) p-nitroaniline

62. Which types of isomerism is shown by 2,3-dichlorobutane ?
(A) Diastereo (B) Optical (C) Geometric (D) Structural

63. The reaction

$$R-C(=O)X + \overset{\ominus}{Nu} \rightarrow R-C(=O)Nu + X^{\ominus}$$

is fastest when X is
(A) Cl (B) NH_2 (C) OC_2H_5 (D) OCOR

64. Elimination of bromine from 2-bromobutane results in the formation of -
(A) equimolar mixture of 1 and 2-butene
(B) predominantly 2-butene
(C) predominantly 1-butene
(D) predominantly 2-butyne

65. Equimolar solutions in the same solvent have
(A) same boiling point but different freezing point
(B) same freezing point but different boiling point
(C) same boiling and same freezing points
(D) different boiling and different freezing points

66. Which of the statements in relation to the hydrogen atom is correct ?
(A) 3s orbital is lower in energy than 3p orbital
(B) 3p orbital is lower in energy than 3d orbital
(C) 3s and 3p orbitals are of lower energy than 3d orbital
(D) 3s, 3p and 3d orbitals all have the same energy

67. The structure of diborane (B_2H_6) contains
(A) four 2c-2e bonds and two 3c-2e bonds
(B) two 2c-2e bonds and four 3c-2e bonds
(C) two 2c-2e bonds and two 3c-2e bonds
(D) four 2c-2e bonds and four 3c-2e bonds

68. The value of the spin only magnetic moment for one of the following configurations is 2.84 BM. The correct one is
(A) d4 (in strong ligand field)
(B) d4 (in weak ligand field)
(C) d3 (in weak as well as in strong fields)
(D) d5 (in strong ligand field)

69. Which of the following factors may be regarded as the main cause of lanthanide contraction?
(A) Poor shielding of one of 4f electron by another in the subshell
(B) Effective shielding of one of 4f electrons by another in the subshell
(C) Poorer shielding of 5d electrons by 4f electrons
(D) Greater shielding of 5d electron by 4f electrons.

70. Reaction of cyclohexanone with dimethylamine in the presence of catalytic amount of an acid forms a compound if water during the reaction is continuously removed. The compound formed is generally know as

(A) a Schiffs base (B) an enamine
(C) an imine (D) an amine

71. p-cresol reacts with chloroform in alkaline medium to give the compound A which adds hydrogen cyanide to form, the compound B. The latter on acidic hydrolysis gives chiral carboxylic acid. The structure of the carboxylic acid is

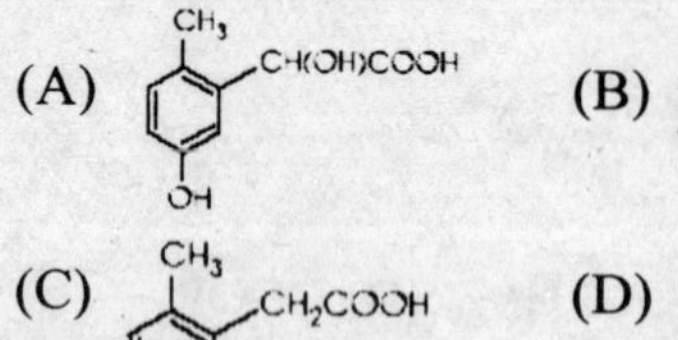

(B) CH_3 / $CH(OH)COOH$ / OH

(C)
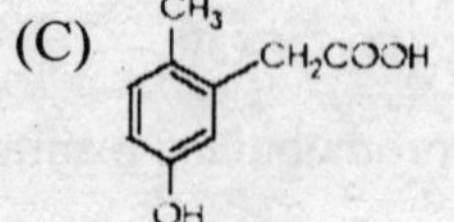

(D)
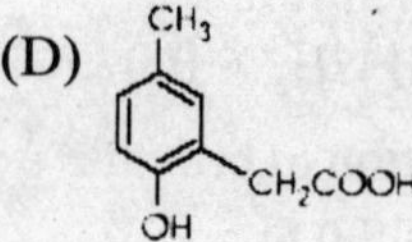

72. An organic compound having molecular mass 60 is found to contain C = 20%, H = 6.67% and N = 46.67% while rest is oxygen. On heating it gives NH_3 alongwith a solid residue. The solid residue give violet colour with alkaline copper sulphate solution. The compound is

(A) CH_3NCO (B) CH_3CONH_2
(C) $(NH_2)_2CO$ (D) $CH_3CH_2CONH_2$

73. If the bond dissociation energies of XY, X_2 and Y_2 (all diatomic molecules) are in the ratio of 1 : 1 : 0.5 and “fH for the formation of XY is –200 kJ mole^{-1}. The bond dissociation energy of X_2 will be

(A) 100 kJ mol^{-1} (B) 200 kJ mol^{-1}
(C) 300 kJ mol^{-1} (D) 400 kJ mole^{-1}

74. $t_{1/4}$ can be taken as the time taken for the concentration of a reactant to drop $t_{3/4}$ of its initial value. If the rate constant for a first order reaction is K, the $t_{1/4}$ can be written as

(A) 0.10/K (B) 0.29/K (C) 0.69/K (D) 0.75/K

75. An amount of solid NH_4HS is placed in a flask already containing ammonia gas at a certain temperature and 0.50 atm. pressure. Ammonium hydrogen sulphide decomposes to yield NH_3 and H_2S gases in the flask. When the decomposition reaction reaches equilibrium, the total pressure in the flask rises to 0.84 atm. The equilibrium constant for NH_4HS decomposition at this temperature is

(A) 0.30 (B) 0.18 (C) 0.17 (D) 0.11

AIEEE - 2002

1.	(D)	39.	(A)
2.	(A)	40.	(D)
3.	(B)	41.	(C)
4.	(A)	42.	(B)
5.	(A)	43.	(D)
6.	(C)	44.	(C)
7.	(C)	45.	(B)
8.	(D)	46.	(B)
9.	(A)	47.	(C)
10.	(A)	48.	(A)
11.	(B)	49.	(C)
12.	(C)	50.	(A)
13.	(D)	51.	(D)
14.	(A)	52.	(A)
15.	(A)	53.	(A)
16.	(B)	54.	(A)
17.	(B)	55.	(B)
18.	(D)	56.	(D)
19.	(C)	57.	(D)
20.	(C)	58.	(D)
21.	(A)	59.	(D)
22.	(D)	60.	(B)
23.	(A)	61.	(B)
24.	(C)	62.	(C)
25.	(D)	63.	(D)
26.	(C)	64.	(A)
27.	(B)	65.	(D)
28.	(C)	66.	(B)
29.	(A)	67.	(B)
30.	(C)	68.	(C)
31.	(C)	69.	(B)
32.	(A)	70.	(C)
33.	(B)	71.	(D)
34.	(B)	72.	(B)
35.	(C)	73.	(A)
36.	(B)	74.	(A)
37.	(C)	75.	(D)
38.	(A)		

AIEEE - 2003

1.	(B)	39.	(D)
2.	(B)	40.	(A)
3.	(B)	41.	(C)
4.	(B)	42.	(D)
5.	(C)	43.	(D)
6.	(A)	44.	(B)
7.	(A)	45.	(A)
8.	(C)	46.	(C)
9.	(B)	47.	(B)
10.	(A)	48.	(B)
11.	(B)	49.	(B)
12.	(D)	50.	(A)
13.	(D)	51.	(B)
14.	(B)	52.	(C)
15.	(A)	53.	(D)
16.	(A)	54.	(C)
17.	(B)	55.	(C)
18.	(B)	56.	(A)
19.	(B)	57.	(D)
20.	(D)	58.	(C)
21.	(C)	59.	(B)
22.	(D)	60.	(D)
23.	(C)	61.	(A)
24.	(C)	62.	(B)
25.	(B)	63.	(D)
26.	(A)	64.	(A)
27.	(A)	65.	(C)
28.	(D)	66.	(C)
29.	(D)	67.	(B)
30.	(D)	68.	(C)
31.	(D)	69.	(B)
32.	(B)	70.	(C)
33.	(A)	71.	(D)
34.	(A)	72.	(C)
35.	(D)	73.	(A)
36.	(A)	74.	(D)
37.	(B)	75.	(D)
38.	(C)		

AIEEE - 2004

1.	(C)	39.	(A)
2.	(B)	40.	(B)
3.	(C)	41.	(C)
4.	(C)	42.	(A)
5.	(C)	43.	(A)
6.	(B)	44.	(C)
7.	(C)	45.	(A)
8.	(B)	46.	(D)
9.	(B)	47.	(A)
10.	(D)	48.	(C)
11.	(B)	49.	(D)
12.	(B)	50.	(B)
13.	(B)	51.	(B)
14.	(D)	52.	(C)
15.	(D)	53.	(C)
16.	(A)	54.	(B)
17.	(D)	55.	(C)
18.	(A)	56.	(A)
19.	(D)	57.	(C)
20.	(C)	58.	(A)
21.	(A)	59.	(B)
22.	(A)	60.	(C)
23.	(B)	61.	(A)
24.	(C)	62.	(A)
25.	(B)	63.	(B)
26.	(D)	64.	(D)
27.	(A)	65.	(D)
28.	(C)	66.	(A)
29.	(D)	67.	(D)
30.	(B)	68.	(B)
31.	(D)	69.	(D)
32.	(C)	70.	(B)
33.	(C)	71.	(D)
34.	(C)	72.	(A)
35.	(B)	73.	(B)
36.	(B)	74.	(B)
37.	(D)	75.	(A)
38.	(D)		

AIEEE - 2005

1. (D)	11. (D)	21. (C)	31. (D)	41. (B)	51. (D)	61. (A)	71. (B)
2. (B)	12. (D)	22. (A)	32. (A)	42. (A)	52. (C)	62. (B)	72. (C)
3. (C)	13. (A)	23. (D)	33. (D)	43. (C)	53. (B)	63. (A)	73. (B)
4. (A)	14. (A)	24. (A)	34. (B)	44. (C)	54. (B)	64. (B)	74. (B)
5. (C)	15. (C)	25. (A)	35. (D)	45. (B)	55. (D)	65. (C)	75. (D)
6. (C)	16. (D)	26. (D)	36. (B)	46. (C)	56. (C)	66. (D)	
7. (A)	17. (D)	27. (C)	37. (D)	47. (D)	57. (B)	67. (A)	
8. (B)	18. (B)	28. (C)	38. (B)	48. (B)	58. (B)	68. (A)	
9. (B)	19. (A)	29. (A)	39. (C)	49. (C)	59. (C)	69. (C)	
10. (A)	20. (D)	30. (C)	40. (B)	50. (C)	60. (D)	70. (B)	

AIIMS - 1993

1. Which of the following statements is wrong ?
(A) Red phosphorus is more active than white phosphorus
(B) Red phosphorus has polymerized structure
(C) White phosphorus has a higher internal energy than red phosphorus
(D) White phosphorus is metastable at all temperatures upto the melting point of red phosphorus

2. Ammonia reacts with excess of chlorine to form
(A) NH_4Cl (B) NCl_2 and HCl
(C) N_2 and NH_4Cl (D) NCl_3 and HCl

3. The empirical formula of an organic compound containing 75% C and 25% H is
(A) C_2H_4 (B) C_3H_8 (C) CH_4 (D) C_2H_6

4. Which type of isomerism among the following, correctly describes the isomeric relationship between the following two isomers (i and ii)
i. $CH_3 - CH(OH)CH_2CH_2 - CH_3$
ii. $CH_3 - CH_2 - O–CH_2 - CH_2 - CH_3$
(A) Chain isomerism (B) Functional isomerism
(C) Position isomerism (D) Carbon-skeleton isomerism

5. Ethene molecule contains
(A) 2 σ & 4 π bonds (B) 4 σ & 2 π bonds
(C) 6 σ bonds (D) 5 σ & 1 π bonds

6. Indicate which olefin amongst the following has the E-(ie. trans) configuration
(A) (F)(Cl)C = C(CH_3)(C_2H_5) (B) (Cl)(H)C = C(H)(Cl)
(C) (H)(Br)C = C(H)(Cl) (D) (Cl)(Br)C = C(H)(Cl)

7. Skin turns yellow in contact with concentrated niric acid because
(A) Skin is burnt
(B) Proteins are converted in xantho proteins
(C) Water is removed from the skin
(D) Nitro cellulose is formed

8. On hydrolysis, P_4O_6 finally gives
(A) H_3PO_4 (B) $H_6P_4O_{12}$
(C) H_3PO_3 (D) $H_4P_2O_7$

9. HBr can be prepared by treating
(A) KBr with sulphuric acid
(B) KBr with nitric acid
(C) KBr with phosphoric acid
(D) Bromine with sulphuric acid

10. With reference to chlorine which one cf the following statements is false ?
(A) It has the outer electronic configuration of ns^2np^5.
(B) It has the highest electronegativity
(C) It has the highest electron affinity in the group
(D) It is a better oxidizing agent than bromine

11. The statement which is not correct for nitrogen is ?
(A) Forms covalent compounds like NH_3, NF_3 etc.
(B) Shows greater stability as a free element
(C) Has a very small atomic radius
(D) Is unable to show greater covalency than 3

12. H_2O has a net dipole moment but BeF_2 has zero dipole moment because
(A) Beryllium has more electronegativity than oxygen
(B) Fluorine has more electrongeativity than oxygen
(C) There is lone pair of electrons in H_2O, but no lone pair in BeF_2.
(D) H_2O molecule is linear while BeF_2 molecule is bent.

13. The anti Markownikoff addition of HBr is not observed in
(A) Propylene (B) 1-butene
(C) 2-butene (D) 2-pentene

14. The IUPAC name of the compound represented by the formula

$$CH_3 - CH_2 - CH_2 - CH = CH - \underset{\displaystyle CH_3}{\underset{|}{CH}} - CH_3 \text{ is}$$

(A) 5-isopropyl-4-pentene (B) 2-methyl-3-heptene
(C) 1-isopropyl-3-pentene (D) 6-methyl-4-heptene

15. Which of the following is the systematic name for allyl alconol ?
(A) Allylol (B) Prop-2-en-1-ol
(C) Rectified spirit (D) Isopropyl alcohol

16. NH_3 is a complexing agent which forms stable silver diammine and copper (ii) tetrammine cations in aqueous solutions of Ag (I) and Cu (II) respectively. In each of the complex ions
(A) NH_3 molecules are present which are held to the metal cations by hydrogen bonding
(B) NH_3 molecules are present which are held to metal cations by ion dipole interaction.
(C) NH_4^+ ions are present which are held to the metal atom by atomdipole intreraction
(D) NH_2^- ions are present which are held to the metal cations by ion-ion interactions

17. The compound B formed in the following sequence of reactions $CH_3CH_2CH_2 - OH \xrightarrow{PCl_5} A \xrightarrow[KOH]{Alcohol} B$ will be
(A) Propane (B) Propene (C) Propyne (D) Propanol

18. Iron forms an oxide Fe_2O_3 according to the equation,
$2Fe + \frac{3}{2} O_2 \rightarrow Fe_2O_3 \quad \Delta H = -822 \text{ kJ mol}^{-1}$
What is the value of ΔH, if one mole of oxygen was formed by the decomposition of Fe_2O_3 ?
(A) +548 kJ (B) –548 kJ (C) –411 kJ (D) +411 kJ

19. Which of the following isoelectronic species has more electrons than neutrons ?
(A) Na^+ (B) Mg^{2+} (C) O^{2-} (D) F^-

20. In the photo-electron emission, the energy of the

emitted electrons is
(A) greater than the incident photon
(B) same as that of the incident photon
(C) smaller than the incident photon
(D) proportional to the intensity of incident proton

21. Which of the following sets (n,l,m,s) of quantum numbers is wrong ?
(A) 2, 1, 0, $+\frac{1}{2}$ (B) 3, 2, 2, $+\frac{1}{2}$
(C) 2, 2, 0, $+\frac{1}{2}$ (D) 3, 2, 2, $-\frac{1}{2}$

22. The fact that BCl_3 is a planar molecule, while NCl_3 is pyramidal can be rationalized in many ways. Of the following statements the best rationalisation is
(A) Nitrogen is more electronegative than boron
(B) The nitrogen atom is smaller than the boron atom
(C) The nitrogen atom in NCl_3 has a lone pair of electrons while the boron atom in BCl_3 does not have
(D) The boron atom in BCl_3 is sp^3 hybridised while nitrogen atom in NCl_3 is sp^2 hybridised.

23. Which of the following statements is wrong ?
(A) The gas constant per molecule is known as Boltzmann constant
(B) The product of pressure and volume of a fixed amount of gas is independent of temperature.
(C) The molecules of different gases have the same kinetic energy at a given temperature
(D) The gas equation is not valid at high pressures and low temperatures

24. The following diagram illustrates a possible electronic configuration of which of the following species ?

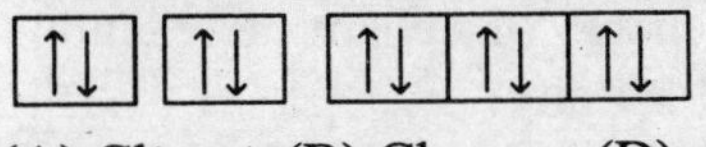

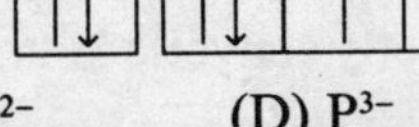

(A) Cl^+ (B) Cl^- (D) Ar^{2-} (D) P^{3-}

25. A mole of N_2H_4 loses ten moles of electrons to form a new compound Y. Assuming that all the nitrogen appears in the new compound, what is the oxidation state of nitrogen in Y ? (There is no change in the oxidation number of hydrogen).
(A) +3 (B) +5 (C) –1 (D) –3

26. In which of the following compounds, the ratio of anion size to the cation size is the lowest ?
(A) CsI (B) LiI (C) LiF (D) CsF

27. Which of the following is isoelectronic with CO ?
(A) CN^- (B) O_2^+ (C) O^{2-} (D) N^{2+}

28. At a certain temperature when 3 moles of hydrogen and one mole of nitrogen are allowed to react, ammonia is formed. When 1 mole of ammonia gas is formed, the total pressure of the vessel is 15 atmospheres. The partial pressure of nitrogen in the vessel is
(A) 9 atm (B) 6 atm (C) 7.5 atm (D) 2.5 atm

29. Which of the following is the strongest acid ?
(A) C_2H_5COOH (B) CH_3COOH
(C) $CH_2BrCOOH$ (D) $CHBr_2COOH$

30. Which of the following compounds would be the most susceptible to electrophilic attack by nitronium ions ?
(A) NO_2, CF_3 (B) OCH_3, CF_3 (C) CH_3, NO_2 (D) CH_3

31. The product of oxidation of benzene with air in the presence of V_2O_5 at 1-2 atm and at 400° C-450°C is
(A) Benzoic acid (B) Adipic acid
(C) Succinic acid (D) Maleic anhydride

32. The function of anhydrous aluminium chloride in Friedel Craft's reaction is to
(A) Produce the electrophile
(B) Absorb water (C) Absorb hydrochloric acid
(D) Produce nucleophile

33. The compound of formula $NH_2 - CH_2 - COOH$ may behave
(A) Only as acid (B) As both acid and base
(C) Only as a base (D) As neither acid nor base

34. The most appropriate method to convert propanoic acid into 3-pentanone is
(A) Heating the dry calcium salt of the propanoic acid in the absence of air
(B) Heating the sodium salt of the acid with ethylchloride
(C) Electrolysis of fused sodium salt of propanoic acid
(D) Heating strogly the sodium salt of propanoic acid

35. Which of the organic structures corresponds to the product expected when excess of benzene react with CH_2Cl_2 in the presence of $AlCl_3$?
(A) $C_6H_5-CHCl-C_6H_5$ (B) $C_6H_5CHCl_2$
(C) $C_6H_5-CH_2-C_6H_5$ (D) $C_6H_5-CCl_2-C_6H_5$

36. Which pair of substances from the following pairs can act as a buffer solution ?
(A) HNO_3 and KNO_3 (B) NaOH and NaCl
(C) Na_2SO_4 and H_2SO_4 (D) NH_4OH and NH_4Cl

37. What is the pH of 1 M sodium acetate ?
(K_a for acetic acid is 1.8×10^{-5} and $K_w = 1 \times 10^{-14}$)
(A) 7 (B) 9.4 (C) 2.4 (D) 4.8

38. Given the reaction $HA(aq) + B^-(aq) \leftrightarrow HB(aq) + A^-(aq)$ Which of the following is correct, if the equilibrium constant value for the above reaction is 10^{-3} ?
(A) B^- is a stronger base than A^-
(B) HB is stronger acid than HA
(C) HA is a stronger acid than H_3O^+
(D) K_c for the reverse reaction is 10^{-3}

39. Which of the following reactions shows that phenol is a weak acid ?

(A) $C_6H_5OH + 3H_2 \rightarrow C_6H_{11}OH$
(B) C_6H_5OH + dil.$HNO_3 \rightarrow$ mono ni.trophenol
(C) $C_6H_5OH + NH_3 \xrightarrow{ZnCl_2}$
(D) C_6H_5OH + aq. NaOH $\rightarrow C_5H_5ONa + H_2O$

40. Which of the following compounds can give haloform test ?
(A) $C_2H_5COC_2H_5$ (B) C_2H_5CHO
(C) C_3H_7CHO (D) $CH_3COC_2H_5$

41. The compound which can be used as an antifreeze in automobile radiators is
(A) Glyceraldehyde (B) Oxalic acid
(C) Glycol (D) Glycerol

42. Glycerol reacts with exces HI togive
(A) 1,2,3, triodopropane (B) Propene
(C) 2-iodopropane (D) Allyl iodide

43. When salicylic acid is distilled with sodalime, we get
(A) Toluene (B) Carbolic acid
(C) Benzene (D) Benzoic acid

44. Ozone is
(A) A compound of oxygen (B) An allotrope of oxygen
(C) An isotope of oxygen (D) An isomer of oxygen

45. It is possible to obtain oxygen from air by fractional distillation because oxygen
(A) Is in a different group of periodic table than nitrogen
(B) Is more reactive than nitrogen
(C) Has higher boiling point than nitrogen
(D) Has lower density than nitrogen

46. The product obtained by heating propanone and the paste of beaching powder is
(A) Trichloromethane (B) Propyl chloride
(C) Ethyl chloride (D) Chloral

47. In which of the following reactions K_p is less than K_c ?
(A) $H_2(g) + Br_2(g) \leftrightarrow 2HBr(g)$
(B) $PCl_5(g) \leftrightarrow PCl_3(g) + Cl_2(g)$
(C) $N_2(g) + 3H_2(g) \leftrightarrow 2NH_3(g)$
(D) $2SO_3(g) \leftrightarrow 2SO_2(g) + O_2(g)$

48. Equilibrium constant for the reaction,
$FeO(s) + CO(g) \leftrightarrow Fe(s) + CO_2(g)$
is 0.5. The equilibrium constant for the reverse reaction
$Fe(s) + CO_2(g) \leftrightarrow FeO(s) + CO_2(g)$ is
(A) –0.5 (B) 0.5 (C) 2 (D) 0.25

49. Which acid does not have a carboxylic group ?
(A) Benzoic acid (B) Salicylic acid
(C) Picric acid (D) Methanoic acid

50. When sodium acetate is added to acetic acid, it
(A) Increases ionization of sodium acetate
(B) Is independent of concentration
(C) Increases ionization of acetic acid
(D) Decreases ionization of acetic acid

AIIMS - 1994

1. Which of the following can differentiate between C_2H_5OH and CH_3OH ?
(A) H_2O (B) I_2 + KOH (C) HCl (D) NH_3

2. Which of the following is a Lewis-acid ?
(A) Cl^- (B) Ag^+ (C) Br^- (D) S^-

3. Which of the following gas is produced in the reaction, $KO_2 + CO_2$?
(A) H_2 (B) N_2 (C) O_2 (D) CO

4. The long form of periodic table is based on
(A) Atomic number (B) Electronegativity
(C) Mass of the atom (D) Shape of the atom

5. When chloroform is exposed to air and sunlight, it gives
(A) Phosgene (B) Lewisite
(C) Mustard gas (D) Carbon tetrachloride

6. Which is an oxidizing substance amongst the following?
(A) SO_4^{2-} (B) SO_3 (C) CO_2 (D) NO_2

7. The number of electrons and neutrons in an element is 18 and 20 respectively. Its mass number is
(A) 17 (B) 22 (C) 37 (D) 38

8. Which of the following is a strong reducing agent ?
(A) Cr (B) Na (C) Cl (D) Ca

9. Aspirin is an acetyation product of
(A) o-hydroxy benzioc acid (B) o-dihydroxy benzene
(C) m-hydroxy benzoic acid (D) p-dihydroxy benzene

10. In which cell, the free energy of a chemical reaction is directly converted into electricity?
(A) Fuel cell (B) Leblanche cell
(C) Concentration cell (D) lead storage battery

11. Which of the following elements will have the highest electron affinity ?
(A) Nitrogen (B) Chlorine
(C) Fluorine (D) Phosphorus

12. Heterolytic fission of an organic covalent bond gives only
(A) Free radicals (B) Cation
(C) Anion (D) both (B) and (C)

13. Which of the following will not undergo aldol condensation ?
(A) Acetone (B) Acetaldehyde
(C) Propionaldehyde (D) Formaldehyde

14. The molecular mass of a volatile substance may be measured by
(A) Liebig's method (B) Hoffmann's method
(C) Victor Meyer's method (D) None of these

15. Number of water molecules in Mohr's salt is
(A) 5 (B) 6 (C) 7 (D) 8

16. Which of the following glasses is the most heat

resistant ?
(A) flint glass (B) Pyrex glass
(C) Bottle glass (D) Water glass

17. The number of electrons in $[_{19}K^{40}]^{-1}$ is
(A) 19 (B) 20 (C) 28 (D) 40

18. Which of the following is an alum ?
(A) $NaAlO_2$ (B) $KCl.MgCl_2.6H_2O$
(C) $FeSO_4(NH_4)_2SO_4.6H_2O$
(D) $Na_2SO_4.Al_2(SO_4)_3.24H_2O$

19. Which of the following metals reacts with water
(A) Copper (B) Sodium (C) Nickel (D) Silver

20. Permanent hardness of water can be removed by adding
(A) K (B) Cl_2 (C) Na_2CO_3 (D) Ca(OCl)Cl

21. $MgCl_2.6H_2O$ when heated gives
(A) Magnesium oxide (B) Magnesium chloride
(C) Magnesium dichloride
(D) Magnesium oxychloride

22. The formation of 2-butene as a major product by dehydration of 2-butanol is in accordance with
(A) Huckel rule (B) Saytzeff's rule
(C) Blanc rule (D) Markiownikoff's rule

23. Which of the following compounds gives Cannizzaro's reaction ?
(A) HCHO (B) CH_3CHO
(C) CH_3CH_2CHO (D) $CH_3CH_2CH_2CHO$

24. The type of isomerism not exhibited by alkenes is
(A) Metamerism (B) Stereoisomerism
(C) Chain isomerism (D) Position isomerism

25. The heat liberated when 1.89 g of benzoic acid is burnt in a bomb clarorimeter at 25°C increases the temperature of 18.94 kg of water by 0.632°C. If the specific heat of water at 25°C is 0.9998 cal/g-deg, the value of the heat of combustion of benzoic acid is
(A) 771.4 kcal (B) 871.2 kcal
(C) 881.1 kcal (D) 981.1 kcal

26. Assertion : Halogens do not occur in free state. Reason : Halogens are highly reactive.
(A) A (B) B (C) C (D) D

27. Assertion : The boiling point of ethanol is much higher than that of diethyl ether. Reason : In ethanol, the molecules are associated by the formation of intermolecular hydrogen bonding, whereas in diethyl ether it is not possible.
(A) A (B) B (C) C (D) D

28. Assertion : Physical adsorption of molecules on surface requires activation energy. Reason : Because the bonds of the absorbed molecules are broken.
(A) A (B) B (C) C (D) D

29. Assertion : Ethers behave as bases in the presence of mineral acids. Reason : Due to the presence of lone pair of electrons on the oxygen.
(A) A (B) B (C) C (D) D

30. Assertion : Reaction of conc. H_2SO_4 on NaBr and NaI does not give HBr and HI. Reason : Both HBr and HI are oxidized by conc. H_2SO_4 to Br_2 and I_2.
(A) A (B) B (C) C (D) D

31. Assertion : Ionic compounds tend to be non-volatile. Reason : The inter molecular forces in these compounds are weak.
(A) A (B) B (C) C (D) D

32. Assertion : Helium and beryllium have similar outer electronic configuration of type ns^2.Reason : Both are chemically intert.
(A) A (B) B (C) C (D) D

33. Assertion : A solution of $KMnO_4$ is decolourised by SO_2. Reason : SO_2 is acidic in character.
(A) A (B) B (C) C (D) D

34. Assertion : Alkanes having more than three carbon atoms exhibit chain isomerism. Reason : All carbon atoms in alkanes are sp-hybridized
(A) A (B) B (C) C (D) D

35. Assertion : The first ionization energy of aluminium is lower than that of magnesium Reason : The ionic radius of aluminium is smaller than that of magnesium
(A) A (B) B (C) C (D) D

36. The compound contaning co-ordinate bond is
(A) O_3 (B) H_2SO_4 (C) SO_3 (D) All of these

37. Ethyl alcohol exhibits acidic character on reacting with
(A) Acetic acid (B) Sodium metal
(C) Hydrogen iodide (D) All of these

38. Which of the following is called an ethanoic acid ?
(A) HCOOH (B) CH_3COOH
(C) CH_3CH_2COOH (D) $CH_3CH_2CH_2COOH$

39. When alkyl halides are heated with dry Ag_2O, they give
(A) Ether (B) Benzene (C) Ester (D) Ketone

40. Which metal is present in brass, bronze and german-silver ?
(A) Mg (B) Zn (C) Al (D) Cu

41. When propyne is treated with aqueous H_2SO_4 in the presence of $HgSO_4$, the product formed is
(A) Ether (B) Propanal (C) Acetone (D) Ethanal

42. In Wolf-Kishner reduction, the carbonyl group of aldehydes and ketones is converted into
(A) $-CH_2-$ (B) $-CHOH-$ (C) $-CH_2OH$ (D) $-CH_3$

43. Which is the strongest of the following acids ?
(A) HCOOH (B) CH_3CCOH
(C) CCl_3COOH (D) $CH_2ClCOOH$

44. Which of the following gases will have the highest rate of diffusion ?
(A) CO_2 (B) NH_3 (C) N_2 (D) O_2

45. If the density ratio of O_2 and H_2 is 16 : 1, then ratio of their V_{rms} will be
(A) 1 : 4 (B) 1 : 16 (C) 4 : 1 (D) 16 :1

46. Which of the following represents bond in silica ?
(A) Si–Si–Si–Si (B) Si–C–Si–O–Si
(C) Si–C–Si–C–Si (D) –Si–O–Si–O–Si–

47. Which of the following compounds will be most easily attacked by an electrophile ?
(A) benzene (B) chlorobenzene (Cl) (C) phenol (OH) (D) toluene (CH_3)

48. Which of the following is not true about e.m.f. of a cell ?
(A) It is responsible for the flow of steady current in the cell
(B) Work calculated from it is not the maximum work obtainable from the cell
(C) It is maximum voltage obtainable from the cell
(D) It is the potential difference between two electrodes when no current is flowing in circuit

49. When ethanal is treated with Fehling's solution, it gives a precipitate of
(A) Cu_2O (B) Cu_3O (C) Cu (D) CuO

50. Which of the following statements regarding entropy is correct ?
(A) At 0°C, the entropy of all crystalline substances is taken to be zero
(B) At 0°C, the entropy of perfect crystalline substances is taken to be zero
(C) At absolute zero temperature, the entropy of all crystalline substances is taken to be zero.
(D) At absolute zero temperature, entropy of a perfect crystalline substance is taken to be zero.

51. Which of the following expressions gives the de Broglie relationship ?
(A) $\frac{h}{mv} = p$ (B) $\lambda = \frac{h}{mv}$ (C) $\lambda = \frac{h}{mp}$ (D) $\lambda m = \frac{v}{p}$

52. According to Dalton's atom theory, the smallest particle in which matter can exit is called
(A) An atom (B) An ion
(C) An electron (D) A molecule

53. The position of both an electron and a helium atom is known within 1.0 nm. The momentum of the electron is known within 5.0×10^{-26} kg ms^{-1}. The minimum uncertainty in the measurement of the momentum of the helium atom is
(A) 5.0×10^{-26} kg ms^{-1} (B) 6.0×10^{-26} kg ms^{-1}
(C) 7.0×10^{-26} kg ms^{-1} (D) 8.0×10^{-26} kg ms^{-1}

54. A bivalent metal has the equivalent weight of 12. The molecular weight to its oxide will be
(A) 36 (B) 40 (C) 24 (D) 32

55. The law of equilibrium was first given by
(A) Goldberg (B) Waage
(C) Boyle (D) Both (A) and (B)

56. A mixture of 40 g of oxygen and 40 g of helium has a total pressure of 0.9 atm. The partial pressure of oxygen is
(A) 0.1 atm (B) 0.2 atm (C) 0.5 atm (D) 0.9 atm

57. The weight of a metal of equivalent weight 12, which will give 0.475 g of its chloride, is
(A) 0.12 g (B) 0.16 g (C) 0.18 g (D) 0.24 g

58. Which of the following is called Baeyer's reagent
(A) Neutral $KMnO_4$ solution
(B) Alkaline $KMnO_4$ solution
(C) Saturated $KMnO_4$ solution
(D) Acidic $KMnO_4$ solution

59. The number of water molecules in gypsum and plaster of pairs is respectively
(A) 2 and 1/2 (B) 2 and 1 (C) 5 and 2 (D) 7 and 2

60. A catalyst is used to
(A) Increase the product
(B) Decrease the product
(C) Minimise the time of rate of reaction
(D) None of these

AIIMS - 1995

1. If an atom is reduced, its oxidation number
(A) Does not change (B) Sharply decreases
(C) Increases (D) Slightly decreases

2. Which of the following has the maximum electonegativity ?
(A) C (B) N (C) F (D) O

3. The property, which can be classified as an intensive property, is
(A) Mass (B) Temperature
(C) Volume (D) Heat capacity

4. Two different gases enclosed in different flasks A and B at same temperature and pressure were found to contain same number of molecules. The ratio of volumes of the flasks A and B must be
(A) 1 : 1 (B) 1 : 2 (C) 1 : 3 (D) 1 : 4

5. Assertion : Formic acid reduces mercuric chloride to mercurous chloride on heating, while acetic acid does not. Reason : Formic acid is a stronger acid than acetic acid.
(A) A (B) B (C) C (D) D

6. Assertion : Sulphur dioxide and chloride are both bleaching agents. Reason : Both are reducing agents.
(A) A (B) B (C) C (D) D

7. Assertion : Enthalpy of graphite is lower than that of diamond. Reason : Entropy of graphite is greater than that of diamond.
(A) A (B) B (C) C (D) D

8. Assertion : Copper liberates hydrogen from a solution of hydrochloric acid. Reason : Hydrogen is below copper in the electro-chemical series.

(A) A (B) B (C) C (D) D

9. Assertion : Phenol undergoes Kolbe reaction whereas ethanol does not. Reason : Their ionization energies are low

(A) A (B) B (C) C (D) D

10. Assertion : Alkali metals impart colour to the flame. Reason : Their ionization energies are low.

(A) A (B) B (C) C (D) D

11. Assertion : Bond order in a molecule can assume any value positive or negative, integral or fractional, including zero. Reason : It depends on the number of electrons in the bonding and antibonding orbitals.

(A) A (B) B (C) C (D) D

12. Assertion : The molecularity of the reaction, $H_2 + Br_2 = 2HBr$ is two. Reason : The order of this reaction is 3/2.

(A) A (B) B (C) C (D) D

13. Assertion : Cyclobutane is less stable than cyclopentane. Reason : Presence of 'bent bonds' causes "loss of orbital overlap ".

(A) A (B) B (C) C (D) D

14. Assertion : The dumas method is more applicable to nitrogen containing organic compounds than the Kjeldahl's method. Reason : The Kjeldahl's method does not give satisfactory results for compounds in which nitrogen is directly linked to oxygen.

(A) A (B) B (C) C (D) D

15. The temperature, at which the density of O_2 at 1 atm. is the same as that of CH_4 at S.T.P. is

(A) 273°C (B) 546°C (C) 100°C (D) 150°C

16. The number of unpaired electrons in $1s^2\ 2s^2\ 2p^3$ is

(A) 1 (B) 2 (C) 3 (D) 5

17. The amount of zinc (at.wt.=65) necessary to produce 224 ml of H_2 by the reaction with an acid will be

(A) 0.065 gm (B) 0.65 gm (C) 6.5 gm (D) 7.5 gm

18. Rutherford's scattering experiment is related to the size of

(A) Nucleus (B) Proton (C) Electron (D) Neutron

19. Which of the following compounds is not soluble in HNO_3 ?

(A) PbS (B) AgCl (C) CuS (D) CdS

20. Which of the following molecule or ions is a bidentate ligand ?

(A) Br_2^+ (B) $CH_3-C \equiv N$ (C) $C_2O_4^{2-}$ (D) CH_3NH_2

21. What is the correct bond angle in dimethyl ether ?

(A) 109° (B) 110° (C) 120° (D) 180°

22. The molality of a solution having 18 gm of glucose (mol. wt. = 180) dissolved in 500 gm of water is

(A) 0.1 M (B) 0.5 M (C) 0.2 M (D) 2.2 M

23. Nessler's reagent is used in the test of

(A) NH_3 (B) NH_4^+ (C) NH_4Cl (D) All of these

24. How many gm of silver will be displaced from a solution of $AgNO_3$ by 4 gm of magnesium ?

(A) 18 gm (B) 36 gm (C) 4 gm (D) 16 gm

25. The IUPAC name of the compound having the formula, CCl_3CH_2CHO is

(A) 1,1, 1- trichloropropanal (B) 1, 2, 1- dichloromethanal
(C) 2, 2, 2- trichloropropanal
(D) 3, 3, 3, - trichloropropanal

26. Which of the following is the least stable and its existence is doubtful ?

(A) Cl_4 (B) GeI_4 (C) SnI_4 (D) PbI_4

27. The number of electrons required to deposit 1gm equivalent aluminium (at. wt. = 27) from a solution of aluminium chloride will be

(A) 1 (B) 2 (C) 3 (D) 4

28. Which of the following complexes has square planar structure ?

(A) $Ni(CO)_4$ (B) $[NiCl_4]^{2-}$
(C) $[Zn(CN)_4]^{2-}$ (D) $[Cu(NH_3)_4]^{2+}$

29. In reaction , $N_2 + 3H_2 \rightleftharpoons 2NH_3$, increase in pressure will favour

(A) Forward direction (B) Backward direction
(C) Reversible reaction (D) Irreversible reaction

30. In order to protect iron from corrosion, iron is coated with

(A) Zinc (B) Tin (C) Nickel (D) Copper

31. By the electrolysis of aqueous solution of $CuSO_4$, the products obtained at both the electrodes are

(A) H_2 at anode and Cu at cathode
(B) $H_2S_2O_8$ at anode and O_2 at cathode
(C) O_2 at anode and H_2 at cathode
(D) O_2 at anode and Cu at cathode

32. A buffer solution contains 0.1 M of acetic acid and 0.1 M of sodium acetate. What will be its pH ? (p_{ka} of acetic acid is 4.75)

(A) 4.00 (B) 4.75 (C) 5.00 (D) 5.25

33. Which of the following compounds turns black on the addition of ammonium hydroxide ?

(A) $PbCl_2$ (B) Hg_2Cl_2 (C) $CuCl_2$ (D) AgCl

34. How many electrons are delivered at the cathode during electrolysis by a current of 1 A in 60 seconds ?

(A) 6.0×10^{23} (B) 6.0×10^{20}
(C) 3.74×10^{20} (D) 7.48×10^{21}

35. Which of the following alkaline earth metal has highest ionic mobility in aqueous solution ?

(A) Ca^{2+} (B) Mg^{2+} (C) Be^{2+} (D) Ba^{2+}

36. In the presence of an acid, hydrolysis of methyl cyanide produces

(A) Acetic acid (B) Methylamine
(C) Methyl alcohol (D) Formic acid

37. Lithopone, a white pigment consists of

(A) PbS and MgO (B) $BaSO_4$ and $PbSO_4$

(C) ZnS and $BaSO_4$ (D) Al_2O_3 and $CaCO_3$

38. Chloroform by reacting with conc. HNO_3 produces
(A) Water gas (B) Laughing gas
(C) Tear gas (D) Producer gas

39. Euchlorine is produced by heating a mixture of
(A) KCl + conc. HCl (B) $KClO_3$ + conc. HCl
(C) KCl + conc. H_2SO_4 (D) K_2ClO_3 + conc.H_2SO_4

40. Which of the following is a condensation polymer ?
(A) Dacron (B) Neoprene (C) Teflon (D) Polystrene

41. Which of the following is the electronic configuration of Cu^{2+} (Z = 29)
(A) [Ar] $4s^1\ 3d^8$ (B) [Ar] $4s^1\ 3d^{10}$
(C) [Ar] $3d^9$ (D) [Ar] $4s^2 3d^{10} 4p^1$

42. Atom bomb is based on the principle of
(A) Radioactivity (B) Nuclear fusion
(C) Nuclear fission (D) Fusion and fission

43. In which of the following reaction $K_P > K_C$?
(A) $H_2 + I_2 \rightarrow 2HI$ (B) $N_2 + 3H_2 \rightarrow 2NH_3$
(C) $PCl_3 + Cl_2 \rightarrow PCl_5$ (D) $2SO_3 \rightarrow O_2 + 2SO_2$

44. The oxidation number of phosphorus and basicity of acid in pyrophosphoric acid respectively are
(A) + one and four (B) + three and one
(C) + four and three (D) + five and four

45. The oxidation number of oxygen atom in O_2^{2-} ion is
(A) –3 (B) –5 (C) –1 (D) –2

46. The important ore of aluminium is
(A) Corundum (B) Ruby (C) Kaolin (D) Bauxite

47. Which of the following oxyacids does not exist
(A) $HBiO_3$ (B) H_3BiO_4 (C) H_3SbO_3 (D) H_3AsO_4

48. The fractional distillation is used in processing of
(A) Crude oil (B) Coal tar
(C) Petroleum (D) All of these

49. If pH value of a solution is 3 and by adding water, it becomes 6, then the dilution is increased by
(A) 10 times (B) 100 times
(C) 500 times (D) 1000 times

50. Which of the following kinds of catalysis can be explained by the adsorption theory ?
(A) Enzyme catalysis (B) Acid base catalysis
(C) Heterogeneous catalysis
(D) Homogeneous catalysis

51. The element, with atomic number 118, will be
(A) Akali metal (B) Noble gas
(C) Transition element (D) Alkaline earth metal

52. Which of the following molecules has regular geometry ?
(A) H_2O (B) XeF_4 (C) PF_3 (D) SF_6

53. The complete combustion of CH_4 gives
(A) CO +H_2 (B) CO +N_2
(C) CO + H_2O (D) CO + N_2O

54. The volume of carbon dioxide gas evolved at S.T.P. by heating 7.3 gm of $Mg(HCO_3)_2$ will be
(A) 1120 ml (B) 2000 ml (C) 2240 ml (D) 2340 ml

55. The maximum covalency of an element of atomic number 7 is
(A) 2 (B) 3 (C) 4 (D) 5

56. The molarity of pure water is
(A) 1.16 M (B) 5.56 M (C) 18.36 M (D) 55.56 M

57. A certain gas diffuses four times as quickly as oxygen. The molecular weight of the gas is
(A) 2 (B) 16 (C) 1 (D) 1.5

58. The equivalent weight of oxygen, when it is converted to oxide is equal to
(A) Molecular weight (B) $\frac{\text{Molecular weight}}{2}$
(C) $\frac{\text{Molecular weight}}{3}$ (D) $\frac{\text{Molecular weight}}{4}$

59. Which of the following is used in photography ?
(A) AgCl (B) AgBr (C) Ag_2S (D) $Ag_2C_2O_4$

60. In the presence of mercuric ion and conc. sulphuric acid, the reaction of acetylene reacts with water to produce
(A) $CH_3 - CHO$ (B) $CH_3 - COOH$
(C) $CH_3 - CO - CH_3$ (D) $CH_3 - CH_2 - OH$

AIIMS-1996

1. In the reaction , $2P_2O_5 + HNO_3 \rightarrow P_4O_5 + X$, the term X is
(A) N_2O_4 (B) N_2O_5 (C) H_2O only (D) PNO_3

2. The aqueous solution of which of the following salts will have the lowest pH ?
(A) NaClO (B) $NaClO_2$ (C) $NaClO_3$ (D) $NaClO_4$

3. Which of the following molecules is paramagnetic ?
(A) NO_2 (B) N_2O_5 (C) P_2O_5 (D) $H_2S_2O_7$

4. If active mass of a 6% solution of a compound is 2, its molecular weight will be
(A) 15 (B) 22 (C) 30 (D) 60

5. Hydrolysis of beryllium carbide produces
(A) Methane (B) Ethene
(C) Acetylene (D) None of these

6. Which of the following species is isotonic with $_{37}Rb^{86}$?
(A) $_{38}Kr^{87}$ (B) $_{19}Mg^{39}$ (C) $_{36}Kr^{87}$ (D) $_{37}Rb^{84}$

7. Which of the following indicators is known as metal indicator ?
(A) Phenol red (B) Methyl orange
(C)Phenolphthalein (D) Erichrome-Black-T

8. Which of the following is the correct sequence of atomic weights of the given elements ?
(A) Fe > Co > Ni (B) Co > Ni > Fe
(C) Ni > Co > Fe (D) Co > Fe > Ni

9. The equilibrium constant of a reaction is 300. If the

volume of a reaction flask is tripled, the equilibrium constant will be.

(A) 300 (B) 600 (C) 100 (D) 150

10. In the reaction sequence,

$$CaC_2 \xrightarrow{H_2O} A \xrightarrow[HgSO_4]{dil\ H_2SO_4} B \xrightarrow{H_2/Ni} C,$$

'C' is

(A) HCHO (B) CH_3OH (C) CH_3CHO (D) C_2H_5OH

11. Avogradro's number of oxygen atoms weight

(A) 32 g (B) 56 g (C) 8g (D) 16 g

12. Which of the following compound is known as white vitriol ?

(A) $CaSO_4.2H_2O$ (B) $MgSO_4.7H_2O$
(C) $ZnSO_4.7H_2O$ (D) $Na_2SO_4.7H_2O$

13. The maximum number of electrons, present in an orbit that is represented by azimuthal quantum number (l) = 3, will be

(A) 2 (B) 6 (C) 8 (D) 14

14. A radioactive nuclide decays at such a rate that after 96 min, only 1/8th of the original amount remains. The value of half-life of the nuclide is

(A) 12 min (B) 24 min (C) 32 min (D) 48 min

15. How many litres of ammonia gas at S.T.P. would be needed to prepare 100 ml of 2.5 M ammonium hydroxide solution ?

(A) 5.6 lit (B) 11.2 lit (C) 0.056 lit (D) 0.56 lit

16. Silicon is an important constituent of

(A) Rocks (B) Animals (C) Alloys (D) Vegetables

17. Which of the following alkanes is optically active ?

(A) Propane (B) 2-methyl butane
(C) 3-methyl hexane (D) 2,3,4-trimethyl pentane

18. Thermite is a mixture of

(A) Fe+Al (B) Cu + Mg (C) Zn + Mg (D) Fe_2O_3 + Al

19. The boiling point of a compound is raised by

(A) Volatility of compound
(B) Non-polarity in the molecules
(C) Intermolecular hydrogen bonding
(D) Intramolecular hydrogen bonding

20. The strongest bronsted base among the following anions is

(A) ClO^- (B) ClO_2^- (C) ClO_3^- (D) ClO_4^-

21. The pH of 0.001-N acetic acid solution, which is 10% dissociated, is

(A) 3 (B) 4 (C) 1 (D) 2

22. Which of the following gas molecules have equal total kinetic energy and translation kinetic energy ?

(A) He (B) N_2 (C) O_2 (D) CH_4

23. Which of the following compounds will undergo hydrolysis ?

(A) $NaNO_3$ (B) NH_4Cl (C) KCl (D) NaCl

24. The maximum energy is possessed by an electron, when it is present

(A) In nucleus
(B) In ground energy state
(C) In first excited state
(D) At infinite distance from the nucleus

25. The rate, at which a substance reacts, depends upon its

(A) Molecular mass (B) Atomic mass
(C) Equivalent mass (D) Active mass

26. The number of molecules in 4.25 g of ammonia is approximately

(A) 0.5×10^{23} (B) 1.5×10^{23}
(C) 3.5×10^{23} (D) 2.5×10^{23}

27. Absolute ether is diethyl ether, which

(A) Is obtained from Williamson's synthesis
(B) Is mixed with absolute alcohol
(C) Contains trace of peroxide
(D) Is totally free from moisture and peroxides

28. In the P^{3-}, S^{2-} and Cl^- ions, the increasing order of size is

(A) $Cl^- < S^{2-} < P^{3-}$ (B) $P^{3-} < S^{2-} < Cl^-$
(C) $S^{2-} < Cl^- < P^{3-}$ (D) $S^{2-} < P^{3-} < Cl^-$

29. The pH value of human blood is about

(A) 2.3 to 3.4 (B) 3.5 to 4.6
(C) 4.5 to 5.6 (D) 7.3 to 7.5

30. Which of the following is a sulphide ore ?

(A) Magnetite (B) Malachite
(C) Carnallite (D) Copper pyrites

31. Which of the following contains cobalt ?

(A) Haemoglobin (B) Vitamin D
(C) Virtamin B_{12} (D) Chlorophyll

32. Ostwald's dilution law is applicable to only

(A) Solute (B) Solvent
(C) Strong electrolytes (D) Weak electrolytes

33. Bismuth chloride $BiCl_3$, on hydrolysis, yields a white precipitate of

(A) Bismuth oxyacid (B) Bismuth oxychloride
(C) Bismuth hydoxide (D) None of these

34. Which of the following is the strongest base ?

(A) H_2^+ (B) HCO_3^- (C) H (D) H^-

35. The volume of 3.0 N-H_2O_2 strength is

(A) 4.2 litres (B) 8.4 litres
(C) 16.8 litres (D) 33.6 litres

36. Which of the following carbonium ions will be most stable ?

(A) CH_3^+ (B) $(CH_3)_3C^+$
(C) $(CH_3)_2CH^+$ (D) $CH_3CH_2^+$

37. Assertion : A special line will be seen for a $2p_x \rightarrow 2p_y$ transition. Reason : Energy is released in the form of waves of light when the electron drops form $2p_x$ to $2p_y$ orbital.

(A) A (B) B (C) C (D) D

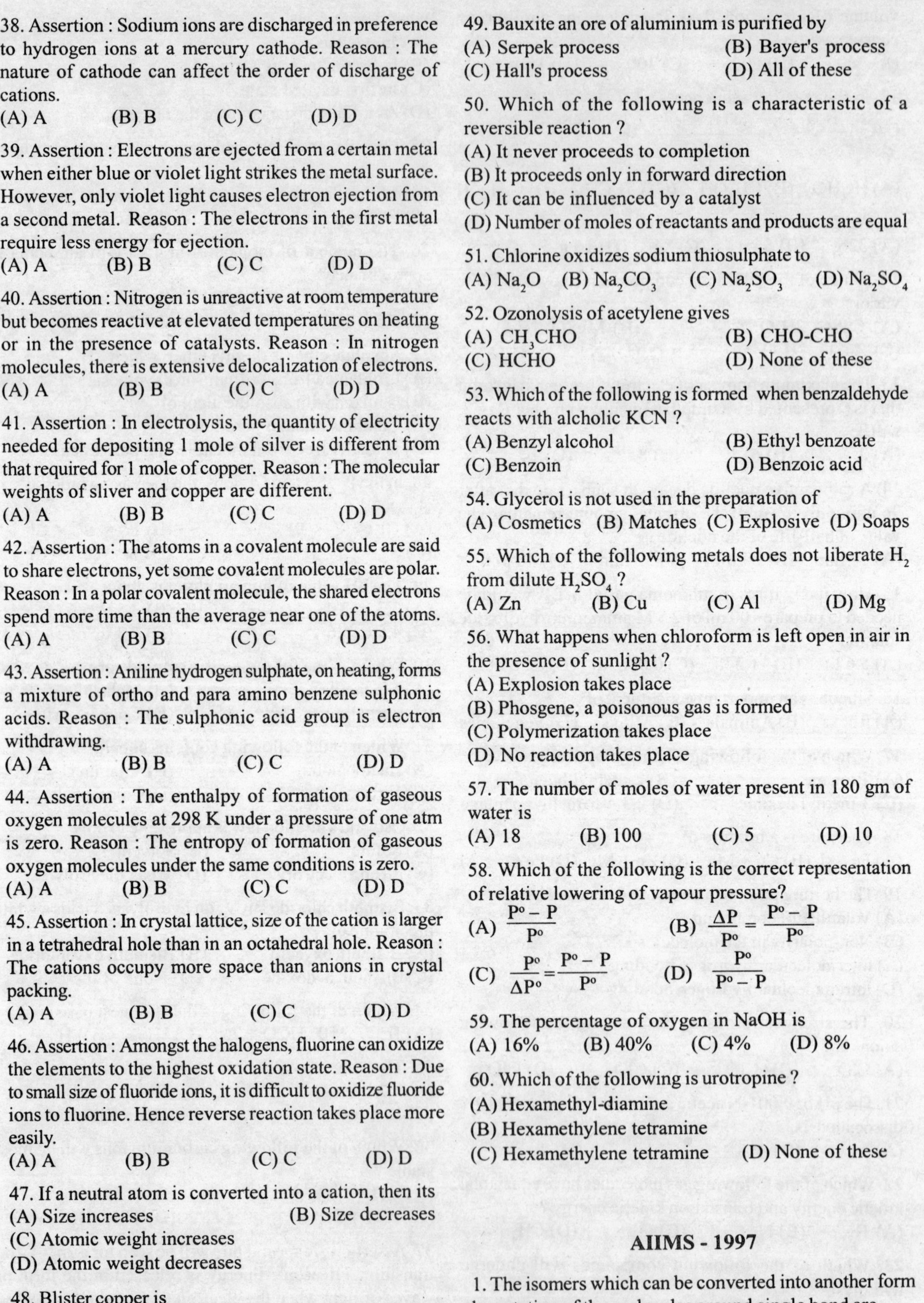

38. Assertion : Sodium ions are discharged in preference to hydrogen ions at a mercury cathode. Reason : The nature of cathode can affect the order of discharge of cations.
(A) A (B) B (C) C (D) D

39. Assertion : Electrons are ejected from a certain metal when either blue or violet light strikes the metal surface. However, only violet light causes electron ejection from a second metal. Reason : The electrons in the first metal require less energy for ejection.
(A) A (B) B (C) C (D) D

40. Assertion : Nitrogen is unreactive at room temperature but becomes reactive at elevated temperatures on heating or in the presence of catalysts. Reason : In nitrogen molecules, there is extensive delocalization of electrons.
(A) A (B) B (C) C (D) D

41. Assertion : In electrolysis, the quantity of electricity needed for depositing 1 mole of silver is different from that required for 1 mole of copper. Reason : The molecular weights of sliver and copper are different.
(A) A (B) B (C) C (D) D

42. Assertion : The atoms in a covalent molecule are said to share electrons, yet some covalent molecules are polar. Reason : In a polar covalent molecule, the shared electrons spend more time than the average near one of the atoms.
(A) A (B) B (C) C (D) D

43. Assertion : Aniline hydrogen sulphate, on heating, forms a mixture of ortho and para amino benzene sulphonic acids. Reason : The sulphonic acid group is electron withdrawing.
(A) A (B) B (C) C (D) D

44. Assertion : The enthalpy of formation of gaseous oxygen molecules at 298 K under a pressure of one atm is zero. Reason : The entropy of formation of gaseous oxygen molecules under the same conditions is zero.
(A) A (B) B (C) C (D) D

45. Assertion : In crystal lattice, size of the cation is larger in a tetrahedral hole than in an octahedral hole. Reason : The cations occupy more space than anions in crystal packing.
(A) A (B) B (C) C (D) D

46. Assertion : Amongst the halogens, fluorine can oxidize the elements to the highest oxidation state. Reason : Due to small size of fluoride ions, it is difficult to oxidize fluoride ions to fluorine. Hence reverse reaction takes place more easily.
(A) A (B) B (C) C (D) D

47. If a neutral atom is converted into a cation, then its
(A) Size increases (B) Size decreases
(C) Atomic weight increases
(D) Atomic weight decreases

48. Blister copper is
(A) Pure copper (B) Ore of copper (C) Alloy of copper
(D) Copper containing 1% impurity

49. Bauxite an ore of aluminium is purified by
(A) Serpek process (B) Bayer's process
(C) Hall's process (D) All of these

50. Which of the following is a characteristic of a reversible reaction ?
(A) It never proceeds to completion
(B) It proceeds only in forward direction
(C) It can be influenced by a catalyst
(D) Number of moles of reactants and products are equal

51. Chlorine oxidizes sodium thiosulphate to
(A) Na_2O (B) Na_2CO_3 (C) Na_2SO_3 (D) Na_2SO_4

52. Ozonolysis of acetylene gives
(A) CH_3CHO (B) CHO-CHO
(C) HCHO (D) None of these

53. Which of the following is formed when benzaldehyde reacts with alcholic KCN ?
(A) Benzyl alcohol (B) Ethyl benzoate
(C) Benzoin (D) Benzoic acid

54. Glycerol is not used in the preparation of
(A) Cosmetics (B) Matches (C) Explosive (D) Soaps

55. Which of the following metals does not liberate H_2 from dilute H_2SO_4 ?
(A) Zn (B) Cu (C) Al (D) Mg

56. What happens when chloroform is left open in air in the presence of sunlight ?
(A) Explosion takes place
(B) Phosgene, a poisonous gas is formed
(C) Polymerization takes place
(D) No reaction takes place

57. The number of moles of water present in 180 gm of water is
(A) 18 (B) 100 (C) 5 (D) 10

58. Which of the following is the correct representation of relative lowering of vapour pressure?
(A) $\frac{P^o - P}{P^o}$ (B) $\frac{\Delta P}{P^o} = \frac{P^o - P}{P^o}$
(C) $\frac{P^o}{\Delta P^o} = \frac{P^o - P}{P^o}$ (D) $\frac{P^o}{P^o - P}$

59. The percentage of oxygen in NaOH is
(A) 16% (B) 40% (C) 4% (D) 8%

60. Which of the following is urotropine ?
(A) Hexamethyl-diamine
(B) Hexamethylene tetramine
(C) Hexamethylene tetramine (D) None of these

AIIMS - 1997

1. The isomers which can be converted into another form by rotation of the molecules around single bond are
(A) Enantiomers (B) Diastereomers
(C) Geometrical isomers (D) Conformers

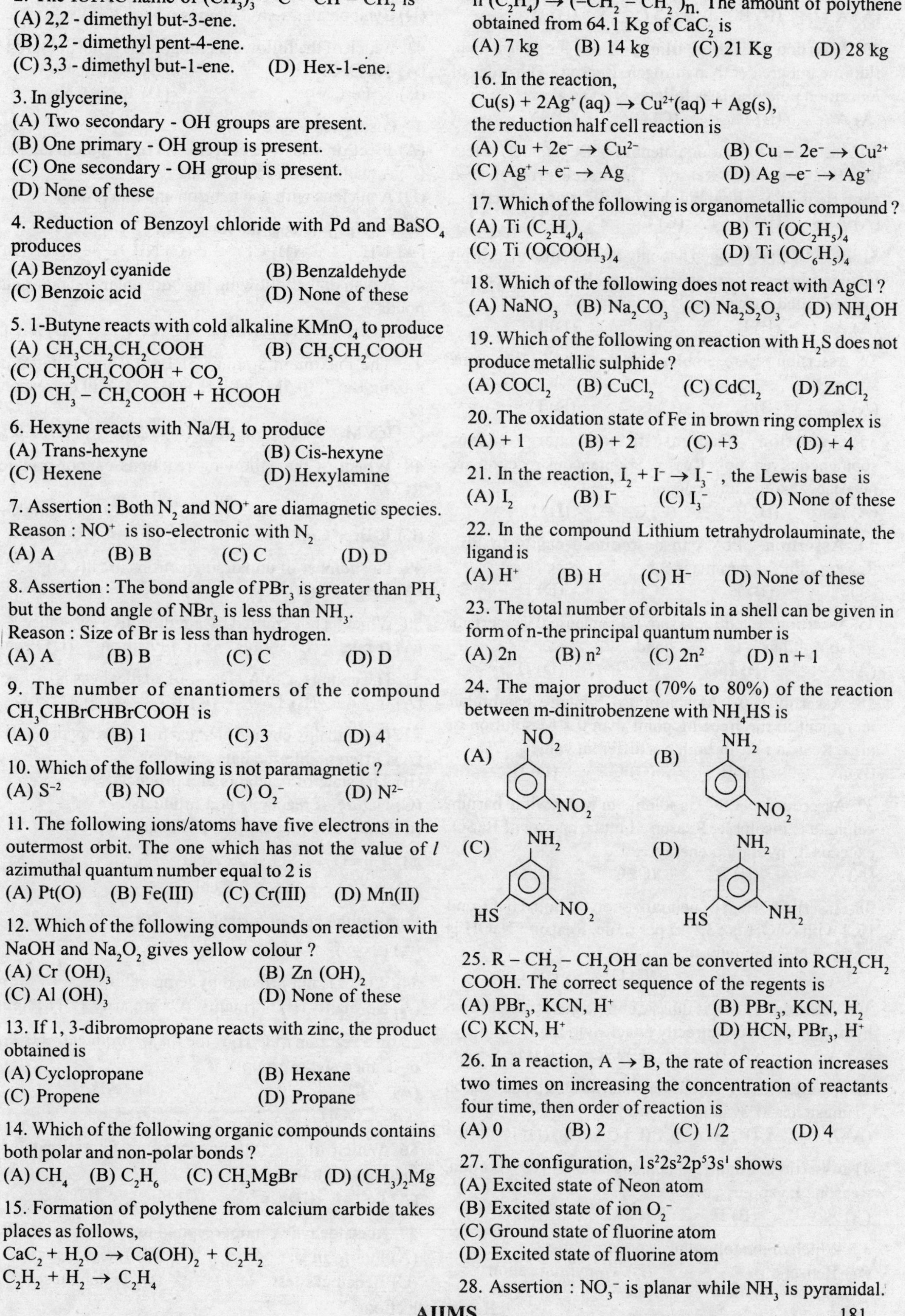

2. The IUPAC name of $(CH_3)_3 - C - CH = CH_2$ is
(A) 2,2 - dimethyl but-3-ene.
(B) 2,2 - dimethyl pent-4-ene.
(C) 3,3 - dimethyl but-1-ene. (D) Hex-1-ene.

3. In glycerine,
(A) Two secondary - OH groups are present.
(B) One primary - OH group is present.
(C) One secondary - OH group is present.
(D) None of these

4. Reduction of Benzoyl chloride with Pd and $BaSO_4$ produces
(A) Benzoyl cyanide (B) Benzaldehyde
(C) Benzoic acid (D) None of these

5. 1-Butyne reacts with cold alkaline $KMnO_4$ to produce
(A) $CH_3CH_2CH_2COOH$ (B) CH_5CH_2COOH
(C) $CH_3CH_2COOH + CO_2$
(D) $CH_3 - CH_2COOH + HCOOH$

6. Hexyne reacts with Na/H_2 to produce
(A) Trans-hexyne (B) Cis-hexyne
(C) Hexene (D) Hexylamine

7. Assertion : Both N_2 and NO^+ are diamagnetic species.
Reason : NO^+ is iso-electronic with N_2
(A) A (B) B (C) C (D) D

8. Assertion : The bond angle of PBr_3 is greater than PH_3 but the bond angle of NBr_3 is less than NH_3.
Reason : Size of Br is less than hydrogen.
(A) A (B) B (C) C (D) D

9. The number of enantiomers of the compound $CH_3CHBrCHBrCOOH$ is
(A) 0 (B) 1 (C) 3 (D) 4

10. Which of the following is not paramagnetic ?
(A) S^{-2} (B) NO (C) O_2^- (D) N^{2-}

11. The following ions/atoms have five electrons in the outermost orbit. The one which has not the value of l azimuthal quantum number equal to 2 is
(A) Pt(O) (B) Fe(III) (C) Cr(III) (D) Mn(II)

12. Which of the following compounds on reaction with NaOH and Na_2O_2 gives yellow colour ?
(A) Cr $(OH)_3$ (B) Zn $(OH)_2$
(C) Al $(OH)_3$ (D) None of these

13. If 1, 3-dibromopropane reacts with zinc, the product obtained is
(A) Cyclopropane (B) Hexane
(C) Propene (D) Propane

14. Which of the following organic compounds contains both polar and non-polar bonds ?
(A) CH_4 (B) C_2H_6 (C) CH_3MgBr (D) $(CH_3)_2Mg$

15. Formation of polythene from calcium carbide takes places as follows,
$CaC_2 + H_2O \rightarrow Ca(OH)_2 + C_2H_2$
$C_2H_2 + H_2 \rightarrow C_2H_4$
$n\,(C_2H_4) \rightarrow (-CH_2 - CH_2-)_n$. The amount of polythene obtained from 64.1 Kg of CaC_2 is
(A) 7 kg (B) 14 kg (C) 21 Kg (D) 28 kg

16. In the reaction,
$Cu(s) + 2Ag^+(aq) \rightarrow Cu^{2+}(aq) + Ag(s)$,
the reduction half cell reaction is
(A) $Cu + 2e^- \rightarrow Cu^{2-}$ (B) $Cu - 2e^- \rightarrow Cu^{2+}$
(C) $Ag^+ + e^- \rightarrow Ag$ (D) $Ag - e^- \rightarrow Ag^+$

17. Which of the following is organometallic compound ?
(A) Ti $(C_2H_4)_4$ (B) Ti $(OC_2H_5)_4$
(C) Ti $(OCOOH_3)_4$ (D) Ti $(OC_6H_5)_4$

18. Which of the following does not react with AgCl ?
(A) $NaNO_3$ (B) Na_2CO_3 (C) $Na_2S_2O_3$ (D) NH_4OH

19. Which of the following on reaction with H_2S does not produce metallic sulphide ?
(A) $COCl_2$ (B) $CuCl_2$ (C) $CdCl_2$ (D) $ZnCl_2$

20. The oxidation state of Fe in brown ring complex is
(A) + 1 (B) + 2 (C) +3 (D) + 4

21. In the reaction, $I_2 + I^- \rightarrow I_3^-$, the Lewis base is
(A) I_2 (B) I^- (C) I_3^- (D) None of these

22. In the compound Lithium tetrahydrolauminate, the ligand is
(A) H^+ (B) H (C) H^- (D) None of these

23. The total number of orbitals in a shell can be given in form of n-the principal quantum number is
(A) 2n (B) n^2 (C) $2n^2$ (D) n + 1

24. The major product (70% to 80%) of the reaction between m-dinitrobenzene with NH_4HS is

(A) NO_2 / NO_2 (B) NH_2 / NO_2
(C) NH_2 / HS / NO_2 (D) NH_2 / HS / NH_2

25. $R - CH_2 - CH_2OH$ can be converted into RCH_2CH_2COOH. The correct sequence of the regents is
(A) PBr_3, KCN, H^+ (B) PBr_3, KCN, H_2
(C) KCN, H^+ (D) HCN, PBr_3, H^+

26. In a reaction, $A \rightarrow B$, the rate of reaction increases two times on increasing the concentration of reactants four time, then order of reaction is
(A) 0 (B) 2 (C) 1/2 (D) 4

27. The configuration, $1s^2 2s^2 2p^5 3s^1$ shows
(A) Excited state of Neon atom
(B) Excited state of ion O_2^-
(C) Ground state of fluorine atom
(D) Excited state of fluorine atom

28. Assertion : NO_3^- is planar while NH_3 is pyramidal.

Reason : N in NO_3^- is sp^2 and in NH_3 is sp^3 hybridisation.
(A) A (B) B (C) C (D) D

29. Assertion : Electron affinity of oxygen is less than fluorine but greater than nitrogen. Reason : The order of ionisation potential is as follows N > O < F.
(A) A (B) B (C) C (D) D

30. Assertion : Ionisation potential of Be (at.no.4) is less than B (atomic no.5). Reason : The first electron released from Be is of p-orbital but that from B is of s-orbital.
(A) A (B) B (C) C (D) D

31. Assertion : Hydrogen has only one electron in its orbit but it produces several spectral lines. Reason : There are many excited energy levels available.
(A) A (B) B (C) C (D) D

32. Assertion : Li resembles Mg. Reason : Li^+ has same size as Mg^{2+}
(A) A (B) B (C) C (D) D

33. Assertion : Decrease in free energy causes spontaneous reaction. Reason Spontaneous reaction are invariably exothermic reaction.
(A) A (B) B (C) C (D) D

34. Assertion : Pb^{4+} can be reduced easily to Pb^{2+} Reason : Pb^{2+} is paramagnetic
(A) A (B) B (C) C (D) D

35. Assertion : Lactic acid shows geometrical isomerism. Reason : It has – C = C – bond
(A) A (B) B (C) C (D) D

36. Assertion : 0.1 M solution of glucose has higher increment in the freezing point than 0.1 M solution of urea. Reason : K_f for both has different value.
(A) A (B) B (C) C (D) D

37. Assertion : Na_2SO_4 is soluble in water while barium sulphate is insoluble. Reason : Lattice energy of $BaSO_4$ exceeds its hydration energy.
(A) A (B) B (C) C (D) D

38. Assertion : Heat of neutralization for both HNO_3 and HCl with NaOH is 53.7 kJ per mole. Reason : NaOH is a strong electrolyte/base.
(A) A (B) B (C) C (D) D

39. Assertion : H_3PO_3 is dibasic compound. Reason : Two hydrogen atoms are directly attached to the P.
(A) A (B) B (C) C (D) D

40. Assertion : Tungsten has very high melting point. Reason : Tungsten is a covalent compound.
(A) A (B) B (C) C (D) D

41. Assertion : Liquid ammonia is used for refrigeration. Reason : It vapourises quickly.
(A) A (B) B (C) C (D) D

42. Which of the following is not aromatic ?
(A) Benzene (B) Tropylium cation
(C) Cyclopentadienylanion
(D) Cyclooctatetra-enyl dianions

43. Which of the following compounds is not coloured ?
(A) Na_2CuCl_4 (B) Na_2CdCl_4
(C) $K_4Fe(CN)_6$ (D) $K_3Fe(CN)_6$

44. Positron is
(A) Electron with +ve charge (B) A helium nucleus
(C) A nucleus with two protons
(D) A nucleus with one neutron and one proton

45. $K_2[HgI_4]$ detects the ion/group
(A) NH_4^+ (B) Cl^- (C) NH_2 (D) NO

46. Which of the following has both polar and non-polar bonds ?
(A) HCl (B) $AlCl_3$ (C) C_2H_6 (D) NH_4Cl

47. The maximum amount of $BaSO_4$ precipitated on mixing $BaCl_2$ (0.5M) with H_2SO_4 (1M) will correspond to
(A) 0.5 M (B) 1.0 M (C) 1.5 M (D) 2.0 M

48. Which of the following reactions cannot produce RCOAr ?
(A) $RCOCl + H - Ar \xrightarrow{AlCl_3}$ (B) $ArCOCl + RMgX \rightarrow$
(C) $R\text{-}Br + CrO_3 \rightarrow$ (D) None of these

49. The number of unpaired electrons in $Ni(CO)_4$
(A) Zero (B) One (C) Three (D) Five

50. Which is not emitted by a radioactive substance ?
(A) α-rays (B) β-rays (C) Positron (D) Proton

51. The compressibility factor of an ideal gas is
(A) 0 (B) 1 (C) 2 (D) 4

52. The enthalpy change of a reaction does not depend on
(A) Different intermediate reactions
(B) The state of reactants and products.
(C) Nature of reactants and products
(D) Initial and final enthalpy change of a reaction

53. $S + \frac{3}{2}O_2 \rightarrow SO_3 + 2x$ Kcal
$SO_2 + \frac{1}{2}O_2 \rightarrow SO_3 + y$ Kcal.
Find out the heat of formation of SO_2
(A) (2x–y) (B) (2x+y) (C) (x+y) (D) 2x/y

54. Which is not affected by temperature ?
(A) Normality (B) Formality (C) Molarity (D) Molality

55. In a reaction of C_6H_5Y, the major product (> 60%) is m-isomer, so the group Y is
(A) –OH (B) $-NH_2$
(C) –COOH (D) –Cl

56. Which of the following cannot be estimated by iodometric titration ?
(A) Fe^{3+} (B) Cu^{2+} (C) Pb^{2+} (D) Ag^+

57. Acetaldehyde cannot respond to
(A) Iodoform test (B) Lucas test
(C) Benedict's test (D) Tollen's test

AIIMS -1998

1. Ratio of the energy of a photon of wavelengths 3000 Å and 6000 Å is
(A) 1 : 3 (B) 1 : 6 (C) 2 : 1 (D) 1 : 2

2. If a compound, on analysis was found to contain C = 18.5%, H = 1.55%, Cl = 55.04% and O = 24.81%, its empirical formula is
(A) CHClO (B) CH_2ClO (C) C_2H_2OCl (D) $ClCH_2O$

3. What is the oxidation number of sulphur in $Na_2S_4O_6$?
(A) $\frac{2}{3}$ (B) $\frac{3}{2}$ (C) $\frac{3}{5}$ (D) $\frac{5}{2}$

4. Which of the following is the most electropositive element ?
(A) Aluminium (B) Magnesium
(C) Phosphorus (D) Sulphur

5. Which of the following informations can be obtained on the basis of Le Chatelier's principle
(A) Equilibrium constant of a chemical reaction
(B) Shift in equilibrium position on changing the value of a constant
(C) Entropy change in a reaction
(D) Dissociation constant of a weak acid

6. The physical adsorption of gases on the solid surface is due to
(A) Hydrogen bonding (B) Covalent bonding
(C) Van der Waalls force (D) All of these

7. Which of the following salts has the same value of Vant Hoff factor as that of $K_3[Fe(CN)_6]_2$?
(A) Na_2SO_4 (B) Fe_3O_4 (C) $Al_2(SO_4)_3$ (D) $Al(NO_3)_3$

8. In an endothermic reaction, the value of change in enthalpy (ΔH) is
(A) Positive (B) Negative (C) Zero (D) Either (b) or (c)

9. Which of the following shows maximum paramagnetic character ?
(A) $[Cr(H_2O)_6]^{3+}$ (B) $[Fe(CN)_6]^{4-}$
(C) $[Fe(CN)_6]^{3-}$ (D) $[Cu(H_2O)_6]^{2+}$

10. Which of the following explains the sequence of filling the electrons in different shells ?
(A) Hund's rule (B) Octet rule
(C) Aufbau principle (D) all of these

11. The specific conductance of a N/10 KCl at 25°C is 0.0112 ohm^{-1} cm^{-1}. The resistance of a cell containing solution at the same temperature was found to be 55 ohms. The cell constant will be
(A) 0.0616 cm^{-1} (B) 0.616 cm^{-1}
(C) 6.16 cm^{-1} (D) 616 cm^{-1}

12. Plank's constant has the same dimesion as that of
(A) Radiant energy (B) Angular momentum
(C) Power (D) Work

13. Which of the following is the most stable alkene ?
(A) $R_2 - C = CR_2$ (B) RCH = CHR
(C) RCH = CH (D) $CH_2 = CH_2$

14. The purification of alumina is carried out by
(A) Castner process (B) Bosch process
(C) Baeyer's process (D) Hoope's process

15. Which of the following substances is used, in the laboratory, for fast drying of neutral gases ?
(A) Sodium phosphate (B) Phosphorus pentoxide
(C) Sodium sulphate (D) Anhydrous calcium chloride

16. The equivalent weight of an acid is equal to
(A) Molecular weight/basicity
(B) Molecular weight x acidity
(C) Molecular weight x basicity
(D) Molecular weight/acidity

17. When the hybridization state of carbon atom changes from sp^3 to sp^2 to sp, the angle between the hybridized orbitals
(A) Increases progressively
(B) Decreases gradually
(C) Decreases considerably (D) All of these

18. The most reactive compound, for the electrophillic nitration, is
(A) Benzene (B) Nitrobenzene
(C) Benzoic acid (D) Toluene

19. Hess law is applicable for the determination of heat of
(A) Reaction (B) Formation
(C) Transition (D) All of these

20. The boiling point of water (100°C) becomes 100.52°C, if 3 grams of a non volatile solute is dissolved in 200 ml of water. The molecular weight of solute is (if K_b for water is = 0.6 K/m)
(A) 12.2 g mol^{-1} (B) 15.4 g. mol^{-1}
(C) 17.3 g. mol^{-1} (D) 20.4 g. mol^{-1}

21. The product obtained when acetic acid is treated with phosphorus trichloride, is
(A) $CH_3 - \overset{\overset{O}{\|}}{C} - O - PCl_2$ (B) $CH_3 - \overset{\overset{O}{\|}}{C} - OCl$
(C) $CH_3 - \overset{\overset{O}{\|}}{C} - Cl$ (D) $CH_3 - CH_2 - \overset{\overset{O}{\|}}{CH_2} - OH$

22. Which of the following is the main cause of late discovery of neutron ?
(A) Neutron is highly ustable particle
(B) Neutron is chargeless particle
(C) Neutron in nucleus moves very fast (D) All of these

23. Which of the following is used to make 'non-stick' cookware ?
(A) Polystyrene (B) Polytetrafluoroethene
(C) Poly-ethylene (D) None of these

24. The reaction,
benzene + methyl halide $\xrightarrow{\text{Anhy. } AlCl_3}$ Toluene, is known as
(A) Kolbe reaction (B) Wurtz reaction

(C) Perkin's reaction (D) Friedal craft reaction

25. Which of the following has been universally accepted as a reference electrode at all temperatures and has been assigned a value of zero volt ?
(A) Graphite electrode (B) Copper electrode
(C) Platinum electrode (D) Standard hydrogen electrode

26. Nitrosoamines ($R_2N-N=O$) are soluble in water. On heating them with concentrated H_2SO_4, they give secondary amines. This reaction is called
(A) Perkin reaction (B) Fittig's reaction
(C) Sandmeyer's reaction
(D) Liebermann niotroso reaction

27. Acetamide is treated separately with the following reagents. Which one of these would give methylamine ?
(A) KOH (B) NaOH
(C) NaOH + Br_2 (D) None of these

28. The ionic product of water at 25°C is 10^{-14}. Its ionic product at 90°C will be
(A) 1 x 10^{-20} (B) 1 x 10^{-12} (C) 1 x 10^{-14} (D) 1 x 10^{-16}

29. Ethanol and dimethyl ether form a pair of functional isomers. The boiling point of ethanol is higher than that of dimethyl ether, due to the presence of
(A) CH_3-group in ethanol
(B) CH_3-group in dimethyl ether
(C) H-bonding in ethanol
(D) H-bonding in dimethyl ether

30. A first order reaction, which is 30% complete in 30 minutes has a half-life period of
(A) 102.2 min (B) 120.2 min
(C) 24.2 min (D) 58.2 min

31. The IUPAC name of the compound,

$$CH_3 - \underset{\underset{OH}{|}}{C} = CH - CH_2 - COOH \quad \text{is}$$

(A) Hydroxy pentenoic acid
(B) 4-hydroxy-3-pentenoic acid
(C) 4-hydroxy-4-pentenoic acid
(D) 4-hydroxy-4-methyl-3-ene-pentanoic acid

32. Baeyer's reagent is used in the laboratory for
(A) Detection of glucose
(B) Oxidation process (C) reduction process
(D) Detection of double bond

33. The reaction of primary amine with chloroform and ethanolic KOH, is called
(A) Carbylamine reaction (B) Kolbe's reaction
(C) Reimer-Tiemann reaction (D) None of these

34. The rate constant of a reaction is 0.69 x 10^{-2} min^{-1} and the initial concentration is 0.2 mol L^{-1}. The half-life period is
(A) 400 sec (B) 600 sec (C) 800 sec (D)-1200 sec

35. One gram mole of a gas at N.T.P. occupies 22.4 litres. This fact was derived from
(A) Avogadro's hypothesis (B) Dalton's atomic theory
(C) Law of gaseous volumes
(D) Law of constant proportions

36. An acyl halide is formed when PCl_5 reacts with an
(A) Amide (B) Ester (C) Acid (D) Alcohol

37. In the reaction, $4Fe + 3O_2 \rightarrow 4Fe^{3+} + 6O_2^{2-}$, which of the following statements is incorrect ?
(A) It is a redox reaction
(B) Fe^{3+} is an oxidizing agent
(C) Metallic iron is a reducing agent
(D) Metallic iron is reduced to Fe^{3+}

38. An equilibrium mixture of the reaction, $2H_2S\ (g) = 2H_2(g) + S_2$, had 0.5 mole H_2S, 0.10 mole H_2 and 0.4 mole S_2 in one litre vessel. the value of K_C is
(A) 0.004 (B) 0.008 (C) 0.016 (D) 0.160

39. Which of the following compounds is 2-bromotoluene?

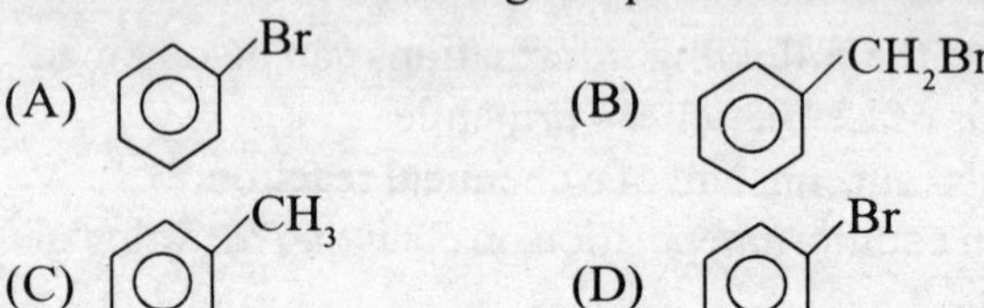

40. At what pressure will a quantity of gas, which occupies 100 ml at a pressure of 720mm, occupy a volume of 84 ml ?
(A) 736.18 mm (B) 784.15 mm
(C) 820.20 mm (D) 857.14 mm

41. Which of the following elements are analogous to the lanthanides ?
(A) Actinides (B) Borides (C) Carbides (D) Hydrides

42. In the gas equation : PV = nRT
(A) V is the volume of one mole of the gas
(B) n is the number of molecules of the gas
(C) n number of moles of the gas have volume V
(D) P is the pressure of one mole of the gas

43. The functional group which is found in amino acid, is
(A) –COOH group (B) $-NH_2$ group
(C) $-CH_3$ group (D) Both (a) and (b)

44. Principle quantum number of an atom is related to the
(A) Size of the orbital (B) Spin angular momentum
(C) Orbital angular momentum
(D) Orientation of the orbital in space

45. A certain current liberated 0.504 g of hydrogen in 2 hours. How many grams of copper can be liberated by the same current flowing for the same time in copper sulphate solution
(A) 12.7 g (B) 16.0 g (C) 31.8 g (D) 63.5 g

46. If enthalpies of fornation for $C_2H_4(g)$, $CO_2(g)$ and H_2O (l) at 25°C and 1 atm pressure are 52, –394 and –286 kJ/mol respectively, the enthalpy of combustion of $C_2H_4(g)$ will be
(A) +14.2 kJ/mol (B) +1412 kJ/mol
(C) –141.2 kJ/mol (D) –1412 kJ/mol

47. A compound with empirical formula CH_2O has a vapour density of 30. Its molecular formula is

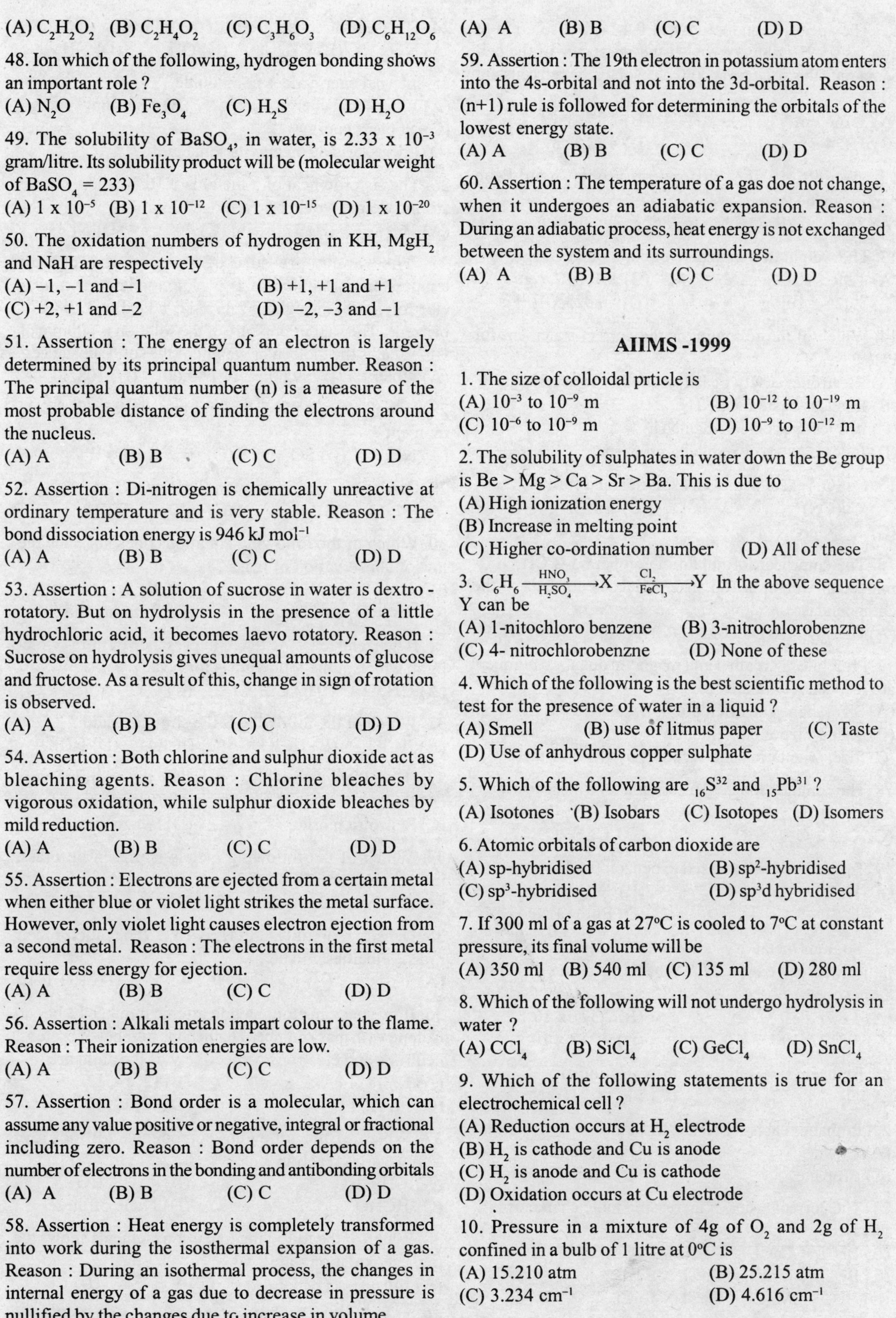

(A) $C_2H_2O_2$ (B) $C_2H_4O_2$ (C) $C_3H_6O_3$ (D) $C_6H_{12}O_6$

48. Ion which of the following, hydrogen bonding shows an important role ?
(A) N_2O (B) Fe_3O_4 (C) H_2S (D) H_2O

49. The solubility of $BaSO_4$, in water, is 2.33 x 10^{-3} gram/litre. Its solubility product will be (molecular weight of $BaSO_4$ = 233)
(A) 1 x 10^{-5} (B) 1 x 10^{-12} (C) 1 x 10^{-15} (D) 1 x 10^{-20}

50. The oxidation numbers of hydrogen in KH, MgH_2 and NaH are respectively
(A) –1, –1 and –1 (B) +1, +1 and +1
(C) +2, +1 and –2 (D) –2, –3 and –1

51. Assertion : The energy of an electron is largely determined by its principal quantum number. Reason : The principal quantum number (n) is a measure of the most probable distance of finding the electrons around the nucleus.
(A) A (B) B (C) C (D) D

52. Assertion : Di-nitrogen is chemically unreactive at ordinary temperature and is very stable. Reason : The bond dissociation energy is 946 kJ mol^{-1}
(A) A (B) B (C) C (D) D

53. Assertion : A solution of sucrose in water is dextro - rotatory. But on hydrolysis in the presence of a little hydrochloric acid, it becomes laevo rotatory. Reason : Sucrose on hydrolysis gives unequal amounts of glucose and fructose. As a result of this, change in sign of rotation is observed.
(A) A (B) B (C) C (D) D

54. Assertion : Both chlorine and sulphur dioxide act as bleaching agents. Reason : Chlorine bleaches by vigorous oxidation, while sulphur dioxide bleaches by mild reduction.
(A) A (B) B (C) C (D) D

55. Assertion : Electrons are ejected from a certain metal when either blue or violet light strikes the metal surface. However, only violet light causes electron ejection from a second metal. Reason : The electrons in the first metal require less energy for ejection.
(A) A (B) B (C) C (D) D

56. Assertion : Alkali metals impart colour to the flame. Reason : Their ionization energies are low.
(A) A (B) B (C) C (D) D

57. Assertion : Bond order is a molecular, which can assume any value positive or negative, integral or fractional including zero. Reason : Bond order depends on the number of electrons in the bonding and antibonding orbitals
(A) A (B) B (C) C (D) D

58. Assertion : Heat energy is completely transformed into work during the isosthermal expansion of a gas. Reason : During an isothermal process, the changes in internal energy of a gas due to decrease in pressure is nullified by the changes due to increase in volume.
(A) A (B) B (C) C (D) D

59. Assertion : The 19th electron in potassium atom enters into the 4s-orbital and not into the 3d-orbital. Reason : (n+1) rule is followed for determining the orbitals of the lowest energy state.
(A) A (B) B (C) C (D) D

60. Assertion : The temperature of a gas doe not change, when it undergoes an adiabatic expansion. Reason : During an adiabatic process, heat energy is not exchanged between the system and its surroundings.
(A) A (B) B (C) C (D) D

AIIMS -1999

1. The size of colloidal prticle is
(A) 10^{-3} to 10^{-9} m (B) 10^{-12} to 10^{-19} m
(C) 10^{-6} to 10^{-9} m (D) 10^{-9} to 10^{-12} m

2. The solubility of sulphates in water down the Be group is Be > Mg > Ca > Sr > Ba. This is due to
(A) High ionization energy
(B) Increase in melting point
(C) Higher co-ordination number (D) All of these

3. $C_6H_6 \xrightarrow[H_2SO_4]{HNO_3} X \xrightarrow[FeCl_3]{Cl_2} Y$ In the above sequence Y can be
(A) 1-nitochloro benzene (B) 3-nitrochlorobenzne
(C) 4- nitrochlorobenzne (D) None of these

4. Which of the following is the best scientific method to test for the presence of water in a liquid ?
(A) Smell (B) use of litmus paper (C) Taste
(D) Use of anhydrous copper sulphate

5. Which of the following are $_{16}S^{32}$ and $_{15}Pb^{31}$?
(A) Isotones (B) Isobars (C) Isotopes (D) Isomers

6. Atomic orbitals of carbon dioxide are
(A) sp-hybridised (B) sp^2-hybridised
(C) sp^3-hybridised (D) sp^3d hybridised

7. If 300 ml of a gas at 27°C is cooled to 7°C at constant pressure, its final volume will be
(A) 350 ml (B) 540 ml (C) 135 ml (D) 280 ml

8. Which of the following will not undergo hydrolysis in water ?
(A) CCl_4 (B) $SiCl_4$ (C) $GeCl_4$ (D) $SnCl_4$

9. Which of the following statements is true for an electrochemical cell ?
(A) Reduction occurs at H_2 electrode
(B) H_2 is cathode and Cu is anode
(C) H_2 is anode and Cu is cathode
(D) Oxidation occurs at Cu electrode

10. Pressure in a mixture of 4g of O_2 and 2g of H_2 confined in a bulb of 1 litre at 0°C is
(A) 15.210 atm (B) 25.215 atm
(C) 3.234 cm^{-1} (D) 4.616 cm^{-1}

11. The specific conductivity of 0.1 N KCl solution at 20°C is 0.0212 ohm^{-1} cm^{-1}. If the resistance of the cell containing this solution at 20°C is 55 ohm, the cell constant is

(A) 1.166 cm^{-1} (B) 2.173 cm^{-1}
(C) 3.324 cm^{-1} (D) 4.616 cm^{-1}

12. Which of the following atoms would be paramagnetic ?

(A) Zn (B) Ca (C) N (D) Be

13. The weight of a molecule of compound $C_{60}H_{122}$ is

(A) 1.09 x 10^{-21} g (B) 1.4 x 10^{-21} g
(C) 5.025 x 10^{23} g (D) 16.023 x 10^{23} g

14. Which of the following is the correct statement for PH_3 ?

(A) Electronegativity of PH_3 is greater than NH_3
(B) It is less basic than NH_3
(C) It is less poisonous than NH_3
(D) It does not show reducing properties

15. $CuSO_4$ reacts with KCN solution and forms

(A) Cu(CN) (B) $Cu(CN)_2$
(C) K_3 $[Cu(CN)_4]$ (D) K_4 $[Cu(CN)_6]$

16. The empirical formula of a compound is CH_2O. Its molecular weight is 180. The molecular formula of the compound is

(A) $C_5H_{10}O_5$ (B) C_4HO_4 (C) $C_3H_6O_3$ (D) $C_6H_{12}O_6$

17. Principle, azimuthal and magnetic quantum numbers are respectively related to

(A) Size, shape and orientation
(B) Shape, size and orientation
(C) Size, orientation and shape (D) None of these

18. The common molecular formula for a disaccharide is

(A) $C_{12}H_{22}O_{11}$ (B) $C_{18}H_{22}O_{11}$
(C) $C_{10}H_{18}O_8$ (D) $C_{10}H_{20}O_{10}$

19. Toluene can be oxidized to benzoic acid by

(A) $KMnO_4$ (B) $K_2Cr_2O_7$
(C) H_2SO_4 (D) Both (a) and (b)

20. Sodium metal crystallizes as a body centred cubic lattice with cell edge 4.29 Å. What is the radius of sodium atom (in cm) ?

(A) 1.857 x 10^{-8} (B) 2.371 x 10^{-7}
(C) 3.817 x 10^{-8} (D) 9.312 x 10^{-7}

21. Beilstein test is used for the detection of

(A) N_2 (B) CO_2 (C) Cl (D) Na

22. Propane reacts with HBr to form

(A) Ethane (B) Hexane
(C) Propane (D) Bromo propane

23. The correct order of increasing ionic character is

(A) $BeCl_2 < BaCl_2 < MgCl_2 < CaCl_2$
(B) $BeCl_2 < CaCl_2 < MgCl_2 < BaCl_2$
(C) $BeCl_2 < MgCl_2 < CaCl_2 < BaCl_2$
(D) $BeCl_2 < MgCl_2 < BaCl_2 < CaCl_2$

24. The meta-directing group of the following is

(A) NO_2 (B) CH_3 (C) OH (D) NH_2

25. Internal energy does not include

(A) Vibrational energy (B) Rotational energy
(C) Nuclear energy
(D) Energy arising by gravitational pull

26. The specific heat of a metal is 0.16 Its approximate atomic weight would be

(A) 16 (B) 32 (C) 40 (D) 64

27. The vapour pressure of benzene at a certain temperature is 640mm of Hg. A non-volatile and non-electrolyte solid weighing 2.175 g is added to 39.08 g of benzene. The vapour pressure of the solution is 600mm of Hg. What is the molecular weight of the solid substance ?

(A) 49.59 (B) 59.60 (C) 69.60 (D) 108.30

28. Which among the following has the largest dipole moment ?

(A) NH_3 (B) SO_3 (C) HI (D) H_2O

29. Oxidation state of osmium (Os) in OsO_4 is

(A) +4 (B) +6 (C) +7 (D) +8

30. Which of the following equations does not obey the law of conservation of mass ?

(A) $H_2 + O \rightarrow 2H_2O$ (B) $2H + O_2 \rightarrow H_2O$
(C) $2H + O_2 \rightarrow 2H_2O$ (D) Both (a) and (b)

31. If benzene reacts with Cl_2 in presence of ultra- violet light, which of the following is formed ?

(A) CH_5Cl (B) $C_6H_6Cl_6$ (C) C_6Cl_6 (D) CCl_4

32. Which of the following is the strongest acid ?

(A) HClO (B) $HClO_2$ (C) $HClO_3$ (D) $HClO_4$

33. Benzaldehyde can be prepared by the hydrolysis of

(A) Benzyl chloride (B) Benzal chloride
(C) Benzotrichloride (D) Benzo nitrite

34. Which of the following metals is present in brass, bronze and german-silver ?

(A) Cu (B) Al (C) Na (D) Mg

35. Density ratio of O_2 and H_2 is 16 :1. The ratio of their r.m.s. velocities will be

(A) 1 : 4 (B) 1 : 16 (C) 4 : 1 (D) 16 : 1

36. If magnesium atom having atomic number 12 has an isotope with mass number 26, then the number of protons, neutrons and electrons respectively of the atom are

(A) 12, 12, 14 (B) 12, 14, 12
(C) 14, 12, 12 (D) 12, 12, 12

37. Which of the following compounds will not give Cannizzaro's reaction ?

(A) CH_3CHO (B) C_6H_5CHO
(C) HCHO (D) None of these

38. Which of the following transition elements shows the highest oxidation state ?

(A) Mn (B) Fe (C) V (D) Cr

39. The equivalent weight of phosphoric acid (H_3PO_4) in

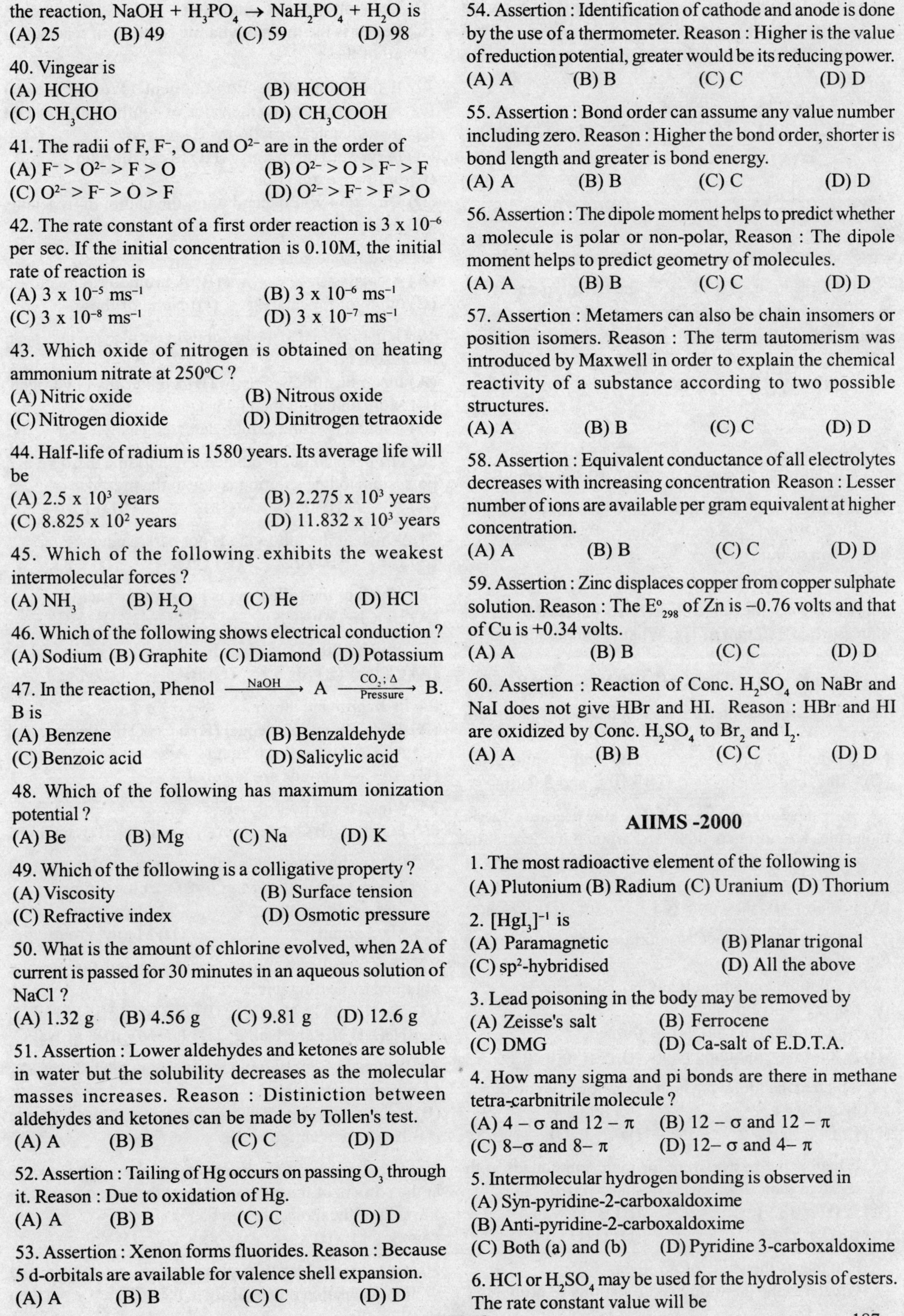

the reaction, $NaOH + H_3PO_4 \rightarrow NaH_2PO_4 + H_2O$ is
(A) 25 (B) 49 (C) 59 (D) 98

40. Vingear is
(A) HCHO (B) HCOOH
(C) CH_3CHO (D) CH_3COOH

41. The radii of F, F^-, O and O^{2-} are in the order of
(A) $F^- > O^{2-} > F > O$ (B) $O^{2-} > O > F^- > F$
(C) $O^{2-} > F^- > O > F$ (D) $O^{2-} > F^- > F > O$

42. The rate constant of a first order reaction is 3 x 10^{-6} per sec. If the initial concentration is 0.10M, the initial rate of reaction is
(A) 3 x 10^{-5} ms^{-1} (B) 3 x 10^{-6} ms^{-1}
(C) 3 x 10^{-8} ms^{-1} (D) 3 x 10^{-7} ms^{-1}

43. Which oxide of nitrogen is obtained on heating ammonium nitrate at 250°C ?
(A) Nitric oxide (B) Nitrous oxide
(C) Nitrogen dioxide (D) Dinitrogen tetraoxide

44. Half-life of radium is 1580 years. Its average life will be
(A) 2.5 x 10^3 years (B) 2.275 x 10^3 years
(C) 8.825 x 10^2 years (D) 11.832 x 10^3 years

45. Which of the following exhibits the weakest intermolecular forces ?
(A) NH_3 (B) H_2O (C) He (D) HCl

46. Which of the following shows electrical conduction ?
(A) Sodium (B) Graphite (C) Diamond (D) Potassium

47. In the reaction, Phenol $\xrightarrow{\text{NaOH}}$ A $\xrightarrow[\text{Pressure}]{CO_2;\ \Delta}$ B. B is
(A) Benzene (B) Benzaldehyde
(C) Benzoic acid (D) Salicylic acid

48. Which of the following has maximum ionization potential ?
(A) Be (B) Mg (C) Na (D) K

49. Which of the following is a colligative property ?
(A) Viscosity (B) Surface tension
(C) Refractive index (D) Osmotic pressure

50. What is the amount of chlorine evolved, when 2A of current is passed for 30 minutes in an aqueous solution of NaCl ?
(A) 1.32 g (B) 4.56 g (C) 9.81 g (D) 12.6 g

51. Assertion : Lower aldehydes and ketones are soluble in water but the solubility decreases as the molecular masses increases. Reason : Distiniction between aldehydes and ketones can be made by Tollen's test.
(A) A (B) B (C) C (D) D

52. Assertion : Tailing of Hg occurs on passing O_3 through it. Reason : Due to oxidation of Hg.
(A) A (B) B (C) C (D) D

53. Assertion : Xenon forms fluorides. Reason : Because 5 d-orbitals are available for valence shell expansion.
(A) A (B) B (C) C (D) D

54. Assertion : Identification of cathode and anode is done by the use of a thermometer. Reason : Higher is the value of reduction potential, greater would be its reducing power.
(A) A (B) B (C) C (D) D

55. Assertion : Bond order can assume any value number including zero. Reason : Higher the bond order, shorter is bond length and greater is bond energy.
(A) A (B) B (C) C (D) D

56. Assertion : The dipole moment helps to predict whether a molecule is polar or non-polar, Reason : The dipole moment helps to predict geometry of molecules.
(A) A (B) B (C) C (D) D

57. Assertion : Metamers can also be chain insomers or position isomers. Reason : The term tautomerism was introduced by Maxwell in order to explain the chemical reactivity of a substance according to two possible structures.
(A) A (B) B (C) C (D) D

58. Assertion : Equivalent conductance of all electrolytes decreases with increasing concentration Reason : Lesser number of ions are available per gram equivalent at higher concentration.
(A) A (B) B (C) C (D) D

59. Assertion : Zinc displaces copper from copper sulphate solution. Reason : The E^o_{298} of Zn is –0.76 volts and that of Cu is +0.34 volts.
(A) A (B) B (C) C (D) D

60. Assertion : Reaction of Conc. H_2SO_4 on NaBr and NaI does not give HBr and HI. Reason : HBr and HI are oxidized by Conc. H_2SO_4 to Br_2 and I_2.
(A) A (B) B (C) C (D) D

AIIMS -2000

1. The most radioactive element of the following is
(A) Plutonium (B) Radium (C) Uranium (D) Thorium

2. $[HgI_3]^{-1}$ is
(A) Paramagnetic (B) Planar trigonal
(C) sp^2-hybridised (D) All the above

3. Lead poisoning in the body may be removed by
(A) Zeisse's salt (B) Ferrocene
(C) DMG (D) Ca-salt of E.D.T.A.

4. How many sigma and pi bonds are there in methane tetra-carbnitrile molecule ?
(A) 4 – σ and 12 – π (B) 12 – σ and 12 – π
(C) 8–σ and 8– π (D) 12– σ and 4– π

5. Intermolecular hydrogen bonding is observed in
(A) Syn-pyridine-2-carboxaldoxime
(B) Anti-pyridine-2-carboxaldoxime
(C) Both (a) and (b) (D) Pyridine 3-carboxaldoxime

6. HCl or H_2SO_4 may be used for the hydrolysis of esters. The rate constant value will be

(A) Same for both acids (B) Greater if HCl is used
(C) Grater if H_2SO_4 is used
(D) Will depend upon the nature of ester

7. $CH_3 - COCl + H_2 \xrightarrow{[Pd]} X$. X is
(A) $CH_3 - CHO$ (B) $CH_3 - CH_2 - OH$
(C) $CH_3 - C(OH)(H) - Cl$ (D) Both (a) and (b)

8. In the reaction, A + B → C, the potential energy as function of reaction coordinate is shown in the figure. What is the value of activation energy E_a ?

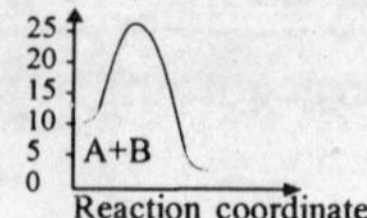

(A) 20 K cal/mole (B) 15 K cal/mole
(C) 5 K cal/mole (D) 10 K cal/mole

9. Sodium chloride crystalises in the face centered cubic structure. What is the total number of sodium ions per unit cell in sodium chloride ?
(A) 2 (B) 4 (C) 6 (D) 8

10. A 14.0 g sample of nitrogen (N_2) occupies 8.0 litres at a pressure of 380 mm of Hg. What is temperature of N_2 under these conditions ?
(A) 100.0 K (B) 297.6 K (C) 197.6 K (D) 97.6 K

11. For a gas, deviation from ideal behaviour is maximum at
(A) 0°C and 1.0 atm (B) 100° and 2.0 atm
(C) –10°C and 2.0 atm (D) –10°C and 5.0 atm

12. In a gaseous mixture at 3.0 atm pressure, 25% molecules are nitrogen, 30% are carbon dioxide and the remaining are oxygen. What is the partial pressure of oxygen ?
(A) 0.75atm (B) 0.90 atm (C) 1.20 atm (D) 1.35 atm

13. When an excess of KCN solution is added to a copper salt solution
(A) A solution containing $[Cu(CN)_2]^-$ is formed
(B) Copper is precipitated
(C) A precipitate of $Cu(CN)_2$ is formed
(D) A solution containing $[Cu(CN)_4]^{2-}$ is formed

14. Which of the following is an example of a π complex ?
(A) $[Cu(NH_3)_4]^{2+}$ (B) $[NiCl_4]^{2-}$
(C) $CoCl_2 . 2NH_3$ (D) $Ni(CO)_4$

15. Which is the best solvent for ionic compounds with reference to their di-electric constants (D) at 25°C
(A) A (D = 78.5) (B) B (D = 24.3)
(C) C (D = 32.6) (D) D (D = 20.7)

16. If higher is the value of K_C, then
(A) Greater is the thermodynamic stability of products
(B) Reaction goes to completion
(C) Lesser is the thermodynamic stability of reactants
(D) all of these

17. If the value of reaction quotient 'Q' of a chemical reaction is greater than the value of equilibrium constant K_C, then chemical reaction will proceed
(A) In forward direction (B) In equilibrium
(C) In reverse direction
(D) Direction will depend upon the nature of reactants and products.

18. Co-enzyme acts as
(A) A catalyst (B) A promoter for catalyst
(C) Inhibitor for catalyst (D) None of these

19. Helium is used in the balloons in the place of hydrogen because it is
(A) Incombustible (B) Lighter than hydrogen
(C) More abundant than hydrogen
(D) Radioactive and can be detected easily

20. The deep colour produced when iodine dissolve in a potassium iodide solution is due to the presence of
(A) I^+ (B) I^- (C) I_2^- (D) I_3^-

21. Which of the following is not paramagnetic ?
(A) ClF_3 (B) BrO_2 (C) ClO_2 (D) ClO_3

22. Which of the following is the strongest acid ?
(A) HOCl (B) $HClO_2$ (C) $HClO_3$ (D) $HClO_4$

23. Which of the following doesn't act as Lewis acid ?
(A) CCl_4 (B) SF_3 (C) BF_3 (D) $SnCl_4$

24. In dichromate dimer
(A) 4 Cr–O bonds are equal (B) 6 Cr–O bonds are equal
(C) All Cr–O bonds are equal
(D) All Cr–O bonds are unequal

25. Which is not hydrolysed ?
(A) PCl_3 (B) NCl_3 (C) $AsCl_3$ (D) $SnCl_3$

26. CO_2 is liberated on adding sodium carbonate to a carboxylic acid. The carbon of CO_2 comes from
(A) Carboxylic group (B) Carbonate
(C) Alkyl group (D) Methyl group

27. The formula of microcosmic salt and the product obtained by heating it is
(A) $Na(NH_4)PO_4$; $NaPO_3$ (B) $Na(NH_4)_2HPO_4$; $NaPO_3$
(C) $Na(NH_4)HPO_4$; $NaPO_4$ (D) $Na(NH_4)HPO_4$; $NaPO_3$

28. Which of the following is true about ClO_2 ?
(A) It is paramagnetic
(B) It dimerises in liquid phase (C) It is an angular molecule
(D) It is very reactive

29. A molecule contains atoms x and y so that x occurs at the corners of the cube while y at the face centre. The formula of the molecule can be
(A) xy_3 (B) x_3y (C) xy_2 (D) x_2y

30. A gas z is bubbled through a solution containing x^- and y^-. If the reduction potential are in the order x > y > z, then

(A) y will oxidize both x and z
(B) y will reduce both x and z
(C) y will oxidize z and not x
(D) y will oxidize x and not z

31. Size of the iodine species increases in the order
(A) $I^+ < I < I^-$ (B) $I^+ < I^- < I$
(C) $I^- < I^+ < I^-$ (D) $I < I^- < I^+$

32. Which of the following molecules has a tetrahedral shape ?
(A) Hg_2Cl_2 (B) CO_3^{-2} (C) NH_4^+ (D) $Ni(CN_4)^{2-}$

33. Fluorine molecule is formed by the overlap of
(A) p–p orbitals by side-to-side manner
(B) p–p orbitals by end-to-end manner
(C) s–p orbitals (D) s–s orbitals

34. Polarizability of halide ions increases in the order
(A) F^-, I^-, Br^-, Cl^- (B) Cl^-, Br^-, I^-, F^-
(C) I^-, Br^-, Cl^-, F^- (D) F^-, Cl^-, Br^-, I^-

35. Which of the following cannot exist on the basis of Molecular Orbital theory ?
(A) He (B) C_2 (C) H_2^+ (D) He_2^+

36. Which of the following has the largest size ?
(A) Br^- (B) I^- (C) I (D) I^+

37. How many grams of sucrose should be dissolved in 500 gms of water so as to get a solution which has a difference of 104°C between boiling point and freezing point ($K_f = 1.86$ K kg mol^{-1}, $K_b = 0.52$ kg mol^{-1})
(A) 1.68 (B) 3.36 (C) 8.40 (D) 0.840

38. Which of the following has a tendency for covalent compound formation ?
(A) Ca (B) Ba (C) Na (D) Mg

39. 12 g of an alkaline earth metal gives 14.8 gm of its nitride. Atomic weight of the metal is
(A) 12 (B) 20 (C) 40 (D) 14.8

40. $CuSO_4$ dissolves in NH_3 due to the formation of
(A) $Cu(OH)_2$ (B) $Cu[(NH_3)_4]SO_4$
(C) $Cu[(NH_3)_4(OH)_2$ (D) CuO

41. Which of the following is paramagnetic ?
(A) NO_2 (B) N_2O (C) N_2O_3 (D) N_2O_5

42. Oxidation state of sulphur in S_8, H_2S and S_2F_2 is
(A) 0, –2, –1 (B) 1, + 2, –1
(C) 1, –2, +1 (D) 0, –2, +1

43. Assertion : Bond order can have positive, negative or fractional value. Reason : bond order depends upon the bonding and antibonding orbital electrons.
(A) A (B) B (C) C (D) D

44. Assertion : Formic acid reduces mercuric chloride to mercurous chloride on heating, while acetic acid does not. Reason : Formic acid is stronger than acetic acid.
(A) A (B) B (C) C (D) D

45. How many unpaired electrons are present in an atom of configuration $1s^2\ 2s^2\ 2p^4$?
(A) 1 (B) 2 (C) 5 (D) 7

46. When the value of azimuthal quantum number is 3, magnetic quantum number can have values
(A) +3, +2, +1, 0, –1, –2, –3 (B) +2, +1, 0, –1, –2,
(C) +1, 0, –1 (D) +1, –1

47. Spectrum given due to transition of an electron form M to L shell is
(A) Absorption (B) Continuous
(C) Emission (D) γ - ray

48. Which of the following has no neutron in its nucleus ?
(A) Tritium (B) Hydrogen (C) Helium (D) Deuterium

49. Which of the following relations is called Heisenberg's uncertainty principle ?
(A) $\Delta p \Delta x \geq \frac{h}{4\pi}$ (B) $\Delta u \Delta x \geq \frac{h}{4\pi m}$
(C) $\Delta E \Delta t \geq \frac{h}{4\pi}$ (D) All of these

50. Hund's rule of maximum multiplicity explains
(A) Arrangement of electrons in orbitals
(B) Filling of electrons in orbitals
(C) Filling of electrons in orbitals of atom
(D) Pairing of electrons in shells

51. The orbital angular momentum for s-orbital may be
(A) 1 (B) zero (C) $\frac{\sqrt{2}h}{2\pi}$ (D) all of these

52. Which relation is called Bragg's equation ?
(A) $mvr = \frac{nh}{2\pi}$ (B) $n\lambda = 2d \sin\theta$ (C) $\lambda = \frac{h}{mv}$ (D) none

53. The boiling point of amines and corresponding alcohols and acids follow the order
(A) $RCOOH < RCH_2NH_2 < RCH_2OH$
(B) $RCH_2NH_2 < RCH_2OH < RCOOH$
(C) $RCH_2OH < RCOOH < RCH_2NH_2$
(D) $RCOOH < RCH_2OH < RCH_2NH_2$

54. Which of the following is paraldehyde ?
(A) [six-membered ring with three O atoms] (B) [six-membered ring with three O atoms bearing H_3C, CH_3, CH_3]
(C) A 40% HCHO solution (D) None of these

55. Identify X in the reaction,
$CH_2 = CH_2 + H_2SO_4 \rightarrow$ intermediate $\xrightarrow[\text{boil}]{H_2O}$ X
(A) CH_3OH (B) CH_3CH_2OH
(C) CH_3COCH_3 (D) CH_3OCH_3

56. Identify X in the reaction,
$HCHO + CH_3MgI \xrightarrow{\text{dry ether}}$ Intermediate $\xrightarrow{H_2O / H^+}$
(A) CH_3COCH_3 (B) CH_3OCH_3
(C) CH_3OH (D) CH_3CH_2OH

57. A gas X decolourises bromine-water and also Bayer's reagent, It burns in air with sooty flame. When it reacts with HCl, it forms finally ethyledene chloride. X is
(A) C_2H_6 (B) C_3H_4 (C) C_2H_4 (D) C_2H_2

58. Ammonical solution of cuprous chloride gives red precipitate with

(A) $H-C \equiv C-CH_3$ (B) $H-C = C-H$ (with H below each C)

(C) $(C_2H_5)_2-C = CH_2$ (D) $CH_3-C \equiv C-C_2H_5$

59. For the given molecular formula, $C_4H_{10}O$, how many isomers are possible ?

(A) 4 (B) 7 (C) 8 (D) 10

60. The compound that shows optical isomerism is

(A) $CH_3-CH(Cl)-CH_3$ (B) 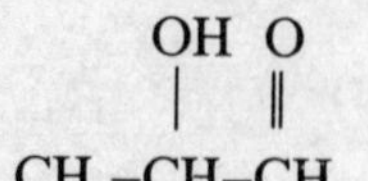 (cyclohexane with CH_2OH and OH)

(C) $CH_3-CH(OH)-CH=O$ (D) $CH_3-CH_2-CH(OH)-CH_2CH_3$

AIIMS - 2001

1. Which α-amino acid can cross link peptide chains ?

(A) Glutamine (B) Tyrosine (C) Serine (D) Cysteine

2. The movement of colloidal particles, under applied electric field, is known as

(A) Dialysis (B) Electrophoresis
(C) Electrodialysis (D) None of these

3. Which of the following proteins destroys the antigen when it enters in body cell ?

(A) Chromoprotein (B) Phosphoprotein
(C) Antibodies (D) Insulin

4. $_{88}Ra^{226}$ is

(A) Lanthanide (B) Rare gas
(C) Radioactive (D) non-radioactive

5. The charge on $[Ag(CN)_2]^-$ complex is

(A) –1 (B) +1 (C) +2 (D) +3

6. Which of the following hydroxides is insoluble in water ?

(A) $Be(OH)_2$ (B) $Mg(OH)_2$ (C) $Ca(OH)_2$ (D) $Ba(OH)_2$

7. The electronic configuration of transition elements is exhibited by

(A) ns^1 (B) ns^2np^5 (C) $ns^2(n-1)d^{10}$ (D) $(n-1)d^{1-10}\ ns^{0-2}$

8. Which of the following statements is correct regarding the drawbacks of raw rubber ?

(A) It is plastic in nature (B) It has little durability
(C) It has large water-absorption capacity
(D) All of these

9. Which of the following is used as a moderator in nuclear reactor ?

(A) H_2O (B) NaOH (C) D_2O (D) N_2O

10. The hardness of water is measured by ?

(A) E.D.T.A. method (B) Distribution method
(C) Conductivity method (D) All of these

11. The mass of copper deposited from a solution of $CuSO_4$ by passage of 5 A current for 965 seconds is (Mol. wt. of Copper = 63.5)

(A) 15.875 g (B) 1.5875 g (C) 4825 g (D) 96500 g

12. Which of the following will not give Cannizzaro's reaction ?

(A) HCHO (B) CCl_3CHO (C) C_6H_5CHO (D) CH_3CHO

13. In KHX_2, X is halogen of the following acid

(A) HBr (B) HI (C) HF (D) HCl

14. Which of the following is explosive ?

(A) KN_3 (B) AgN_3 (C) Ba_3N_2 (D) NaN_3

15. Increasing order of base strength is SH^-, I , OH^-, NH_2^- is (proton affinity order)

(A) $SH^- < I^- < OH^- < NH_2^-$
(B) $I^- < SH^- < OH^- < NH_2^-$
(C) $NH_2^- < OH^- < SH^- < I^-$ (D) None of these

16. Given for $M \rightarrow M^{+2} + 2e^-$, $E^o_{OP} = 0.44V$; $M^{+2} \rightarrow M^{+3} + e^-$, $E^o_{OP} = -0.77$ V. Determine the value of E^o_{OP} for $M \rightarrow M^{+3} + 3e^-$,

(A) +0.33 V (B) 0.036 V (C) –0.33 V (D) 1.21 V

17. Which carbocation is more stable?

(A) $\overset{\oplus}{C}H_3$ (B) $CH_3\overset{\oplus}{C}H_2$ (C) $(CH_3)_2\overset{\oplus}{C}H$ (D) $(CH_3)_3\overset{\oplus}{C}$

18. Which is strongest acid among the following ?

(A) $CH_2(COOH)_2$ (B) $HO_2C-(CH_2)_2-CO_2H$

(C) $H-C-COOH \ \|\ H-C-COOH$ (D) $H-C-COOH \ \|\ HOOC-C-H$

19. Which ion shows aromatic character ?

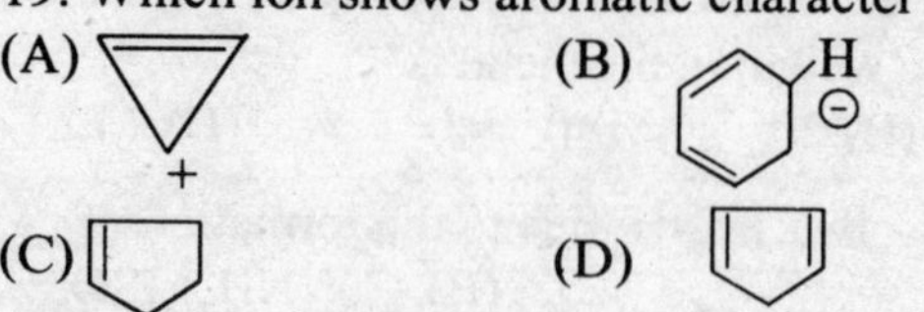

20. Given, $\frac{_{92}U^{238}}{_{82}Pb^{206}} = \frac{4}{1}$,half life of $_{92}U^{238} = 1.35x10^9$, then the age of crystal will be

(A) 4.2 x 10^8 (B) 4.2 x 10^9 (C) 4.2 x 10^{10} (D) 4.2 x 10^{11}

21. In NaCl if $r_{Na^+} = 100$ pm, maximum size of r_{cl^-} will be

(A) 241.5 pm (B) 261.5 pm (C) 251.55 pm (D) 271.5 pm

22. Pthalic anhydride on treatment with ammonia forms

(A) Pthalimide (B) Benzaldehyde
(C) Pthalic acid (D) Benzoic acid

23. Which of the following is hardest ?

(A) Boron nitirde (B) Boron carbide
(C) Silicon boride (D) Aluminium boride

24. In which $p\pi$-$d\pi$ overlapping occurs

(A) O_2 (B) N_2 (C) CO_3^{-2} (D) PO_4^{-3}

25. In which of the pair same structure is not possible ?

(A) SF_4, SIF_4 (B) $BeCl_2$, C_2H_2
(C) BF_3, SO_3 (D) H_2O, H_2S

26. In diborane formation of which intermediate is not possible
(A) $B_2H_3(CH_3)_3$ (B) $B_2H_4(CH_3)_2$
(C) $B_2H(CH_3)_5$ (D) $B_2H_2(CH_3)_4$

27. Which compound is not possible ?
(A) $[Be(OH)_6]^{-4}$ (B) $[Sn(OH)_6]^{-4}$ (C) SF_6 (D) IF_7

28. Hybridisation respectively in CH_4, $\overset{+}{C}H_3$, $\dot{C}H_3$ are
(A) sp^3, sp^2, sp^2 (B) sp^3, sp^3, sp^3
(C) sp^2, sp^2, sp^2 (D) sp^2, sp^3, sp^2

29. Correct order of bond energy is
(A) $He_2^+ < NO < O_2^+ < C_2^{-2}$
(B) $C_2^{-2} < NO < O_2^+ < He_2^+$
(C) $C_2^{-2} < NO < He_2^+ < O_2^+$ (D) None of these

30. Which one has minimum bond length ?
(A) O_2^+ (B) O_2 (C) O_2^- (D) F_2

31. Which of the sets is correctly matched ?
(A) n = 5, l = 2, m = 2, s = –1/2
(B) n = 5, l = 5, m = 2, s = +1/2
(C) n = 3, l = 2, m = 2, s = 0
(D) n = 3, l = 2, m = 3, s = +1/2

32. Which alkene on ozonolysis gives the following compounds, (i) methanal (ii) Ethanal (iii) 2-oxo propanal?
(A) 3 - methyl - 1,3 - hexadiene
(B) 3 - methyl - 1,2 - butadiene
(C) 3 - methyl - 1,3 - pentadiene
(D) 2 - methyl - 1,3 - hexadiene

33. What is the correct order of base strength ?
(A) $NH_3 > NH_2 - NH_2 > NH_2OH > N_3H$
(B) $NH_2OH > N_3H > NH_3 > NH_2 - NH_2$
(C) $NH_2 - NH_2 > NH_3 > NH_2OH > N_3H$
(D) None of these

34. Strongest conjugate base of the following is
(A) ClO_3^- (B) ClO_4^- (C) ClO^- (D) ClO_2^-

35. If for ammonium formate pK_a = 3.78, pK_b = 4.78, pH will be equal to
(A) 6.9 (B) 6.5 (C) 7.3 (D) 6.2

36. $2SO_3\ (g) \rightleftharpoons 2SO_2\ (g) + O_2\ (g)$. If at 700 K, $K_p = 1.3 \times 10^{-3}\ kP_a$, then K_c is
(A) 2.33×10^{-5} mole lit^{-1} (B) 2.33×10^{-4} mole lit^{-1}
(C) 2.33×10^{-7} mole lit^{-1} (D) 2.33×10^{-6} mole lit^{-1}

37. In crystalline solid, which is true ?
(A) Irregular arrangment
(B) With irregular arrangement, some regularity
(C) Regular arrangement
(D) with regular arrangement some irregularity

38. C (Graphite) + $O_2 \rightarrow CO_2$, ΔH = –a, $2CO + O_2 \rightarrow 2CO_2$, ΔH = –b. Determine heat of formation of CO
(A) $\frac{b-a}{2}$ (B) $\frac{b-2a}{2}$ (C) $\frac{a-b}{2}$ (D) $\frac{a-2b}{2}$

39. Which of the following will be hydrolysed fastest with alc. KOH ?
(A) $CH_3CH_2CONHCH_3$ (B) $(CH_3CO)_2O$
(C) $CH_3COOC_2H_5$ (D) C_2H_5CN

40. $CH_3CH_2CO_2H \xrightarrow{\text{P and } Br_2} [A] \xrightarrow{\text{(i) alc KOH (ii) } H^+} [B]$; [B] is
(A) $CH_3CH_2CO_2H$ (B) CH_3CH_2CHO
(C) $CH_2 = CH - CHO$ (D) $CH_2 = CH - CO_2H$

41. In colloidal solution, the movement of dispersion medium under the influence of electric field is called
(A) Electro- osmosis (B) Cataphoresis
(C) Coagulation (D) Brownian movement

42. Freundlich equation is
(A) $\log \frac{x}{m} = \log K + \frac{1}{n} \log P$
(B) $\log \frac{x}{m} = \log K - \frac{1}{n} \log P$
(C) $\log \frac{x}{m} = \log K + \frac{1}{n} \log P$
(D) $\log \frac{x}{m} = \log K + \frac{1}{P} \log n$

43. In the reaction of benzene with conc. $HNO_3.H_2SO_4$ which is velocity determining step ?
(A) $\overset{\oplus}{N}O_2$ formation (B) Carbocation formation
(C) Removal of $H^{\oplus}$ (D) All of these

44. In which of the spcies dsp^2 hybridisation occurs ?
(A) $[CuCl_4]^{-2}$ (B) $[PtCl_4]^{-2}$
(C) $[NiCl_4]^{-2}$ (D) $[CoCl_4]^{-2}$

45. Which of the complex compounds forms a maximum number of isomers ?
(A) $[Ru(NH_3)_4Cl_2]^{+2}$ (B) $[Co(en)_2Cl_2]^{+2}$
(C) $[Cu(NH_3)_4]^{+2}$ (D) $[PtCl_4]^{-2}$

46. Amongst the compounds, which is least soluble ?
(A) SrF_2 (B) BaF_2 (C) MgF_2 (D) CaF_2

47. In the presence of reductant Na / liq. NH_3, an alkyne, forms .... alkene
(A) Trans (B) Cis (C) Both of these (D) none of these

48. Fe^{+3} and Sn^{2+} cannot stay together because
(A) Fe^{+3} reduces into Fe^{+2} and Sn^{+2} oxidises into Sn^{+4}
(B) No reaction occurs (C)Precipitation occurs
(D) Fe^{+3} get oxidizes and Sn^{+2} gets reduced

49. If energy of the e^- in H-atom in some excited is –3.4 eV, what will be its angular momentum
(A) 1.8×10^{-30} Kgm^2s^{-1} (B) 2.1×10^{-34} Kgm^2s^{-1}
(C) 9.2×10^{-30} Kgm^2s^{-1} (D) 1.2×10^{-32} Kgm^2s^{-1}

50. A dye absorbs strongly in the region, 605-750 nm. The colour of the reflected light will be
(A) Blue – green (B) Blue – red
(C) Red – green (D) Green – blue

51. Yeast cells derive their energy from glucose by
(A) Glycolysis (B) Respiration
(C) fermentation (D) None of these

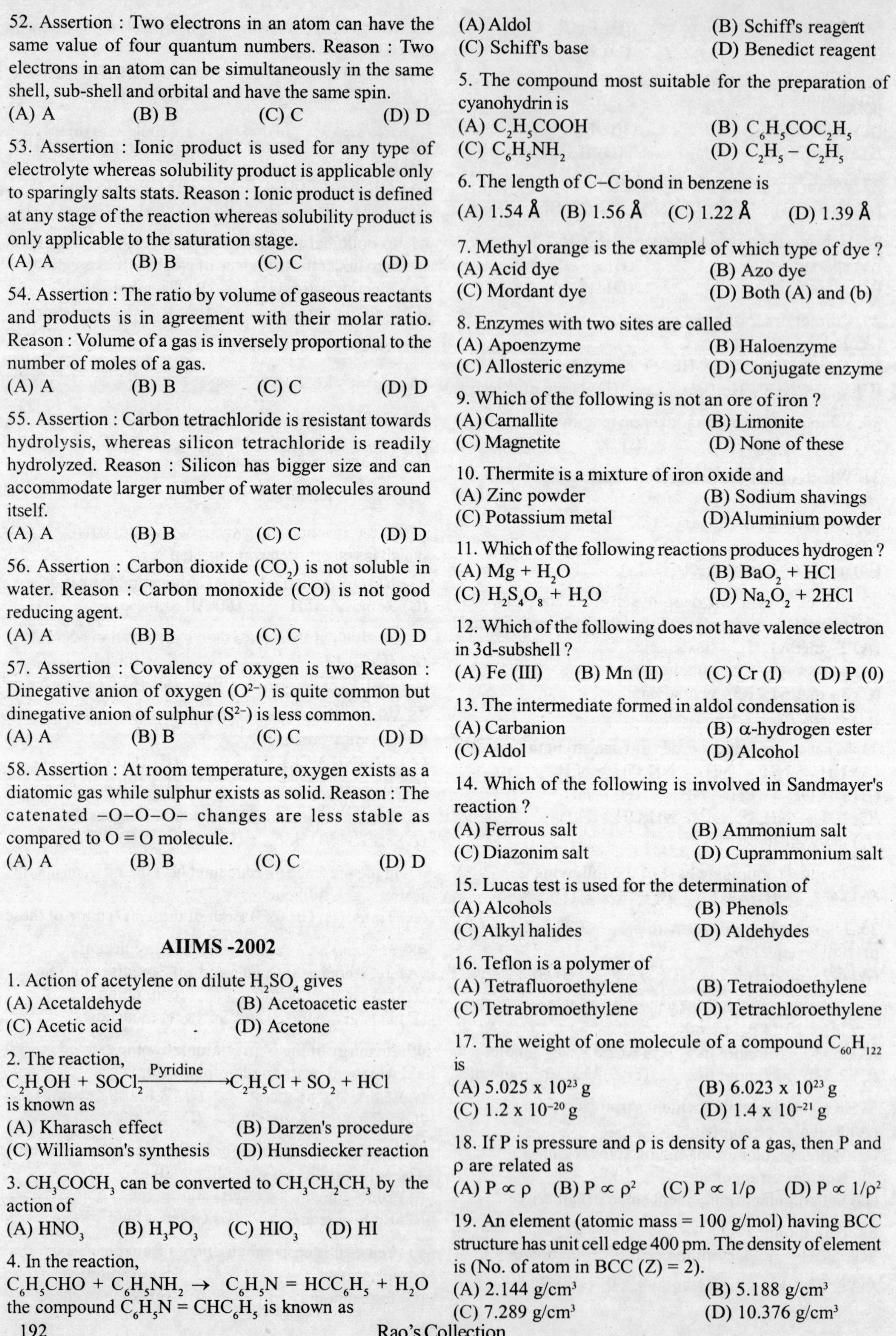

52. Assertion : Two electrons in an atom can have the same value of four quantum numbers. Reason : Two electrons in an atom can be simultaneously in the same shell, sub-shell and orbital and have the same spin.
(A) A (B) B (C) C (D) D

53. Assertion : Ionic product is used for any type of electrolyte whereas solubility product is applicable only to sparingly salts stats. Reason : Ionic product is defined at any stage of the reaction whereas solubility product is only applicable to the saturation stage.
(A) A (B) B (C) C (D) D

54. Assertion : The ratio by volume of gaseous reactants and products is in agreement with their molar ratio. Reason : Volume of a gas is inversely proportional to the number of moles of a gas.
(A) A (B) B (C) C (D) D

55. Assertion : Carbon tetrachloride is resistant towards hydrolysis, whereas silicon tetrachloride is readily hydrolyzed. Reason : Silicon has bigger size and can accommodate larger number of water molecules around itself.
(A) A (B) B (C) C (D) D

56. Assertion : Carbon dioxide (CO_2) is not soluble in water. Reason : Carbon monoxide (CO) is not good reducing agent.
(A) A (B) B (C) C (D) D

57. Assertion : Covalency of oxygen is two Reason : Dinegative anion of oxygen (O^{2-}) is quite common but dinegative anion of sulphur (S^{2-}) is less common.
(A) A (B) B (C) C (D) D

58. Assertion : At room temperature, oxygen exists as a diatomic gas while sulphur exists as solid. Reason : The catenated –O–O–O– changes are less stable as compared to O = O molecule.
(A) A (B) B (C) C (D) D

AIIMS -2002

1. Action of acetylene on dilute H_2SO_4 gives
(A) Acetaldehyde (B) Acetoacetic easter
(C) Acetic acid (D) Acetone

2. The reaction,
$C_2H_5OH + SOCl_2 \xrightarrow{\text{Pyridine}} C_2H_5Cl + SO_2 + HCl$
is known as
(A) Kharasch effect (B) Darzen's procedure
(C) Williamson's synthesis (D) Hunsdiecker reaction

3. CH_3COCH_3 can be converted to $CH_3CH_2CH_3$ by the action of
(A) HNO_3 (B) H_3PO_3 (C) HIO_3 (D) HI

4. In the reaction,
$C_6H_5CHO + C_6H_5NH_2 \rightarrow C_6H_5N = HCC_6H_5 + H_2O$
the compound $C_6H_5N = CHC_6H_5$ is known as
(A) Aldol (B) Schiff's reagent
(C) Schiff's base (D) Benedict reagent

5. The compound most suitable for the preparation of cyanohydrin is
(A) C_2H_5COOH (B) $C_6H_5COC_2H_5$
(C) $C_6H_5NH_2$ (D) $C_2H_5 - C_2H_5$

6. The length of C–C bond in benzene is
(A) 1.54 Å (B) 1.56 Å (C) 1.22 Å (D) 1.39 Å

7. Methyl orange is the example of which type of dye ?
(A) Acid dye (B) Azo dye
(C) Mordant dye (D) Both (A) and (b)

8. Enzymes with two sites are called
(A) Apoenzyme (B) Haloenzyme
(C) Allosteric enzyme (D) Conjugate enzyme

9. Which of the following is not an ore of iron ?
(A) Carnallite (B) Limonite
(C) Magnetite (D) None of these

10. Thermite is a mixture of iron oxide and
(A) Zinc powder (B) Sodium shavings
(C) Potassium metal (D)Aluminium powder

11. Which of the following reactions produces hydrogen ?
(A) $Mg + H_2O$ (B) $BaO_2 + HCl$
(C) $H_2S_4O_8 + H_2O$ (D) $Na_2O_2 + 2HCl$

12. Which of the following does not have valence electron in 3d-subshell ?
(A) Fe (III) (B) Mn (II) (C) Cr (I) (D) P (0)

13. The intermediate formed in aldol condensation is
(A) Carbanion (B) α-hydrogen ester
(C) Aldol (D) Alcohol

14. Which of the following is involved in Sandmayer's reaction ?
(A) Ferrous salt (B) Ammonium salt
(C) Diazonim salt (D) Cuprammonium salt

15. Lucas test is used for the determination of
(A) Alcohols (B) Phenols
(C) Alkyl halides (D) Aldehydes

16. Teflon is a polymer of
(A) Tetrafluoroethylene (B) Tetraiodoethylene
(C) Tetrabromoethylene (D) Tetrachloroethylene

17. The weight of one molecule of a compound $C_{60}H_{122}$ is
(A) 5.025×10^{23} g (B) 6.023×10^{23} g
(C) 1.2×10^{-20} g (D) 1.4×10^{-21} g

18. If P is pressure and ρ is density of a gas, then P and ρ are related as
(A) $P \propto \rho$ (B) $P \propto \rho^2$ (C) $P \propto 1/\rho$ (D) $P \propto 1/\rho^2$

19. An element (atomic mass = 100 g/mol) having BCC structure has unit cell edge 400 pm. The density of element is (No. of atom in BCC (Z) = 2).
(A) 2.144 g/cm³ (B) 5.188 g/cm³
(C) 7.289 g/cm³ (D) 10.376 g/cm³

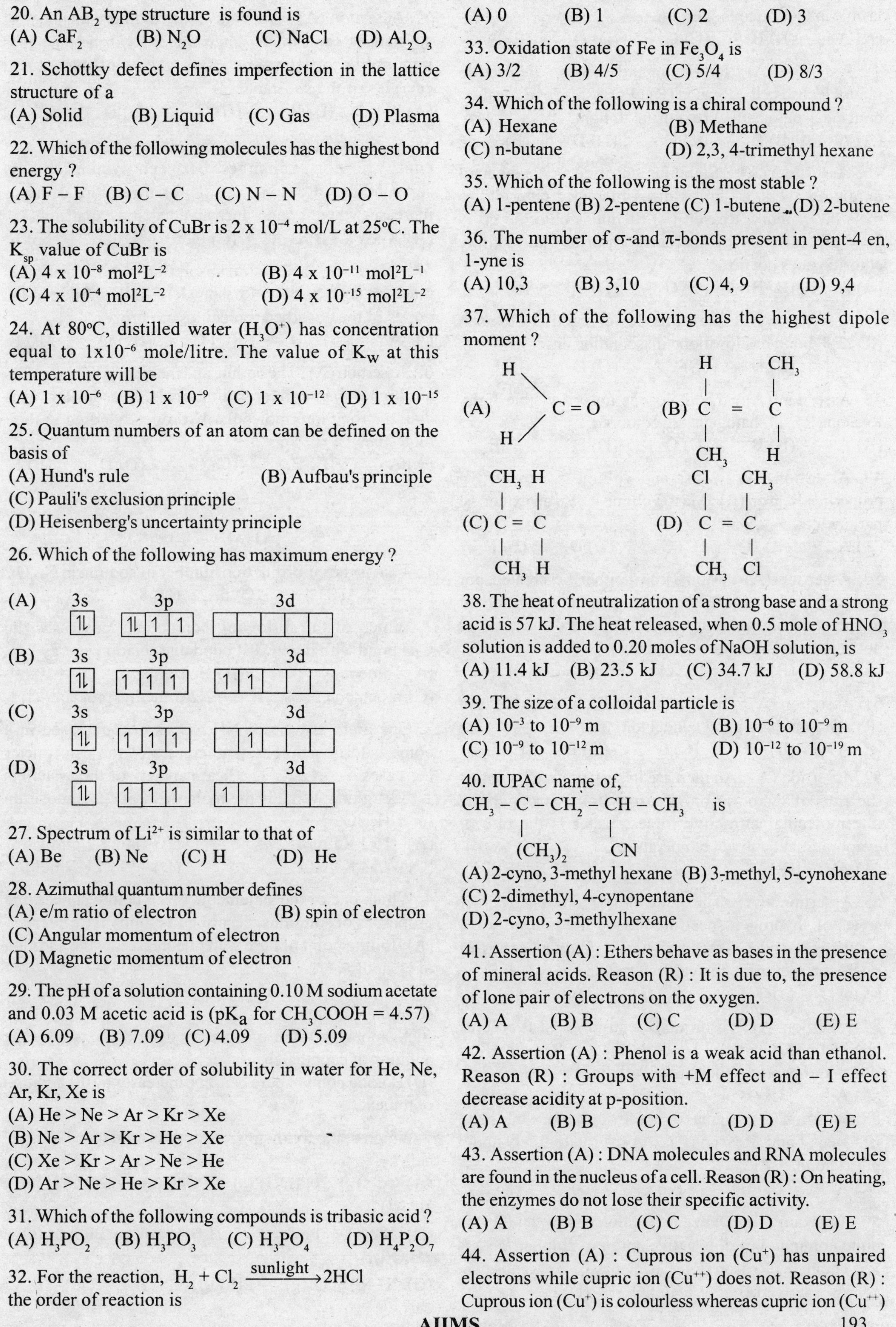

20. An AB_2 type structure is found is
(A) CaF_2 (B) N_2O (C) NaCl (D) Al_2O_3

21. Schottky defect defines imperfection in the lattice structure of a
(A) Solid (B) Liquid (C) Gas (D) Plasma

22. Which of the following molecules has the highest bond energy ?
(A) F – F (B) C – C (C) N – N (D) O – O

23. The solubility of CuBr is 2×10^{-4} mol/L at 25°C. The K_{sp} value of CuBr is
(A) 4×10^{-8} mol^{2}L^{-2} (B) 4×10^{-11} mol^{2}L^{-1}
(C) 4×10^{-4} mol^{2}L^{-2} (D) 4×10^{-15} mol^{2}L^{-2}

24. At 80°C, distilled water (H_3O^+) has concentration equal to 1×10^{-6} mole/litre. The value of K_w at this temperature will be
(A) 1×10^{-6} (B) 1×10^{-9} (C) 1×10^{-12} (D) 1×10^{-15}

25. Quantum numbers of an atom can be defined on the basis of
(A) Hund's rule (B) Aufbau's principle
(C) Pauli's exclusion principle
(D) Heisenberg's uncertainty principle

26. Which of the following has maximum energy ?

(A) 3s [⇅] 3p [⇅|1|1] 3d [| | | |]

(B) 3s [⇅] 3p [1|1|1] 3d [1|1| | |]

(C) 3s [⇅] 3p [1|1|1] 3d [|1| | |]

(D) 3s [⇅] 3p [1|1|1] 3d [1| | | |]

27. Spectrum of Li^{2+} is similar to that of
(A) Be (B) Ne (C) H (D) He

28. Azimuthal quantum number defines
(A) e/m ratio of electron (B) spin of electron
(C) Angular momentum of electron
(D) Magnetic momentum of electron

29. The pH of a solution containing 0.10 M sodium acetate and 0.03 M acetic acid is (pK_a for CH_3COOH = 4.57)
(A) 6.09 (B) 7.09 (C) 4.09 (D) 5.09

30. The correct order of solubility in water for He, Ne, Ar, Kr, Xe is
(A) He > Ne > Ar > Kr > Xe
(B) Ne > Ar > Kr > He > Xe
(C) Xe > Kr > Ar > Ne > He
(D) Ar > Ne > He > Kr > Xe

31. Which of the following compounds is tribasic acid ?
(A) H_3PO_2 (B) H_3PO_3 (C) H_3PO_4 (D) $H_4P_2O_7$

32. For the reaction, $H_2 + Cl_2 \xrightarrow{\text{sunlight}} 2HCl$ the order of reaction is
(A) 0 (B) 1 (C) 2 (D) 3

33. Oxidation state of Fe in Fe_3O_4 is
(A) 3/2 (B) 4/5 (C) 5/4 (D) 8/3

34. Which of the following is a chiral compound ?
(A) Hexane (B) Methane
(C) n-butane (D) 2,3, 4-trimethyl hexane

35. Which of the following is the most stable ?
(A) 1-pentene (B) 2-pentene (C) 1-butene (D) 2-butene

36. The number of σ-and π-bonds present in pent-4 en, 1-yne is
(A) 10,3 (B) 3,10 (C) 4, 9 (D) 9,4

37. Which of the following has the highest dipole moment ?

(A) $(H)(H)C = O$ (B) $(H)(CH_3)C = C(CH_3)(H)$

(C) $(CH_3)(CH_3)C = C(H)(H)$ (D) $(Cl)(CH_3)C = C(CH_3)(Cl)$

38. The heat of neutralization of a strong base and a strong acid is 57 kJ. The heat released, when 0.5 mole of HNO_3 solution is added to 0.20 moles of NaOH solution, is
(A) 11.4 kJ (B) 23.5 kJ (C) 34.7 kJ (D) 58.8 kJ

39. The size of a colloidal particle is
(A) 10^{-3} to 10^{-9} m (B) 10^{-6} to 10^{-9} m
(C) 10^{-9} to 10^{-12} m (D) 10^{-12} to 10^{-19} m

40. IUPAC name of,
$CH_3 - C(CH_3)_2 - CH_2 - CH(CN) - CH_3$ is
(A) 2-cyno, 3-methyl hexane (B) 3-methyl, 5-cynohexane
(C) 2-dimethyl, 4-cynopentane
(D) 2-cyno, 3-methylhexane

41. Assertion (A) : Ethers behave as bases in the presence of mineral acids. Reason (R) : It is due to, the presence of lone pair of electrons on the oxygen.
(A) A (B) B (C) C (D) D (E) E

42. Assertion (A) : Phenol is a weak acid than ethanol. Reason (R) : Groups with +M effect and – I effect decrease acidity at p-position.
(A) A (B) B (C) C (D) D (E) E

43. Assertion (A) : DNA molecules and RNA molecules are found in the nucleus of a cell. Reason (R) : On heating, the enzymes do not lose their specific activity.
(A) A (B) B (C) C (D) D (E) E

44. Assertion (A) : Cuprous ion (Cu^+) has unpaired electrons while cupric ion (Cu^{++}) does not. Reason (R) : Cuprous ion (Cu^+) is colourless whereas cupric ion (Cu^{++})

is blue in the aqueous solution.

(A) A (B) B (C) C (D) D (E) E

45. Assertion (A) : Potassium and caesium are used in photoelectric cells. Reason (R) : Potassium and caesium emit electrons easily on exposure to light.

(A) A (B) B (C) C (D) D (E) E

46. Assertion (A) : Stannous chloride is a powerful oxidizing agent which oxidizes mercuric chloride, to metallic mercury. Reason (R) Stannous chloride gives grey precipitate with mercuric chloride, but stannic chloride does not do so.

(A) A (B) B (C) C (D) D (E) E

47. Assertion (A) : Fluorine has lower reactivity. Reason (R) : F-F bond has low bond dissociation energy.

(A) A (B) B (C) C (D) D (E) E

48. Assertion (A) : The halogens absorb visible light. Reason (R) : all halogens are coloured.

(A) A (B) B (C) C (D) D (E) E

49. Assertion (A) : Mass and volume are extensive properties.Reason (R) : Mass/Volume is also an extensive parameter.

(A) A (B) B (C) C (D) D (E) E

50. Assertion (A) : Atoms can neither be created nor destroyed. Reason (R) : Under similar condition of temperature and pressure, equal volume of gases does not contain equal number of atoms.

(A) A (B) B (C) C (D) D (E) E

51. Assertion (A) : Diamond is a bad conductor. Reason (R) : Graphite is a good conductor.

(A) A (B) B (C) C (D) D (E) E

52. Assertion (A) : An increase in surface area increases the rate of evaporation. Reason (R) : Stronger the intermolecular attractive forces, faster is the rate of evaporation at a given temperature

(A) A (B) B (C) C (D) D (E) E

53. Assertion : (A) : The value of n for a line in Balmer series of hydrogen spectrum having the highest wave length is 4 and 6.Reason (R) : For Balmer series of hydrogen spectrum, the value $n_1 = 2$ and $n_2 = 3, 4, 5$

(A) A (B) B (C) C (D) D (E) E

54. Assertion (A) : Absorption spectrum consists of some bright lines separated by dark spaces. Reason (R) : Emission spectrum consists of dark lines.

(A) A (B) B (C) C (D) D (E) E

55. Assertion (A) : Sigma (σ) is a strong bond, while pi(π) is a weak bond. Reason (R) : Atoms rotate freely about pi (π) bond.

(A) A (B) B (C) C (D) D (E) E

56. Assertion (A) : Di negative anion of oxygen (O^{2-}) is quite common but di negative anion of sulphur (S^{2-}) is less common. Reason (R) : Covalency of oxygen is two.

(A) A (B) B (C) C (D) D (E) E

57. Assertion (A) : Absolute value of internal energy of substances can not be determined. Reason (R) : It is impossible to determine exact values of constituent energies of the substances.

(A) A (B) B (C) C (D) D (E) E

58. Assertion (A) : During an adiabatic process, heat energy is not exchanged between system and its surroundings. Reason (R) : The temperature of a gas increases when it undergoes an adiabatic expansion.

(A) A (B) B (C) C (D) D (E) E

59. Assertion (A) : Physical absorption of molecules takes place on surface only. Reason (R) : In this process, the bonds of the absorbed molecules are broken.

(A) A (B) B (C) C (D) D (E) E

60. Assertion (A) : The boiling and melting points of amides are higher than corresponding acids. Reason (R) : It is due to strong intermolecular hydrogen bonding in their molecules

(A) A (B) B (C) C (D) D (E) E

AIIMS -2003

1. What is the co-ordination number of sodium in Na_2O ?

(A) 2 (B) 4 (C) 6 (D) 8

2. Which of the following compounds possesses the C-H bond with the lowest bond dissociation energy ?

(A) Toluene (B) Benzene

(C) n-pentane (D) 2,2-dimethyl propane

3. One gram sample of NH_4NO_3 is decopmposed in a bomb calorimeter. The temperature of the calorimeter increases by 6.12 K. The heat capacity of the system is 1.23 kJ/g/deg. What is the molar heat of decomposition for NH_4NO_3 ?

(A) –16.1 KJ/mol (B) –602 KJ/mol

(C) –7.53 KJ/mol (D) –398.1 KJ/mol

4. Which one of the statements given below concerning properties of solutions, describes a colligative effect ?

(A) Boiling point of pure water decreases by the addition of ethanol

(B) Vapour pressure of pure water decreases by the addition of nitric acid

(C) Vapour pressure of pure benzene decreases by the addition of naphthalene

(D) Boiling point of pure benzene increases by the addition of toluene

5. Which of the following reactions is used to make a fuel cell ?

(A) $Cd(s) + 2Ni(OH)_3(s) \rightarrow CdO(s) + 2Ni(OH)_2(s) + H_2O(l)$

(B) $2Pb(s) + PbO_2(s) + 2H_2SO_4(aq) \rightarrow 2PbSO_4(s) + 2H_2O(l)$

(C) $2H_2(g) + O_2(g) \rightarrow 2H_2O(l)$

(D) $2Fe(s) + O_2(g) + 4H^+(aq) \rightarrow 2Fe^{2+}(aq) + 2H_2O(l)$

6. Which of the following is not a buffer solution ?
(A) 3M H_2CO_3 + 3M $KHCO_3$
(B) 0.8M H_2S + 0.8M KHS_+
(C) 2M $C_6H_5NH_2$ + 2M $C_6H_5NH_3Br$
(D) 0.05M $KClO_4$ + 0.05M $HClO_4$

7. Which one of the following has ΔS^o greater than zero ?
(A) $CaO(s) + CO_2(g) \rightleftharpoons CaCO_3(s)$
(B) $NaCl(aq) \rightleftharpoons NaCl(s)$
(C) $NaNO_3(s) \rightleftharpoons Na^+(aq) + NO_3^-(aq)$
(D) $CaO(s) + CO_2(g) \rightleftharpoons CaCO_3(s)$

8. The quantum number 'm' of a free gaseous atom is associated with
(A) The effective volume of the orbital
(B) The shape of the orbital
(C) The spatial orientation of the orbital
(D) The energy of the orbital in the absence of a magnetic field

9. Among the following the strongest acid is
(A) CH_3COOH (B) C_6H_5COOH
(C) m-$CH_3OC_6H_4COOH$ (D) p-$CH_3OC_6H_4COOH$

10. Among the following the weakest base is
(A) $O_2NCH_2NH_2$ (B) CH_3NHCHO
(C) $C_6H_5CH_2NH_2$ (D) $C_6H_5CH_2NHCH_3$

11. IUPAC name of [structure: cyclohexene ring with CH_3] is
(A) 3-methyl cyclohexene
(B) 1-methyl cyclohex-2-ene
(C) 6-methyl cyclohexene
(D) 1-methyl cyclohex-5-ene

12. Intermolecular hydrogen bonding is strongest in
(A) Methylamine (B) Phenol
(C) Formaldehyde (D) Methanol

13. The ortho/para directing group among the following is
(A) $COCH_3$ (B) $NHCOCH_3$ (C) COOH (D) CN

14. Treatment of benzene with isobutene in the presence of sulphuric acid gives
(A) Isobutyl benzene (B) Tert-butyl benzene
(C) n-butyl benzene (D) No reaction

15. The most reactive nucleophile among the following is
(A) CH_3O^- (B) $C_6H_5O^-$
(C) $(CH_3)_2CHO$ (D) $(CH_3)_3CO^-$

16. The absolute configuration of the following compound is

CH_3
H — C — Cl
Cl — C — H
C_2H_5

(A) 2R,3S (B) 2R,3R (C) 2 S,3R (D) 2S,3S

17. Subunits present in haemoglobin are-
(A) 2 (B) 3 (C) 4 (D) 5

18. At higher temperature, iodoform reaction is given by
(A) $CH_3CO_2CH_3$ (B) $CH_3CO_2C_2H_5$
(C) $C_6H_5CO_2CH_3$ (D) $CH_3CO_2C_6H_5$

19. Among the following the achiral amino acid is
(A) 2-ethylalanine (B) 2-methylglycine
(C) 2-hydroxymethyl serine (D) Tryptophan

20. The paramagnetic species is
(A) TiO_2 (B) BaO_2 (C) KO_2 (D) SiO_2

21. The reagent commonly used to determine the hardness of water titrimetrically is
(A) Oxalic acid (B) Disodium salt of EDTA
(C) Sodium citrate (D) Sodium thiosulphate

22. The true statement for the acids of phosphorus H_3PO_2, H_3PO_3 and H_3PO_4 is
(A) The order of their acidity is $H_3PO_4 > H_3PO_3 > H_3PO_2$
(B) All of them are reducing in nature
(C) All of them are tribasic acids
(D) The geometry of phosphorus is tetrahedral in the three

23. The ion which is not tetrahedral in shape is
(A) $Cu(NH_3)_4^{2+}$ (B) $NiCl_4^{2-}$ (C) BF_4 (D) NH_4^+

24. The complex used as an anticancer agent is
(A) $[Co(NH_3)_3 Cl_3]$ (B) cis-$[Pt\ Cl_2(NH_3)_2]$
(C) cis-$K_2[PtCl_2Br_2]$ (D) Na_2CoCl_4

25. The coloureless species is -
(A) VCl_3 (B) $VOSO_4$
(C) Na_3VO_4 (D) $[V(H_2O)_6]\ SO_4.H_2O$

26. The ligand called π–acid is
(A) CO (B) NH_3 (C) $C_2O_4^{2-}$ (D) Ethylene diamine

27. The compound used for gravimetric estimation of copper (II) is
(A) $Cu_2(SCN)_2$ (B) Cu_2O (C) Cu_2I_2 (D) Cu_2CO_3

28. In the extraction of copper from its sulphide ore, the metal is formed by reduction of Cu_2O with
(A) Cu_2S (B) SO_2 (C) FeS (D) CO

29. Nitrobenzene gives N-phenylhydroxylamine by
(A) Sn/HCl (B) H_2/Pd-c (C) Zn/NaOH (D) Zn/NH_4Cl

30. Propan-1-ol can be prepared from propene with
(A) B_2H_6 following by H_2O_2
(B) H_2O/H_2SO_4
(C) $Hg(OAc)_2/H_2O$ following by $NaBH_4$
(D) CH_3CO_2H/H_2SO_4

31. Which of the following are arranged in the decreasing order of dipole moment ?
(A) CH_3Cl, CH_3Br, CH_3F (B) CH_3F, CH_3Cl, CH_3Br
(C) CH_3Br, CH_3Cl, CH_3F (D) CH_3Br, CH_3F, CH_3Cl

32. MnO_4^- (1mole) in neutral aqueous medium is disproportionate to
(A) 1/3 mole of Mn_2O_7 and 1/3 mole of MnO_2

(B) 2/3 mole of MnO^-_4 and 1/3 mole of MnO_2
(C) 1/3 mole of MnO^-_4 and 2/3 mole of MnO_2
(D) 2/3 mole of Mn_2O_7 and 1/3 mole of MnO_2

33. Lanthanide for which +II and +III oxidation states are common is
(A) La (B) Nd (C) Ce (D) Eu

34. The mixture of concentrated HCl and HNO_3 made in 3:1 ratio contains
(A) NCl_3 (B) N_2O_4 (C) ClO_2 (D) NOCl

35. On dissolving moderate amount of sodium metal in liquid NH_3 at low temperature, which of the following does not occur ?
(A) Blue coloured solution is obtained
(B) Na+ ions are formed in the solution
(C) Liquid NH_3 becomes good conductor of electricity
(D) Liquid ammonia remains diamagnetic

36. Which one of the following is not a surfactant ?
(A) $CH_3-(CH_2)_{15}-N^+(CH_3)_2-CH_3Br^-$
(B) $CH_3-(CH_2)_{14}-CH_2-NH_2$
(C) $CH_3-(CH_2)_{16}-CH_2OSO_2^-\ Na^+$
(D) $OHC-(CH_2)_{14}-CH_2COO^-\ Na^+$

37. Time required to deposit one milli mole of aluminium metal by the passage of 9.65 amperes through aqueous solution of aluminium ion is
(A) 30s (B) 10s (C) 30,000s (D) 10,000s

38. In which of the following acid-base titrations, pH is greater than 8 at equivalence point ?
(A) Hydrochloric acid versus ammonia
(B) Acetic acid versus ammonia
(C) Acetic acid versus sodium hydroxide
(D) Hydrochloric acid versus sodium hydroxide

39. Which of the following is not a green house gas ?
(A) Carbon dioxide (B) Water vapour
(C) Methane (D) Oxygen

40. The potential energy diagram for a reaction $R \rightarrow P$ is given,

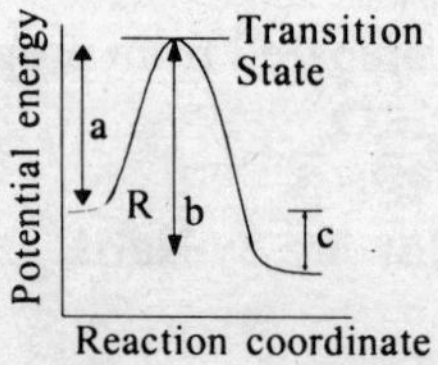

Potential energy
Transition State
a
R
b
c
Reaction coordinate

ΔH^o of the reaction corresponds to the energy
(A) a (B) b (C) c (D) a + b

41. Assertion: Increasing pressure on pure water decreases its freezing point.
Reason : Density of water is maximum at 273 K.
(A) (B) (C) (D)

42. Assertion: The micelle formed by sodium stearate in water has $-COO^-$ groups at the surface.
Reason : Surface tension of water is reduced by the addition of stearate.
(A) (B) (C) (D)

43. Assertion: The O–Obond length in H_2O_2 is shorter than that of O_2F_2.
Reason : H_2O_2 is an ionic compound.
(A) (B) (C) (D)

44. (A) Hemoglobin is an oxygen carrier.
Reason : Oxygen binds as O^{2-} to Fe of hemoglobin.
(A) (B) (C) (D)

45. Assertion: Solution of Na_2CrO_4 in water is intensely coloured.
Reason : Oxidation state of Cr in Na_2CrO_4 is +VI.
(A) (B) (C) (D)

46. Assertion: NF_3 is a weaker ligand than $N(CH_3)_3$.
Reason : NF_3 ionizes to give F^- ions in aqueous solution.
(A) (B) (C) (D)

47. Assertion: PbI_4 is a stable compound .
Reason : Iodide stabilizes higher oxidation state.
(A) (B) (C) (D)

48. Assertion: ${}_{11}Na^{22}$ emits a positron giving ${}_{12}Mg^{22}$.
Reason : In β^+ emission neutron is transformed into proton.
(A) (B) (C) (D)

49. Assertion: Glycosides are hydrolyzed in acidic conditions.
Reason : Glycosides are acetals.
(A) (B) (C) (D)

50. Assertion: Benzyl bromide when kept in acetone water it produces benzyl alcohol.
Reason : The reaction follows $S_N{}^2$ mechanism.
(A) (B) (C) (D)

51. Assertion: Activity of an enzyme is pH dependent.
Reason : Change in pH affects the solubility of the enzyme in water.
(A) (B) (C) (D)

52. Assertion: Alkyl benzene is not prepared by Friedel-Crafts alkylation of benzene.
Reason : Alkylhalides are less reactive than acylhalides.
(A) (B) (C) (D)

53. Assertion: Hydroxyketones are not directly used in Grignard reaction.
Reason : Grignard reagents react with hydroxyl group.
(A) (B) (C) (D)

54. Assertion: Trans-2-butene on reaction with Br_2 gives meso-2, 3-dibromobutane.
Reason : The reaction involves syn-addition of bromine.
(A) (B) (C) (D)

55. Assertion: Cis-1, 3-dihydroxy cyclohexane exists in boat conformation.
Reason : In the chair form, there will not be hydrogen bonding beyween the two hydroxyl groups.

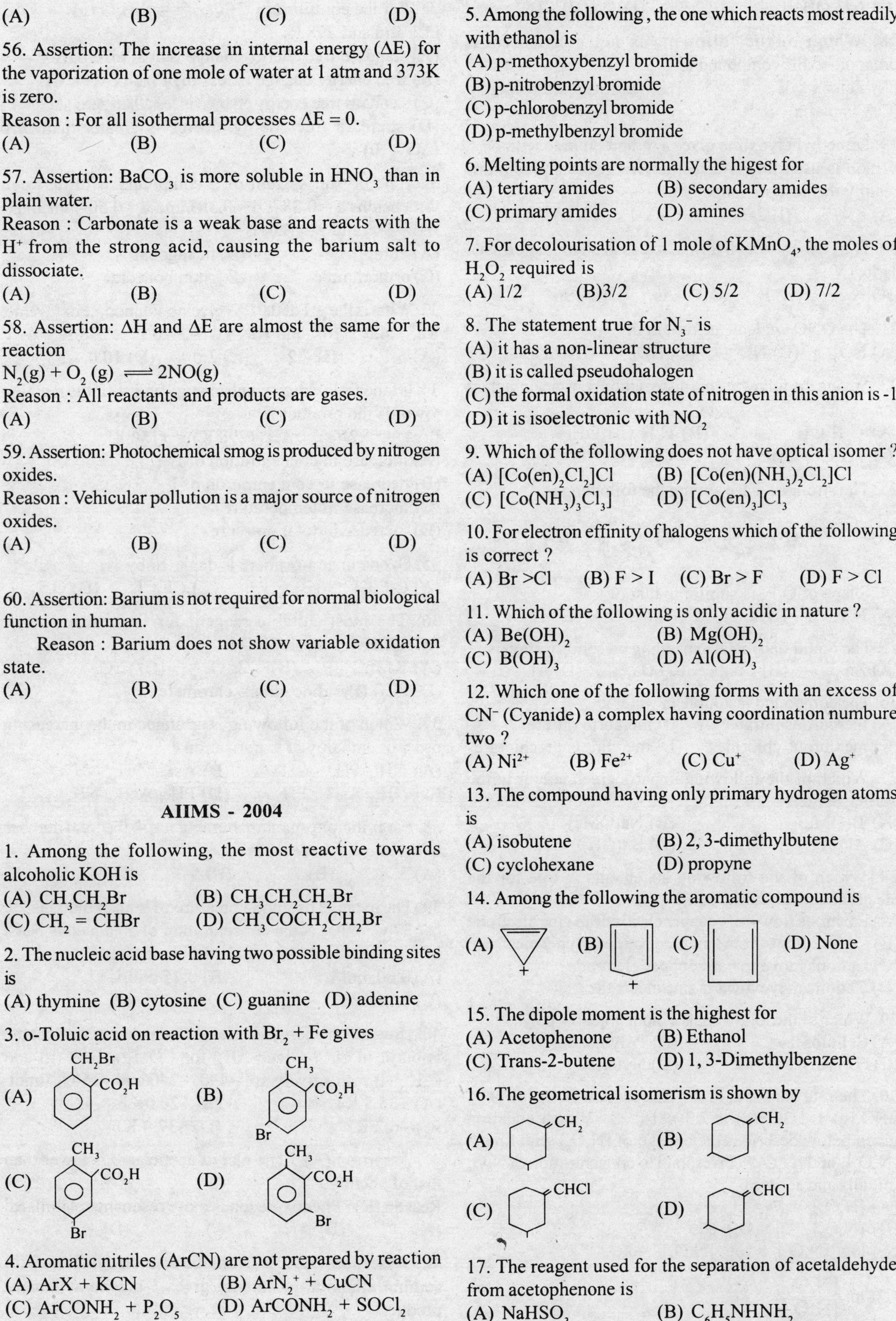

(A) (B) (C) (D)

56. Assertion: The increase in internal energy (ΔE) for the vaporization of one mole of water at 1 atm and 373K is zero.
Reason : For all isothermal processes $\Delta E = 0$.
(A) (B) (C) (D)

57. Assertion: $BaCO_3$ is more soluble in HNO_3 than in plain water.
Reason : Carbonate is a weak base and reacts with the H^+ from the strong acid, causing the barium salt to dissociate.
(A) (B) (C) (D)

58. Assertion: ΔH and ΔE are almost the same for the reaction
$N_2(g) + O_2(g) \rightleftharpoons 2NO(g)$
Reason : All reactants and products are gases.
(A) (B) (C) (D)

59. Assertion: Photochemical smog is produced by nitrogen oxides.
Reason : Vehicular pollution is a major source of nitrogen oxides.
(A) (B) (C) (D)

60. Assertion: Barium is not required for normal biological function in human.
Reason : Barium does not show variable oxidation state.
(A) (B) (C) (D)

AIIMS - 2004

1. Among the following, the most reactive towards alcoholic KOH is
(A) CH_3CH_2Br (B) $CH_3CH_2CH_2Br$
(C) $CH_2 = CHBr$ (D) $CH_3COCH_2CH_2Br$

2. The nucleic acid base having two possible binding sites is
(A) thymine (B) cytosine (C) guanine (D) adenine

3. o-Toluic acid on reaction with Br_2 + Fe gives
(A) [ring with CH_2Br, CO_2H] (B) [ring with CH_3, CO_2H, Br]
(C) [ring with CH_3, CO_2H, Br] (D) [ring with CH_3, CO_2H, Br]

4. Aromatic nitriles (ArCN) are not prepared by reaction
(A) ArX + KCN (B) ArN_2^+ + CuCN
(C) $ArCONH_2 + P_2O_5$ (D) $ArCONH_2 + SOCl_2$

5. Among the following, the one which reacts most readily with ethanol is
(A) p-methoxybenzyl bromide
(B) p-nitrobenzyl bromide
(C) p-chlorobenzyl bromide
(D) p-methylbenzyl bromide

6. Melting points are normally the higest for
(A) tertiary amides (B) secondary amides
(C) primary amides (D) amines

7. For decolourisation of 1 mole of $KMnO_4$, the moles of H_2O_2 required is
(A) 1/2 (B) 3/2 (C) 5/2 (D) 7/2

8. The statement true for N_3^- is
(A) it has a non-linear structure
(B) it is called pseudohalogen
(C) the formal oxidation state of nitrogen in this anion is -1
(D) it is isoelectronic with NO_2

9. Which of the following does not have optical isomer ?
(A) $[Co(en)_2Cl_2]Cl$ (B) $[Co(en)(NH_3)_2Cl_2]Cl$
(C) $[Co(NH_3)_3Cl_3]$ (D) $[Co(en)_3]Cl_3$

10. For electron effinity of halogens which of the following is correct ?
(A) Br > Cl (B) F > I (C) Br > F (D) F > Cl

11. Which of the following is only acidic in nature ?
(A) $Be(OH)_2$ (B) $Mg(OH)_2$
(C) $B(OH)_3$ (D) $Al(OH)_3$

12. Which one of the following forms with an excess of CN^- (Cyanide) a complex having coordination numbure two ?
(A) Ni^{2+} (B) Fe^{2+} (C) Cu^+ (D) Ag^+

13. The compound having only primary hydrogen atoms is
(A) isobutene (B) 2, 3-dimethylbutene
(C) cyclohexane (D) propyne

14. Among the following the aromatic compound is
(A) [cyclopropenyl cation, +] (B) [cyclopentadienyl cation, +] (C) [cyclobutadiene] (D) None

15. The dipole moment is the highest for
(A) Acetophenone (B) Ethanol
(C) Trans-2-butene (D) 1, 3-Dimethylbenzene

16. The geometrical isomerism is shown by
(A) [ring $=CH_2$] (B) [ring $=CH_2$]
(C) [ring $=CHCl$] (D) [ring $=CHCl$]

17. The reagent used for the separation of acetaldehyde from acetophenone is
(A) $NaHSO_3$ (B) $C_6H_5NHNH_2$

(C) NH_2OH (D) NaOH - I_2

18. Which of the following is not considered as organometallic compound ?

(A) Zeise's salt (B) Grignard reagent
(C) Cis-platin (D) Ferrocene

19. Dimethyl glyoxime gives a red precipitate with Ni^{2+}, Which is used for its detection. To get this precipitate readily the best pH range is

(A) <1 (B) 2-3 (C) 3-4 (D) 9-11

20. The element which forms oxides in all oxidation state +I to +V is

(A) N (B) P (C) As (D) Sb

21. The ONO angle is maximum in

(A) NO_2 (B) NO_2^+ (C) NO_3^- (D) NO_2^-

22. Among the following the dissociation constant is highest for

(A) C_6H_5OH (B) $C_6H_5CH_2OH$
(C) $CH_3C \equiv CH$ (D) $CH_3NH_3^+Cl^-$

23. The strongest base among the following is

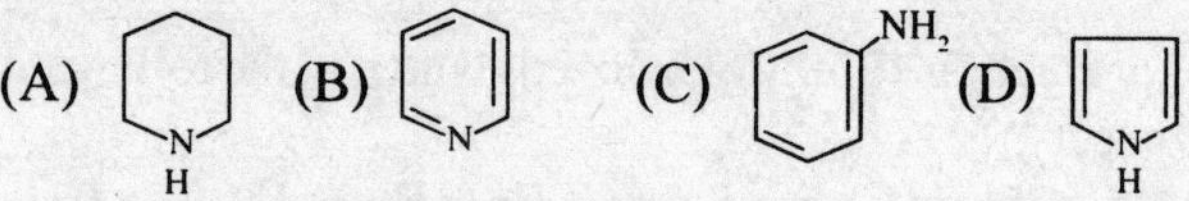

24. Shape of O_2F_2 is similar to that of

(A) H_2F_2 (B) C_2H_2 (C) C_2F_2 (D) H_2O_2

25. The liquid fied metal expanding on solidification is

(A) Zn (B) Cu (C) Ga (D) Al

26. The compound insoluble in water is

(A) mercurous nitrate (B) mercuric nitrate
(C) mercurous chloride (D) mercurous perchlorate

27. Which of the following imparts green colour to the burner flame ?

(A) $B(OMe)_3$ (B) Na(OMe)
(C) $Al(OPh)_3$ (D) $Sn(OH)_2$

28. Which of the following statements is true for the electrochemical Daniel cell ?

(A) Electron's flow from copper electrode to zinc electrode
(B) Current flows from zinc electrode to copper electrode
(C) Cations move toward copper electrode
(D) Cations move toward zinc electrode

29. Which of the following is a biodegradable polymer?

(A) Cellulose (B) Polythene
(C) Polyvinyl chloride (D) Nylon-6

30. The rate constant, k, for the reaction $N_2O_5\ (g) \rightarrow 2NO_2(g) + 1/2O_2\ (g)$ is $2.3 \times 10^{-2}\ S^{-1}$.Which equation given below describes the change of $[N_2O_5]$ with time ? $[N_2O_5]_0$ and $[N_2O_5]_t$ correspond to concentration of N_2O_5 initially and at time t

(A) $[N_2O_5]_t = [N_2O_5]_0 + kt$
(B) $[N_2O_5]_0 = [N_2O_5]_t\ e^{kt}$
(C) $\log_{10}[N_2O_5]_t = \log_{10}[N_2O_5]_0 - kt$
(D) In $\frac{[N_2O_5]_0}{[N_2O_5]_t} = kt$

31. For the equilibrium $H_2O \rightleftharpoons H_2O\ (g)$ at 1 atm and 298 K,

(A) standard free energy change is qual tozero ($\Delta G^o = 0$)
(B) free energy change is less than zero ($\Delta G < 0$)
(C) standard free energy change is less than zero ($\Delta G^o < 0$)
(D) standard free energy change is greater than zero ($\Delta G^o > 0$)

32. The crystal system of a compound with unit cell dimensions a = 0.387, b = 0.387 and c = 0.504 nm and $\alpha = \beta = 90^o$ and $\gamma = 120^o$ is

(A) cubic (B) hexagonal
(C) orthorhmbic (D) rhombohedral

33. What is the pH of 0.01 M glycine solution ? For glycine, $Ka_1 = 4.5 \times 10^{-3}$ and $Ka_2 = 1.7 \times 10^{-10}$ at 298 K

(A) 6.1 (B) 7.2 (C) 3.0 (D) 10.0

34. Of the following, which change will shift the reaction towards the product ?

$I_2(g) \rightleftharpoons 2I(g)$, ΔH^o (298 K) = +150 kJ

(A) increase in concentration of I
(B) decrease in concentration of I_2
(C) increase in temperature
(D) increase in total pressure

35. Ozone in stratosphere is depleted by

(A) $C_6F_6Cl_6$ (B) C_6F_6 (C) CF_2Cl_2 (D) C_7F_{16}

36. The most suitable reagent for the conversion of $RCH_2OH \longrightarrow RCHO$ is

(A) $KMnO_4$ (B)$K_2Cr_2O_7$ (C) CrO_3
(D) PCC (Pyridine chloro chromate)

37. Which of the following is arranged in the increasing order of enthalpy of vaporisation ?

(A) NH_3, PH_3, AsH_3 (B) AsH_3, PH_3, NH_3
(C) NH_3, AsH_3, PH_3 (D) PH_3, AsH_3, NH_3

38. For principle quantum number n = 4, the total number of orbitals having $l = 3$ is

(A) 3 (B) 7 (C) 5 (D) 9

39. The average osomotic pressure of human blood is 7.8 bar 37°C. What is the concentration of an aqueous NaCl solution that could be used in the blood stream ?

(A) 0.60 mol/L (B) 0.45 mol/L
(C) 0.16 mol/L (D) 0.32 mol/L

40. How much energy is released when 6 mole of octane is burnt in air ? Given ΔH_f^o for CO_2 (g), H_2O (g) and C_8H_{18} (l) respectively are –490, –240 and +160 KJ/mol

(A) –35.5 KJ (B) –20.0KJ
(C) –6.2 KJ (D) –37.4 KJ

41. Assertion (A) : The pka of acetic acid is lower than that of phenol.
Reason (R) : Phenoxide ion is more resonance stabilised.

(A) (B) (C) (D)

42. Assertion (A) : 2-Bromobutane on reaction with sodium ethoxide in ethanol gives 1-butene as amajor product.

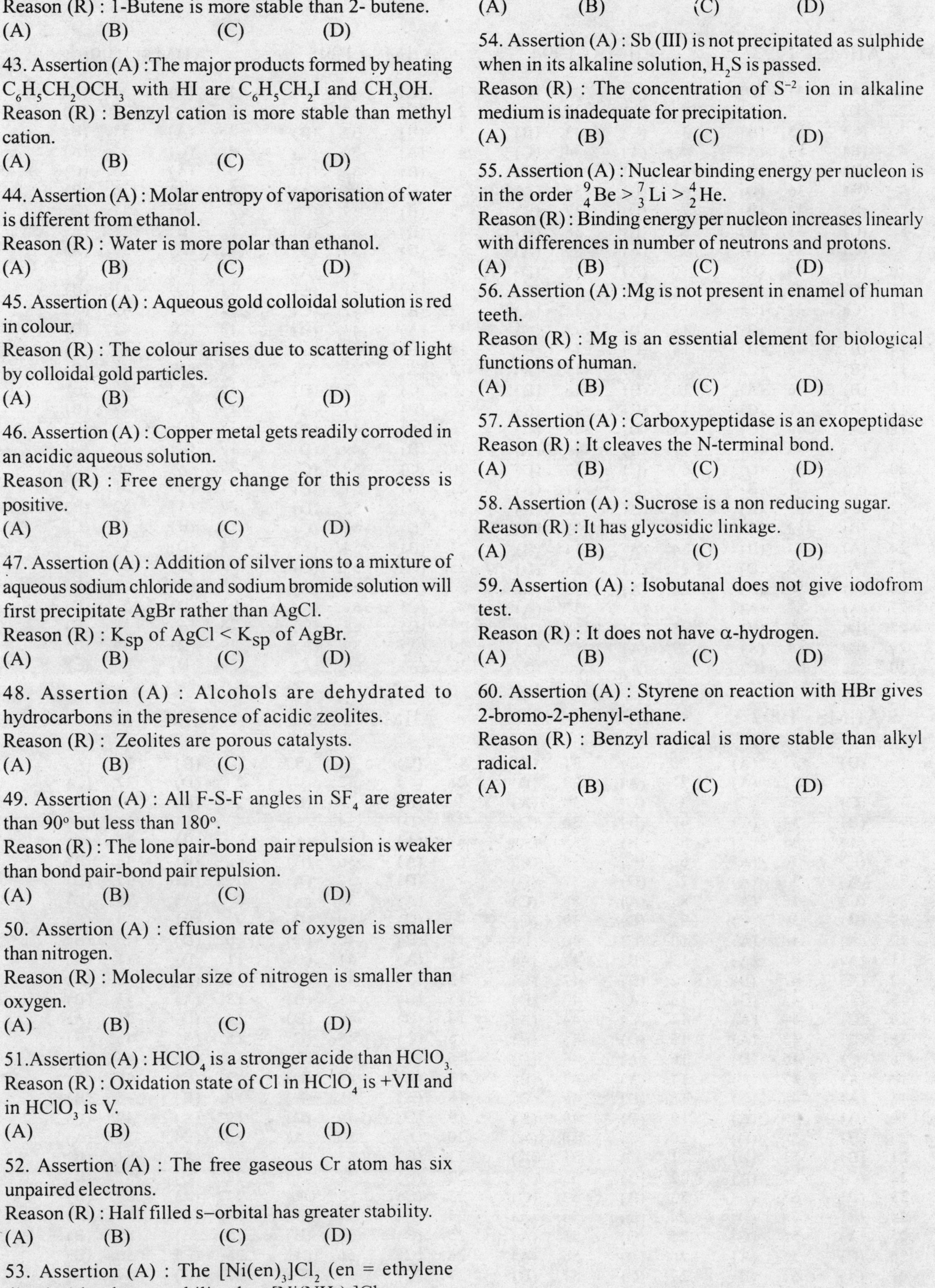

Reason (R) : 1-Butene is more stable than 2- butene.
(A) (B) (C) (D)

43. Assertion (A) :The major products formed by heating $C_6H_5CH_2OCH_3$ with HI are $C_6H_5CH_2I$ and CH_3OH.
Reason (R) : Benzyl cation is more stable than methyl cation.
(A) (B) (C) (D)

44. Assertion (A) : Molar entropy of vaporisation of water is different from ethanol.
Reason (R) : Water is more polar than ethanol.
(A) (B) (C) (D)

45. Assertion (A) : Aqueous gold colloidal solution is red in colour.
Reason (R) : The colour arises due to scattering of light by colloidal gold particles.
(A) (B) (C) (D)

46. Assertion (A) : Copper metal gets readily corroded in an acidic aqueous solution.
Reason (R) : Free energy change for this process is positive.
(A) (B) (C) (D)

47. Assertion (A) : Addition of silver ions to a mixture of aqueous sodium chloride and sodium bromide solution will first precipitate AgBr rather than AgCl.
Reason (R) : K_{sp} of AgCl < K_{sp} of AgBr.
(A) (B) (C) (D)

48. Assertion (A) : Alcohols are dehydrated to hydrocarbons in the presence of acidic zeolites.
Reason (R) : Zeolites are porous catalysts.
(A) (B) (C) (D)

49. Assertion (A) : All F-S-F angles in SF_4 are greater than 90° but less than 180°.
Reason (R) : The lone pair-bond pair repulsion is weaker than bond pair-bond pair repulsion.
(A) (B) (C) (D)

50. Assertion (A) : effusion rate of oxygen is smaller than nitrogen.
Reason (R) : Molecular size of nitrogen is smaller than oxygen.
(A) (B) (C) (D)

51.Assertion (A) : $HClO_4$ is a stronger acide than $HClO_3$.
Reason (R) : Oxidation state of Cl in $HClO_4$ is +VII and in $HClO_3$ is V.
(A) (B) (C) (D)

52. Assertion (A) : The free gaseous Cr atom has six unpaired electrons.
Reason (R) : Half filled s–orbital has greater stability.
(A) (B) (C) (D)

53. Assertion (A) : The $[Ni(en)_3]Cl_2$ (en = ethylene diamine) has lower stability than $[Ni(NH_3)_6]Cl_2$.
Reason (R) : In $[Ni(en)_3]Cl_2$ the geometry of Ni is trigonal bipyramidal.
(A) (B) (C) (D)

54. Assertion (A) : Sb (III) is not precipitated as sulphide when in its alkaline solution, H_2S is passed.
Reason (R) : The concentration of S^{-2} ion in alkaline medium is inadequate for precipitation.
(A) (B) (C) (D)

55. Assertion (A) : Nuclear binding energy per nucleon is in the order ${}^{9}_{4}Be > {}^{7}_{3}Li > {}^{4}_{2}He$.
Reason (R) : Binding energy per nucleon increases linearly with differences in number of neutrons and protons.
(A) (B) (C) (D)

56. Assertion (A) :Mg is not present in enamel of human teeth.
Reason (R) : Mg is an essential element for biological functions of human.
(A) (B) (C) (D)

57. Assertion (A) : Carboxypeptidase is an exopeptidase
Reason (R) : It cleaves the N-terminal bond.
(A) (B) (C) (D)

58. Assertion (A) : Sucrose is a non reducing sugar.
Reason (R) : It has glycosidic linkage.
(A) (B) (C) (D)

59. Assertion (A) : Isobutanal does not give iodofrom test.
Reason (R) : It does not have α-hydrogen.
(A) (B) (C) (D)

60. Assertion (A) : Styrene on reaction with HBr gives 2-bromo-2-phenyl-ethane.
Reason (R) : Benzyl radical is more stable than alkyl radical.
(A) (B) (C) (D)

AIIMS - 1993

1.	(A)	31.	(D)
2.	(D)	32.	(A)
3.	(C)	33.	(B)
4.	(B)	34.	(A)
5.	(D)	35.	(B)
6.	(B)	36.	(D)
7.	(B)	37.	(B)
8.	(C)	38.	(B)
9.	(C)	39.	(D)
10.	(B)	40.	(D)
11.	(D)	41.	(C)
12.	(C)	42.	(C)
13.	(C)	43.	(B)
14.	(B)	44.	(B)
15.	(B)	45.	(C)
16.	(B)	46.	(A)
17.	(B)	47.	(C)
18.	(A)	48.	(C)
19.	(C)	49.	(C)
20.	(C)	50.	(D)
21.	(C)	51.	(B)
22.	(C)	52.	(A)
23.	(B)	53.	(A)
24.	(A)	54.	(B)
25.	(A)	55.	(D)
26.	(C)	56.	(A)
27.	(A)	57.	(A)
28.	(D)	58.	(B)
29.	(D)	59.	(A)
30.	(B)	60.	(C)

AIIMS - 1994

1.	(A)	31.	(C)
2.	(B)	32.	(C)
3.	(C)	33.	(B)
4.	(A)	34.	(C)
5.	(A)	35.	(B)
6.	(B)	36.	(D)
7.	(D)	37.	(B)
8.	(B)	38.	(B)
9.	(A)	39.	(A)
10.	(A)	40.	(D)
11.	(B)	41.	(C)
12.	(D)	42.	(A)
13.	(D)	43.	(C)
14.	(C)	44.	(B)
15.	(B)	45.	(A)
16.	(B)	46.	(D)
17.	(B)	47.	(C)
18.	(D)	48.	(B)
19.	(B)	49.	(A)
20.	(C)	50.	(D)
21.	(C)	51.	(B)
22.	(B)	52.	(A)
23.	(A)	53.	(A)
24.	(A)	54.	(B)
25.	(A)	55.	(D)
26.	(A)	56.	(A)
27.	(A)	57.	(A)
28.	(D)	58.	(B)
29.	(A)	59.	(A)
30.	(A)	60.	(C)

AIIMS - 1995

1.	(C)	31.	(D)
2.	(C)	32.	(B)
3.	(B)	33.	(B)
4.	(A)	34.	(C)
5.	(B)	35.	(D)
6.	(C)	36.	(A)
7.	(B)	37.	(C)
8.	(D)	38.	(C)
9.	(C)	39.	(B)
10.	(A)	40.	(A)
11.	(A)	41.	(C)
12.	(B)	42.	(C)
13.	(A)	43.	(D)
14.	(B)	44.	(D)
15.	(A)	45.	(D)
16.	(C)	46.	(D)
17.	(B)	47.	(B)
18.	(A)	48.	(D)
19.	(B)	49.	(D)
20.	(C)	50.	(C)
21.	(B)	51.	(B)
22.	(C)	52.	(D)
23.	(D)	53.	(C)
24.	(B)	54.	(A)
25.	(D)	55.	(C)
26.	(D)	56.	(D)
27.	(C)	57.	(A)
28.	(D)	58.	(D)
29.	(A)	59.	(B)
30.	(A)	60.	(A)

AIIMS - 1996

1.	(B)	31.	(C)
2.	(D)	32.	(D)
3.	(A)	33.	(B)
4.	(C)	34.	(A)
5.	(A)	35.	(C)
6.	(A)	36.	(B)
7.	(D)	37.	(D)
8.	(B)	38.	(A)
9.	(A)	39.	(A)
10.	(D)	40.	(C)
11.	(D)	41.	(B)
12.	(C)	42.	(A)
13.	(D)	43.	(D)
14.	(C)	44.	(C)
15.	(A)	45.	(D)
16.	(A)	46.	(B)
17.	(C)	47.	(B)
18.	(D)	48.	(D)
19.	(C)	49.	(D)
20.	(A)	50.	(C)
21.	(B)	51.	(D)
22.	(A)	52.	(B)
23.	(B)	53.	(C)
24.	(D)	54.	(B)
25.	(D)	55.	(B)
26.	(D)	56.	(B)
27.	(D)	57.	(D)
28.	(A)	58.	(B)
29.	(D)	59.	(B)
30.	(D)	60.	(C)

AIIMS - 1997

1.	(D)	31.	(A)
2.	(C)	32.	(A)
3.	(C)	33.	(C)
4.	(B)	34.	(C)
5.	(A)	35.	(D)
6.	(C)	36.	(A)
7.	(A)	37.	(A)
8.	(C)	38.	(A)
9.	(B)	39.	(C)
10.	(A)	40.	(A)
11.	(A)	41.	(A)
12.	(A)	42.	(D)
13.	(A)	43.	(B)
14.	(C)	44.	(A)
15.	(D)	45.	(A)
16.	(C)	46.	(D)
17.	(A)	47.	(A)
18.	(A)	48.	(C)
19.	(A)	49.	(A)
20.	(B)	50.	(D)
21.	(B)	51.	(B)
22.	(C)	52.	(A)
23.	(B)	53.	(A)
24.	(B)	54.	(D)
25.	(A)	55.	(C)
26.	(C)	56.	(A)
27.	(A)	57.	(B)
28.	(A)		
29.	(C)		
30.	(D)		

AIIMS - 1998

1.	(C)	31.	(B)
2.	(A)	32.	(D)
3.	(D)	33.	(A)
4.	(B)	34.	(D)
5.	(B)	35.	(A)
6.	(C)	36.	(C)
7.	(D)	37.	(D)
8.	(A)	38.	(C)
9.	(C)	39.	(C)
10.	(C)	40.	(D)
11.	(B)	41.	(A)
12.	(B)	42.	(C)
13.	(A)	43.	(D)
14.	(C)	44.	(A)
15.	(D)	45.	(B)
16.	(A)	46.	(D)
17.	(A)	47.	(B)
18.	(D)	48.	(C)
19.	(D)	49.	(B)
20.	(C)	50.	(A)
21.	(C)	51.	(A)
22.	(B)	52.	(A)
23.	(B)	53.	(C)
24.	(D)	54.	(A)
25.	(D)	55.	(A)
26.	(D)	56.	(A)
27.	(C)	57.	(D)
28.	(B)	58.	(D)
29.	(C)	59.	(A)
30.	(D)	60.	(D)

AIIMS - 1999

1.	(C)	31.	(B)
2.	(C)	32.	(D)
3.	(B)	33.	(B)
4.	(D)	34.	(A)
5.	(A)	35.	(A)
6.	(A)	36.	(B)
7.	(D)	37.	(A)
8.	(A)	38.	(A)
9.	(C)	39.	(D)
10.	(B)	40.	(D)
11.	(A)	41.	(C)
12.	(C)	42.	(D)
13.	(B)	43.	(B)
14.	(B)	44.	(B)
15.	(C)	45.	(C)
16.	(D)	46.	(B)
17.	(A)	47.	(D)
18.	(A)	48.	(A)
19.	(D)	49.	(D)
20.	(A)	50.	(A)
21.	(C)	51.	(B)
22.	(D)	52.	(A)
23.	(C)	53.	(A)
24.	(A)	54.	(D)
25.	(D)	55.	(B)
26.	(C)	56.	(A)
27.	(C)	57.	(C)
28.	(D)	58.	(A)
29.	(D)	59.	(A)
30.	(D)	60.	(A)

AIIMS - 2000

1.	(B)	31.	(A)
2.	(D)	32.	(C)
3.	(D)	33.	(B)
4.	(C)	34.	(D)
5.	(B)	35.	(A)
6.	(B)	36.	(B)
7.	(B)	37.	(D)
8.	(A)	38.	(D)
9.	(B)	39.	(C)
10.	(D)	40.	(B)
11.	(D)	41.	(A)
12.	(D)	42.	(D)
13.	(A)	43.	(B)
14.	(D)	44.	(A)
15.	(A)	45.	(B)
16.	(D)	46.	(A)
17.	(C)	47.	(C)
18.	(B)	48.	(B)
19.	(A)	49.	(D)
20.	(D)	50.	(A)
21.	(A)	51.	(B)
22.	(D)	52.	(B)
23.	(A)	53.	(B)
24.	(C)	54.	(B)
25.	(B)	55.	(B)
26.	(B)	56.	(D)
27.	(D)	57.	(D)
28.	(A)	58.	(A)
29.	(A)	59.	(B)
30.	(C)	60.	(C)

AIIMS - 2001

Q.	Ans.	Q.	Ans.
1.	(D)	31.	(A)
2.	(B)	32.	(C)
3.	(C)	33.	(C)
4.	(C)	34.	(C)
5.	(A)	35.	(B)
6.	(A)	36.	(C)
7.	(D)	37.	(D)
8.	(D)	38.	(B)
9.	(C)	39.	(B)
10.	(A)	40.	(D)
11.	(B)	41.	(A)
12.	(D)	42.	(C)
13.	(C)	43.	(B)
14.	(C)	44.	(A)
15.	(B)	45.	(B)
16.	(B)	46.	(C)
17.	(D)	47.	(A)
18.	(C)	48.	(A)
19.	(D)	49.	(B)
20.	(A)	50.	(A)
21.	(A)	51.	(A)
22.	(A)	52.	(D)
23.	(A)	53.	(B)
24.	(D)	54.	(C)
25.	(A)	55.	(B)
26.	(C)	56.	(D)
27.	(A)	57.	(A)
28.	(A)	58.	(A)
29.	(A)		
30.	(A)		

AIIMS - 2002

Q.	Ans.	Q.	Ans.
1.	(A)	31.	(C)
2.	(B)	32.	(A)
3.	(D)	33.	(D)
4.	(C)	34.	(D)
5.	(B)	35.	(B)
6.	(D)	36.	(A)
7.	(B)	37.	(A)
8.	(C)	38.	(A)
9.	(A)	39.	(B)
10.	(D)	40.	(C)
11.	(A)	41.	(A)
12.	(D)	42.	(D)
13.	(A)	43.	(D)
14.	(C)	44.	(E)
15.	(A)	45.	(A)
16.	(A)	46.	(E)
17.	(D)	47.	(E)
18.	(A)	48.	(A)
19.	(B)	49.	(B)
20.	(A)	50.	(C)
21.	(A)	51.	(B)
22.	(B)	52.	(C)
23.	(A)	53.	(E)
24.	(C)	54.	(D)
25.	(C)	55.	(C)
26.	(B)	56.	(B)
27.	(C)	57.	(A)
28.	(C)	58.	(C)
29.	(D)	59.	(C)
30.	(C)	60.	(A)

AIIMS - 2003

Q.	Ans.	Q.	Ans.
1.	(B)	31.	(B)
2.	(A)	32.	(B)
3.	(B)	33.	(D)
4.	(C)	34.	(D)
5.	(C)	35.	(D)
6.	(D)	36.	(B)
7.	(C)	37.	(A)
8.	(C)	38.	(C)
9.	(C)	39.	(D)
10.	(A)	40.	(C)
11.	(A)	41.	(C)
12.	(B)	42.	(A)
13.	(B)	43.	(C)
14.	(B)	44.	(C)
15.	(D)	45.	(B)
16.	(D)	46.	(C)
17.	(C)	47.	(D)
18.	(B)	48.	(D)
19.	(C)	49.	(A)
20.	(C)	50.	(A)
21.	(B)	51.	(A)
22.	(D)	52.	(C)
23.	(A)	53.	(A)
24.	(B)	54.	(C)
25.	(C)	55.	(D)
26.	(A)	56.	(A)
27.	(A)	57.	(A)
28.	(A)	58.	(C)
29.	(D)	59.	(B)
30.	(A)	60.	(B)

AIIMS - 2004

Q.	Ans.	Q.	Ans.
1.	(D)	31.	(D)
2.	(D)	32.	(B)
3.	(C)	33.	(C)
4.	(A)	34.	(C)
5.	(A)	35.	(C)
6.	(A)	36.	(D)
7.	(C)	37.	(D)
8.	(B)	38.	(B)
9.	(C)	39.	(C)
10.	(B)	40.	(D)
11.	(C)	41.	(C)
12.	(D)	42.	(D)
13.	(A)	43.	(A)
14.	(D)	44.	(B)
15.	(A)	45.	(B)
16.	(C)	46.	(D)
17.	(A)	47.	(C)
18.	(C)	48.	(B)
19.	(D)	49.	(C)
20.	(A)	50.	(B)
21.	(B)	51.	(B)
22.	(A)	52.	(B)
23.	(A)	53.	(D)
24.	(D)	54.	(D)
25.	(C)	55.	(C)
26.	(C)	56.	(B)
27.	(A)	57.	(C)
28.	(C)	58.	(B)
29.	(A)	59.	(C)
30.	(D)	60.	(D)

BHU - 1995

1. The osmotic pressure of a 5% solution of cane sugar at 15^0C is (Mol. wt of cane sugar = 342)
(A) 4 atm (B) 3.57 atm (C) 3.45 atm (D) 2.35 atm

2. A colloidal solution is subjected to an electric field. The colloidal particles move towards the anode. Coagulation of the same colloidal solution is studied using NaCl, $BaCl_2$ and $AlCl_3$ solution. The order of coagulating power is
(A) $NaCl > AlCl_3 > BaCl_2$ (B) $AlCl_3 > BaCl_2 > NaCl_3$
(C) $NaCl > BaCl_2 > AlCl_3$ (D) $BaCl_2 > NaCl > AlCl_3$

3. $[Zn^{2+}] = 0.1$ M and $E^o = -0.76$ V, then half cell potential at 298 K for the reaction $Zn^{2+}(aq) + 2e^- \rightarrow Zn(s)$ is
(A) –0.698 V (B) 0.698 V (C) 0.789 V (D) –0.789 V

4. Which of the following statements is correct ?
(A) $SnCl_4$ can not be reduced
(B) Mosaic gold has formula SnS_2
(C) Xenon has configuration (outer) ns^2p^6
(D) None of these

5. In which of the following reactions, hydrogen peroxide is a reducing agent ?
(A) $2FeCl_2 + 2HCl + H_2O_2 \rightarrow 2FeCl_3 + 2H_2O$
(B) $Cl_2 + H_2O_2 \rightarrow 2HCl + O_2$
(C) $2H_1 + H_2O_2 \rightarrow 2H_2O + I_2$
(D) $H_2SO_3 + H_2O_2 \rightarrow H_2SO_4 + H_2O$

6. Which of the following is a Lewis base ?
(A) Acetone (B) Amine (C) CH_4 (D) C_2H_5OH

7. The one with square planar shape is
(A) BF_4 (B) $[Cu(NH_3)_4]^{2+}$
(C) $[CrF_6]^{3-}$ (D) None of these

8. A radioactive substance has half-life 5760 years. A sample of 100 mg containing the substance is reduced to 25 mg in
(A) 1440 years (B) 280 years
(C) 11520 years (D) 2880 years

9. The positron has mass equal to
(A) Electron (B) Proton (C) Deutron (D) α-particle

10. The electronic configuration of Fe^{3+} is
(A) $[Ar]3d^6$ (B) $[Ar]3d^5$ (C) $[Ar]2d^4$ (D) $3d^64s^2$

11. The bond angle of sp^2 - hydrid orbital is
(A) 105° (B) 109°8" (C) 120° (D) 180°

12. Rutherford's experiment led to the discovery of
(A) Nucleus (B) α-particle (C) β-particle (D) Electron

13. The order of stability of carbonium ion is
(A) $3^o > 1^o > 2^o > CH_3^+$ (B) $2^o > 3^o > 1^o > CH_3$
(C) $CH_3^+ > 1^o > 2^o > 3^o$ (D) $3^o > 2^o > 1^o > CH_3^+$

14. The complete combustion of CH_4 gives
(A) $CO_2 + H_2O$ (B) $CO_2 + H_2$
(C) $CO + H_2$ (D) $CO + H_2O$

15. The Grignard reagent, on reaction with acetone forms
(A) Tertiary alcohol (B) Secondary alcohol
(C) Acetic acid (D) Acetaldehyde

16. Zn, on reacting with cold dil. HNO_3, gives
(A) ZnNO (B) NO (C) NO_2 (D) NH_4NO_3

17. Which of the following is used in cancer therapy ?
(A) Ni (B) Rn (C) Po (D) Co

18. A radioactive substance has half-life of 60 minutes. After 3 hours, what percentage of radioactivity will remain ?
(A) 50% (B) 25% (C) 175% (D) 12.5%

19. The number of atoms in 4.25 grams of NH_3 is approximately
(A) 1×10^{23} (B) 1.5×10^{23} (C) 2×10^{23} (D) 6×10^{23}

20. Tetraethyl lead is a
(A) Solvent (B) Petroleum additive
(C) Oxidising agent (D) Fire extinguisher.

21. The chemical formula of plaster of paris is
(A) $CaSO_4.2H_2O$ (B) $CaSO_4.H_2O$
(C) $CaSO_4.\frac{1}{2}H_2O$ (D) $CaSO_4.\frac{1}{4}H_2O$

22. The azimuthal quantum number for the last electron of sodium atom is
(A) 1 (B) 2 (C) 3 (D) 0

23. $PbCl_4$ exists but $PbBr_4$ and PbI_4 do not, because of
(A) More electropositive nature of bromine and iodine
(B) Inability of bromine and iodine to oxidize Pb^{+2} to Pb^{+4}
(C) Br^- and I^- ions are bigger in size
(D) Chlorine is a gas

24. Which of the following is a microcosmic salt ?
(A) $Na(NH_4).H_2O$ (B) $Na(NH_4)HPO_4.4H_2O$
(C) $Na(NH_3)HPO_4.4H_2O$ (D) $K(NH_4)HPO_3.2H_2O$

25. Which of the following represents the correct pattern of electron filling in chromium ?
(A) [↑↓][↑↓][↑][ ][ ] [↑] (B) [↑][↑][↑][↑][ ] [↑↓]
(C) [↑][↑][↑][↑][↑] [↑] (D) [↑↓][↑][↑][↑][↑] [ ]

26. Which of the following contains amino group ?
(A) Basic days (C) Mordant dyes
(B) Vat dyes (D) Acid dyes

27. The chemical formula of lithopone is
(A) $BaS + ZnSO_4$ (B) $ZnS + BaSO_4$
(C) $BaSO_4 + ZnCO_3$ (D) $BaCO_3 + ZnSO_4$

28. Which of the following is called Baeyer's reagent ?
(A) Saturated $KMnO_4$ solution
(B) Neutral $KMnO_4$ solution
(C) Alkaline $KMnO_4$ solution
(D) Acidic $KMnO_4$ solution

29. The ratio of π to σ bonds in benzene is
(A) 3 : 1 (B) 1 : 6 (C) 1 : 4 (D) 1 :2

30. A war gas is prepared from
(A) PH_3 (B) C_2H_2 (C) Zinc phosphate (D) Chloropicrin

31. The sides of match box have coating of
(A) Potassium chlorate and red lead

(B) Potassium chlorate and antimony sulphide
(C) Antimony sulphide and red phosphorus
(D) Antimony shlphide and red lead

32. The crystal structure of NaCl is
(A) fcc (B) bcc (C) scc (D) Either 'B' or 'C'

33. The cynamide process is used in the formation of
(A) NH_3 (B) HNO_3 (C) N_2 (D) PH_3

34. Which of the following is an n, p reaction ?
(A) ${}_5C^{13} + {}_1H^1 \rightarrow {}_6C^{14}$ (B) ${}_7N^{14} + {}_1H^1 \rightarrow {}_8O^{15}$
(C) ${}_{13}Al^{27} + {}_0n^1 \rightarrow {}_{12}Mg^{27} + {}_1H^1$
(D) ${}_{92}U^{235} + {}_0n^1 \rightarrow {}_{54}Xe^{140} + {}_{88}Sr^{94} + {}_{30}n^1$

35. If $E^0 = 1.10$ volt for the cell reaction $Zn(s) + Cu^{2+}(aq) \rightarrow Cu(s) + Zn^{2+}(aq)$, the value of ΔG^0 is
(A) 212.3 kJ (B) 106.05 kJ (C) –212.3 kJ (D) 424.6 kJ

36. Which of the following represents a noble gas electronic configuration ?
(A) $1s^2 2s^2 2p^6 3s^2 3p^6 3d^{10} 4s^2 4p^6 4d^{10} 5s^2 5p^6 5d^1 6s^2$
(B) $1s^2 2s^2 2p^6 3s^2 3p^6 3d^{10} 4s^2 4p^6 4d^{10} 5s^2 5p^6 5d^5 6s^2$
(C) $1s^2 2s^2 2p^6 2s^2 3p^6 3d^{10} 4s^2 4p^6 4d^{10} 5s^2 5p^6$
(D) $1s^2 2s^2 2p^6 3s^2 3p^6 3d^{10} 4s^2 4f^{14} 5s^2 5p^6 5p^1$

37. For the reaction $C + O_2 \rightarrow CO_2$
(A) $\Delta H > \Delta E$ (B) $\Delta H < \Delta E$
(C) $\Delta H = \Delta E$ (D) Either 'A' or 'B'

38. Silver chloride is soluble in methylamine due to the formation of
(A) AgOH (B) $Ag(CH_3NH_2)Cl$
(C) $[Ag(CH_3NH_2)_2]Cl$ (D) $Ag + CH_3Cl + NH_4Cl$

39. For a process to be spontaneous
(A) ΔG should be –ve (B) ΔG should be +ve
(C) ΔH should be –ve (D) ΔH should be +ve

40. Which of the following type of metals acts as the most efficient catalyst ?
(A) Alkali (B) Transition
(C) Alkaline earth (D) None of these

41. The solubility product of $BaSO_4$ is 1.5×10^{-9} at 18°C. Its solubility in water at 18°C is
(A) 1.5×10^{-9} mole litre^{-1} (B) 1.5×10^{-5} mole litre^{-1}
(C) 3.9×10^{-9} mole litre^{-1} (D) 3.9×10^{-5} mole litre^{-1}

42. The pH value of a solution whose hydronium ion concentration is 6.2×10^{-9} mole L^{-1} is
(A) 6.21 (B) 7.21 (C) 7.75 (D) 8.21

43. Which of the following is a buffer solution ?
(A) $K_2SO_4 + H_2SO_4$ (B) $NH_4OH + NH_4Cl$
(C) $NaOH + CH_3COONa$ (D) $NaOH + Na_2SO_4$

44. Which of the following gives aldol condensation ?
(A) Acetaldehyde (B) Formaldehyde
(C) Benzaldehyde (D) Benzophenone

45. Which of the following composition for an alloy is correct ?
(A) Gun metal - Cu + Zn + Sn
(B) Duralumin - Al + Cu + Mg + Ag
(C) German silver - Cu + Zn + C (D) Solder-Pb + Al

46. Mac Arther process is used for
(A) Hg (B) Fe (C) Cl (D) O_2

47. The catalyst $SnCl_2$ / HCl is used in
(A) Clemmensen reduction
(B) Rosen mund's reduction
(C) Stephen's reduction (D) Cannizzaro's reduction

48. The standard reduction potentials for the following half-cell reactions are $Zn \rightarrow Zn^{2+} + 2e^+$; $E^o = -0.76$ V ; $Fe \rightarrow Fe^{2+} + 2e^-$; $E^o = -0.44$ V The e.m.f for the cell reaction $Fe^{2+}(aq) + Zn(s) \rightarrow Zn^{2+}(aq) + Fe(s)$ will be
(A) –0.32 V (B) +0.32 V (C) +1.20 V (D) –1.20 V

49. Iodoform test is not given by
(A) Acetic acid (B) Ethyl alcohol
(C) Acetone (D) None of these

50. The IUPAC name of $C_2H_5CONH_2$ is
(A) Propanamide (B) Ethanamide
(C) Methanamide (D) Benzamide

BHU - 1996

1. Iodoform CHI_3 reacts with Ag powder to produce
(A) Methane (B) butane (C) Ethene (D) Acetylene

2. The partial pressure of hydrogen in a flask containing 2g hydrogen 16g oxygen and 14g nitrogen is
(A) Half of the total pressure
(B) Equal to total pressure (C) 1/8th of the total pressure
(D) 1/3 of the total pressure

3. A closed vessel contains equal number of oxygen and hydrogen molecules at a total pressure of 740mm. If oxygen is removed from the system, the pressure will.
(A) Become half of 740 mm (B) Remain uncharged
(C) Become 1/4th of 740 mm
(D) Become two times of 740 mm

4. In a diamond, the bond present is
(A) Covalent (B) Electrovalent
(C) Hydrogen (D) Coordinate bond

5. Which of the following is a part of Nessler's reagent ?
(A) HgI_4 (B) Hg_2I_2 (C) HgI_2^{-4} (D) HgI_4^{-2}

6. Ester and acetamide are distinguished by
(A) Hydrolysis with a strong acid or an alkali
(B) Derivatives of fatty acids
(C) Grignard reagent (D) Both 'A' and 'B'

7. The smell of agarbatti mixed in the atmosphere, is due to
(A) Osmosis (B) Diffusion
(C) Evaporation (D) Condensation

8. Oxygen is paramagnetic. This is shown by
(A) Valency theory (B) Molecular orbital theory
(C) Quantum theory (D) Gravitational theory

9. Half-life period of a metal is 20 days. How much

fraction of the metal remains after 80 days ?

(A) $\frac{1}{16}$ (B) $\frac{1}{4}$ (C) $\frac{1}{8}$ (D) 1

10. When 0.04 Faraday of electricity is passed through a solution of $CaSO_4$, the weight of Ca^{+2} metal deposited at the cathode is

(A) 0.2 g (B) 0.4 g (C) 0.6 g (D) 0.8 g

11. ${}_7N^{14} + {}_2He^4 \rightarrow {}_1H^1 + x$. The value of x is

(A) ${}_8O^{18}$ (B) ${}_6O^{14}$ (C) ${}_8O^{16}$ (D) ${}_8O^{17}$

12. The amount of energy which is released due to the addition of extra electron to the outermost orbit of gaseous atom is called

(A) electron capacity (B) Electron affinity
(C) Ionisation potential (D) Electronegativity

13. 15 c.c.of a gaseous hydrocarbon required 45 c.c. of oxygen for complete combustion and 30 c.c. of CO_2 is formed. The formula of the hydrocarbon is

(A) C_3H_6 (B) C_2H_6 (C) C_4H_{10} (D) C_2H_4

14. Milk is an example of

(A) Gel (B)Sol (C) Suspension (D) Emulsion

15. When a strong beam of light is passed through a colloidal solution, the light will

(A) Give a rainbow (B) Be scattered
(C) Be reflected (D) Pass unchanged

16. Phenolpthalein is obtained by heating phthalic anhydride with concentrated H_2SO_4 and

(A) Phenol (B) Benzene
(C) Benzyl alcohol (D) Benzoic acid

17. The reaction
$C_6H_5CHO + (CH_3CO)_2O \xrightarrow{CH_3COONa} C_6H_5CH = CH-COOH + CH_3COOH$ is known is

(A) Perkin's reaction (B) Claisen condensation
(C) Benzoin condensation (D) Cannizaro's reaction

18. The four bonds in methane are directed in space with an angle of

(A) 109°28' (B) 180° (C) 120° (D) 90°

19. Moist ammonia gas is dried over

(A) P_2O_5 (B) Soda lime
(C) Conc.H_2SO_4 (D) Anhydrous $CaCl_2$

20. Molecular weight of urea is 60. A solution of urea containing 6g urea in one litre is

(A) 1 molar (B) 1.5 molar (C) 0.1 molar (D) 0.01 molar

21. Which of the following is contained in alcoholic beverages ?

(A) 2-propanol (B) Methanol (C) Glycerol (D) Ethanol

22. Maltose on hydrolysis, gives

(A) Glucose (B) Galactose + glucose
(C) Mannose + glucose (D) Mannose + fructose

23. Which of the following is an antibiotic ?

(A) Penicillin (B) Albumin (C) Insulin (D) Ascorbic acid

24. What is formed, when nitrobenzene is reduced using zinc and alkali ?

(A) Phenol (B) Aniline
(C) Nitrosobenzene (D) Hydrazobenzene

25. What is obtained when benzoyl chloride reacts with aniline in the presence of sodium hydroxide ?

(A) Benzoic acid (B) Azobenzene
(C) Acetanilide (D) Benzanilide

26. Carbylamine reaction is given by

(A) Primary-amine (B) Secondary-amine
(C) Tertiary-amine (D) Quaternary salts

27. Iodine dissolves readily in

(A) Water (B) Alcohol
(C) Potassium iodide (D) Carbon tetrachloride

28. Aqueous solution of Cu (II) gives an intense blue solution with ammonia, which is due to the formation of

(A) $[Cu(H_2O)_6]^{2+}$ (B) $Cu(OH)_2$
(C) $[Cu(NH_3)_4]^{2+}$ (D) $Cu(NH_3)_2$

29. Gold is not used in

(A) Medicine (B) Coins (C) Photography (D) Jewellery

30. Ethene can be produced from ethanol by

(A) Dehydrohalogenation (B) Dehydrogenation
(C) Dehydration with conc. H_2SO_4 at 170°C
(D) Reduction with hydrogen iodide

31. With oxalic acid, glycerol at 530 K gives

(A) Allyl alcohol (B) Formic acid
(C) Glyceraldehyde (D) Glycerol mono-oxalate

32. Picric acid is

(A) Phenol (B) Nitrolakane
(C) Trinitrophenol (D) Carboxylic acid

33. Cresol is

(A) Greenish yellow oil (B) Heavier than water
(C) Made up of mainly cresols (D) All of theses

34. Triamino benzene is a

(A) 2°-amine (B) 3°-amine
(C) 1°-amine (D) Quaternary salt

35. Vitamin E is

(A) Fat soluble (B) Water soluble
(C) Soluble in ether (D) Soluble in alcohol

36. Which one of the following vitamins is related to ergosterol ?

(A) Vitamin E (B) Vitamin K
(C) Vitamin C (D) Vitamin D

37. The absorption of hydrogen by palladium is called

(A) Hydration (B) Reduction
(C) Occlusion (D) Hydrogenation

38. The reaction $C_6H_5NH_2 + CHCl_3 + 3KOH \rightarrow C_6H_5NC + 3KCl + 3H_2O$ is known is

(A) Reimer-Tiemann reaction (B) Kolbe reaction
(C) Carbylamine reaction
(D) Hofmann's degradation

39. The rate of diffusion of hydrogen gas is

(A) Same the He gas (B) 1.4 times to He gas

(C) two times to He gas (D) Five times to he gas

40. The relative rate of diffusion of a gas (molecular weight = 128) as compared to oxygen is
(A) 2 times (B) 0.25 times
(C) 0.125 times (D) 0.5 times

41. Which of the following processes is not used for the purification of solid impurities ?
(A) Distillation (B) Sublimation
(C) Crystallisation (D) Steam distillation

42. Which of the following is metalloid ?
(A) Pb (B) Zn (C) As (D) Na

43. The best quality of steel is obtained by
(A) L–D process (B) Decon's process
(C) Bessemer process (D) Open Hearth process

44. The number of molecules present in 1 dm^3 of hydrogen at S.T.P. is
(A) 3.69×10^{22} (B) 2.69×10^{22}
(C) 1.69×10^{22} (D) 1.09×10^{22}

45. Four litres of CO_2 is kept at 27°C. What should be the volume if the temperature is lowered to 150 K at the same pressure ?
(A) 0.5 litre (B) 1 litre (C) 2 litres (D) 4 litres

46. 100 ml of 0.2 M H_2SO_4 is added to 100 ml of 0.2 M NaOH. The resulting solution will be.
(A) Acidic (B) Basic (C) Neutral (D) Slightly basic

47. Element boron is obtained by the reduction of oxides of
(A) Non-metal (B) Noble gas
(C) Electronegative metal (D)Electropositive metal

48. Benzene show mainly........... reaction
(A) Electrophilic substitution (B) Electrophilic addition
(C) Nucleophilic substitution (D) Nucleophilic addition

49. Which of the following is the most electronegative element ?
(A) Fluorine (B) Chlorine (C) Bromine (D) Iodine

50. Which of the following solutions in water possesses the lowest vapour pressure ?
(A) 0.1 M NaCl (B) 0.1 M $BaCl_2$
(C) 0.1 M KCl (D) None of these

BHU - 1997

1. Calculate the standard heat of formation of carbon disulphide (l) given that the standard heat of combustion of carbon (s), sulphur (s) and carbon disulphide (l) are –393.3, –293.72 and –1108.76 kJ mol^{-1} respectively.
(A) –128.02 kJ mol^{-1} (B) +12.802 kJ mol^{-1}
(C) +128.02 kJ mol^{-1} (D) –12.802 kJ mol^{-1}

2. The enthalpy of water is 386 kJ. What is the entropy of water ?
(A) 1.5 kJ (B) 22.05 kJ (C) 0.5 kJ (D) 1.03 kJ

3. The solubility of AgCl at 20°C is 1.35×10^{-5} gm/lit. The solubility product of AgCl is
(A) 2×10^{-16} (B) 1.035×10^{-5}
(C) 1.0×10^{-14} (D) 108×10^{-3}

4. The solubility product of Ag_2CrO_4 is 32×16^{-12}. What is the concentration of CrO_4^- ions in that solution ?
(A) 2×10^{-4} m (B) 16×10^{-4} m
(C) 8×10^{-4} m (D) 8×10^{-8} m

5. The formula of green vitriol is
(A) $FeSO_4.7H_2O$ (B) $MgSO_4.7H_2O$
(C) $ZnSO_4.7H_2O$ (D) $CuSO_4.5H_2O$

6. Which of the following statements is correct for covalent bond ?
(A) It may be polar or non-polar
(B) Direction is non-polar
(C) Electrons are shared between two atoms
(D) Valency electrons are attracted.

7. I.U.P.A.C. name of the compound
$CH_3– CH = C– CH_3$
with CH_2CH_3 on the third carbon, is
(A) 3-methyl-3 pentene (B) 2-ethyl-2 butene
(C) 3-ethyl-2 butene (D) 3-methyl-2 pentene

8. The change in optical rotation with time of freshly prepared solution of sugar is known as
(A) Inversion (B) Specific rotation
(C) Rotatory motion (D) Mutarotation

9. Hydroysis of sucrose is called
(A) Saponification (B) Hydration
(C) Esterification (D) Inversion

10. Sweetest sugar is
(A) Fructose (B) Maltose (C) Sucrose (D) Glucose

11. The metal present in vitamin B_{12} is
(A) Magnesium (B) Iron (C) Cobalt (D) Manganese

12. Which of the following statements is correct ?
(A) Phenol is more acidic than ethyl alcohol
(B) Phenol is more acidic than carboxylic acid
(C) Phenol is less acidic than ethyl alcohol.
(D) Phenol is more acidic than carbonic acid

13. The colour of P-amino azobenzene is
(A) Orange (B) Congo red (C) Bismark brown (D) Indigo

14. Oxidation number of N in HNO_3 is
(A) –3, +5 (B) +5 (C) –3.5 (D) +3.5

15. If HNO_3 changes into N_2O, the oxidation number is changed by
(A) 1 (B) +4 (C) +2 (D) 6

16. When $CuSO_4$ reacts with KI, the oxidation number of Cu changes by
(A) 0 (B) –1 (C) 1 (D) 2

17. Which of the following elements has the maximum electron affinity?
(A) Cl (B) F (C) Br (D) I

18. 0.15 gm of a substance dissolved in 15g of solvent is boiled at a temperature higher by 0.216°C than that of the

pure solvent. Find out the molecular weight of the substance Molal elevation constant for the solvent is 2.16°C.

(A) 10 (B) 10.1 (C) 100 (D) 1.01

19. The size of particles in suspension, true solution and colloidal solution varies in the order

(A) Suspension > colloidal + true solution

(B) True solution > suspension > colloidal

(C) Suspension > colloidal > true solution

(D) None of these

20. Froth flotation process is used for the metallurgy of

(A) Oxide ores (B) Chloride ores

(C) Sulphide ores (D) Amalgams

21. D_2O is used more in

(A) Pharmaceutical preparations

(B) Chemical industry (C) Nuclear reactor

(D) insecticide preparation

22. What is lye ?

(A) 10% solution of NaCl (B) 10% solution of KOH

(C) 10% solution of NaOH (D) 10% solution of Na_2CO_3

23. Which of the following ions is coloured ?

(A) Ti^{4+} (B) V^{3+} (C) Cu^{+} (D) Cu^{2+}

24. The largest number of molecules is present in

(A) 34 gm of water (B) 54 gm of N_2O_5

(C) 28 gm of CO_2 (D) 46 gm of CH_3OH

25. Volume of a gas at N.T.P. is 1.12 x 10^{-7} c.c. Calculate the number of molecules in it.

(A) 30.1 x 10^{23} (B) 3.01 x 10^{20}

(C) 3.01 x 10^{12} (D) 3.01 x 10^{24}

26. Lowering of vapour pressure is the highest for

(A) Urea (B) 0.1 M glucose

(C) 0.1 M $MgSO_4$ (D) 0.1 M $BaCl_2$

27. $C + O_2 \rightarrow CO_2$; $\Delta H = X$

$CO + \frac{1}{2}O_2 \rightarrow CO_2$; $\Delta H = Y$, then the heat of formation of CO is

(A) X+Y (B) 2X−Y (C) X−Y (D) Y−2X

28. pH of .01 M NaOH solution is

(A) 12 (B) 1 (C) 2 (D) 14

29. What is the correct sequence of bond order ?

(A) $O_2^+ > O_2^- > O_2$ (B) $O_2^+ > O_2 > O_2^-$

(C) $O_2^+ > O_2^{-1} > O_2^{+1}$ (D) $O_2^{-1} > O_2^{+1} > O_2$

30. Which of the following is not a Lewis acid ?

(A) $FeCl_3$ (B) PH_3 (C) BF_3 (D) $AlCl_3$

31. Gibbs free energy for reaction at equilibrium is

(A) Zero (B) −ve (C) +ve (D) +ve or −ve

32. Which of the following is optically active ?

(A) Ethylene glycol (B) Oxalic acid

(C) Glycerol (D) Tartaric acid

33. The half-life of an isotope is 10 hrs. How much will be left behind after 4 hrs in 1 gm sample?

(A) 45.6 x 10^{23} atoms (B) 4.56 x 10^{23} atoms

(C) 4.56 x 10^{21} atoms (D) 45.6 x 10^{20} atoms

34. A sample of $Na_2CO_3.H_2O$ weighing 0.62g is added to 100 ml of 0.1 N H_2SO_4 solution. What will the resulting solution ?

(A) Basic (B) Acidic (C) Neutral (D) None of these

35. Which of the following statements is correct ?

(A) All halogens form oxyacid

(B) All halogens except fluorine form oxyacids

(C) Only chlorine and bromine form oxyacid

(D) Only iodine form oxyacid

36. When ethyl iodide and propyl iodide react with Na in the presence of ether, they form

(A) One alkane (B) Two alkanes

(C) Four alkanes (D) Three alkanes

37. On passing 0.5 Faraday of electricity through NaCl, the amount of Cl liberated at cathode is

(A) 35.5 gm (B) 17.75 gm (C) 71 gm (D) 142 gm

38. The enthalpy of neutralization of HCN by NaOH is −12.13 kJ mol^{-1}. The enthalpy of ionisation of HCN will be

(A) 4.519 kJ (B) 45.10 kJ (C) 451.9 kJ (D) 45.19 kJ

39. Km value of an enzyme is the subsrate concentration of which

(A) Reaction attains half of its maximum velocity

(B) Reaction attains 8 times maximum velocity

(C) Reaction attains two times of its maximum velocity

(D) Reaction attains same as maximum velocity

40. The values of heat of formation of SO_2 and SO_3 are −298.2 kJ and −98.2 kJ. The heat of reaction of the reaction, $SO_2 + \frac{1}{2} O_2 \rightarrow SO_3$ will be

(A) −200 kJ (B) −356.2 kJ (C) +200kJ (D) −396.2 kJ

41. The pH of 0.0001 M NaOH is

(A) 12 (B) 11 (C) 10 (D) 4

42. Ethyl alcohol gives ethyl chloride with

(A) $SOCl_2$ (B) KCl (C) Cl_2 (D) NaCl

43. Which of the following is not a Bronsted acid ?

(A) NH_4^+ (B) CH_3COO^- (C) H_2O (D) HSO_4^-

44. Electronic configuration of Sc^{21} is

(A) $1s^2, 2s^2p^63s^2, 3p^64s^2, 3d^1$ (B) $1s^2, 2s^2, 2p^63s^2, 3p^64s^1, 3d^2$

(C) $1s^2, 2s^2, 2p^63s^2, 3p^64s^0, 3d^3$ (D) $1s^2, 2s^2p^63s^2, 3p^64s^2, 3d^2$

45. Isopropyl alcohol on oxidation gives

(A) Acetone (B) Acetaldehyde (C) Ether (D) Ethylene

46. Toluene on oxidation with chromyl chloride gives

(A) Benzoic acid (B) Benzaldelyde

(C) Chlorobenzene (D) None of these

47. The main product (X) in the reaction

$CH_3CONH_2 \xrightarrow{Br_2 + NaOH} X$ is

(A) CH_3COOH (B) $CH_3CH_2NH_2$ (C) CH_3NH_2 (D) CH_3Br

48. Palmitic acid is

(A) $C_{16}H_{31}COOH$ (B) $C_{17}H_{35}COOH$

(C) $C_{15}H_{31}COOH$ (D) $C_{17}H_{31}COOH$

49. Chloroform reacts with NaOH to give

(A) Soidium oxalate (B) CH_3COONa
(C) CH_3OH (D) HCOONa

50. Which of the following on hydrolysis forms acetic acid ?
(A) C_2H_5OH (B) $C_2H_5NH_2$ (C) CH_3CN (D) CH_3OH

BHU - 1998

1. A reaction is A+B $\rightarrow$ C + D. Initially it starts with equal concentration of A and B. At equilibrium the C is two times of moles of A. What is the equilibrium constant of the reaction ?
(A) 1/4 (B) 1/2 (C) 4 (D) 2

2. What is the amount of chlorine evolved when 2 amperes of current is passed for 30 minutes in an aqueous solution of NaCl ?
(A) 33 g (B) 99 g (C) 66 g (D) 1.32 g

3. H_2S is passed through an acidified solution containing Ag, Cu and Zn. Which forms precipitate ?
(A) Ag (B) Zn (C) Cu (D) None of these

4. Cis and trans 2-butenes are
(A) Conformational isomers (B) Optical isomers
(C) Position isomers (D) Geometrical isomers

5. Which of the following is a chiral molecule ?
(A) $CHBr_3$ (B) CHClBrI (C) CH_3Cl (D) CH_2Cl_2

6. 0.2 gm of fine animal charcoal is mixed with half litre of acetic acid solution and shaken for 30 minutes. Then
(A) Concentration increases
(B) The concentration of the solution decreases
(C) Concentration remains same (D) None of these

7. Which of the following has highest chlorine content ?
(A) Pyrene (B) DDT (C) Chloral (D) Gammexane

8. Propene reacts with HBr in presence of peroxides to give
(A) 2-bromopropane (B) 1-bromopropane
(C) 3-bromopropane (D) None of these

9. DDT is prepared by reacting chlorobenzene with
(A) CCl_4 (B) CCl_3-CHO (C) $CHCl_3$ (D) Ethane

10. IUPAC name of

$$CH_3-\underset{C_2H_5}{\overset{H}{\underset{|}{\overset{|}{C}}}}-\underset{CH_3}{\overset{C_4H_9}{\underset{|}{\overset{|}{C}}}}-CH_3 \text{ is :}$$

(A) 2-ethyl, 3,3-dimethyl-heptane
(B) 2-butyl, 2-methyl, 3-ethyl-butane
(C) 3,4,4-trimethyl-heptane (D) 3,4,4-trimethyloctane

11. Which of the following is an electrophile ?
(A) H_2O (B) NH_3 (C) $AlCl_3$ (D) $C_2H_5NH_2$

12. In which reaction aromatic aldehyde is treated with acid anhydride in presence of corresponding salt of the acid to give unsaturated aromatic acid ?
(A) Friedel-Crafts reaction (B) Perkin's reaction
(C) Wurtz's reaction (D) None of these

13. The gas molecules have r.m.s. velocity of its molecules as 1000 m/s. What is its average velocity ?
(A) 546 m/s (B) 960 m/s (C) 1012 m/s (D) 921.58 m/s

14. A container has hydrogen and oxygen mixture in the ratio of 4:1 by weight, then
(A) Internal energy increases
(B) Entropy of the gases increases
(C) Internal energy of the gases decreases
(D) Entropy of the gases decreases.

15. Specific heat of metal is 2.3 and its chloride contains 87% of chlorine. What is the molecular weight of the metal ?
(A) 24 (B) 54 (C) 62 (D) 27

16. A container contains 1 mole of a gas at 1 atm. pressure 27°C temperature and volume is 24.6 lit. If pressure is 10 atm and temperature 327°C, then the volume approximately is
(A) 2 litre (B) 48 litre (C) 15 litre (D) 10 litre

17. Increase in boiling point of a sucrose solution is 0.1 K, then what is increase in boiling point of the same concentration of NaCl solution ?
(A) 0.4 K (B) 0.58 (C) 0.1K (D) 0.2 K

18. Dialysis can separate
(A) Glucose and fructose (B) Glucose and sucrose
(C) Glucose and NaCl (D) Glucose and proteins

19. Which of the following is not a macromolecule ?
(A) DNA (B) Starch (C) Palmitate (D) Insulin

20. If a gas at constant temperature and pressure expands, its
(A) Internal energy decreases
(B) Internal energy increases
(C) Internal energy remains the same
(D) Its entropy increases and then decreases

21. Equal volumes of methanoic acid and sodium hydroxide are mixed. if x is the heat of formation of water, then heat evolved on neutralisation is
(A) More than x (B) Equal to x
(C) Twice of x (D) Less than x

22. If 900 J/g of heat is exchanged at the boiling point of water, then what is the increase in entropy ?
(A) 43.4 J/mole (B) 87.2 J/mole
(C) 900 J/mole (D) Zero

23. A catalyst in a chemical reaction
(A) Increases the activation energy of the reaction
(B) Changes the equilibrium constant of a reaction
(C) Does not initiate a reaction
(D) Does not change the rate of the reaction

24. If initial concentration is reduced to 1/4th in a zero order reaction, the time taken for half the reaction to complete
(A) remains same (B) Becomes 4 times

(C) Becomes one-fourth (D) Doubles

25. CH_3COOH is reacted with $CH \equiv CH$ in presence of Hg^{++}, the product is

(A) CH_3
|
HC–$(OOC - CH_3)_2$

(B) CH_3
|
$COO - CH_3$

(C) CH_2OOCCH_3
|
CH_2OOCCH_3

(D) None of these

26. Formaldehyde when reacts with methyl magnesium bromide gives
(A) CH_3CHO (B) HCHO (C) C_2H_5OH (D) CH_3COOH

27. Aromatic compounds undergo mostly
(A) Nucleophilic addition (B) Electrophilic addition
(C) Electrophilic substitution (D) None of these

28. What is the oxidation number of Cr in $K_2Cr_2O_7$?
(A) +6 (B) –6 (C) +3 (D) –5

29. Ratio of masses of proton and electron is
(A) Infinite (B) $1.8x10^{+3}$ (C) 1.8 (D) None of these

30. An electron has magnetic quantum number as - 3, what is its principal quantum number ?
(A) 1 (B) 2 (C) 3 (D) 4

31. Wavelength associated with electron motion
(A) Remains same irrespective of speed of electron
(B) Increases with increase in speed of electron
(C) Decreases with speed increase of e^- (D) Is zero

32. Which compound volatiles easily on heating ?
(A) $MgCl_2$ (B) $HgCl_2$ (C) $ZnCl_2$ (D) None of these

33. One of the following is false for Hg
(A) It is a metal (B) It has high specific heat
(C) It can evolve hydrogen from H_2S
(D) It is less reactive than hydrogen.

34. Which among the following is the strongest reducing agent ? Given $Fe^{2+} + 2e^- \rightarrow Fe(-0.44\ V)$; $Ni^{2+} + 327°C, e^- \rightarrow Ni\ (-0.25V)$; $Sn^{2+} + 2e^- \rightarrow Sn(-0.14V)$ and $Fe^{3+} + e^- \rightarrow Fe^{2+}\ (-77\ V)$?
(A) Fe (B) Fe^{2+} (C) Ni (D) Sn

35. "Reference protein" is obtained from
(A) Meat (B) Milk (C) Egg (D) Soyabean

36. Carborundum is
(A) SiC (B) CS_2 (C) BN (D) SiO_2

37. Units of mole fraction are
(A) Moles/lit. (B) Moles/lit^2 (C) Moles lit (D) Dimensionless.

38. The neutralisation of a strong acid by a strong base liberates an amount of energy per mole of H^+ that
(A) Depends upon the temperature at which the reaction takes place
(B) Depends upon which acid and base are involved
(C) Depends upon which catalyst is used
(D) Is always the same

39. Heat exchanged in a chemical reaction at constant temperature and constant pressure is called
(A) Internal energy (B) Enthalpy
(C) Entropy (D) Free energy

40. When acetylene is passed through dil H_2SO_4 in the presence of $HgSO_4$ the compound formed is
(A) C_2H_5OH (B) Acetone
(C) Carbide of Hg (D) Acetadehyde

41. Carbon percentage is maximum in
(A) Pyrene (B) Gammaxine(C) Ethylene glycol (D) PVC

42. Which of the following oxides reacts with HCl and NaOH ?
(A) CaO (B) ZnO (C) N_2O_5 (D) CO_2

43. Which of the following is least ionic ?
(A) AgCl (B) KCl (C) $BaCl_2$ (D) $COCl_2$

44. Alkyl halides react with metallic sodium in the presence of anhydrous ether in
(A) Wurtz's reaction (B) Aldol's reaction
(C) Perkin's reaction (D) None of these

45. Aspirin is a/an
(A) Antibiotic (B) Antipyretic
(C) Antiseptic (D) None of these

46. 240g of urea are present in 10 litre solution. The active mass of urea will be
(A) 0.2 (B) 0.06 (C) 0.4 (D) 0.08

47. Degree of ionsation of a solution depends upon
(A) Temperature (B) Nature of the electrolyte
(C) Nature of the solvent (D) All of these

48. Which of the following is most acidic ?
(A) Na_2O (B) MgO (C) Al_2O_3 (D) CaO

49. Which metal is protected by a layer of its own oxide ?
(A) Aluminium (B) Silver (C) Gold (D) Iron

50. Which of the following combines with Fe(II) ions to from a brown complex ?
(A) N_2O (B) NO (C) N_2O_3 (D) N_2O_5

BHU - 1999

1. The rate constant of a first order reaction is 3×10^{-6} per sec. If the initial concentration is 0.10 M, the initial rate of reaction is
(A) $3 \times 10^{-7}\ ms^{-1}$ (B) $3 \times 10^{-8}\ ms^{-1}$
(C) $3 \times 10^{-5}\ ms^{-1}$ (D) $3 \times 10^{-6}\ ms^{-1}$

2. Heat of neutralistion of a strong acid against a strong base is constant and equal to
(A) 57.3 kcal (B) 59.3 kcal (C) 13.7 kcal (D) 53.7 kcal

3. The metal that forms a self protecting film of oxide to prevent corrosion, is
(A) Cu (B) Al (C) Na (D) Au

4. The molarity of pure water is
(A) 34.3 (B) 55.3 (C) 1 (D) 18

5. Which of the following ions is the largest in size
(A) K+ (B) S^{2-} (C) Cl^- (D) Ca^{2+}

6. Which of the following compounds is the strongest base ?
(A) Ammonia (B) Aniline
(C) Methylamine (D) N-methyl aniline

7. The conjugate acid of NH_2^- ion is
(A) NH_3 (B) NH_4^+ (C) N_2H_4 (D) NH_2OH

8. Partial pressure of hydrogen in a flask containing 2 g hydrogen, 16 g oxygen and 14 g nitrogen is
(A) Equal to the total pressure(B) 1/8 the of the total pressure
(C) Half of the total pressure(D) 1/3rd of the total pressure

9. Vapour density of a metal chloride is 77. If its equivalent weight is 3, its atomic weight will be
(A) 9 (B) 12 (C) 3 (D) 6

10. The average kinetic energy of an ideal gas per molecule in S.I. units at 25°C will be
(A) 61.7×10^{-21}J (B) 6.17×10^{-21} J
(C) 6.17×10^{-20} J (D) 7.16×10^{-20} J

11. Ethyl amine reacts with chloroform and potassium hydroxide, to form
(A) Amine (B) Alcohol (C) Aldehyde (D) Ethyl isocyanide

12. Which of the following explains the sequence of filling the electrons in different shells ?
(A) Octete rule (B) Hund's rule
(C) Aufbau principle (D) All of these

13. The tetrahedral structure is possessed by
(A) CH_4 (B) NH_4^+ (C) BF_4^- (D) All of these

14. The type of hybridisation of boron in diborane is
(A) sp^3-hybridisation (B) sp^2d^2-hybridisation
(C) sp-hybridisation (D) sp^2-hybridisation

15. Copper turnings, when heated with concentrated sulphuric acid, give
(A) O_2 (B) SO_2 (C) SO_3 (D) H_2S

16. Electronic configuration of calcium atom may be written as
(A) [Ar] $4s^2$ (B) [Xe] sp^2 (C) [Ne] $2s^2$ (D) [Ne] $3s^2$

17. Which one of the following statements is wrong for gases ?
(A) Volume of the gas is equal to volume of the container confining the gas
(B) Confined gas exerts uniform pressure on the walls of its container in all directions
(C) Gases do not have a definite shape and volume
(D) Mass of the gas can not be determined by weighing a container in which gas is enclosed

18. If HNO_3 changes into N_2O, the oxidation number is changed by
(A) 2 (B) 6 (C) 0 (D) +4

19. The values of heat of formation of SO_2 and SO_3 are −298.2 kJ and −98.2 kJ respectively. The heat of formation of the reaction $SO_2 + 1/2O_2 \rightarrow SO_3$ will be
(A) −200 kJ (B) −356.2 kJ
(C) +200 kJ (D) −396.2 kJ

20. The high density of water as compared to ice is due to
(A) Dipole-dipole interactions
(B) Dipole-induced dipole interactions
(C) Hydrogen bonding interactions
(D) Induced dipole induced dipole interactions

21. The number of atoms in 4.25 g of NH_3 is approximately
(A) $1.0^{21} \times 10^{23}$ (B) $1.0^{52} \times 10^{23}$
(C) $2.0^{23} \times 10^{23}$ (D) $6.0^{23} \times 10^{23}$

22. What is the oxidation number of Co in $[Co(NH_3)_4ClNO_2]$
(A) +2 (B) + 3 (C) +4 (D) +5

23. 0.24 g of volatile gas upon vaporisation gives 45 ml vapour at STP. What will be the vapour density of the substance ? (Density of H_2 = 0.089)
(A) 5.992 (B) 59.93 (C) 95.39 (D) 95.93

24. The Bohr's orbit radius for the hydrogen atom (n=1) is approximately 0.530 Å. The radius for the first excited state (n = 2) orbit is
(A) 0.13Å (B) 1.06Å (C) 4.77Å (D) 2.22Å

25. A certain mass of gas occupies a volume of 300 c.c. at 27°C and 620 mm pressure. The volume of this gas at 47°C and 640 mm pressure will be
(A) 310 c.c. (B) 350 c.c. (C) 400 c.c. (D) 510 c.c.

26. What will be the molarity of a solution containing 5 g of sodium hydroxide in 250 ml solution ?
(A) 0.1 (B) 0.5 (C) 1 (D) 2

27. In which of the following reactions, H_2O_2 is acting as a reducing agent ?
(A) $H_2O_2 + SO_2 \rightarrow H_2SO_4$ (B) $2KI + H_2O_2 \rightarrow 2KOH + I_2$
(C) $PbS + 4H_2O_2 \rightarrow PbSO_4 + 4H_2O$
(D) $Ag_2O + H_2O_2 \rightarrow 2Ag + H_2O + O_2$

28. Gold number is associated with
(A) Protective colloids (B) Amount of pure gold
(C) Electrophoresis (D) Purple of cassius

29. To prepare a pure sample of n-hexane using sodium metal as one reactant, the other reactant or reactants will be
(A) n-propyl bromide
(B) Ethyl bromide and n-butyl bromide
(C) Ethyl chloride and n-butyl bromide
(D) Methyl bromide and n-pentyl chloride

30. The decreasing order of boiling points is
(A) iso-pentane > n-pentane > neo-pentane
(B) neo-pentane > iso-pentane > n-pentane
(C) n-pentane > iso-pentane > neo-pentane
(D) n-pentane > neo-pentane > iso-pentane

31. Molecular weight of urea is 60. The number of moles in the solution of urea containing 6 g urea in one litre is
(A) 0.1 (B) 0.2 (C) 0.4 (D) 0.8

32. If a gas at constant temperature and pressure expands, then its
(A) Internal energy decreases
(B) Internal energy remains the same
(C) Internal energy increases

(D) entropy increases first and then decreases

33. Which pair of species have same percentage of carbon ?
(A) CH_3COOH and C_2H_5OH
(B) CH_3COOH and $C_6H_{12}O_6$
(C) $HCOOCH_3$ and $C_{12}H_{22}O_{11}$ (D) $C_6H_{12}O_6$ and $C_{12}H_{22}O_{11}$

34. A biological catalyst is essentially a/an
(A) Enzyme (B) Amino acid
(C) Carbohydrate (D) Nitrogen molecule

35. Heating a mixture of sodium benzoate and soda lime gives
(A) Benzene (B) Aniline (C) Toluene (D) None of these

36. Haemoglobin contains 0.334% of iron by weight. If the molecular weight of haemoglobin is approximately 67200, then the number of iron atoms (atomic weight of Fe is 56) present in one molecule of haemoglobin are
(A) 1 (B) 4 (C) 5 (D) 8

37. Nitrobenzene combines with hydrogen in the presence of platinum to yield
(A) Toluene (B) Benzene (C) Aniline (D) Azobenzene

38. In the following nuclear reaction :
$_4Be + {_2He^4} \rightarrow {_6C^{12}} + {_0n^1}$; the mass no of Be atom is
(A) 4 (B) 6 (C) 7 (D) 9

39. Which ion is detected by Nesseler's reagent ?
(A) MnO_4 (B) CrO_4^{2-} (C) NH_4^+ (D) PO_4^{3-}

40. Standard state Gibbs free energy change for isomerization reaction, cis-2-pentene $\rightleftharpoons$ trams-2-pentene, is –3.67 kJ/mol at K. If more trans-2- pentene is added to the reaction vessel, then
(A) More cis-2- pentene is formed
(B) Additional trans-2- penetene is formed
(C) Equilibrium remains unaffected
(D) Equilibrium is shifted in forward direction

41. When diethyl ether is treated with Cl_2, in the presence of sunlight, the product formed is
(A) $CH_3CHCl–O–CH_2CH_3$ (B) $CH_3CHCl–O–CHClCH_3$
(C) $CCl_3CCl_2OCCl_2CCl_3$ (D) $CH_3CHCl–O–CH_2CH_2Cl$

42. Chloramphenicol is an axample of
(A) Antibiotic (B) Antipyretic
(C) Antiseptic (D) Sulpha drug

43. Terylene is used for making
(A) Sails (B) Fabrics (C) Seat belts (D) All of these

44. Natural rubber is a polymer of
(A) Ethylene (B) Isoprene
(C) Acrylic acid (D) 1,4 butadiene

45. Starch is converted into maltose by
(A) Maltase (B) Invertase (C) Zymase (D) Diastase

46. Choroform on treatment with concentrated HNO_3 gives
(A) CCl_3NO_2 (B) $CHCl_2HNO_3$
(C) $CHCl_2NO_2$ (D) $CHCl_2NO_3$

47. The waxes are long chain compounds of fatty acids, which belong to the class of
(A) Esters (B) Ethers (C) Alcohols (D) Aldelydes

48. Ehylamine can be prepared by the action of bromine and caustic potash on
(A) Acetamide (B) Formamide
(C) Propionamide (D) Methyl cyanide

49. Ethyl alcohol is soluble in water in all proportions, because it
(A) Is basic in nature (B) Is acidic in nature
(C) Dissociates in water
(D) Forms hydrogen bond with water

50. In Liebermann's nitroso reaction, sequential changes in the colour of phenol occurs as
(A) Red → green → white (B) Green → yellow → white
(C) Red → deep blue → green
(D) Brown or red → green → red → deep blue

BHU - 2000

1. The number of unpaired electrons in $Ni(CO)_4$ is
(A) Three (B) Four (C) Zero (D) One

2. Which of the following has square planar structure ?
(A) $[Ni](CO_4)]$ (B) $[Ni(Cl_4)]^{-2}$
(C) $[Ni(CN)_4]^{2+}$ (D) All of these

3. Which one of the following pairs of substances will not evolve H_2 gas ?
(A) Cu and HCl (g) (B) Sodium and ethyl alcohol
(C) Iron and H_2SO_4 (aq) (D) Iron and steam

4. Optical glass has high percentage of
(A) MgO (B) Al_2O_3 (C) B_2O_3 (D) PbO

5. Cyanide process is used for extraction of
(A) Ag (B) Pt (C) Zn (D) Ni

6. Epsom salt is
(A) $MgSO_4$ (B) $MgSO_4.2H_2O$
(C) $MgSO_4.7H_2O$ (D) $MgSO_4.\frac{1}{2}H_2O$

7. Iodine test is shown by
(A) Polypeptides (B) glycogen (C) Starch (D) glucose

8. Na when reacts with phenol,
(A) No reaction occurs (B) H_2 is evolved
(C) There is oxidation of benzene involving release of CO_2
(D) Benzene is formed

9. Tetrabromoethane on heating with Zn gives
(A) Ethyl Bromide (B) Ethane (C) Ethene (D) Ethyne

10. Total number of acyclic isomers including stereoisomers of C_4H_7Cl are
(A) 9 (B) 10 (C) 12 (D) 13

11. A fruity smell is produced by reaction of C_2H_5OH with
(A) PCl_5 (B) $CH_3–\overset{\overset{O}{\|}}{C}–CH_3$ (C) $CH_3–\overset{\overset{O}{\|}}{C}–OH$
(D) None of these

12. On reaction with HNO_2 , $C_2H_5NH_2$ produces
(A) C_2H_5OH (B) $C_2H_5–NO_2$
(C) $CH_3–CHO$ (D) CH_3COOH

13. Type of forces between the molecules of C_6H_6 are
(A) Dipole-Dipole interactions (B) Dispersive forces
(C) H-bonding (D) van der Waal's forces

14. The material used in warfares for making smoke screens for military purposes is
(A) $SiCl_4$ (B) PCl_5 (C) PH_3 (D) CCl_4

15. Lithopone is
(A) $ZnSO_4 + PbS$ (B) $ZnS + BaSO_4$
(C) $ZnS + Ba(OH)_2$ (D) $ZnSO_4 + BaSO_4$

16. Clemmensen reduction of a ketone is carried out in presence of
(A) $LiAlH_4$ in ether (B) Zn-Hg with HCl
(C) H_2 with Pd as catalyst (D) Glycol with KOH

17. C_3H_9N cannot represent
(A) 3° amines (B) Quarternary salt
(C) 1° amines (D) 2° amines

18.
$$C_6H_5OH \xrightarrow[NaOH(aq)]{ClCOCH_3(aq)} C_6H_5\,O{-}\overset{\overset{O}{\|}}{C}{-}CH_3$$ is an example of
(A) Schotten-Baumann reaction (B) Acetlyation
(C) Reimer Tiemen reaction (D) Benzoylation

19. Sides of safety match boxes contain
(A) P_4S_3 (B) $KClO_3$, KNO_3, Sulphur + Antimony
(C) Red phosphorus + Sand powder
(D) $Ca_3(PO_4)_2$ + Glass pieces

20. The formula of borax is
(A) $Na_2[B_4O_5(OH)_4]\ 8H_2O$ (B) $Na_2B_4O_7.4H_2O$
(C) $Na_2[B_4O_5(OH)_4]\ 10H_2O$ (D) $Na_2B_4O_7.8H_2O$

21. Maximum value of paramagnetic behaviour is shown by
(A) $[Fe(CN)_6]^{3-}$ (B) $[Sc(CN_6]^{3-}$
(C) $[Cr(CN)_6]^{3-}$ (D) $[Co(CN)_6]^{3-}$

22. β-particles are
(A) Neutral particles orbiting nucleus
(B) A particle produced by splitting of e- into H^+ and ${}_0n^1$
(C) Charged particles emitted from nucleus
(D) High energy positively charged particles emitted from nucleus

23. Gas equation, PV = nRT is obeyed by
(A) Only isothermal process (B) Only adiabatic process
(C) Both (A) and (B) (D) None of these

24. For a redox reaction,
$HNO_4^- + C_2O_4^{2-} + H^+ \rightarrow Mn^{2+} + CO_2 + H_2O$, Correct stichometric coefficients of HNO_4^- + $C_2O_4^{2-}$ OH^+ are
(A) 2,5,16 (B) 16,5,2 (C) 5,11,2 (D) 2,16,5

25. Densities of two gases are in ratio 1:2 and their temperatures are in ratio 2:1, then ratio of their respective pressures is
(A) 1 : 1 (B) 1 : 2 (C) 2 : 1 (D) 4 : 1

26. Which of the following statements is true ?
(A) All metal nitrates are insoluble in H_2O
(B) Solubility depends upon temperature
(C) All metal nitrates are soluble in H_2O
(D) Either (A) or (B)

27. BF_3 molecule has planar structure. Correct explanation for this is
(A) BF_3 is sp^3 hybridised (B) BF^3 is sp^2 hybridised
(C) BF_3 is sp hybridised (D) BF_3 is sp^3d hybridised

28. Electronic configuration of most electronegative element is
(A) $1s^22s^22p^43s^1$ (B) $1s^22s^22p^63s^23p^5$
(C) $1s^22s^22p^5$ (D) $1s^22s^22p^63s^23p^6$

29. The reaction of aniline with acetyl chloride in presence of NaOH gives
(A) Acetamide (B) Aniline hydrochloride
(B) P- chloroaniline (D) None of these

30. Which of the following represents a noble gas configuration ?
(A) $1s^2,2s^22p^6,3s^23p^63d^{10},4s^24f^{14},5s^2,\ 5p^65d^1$
(B) $1s^2,2s^22p^6,3s^23p^63d^{10},4s^24p^64d^{10},5s^25p^65d^1,6s^2$
(C) $1s^2,2s^22p^6,3s^23p^63d^{10},\ 4s^24p^64d^{10},\ 5s^2,\ 5p^6$
(D) $1s^2,2s^22p^6,3s^23p^63d^{10},4s^24p^64d^{10},5s^25p^65d^5,6s^2$

31. KI and $CuSO_4$ solutions when mixed give
(A) $CuI_2 + K_2SO_4$ (B) $Cu_2I_2 + K_2SO_4$
(C) $Cu_2I_2 + I_2 + K_2SO_4$ (D) $CuI_2 + I_2 + K_2SO_4$

32. Which of the following elements is used in cancer therapy ?
(A) Rn (B) Th (C) Co (D) U

33. Partial pressure of a solution component is directly proportional to its mole fraction. This statement is known as
(A) Henry's law (B) Raoult's law
(C) Distribution law (D) Ostwald's dilution law

34. The relationship which describes the variation of vapour pressure with temperature is called
(A) Hess' Law (B) Arrhenius equation
(C) Kirchhoff's law (D) Clausius-Clapeyron equation

35. Which of the following is not a colligative property ?
(A) Depression is freezing point
(B) Osmotic pressure (C) Elevation of boiling point
(D) Increase in freezing point

36. The bond length between carbon and carbon in sp^2 hybridised molecule is.
(A) 1.2Å (B) 1.32Å (C) 1.54Å (D) 1.4 Å

37. Carbon atoms in benzene molecule is inclined at an angle of
(A) 120° (B) 180° (C) 109°28' (D) 60°

38. Bakellite is prepared by the reaction between
(A) Ethylene glycol and dimethyl terepthalate
(B) Urea and formaldehyde
(C) Phenol and formaldehyde
(D) Tetramethylene glycol and hexamethylene diisocyanate

39. Aspirin is
(A) Salicyaldehyde (B) Salicyclic acid

(C) Both (A) and (B) (D) 2 Acetoxy benzoic acid

40. Orlon is a polymer of
(A) PVC (B) Bakellite (C) Acrylonitrile (D) Nylon

41.In the given reaction $K_2Cr_2O_7 + XH_2SO_4 + YSO_2 \rightarrow K_2SO_4 + Cr_2(SO_4)_3 + ZH_2O$, X, Y, Z, are
(A) 1,3,1 (B) 4,1,4 (C) 3,2,3 (D) 2,1,2

42. For an endothermic reaction ΔS is (+) ve. The reaction is (A) Feasible when $T\Delta S > \Delta H$
(B) Feasible when $\Delta H > T\Delta S$
(C) Feasible at all temperatures (D) Not feasible at all

43. Which solutions are mixed to form a buffer solution ?
(A) A strong acid and its salt of strong base
(B) Strong acid and its salt of weak base
(C) Weak acid and its salt of strong base
(D) Weak acid and its salt of weak base

44. Which of the following is not possible
(A) n = 3, l = 0, m = 0 (B) n = 3, l = 1, m = –1
(C) n = 2, l = 0, m = –1 (D) n = 2, l = 1, m = 0

45. Bayer's Reagent is
(A) Acidified $KMnO_4$ (B) Alkaline $KMnO_4$
(C) $ZnSO_4$ (D) $CuSO_4.2H_2O$

46. The ratio of π and σ bonds in benzene is
(A) 1 : 4 (B) 2 : 1 (C) 1 : 1 (D) 2 : 2

47. P_2O_5 is a/an
(A) Reducing agent (B) Oxidzing agent
(C) Buffer (D) None of these

48. Which of the following is formed by the oxidation of chloroform ?
(A) CH_3CO_2 (B) $COCl_2$ (C) CO_2 (D) O_2

49. Electrolysis of potassium sulphate yields
(A) $K_2S_2O_8$ (B) K_2SO_8 (C) KS_2O_4 (D) KSO_4

50. What is plaster of Paris ?
(A) $CaSO_4.H_2O$ (B) $CaSO_4.2H_2O$
(C) $CaSO_4$.½H_2O (D) None of these

BHU - 2001

1. Wavelength of spectral line emitted is inversely proportional to
(A) Radius (B) Velocity
(C) Quantum number (D) Energy

2. Which one of the following represents an impossible arrangements ?

	n	ℓ	m	n
(A)	4	0	0	1/2
(B)	3	2	-3	1/2
(C)	3	2	-2	1/2
(D)	5	3	0	1/2

3. Electronic configuration of chromium is
(A) $[Ne]3s^23p^63d^44s^2$ (B) $[Ne]3s^23p^63d^54s^1$
(C) $[Ne]3s^23p^63d^14s^2$ (D) $[Ne]3s^23p^64s^24d^4$

4. Which of the following relates to photons having both wave as well as particle nature
(A) Interference (B) Diffraction (C) $E = mc^2$ (D) $\lambda = c/v$

5. Which of the following is isotonic with $_{37}Rb^{86}$?
(A) $_{36}Kr^{84}$ (B) $_{37}Rb^{85}$ (C) $_{38}Sr^{87}$ (D) $_{39}Y^{89}$

6. The type of bonding present in N_2O_5 is
(A) Covalent only (B) Covalent and electrovalent
(C) Coordinate (D) Covalent and coordinate

7. During formation of a chemical bond between two atoms ?
(A) Only forces repulsion are operative
(B) Only forces attraction are operative
(C) Neither of the forces are operative
(D) Both of the above two forces are operative

8. Among the alkaline earth metals the element forming predominantly covalent compounds is
(A) Be (B) Mg (C) Sr (D) Ca

9. O_2^{2+} has a bond order of
(A) 1.5 (B) 1 (C) 2.5 (D) 3

10. Hydrogen bonding is maximum in
(A) Ethanol (B) Diethyl ether
(C) Ethyl chloride (D) Triethyl amine

11. Which of the following pair posses the same structure
(A) PH_3 and BCl_3 (B) SO_2 and NH_3
(C) PCl_5 and SF_6 (D) NH_4^+ and SO_4^{2-}

12. Which one has dipole moment ?
(A) CCl_4 (B) 2 Methyl propane
(C) Cis-Butene-2 (D) Trans butene -2

13. In which of the following reactions no change in valency occurs ?
(A) $Cl_2 + NaOH \rightarrow NaClO + NaCl + H_2O$
(B) $SnCl_2 + Cl_2 \rightarrow SnCl_4$ (C) $2H_2O_2 \rightarrow 2H_2O_2 + O_2$
(D) $AgNO_3 + KCl \rightarrow AgCl + KNO_3$

14. Given :
$C(s) + O_2(g) \rightarrow CO_2(g) + 94.2$ kcal
$H_2(g) + 1/2O_2(g) \rightarrow H_2O(l) + 68.3$ kcal
$CH_4(g) + 2O_2(g) \rightarrow CO_2(g) + 2H_2O(l) + 210.8$ kcal
The heat of formation of methane in Kcal will be
(A) 45.9 (B) 47.8 (C) 20.0 (D) 47.3

15. Faraday constant
(A) Depends on the number of electrons
(B) Is a numerical constant
(C) Depends on equivalent mass
(D) Depends upon the current passed

16. Molten sodium chloride conducts electricity due to the presence of
(A) Free electrons (B) Free ions
(C) Free molecules (D) Atoms of sodium and chlorine

17. A catalyst
(A) Is not used up in the reaction
(B) Changes in composition during the reaction

(C) Is used up in reaction
(D) Increases equilibrium constant

18. The IUPAC name of the compound
$CH_3 - \underset{\displaystyle CH_3}{\underset{|}{C}} = CH_2CH_2OH$ is
(A) 2-methyl-2-butenol (B) 2-methyl-3-butenol
(C) 3-methyl-2-butenol (D) 3-methyl-but-2-enal

19. IUPAC name for
$CH_3CH_2 - \overset{\displaystyle CH_3}{\overset{|}{CH}} - CH_2COCl$ is
(A) 3-methyl butanoyl chloride
(B) 1-chloro-3ethyl butanone
(C) 3-methyl pentanoyl chloride
(D) 1-chloro-3methyl pentanone

20. Which of the following alkanes is optically active ?
(A) Propane (B) 2-methyl butane
(C) 3-methyl hexane (D) 2,3,4-trimethyl pentane

21. The C–H bond distance is longest in
(A) C_2H_2 (B) C_2H_4 (C) C_2H_6 (D) C_6H_6

22. The type of hybridization of carbon atom in 1,2 butadine is
(A) sp (B) sp^3 and sp^2 (C) sp^2 and sp (D) sp, sp^2 and sp^3

23. Isotopes are elements which have the same
(A) Number of neutrons (B) Mass number
(C) Atomic number (D) Number of atoms

24. Electrons in an orbital give rise to magnetic field
(A) In the opposite direction (B) In either direction
(C) In the same direction
(D) No magnetic field is created due to spin

25. The function of anhydrous $AlCl_3$ in Friedelcraft's reaction is
(A) Produce attacking electrophile
(B) Absorb H_2O (C) Absorb HCl
(D) Produce nucleophile

26. Phenol is prepared by
(A) Cumene process (B) From $C_6H_5N_2Cl$
(C) From $C_6H_5SO_3Na$ (D) All of these

27. Aldehydes are oxidized to acids by
(A) $K_2Cr_2O_7/H^+$ (B) Tollen's reagent
(C) Fehling's solution (D) All of these

28. Steam distillation is used for purification of
(A) Covalent compounds (B) Volatile liquids
(C) Inorganic liquids (D) Organic liquids

29. In a periodic table there are
(A) 7 periods and 9 groups (B) 18 periods and 9 groups
(C) 7 periods and 18 groups (D) 7 periods and 10 groups

30. The effective nuclear charge on $[Co(NH_3)Cl_3$ is
(A) 3 (B) 2 (D) 6 (D) 0

31. Essential condition for the formation of coordination compounds is
(A) Small size (B) High charge density
(C) Presence of empty orbitals
(D) All the above

32. During extraction of copper, the walls of blast furnace is covered with
(A) CaO (B) SiO_2 (C) Graphite (D) NaCl

33. Which of the following is not hydrolyzed ?
(A) $SiCl_4$ (B) $PbCl_2$ (C) NaCl (D) CCl_4

34. Hybridization of carbon atoms in benzene is
(A) sp^3 (B) sp (C) sp^2 (D) dsp^2

35. Which of the following statements is correct ?
(A) Lower alcohols are soluble in water and possess characteristic odour
(B) M.P. and B.P. of alcohol increases with increase in molecular weight and so is their solubility in water
(C) Solubility of alcohol in water increases with increase in molecular weight
(D) Higher alcohols are insoluble in water and possess characteristic odour

36. When chloroform is exposed to air and sunlight, it gives
(A) Phosgene (B) Lewistie
(C) Mustard gas (D) Carbon tetrachloride

37. Meso-tartaric acid is
(A) Optically active because of asymmetric carbon atom
(B) Optically inactive
(C) Optically inactive due to external compensation
(D) Optically active because of molecular symmetry

38. In which of the following reactions H_2O_2 acts as a reducing agent ?
(A) $Na_2AsO_3 + H_2O_2 \rightarrow Na_3AsO_4 + H_2O$
(B) $K_2Cr_2O_7 + H_2SO_4 + 4H_2O_2 \rightarrow 2CrO_5 + K_2SO_4 + 5H_2O$
(C) $2KI + H_2O \rightarrow 2KOH + I_2$
(D) $PbO_2 + H_2O_2 \rightarrow PbO + H_2O + O_2$

39. The most stable oxidation state of Nickel is
(A) +1 (B) +2 (C) +3 (D) 0

40. The average speed of an ideal gas molecule at 27°C is 0.3 ms^{-1}. Calculate the average speed at 927°C.
(A) 0.4ms^{-1} (B) 0.6 ms^{-1} (C) 0.9 ms^{-1} (D) 0.2 ms^{-1}

41. The yield of the product can be increased by
(A) Carrying the reaction in a closed container vessel
(B) By escaping out the gaseous product
(C) Decreasing the concentration of reactant
(D) Increasing the concentration of product

42. For a reaction to occur spontaneously
(A) ΔH and ΔS must be negative with $T\Delta S > \Delta H$
(B) ΔH must be negative and ΔS + ve
(C) ΔS must be negative and ΔH + ve
(D) ΔH and ΔS both +ve, $\Delta H > T\Delta S$

43. The reaction
$$C_6H_6 + CH_3Cl \xrightarrow{AlCl_3\ (anhyd.)} C_6H_5CH_3 + HCl$$

(A) Friedel Craft's reaction (B) Wurtz reaction
(C) Sandmeyer's reaction (D) Wurtz-Fittig reaction

44. 1, 3-butadiene reacts with ethene to form
(A) Benzene (B) Cyclohexane
(C) Cyclohexene (D) 2,3 dimethyl butane

45. Carboxylic acid is formed by the hydrolysis of an addition compound formed by reaction of Grignard's reagent with
(A) Methyl alcohol (B) Methanal
(C) Carbon dioxide (D) Methane

46. "The electronegative part of the addendum goes to less protonated carbon atom". It is the statement of
(A) Saytzeff's Rule (B) Markownikoff's Rule
(C) Anti-Markownikoff's Ruls (D) None of these

47. When sodium derivative of phenol reacts with methyl chloride, it forms
(A) Anisole (B) Chlorobenzene
(C) Benzoic acid (D) Benoyl chloride

48. With excess of bromine, phenol reacts to form

(A) (B) (C)

(D) mixture of A and B

49. Formation of cyanohydrin from aldehydes is
(A) Nucleophilic substitution reaction
(B) Nucleophilic addition reaction
(C) Electrophilic substitution reaction
(D) Electrophilic addition reaction

50. In the nitration of benzene with a mix of conc. HNO_3 and conc. H_2SO_4, the reactive species involved is
(A) NO_3 (B) NO_2 (C) NO_2 (D) NO_2^+

BHU - 2002

1. For a reaction to be non-spontaneous at all temperatures, the signs of enthalpy; entropy and free-energy changes must be
(A) +, – and + (B) +, + and – (C) –, + and – (D) –, – and +

2. The incorrect statement of the following is
(A) Entropy is a measure of disorder.
(B) Entropy is expressed in the units of $JK^{-1}\ mol^{-1}$.
(C) The spontaneous process always accompanies the decrease in free energy.
(D) Entropy is not a state function.

3. Calculate the standard enthalpy change (ΔH^o) for the reaction $OF_2(g) + H_2O(g) \rightarrow O_2(g) + 2HF(g)$
given that standard enthalpies of formation ΔH_f^o of $OF_2(g)$, $H_2O(g)$ and $HF(g)$ are +23.0, –241.8 and –268.6 kJ $mole^{-1}$ respectively
(A) –318.4 kJ (B) 318.4 kJ (C) –418.4 kJ (D) 418.4 kJ

4. Calculate the number of coulombs required to deposit 40.5 g of Al when the electrode reaction is as follows :
$Al^{3+} + 3e^- \rightarrow Al$
(A) 4.34×10^5 C (B) 434×10^5 C
(C) 4.43×10^5 C (D) 43.40×10^5 C

5. ${}_{95}Am^{241}$ and ${}_{90}Th^{234}$ respectively belong to
(A) (4n+1) & (4n+3) radioactive disintegration series
(B) (4n+1) & (4n+2) radioactive disintegration series
(C) (4 n+1) & 4n radioactive disintegration series
(D) 4n & (4n +1) radioactive disintegration series

6. Volume strength of H_2O_2 in a 3.4% (w/v) solution of H_2O_2 is
(A) 11.2 ml (B) 5.6 ml (C) 2.8 ml (D) 22.4 ml

7. The temporary hardness of water due to calcium bicarbonate can be removed by adding
(A) $CaCO_3$ (B) $Ca(OH)_2$ (C) $CaCl_2$ (D) HCl

8. Which of the following is a colligative property ?
(A) Elevation in boiling point
(B) Relative lowering of vapour pressure
(C) Osmotic pressure (D) All the above

9. The Osmotic pressure of equimolar solutions of glucose, sodium chloride and barium chloride will be in the order
(A) $BaCl_2$>NaCl>Glucose (B) $BaCl_3$>Glucose>NaCl
(C) Glucose>$BaCl_2$>NaCl (D) NaCl>$BaCl_2$>Glucose

10. The IUPAC name of

$$CH_3-\underset{\underset{CH_3}{|}}{C}=CH-\overset{\overset{O}{||}}{C}-CH_3$$ is

(A) 2-Methylpent-3-en-2one
(B) 4-Methylpent-3-en-2one
(C) 3-Methylpent-2-en-1one (D) None of these

11. What is the mole fraction of a solute in 2.5 M aqueous solution ?
(A) 4.3 (B) 43 (C) 0.043 (D) 0.43

12. One molecule of an alkane requires how many molecules of oxygen for complete combustion ?
(A) (n + 1) x 2 (B) (3n + 1)/2 (C) (3n + 1) (D) (n + 1)/2

13. A compound formed by elements A and B crystallizes in the cubic arrangement in which A atoms are at the corner and B atoms are at the face centre of a cube. What is the formula of the compound
(A) A_2B (B) A_2B_2 (C) AB_3 (D) AB_3

14. For electronic transition from n = 2 → n = 1. Which of the following produces shortest wavelength ?
(A) H-atom (B) D-atom (C) He^+ion (D) Li^{2+} ion

15. If uncertainty in position of an electron is zero, the uncertainty in its momentum will be
(A) Zero (B) Infinite
(C) Less than h/4π (D) grater than h/4π

16. Which of the following is correct statement for

$3d_{xy}, 3d_{yz}, 3d_{zx}, 3d_{x^2-y^2}, 3d_{z^2}$?
(A) All orbitals have same energy
(B) $3d_{yz}$ has maximum energy (C) $3d_{z^2}$ has lowest energy
(D) All orbitals have different energies

17. The expected outermost electronic configuration of Pt atomic number 78 is
(A) $[Xe]6s^2\ 4f^{14}5d^8$ (B) $[Xe]6s^2\ 4f^{14}5d^7$
(C) $[Xe]6s^1\ 4f^{14}5d^9$ (D) $[Xe]6s^1\ 4f^{14}5d^{10}$

18. Magnetic moment of sodium in ground state is
(A) $\sqrt{6}$ (B) 2 (C) $\sqrt{3}$ (D) None of these

19. The order of O–O bond length in H_2O_2, O_3 and O_2 is
(A) $O_2 > O_3 > H_2O_2$ (B) $O_3 > H_2O_2 > O_2$
(C) $H_2O_2 > O_3 > O_2$ (D) $O_2 > H_2O_2 > O_3$

20. The correct order of increasing basicity is
(A) $CH_4 < NH_3 < H_2O < HF$ (B) $HF < H_2O < NH_3 < CH_4$
(C) $HF < NH_3 < N_2O < CH_4$ (D) $HF < NH_3 < H_2O = CH_4$

21. Which of the following is the strongest base ?
(A) ClO^- (B) CHO_2^- (C) CHO_3^- (D) ClO_4^-

22. The strongest reducing agent amongst the following is
(A) $H_2PO_2^-$ (B) $P_2O_7^{4-}$ (C) $P_2O_4^{4-}$ (D) H_3PO_4

23. The pH of a 0.05 M solution of a weak monobasic acid which is 20% ionized is
(A) 2.5 (B) 1.5 (C) 5.2 (D) 2.0

24. For a reversible process at equilibrium, the change in entropy many be expressed as
(A) $\Delta S = Tq_{rev}$ (B) $\Delta S = \frac{\Delta H}{T}$
(C) $\Delta S = \frac{q_{rev}}{T}$ (D) $\Delta S = T\Delta H$

25. The most appropriate reagent (R) for the reaction

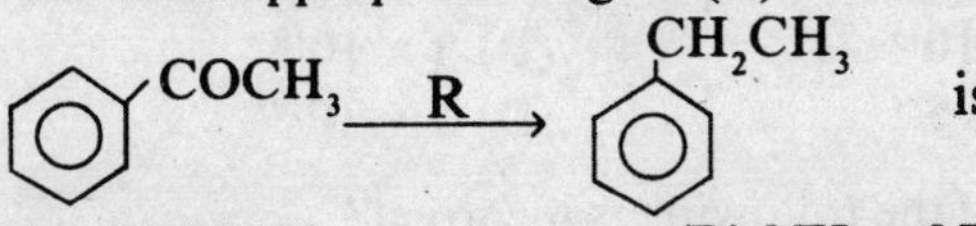

$$C_6H_5COCH_3 \xrightarrow{R} C_6H_5CH_2CH_3$$ is

(A) Zn (Hg)/HCl (B) $NH_2 - NH_2/OH^-$
(C) H_2/Ni (D) $NaBH_4$

26. $$C_6H_6 + CH_3 - \underset{\underset{CH_3}{|}}{CH} - CH = CH_2 \xrightarrow[\text{or } H_2SO_4]{AlCl_3} P.$$

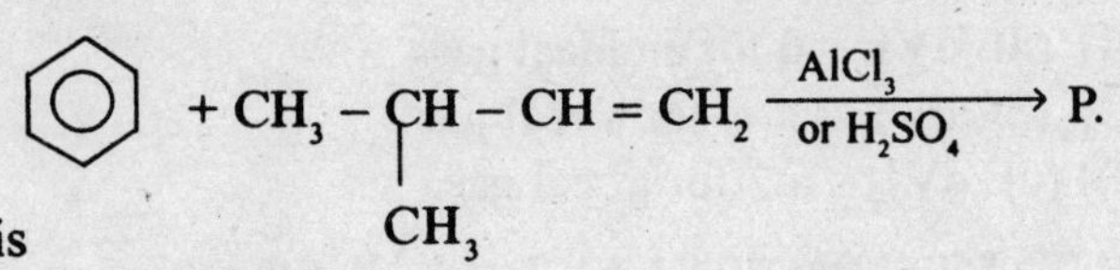

P is

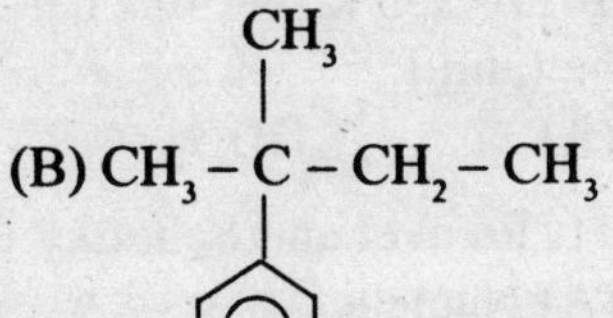

(A) 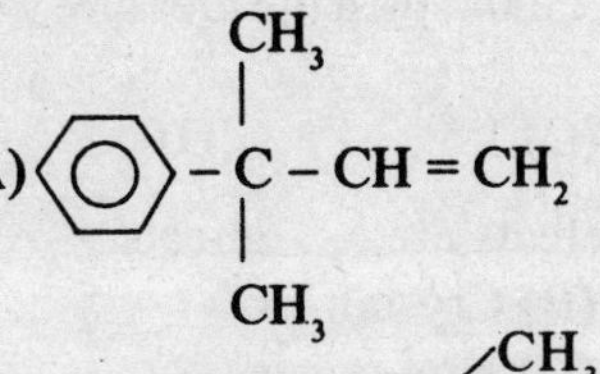 $C_6H_5 - \underset{\underset{CH_3}{|}}{\overset{\overset{CH_3}{|}}{C}} - CH = CH_2$ (B) $CH_3 - \underset{\underset{C_6H_5}{|}}{\overset{\overset{CH_3}{|}}{C}} - CH_2 - CH_3$

(C) $CH_3 - \underset{\underset{C_6H_5}{|}}{CH} - CH(CH_3)_2$ (D) Both (B) and (C)

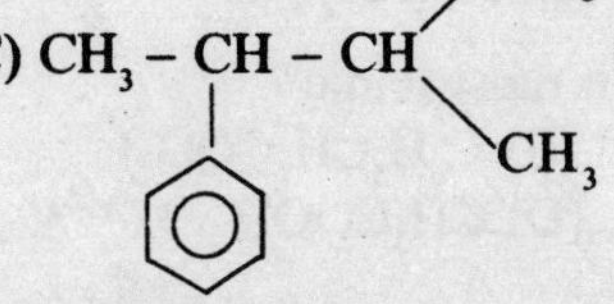

27. Which of the amino acids contains sulphur ?
(A) Valine (B) Alanine (C) Cystein (D) Threonine

28. Herculon is the polymer of
(A) Polyisobutylene (B) Polyvinylidene chloride
(C) Polymethyl methacrylate (D) Polypropylene

29. Which of the following has largest anionic size ?
(A) F (B) Cl^- (C) O_2^- (D) I^-

30. Element having maximum no. of stable isotopes is
(A) Sn (B) Fe (C) Pb (D) Os

31. Which element of 2nd period forms most acidic oxide ?
(A) B (B) C (C) F (D) N

32. The comman oxidation states of 14th group elements is/are
(A) +4 (B) +2, +4 (C) $Ni(OH)_2$ (D) +3

33. Which of the following compounds is least soluble in acidic solution than in pure water ?
(A) AgBr (B) $CaCO_3$ (C) $Ni(OH)_2$ (D) $Ca(PO_4)_2$

34. The chemical process in production of steel from haematite involves
(A) Reduction followed by oxidation
(B) Reduction (C) Oxidation
(D) Oxidation followed by reduction

35. Which of the following transition metal oxides is not paramagnetic ?
(A) TiO (B) V_2O_5 (C) CuO (D) VO_2

36. When copper sulphate solution is treated with potassium iodide and excess of hypo solution in the resulting solution, a white precipitate is formed. The white ppt. is due to
(A) NaI (B) Cu_2I_2 (C) CuI_2 (D) $Na_2S_4O_6$

37. How many isomeric amines with formula C_3H_4N are possible ?
(A) 2 (B) 3 (C) 4 (D) 5

38. Conversion of CH_4 to CH_3Cl is an example of
(A) Free radical substitution reaction
(B) Nucleophilic substitution reaction
(C) Electrophonic substitution reaction
(D) Free radical addition reaction

39. Tertiary alkyl free radical
(A) Have six hyperconjugative forms
(B) Have seven hyperconjugative forms
(C) Have three hyperconjugative forms
(D) Have nine hyperconjugative forms

40. Which of the following groups has maximum +ve inductive effect ?
(A) $(CH_3)_3C-$ (B) CH_3CH_2- (C) $(CH_3)_2CH-$ (D) $-CH_3$

41. Formic acid is considered as a resonance hybrid of

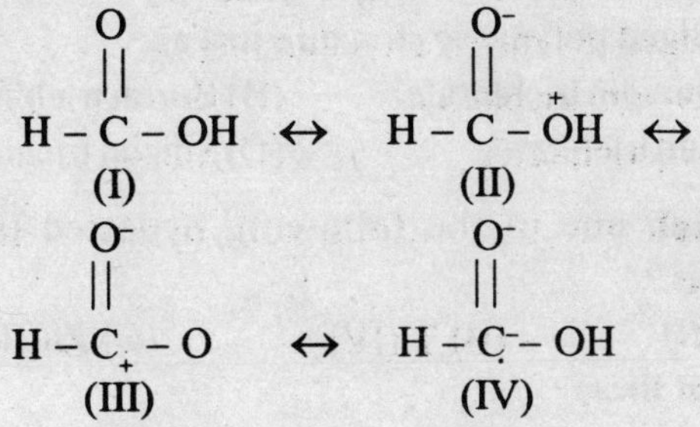

$$\underset{(I)}{H - \overset{\overset{O}{||}}{C} - OH} \leftrightarrow \underset{(II)}{H - \overset{\overset{O^-}{||}}{C} - \overset{+}{O}H} \leftrightarrow$$

$$\underset{(III)}{H - \overset{\overset{O^-}{||}}{\underset{+}{C}} - O} \leftrightarrow \underset{(IV)}{H - \overset{\overset{O^-}{||}}{C^-} - OH}$$

Which of the following order is correct for the stability of the four contributing structures ?

(A) I > III > II > IV (B) I > IV > III > II
(C) 1 > II > III > IV (D) I > II > IV > III

42. Zeigler-Natta catalyst is
(A) Mixture of titanium (III) chloride and triethyl aluminum.
(B) Mixture of zirconium (II) chloride and triethyl aluminum.
(C) Mixture of zirconium (III) chloride and dimethyl aluminum.
(D) Mixture of titanium (III) chloride and dimethyl aluminum.

43. In the reaction

$$CH_3-\underset{\displaystyle Cl}{\overset{}{CH}}-CH_3 \xrightarrow[ether]{Mg} A \xrightarrow{D_2O} B$$

identify B
(A) $CH_3-CH_2-CH_2-D$ (B) $CH_3-CH_2-CH_2-Mg(OD)$
(C) $CH_3-\underset{\displaystyle D}{CH}-CH_3$ (D) no reaction from A to B.

44. Which of the following does not give iodoform test ?
(A) $CH_3-\underset{\displaystyle CH_3}{CHOH}$ (B) $CH_3-\underset{\displaystyle OH}{CH}-CH_2-CH_2-CH_3$
(C) Acetophenone (D) Benzophenone

45. Zeolite is
(A) Hydrated aluminium silicate
(B) Hydrated sodium silicate
(C) Hydrated sodium potassium silicate
(D) Hydrated sodium aluminium silicate

46. Bell metal is an alloy of
(A) Cu and Zn (B) Cu and Sn
(C) Al and Zn (D) Al and Sn

47. The hydrolysis of trimethyl chlorosilane $(CH_3)_3SiCl$ yields
(A) $[Si(CH_3)_3]_2O$ (B) $(CH_3)_3SiOH$
(C) $(CH_3)_3SiH$ (D) Does not hydrolyse

48. Which of the following compounds of platinum is used in cancer chemotherapy ?
(A) Tetrammine platinum (II) chloride
(B) Potassium tetrachloroplatinate (II)
(C) Cis-diammine dichloroplatinum (II)
(D) Trans-diammine dichloro platinum (II)

49. Beryllium chloride, $BeCl_2$ is essentially a covalent and has a bridged polymeric structure just as
(A) Aluminium trichloride (B) Boron trichloride
(C) Galium trichloride (D) Silicon trichloride

50. Which one of the following hydrated ions is colourless ?
(A) Sc(III) (B) Ti(IV) (C) Zn(II)
(D) All of these

BHU - 2003

1. Radius of second Bohr's orbit of H atom is .... the first Bohr orbit of H atom.
(A) 4 times (B) 2 times (C)5 times (D) 3 times

2. Maximum number of orbitals in *l*-subshell is
(A) 16 (B) 11 (C) 13 (D) 14

3. a/V^2 is given for in van der Waal's equation
(A) internal pressure (B) intermolecular attraction
(C) both (A) and (B) (D) temperature correction.

4. H_2O is liquid, H_2S is gas due to
(A) H- bonding (B) Molecular attraction
(C) H-bond and three dimensional ice structure
(D) H-bonding and molecular attraction

5. H-bonding is maximum in
(A) CH_3COCH_3 (B) C_6H_5COOH
(C) CH_3CH_2OH (D) C_6H_5OH

6. Graphite is soft while diamond is hard because
(A) Graphite is in planar form while diamond is in tetrahedral form
(B) Diamond has sp^2 hybridisation but graphite has sp^3 hybridisation
(C) Graphite is in powder form
(D) Graphite is covalent and diamond is ionic

7. A substance has half life period of 4 sec. After t = 20 sec, how much percentage of the substance has decayed?
(A) 96.875 (B) 3 (C) 40 (D) 50

8. In element has half life period of 6.5 hrs. Initial number of atoms is 48×10^{19}, then after 26hrs, the number of atoms will be
(A) 1.6×10^{19} (B) 3×10^{19}
(C) 2.4×10^{19} (D) 5×10^{19}

9. Which of the following is incorrect?
(A) Density of liquid = density of vapour at critical temperature
(B) $(dE/dV)_T = 0$ for an ideal gas
(C) $(dE/dV)_T \neq a/v^2$ for a real gas
(D) $(dE/dV)_T = a/2$ for a real gas.

10. T= 550 K, V= 49.8 l, n = 2 mol., then the pressure will be (atm.)
(A) 2 (B) 4 (C) 1.5 (D) 2.5

11. Frenkel and Schottky defects are...... defects.
(A) Nucleus (B) Crystal
(C) Nuclear (D) Non-crystal

12. Which of the following is most acidic ?
(A) HCOOH (B) $C_6H_5CH_2COOH$
(C) $ClCH_2COOH$ (D) CH_3COOH

13. Diamond exists as
(A) Tetrahedral (B) Octahedral
(C) Planar (D) Hectahedral

14. Most stable carbanion is

(A) CH_2^- (p-nitrophenyl, $NO2$) (B) CH_2^- (p-methylphenyl, CH_3) (C) CH_3^- (D) CH_4^-

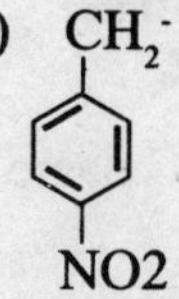

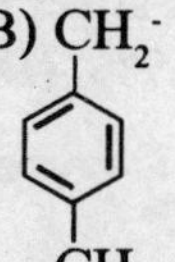

15. Which of the following is the strongest nucleophile ?

(A) ROH (B) CH_3O^- (C) RNH_2 (D) $C_6H_5O^-$

16. $(CH_3)_2C = CHCH_3 \xrightarrow{catalyst/H_2}$

Number of optical isomers possible are

(A) 0 (B) 4 (C) 2 (D) 3

17. Propene and propyne are distinguished by using

(A) $K_2Cr_2O_7$ (B) H_2SO_4 (C) $AgNO_3$ (D) NH_4OH

18. Isobutene +HBr $\xrightarrow{peroxide}$ X. X is

(A) iso-butyl bromide (B) tertiary butyl bromide

(C) tertiary butyl alcohol (D) iso-butyl alcohol

19. $CH_3CH_2C{\equiv}CH \xrightarrow{R}$ Butanone. R is

(A) $KMnO_4$ (B) Hg^{2+}/H^+ (C) $K_2Cr_2O_7$ (D) H_2SO_4

20. Which of the following substances doesnot give iodoform test ?

(A) CH_3COCH_3 (B) $CH_3CHOHCH_3$

(C) CH_3OH (D) C_2H_5OH

21. Correct order of polarising power is

(A) $Cs^+<K^+<Mg^{2+}<Al^{3+}$ (B) $K^+<Cs^+<Mg^{2+}<Al^{3+}$

(C) $Cs^+<K^+<Al^{3+}<Mg^{2+}$ (D) $K^+<Cs^+<Al^{3+}<Mg^{2+}$

22. Carborundum is

(A) Al_2O_3 (B) SiC (C) $AlCl_3$ (D) $CaCO_3$

23.Correct statement is

(A) $K=4s^1$, $Cr=3d^54s^1$, $Cu=3d^{10}4s^1$

(B) $K=4s^2$, $Cr=3d^54s^1$, $Cu=3d^{10}4s^1$

(C) $K=4s^2$, $Cr=3d^44s^2$, $Cu=3d^94s^2$

(D) $K=4s^1$, $Cr=3d^44s^2$, $Cu=3d^94s^2$

24. $Na_2S_2O_3$ + I_2 $\rightarrow$ Product is

(A) $Na_2S_4O_6$ (B) NaI (C) Na_2S (D) S_2

25. Which of the following statements is wrong ?

(A) $Ni(CO)_4$ has oxidation number +4 for Ni

(B) $Ni(CO)_4$ has oxidation number 0 for Ni

(C) Both (A) & (B) (D) None of these

26. In $[NiCl_4]^{2-}$, the number of unpaired electrons is

(A) 2 (B) 3 (C) 4 (D) 4.5

27. Phenol and benzoic acid are distinguished by

(A) $NaHCO_3$ (B) NaOH (C) Na_2CO_3 (D) H_2SO_4

28. $NaOH/H^+$ reacts with

(A) $C_6H_5OCH_3$ (B) CH_3OH

(C) C_2H_5OH (D) CH_3COCH_3

29. Benzophenone doesnot react with

(A) RNH_2 (B) Na_2CO_3 (C) NaOH (D) SO_3

30.CH3COOH is formed by hydrolysis of

(A) CH_3CN (B) CH_3CH_2CN

(C) CH_3NH_2 (D) $CH_3CH_2NH_2$

31. Which of the following has ester linkage ?

(A) Terylene (B) PVC (C) Bakelite (D) Polythene

32. $\delta H_f(H_2O)$ =X. Heat of neutralisation of CH_3COOH and NaOH will be

(A) less than X (B) less than 2X

(C) between X and 2X (D) X

33. NaOH (aq) , HCl (aq) ,NaCl (aq) , concentration of each is 10^{-3}M. The pH will be respectively-

(A) 11 , 3 , 7 (B) 10 , 2 , 6 (C) 10 , 6 , 2 (D) 3 , 4 , 7

34. K_{sp} (AB)= 1× 10^{-10}. If [A+] = $1{\times}10^{-5}$, which of the following solutions of [B-] cannot be precipitated ?

(A) $1{\times}10^{-5}$ (B) 5×10^{-6} (C) 2×10^{-5} (D) 5×10^{-5}

35. C (diamond) + $O_2 \rightarrow CO_2$; ΔH = -395.4 kJ/mol.

C (graphite) + $O_2 \rightarrow CO_2$; ΔH = -393.5 kJ/mol.

C (graphite) $\rightarrow$ C (diamond) ; Δ H = ?

(A) + 1.9 (B) -1.9 (C) +3.8 (D) -3.8

36. A_2 (g) + B_2 (g) $\leftrightarrow$ 2AB (g) ; ΔH = +ve.

(A) it ocuurs at 1000 atm. pressure

(B) unaffected by pressure

(C) it occurs at high temperature

(D) it occurs at high temperature and high pressure

37. A + B $\rightarrow$ Product. The value of rate constant depends on

(A) conc. of A and B (B) temperature

(C) pressure (D) all of these

38. r = $k[NO]^2$ $[O_2]$. Rate constant can be increased by

(A) increasing [NO] (B) increasing $[O_2]$

(C) increasing temperature (D) decreasing temperature

39. 1 A current is supplied for 30 minutes through NaCl. How much chlorine will be liberated ?

(A) 0.66g (B) 0.66 mole (C) 0.33g (D) 0.33 mole

40. $KMnO_4$+$H_2C_2O_4$ $\xrightarrow{H^+}$. Change in the oxidation number of Mn is

(A) 7 to 2 (B) 7 to 5 (C) 2 to 7 (D) 5 to 7

41. Zn^{2+} + 2 $e^- \rightarrow$ Zn(s) ; E^0 = 0.76

Fe^{3+} + $e^- \rightarrow$ Fe^{2+} ; E^0 = +0.77

Cr^{3+} + $3e^- \rightarrow$ Cr ; E^0 = -0.79

H+ + e-$\rightarrow$ 1/2 H_2 ; E^0 = 0.00

Stronger reducing agent is

(A) Fe^{2+} (B) Zn (C) Cr (D) H_2

42. Cryolite helps in

(A) increasing the electrical conductivity

(B) decreasing the electrical conductivity

(C) lowering the melting point

(D) increasing the melting point

43. Which of the folllowing has the least solubility

(A) CoS (B) Bi_2S_3 (C) PbS (D)Ag_2S

BHU - 2004

1. Pyrosilicate ion is
(A) SiO_2^{2-} (B) SiO_4^{2-} (C) $Si_2O_6^{7-}$ (D) $Si_2O_7^{6-}$

2. Which of the following statement is correct
(A) BCl_3 and AlC_3 are both strong Lewis acids and BCl_3 is stronger than AlC_3
(B) BCl_3 and $AlCl_3$ are both Lewis acids and $AlCl_3$ is stronger then BCl_3
(C) BCl_3 and $AlCl_3$ are both equally strong Lewis acids
(D) Both BCl_3 and $AlCl_3$ are not Lewis acids

3. Due to reference in biological role, Ca^{2+} ions are imporant in
(A) Defence mechanism
(B) Triggering the contraction of muscles
(C) Formation of ATP
(D) Generating right electrical potential across cell membrane

4. The volume strength of 1.5 N H_2O_2 solution is
(A) 4.8 (B) 5.2 (C) 8.4 (D) 8.8

5. Excess of Sodium hydroxide reacts with zine to form.
(A) ZnH_2 (B) Na_2ZnO_2 (C) ZnO (D) $Zn(OH)_2$

6. Which of the following halogen oxides is ionic?
(A) ClO_2 (B) BrO_2 (C) I_2O_3 (D) I_4O_9

7. The sides of match box have coating of
(A) Antimony sulphide and red phosphorus
(B) Potassium chlorate and antimony sulphide
(C) Potassium chlorate and red lead
(D) Antimony sulphide and red lead

8. The de Broglie wavelength of a tennis ball of mass 60 g moving with a velocity of 10 metres per second is approximately
(A) 10^{-31}metres (B) 10^{-16} metres
(C) 10^{-25}metres (D) 10^{-33}metres

9. What is correct sequence of bond order ?
(A) $O_2^+ > O_2^- > O_2^+$ (B) $O_2^+ > O_2 > O_2^-$
(C) $O_2^+ > O_2^- > O_2$ (D) $O_2^- > O_2^+ > O_2$

10. ${}_{92}U^{215} + {}_0n^1 \rightarrow {}_{92}U^{216} \rightarrow$ fission products + neutrons + 3.20 x10^{-11}. The engery released when 1 g of ${}_{92}U^{235}$ finally undergoes fission is
(A) 12.75 x 10^8 KJ (B) 16.40 x 10^7 KJ
(C) 8.20 x 10^7 KJ (D) 6.50 x 10^6 KJ

11. Which of the following is an n, p reaction ?
(A) ${}_{13}Al^{27} + {}_0n^1 \rightarrow {}_{12}Mg^{27} + {}_1H^1$
(B) ${}_7Al^{14} + {}_1H^1 \rightarrow {}_8O^{15}$
(C) ${}_6C^{15} + {}_1H^1 \rightarrow {}_6C^{14} + {}_1e^0$
(D) ${}_{92}U^{215} + {}_0n^1 \rightarrow {}_{54}Xe^{140} + {}_{38}SR^{94} + 3\,{}_0n^1$

12. A wooden piece is 11460 years old. What is the fraction of ^{14}C activity left in the piece? (Half- life period of ^{14}C is 5730 years)
(A) 0.12 (B) 0.25 (C) 0.50 (C) 0.75

13. The values of observed and calculated molecular weights of silver nitrate are 92.64 and 170 respectively. The degree of dissociation of silver nitrate will be
(A) 60% (B) 83.5% (C) 46.7% (D) 60.23%

14. Which of the following is correct IUPAC name for the following compound ?

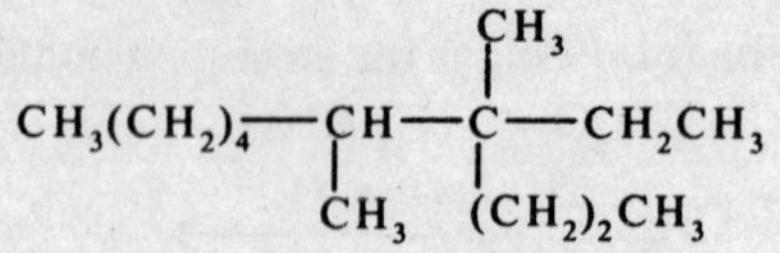

$$CH_3(CH_2)_4-CH(CH_3)-C(CH_3)(CH_2CH_3)-(CH_2)_2CH_3$$

(A) 6, 7-dimethyl-7-n-propylonane
(B) 4, 5-dimethyl-4-ethyldecane
(C) 3, 4 dimethyl-3-n-propylonone
(D) 6, 7-dimethyl-7-ethyldecane

15. By Friedel Craft's acylation reaction, the effective electrophile be
(A) $AlCl_3$ (B) $RCOCl^{(+)}$ (C) $RCO^{(+)}$ (D) RCOCl

16. Consider the following reaction,

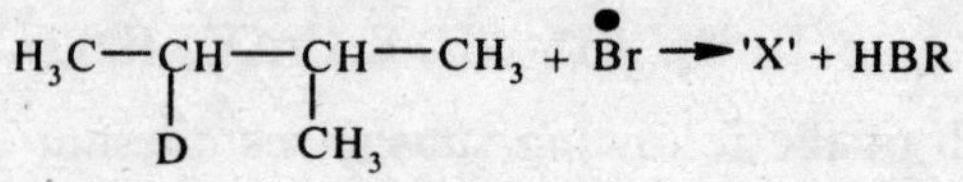

$$H_3C-CH(D)-CH(CH_3)-CH_3 + \dot{B}r \rightarrow 'X' + HBR$$

Identify the structure of the major product 'X'

(A) 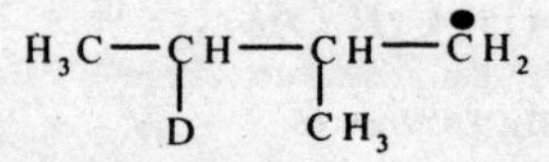 $H_3C-CH(D)-CH(CH_3)-\dot{C}H_2$ (B) 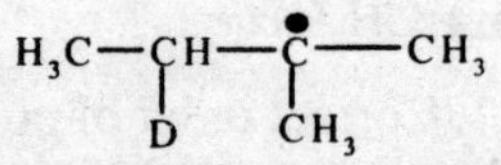 $H_3C-CH(D)-\dot{C}(CH_3)-CH_3$

(C) $H_3C-\dot{C}(D)-CH(CH_3)-CH_3$ (D) $H_3C-\dot{C}H-CH(CH_3)-CH_3$

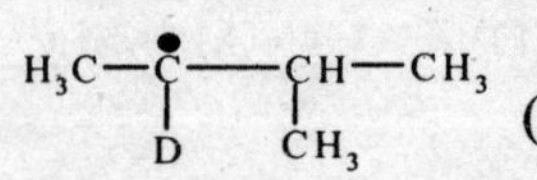

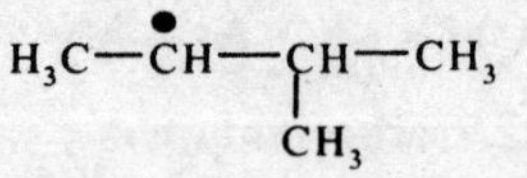

17. In the reaction,

$$C_6H_5CH_3 \xrightarrow{\text{oxidation}} A \xrightarrow{\text{NaOH}} B \xrightarrow[\Delta]{\text{Soda-lime}} C,$$ the product C is
(A) C_6H_5OH (B) C_6H_6
(C) C_6H_5COONa (D) C_6H_5ONa

18. Which of the following will convert $CH\equiv C-CH_2-CH_3$ to $CH_3COCH_2CH_3$?
(A) H_2O/H^+ (B) Hg^{2+}/H_2SO_4
(C) H_2SO_4(conc.)/H_3PO_4 (D) $K_2Cr_2O_7$ /$KMnO_4$

19. The compound benzophenone $C_6H_5COC_6H_5$ does not react with
(A) $NaHSO_3$ (B) CH_3OH (C) HCN (D) NH_2OH

20. With oxalic acid, glycerol at 530°C gives
(A) Glyceraldehyde (B) Formic acid
(C) Allyl alcohol (D) Glycerol mono-oxalate

21. Orthonitrophenol is steam volatile while para isomer is not. This is because of
(A) More solubility of ortho form
(B) Intramolecular hydrogen bonding in ortho nitrophenol
(C) More lattice engergy or ortho form
(D) All of the above

22. Silver is present in

(A) Limonite (B) Magnetite (C) Galena (D) Dolomite

23. Method used for obtaining highly pure silicon used as a semiconductor material, is
(A) Oxidation (B) Electrochemical
(C) Crystallization (D) Zone refining

24. van Arkel method of purification of metals involves converting the metal into a
(A) Volatile stable compound
(B) Volatile unstable compound
(C) Non volatile stable compound
(D) None of the above

25. Which one of the following reactions is an example for calcination process ?
(A) $2Ag + 2HCl + (O) \rightarrow 2AgCl + H_2O$
(B) $2Zn + O_2 \rightarrow 2\, ZnO$
(C) $2ZnS + 3O_2 \rightarrow 2ZnO + 2SO_2$
(D) $MgCO_3 \rightarrow MgO + CO_2$

26. The method of Zone refining of metals is based on the principle of
(A) Greater mobility of the pure metal than that of the impurity
(B) Higher melting point of the impurity than that of the pure metal
(C) Greater noble character of the solid than that of the impurity
(D) Greater solubility of the impurity in the molten state than in the solid

27. Calcium is obtained by the
(A) Electrolysis of solution of calcium chloride in water
(B) Electrolysis of molten anhydrous calcium chloride
(C) Roasting of limestone
(D) Reduction of calcium chloride with carbon.

28. The reactivity towards nucleophile in increasing order to addition to carbonyl group among cyclopentanone, 3-pentanone and n-pentanal is
(A) Cyclopentanone, 3-pentanone, n-pentanal
(B) 3-Pentanone, cyclopentanone, n-pentanal
(C) n-Pentanal, 3-pentanone, cyclopentanone
(D) n-Pentanal, cyclopentanone, 3-pentanone

29. When diethyl ether is treated with exvess of Cl_2 in the presence of sunlight, then the product formed is
(A) $CH_3CHCl - O - CH_2CH_3$
(B) $CH_3CHCl - O - CHClCH_3$
(C) $CCl_3CCl_2 - O - CCl_2CCl_3$
(D) $CH_3CCl_2 - O - CHClCH_3$

30. The end product 'C' in the following sequence of chemical reactions is

$CH_3COOH \xrightarrow{CaCO_3} A \xrightarrow{heat} B \xrightarrow{NH_2OH} C$

(A) Acetaldehyde oxime (B) Formaldehyde oxime
(C) Methyl nitrate (D) Acetoxime

31. The reaction $C_6H_5CHO + (CH_3CO)_2O \xrightarrow{CH_3COONa}$ $C_6H_5CH = CH\text{-}COOH + CH_3COOH$ is known as
(A) Benzoin condensation (B) Claisen condensation
(C) Perkin's reaction (D) Cannizaro's reaction

32. Principal product (X) in the reaction

$CH_3CONH_2 \xrightarrow{Br_2 + NaOH} X$ is

(A) CH_3NH_2 (B) $CH_2CH_2NH_2$
(C) CH_3COOH (D) CH_3Br

33. The common acid used in the manufacture of rayon and plastics is
(A) Methanoic acid (B) Ethanoic acid
(C) Propanoic acid (D) Butanoic acid

34. The change in optical rotation (with time) of freshly prepared solution of sugar is known as
(A) Rotatory (B) Specific rotation
(C) Inversion (D) Mutarotation

35. For the reaction $Pt/H_2(1\ atm)/H^+(aq)/Cl^-(aq)/AgCl/Ag$ K_c equilibrium constant is represented as

(A) $K_c = \frac{[Cl^-][AgCl]}{[H^+][H_2]}$ (B) $K_c = [H^+][Cl^-]$

(C) $K_c = \frac{[H^+][H_2]}{[Cl^-][AgCl]}$ (D) $K_c = \frac{[H_2]}{[Ag]}$

36. In the equilibrium $PCl_5 \rightleftharpoons Cl_2 + 22$ Kcal, the dissociation of PCl_5 will be more on
(A) Decreasing pressure (B) Decreasing temperature
(C) Increasing temperature
(D) Increasing the concentration of chlorine

37. The solubility product of $BaSO_4$ is 1.5×10^{-9} at 18°C. Its solubility in water at 18°C is
(A) 3.9×10^{-9} mole litre^{-1} (B) 1.5×10^{-5} mole litre^{-1}
(C) 1.5×10^{-9} mole litre^{-1} (D) 3.9×10^{-5} mole litre^{-1}

38. The plot of concentration of the reactant Vs. time for a reaction is a straight line with a negative slope. This reaction follows a
(A) First order rate equation
(B) Zero order rate equation
(C) Second order rate equation
(D) Third order rate equation

39. The standard electorde potential for $Pb^{2}+$ /Pb and Zn^{2+}/Zn are- 0.126 V and -0.763 respectively. The e.m.f. of the cell Zn/zn^{2+} (1.0M)/Pb^{2}+(1.0M)Pb is
(A) - 0.889V (B) - 0.637V (C) 0.637 V (D) 0.889V

40. When 0.04 Faraday of elctricity is passed through a solution of $CaSO_4$, then the weight of Ca^{2+} metal deposited at the cathode is
(A) 0.6 g (B) 0.4 g (C) 0.2 g (D) 0.8 g

41. The molar conductances of NaCl, HCl and CH_3COONa at infinite dilution are 126.45, 426.16 and 91 ohm^{-1} cm^2 mol^{-1} respectively. The molar conductance of CH_3COOH at infinite dilution is
(A) 201.28 ohm^{-1} cm^2 mol^{-1} (B) 698.28 ohm^{-1} cm^2 mol^{-1}

(C) 390.71 ohm^{-1} cm^2 mol^{-1} (D) 540.48 ohm^{-1} cm^2 mol^{-1}

42. A catalyst is a substance which
(A) Changes the rate of a chemical reaction
(B) Changes the equilibrium constant of a chemical reaction
(C) Increase the equilibrium concentration of the products.
(D) Supplies energy to a chemical reaction.

43. A colloidal solution is subjected to an electrical field. The particles move towards anode. The coaulation of same sol is studied using NaCl,$BaCl_2$ and $AlCl_3$ solutions. Their coagulating power should be
(A) NaCl > $BaCl_2$ > AlC_3 (B) $BaCl_2$> $AlCl_3$ > NaCl
(C) $AlCl_3$ > $BaCl_2$ > NaCl (D) $BaCl_2$ > NaCl > $AlCl_3$

44. The relationship which describes the variation of vapour pressure with temperature is called
(A) Hess's law (B) Arrhenius equation
(C) Kirchoff's law (D) Clausius-Clapeyron

45. The orbitals are called degenerate when
(A) They have same wave functions but different engergy
(B) They have different wave functions but same engery
(C) They have the same engergy
(D) They have the same wave functions.

46. Fibre reactive dyes ae derivatives of
(A) Anthraquinone (B) 2, 4-dichloro-1, 3, 5-triazone
(C) 2, 4, 6-triamino-1, 3, 5-triazine (D) Naphthalene

47. In the following four elements, the ionization potential of which one is the highest ?
(A) Argon (B) Barium (C) Cesium (D) Oxygen

48. In the commercial electrochemical process for aluminium extraction the electrolyte used is
(A) $Al(OH)_3$ in NaOH solution
(B) An aqueous solution of $Al_2(SO_4)_3$
(C) A molten mixture of Al_2O_3 and Na_3AlF_6
(D) A molten mixture of Al_2O_3 and $Al(OH)_3$

49. Which of the following is an azo dye?
(A) Orange-1 (B) Malachite green
(C) Indigo (D) Martius yellow

50. In Solvay process of manufacture of Na_2CO_3, the by-products are
(A) NH4Cl.CaO (B) CaO, Na_2CO_3
(C) $CaCl_2$, CO_2, NH_3 (D) Na_2CO_3,CO_2

BHU - 1995

1. (C)	27. (B)
2. (B)	28. (C)
3. (D)	29. (C)
4. (B)	30. (D)
5. (B)	31. (C)
6. (B)	32. (A)
7.	33. (A)
8. (C)	34. (C)
9. (A)	35. (C)
10. (B)	36. (C)
11. (C)	37. (C)
12. (A)	38. (C)
13. (D)	39. (A)
14. (A)	40. (B)
15. (A)	41. (D)
16. (D)	42. (D)
17. (D)	43. (B)
18. (D)	44. (A)
19. (D)	45. (A)
20. (B)	46. (A)
21. (C)	47. (C)
22. (D)	48. (B)
23. (B)	49. (A)
24. (B)	50. (A)
25. (C)	
26. (A)	

BHU - 1996

1. (D)	27. (C)
2. (A)	28. (C)
3. (A)	29. (C)
4. (A)	30. (C)
5. (D)	31. (A)
6. (A)	32. (C)
7. (B)	33. (A)
8. (B)	34. (C)
9. (A)	35. (A)
10. (D)	36. (D)
11. (D)	37. (C)
12. (B)	38. (C)
13. (D)	39. (B)
14. (D)	40. (D)
15. (B)	41. (A)
16. (A)	42. (C)
17. (A)	43. (D)
18. (A)	44. (B)
19. (B)	45. (C)
20. (C)	46. (A)
21. (D)	47. (D)
22. (A)	48. (A)
23. (A)	49. (A)
24. (D)	50. (B)
25. (D)	
26. (A)	

BHU - 1997

1. (B)	27. (C)
2. (D)	28. (A)
3. (C)	29. (B)
4. (A)	30. (B)
5. (A)	31. (A)
6. (C)	32. (D)
7. (D)	33. (B)
8. (D)	34. (C)
9. (D)	35. (B)
10. (A)	36. (D)
11. (C)	37. (B)
12. (A)	38. (D)
13. (B)	39. (A)
14. (B)	40. (A)
15. (B)	41. (C)
16. (A)	42. (A)
17. (A)	43. (B)
18. (C)	44. (A)
19. (C)	45. (A)
20. (C)	46. (B)
21. (C)	47. (C)
22. (B)	48. (C)
23. (D)	49. (D)
24. (A)	50. (C)
25. (B)	
26. (D)	

BHU - 1998

1. (C)	27. (C)
2. (A)	28. (A)
3. (C)	29. (B)
4. (D)	30. (D)
5. (B)	31. (C)
6. (B)	32. (B)
7. (A)	33. (C)
8. (B)	34. (A)
9. (B)	35. (C)
10. (D)	36. (A)
11. (C)	37. (D)
12. (B)	38. (D)
13. (D)	39. (B)
14. (B)	40. (D)
15. (D)	41. (A)
16. (C)	42. (B)
17. (B)	43. (A)
18. (D)	44. (A)
19. (C)	45. (B)
20. (C)	46. (C)
21. (D)	47. (D)
22. (A)	48. (D)
23. (C)	49. (A)
24. (A)	50. (B)
25. (A)	
26. (C)	

BHU - 1999

1. (A)	27. (D)
2. (C)	28. (A)
3. (B)	29. (A)
4. (B)	30. (C)
5. (C)	31. (A)
6. (C)	32. (B)
7. (A)	33. (B)
8. (C)	34. (A)
9. (B)	35. (A)
10. (B)	36. (B)
11. (D)	37. (C)
12. (C)	38. (D)
13. (D)	39. (C)
14. (A)	40. (A)
15. (B)	41. (C)
16. (A)	42. (A)
17. (D)	43. (D)
18. (D)	44. (B)
19. (C)	45. (D)
20. (C)	46. (A)
21. (D)	47. (A)
22. (A)	48. (C)
23. (B)	49. (D)
24. (D)	50. (D)
25. (A)	
26. (B)	

BHU - 2000

1. (C)	27. (B)
2. (C)	28. (C)
3. (A)	29. (D)
4. (D)	30. (C)
5. (A)	31. (C)
6. (A)	32. (C)
7. (C)	33. (B)
8. (B)	34. (D)
9. (D)	35. (D)
10. (A)	36. (B)
11. (C)	37. (A)
12. (A)	38. (C)
13. (B)	39. (D)
14. (A)	40. (C)
15. (B)	41. (A)
16. (B)	42. (A)
17. (B)	43. (C)
18. (B)	44. (C)
19. (C)	45. (B)
20. (A)	46. (A)
21. (C)	47. (B)
22. (C)	48. (B)
23. (A)	49. (A)
24. (A)	50. (C)
25. (A)	
26. (C)	

BHU - 2001

1. (D)	27. (D)
2. (B)	28. (B)
3. (B)	29. (C)
4. (D)	30. (A)
5. (C)	31. (D)
6. (D)	32. (B)
7. (D)	33. (D)
8. (A)	34. (C)
9. (D)	35. (A)
10. (A)	36. (A)
11. (D)	37. (B)
12. (C)	38. (D)
13. (D)	39. (C)
14. (C)	40. (B)
15. (B)	41. (B)
16. (B)	42. (B)
17. (A)	43. (A)
18. (C)	44. (C)
19. (C)	45. (C)
20. (C)	46. (B)
21. (C)	47. (A)
22. (D)	48.
23. (C)	49. (B)
24. (A)	50. (D)
25. (A)	
26. (D)	

BHU - 2002

1. (C)	27. (C)
2. (D)	28. (D)
3. (A)	29. (D)
4. (A)	30. (B)
5. (B)	31. (C)
6. (A)	32. (B)
7. (B)	33. (A)
8. (D)	34. (B)
9. (A)	35. (B)
10. (B)	36. (B)
11. (C)	37, (C)
12. (B)	38. (A)
13. (C)	39. (D)
14. (D)	40. (A)
15. (B)	41. (C)
16. (A)	42. (A)
17. (C)	43. (C)
18. (C)	44. (D)
19. (C)	45. (D)
20. (B)	46. (A)
21. (A)	47. (A)
22. (A)	48. (D)
23. (D)	49. (A)
24. (C)	50. (C)
25. (B)	
26. (B)	

BHU - 2003

1. (A)	27. (A)
2. (B)	28. (D)
3. (B)	29. (C)
4. (A)	30. (A)
5. (D)	31. (A)
6. (A)	32. (A)
7. (A)	33. (A)
8. (B)	34. (B)
9. (D)	35. (A)
10. (C)	36. (B)
11. (B)	37. (B)
12. (C)	38. (C)
13. (A)	39. (A)
14. (A)	40. (A)
15. (D)	41. (C)
16. (A)	42. (C)
17. (C)	43. (D)
18. (A)	
19. (B)	
20. (C)	
21. (A)	
22. (A)	
23. (A)	
24. (A)	
25. (A)	
26. (A)	

BHU - 2004

1. (D)	26. (D)
2. (A)	27. (B)
3. (B)	28. (A)
4. (C)	29. (C)
5. (B)	30. (D)
6. (D)	31. (C)
7. (A)	32. (A)
8. (D)	33. (B)
9. (B)	34. (D)
10. (C)	35. (B)
11. (A)	36. (B)
12. (B)	37. (D)
13. (B)	38. (C)
14. (B)	39. (C)
15. (C)	40. (D)
16. (B)	41. (C)
17. (B)	42. (A)
18. (B)	43. (C)
19. (B)	44. (D)
20. (C)	45. (C)
21. (B)	46. (B)
22. (C)	47. (A)
23. (D)	48. (C)
24. (A)	49. (A)
25. (D)	50. (C)

CBSE -1990

1. Among the following compounds the one which exhibits geometrical isomerism is

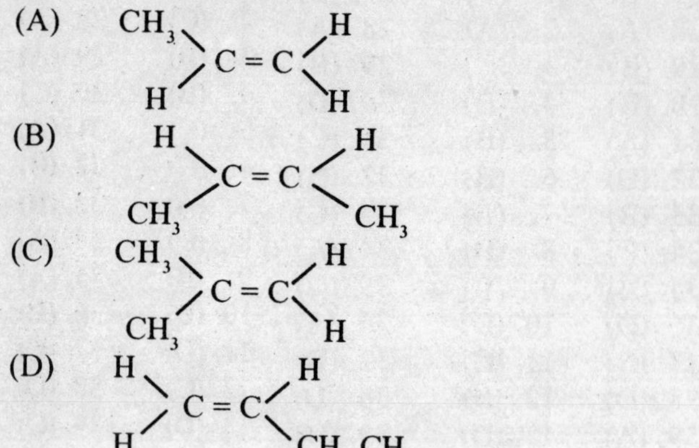

2. Which of the following compounds can exist as optical isomers ?

(A) $(CH_3)(CH_3)C(H)(COOH)$ (B) $(CH_3)(C_2H_5)C(H)(COOH)$

(C) $(CH_3)(CH_3)C(OH)(COOH)$ (D) $(H)(OH)C(H)(COOH)$

3. Identify Z in

$C_6H_5NH_2 \xrightarrow[2.H_2O\ boil]{1.HNO_2(250K)} X \xrightarrow{NaOH} Y \xrightarrow{CH_3I} Z$

(A) 2,6-dimethyl-4-methyl-N-methylaniline ($NH\text{-}CH_3$, H_3C, CH_3, CH_3 on benzene ring) (B) OCH_3, H_3C, CH_3, CH_3 on benzene ring

(C) OCH_3, HO, OH, OH on benzene ring (D) $C_6H_5OCH_3$

4. An aromatic amine (A) was treated with alcoholic potash and another compound (Y) when foul semlling gas was formed with formula C_6H_5NC. Y was formed by reacting a compound (Z) with Cl_2 in the presence of slaked lime. The compound (Z) is

(A) CH_3OCH_3 (B) $CHCl_3$

(C) $C_6H_5NH_2$ (D) CH_3CH_2OH

5. 28 gm of N_2 and 6gm of H_2 were kept at 400°C in 1-liltre vessel, the equilibrium mixture contained 27.54 gm of NH_3.The approximate value of K_C;for the above reaction can be in (mol^{-2} litre^2).

(A) 25 (B) 100 (C) 75 (D) 50

6. On heating glucose with Fehling's solution; we get a precipitate whose colour is

(A) black (B) white (C) yellow (D) red

7. When NH_4Cl is added to a solution of NH_4OH

(A) pH will increase (B) pH will decrease

(C) the dissociation of NH_4OH is increased.

(D) It has no effect on pH at all.

8. Identify B in

$C_6H_4(CH_2OH)_2 \xrightarrow[oxidation]{vigorous} A \xrightarrow[heating]{dry} B$

(A) $C_6H_4(COOH)_2$ (B) C_6H_5COOH

(C) naphthalene-fused anhydride (C(=O)–O–C(=O)) (D) $[C_6H_4(COOH)(CO)]^-$

9. Identify Y in reaction

$C_6H_5OH \xrightarrow{Reimer\text{-}Tiemann} X \xrightarrow{NaOH} Y$

(A) OH, OH on benzene ring with CH(OH)–cyclohexyl bearing OH

(B) OH, OH on benzene ring with CH_2OH + $C_6H_4(OH)COOH$

(C) OH, OH on benzene ring –C(=O)– C_6H_4OH

(D) $C_6H_4(OH)$–C(=O)–CH(OH)–$C_6H_4(OH)$

10. The boiling point of a solvent containing non-volatile solute is

(A) depressed (B) elevated

(C) does not change (D) none of these

11. Number of moles of the solute dissolved per litre of the solutions is known as

(A) formality (B) normality

(C) molality (D) molarity

12. Given that

$2C(s) + 2O_2(g) \longrightarrow 2CO_2(g),\ \Delta H = -787KJ$

$H_2(g) + \frac{1}{2}O_2(g) \longrightarrow H_2O(l),\ \Delta H = -286KJ$

$C_2H_2(g) + 2\frac{1}{2}O_2(g) \longrightarrow 2CO_2(g) + H_2O(l), \Delta H = -1310KJ$

Heat of formation of acetylene is
(A) –800KJ (B) +237KJ (C) +1802 KJ (D) –1802KJ

13. For the reaction $H_2(g) + I_2(g) = 2HI(g)$ at 721K, K_C the value of equilibrium constant is 50. When the equilibrum concentration of both is 0.5N, value of K_P under the same conditions will be
(A) 0.02 (B) 0.2 (C) 50.0 (D) 50/RT

14. The half-life of a first order reaction is 69.35. The value of the rate constant of the reaction is
(A) 0.01 s^{-1} (B) 0.001s^{-1} (C) 1.0 s^{-1} (D) 0.1 s^{-1}

15. The number of sigma (σ) and pi (π)bonds in one molecule of benzene is
(A) 6 σ and 6π (B) 12 σ and 6π
(C) 12 σ and 12π (D) 12 σ and 3π

16. 1 mole of ice is converted into liquid stage at 273K. $H_2O(s)$ and $H_2O(l)$ have entropies 38.20 and 60.01 J $mole^{-1}$ deg^{-1}. Enthalpy change in conversion will be
(A) 59.85 J/mole (B) 5958.5 J/mole
(C) 595.85 J/mole (D) 59585.0 J/mole

17. The reaction equation

$$C_6H_5Cl + NaOH \xrightarrow[200\ atm.]{200°C} C_6H_5ONa + NaCl + H_2O$$

illustrates
(A) Carbylamine test (B) Holoform reaction
(C) Down's process (D) Kolbe's reaction

18. A mixture of two completely miscible non-ideal liquids which distil as such without change in its compositon at a constant temperature as if it were a pure liquid. This mixture is known as
(A) binary liquid mixture (B) azeotropic mixture
(C) eutectic mixture (D) ideal mixture

19. By which of the following processes can acetic acid be converted into methane ?
(A) Dehydrogenation (B) Dehydration
(C) Decarboxylation (D) Polymerisation

20. Heredity traits are maintained by
(A) DDT (B) TNT (C) RNA (D) DNA

21. In Faraday-Tyndall effect the colloidally suspended particles
(A) show Brownian movement
(B) trace out the path of a strong beam of light
(C) show electrophoresis (D) coagulate

22. Which of the following has both ionic and cavalent bonds
(A) CO_2 (B) H_2O (C) $CaCl_2$ (D) NH_4Cl

23. Absolute zero is defined as
(A) at which liquid helium boils
(B) at which water boils at 298 K
(C) the temperature at which all molecular motion ceases
(D) All of the above

24. A catalyst is a substance which
(A) alters the chemical equilbrium
(B) lowers the activation energy which the reactant molecules must aquire in order to form the intermediate activated complex
(C) changes the equilbrium of a reaction so that the concentration of product increases
(D) None of these

25. Which one of the following is not found in alkenes ?
(A) Chain isomerism (B) Geometrical isomerism
(C) Mesomerism (D) Position isomerism

26. $K_4Fe(CN)_6$ is called
(A) potassium hexacyanoferrate (III)
(B) prussian blue
(C) potassium hexacyanoferrate (II)
(D) potassium ferricyanide

27. Root mean square velocity of a gas molecule is proportional to
(A) $m^{-1/2}$ (B) m (C) $m^{1/2}$ (D) M^o

28. The ionisation of hydrogen atom would give rise to
(A) hydride ion (B) hydronium ion
(C) proton (D) hydroxyl ion

29. Which of the following shows hydrogen bonding ?
(A) HCl (B) H_2S (C) HBr (D) HF

30. Which of the following has the least bond angle ?
(A) NH_3 (B) H_2O (C) BeF_2 (D) CH_4

31. Heat absorbed or evolved in a reaction at constant temperature and volume is
(A) P ΔV (B) ΔnRT (C) ΔH (D) ΔE

32. If the system undergoes contraction of volume the work done by system is
(A) positive (B) negative (C) negligible (D) zero

33. Conjugate base of HCO_3^- is
(A) H_2CO_3 (B) CO_3^{2-} (C) H (D) CO_2

34. Which of the following acts as both Bronsted acid and base ?
(A) OH^- (B) H_3PO_4 (C) H_2O (D) NH_3

35. Conjugate base for HSO_4^- is
(A) H^+ (B) H_2O (C) SO_4^{-2} (D) H_2SO_4

36. The vapour density of ozone would be
(A) 24 (B) 96 (C) 48 (D) 16

37. Identify X in the reaction

$$p\text{-}ClC_6H_4NH_2 \xrightarrow[\Delta]{(CHCl_3 / KOH)} \text{Intermediate} \xrightarrow[300\ K]{HCl} X$$

(A) $p\text{-}ClC_6H_4NC$ (B) $p\text{-}ClC_6H_4NHCH_3$

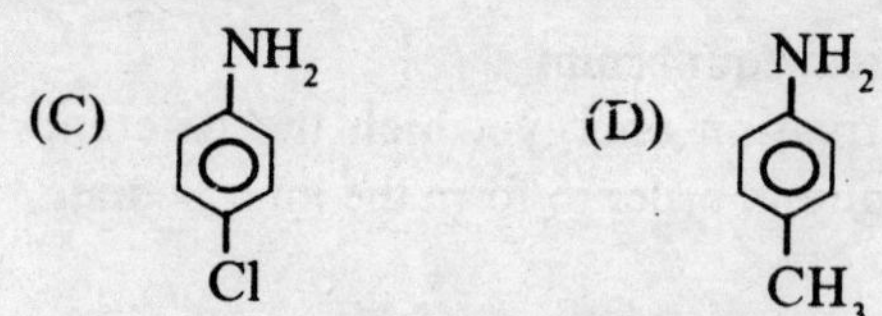

38. The ability of a substance, to assume two or more crystalline structure is called
(A) isomorphism. (B) amorphism
(C) Isomerism. (D) polymorphism.

39. The compound formed when malonic ester reacts with urea is
(A) cinnamic acid. (B) butyric acid.
(C) barbiatyric acid. (D) crotonic acid .

40. Which acid is present is vinegar ?
(A) HCl (B) CH_3COOH
(C) Tartaric acid (D) Citric acid

41. Identify X in the series :

$NH-CO-CH_3$ (on benzene ring) $\xrightarrow{HNO_3 / H_2SO_4}$ Intermediate $\xrightarrow[\Delta]{H_2O / H^+}$ X

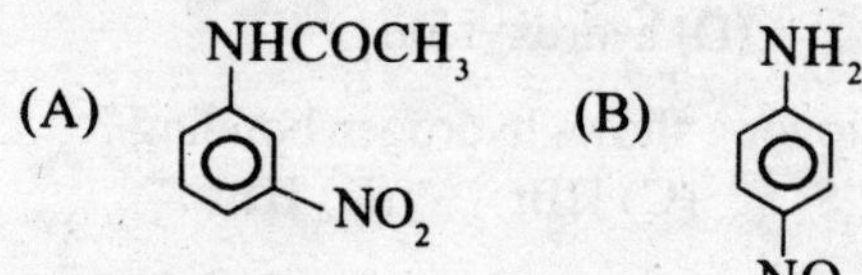

(C) NH_2 with NO_2 (meta) on benzene ring (D) NH_2 with O_2N, NO_2, NO_2 on benzene ring

42. The reagents which can be used to distinguish acetophenone from benzophenone is/are
(A) aqueous solution of $NaHSO_3$
(B) 2, 4-dinitrophenylhydrazine
(C) Benedict's reagent. (D) $I_2$and Na_2CO_3

43. when two atoms co-axially overlap to form a molecule then which of the following bond will be fromed
(A) 1π bond (B) $2\ \sigma$ bonds (C) $1\ \sigma$ bond (D) $2\ \pi$ bonds

44. In a compound with atomic number $_{29}X^{60}$; the number of electrons is
(A) 40 (B) 59 (C) 29 (D) 30

45. which of the following undergoes substitution reactions ?
(A) Alkanes (B) Akenes
(C) Alkynes (D) Any of the above

46. $H_2 + I_2 \rightleftharpoons 2HI$
in the above equilibrium system if the concentration of the reactants at 25°C is increased, the value of K_C will
(A) depend on the nature of the reactants.
(B) Increase (C) decrease (D) remain same

47. The solution which has the highest bolling point is
(A) 0.1 N Na_2SO_4 (B) 0.1 N $MgSO_4$
(C) 0.1M$Al_2(SO_4)_3$ (D) 0.1 M $BaSO_4$

48. 2.0 litres of air formed 1.715 l ozonised air when passed through Brodiel apparatus. The volume of ozone formed is
(A) 225.0ml. (B) 340.0ml (C) 85.0ml (D) 170.0ml

49. 50.0ml of 0.2N HCI, 50.0 ml of 0.1 N H_2SO_4 and 100.0ml of 0.2 N HNO_3 are mixed. The normally of resulting solution is
(A) 0.10N (B) 0.15N (C) 0.175N (D) 0.20N

50. The shape of the sp^3d^2 hybrld orbital is
(A) square planar (B) trigonal bipyramidal
(C) octahedral (D) pentagonal-bi-pyramidal

51. Different lines in Balmer series of hydrogen spectrum lies in
(A) UV region (B) Far Infrared region
(C) IR region (D) visible region

52. Amide is converted into primary amine when it is heated with
(A) H_2SO_4 (B) NaOH + Br_2
(C) KOH (alkaline) (D) KOH (aqueous)

53. The minimum bond distance is in
(A) $C-C$ (B) $C=C$ (C) $C\equiv C$ (D) $C-C=C-C$

54. If ΔH is the change in enthalpy and ΔE, the change in the internal energy accompanying a gaseous reaction
(A) $\Delta H < \Delta E$ only if the number of moles of the products is greater than the number of the reactants.
(B) ΔH is always less than ΔE
(C) ΔH is always greater than ΔE
(D) $\Delta H < \Delta E$ only if the number of moles of products is less than the number of moles of the reactants.

55. Given that for the reaction
$C_2H_4(g) + 3O_2(g) \longrightarrow 2CO_2(g) + 2H_2O\ (l)$ at 298°K
$\Delta E = -1415KJ$
If R = 0.0084 KJ per degree, ΔH =
(A) –1400 KJ (B) –1420 KJ
(C) –1410 KJ (D) –1430 KJ

56. Ethoxy ethane and methoxy propane are
(A) geometrical isomers (B) optical isomers
(C) functional group isomers (D) metamers

57. An aqueous solsution. whose pH = 0 is
(A) neutral (B) amphoteric
(C) acidic (D) alkaline

58. Washing soda has the formula
(A) Na_2CO_3 , $7H_2O$ (B) Na_2CO_3 , $10H_2O$
(C) Na_2CO_3 , H_2O (D) Na_2CO_3

59. Aspirin is
(A) antipyretic (B) psychedelic
(C) antibiotic (D) sedative

60. Lemon is sour due to the presence of
(A) tartaric acid (B) citric acid
(C) sulphuric acid (D) oxalic acid

61. Poisonous gas present in the exhaust fumes of a jet plane is
(A) methane (B) C_2H_2

(C) fluorocarbon (D) carbon monoxide

62. Calomel is
(A) $HgCl_2$ (B) Hg_2Cl_2
(C) Hg+$HgCl_2$ (D) Hg_2Cl_2+Hg

63. The general formula $(RCO)_2O$ represents
(A) an ether (B) ketone
(C) anester (D) an acid anhydride

64. Gypsum is
(A) $CaSO_4$, $2H_2O$ (B) $CaSO_4$, $3H_2O$
(C) $CaSO_4$ (D) $CaSO_4$, $1/2H_2O$

65. In which of the following the bond angle between the two bonds is minimum ?
(A) H_2O (B) NH_3 (C) CO_2 (D) CH_4

66. Alkali metals lose electrons in
(A) s-orbitals (B) p-orbitals
(C) d-orbitals (D) f-orbitals

67. Which of the following is strongest acid ?
(A) HF (B) HI (C) HBr (D) HCl

68. Cellophane used for wrapping is
(A) cellulose nitrate (B) synthetic cellulose
(C) cellulose acetate (D) regenerated cellulose

69. The composition of 'Golden spongles' is
(A) As_2S_2 (B) $BaCrO_4$ (C) $PbCrO_4$ (D) PbI_2

70. Lucas test is for
(A) alcohols (B) amines
(C) diethylether (D) glacial acetic acid

71. The first synthetic organic compound is
(A) benzene (B) urea (C) glucose (D) methane

72. Avogadro's number being $6x10^{23}$, the number of atoms presents in 6 gms of carbon is
(A) $6 \times 10^{23} \times 1$ (B) $6 \times 10^{23} \times 6$
(C) $\frac{6 \times 10^{23}}{6}$ (D) $\frac{6 \times 10^{23}}{2}$

73. $CHCl_3$ on oxidation gives
(A) phosgene (B) formic acid
(C) chloropicrin (D) CCl_4

74. Which of the following is monatomic ?
(A) Oxygen (B) Hydrogen
(C) Helium (D) Fluorine

75. In ethylene molecules each carbon atom involves
(A) sp-hybridisation (B) sp^2-hybridisation
(C) sp^3-hybridisation (D) none of these

CBSE - 1991

1. The volume of CO_2 at STP obtained by heating one gram of pure $CaCO_3$ will be
(A) 0.224 litre (B) 11.2 litre
(C) 1 litre (D) 22.4 litre

2. Which of the following solutions would have the highest Osmotic pressure ?
(A) M/10 $BaCl_2$ (B) M/10 Glucose
(C) M/10 NaCl (D) M/10 Urea

3. Sodium phenoxide reacts with CO_2 to yield a product which on hydrolysis yields a new compounds. The new compound is
(A) Phthalic acid (B) Salicylic acid
(C) Salicylaldehyde (D) Benzoic acid

4. Oleum is
(A) fuming H_2SO_4 (B) Oil of vitriol
(C) Castor oil (D) None of these

5. $CH_2 = CH_2$ is a
(A) monomer (B) polymer (C) isomer (D) equimer

6. The reaction ,
$2SO_2 + O_2 + 2H_2O \longrightarrow 2H_2SO_4$
is an example of
(A) synthesis (B) analysis
(C) displacement (D) double decomposition

7. The Hydrogen bond is the strongest in
(A) F – H.....F (B) F – H......O
(C) O – H .....S (D) S – H......O

8. Father of Chemistry is
(A) Lavosier (B) Priestlay
(C) Berthold (D) Aristotle

9. I.U.P.A.C. name of the compound,

$$CH_3 - \overset{\overset{\displaystyle CH_3}{|}}{\underset{\underset{\displaystyle CH_3}{|}}{C}} - CH = CH_2$$ is

(A) 1, 1, 1 trimethyl-3 propane
(B) 3, 3 dimethyl-1-butene (C) 3, 3, 3 trimethyl-1-propane
(D) 1, 1-dimethyl-3-butene

10. $C_2H_5Br + NaOH$ ------→ $C_2H_5OH + NaBr$
This reaction is known as .... substitution
(A) electrophilic (B) nucleophilic
(C) Both (A) and (B) (D) None of these

11. An sp^3 hybrid orbital contains
(A) 1/4th s-character (B) 1/2th s-character
(C) 2/3rd s-character (D) 3/4th s-character

12. Chloride of metal is MCl_2. The formula of its phosphate will be
(A) M_2PO_4 (B) $M_3(PO_4)_2$
(C) $M(PO_4)_3$ (D) MPO_4

13. The charge on a cation M is +2 and on an anion A is -3. The compound formed by these has the formula,
(A) M_3A_2 (B) M_2A_3 (C) MA_2 (D) M_2A

14. Which is not present in Grignard reagent
(A) Methyl group (B) Magnesium
(C) Halogen (D) -COOH group

15. Grignard reagent reacts with sulphur and forms
(A) hypo (B) thio alcohol
(C) methyl alcohol (D) formic acid

16. Glycerol reacts with $P_4 + I_2$ to form
(A) aldehyde (B) allyl Iodide
(C) allyl acohol (D) acetylene

17. $CH_3COCl \xrightarrow[BaSO_4]{Pd/H_2} X$. Here X is
(A) acetaldehyde (B) propionaldehyde
(C) acetone (D) acetic anhydride

18. The weakest acid among the following is
(A) CH_3COOH (B) $CH_2ClCOOH$
(C) $CHCl_2COOH$ (D) Cl_3CCOOH

19. Which of the following is the most stable ?
(A) $(CH_3)_3\ C^+$ (B) $C_6H_5CH_2^+$
(C) $CH_3CH_2^+$ (D) $(CH_3)_2\ CH^+$

20. The vapour density of a gas is 11.2. The volume occupied by 11.2 gms of this gas at NTP is
(A)1 litre (B) 11.2 litre
(C) 22.4 litre (D) 20 litre

21. On hydrolysis of starch we get finally
(A) glucose and fluctose (B) sucrose
(C) glucose (D) fructose

22. Insulin, a hormone, is chemically a
(A) fat (B) Steroid (C) protein (D) Carbohydrate

23. Natural rubber is polymer of
(A) polyethylene (B) PVC
(C) Isoprene (D) Insulin

24. NH_4OH is a weak base because
(A) It is only slightly ionised
(B) It is not hydroxide of metal
(C) It has low vapour pressure
(D) It has low density

25. Glass is a
(A) Liquid (B) transparent organic polymer
(C) solid (D) Super-cooled liquid

26. Basicity of orthophosphoric acid is
(A)2 (B) 3 (C) 4 (D) 5

27. Which is more electronegative ?
(A) F (B) Cl (C) I (D) None of these

28. Maximum number of electrons in any orbit is
(A) n^2 (B) $2n^2$ (C) $1/2\ n^2$ (D) None of these

29. No two electrons in an atom can have the same set of four quantum numbers. This principle was evolved by
(A) Helsenberg (B) Pauli
(C) Maxwell (D) De-Broglle

30. In a potassium atom, electronic energy levels are in the following order
(A) 4s < 3d (B) 4s<3p (C) 4s > 3d (D) 4s > 4p

31. For orthorhombic axial distance $a \neq b \neq c$, the Axial angle is
(A) $\alpha = \beta = \gamma \neq 90^\circ$ (B) $\alpha = \beta = \gamma = 90^\circ$
(C) $\alpha = \gamma = 90^\circ, \beta \neq 90^\circ$ (D) $\alpha \neq \beta \neq \gamma \neq 90^\circ$

32. In which bond, angle is the highest ?
(A) sp^3d (B) sp (C) sp^3 (D) sp^2

33. Which type of hybridization occurs in ethylene ?
(A) sp (B) sp^2 (C) sp^3d (D) sp^3d^2

34. dsp^2 hybridization has arrangement which is
(A) trigonal planar (B) tetrahedral
(C) square planar (D) trigonal blpyramid.

35. Strength of hydrogen bond is intermediate between
(A) lonic and covalent (B) lonic and metallic
(C) van der Waal and covalent
(D) metallic and covalent

36. Both lonic and covalent bonds are present in
(A) SO_2 (B) NaOH (C) CH_4 (D) KCl

37. The shortest carbon-carbon bond distance is found in
(A) diamond (B) ethane (C) benzene (D) acetylene

38. In the following molecule the two carbon atoms marked by asterisk * possess the following type of hybridization orbitals.

$$H_3C - {}^*C \equiv {}^*C - CH_3$$

(A) sp orbitals (B) s orbitals
(C) sp^3 orbitals (D) sp^2 orbitals

39. Which one of following is paramagnetic?
(A) O_2 (B) CO_2 (C) H_2O (D) NO_2

40. Hydrogen bonding is maximum in
(A) ethanol (B) dlethyl ether
(C) ethyl chloride (D) triethyl amine

41. The transition elements have a general electronic configuration
(A) $ns^2\ np^6\ nd^{1-10}$ (B) $(n-1)d^{1-10}\ ns^{0-2}\ np^{0-6}$
(C) $(n-1)d^{1-10}\ ns^{0-2}$ (D) $nd^{1-10}\ ns^{0-2}$

42. The correct order of second lonisation potential of C,N,O and F is
(A) C>N>O>F (B) O>N>F>C
(C) O>F>N>C (D) F>O>N> C

43. Energy of third orbit of Bohr's atom is
(A) −13.6 eV (B) −3.4 eV
(C) −1.5 eV (D) None of these

44. Which of following has the lowest ionisation potential ?
(A) $3p^6$ (B) $2p^6$ (C) $3d^2$ (D) $4s^1$

45. How do we form pure N_2 gas ?
(A) $N_2O + Cu$ (B) $(NH_4)_2Cr_2\ O_7$
(C) $NH_3 + NaNO_2$ (D) $NH_4CI + NaNO_2$

46. P_4O_6 is heated with H_2O to give
(A) hypophosphoric acid
(B) hypophosphorus acid
(C) orthophosphorus acid
(D) orthophosphoric acid

47. Which of the following is formed by sea salt ?
(A) Halogen (B) Chlorine
(C) Oxygen (D) Sulphur

48. For the reaction $N_2 + 3H_2 \rightleftharpoons 2NH_3$, $\Delta H=$
(A) $\Delta E + 2RT$ (B) $\Delta E - 2RT$
(C) $\Delta E + RT$ (D) $\Delta E - 2RT$

49.Cr has electronic formation
(A) $3s^2 3p^6 3d^6$ (B) None of these
(C) $3s^2 3p^6 3d^4 4s^2$ (D) $3s^2 3p^6 3d^5 4s^1$

50. Oxidation number of Mn in K_2MnO_4 is
(A) +6 (B) -6 (C) +7 (D) -5

51. In reversible reaction both side rates of reaction are
(A) Same (B) different
(C) One side more (D) reactant rate less

52. If A+B→C reactant takes some heat then reaction is completed at the
(A) high temperature and low pressure
(B) low temperature and high pressure
(C) high temperature and high pressure
(D) low temperature and low pressure

53. PH_4I + NaOH forms
(A) PH_3 (B) NH_3 (C) P_4O_6 (D) P_4O_{10}

54. The pH of 0.001M HCl is
(A) 2 (B) 10 (C) 3 (D) 11

55. Which of the following has more oxidising power ?
(A) Br (B) Cl (C) I (D) As

56. Cinnabar is an ore of
(A) gold (B) silver (C) mercury (D) zinc

57. Calculate the OH^- ion concentration of NaOH of whose pH is 2
(A) 10^{-2} (B) 10^{-9} (C) 10^{-12} (D) 10^{-7}

58. Action of water on carbide forms
(A) ethane (B) ethene (C) ethyne (D) ethanol

59. $C_2H_2 \xrightarrow[H_2SO_4]{HgOH\ 1\%} \Delta \xrightarrow{O} X.$
X is
(A) Acid (B) aldehyde (C) Ketone (D) ethanol

60. Nitrophenol form phend is formed by ...... substitution
(A) Nucleophilic (B) Electrophilic
(C) Both (A) & (B) (D) None of these

61. Benzene reacts with CH_3COCl in the presence of $AlCl_3$ to give
(A) $C_6H_5CO_2Cl$ (B) $C_6H_5COCH_3$
(C) C_6H_5Cl (D) $C_6H_5CH_3$

62. Which one of the following solutions has the highest boiling point ?
(A) 0.1M glucose (B) 0.1M $BaCl_2$
(C) 0.1M NaCl (D) 0.1M Urea

63. Ce 58 is a member of
(A) d-block (B) f-block (C) s-block (D) p-block

64. I.U.P.A.C. name for CH_3CHO is
(A) acetaldehyde (B) methyl aldehyde
(C) formyl methane (D) ethanal

65. $_{20}Ca^{40}$ and $_{19}K^{40}$ are
(A) Isobars (B) Isotones (C) Isomers (D) Isotopes

66. When sucrose is heated with conc.HNO_3, the product is
(A) Sucrose nitrate (B) Formic acid
(C) Oxalic acid (D) Citric acid

67. Kjeldahl's method is used in the estimation of
(A) Nitrogen (B) Halogen
(C) Sulphur (D) Oxygen

68. If formaldehyde and KOH are heated together, we get
(A) methane (B) acetylene
(C) methanol (D) ethylacetate

69. The compound whose solution has the highest pH is
(A) NaCl (B) $NaHCO_3$ (C) NH_4Cl (D) Na_2CO_3

70. What makes a lemon sour ?
(A) Oxalic acid (B) Tartaric acid
(C) Citric acid (D) Hydrochloric acid

71. The Isomers must have the same
(A) structural formulae (B) molecular formulae
(C) chemical properties (D) physical properties

72. Which is of the following is obtained when chloral is boiled with NaOH ?
(A) CH_3Cl (B) $CHCl_3$ (C) CCl_4 (D) All of the above

73. Kolbe-Schmldt's reaction is used for preparing
(A) salicyclic acid (B) Salicylaldehyde
(C) Phenol (D) Hydrocarbon

74. Aqueous solution of ammonia consists of
(A) NH_4^+ (B) NH_4^+ and OH^- (C) H^+ (D) CH^-

75. Isocyanide test is used for the detection of
(A) primary alcohols (B) primary amines
(C) secondary amines (D) secondary alcohols

CBSE - 1992

1. 2-Methyl, -2-butene will be represented as
(A) $CH_3-\underset{\displaystyle CH_3}{\underset{|}{CH}}-CH_2-CH_3$ (B) $CH_3-\underset{\displaystyle H}{\underset{|}{C}}=\underset{\displaystyle CH_3}{\underset{|}{C}}-CH_3$
(C) $CH_3-CH_2-\overset{\displaystyle CH_3}{\overset{|}{C}}=CH_2$ (D) $CH_3-\overset{\displaystyle CH_3}{\overset{|}{CH}}-CH-CH_3$

2. Select the true statement about benzene from amongst the following
(A) There is cyclic delocalisation of pi eletrons in benzene
(B) Because of unsaturation benzene easily undergoes addition reactions
(C) There are two types of C-C bonds in benzene molecule
(D) Monosubstitution of benzene gives three isomeric products

3. A. straight chain hydrcarbon has the molecular formula C_8H_{10}. The hybridisation for the carbon atoms from one

end of the chain to the other are respectively sp^3, sp^2, sp^2, sp^3, sp^2, sp^2, sp and sp. The structural formula of hydrocarbon would be
(A) $CH_3–CH_2–CH = CH–CH = CH–C≡CH$
(B) $CH_3–CH = CH–CH_2–C ≡ C–CH = CH_2$
(C) $CH_3–C≡C–CH_2–CH = CH–CH = CH_2$
(D) $CH_3–CH = CH–CH_2–CH = CH–C≡CH$

4. On passing acetylene through a dilute solution of sulphuric acid containing mercuric sulphate, the major product obtained will be
(A) ethylene (B)ethylene glycol
(C) acetaldehyde (D) acetic acid

5. To prepare a pure sample of n-hexane using sodium metal as one reactant, the other reactant or reactants will be
(A) ethyl chloride and n-butyl chloride
(B) ethyl bromide and n-butyl bromide
(C) methylbromide and n-pentyl bromide
(D) n-propyl bromide

6. Select from amongst the following the correct statement
(A) Sodium metal is stored under kerosene oil
(B) One of the oxides of carbon is a basic oxide
(C) Metals can form only basic oxides
(D) To prevent combination of white phosphorus wih oxygen it is kept in kerosene oil.

7. A solution containing ammonium chloride and sodium nitrite on boiling produces.
(A) ammoina (B) nitric oxide
(C) Nitrogen (D) nitrous oxide

8. On electrolysis of dilute sulphuric acid using platinum electrodes, the product obtained at the anode will be
(A) Hydrogen (B) oxygen
(C) Hydrogen sulphide (D) Sulphur dioxide

9. The IUPAC name of

$$\begin{array}{c} \quad OH \quad\quad CH_3 \\ \quad | \quad\quad\quad | \\ CH_3–CH–CH_2–CH–CHO \end{array}$$

Will be
(A) 4-Hydroxy-2-methyl-pentanol
(B) 4-Hydroxy-2, methyl pentenal
(C) 4-Hydroxy-2- methyl pentanal
(D) 2-Hydroxy-4-methyl pentanal

10. How many isomers of $C_5H_{11}OH$ will be primary alhocols ?
(A) 5 (B) 4 (C) 2 (D) 3

11. Methanol is industrially prepared by
(A) reacting HCHO with a solution of NaOH
(B) reduction of CO using H_2 and $ZnO\text{-}Cr_2O_3$
(C) Oxidation of CH_4 by steam at 900°C
(D) reduction of HCHO using $LiAlH_4$

12. When phenol is treated with excess of bromine water, the product formed would be
(A) a mixture of o- and p- bromophenol
(B) 2, 4, 6- tribromophenol
(C) m-bromophenol (D) 2, 4 dibromophenol

13. Which one of the following on oxidation will not give a carboxylic acid with the same number of carbon atoms ?
(A) CH_3COCH_3 (B) CCl_3CH_2CHO
(C) $CH_3CH_2CH_2OH$ (D) CH_3CH_2CHO

14. For the carbylamine reaction we need hot alcoholic KOH and
(A) a primary amine and an alkyl halide
(B) any amine and chloroform
(C) chloroform and silver powder
(D) a monoalkylamine and trichloromethane

15. HBr will react fastest with
(A) 2-methylpropan -1-o1 (B) 2-methylpropan -2-o1
(C) propan-2-o1 (D) propan-1-o1

16. The reaction of toluene with chlorine in presence of ferric chloride gives mainly
(A) o-and p-chlorotoluene (B) benzoyl chloride
(C) m-chlorotoluene (D) benzyl chloride

17. Geometrical isomerism will be shown by
(A) 2-butene (B) 1-butene
(C) ethylene (D) propylene

18. The most reactive compound for electrophilic nitration will be
(A) toluene (B) benzoic acid
(C) nitrobenzene (D) benzene

19. Which one of these will react with sodium metal ?
(A) Ethene (B) Ethane (C) Ether (D) Ethyne

20. Which one of the following bonds will be the most polar ?
(A) C—F (B) N—N (C) N—Cl (D) O—F

21. In a molecule of carbon tetrachloride the four C—Cl bonds are directed towards the corners of a
(A) square (B) tetrahedron
(C) prism (D) cube

22. Number of electrons shared in the formation of nitrogen molecule is
(A) 2 (B) 8 (C) 6 (D) 10

23. Elements of which one of the following groups will form anions most readily ?
(A) Oxygen group (B) Nitrogen group
(C) Halogens (D) Alkali metals

24. Strongest hydrogen bonding is shown by
(A) Hydrogen fluoride (B) Hydrogen sulphide
(C) Water (D) ammonia

25. A co-ordinate bond is formed when an atom in a molecule has
(A) all its valency shell electrons shared
(B) a single unshared electron
(C) electric charge on it

(D) one or more unshared electron pairs

26. If an isotope of hydrogen has two neutrons in its atom, its atomic number and atomic mass number will respectively be.

(A) 2 and 1 (B) 3 and 1
(C) 1 and 1 (D) 1 and 3

27. If the valency shell electronic structure for an elements is $ns^2\ np^5$ This element will belong to the group of

(A) alkali metals (B) unert metals
(C) noble gases (D) halogens

28. A sudden large jump between the values of second and third ionisation energies of an element would be associated with the electronic configuration

(A) $1s^2\ 2s^2\ 2p^6\ 3s^2\ 3p^2$ (B) $1s^2\ 2s^2\ 2p^6\ 3s^2$
(C) $1s^2\ 2s^2\ 2p^6\ 3s^1$ (D) $1s^2\ 2s^2\ 2p^6\ 3s^2\ 3p^1$

29. The lustre of a metal is on account of

(A) high polish of metals
(B) chemical inertness of metals
(C) reflection of light due to presence of free electrons
(D) high density of metals

30. The binding energy of an electron in the nth Bohr orbit of hydrogen atom is

(A) $13.6/n^4$ eV (B) $13.6/n^3$ eV
(C) $13.6/n^2$ eV (D) $13.6/n$ eV

31. Electronic configuration of calcium atom may be written as

(A) Ne, $4s^2$ (B) Kr, $4p^2$ (C) Ne, $4p^2$ (D) Ar, $4s^2$

32. H_3PO_2 is the formula for one of the phosphorus acids. Its name and basicity are respectively.

(A) hypophosphoric acid and one
(B) hypophosphoric acid and two
(C) phosphorus acid and two
(D) hypophosphorus acid and two

33. Which one of the following would quickly absorb oxygen ?

(A) Alkaline solution of pyrogallic acid
(B) Concentrated sulphuric acid
(C) Lime water.
(D) Alkaline solution of copper sulphate

34. When chlorine is passed over dry slaked lime at room temperature, the main reaction procuct is

(A) $Ca(ClO_2)_2$ (B) $CaCl_2$
(C) $CaOCl_2$ (D) $Ca\ (OCl)_2$

35. In the manufacture of bromine from sea water, the mother liquor containing bromide is treated with

(A) iodine (B) Sulphor dioxide
(C) carbon dioxide (D) chlorine

36. Water gas is produced by

(A) saturating hydrogen with moisture
(B) mixing oxygen and hydrogen in the ratio of 1:2
(C) passing steam through a red hot coke bed
(D) heating a mixture of CO_2 and CH_4 in petroleum refineries

37. Which one of the following species is likely to have a linear structure.

(A) SO_2 (B) CO_2 (C) CO_3^{2-} (D) SO_4^{2-}

38. Which one of the following pairs is isoelectronic ?

(A) Ar and Cl (B) Na^+ and Ne
(C) Na^+ and Mg (D) Mg and Ne

39. In a group of the periodic table the ionisation potentials of elements decrease from the top to the bottom because of

(A) decrease in chemical reactivities
(B) decrease in electronegativitie
(C) increase of densities
(D) increase in atomic size

40. Which one of the following substances is used in the laboratory for fast drying of neutral gases

(A) Phoshorus pentoxide (B) Active charcoal
(C) Anhydrous calcium chloride (D) Na_3PO_4

41. Nitrogen is a relatively inactive element because

(A) its electronegativity is fairly high
(B) its atom has a stable electronic configuration
(C) it has a low atomic radius
(D) dissociation energy of its molecule is fairly high

42. Baeyer's reagent is used in the laboratory for

(A) detection of double bonds (B) detection of glucose
(C) reduction (D) oxidation

43. Select the correct statement. In the gas equation, PV = nRT

(A) V denotes volume of one mole of the gas
(B) n is the number of molecules of the gas
(C) n moles of the gas have a volume V
(D) p is the pressure of the gas when only one mole of gas is present

44. 3.2g of oxygen (At.wt.=16) and 0.2 g of hydrogen (At.wt.=1) are placed in a 1.12 litre flask at 0°C. The total pressure of the gas mixture will be.

(A)3 atm (B) 2 atm (C) 1 atm (D) 4 atm

45. 50 ml of a gas A diffuse through a membrane in the same time as for the diffusion of 40 ml of gas B under identical pressure temperature conditions.If the molecular weigh of A is 64 that of B would be.

(A) 200 (B) 80 (C) 100 (D) 250

46. 0.1 M solution of which one of these substance will react basic ?

(A) Sodium borate (B) Ammonium chloride
(C) Calcium nitrate (D) Sodium sulphate

47. If a 0.1 molar solution of glucose (mol.wt.=180) and 0.1 molar solution of urea(mol.wt.60) are placed on the two sides of a semipermeable membrane to equal heights, then it will be correct to say that

(A) there will be no net movement across the membrane
(B) glucose will flow across the membrane into urea solution
(C) urea will flow across the membrane into glucose solution

(D) water will flow from urea solution into glucose solution

48. In reversible reaction a catalyst is the substance which
(A) increase rate of the forward reaction
(B) decreases the value of enthalpy change in the reaction
(C) reduces the time required for reaching the equilibrium state in the reaction
(D) decrease the rate of the reverse reaction

49. In the reversible reaction of the type $A+B \rightleftharpoons AB$, in general
(A) both the reactions will be exothermic
(B) the combination reaction will be endothermic and the dissociation reaction will be exothermic
(C) neither of the reactions will be endothermic
(D) the combination reaction will be exothermic and the dissociation reaction will be endothermic

50. A closed flask contains water in all its three states, solid, liquid and vapour, at 0°C. In this situation the average kinetic energy of water molecule will be
(A) the greatest in the vapour state
(B) the same in all the three states
(C) the greatest in the solid state
(D) greater in the liquid than in the vapour state

51. Natural rubber is a polymer of
(A) n-pentene-1 (B) propylene
(C) butadiene (D) isoprene

52. On bromination propionic acid yields two isomeric 2-bromopropionic acids. This pair will be an example of
(A) optical isomers (B) cis-trans isomers
(C) chain isomers (D) position isomers

53. The correct value of the gas constant R is close to
(A) 0.082 litre–atmosphere K
(B) 0.082 litre-atmosphere-K^{-1} mol^{-1}
(C) 0.082 $litre^{-1}$ $atmosphere^{-1}$ –K–mol
(D) 0.082 litre $atmosphere^{-1}$– K–mol^{-1}

54. One litre of a gas weighs 2g at 300K and 1 atm pressure. If the pressure is made 0.75 atm., at which of the following temperatures will one litre of the same gas weigh one gram.
(A) 450 K (B) 600K (C) 800 K (D) 900K

55. The internal energy of one mole of an ideal gas is given by
(A) 1/2 RT (B) 3/2 kT (C) 3/2 RT (D) 1/2 kT

56. Based on the following thermochemical equations
$H_2O(g) + C(s) = CO(g) + H_2(g)$; $\Delta H = 131$ kJ
$CO(g) + 1/2\ O_2(g) = CO_2(g)$; $\Delta H = -282$ kJ
$H_2(g) + 1/2O_2(g) = H_2O(g)$; $\Delta H = -242$ kJ
$C(s) + O_2(g) = CO_2(g)$; $\Delta H = x$ kJ
The value of x will be
(A) –393 kJ (B) –655 kJ (C) 393 kJ (D) 655 kJ

57. Which one of the following modes of expressing concentration of solutions is independent of temperature ?
(A) Normality (B) Grams per litre
(C) Molarity (D) Molality

58. The dimer of methanol will have the structure

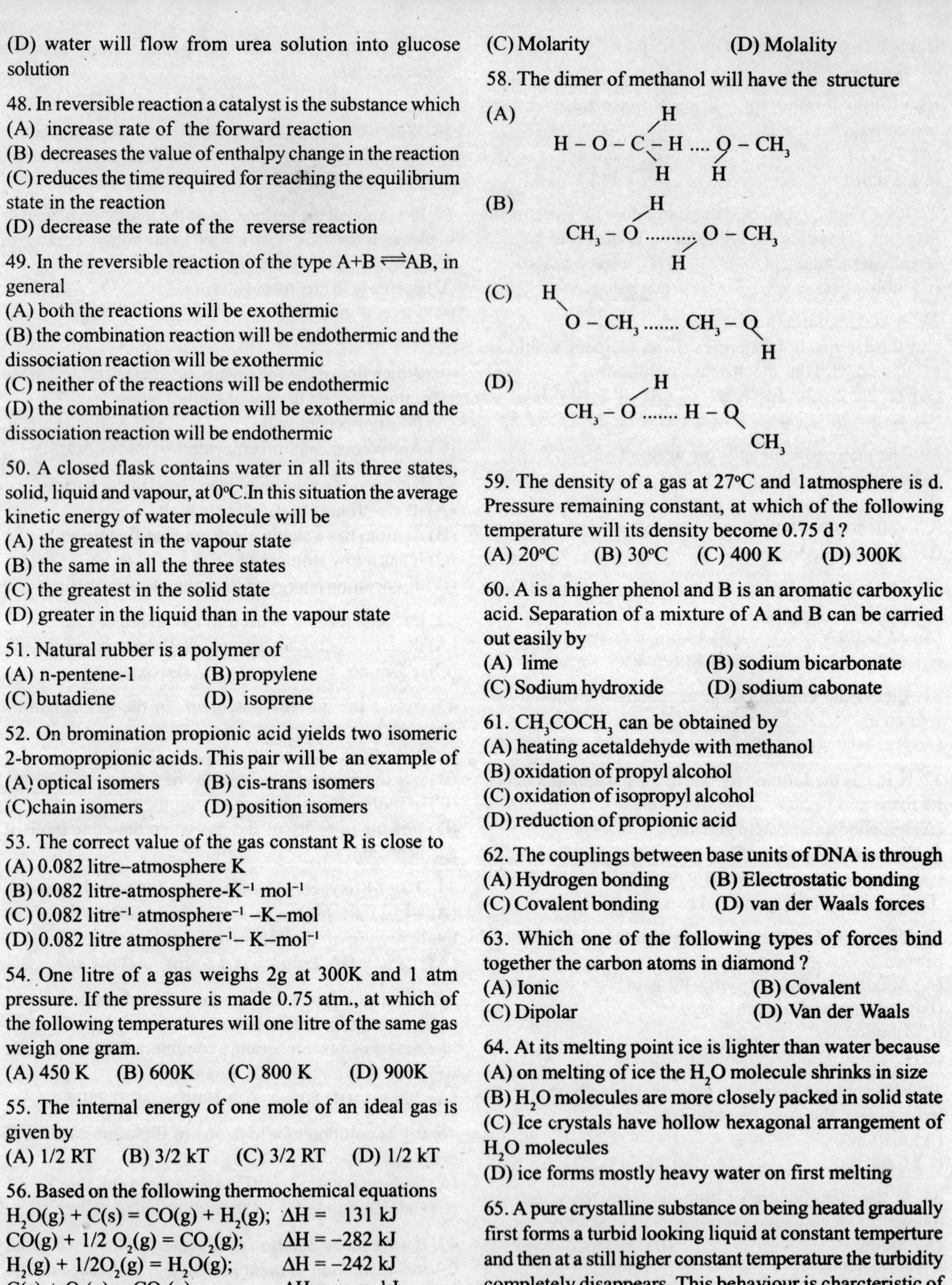

(A) $H-O-C(H)(H)-H \;....\; O(H)-CH_3$
(B) $CH_3-O(H) \;.........\; O(H)-CH_3$
(C) $H-O-CH_3 \;.......\; CH_3-O-H$
(D) $CH_3-O(H) \;.......\; H-O-CH_3$

59. The density of a gas at 27°C and 1atmosphere is d. Pressure remaining constant, at which of the following temperature will its density become 0.75 d ?
(A) 20°C (B) 30°C (C) 400 K (D) 300K

60. A is a higher phenol and B is an aromatic carboxylic acid. Separation of a mixture of A and B can be carried out easily by
(A) lime (B) sodium bicarbonate
(C) Sodium hydroxide (D) sodium cabonate

61. CH_3COCH_3 can be obtained by
(A) heating acetaldehyde with methanol
(B) oxidation of propyl alcohol
(C) oxidation of isopropyl alcohol
(D) reduction of propionic acid

62. The couplings between base units of DNA is through
(A) Hydrogen bonding (B) Electrostatic bonding
(C) Covalent bonding (D) van der Waals forces

63. Which one of the following types of forces bind together the carbon atoms in diamond ?
(A) Ionic (B) Covalent
(C) Dipolar (D) Van der Waals

64. At its melting point ice is lighter than water because
(A) on melting of ice the H_2O molecule shrinks in size
(B) H_2O molecules are more closely packed in solid state
(C) Ice crystals have hollow hexagonal arrangement of H_2O molecules
(D) ice forms mostly heavy water on first melting

65. A pure crystalline substance on being heated gradually first forms a turbid looking liquid at constant temperture and then at a still higher constant temperature the turbidity completely disappears. This behaviour is charcteristic of substances forming
(A) isomeric crystals (B) liquid crystals
(C) isomorphous crystals (D) allotropic crystals

66. Bakelite is a product of the reaction between
(A) Phenol and methanal

(B) Formaldehyde and NaOH
(C) Aniline and urea
(D) Phenol and chloroform

67. Which one of the following compounds will give a characteristic deep colour with ferric chloride solution ?
(A) Acetone (B) Ethanol (C) Phenol (D) Acetic acid

68. Find the strongest acid from amongst the following
(A) CF_3COOH (B) CBr_3COOH
(B) CH_3 COOH (D) CCl_3COOH

69. Benzoic acid gives benzene on being heated with X and phenol gives benzene on being heated with Y. Therefore X and Y are respectively
(A) Soda lime and copper
(B) Zinc dust and sodium hydroxide
(C) Zinc dust and soda lime
(D) Soda lime and zinc dust

70. The solubility product of AgCl is 1.8×10^{-10} mol^2–$litre^{-2}$. On mixing the following solutions in equal volumes, in which case the precipitation of AgCl could be expected ?
(A) 10^{-10} M of Ag^+ and 10^{-10} M of Cl^-
(B) 10^{-4} M of Ag^+ and 10^{-4} M of Cl^-
(C) 10^{-5} M of Ag^+ and 10^{-5} M of Cl^-
(D) 10^{-6} M of Ag^+ and 10^{-6} M of Cl^-

71. In which one of the following solvents will AgBr have the highest solubility ?
(A) 10^{-3} MNaBr (B) 10^{-3} M NH_4 OH
(C) Pure water (D) 10^{-3} M HBr

72. In the gas phase reaction $C_2H_4 + H_2 \rightleftharpoons C_2H_6$ the equilibrium constant can be expressed in units of
(A) $litre^{-1}$ - mol^{-1} (B) litre- mol^{-1}
(C) mol^2 - $litre^{-2}$ (D) mol - $litre^{-1}$

73. 10 ml of a strong acid solution of pH = 2.000 are mixed with 990 ml of a buffer solution of pH = 4.000. The pH of the resulting solution will be
(A) 4.002 (B) 4.000 (C) 4.200 (D) 3.800

74. Which one of the following information can be obtained on the basis of Chatelier principle ?
(A) Dissociation constant of a weak acid
(B) Entropy change in a reaction
(C) Equilibrium constant of a chemical reaction
(D) shift in equilibrium position on changing value of a constraint

75. An ideal gas cannot be liquefied because
(A) its critical temperature is always 0°C
(B) its molecules are relatively small in size
(C) it solidifies before becoming a liquid
(D) forces operative between its molecules are negligible.

CBSE - 1993

1. Among aceticacid, phenol and n-hexanol, which of the compounds will react with $NaHCO_3$ solution to give sodium salt and carbon dioxide ?
(A) Acetic acid and phenol (B) Acetic acid
(C) Phenol (D) n-hexanol

2. Reduction of 2-butyne with sodium in liquid ammonia gives predominantly.
(A) Cis 2- Butene (B) n-Butene
(C) Trans-2-Butene (D) No reaction

3. In which of the following the solubility of AgCl will be maximum ?
(A) 0.1 M NaBr (B) 0.1 $NaNO_3$
(C) H_2O (D) 0.1 M NaCl

4. One mole of CO_2 contains
(A) 3 gm atoms of CO_2
(B) 6.023×10^{23} atoms of carbon
(C) 6.023×10^{23} atoms of oxygen
(D) 18.1×10^{23} molecules of CO_2

5. In graphite, electrons are
(A) present in anti-bonding orbitals
(B) localised on each C-atom
(C) localised on every third C- atom
(D) spread out between the structure

6. Which of the following fertilizer has the highest Nitrogen percentage ?
(A) Ammonium sulphate (B) Calcium cynamide
(C) Urea (D) Ammonium nitrate

7. Which of the following can distinguish propyne from propene ?
(A) Ammonical $AgNO_3$ (B) Dil. H_2SO_4
(C) Br_2–H_2O (D) Aqueous $KMnO_4$

8. If ionization potential for hydrogen atom is 13.6 eV, then ionization potential for He^+ will be
(A) 54.4 eV (B) 6.8 eV (C) 13.6 eV (D) 24.5 eV

9. Which kind of valency exists in CaH_2 and C_2H_2 ?
(A) Covalency in both
(B) Electrovalency in CaH_2 and covalency in C_2H_2
(C) Electrovalency in both
(D) Covalency in CaH_2 and electrovalency in C_2H_2

10. Which of the following statements is not correct ?
(A) Double bond is shorter than single bond
(B) Sigma bond is weaker than π(Pi) bond
(C) Double bond is stronger than single bond
(D) Covalent bond is stronger than hydrogen bond

11. A compound is treated with $NaNH_2$ to give sodium salt. Identify the compound
(A) C_2H_6 (B) C_2H_4 (C) C_2H_2 (D) C_6H_6

12. Which is the correct symbol relating the two Kekule benzene ?
(A) $\rightleftarrows$ (B) $\rightarrow$ (C) $\rightleftharpoons$ (D) $\leftrightarrow$

13. Number of atoms in one unit cell of a face centred cubic lattice is
(A) 4 (B) 1 (C) 2 (D) 6

14. Which compound will show the highest lattice energy ?
(A) RbF (B) CsF (C) NaF (D) KF

15. Which is an extensive property of the system ?
(A) Temperature (B) Refractive index
(C) Volume (D) Viscosity

16. Which of the following is not a state function of thermodynamic system ?
(A) Work (B) enthalpy
(C) Free energy (D) Internal Energy

17. One would expect proton to have very large
(A) charge (B) ionization potential
(C) hydration energy (D) radius

18. Conjugate base of NH_4^+ ion is
(A) NH_3 (B) OH^- (C) KOH (D) NH_4OH

19. Ratio of the energy of a photon of wavelength 3000 Å and 6000Å is
(A) 2 (B) 1/6 (C) 1/2 (D) 1/3

20. The number of atoms is greatest in
(A) 71 gm of chlorine (B) 48 grm of magnesium
(C) 127 gm of iodine (D) 4 gm of hydrogen

21. Which of the following compounds does not contain double bond or triple bond ?
(A) C_2H_4 (B) H_2O (C) N_2 (D) HCN

22. Which of the following species has four lone pairs of electrons ?
(A) I^- (B) O^- (C) Cl^- (D) He

23. Which of the following sets have strongest tendency to form anions ?
(A) Ga, In, T (B) Na, Mg , AI
(C) N, O, F (D) V, Cr, Mn

24. What is the oxidation number of S in $Na_2S_4O_6$?
(A) 1½ (B) 2½ (C) 2 (D) 3

25. In which of the following cases does the reaction go to farthest to completion ?
(A) K = 10 (B) K = 1 (C) $K = 10^3$ (D) $K=10^{-2}$

26. Bromoethane when undergoes Wurtz reaction gives
(A) 2-3 dimethyl butane (B) 2-methyl propane
(C) n-butane (D) n-propane

27. Internal energy and pressure of a gas of unit volume are related as
(A) $P = \frac{2}{3} E$ (B) $P = \frac{3}{2} E$
(C) $P = \frac{1}{2} E$ (D) P = 2 E

28. The electronic configuration of an ideal gas is
(A) $1s^2\ 2s^2\ 2p^6$ (B) $1s^2\ 2s^2\ 2p^6\ 3s^1$
(C) $1s^2\ 2s^2\ 2p^6\ 3s^2\ 3p^5$ (D) $1s^2\ 2s^2\ 2p^1$

29. Propene and propyne may be differentiated by
(A) Br_2-H_2O (B) Alkaline–$KMnO_4$
(C) Acidic–$K_2Cr_2O_7$ (D) Ammoniacal–Cu_2Cl_2

30. Some Cl_2 is liberated when we heat
(A) $K_2Cr_2O_7$ and MnO_2 (B) $K_2Cr_2O_7 + HCl + MnO_2$
(C) $KMnO_4$+ NaCl (D) $Pb_2(NO_3)_4 + MnO_2$

31. What is formed when primary alcohol undergoes catalytic dehydrogenation ?
(A) Alkene (B) Acid (C) Aldehyde (D) Ketone

32. Which one of the following on oxidation gives a ketone ?
(A) Primary alcohol (B) Secondary alcohol
(C) Tertiary alcohol (D) All of these

33. Half-life period of a substance is 1600 minutes. How much fraction of the substance will remain after 6400 minutes ?
(A) 1/16 (B) 1/2 (C) 1/4 (D) 1/8

34. Which of the following has the greatest electron affinity ?
(A) Fluorine (B) Bromine
(C) Chlorine (D) Iodine

35. The weakest bond is
(A) Hydrogen (B) van der Waal's
(C) covalent (D) coordinate

36. When $CuSO_4$ is electrolysed by platinum electrode,
(A) sulphur is liberated at cathode, oxygen at anode
(B) copper is liberated at cathode, sulphur at anode
(C) copper is liberated at cathode, oxygen at anode
(D) oxygen is liberated at cathode, copper at anode

37. Atomic size of an element is 7×10^{-10}, then size of its molecules is
(A) 7×10^{-8} (B) 7×10^{-10}
(C) 7×10^{-12} (D) 7×10^{-15}

38. Impossible configuration is represented by
(A) $1s^2, 2s^2, 2p^2, 3s^1$ (B) $1s^2, 2s^2, 2p^6$
(C) $1s^2, 2s^2$ (D) None

39. Bohr model postulate
(A) Angular momentum = $\frac{mn}{2\pi}$
(B) Angular momentum = $\frac{hm}{2\pi}$
(C) Angular momentum = $\frac{nh}{2\pi}$
(D) None of the above

40. Ionic bond is formed between
(A) CH_4 (B) O_3 (C) Both (A) and (B) (D) NaCl

41. Strongest bond is present in
(A) CsF (B) NaCl
(C) Both (A) and (B) (D) None of the above

42. When we move from left to right in second period, the atomic volume of the element.
(A) increases (B) increase with constant state
(C) decreases (D) None of the above

43. According to Bohr's model of atom
(A) total energy of electron is quantised.
(B) angular momentum of electron is quantised
(C) Both (A) and (B)
(D) None of the above

44. Reactivity of hydrogen atoms attached to different atoms in alkanes has order.
(A) Primary > Secondary > Tertiary
(B) Tertiary > Primary > Secondary
(C) Both (A) and (B)
(D) Tertiary > Secondary > Primary

45. Enzyme takes part in reaction and
(A) decreases the rate of chemical reaction
(B) increases the rate of chemical reaction
(C) Both (A) and (B) (D) None of the above

46. A gas "disobeys" gas law at
(A) low pressure and high temperature
(B) both low pressure and temperature
(C) high pressure and low temperature
(D) None of the above

47. 4th electron of Be will have the four quantum numbers

	n	I	m	s
(A)	2	1	0	+½
(B)	2	0	0	−½
(C)	1	0	0	+½
(D)	1	1	1	−½

48. Which of the following has largest size ?
(A) Na (B) Na+ (C) Na^- (D) None of the above

49. Oxidation number of iron in $[K_4Fe(CN)_6]$ is
(A) +3 (B) +2 (C) +10 (D) +1

50. Sodium formate on heating yields.
(A) oxalic acid and H_2 (B) sodium oxalate and H_2
(C) CO_2 and NaOH (D) sodium oxalate

51. Which of the following sequences shows decreasing order of strength of bonds ?
(A) Electrovalent > Covalent > van der Waal's
(B) Electrovalent > van der Waal's > Covalent
(C) Covalent > Electrovalent > van der Waal's
(D) Covalent > van der Waal's > Electrovalent

52. When SO_2 solution is passed over which of the following it turns green ?
(A) $K_2Cr_2O_7$ (B) H_2SO_4 (C) SO_3 (D) P_2O_5

53. An atom with atomic number 20 is most likely to combine chemically with the atom whose atomic number is
(A) 16 (B) 10 (C) 11 (D) 14

54. Which of the following elements is extracted commercially by the electrolysis of an aqueous solution of its compound ?
(A) Al (B) Na (C) Cl (D) Br

55. Ammonium dichromate on heating gives
(A) N_2 (B) N_2O (C) NO (D) NO_2

56. A solution of potassium bromide is treated with each of the following. Which one would liberate bromine ?
(A) Hydrogen iodide (B) Sulphur dioxide
(C) Chlorine (D) Iodine

57. When propene is heated at 400°C in presence of chlorine. It gives the following compound
(A) Polyvinyl chloride (B) Allyl chloride
(C) 1,2-Dichloro propene (D) No reaction

58. At 90°C, pure water $[H_3O^+] = 10^{-6}$, what is the value of K_W at 90°C ?
(A) 10^{-8} (B) 10^{-6} (C) 10^{-12} (D) 10^{-11}

59. When hydrochloric gas is treated with propene in presence of benzoyl peroxide, It gives
(A) 2-chloro propane (B) allyl chloride
(C) n-propyl chloride (D) No reaction

60. Industrial preparation of chloroform employs acetone and
(A) phosgene (B) calcium hypochloride
(C) chlorine gas (D) sodium chloride

61. On doping Ge metal with a little amount of indium one gets
(A) rectifier (B) insulator
(C) n-type semiconductor (D) p-type semi conductor

62. The ratio amongst most most probable velocity, mean velocity and r.m.s. velocity is in order of
(A) $1 : \sqrt{2} : \sqrt{3}$ (B) $\sqrt{2} : \sqrt{3} : \sqrt{8/\pi}$
(C) $\sqrt{2} : \sqrt{8/\pi} : \sqrt{3}$ (D) 1 : 2 : 3

63. Na^+, Mg^{+2}, Al^{+3}, Si^{+4} are isoelectronic, their ionic size will follow the order of
(A) $Na^+ > Mg^{+2} < Al^{+3} < Si^{+4}$
(B) $Na^+ < Mg^{+2} > Al^{+2} > Si^{+4}$
(C) $Na^+ > Mg^{+2} > Al^{+3} > Si^{+4}$
(D) $Na^+ < Mg^{+2} < Al^{+3} < Si^{+4}$

64. A solution containing NH_4Cl and sodium nitrite on boiling produces
(A) N-peroxide (B) Ammonia
(C) Nitrogen (D) Nitrous oxide

65. For covalent solid, the units which occupy lattice point are
(A) electrons (B) atoms (C) ions (D) molecules

66. Nitrogen can be purified from the impurities of oxides of nitrogen and ammonia by passing through.
(A) a solution of KOH
(B) alkaline solution of pyrogallol
(C) a solution of $K_2Cr_2O_7$, acidified with H_2SO_4
(D) conc. HCl

67. $[CH_3]_2$ C = $CHCOCH_3$ can be oxidised to $[CH_3]_2$ C = CHCOOH by
(A) chromic acid (B) NaOI
(C) Cu at 300°C (D) $KMnO_4$

68. Ethanol and dimethyl ether are a pair of functional isomers. The boiling point of ethanol is higher than that of dimethyl ether due to the presence of
(A) H-bonding in dimethyl ether
(B) CH_3 group in dimethyl ether
(C) H-bonding in ethanol
(D) CH_3 group in ethanol

69. Increasing order of acid strength among p-Methoxyphenol, p-Methylphenol and p-Nitrophenol are as

(A) p-Nitrophenol, p-Methoxyphenol, p-Methylphenol
(B) p-Methylphenol, p-Methyoxyphenol, p-Nitroxyphenol
(C) p-Nitrophenol, p-Methylphenol, p-Methoxyphenol
(D) p-Methoxyphenol, p-Methylphenol, p-Nitrophenol

70. When the hybridization state of carbon atom changes from sp^3 to sp^2 and finally to sp, the angle between the hybridized orbitals

(A) decreases gradually (B) decreases considerably
(C) is not affected (D) increases progressively

71. Benzene reacts with N-propyl chloride in the presence of anhydrous $AlCl_3$ to give

(A) 3-propyl-1-chlorobenzene (B) N-propylbenzene
(C) iso-propylbenzene (D) no reaction

72. Which of the following reagents is useful in separating benzoic acid from phenol ?

(A) 5% aqueous NaOH (B) 5% aqueous $NaHCO_3$
(C) Dilute HCl (D) Dilute H_2SO_4

73. When acetylene reacts with water in the presence of $HgSO_4$, the compound isolated is

(A) CH_3COCH_3 (B) CH_3-CHO
(C) $CH_2 = CH_2$ (D) $CH_2 = CH(OH)$

74. The restricted rotation about carbon-carbon double bond in 2-butene is due to

(A) overlap of one s and one sp^2 hybridized orbitals.
(B) overlap of two sp^2 hybridized orbitals
(C) overlap of one p and one sp^2 hybridized orbitals
(D) overlap of two p orbitals

75. chlorobenzene reacts with Mg in dry ether to give a compound (A) which further reacts with ethanol to yield

(A) phenol (B) benzene
(C) ethyl benzene (D) phenyl ether

CBSE -1994

1. For the adsorption of a gas on a solid, the plot of log x/m versus log P is linear with slope equal to

(A) n (B) 1/n (C) k (D) log k

2. Select the rate law that corresponds to the data shown for the following reaction ,

$$A + B \rightarrow C$$

Exp.	$[A]_0$	$[B]_0$	Initial Rate
1.	0.012	0.035	0.10
2.	0.024	0.070	0.80
3.	0.024	0.035	0.10
4.	0.012	0.070	0.80

(A) Rate = $k [B]^3$ (B) Rate = $k [B]^4$
(C) Rate = $k[A][B]^3$ (D) Rate = $k[A]^2[B]^2$

3. The resistance of a 0.1N solution of acetic acid is 250 ohm, when measured in a cell of cell constant 1.15 cm^{-1}. The equivalent conductance (in ohm^{-1} cm^2, $equiv^{-1}$) of 0.1 N acetic acid is.

(A) 18.4 (B) 2.3 (C) 4.6 (D) 9.2

4. Which one of the following salts would have the same value of the Van't Hoff factor (I) as that of $K_3[Fe(CN)_6]$?

(A) $Al_2(SO_4)_3$ (B) NaCl
(C) $Al(NO_3)_3$ (D) Na_2SO_4

5. Molecular species having highest bond order from the following is

(A) O^+_2 (B) O_2^{2-} (C) O_2 (D) O^-_2

6. Which of the following ions is not isoelctronic with O^{2-} ?

(A) Tl^{++} (B) Na^+ (C) N^{3-} (D) F^-

7. A solid with high electrical and thermal conductivity from the following is

(A) sodium chloride (B) ice (C) silicon (D) lithium

8. Ionic solids with Schottky defects contain in their structure

(A) equal number of cation and anion vacancies
(B) anion vacancies and interstitial anions
(C) cation vacancies
(D) cation vacancies and interstitial cations

9. If a molecule MX_3 has zero dipole moment, the sigma bonding orbitals used by M (at. no.<21) are

(A) sp^2 hybrid (B) pure p-orbitals
(C) sp^3 hybrid (D) sp hybrid

10. Which of the following molecules/ions has a triangular pyramidal shape?

(A) NO_3 (B) CO_3^{2-} (C) BF_3 (D) H_3O^+

11. Strongest acid from the following is

(A) HBr (B) H_2S (C) HF (D) H_3PO_3

12. The most powerful oxidizing agent from the following is

(A) H_3PO_3 (B) H_3PO_4 (C) H_2SO_4 (D) H_3BO_3

13. A 10 volume solution of H_2O_2 means

(A) 10 grams of H_2O_2 solution liberate 10 mL of O_2 at STP
(B) 10 gram equivalents of H_2O_2 liberates 10 mL of O_2 at STP
(C) 10 litres of H_2O_2 liberates 10 mL of O_2 at STP
(D) 1 mL of solution of H_2O_2 liberates 10 mL of O_2 at STP.

14. Which of the following is a false statement about boric acid, H_3BO_3 ?

(A) It is prepared by acidifying an aqueous solution of borax.
(B) It is a strong tribasic acid.
(C) It does not act as proton donor but acts as a Lewis acid by accepting hydroxyl ion.
(D) It has a layer structure in which planar BO_3 units are joined by hydrogen bonds.

15. Carbon and silicon belong to Gr.IV. The maximum coordination number of carbon in commonly occurring compounds is four whereas that of Si is six. This is due to

(A) larger size of silicon
(B) availability of low-lying d-orbitals in silicon.

(C) silicon being vulnerable to attack by nucleophilic reagents
(D) more electropositive nature of silicon.

16. Which one of the following oxides of nitrogen is paramagnetic.
(A) NO_2 (B) N_2O_2 (C) N_2O (D) N_2O_4

17. All the following substances react with water. The pair that yields the same gaseous product is.
(A) K and KO_2 (B) Ca and CaH_2
(C) Na and Na_2O_2 (D) Ba and BaO_2

18. Which of the products is formed when $SiCl_4$ vapour is passed over hot Mg ?
(A) $SiCl_2 + MgCl_2$ (B) $Si + MgCl_2$
(C) $Mg_2Si + Cl_2$ (D) $MgSiCl_6$

19. Which of the following statements is false ?
(A) Carbon monoxide reduces iron (III) oxide to iron
(B) Mercury (II) iodide dissolves in excess of KI solution.
(C) Zinc disolves in hot NaOH solution.
(D) Tin (IV) chloride is made by dissolving tin in conc. HCl

20. Which of the following will displace Br_2 from an acqueous solution containingbromideions ?
(A) Cl_2 (B) I_3^- (C) I_2 (D) Cl^-

21. Suggest a purification method for obtaining highly pure silicon used as semiconductor from the following.
(A) electrochemical refining method
(B) reduction method (C) zone refining method
(D) oxidation method.

22. The most durable metal plating on iron to protect against corrosion is.
(A) tin plating (B) zinc plating.
(C) nickel plating (D) copper plating

23. Which of the following ligands is expected to be bidentate ?
(A) $C_2O_4^{2-}$ (B) $CH_3C \equiv N$ (C) Br^- (D) CH_3NH_2

24. IUPAC name for the complex compound $K_3[Fe(CN)_6]$ is
(A) potassiumhexacyanoferrate (II)
(B) potassiumhexacyanoferrate (III)
(C) potassiumferrocyanideiron (III)
(D) potassiumcyanohexaferrate (II)

25. Which one of the following transition elements shows the highest oxidation state ?
(A) Mn (B) Fe (C) V (D) Cr

26. Which one one of the following has tetrahedral geometry ?
(A) $[CO(NH_3)_6]^{3+}$ (B) $[Ni(CN)_4]^{2-}$
(C) $Fe(CO)_5$ (D) $[NiCl_4]^{2-}$

27. Which of the following complexes is likely to show optical activity ?
(A) Trans-$[Co(NH_3)_4Cl_2]^+$ (B) $[Cr(H_2O)_6]^{3+}$
(C) Cis-$[Co(NH_3)_2en_2]^{3+}$ (where en is ethylenediamine)
(D) Trans-$[Co(NH_3)_2\ en_2]^{3+}$

28. Which one of the following metal ions plays an important role in muscle contraction ?
(A) K^+ (B) Na^+ (C) Mg^{2+} (D) Ca^{2+}

29. Which of the following statement concerning Lanthanide elements is false.
(A) All lanthanides are highly dense metals.
(B) Most characteristic oxidation state of lanthanide elements is +3.
(C) Lanthanides are separated from one another by ion exchange method.
(D) The ionic radii of trivalent lanthanides steadily increase with increase in the atomic number.

30. Which one of the following IUPAC names is correct ?
(A) 2-methyl-3 ethylpentane
(B) 2-ethyl-3-methylpentane
(C) 3-ethyl-2methylpentane
(D) 3-methyl-2-ethylpentane

31. The compound which will not react with sodium is.
(A) $CH_3–CHOH–CH_3$ (B) $CH_3–O–CH_3$
(C) $CH_3–COOH$ (D) C_2H_5OH

32. Reaction of t-butyl bromide with sodium methoxide produces.
(A) isobutane (B) isobutylene
(C) t-buty methyl ether (D) sodium t-butoxide

33. Chiral molecules are those which are
(A) unstablemolecules
(B) show geometrical somerism
(C) superimposable on their mirror images
(D) not superimposable on their mirror images

34. Which of the following does not observe the Anti-Markownikoff's addition of HBr ?
(A) Pent-2-ene (B) But-1-ne
(C) but-2-ene (D) Propene

35. Grignard reagent is prepared by the reaction between.
(A) magnesium and alkyl halide.
(B) zinc and alkyl halide
(C) magnesium and alkane.
(D) magnesium and aromatic hydrocarbon.

36. The most suitable method of separation of a 1:1 mixture of ortho- and para-nitrophenols is.
(A) sublimation (B) chromatography
(C) crystallisation (D) distillation

37. The disaccharide present in milk is
(A) lactose (B) maltose
(C) cellobiose (D) sucrose

38. The action of nitrous acid on an aliphatic primary amine gives.
(A) alkyl nitrite (B) secondary amine
(C) alcohol (D) nitro alkane

39. Which of the following statements is false about resonance contributing structures ?
(A) Contributing structures represent hypothetical molecules having no real existence.
(B) Contributing structures are less stable than the

resonance hybrid.

(C) Contributing structures contribute to the resonance hybrid in proportion of their energies

(D) Equivalent contributing structures make the resonance very important.

40. Which one of the following compounds gives positive test with Tollen's reagent ?

(A) Acetone (B) Acetic acid (C) sucrose (D) Acetaldehyde

41. For which of the following compounds the Lassaigne's test for the detection of nitrogen will fail ?

(A) $H_2NCONHNH_2 + HCl$ (B) $H_2NNH_2 + 2HCl$

(C) H_2NCONH_2 (D) $C_6H_5NHNH_2.2HCl$

42. The peptide bond is a key feature in

(A) Polysaccharides (B) proteins

(C) nucleotides (D) vitamins.

43. The reactive species in nitration of benzene is

(A) NO_3^- (B) NO_2^- (C) NO_2^+ (D) HNO_3

44. Which of the following compounds reacts slower than benzene in electrophilic bromination

(A) 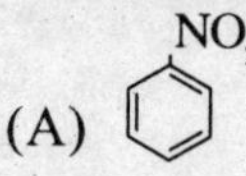

(B) 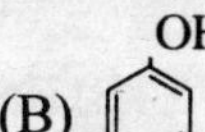

(C) CH_3 (benzene ring)

(D) NH_2 (benzene ring)

45. The total number of valence electrons in 4.2 g. of N_3^- ion is

(A) 1.6 N_A (B) 3.2N_A (C) 2.1 N_A (D) 4.2 N_A

(where N_A is the Avogadro number)

46. The table shown below lists the bond dissociation energies (E_{diss})for single covalent bonds formed between carbon atoms (C) and atoms of elements 'A', 'B', 'C' and 'D'. which element has the smallest atoms ?

Bond	E_{diss} (kJ mol^{-1})
C-A	240
C-B	328
C-C	276
C-D	485

(A) A (B) B (C) C (D) D

47. For which one of the following sets of four quantum numbers an electron will have the highest energy ?

	n	ℓ	m	s
(A)	4	1	0	−1/2
(B)	5	0	0	−1/2
(C)	3	2	1	1/2
(D)	4	2	−1	1/2

48. In a radioactive decay an emitted electron comes from

(A) the nucleus of the atom

(B) the orbit with principal quantum number 1

(C) the inner orbital of the atom

(D) the outermost orbit of the atom.

49. Which one of the following pairs of atoms or ions will have identical ground state electronic configuration ?

(A) Li^+ and He^- (B) Cl^- and Ar

(C) Na and K (D) $F^+$and Ne

50. Which of the following is most soluble?

(A) Bi_2S_3(K_{sp} = 1 x 10^{-70}) (B) MnS (K_{sp} = 7 x10^{-16})

(C) CuS(K_{sp} = 8 x 10^{-37}) (D) Ag_2S (K_{sp} = 6 x 10^{-51})

51. At 80° C pure distilled water has [H_3O^+] concentration equal to 1 x10^{-6} mol/L.The value of K_w at this temperature will be

(A) 1 x 10^{-14} (B) 1 x 10^{-6}

(C) 1 x 10^{-8} (D) 1 x 10^{-12}

52. 50 mL of hydrogen diffuses out through a small opening in a vessel in 20 minutes. The time needed for 40 mL of oxygen to diffuse out is

(A) 12 min (B) 64 min (C) 8 min (D) 32 min

53. The oxidation state of I in $H_4IO_6^-$ is

(A) +7 (B) −1 (C) +5 (D) +1

54. The formula for Calcium chlorite is

(A) $CaClO_2$ (B) $Ca(ClO_2)_2$

(C) $Ca(ClO_4)_2$ (D) $Ca(ClO_3)_2$

55. The number of atoms in 100 g of an FCC crystal with density d = 10.0g/cm^3 and cell-edge equal 100 pm, is equal to

(A) 3 x 10^{25} (B) 4 x 10^{25}

(C) 1 x 10^{25} (D) 2 x 10^{25}

56. A molecule in which sp^2 hybrid orbitals are used by the central atom in forming covalent bonds is

(A) PCl_5 (B) N_2 (C) He_2 (D) SO_2

57. Which one of the following molecules does not possess a permanent dipole moment ?

(A) CS_2 (B) SO_3 (C)H_2S (D) SO_2

58. How many grams of CH_3OH would have to be added to water to prepare 150 mL of solution that is 2.0 M CH_3OH ?

(A) 9.6 (B) 2.4 (C) 9.6 x 10^3 (D) 4.3 x 10^2

59. Consider the following reaction occurring in an automobile:

$2 C_8H_{18}(g) + 25O_2(g) \rightarrow 16 CO_2(g) + 18H_2O(g)$

The sign of ΔH, ΔS and ΔG would be

(A) +, − , + (B) −, +, −

(C) −, +, + (D) +, +, −

60. 1.1 mol of A is mixed with 2.2 mol of B and the mixture is kept in a one litre flask till the equilibrium, $A + 2B \rightleftharpoons 2C + D$ is reached. At equilibrium 0.2 mol of C is formed. The equilibrium constant of the above reaction is

(A) 0.002 (B) 0.004 (C) 0.001 (D) 0.003

61. Which substance is serving as a reducing agent in the following reaction?

$14H^+ + Cr_2O_7^{2-} + 3Ni \rightarrow 2Cr^{3+} + 7H_2O + 3Ni^{2+}$

(A) H^+ (B) Cr_2O_7 (C) H_2O (D) Ni

62. The standard reduction potentials at 25°C of Li^+ / Li, Ba^{2+} / Ba, Na^+ / Na and Mg^{2+} / Mg are –3.05, –2.73, –2.7 and –2.37 volt respectively. Which one of the following is the strongest oxidising agent ?
(a) Na^+ (B) Li^+ (C) Ba^{2+} (D) Mg^{2+}

63. The latent heat of vaporisation of a liquid at 500 Kand 1 atm pressure is 10.0 Kcal/mol. What will be the change in internal energy (ΔE) of 3 moles of liquid at the same temperature and pressure ?
(A) 13.0 kcal/mol (B) –13.0 kcal/mol
(C) 7.0 kcal/mol (D) –7.0 kcal/mol

64. How many significant figures should there be in the answer to the following problem ?

$$\frac{(29.2 - 20.2)(1.79 \times 10^5)}{1.39}$$

(A) 3 (B)1 (C) 4 (D) 2

65. At 25°C, the highest osmotic pressure is exhibited by 0.1 M solution of
(A) $CaCl_2$ (B) KCl (C) glucose (D) Urea

66. Which of the following orders is incorrect with respect to the property indicated ?
(A) Benzamide < Cyclohexylamine < Aniline (Base-strength)
(B) Formicacid >Aceticacid > Propanoicacid (Acid-strength)
(C) Cyclohexanol < Phenol < Benzoic acid (Acid-strength)
(D) Fluoroacetic acid>Chloroacetic acid>Bromoacetic acid (Acid-strength)

67. An acyl halide is formed by reacting PCl_5 with an
(A) ester (B) amide (C) acid (D) alcohol

68. Diazo coupling is useful to prepare some.
(A) pesticides (B) dyes
(C) proteins (D) vitamins

69. On heating glycerol with conc. H_2SO_4, a compound is obtained which has bad odour. The compound is
(A) allyl alcohol (B) formic acid
(C) acrolein (D) glycerol sulphate

70. A compound 'A' has a molecular formula C_2Cl_3OH. It reduces Fehling's solution and on oxidation gives a monocarboxylic acid 'B'.'A' is obtained by the action of chlorine on ethyl alcohol, 'A' is
(A) methyl chloride (B) monochloroacetic acid
(C) chloral (D) chloroform

71. The function of enzymes in the living system is to
(A) transport oxygen
(B) provide immunity
(C) catalyse biochemical reactions
(D) provide energy..

72. Which of the following statements is false ?
(A) Some disinfectants can be used as anti-septics in lower concentrations.
(B) Sulphadiazine is a synthetic antibacterial
(C) Ampicillin is a natural antibiotic
(D) Aspirin is an analgesic and antipyretic

73. Which of the following compounds will undergo self aldol condensation in the presence of cold, dilute alkali ?
(A) $CH_2 = CH–CHO$ (B) $CH \equiv C–CHO$
(C) C_6H_5–CHO (D) CH_3CH_2CHO

74. Meso-tartaric acid is optically inactive due to the presence of
(C) molecular symmetry
(D) external compensation.
(A) geometrical isomerism.
(B) two chiral carbon atoms

75. An example of a natural biopolymer is
(A) teflon (B) rubber (C) nylon-66 (D) DNA

CBSE -1995

1. BCl_3 is a planar molecule whereas NCl_3 is pyramidal because.
(A) nitrogen atom is smaller than boron atom.
(B) BCl_3 has no lone pair of electrons but NCl_3 has a lone pair of electrons.
(C) B-Cl bond is more polar than N-Cl bond
(D) N-Cl bond is more covalent than B-Cl bond.

2. Identify the correct statement:
(A) Gypsum contains a lower percentage of plaster of calcium than plaster of Paris
(B) Gypsum is obtained by heating plaster of Paris.
(C) Plaster of Paris can be obtained by hydration of gypsum.
(D) Plaster of Paris is obtained by partial oxidation of gypsum.

3. If one end of a piece of a metal is heated the other end becomes hot after some time. This is due to
(A) energised electrons moving to the other part of the metal
(B) resistance of the metal
(C) mobility of atoms in the metal
(D) minor perturbation in the energy of atoms.

4. Sodium is made by the electrolysis of a molten mixture of about 40% NaCl and 60% $CaCl_2$ because
(A) Ca^{++} can displace Na from NaCl
(B) Ca^{++} can reduce NaCl to Na
(C) $CaCl_2$ helps in conduction of electricity
(D) This mixture has a lower melting point than NaCl

5. The solubility in water of sulphates down the Be group is Be>Mg>Ca>Sr>Ba. This is due to
(A) high heat of solvation for smaller ions like Be^{2+}
(B) increasing molecular weight.
(C) decreasing lattice energy
(D) increase in melting points

6. Aluminium (III) chloride forms a dimer because
(A) higher coordination number can be achieved by aluminium.
(B) aluminium has high ionization energy

(C) aluminium belongs to 3rd group
(D) it cannot form a trimer.

7. Mercury is the only metal which is liquid at 0^oC. This is due to its
(A) low ionisation potential
(B) very high ionisation energy and weak metallic bond.
(C) high atomic weight (D) high vapour pressure.

8. In the metal carbonyls of the general formula M (CO) where M = metal: x = 4, the metal is bonded to
(A) carbon and oxygen (B) Carbon
(C) Oxygen (D) $C \equiv O$ triple bond

9. The number of geometrical isomers for $[Pt(NH_3)_2Cl_2]$ is
(A) Three (B) four (C) Two (D) one

10. Coordination number and oxidation state of Cr in $K_3Cr(C_2O_4)_3$ are repsectively
(A) 4 and +2 (B) 6 and +3
(C) 3 and +3 (D) 3 and 0

11. Amongst TiF_6^{2-}, CoF_6^{3-}, Cu_2Cl_2 and $NiCl_4^{2-}$(atomic numbers Ti = 22, Co = 27, Cu = 29, Ni = 28). The colourless species are
(A) CoF_6^{3-} and $NiCl_4^{2-}$ (B) TiF_6^{2-} and CoF_6^{3-}
(C) Cu_2Cl_2 and $NiCl_4^{2-}$ (D) TiF_6^{2-} and Cu_2Cl_2

12. The IUPAC name for
$CH_3CH = CHCH_2\underset{\displaystyle NH_2}{\underset{|}{C}H}CH_2COOH$ is
(A) 5-amino hex-2-ene-carboxylic acid.
(B) 5-amino-2 heptenoic acid.
(C) 3- amino-5- heptenoic acid.
(D) β-amino-δ- heptenoic acid.

13. In which of the compounds below is there more than one kind of hybridization (sp, sp^2, sp^3) for carbon?
(A) $CH_3CH_2CH_2CH_3$ (B) CH_3- CH = CH -CH_3
(C) CH_2 = CH-CH= CH_2 (D) H-C $\equiv$ C-H

14. The standard state Gibbs Free energy change for the isomerization reaction :
cis. 2 pentene = trans-2 pentene
is -3.67kJ/mol at 400K. If more trans-2 pentene is added to the reaction vessel
(A) Equilibrium remains unaffected.
(B) more cis-2 pentene is formed.
(C) equilibrium is shifted in the forward direction.
(D) additional trans-2 pentene is formed.

15. An ideal gas obeying kinetic theory of gases can be liquefied if
(A) its temperature is more than critical temperature T.
(B) it can not be liquefied at any P and T.
(C) its pressure is more than criticial pressure P.
(D) its pressure is more than P_C at a temperature less than T_C

16. A chemical reaction is catalysed by a catalyst. X. Hence X
(A) does not affect equilibrium constant of the reaction
(B) decreases rate constant of the reaction.
(C) reduces enthalpy of the reaction.
(D) increases activation energy of the reaction.

17. pH of a 10M solution of HCl is
(A) 0 (B) 1 (C) less than 0 (D) 2

18. An electrochemical cell is set up as follows:
Pt (H_2.1atm)/0.1MHCl/0.1M acetic acid/ (H_2,1atm) Pt.
E.M.F. of this cell will not be zero because
(A) the pH of 0.1 M HCl and 0.1 M acetic acid is not the same.
(B) acids used in the two compartments are different.
(C) E.M.F. of a cell depends on the molarities of acids used .
(D) the temperature is constant.

19. If enthalpies of formation for $C_2H_4(g)$, $CO_2(g)$ and $H_2O(l)$at 25^oC and 1 atm pressure be 52,-394 and 286 kJ mol^{-1} respectively, enthalpy of combustion of C_2H_4 (g) will be
(A) +1412 kJ mol^{-1} (B) -1412 kJ mol^{-1}
(C) +14.2 kJ mol^{-1} (D) -141.2 kJ mol^{-1}

20. Half life for radioactive ^{14}C is 5760 years. In how many years 200 mg of ^{14}C sample will be reduced to 25 mg?
(A) 5760 year (B) 17280 year
(C) 11520 year (D) 23040 year

21. According to Raoult's law the relative lowering of vapour pressure for a solution is equal to the
(A) mole fraction of solute. (B) moles of solvent
(C) mole fraction of solvent (D) moles of solute.

22. The oxide which cannot act as a reducing agent is
(A) SO_2 (B) NO_2 (C) CO_2 (D) ClO_2

23. About 20km above the earth there is an ozone layer, each one of the following statements about ozone and ozone layer is true?
(A) Ozone has a triatomic linear molecule.
(B) Ozone layer is harmful to us because it cuts out a radiation useful for photosynthesis.
(C) Ozone layer is beneficial to us because ozone cuts out the ultraviolet radiation of the sun.
(D) The conversion of ozone to oxygen is an endothermic reaction

24. Ground state electronic configuration of valence shell electrons in nitrogen molecule (N_2) is written as $(2s\sigma)^2 (2s\sigma^*)^2(2p\sigma)^2 (2p\pi)^4$. Hence the bond order in nitrogen molecule is
(A) 2 (B) 3 (C) 0 (D) 1

25. After the emission of one α-particle followed by one, β-particle from the atom of X_{92}^{238}, the number of neutrons in the atom will be
(A) 142 (B) 146 (C) 144 (D) 143

26. The electronic configuration of an element is $1s^2$ $2s^2$ $2p^6$ $3s^2$ $3p^3$, What is the atomic number of the element which is just- below the above element in the periodic

table?
(A) 31 (B) 49 (C) 33 (D) 34

27. The correct order of the O-O bond length in O_2, H_2O_2 and O_3 is
(A) $O_2 > O_3 > H_2O_2$ (B) $O_3 > H_2O_2 > O_2$
(C) $H_2O_2 > O_3 > O_2$ (D) $O_2 > H_2O_2 > O_3$

28. The radius of hydrogen atom in the ground state is 0.53 Å. The radius of Li^{2+} ion in a similar state is
(A) 1.06 Å (B) 0.265 Å
(C) 0.17 Å (D) 0.53 Å

29. Which of the following species is para-magnetic?
(A) CO (B) NO^- (C) O_2^{2-} (D) CN^-

30. The uncertainty in the position of an electron (mass=9.1×10^{-28}g) moving with a velocity of 3.0×10^4 cms^{-1} accurate upto 0.011% will be
(A) 1.92 cm (B) 7.68 cm
(C) 5.76 cm (D) 3.84 cm

31. The number of moles of oxygen in one litre of air containing 21% oxygen by volume, in standard conditions is
(A) 0.186 mole (B) 0.21 mole
(C) 2.10 mole (D) 0.093 mole

32. Normal boiling point of water is 373 k (at 760mm),. Vapour pressure of water at 298 K is 23 mm. If the enthalpy of evaporation is 40.656kJ/mol, the boiling point of water at 23 mm pressure will be
(A) 51.6 K (B) 12.5 K (C) 25 K (D) 75.7 K

33. Dimensions of pressure are the same as that of
(A) Energy (B) Force
(C) Energy per unit volume
(D) Force per unit volume

34. Concentration units independent of temeperature would be
(A) Normality (B) Weight volume percent
(C) Molality (D) Molarity

35. Solubility of AgCl will be minimum in
(A) 0.001 M $AgNO_3$ (B) Pure water
(C) 0.01 M $CaCl_2$ (D) 0.01 M NaCl

36. For a reaction to occur spontaneously
(A) (Δ H+TΔS) must be negative.
(B) (Δ H-TΔS) must be negative
(C) Δ H must be negative (D) ΔS must be negative

37. Examine the following common chemical structures to which simple functional groups are often attached.

(i) 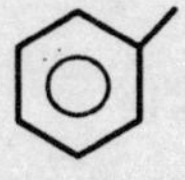 (ii) 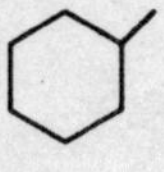 (iii)

(iv) $CH_3CH_2CH_2CH_2-$ (v) $H_2C=C\langle^{H}_{H}$

Which of these systems have essentially planar geometry ?
(A) (i) and (v) (B) (ii) and (iii)
(C) (ii), (iii) and (iv) (D) (iv)

38. Alkene $RCH=CH_2$ reacts readily with B_2H_6 and the product on oxidation with alkaline hydrogen peroxides produces
(A) $R-CH_2-CHO$ (B) RCH_2CH_2OH
(C) $R-\underset{\underset{O}{\|}}{C}-CH_3$ (D) $R-\underset{\underset{OH}{|}}{CH}-\underset{\underset{OH}{|}}{CH_2}$

39. When 3.3- dimethyl, 2 butanol is heated with H_2SO_4, the major product obtained is
(A) cis and trans isomers of product
(B) 3,3 dimethyl-l butene (C) 2,3 dimethyl 1-2 butene
(D) 2,3 -dimethyl-1 butene

40. In the reaction
$CH_3CHO + HCN \rightarrow CH_3CH(OH)CN$
a chiral centre is produced. This product would be
(A) laevorotatory. (B) meso compound.
(C) dextrorotatory. (D) racemic mixture.

41. Oxidation of toluene with CrO_3 in the presence of $(CH_3CO)_2O$ gives a product 'A' which on treatment with aqueous NaOH produces
(A) C_5H_5CHO (B) $(C_6H_5CO)_2O$
(C) C_6H_5COONa (D) 2,4 diacetyl toluene

42. The general order of reactivity of carbonyl compounds for nucleophilic addition reactions is
(A) $Ar_2C=O > R_2C=O > ArCHO > RCHO > H_2C=O$
(B) $H_2C=O > RCHO > ArCHO > R_2C=O > Ar_2C=O$
(C) $ArCHO > Ar_2C=O > RCHO > R_2C=O > H_2C=O$
(D) $H_2C=O > R_2C=O > Ar_2C=O > RCHO > ArCHO$

43. α-D Glucose and β-D Glucose differ from each other due to difference in one of carbons with respect to its
(A) Size of hemiacetal ring. (B) Number of OH groups.
(C) Configuration (D) Conformation.

44. Predict which one of the following molecules is most likely to organize and form a bilayer membrane structure.

I. CH_2OH | $CHOH$ | $CH_2OP(=O)(O^-)-O(CH_2)_2-N-(CH_3)_2$

II. $CH_2OCO(CH_2)_{14}CH_3$ | $CHOCO(CH_2)_{14}CH_3$ | $CH_2OP(=O)(O^-)-O(CH_2)_2-N-(CH_3)_2$

III. $CH_2OCO(CH_2)_{14}CH_3$ | $CHOCO(CH_2)_{14}CH_3$ | CH_2OCH_3

IV. 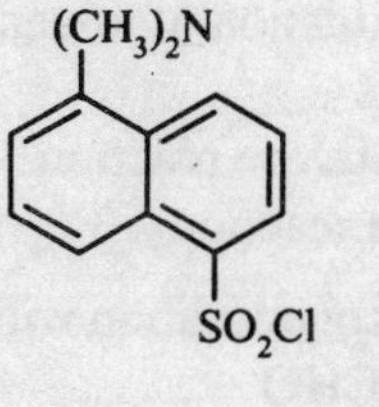 $(CH_3)_2N$ ... SO_2Cl (naphthalene)

(A) II (B) I (C) III (D) IV

45. The increasing order of acidity among phenol, p-methylphenol, m-nitrophenol and p-nitropnenol is
(A) m-nitrophenol, p-nitrophenol, phenol, p-methylphenol
(B) p-methyl phenol, m-nitrophenol, phenol, p-nitrophenol
(C) p-methyl phenol, phenol, m-nitrophenol, p-nitrophenol
(D) phenol, p-methylphenol, p-nitrophenol, m-nitrophenol

46. Secondary structure of a protein refers to
(A) mainly denatured proteins and structures of prosthetic groups
(B) three-dimensional structure, specially the bond between amino acid residues that are distinct from each other in the polypeptide chain
(C) linear sequence of amino acid residues in the polypeptide chain
(D) regular folding patterns of continuous portions of the polypeptide chain

47. Bakelite is prepared by the reaction between
(A) tetramethylene glycol and hexamethylenediisocyanate.
(B) phenol and formaldehyde.
(C) urea and formaldehyde.
(D) ethylene glycol and dimethyl terephthalate.

48. The pH of blood does not appreciably change by a small addition of an acid or a base because blood
(A) contains serum protein which acts as buffer.
(B) contains iron as a part of the molecule.
(C) can be easily coagulated. (D) it is body fluid.

49. Of the following statements about enzymes which ones are true?
(i) Enzymes lack in nucleophilic groups
(ii) Enzymes are highly specific both in binding chiral substrates and in catalyzing their reactions
(iii) Enzymes catalyze chemical reactions by lowering the activation energy
(iv) Pepsin is a proteolytic enzyme
(A) (i) and (iv) (B) (i) and (iii)
(C) (ii), (iii) and (iv) (D) (i)

50. Oxidation of glucose is one of the most important reactions in a living cell. What is the number of ATP molecules generated in cells from one molecule of glucose?
(A) 18 (B) 28 (C) 38 (D) 12

CBSE - 1996

1. The structure and hybridisation of $Si(CH_3)_4$ is
(A) octahedral, sp^3d (B) tetrahedral, sp^3
(C) bent, sp (D) trigonal, sp^2

2. The oxidation of toluene to benzaldehyde by chromyl chloride is
(A) Cannizzaro reaction (B) Wurtz reaction
(C) Etard reaction (D) Riemer-Tiemann reaction

3. Aldol condensation will not take place in
(A) CH_3CHO (B) CH_3COCH_3
(C) HCHO (D) CH_3CH_2CHO

4. The ionic prouct of water at 25°C is 10^{-14} the ionic product at 900°C will be
(A) 1×10^{-14} (B) 1×10^{-16} (C) 1×10^{-20} (D) 1×10^{-12}

5. Which will be the longest wavelength line in Balmer series of spectrum ?
(A) 566 nm (B) 556nm (C) 546 nm (D) 656nm

6. Which of the following mixtures of gases does not obey Dalton's Law of partial pressure?
(A) O_2 amd CO_2 (B) N_2 amd O_2
(C) Cl_2 and SO_2 (D) CO_2 and He

7. Which of the following compounds gives benzoic acid on oxidation?
(A) Chlorophenol (B) Chlorotolune
(C) Chlorobenzene (D) Benzylchloride

8. Among the following oxides, the least acidic is
(A) As_4O_6 (B) As_4O_{10} (C) P_4O_6 (D) P_4O_{10}

9. Reducton of thiosulphate with iodine gives
(A) sulphate ion (B) sulphite ion.
(C) tetrathionate ion (D) sulhide ion.

10. Which of the following is largest in size ?
(A) Cl^- (B) S^{2-} (C) Na^+ (D) F^-

11. The relationship between osmotic pressure at 273 k when 10 g glucose(P_1), 10 g urea (P_2) and 10 g sucrose (P_3) are dissolved in 250 ml of water is
(A) $P_2>P_1>P_3$ (B) $P_2>P_3>P_1$
(C) $P_1>P_2>P_3$ (D) $P_3>P_1>P_2$

12. The equilibrium constant for the reaction $N_2+3H_2=2NH_3$ is K, then the equilibrium constant for the equilibrium $NH_3 = 1/2\ N_2 + 3/2 H_2$ is
(A) $\sqrt{K}$ (B) $\sqrt{1/K}$ (C) 1/K (D) $1/K^2$

13. When calomel reacts with NH_4OH, we get
(A) $HgNH_2Cl$ (B) NH_2-Hg-Hg-Cl
(C) Hg_2O (D) HgO

14. Boron compounds behave as lewis acids because of their
(A) acidic nature. (B) covalent nature.
(C) electron deficiency (D) ionisation property.

15. The basic character of hydrides of the 15th group elements decreases in the order:
(A) $NH_3>PH_3>AsH_3>SbH_3$
(B) $SbH_3>AsH_3>PH_3>NH_3$
(C) $SbH_3>PH_3>AsH_3>NH_3$
(D) $NH_3>SbH_3>PH_3>AsH_3$

16. Which of the following bonds has the highest energy?
(A) S-S (B) O-O (C) Se-Se (D) Te-Te

17. A transition element x has a configuration $(Ar)3d^4$ in its +3 oxidation state. Its atomic number is
(A) 25 (B) 26 (C) 22 (D) 19

18. Which of the following has zero dipole moment?
(A) cis 2-butene (B) trans 2-butene
(C) 1-butene (D) 2-methyl-1-propene.

19. The reaction

$CH_2 = CH - CH_3 + HBr \rightarrow CH_3CHBr - CH_3$ is

(A) electrophilic addition

(B) nucleophilic addition

(C) electrophilic substitution (D) free radical addition.

20. p-Nitrotoluene on further nitration gives

(A) 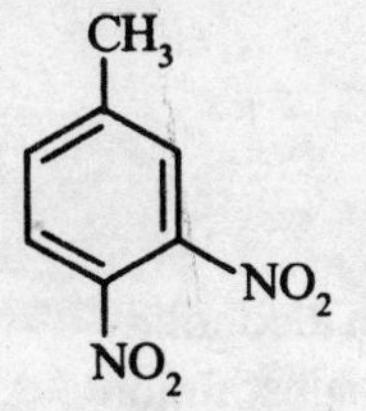 CH_3, NO_2, NO_2 (B) 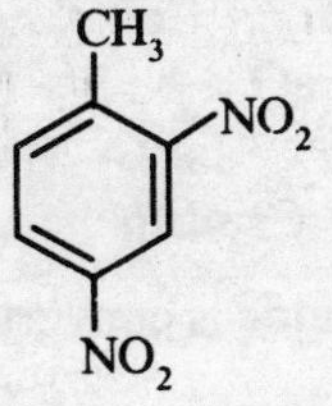 CH_3, NO_2, NO_2

(C) 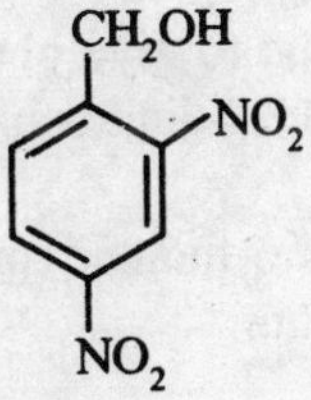 CH_2OH, NO_2, NO_2 (D) 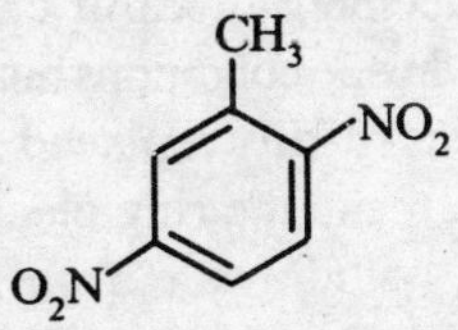 CH_3, NO_2, O_2N

21. The bond length between sp^3 hybridised carbon and other carbon atom is minimum in

(A) propane (B) butane

(C) propene (D) propyne

22. Electrophile in the case of chlorination of benzene in the presence of $FeCl_3$ is

(A) Cl (B) $FeCl_3$ (C) Cl^+ (D) Cl^-

23. Which of the following has the smallest size?

(A) Mg^{+2} (B) Cl^- (C) Na^+ (D) F^-

24. In a Bohr's model of atom when an electron jumps from or falls ton =1 to n = 3, how much energy will be emitted or absobed ?

(A) 2.15×10^{-11} ergs (B) 0.1911×10^{-10} ergs

(C) 2.389×10^{-10} ergs (D) 0.239×10^{-10} ergs

25. Which of the following is used as an anti-knocking agent ?

(A) T.E.L. (B) C_2H_5OH (C) Glyoxal (D) Freon

26. The electronic configuration of transition elements is exhibited by

(A) $ns^2(n-1)d^{1-10}$ (B) $ns^2(n-1)d^{10}$

(C) ns^1 (D) $ns^2\ np^5$

27. Which of the following does not have sp^2 hybridisation

(A) C_3H_6 (B) $BeCl_2$ (C) C_2H_4 (D) C_2H_2

28. At what temperature the rate of effusion of N_2 would be 1.625 times than the rate of SO_2 at 500°C ?

(A) 373K (B) 273K (C) 110 K (D) 173K

29. How many atoms are present in each unit cell of a F.C.C. crystal contains ?

(A) 4 (B) 6 (C) 3 (D) 8

30. The average kinetic energy of an ideal gas per molecule in S.I. units at 250 C will be

(A) 61.7×10^{-21} (B) 6.17×10^{-21}

(C) 6.17×10^{-20} (D) 7.16×10^{-20}

31. 0.24 gm of a volatile gas upon vapourisation gives 45 ml vapour at NTP. What will be the vapour density of the substance ? (Density of H_2 = 0.089)

(A) 95.93 (B) 59.93 (C) 95.39 (D) 5.993

32. If we mix a pentavalent impurity in a crystal lattice of germanium, what type of semiconductor formation will occur ?

(A) P-type semiconductor (B) N-type smiconductor

(C) Both (A) and (B) (D) None of these.

33. What is the molarity of H_2SO_4 solution that has a density 1.84gm/cc at 35°C and contains 98% by weight?

(A) 4.18 M (B) 8.14 M (C) 18.4 M (D) 18 M

34. Vapour pressure of CCl_4 at 25°C is 143mm Hg. 0.5gm of a non-volatile solute (mol.weight = 65) is dissolved in 100 ml CCl_4. Find the vapour pressure of the solution. (Density of CCl_4 = 1.58g/cm²)

(A) 199.34 mm (B) 143.99 mm

(C) 149.93 mm (D) 94.39 mm

35. A colloidal system has particles of size

(A) 10^{-4} m to 10^{-10} m (B) 10^{-5} m to 10^{-7} m

(C) 10^{-9} m to 10^{-12} m (D) 10^{-6} m to 10^{-9} m

36. The given reaction

$2FeCl_3 + SnCl_2 \rightarrow 2FeCl_2 + SnCl_4$

is an example of

(A) first order reaction. (B) Second order reaction.

(C) Third order reacton. (D) None of these.

37. When a few typical solutes such as protein particles, blood corpuscles are separated by a particular selective membrane the process is called

(A) dialysis. (B) diffusion.

(C) transpiration. (D) endosmosis.

38. A 5 ampere of current is passed through a solution of zinc sulphate for 40 minutes. Find the amount of zinc deposited at the cathode.

(A) 40.65 gm (B) 4.065 gm

(C) 0.4065 gm (D) 65.04 gm

39. Which of the following is not a Lewis acid ?

(A) SiF_4 (B) C_2H_4 (C) BF_3 (D) $FeCl_3$

40. pH value of N/10 NaOH solution is

(A) 10 (B) 11 (C) 12 (D) 13

41. The amount of zinc required to produce 224 ml of H_2 at STP on treatment with dilute H_2SO_4 will be

(A) 0.65 gm (B) 6.5 gm (C) 65 gm (D) 0.065 gm

42. IUPAC name of $(CH_3)_2$ CH-CH_2-CH_2 Br is

(A) 1-bromo-pentane (B) 2-methy1-4bromo-butane

(C) 1-bromo-3-methyl-butane

(D) 2-methyl-3-bromo-propane.

43. Total number of moles for the reaction ,

$2\ HI = H_2 + I_2$ if α is dissociation constant, is

(A) 1 (B) $1-\alpha$ (C) 2 (D) $2-\alpha$

44. Which of the following represents calcium chlorate?
(A) $CaClO_2$ (B) $Ca(ClO_4)_2$
(C) $Ca(ClO_3)_2$ (D) $Ca(ClO_2)_2$

45. An element (atomic mass = 100g/mol) having bcc structure has unit cell edge 400 pm. Then, density of the element is
(A) 10.376 g/cm³ (B) 5.188 g/cm³
(C) 7.289 g/cm³ (D) 2.144 g/cm³

46. $KMnO_4$ reacts with oxalic acid according to the equation :
$2MnO_4 + 5C_2O_4^{-2} + 16H^+ \text{--->} 2Mn^{2+} + 10CO_2 + 8H_2O$
Here 20 ml of 0.1 M $KMnO_4$ is equivalent to
(A) 20ml of 0.5 M $C_2H_2O_4$ (B) 50 ml of 0.1 M $C_2H_2O_4$
(C) 50 ml of 0.5 M $C_2H_2O_4$ (D) 20 ml if 0.1 M $C_2H_2O_4$

47. The number of isomers in C_5H_{12} will be
(A) 6 (B) 3 (C) 5 (D) 7

48. Which of the following is the correct equation ?
(A) $\Delta U = \Delta Q - W$ (B) $\Delta W = \Delta U + \Delta Q$
(C) $\Delta U = \Delta W + \Delta Q$ (D) None of these

49. One of the following has greatest electron affinity. Identify it.
(A) F (B) Cl (C) I (D) Br

50. Reducton potentials for the following half-cell reactions are ,
$Zn = Zn^{2+} + 2e^-$; $E_0 = 0.76$ V
$Fe = Fe^{2+} + 2e^-$; $E_0 = +0.44$V
The EMF for the cell reaction $Fe^{2+} + Zn = Zn^{2+} + Fe$ will be
(A) -0.32 V (B) +1.20 V (C) -1.20 V (D) +0.32 V

CBSE -1997

1. Which of the following techniques is most suitable for identification of cyclohexanone in a mixture containing benzoic acid, isoamylalcohol, cyclohexane and cyclohexanone ?
(A) Sublimation (B) Evaporation
(C) Crystallisation (D) IR spectroscopy

2. In graphite, electrons are
(A) localised on every third C-atom
(B) present in anti-bonding orbital
(C) localised on each C-atom
(D) spread out between the structure

3. The lanthanide contraction is responsible for the fact that
(A) Zr and Hf have about the same radius
(B) Zr and Zn have the same oxidation state
(C) Zr and Y have about the same radius
(D) Zr and Nb have similar oxidation state

4. Carbon-14 dating method is that the
(A) Carbon-14 fraction is the same in all objects.
(B) Carbon-14 highly insoluble
(C) Ratio of carbon-14 and carbon-12 are constant.
(D) all the above

5. The intermetallic compound LiAg crystallizes in cubic lattice in which both lithium and silver have co-ordination number of eight. The crystal class is
(A) Simple cube (B) Body centre cube
(C) Face centre cube (D) None of these

6. Which of the following does not represent the correct of the property indicated ?
(A) $Mn^{2+} < Ni^{+2} < Co^{2+} < Fe^{2+}$
(B) $FeO > CaO > NiO > CuO$
(C) $Sc^{3+} > Cr^{3+} > Fe^{3+} > Mn^{3+}$
(D) $Sc < Ti < Cr < Mn$

7. The alkylhalde is converted into an alcohol by
(A) addition (B) substitution
(C) dehydrohalogenation (D) elimination

8. Replacement of Cl of chlorobenzene to give phenol requires drastic conditions but chlorine of 2,4-dinitrochlorobenzene is readily replaced since.
(A) NO_2 make the ring electron rich at ortho and para
(B) NO_2 withdraws e^- from meta position.
(C) NO_2 denotes e^- at meta position
(D) NO_2 withdraws e^- from ortho/para positions

9. Reaction of O_2 and CO with Haemoglobin gives
(A) only oxygen-haeme complex
(B) only CO-haeme complex
(C) both but oxygen-haeme is more stable
(D) both but CO-haeme complex is more stable

10. A physician wishes to prepare a buffer solution at pH = 3.85 that efficiently resists changes in pH yet contains only small concentration of the buffering agent. Which one of the following weak acids together with its sodium salt would be best to use
(A) 2,5-dihydroxy benzoic aci (pka =2.97)
(B) Acetoacetic acid (pka =3.58)
(C) m-chlorobenzoic acid (pka=3.98)
(D) p-chlorocinnamic acid (pka-4.41)

11. The most stable conformations of n-butane is
(A) skew boat (B) eclipsed
(C) gauche (D) staggered-anti

12. Which one of the following chemical units is certainly to be found in enzyme ?
(A) (B)
(C) (D)

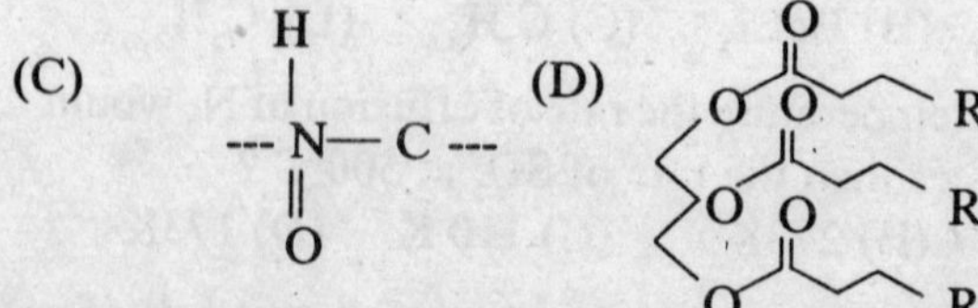

13. Ketones [R—C—R1],

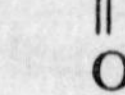

(Where R = R1 = alkyl group) can be obtained in one

step by

(A) hydrolysis of ester (B) oxidation of primary alcohol
(C) oxidation of tertiary alcohol
(D) reaction of acid halide with alcohol

14. In the commercial gasolines, the type of hydrocarbons which are more desirable is
(A) linear unsaturated hydrocarbon (B) toluene
(C) branched hydrocarbon
(D) straight-chain hydrocarbon

15. The volume strength of 1.5 N H_2O_2 solution is
(A) 4.8 (B) 5.2 (C) 8.8 (D) 8.4

16. Calcium is obtained by the
(A) roasting of limestone
(B) electrolysis of solution of calcium chloride in H_2O
(C) reduction of calcium chloride with carbon
(D) electrolysis of molten anhydrous calcium chloride

17. The number of geometrical isomers of the complex $[Co(NO_2)_3(NH_3)_3]$ is
(A) 2 (B) 4 (C) 3 (D) 0

18. The formula of dichlorobis(urea)copper (II) is
(A) $[Cu\{O = C(NH_2)_2\}]\ Cl_2$
(B) $[CuCl_2]\ \{O = C(NH_2)_2H_2\}$
(C) $[CuCl_2\ \{O = C(NH_2)_2\}_2]$
(D) $[Cu\ \{O = C(NH_2)_2\}Cl]Cl$

19. the cylindrical shape of an alkyne is due to
(A) three sigma C-C bonds (B) three π C-C bonds
(C) two sigma C-C and one π C-C bonds
(D) one sigma C-C and two π C-C bonds

20. The electronic configuration of gadolinium (At.no.64) is
(A) $[Xe]\ 4f^8\ 5d^9\ 6s^2$, (B) $[Xe]\ 4f^7\ 5d^1\ 6s^2$,
(C) $[Xe]\ 4f^3\ 5d^5\ 6s^2$, (D) $[Xe]\ 4f^6\ 5d^2\ 6s^2$,

21. The structural formula of hypophosphorus acid is

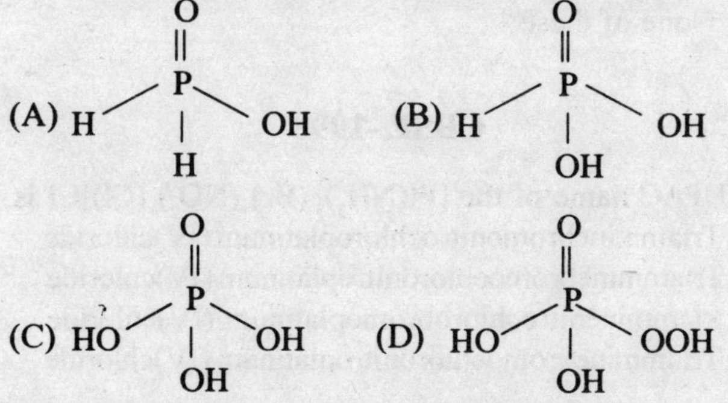

22. CrO_3 dissolves in aqueous NaOH to give
(A) CrO_4^{2-} (B) $Cr(OH)_3$ (C) CrO_7^{2-} (D) $Cr(OH)_2$

23. The ion that is isoelectronic with CO is
(A) CN^- (B) O_2^+ (C) O_2^- (D) N_2^+

24. The reaction is described as

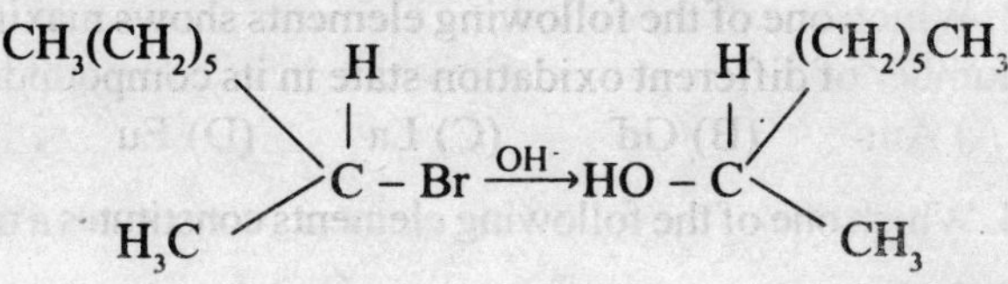

(A) S_E^2 (B) S_N^1 (C) S_N^2 (D) S_N^0

25. Tautomerism will be exhibited by
(A) $(CH_3)_3CNO$ (B) $(CH_3)_2NH$
(C) R_3CNO_2 (D) RCH_2NO_2

26. 1-phenylethanol can be prepared by reaction of benzaldehyde with
(A) Methyl bromide (B) ethyl iodide and magnesium
(C) methyl bromide and aluminium bromide
(D) methyl iodide and magnesium

27. Which one of the following is used to make 'non-stick' cookware ?
(A) PVC (B) Polystyrene
(C) Poly ethylene terephtalate
(D) Polyterafluoroethylene

28. Which one of the following is correct order of the size of iodine species ?
(A) $I > I^- > I^+$ (B) $I > I^+ > I^-$
(C) $I^+ > I^- > I$ (D) $I^- > I > I^+$

29. The high density of water compared to ice is due to
(A) dipole-induced dipole interactions
(B) induced dipole induced dipole interactions
(C) hydrogen bonding interactions
(D) dipole-dipole interactions

30. Given that
$C + O_2 \rightarrow CO_2,\ \Delta H^o = -x$ kJ
$2CO + O_2 \rightarrow 2CO_2,\ \Delta H^o = -y$ kJ
The enthalpy of formation of carbon monoxide can be represented as
(A) Y – 2x (B) $\frac{2x-y}{2}$ (C) $\frac{y-2x}{2}$ (D) 2x – y

31. N_2 and O_2 are converted into monocations, N_2^+ and O_2^+ respectively. Which is wrong ?
(A) In N_2, the N-N bond weakens.
(B) In O_2, the O-O bond order increases.
(C) In O_2 paramagnetism decreases.
(D) N_2^+ becomes diamagnetic.

32. The compound obtained by heating a mixture of a primary amine and chlorofrom with ethanolic potassium hydroxide (KOH) is
(A) an alkyl isocyanide (B) an alkyl halide (C) an amide
(D) an amide and nitro compound

33. E^0 for the cell, $Zn / Zn^{2+}(aq) // Cu^{2+}(aq) / Cu$ is 1.10 V at 25°C, the equilibrium constant for the reaction $Zn + Cu^{2+}(aq) \rightleftharpoons Cu + Zn^{2+}(aq)$ is of the order of
(A) 10^{-18} (B) 10^{17} (C) 10^{-28} (D) 10^{-37}

34. which of the following is redox reaction ?
(A) H_2SO_4 with NaOH
(B) formation of O_3 from O_2 in atmosphere by lightning
(C) formation of Nitrogen oxides from nitrogen and oxygen by lightning
(D) Evaporation of H_2O.

35. Among the following compounds (I-III) the correct case of reaction with electrophilie is

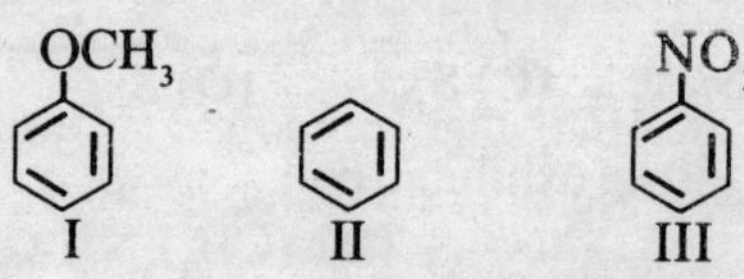

(A) II>III>I (B) III<I<II (C) I>II>III (D) I=II>III

36. Which of the following reactions is expected to readily give a hydrocarbon product in good yields ?

(A) $RCOOK \xrightarrow[\text{Oxidation}]{\text{Electrolytic}}$

(B) $RCOO\,Ag \xrightarrow{I_2}$

(C) $CH_3CH_3 \xrightarrow[hv]{Cl_2}$

(D) $(CH_3)_3CCl \xrightarrow{C_2H_5OH}$

37. The function of enzymes in the living system is to
(A) Transport oxygen (B) provide immunity
(C) provide energy
(D) catalyse biochemical reactions

38. Which of the following 0.10 M aqueous solution will have the lowest freezing point ?
(A) $Al_2(SO_4)_3$ (B) $C_5H_{10}O_5$ (C) KI (D) $C_{12}H_{22}O_{11}$

39. The AsF_5 molecule is trigonal pyramidal. The hybride orbitals used by the As atoms for bonding are
(A) $d_{x^2-y^2}$, d_{z^2}, S, P_x, P_y (B) dxy, s, Px, Py, Pz
(C) s, P_x, P_y, P_z, P_{z^2} (D) $d_{x^2-y^2}$, s, P_x, P, P_z

40. Equivalent conductance of NaCl, HCl and C_2H_5COONa at infinite dilution are 126.45, 426.16 and 91 $ohm^{-1}\,cm^2$ respectivly. The equivalent conductance of C_2H_5COONa is
(A) 201.28 ohm^{-1}/cm^{-2} (B) 390.71/ohm cm^2
(C) 698.28/ohm cm^2 (D) 540.48/ohm cm^2

41. $U^{235} + n \rightarrow U^{235} \rightarrow$ fission product + neutron + 3.20×10^{-11} J
The energy released when 1gm of U^{235} finally undergoes fission is
(A) 12.75×10^5 KJ (B) 18.60×10^9 KJ
(C) 8.21×10^7 KJ (D) 6.55×10^6 KJ

42. Which one of the following has the highest dipole moment ?
(A) SbH_3 (B) AsH_3 (C) NH_3 (D) PH_3

43. The solubility product of CuS Ag_2S and HgS are 10^{-31}, 10^{-44}, 10^{-54} respectively. The solubilities of these sulphides are in the order
(A) Ag_2S>CuS>HgS (B) AgS>HgS>CuS
(C) HgS>Ag_2S>CuS (D) CuS>Ag_2S>HgS

44. Which of the following elements is responsible for oxidation of water to O_2 in biological processes ?
(A) Fe (B) Mn (C) Cu (D) Mo

45. The ability to bring about coagulation of a given colloidal solution depends upon
(A) its charge. (B) the sign of the charge alone.
(C) the magnitude of charge.
(D) both magnitude and charge

46. H_2 and O_2 are converted into monoanions N_2^- and O_2^- respectively, which of the following statements is wrong ?
(A) In NH_3 decreases
(B) N_2^+ becomes diamagnetic
(C) In H_2 the N-N bond weakens
(D) In O_2 the O-O bond order increases

47. For two ionic solids CaO and KI, identify the wrong statement among the following
(A) Lattice energy of CaO is much larger than that of KI
(B) KI is soluble in benzene
(C) CaO has high m.p. (D) KI has high m.p.

48. Which of the following can possibly be used as analgesic without causing addiction and mood modification ?
(A) Morphine (B) N-acetyl- para- aminophenol
(C) Diazepam (D) tetrahydrolumional

49. In a reaction

$CH_2 = CH_2 \xrightarrow[\text{Acid}]{\text{Hypochlorous}} M \xrightarrow{R}$ $\begin{matrix} CH_2OH \\ | \\ CH_2OH \end{matrix}$

M = molecules
R = reagent

M and R are
(A) CH_3CH_2Cl and NaOH
(B) CH_2Cl, CH_2OH and aq. $NaHCO_3$
(C) CH_3CH_2OH and HCl
(D) $CH_2 - CH_2$ (bridged by O) and Heat.

50. The hydride ion H^- is stronger base than its hydroxide ion OH^-. Which of the following reaction will occur if sodium hydride (NaH) is dissolved in water ?
(A) $H^-(aq) + H_2O \rightarrow H_2O$
(B) $H^- + H_2O \rightarrow$ No reaction
(C) $H^-(aq) + H_2O \rightarrow OH^- + H_2$
(D) None of these

CBSE -1998

1. IUPAC name of the $[Pt(NH_3)_3(Br)(NO_2)(Cl)]Cl$ is
(A) Triamminebromonitrochloroplatinum (IV)chloride
(B) Triamminebromochloronitroplatinum (IV)chloride
(C) triamminenitrochlorobromoplatinum (IV)chloride
(D) Triamminebromochloronitroplatinum (IV)chloride

2. Which one of the following pairs of substances on reaction will not evolve H_2 gas ?
(A) Iron and H_2SO_4
(B) Iron and steam
(C) Copper and HCl (aqueous)
(D) Sodium and Ethyl alcohol

3. Which one of the following elements shows maximum number of different oxidation state in its compounds ?
(A) Am (B) Gd (C) La (D) Eu

4. Which one of the following elements constitutes a major

impurity in pig iron ?
(A) Oxygen (B) Silicon
(C) Graphite (D) sulphur

5. Activation energy of a chemical reaction can be determined by.
(A) evaluating rate constant at standard temperature
(B) evaluating rate constant at two different temperatures
(C) changing concentrations of reactants
(D) evaluating velocities of reaction at two different temperatures

6. For the cell reaction Cu^{2+} [C_1(aq)]+Zn(s) = Zn^{2+}[C_2(aq)]+ Cu(s) of an electrochemical cell, the change in free energy, ΔG, at a given temperature is a function of
(A) $\ln(C_1)$ (B) $\ln(C_2/C_1)$ (C) $\ln C_1 + C_2$ (D) $\ln(C_2)$

7. At the critical micelle concentration (cmc) the surfactant molecules
(A) decompose (B) becomes completely soluble
(C) associate (D) dissociate

8. Without losing its concentration $ZnCl_2$ solution cannot be kept in contact with
(A) Pb (B) Ag (C) Al (D)Au

9. One mole of an ideal gas at 300K is expanded isothermally from an initial volume of 1 litre to 10 litre. The ΔE for this process is
(R = 2 cal. mol^{-1} K^{-1})
(A) 163.7 cal (B) 1381.1 cal
(C) 9 lit. atm (D) zero

10. The Bohr orbit radius for hydrogen atom (n=1) is approximately 0.530Å. The radius for the first excited state (n=2) orbit is (in Å)
(A) 1.06 (B) 0.13 (C) 4.77 (D) 2.12

11. Among the following which one is paramagnetic? (At.no.: Be=4, Ne=10, As=33, Cl=17)
(A) Be^- (B) Ne^{+2} (C) Cl^- (D) As^+

12. In PO_4^{3-} ion, the formal charge on each oxygen atom and P-O bond order respectively are
(A) –0.75, 1.25 (B) –3,1.25
(C) –0.75, 1.0 (D) –0.75, 0.6

13. Schottky defect in crystals is observed when.
(A) unequal number of cations and anions are missing from the lattice.
(B) an ion leaves its normal sites and occupies an interstitial site.
(C) equal number of cations and and anions are missing from the lattice
(D) density of crystal is increased.

14. The second order Bragg diffraction of of X ray with λ = 1.0 Å from a set of parallel planes in a metal occurs at an angle 60°. The distance between the scattering planes in the crystal is
(A) 0.57 Å (B) 1.00 Å (C) 1.15 Å (D) 2.00 Å

15. Aniline is reacted with bromine water and the resulting product is treated with an aqueous solution of sodium nitrite in presence of dilute HCl. The compound so formed is converted into a tetrafluoroborate which is subsequently heated dry. The final product is
(A) p-bromoaniline
(B) 2,4,6-tribromofluorobenzene
(C) p-bromofluorobenzene
(D) 1, 3, 5-tribromobenzene

16. The total number of possible isomers for the complex compound $[Cu^{II}(NH_3)_4]$ $[Pt^{II}Cl_4]$ is
(A) 4 (B) 3 (C) 6 (D) 5

17. A coordination complex compound of cobalt has the molecular formula containing five ammonia molecules, one nitro-group and two chlorine atoms for one cobalt atom. One mole of this compound produce three ions in an aqueous solution. On reacting this solution with excess of $AgNO_3$ solution, we get two moles of AgCl precipitate. The ionic formula for this complex would be.
(A) $[Co(NH_3)_5]$ $[(NO_2)_2Cl_2]$
(B) $[Co(NH_3)_4(NO_2)$ Cl] $[(NH_3)$ Cl]
(C) $[Co(NH_3)Cl]$ $[Cl(NO_2)]$
(D) $[Co(NH_3)_5(NO_2)]Cl_2$

18. When a substance A reacts with water it produces a combustible gas B and a solution of substance C in water. When another substance D reacts with the solution of C, it also produces the same gas B on warming but D can produce B on reaction with dilute sulphuric acid at room temp. A imparts a deep golden yellow colour to a smokeless flame of Bunsen burner. A,B, C and D are respectively
(A) K, H_2, KOH, Al (B) Na, H_2, NaOH, Zn
(C) CaC_2, C_2H_2, $Ca(OH)_2$, Fe (D) Ca, H_2, $Ca(OH)_2$,Sn

19. Given the numbers: 161cm., 0.161cm, 0.0161cm. The number of significant figures for the three numbers is
(A) 3, 4 and 5 respectively
(B) 3, 4 and 4 respectively
(C) 3, 3 and 3 respectively
(D) 3, 3 and 4 respectively

20. Haemoglobin contains 33% of iron by weight. The molecular weight of haemoglobin is approximately 67200. The number of iron atoms (at. wt. of Fe = 56) present in one molecule of haemoglobin are
(A) 6 (B) 1 (C) 4 (D) 2

21. The position of both, an electron and a helium atom is known within 1.0 nm. Further the mementum of the electron is known within 5.0 x 10^{-26} kg. ms^{-1}. The minimum uncertainty in the measurement of momentum of helium atom is.
(A) 50 kg ms^{-1} (B) 5.0 x10^{-26} kg ms^{-1}
(C) 80 kg ms^{-1} (D) 80 x 10^{-4}

22. The number of neutrons in a parent nucleus X which give $^{14}_{7}N$ after two successive β emissions would be
(A) 8 (B) 9 (C) 6 (D) 7

23. The number of antibonding electron pairs in O_2^{2-} molecular ion on the basis of molecular orbital theory is (at. no .of O = 8).

(A) 4 (B) 3 (C) 2 (D) 5

24. The edge length of face centred unit cubic cell is 508 pm. if the radius of the cation is 110 pm, the radius of anion is

(A) 110 pm (B) 288 pm (C) 618 pm (D) 398 pm

25. In crystals of which of the following ionic compounds would expect maximum distance between centres of cations and anions ?

(A) LiI (B) LiF (C) CsF (D) CsI

26. A 5% solution of cane sugar (mol.wt.=342) is isotonic with1% solution of substance X. The molecular weight of X is

(A) 171.2 (B) 68.4 (C) 34.2 (D) 136.2

27. The vapour pressure of a solvent decreased by 10 mm of mercury when a non-volatile solute was added to the solvent. The mole fraction of the solute in the solution is 0.2. What should be the mole fraction of solvent if the decrease in the vapour pressure is to be 20 mm of mercury ?

(A) 0.4 (B) 0.2 (C) 0.8 (D) 0.6

28. In the reaction $4NH_3(g) + 5O_2(g) \rightarrow 4NO(g) + 6H_2O(l)$ when 1 mole of ammonia and 1 mole of O_2 are made to react to completion.

(A) 1.0 mole of NO will be produced.
(B) All the oxygen will be consumed.
(C) All the ammonia will be Consumed.
(D) 1.0 mole of H_2O is produced.

29. A one litre flask is full of bromine vapour. The intensity of brown colour of vapour will not decrease appreciably on adding to the flask some.

(A) pieces of marble (B) carbon tetrachloride
(C) animal charcoal powder (D) carbon disulphide

30. Which one of the following species will impart colour to an aqueous solution ?

(A) Cr^{3+} (B) Ti^{4+} (C) Cu^{+} (D) Zn^{2+}

31. Repeated use of which of the following fertiliser would increase the acidity of soil ?

(A) Urea (B) Potassium nitrate
(C) Ammonium sulphate (D) Superphosphate of lime

32. The first ionization potential (eV) of Be and B respectively are.

(A) 8.29, 9.32 (B) 9.32, 8.29
(C) 9.32, 9.32 (D) 8.29, 8.29

33. An organic compound containing C, H and N gave the following analysis what would be its emipirical formula ? C = 40%, H=13.33%, N = 46.67%

(A) C_2H_7N (B) $C_2H_7N_2$ (C) CH_4N (D) CH_3N

34. If K_1 and K_2 are the respective equilibrium constants for the two reactions

Xe_2F_6 (g) $+H_2O \rightleftharpoons XeOF_4(g) + 2HF(g)$
$XeO_4(g) + XeF_6(g) \rightleftharpoons XeOF_4(g) + XeO_3F(g)$

the equilibrium constant for the reaction

$XeO_4(g) + 2HF(g) \rightleftharpoons XeO_3F_2(g) + H_2O(g)$ will be

(A) $K_1.K_2$ (B) $K_1/(K_2)^2$
(C) K_1 / K_2 (D) K_2/K_1

35. Identity the correct statement regarding entropy

(A) At absolute zero entropy of perfectly crystalline substance is taken to be zero.
(B) At 0°C the entropy of a perfectly crystalline substance is taken to be zero
(C) At absolute zero of temp., the entropy of perfectly crystalline substance is +ve
(D) At absolute zero of temperature, the entropy of all crystalline substance is to be zero.

36. Which one of the following compounds will be most easily attacked by an electrophile

(A) $DCH_2CH_2CH_2Cl$ (B) CH_3CHDCH_2Cl
(C) $CH_3CHClCH_2D$ (D) CH_3CH_2CHDCl

37. Which of the following compounds is resistant to nucleophilic attack by hydroxyl ions ?

(A) Methyl acetate (B) Acetonitrile
(C) Acetamide (D) Dimethyl ether

38. Reaction of $CH_2 - CH_2$ (with O bridging, epoxide) with RMgX leads to formation of

(A) RCHOHR (B) RCHOHCH3
(C) R_2CHCH_2OH (R and R on CH) (D) RCH_2CH_2OH

39. Iodoform test is not given by

(A) 2-Pentanone (B) 3-Pentanone
(C) Ethanal (D) Ethanol

40. Which one of the following esters cannot give Claisen self-condensation ?

(A) $CH_3-CH_2-CH_2-CH_2COOC_2H_5$
(B) $C_6H_5CH_2COOC_2H_5$
(C) $C_6H_5COOC_2H_5$
(D) $C_6H_{11}CH_2COOC_2H_5$

41. Which one of the following compounds will be most easily attacked by an electrophile

(A) benzene ring with –Cl (B) benzene ring
(C) benzene ring with –OH (D) benzene ring with –CH_3

42. Aspirin is an acetylation product of

(A) p-dihydroxybenzene.
(B) o-hydroxy benzoic acid.
(C) o-dihydroxybenzene.
(D) m-hydroxy benzoic acid.

43. In DNA the complementary bases are

(A) uracil and adenine: cytosine and guanine

(B) adenine and thymine: guanine and cytosine
(C) adenine and thymine: guanine and uracil
(D) adenine and guanine: thymine and cytosine

44. The number of molecules of ATP produced in the lipid metabolism of a molecule of palmitic acid is
(A) 56 (B) 36 (C) 86 (D) 130

45. Which of the following is not compatible with arenes ?
(A) Greater stability
(B) Resonance
(C) Delocalisation of π–electrons
(D) Electrophilic addition

46. IUPAC name for the compound is

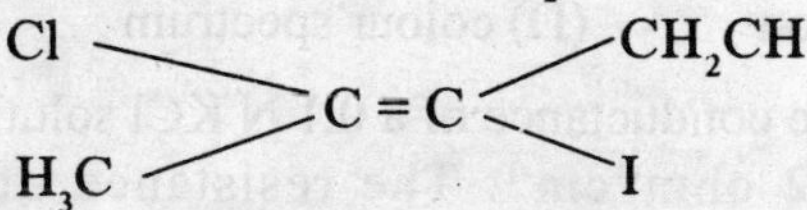

(A) cis-2-chloro-3-iodo-2-pentene
(B) trans-3-iodo-4-chloro-3-pentene
(C) trans-2-chloro-3-iodo-2-pentene
(D) cis-3-iodo,4-chloro,3-pentene

47. Which one of the following orders is correct regarding the -I effect of the substituents ?
(A) $-NR_2 < -OR < -F$ (B) $-NR_2 > -OR < -F$
(C) $-NR_2 < -OR > -F$ (D) $-NR_2 > -OR > -F$

48. 2–Bromopentane is heated with potassium ethoxide in ethanol. The major product obtained is
(A) 2-ethoxypentane (B) Pentene-1
(C) Cis-pentene-2 (D) Trans-pentene-2

49. Glucose molecule reacts with X number of molecules of phenylhydrazine to yield osazone. The value of 'X' is
(A) three (B) two (C) one (D) four

50. An ester (A) with molecular formula $C_9H_{10}O_2$ was treatred with excess of CH_3 MgBr and the complex so formed was treated with H_2SO_4 to give an olefin (B). Ozonolysis of (B) gave a ketone with molecular formula C_8H_8O which shows +ve iodoform test. The structure of (A) is
(A) C_6H_5 $COOC_2H_5$
(B) $H_3COCH_2COC_6H_5$
(C) P–H_3CO–C_6H_4–$COCH_3$
(D) $C_6H_5COOC_6H_5$

CBSE -1999

1. When acetylene is passed through dil. H_2SO_4 in the presence of $HgSO_4$, the compound formed is
(A) acetic acid (B) acetaldehyde
(C) ether (D) ketone

2. The type of hybridisation of boron in diborane is
(A) sp-hybridisation (B) sp^2-hybridisation
(C) sp^3-hybridisation (D) sp^3d^2-hybridisation

3. Bell metal is an alloy of
(A) Cu + Pb (B) Cu + Sn
(C) Cu + Zn (D) Cu + Ni

4. Which of the following phosphorus is the most reactive ?
(A) Red phosphorus (B) White phosphorus
(C) Scarlet phosphorus (D) Violet phosphorus

5. The vapour pressure of benzene at a certain temperature is 640 mm of Hg. A non-volatile and non-electrolyte solid, weighing 2.175g is added to 39.08 of benzene. If the vapour pressure of the solution is 600 nm of Hg, what is the molecular weight of solid substance ?
(A) 69.60 (B) 79.82 (C) 49.50 (D) 59.60

6. Which of the following shows electrical conduction?
(A) Potassium (B) Graphite
(C) Diamond (D) Sodium

7. Ethyl chloride is converted into diethyl ether by
(A) Wurtz synthesis (B) Grignard reaction
(C) Perkin's reaction (D) Williamson's synthesis

8. In the reaction

$CH_3CN + 2H \xrightarrow[\text{Ether}]{\text{HCl}} X \xrightarrow{\text{Boiling } H_2O} Y$; the term Y is

(A) Acetaldehyde (B) Dimethylamine
(C) Acetone (D) Ethylamine

9. Which of the following configuration is correct for iron?
(A) $1s^2\ 2s^2\ 2p^6\ 3s^2\ 3p^6\ 3d^5$
(B) $1s^2\ 2s^2\ 2p^6\ 3s^2\ 3p^6\ 4s^2\ 3d^5$
(C) $1s^2\ 2s^2\ 2p^6\ 3s^2\ 3p^6\ 4s^2\ 3d^7$
(D) $1s^2\ 2s^2\ 2p^6\ 3s^2\ 3p^6\ 4s^2\ 3d^6$

10. Which of the following has more unpaired d-electrons?
(A) Zn^+ (B) Fe^{2+} (C) N^{3+} (D) Cu^+

11. At 25°C and 730 mm pressure, 380 ml of dry oxygen was collected. If the temperature is constant, what volume will the oxygen occupy at 760 mm pressure?
(A) 569 ml (B) 621 ml (C) 365 ml (D) 449 ml

12. How many g of dibasic acid (mol. wt. 200) should be present in 100 ml of the aqueous solution to give 0.1 normallity ?
(A) 1g (B) 2g (C) 10g (D) 20 g

13. If 0.15 g of a solute dissolved in 15g of solvent boiled at a temperature higher by 0.216°C than that of the pure solvent, The mol. wt. of the solute is (Molal elevation constant for the solvent is 2.16°C)
(A) 10.1 (B) 100 (C) 1.01 (D) 10

14. Oxidation state of Fe in Fe_3O_4 is
(A) 3/2 (B) 4/5 (C) 5/4 (D) 8/3

15. Who modified Bohr's theory by introducing elliptical orbits for electron path?
(A) Hund (B) Thomson
(C) Rutherford (D) Sommerfield

16. The uncertainty in momentum of an electron is 1×10^{-5} kg-m/s. The uncertainty in its position will be ($h = 6.62 \times 10^{-34}$ kg-m^2/s)

(A) 1.05×10^{-26} m (B) 1.05×10^{-25} m
(C) 5.27×10^{-30} m (D) 5.25×10^{-30} m

17.For a first-order reaction, the half life period is independent of
(A) first power of final concentration
(B) square root of final concentration
(C) initial concentration
(D) cube root of initial concentration

18. The concentration of $[H^+]$ and concentration of $[OH^-]$ of a 0.1 aqueous solution of 2% ionised weak acid is [ionic product of water = 1×10^{-14}]
(A) 0.02×10^{-3} M and 5×10^{-11} M
(B) 1×10^{-3} M and 3×10^{-11} M
(C) 2×10^{-3} M and 5×10^{-12} M
(D) 3×10^{-2} M and 4×10^{-13} M

19. In which of the following compounds transition metal has zero oxidation state?
(A) $[Fe(CO)_5]$ (B) $NH_2.NH_2$
(C) NOCl (D) CrO_5

20. An organic compound containing C, H and N gave the following analysis : C = 40%, H = 13.33% and N= 46.67% its empirical formula would be
(A) CH_4N (B) C_2H_7N (C) CHN (D) C_2H_2N

21. The solubility of a saturated solution of calcium fluoride is 2×10^{-4} moles per litre. Its solubility product is
(A) 12×10^{-2} (B) 14×10^{-4}
(C) 22×10^{-11} (D) 32×10^{-12}

22. Which of the following is used in the preparation of chlorine?
(A) Both MnO_2 and $KMnO_4$
(B) Only MnO_2 (C) Only $KMnO_4$
(D) MnO_2 or $KMnO_4$

23. The strongest conjugate base is
(A) NO_3^- (B) Cl^- (C) SO_4^{2-} (D) CH_3COO^-

24.Which of the following oxides is the most acidic?
(A) As_2O_3 (B) Sb_2O_5 (C) N_2O_5 (D) P_2O_3

25. The de Broglie wavelength of a particle with mass 1g and velocity 100 m/s is
(A) 6.63×10^{-33} m (B) 6.63×10^{-34} m
(C) 6.63×10^{-35} m (D) 6.65×10^{-35} m

26.Which one of the following statements is wrong for gases?
(A) Gases do not have a definite shape and volume
(B) Mass of a gas cannot be determined by weighing a container in which it is enclosed
(C) Volume of the gas is equal to volume of container confining the gas
(D) Confined gas exerts uniform pressure on the walls of its container in all directions

27. The structure of H_2O_2 is
(A) Planar (B) Non-planar
(C) Spherical (D) Linear

28. In an endothermic reaction, the value of ΔH is
(A) Zero (B) Positive
(C) Negative (D) Constant

29.Which of the following elements has maximum electron affinity?
(A) I (B) F (C) Cl (D) Br

30.The most suitable method of the separation of a mixture of ortho and para-nitrophenol mixed in the ratio of 1 : 1 is
(A) distillation (B) crystallisation
(C) vapourisation (D) colour spectrum

31. The specific conductance of a 0.1 N KCl solution at 23°C is 0.012 $ohm^{-1}cm^{-1}$. The resistance of cell containing the solution at the same temperature was found to be 55 ohm. The cell constant will be
(A) 0.142 cm^{-1} (B) 0.616 cm^{-1}
(C) 0.918 cm^{-1} (D) 1.12 cm^{-1}

32. In the reaction : $S + 3/2\, O_2 \rightarrow SO_3 + 2x$ kcal and $SO_2 + 1/2\, O_2 \rightarrow SO_3 + y$ kcal, the heat of formation of SO_2 is
(A) (x + y) (B) (x - y) (C) (2x + y) (D) (2x - y)

33. Purification of aluminium, by electrolyte refining, is known is
(A) Hall's process (B) Bayer's process
(C) Hoope's process (D) Serpeck's process

34. Sulphur-35 (34.96903 a.m.u.) emits a β-particle but no γ-ray. the product is chlorine-35 (34.96885 a.m.u.).the maximum energy emitted by the β-particle is
(A) 16.758 MeV (B) 1.6758 MeV
(C) 0.16758 MeV (D) 0.016758 MeV

35. Which of the following compounds will be most easily attacked by an electrophile ?

(A) 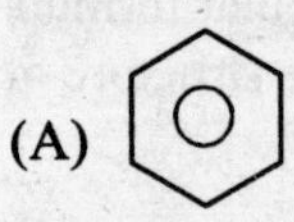
(B) 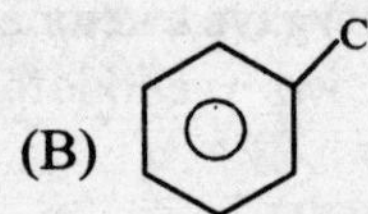

(C) 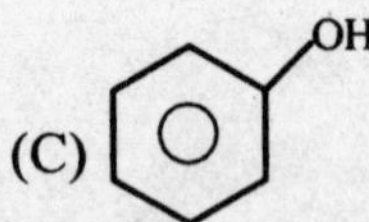

(D) 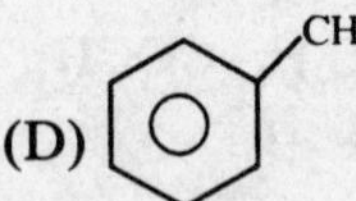

36. Natural rubber is a polymer of
(A) butadiene (B) ethyne (C) styrene (D) isoprene

37. Terylene is a condensation polymer of ethylene glycol and
(A) salicylic acid (B) terphthalic acid
(C) benzoic acid (D) phthallic acid

38. In the following radioactive decay, ${}_{92}X^{232} \rightarrow {}_{89}Y^{220}$, how many α and β particles are ejected from X to form Y ?
(A) 3α and 3β (B) 5α and 3β
(C) 3α and 5β (D) 5α and 5β

39. Cassiterite is an ore of
(A) Sb (B) Sn (C) Mn (D) Ni

40. In Friedel-Craft's alkylation, besides $AlCl_3$ the other reactants are
(A) C_6H_6+NH_3 (B) C_6H_6+CH_4
(C) C_6H_6+CH_3Cl (D) C_6H_6+CH_3COCl

41. Phenyl isocyanides are prepared from which of the following reaction?
(A) Rosenmund's reaction (B) Carbylamine reaction
(C) Reimer-Tieman reaction (D) Wurtz reaction

42. The decomposition of organic compounds in the presence of oxygen and without the development of odoriferous substances is called.
(A) decay (B) N_2-fixation
(C) nitrification (D) denitrification

43. Aldol condensation will not take place in
(A) HCHO (B) CH_3CHO
(C) CH_3COCH_3 (D) CH_3CH_2CHO

44. Which of the following is the sweetest sugar?
(A) Fructose (B) Maltose (C) Sucrose (D) Glucose

45. The number of atoms in 4.25g of NH_3 is approximately
(A) 1 x 10^{23} (B) 2 X 10^{23}
(C) 4 x 10^{23} (D) 6 x 10^{23}

46. Percentage of lead in lead pencil is
(A) 20 (B) 80 (C) zero (D) 70

47. Gammexane is
(A) chlorobenzene (B) benzyl chloride
(C) bromobenzene (D) benzene hexachloride

48. The half-life of ${}_6C^{14}$ if its λ is 2.31 x 10^{-4}, is
(A) 2 x 10^2 years (B) 3 x 10^2 years
(C) 3.5 x 10^4 years (D) 4 x 10^2 years

49. Which one of the following compounds will react with $NaHCO_3$ solution to give sodium salt and carbon dioxide?
(A) Acetic acid (B) n-hexanol
(C) Phenol (D) Both (B) and (C)

50. Amides may be converted into amines by a reaction named after
(A) Hoffmann (B) Kekule
(C) Perkin (D) Claisen

CBSE - 2000

1. Calculate the wavelength of light
Given Energy = 3.03 x 10^{-19} J, h = 6.36 x 10^{-34} J.S., c = 3.00 10^8 m/s
(A) 0.656 nm (B) 65.6 nm (C) 6.56 nm (D) 656 nm

2. Which of the following is correct about enzymes
(A) It changes the equilibrium constant.
(B) It increases the activation energy.
(C) It reduces the activation energy.
(D) None of these

3. Which of the following will not give coloured ion
(A) V^{+3} (B) Cr^{+3} (C) Sc^{+3} (D) Ti^{+3}

4. During Friedel-Craft reaction which of the following reacts in presence of $AlCl_3$?
(A) Toluene ($C_6H_5CH_3$) + HCl (B) p-cresol (OH, CH_3 on benzene ring) + CH_2Cl_2
(C) Benzene + $CHCl_3$ (D) Benzene + CH_3Cl

5. Arrange X-rays, γ-rays, UV rays in decreasing order of energy value for photocells.
(A) X < UV < γ (B) γ < X < UV
(C) UV < X < γ (D) γ > x > UV

6. At infinite dilution for an aq. solution of $BaCl_2$, molar conductivity for Ba^{2+} and Cl^- ions are Λ_m^∞ = 127.32 s cm²/ mol, Λ_m^∞ = 76.34 s cm²/mol respectively. What is Λ_m^∞ for $BaCl_2$ at the same dilution ?
(A) 280 s cm² mol⁻¹ (B) 330.98 s cm² mol⁻¹
(C) 90.98 s cm² mol⁻¹ (D) 203.6 s cm² mol⁻¹

7. What is E_0 for the cell involving disproportionation of unstable Cu^+ ion into its constituent oxidized-reduced states, if E_0 Cu^{++}/Cu^+ = 0.15V, E_0 Cu^+/Cu = 0.340V ?
(A) –0.19 V (B) –0.38V (C) +0.38V (D) 0.49 V

8. A container contains CO and another H_2 in equal molar concentrations at the same temperature. What is the ratio of avg. molar K.E.'s of the two gases ?
(A) Depends on volume (B) 1 : 1
(C) KE_{CO} > KE_{n_2} (D) KE_{H_2} > KE_{CO}

9. Set containing iso-electric species is
(A) C_2^{2-}, NO^+, CN^-, O_2^{+2} (B) CO, NO, O_2, CN
(C) CO_2, NO_2, O_2, N_2O_5 (D) CO, CO_2, NO, NO_2

10. For a reversible reaction if the conc. of reactants is doubled, how will the rate constant and to what extent is affected at NTP. Reaction depends only on concentration of one of the reactants.
(A) Remains same (B) Nothing can be inferred
(C) Doubles (D) Halves

11. The correct relation between decay constant and half-life of a radioactive decay reaction.
(A) $t_{1/2} = \frac{\log_e 2}{\lambda}$ (B) $t_{1/2} = \frac{2.303 \log_e 2}{\lambda}$
(C) $t_{1/2} = \lambda \log_e 2$ (D) $t_{1/2} = 2.303 \log_{10} 2\lambda$

12. Starting with 300g of a radioactive sample what amount would remain radioactive after 1 hr, if half-life of initial sample is 15 min ?
(A) No activity remains (B) No decay occurs
(C) 150 gm (D) 18.75 gm

13. Mg is an important component of which bimolecule occurring extensively in living world ?
(A) Haemoglibin (B) Chlorophyll
(C) Florigen (D) ATP

14. ${}^{14}_{6}C \rightarrow {}^{14}_{5}C + {}^{0}_{+1}\beta + X$. Identify X.
(A) Electron (B) Neutrino (C) Proton (D) Neutron

15. α–D(+) glucose and β–D(+ve) glucose are.
(A) Enantiomers (B) Geometrical isomers
(C) Epimeres (D) Anomers

16. For every diprotic acid of the type H_2X. How would you relate ionization constants ka_1 and Ka_2 ?
(A) $Ka_1 = ka_2$ (B) $Ka_1 < ka_2$
(C) $ka_2 < ka_1$ (D) data insufficient

17. $ZnCO_3(s) \longrightarrow ZnO(s) + CO_2(g)$ Expression for the partial pressure constant for the above reaction would be.
(A) $K_p = \dfrac{[ZnO][CO_2]}{[ZnCO_3]}$ (B) $K_p = \dfrac{P(ZnO)P(CO_2)}{P(ZnCO_3)}$
(C) $K_p = P(ZnO)^2\, P(CO_2)$ (D) $K_p = P(CO_2)$

18. Which of the following relations holds good to the laws of thermodynamics for the reaction.
$C_2H_4(g) + 3O_2(g) \longrightarrow 2CO_2(g) + 2H_2O(l)$
(A) $\Delta H = \Delta E + RT$ (B) $\Delta H = \Delta E - RT$
(C) $\Delta H = \Delta E + 2RT$ (D) $\Delta H = \Delta E - 2RT$

19. Compounds consisting of equal number of cations and anions in a 3-d lattice separated by equal distances in space are.
(A) Molecular compounds. (B) Ionic compounds
(C) Covalent (D) Amorphous

20. Colloidal solutions can be destroyed by.
(A) Dialysis
(B) Separation by animal parchment.
(C) Electrophoresis
(D) Addition of electrolytes

21. For $BaCO_3(s) \rightarrow BaO(s) + CO_2(g)$, if 98.7g of $BaCO_3$ is decomposed at NTP, the reaction on completion gives ....amount of CO_2 (At. wt. of Ba = 137.4)
(A) 11.2*l* (B) 44.8*l* (C) 22.4*l* (D) 0.56*l*

22. For a reaction to proceed spontaneously.
(A) ΔH is +ve (B) ΔE is +ve
(C) ΔG is –ve (D) G is +ve

23. For the melting of a solid at 25°C, the fusion process requires energy equivalent to 2906 Joules to be added to the system. Considering the process to be reversible at fusion point, what is the entropy value of the process ?
(A) 9.76 J/K (B) 11.272 J/K
(C) 2.33 J/K (D) Insufficient data

24. Which of the following at ordinary conditions of temperature and pressure exists as linear polymer molecule due to extensive intermolecular force ?
(A) HF (B) NH_3 (C) H_2O (D) HCl

25. Which is planar molecule ?
(A) XeO_4 (B) XeF_4 (C) $XeOF_4$ (D) XeO_2F_2

26. What is the shape of $Fe(CO)_5$ molecule ?
(A) Tetrahedral (B) Octahedral
(C) Trigonal bi-pyramidal (D) Square pyramidal

27. Half-life of active hydrogen is.
(A) Less than a second (B) 1 sec
(C) 1 min (D) 1 hour

28. Which of the following is incorrect with respect to property indicated ?
(A) EN → $F_2 > Cl_2 > Br_2$ (B) EA → $Cl_2 > F_2 > Br_2$
(C) Oxidising power → $F_2 > Cl_2 > Br_2$
(D) Bond energy → $F_2 > Cl_2 > Br_2$

29. Find out ΔG for the following reaction.
$2ZnS + 3O_2 \rightarrow 2ZnO + 2SO_2$, ΔG = ?
Data given :
$2Zn + O_2 \rightarrow 2ZnO$, ΔG = – 616 J
$2Zn + S_2 \rightarrow 2ZnS$, ΔG = – 293 J
$S_2 + 2O_2 \rightarrow 2SO_2$, ΔG = – 408 J
(A) –731 J (B) –1317 J (C) –501 J (D) +731 J

30. Bond angle (dihydral H–C–H) in staggered conformation of C_2H_6 is.
(A) 0° (B) 90° (C) 120° (D) 60°

31. Which of the following exhibits highest oxidation state?
(A) Sc (B) Fe (C) Zn (D) Mn

32. Which of the following will give 4 isomers ?
(A) $[Co(en)_2(H_2C)_2]Cl_3$ (B) $[Co(en)_3]Cl_2$
(C) $[Co(en)_3]Cl_3$ (D) $[Co(en)_2(NO_2)_2]Cl$

33. Which of the following is electron deficient ?
(A) NH_3 (B) BCl_3 (C) PCl_3 (D) None of these

34. Polarization of acrolein may be written as.
(A) $\overset{-\delta}{CH_2} = CH - \overset{\delta+}{CH} = O$ (B) $\overset{-\delta}{CH_2} = CH - CH = \overset{\delta+}{O}$
(C) $\overset{\delta-}{CH_2} = \overset{\delta+}{CH} - CH = O$ (D) $\overset{\delta-}{CH_2} = CH - CH = \overset{-\delta}{O}$

35. A compound formed by elements A & B. crystallizes in the cubic structure where the A atoms are at the corners of the cube and B atoms are at the body centres. The simplest formula of the compound is.
(A) AB (B) A_6B (C) A_8B_4 (D) AB_6

36. What will happen if $LiAlH_4$ is added to an ester ?
(A) Two units of alcohol are obtained.
(B) One unit of alcohol and one unit of acid is obtained.
(C) Two units of acid are obtained.
(D) None of these

37. $CF_2 - CF_2$ is a unit of
(A) Teflon (B) Polyethene (C) Bakelite (D) Buna-S

38. Ethyl alcohol is heated with conc. H_2SO_4. The product formed is.
(A) $H_3C-\underset{\underset{O}{\|}}{C}-OC_2H_5$ (B) C_2H_6 (C) C_2H_4 (D) C_2H_2

39. In the following compounds

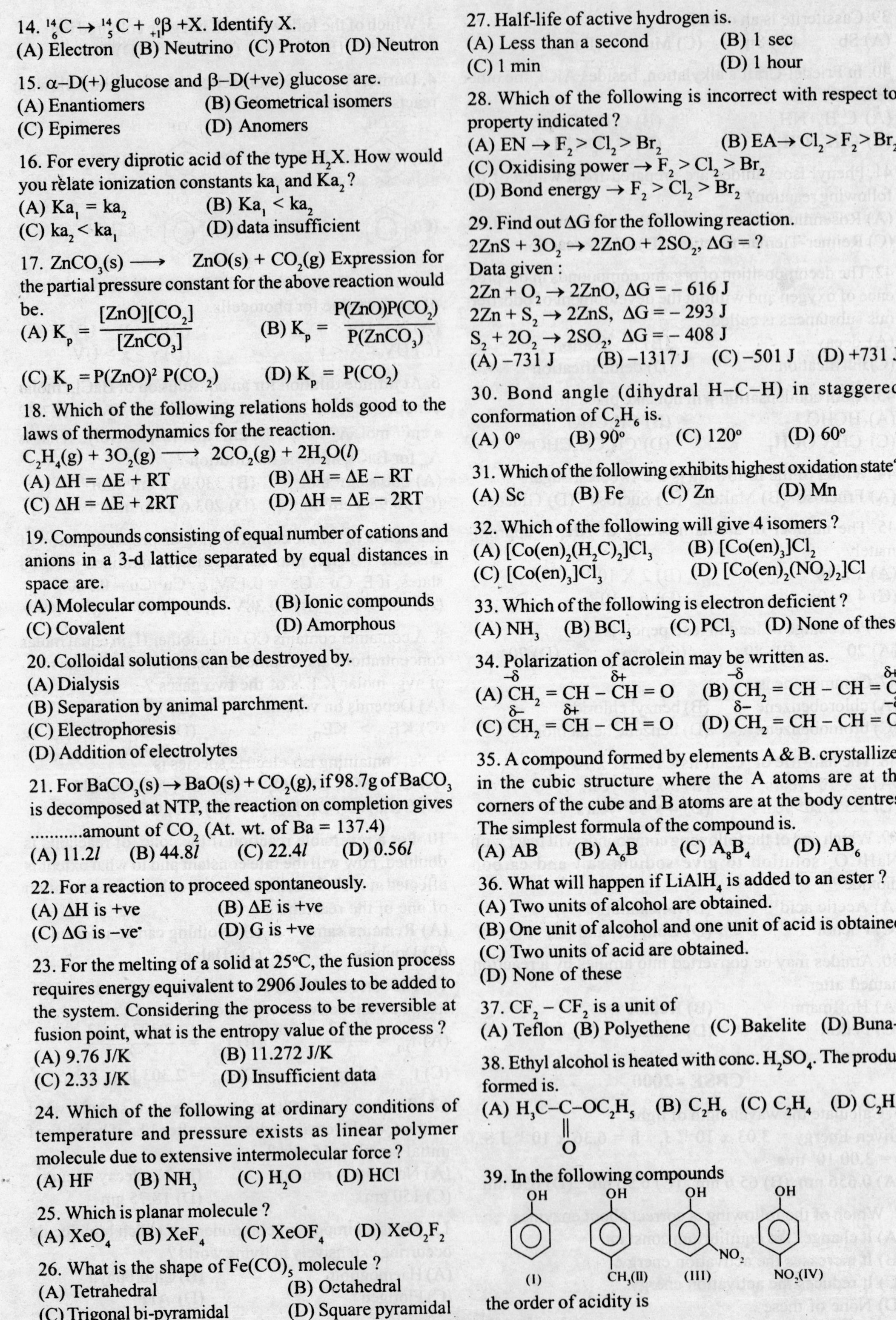

the order of acidity is

(A) III > IV > I > II (B) I > IV > III > II
(C) II > I > III > IV (D) IV > III > I > II

40. R & S forms of a pair of enantiomers differ from each other in.
(A) Solubility in racemic mixture.
(B) Optical rotation of polarized light.
(C) Reaction with racemic mixture (D) None of these

41. Conjugate acid of NH_2^-
(A) NH_3 (B) NH_2^+ (C) NH_3^- (D) NH_4^+

42. The frequency emitted for which transition has higher energy in terms of light radiation ?
(A) $6 \rightarrow 1$ (B) $2 \rightarrow 1$ (C) $1 \rightarrow 2$ (D) $3 \rightarrow 4$

43. Which of the following solutions have lowest freezing point ?
(A) 0.1 M $NaNO_3$ (B) 0.01 M glucose
(C) 0.01 M BCl_3 (D) 0.2 M Na_2SO_4

44. A, B, C are three elements forming a part of compound in oxidation states of +2, +5 and -2 respectively. What could be the compound ?
(A) $A_2(BC)_2$ (B) $A_2(BC_4)_3$
(C) $A_3(BC_4)_2$ (D) ABC

45. Phenols are more acidic than aliphatic alcohols because.
(A) Phenoxide ion does not show resonance
(B) Alcohols do not loose H-atom at all.
(C) Phenoxide ion is stabilized by resonance.
(D) Phenols are more soluble in polar solvents

46. Which is the order of stability for I 1-butene, II cis 2-butene, III trans 2-butene
(A) I < II < III (B) I > II > III
(C) III > I > II (D) I > II < III

47.

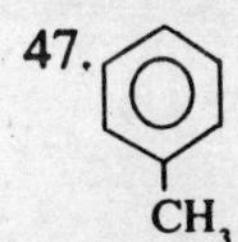

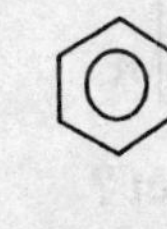

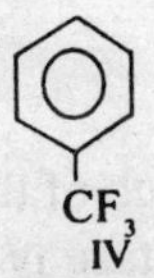

Arrange in order of electrophilic substitution reactions.
(A) I > II > III > IV (B) IV > I > II > III
(C) III > IV > II > I (D) II > IV > III > I

48. 1-hydroxy propane can be obtained from Propene treating with.
(A) B_2H_6 + NaOH (B) $B_2H_6 + H_2O_2 / OH^-$
(C) $LiAlH_4 + H_2O / H^+$ (D) $HCl + H_2O / H^+$

49. A compound 'A' on reduction gives 'B' which on further reaction with $CHCl_3$ and alc. KOH gives compound C which on hydrolysis gives aniline. Identify A.
(A) Nitrobenzene (B) Methylamine
(C) Nitromethane (D) Nitrosobenzene

50. Which reagent bring about the conversion of carboxylic acids into esters.
(A) $LiAlH_4$ (B) $Al(OC_2H_5)_3$
(C) C_2H_5OH (D) dry HCl + C_2H_5OH

CBSE -2001

1. In steam distillation of toluene, the pressure of toluene vapour is
(A) Equal to vapour pressure to that of toluene in simple distillation
(B) Equal to pressure of barometer
(C) Less than pressure of barometer
(D) More than vapour pressure of toluene in simple distillation.

2. A compound of molecular formula C_7H_{16} shows optical isomerism the compound will be
(A) 2,3- Dimethyl pentane (B) 2,2- Dimethyl butane
(C) 2-Methyl hexane (D) none of these

3. Change in enthalpy for reaction
$2H_2O_2(l) \rightarrow 2H_2O(l) + O_2(g)$, if heats of formation of $H_2O_2(l)$ and $H_2O(l)$ are -188 and -288 kJ/mol respectively is
(A) -196 kJ/mol (B) +196 kJ/mol
(C) +948 kJ/mol (D) -948 kJ/mol

4. When I mole of a gas is heated at constant volume, the temperature is raised from 298 to 308K. Heat supplied to the gas is 500J. Then which statement is correct
(A) $q = W = 500$ J, $\Delta U = 0$
(B) $\Delta U = 0$. $q = W = -500$ J
(C) $q = W = 500$ J, $\Delta U = 0$
(D) $q = \Delta U = 500$ J, $W = 0$

5. Enthalpy of $CH_4 + \frac{1}{2} O_2 \rightarrow CH_3OH$ is negative. If enthalpies of combustion of CH_4 and CH_3OH are x and y respectively, then which relation is correct ?
(A) $x = y$ (B) $x \geq y$ (C) $x > y$ (D) $x < y$

6. For reaction $2N_2O_5 \rightarrow 4NO_2 + O_2$, rate and rate constant are 1.02×10^{-4} and 3.4×10^{-5} sec^{-1} respectively . Then conc. of N_2O_5 at that time will be -
(A) 1.732 (B) 3 (C) 1.02×10^{-4} (D) 3.4×10^5

7. A human body required 0.01 activity of radioactive substance after 24 hrs. Half-life of radioactive substance is 6 hrs. Then the max activity of radioactive substance that can be injected is -
(A) 0.16 (B) 0.32 (C) 0.08 (D) 0.04

8. When a bio-chemical reaction is carried out in laboratory (outside the human body) in absence of enzyme, the rate of reaction obtained is 10^{-6} times. The activation energy of reaction in presence of enzyme is
(A) $\frac{6}{RT}$ (B) P is required
(C) Different from E_a obtained in laboratory (D) None

9. Molarity of liquid HCl, if density of solution is 1.17 gm/c.c is -
(A) 36.5 (B) 18.25 (C) 32.05 (D) 42.10

10. Percentage of Se in peroxidase anhydrous enzyme is 0.5% by weight (atomic weight of Se=78.4), then minimum molecular weight of peroxidase anhydrous enzyme is -
(A) 1.568×10^4 (B) 1.568×10^3

(C) 15.68 (D) 2.136×10^4

11. Specific volume of a cylindrical virus particle is 6.02×10^{-2} c.c/gm whose radius and length are 7Å and 10Å respectively. If $N_A = 6.02 \times 10^{23}$, find the molecular weight of virus.
(A) 15.4 kg/mol (B) 1.54×10^4 kg/mol
(C) 3.08×10^4 kg/mol (D) 3.08×10^3 kg/mol

12. Pure water can be obtained from sea water by
(A) Centrifugation (B) Plasmolysis
(C) Reverse osmosis (D) Sedimentation

13. Std. electrode poetntials are.
Fe^{+2}/Fe, $E^0 = -0.44$ and Fe^{+3}/Fe^{+2} $E^0 = 0.77$
Now an iron piece is dipped in aqueous solution of a mixture of Fe^{3+} and Fe^{2+} ions then -
(A) Fe^{+2}/Fe^{3+} remains unchanged
(B) Fe^{3+} increases (C) Fe^{3+} decreases
(D) Fe^{+2} decreases

14. Which is not correct regarding adsorption of a gas on surface of solid ?
(A) On increasing temperature adsorptiion increases continuously
(B) Enthalpy and entropy changes are -ve
(C) Adsorption is more for some specific substance
(D) Reversible

15. $PbO_2 \longrightarrow PbO$ $\Delta G_{298} < 0$
$SnO_2 \longrightarrow SnO$ $\Delta G_{298} > 0$
Most probable oxidation states of Pb and Sn are-
(A) Pb^{+4}, Sn^{+4} (B) Pb^{+4}, Sn^{+2}
(C) Pb^{+2}, Sn^{+2} (D) Pb^{+2}, Sn^{+4}

16. The correct order of acidity of the following is

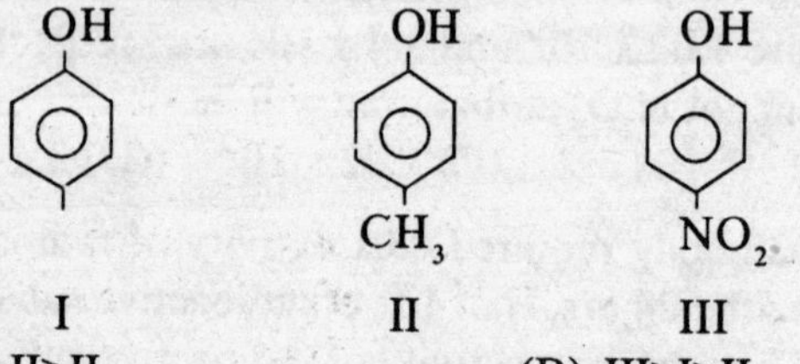

(A) I>II>II (B) III>I>II
(C) II>III>I (D) I>III>II

17. $CH_3 - CH_2 - CH(Cl) - CH_3$ obtained by chlorination of n-butane will be -
(A) Meso form (B) Racemic mixture
(C) d-form (D) l-form

18. Which alkene on ozonolysis gives CH_3CH_2CHO and CH_3COCH_3 ?
(A) $CH_3CH_2CH = CHCH_3$
(B) $CH_3 - C(CH_3) = CHCH_3$
(C) $CH_3CH_2CH = C(CH_3)_2$
(D) $CH_3CH_2CH = CHCH_2CH_3$

19. Intermediates formed during reaction of $RCONH_2$ with Br_2 and KOH are -
(A) RCONHBr and RCNO
(B) RNHCOBr and RNCO
(C) RNHBr and RCONHBr (D) $RCONBr_2$

20. An organic compound $A(C_4H_9Cl)$ on reacting with Na/diethyl ether gives a hydrocarbon which on monochlorination gives only one chloro derivative. The compound A is-
(A) t- Butyl chloride (B) Sec-butyl chloride
(C) Iso- butyl chloride (D) n-Butyl chloride

21. Which of the following is incorrect ?
(A) Fehling's solution is used in detection of glucose
(B) Tollen's reagent is used to detect unsaturation
(C) $FeCl_3$ is used in detection of phenolic group
(D) $NaHSO_3$ is used in detection of carbonyl group

22. Which of the following gives positive Fehling's solution test ?
(A) Surcose (B) Glucose (C) Fats (D) Proteins

23. Which of the following is not correctly matched ?
(A) Neoprene $\left[-CH-\underset{Cl}{C} = CH-CH_2- \right]_n$
(B) Nylon 6,6 $\left[-NH-(CH_2)_6-NH-CO-(CH_2)_4-\overset{O}{\overset{\|}{C}}- \right]_n$
(C) Terylene $\left[-OCH_2-CH_2-\overset{O}{\overset{\|}{C}}-C_6H_4-\overset{O}{\overset{\|}{C}}- \right]_n$
(D) PMMA $\left[-CH_2-\overset{CH_3}{\underset{COOH_3}{C}}- \right]_n$

24. Which of the following is correct ?
(A) diastase is an enzyme
(B) Acetophenone is an ether
(C) Cycloheptane is an aromatic compound
(D) All the above

25. The incorrect IUPAC name is –
(A) $CH_3COCH(CH_3)_2$ 2 menthyl-3butanone
(B) $(CH_3)_2CH-CH(CH_2CH_3)-CH_3$ 2, 3-dimethyl pentane
(C) $CH_3 - C \equiv CCH(CH_3)_2$ 4-methyl-2-pentyne
(D) CH_3-CHCl- CHBr - CH_3 2-bromo-3chloro butane

26. In the preparation of an alkene from an alcohol using Al_2O_3, which is effective factor ?
(A) Porosity of Al_2O_3 (B) Temperature
(C) Concentration (D) Surface area of Al_2O_3

27. Which of the following is correct ?
(A) Reduction of any aldehyde give secondary alcohol
(B) Reaction of vegetable oil with H_2SO_4 give glycerine

(C) Alcoholic idonie with NaOH give iodoform

(D) Surcose on reaction with NaCl give invert sugar

28. Which of the following is correct about H-bonding in nucleotide ?

(A) A - T , G - C (B) A - G, T - C

(C) G - T, A - C (D) A - A, T - T

29. Which is correct statement ?

(A) Proteins are composed of only one type of amino acid

(B) Starch is polymer of α-glucose

(C) Amylose is a component of cellulose

(D) In cyclic structure of fructose there are 4C and one oxygen atom

30. Which statement is incorrect about peptide $\left[\begin{array}{c} O \\ \| \\ -C-NH \end{array}\right]$ bond ?

(A) C - N bond length in proteins is longer than usual bond length of C - N bond

(B) Spectroscopic analysis shows planar structure of - CO - NH - group

(C) C - N bond length in proteins is smaller than usual bond lenght of C - N bond

(D) None of these

31. Which of the following will exhibit maximum molar conducitivity ?

(A) $K_4[Fe(CN)_6]$ (B) $[Co(NH_3)_6]Cl_3$

(C) $[Cu(NH_3)_4]Cl_2$ (D) $[Ni(CO)_4]$

32. The number of orbitals represented by following quantum numbers is

n = 3, I = 2, m + 2 is

(A) 1 (B) 2 (C) 3 (D)4

33. In HS^-, I^-, $R-NH_2$, NH_3 order of proton accepting tendency will be -

(A) $I^- > NH_3 > RNH_2 > HS^-$

(B) $NH_3 > R-NH_2 > HS^- > I^-$

(C) $RNH_2 > NH_3 > HS^- > I^-$

(D) $HS^- > RNH_2 > NH_3 > I^-$

34. Beans are cooked earlier in pressure cooker because -

(A) Boiling point increases with increasing pressure

(B) Boiling point decrease with increasing pressure

(C) Extra pressure of pressure cooker, softens the beans

(D) Internal energy is not lost while cooking in pressure cooker.

35. Most convenient method to protect bottom of ship made of iron is

(A) Connecting it with Mg block

(B) Connecting it with Pb block

(C) Coating it with red led oxide

(D) white tin plating

36. When molten zinc is converted into solid state, it acquires HCP structure. The number of nearest neighbours will be -

(A) 6 (B) 8 (C) 12 (D) 4

37. Nitrogen forms N_2, but as soon as phosphorus forms P_2 it is immediaetly converted into P_4. The reason is -

(A) Triple bond present between phosphorus atoms

(B) $p\pi - p\pi$ bonding is weak

(C) $p\pi - p\pi$ bonding is strong

(D)Multiple bonds are formed easily

38. If ionisation constant of CH_3COOH is 1.7 x 10^{-5} and conc. of H^+ ions is 3.4 x 10^{-4}, find out initial conc. of CH_3COOH molecules

(A) 3.4 x 10^{-4} (B) 3.4 x 10^{-4}

(C) 6.9 10^{-4} (D) 7.24 x 10^{-3}

39. Solubility of a M_2S salt is 3.5 x 10^{-16}. Find out its solubility product

(A) 1.7×10^{-6} (B) 1.7×10^{-16}

(C)1.7×10^{-18} (D) 1.7×10^{-12}

40. If $^{a}_{b}X$ species emits firstly a positron, then two α - and two - β particles and in the last one α- particle is also emitted and is finally converted into $^{d}_{c}Y$ species, then the correct relation is

(A) a = c + 12, d = b - 5 (B) a = c - 8, d = b -1

(C) a = c - 6, d = b (D) a = c - 4, d = b - 2

41. Which one of the following pairs has the isostructural species ?

(A) XeF_2, IF_2^- (B) NH_3+ BF_3

(C) CO_3^{-2} , SO_3^{-2} (D) PCl_5, ICl_5

42. In which of the following, bond angle is maximum ?

(A) PCl_3 (B) SCl_2 (C) NH_3 (D) NH_4^+

43. Which of the following statements is not correct ?

(A) $La(OH)_3$ is less basic than $Lu(OH)_3$

(B) In lanthanide series ionic radius of Ln^{+3} ions decreases

(C) La is actually an element of transition series rather than a lanthanide.

(D) Atomic radius of Zr and Hf are same because of Ianthanide contraction.

44. Correct order of Ist IP among following elements Be, B, C, N, O is

(A) B<Be<C<O<N (B) B<Be<C<N<O

(C) Be<B<C<N<O (D) Be<B<C<O<N

45. Which of the following will give maximum number of isomers ?

(A) [$Co(NH_3)_4Cl_2$] (B) $Ni[(en)(NH_3)_4]^{+2}$

(C) $[Ni(C_2O_4)(en)_2]^{-2}$ (D) $[Cr(SCN)_2(NH_3)_4]^+$

46. Coordinatiion no. of Ni in $[Ni(C_2O_4)_3]^{-4}$ is

(A) 3 (B) 6 (C) 4 (D) 2

47. Which of the following organometallic compounds is both·δ and π bonded ?

(A) $[Fe(n^5\text{-}C_5H_5)_2]$ (B) $K[PtCl_3\ (n^2\text{-}C_2H_4)]$

(C) $[Co(CO)_5NH_3]^{+2}$ (D) $Fe\ (CH_3)_3$

48. Which statement is incorrect ?

(A) $Ni(CO)_4$ Tetrahedral, paramagnetic

(B) $Ni(CN)_4^{-2}$ -Square planar diamagnetic
(C) $Ni(CO)_4$ - Tetrahedral, diamagnetic
(D) $[Ni(Cl)_4]^{-2}$ - Tetrahedral, paramagnetic

49. In X - H ....Y, X and Y both are electronegative elements -
(A) Electron density on X will increase and on H will decrease
(B) On both X and H electron density will increase
(C) On both X and H electron density will decrease
(D) On X electron density will decrease and on H it increases

50. Main axis of a diatomic molecule is Z. Atomic orbitals p_X and p_Y overlap to form-
(A) δ Molecular orbital
(B) σ Molecular orbital
(C) π Molecular orbital
(D) No Molecular orbital will be formed.

CBSE -2002

1. In the following reaction product 'P' is

$$R-\underset{\underset{O}{\|}}{C}-Cl \xrightarrow[\text{Pd-BaSO}_4]{H_2} P$$

(A) RCHO (B) RCH_3 (C) RCH_2OH (D) RCOOH

2. $\overset{\ominus}{\ddot{C}}H_2-\underset{\underset{O}{\|}}{C}-CH_3$ and $CH_2=\underset{\underset{\ominus}{:\ddot{O}:}}{C}-CH_3$ are

(A) Resonating structures (B) Tautomers
(C) Geometrical isomers (D) Optical isomers

3. Reactivity order of halides for dehydrohalogenation is
(A) R-I > R-Cl > R-Br > R-F
(B) R-F > R - Cl > R -Br > R -I
(C) R-I > R-Br > R-Cl > R-F
(D) R-F > R-I > R-Br > R-Cl

4. Monomer of $\left[-\underset{\underset{CH_3}{|}}{\overset{\overset{CH_3}{|}}{C}}-CH_2-\right]_n$ is

(A) 2-methyl propene (B) Styrene
(C) Propylene (D) Ethene

5. $C_6H_5MgBr \xrightarrow[\text{(ii) } H_3O^{\oplus}]{\text{(i) } CO_2} P$, In the above reaction product 'P' is

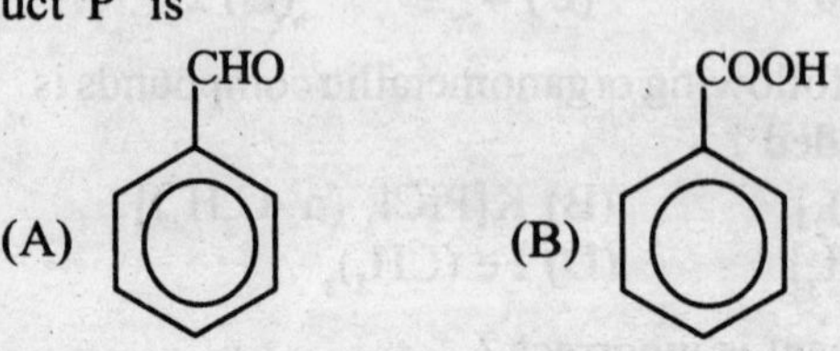

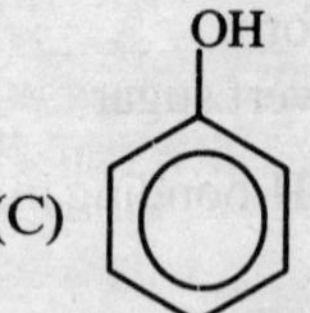

(C) (phenol) (D) $C_6H_5-\underset{}{\overset{\overset{O}{\|}}{C}}-C_6H_5$

6. Cellulose is polymer of
(A) Glucose (B) Fructose (C) Ribose (D) Sucrose

7. $CH_3CH_2Cl \xrightarrow{NaCN} X \xrightarrow{Ni/H_2} Y \xrightarrow{\text{Acetic Anhydride}} Z$

Z in the above reaction sequence is
(A) $CH_3CH_2CH_2NH_2$
(B) $CH_3CH_2CH_2CONHCH_3$
(C) $CH_3CH_2CH_2NHCOCH_3$
(D) $CH_3CH_2CH_2CONHCOCH_3$

8. When phenol is treated with $CHCl_3$ and NaOH the product formed is
(A) Benzaldehyde (B) Salicylaldehyde
(C) Salicylic acid (D) Benzoic acid

9. The percentage of C, H and N in an organic compound are 40%, 13.3% and 46.7% respectively then empirical formula is
(A) $C_3H_{13}N_3$ (B) CH_2N (C) CH_4N (D) CH_6N

10. Enzymes are made up of
(A) Edible proteins
(B) Proteins with specific structure
(C) Nitrogen containing carbohydrates
(D) Carbohydrates

11. Geometrical isomers differ in
(A) Position of functional group
(B) Position of atoms
(C) Spatial arrangement of atoms
(D) Length of carbon chain

12. When $CH_3CH_2CHCl_2$ is treated with $NaNH_2$, the product formed is
(A) CH_3-CH = CH_2 (B) CH_3-C ≡ CH
(C) $CH_3CH_2CH\langle^{NH_2}_{NH_2}$ (D) $CH_3CH_2CH\langle^{Cl}_{NH_2}$

13. ${}_{92}U^{235}$, nucleus absorb a neutron and disintigrates into ${}_{54}Xe^{139}$, ${}_{38}Sr^{94}$ and x. What will be the product x :-
(A) 3 -neutrons (B) 2-neutrons
(C) α -particle (D) β–particle

14. In hydrogen atom, energy of first excited state is -3.4 eV. Find out the KE of same orbit of Hydrogen atom
(A) + 3.4 eV (B) + 6.8 eV (C) -13.6 eV (4) + 13.6 eV

15. For the reaction $BaO_2(s) \rightleftharpoons BaO(s) + 1/2\ O_2(g)$; ΔH= + ve. In equilibrium condition Pressure of O_2 depends on
(A) Increase of mass of BaO_2
(B) Increase of mass of BaO

(C) Increase of temp. on Equilibrium
(D) Increase of mass of BaO_2 and BaO both

16. Solubility of MX_2 -type electrolytes is 0.5×10^{-4} Mole/lit. Then find out K_{sp} of electrolytes
(A) 5×10^{-12} (B) 25×10^{-10}
(C) 1×10^{-13} (D) 5×10^{-13}

17. 1 M and 2.5 litre NaOH solution mixed with another 0.5 M and 3 litre NaOH solution. Find out molarity of resultant solution
(A) 0.73 M (B) 0.50 M (C) 0.80 M (D) 1.0 M

18. Which has highest pH
(A) CH_3COOK (B) Na_2CO_3
(C) NH_4Cl (D) $NaNO_3$

19. Solution of 0.1 N NH_4OH and 0.1 N NH_4Cl has pH 9.25. Find out pk_b of NH_4OH
(A) 9.25 (B) 4.75 (C) 3.75 (4) 8.25

20. At which conditions van der Waal's real gas acts as an ideal gas,
(A) High temp., High pressure
(B) Low temp., Low pressure
(C) High temp., Low pressure
(D) Low temp., High pressure

21. Unit of entropy is :
(A) $JK^{-1} mol^{-1}$ (B) $J mol^{-1}$ (C) $J^{-1}K^{-1}mol^{-1}$ (D) $JK mol^{-1}$

22. In a closed insulated container a liquid is stirred with a paddle to increase the temperature, which of the following is true
(A) $\Delta E = W \neq 0, q = 0$ (B) $\Delta E = W = q \neq 0$
(C) $\Delta E = 0, W = q \neq 0$ (D) $W = 0\ \Delta E = q \neq 0$

23. 2 mole of ideal gas at 27°C temp, is expanded reversibly from 2 lit. to 20 lit. Find out the entropy change (R = 2 cal/mol K)
(A) 92.1 (B) 0 (C) 4 (D) 9.2

24. Which is not a true statement
(A) α-carbon of α-amino acid is asymmetric
(B) All proteins are found in L-form
(C) Human body can synthesize all proteins they need.
(D) At pH = 7 both amino and carboxylic groups exist in ionised form.

25. Heat of combustion $\Delta H°$ for C(s), $H_2(g)$ and $CH_4(g)$ are -94, -68 and -213 Kcal/mol, then $\Delta H°$ for $C(s) + 2H_2(g) \rightarrow CH_4(g)$ is
(A) -17 Kcal (B) -111 Kcal
(C) -170 Kcal (D) -85 Kcal

26. $3A \rightarrow 2B$, rate of reaction $\frac{+d(B)}{dt}$ is equal to
(A) $-\frac{3}{2}\frac{d(A)}{dt}$ (B) $-\frac{2}{3}\frac{d(A)}{dt}$
(C) $-\frac{1}{3}\frac{d(A)}{dt}$ (D) $+2\frac{d(A)}{dt}$

27. $2A \rightarrow B + C$ would be a zero order reaction when
(A) The rate of reaction remains same at any concentration of A
(B) The rate remains unchanged at any conc. of B and C
(C) The rate of reaction is proportional to square of concentration of A
(D) The rate of reaction doubles if concentration of B is increased to double

28. Which has maximum molecules
(A) 7 gm N_2 (B) 2 gm H_2
(C) 16 gm NO_2 (D) 16 gm O_2

29. A solution contains non volatile solute of molecular mass M_2. Which of the following can be used to calculate the molecular mass of solute in, terms of osmotic pressure
(A) $M_2 = \left[\frac{m_2}{\pi}\right]VRT$ (B) $M_2 = \left[\frac{m_2}{V}\right]\frac{RT}{\pi}$
(C) $M_2 = \left[\frac{m_2}{V}\right]\pi RT$ (D) $M_2 = \left[\frac{m_2}{V}\right]\frac{\pi}{RT}$

Note: $m_2 \rightarrow$ Mass of solute
$V \rightarrow$ Volume of solution
$\pi \rightarrow$ Osmotic pressure

30. A solution containing components A and B follows Raoult's law
(A) Volume of the solution is different from the sum of the volumes of solute and solvent
(B) A-B attraction force is greater than A -A and B-B
(C) A-B attraction force is less than A-A and B-B
(D) A-B attraction force remains same as A-A and B-B

31. Which reaction is not feasible
(A) $2KI + Br \rightarrow 2 KBr + I_2$
(B) $2 KBr + I_2 \rightarrow 2 KI + Br_2$
(C) $2 KBr + Cl_2 \rightarrow 2 KCl + Br_2$
(D) $2H_2O + 2F_2 \rightarrow 4 HF + O_2$

32. In electrolysis of NaCl when Pt electrode is taken then H_2 is liberated at cathode while with Hg cathode it forms sodium amalgam
(A) Hg is more inert than Pt
(B) More voltage is required to reduce H^+ at Hg than at Pt
(C) Na is dissolved in Hg while it does not dissolve in Pt
(D) Conc. of H^+ ions is larger when Pt electrode is taken.

33. Which of the following statements is ture
(A) bond energy of F_2 is less than Cl_2
(B) silicon exibits 4 coordination number in its compound
(C) Mn(III) oxidation state is more stable than Mn(II)in aqueous state
(D) elements of 15th gp shows only +3 and +5 oxidation states.

34. Which of the following orders is wrong
(A) $NH_3 < PH_3 < AsH_3$ - Acidic
(B) Li < Be < B < C - I[st] IP
(C) $Al_2O_3 < MgO < Na_2O < K_2O$ - Basic
(D) $Li^+ < Na^+ < K^+ < Cs^+$ - Ionic radius

35. General electronic configuration of lanthanides is
(A) $(n-2)f^{1-14}$ $(n-1)s^2p^6d^{0-1}$ ns^2
(B) $(n-2)f^{10-14}$ $(n-1)$ d^{0-1} ns^2
(C) $(n-2)f^{0-14}(n-1)$ d^{10} ns^2 (D) $(n-2)d^{0-1}(n-1)f^{1-14}$ ns^2

36. An atom has elcctronic configuration $1s^2 2s^2 2p^6 3s^2 3p^6 3d^3 4s^2$, it is
(A) Ti (B) Mn (C) V (D) Sc

37. Which of the following is iso-electronic
(A) CO_2, NO_2 (B) NO_2^-, CO_2
(C) CN^-, CO (D) SO_2, CO_2

38. Which of the following has $p_\pi - d_\pi$ bonding
(A) NO_3^- (B) SO_3^{-2} (C) BO_3^{-3} (D) CO_3^{-2}

39. In NO_3^- ion number of bond pair and lone pair of electrons on nitrogen atom are
(A) 2, 2 (B) 3, 1 (C) 1, 3 (D) 4, 0

40. Which of the following shows maximum number of oxidation states
(A) Mn (B) V (C) Cr (D) Fe

41. Atomic number of Cr and Fe are respectively 25 and 26. Which of the following complexes shows paramagnetic character
(A) $[Cr(CO)_6]$ (B) $[Fe(CO)_5]$
(C) $[Fe(CN)_6]^{-4}$ (D) $[Cr(NH_3)_6]^{+3}$

42. The hypothetical complex chlorodiaquotriammine cobalt (III) chloride can be represented as
(A) $[CoCl(NH_3)_3(H_2O)_2]Cl_2$ (B) $[Co(NH_3)_3(H_2O)Cl_3]$
(C) $[Co(NH_2)_3(H_2O)_2Cl]$ (D) $[Co(NH_3)_3(H_2O)_3]Cl_3$

43. In the silver plating of copper, $K[Ag(CN)_2)]$ is used instead of $AgNO_3$. The reason is
(A) A thin layer of Ag is formed on Cu
(B) More voltage is required
(C) Ag^+ ions are completely removed from solution
(D) Less availability of Ag^+ ions, as Cu can not displace Ag from $[Ag(CN)_2]^-$ ion

44. $CuSO_4$ when reacts with KCN forms CuCN which is insoluble in water. It is soluble in excess of KCN due to formation of the complex
(A) $K_2[Cu(CN)_4]$ (B) $K_3[Cu(CN)_4]$
(C) $CuCN_2$ (D) $Cu[KCu(CN)_4]$

45. Position of non polar & polar part in miscell is
(A) Polar at outer surface but non polar at inner surface.
(B) Polar at inner surface non polar at outer surface.
(C) Distributed over all the surface.
(D) Present in the surface only.

46. In borax bead test which compound is formed
(A) Ortho borate (B) Meta borate
(C) Double oxide (D) Tetra borate

47. Zn gives H_2 gas with H_2SO_4 & HCl but not with HNO_3 because
(A) Zn acts as oxidising agent when reacts with HNO_3
(B) HNO_3 is weaker acid than H_2SO_4 & HCl
(C) In electrochemical series Zn is above hydrogen
(D) NO_3^- is reduced in preference to hydronium ion

48. IUPAC name of the following is
$CH_2 = CH - CH_2 - CH_2 - C \equiv CH$
(A) 1, 5-hexenyne (B) 1-hexene-5-yne
(C) 1-hexyne-5-ene (D) 1, 5-hexynene

49. [Benzene ring with C≡N and OCH_3] + $CH_3MgBr \xrightarrow{H_3O^{\oplus}} P$

Product 'P' in the above reaction is
(A) [Benzene ring with $CH(OH)-CH_3$ and OCH_3] (B) [Benzene ring with $C(=O)-CH_3$ and OCH_3]
(C) [Benzene ring with CHO and OCH_3] (D) [Benzene ring with COOH and OCH_3]

50. n–propyl alcohol and isopropyl alcohol can be chemically distinguished by which reagent
(A) PCl_5 (B) Reduction
(C) Oxidation with Pot.dichromate (D) Ozonolysis

CBSE -2003

1. Which one of the following characteristics of the transition metals is associated with their catalytic activity?
(A) Paramagnetic behaviour
(B) Variable oxidation states
(C) High enthalpy of atomisation
(D) Colour of hydrated ions

2. The number of unpaired electrons in the complex ion $[CoF_6]^{3-}$ is
(A) Zero (B) 2 (C) 3 (D) 4
(Atomic no. of Co = 27)

3. The correct order of ionic radii of Y^{3+}, La^{3+}, Eu^{3+} and Lu^{3+} is
(A) $La^{3+} < Eu^{3+} < Lu^{3+} < Y^{3+}$
(B) $Y^{3+} < La^{3+} < Eu^{3+} < Lu^{3+}$
(C) $Y^{3+} < Lu^{3+} < Eu^{3+} < La^{3+}$
(D) $Lu^{3+} < Eu^{3+} < La^{3+} < Y^{3+}$
(Atomic nos. Y = 39, La = 57, Eu = 63, Lu = 71)

4. Chargaff's rule states that in an organism
(A) amounts of all bases are equal
(B) amount of adenine (A) is equal to that of thymine (T) and the amount of guanine (G) is equal to that of cytosine (C)
(C) amount of adenine (A) is equal to that of guanine (G) and the amount of thymine(T) is equal to that of

cytosine (C)
(D) amount of adenine (A) is equal to that of cytosine (C) and the amount of thymine (T) is equal to that of guanine (G)

5. The activation energy for a simple chemical reaction $A \longrightarrow B$ is E_a in forward direction. The activation energy for reverse reaction
(A) is always double of E_a
(B) is negative of E_a
(C) is always less than E_a
(D) can be less or more than E_a

6. In the reaction
$CH_3CHO + HCN \longrightarrow CH_3CH(OH)CN \xrightarrow{H.\ OH} CH_3CH(OH)COOH$
an asymmetric centre is generated. The acid obtained would be
(A) L-isomer
(B) 50% D + 50% L-isomer
(C) 20% D + 80% L-isomer
(D) D - isomer

7. Name of the compound given below is

CH_3, H_3C, CH_3, CH_3

(A) 5-ethyl-6-methyloctane
(B) 4-ethyl-3-methyloctane
(C) 3-methyl-4-ethyloctane
(D) 2, 3-diethylheptane

8. According to the adsorption theory of catalysis, the speed of the reaction increases because
(A) adsorption lowers the activation energy of the reaction
(B) the concentration of reactant molecules at the active centres of the catalyst becomes high due to adsorption.
(C) in the process of adsorption, the activation energy of the molecules becomes large.
(D) adsorption produces heat which increases the speed of the reaction

9. The emf of a Daniell cell at 298 K is E_1.
$Zn \,|\, ZnSO_4\ (0.01M) \,||\, CuSO_4\ (1.0M) \,|\, Cu$
When the concentration of $ZnSO_4$ is 1.0M and that of $CuSO_4$ is 0.01 M, the emf changed to E_2. What is the relationship between E_1 and E_2 ?
(A) $E_2 = 0 \neq E_1$ (B) $E_1 > E_2$ (C) $E_1 < E_2$ (D) $E_1 = E_2$

10. The correct order of reactivity towards the electrophilic substitution of the compounds, aniline (I), benzene(II) and nitrobenzene (III) is
(A) I > II > III (B) III > II > I
(C) II > III > I (D) I < II > III

11. Which one of the following is a free-radical substitution reaction ?
(A) $CH_3CHO + HCN \rightarrow CH_3CH(OH)CN$
(B) $C_6H_5CH_3 + Cl_2 \xrightarrow{Boiling} C_6H_5CH_2Cl$
(C) $C_6H_6 + CH_3Cl \xrightarrow{Anhy\ AlCl_3} C_6H_5CH_3$
(D) $C_6H_5CH_2Cl + AgNO_2 \longrightarrow C_6H_5CH_2NO_2$

12. The radio isotope, tritium (^3_1H) has a half-life of 12.3 years. If the initial amount of tritium is 32 mg., how many milligrams of it would remain after 49.2 years ?
(A) 2 mg (B) 4 mg (C) 8 mg (D) 1 mg

13. The densities of graphite and diamond at 298K are 2.25 and 3.31 g cm^{-3}, respectively. If the standard free energy difference (ΔG^o) is equal to 1895 J mol^{-1}, the pressure at which graphite will be transformed into diamond at 298 K is
(A) 9.92×10^5 Pa (B) 9.92×10^8 Pa
(C) 9.92×10^7 Pa (D) 9.92×10^5 Pa

14. The reaction $A \rightarrow B$ follows first order kinetics. The time taken for 0.8 mole of A to produce 0.6 mole of B is 1 hour. What is the time taken for conversion of 0.9 mole of A to produce 0.675 mole of B ?
(A) 2 hours (B) 1 hour (C) 0.5 hour (D) 0.25 hour

15. On the basis of the information available from the reaction
$4/3\ Al + O_2 \rightarrow 2/3\ Al_2O_3$, $\Delta G = -827$ KJ mol^{-1} of O_2
the minimum emf required to carry out an electrolysis of Al_2O_3 is (F = 96500 C mol^{-1})
(A) 8.56 V (B) 2.14 V (C) 4.28 V (D) 6.42 V

16. Which of the following statements is not correct for sigma- and pi- bonds formed between two carbon atoms ?
(A) Sigma-bond determines the direction between carbon atoms but a pi-bond has no primary effect in this regard.
(B) Sigma-bond is stronger than a pi-bond
(C) Bond energies of sigma- and pi-bonds are of the order of 264 KJ/mol and 347 KJ/mol, respectively.
(D) Free rotation of atoms about a sigma-bond is allowed but not in case of a pi-bond

17. Among the following, which is not the π-bonded organometallic compound ?
(A) $(CH_3)_4Sn$ (B) $K[PtCl_3(\eta^2-C_2H_4)]$
(C) $Fe(\eta^2-C_5H_5)_2$ (D) $Cr(\eta^6-C_6H_6)_2$

18. For the reaction $C_3H_8(g) + 5O_2(g) \rightarrow 3CO_2(g) + 4H_2O(l)$
at constant temperature, $\Delta H - \Delta E$ is
(A) – RT (B) +RT (C) – 3RT (D) + 3RT

19. The pyknometric density of sodium chloride crystal is 2.165×10^3 kg m^{-3} while its X-ray density is 2.178 x 10

$kg\ m^{-3}$. The fraction of unoccupied sites in sodium chloride crystal is

(A) 5.96×10^{-3} (B) 5.96

(C) 5.96×10^{-2} (D) 5.96×10^{-1}

20. Formation of a solution from two components can be considered as

I. Pure solvent $\rightarrow$ separated solvent molecules, ΔH_1

II. Pure solute $\rightarrow$ separated solute molecules, ΔH_2

III. Separated solvent and solute molecules $\rightarrow$ solution, ΔH_3

solution formed will be ideal at

(A) $\Delta H_{soln} = \Delta H_1 - \Delta H_2 - \Delta H_3$

(B) $\Delta H_{soln} = \Delta H_3 - \Delta H_1 - \Delta H_2$

(C) $\Delta H_{soln} = \Delta H_1 - \Delta H_2 - \Delta H_3$

(D) $\Delta H_{soln} = \Delta H_1 - \Delta H_2 - \Delta H_3$

21. The solubility product of AgI at 25°C is $1.0 \times 10^{-16}\ mol^2\ L^{-2}$. The solubility of AgI is 10^{-4} N solution of KI at 25°C is approximately (in $mol\ L^{-1}$)

(A) 1.0×10^{-8} (B) 1.0×10^{-16}

(C) 1.0×10^{-12} (D) 1.0×10^{-10}

22. Which of the follwing pairs of compounds are enantiomers ?

(A)
```
     CH3              CH3
 HO ─┼─ H         HO ─┼─ H
  H ─┼─ OH  and   HO ─┼─ H
     CH3              CH3
```

(B)
```
     CH3              CH3
  H ─┼─ OH        HO ─┼─ H
 HO ─┼─ H   and    H ─┼─ OH
     CH3              CH3
```

(C)
```
     CH3              CH3
  H ─┼─ OH        HO ─┼─ H
 HO ─┼─ H   and   HO ─┼─ H
     CH3              CH3
```

(D)
```
     CH3              CH3
  H ─┼─ OH         H ─┼─ OH
 HO ─┼─ H   and    H ─┼─ OH
     CH3              CH3
```

23. The following equilibria are given :

$N_2 + 3H_2 \rightleftharpoons 2NH_3 \quad K_1$

$N_2 + O_2 \rightleftharpoons 2NO \quad K_2$

$H_2 + \frac{1}{2} O_2 \rightleftharpoons H_2O \quad K_3$

The equilibrium constant of the reaction

$2NH_3 + \frac{5}{2} O_2 \rightleftharpoons 2NO + 3H_2O$

in terms of K_1, K_2 and K_3 is

(A) $K_2K_3^3/K_1$ (B) $K_1K_2K_3$

(C) K_1K_2/K_3 (D) $K_1K_3^2/K_2$

24. For which one of the following equations is ΔH°_{react} equal to ΔH°_f for the product ?

(A) $2\ CO(g) + O_2(g) \rightarrow 2CO_2\ (g)$

(B) $N_2(g) + O_3(g) \rightarrow N_2O_3(g)$

(C) $CH_4(g) + 2Cl_2(g) \rightarrow CH_2Cl_2(l) + 2HCl(g)$

(D) $Xe(g) + 2F_2(g) \rightarrow XeF_4(g)$

25. Which one of the following compounds is not a protonic acid ?

(A) $SO_2(OH)_2$ (B) $B(OH)_3$

(C) $PO(OH)_3$ (D) $SO(OH)_2$

26. Vitamin B_{12} contains

(A) Ca (II) (B) Fe (II) (C) Co (III) (D) Zn (II)

27. Which one of the following octahedral complexes will not show geometric isomerism ? (A and B are monodentate ligands)

(A) $[MA_5B]$ (B) $[MA_2B_4]$

(C) $[MA_3B_3]$ (D) $[MA_4B_2]$

28. The molar heat capacity of water at constant pressure, C is $75\ JK^{-1}\ mol^{-1}$. When 1.0 kJ of heat is supplied to 100g of water which is free to expand, the increase in temperature of water is

(A) 2.4 K (B) 4.8 K (C) 6.6 K (D) 1.2 K

29. Acrilon is a hard, horny and a high melting material. Which of the following represents its sturcture.

(A) $\left[-CH_2-\underset{\displaystyle Cl}{\underset{|}{CH}}- \right]_n$

(B) $\left[-CH_2-\underset{\displaystyle CN}{\underset{|}{CH}}- \right]_n$

(C) $\left[-CH_2-\overset{\displaystyle CH_3}{\overset{|}{\underset{\displaystyle COOHCH_3}{\underset{|}{CH}}}}- \right]_n$

(D) $\left[-CH_2-\underset{\displaystyle COOC_2H_5}{\underset{|}{CH}}- \right]_n$

30. Glycolysis is

(A) conversion of glucose to haem.

(B) oxidation of glucose to glutamate

(C) conversion of pyruvate to citrate

(D) oxidation of glucose to pyruvate

31. Which of the following statements is not true ?

(A) HF is a stronger acid than HCl

(B) Among halide ions, iodide is the most powerful reducing agent

(C) Fluorine is the only halogen that does not show a variable oxidation state

(D) HOCl is a stronger acid than HOBr.

32. The basic character of the transition metal monoxides follows the order

(A) TiO > VO > CrO > FeO

(B) VO > CrO > TiO > FeO

(C) CrO > VO > FeO > TiO

(D) TiO > FeO > VO > CrO
(Atomic nos. Ti = 22, V = 23, Cr = 24, Fe = 26)

33. Which one of the following monomers gives the polymer neoprene on polymerisation ?

(A) $CF_2 = CF_2$ (B) $CH_2 = CHCl$

(C) $CCl = CCl_2$ (D) $CH_2 = C(Cl) - CH = CH_2$

34. A and B in the following reactions are :

$$R - \underset{\underset{O}{\parallel}}{C} - R' \xrightarrow{HCN/KCN} A \xrightarrow{B} R - \overset{R'}{C}\begin{smallmatrix} OH \\ CH_2NH_2 \end{smallmatrix}$$

(A) A = $RR'C\begin{smallmatrix} CN \\ OH \end{smallmatrix}$, B = $LiAlH_4$

(B) A = $RR'C\begin{smallmatrix} OH \\ COOH \end{smallmatrix}$, B = NH_2

(C) A = $RR'C\begin{smallmatrix} CN \\ OH \end{smallmatrix}$, B = H3O

(D) A = $RR'CH_2CN$, B = NaOH

35. The compound,

$$CH_3 - \overset{\overset{CH_3}{|}}{C} = CH - CH_3$$

on reaction with $NaIO_4$ in the presence of $KMnO_4$ gives

(A) $CH_3CHO + CO_2$ (B) CH_3COCH_3

(C) $CH_3COCH_3 + CH_3COOH$

(D) $CH_3COCH_3 + CH_3CHO$

36. In a set of the given reactions, acetic acid yielded a product C.

$$CH_3COOH + PCl_5 \longrightarrow A \xrightarrow[\text{Anh. AlCl}_3]{C_6H_6} B \xrightarrow[\text{ether}]{C_2H_5MgBr} C$$

product C would be

(A) $CH_3-\overset{\overset{C_2H_5}{|}}{C}(OH)C_6H_5$ (B) $CH_3CH(OH)C_2H_5$

(C) $CH_3COC_6H_5$ (D) $CH_3-CH(OH)C_6H_5$

37. The final product C, obtained in the reaction

$$\text{(p-toluidine: } NH_2, CH_3) \xrightarrow{AC_2O} A \xrightarrow[CH_3COOH]{Br_2} B \xrightarrow[H^+]{H_2O} C$$ would be

(A) ring with $NHCOCH_3$, Br, CH_3 (B) ring with NH_2, $COCH_3$, CH_3 (C) ring with $COCH_3$, Br, CH_3 (D) ring with NH_2, Br, CH_3

38. The value of planck's constant is 6.63×10^{-34} Js. The velocity of light is $3.0 \times 10^8 ms^{-1}$. Which value is closest to the wavelenght in nanometers of a quantum of light with frequency of $8 \times 10^{15} s^{-1}$?

(A) 3×10^7 (B) 2×10^{-25} (C) 5×10^{-18} (D) 4×10^1

39. The ions O^{2-}, F^-, Mg^+, Mg^{2+} and Al^{3+} are isoelectronic. Their ionic radii show.

(A) a decrease from O^{2-} to F^- and then increase from Na^+ to Al^{3+}

(B) a significant increase from O^{2-} to Al^{3+}

(C) a significant decrease from O^{2-} to Al^{3+}

(D) an increase from O^{2-} to F^- and then decrease from Na^+ to Al^{3+}

40. The temperature dependence of the rate constant (k) of a chemical reaction is written in terms of Arrhenius equation, $k = A.e^{-E^*/RT}$. Activation energy (E*) of the reaction can be calculated by plotting.

(A) log k vs 1/log T (B) k vs T

(C) k vs 1/ log T (D) log k vs 1/T

41. If the rate of the reaction is equal to the rate constant, the order of the reaction is

(A) 3 (B) 0 (C) 1 (D) 2

42. The oxidation state of sulphur in the anions SO_3^{2-}, $S_2O_4^{2-}$ and $S_2O_6^{2-}$ follow the order

(A) $S_2O_6^{2-} < S_2O_4^{2-} < SO_3^{2-}$

(B) $S_2O_4^{2-} < SO_3^{2-} < S_2O_6^{2-}$

(C) $S_3^{2-} < S_2O_4^{2-} < S_2O_6^{2-}$

(D) $S_2O_4^{2-} < S_2O_6^{2-} < SO_3^{2-}$

43. The reaction quotient (Q) for the reaction $N_2(g) + 3H_2(g) \rightleftharpoons 2NH_3(g)$ is given by $Q = \frac{[NH_3]^2}{[N_2][H_2]^3}$

The reaction will proceed from right to left if

(A) Q = 0 (B) Q = K_C (C) Q < K_C (D) Q > K_C

Where K_c is the equlibrium constant.

44. What is the entropy change (in $JK^{-1} mol^{-1}$) when one mole of ice is converted into water at 0°C ? (The enthalpy change for the conversion of ice to liquid water is 6.0 KJ mol^{-1} at 0°C)

(A) 21.98 (B) 20.13 (C) 2.013 (D) 2.198

45. The method of zone refining of metals is based on the principle of

(A) greater solubility of the impurity in the molten state than in the solid

(B) greater mobility of the pure metal than that of the impurity

(C) higher melting point of the impurity than that of the pure metal

(D) greater noble character of the solid metal than that of the impurity

46. According to IUPAC nomenclature sodium nitroprusside is named as

(A) sodium pentacyanonitrosyl ferrate (III)

(B) sodium nitroferricyanide

(C) sodium nitroferrocyanide

(D) sodium pentacyanonitrosyl ferrate (II)

47. Phospholipids are esters of glycerol with
(A) three phosphate groups
(B) three carboxylic acid residues
(C) two carboxylic acid residues and one phosphate group
(D) one carboxylic acid residue and two phosphate groups.

48. In Haber process 30 litres of dihydrogen and 30 litres of dinitrogen were taken for reaction which yielded only 50% of the expected product. What will be the composition of gaseous mixture under the aforesaid condition in the end ?
(A) 20 litres ammonia, 25 litres nitrogen, 15 litres hydrogen
(B) 20 litres ammonia, 20 litres nitrogen, 20 litres hydrogen
(C) 10 litres ammonia, 25 litres nitrogen, 15 litres hydrogen
(D) 20 litres ammonia, 10 litres nitrogen, 30 litres hydrogen

49. Which one of the following orders of acid strength is correct ?
(A) RCOOH > HC ≡ CH > HOH > ROH
(B) RCOOH > ROH > HOH > HC ≡ CH
(C) RCOOH > HOH > ROH > HC ≡ CH
(D) RCOOH > HOH > HC ≡ CH > ROH

50. When m-chlorobenzaldehyde is treated with 50% KOH solutiion, the product (s) obtained is (are)

(A) (m-ClC₆H₄)–CH(OH) — CH(OH)–(C₆H₄Cl-m)

(B) (m-HOC₆H₄)–COO⁻ + (m-HOC₆H₄)–CH₂OH

(C) (m-ClC₆H₄)–COO⁻ + (m-ClC₆H₄)–CH₂OH

(D) (m-HOC₆H₄)–CH(OH) — CH(OH)–(C₆H₄OH-m)

CBSE - 2004

1. The frequency of radiation emitted when the electron falls from n = 4 to n = 1 in a hydrogen atom will be (Given ionization energy of H = 2.18×10^{-18} J atom^{-1} and h = 6.625×10^{-34} Js)
(A) 1.54×10^{15} s^{-1} (B) 1.03×10^{15} s^{-1}
(C) 3.08×10^{15} s^{-1} (D) 2.00×10^{15} s^{-1}

2. The helical structure of protein is stabilized by
(A) dipeptide bonds (B) hydrogen bonds
(C) ether bonds (D) peptide bonds

3. In BrF_3 molecule, the lone pairs occupy equatorial positions to minimize
(A) lone pair - bond pair repulsion only
(B) bond pair - bond pair repulsion only
(C) lone pair - lone pair repulsion and lone pair - bond pair repulsion
(D) lone pair - lone pair repulsion only

4. The radioactive isotope $_{27}Co^{60}$ which is used in the treatment of cancer can be made by (n, p)reaction. For this reaction the target nucleus is
(A) $_{28}Ni^{59}$ (B) $_{27}Co^{59}$ (C) $_{28}Ni^{60}$ (D) $_{27}Co^{60}$

5. The maximum number of molecules is present in
(A) 15 L of H_2 gas at STP (B) 5 L of N_2 gas at STP
(C) 0.5 g of H_2 gas (D) 10 g of O_2 gas

6. A compound formed by elements X and Y crystallizes in a cubic structure in which the X atoms are at the corners of a cube and the Y atoms are at the facecentres. The formula of the compound is
(A) XY_3 (B) X_3Y (C) XY (D) XY_2

7. Camphor is often used in molecular mass determination because
(A) it is readily available
(B) it has a very high cryoscopic constant
(C) it is volatile (D) it is solvent for organic substances

8. The molecular formula of diphenyl methane, C_6H_5–CH_2–C_6H_5, is $C_{13}H_{12}$. How many structural isomers are possible when one of the How many structural isomers are possible when one of the hydrogens is replaced by a chlorine atom ?
(A) 6 (B) 4 (C) 8 (D) 72

9. Which of the following is least reactive in a nucleophilic substitution reaction ?
(A) $(CH_3)_3C$ - Cl (B) CH_2 = CHCl
(C) CH_3CH_2Cl (D) CH_2=$CHCH_2Cl$

10. Using anhydrous $AlCl_3$ as catalyst, which one of the following reactions produces ethylbenzene (PhEt) ?
(A) H_3C - CH_2OH + C_6H_6 (B) CH_3 - CH = CH_2 + C_6H_6
(C) H_2C = CH_2 + C_6H_6 (D) H_3C - CH_3 + C_6H_6

11. Reaction of HBr with propene in the presence of peroxide gives
(A) isopropyl bromide (B) 3-bromo propane
(C) allyl bromide (D) n-propyl bromide

12. Which of the following is responsible for depletion of the ozone layer in the upper strata of the atmosphere ?
(A) Polyhalogens (B) Ferrocene (C) Fullerenes (D) Freon

13. Chloropicrin is obtained by the reaction of
(A) steam on carbon tetrachloride
(B) nitric acid on chlorobenzene
(C) chlorine on picric acid (D) nitric acid on chloroform

14. Which one of the following can be oxidised to the

corresponding carbonyl compound ?
(A) 2-hydroxy-propane (B) Ortho-nitrophenol
(C) 2-methyl-2-hydroxy-propane (D) Phenol

15. The OH group of an alcohol or the - COOH group of a carboxylic acid can be replaced by - Cl using
(A) phosphorus pentachloride (B) hypochlorous acid
(C) chlorine (D) hydrochloric acid

16. Which of the following will not form a yellow precipitate on heating with an alkaline solution of iodine?
(A) $CH_3CH(OH)CH_3$ (B) $CH_3CH_2CH(OH)CH_3$
(C) CH_3OH (D) CH_3CH_2OH

17. Among the following, the pair in which the two species are not isostructural is
(A) SiF_4 and SF_4 (B) IO_3^- and XeO_3
(C) BH_4^- and NH_4^+ (D) PF_6^- and SF_6

18. Which one of the following statements about the zeolites is false ?
(A) They are used as cation exchangers
(B) They have open structure which enables them to take up small molecules
(C) Zeolites are aluminosilicates having three dimensional network
(D) Some of the SiO_4^{4-} units are replaced by AlO_4^{5-} and AlO_6^{9-} ions in zeolites

19. Ionic radii are
(A) inversely proportional to effective nuclear charge
(B) inversely proportional to square of effective nuclear charge
(C) directly proportional to effective nuclear charge
(D) directly proportional to square of effective nuclear charge

20. H_2O is dipolar, whereas BeF_2 is not. It is because
(A) the electronegativity of F is greater than that of O
(B) H_2O involves hydrogen bonding whereas BeF_2 is a discrete molecule
(C) H_2O is linear and BeF_2 is angular
(D) H_2O is angular and BeF_2 is line]

21.Which is the best description of the behaviour of bromine in the reaction given below ?
$H_2O + Br_2 \longrightarrow HOBr + HBr$
(A) Proton acceptor only (B) Both oxidized and reduced
(C) Oxidized only (D) Reduced only

22. In a regular octahedral molecule, MX_6 the number of X - M - X bonds at 180° is
(A) three (B) two (C) six (D) four

23. Among K, Ca, Fe and Zn, the element which can form more than one binary compound with chlorine is
(A) Fe (B) Zn (C) K (D) Ca

24. Which of the following is considered to be an anticancer species ?
(A) $[PtCl_3(CH_2{=}CH_2)]$ (B) $[PtCl_4]$
(C) $[Pt(H_3N)_2Cl_2]$ (cis) (D) $[Pt(H_3N)_2Cl_2]$ (trans)

25. Which of the following coordination compounds would exhibit optical isomerism ?
(A) pentaamminenitrocobalt(III) iodide
(B) diamminedichloroplatinum(II)
(C) trans-dicyanobis (ethylenediamine) chromium(III) chloride
(D) tris-(ethylenediamine)cobalt(III) bromide

26. Among $[Ni(CO)_4]$, $[Ni(CN)_4]^{2-}$, $[NiCl_4]^{2-}$ species, the hybridization states at the Ni atom are, respectively
(A) sp^3 ,dsp^2, dsp^2 (B) sp^3, dsp^2, sp^3
(C) sp^3, sp^3, dsp^2 (D) dsp^2, sp^3, sp|
(At. no. of Ni = 28)

27. Among the following series of transition metal ions, the one where all metal ions have $3d^2$ electronic configuration is
(A) Ti^{3+}, V^{2+}, Cr^{3+}, Mn^{4+} (B) Ti^+, V^{4+}, Cr^{6+}, Mn^{7+}
(C) Ti^{4+}, V^{3+}, Cr^{2+} , Mn^{3+} (D) Ti^{2+}, V^{3+}, Cr^{4+}, Mn^{5+}
(At. nos. Ti = 22; V = 23; Cr = 24; Mn = 25)

28. CN^- is a strong field ligand. This is due to the fact that
(A) it carries negative charge (B) it is a pseudohalide
(C) it can accept electrons from metal species
(D) it forms high spin complexes with metal species

29. Considering H_2O as a weak field ligand, the number of unpaired electrons in $[Mn(H_2O)_6]^{2+}$ will be (At. no. of Mn = 25)
(A) three (B) five (C) two (D) four

30. Which of the following does not have a metal - carbon bond ?
(A) $Al(OC_2H_5)_3$ (B) C_2H_5MgBr
(C) $K[Pt(C_2H_4)Cl_3]$ (D) $Ni(CO)_4$

31. Lanthanoids are
(A) 14 elements in the sixth period (atomic no. = 90 to 103) that are filling 4f sublevel
(B) 14 elements in the seventh period (atomic no. = 90 to 103) that are filling 5f sublevel
(C) 14 elements in the sixth period (atomic no. = 58 to 71) that are filling 4f sublevel
(D) 14 elements in the seventh period (atomic no. = 58 to 71) that are filling 4f sublevel

32.In an octahedral structure, the pair of d orbitals involved in d2sp3 hybridization is
(A) $d_{x^2-y^2}$, d_{z^2} (B) d_{xz}, $d_{x^2-y^2}$
(C) d_{z^2}, d_{xz} (D) d_{xy}, d_{yz}

33. Aniline when diazotized in cold and then treated with dimethyl aniline gives a coloured product. Its structure would be

(A) $CH_3NH-C_6H_4-N{=}N-C_6H_4-NHCH_3$

(B) $CH_3-C_6H_4-N=N-C_6H_4-NH_2$

(C) $(CH_3)_2N-C_6H_4-N=N-C_6H_5$

(D) $(CH_3)_2N-C_6H_4-NH-C_6H_5$

34. Which one of the following is a chain growth polymer ?
(A) Starch (B) Nucleic acid (C) Polystyrene (D) Protein

35. Which one of the following structures represents the peptide chain?

(A) $-NH-C(=O)-NH-C-NH-C(=O)-NH-$ (B) $-NH-C(=O)-C-C-C-NH-C-C-C-$

(C) $-NH-C-C(=O)-NH-C-C(=O)-NH-C-C(=O)-$ (D) $-NH-C-C-C(=O)-NH-C-C-NH-C(=O)-C-C-$

36. A sequence of how many nucleotides in messenger RNA makes a codon for an amino acid ?
(A) Three (B) Four (C) One (D) Two

37. The hormone that helps in the conversion of glucose to glycogen is
(A) Cortisone (B) Bile acids (C) Adrenaline (D) Insulin

38. The enzyme which hydrolyses triglycerides to fatty acids and glycerol is called
(A) Maltase (B) Lipase (C) Zymase (D) Pepsin

39. The correct statement in respect of protein haemoglobin is that it
(A) functions as a catalyst for biological reactions
(B) maintains blood sugar level
(C) acts as an oxygen carrier in the blood
(D) forms antibodies and offers resistance to diseases

40. Number of chiral carbons in β-D-(+)-glucose is
(A) five (B) six (C) three (D) four

41. The solubility product of a sparingly soluble salt AX_2 is 3.2×10^{-11}. Its solubility (in moles/litre) is
(A) 5.6×10^{-6} (B) 3.1×10^{-4} (C) 2×10^{-4} (D) 4×10^{-4}

42. The rapid change of pH near the stoichiometric point of an acid-base titration is the basis of indicator detection. pH of the solution is related to ratio of the concentrations of the conjugate acid (HIn) and base (In-) forms of the indicator by the expression

(A) $\log In^- / HIn = pK_{In} - pH$
(B) $\log HIn / In^- = pK_{In} - pH$
(C) $\log HIn / In^- = pH - pK_{In}$
(D) $\log In^- / HIn = pH - pK_{In}$

43. If the bond energies of H - H, Br - Br, and H - Br are 433, 192 and 364 kJ mol^{-1} respectively, the DH° for the reaction
$H_2(g) + Br_2(g) \longrightarrow 2HBr(g)$ is
(A) -261 kJ (B) + 103 kJ (C) + 261 kJ (D) - 103 kJ

44. Standard enthalpy and standard entropy changes for the oxidation of ammonia at 298 K are -382.64 kJ mol^{-1} and -145.6 J K^{-1} mol^{-1}, respectively. Standard Gibbs energy change for the same reaction at 298 K is
(A) - 221.1 kJ mol^{-1} (B) - 339.3 kJ mol^{-1}
(C) - 439.3 kJ mol^{-1} (D) - 523.2 kJ mol^{-1}

45. The work done during the expansion of a gas from a volume of 4 dm^3 to 6 dm^3 against a constant external pressure of 3 atm is (1 L atm = 101.32 J)
(A) - 6 J (B) - 608 J (C) + 304 J (D) - 304 J

46. Considering entropy (S) as a thermodynamic parameter, the criterion for the spontaneity of any process is
(A) $\Delta S_{system} + \Delta S_{surroundings} > 0$
(B) $\Delta S_{system} - \Delta S_{surroundings} > 0$
(C) $\Delta S_{system} > 0$ only (D) $\Delta S_{surroundings} > 0$ only

47. The rate of a first order reaction is 1.5×10^{-2} mol L^{-1} min^{-1} at 0.5 M concentration of the reactant. The half-life of the reaction is
(A) 0.383 min (B) 23.1 min (C) 8.73 min (D) 7.53 min

48. The standard e.m.f. of a galvanic cell involving cell reaction with n = 2 is found to be 0.295 V at 25°C. The equilibrium constant of the reaction would be
(Given F = 96500 C mol^{-1}; R = 8.314 J K^{-1} mol^{-1})
(A) 2.0×10^{11} (B) 4.0×10^{12} (C) 1.0×10^{2} (D) 1.0×10^{10}

49. Which of the following forms cationic micelles above certain concentration ?
(A) Sodium dodecyl sulphate (B) Sodium acetate
(C) Urea (D) Cetyltrimethylammonium bromide

50. A solid compound 'X' on heating gives CO_2 gas and a residue. The residue mixed with water forms 'Y'. On passing an excess of CO_2 through 'Y' in water, a clear solution, 'Z', is obtained. On boiling 'Z', compound 'X' is reformed. The compound 'X' is
(A) $Ca(HCO_3)_2$ (B) $CaCO_3$ (C) Na_2CO_3 (D) K_2CO_3

CBSE - 2005

1. The monomer of the polymer;

$$\sim\!\sim CH_2-C(CH_3)_2-CH_2-\overset{\oplus}{C}(CH_3)_2$$ is:

(A) $H_2C=C(CH_3)_2$ (B) $CH_3CH = CHCH_3$
(C) $CH_3CH = CH_2$ (D) $(CH_3)_2 C = C(CH_3)_2$

2. Aniline in a set of reactions yielded a product D.

$$C_6H_5NH_2 \xrightarrow[HCl]{NaNO_2} A \xrightarrow{CuCN} B \xrightarrow[Ni]{H_2} C \xrightarrow{HNO_2} D$$

The structure of the product D would be:
(A) C_6H_5NHOH (B) $C_6H_5NHCH_2CH_3$
(C) $C_6H_5CH_2NH_2$ (D) $C_6H_5CH_2OH$

3. The best method for the separation of naphthalene

and benzoic acid from their mixture is:
(A) distillation (B) sublimation
(C) chromatography (D) crystallisation

4. Names of some compounds are given. Which one is not in IUPAC system?

(A) $CH_3 - CH_2 - CH_2 - CH(CH_2CH_3) - CH(CH_3) - CH_2CH_3$
3-Methyl-4-ethyl heptane

(B) $CH_3 - CH(OH) - CH(CH_3) - CH_3$
3-Methyl-2-butanol

(C) $CH_3 - CH_2 - C(=CH_2) - CH(CH_3) - CH_3$
2-Ethyl-3-methyl-but-1-ene

(D) $CH_3 - C \equiv C - CH - (CH_3)_2$
4-Methyl-2-pentyne

5. Which of the following undergoes nucleophilic substitution exclusively by S_N1 mechanism ?
(A) Ethyl chloride (B) Isopropyl chloride
(C) Chlorobenzene (D) Benzyl chloride

6. Which one of the following compounds is most acidic?
(A) $Cl-CH_2-CH_2-OH$ (B) 

(C) 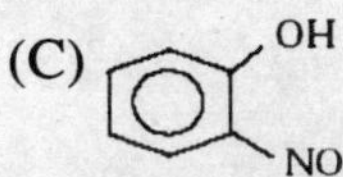 (D) 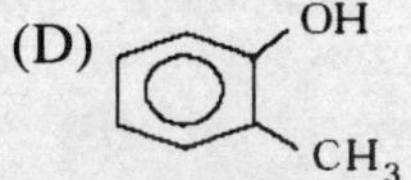

7. Which one of the following arrangements represents the correct order of electron gain enthalpy (with negative sign) of the given atomic species ?
(A) S < O < Cl < F (B) Cl < F < S < O
(C) F < Cl < O < S (D) O < S < F < Cl

8. The main reason for larger number of oxidation states exhibited by the actinoids than the corresponding lanthanoids, is
(A) more energy difference between 5f and 6d orbitals than between 4f and 5d orbitals.
(B) lesser energy difference between 5f and 6d orbitals than between 4f and 5d orbitals.
(C) larger atomic size of actinoids than the lanthanoids.
(D) greater reactive nature of the actinoids than the lanthanoids.

9. What is the correct relationship between the pHs of isomolar solutions of sodium oxide (pH_1), sodium sulphide (pH_2), sodium selenide (pH_3) and sodium telluride (pH_4)?
(A) $pH_1 > pH_2 > pH_3 > pH_4$ (B) $pH_1 > pH_2 \approx pH_3 > pH_4$
(C) $pH_1 < pH_2 < pH_3 < pH_4$ (D) $pH_1 < pH_2 < pH_3 \approx pH_4$

10. Which one of the following alkenes will react faster with H_2 under catalytic hydrogenation conditions ?

(A) $RHC=CHR$ (R, R cis; H, H) (B) $R_2C=CH_2$ (R, R on one carbon; H, H on the other)

(C) $R_2C=CHR$ (D) $R_2C=CR_2$

11. The mass of carbon anode consumed (giving only carbondioxide) in the production of 270 kg of aluminium metal from bauxite by the Hall process is

(A) 270 kg (B) 540 kg (C) 90 kg (D) 180 kg

12. Which one of the following forms micelles in aqueous solution above certain concentration?
(A) Dodecyl trimethyl ammonium chloride
(B) Glucose (C) Urea (D) Pyridinium chloride

13. Which of the following pairs of a chemical reaction is certain to result in a spontaneous reaction ?
(A) Exothermic and increasing disorder
(B) Exothermic and decreasing disorder
(C) Endothermic and increasing disorder
(D) Endothermic and decreasing disorder

14. Which one of the following oxides is expected to exhibit paramagnetic behaviour?
(A) CO_2 (B) SiO_2 (C) SO_2 (D) ClO_2

15. Which of the following molecules has trigonal planar geometry ?
(A) BF_3 (B) NH_3 (C) PCl_3 (D) IF_3

16. A reaction occurs spontaneously if
(A) $T\Delta S < \Delta H$ and both ΔH and ΔS are + ve
(B) $T\Delta S > \Delta H$ and ΔH is + ve and ΔS is – ve
(C) $T\Delta S > \Delta H$ and both ΔH and ΔS are + ve
(D) $T\Delta S = \Delta H$ and both ΔH and ΔS are + ve

17. In face-centred cubic lattice, a unit cell is shared equally by how many unit cells ?
(A) 2 (B) 4 (C) 6 (D) 8

18. A nuclide of an alkaline earth metal undergoes radioactive decay by emission of the ±-particles in succession. The group of the periodic table to which the resulting daughter element would belong is
(A) Gr. 4 (B) Gr. 6 (C) Gr. 14 (D) Gr. 16

19. For a first order reaction A → B the reaction rate at reactant concentration of 0.01 M is found to be 2.0×10^{-5} mol L^{-1} S^{-1}. The half life period of the reaction is
(A) 30 s (B) 220 s (C) 300 s (D) 347 s

20. The major organic product formed from the following reaction

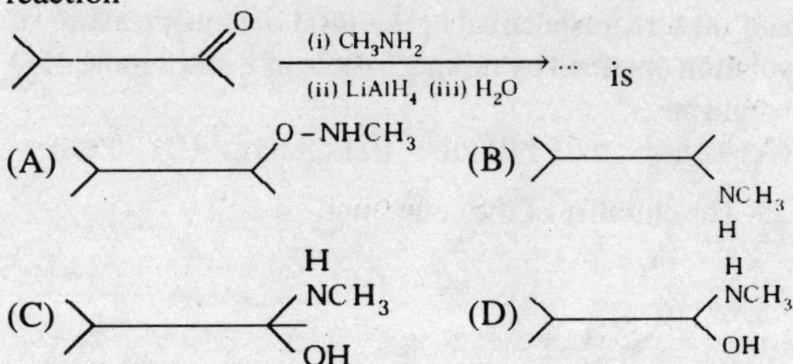

21. Electrolytic reduction of nitrobenzene in weakly acidic medium gives
(A) N-Penylhydroxylamine (B) Nitrosobenzene
(C) Aniline (D) p-Hydroxyaniline

22. 4.5 g of aluminium (at. mass 27 amu) is deposited at

cathode from Al3+ solution by a certain quantity of electric charge. The volume of hydrogen produced at STP from H+ ions in solution by the same quantity of electric charge will be

(A) 44.8 L (B) 22.4 L (C) 11.2 L (D) 5.6 L

23. At 25°C, the dissociation constant of a base, BOH, is 1.0×10^{-2}. The concentration of hydroxyl ions in 0.01 M aqueous solution of the base would be

(A) 1.0×10^{-5} mol L^{-1} (B) 1.0×10^{-6} mol L^{-1}
(C) 2.0×10^{-6} mol L^{-1} (D) 1.0×10^{-7} mol L^{-1}

24. A solution of urea (mol. mass 56 g mol^{-1}) boils at 100.18°C at the atmospheric pressure. If Kf and Kb for water are 1.86 and 0.512 K kg mol^{-1} respectivley, the above solution will freeze at

(A) 0.654°C (B) –0.654°C (C) 6.54°C (D) –6.54°C

25. Which one of the following is an inner orbital complex as well as diamagnetic in behaviour?

(A) $[Zn(NH_3)_6]^{2+}$ (B) $[Cr(NH_3)_6]^{3+}$
(C) $[Co(NH_3)_6]^{3+}$ (D) $[Ni(NH_3)_6]^{2+}$

26. A solution has a 1 : 4 mole ratio of pentane to hexane. The vapour pressures of the pure hydrocarbons at 20°C are 400 mm Hg for pentane and 120 mm Hg for hexane. The mole fraction of pentane in the vapour phase would be:

(A) 0.200 (B) 0.549 (C) 0.786 (D) 0.478

27. H2S gas when passed through a solution of cations containing HCl precipitates the cations of second group of qualitative analysis but not those belonging to the fourth group. It is because

(A) presence of HCl decreases the sulphide ion concentration.
(B) solubility product of group II sulphides is more than that of group IV sulphides.
(C) presence of HCl increases the sulphide ion concentration.
(D) sulphides of group IV cations are unstable in HCl.

28. The vapour pressure of two liquids P and Q are 80 and 60 torr, respectively. The total vapour pressure of solution obtained by mixing 3 mole of P and 2 mole of Q would be

(A) 72 torr (B) 140 torr (C) 68 torr (D) 20 torr

29. The chirality of the compound

Br, H, Cl, H_3C on C (wedge/dash structure) is:

(A) R (B) S (C) E (D) Z

30. Which functional group participates in disulphide bond formation in proteins ?

(A) Thioester (B) Thioether (C) Thiol (D) Thiolactone

31. Which of the following is the electron deficient molecule?

(A) C_2H_6 (B) B_2H_6 (C) SiH_4 (D) PH_3

32. The correct sequence of increasing covalent character is represented by

(A) LiCl < NaCl < $BeCl_2$ (B) $BeCl_2$ < LiCl < NaCl
(C) NaCl < LiCl < $BeCl_2$ (D) $BeCl_2$ < NaCl < LiCl

33. The correct order in which the O–O bond length increases in the following is:

(A) $O_2 < H_2O_2 < O_3$ (B) $O_3 < H_2O_2 < O_2$
(C) $H_2O_2 < O_2 < O_3$ (D) $O_2 < O_3 < H_2O_2$

34. The energy of second Bohr orbit of the hydrogen atom is –328 kJ mol^{-1}; hence the energy of fourth Bohr orbit would be:

(A) – 41 kJ mol^{-1} (B) –82 kJ mol^{-1}
(C) –164 kJ mol^{-1} (D) –1312 kJ mol^{-1}

35. In a set of reactions acetic acid yielded a product

$$CH_3COOH \xrightarrow{SOCl_2} A \xrightarrow[\text{Anhy. } AlCl_3]{\text{Benzene}} B \xrightarrow{HCN} C \xrightarrow{H \cdot OH} D$$

The structure of D would be:

(A) $C_6H_5-CH_2-C(COOH)(OH)-CH_3$ (B) $C_6H_5-C(CN)(OH)-CH_3$

(C) $C_6H_5-CH_2-C(OH)(CN)-CH_3$ (D) $C_6H_5-C(OH)(CH_3)-COOH$

36. The mole fraction of the solute in one molal aqueous solution is :

(A) 0.009 (B) 0.018 (C) 0.027 (D) 0.036

37. Which amongst the following is the most stable carbocation?

(A) $\overset{+}{C}H_3$ (B) $CH_3\overset{+}{C}H_2$
(C) $CH_3-\overset{+}{C}H-CH_3$ (D) $(CH_3)_3C^+$

38. The rate of reaction between two reactants A and B decreases by a factor of 4 if the concentration of reactant B is doubled. The order of this reaction with respect to reactant B is:

(A) 2 (B) -2 (C) 1 (D) -1

39. Products of the following reaction:

$$CH_3C \equiv C \cdot CH_2CH_3 \xrightarrow[(2)\ \text{Hydrolysis}]{(1)\ O_3} \text{...are:}$$

(A) $CH_3COOH+CO_2$
(B) CH_3COOH + HOOC $.CH_2CH_3$
(C) $CH_3CHO + CH_3CH_2CHO$
(D) $CH_3COOH + CH_3COCH_3$

40. Which one of the following is expected to exhibit optical isomerism?

(en = ethylenediamine)

(A) cis-$[Pt(NH_3)_2 Cl_2]$ (B) trans-$[Pt(NH_3)_2Cl_2]$

(C) cis-$[Co(en)_2Cl_2]$ (D) trans-$[Co(en)_2 Cl_2]$

41. The number of moles of $KMnO_4$ reduced by one mole of KI in alkaline medium is:

(A) one (B) two (C) five (D) one fifth

42. The cell membranes are mainly composed of

(A) fats (B) proteins (C) phospholipids (D) carbohydrates

43. Which of the following would have a permanent dipole moment?

(A) SiF_4 (B) SF_4 (C) XeF_4 (D) BF_3

44. Four successive members of the first row transition elements are listed below with their atomic numbers. Which one of them is expected to have the highest third ionization enthalpy?

(A) Vanadium (Z = 23) (B) Chromium (Z = 24)

(C) Manganese (Z = 25) (D) Iron (Z = 26)

45. The absolute enthalpy of neutralisation of the reaction:

$MgO (s) + 2HCl (aq) \rightarrow MgCl_2 (aq) + H_2O (l)$ will be:

(A) -57.33 kJ mol^{-1} (B) > -57.33 kJ mol^{-1}

(C) < -57.33 kJ mol^{-1} (D) 57.33 kJ mol^{-1}

46. Which one of the following pairs represents stereoisomerism?

(A) Structural isomerism and Geometrical isomerism.

(B) Optical isomerism and Geometrical isomerism.

(C) Chain isomerism and Rotational isomerism.

(D) Linkage isomerism and Geometrical isomerism.

47. Equilibrium constants K_1 and K_2 for the following equilibria:

$$NO (g) + \frac{1}{2} O_2 \overset{K_1}{\rightleftharpoons} NO_2 (g) \text{ and}$$

$$2NO_2 (g) \overset{K_2}{\rightleftharpoons} 2NO (g) + O_2 (g)$$

(A) $K_2 = \frac{1}{K_1^2}$ (B) $K_2 = K_1^2$

(C) $K_2 = \frac{1}{K_1}$ (D) $K_2 = \frac{K_1}{2}$

48. The surface tension of which of the following liquid is maximum ?

(A) C_2H_5OH (B) CH_3OH (C) H_2O (D) C_6H_6

49. The aqueous solution containing which one of the following ions will be colourless ?

(A) Sc^{3+} (B) Fe^{2+} (C) Ti^{3+} (D) Mn^{2+}

50. The correct order of acid strength is:

(A) $HClO_4 < HClO_3 < HClO_2 < HClO$

(B) $HClO < HClO_2 < HClO_3 < HClO_4$

(C) $HClO_4 < HClO < HClO_2 < HClO_3$

(D) $HClO_2 < HClO_3 < HClO_4 < HClO$

CBSE -1990

1.	(B)	39.	(C)
2.	(B)	40.	(B)
3.	(D)	41.	(B)
4.	(D)	42.	(D)
5.	(C)	43.	(C)
6.	(D)	44.	(C)
7.	(B)	45.	(A)
8.	(C)	46.	(D)
9.	(B)	47.	(C)
10.	(B)	48.	(A)
11.	(D)	49.	(C)
12.	(B)	50.	(C)
13.	(C)	51.	(D)
14.	(B)	52.	(B)
15.	(D)	53.	(C)
16.	(C)	54.	(D)
17.	(B)	55.	(B)
18.	(B)	56.	(D)
19.	(C)	57.	(C)
20.	(D)	58.	(B)
21.	(A)	59.	(A)
22.	(D)	60.	(B)
23.	(C)	61.	(D)
24.	(B)	62.	(B)
25.	(C)	63.	(D)
26.	(C)	64.	(A)
27.	(A)	65.	(A)
28.	(C)	66.	(A)
29.	(D)	67.	(B)
30.	(B)	68.	(C)
31.	(D)	69.	(C)
32.	(B)	70.	(A)
33.	(B)	71.	(B)
34.	(C)	72.	(D)
35.	(C)	73.	(A)
36.	(A)	74.	(C)
37.	(A)	75.	(B)
38.	(D)		

CBSE -1991

1.	(A)	39.	(A)
2.	(A)	40.	(A)
3.	(B)	41.	(C)
4.	(A)	42.	(C)
5.	(A)	43.	(C)
6.	(A)	44.	(D)
7.	(A)	45.	(D)
8.	(A)	46.	(C)
9.	(B)	47.	(A)
10.	(B)	48.	(B)
11.	(A)	49.	(D)
12.	(B)	50.	(A)
13.	(A)	51.	(A)
14.	(D)	52.	(C)
15.	(B)	53.	(A)
16.	(B)	54.	(C)
17.	(A)	55.	(B)
18.	(A)	56.	(C)
19.	(B)	57.	(C)
20.	(B)	58.	(C)
21.	(C)	59.	(A)
22.	(C)	60.	(B)
23.	(C)	61.	(B)
24.	(A)	62.	(B)
25.	(D)	63.	(B)
26.	(B)	64.	(D)
27.	(A)	65.	(A)
28.	(B)	66.	(C)
29.	(B)	67.	(A)
30.	(A)	68.	(C)
31.	(C)	69.	(D)
32.	(B)	70.	(C)
33.	(B)	71.	(B)
34.	(C)	72.	(B)
35.	(C)	73.	(A)
36.	(B)	74.	(B)
37.	(D)	75.	(B)
38.	(A)		

CBSE -1992

1.	(B)	39.	(D)
2.	(A)	40.	(C)
3.	(D)	41.	(D)
4.	(C)	42.	(A)
5.	(D)	43.	(C)
6.	(A)	44.	(D)
7.	(C)	45.	(C)
8.	(B)	46.	(A)
9.	(C)	47.	(A)
10.	(B)	48.	(C)
11.	(B)	49.	(D)
12.	(B)	50.	(A)
13.	(A)	51.	(D)
14.	(D)	52.	(A)
15.	(B)	53.	(B)
16.	(A)	54.	(A)
17.	(A)	55.	(C)
18.	(A)	56.	(A)
19.	(D)	57.	(D)
20.	(A)	58.	(A)
21.	(B)	59.	(C)
22.	(C)	60.	(B)
23.	(C)	61.	(C)
24.	(A)	62.	(A)
25.	(D)	63.	(B)
26.	(D)	64.	(C)
27.	(D)	65.	(B)
28.	(B)	66.	(A)
29.	(C)	67.	(C)
30.	(C)	68.	(A)
31.	(D)	69.	(D)
32.	(B)	70.	(B)
33.	(A)	71.	(B)
34.	(C)	72.	(B)
35.	(D)	73.	(B)
36.	(C)	74.	(D)
37.	(B)	75.	(A)
38.	(B)		

CBSE -1993

1.	(B)	39.	(C)
2.	(B)	40.	(D)
3.	(D)	41.	(B)
4.	(B)	42.	(C)
5.	(D)	43.	(B)
6.	(C)	44.	(D)
7.	(A)	45.	(B)
8.	(A)	46.	(C)
9.	(B)	47.	(B)
10.	(B)	48.	(C)
11.	(C)	49.	(B)
12.	(A)	50.	(B)
13.	(A)	51.	(A)
14.	(C)	52.	(A)
15.	(C)	53.	(A)
16.	(A)	54.	(C)
17.	(C)	55.	(A)
18.	(A)	56.	(C)
19.	(A)	57.	(B)
20.	(D)	58.	(C)
21.	(B)	59.	(C)
22.	(C)	60.	(B)
23.	(C)	61.	(D)
24.	(B)	62.	(C)
25.	(C)	63.	(C)
26.	(C)	64.	(A)
27.	(A)	65.	(D)
28.	(A)	66.	(A)
29.	(D)	67.	(A)
30.	(B)	68.	(C)
31.	(C)	69.	(C)
32.	(B)	70.	(D)
33.	(A)	71.	(B)
34.	(B)	72.	(B)
35.	(B)	73.	(B)
36.	(B)	74.	(D)
37.	(D)	75.	(B)
38.	(A)		

CBSE -1994

1.	(A)	39.	(D)
2.	(A)	40.	(D)
3.	(C)	41.	(B)
4.	(C)	42.	(B)
5.	(A)	43.	(C)
6.	(A)	44.	(A)
7.	(D)	45.	(A)
8.	(A)	46.	(D)
9.	(A)	47.	(D)
10.	(C)	48.	(A)
11.	(A)	49.	(B)
12.	(C)	50.	(B)
13.	(D)	51.	(D)
14.	(B)	52.	(B)
15.	(B)	53.	(A)
16.	(A)	54.	(B)
17.	(B)	55.	(B)
18.	(B)	56.	(D)
19.	(D)	57.	(A)
20.	(A)	58.	(A)
21.	(C)	59.	(B)
22.	(B)	60.	(C)
23.	(A)	61.	(D)
24.	(B)	62.	(D)
25.	(A)	63.	(B)
26.	(B)	64.	(D)
27.	(C)	65.	(A)
28.	(D)	66.	(A)
29.	(D)	67.	(C)
30.	(C)	68.	(B)
31.	(B)	69.	(C)
32.	(B)	70.	(C)
33.	(D)	71.	(C)
34.	(C)	72.	(C)
35.	(A)	73.	(D)
36.	(D)	74.	(C)
37.	(A)	75.	(D)
38.	(C)		

CBSE -1995

1.	(B)	27.	(C)
2.	(A)	28.	(C)
3.	(A)	29.	(B)
4.	(D)	30.	(D)
5.	(A)	31.	(D)
6.	(A)	32.	(D)
7.	(B)	33.	(D)
8.	(D)	34.	(C)
9.	(C)	35.	(C)
10.	(B)	36.	(B)
11.	(D)	37.	(A)
12.	(C)	38.	(B)
13.	(B)	39.	(C)
14.	(B)	40.	(D)
15.	(B)	41.	(A)
16.	(A)	42.	(B)
17.	(C)	43.	(C)
18.	(A)	44.	(B)
19.	(B)	45.	(C)
20.	(B)	46.	(D)
21.	(A)	47.	(B)
22.	(C)	48.	(A)
23.	(C)	49.	(C)
24.	(B)	50.	(A)
25.	(D)		
26.	(C)		

CBSE -1996

1.	(B)	27.	(D)
2.	(C)	28.	(A)
3.	(C)	29.	(A)
4.	(D)	30.	(B)
5.	(D)	31.	(B)
6.	(D)	32.	(B)
7.	(D)	33.	(C)
8.	(A)	34.	(C)
9.	(C)	35.	(D)
10.	(B)	36.	(C)
11.	(A)	37.	(A)
12.	(B)	38.	(B)
13.	(A)	39.	(A)
14.	(C)	40.	(D)
15.	(A)	41.	(A)
16.	(A)	42.	(D)
17.	(A)	43.	(C)
18.	(B)	44.	(C)
19.	(A)	45.	(B)
20.	(B)	46.	(B)
21.	(D)	47.	(B)
22.	(C)	48.	(A)
23.	(A)	49.	(B)
24.	(B)	50.	(D)
25.	(A)		
26.	(A)		

CBSE -1997

1.	(D)	27.	(D)
2.	(D)	28.	(D)
3.	(A)	29.	(C)
4.	(C)	30.	(C)
5.	(B)	31.	(D)
6.	(C)	32.	(A)
7.	(B)	33.	(D)
8.	(D)	34.	(C)
9.	(D)	35.	(C)
10.	(A)	36.	(A)
11.	(D)	37.	(D)
12.	(C)	38.	(A)
13.	(C)	39.	(A)
14.	(B)	40.	(B)
15.	(D)	41.	(C)
16.	(D)	42.	(C)
17.	(A)	43.	(A)
18.	(C)	44.	(A)
19.	(D)	45.	(D)
20.	(B)	46.	(B)
21.	(A)	47.	(B)
22.	(A)	48.	(B)
23.	(A)	49.	(B)
24.	(C)	50.	(C)
25.	(D)		
26.	(B)		

CBSE -1998

1.	(B)	27.	(D)
2.	(C)	28.	(B)
3.	(A)	29.	(A)
4.	(C)	30.	(A)
5.	(B)	31.	(C)
6.	(B)	32.	(B)
7.	(C)	33.	(C)
8.	(A)	34.	(D)
9.	(B)	35.	(A)
10.	(D)	36.	(A)
11.	(B)	37.	(D)
12.	(A)	38.	(D)
13.	(C)	39.	(B)
14.	(D)	40.	(C)
15.	(B)	41.	(C)
16.	(C)	42.	(B)
17.	(D)	43.	(D)
18.	(B)	44.	(D)
19.	(C)	45.	(D)
20.	(C)	46.	(C)
21.	(B)	47.	(A)
22.	(B)	48.	(A)
23.	(B)	49.	(A)
24.	(A)	50.	(A)
25.	(C)		
26.	(B)		

CBSE -1999

1.	(B)	27.	(B)
2.	(C)	28.	(B)
3.	(B)	29.	(C)
4.	(B)	30.	(A)
5.	(A)	31.	(B)
6.	(B)	32.	(D)
7.	(D)	33.	(C)
8.	(A)	34.	(C)
9.	(D)	35.	(C)
10.	(B)	36.	(D)
11.	(C)	37.	(B)
12.	(A)	38.	(A)
13.	(B)	39.	(B)
14.	(D)	40.	(C)
15.	(D)	41.	(B)
16.	(C)	42.	(C)
17.	(C)	43.	(A)
18.	(C)	44.	(A)
19.	(A)	45.	(D)
20.	(A)	46.	(C)
21.	(D)	47.	(D)
22.	(A)	48.	(B)
23.	(D)	49.	(A)
24.	(C)	50.	(A)
25.	(A)		
26.	(B)		

CBSE -2000

1.	(D)	27.	(A)
2.	(C)	28.	(D)
3.	(C)	29.	(A)
4.	(D)	30.	(D)
5.	(D)	31.	(D)
6.	(B)	32.	(D)
7.	(D)	33.	(B)
8.	(B)	34.	(D)
9.	(A)	35.	(A)
10.	(A)	36.	(A)
11.	(A)	37.	(A)
12.	(D)	38.	(C)
13.	(B)	39.	(D)
14.	(B)	40.	(B)
15.	(D)	41.	(A)
16.	(C)	42.	(B)
17.	(D)	43.	(C)
18.	(A)	44.	(C)
19.	(B)	45.	(C)
20.	(D)	46.	(A)
21.	(A)	47.	(A)
22.	(C)	48.	(B)
23.	(A)	49.	(A)
24.	(A)	50.	(D)
25.	(B)		
26.	(D)		

CBSE -2001

1.	(D)	27.	(C)
2.	(A)	28.	(A)
3.	(A)	29.	(B)
4.	(D)	30.	(A)
5.	(D)	31.	(A)
6.	(B)	32.	(A)
7.	(A)	33.	(C)
8.	(C)	34.	(A)
9.	(C)	35.	(A)
10.	(A)	36.	(C)
11.	(A)	37.	(B)
12.	(C)	38.	(D)
13.	(C)	39.	(B)
14.	(A)	40.	(A)
15.	(D)	41.	(A)
16.	(B)	42.	(D)
17.	(B)	43.	(A)
18.	(C)	44.	(A)
19.	(A)	45.	(D)
20.	(A)	46.	(B)
21.	(B)	47.	(C)
22.	(B)	48.	(A)
23.	(C)	49.	(A)
24.	(A)	50.	(C)
25.	(A)		
26.	(B)		

CBSE -2002

1. (A)
2. (A)
3. (C)
4. (A)
5. (B)
6. (A)
7. (C)
8. (B)
9. (C)
10. (B)
11. (C)
12. (B)
13. (A)
14. (A)
15. (C)
16. (D)
17. (A)
18. (B)
19. (B)
20. (C)
21. (A)
22. (A)
23. (D)
24. (B)
25. (A)
26. (B)
27. (A)
28. (B)
29. (B)
30. (D)
31. (B)
32. (B)
33. (A)
34. (B)
35. (A)
36. (C)
37. (C)
38. (B)
39. (D)
40. (A)
41. (D)
42. (A)
43. (D)
44. (B)
45. (A)
46. (B)
47. (D)
48. (B)
49. (B)
50. (C)

CBSE -2003

1. (B)
2. (D)
3. (C)
4. (B)
5. (D)
6. (B)
7. (B)
8. (A)
9. (B)
10. (A)
11. (B)
12. (A)
13. (A)
14. (B)
15. (B)
16. (C)
17. (A)
18. (C)
19. (A)
20. (C)
21. (C)
22. (B)
23. (A)
24. (D)
25. (B)
26. (C)
27. (A)
28. (A)
29. (B)
30. (D)
31. (A)
32. (A)
33. (D)
34. (A)
35. (C)
36. (A)
37. (D)
38. (D)
39. (C)
40. (D)
41. (B)
42. (B)
43. (D)
44. (A)
45. (A)
46. (D)
47. (C)
48. (C)
49. (C)
50. (C)

CBSE -2004

1. (C)
2. (B)
3. (D)
4. (C)
5. (A)
6. (B)
7. (A)
8. (B)
9. (B)
10. (A)
11. (D)
12. (B)
13. (D)
14. (A)
15. (A)
16. (C)
17. (A)
18. (D)
19. (A)
20. (D)
21. (B)
22. (A)
23. (A)
24. (C)
25. (C)
26. (B)
27. (D)
28. (D)
29. (B)
30. (A)
31. (C)
32. (A)
33. (C)
34. (C)
35. (C)
36. (A)
37. (D)
38. (B)
39. (C)
40. (A)
41. (C)
42. (C)
43. (D)
44. (B)
45. (B)
46. (A)
47. (B)
48. (D)
49. (D)
50. (B)

CBSE - 2005

1. (A)
2. (D)
3. (D)
4. (A)
5. (D)
6. (C)
7. (B)
8. (B)
9. (A)
10. (A)
11. (D)
12. (A)
13. (A)
14. (D)
15. (A)
16. (C)
17. (C)
18. (C)
19. (D)
20. (B)
21. (C)
22. (D)
23. (D)
24. (B)
25. (C)
26. (D)
27. (A)
28. (A)
29. (A)
30. (C)
31. (B)
32. (C)
33. (D)
34. (B)
35. (D)
36. (B)
37. (D)
38. (B)
39. (B)
40. (C)
41. (B)
42. (C)
43. (B)
44. (C)
45. (C)
46. (B)
47. (A)
48. (C)
49. (A)
50. (B)

CPMT - 1996

1. A catalyst
(A) does not take part in reaction
(B) is specific in action (C) Takes part in reaction
(D) None of these

2. Night-blindness is due to
(A) Vitamin A (B) Vitamin B_5
(C) Vitamin C (D) Vitamin D

3. HCOOH shows all tests of aldehyde because
(A) It has aldehyde group
(B) It is member of aldehydes
(C) All acids show test of aldehyde
(D) Does not show any test

4. Which of the following is the sweetest sugar ?
(A) Sucrose (B) Fructose (C) Glucose (D) Lactose

5. Which of the following does not show any reducing test for aldehyde ?
(A) Sucrose (B) Fructose (C) Ketose (D) Lactose

6. Benzene is obtained from
(A) Coaltar (B) Plants (C) Animals (D) Charcoal

7. In Dumas method which of the following is measured ?
(A) Nitrogen (B) Ammonium Nitrite
(C) Cynogen (D) Cyno Benzene

8. Alumino thermic process is used for metallurgy of.
(A) Pb (B) Ag (C) Al (D) None of these

9. In $CH_3CH_2CH_2OH$ which bond is most easily broken ?
(A) C—C (B) C—H (C) O—H (D) C—O

10. When CH_3COCH_3 reacts with Cl_2, which of the following is formed ?
(A) $CHCl_3$ (B) CCl_4 (C) CCl_2H_2 (D) CH_3Cl

11. When sodium propanoate is heated with soda lime, which of the following is formed ?
(A) CH_4 (B) CH_3CH_3 (C) Butane (D) Propane

12. Half-life of 10 gm of a radioactive substance is 10 days. The half-life of 20 gm is
(A) 10 days (B) 20 days (C) 25 days (D) Infinite

13. In qualitative analysis NH_4Cl is added before NH_4OH
(A) To decrease $[OH]^-$ conc.
(B) To increase $[OH^-]$ conc.
(C) For making HCl (D) None of these

14. If HBr is added to $CH_2 = C(CH_3) — CH_3$, the compound formed is

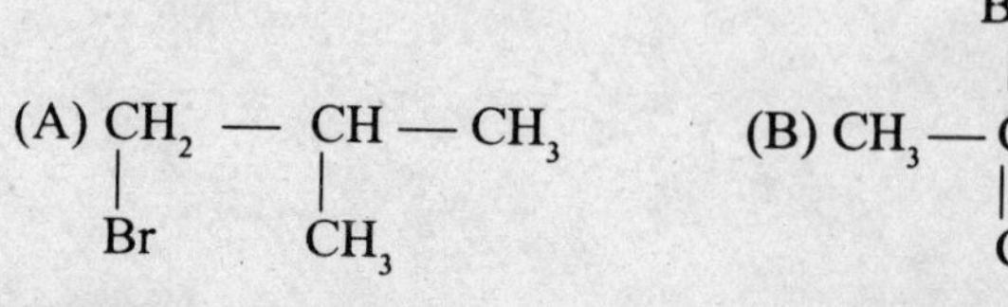

(A) $CH_2(Br) — CH(CH_3) — CH_3$ (B) $CH_3 — C(Br)(CH_3) — CH_3$
(C) $CH_2 = C(CH_2Br) — CH_3$ (D) None of these

15. Electrovalent compounds are
(A) Good conductor of electricity
(B) Unploar in nature (C) low mp & bp
(D) Easily available

16. Which of the following compounds shows highest m.p.?
(A) Coordinate (B) Electrovalent
(C) Covalent (D) None of these

17. At. No. of an element is 26. The element shows
(A) Ferromagnetism (B) Diamagnetism
(C) Paramagnetism (D) None of these

18. Which of the following is not obtained when C_2H_5OH is heated with conc. H_2SO_4 ?
(A) $C_2H_5SO_4H$ (B) $CH_3CH_2CH_2CH_3$
(C) $CH_2 = CH_2$ (D) $C_2H_5OC_2H_5$

19. $2\ SO_3 \rightleftharpoons 2\ SO_2 + O_2$. If $K_C = 100$, $\alpha = 1$ when half of the reaction is completed the concentration of SO_3 and SO_2 are equal, the concentration of O_2 is
(A) 0.001 M (B) $\frac{1}{2}$ of SO_2
(C) 2 time of SO_2 (D) Data incomplete

20. Solubility product of $PbCl_2$ is 1.0×10^{-6}. The solubility is
(A) 6.0×10^{-7} (B) 6.3×10^{-6}
(C) 7.5×10^{-5} (D) None of these

21. $CH_3 — C(OH)(CH_3) — CH_2 — CH(CH_3) — CH_3$, IUPAC name is
(A) 2,4 dimethyl pentanol-2
(B) 2,4 dimethyl pentanol-2
(C) 2,2 dimethyl butanol-2 (D) Butanol-2

22. Which connot displace H_2 from its compound ?
(A) Zn (B) Cu (C) Hg (D) None of these

23. Electromagnetic radiation with maximum frequency is
(A) Ultraviolet (B) Radiowave
(C) X-rays (D) Infrared

24. Which of the following contains maximum number of unpaired electron?
(A) CO^{+3} (B) Fe^{+2} (C) Mn^{+2} (D) Co^{+2}

25. Which of the following conditions is used to find atomic Cl from molecular Cl_2?
(A) Low temp., high pressure
(B) High temp., low pressure
(C) High temp., high pressure
(D) Low temp, low pressure

26. Zn does not show variable valency because
(A) Of complete 'd' shell (B) Intertpair effect
(C) Of complete 4s subshell
(D) None of these

27. $_{20}X^{40}$ and $_{21}X^{40}$ are
(A) Isotones (B) Isostereomers

(C) Isobars (D) Isotopes

28. 1 AMU is equal to
(A) 1 gm of H_2 (B) 1.66×10^{-23} kg
(C) $\frac{1}{12}$ of C –12 (D) $\frac{1}{16}$ of O – 16

29. The property of attracting electrons by the halogen atom is called
(A) Electronegativity (B) Electronic attraction
(C) Ionisation potential (D) Electron affinity

30. Ionization of Hydrogen requires .... energy
(A) < 13.6 eV (B) 1.5 eV
(C) 13.6 eV (D) > 13.6 eV

31. Separation of Pentene-1 and Pentyne-1 is done by
(A) $Ag(NH_3)^+$ (B) Conc. H_2SO_4
(C) HCl (D) Br_2

32. $CH_2 = CH — (CH_2)_5COOH \xrightarrow[HBr]{Peroxide}$ Z, Z is
(A) $CH_2 = CH — (CH_3)_5$ COOH
|
Br
(B) $BrCH_2 — (CH_2)_6COOH$
(C) C_6H_5COOH
(D) $CH_2 = CH — (CH_3)_5 CH_2OH$

33. Which of the following has highest density ?
(A) B (B) Pb (C) Fe (D) Cu

34. Heat of combustion is
(A) Always -ve (B) +ve (C) Both A and B (D) 0

35. Same mass of CH_4, H_2 is taken in a container, the partial pressure caused by H_2 is
(A) $\frac{1}{2}$ (B) 1 (C) $\frac{8}{9}$ (D) $\frac{1}{9}$

36. Which of the following contains asymmetric centre ?
(A) Butene-2 (B) 2,2-dimethyl propane
(C) Hexyne-2 (D) None of these

37. Which of the following shows only one Brominated compound ?
(A) Butene-2 (B) 2,2-dimethyl propane
(C) Butyne-1 (D) Butanol-3

38. When NaOH is prepared, the gas released is
(A) Cl_2 (B) H_2 (C) O_2 (D) H_2O

39. Which gas is obtained when urea is heated with HNO_2 ?
(A) N_2 (B) H_2 (C) O_2 (D) NH_3

40. Kerosene is used as fuel because it is
(A) Less volatile (B) More volatile
(C) Cheap (D) Abundantly available

41. Which of the following is used to flow electricity ?
(A) Cu (B) Ag (C) Fe (D) Al

42. If the reduction potential is more
(A) Act as oxidizing agent (B) Redox nature
(C) It is easily oxidized (D) Easily reduced

43. Which of the following cannot work as oxidizing agent ?
(A) H_2O_2 (B) I_2 (C) O_2 (D) $KMnO_4$

44. When sulphuric ether is reacted in excess of HI, what is formed ?
(A) C_2H_5OI (B) $C_2H_5I + H_2O$
(C) C_2H_5I (D) $C_2H_5OC_2H_4I$

45. When ether is reacted with O_2 it forms
(A) Peroxide (B) Acid (C) Ketone (D) TNT

46. To prevent rancidification of food material, which of the following is added ?
(A) Reducing agent (B) Anti-oxidant
(C) Oxidising agent (D) None of these

47. For the separation of 2 immiscible liquids which method is used ?
(A) Chromatography (B) Fractional column
(C) Fractional distillation (D) Separating funnel

48. Oxidation state of N in $(NH_4)_2$ SO_4 is
(A) +1 (B) 3 (C) –1/3 (D) –1

49. Atomic number of N is 7. The at. no. of IIIrd member of Nitrogen family is
(A) 33 (B) 43 (C) 23 (D) 15

50. In esterification OH^- ion for making H_2O comes from
(A) Acid (B) Alcohol
(C) Ketone (D) Carbohydrate

51. If the concentration of HX and X^- are equal and the K_b of the solution is 10^{-10}, pH of the solution is
(A) 14 (B) 8 (C) 4 (D) 10

52. Number of unpaired electrons in aninert gas is
(A) Zero (B) 8 (C) 4 (D) 18

53. Which of the following is strongest base ?
(A) $Al(OH)_3$ (B) $Mg(OH)_2$
(C) Ca $(OH)_2$ (D) $Ba(OH)_2$

54. What is the pH of Ba $(OH)_2$ of normality 10 ?
(A) 7 (B) 9 (C) 4 (D) 10

55. Which of the following has least covalent P-H bond ?
(A) PH_3 (B) $P_2H_6^-$ (C) P_2H_5 (D) PH_6^+

56. $C + \frac{1}{2}O_2 \rightarrow CO, \ \Delta H = 42; CO + \frac{1}{2}O_2 \rightarrow CO_2, \Delta H = +24.$
The heat of formation of CO_2 is
(A) –16 Kcal (B) +66 Kcal
(C) + 16 Kcal (D) –66 Kcal

57. Which of the following is not linear ?
(A) CO_2 (B) Hg_2Cl_2 (C) $SnCl_2$ (D) HCl

58. During smelting an additional substance is added which combines with impurities to form a fusible product which is known as -
(A) Gangue (B) Flux (C) Slag (D) Mud

59. When calcium acetate is heated, the product formed is ?
(A) CH_3COCH_3 (B) CH_3COOH
(C) CH_3CHO (D) HCOOH

60. When silver nitrate is heated what is formed ?

(A) Ag_2O (B) Ag_2O_3 (C) Ag (D) AgO_2

61. Which of the following metals will not react with a solution of $CuSO_4$?
(A) Fe (B) Zn (C) Mg (D) Hg

62. Calcium is obtained by the-
(A) Electrolysis of solution of $CaCl_2$ in water
(B) Reduction of $CaCl_2$ with carbon
(C) Electrolysis of molten $CaCl_2$
(D) Roasting of lime stone

63. Which statement is not correct for n = 5, m = 3 ?
(A) $l = 4$ (B) $l = 0, 1, 2, 3; s = +\frac{1}{2}$
(C) $l = 3; s = +\frac{1}{2}$ (D) $l = 2; s = +\frac{1}{2}$

64. $1s^2, 2s^2, 2p^6, 3s^1$ shows configuration of -
(A) Al^{+3} in ground state (B) Ne in excited state
(C) Mg^{+1} in excited state (D) None of these

65. What is the name of reaction when Benzaldehyde changes into Benzyl alcohol?
(A) Friedel-Craft reaction (B) Kolbe's reaction
(C) Wurtz reaction (D) Cannizaro reaction

66. Which of the following reactions shows Industrial method of CH_3OH ?
(A) $CO + H_2$ reduction (B) $CH_3Cl + H_2O$
(C) $CH_3NH_2 + HNO_2$ (D) CH_3Br + aqKOH

67. Which of the following is an example of Friedal-Craft's reaction ?
(A) $C_6H_6 + Cl_2 \xrightarrow{AlCl_3} C_6H_5Cl + HCl$
(B) $C_6H_6 + C_6H_5COCl \xrightarrow{AlCl_3} C_6H_5COC_6H_5 + HCl$
(C) $CH_3COOAg \longrightarrow CH_3Br + CO_2 + AgBr$
(D) $C_2H_5OH \xrightarrow{H_2SO_4} C_2H_5OC_2H_5$

68. $CH_3CONH_2 \xrightarrow{Na + ROH} Z + H_2O$, what is Z ?
(A) $CH_3CH_2NH_2$ (B) CH_3CH_2NC
(C) $CH_3CH_2CH_3$ (D) NH_2CONH_2

69. In which of the following reactions Phenol or Sodium Phenoxide is not formed ?
(A) $C_6H_5N_2Cl$ + Alc KOH (B) C_6H_5OCl + NaOH
(C) $C_6H_5N_2Cl$ + Aq NaOH (D) $C_6H_5NNCl \xrightarrow[\Delta]{H_2O}$

70. What is formed when HNO_3 is heated ?
(A) NO_2 (B) NO (C) N_2O_5 (D) N_2O_4

71. NaCl and CH_3COOH kept in separate containers. If respectively pressure is P_1 and P_2, then what is the correct statement ?
(A) $P_1 > P_2$ (B) $P_1 = P_2$
(C) $P_1 < P_2$ (D) $P_1 = P_2 = 0$ atm

72. If one mole of substance is present in 1 kg of solvent, it is
(A) Molar concentration (B) Molal concentration
(C) Normality (D) Strength gm/gm

73. Gas deviates form ideal gas nature as the molecules
(A) Are colourless (B) Attract each other
(C) Contain covalent bond (D) Show Brownian motion

74. Which of the following is employed in metal welding ?
(C) Butane (D) Iso-Butene
(A) Acetylene (B) Ethane

75. 2N-HCl will have the same molar concentration as-
(A) $0.5N - H_2SO_4$ (B) $1.0N - H_2SO_4$
(C) $2N - H_2SO_4$ (D) $4N - H_2SO_4$

76. Which of the following pairs can not be distinguished by passing H_2S ?
(A) Hg, Pb (B) Cd, Pb
(C) As, Cd (D) Zn, Mn

77. $2NO_2 \rightleftharpoons 2NO + O_2$, K= 1.6×10^{-12}
$2NO + O_2 \rightleftharpoons 2NO_2$, $K^1 = ?$
(A) $K^1 = \frac{1}{K^2}$ (B) $K^1 = \frac{1}{K}$
(C) $K^1 = \frac{1}{\sqrt{K}}$ (D) None of these

78. Colour of colloidal solutions is due to-
(A) Due to formation of hydrated crystal
(B) Different size of colloidal particles
(C) Due to formation of complex
(D) None of these

79. Which of the following is a property of colloid ?
(A) Scattering of light (B) They show attraction
(C) Dialysis (D) Emulsion

80. The stability of lyophilic colloid is due to -
(A) Charge on their particles
(B) a layer of medium of dispersion on their particles
(C) Small-size of particles
(D) The large-size of their particles

81. Candlles (Parafin wax) contain
(A) Higher unsaturated Hydrocarbons
(B) Lower unsaturated Hydrocarbons
(C) Higher saturated Hydrocarbons
(D) Lower saturated Hydrocarbo

82. 10 volume of H_2O_2 is a solution with a strength of approximately -
(A) 30% (B) 3% (C) 1% (D) 10%

83. Repeated use of which of the following fertiliser would increase the acidity of soil ?
(A) Urea (B) Potassium nitrate
(C) Ammonium sulphate (D) Superphosphate of lime

84. Which of the following species will impart colour to an aqueous solution ?
(A) Cr^{3+} (B) Ti^{4+} (C) Cu^+ (D) Zn^{2+}

CPMT - 1997

1. Alkylhalide heated with dry Ag_2O gives
(A) Ketone (B) AgCl + Hydrocarbon
(C) Ester (D) Ether

2. If glycerol is heated with oxalic acid at 100-110°C, it gives an acid. It is
(A) Acetic acid (B) Formic acid
(C) Propanoic acid (D) Benzoic acid

3. Metal that causes tears is
(A) Ni (B) Sb (C) Fe (D) Sn

4. Aluminium is most abundant is earht's crust, yet it is obtained from bauxite because
(A) Bauxite contains maximum aluminium
(B) Bauxite is available in large quantity
(C) Easy extraction of aluminium from it
(D) Bauxite is less impure

5. Aldehydes and ketones can be distinguished by
(A) Fehling's solution (B) H_2SO_4 solution
(C) Alkaline $KMnO_4$ (D) NH_3

6. Which metal can't be obtained by electrolysis ?
(A) Cu (B) Mg (C) Cr (D) Ni

7. $MgCl_2 . 6H_2O$ on heating gives
(A) Magnesium oxychloride (B) Magnesium dichloride
(C) Magnesium oxide (D) Magnesium chloride

8. Wool when heated with BaO at 1100°C, it gives a compound. The compound is
(A) $BaZnO_2$ (B) Ba + ZnO_2 (C) $BaCdO_2$ (D) BaO_2 + Zn

9. The number of electrons and neutrons of an element is 18 and 20 respectively. Its mass number is
(A) 2 (B) 38 (C) 17 (D) 37

10. The number of electrons in $[{}^{40}_{19}K]^{-1}$ is
(A) 19 (B) 20 (C) 18 (D) 40

11. Which of the following has zero dipole moment ?
(A) CH_2Cl_2 (B) CH_4 (C) PH_4 (D) PH_3

12. Phenol + Zn (powder) .... ?
(A) Benzene (B) Benzaldehyde
(C) Anisole (D) Phenyl acetate

13. Number of π (pi) and σ (sigma) bonds in benzene is
(A) $3\pi + 12\sigma$ (B) $12\pi + 12\sigma$ (C) $6\pi + 12\sigma$ (D) $6\pi + 6\sigma$

14. When primary alcohol is oxidized with chlorine, it gives
(A) CH_3CHO (B) CCl_3CHO
(C) CH_3COCH_3 (D) Cl_3CHO

15. IUPAC name of $CH_3 - CH = CH - C \equiv CH$ is
(A) Pent-2-en-4-yne (B) Pent-3-en-1-yne
(C) Pent-3-yne-1-en (D) Pent-2-yne-1-en

16. When HNO_3 is dropped into the palm and washed with water it turns into yellow. It shows the prescnce of
(A) NO (B) N_2O_5 (C) NO_2 (D) N_2O

17. Which of the following is a polymer ?
(A) CH_4 (B) PVC (C) SO_2 (D) CO_2

18. The size of colloidal particle is
(A) 10^{-6} to 10^{-9} m (B) 10^{-9} to 10^{-12} m
(C) 10^{-3} to 10^{-9} m (D) 10^{-12} to 10^{-19} m

19. Fluorine has lower electron affinity than chlorine because of
(A) Bigger radius of fluorine, less density
(B) Smaller radius of chlorine, high density
(C) Smaller radius of fluorine, high density
(D) Smaller radius of chlorine, high density

20. Which one of the following shows bond in silica?
(A) –Si–O–Si–O–Si– (B) Si–C–Si–O–Si
(C) Si–C–Si–C–Si (D) Si–Si–Si–Si

21. E_0 of a cell a A + bB $\rightarrow$ Cc +Dd is-
(A) $-\frac{RT}{nF}\log\frac{[C]^c[D]^d}{[A]^a[B]^b}$ (B) $-\frac{RT}{nF}\log\frac{[a]^A[b]^B}{[c]^C[d]^D}$
(C) $-\frac{RT}{nF}\log\frac{[C]^c[d]^D}{[A]^a[B]^B}$ (D) $-\frac{RT}{nF}\log\frac{[C]^c[d]^D}{[a]^A[B]^b}$

22. Oxidation state of Ni in Ni $(CO)_4$ is
(A) 2 (B) 3 (C) 1 (D) 0

23. Oxidation number of C in HCHO is
(A) 0 (B) +4 (C) -2 (D) +2

24. van der Waal's equation is true for
(A) Ideal gas (B) Real gas
(C) Gaseous substance (D) None of these

25. Number of water molecules in Mohr's salt is
(A) 7 (B) 6 (C) 5 (D) 8

26. The difference of water molecules in gypsum and plaster of Paris is
(A) $\frac{1}{2}$ (B) $1\frac{1}{2}$ (C) 5/2 (D) 2

27. The formula of corrosive sublimate is
(A) $HgCl_2$ (B) Hg_2Cl_2 (C) Hg_2O (D) Hg

28. Which one has maximum alkaline nature ?
(A) PH_3 (B) NH_3 (C) AsH_3 (D) SbH_3

29. Example of sp^2 hybridisation is
(A) $C_2H_5^+$ (B) $C_2H_5^-$ (C) CH_3^+ (D) CH_3^-

30. Which one has large radius ?
(A) Na+ (B) F (C) F- (D) Na

31. Nessler's reagent is
(A) $KHgI_4$ (B) K_2HgI_4 + NH_4OH
(C) K_2HgI_4 + KOH (D) $KHgI_4$ + NH_4OH

32. Electronic configuration of a transition element is
(A) $1s^2 2s^2 2p^6 3s^2 3p^6 4s^2$
(B) $1s^2 2s^2 2p^6 3s^4 3p^6 3d^{10} 4s^2 4p^1$
(C) $1s^2 2s^2 2p^6 3s^2 3p^6 3d^{10} 4s^2 4p^6$
(D) $1s^2 2s^2 2p^6 3s^2 3p^6 3d^2 4s^2$

33. $yMnO_4^- + xH^+ + 5C_2O_4^{2-} \rightarrow yMn^{++} + 10CO_2 + \frac{x}{2}H_2O$.
X and y is -
(A) 2 and 16 (B) 16 and 2 (C) 8 and 16 (D) 5 and 2

34. Which oxychloride has maximim pH ?

(A) NaClO (B) $NaClO_2$
(C) $NaClO_3$ (D) $NaClO_4$

35. The b.p of glycerol is more than propanol because of
(A) Hydrogen bonding (B) Hybridisation
(C) Resonance (D) All of these

36. Identify the correct statement in the following compounds HF, HBr, H_2Te, H_3P
(A) HBr (Strong acid) (B) H_3P (strong base)
(C) HF (strong acid) (D) H_2Te (strong base)

37. Vant Hoff factor of $Ca(NO_3)_2$ is
(A) 1 (B) 2 (C) 3 (D) 4

38. Which acid will not form hydrocarbon ?
(A) Cinnamic acid (B) Isothallic
(C) Salicylic acid (D) Picric acid

39. $R-OH+HX \rightarrow R-X+H_2O$ In the above reaction, the reactivity of different alcohols is
(A) Tertiary<Secondary>Primary
(B) Secondary<Primary<Tertiary
(C) Tertiary>Secondary>Primary
(D) Tertiary<Secondary<Primary

40. Constitution of petrol is-
(A) C_3-C_5 (B) $C_{12}-C_{18}$ (C) C_5-C_{12} (D) C_8-C_{10}

41. Density of methane at 25°C and 6 atm, pressure is (Given R = 0.0821 atm)
(A) 4 gm/L (B) 8 gm/L
(C) 12 gm/L (D) 16 gm/L

42. The pH value of 0.1M NaOH solution is (when there is a given reaction $[H^+][OH^-] = 10^{-14}$
(A) 5 (B) 18 (C) 13 (D) 12

43. Which of the following relationships is true ?
(A) $K_C = K_P(RT)^{\Delta n}$ (B) $K_P = K_C(RT)^{\Delta n}$
(C) $\frac{K_c}{RT} = (KP)^{\Delta n}$ (C) $\frac{K_p}{RT} = (KC)^{\Delta n}$

44. Density ratio of O_2 and H_2 is 16:1. The ratio of its V_{rms} is -
(A) 4 : 1 (B) 16 : 1 (C) 1 : 4 (D) 1 : 16

45. Metal present in blood is
(A) Al (B) Mg (C) Cu (D) Fe

46. Which orbital of carbon will not emit photon after absorbing it ?
(A) 3p (B) 2p (C) 1s (D) 2s

47. Number of isomers of C_7H_9N is
(A) 3 (B) 5 (C) 4 (D) 6

48. The long form of periodic table is based on
(A) Shape of the atom (B) Mass of the atom
(C) atomic number (D) Electronegativity

49. X in the given reaction is $HNO_3 + 2P_2O_5 \rightarrow P_4O_{10} + X$
(A) N_2O_4 (B) N_2O_5 (C) NO_2 (D) H_2O

50. White phosphorus + Caustic soda → Phosphine+ NaH_2PO_2. The above reaction is
(A) Oxidation reaction (B) Reduction reaction
(C) Neutralisation reaction (D) Redox reaction

51. Which one is an oxidizing agent in the reaction
$2CrO_4^{2-} + 2H^+ \rightarrow Cr_2O_7^{2-} + H_2O$
(A) H^+ (B) $Cr_2O_4^-$ (C) Cr^{++} (D) None of these

52. Which one of the following is a strong reducing agent ?
(A) Cl (B) Cr (C) HNO_3 (D) Na^+

53. Which one is an oxidising substance ?
(A) CH_2O_2 (B) CO (C) H_2S (D) CO_2

54. When iodine reacts with NaF, NaBr and NaCl -
(A) It gives mixture of F_2, Cl_2 and Br_2
(B) It gives chlorine (C) It gives bromine
(D) None of these

55. Which statement for equilibrium constant is true for the reaction $A+B \rightleftharpoons C$?
(A) Increases with temperature
(B) Does not change with temperature
(C) Changes when catalyst is added
(D) Changes with temperature

56. When an acid or alkali is mixed with buffer solution, then pH of buffer solution
(A) Remains unchanged (B) Changes slightly
(C) Increases (D) Decreases

57. The number of atoms in Cr and O are 4.8×10^{10} and 9.6×10^{10} respectively. Its empirical formula is
(A) Cr_2O_3 (B) CrO_2 (C) Cr_2O_4 (D) None of these

58. Propane is obtained from propene by
(A) Catalytic hydrogenation (B) Wurtz reaction
(C) Dehydrogenation (D) Frakland reaction

59. Which substance will give amide when heated with NH_3 ?
(A) Potassium (B) Hydrogen
(C) Ethane (D) Nitrogen

60. The best method to separate the mixture of ortho-and paranitrophenol (1:1) is
(A) Distillation (B) Crystallization
(C) Vaporisation (D) None of these

61. If N and S are present in an organic compound, in Lassaigne test, they are converted into
(A) Na_2S and NaCN (B) NaSCN
(C) Na_2SO_3 and NaCN (D) Na_2S and NaCNO

62. The indicator that is used in the titration of Na_2CO_3 solution with HCl is
(A) Methyl organe (B) Dil. H_2SO_4
(C) Phenolphthalein (D) None of these

63. What is the solubility of calcium fluoride in a saturated solution if its solubility product is 3.2×10^{-11} ?
(A) 2.0×10^{-4} mole/litre (B) 12.0×10^{-3} mole/litre
(C) 0.2×10^{-4} mole/litre (D) 2×10^{-3} mole/litre

64. Which of the following order is correct ?

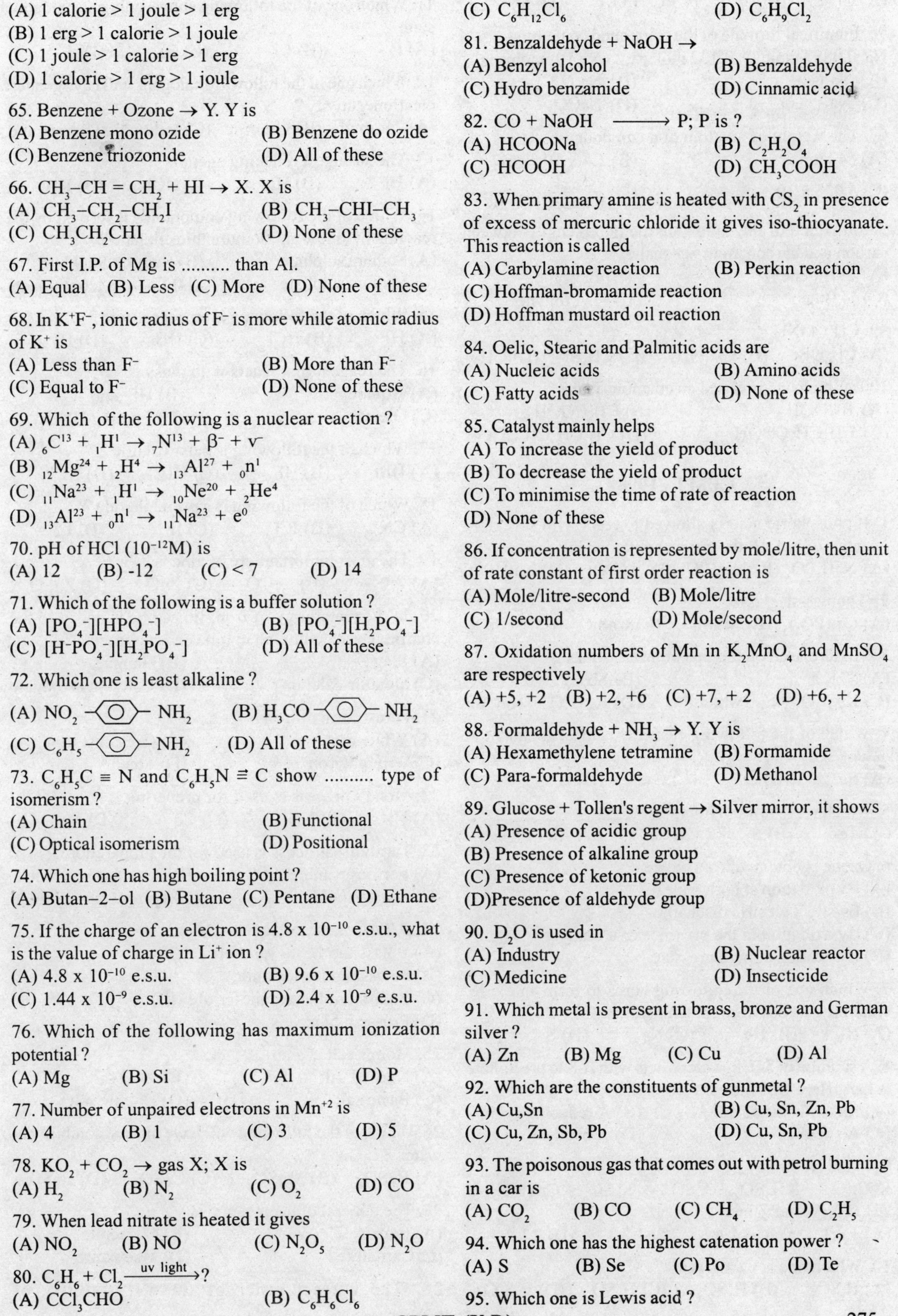

(A) 1 calorie > 1 joule > 1 erg
(B) 1 erg > 1 calorie > 1 joule
(C) 1 joule > 1 calorie > 1 erg
(D) 1 calorie > 1 erg > 1 joule

65. Benzene + Ozone $\rightarrow$ Y. Y is
(A) Benzene mono ozide (B) Benzene do ozide
(C) Benzene triozonide (D) All of these

66. $CH_3{-}CH = CH_2 + HI \rightarrow X$. X is
(A) $CH_3{-}CH_2{-}CH_2I$ (B) $CH_3{-}CHI{-}CH_3$
(C) CH_3CH_2CHI (D) None of these

67. First I.P. of Mg is .... than Al.
(A) Equal (B) Less (C) More (D) None of these

68. In K^+F^-, ionic radius of F^- is more while atomic radius of K^+ is
(A) Less than F^- (B) More than F^-
(C) Equal to F^- (D) None of these

69. Which of the following is a nuclear reaction ?
(A) ${}_6C^{13} + {}_1H^1 \rightarrow {}_7N^{13} + \beta^- + v^-$
(B) ${}_{12}Mg^{24} + {}_2H^4 \rightarrow {}_{13}Al^{27} + {}_0n^1$
(C) ${}_{11}Na^{23} + {}_1H^1 \rightarrow {}_{10}Ne^{20} + {}_2He^4$
(D) ${}_{13}Al^{23} + {}_0n^1 \rightarrow {}_{11}Na^{23} + e^0$

70. pH of HCl (10^{-12}M) is
(A) 12 (B) -12 (C) < 7 (D) 14

71. Which of the following is a buffer solution ?
(A) $[PO_4^-][HPO_4^-]$ (B) $[PO_4^-][H_2PO_4^-]$
(C) $[H^-PO_4^-][H_2PO_4^-]$ (D) All of these

72. Which one is least alkaline ?
(A) $NO_2{-}C_6H_4{-}NH_2$ (B) $H_3CO{-}C_6H_4{-}NH_2$
(C) $C_6H_5{-}C_6H_4{-}NH_2$ (D) All of these

73. $C_6H_5C \equiv N$ and $C_6H_5N \equiv C$ show .... type of isomerism ?
(A) Chain (B) Functional
(C) Optical isomerism (D) Positional

74. Which one has high boiling point ?
(A) Butan–2–ol (B) Butane (C) Pentane (D) Ethane

75. If the charge of an electron is 4.8×10^{-10} e.s.u., what is the value of charge in Li^+ ion ?
(A) 4.8×10^{-10} e.s.u. (B) 9.6×10^{-10} e.s.u.
(C) 1.44×10^{-9} e.s.u. (D) 2.4×10^{-9} e.s.u.

76. Which of the following has maximum ionization potential ?
(A) Mg (B) Si (C) Al (D) P

77. Number of unpaired electrons in Mn^{+2} is
(A) 4 (B) 1 (C) 3 (D) 5

78. $KO_2 + CO_2 \rightarrow$ gas X; X is
(A) H_2 (B) N_2 (C) O_2 (D) CO

79. When lead nitrate is heated it gives
(A) NO_2 (B) NO (C) N_2O_5 (D) N_2O

80. $C_6H_6 + Cl_2 \xrightarrow{\text{uv light}} ?$
(A) CCl_3CHO (B) $C_6H_6Cl_6$
(C) $C_6H_{12}Cl_6$ (D) $C_6H_9Cl_2$

81. Benzaldehyde + NaOH $\rightarrow$
(A) Benzyl alcohol (B) Benzaldehyde
(C) Hydro benzamide (D) Cinnamic acid

82. CO + NaOH $\longrightarrow$ P; P is ?
(A) HCOONa (B) $C_2H_2O_4$
(C) HCOOH (D) CH_3COOH

83. When primary amine is heated with CS_2 in presence of excess of mercuric chloride it gives iso-thiocyanate. This reaction is called
(A) Carbylamine reaction (B) Perkin reaction
(C) Hoffman-bromamide reaction
(D) Hoffman mustard oil reaction

84. Oelic, Stearic and Palmitic acids are
(A) Nucleic acids (B) Amino acids
(C) Fatty acids (D) None of these

85. Catalyst mainly helps
(A) To increase the yield of product
(B) To decrease the yield of product
(C) To minimise the time of rate of reaction
(D) None of these

86. If concentration is represented by mole/litre, then unit of rate constant of first order reaction is
(A) Mole/litre-second (B) Mole/litre
(C) 1/second (D) Mole/second

87. Oxidation numbers of Mn in K_2MnO_4 and $MnSO_4$ are respectively
(A) +5, +2 (B) +2, +6 (C) +7, + 2 (D) +6, + 2

88. Formaldehyde + $NH_3 \rightarrow$ Y. Y is
(A) Hexamethylene tetramine (B) Formamide
(C) Para-formaldehyde (D) Methanol

89. Glucose + Tollen's regent $\rightarrow$ Silver mirror, it shows
(A) Presence of acidic group
(B) Presence of alkaline group
(C) Presence of ketonic group
(D)Presence of aldehyde group

90. D_2O is used in
(A) Industry (B) Nuclear reactor
(C) Medicine (D) Insecticide

91. Which metal is present in brass, bronze and German silver ?
(A) Zn (B) Mg (C) Cu (D) Al

92. Which are the constituents of gunmetal ?
(A) Cu,Sn (B) Cu, Sn, Zn, Pb
(C) Cu, Zn, Sb, Pb (D) Cu, Sn, Pb

93. The poisonous gas that comes out with petrol burning in a car is
(A) CO_2 (B) CO (C) CH_4 (D) C_2H_5

94. Which one has the highest catenation power ?
(A) S (B) Se (C) Po (D) Te

95. Which one is Lewis acid ?

(A) Cl^- (B) Ag^+ (C) C_2H_5OH (D) S^-

96. Empirical formula of the compound containing Na, 0.0887 mole;O, 0.132 mole;C,2.65 x 10^{22}atoms is
(A) Na_2CO_3 (B) $Na_3O_2C_5$
(C) $Na_{0.0877}O_{0.132}C_{2.65x10^{22}}$ (D) NaCO

97. The weight of an atom of a compound $C_{60}H_{22}$ is
(A) 842 gm (B) 1.4 x 10^{-21}gm
(C) 5.025 x 10^{23} gm (D) $\frac{1}{6.023 \times 10^{23}}$ gm

98. In which of the following the bond length between carbon and carbon atom is equal ?
(A) 2–Butene (B) Benzene
(C) 1–Butene (D) 1–Propyne

99. $CH_3CONH_2 \xrightarrow{NaOBr} Z$. Z is
(A) CH_3OBr (B) CH_3NH_2 (C) CH_3Br (D) CH_4

100. Which one is called an ethanoic acid ?
(A) HCOOH (B) CH_3COOH
(C) CH_3CH_2COOH (D) $CH_3CH_2CH_2COOH$

CPMT - 1998

1. If phosphoric acid is allowed to react with sufficient quantity of NaOH, the product obtained is-
(A) NaH_2PO_4 (B) $NaHPO_3$ (C) Na_3PO_4 (D) Na_2HPO_4

2. 'Thomas-slag' is-
(A) $Ca_3(PO_4)_2$ (B) $MnSiO_3$ (C) $CaSiO_3$ (D) $FeSiO_3$

3. Mixture used on the tips of match sticks is-
(A) S+K (B) Sb_2S_3
(C) $K_2Cr_2O_7$ + S + white P (D) $K_2Cr_2O_7$ + K + S

4. Which of the following elements is most abundant in nature ?
(A) Si (B) O (C) Ca (D) C

5. Which will show polymorphism?
(A) Te (B) S (C) Se (D) All of these

6. Ozone is obtained from oxygen
(A) By oxidation at high temp.
(B) By silent electric discharge
(C) By oxidation in the presence of a catalyst
(D) By heating at high pressure

7. Which one of the following burns to form an oxide, which is gaseous at room temperature ?
(A) H_2 (B) He (C) Na (D) S

8. A solution of SO_2 in water reacts with H_2S to precipitate sulpur. Here SO_2 acts as-
(A) An oxidising agent (B) A reducing agent
(C) An acid (D) A catalyst

9. In the reaction-
$K_2Cr_2O_7 + XH_2SO_4 + YSO_2 \rightarrow K_2SO_4 + Cr_2(SO_4)_3 + ZH_2O$ X,Y and Z are-
(A) 1,3,1 (B) 4,1,4 (C) 3,2,3 (D) 2,1,2

10. Which one is Caro's acid ?
(A) H_2SO_3 (B) H_2SO_4 (C) H_2SO_5 (D) $H_2S_2O_8$

11. Which one of the following shows only one oxidation state ?
(A) F (B) Cl (C) Br (D) I

12. Which one of the following halogens will have greatest electronegativity ?
(A)Chlorine (B) Fluorine (C) Iodine (D) Bromine

13. The strongest reducing agent is
(A) HF (B) HI (C) HCl (D) HBr

14. Which of the following compounds is formed by the reaction of iodine and sodium thiosulphate ?
(A) Sodium sulphide (B) Sodium Suphhite
(C) Sodium sulphate (D) Sodium tetrathionate

15. Which of the following is strongest acid ?
(A) HF (B) HCl (C) HBr (D) HI

16. The reagent that reacts with glass is
(A) Aquaregia (B) HF acid
(C) Oleum (D) Fuming HNO_3

17. Which of the following is most volatile ?
(A) HBr (B) HF (C) HI (D) HCl

18. Which of the following is a pseudohalide ?
(A) CN^- (B) ICl (C) IF_5 (D) I_3^-

19. The most important ore of zinc is
(A) ZnS (B) $ZnCO_3$ (C) ZnO (D) $ZnSO_4$

20. In the metallurgy of iron, limestone is added to the charge. The calcium ion in lime ends up in
(A) Slag (B) Gangue
(C) Metallic calcium (D) Calcium carbonate

21. Purest form of iron is
(A) White cast iron (B) Grey cast iron
(C) Wrought iron (D) Steel

22. Mond's process is used for preparing
(A) Ni (B) NH_3 (C) Co (D) Mn

23. Turnbull's process is used for the preparation of
(A) Ferrouscyanide (B) Ferri-ferrocyanide
(C) Ferricyanide (D) Ferrous ferricyanide

24. IUPAC name of $K_4[Fe(CN)_6]$ is
(A) Potassium hexacyanoferrate (II)
(B) Potassium ferricyanide
(C) Potassium hexacyanoferrate (III)
(D) Prussian blue

25. Mohr's salt is a/an
(A) Normal salt (B) Acid salt
(C) Basic salt (D) Double salt

26. Which of the following will have highest solubility in water ?
(A) $BeSO_4$ (B) $MgSO_4$ (C) $CaSO_4$ (D) $BaSO_4$

27. The mineral of magnesium is
(A) Bauxite (B) Malachite
(C) Carnallite (D) Haematite

28. The correct order of thermal stability of

$BaCO_3$ (a), $CaCO_3$ (b) and $MgCO_3$ (c) will be
(A) a > b > c (B) a < b < c
(C) a > b < c (D) a < b > c

29. A piece of magnesium ribbon is heated to redness in the atmosphere of nitrogen and cooled with water. The evolved gas is
(A) O_2 (B) NH_3 (C) H_2 (D) N_2

30. Which mineral is used in the isolation of radium ?
(A) Limestone (B) Pitchblende
(C) Rutile (D) Haematite

31. Magnesium burns in
(A) CO (B) NO_2 (C) N_2 (D) None of these

32. Bleaching powder is a compound having a formula-
(A) $CaCl_2$ (B) CaClO (C) $CaOCl_2$ (D) $CaOCl_3$

33. The formula for Plaster of Paris is
(A) $CaSO_4$ (B) 2 $CaSO_4$ H_2O
(C) $CaSO_4$ $\frac{3}{2}$ H_2O (D) $CaSO_4$ H_2O

34. Chemically quick lime is
(A) CaO (B) $Ca(OH)_2$ (C) $CaCO_3$ (D) $CaSO_4$ $2H_2O$

35. Mixture of $MgCl_2$ and MgO is called-
(A) Complex salt (B) Double salt
(C) Portland cement (D) Sorel's cement

36. Which of the following has hexagonal close packed crystal structure ?
(A) NaCl (B) CsCl (C) RbCl (D) Mg

37. Lucas reagent is-
(A) $ZnCl_2$ + conc. HCl (B) $MnO_2 + H_2O$
(C) H_2SO_4 + HCl (D) Na + H_2O

38. Cinnabar is an ore of-
(A) Copper (B) Nickel (C) Mercury (D) Tungsten

39. Calomel is-
(A) Hg_2Cl_2 and Hg (B) Hg_2Cl_2
(C) Hg and $HgCl_2$ (D) $HgCl_3$

40. Which one is called inorganic benzene ?
(A) B_6H_6 (B) N_6H_6 (C) C_6H_6 (D) $B_3N_3H_6$

41. Which of the following is the most abundant in the earth's crust?
(A) In (B) Ga (C) Al (D) B

42. Aluminium appears like gold, when it is mixed with-
(A) 90% Cu (B) 50% Ni
(C) 90% Sn (D) 50% Co

43. Which of the following is a true peroxide ?
(A) BaO_2 (B) SO_2 (C) NO_2 (D) MnO_2

44. Structure of H_2O_2 is
(A) Linear (B) Tetrahedral
(C) Square planar (D) Unsymmetrical

45. Froth flotation process is used for the concentration of
(A) Oxide ores (B) Sulphide ores
(C) Chloride ores (D) Amalgams

46. The heating of pyrites to remove sulphur is known as
(A) Bessemerisation (B) Calcination
(C) Roasting (D) Smelting

47. Which of these, I-Mg , II-Sn, III-Sulphur, IV-F are prepared by electrolytic method ?
(A) I and II (B) II and III (C) III and IV (D) I and IV

48. In electrorefining process of a metal, the impure metal is made the anode and a strip of pure metal the cathode during the electrolysis of an aqueous solution of a complex metal salt. This method cannot be used for refining of
(A) Cu (B) Zn (C) Al (D) Ag

49. Which one of the following fluxes is used to remove acidic impurities in metallurgical process ?
(A) Silica (B) Sodium chloride
(C) Lime-Stone (D) Radium carbonate

50. The correct representation of sulphur is
(A) S (B) S_2 (C) S_4 (D) S_8

51. Which process of purification is represented by the following equation ?

$$\underset{\text{(Impure)}}{Ti + 2I_2} \xrightarrow{250^\circ C} TiI_4 \xrightarrow{1400^\circ C} \underset{\text{(Pure)}}{Ti + 2I_2}$$

(A) Cupellation (B) Poling
(C) Van Arkel's process (D) Zone refining

52. Sulphide ore of zinc is concentrated by
(A) Distillation (B) Electromagnetic process
(C) Froth floatation process (D) Gravity process

53.Electrometllurgical process of the fused salt is employed to extract
(A) Na (B) Ag (C) Fe (D) Pb

54. An important ore of iron is
(A) Pitchblende (B) Galena
(C) Malachite (D) Haematite

55. Which of the following is most basic?
(A) Na_2O (B) BaO (C) As_2O_3 (D) Al_2O_3

56. Which of the following imparts violet colouration to the Bunsen-burner non-luminous flame ?
(A) $CaCl_2$ (B) KCl (C) NaCl (D) $BaCl_2$

57. Chile saltpetre is
(A) KNO_2 (B) $NaNO_3$ (C) Na_2SO_4 (D) K_2SO_4

58. The characteristic property of elements in a group is
(A) Ionisation energy (B) Electronegativity
(C) Number of electrons in outermost orbit
(D) Electron affinity

59. The basis of modern periodic law is
(A) Atomic number (B) Atomic size
(C) Atomic volume (D) Atomic mass

60. The long form of periodic table consists of
(A) 7 Horizontal and 7 vertical columns
(B) 8 Horizontal and 7 vertical columns
(C) 7 Horizontal and 18 vertical columns

(D) 8 Horizontal and 8 vertical columns

61. Which of the following indicates the correct variation in electronegativity ?
(A) F > O > N > C (B) F > N > O > C
(C) F > N > O > C (D) F > N > O > C

62. Na^+, Mg^{++}, Al^{+++}, and Si^{++++} are isoelectronic. The order of their ionic size is
(A) $Na^+ > Mg^{++} < Al^{+++} < Si^{++++}$
(B) $Na^+ < Mg^{++} > Al^{+++} > Si^{++++}$
(C) $Na^+ > Mg^{++} > Al^{+++} > Si^{++++}$
(D) $Na^+ < Mg^{++} > Al^{+++} < Si^{++++}$

63. Which has the highest electron affinity ?
(A) Br (B) I (C) F (D) Cl

64. The correct order of electron affinity of the elements of oxygen family in the periodic table is
(A) O > S > Se (B) S > O > Se
(C) S > Se > O (D) Se > O > S

65. Which of the following elements has the highest electronegativity ?
(A) Li (B) Cr (C) Ac (D) F

66. The correct IUPAC name of $Fe_4[Fe(CN)_6]_3$ is
(A) Ferroso ferric cyanide
(B) Ferric ferrous hexacyanate
(C) Iron (III) hexacyanoferrate (II)
(D) Hexacyanoferrate (III, II)

67. Which of the following metals absorb hydrogen to the highest extent ?
(A) Zn (B) Pd (C) Al (D) K

68. Change of ordinary hydrogen into atomic hydrogen is a
(A) Nuclear reaction (B) Decomposition reaction
(C) Exothermic reaction (D) Endothermic reaction

69. Tritium is an isotope of
(A) Hydrogen (B) Tellurium
(C) Titanium (D) Tantalum

70. Heavy water is used as
(A) Washing water (B) Detergent
(C) Drinking water (D) A moderator

71. Duralumin alloy consists of
(A) Al and Mg (B) Al and Ni
(C) Al, Mg and Ni (D) Al, Mg, Mn and Cu

72. The blue coloured mineral 'Lapis-lazuli,' which is used as a semi-precious stone is a mineral of the following class-
(A) Basic copper carbonate (B) Prussian blue
(C) Sodium-alumino silicate (D) Zinc Cobaltate

73. Which of the following processes does not involve a catalyst ?
(A) Thermite process (B) Ostwald's process
(C) Haber's process (D) Contact process

74. Which one of the following is not a characteristic property of carbon ?
(A) Catenation (B) Multiple bond formation
(C) Availability of d-orbitals for bonding
(D) Highest electronegative element of the group

75. Poisonous gas present in fumes of motor car is-
(A) CO (B) CO_2 (C) CH_4 (D) C_2H_2

76. 'Activated charcoal' is employed to remove colouring matter from pure substance, it does so by
(A) Bleaching (B) Oxidation
(C) Reduction (D) Adsorption

77. Both heat and light are both produced when coal gas is heated. This is due to-
(A) CO and H_2 (B) CH_4 and H_2
(C) Hydrocarbons (D) CO_2 and CH_4

78. Dry ice is-
(A) Ice which is completely free from the water on the suface
(B) Silid SO_2 (C) Solid CO_2 (D) Solid benzene

79. Water gas is a mixture of
(A) CO and H_2 (B) CO and CO_2
(C) CO_2 and H_2O (D) CO and steam

80. Which glass has the highest percentage of lead ?
(A) Jena glass (B) Pyrex glass
(C) Soda glass (D) Flint glass

81. White lead is
(A) $PbCO_3$ (B) $PbCO_3$. PbO
(C) $2PbCO_3$.Pb (OH_2) (D) $2PbSO_4$. PbO

82. Ammonia is dried over
(A) Slaked lime (B) Quick-lime
(C) Calcium chloride (D) Phosphorus pentachloride

83. Which of the following is amphoteric ?
(A) P_2O_5 (B) As_2O_3 (C) N_2O_3 (D) N_2O_5

84. Which of the following oxides of nitrogen is the anhydride of HNO_3 ?
(A) N_2O_4 (B) N_2O_5 (C) NO (D) N_2O_3

85. 'Aqua regia' is a mixture of
(A) $3HCl^- + HNO_3$ (B) $H_3PO_4 + H_2SO_4$
(C) $3HNO_3 + HCl$ (D) $HCl + CH_3COOH$

89. The correct sequence of alkali metals down the group is
(A) Na, K, Li, Rb, Cs, Fr (B) Fr, Cs, Rb, K, Na, Li
(C) Li, Na, K, Cs, Fr, Rb (D) Li, Na, K, Rb, Cs, Fr

87. Baking soda is
(A) Na_2CO_3. $10H_2O$ (B) Na_2SO_4. $10H_2O$
(C) Na_2SO_4 (D) $NaHCO_3$

88. On heating sodium metal in a current of dry ammonia, the compound formed is
(A) Sodium nitride (B) Sodium hydride
(C) Sodium amide (D) Sodium azide

89. Microcosmic salt is
(A) Na_2HPO_4 . $2H_2O$ (B) $(NH_4)_2HPO_4.2H_2O$

(C) $Na(NH_4)HPO_3$ (D) $Na(NH_4)HPO_4.4H_2O$

90. In the extraction of copper, we use
(A) Cu_2S (B) Cuprite
(C) Silver argentocyanide (D) $CuFeS_2$

91. Brass is an alloy of Cu and-
(A) Zn (B) Sb (C) Pb (D) Bi

92. Bell metal is an alloy of
(A) Zn and Cu (B) Zn and Pb
(C) Cu and Sn (D) Cu and Ni

93. The electronic configuration of $_{29}Cu$ is
(A) $1s^22s^22p^63s^23p^63d^94s^2$
(B) $1s^22s^22p^63s^23p^63d^{10}4s^1$
(C) $1s^22s^22p^63s^23p^63d^{10}4s^24p^4$ (D) None of these

94. Blue complex formed on addition of conc. NH_4OH solution to a Cu (II) salt solution has following ion
(A) $[Cu(NH_4)_4]^{2+}$ (B) $[Cu(NH_3)_2]^{2+}$
(C) $[Cu(NH_3)_4]^{2+}$ (D) $[Cu(NH_4)_2]^{2+}$

95. The metal hydroxide which is soluble in excess of ammonium hydroxide is
(A) $Fe(OH)_2$ (B) $Fe(OH)_3$
(C) $Al(OH)_3$ (D) $Cu(OH)_2$

96. In solid $CuSO_4$ $5H_2O$, copper is coordinated to -
(A) Four water molecules (B) Five water molecules
(C) One suphate ion (D) One water molecule

97. AgCl on fusion with Na_2CO_3 gives
(A) Ag (B) Silver carbide (C) Ag_2CO_3 (D) Ag_2O

98. When $AgNO_3$ (solid) is strongly heated, the products are
(A) Ag, NO_2 and O_2 (B) Ag, N_2 and O_2
(C) $AgNO_2$ and O_2 (D) Ag, NO and O_2

99. Cyanide process is used for the extraction of-
(A) Cu (B) Na (C) Ag (D) Al

100. Blister copper is
(A) Ore of copper
(B) Copper containing 1%impurity
(C) Pure copper (D) Alloy of copper

CPMT - 1999

1. Which of the following ions does not have S-S linkage ?
(A) $S_2O_5^{2-}$ (B) $S_2O_3^{2-}$ (C) $S_2O_8^{2-}$ (D) $S_2O_6^{2-}$

2. The configuration of chromium atom in the ground state is
(A) $[Ar]3d^44s^2$ (B) $[Ar]3d^54s^1$
(C) $[Ar]3d^64s^2$ (D) $[Ar]3d^14s^2$

3. $KMnO_4$ will oxidise acetylene to
(A) Ethylene glycol (B) Ethyl alcohol
(C) Oxalic acid (D) Acetic acid

4. Amino acids are the building blocks of
(A) Carbohydrates (B) Vitamins
(C) Fats (D) Proteins

5. Small liquid droplets dispersed in another liquid is called
(A) Suspension (B) Emulsion
(C) Gel (D) True solution

6. The reaction $2SO_2 + 2O_2 \rightleftharpoons 2SO_3$ will be favoured by
(A) Low temperature and low pressure
(B) High temperature and high pressure
(C) High temperature and low pressure
(D) Low temperature and high pressure

7. In the extraction of Cu, the reaction which takes place in Bessemer converter is
(A) $2CuFeS_2 + O_2 \rightarrow Cu_2S + FeS + SO_2$
(B) $2Cu_2S + 3O_2 \rightarrow 2Cu_2O + 2SO_2$
(C) $2Cu_2O + Cu_2S \rightarrow 6Cu + SO_2$
(D) $2FeS + 3O_2 \rightarrow 2FeO + 2SO_2$

8. Maximum electrons in a d-orbital can be
(A) 6 (B) 14 (C) 2 (D) 10

9. Value of universal gas constant 'R' depends upon
(A) Temperature of the gas (B) Volume of gas
(C) Number of moles of gas (D) None of these

10. Which quantum number determines the shape of the subshell ?
(A) Azimuthal quantum number
(B) Magnetic quantum number
(C) Principal quantum number
(D) Spin quantum number

11. $KMnO_4$ reacts with Ferrous Sulphate according to the equation ,
$MnO_4^- + 5Fe^{2+} + 8H^+ \rightarrow Mn^{2+} + 5Fe^{3+} + 4H_2O$
Here 10 ml of 0.1 M $KMnO_4$ is equivalent to
(A) 20ml of 0.1 M $FeSO_4$ (B) 30ml of 0.1 M $FeSO_4$
(C) 40ml of 0.1 M $FeSO_4$ (D) 50ml of 0.1 M $FeSO_4$

12. Which of the following are isoelectronic species ?
I - $\overset{+}{C}H_3$, II-NH_2^-, III NH_4^+, IV-NH_3
(A) I, II, III (B) II, III, IV (C) I, II, IV (D) II and I

13. Which of the following species is responsible for sulphonation of benzene ring ?
(A) SO_3 (B) SO_2^- (C) H_2SO_4 (D) HSO_3^-

14. Which of the following is an example of S_N2 reaction ?
(A) $CH_3Br + OH \longrightarrow CH_3OH + Br$
(B) $CH_3CH(Br)CH_3 + OH^- \longrightarrow CH_3CH(OH)CH_3 + Br^-$
(C) $CH_3CH_2OH \xrightarrow{-H_2O} CH_2 = CH_2$
(D) $CH_3-CH(CH_3)-CH(Br)-CH_3 + OH^- \longrightarrow CH_3-C(CH_3)(OH)-CH_3 + Br^-$

15. Which of the following is not true in case of reaction with heated copper at 300°C ?
(A) Phenol → Benzyl alcohol

(B) Primary alcohol → Aldehyde
(C) Secondary alcohol → Ketone
(D) Tertiary alcohol → Olefin

16. Among Cl^-, Br^-, I^- the correct order for being oxidised to dihalogen is
(A) $I^- > Cl^- > Br^-$ (B) $Cl^- > Br^- > I^-$
(C) $I^- > Br^- > Cl^-$ (D) $Br^- > I^- > Cl^-$

17. In the following reaction, X is

$$X \xrightarrow{\text{Bromination}} Y \xrightarrow{NaNO_2+HCl} Z \xrightarrow[C_2H_5OH]{\text{Boiling}} \text{Tribromo Benzene}$$

(A) Benzoic acid (B) Salicylic acid
(C) Phenol (D) Aniline

18. Bond length of Ethane (I), Ethene (II), Acetylene (III) and Benzene (IV) follows the order
(A) I > II > III > IV (B) I > II > IV > III
(C) I > IV > II> III (D) III > IV > II > I

19. Which of the following possesses highest melting point ?
(A) m- dichlorobenzene (B) p-dichlorobenzene
(C) Chlorobenzene (D) o-dichlorobenzene

20. Which of the following carbon atoms is likely to possess geometry ? $H_2\overset{1}{C} = \overset{2}{C}H-\overset{3}{C}H_2-\overset{4}{C}OOH$
(A) 1 (B) 2 (C) 3 (D) 4

21. Which of the following will not give iodoform test ?
(A) Ethanol (B) Ethanal
(C) Isopropyl alcohol (D) Benzyl alcohol

22. The correct order of boiling point for primary (1°), secondary (2°) and tertiary (3°) alcohols is
(A) 1° > 2° > 3° (B) 3° > 2° > 1°
(C) 2° > 1° > 3° (D) 2° > 3° > 1°

23. Which of the following set of quantum numbers belong to highest energy ?
(A) n = 3, l = 1, m = 1, s = + 12
(B) n = 4, l = 0,m = 0, s = + 1/2
(C) n = 3, l = 0, m = 0, s = + 1/2
(D) n = 3, l = 2, m = 1, s = + 1/2

24. O_2 molecule is
(A) Paramagnetic (B) Diamagnetic
(C) Ferromagnetic (D) None of these

25. The rise in the boiling point of a solution containing 1.8 gram of glucose in 100g of a solvent is 0.1°C. The molal elevation constant of the liquid is
(A) 0.01 k/m (B) 0.1 k/m (C) 1 k/m (D) 10 k/m

26. Number of moles of a solute per kilogram of a solvent is called
(A) Molarity (B) Molality (C) Normality (D) Formality

27. Enthalpy of an exothermic reaction is always :
(A) Positive (B) Negative
(C) Zero (D) May be positive or negative

28. The osmotic pressure of 1 m solution at 27°C is
(A) 2.46 atm (B) 24.6 atm
(C) 1.21 atm (D) 12.1 atm

29. The unit of equivalent conductivity is
(A) Ohm cm^2 (g-equivalent)
(B) S cm^{-2} (C) Ohm cm
(D) $Ohm^{-1}cm^2$ (g equivalent)$^{-1}$

30. At infinite dilution, the percentage ionisation for both strong and weak electrolytes is
(A) 1% (B) 20% (C) 50% (D) 100%

31. The standard reduction potentials of 4 elements are given below. Which of the following will be the most suitable reducing agent ? I = – 3.04 V, II = –1.90 V, III = 0 V, IV = 1.90 V
(A) I (B) II (C) III (D) IV

32. A catalyst increases rate of a reaction by
(A) Decreasing enthalpy (B) Decreasing internal energy
(C) Decreasing activation energy
(D) Increasing activation energy

33. pH of a solution can be expressed as
(A) $-\log_e (H^+)$ (B) $-\log_{10} (H^+)$
(C) $\log_e (H^+)$ (D) $\log_{10} (H^+)$

34. IUPAC name of the following compound is

$$\begin{array}{l} CH_3-CH = C-CH_2-CH_3 \\ \qquad\qquad\quad | \\ \qquad\quad CH_2-CH_2-CH_3 \end{array}$$

(A) 3-ethyl 2-hexene (B) 3-propyl 2-hexene
(C) 3-propyl 3-hexene (D) 4-ethyl 4-hexene

35. Which of the following is isoelectronic as well as has same structure as that of N_2O ?
(A) NO_2 (B) CO_2 (C) N_3H (D) H_2O

36. Which of the following has maximum hydrogen bonding ?
(A) Ethylamine (B) Ammonia
(C) Ethyl alcohol (D) Diethyl ether

37. Which of the given bonds is most polar ?
(A) C - O (B) C - F (C) O - F (D) N - F

38. Which of the following is the most suitable method for removing traces of water from ethanol ?
(A) Heating with Na metal
(B) Reacting with Mg
(C) Passing dry HCl through it
(D) Distilling it

39. Which of the following is most acidic ?
(A) Phenol (B) Benzyl alcohol
(C) m-chlorophenol (D) Cyclohexanol

40. Propyne on polymerisation gives
(A) Mesitylene (B) Benzene
(C) Ethyl benzene (D) Propyl benzene

41. Which of the following catalyst is used in the polymerisation of HC ≡ CH to C_6H_6 ?
(A) $NbCl_5$ (B) HCl (C) $AlCl_3$ (D) $HgSO_4$

42. Which of the following reactions will not give primary

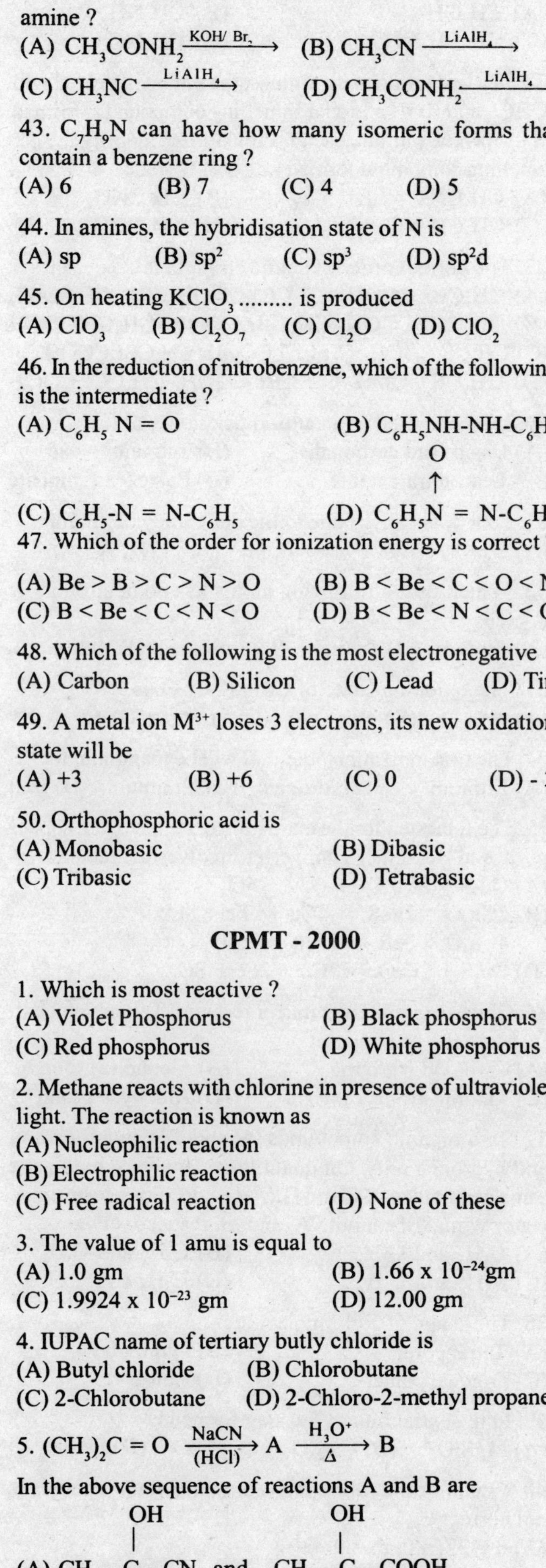

amine ?

(A) $CH_3CONH_2 \xrightarrow{KOH/\ Br_2}$ (B) $CH_3CN \xrightarrow{LiAlH_4}$

(C) $CH_3NC \xrightarrow{LiAlH_4}$ (D) $CH_3CONH_2 \xrightarrow{LiAlH_4}$

43. C_7H_9N can have how many isomeric forms that contain a benzene ring ?

(A) 6 (B) 7 (C) 4 (D) 5

44. In amines, the hybridisation state of N is

(A) sp (B) sp^2 (C) sp^3 (D) sp^2d

45. On heating $KClO_3$ .... is produced

(A) ClO_3 (B) Cl_2O_7 (C) Cl_2O (D) ClO_2

46. In the reduction of nitrobenzene, which of the following is the intermediate ?

(A) $C_6H_5\ N = O$ (B) $C_6H_5NH\text{-}NH\text{-}C_6H_5$

(C) $C_6H_5\text{-}N = N\text{-}C_6H_5$ (D) $C_6H_5\overset{\overset{O}{\uparrow}}{N} = N\text{-}C_6H_5$

47. Which of the order for ionization energy is correct ?

(A) Be > B > C > N > O (B) B < Be < C < O < N

(C) B < Be < C < N < O (D) B < Be < N < C < O

48. Which of the following is the most electronegative

(A) Carbon (B) Silicon (C) Lead (D) Tin

49. A metal ion M^{3+} loses 3 electrons, its new oxidation state will be

(A) +3 (B) +6 (C) 0 (D) - 3

50. Orthophosphoric acid is

(A) Monobasic (B) Dibasic

(C) Tribasic (D) Tetrabasic

CPMT - 2000

1. Which is most reactive ?

(A) Violet Phosphorus (B) Black phosphorus

(C) Red phosphorus (D) White phosphorus

2. Methane reacts with chlorine in presence of ultraviolet light. The reaction is known as

(A) Nucleophilic reaction

(B) Electrophilic reaction

(C) Free radical reaction (D) None of these

3. The value of 1 amu is equal to

(A) 1.0 gm (B) 1.66×10^{-24}gm

(C) 1.9924×10^{-23} gm (D) 12.00 gm

4. IUPAC name of tertiary butly chloride is

(A) Butyl chloride (B) Chlorobutane

(C) 2-Chlorobutane (D) 2-Chloro-2-methyl propane

5. $(CH_3)_2C = O \xrightarrow[(HCl)]{NaCN} A \xrightarrow[\Delta]{H_3O^+} B$

In the above sequence of reactions A and B are

(A) $CH_3-\overset{OH}{\underset{CH_3}{C}}-CN$ and $CH_3-\overset{OH}{\underset{CH_3}{C}}-COOH$

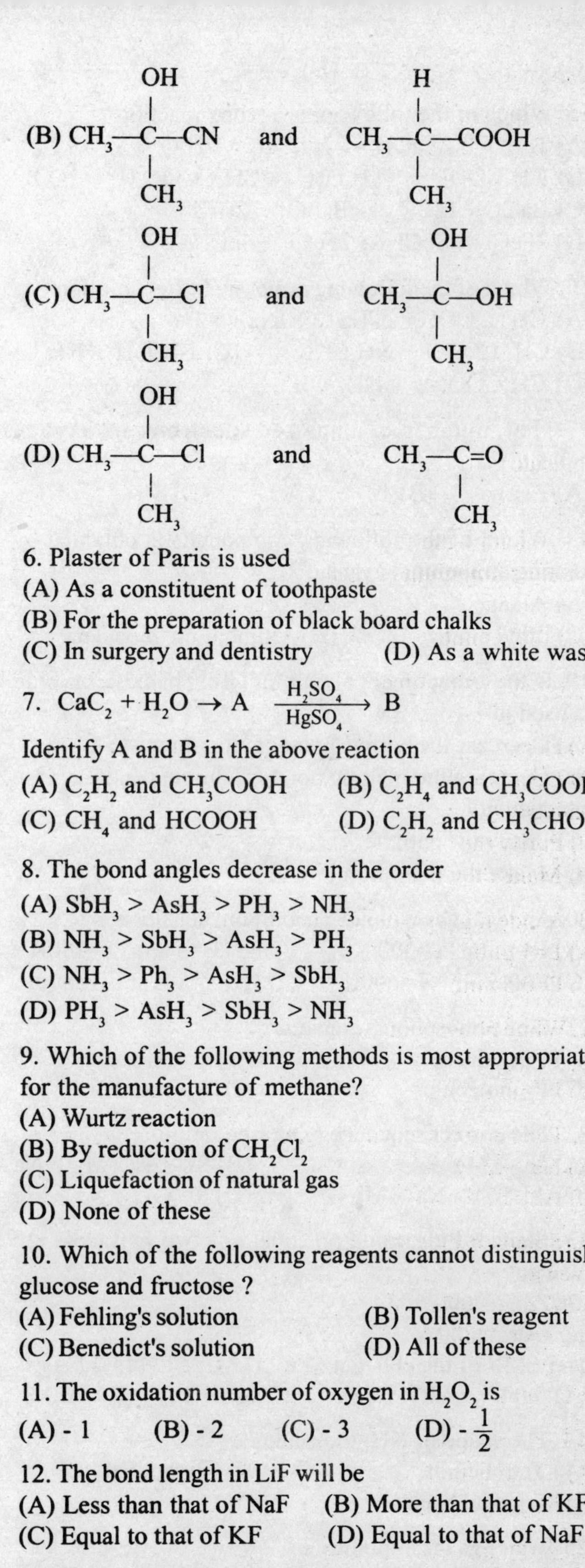

(B) $CH_3-\overset{OH}{\underset{CH_3}{C}}-CN$ and $CH_3-\overset{H}{\underset{CH_3}{C}}-COOH$

(C) $CH_3-\overset{OH}{\underset{CH_3}{C}}-Cl$ and $CH_3-\overset{OH}{\underset{CH_3}{C}}-OH$

(D) $CH_3-\overset{OH}{\underset{CH_3}{C}}-Cl$ and $CH_3-\underset{CH_3}{C}=O$

6. Plaster of Paris is used

(A) As a constituent of toothpaste

(B) For the preparation of black board chalks

(C) In surgery and dentistry (D) As a white wash

7. $CaC_2 + H_2O \rightarrow A \xrightarrow[HgSO_4]{H_2SO_4} B$

Identify A and B in the above reaction

(A) C_2H_2 and CH_3COOH (B) C_2H_4 and CH_3COOH

(C) CH_4 and HCOOH (D) C_2H_2 and CH_3CHO

8. The bond angles decrease in the order

(A) $SbH_3 > AsH_3 > PH_3 > NH_3$

(B) $NH_3 > SbH_3 > AsH_3 > PH_3$

(C) $NH_3 > Ph_3 > AsH_3 > SbH_3$

(D) $PH_3 > AsH_3 > SbH_3 > NH_3$

9. Which of the following methods is most appropriate for the manufacture of methane?

(A) Wurtz reaction

(B) By reduction of CH_2Cl_2

(C) Liquefaction of natural gas

(D) None of these

10. Which of the following reagents cannot distinguish glucose and fructose ?

(A) Fehling's solution (B) Tollen's reagent

(C) Benedict's solution (D) All of these

11. The oxidation number of oxygen in H_2O_2 is

(A) - 1 (B) - 2 (C) - 3 (D) $-\frac{1}{2}$

12. The bond length in LiF will be

(A) Less than that of NaF (B) More than that of KF

(C) Equal to that of KF (D) Equal to that of NaF

13. The rate of diffusion of a gas having molecular weight just double of that of nitrogen gas is 56 ml s^{-1}. The rate of diffusion of nitrogen gas will be

(A) 112.0 ml s^{-1} (B) 90.00 ml s^{-1}

(C) 79.19 ml s^{-2} (D) 56 ml s^{-1}

14. The values of x and y in the following redox reaction $x\,Cl_2 + 6OH^- \rightarrow ClO_3^- + y\,Cl^- + 3H_2O$ are

(A) x = 5, y = 3 (B) x = 2, y = 4

(C) x = 3, y = 5 (D) x = 4, y = 2

15. Which of the following is a redox reaction ?
(A) $H_2SO_4 + 2NaOH \rightarrow Na_2SO_4 + 2H_2O$
(B) $CH_3COOH + C_2H_5OH \rightarrow CH_3COOC_2H_5 + H_2O$
(C) $BaCl_2 + H_2SO_4 \rightarrow BaSO_4 + 2HCl$
(D) $2FeCl_3 + SnCl_2 \rightarrow 2FeCl_2 + SnCl_4$

16. Which of the following is a basic buffer solution ?
(A) CH_3 COOH + CH_3 COONa
(B) CH_3 COOH + NH_4Cl (C) $NH_4OH + NH_4Cl$
(D) CH_3COONa + NH_4 OH

17. The number of unpaired electrons in oxygen molecule is
(A) Zero (B) 1 (C) 2 (D) 3

18. Which of the following compounds is obtained by heating ammonium cyanate ?
(A) Alkane (B) Urea
(C) Ethyl amine (D) Ammonium thiocyanate

19. In the extraction of aluminium from bauxite, cryolite is used to -
(A) Decrease the boiling point of aluminium
(B) Decrease the melting point of alumina and increase conductivity
(C) Purify the bauxite
(D) Make the mass lighter

20. An ideal gas will have maximum density at
(A) P=1 atm, T=300 K (B) P=2 atm, T=150K
(C) P=0.5 atm, T=600K (D) P=1.0 atm, t=500K

21. White phosphorus contains
(A) P_2 molecules (B) P_6 molecules
(C) P_4 molecules (D) P_8 molecules

22. The correct sequence of atomic radii is
(A) Na > Mg > Al > Si (B) Si > Al > Mg > Na
(C) Al > Si > Na > Mg (D) Si > Al > Na > Mg

23. Standard electrode potentials of two half cells are given as
$Zn^{2+}/Zn = -0.76$ V ;
$Ag^+/Ag = 0.80$ V
The EMF of the cell Zn / Zn^{2+} (1M)//Ag^+ (1M) / Ag is
(A) 0.04 V (B) -1.56V (C) 1.56 V (D) 1.20 V

24. The shape of NH_3 molecule is
(A) Tetrahedral (B) Pyramidal
(C) Trigonal planar (D) Linear

25. Water gas is a mixture of
(A) CO and H_2 (B) CO_2 and H_2O
(C) CO and N_2 (D) CO, H_2 and N_2

26. Heavy water is
(A) Water containing sulphates of Ca and Mg
(B) Water containing heavy ions
(C) Water at 4° (D) D_2O

27. When acetamide is treated with Br_2 in the presence of a base, the product obtained is
(A) CH_3OH (B) CH_3NH_2
(C) $CH_3CH_2NH_2$ (D) CH_3NC

28. In a test for primary amines, the amine is treated with $CHCl_3$ and KOH and a bad smelling compound is formed. If the primary amine used is ethylamine, identify the bad smelling compound formed
(A) CH_3CN (B) CH_3CNO
(C) CH_3CH_2NC (D) CH_3NCO

29. The correct order of acidic strength is
(A) $CH_3CBr_2COOH > CH_3CCl_2COOH > CH_3CF_2COOH$
(B) $CH_3CCl_2COOH > CH_3CBr_2COOH > CH_3CF_2COOH$
(C) $CH_3CF_2COOH > CH_3CCl_2COOH > CH_3CBr_2COOH$
(D) $CH_3CF_2COOH > CH_3CBr_2COOH > CH_3CCl_2COOH$

30. Which is used as an anti-knock in petrol fuel ?
(A) Basic lead carbonate (B) Tetraethyl lead
(C) Lead tetra acetate (D) Basic lead sulphate

31. Scurvy is caused due to the deficiency of vitamin
(A) B_1 (B) C (C) K (D) A

32. Which of the following atoms has no neutron in its nucleus ?
(A) Protium (B) Tritium (C) Lithium (D) Helium

33. The oxidation state of Cr in $K_2Cr_2O_7$ is
(A) -5 (B) +6 (C) - 6 (D) +5

34. The first ionization potential will be maximum for
(A) Lithium (B) Hydrogen (C) Uranium (D) Iron

35. The final step for the extraction of copper from copper pyrites in Bessemer's converter involves the reaction
(A) $Cu_2S + 2Cu_2O \rightarrow 6Cu + SO_4$
(B) $2Cu_2O + 2FeS \rightarrow 4Cu + 2Fe + SO_2$
(C) $4Cu_2O + FeS \rightarrow 8\ Cu + FeSO_4$
(D) $Cu_2S + 2FeO \rightarrow 2Cu + 2Fe + SO_2$

36. Glucose gives silver mirror test with Tollen's reagent. It shows the presence of
(A) Carboxylic group (B) Alcoholic group
(C) Ketonic group (D) Aldehyde group

37. Two organic compounds 'A' and 'B' contain carbon and hydrogen only. On quantitative analysis, both gave same percentage of C and H. 'A' can decolourise bromine water while 'B' cannot. 'A' and 'B' respectively are
(A) C_2H_2 and C_6H_6 (B) C_2H_4 and C_2H_6
(C) C_2H_6 and C_6H_6 (D) C_2H_2 and C_2H_6

38. The shape of XeF_2 molecule is
(A) Tetrahedral (B) Pyramidal
(C) Trigonal planar (D) Linear

39. In the extraction of Fe, slag formed is
(A) $MgSiO_3$ (B) $CaSiO_3$ (C) FeO (D) $FeSiO_3$

40. Which of the following sets of quantum numbers is not correct ?
(A) n = 2, ℓ= 0, m = 0, s+$\frac{1}{2}$
(B) n = 4, ℓ= 3, m = 2, s= + $\frac{1}{2}$

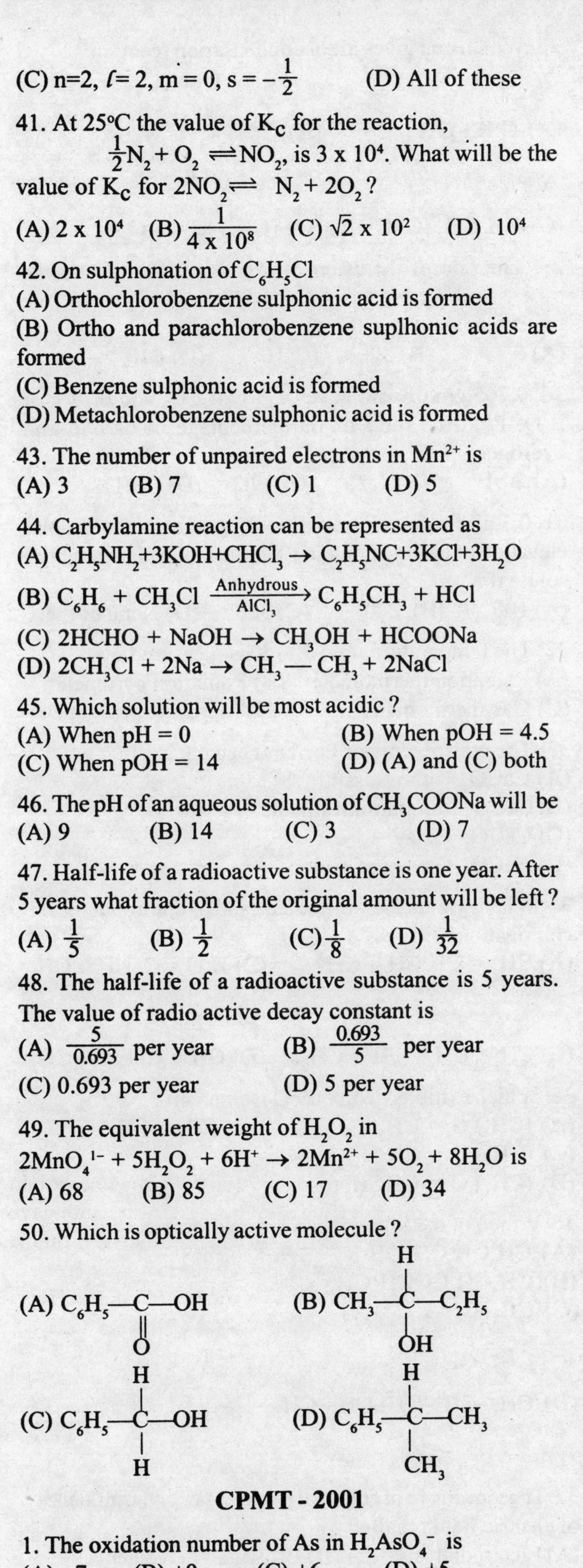

(C) n=2, l = 2, m = 0, s = $-\frac{1}{2}$ (D) All of these

41. At 25°C the value of K_C for the reaction, $\frac{1}{2}N_2 + O_2 \rightleftharpoons NO_2$, is 3 x 10^4. What will be the value of K_C for $2NO_2 \rightleftharpoons N_2 + 2O_2$?

(A) 2 x 10^4 (B) $\frac{1}{4 \times 10^8}$ (C) $\sqrt{2}$ x 10^2 (D) 10^4

42. On sulphonation of C_6H_5Cl

(A) Orthochlorobenzene sulphonic acid is formed

(B) Ortho and parachlorobenzene suplhonic acids are formed

(C) Benzene sulphonic acid is formed

(D) Metachlorobenzene sulphonic acid is formed

43. The number of unpaired electrons in Mn^{2+} is

(A) 3 (B) 7 (C) 1 (D) 5

44. Carbylamine reaction can be represented as

(A) $C_2H_5NH_2$+3KOH+$CHCl_3$ $\rightarrow$ C_2H_5NC+3KCl+$3H_2O$

(B) $C_6H_6 + CH_3Cl \xrightarrow[AlCl_3]{Anhydrous} C_6H_5CH_3 + HCl$

(C) 2HCHO + NaOH $\rightarrow$ CH_3OH + HCOONa

(D) $2CH_3Cl + 2Na \rightarrow CH_3—CH_3 + 2NaCl$

45. Which solution will be most acidic ?

(A) When pH = 0 (B) When pOH = 4.5

(C) When pOH = 14 (D) (A) and (C) both

46. The pH of an aqueous solution of CH_3COONa will be

(A) 9 (B) 14 (C) 3 (D) 7

47. Half-life of a radioactive substance is one year. After 5 years what fraction of the original amount will be left ?

(A) $\frac{1}{5}$ (B) $\frac{1}{2}$ (C) $\frac{1}{8}$ (D) $\frac{1}{32}$

48. The half-life of a radioactive substance is 5 years. The value of radio active decay constant is

(A) $\frac{5}{0.693}$ per year (B) $\frac{0.693}{5}$ per year

(C) 0.693 per year (D) 5 per year

49. The equivalent weight of H_2O_2 in

$2MnO_4^{1-} + 5H_2O_2 + 6H^+ \rightarrow 2Mn^{2+} + 5O_2 + 8H_2O$ is

(A) 68 (B) 85 (C) 17 (D) 34

50. Which is optically active molecule ?

(A) $C_6H_5—C(=O)—OH$

(B) $CH_3—C(H)(OH)—C_2H_5$

(C) $C_6H_5—C(H)(H)—OH$

(D) $C_6H_5—C(H)(CH_3)—CH_3$

CPMT - 2001

1. The oxidation number of As in $H_2AsO_4^-$ is

(A) –7 (B) +9 (C) +6 (D) +5

2. The pH of 0.01 M NaOH is

(A) 5 (B) 7 (C) 9 (D) 12

3. Bauxite is an ore of

(A) Aluminium (B) Iron (C) Zine (D) Copper

4. The weakest acid among the following is

(A) HI (B) HBr (C) HCl (D) HF

5. Which of the following set of quantum numbers is not possible ?

(A) n = 3, l = 2, m = – 3, s = + 1/2

(B) n = 2, l = O, m = O, s = + 1/2

(C) n = 2, l = 1, m = – 1, s = – 1/2

(D) n = 3, l = 2, m = – 2, s = + 1/2

6. The pH of 0.1 M solution of the following salts increases in the order

(A) NaCl < NH_4Cl < NaCN < HCl

(B) HCl < NH_4Cl < NaCl < NaCN

(C) NaCN < NH_4Cl < NaCl < HCl

(D) HCl < NaCl < NaCN < NH_4Cl

7. A mixture of carbonates of two bivalent metals contains 13.5% of one metal. If 2.5 g of the mixture liberates 1.33 g CO_2, the % by weight of the other metal is

(A) 13.6 (B) 12 (C) 13.44 (D) 15

8. If the ratio of co-ordination number of P to that of Q be Y : Z, then the formula of the soild is

(A) P_yQ_x (B) P_xQ_y (C) $P_{1/y}\ Q_{1/z}$ (D) None of these

9. The addition product of compound A with excess of methyl magnesium iodide followed by acidification yields t-butanol. The compound A is

(A) Methanol (B) Ethanol

(C) Propanol (D) Methyl ethanoate

10. When 1x10^{-3} mole of the chloride of an element Y was completely hydrolysed. it was found that the resulting solution required 20 ml of 0.1 M aqueous silver nitrate for complete precipitation of the chloride ion. The element Y could be

(A) Aluminium (B) Phosphorous

(C) Silicon (D) Sulphur

11. A.G.M. counter is used to study the radioactive process. In the absence of radioactive substance A, it counts 3 disintegration per second (dps). At the start in the presence of A, it records 23 dps and after 10 minutes 13 dps, (1) What does it count after 20 minutes? (2) What is the half-life of A ?

(A) 8 dps. 10 min (B) 5 dps, 10 min

(C) 5 dps, 20 min (D) 5 dps, 5 min

12. 2 mol. N_2 and 3 mol. H_2 gas are allowed to react in a 20 liters flask at 400 K and after complete conversion of H_2 into NH_3, 10 litres H_2O was added and temperature reduced to 300 K. Pressure of the gas after reaction is :

$N_2 + 3H_2 \rightarrow 2NH_3$

(A) R x 300/20 (B) R x 300/10

(C) 3R x 300/20 (D) 3R x 300/10

13. Which of the following dissolves in $NaHCO_3$?

(A) CH_3CHO (B) C_2H_5OH

(C) CH_3COOH (D) CH_3COCH_3

14. IUPAC name of the compound,

$CH_3C(CH_3)_2\ CH_2CH = CH_2$ is

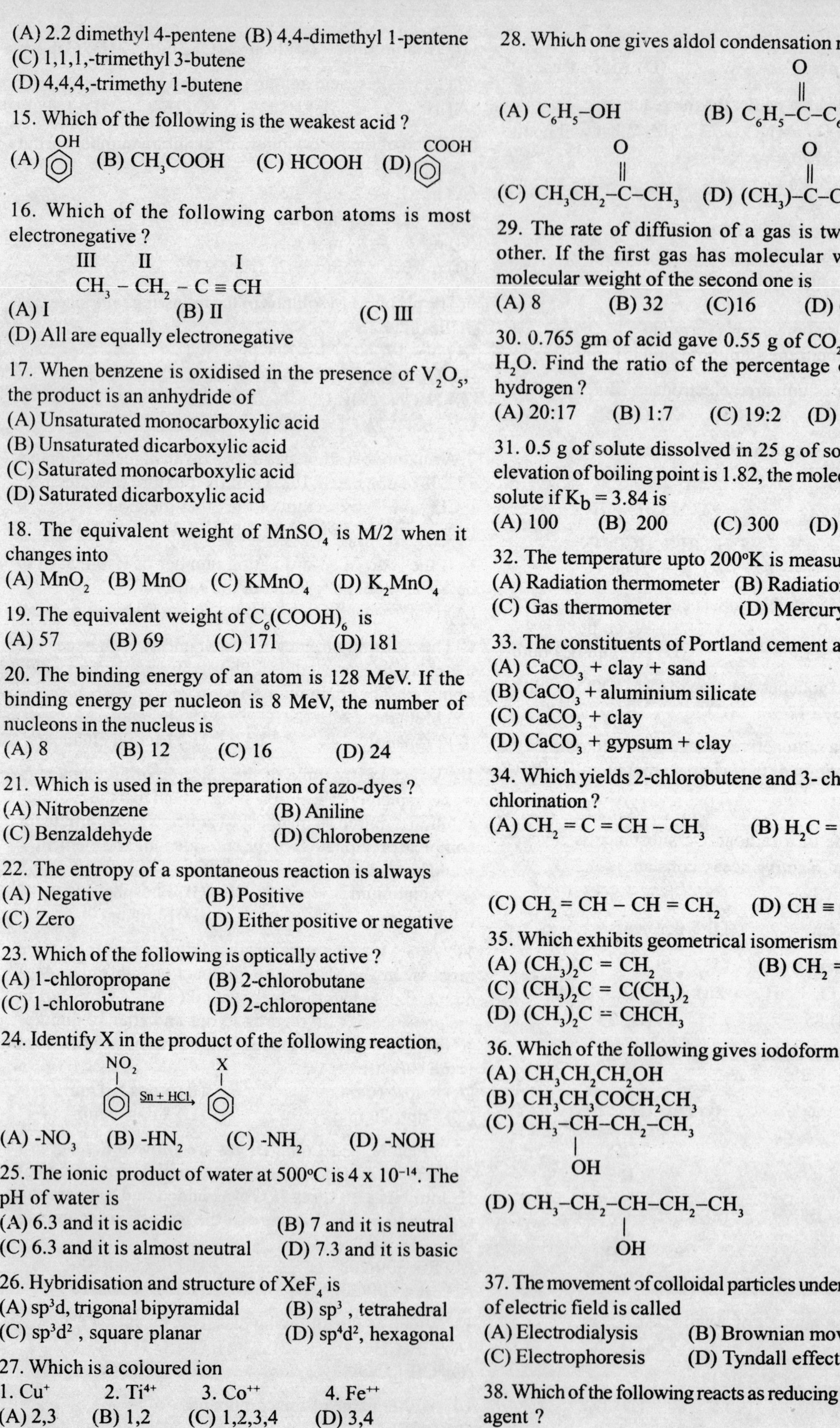

(A) 2.2 dimethyl 4-pentene (B) 4,4-dimethyl 1-pentene
(C) 1,1,1,-trimethyl 3-butene
(D) 4,4,4,-trimethy 1-butene

15. Which of the following is the weakest acid ?
(A) Phenol (OH on benzene ring) (B) CH_3COOH (C) HCOOH (D) Benzoic acid (COOH on benzene ring)

16. Which of the following carbon atoms is most electronegative ?

$$\overset{III}{CH_3} - \overset{II}{CH_2} - \overset{I}{C} \equiv CH$$

(A) I (B) II (C) III
(D) All are equally electronegative

17. When benzene is oxidised in the presence of V_2O_5, the product is an anhydride of
(A) Unsaturated monocarboxylic acid
(B) Unsaturated dicarboxylic acid
(C) Saturated monocarboxylic acid
(D) Saturated dicarboxylic acid

18. The equivalent weight of $MnSO_4$ is M/2 when it changes into
(A) MnO_2 (B) MnO (C) $KMnO_4$ (D) K_2MnO_4

19. The equivalent weight of $C_6(COOH)_6$ is
(A) 57 (B) 69 (C) 171 (D) 181

20. The binding energy of an atom is 128 MeV. If the binding energy per nucleon is 8 MeV, the number of nucleons in the nucleus is
(A) 8 (B) 12 (C) 16 (D) 24

21. Which is used in the preparation of azo-dyes ?
(A) Nitrobenzene (B) Aniline
(C) Benzaldehyde (D) Chlorobenzene

22. The entropy of a spontaneous reaction is always
(A) Negative (B) Positive
(C) Zero (D) Either positive or negative

23. Which of the following is optically active ?
(A) 1-chloropropane (B) 2-chlorobutane
(C) 1-chlorobutrane (D) 2-chloropentane

24. Identify X in the product of the following reaction,

$$C_6H_5NO_2 \xrightarrow{Sn + HCl} C_6H_5X$$

(A) $-NO_3$ (B) $-HN_2$ (C) $-NH_2$ (D) -NOH

25. The ionic product of water at 500°C is 4×10^{-14}. The pH of water is
(A) 6.3 and it is acidic (B) 7 and it is neutral
(C) 6.3 and it is almost neutral (D) 7.3 and it is basic

26. Hybridisation and structure of XeF_4 is
(A) sp^3d, trigonal bipyramidal (B) sp^3 , tetrahedral
(C) sp^3d^2 , square planar (D) sp^4d^2, hexagonal

27. Which is a coloured ion
1. Cu^+ 2. Ti^{4+} 3. Co^{++} 4. Fe^{++}
(A) 2,3 (B) 1,2 (C) 1,2,3,4 (D) 3,4

28. Which one gives aldol condensation reaction ?
(A) C_6H_5-OH (B) $C_6H_5-\overset{O}{\overset{||}{C}}-C_6H_5$
(C) $CH_3CH_2-\overset{O}{\overset{||}{C}}-CH_3$ (D) $(CH_3)-\overset{O}{\overset{||}{C}}-C-C_6H_5$

29. The rate of diffusion of a gas is twice that of the other. If the first gas has molecular weight 16, the molecular weight of the second one is
(A) 8 (B) 32 (C)16 (D) 64

30. 0.765 gm of acid gave 0.55 g of CO_2 and 0.12 g of H_2O. Find the ratio of the percentage of carbon and hydrogen ?
(A) 20:17 (B) 1:7 (C) 19:2 (D) 18:11

31. 0.5 g of solute dissolved in 25 g of solvent. If molal elevation of boiling point is 1.82, the molecular weight of solute if $K_b = 3.84$ is
(A) 100 (B) 200 (C) 300 (D) None of these

32. The temperature upto 200°K is measured by
(A) Radiation thermometer (B) Radiation pyrometer
(C) Gas thermometer (D) Mercury thermometer

33. The constituents of Portland cement are
(A) $CaCO_3$ + clay + sand
(B) $CaCO_3$ + aluminium silicate
(C) $CaCO_3$ + clay
(D) $CaCO_3$ + gypsum + clay

34. Which yields 2-chlorobutene and 3- chlorobutene on chlorination ?
(A) $CH_2 = C = CH - CH_3$ (B) $H_2C = \underset{CH_3}{\underset{|}{C}}-CH = CH_2$
(C) $CH_2 = CH - CH = CH_2$ (D) $CH \equiv C-CH = CH_2$

35. Which exhibits geometrical isomerism ?
(A) $(CH_3)_2C = CH_2$ (B) $CH_2 = CH_2$
(C) $(CH_3)_2C = C(CH_3)_2$
(D) $(CH_3)_2C = CHCH_3$

36. Which of the following gives iodoform test
(A) $CH_3CH_2CH_2OH$
(B) $CH_3CH_2COCH_2CH_3$
(C) $CH_3-\underset{OH}{\underset{|}{CH}}-CH_2-CH_3$
(D) $CH_3-CH_2-\underset{OH}{\underset{|}{CH}}-CH_2-CH_3$

37. The movement of colloidal particles under the influence of electric field is called
(A) Electrodialysis (B) Brownian movement
(C) Electrophoresis (D) Tyndall effect

38. Which of the following reacts as reducing and oxidizing agent ?

(A) KOH (B) $KMnO_4$ (C) H_2SO_4 (D) H_2O_2

39. Which of the following about fluorine is not correct ?
(A) The electron affinity of chlorine is greater than fluorine
(B) The bond energy of fluorine is less than chlorine
(C) Fluorine does not from oxyacids
(D) Fluorine cannot be prepared by electrolysis of metal fluorides

40. A flask contains 36% nitrogen, 39% oxygen and 26% hydrogen. If the total pressure of the mixture of gases is 500 mm, the partial pressure of oxygen is
(A) 120 mm (B) 145 mm (C) 195 mm (D) 210 mm

41. If, in the reaction $N_2 + O_2 \rightleftharpoons 2NO$, the production of NO is 40% at equilibrium, the K_C for the reaction is
(A) 0.26 (B) 0.44 (C) 0.2 (D) 0.25

42. Which of the following is a fundamental particle ?
(A) Water molecule (B) Oxygen molecule
(C) Alpha particle (D) Proton

43. The radioactive disintegration follows .... type of reaction.
(A) First order (B) Zero order
(C) Pseudo first order (D) Second order

44. The presence of 3 unpaired electrons in nitrogen can be explained by
(A) Heisenberg's uncertainty principle (B) Afbau's rule
(C) Pauli's exclusion law (D) Hund's rule

45. The maximum valency that can be shown by nitrogen and phosphorus is
(A) 3 and 5 (B) 4 and 3 (C) 5 and 5 (D) 5 and 3

46. The two adjacent sheets are bonded together in graphite with
(A) Van der Wall's forces (B) Ionic bonds
(C) Metallic bonds (D) Covalent bonds

47. Which of the following is most stable ?
(A) Si^{++} (B) Pb^{++} (C) Sn^{++} (D) Ge^{++}

48. The metal ion, which is precipitated when H_2S is passed in acidic medium is
(A) Zn^{++} (B) Ni^{++} (C) Cd^{++} (D) Mn^{++}

49. Which enzyme breaks sucrose into glucose and fructose ?
(A) Zymase (B) Maltase (C) Diastase (D) Invertase

50. When sodium oxalate is treated with dilute H_2SO_4, the gases produced are
(A) CO (B) CO_2
(C) Both CO and CO_2 (D) Both CO_2 and SO_2

CPMT - 2002

1. Strongest Bronsted base of the following is
(A) ClO_2^- (B) ClO_4^- (C) ClO^- (D) ClO_3^-

2. Nessler's reagent is
(A) potassium in mercuric iodide
(B) $TiCl_4$ (C) Al_2O_3/Cr_2O_3 (D) anhydrous $AlCl_3$

3. The reaction $2SO_2 + O_2 \rightarrow 2SO_3$; ΔH = -ve is favoured by
(A) high temperature, high pressure
(B) low temperature, low pressure
(C) low temperature, high pressure
(D) high temperature, low pressure

4. To prevent corrosion of iron by zinc coating is called
(A) galvanisation (B) rusting
(C) protection (D) none of these

5. The energy that opposes dissolution of a solvent is
(A) hydration energy (B) lattice energy
(C) internal energy (D) bond energy

6. In a laboratory, H_2O_2 is prepared by
(A) conc $H_2SO_4 + Na_2O_2$ (B) $H_2 + O_2$
(C) cold $H_2SO_4 + BaO_2$ (D) $HCl + BaO_2$

7. Ammonia is dried over
(A) Quick lime (B) slaked lime
(C) anhydrous $CaCl_2$ (D) none of these

8. Fusion mixture is
(A) $Na_2CO_3 + K_2CO_3$ (B) $Na_2CO_3 + NaHCO_3$
(C) $Na_2CO_3 + NaOH$ (D) $Na_2CO_3 + K_2SO_4$

9. Wood charcoal is used to decolourise sugar because it
(A) reduces coloured material
(B) adsorbs coloured material
(C) absorbs decolourised material
(D) none of these

10. The general electronic configuration of transition elements is
(A) $(n-1)d^{1-10} ns^{1-2}$ (B) $np^{1-6} ns^{1-2}$
(C) $(n-1)d^{1-10} ns^2$ (D) $nd^{1-10} (n-1)s^2$

11. Sea weeds are an important source of
(A) Cl_2 (B) I_2 (C) F_2 (D) Br_2

12. Which indicator is suitable for a strong acid and a weak base titration ?
(A) phenolphthalein (B) methyl red
(C) methyl orange (D) litmus

13. Which of the following salts undergoes hydrolysis?
(A) NaCl (B) CH_3COONa (C) K_2SO_4 (D) KNO_3

14. Which of the following is an ore of aluminium ?
(A) dolomite (B) azurite (C) bauxite (D) malachite

15. white lead is
(A) Pb_3O_4 (B) PbO
(C) $2PbCO_3\ Pb(OH)_2$ (D) $Pb(CH_3COO)_2\ Pb(OH)_2$

16. Bleaching powder is obtained by the interaction of Cl_2 and
(A) dry slaked lime (B) hot $Ca(OH)_2$
(C) cold $Ca(OH)_2$ (D) conc $CaCl_2$

17. In graphite C atom is in .... state
(A) sp (B) sp^2 (C) sp^3 (D) none of these

18. O_2 molocule is
(A) diamagnetic (B) ferromagnetic

(C) paramagnetic (D) none of these

19. If the atomic distance in H_2 and Cl_2 are 74 and 198 pm. respectively, the bond length of HCl is
(A) 272 pm (B) 363 pm (C) 136 pm (D) 68 pm

20. Which is not a property of transition elements ?
(A) coloured salts (B) variable valency
(C) fixed valency (D) complex formation

21. Which noble gas is most soluble in water ?
(A) He (B) Ar (C) Ne (D) Xe

22. Solubility of BaF_2 in a solution of $Ba(NO_3)_2$ will be represented by
(A) $[Ba^{2+}]$ (B) $[NO_3]^{-2}$ (C) $\frac{1}{2}[F^-]$ (D) $[Ba]^2$

23. On dissolving KCN in H_2O_2
(A) pH increases (B) pH remains constant
(C) pH decreases (D) none of these

24. Butter is a colloid formed when
(A) fat is dispersed in water
(B) milk is dispersed in water
(C) water is disperesed in fat
(D) milk is dispersed in fat

25. A catalyst can affect the reversible reaction by
(A) changing equilibrium constant
(B) slowing forward reaction
(C) attaining equilibria in both directions
(D) none of these

26. Increasing order of rms velocities of H_2, O_2, N_2 and HBr is
(A) H_2, O_2, N_2, HBr (B) HBr, O_2, N_2, H_2
(C) H_2, N_2, O_2, HBr (D) HBr, N_2, O_2, H_2

27. Claisen condensation is not given by
(A) $C_6H_5-COO-C_6H_5$
(B) $CH_3CH_2COO-C_6H_5$
(C) $C_6H_5-COOH_3$ (D) $C_6H_5-COOCH_2CH_2CH_3$

28. The orbital with maximum energy is
(A) 4s (B) 6d (C) 3d (D) 5p

29. $_{90}Th^{228}$ Changes to $_{83}Bi^{212}$ by the emission of
(A) 4α, 1β (B) 4α, 2β
(C) 5α, 1β (D) 5α, 2β

30. Acetone, formaladehyde and acetaldehyde can be distinguished by
(A) NH_2NH_2 (B) NH_3
(C) NH_2OH (D) phenyl hydrazine

31. How many σ and π bonds are present in the molecule, $C-C\equiv C-C=C$
(A) 10σ, 3π (B) 4σ, 2π
(C) 7σ, 3π (D) 6σ, 7π

32. If M^{+3} ion loses $3e^-$, its oxidation number will be
(A) +6 (B) -3 (C) 0 (D) ÷3

33. Blister copper is
(A) Impure Cu (B) Cu alloy
(C) pure Cu (D) Cu having 1% impurity

34. Refractory metals are used in the construction of furnaces because
(A) they can withstand high temperatures
(B) they are chemically inert
(C) their high melting point (D) none of these

35. Fusion of Na with aniline gives mainly
(A) $Na_2NC_6H_5$ (B) $NH_2NH_2 + C_6H_6$
(C) $NaNH_2C_6H_5$ (D) $N_2 + C_6H_5Na$

36. I.U.P.A.C. name of the compound

$$CH_3-CH=\underset{\displaystyle CH_2-CH_2-CH_3}{\underset{|}{C}}-CH_2-CH_3$$

is
(A) 3 ethyl 2 hexene (B) 4 ethyl 2 hexene
(C) 4 ethyl 4 hexene (D) 3 propyl 2 hexene

37. The compound that shows cis-trans isomerism is
(A) 1 butene (B) 2 propene (C) 2 butene (D) benzene

38. Propyne on polymerisation gives
(A) mesitylene (B) benzene
(C) propylene (D) toluene

39. Which of the following species participate in sulphonation of benzene ?
(A) SO_3 (B) SO_3H^- (C) SO_2 (D) SO_3H^+

40. $AgNO_3$ does not give precipitate with $CHCl_3$ because
(A) $CHCl_3$ does not ionise in water
(B) $CHCl_3$ does not react with $AgNO_3$
(C) $CHCl_3$ is chemically inerrt (D) none of these

41. Picric acid and benzoic acid can be distinguished by
(A) aq. $NaHCO_3$ (B) aq. $FeCl_3$
(C) aq. Na_2CO_3 (D) aq. NaOH

42. Which is known as oil of winter green ?
(A) phenyl salicylate (B) ethyl salicylate
(C) methyl salicylate (D) salicylic acid

43. Which carbohydrate is an essential constituent of plant cell ?
(A) cellulose (B) sucrose (C) starch (D) glucose

44. Function of $AlCl_3$ in Friedel Craft's reaction is to
(A) produce nucleophile (B) make a new bond
(C) produce electrophile (D) break a bond

45. The boiling point of glycerol is more than propanol because of
(A) hybridisation (B) arrangement of molecules
(C) H-bonding (D) all of these

CPMT - 2003

1. On oxidation, an aldehyde produces
(A) an acid (B) an alcohol
(C) a ketone (D) none of these

2. A monoprotic acid in o.1M solution ionises to 0.001%. Its ionisation constant is

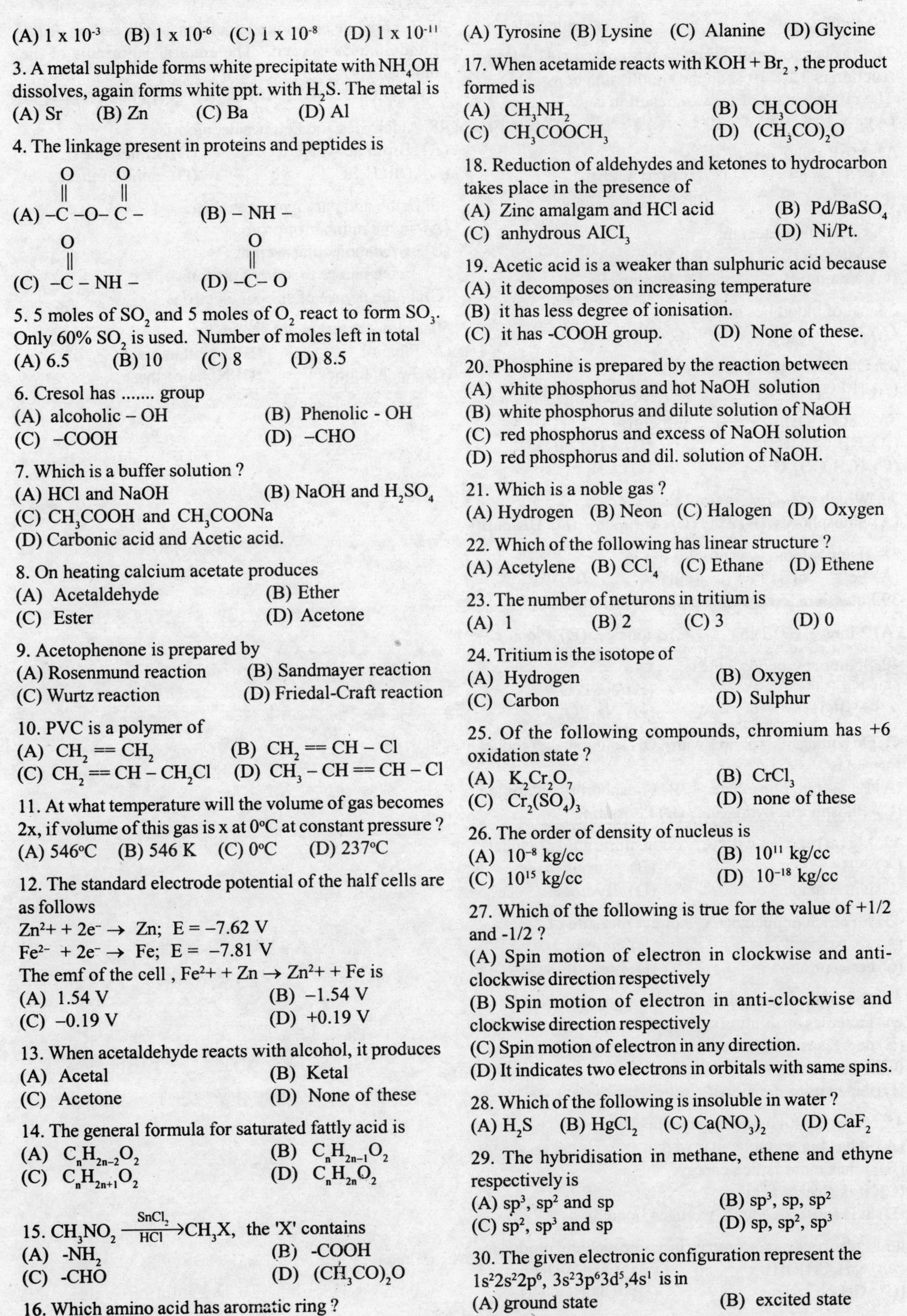

(A) 1×10^{-3} (B) 1×10^{-6} (C) 1×10^{-8} (D) 1×10^{-11}

3. A metal sulphide forms white precipitate with NH_4OH dissolves, again forms white ppt. with H_2S. The metal is
(A) Sr (B) Zn (C) Ba (D) Al

4. The linkage present in proteins and peptides is

(A) $-\overset{\overset{O}{\|}}{C}-O-\overset{\overset{O}{\|}}{C}-$ (B) $-NH-$

(C) $-\overset{\overset{O}{\|}}{C}-NH-$ (D) $-\overset{\overset{O}{\|}}{C}-O$

5. 5 moles of SO_2 and 5 moles of O_2 react to form SO_3. Only 60% SO_2 is used. Number of moles left in total
(A) 6.5 (B) 10 (C) 8 (D) 8.5

6. Cresol has .... group
(A) alcoholic – OH (B) Phenolic - OH
(C) –COOH (D) –CHO

7. Which is a buffer solution ?
(A) HCl and NaOH (B) NaOH and H_2SO_4
(C) CH_3COOH and CH_3COONa
(D) Carbonic acid and Acetic acid.

8. On heating calcium acetate produces
(A) Acetaldehyde (B) Ether
(C) Ester (D) Acetone

9. Acetophenone is prepared by
(A) Rosenmund reaction (B) Sandmayer reaction
(C) Wurtz reaction (D) Friedal-Craft reaction

10. PVC is a polymer of
(A) $CH_2 == CH_2$ (B) $CH_2 == CH - Cl$
(C) $CH_2 == CH - CH_2Cl$ (D) $CH_3 - CH == CH - Cl$

11. At what temperature will the volume of gas becomes 2x, if volume of this gas is x at 0°C at constant pressure ?
(A) 546°C (B) 546 K (C) 0°C (D) 237°C

12. The standard electrode potential of the half cells are as follows
$Zn^2+ + 2e^- \rightarrow Zn$; E = –7.62 V
$Fe^{2-} + 2e^- \rightarrow Fe$; E = –7.81 V
The emf of the cell , $Fe^2+ + Zn \rightarrow Zn^2+ + Fe$ is
(A) 1.54 V (B) –1.54 V
(C) –0.19 V (D) +0.19 V

13. When acetaldehyde reacts with alcohol, it produces
(A) Acetal (B) Ketal
(C) Acetone (D) None of these

14. The general formula for saturated fattly acid is
(A) $C_nH_{2n-2}O_2$ (B) $C_nH_{2n-1}O_2$
(C) $C_nH_{2n+1}O_2$ (D) $C_nH_{2n}O_2$

15. $CH_3NO_2 \xrightarrow[HCl]{SnCl_2} CH_3X$, the 'X' contains
(A) $-NH_2$ (B) -COOH
(C) -CHO (D) $(CH_3CO)_2O$

16. Which amino acid has aromatic ring ?
(A) Tyrosine (B) Lysine (C) Alanine (D) Glycine

17. When acetamide reacts with $KOH + Br_2$, the product formed is
(A) CH_3NH_2 (B) CH_3COOH
(C) CH_3COOCH_3 (D) $(CH_3CO)_2O$

18. Reduction of aldehydes and ketones to hydrocarbon takes place in the presence of
(A) Zinc amalgam and HCl acid (B) $Pd/BaSO_4$
(C) anhydrous $AICI_3$ (D) Ni/Pt.

19. Acetic acid is a weaker than sulphuric acid because
(A) it decomposes on increasing temperature
(B) it has less degree of ionisation.
(C) it has -COOH group. (D) None of these.

20. Phosphine is prepared by the reaction between
(A) white phosphorus and hot NaOH solution
(B) white phosphorus and dilute solution of NaOH
(C) red phosphorus and excess of NaOH solution
(D) red phosphorus and dil. solution of NaOH.

21. Which is a noble gas ?
(A) Hydrogen (B) Neon (C) Halogen (D) Oxygen

22. Which of the following has linear structure ?
(A) Acetylene (B) CCl_4 (C) Ethane (D) Ethene

23. The number of neturons in tritium is
(A) 1 (B) 2 (C) 3 (D) 0

24. Tritium is the isotope of
(A) Hydrogen (B) Oxygen
(C) Carbon (D) Sulphur

25. Of the following compounds, chromium has +6 oxidation state ?
(A) $K_2Cr_2O_7$ (B) $CrCl_3$
(C) $Cr_2(SO_4)_3$ (D) none of these

26. The order of density of nucleus is
(A) 10^{-8} kg/cc (B) 10^{11} kg/cc
(C) 10^{15} kg/cc (D) 10^{-18} kg/cc

27. Which of the following is true for the value of +1/2 and -1/2 ?
(A) Spin motion of electron in clockwise and anti-clockwise direction respectively
(B) Spin motion of electron in anti-clockwise and clockwise direction respectively
(C) Spin motion of electron in any direction.
(D) It indicates two electrons in orbitals with same spins.

28. Which of the following is insoluble in water ?
(A) H_2S (B) $HgCl_2$ (C) $Ca(NO_3)_2$ (D) CaF_2

29. The hybridisation in methane, ethene and ethyne respectively is
(A) sp^3, sp^2 and sp (B) sp^3, sp, sp^2
(C) sp^2, sp^3 and sp (D) sp, sp^2, sp^3

30. The given electronic configuration represent the $1s^22s^22p^6, 3s^23p^63d^5,4s^1$ is in
(A) ground state (B) excited state

(C) anionic state (D) cationic state

31. In a chemical equilibrium the rate constant of backward reaction is 3.2×10^{-2} and the equilibrium constant is 2^{-5}. The rate constant of forward reaction is
(A) 8×10^{-2} (B) 4×10^{-2} (C) 1×10^{-3} (D) 2×10^{-2}

32. Most stable alkene is
(A) $CH_2 = CH_2$ (B) $RCH = CHR$
(C) $R_2C = CR_2$ (D) $RCH = CH_2$

33. Cobalt is present in
(A) Vitamin B_2 (B) Vitamin B_1
(C) Vitamin B_5 (D) Vitamin B_{12}

34. pH of blood lies between
(A) 7–8 (B) 6–7 (C) 8–9 (D) 13–14

35. On hydrrolysis PCl_3 gives
(A) H_3PO_4 (B) $POCl_3$ (C) H_3PO_3 (D) PH_3

36. $CH_3COOH \xrightarrow[P_2O_5]{\Delta}$ X. identify X.
(A) CH_3COCH_3 (B) CH_3CHO
(C) $(CH_3CO)_2O$ (D) CH_4

37. Which show maximum valency ?
(A) Phosphorus (B) Tin (C) Antimony (D) Bismuth

38. Haemoglobin is a complex of
(A) Fe^{3+} (B) Fe^{2+} (C)Fe^{4+} (D) Cu^{2+}

39.Potassium ferricyanide produces on ionisation
(A) 2 ions (B) 1 ion (C) 3 ions (D) 4 ions

40. Sodium pyrophosphate is
(A) $Na_2P_2O_7$ (B) $Na_4P_2O_7$
(C) $NaPO_4$ (D) Na_2PO_2

41. On boiling, phosphorus with KOH solution, the product formed is
(A) Potassium phosphate (B) Phosphorus pentaoxide
(C) Phosphorus hydroxide (D) Phosphine

42. Mg on reaction with very dilute nitric acid produces
(A) NH_3 (B) Nitrous oxide
(C) Nitric oxide (D) Hydrogen

43. Electric conductance is a direct measure of
(A) resistance (B) current
(C) concentration (D) none of these

44. The conductivity of strong electrolyte
(A) increases on dilution slightly
(B) decreases on dilution
(C) does not change with dilution
(D) depen upon density of electrolytes itself

45. Water has high boiling point because
(A) it has higher molecular weight
(B) it has more lattice energy
(C) it is weak a acid
(D) it is associated with hydrogen bonding

46. Hydrolysis of acetonitrile in acidic medium produces
(A) CH_3CH_2OH (B) CH_3COOH
(C) CH_3NC (D) CH_3COOCH_3

47. An organic compound contains 54.6% carbon, 9.1% H and 36.56% oxygen. The empirical formula of the compound is
(A) $C_2H_2O_2$ (B) C_2H_4O (C) $C_2H_4O_2$ (D) CH_2O

48. Aldehydes and ketones decolourise
(A) Bromine water (B) Limestone
(C) Dil, H_2SO_4 (D) none of these

49. Ortho and para hydrogen differ
(A) in the number of protons
(B) in the molecular weight
(C) in the nature of spin of protons
(C) in the nature of spin of electrons.

50. Iodoform test is not shown by
(A) Ethanol (B) 2-pentanone
(C) 3-pentanone (D) None of these

CPMT - 1996

1. (A) 27. (C) 53. (D) 79. (A)
2. (A) 28. (C) 54. (D) 80. (B)
3. (A) 29. (A) 55. (D) 81. (A)
4. (B) 30. (D) 56. (D) 82. (B)
5. (A) 31. (A) 57. (C) 83. (C)
6. (A) 32. (B) 58. (C) 84. (A)
7. (A) 33. (B) 59. (A)
8. (D) 34. (A) 60. (C)
9. (C) 35. (C) 61. (D)
10. (A) 36. (D) 62. (C)
11. (B) 37. (D) 63. (B)
12. (A) 38. (B) 64. (C)
13. (A) 39. (A) 65. (D)
14. (B) 40. (B) 66. (A)
15. (A) 41. (A) 67. (B)
16. (B) 42. (B) 68. (A)
17. (A) 43. (B) 69. (B)
18. (B) 44. (B) 70. (A)
19. (D) 45. (A) 71. (C)
20. (D) 46. (B) 72. (B)
21. (A) 47. (D) 73. (B)
22. (C) 48. (B) 74. (A)
23. (C) 49. (A) 75. (C)
24. (C) 50. (A) 76. (A)
25. (A) 51. (D) 77. (C)
26. (A) 52. (A) 78. (B)

CPMT - 1997

1. (D) 27. (A) 53. (C) 79. (A)
2. (B) 28. (B) 54. (D) 80. (B)
3. (D) 29. (C) 55. (D) 81. (A)
4. (B) 30. (D) 56. (A) 82. (B)
5. (A) 31. (C) 57. (B) 83. (D)
6. (B) 32. (D) 58. (A) 84. (C)
7. (C) 33. (B) 59. (A) 85. (C)
8. (A) 34. (A) 60. (A) 86. (C)
9. (B) 35. (A) 61. (B) 87. (D)
10. (B) 36. (A) 62. (A) 88. (A)
11. (B) 37. (C) 63. (A) 89. (D)
12. (A) 38. (D) 64. (A) 90. (B)
13. (A) 39. (C) 65. (C) 91. (C)
14. (B) 40. (C) 66. (B) 92. (B)
15. (B) 41. (A) 67. (C) 93. (B)
16. (C) 42. (C) 68. (B) 94. (A)
17. (B) 43. (B) 69. (C) 95. (B)
18. (A) 44. (C) 70. (C) 96. (A)
19. (C) 45. (D) 71. (C) 97. (B)
20. (A) 46. (C) 72. (A) 98. (B)
21. (A) 47. (B) 73. (B) 99. (B)
22. (D) 48. (C) 74. (A) 100.(B)
23. (A) 49. (C) 75. (A)
24. (B) 50. (D) 76. (D)
25. (B) 51. (D) 77. (D)
26. (B) 52. (D) 78. (C)

CPMT - 1998

1. (C) 27. (C) 53. (A) 79. (A)
2. (A) 28. (A) 54. (D) 80. (D)
3. (C) 29. (B) 55. (A) 81. (C)
4. (B) 30. (B) 56. (B) 82. (B)
5. (D) 31. (C) 57. (B) 83. (B)
6. (B) 32. (C) 58. (C) 84. (B)
7. (D) 33. (B) 59. (A) 85. (A)
8. (A) 34. (A) 60. (C)
9. (A) 35. (D) 61. (A)
10. (C) 36. (D) 62. (C)
11. (A) 37. (A) 63. (D)
12. (B) 38. (C) 64. (C)
13. (B) 39. (B) 65. (D)
14. (D) 40. (D) 66. (C)
15. (D) 41. (C) 67. (B)
16. (B) 42. (A) 68. (D)
17. (D) 43. (A) 69. (A)
18. (A) 44. (D) 70. (D)
19. (A) 45. (B) 71. (D)
20. (A) 46. (C) 72. (C)
21. (C) 47. (D) 73. (A)
22. (A) 48. (C) 74. (C)
23. (D) 49. (C) 75. (A)
24. (A) 50. (D) 76. (D)
25. (D) 51. (C) 77. (C)
26. (A) 52. (C) 78. (C)

CPMT - 1999

1. (C) 27. (B)
2. (B) 28. (B)
3. (C) 29. (D)
4. (D) 30. (D)
5. (B) 31. (A)
6. (D) 32. (C)
7. (C) 33. (B)
8. (C) 34. (A)
9. (D) 35. (B)
10. (A) 36. (C)
11. (D) 37. (B)
12. (B) 38. (B)
13. (A) 39. (C)
14. (A) 40. (A)
15. (A) 41. (C)
16. (C) 42. (C)
17. (D) 43. (D)
18. (C) 44. (C)
19. (B) 45. (D)
20. (C) 46. (A)
21. (D) 47. (B)
22. (A) 48. (A)
23. (D) 49. (B)
24. (A) 50. (C)
25. (C)
26. (B)

CPMT - 2000

1. (D) 27. (B)
2. (C) 28. (C)
3. (B) 29. (C)
4. (D) 30. (B)
5. (A) 31. (B)
6. (C) 32. (A)
7. (D) 33. (B)
8. (C) 34. (B)
9. (C) 35. (A)
10. (D) 36. (D)
11. (A) 37. (A)
12. (B) 38. (D)
13. (C) 39. (B)
14. (C) 40. (C)
15. (D) 41. (B)
16. (C) 42. (B)
17. (C) 43. (D)
18. (B) 44. (A)
19. (B) 45. (A)
20. (B) 46. (A)
21. (C) 47. (D)
22. (A) 48. (B)
23. (C) 49. (C)
24. (B) 50. (B)
25. (A)
26. (D)

CPMT - 2001

1. (A) 27. (D)
2. (D) 28. (C)
3. (A) 29. (A)
4. (D) 30. (B)
5. (A) 31. (B)
6. (B) 32. (A)
7. (C) 33. (A)
8. (C) 34. (B)
9. (D) 35. (D)
10. (D) 36. (C)
11. (A) 37. (C)
12. (B) 38. (D)
13. (C) 39. (D)
14. (B) 40. (D)
15. (A) 41. (B)
16. (A) 42. (D)
17. (B) 43. (A)
18. (C) 44. (B)
19. (A) 45. (A)
20. (C) 46. (D)
21. (D) 47. (B)
22. (A) 48. (C)
23. (B) 49. (D)
24. (D) 50. (D)
25. (C)
26. (C)

CPMT - 2002

1. (C) 27. (A)
2. (A) 28. (B)
3. (C) 29. (A)
4. (A) 30. (B)
5. (B) 31. (A)
6. (C) 32. (A)
7. (A) 33. (D)
8. (A) 34. (A)
9. (B) 35. (A)
10. (A) 36. (A)
11. (B) 37. (C)
12. (C) 38. (A)
13. (B) 39. (A)
14. (C) 40. (A)
15. (C) 41. (B)
16. (C) 42. (C)
17. (B) 43. (A)
18. (C) 44. (C)
19. (C) 45. (C)
20. (C)
21. (D)
22. (C)
23. (A)
24. (C)
25. (C)
26. (B)

CPMT - 2003

1. (A) 27. (A)
2. (A) 28. (D)
3. (B) 29. (A)
4. (C) 30. (A)
5. (C) 31. (C)
6. (B) 32. (C)
7. (C) 33. (D)
8. (D) 34. (A)
9. (D) 35. (C)
10. (B) 36. (C)
11. (B) 37. (A)
12. (C) 38. (B)
13. (A) 39. (D)
14. (D) 40. (B)
15. (A) 41. (D)
16. (A) 42. (D)
17. (A) 43. (A)
18. (A) 44. (A)
19. (B) 45. (D)
20. (A) 46. (B)
21. (B) 47. (B)
22. (A) 48. (D)
23. (B) 49. (C)
24. (A)
25. (A)
26. (B)

EAMCET - 1990

1. An element 'R' is in group III. Which is true in respect of 'R'

(A) It forms an oxide of type R_2O_3
(B) It forms a halide of the type RX_2
(C) It is a gas at room temperature
(D) It has an oxidation State of +4

2. The components present in producer gas is

(A) $CO + N_2$ (B) $CO + H_2$
(C) $CO_2 + N_2$ (D) $CO_2 + H_2$

3. Which of the following indicates the correct variation in electronegativities ?

(A) F > N < O > C (B) F < N > O > C
(C) F < N < O < C (D) F > N > O < C

4. The first Ionisation energy of lithium will be

(A) Greater to Berelium (B) Lesser to Berelium
(C) Equal to sodium (D) Equal to Fluorine

5. Which is correct with respect to 'P' orbits ?

(A) They have a strong directional character
(B) They are five fold degenerate
(C) They are spherical
(D) They have no directional character.

6. The nitride ion in lithium nitride is composed of

(A) 7 protons + 7 electrons
(B) 10 protons + 7 electrons
(C) 7 protons + 10 electrons
(D) 10 protons + 10 electrons

7. The chemical reaction between $Na_2S_2O_3$ and I_2 in aqueous medium produces

(A) $Na_2S + NaI + H_2S$ (B) $Na_2SO_3 + NaI + H_2O$
(C) $Na_2SO_3 + NaI + H_2O$ (D) $Na_2S_4O_6 + NaI$

8. The Root Mean square velocity at NTP of the species can not be calculated from the expression

(A) $\sqrt{3p/d}$ (B) $\sqrt{3PV/M}$ (C) $\sqrt{3RT/M}$ (D) $\sqrt{3RT/d}$

9. The ion that is isoelectronic with carbon monoxide is

(A) O_2^- (B) N_2^+ (C) CN^- (D) O_2^{2+}

10. An element 'Y' has a ground state configuration 2, 8, 8. The type of bond that exists between the atoms 'Y' is

(A) Ionic (B) Covalent
(C) Metallic (D) van der Waals

11. Which of the following mixtures in aqueous solution acts as a buffer ?

(A) $HNO_3 + KNO_3$ (B) $H_2SO_4 + K_2SO_4$
(C) $NH_4OH + NH_4Cl$ (D) $CH_3COOH + KaCl$

12. How many moles of KI are required to produce 0.4 moles of K_2HgI_4 ?

(A) 3.2 (B) 1.6 (C) 0.4 (D) 0.8

13. Which of the following pairs of the gaseous species diffuse through a porous plug with the same rate of diffusion ?

(A) N_2, CO (B) NO, CO_2
(C) NH_3, PH_3 (D) NO, C_2H_6

14. Which pair of the following atomic numbers represent elements which are both 's' block elements ?

(A) 7, 15 (B) 6, 12 (C) 9, 17 (D) 3, 12

15. A radioactive element has a half life of 20 minutes. How much time should elapse before the elements is reduced to 1/8 of its original mass

(A) 80 min. (B) 160 min. (C) 40 min. (D) 60 min.

16. A transition metal 'X' has a configuration $[Ar]3d^4$ in its +3 oxidation state. Its atomic number is

(A) 25 (B) 26 (C) 22 (D) 19

17. A molal solution is one that contains one mole of solute in

(A) 1000 gms of the solvent (B) One litre of the solution
(C) One litre of the solvent (D) 22.4 litres of solution

18. One mole of Diethyl ether on heating with con. HI. gives two moles of

(A) CH_3I (B) C_2H_5I (C) C_2H_5OH (D) CHI_3

19. A silver cup is plated with silver by passing 965 coloumbs of electricity, the amount of silver deposited is

(A) 9.89 gm (B) 107.87gm (C) 1.0787 gm (D) 1.002 gm

20. Which one of the following gives carbylamine reaction?

(A) H_5C_6 / H_3C > NH (B) CH_3HN_2
(C) $(CH_3)_2NH$ (D) $(CH_3)_3N$

Fill in the blanks

1. An organic compound containing H_2 and O_2 but a single carbon atom responds positively to Tollen's reagent. The formula of the compound is ....

2. The formula of the main product obtained when acetic acid reacts with PCl_5 is ....

3. The Molarity of a solution containing 1.0 gms. NaOH in 250 ml of water is ....

4. In the conversion of $Br_2 \longrightarrow BrO_3^-$ the oxidation state of bromine changes from .... to +5.

5. As per the reaction $N_2(g) + 2O_2(g) \rightleftharpoons 2NO_2(g)$, $\Delta H = 66kJ$. The heat of formation of nitrogendixoide is ....

6. Which one of the following $CuCl_2$, or $CaCl_2$ or $CdCl_2$ is paramagnetic in character ....

7. The single electrode potential E of a 0.1M solution M^+ ions ($E^\circ = 2.36V$) is ....

8. The number of moles of water present in 90.0 gms. of water is ....

9. The ratio between the neutrons, present in carbon and silicon with respective atomic masses of 12 and 28 is .....

10. Complete the reaction

$^{14}_{7}N + {}_0n^1 \longrightarrow$ .... $+ {}_1H^1$

11. The equilibrium concentration of x, y and yx_2 are 4, 2 and 2 moles respectively for the equilibrium

$2x + y \leftrightarrow yx_2$. The value of K_C is ....

12. The class of the compound that get reduced to primary alcohols and also respond to Fehlings test is known as ....

13. 0.2 mole of a saturated hydrocarbon 'X' on complete combustion produces 26.4 gm of CO_2. The molecular weight of 'X' is ....

14. The transition metal present in the alloy gun metal is ....

15. The IUPAC name of the compound

$$CH_3 - \overset{\overset{H}{|}}{\underset{\underset{CH_3}{|}}{C}} - CH_2Br$$ is ....

16. Among the ions Sc^{3+}, V^{3+} and Cr^{3+}, the ion that gives colourless compounds is ....

17. Which order reaction obeys the expression $t_{1/2} = 1/KA$ in chemical kinetics ....

18. Maleic and fumaric acids are a pair of .... Isomers.

19. Among the ions Cl^-, S^{2-} and Na^+, the largest ion is ....

20. The number of Hydrogen atoms bridging the boron atoms in diborane is ....

21. The trivial name of 4-methyl aniline is ....

22. The number of ions produced from one molecule of potassium Ferricyanide in aqueous solution are ....

23. Among the metals Na, Mg and Al. the metal with highest melting point is ....

24. A simple oxide of carbon and that of another non-metal has the same C_P/C_V ratio, but a difference of 2 in their molecular weights. The other non-metal could be ....

25. In the equilibrium reaction $N_2 + 3H_2 \rightleftharpoons 2NH_3$ the sign of ΔH accompanying the reaction is ....

26. An element 'X' belongs to the group I or II or V. Its oxide reacts with water to produce a highly acidic solution. The element 'X' belongs to group ....

27. A metallic oxide contains 60% of the metal. The equivalent weight of the metal is ....

28. A solution of P^H 9.0 is one thousand times as basic as a solution of ....

29. The molecular weight of a gas which diffuses through a porous plug at 1/6 th of the speed of hydrogen under identical condition is ....

30. The type of Hybridisation of orbitals employed in the formation of SF_6 molecule is ....

1. Which one of the following series of lines is found in the U.V. region of atomic spectrum of Hydrogen ?
(A) Bracket (B) Lyman (C) Balmer (D) Paschen

2. Aniline is treated with $NaNO_2$ and HCl at 0°C, the product formed is
(A) Benzenediazonium chloride salt
(B) Nitroaniline (C) Chloroaniline
(D) Aminophenol

3. A gaseous mixture contains 56 grams of N_2, 44 grams CO_2 and 16 grams of CH_4. The total pressure of the mixture is 720 mm Hg. The partial pressure of CH_4 is
(A) 180 mm (B) 360 mm (C) 540 mm (D)720 mm

4. The energy of the electron in the Hydrogen atom is given by the expression
(A) $\frac{-e^2}{r^2}$ (B) $\frac{-n^2h^2}{2\pi^2z^2e^4m}$ (C) $\frac{-2p^2z^2e^4m}{n^2h^2}$ (D) $\frac{nh}{2p}$

5. The R.M.S. velocity of an ideal gas at 27°C is 0.3 m/sec. Its R.M.S. velocity at 927°C in m/sec. is
(A) 3.6 (B) 2.4 (C) 0.9 (D) 0.6

6. Three Faradays of electricity was passed through an aqueous solution of Iron (II) bromide. The weight of iron metal (at. wt.= 56) deposited at the cathode in grams is
(A) 112 (B) 168 (C) 56 (D) 84

7. Which of the following gases contains the same number of molecules as that of 16 grams of Oxygen ?
(A) 16 grams of O_3 (B) 32 grams of SO_2
(C) 16 grams of SO_2 (D) All

8. An organic compound has an empirical formula CH_2O. Its V.D. is 45. The molecular formula of compound is
(A) CH_2O (B) C_2H_5O (C) $C_2H_4O_2$ (D) $C_3H_6O_3$

9. When 4p orbitals in any atom are filled completely, the next electron enters in
(A) 4 d (B) 4 f (C) 5s (D) 3d

10. A radio isotope has a half life of 10 days. If 125 grams of it is left, what was its weight 40 days earlier ?
(A) 600 gms (B) 1000 gms
(C) 1250 gms (D) 2000 gms

11. The relative vapour pressure is equal to the
(A) Ratio of the number of solute molecules to the number of solvent molecules
(B) Ratio of the number of solute molecules to the total number of molecules in solution
(C) Ratio of the number of solvent molecules to the number of solute molecules
(D) Ratio of the number of solvent molecules to the total number of molecules in solution

12. In which of the following reactions does the heat change represent the heat of formation of water ?
(A) $2H_2 + O_2 \rightarrow 2H_2O$ $\Delta H = -116$ K.Cal
(B) $H_2 + \frac{1}{2}O_2 \rightarrow H_2O$ $\Delta H = -58$ K. Cal
(C) $H^+ + OH^- \rightarrow H_2O$ $\Delta H = -13.7$ K. Cal
(D) $C_2H_2 + 2\frac{1}{2}O_2 \rightarrow 2CO_2 + H_2$; $\Delta H = -310$ K.Cal

13. Which of the following statements about the halogens is correct ?
(A) They are all diatomic and forms univalent ions
(B) They are all diatomic and forms divalent ions
(C) They all capable of exhibiting several oxidation states
(D) They can mutually displace each other from the solution of their compound with metals.

14. The pK_a of a certain weak acid is 4.0 what should be the salt to acid ratio if we have to prepare a buffer with a pH = 5 using the acid and one of the slats
(A) 10 1 (B) 1 10 (C) 4 5 (D) 5 4

15. The type of bonds present in $K_4[Fe(CN)_6]$ are
(A) Ionic (B) Covalent (C) Ionic & covalent
(D) Ionic, Covalent and Co-ordinate covalent bond.

16. Which of the following oxides of nitrogen is the anhydride of HNO_3
(A) N_2O_5 (B) N_2O (C) N_2O_3 (D) N_2O_4

17. 9.8 grams of H_2SO_4 is present in two litres of a solution. The molarity of the solution is
(A) 0.1 m (B) 0.05 m (C) 0.2m (D) 0.01m

18. One atom of $^{39}_{19}K$ contains
(A) 19 P; 20 n and 19 e (B) 19 P; 20 n and 20 e
(C) 20 P, 19 n and 20 e (D) 20 P, 19 n and 19 e

19. Chloroform reacts with HNO_3 to give
(A) Methane (B) Chloropicrin
(C) Nitro methane (D) Nitrogen

20. The reagent that has action on glass is
(A) Aquaregia (B) Hydrofluoric acid
(C) Oleum (D) Fuming HNO_3

Fill in the blanks

1. If the pressure and the absolute temperature of a given mass of a gas are doubled, the new volume will be .... the initial volume.

2. Phosgene is formed when .... is exposed to air and sunlight.

3. The process of cooling of glass slowly is called ....

4. In the raction $C_2O_4^{2-} + MnO_4^- + H^+ \rightarrow Mn^{+2} + CO_2$ the reductant is ....

5. The non metallic element present in the mineral cryolite is ....

6. Among SO_2, H_2SO_4 and sodium thiosulphate, sulphur has the highest oxidation state in ....

7. Reduction of acetaldehyde in the presence of $LiAlH_4$ gives ....

8. The maximum number of electrons that can be accomodated in all the orbitals for which $l = 3$ is ....

9. The magnetic quantum number "m" for the outer most electron in the Na atom is ....

10. The normal b.p. of CS_2, H_2O and CCl_4 are 41.3°C, 100°C and 77°C respectively. The liquid in which the intermolecular forces are the weakest is ....

11. Water has a higher b.p. than the corresponding hydrides H_2S, H_2Se and H_2Te. This is because of .... in water.

12. The reaction between .... and Bayer's reagent gives ethylene glycol.

13. The mass number of a nuclide is 125. Its nuclear radius is .... Fermi.

14.
$$\begin{array}{c} \quad\quad\quad CH_3 \\ \quad\quad\quad | \\ CH_3-CH_2-CH-CH_2-CH_2-CH_2-CH_3 \end{array}$$
IUPAC name of the compound is ....

15. The number of σ-bonds in the ethene molecule is .....

16. Along with SiO_2 and Al_2O_3 the other major ingredient in portland cement is ....

17. In the reaction
$^{19}_{8}O + ^{2}_{1}H \rightarrow ^{18}_{8}O + ^{3}_{1}X$, X is ....

18. The half life of a radioactive substance is 100 sec. It's disintegration constant is ....

19. Aqueous solution of $FeCl_3$ is acidic because of ....

20. The weight of oxalic acid $H_2C_2O_4.2H_2O$ required to prepare 500ml of 0.2N solution is ....

21. The K.E. of 8 grams of H_2 at 27°C is .... times more than the K.E. of 8grams of O_2 at the same temperature.

22. 0.1 M HCl solution is diluted by 100 times. The pH of the solution so formed is ....

23. A few drops of an organic compound (liq.) are added to schiff's reagent. A pink colour is produced. The organic compound is a/an ....

24. The co-ordination number of the ions in the CsCl crystal lattice is ....

25. Nuclides having different mass number and different atomic numbers, but the same number of neutrons are called ....

26. When an electron in an excited hydrogen atom jumps from an energy level for which n = 5 to a lower level for which n = 2 the spectral line is observed in the .... series of the hydrogen spectrum.

27. The number of moles of barium carbonate which contains 1.5 moles of O_2 atoms is ....

28. The rate of diffusion of He is .... the rate of diffusion of CH_4.

29. The conjugate base of HSO_4^- ion is ....

30. Among the alkali metals, the metal with the highest I.P. is ....

EAMCET - 1992

1. The degree of dissociation of an acid, HA in 0.1 M solution is 0.1%, its dissociation constant is
(A) 1×10^{-10} (B) 1×10^{-14}
(C) 1×10^{-3} (D) 1×10^{-7}

2. When ethyl alcohol reacts with bromine in the presence of red phosphorus the compound formed is
(A) C_2H_6 (B) PBr_3 (C) CH_3Br (D) C_2H_5Br

3. The half-life period of a radioactive element is 140 days. After 560 days, 1 gm of the element will be reduced to
(A) 1/2 gm (B) 1/4 gm (C) 1/8 gm (D) 1/16 gm

4. Bauxite is an ore of
(A) Al (B) Fe (C) Ca (D) Cu

5. The relation between the enthalpy change and internal energy change at constant temperature is
(A) $H = E + PV$ (B) $\Delta H = \Delta E + \Delta n\ RT$
(C) $\Delta H = \Delta E + P(\Delta V)$ (D) $\Delta H = \Delta G + T\Delta S$

6. At 27°C, the ratio of root mean squre velocities of Ozone to Oxygen is
(A) $\sqrt{2/3}$ (B) 0.25 (C) $\sqrt{3/5}$ (D) $\sqrt{4/3}$

7.The bond energies of $N \equiv N$, H–H and N – H bonds are 945, 436 and 391 kJ.mol^{-1} respectively. The enthalpy of the reaction $N_2(g) + 3H_2(g) \rightarrow 2NH_3(g)$ is
(A) –93 kJ (B) 102 kJ (C) 90 kJ (D) 105kJ

8. A catalyst increases the rate of a reaction, because it
(A) Decreases the energy barrier for reaction
(B) Decreases the collision diameter
(C) Increases the activation energy
(D) Increases the temperature coefficient

9. In nuclear reactions, heavy water is used as a
(A) Fuel (B) Projectile (C) Moderator (D) Arrestor

10. Cuprous ion is colourless while cupric ion is coloured because
(A) Cuprous ion has a completely filled d-orbitals and cupric ion has incomplete d-orbitals
(B) Both have half filled p and d-orbitals
(C) Both have unpaired electrons in the d-orbitals
(D) Cuprous ion has incomplete d-orbitals and cupric ion has a completely filled d-orbitals

11. In the reaction $H_2(g) + Cl_2(g) \rightleftharpoons 2HCl(g)$
(A) $K_P = K_C$ (B) $K_P \neq K_C$
(C) $K_P > K_C$ (D) $K_P < K_C$

12. When 9.65 coulombs of electricity is passed through a solution of silver nitrate (atomic weight of Ag=107.87g taking as 108.0) the amount of silver deposited is
(A) 16.2mg (B) 21.2mg (C) 10.8mg (D) 5.4mg

13. When aniline is treated with chloroform in alcoholic KOH, the product formed is
(A) Benzene diazonium chloride (B) Benzanilide
(C) Schiff base (D) Phenyl isocyanide

14. If 0.84 gm of a metal carbonate reacts with 40ml of N/2 H_2SO_4. The equivalent weight of the metal carbonate is
(A) 42 gm (B) 38 gm (C) 84gm (D) 64 gm

15. Al^{3+} has a lower ionic radius than Mg^{2+} ion because
(A) Mg atom has less number of neutrons than Al.
(B) Al^{3+} has a higher nuclear charge than Mg^{2+}
(C) Their electronegativities are different
(D) Al has a lower ionisation potential than Mg atom.

16. Atoms of different elements having identical mass are known as
(A) Isotones (B) Isomers (C) Isotopes (D) Isobars

17. In the long form of periodic table, the elements having lowest ionisation potential are present in
(A) I group (B) IV group
(C) VII group (D) Zero group

18. If 20 ml of 0.4 N NaOH solution completely neutralise 40 ml of a dibasic acid, the molarity of the acid solution is
(A) 0.1 M (B) 0.2 M (C) 0.3 M (D) 0.4 M

19. A sample of a gas has a volume of 0.2 lit. measured at 1 atm. pressure and 0°C. At the same pressure, but at 273°C, its volume will become
(A) 0.1 litre (B) 0.4 litre (C) 0.8 litre (D) 0.6 litre

20. If in a 100 ml of an aqueous hydrochloric acid of pH 1.0, 900 ml of distilled water is added, the pH of the resultant solution will become
(A) 4.0 (B) 7.0 (C) 1.0 (D) 2.0

Fill up the blanks

1. Electronegativity of an atom is given by the average of electron affinity and ....

2. The energy of red stars is due to nuclear ....

3. In diborane, boron atoms undergo .... hybridisation.

4. $C_6H_5NO_2 \xrightarrow{Sn/HCl}$ ....

5. Ethyl bromide on boiling with alcoholic soluton of sodium hydroxide forms ....

6. Ethyl alcohol when passed over copper at 300°C, it is dehydrogenated to give ....

7. ${}_{92}U^{235} + {}_0n^1 \rightarrow 3\,{}_0n^1 +$........................

8. Magnesium is mainly extracted from ....

9. The formula of plaster of paris is ....

10. The correct electronic configuration of potassium is ....

11. The most electronegative element is ....

12. The nuclei which have the same number of neutrons but different number of protons are called ....

13. The pH of 10^{-2} M NaOH solution is ....

14. The oxidation number of manganese in $KMnO_4$ is ....

15. At constant temperature and pressure if $\Delta G < O$ the precess is ....

16. The solution made by mixing appropriate amounts of a weak acid and its salt with a strong base is called ....

17. Chloroform slowly reacts with oxygen of the air in the presence of light to give ....

18. When NH_4NO_3 is heated, the gas formed is ....

19. Ammoniacal solution of silver nitrate is called as ....

20. 3.5 gm of carbon monoxide at O°C and 760 mm pressure contains ....

21. The rate of reaction is generally doubled for every .... degree rise in temperature

22. A radioactive element has a decay constant of $6.93 \times 10^{-3} sec^{-1}$. Its half life is .... minutes.

23.
$$CH_3-CH_2-\underset{\displaystyle |}{\overset{\displaystyle CH_3}{C}}=CH_2$$
The IUPAC name of the compound is ....

24. Dilute $KMnO_4$ oxidises acetylene to ....

25. IP of 1s electron is .... than 2s electrons.

26. Krypton nucleus has a .... radius than that of barium.

27. A "magic number" nucleus contains .... number of protons and .... number of neutrons.

28. The mole fraction of water in 20% aqueous hydrogen peroxide solution is ....

29. The rate of diffusion of nitrogen gas in a diffusion tube is 9 ml sec^{-1}. An unknown gas X diffuses at a rate of 6 ml sec^{-1} in the same diffusion tube. The molecular weight of X is .... gm mol^{-1}

30. d^2sp^3 hybridisation leads to .... geometry.

EAMCET - 1993

1. The oxidation number of manganese in $KMnO_4$ is
(A) 7 (B) 9 (C) 3 (D) 5

2. Silver is extracted from its .... ore.
(A) Sulphide (B) Oxide (C) Nitrate (D) Halide

3. A certain sample of a gas has a volume of 0.2 litres measured at 1 atm. pressure and 0°C. At the same pressure but at 273°C, its volume will be
(A) 0.4 litres (B) 0.8 litres
(C) 27.8 litres (D) 55.6 litres

4. A solution that obeys Raoult's law is called
(A) Ideal solution (B) Saturated solution
(C) Normal solution (D) Molar solution

5. Quicklime is represented by the formula
(A) $Ca(OH)_2$ (B) CaO (C) $CaCO_3$ (D) $CaHCO_3$

6. The catalyst used in the manufacture of sulphuric acid by contact process is
(A) V_2O_5 (B) MnO_2 (C) Al_2O_3 (D) Cr_2O_3

7. An isomer of ethanol is
(A) Methanol (B) Diethly ether
(C) Acetone (D) Dimethyl ether

8. The pH of a 0.001 M aqueous solution of sodium hydroxide will be
(A) 5.0 (B) 7.5 (C) 9.0 (D) 11.0

9. Which of the following poisonous gases is formed when chloroform is exposed to light and moist air ?
(A) Mustard gas (B) Phosgene
(C) Chlorine (D) Carbon monoxide

10. The half-life period of a radioactive element is 120 days. After 480 days, 4.0 gm of the element will be reduced to .... gms.
(A) 0.5 (B) 0.25 (C) 2.0 (D) 1.0

11. In which of the following compounds is hydrogen bonding strongest in the liquid phase ?
(A) HF (B) CH_4 (C) HI (D) PH_3

12. Which of the following contains a coordinate covalent bond ?
(A) Water (B) Ammonia
(C) Ammonium ion (D) Ethylene

13. The hybridization involved in acetylene is
(A) sp (B) sp^2 (C) sp^3 (D) dsp^2

14. The cobalt isotope of mass 60 decays with a half-life of 5.3 years. How many years does it take for 7/8 of a sample of 500 g of cobalt-60 to disintegrate ?
(A) 15.9 (B) 9.2 (C) 10.6 (D) 4.6

15. The reaction of I_2 with $Na_2S_2O_3$ gives NaI and
(A) SO_2 (B) $Na_2S_2O_6$ (C) $Na_2S_4O_6$ (D) $Na_2S_2O_8$

16. If a reaction obeys the following equation.
$K = \frac{2.303}{t} \log_{10} \frac{a}{a-x}$ the order of the reaction will be.
(A) Zero (B) First (C) Second (D) Third

17. Which of the following electronic configurations corresponds to an inert gas ?
(A) $1s^2 2s^2 2p^5$ (B) $1s^2 2s^2 2p^6$
(C) $1s^2 2s^1$ (D) $1s^2 2s^2 2p^6 3s^1$

18. Which of the following alloys contains Cu & Zn ?
(A) Bronze (B) Brass
(C) Gunmetal (D) Type metal

19. The 4f level is successively filled up in
(A) Alkali metals (B) Rare gases
(C) Lanthanides (D) Actinides

20. In which of the following , sulphur has the lowest oxidation number ?
(A) H_2SO_3 (B) H_2S (C) H_2SO_4 (D) SO_2

Fill in the blanks

1. Glass is soluble in a soloution of .... acid.

2. A solution of $Cr(NO_3)_3$ slowly turns green when concentrated hydrochloric acid is added due to formation of Chromium ....

3. The deep colour produced when iodine is dissolved in a solution of potassium iodide, is caused by the presence of .... ion.

4. Old bottles of ether are dangerous because they frequently contain high concentration of ....

5. Charcoal is often used in gas masks because of its high capacity for absorbing .... molecules.

6. Hydrogen gas may be readily prepared in the laboratory by reacting water with ....

7. The use of a catalyst in a chemical reaction .... activation energy of the reaction.

8. The reaction which proceeds with evolution of heat is called .... reaction.

9. Ethyl bromide on treatment with KOH(aq) gives rise to ....

10. The IUPAC name for $CH_3-\underset{CH_3}{\underset{|}{CH}}-\underset{OH}{\underset{|}{CH}}-CH_3$ is ....

11. Valence bond theory of L. Pauling and J.C. Slater accounts for .... characteristics of covalent bond.

12. The ionic product of water $[H^+][OH^-]$ is ....

13. The oxidation number of chlorine in $KClO_3$ is ....

14. According to Aufbau principle, the electron has a tendency to occupy that subshell which has .... energy.

15. The emission of an alpha particle from a radioactive element causes its atomic number to decrease by ....

16. The solutions which tend to keep the concentration of hydrogen ions constant even when small amounts of strong acid or strong base are added to them are known as .... solutions.

17. When hydrogen bond is formed between two molecules it is called .... hydrogen bond.

18. The chemical formula of Plaster of Paris is ....

19. The melting point of diamond is very high, since the type of bonding in it is ....

20. The chemical formula of bleaching powder is ....

21. Oxides of iron in the blast furnace get reduced due to the action of .... on haematite.

22. The reaction of magnesium bromide with acid solution produces ....

23. The rare gas which is more abundant in the atmosphere is ....

24. A 0.1 M solution of sodium bisulphate is .... in nature .

25. The halogen with highest ionisation potential is ....

26. The reaction of acetaldehyde by .... leads to the formation of ethanol .

27. Unpaired electrons in .... in transition metals leads to paramagnetism .

28. Transition elements exhibit variable valency on account of .... d-orbitals .

29. Inner transition elements form different coloured compounds on account of the presence unfilled ....

30. Temporary hardness of water is due to the presence of .... of calcium & magnesium in water .

EAMCET- 1994

1. Given that

$CH_3CHO + \frac{5}{2}O_2 \rightarrow 2CO_2 + 2H_2O : \Delta H = -1168$ KJ mol^{-1}

$CH_3COOH + 2O_2 \rightarrow 2CO_2 + 2H_2O : \Delta H = -876$ KJmol^{-1}

ΔH for the reaction $CH_3CHO + \frac{1}{2} O_2 \rightarrow CH_3COOH$ is

(A) –292 KJ mol^{-1} (B) 378 KJ mol^{-1}
(C) 195 KJ mol^{-1} (D) –2044 KJ mol^{-1}

2. How many grams of copper will be deposited from a solution of $CuSO_4$ by passing 0.5 Faraday of electricity ? (At. wt. of Cu = 63.5)

(A) 15.875 (B) 127 (C) 31.75 (D) 63.5

3. The single electrode reactions,

$Cd \rightleftharpoons Cd^{2+} + 2e^- : E^o = 0.4030V$

$Cu^{2+} + 2e^- \rightleftharpoons Cu : E^o = 0.337V$

can be combined to give the cell reaction

$Cd + Cu^{2+} \rightleftharpoons Cd^{2+} + Cu$, the e.m.f. of such cell is

(A) 0.74V (B) 0.1354V (C) 0.66 (D) 0.066V

4. The ionic product of water is 14. What is the H^+ ion concentration of a 0.1 M NaOH solution ?

(A) 13M (B) 10^{-13} (C) 10^{-14}M (D) 14M

5. The pH of a 0.05M solution of H_2SO_4 is

(A) 5 (B) 1 (C) 3 (D) 0.1

6. When the same amount of Zn is treated with excess H_2SO_4 and with excess NaOH, the ratio of the volumes of H_2 evolved is

(A) 2 :1 (B) 1 :1 (C) 3 : 2 (D) 1 : 2

7. 5.85 gms of NaCl (Mol. wt 58.5) is dissolved in water and the solution made up to 500 ml. The molarity of the solution will be

(A) 0.1 (B) 0.2 (C) 1.0 (D) 0.117

8. The gas that is used for sterilizing water is

(A) O_3 (B) NH_3 (C) SO_2 (D) NO_2

9. The crystalline form of sulphur that is stable at room temperature is
(A) Engel's sulphur (B) Rhombic sulphur
(C) Monoclinic sulphur (D) Plastic sulphur

10. Bleaching action of chlorine in the presence of moisture is an example of
(A) Oxidation (B) Reduction
(C) Displacement (D) Substitution

11. In the electrolysis of HF+KF mixture for the production of fluorine, the role of KF is.
(A) To minimize corrosion
(B) To lower the oxidation potential of fluoride
(C) To make HF a conducting solution
(D) To lower the solubility of fluorine

12. When dilute H_2SO_4 is electrolyzed at room temperature between pt electrodes, which of the following is liberated at the anode ?
(A) S (B) SO_3 (C) H_2 (D) O_2

13. In the reaction $NO_2^- + OCl^- \rightarrow NO_3^- + Cl^-$ the oxidation state of chlorine
(A) Changes from –2 to–1
(B) Changes from 0 to –1
(C) Does not change
(D) Changes from +1 to –1

14. Meta phosphoric acid is represented by
(A) H_3PO_3 (B) H_2PO_3 (C) HPO_3 (D) $H_4P_2O_7$

15. White phosphorous forms PCl_3 , nitrogen does not form NCl_3 because
(A) Nitrogen is more electronegative than Phosphorous
(B) Nitrogen atom is smaller than Phosphorous atom
(C) The ionization energy of nitrogen is greater than that of phosphorous
(D) Unlike phosphorous, nitrogen has no d-orbital in its valence shell.

16. The reagent used for concentrating silver ore is
(A) KCN (B) NH_3 (C) HI (D) HNO_3

17. The metal that has highest electrical conductivity is
(A) Al (B) Cu (C) Ag (D) Au

18. Which of the following is not an allotrope of carbon ?
(A) Graphite (B) Diamond (C) Soot (D) Carborundum

19. The alcohol manufactured from water gas is
(A) Ethanol (B) Butanol (C) Methanol (D) Isobutanol

20. Thermite is a mixture of iron oxide and
(A) Sodium shavings (B) Aluminium powder
(C) Zinc powder (D) Potassium metal

21. The formula of a metal chloride is MCl_3, if it contains 20% of the metal, the atomic weight of the metal is approximately.
(A) 21.3 (B) 106.5 (C) 25.5 (D) 11.8

22. The metal that forms a self protecting film of oxide to prevent corrosion is
(A) Cu (B) Al (C) Pt (D) Au

23. Which one of the following statements is not correct with respect of BF_3 ?
(A) It is an electron deficient compound
(B) It is a Lewis acid
(C) It is an ionic compound (D) It forms adducts

24. The formula of Dolamite is
(A) $MgCO_3$, $CaCl_2$ (B) $MgCO_3$, $CaCO_3$
(C) $MgCO_3$, $CaSO_4$ (D) $MgCl_2$, $CaCO_3$

25. Permanent hardness of water is due to
(A) The sulphates of Ca & Mg
(B) Both chlorides and sulphates of Ca & Mg
(C) The chlorides of Ca and Mg
(D) The bicarbonates of Ca & Mg

26. In aqueous solution, H_2O_2 oxidises H_2S to
(A) Sulphur (B) Sulphuric acid
(C) Caro's acid (D) Marshall's acid

27. A flask contains 3g of H_2 and 24g of O_2. The partial pressure of hydrogen is
(A) 1/8 of the total pressure
(B) 1/9 of the total pressure
(C) 1/2 of the total pressure
(D) 2/3 of the total pressure

28. Which of the following is a tertiary amine ?
(A) t-butyl amine (B) Methyl, ethyl amine
(C) Dimethyl secondary butyl amine
(D) N-methly amine

29. Benzene molecule contains
(A) 6π and 6 σ bonds
(B) 3π and 9 σ bonds
(C) 6π and 12 σ bonds
(D) 3π and 12 σ bonds

30. Aniline on treatment with bromine water gives a white precipitate of
(A) Mono bromo aniline (B) Dibromo aniline
(C) 2, 4, 6-tribromo aniline (D) Tetrabromo anline

31. Among the following, the hydrocarbon which decolourises bromine water but does not give precipitate with ammonical silver nitrate is
(A) C_2H_2 (B) C_6H_6 (C) CH_4 (D) C_2H_4

32. The method of converting high boiling hydrocarbons into low boiling hydrocarbons is known as
(A) Pyrolysis (B) Isomerization
(C) Cracking (D) Inversion

33. On heating with bleaching powder, ethyl alcohol is converted into
(A) CH_3COOH (B) CH_3CHO
(C) CH_3COOH_3 (D) $CHCl_3$

34. Which one of the following compounds is acidic

(A) C_3H_8 (B) C_2H_6 (C) C_2H_4 (D) C_2H_2

35. The equilibrium constant k for the reaction $HA + B \rightleftharpoons BH^+ + A^-$ is 100. If the rate constant for the forward reaction is 10^5, the rate constant for the reverse reaction is

(A) 10^7 (B) 10^3 (C) 10^{-3} (D) 10^{-5}

36. The role of a catalyst in a reversible reaction is

(A) It increases the rate of forward reaction

(B) It decreases the rate of backward reaction

(C) It alters the equilibrium constant

(D) It allows the equilibrium to be achieved quickly

37. In the equilibrium reaction $A(g)+2B(g) \rightleftharpoons C(g)+D(g)$ the equilibrium constant K_C is given by the expression

(A) $K_C = \frac{[C][D]}{[A][B]}$ (B) $K_C = \frac{[C][D]^2}{[A][B]}$

(C) $K_C = \frac{[A][B]^2}{[C][D]}$ (D) $K_C = \frac{[C][D]}{[A][B]^2}$

38. Considering the following reactions

$C + O_2 \rightarrow CO_2$; $\Delta H = -400 KJ\ mol^{-1}$

$2H_2 + O_2 \rightarrow 2H_2O$; $\Delta H = -570\ KJ\ mol^{-1}$

$CO_2 + 2H_2O \rightarrow CH_4 + 2O_2$; $\Delta H = 890\ KJ\ mol^{-1}$

the heat of formation of methane is

(A) -1860 Kj mol^{-1} (B) -1060 KJ mol^{-1}

(C) -720 KJ mol^{-1} (D) -80 KJ mol^{-1}

39. Which of the following gases contains the least number of molecules ?

(At. wt of P = 31, Cl = 35.5, N = 14, O = 16, C = 12, H = 1)

(A) 4.0g of laughing gas (B) 3.0g of Phosphine

(C) 2.0g of marsh gas (D) 10.0g of marsh gas.

40. The molecular weight of a gas that diffuses twice as rapidly as a gas with molecular weight 64 is

(A) 5.4 (B) 8 (C) 128 (D) 16

41. An example of a non-transitional element is

(A) Cerium (B) Silver (C) Cobalt (D) Lead

42. Which of the following is ions is colourless in aqueous solution

(A) Ti^{+2} (B) V^{+3} (C) Mn^{+2} (D) Zn^{+2}

43. The oxidation state of Mo in $Mo_2Cl_8^{-4}$ ion is

(A) -4 (B) -2 (C) $+6$ (D) $+2$

44. Which of the following is the largest ion ?

(A) Na^+ (B) Mg^{2+} (C) O^{2-} (D) F^-

45. Among the following, the element that has the lowest first ionization potential is

(A) Flurine (B) Neon (C) Nitrogen (D) Oxygen

46. BF_3 forms an adduct with NH_3 because

(A) Nitrogen has high electronegativity

(B) Boron has smaller atomic radius than nitrogen

(C) Boron has an empty p-orbital and nitrogen has a lone pair of electrons

(D) Boron has electropositive character

47. The number, of β-particles emitted in the radio active change ${}_{92}U^{238} \rightarrow {}_{82}Pb^{206} + 8\,{}_2He^4$ is

(A) 2 (B) 4 (C) 6 (D) 8

48. The half life period of a radioactive substance is 150 days. Starting with an initial quantity of 160 mg. of the substance, the fraction of it that has disintegrated in 600 days is

(A) 1/2 (B) 1/4 (C) 1/16 (D) 15/16

49. The constancy of e/m ratio for electrons inspite of variations of gas present in the discharge tube or of the material of the cathode shows that

(A) Electrons are - vely charged

(B) Electrons are universal constituents of matter

(C) Electrons are lightest of all particles

(D) Mass of the electron is 1/1838 of the mass of H-atom

50. The basic assumption of Bohr's model of hydrogen atom is that

(A) The radial distance of the electron is quantised

(B) The orbital velocity of the electron is quantised

(C) The energy of the electron is quantised

(D) The angular momentum of the electron is quantised

EAMCET - 1995

1. Ordinary glass is a combination of

(A) SiO_2, Na_2SiO_3, $CaSiO_3$ (B) SiO_2, Na_2CO_3, $CaCO_3$

(C) Na_2CO_3, K_2CO_3, SiO_2 (D) SiO_2, $PbCO_3$, $CaSiO_3$

2. Water gas essentially is a mixture of

(A) CO, N_2, H_2 (B) CO_2, N_2 (C) CO_2, H_2 (D) CO, H_2

3. Which of the following is correct with respect to the 15th group elements ?

(A) Except Nitrogen, all elements exhibit allotropy

(B) All elements except Nitrogen and Bismuth exhibit allotropy

(C) All the elements exhibit allotropy

(D) All elements except Nitrogen and Arsenic exhibit allotropy

4. Catalyst used in the manufacture of HNO_3 by Ostwald process is

(A) Platinum gauge (B) V_2O_5 (C) FeMO (D) MnO_2

5. Which of the following is paramagnetic ?

(A) N_2O_3 (B) N_2 (C) N_2O_4 (D) NO

6. Preparation of fluorine by direct electrolysis of liquid HF is not practical because

(A) Very high temperature is needed

(B) Electrolytic products are different from fluorine

(C) Very low temperature is required

(D) HF is a non electrolyte

7. The oxidation state of Iodine in ICl_2^- is

(A) +1 (B) -1 (C) +2 (D) -3

8. When 6 volumes of Oxygen undergoes complete reaction to ozone, the number of moles of ozone formed are
(A) 4 (B) 2 (C) 6 (D) 3

9. Which one of the following would you expect to have highest electonegativity ?
(A) Mg (B) S (C) B (D) Te

10. The mole percentage of oxygen in a mixture of 7.0 gm of nitrogen and 8.0 gm of oxygen is
(A) 24 (B) 50 (C) 8 (D) 16

11. Which of the following can act as both Bronsted acid and a Bronsted base ?
(i) HCO_3^- (ii) NH_3 (iii) O_2^- (iv) HS^-
(A) (i) & (ii) (B) (ii) & (iii)
(C) None of them (D) All of them

12. Which of the following is not a Lewis acid ?
(A) BF_3 (B) $AlCl_3$ (C) $BeCl_2$ (D) $SnCl_2$

13. Which of the following is least likely to act as a Lewis base ?
(At. No. of P = 15, S = 16, Cl = 17, I = 53)
(A) PCl_3 (B) SCl_2 (C) I^- (D) I^+

14. When a Cu wire is dipped in an aqueous $AgNO_3$ solution, the solution turns blue. This is due to
(A) Oxidation of silver (B) Reduction of copper
(C) Oxidation of copper (D) Reduction of silver

15. How many grammes of copper would be deposited if 3.00 amperes of current is passed through a solution of copper sulphate for 4.0 hours ? (At. Wt. of Cu = 63.54)
(A) 28.44 (B) 56.88 (C) 7.11 (D) 14.22

16. With respect to the figure given, which of the following statements is correct ?

P.E. A C B
Reaction Path

(A) ΔE for the forward reaction is C – B
(B) ΔE for the forward reaction is B – A
(C) $E_{(forward)} > E_{(backward)}$
(D) E (for reverse reaction) = C – A

17. The rate of a reaction, 2A + 3B + 4C $\rightarrow$ products, is equal to $R = [A]^2 [B]^1 [C]^0$. The overall order of the reaction is
(A) 9 (B) 5 (C) 2 (D) 3

18. For a reaction A + B $\rightarrow$ products, the rate of the reaction doubled when the concentration of A was doubled. When the concentration of A & B were doubled, the rate again doubled, the order of the reaction with respect to A & B are
(A) 1, 0 (B) 0, 1 (C) 1, 1 (D) 2, 0

19. The units of rate of a first order reaction are
(A) sec^{-1} (B) $lit.mol^{-1}$
(C) $mole\text{-}lit^{-1}\ sec^{-1}$ (D) No units

20. The time for half life of a first order reaction is 1 Hr. What is the time taken for 87.5% completion of the reaction ?
(A) 1 Hour (B) 2 Hours (C) 3 Hours (D) 4 Hours

21. The compound formed when 2-butene is treated with alkaline $KMnO_4$ is
(A) Acetaldehyde (B) Acetic acid
(C) CH_2OH | CH_2OH (D) CH_2-CH_3 | $CH_3C=O$

22. The products formed when ethyl chloride is treated with silver cyanide
(A) $C_2H_5CN + AgCl$ (B) $C_2H_5NC + AgCl$
(C) $C_2H_4 + AgCl$ (D) None of these

23. Which of the following compounds does not give positive iodoform test ?
(A) Ethanol (B) Acetone
(C) Acetaldehyde (D) Methanol

24. The number of structural isomers possible for $C_3H_5Cl_3$ is
(A) 5 (B) 6 (C) 11 (D) 8

25. A compound has an empirical formula CH_2O. Its vapour density is 45. Its molecular formula is
(A) $C_8H_6O_8$ (B) $C_3H_6O_3$ (C) $C_4H_8O_4$ (D) $C_5H_{10}O_5$

26. The compound which gives n-hexane on heating with sodium in dry ether is
(A) Ethyl bromide (B) n-propyl bromide
(C) Methyl bromide (D) n-Butyl bromide

27. Equal masses of methane and oxygen are introduced into a vessel at 27°C. What fraction of the total pressure is due to Oxygen ?
(A) 1/10 (B) 1/2 (C) 1/3 (D) 1/4

28. A given mass of gas obeys Boyle's law at certain temperature. Which one of the plots will not give a straight line ?
(A) V Vs 1/P (B) P Vs 1/V
(C) log P Vs. log V (D) P Vs. V

29. Which of the following is not a peroxide ?
(A) Na_2O_2 (B) PbO_2 (C) $NaHO_2$ (D) BaO_2

30. In the following two reactions,
(i) $2H_2O_2 \rightarrow 2H_2O + O_2$
(ii) $H_2O_2 + 2I^- + 2H^+ \rightarrow 2H_2O + I_2$; H_2O_2 acts as
(A) An oxidizing agent (B) A reducing agent
(C) Both oxidizing and reducing agent respectively
(D) Both reducing and oxidizing agent respectively

31. Which of the following alkali metals has the biggest tendency for the half reaction ?
$M(g) \rightarrow M^+(aq) + e^-$
(A) Sodium (B) Lithium (C) Potassium (D) Cesium

32. The glass which contains PbO mainly is

(A) Soda glass (B) Flint glass
(C) Pyrex glass (D) Vitreosil

33. Beryllium shows diagonal relationship with aluminium. Which of the following similarity is not correct ?
(A) $Be(OH)_2$ like $Al(OH)_3$ is basic
(B) Be forms beryllates and Al forms aluminates
(C) Be_2C like Al_4C_3 yields methane on hydrolysis
(D) Be like Al is rendered passive by HNO_3

34. Which of the following nitrogen halides is most stable ?
(A) NI_3 (B) NBr_3 (C) NCl_3 (D) NF_3

35. The compound formed during the setting of the plaster of paris is
(A) Cement (B) Gypsum (C) Anhydride (D) Mortar

36. Which of the following statements is not true regarding BF_3 molecule
(A) It is an electron deficient compound
(B) It is Lewis acid (C) It is an ionic compound
(D) It is a covalent compound

37. The oxy acid of phosphorus in which phosphorus has the lowest oxidation state is
(A) Hypohosphorus-acid (B) Orthophosphoric acid
(C) Pyrophosphoric acid (D) Metaphosphoric acid

38. The chief product of electrolysis of Al_2O_3 at the anode in the Hall's process is
(A) AlF_3 (B) CO_2 (C) H_2 (D) Na

39. In the following reaction, the product "R" is

$$CaC_2 \xrightarrow{H_2O} P \xrightarrow[\text{Tube}]{\text{Hot Iron}} Q \xrightarrow[AlCl_3]{CH_3Cl} R$$

(A) Benzene (B) Toluene
(C) Ethyl benzene (D) n-propyl benzene

40. Rutherford's experiment on scattering of α-particles showed for the first time that the atom has
(A) Nucleus (B) Electron (C) Proton (D) Neutron

41. Which of the following molecules has the highest dipole moment ?
(A) BCl_3 (B) CCl_4 (C) CO_2 (D) NF_3

42. The radius of second Bohr's orbit is
(A) 0.053nm (B) 0.106nm (C) 0.212nm (D) 0.0265nm

43. Nuclides having equal number of neutrons but differing in mass numbers and atomic numbers are called
(A) Isomers (B) Isotopes (C) Isotones (D) Isobars

44. In the following reaction
${}_{95}Am^{241} + X \rightarrow {}_{97}Bk^{243} + Y$; X and Y should be
(A) $X = 2{}_0n^1$ and $Y = 2\alpha^4$ (B) $X = \alpha$ and $Y = 2e$
(C) $X = 2e$ and $Y = \alpha$ (D) $X = \alpha$ and $Y = 2n^1_0$

45. The decreasing order of atomic radii for Be, Na and Mg is
(A) Na > Mg > Be (B) Mg > Na > Be
(C) Be > Na > Mg (D) Be > Mg > Na

46. The metal which is not ferromagnetic is
(A) Iron (B) Manganese (B) Cobalt (D) Nickel

47. The name of the element with atomic number 100 was adopted in honour of
(A) Alfred Nobel (B) Enrico Fermi
(C) Dimitri Mendeleev (D) Albert Einstein

48. Blue prints are based on the formation of a transition metal cyanide complex. The metal employed is
(A) Manganese (B) Cobalt (C) Chromium (D) Iron

49. For an ideal gas a plot of $\frac{PV}{RT}$ Vs T looks like

(A) 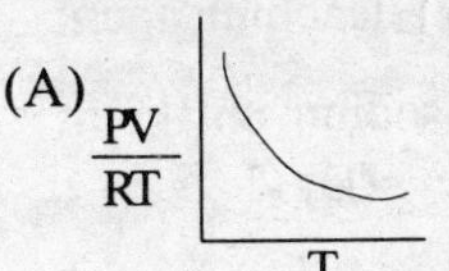
(B) 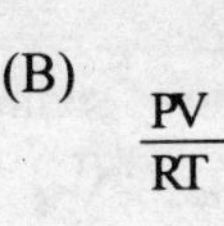 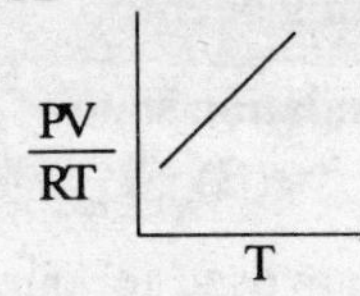
(C) 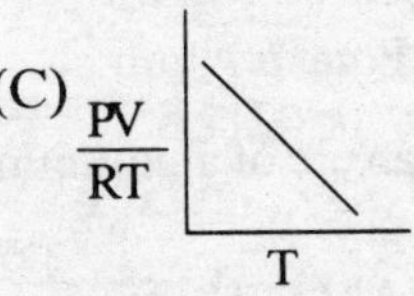
(D) 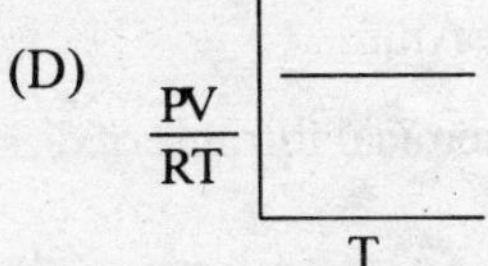

50 Which of the following is a representation of Gay-Lussac's law ?
(A) $P_1T_1 = P_2T_2$ (B) $P_1T_2 = P_2T_1$
(C) $P_1V_1 = P_2V_2$ (D) $V_1T_2 = V_2T_1$

EAMCET - 1996

1. Bleaching action of SO_2 is due to its
(A) Acidic nature (B) Reducing action
(C) Oxidising action (D) None of these

2. The element that can exhibit highest number of oxidation states amongst the following is
(A) V (B) Mn (C) Ni (D) CO

3. The number of moles of AgCl precipitated when excess $AgNO_3$ is mixed with one mole of $[Cr(NH_3)_4Cl_2]Cl$ is
(A) 2 (B) 3 (C) 0 (D) 1

4. The molecular formula of cryolite is
(A) $3NaF\ AlF_3$ (B) AlF_3 (C) $2NaFAlF_3$ (D) $NaFAlF_3$

5. Silver containing lead as an impurity is removed by
(A) Poling (B) Cupellation
(C) Liquation (D) Distillation

6. Diethyl ether is prepared by passing ethyl alcohol vapours over a catalyst under high pressure and temperature. The catalyst employed is
(A) Al_2O_3 (B) Ag_2O (C) SiO_2 (D) CuO

7. Acetone on reaction with chlorine gives normally
(A) Mono chloro acetone (B) Dichloro acetone
(C) Tri chloro acetone (D) Hexa chloro acetone

8. Reagent used for converting ethanoic acid to ethanol is
(A) $LiAlH_4$ (B) BH_3 (C) PCl_5 (D) $K_2Cr_2O_7$

9. Aniline when heated with $NaNO_2$ and HCl at 0-5 °C

the product formed is

(A) Chloro aniline (B) Benzene diazonium Chloride
(C) Chloro benzene (D) Dichloro benzene

10. Oxidation of aniline with potassium dichromate and sulphuric acid gives

(A) o-nitro aniline (B) N-nitroso aniline
(C) p-benzoquinone (D) o-benzoquinone

11. Sodium amalgam is useful as

(A) Oxidising agent (B) Catalyst
(C) Reducing agent (D) Bleaching agent

12. The oxidation state of sodium in sodium amalgam is

(A) zero (B) +2 (C) + 1 (D) -1

13. The alum used for purifying water is

(A) Ammonium alum (B) Chrome alum
(C) Ferric Alum (D) Potash Alum

14. The chemical formula of diaspore, an ore of aluminium is

(A) $Al_2O_3 3H_2O$ (B) $Al_2O_3 2H_2O$
(C) $Al_2O_3 H_2O$ (D) Al_2O_3

15. The inert form of carbon is

(A) Diamond (B) Graphite (C) Coal (D) Charcoal

16. When sodium argentocyanide is treated with zinc dust, silver precipitates because

(A) Zinc forms a complex readily with cyanide
(B) Silver is more electropositive than zinc
(C) Zinc is more electropositive than silver
(D) Both Zn^{+2} and Ag^{+} ions have d^{10} eletronic configuration.

17. The molecular weight of an organic compound is 180. Its empirical formula is CH_2O. The molecular formula is

(A) $C_6H_{12}O_6$ (B) $C_7H_{16}O_5$ (C) $C_8H_4O_5$ (D) $C_5H_8O_7$

18. Action of Zinc dust on tetra bromo ethane gives

(A) CH_3OH (B) $CH \equiv CH$
(C) $CH_2 = CH–CH_2OH$ (D) $CH_3 - CH_3$

19. The number of isomeric structures possible for a molecule having molecular formula C_5H_{12} is

(A) 2 (B) 3 (C) 4 (D) 5

20. An isomer of ethanol is

(A) Methanol (B) Dimethyl ether
(C) Diethyl ether (D) Ethylene glycol

21. At STP the density of a gas (molecular weight 45) in g/l is

(A) 11.2 (B) 1000 (C) 2 (D) 22.4

22. Gases show ideal behaviour at

(A) High pressure and High temperature
(B) Low pressure and High temperature
(C) Low pressure and Low temperature
(D) High pressure and Low temperature

23. The weight in grams of KCl (mol. wt. = 74.5) in 100ml of a 0.1 M KCl solution is

(A) 74.5 (B) 7.45 (C) 0.745 (D) 0.0745

24. The reaction taking place at the anode when an aqueous solution of $CuSO_4$ is electrolysed using inert Pt electrodes is represented as

(A) $2SO_4^{2-} \rightarrow S_2O_4^{2-} + 2e^-$ (B) $Cu^{2+} + 2e^- \rightarrow Cu$
(C) $2H_2O \rightarrow O_2 + 4H^+ + 4e^-$ (D) $2H^+ + 2e^- \rightarrow H_2$

25. If the standard reduction potentials at 298 K for the following half cell reactions

$Zn^{++}(aq) + 2e^- \rightarrow Zn(s)$ –0.762
$Cr^{+++}(aq) + 3e^- \rightarrow Cr(s)$ –0.740
$2H^+(aq) + 2e^- \rightarrow H_2(g)$ 0.000
$Fe^{+++}(aq) + e^- \rightarrow Fe^{++}(aq)$ 0.770

which one is the strongest reducing agent ?

(A) Zn(s) (B) Cr(s) (C) $H_2(g)$ (D) $Fe^{++}(aq)$

26. Which of the following compounds is most acidic ?

(A) CH_3Cl (B) CH_3OH
(C) $H_2C = CH_2$ (D) $HC \equiv CH$

27. Which one of the following reagents and conditions convert benzene to chlorobenzene ?

(A) Cl_2, sunlight, heat (B) HCl, heat
(C) HCl, sunlight, heat (D) Cl_2, $AlCl_3$

28. Hg sticks to the surface of the glass when it comes in contact with

(A) H_2O (B) HNO_3 (C) O_3 (D) I_2

29. The temperature at which heavy water has maximum density is

(A) 4.6°C (B) 11.6°C (C) 1.16°C (D) 4°C

30. Poisonous gas present in the exhaust fumes of an automobile is

(A) C_2H_6 (B) CO_2 (C) CO (D) CH_4

31. Superphosphate is a mixture of

(A) $Ca(H_2PO_4)\ H_2O + CaCl_2\ 2H_2O$
(B) $Ca(H_2PO_4)_2\ H_2O + 2CaSO_4\ 2H_2O$
(C) $Ca_3(PO_4)_2\ H_2O + 2CaSO_4\ 2H_2O$
(D) $Ca_3(PO_4)_2\ H_2O + CaCl_2\ 2H_2O$

32. The bond angle of H_2X (where X is a 16th group element) as one goes down the group

(A) Increases (B) Decreases
(C) Does not change (D) Changes irregularly

33. With respect to oxygen and ozone which one of the following statements is NOT CORRECT

(A) Oxygen valency is 2 in both
(B) Oxygen has 2 bonds and ozone has 3 bonds
(C) Both have similar reactivity with water
(D) Oxygen is colourless and ozone is coloured

34. The element which can displace all other halogens from their compounds is

(A) Cl (B) Br (C) I (D) F

35. Which of the following is used to bombard $_{13}Al^{27}$ to give $_{15}P^{30}$ and a neutron ?

(A) Proton (B) $\alpha-$ particle
(C) $\beta-$ particle (D) deuteron

36. Urine normally has a pH of 6. If a patient eliminates 1.3 litres of urine per day how many moles of H^+ ions does he urinate ?
(A) 1.3 x 10^{-3} (B) 1.3 x 10^{-6}
(C) 1.3 x 10^{-7} (D) 1.3 x 10^{6}

37. If the pH of a solution is increased from 3 to 6, its hydrogen ion concentration will be
(A) Reduced to half (B) Doubled
(C) Reduced by 1000 times
(D) Increases by 1000 times

38. When Zn metal is added to $CuSO_4$ solution, Cu is precipitated it is due to
(A) Oxidation of Cu^{2+} (B) reduction of Cu^{2+}
(C) Hydrolysis of $CuSO_4$ (D) Ionisation of $CuSO_4$

39. The volume in litres of CO_2 liberated at STP when 10g of 90% pure lime stone is heated completely is
(A) 2.24 (B) 22.4 (C) 2.016 (D) 20.16

40. Alkyl halide reacting with metallic sodium in dry ether is called
(A) Friedel-Craft's reaction (B) Sandmeyer's reaction
(C) Wurtz reaction (D) Gabriel's reaction

41. When an alkyl halide reacts with an alkoxide the product formed is ?
(A) Hydrocarbon (B) Unsaturated hydrocarbon
(C) Ether (D) Alcohol

42. Compound 'A' reacts with PCl_5 to give 'B' which on treatment with KCN followed by hydrolysis gave propanoic acid as the product. Compound 'A' is
(A) Propane (B) Ethane
(C) Ethyl choride (D) Ethyl alcohol

43. The de-Broglie wavelength associated with a particle of mass 1 mg moving with a velocity of 10 ms^{-1} is
(A) 6.63 x 10^{-29} m (B) 6.63 x 10^{-31} m
(C) 6.63 x 10^{-34} m (D) 6.63 x 10^{-22} m

44. The rule that explains the reason for chromium to have $[Ar]3d^54s^1$ configuration instead of $[Ar]\ 3d^4s^2$ is
(A) Pauli's exclusion principle (B) Aufbau Principle
(C) Hund's rule (D) Heisenberg principle

45. The State of hybridization of Boron in BCl_3 is
(A) sp (B) sp^3d (C) sp^3 (D) sp^2

46. The effect of repulsion between the two lone pairs of electrons present on oxygen in water molecule is
(A) No change in H-O-H bond angle
(B) Increase in H-O-H bond angle
(C) Decrease in H-O-H bond angle
(D) All atoms will be in one plane

47. Which pair of the following elements have similar chemical properties ?
(A) Beryllium and Boron
(B) Aluminium and Magnesium
(C) Carbon and Nitrogen
(D) Lithium and Magnesium

48. In the equilibrium reaction $A + B \rightleftharpoons C + D$, the activation energy for the forward reaction is 25 k.cals mol^{-1} and that of the backward reaction is 15k.cals mol^{-1}. Which one of the following statements is correct ?
(A) It is a reaction for which $\Delta H = 0$
(B) It is a sublimation process
(C) It is an exothermic process
(D) It is an endothermic process

49. The equilibrium concentration of C_2H_4 in the following gas phase reaction can be increased by
$C_2H_4(g) + H_2(g) \rightleftharpoons C_2H_6(g)$; $\Delta H = -32.7$ Kcal.
(A) Removal of C_2H_6 (B) Removing H_2
(C) Increase in temperature (D) Increase in pressure

50. Liquid benzene burns in oxygen according to $2C_6H_6(l) + 15O_2(g) \rightarrow 12CO_2(g) + 6H_2O(g)$. How many litres of oxygen are required for complete combustion of 39 gm of liquid C_6H_6 (At.Wt. of C = 12, O = 16)
(A) 12.2 (B) 22,4 (C) 42 (D) 84

EAMCET - 1997

1. The hybridization state of carbon atoms in the product formed by the reaction of ethyl chloride with aqueous KOH is
(A) Sp^3 (B) Sp^3d (C) Sp (D) Sp^2

2. The atomic number of a radioactive element increases by one unit in
(A) Alpha emission (B) Beta emission
(C) gamma emission (D) Electron capture

3. Which of the following compounds is a Lewis acid
(A) NCl_3 (B) $CHCl_3$ (C) PCl_3 (D) BCl_3

4. The alkane that yields two isomeric monobromo derivaties is
(A) Propene (B) Ethane
(C) Methane (D) Propane

5. The alcohol that produces turbidity immediately with $ZnCl_2$/conc. HCl at room temperature is
(A) 1-Hydroxybutane (B) 2-Hydroxybutane
(C) 2-Hydroxy - 2- methyl propane
(D) 1-Hydroxy - 2 - methyl propane

6. Ethyl chloride on heating with silver cyanide forms a compound X. The functional isomer of X is
(A) C_2H_5NC (B) C_2H_5CN
(C) $H_3C-NH-CH_3$ (D) $C_2H_5NH_2$

7. The number of π-ibonds in the product formed by passing acetylene through dil H_2SO_4 containing mercuric sulphate is
(A) Zero (B) One (C) Two (D) Three

8. In the complete combustion of butanol C_4H_9OH, if ΔH is enthalphy of combustion and ΔE is the heat of

Combustion at constant volume, then
(A) $\Delta H < \Delta E$ (B) $\Delta H = \Delta E$ (C) $\Delta H > \Delta E$
(D) ΔH, ΔE relation can not to be predicted.

9. Which of the following is not a base ?
(A) $(CH_3)_3N$ (B) HN_3 (C) N_2H_4 (D) NH_2OH

10. Which of the following is reduced by hydrogen peroxide in acidic medium ?
(A) $KMnO_4$ (B) KI
(C) $FeSO_4$ (D) $K_4[Fe(CN)_6]$

11. The electronic configuration of chromium (Z = 24) is
(A) $[Ar]4d^4 4s^2$ (B) $[Ar]3d^5 4s^1$
(C) $[Ar]3d^4 3s^2$ (D) $[Ar]4d^5 4s^1$

12. Hydrolysis of an ester gives a carboxylic acid which on Kolbe's electrolysis yields ethane. The ester is
(A) Methyl ethanoate (B) Methyl methanoate
(C) Ethyl methanoate (D) Methyl propanoate

13. The strongest conjugate base results from
(A) Formic acid (B) Benzoic acid
(C) Acetic acid (D) Acetylene

14. ΔH value for the reaction $H_2 + Cl_2 = 2HCl$ is -4.12 K. Cal. If E_1 is the activation energy of the reactants and E_2 is the activation energy of the products, then E_1 and E_2 for the above reaction are related as
(A) $E_1 > E_2$ (B) $E_1 < E_2$ (C) $E_1 = E_2$
(D) ΔH is not related to E_1 & E_2

15. The value of K_p for the reaction $2H_2S(g) = 2H_2(g) + S_2(g)$ is 1.2×10^{-2} at 1065°C. The value of K_c for this reaction is
(A) 1.2×10^{-2} (B) $< 1.2 \times 10^{-2}$ (C) 83 (D) $> 1.2 \times 10^{-2}$

16. Which of the following metals does not reduce H_2O ?
(A) Cu (B) Li (C) Ca (D) Fe

17. In the Bohr's hydrogen atom, the electronic transition emitting light of longest wave length is
(A) n = 2 to n = 3 (B) n = 4 to n = 3
(C) n = 3 to n = 2 (D) n = 2 to n = 1

18. Which of the following halogens does not exhibit positive oxidation state in its compounds ?
(A) I (B) Br (C) Cl (D) F

19. Which of the following ores is a chloride ?
(A) Horn Silver (B) Zincite (C) Bauxite (D) Felspar

20. In the electro chemical cell $H_2(g)$ 1 atm. / H^+ (1M) // Cu^{2+} (1M) / Cu(s), which one of the following statements is true ?
(A) H_2 is cathode ; Cu is anode
(B) Oxidation occurs at Cu electrode
(C) Reduction occurs at H_2 electrode
(D) H_2 is anode ; Cu is cathode

21. On passing a current through a KCl solution, 19.5 gm. of potassium is deposited. If the same quantity of electricity is passed through a solution of aluminium chloride, the amount of aluminium deposited is
(A) 4.5g (B) 9.0 g (C) 13.5 g (D) 27 g

22. The de Broglie wave length of a particle with mass 1 gram and velocity 100 ms^{-1} is
(A) 6.63×10^{-35}m (B) 6.63×10^{-34}m
(C) 6.65×10^{-33}m (D) 6.63×10^{-32}m

23. Which of the following statements is true for ammonium ion ?
(A) All bonds are ionic
(B) All bonds are coordinate covalent
(C) H atoms are situated at the corners of a square
(D) H atoms are situated at the corners of tertrahedron.

24. Which of the following compounds in aqueous solution gives a white ppt. with perchloric acid ?
(A) $MgCl_2$ (B) $FeCl_3$ (C) NaCl (D) KCl

25. Which of the following sequence is correct with reference to the oxidation number of iodine ?
(A) I_2 < ICl < HI < HIO_4 (B) HIC_4 < ICl < I_2 < HI
(C) I_2 < HI < ICl < HIO_4
(D) HI < I_4 < ICl < I_2 < HIO_4

26. The root mean square velocity of a gas is doubled when the temperature is
(A) Reduced to half (B) Reduced to one-fourth
(C) Increased four times (D) Increased two times

27. Quartz is a crystalline variety of
(A) Sodium silicate (B) Silicon carbide
(C) Silicon (D) Silica

28. Which of the following compounds does not form an ozonide ?
(A) Ethene (B) Propyne (C) Propene (D) Propane

29. The p^H of the aqueous solution containing 0.49 gram. of H_2SO_4 in one litre is
(A) 2 (B) 1 (C) 1.7 (D) 0.3

30. The conjugate acid of NH_2^- ion is
(A) NH_3 (B) NH_4^+ (C) N_2H_4 (D) NH_2OH

31. Which of the following compounds liberates CO_2 from aqueous $NaHCO_3$?
(A) Anilinium Chloride (B) $CHCl_3$
(C) CCl_4 (D) CH_3Cl

32. In which of the following conversions phosphorus pentachloride is used as the reagent ?
(A) $H_3C - O\,CH_3 \rightarrow CH_3\,Cl$
(B) $CH_3\,CH_2\,OH \rightarrow CH_3CH_2\,Cl$
(C) $H_2C = CH_2 \rightarrow CH_3\,CH_2\,Cl$
(D) $CH \equiv CH \rightarrow CH_2 = CHCl$

33. Which of the following elements has the highest first ionization potential ?
(A) Boron (B) Carbon (C) Nitrogen (D) Oxygen

34. Of the following, the one with largest size is
(A) Cl^- (B) Ar (C) K^+ (D) Ca^{2+}

35. The reagent that gives an orange coloured precipitate with acetaldehyde is
(A) NH_2OH (B) $NaHSO_3$ (C) Iodine (D) 2,4–DNP

EAMCET - 1998 (Engineering)

1. Be^{2+} is isoelectronic with
(A) Li^+ (B) H^+ (C) Mg^{2+} (D) Na^+

2. Hg sticks to the surface of the glass when it comes in contact with
(A) H_2O (B) HNO_3 (C) Grease (D) O_3

3. The conjugate base of NH_2^- is
(A) NH_3 (B) NH^{2-} (C) NH_4^+ (D) N_3^-

4. The effective atomic number of Cr in $[Cr(NH_3)_6]\,Cl_3$ is
(A) 35 (B) 36 (C) 27 (D) 33

5. An organic compound having carbon and hydrogen has 80% carbon. The empirical formula of the hydrocarbon is
(A) CH_3 (B) CH_4 (C) CH (D) CH_2

6. The first emission line of hydrogen atomic spectrum in the Balmer series appears at (R = Rydberg constant)
(A) $\frac{5R}{36}\,cm^{-1}$ (B) $\frac{3R}{4}\,cm^{-1}$ (C) $\frac{7R}{144}\,cm^{-1}$ (D) $\frac{9R}{400}\,cm^{-1}$

7. The cell reaction of a cell is $Mg(s) + Cu^{2+}(aq) \rightarrow Cu(s) + Mg^{2+}(aq)$. If the standard reduction potential of Mg & Cu are –2.37 & 0.34 V respectively, the emf of the cell is
(A) 2.03 (B) –2.03 (C) 2.71 (D) –2.71

8. 4.5 moles each of hydrogen and iodine were heated in a sealed 10 litre vessel. At equilibrium 3 moles of HI were found. The equilibrium constant for the reaction $H_2(g) + I_2(g) \rightleftharpoons 2HI(g)$ is
(A) 1 (B) 10 (C) 5 (D) 0.33

9. Which of the following sets of quantum numbers is not possible for a 4p electron ?
(A) $n = 4,\ l = 1,\ m = +1,\ m_S = +\frac{1}{2}$
(B) $n = 4,\ l = 1,\ m = 0,\ m_S = +\frac{1}{2}$
(C) $n = 4,\ l = 1,\ m = 2,\ m_S = +\frac{1}{2}$
(D) $n = 4,\ l = 1,\ m = -1,\ m_S = -\frac{1}{2}$

10. The radius of the nucleus is related to the mass number 'A' by
(A) $R = R_0A^{1/2}$ (B) $R = R_0A$
(C) $R = R_0A^2$ (D) $R = R_0A^{1/3}$

11. 10.6 g of a substance of mol. wt. 106 was dissolved in 100 ml. 10 ml. of this solution was pipetted out into a 100 ml. flask and made up to the mark with distilled water. The molarity of the resulting solution is
(A) 1.0 M (B) 10^{-2}M (C) 10^{-3}M (D) 10^{-4}M

12. Which of the following species has the highest ionization potential ?
(A) Al^+ (B) Ne (C) Li^+ (D) Mg^+

13. I mole of 'Ar' gas will have least density at
(A) STP (B) 0°C & 2 atm.
(C) 273°C & 2 atm. (D) 273°C & 1 atm.

14. As per the modern periodic law the physical and chemical properties of elements are the periodic functions of their
(A) Atomic number (B) Electronic configuration
(C) Atomic weight (D) Atomic size

15. How many litres of CO_2 at STP will be formed when 100 ml of 0.1 M H_2SO_4 reacts with excess of Na_2CO_3 ?
(A) 22.4 (B) 2.24 (C) 0.224 (D) 5.6

16. The pH of a solution is increased from 3 to 6. Its H^+ ion concentration will be
(A) Reduced to half (B) Doubled
(C) Reduced by 1000 times
(D) Increased by 1000 times

17. Liquid benzene burns in oxygen according to $2C_6H_6(l) + 15O_2(g) \rightarrow 12CO_2(g) + 6H_2O(g)$. How many litres of oxygen are required for complete combustion of 39gm of liquid C_6H_6 ? (At. wt. of C = 12, O = 16)
(A) 11.2 (B) 22.4 (C) 42 (D) 84

18. Silver chloride dissolves in excess of NH_4OH. The cation present in this solution is.
(A) Ag^+ (B) $[Ag\ (NH_3)_2]^+$
(C) $[Ag\ (NH_3)_4]^+$ (D) $[Ag\ (NH_3)_6]^+$

19. Which of the following can be used as an anaesthetic
(A) NCl_3 (B) NO_2 (C) N_2O (D) NO

20. Reaction of benzene with alkyl halide in the presence of anhydrous $AlCl_3$ is called.
(A) Friedel - Craft's reaction (B) Wurtz reaction
(C) Williamson's synthesis (D) Bayer's reaction

21. 75% of a first order reaction is completed in 32 min. 50% of the reaction would have been completed in.
(A) 24min (B) 16min (C) 18min (D) 23min

22. Which of the following solutions cannot act as a buffer ?
(A) $NaH_2PO_4 + H_3PO_4$ (B) $CH_3COOH + CH_3COONa$
(C) $HCl + NH_4Cl$ (D) $H_3PO_4 + Na_2HPO_4$

23. The heat of combustion of CH_4, C (graphite) and H_2 (g) are respectively 20 k.cals –40k. cals & –10 k. cals. The heat of formation of CH_4 is
(A) –40 k.cals (B) +40 k.cals
(C) –80 k.cals (D) +80 k.cals

24. A gas decolorizes Br_2 in CCl_4 and forms a precipitate with ammoniacal silver nitrate. The gas is
(A) C_2H_2 (B) C_2H_4 (C) C_2H_6 (D) CH_4

25. Mortar is a mixture of
(A) Slaked lime + plaster of Paris + H_2O
(B) $CaCO_3$ + silica + H_2O
(C) Plaster of Paris + silica
(D) Slaked lime + silica + H_2O

26. Which of the following reacts with concentrated sulphuric acid ?
(A) Au (B) Ag (C) Pt (D) Pb

27. By which of the following reactions can one get N-methyl aniline from aniline ?
(A) Alkylation (B) Acetylation

(C) Benzoylation (D) Bromination

28. Compound 'A' reacts with PCl_5 to give 'B' which on treatment with KCN followed by hydrolysis gave propanoic acid. What are A & B respectively ?
(A) C_3H_8 & C_3H_7Cl (B) C_2H_8 & C_2H_5Cl
(C) C_2H_5Cl & C_2H_5Cl (D) C_2H_5OH & C_2H_5Cl

29. When an alkyl halide reacts with an alkoxide the product is
(A) Hydrocarbon (B) Unsaturated hydrocarbon
(C) Ether (D) Alcohol

30. The reagent used for converting ethanoic acid to ethanol is
(A) $LiAlH_4$ (B) BH_3 (C) PCl_3 (D) $K_2Cr_2O_7/H^+$

31. Acetaldehyde when treated with dilute NaOH gives.
(A) CH_3CH_2OH (B) CH_3COOH
(C) $CH_3-\underset{\underset{OH}{|}}{CH}-CH_2-CHO$ (D) CH_3-CH_3

32. If the equilibrium constant for the reaction, $2AB \rightleftharpoons A_2 + B_2$ is 49, what is the value of equilibrium constant for $AB \rightleftharpoons \frac{1}{2} A_2 + \frac{1}{2} B_2$?
(A) 49 (B) 2401 (C) 7 (D) 0.02

33. C_2H_5 CHO and $(CH_3)_2$ CO can be distinguished by testing with
(A) Phenylhydrazine (B) Hydroxylamine
(C) Fehling's solution (D) Sodium bisulphite

34. Among the following compounds, which have more than one type of hybridization for carbon atom ?
(A) $CH_3\ CH_2\ CH_2\ CH_3$ (B) $CH_3\ CH = CH\ CH_3$
(C) $CH_2 = CH–C \equiv CH$ (D) $H–C \equiv C–H$
(A) b & c (B) b (C) c & d (D) c

35. Which of the following does not participate in SOLVEY'S process for the manufacture of Na_2CO_3 ?
(A) NH_3 (B) NaCl solution (C) CO_2 (D) H_2SO_4

36. Silica is soluble in
(A) H_2SO_4 (B) HF (C) HCl (D) HNO_3

37. Which of the following reactions does not yield an alkyl halide ?
(A) Diethyl ether + Cl_2 (B) Diethyl ether + HI
(C) Diethyl ether + PCl_5 (D) None of these

38. The element which can displace other halogens from their compounds is
(A) Cl (B) F (C) Br (D) I

39. The number of moles of AgCl precipitated when excess of $AgNO_3$ is added to one mole of $[Cr(NH_3)_4Cl_2]Cl$ is
(A) Zero (B) 1.0 (C) 2.0 (D) 3.0

40. Cassiterite is concentrated by
(A) Leaching (B) Electromagnetic separation
(C) Floatation (D) Liquation

41. A mixture contains solid organic compounds A, B, C and D. On heating only 'C' changes from solid to vapour state. 'C' can be separated from the rest in the mixture by
(A) Distillation (B) Sublimation
(C) Fractional distillation (D) Crystallization

42. Which of the following is the strongest Lewis acid ?
(A) BCl_3 (B) BF_3 (C) BI_3 (D) BBr_3

43. Which of the following reactions is an example of calcination process ?
(A) $2Ag + 2HCl + (O) \rightarrow AgCl + H_2O$
(B) $2\ Zn + O_2 \rightarrow 2ZnO$
(C) $2ZnS + 3O_2 \rightarrow 2ZnO + 2SO_2$
(D) $MgCO_3 \rightarrow MgO + CO_2$

44. Acetic acid reacts with PCl_5 to form
(A) $CH_2ClCOOH$ (B) $CHCl_2COOH$
(C) CH_3COCl (D) CCl_3COOH

45. Silver containing lead as an impurity is removed by
(A) Poling (B) Cupellation
(C) Bessemerisation (D) Distillation

46. Which of the following sets of elements does not belong to transitional elements ?
(A) Fe, CO, Ni (B) Cu, Ag, Au
(C) Ti, Zr, Hf (D) Ga, In, Tl

47. CARBOGEN is
(A) Pure form of carbon (B) $COCl_2$
(C) Mixture of CO & CO_2 (D) Mixture of O_2 & CO_2

48. Hydrolysis of trichloromethane with aqueous KOH gives
(A) Methanol (B) Chloral
(C) Acetylene (D) Pot. formate

49. The homologue of ethyne is
(A) C_3H_8 (B) C_3H_4 (C) C_2H_2 (D) C_2H_6

50. In Wurtz reaction the reagent used is
(A) Na (B) Na/liq. NH_3
(C) Na/dry ether (D) Na/dry ethanol

EAMCET - 1998 (Medical)

1. Hydrolysis of acetonitrile in acid medium gives
(A) $CH_3COOC_2H_5$ (B) CH_3CH_2OH
(C) CH_3COOH (D) CH_3NC

2. C_2H_5CHO and CH_3COCH_3 can be distinguished from one another by testing with
(A) Phenylhydrazine (B) 2,4 dinitrophenylhydrazine
(C) Fehling's solution (D) Sodium bisulphate

3. The gaseous substance which decolorizes bromine but does not give a precipitate with ammonical silver nitrate is
(A) C_2H_2 (B) CH_4 (C) C_3H_6 (D) C_2H_4

4. The hybridization involved in the six carbon atoms of benzene is

(A) Three sp^3 & three sp^2 (B) Three sp^3 & three sp
(C) All six sp (D) All six sp^2

5. Which of the following hydrides is more acidic ?
(A) H_2Se (B) H_2Te (C) H_2S (D) H_2O

6. IUPAC name of the compound having the formula $(CH_3)_3$ C–CH = CH_2 is
(A) 3, 3, 3 trimethyl 1-propene
(B) 3, 3 dimethyl 1-butene
(C) 1, 1, 1 trimethyl 3-propane
(D) 1, 1 dimethyl 3 - butene

7. What is the product obtained when chlorine reacts with ethyl alcohol in the presence of NaOH ?
(A) $CHCl_3$ (B) CCl_3CHO (C) C_2H_5Cl (D) CH_3Cl

8. Alkyl halide is not obtained from one of the following reactions. Which one is that ?
(A) Diethyl ether $\xrightarrow{\text{Reduction}}$ X $\xrightarrow{SOCl_2}$
(B) Diethyl ether + HI →
(C) Diethyl ether + PCl_5 →
(D) Diethyl ether + Cl_2 →

9. The oxide of chlorine which is a colourless oily liquid is
(A) ClO_2 (B) Cl_2O_7 (C) Cl_2O (D) Cl_2O_6

10. Which of the following species has the highest ionization potential ?
(A) Li^+ (B) Mg^+ (C) Al^+ (D) Ne

11. In which of the following transition metal complexes does the metal exhibit zero oxidation state ?
(A) $[Co(NH_3)_6]Cl_3$ (B) $[Fe(H_2O)_6]SO_4$
(C) $[Ni(CO)_4]$ (D) $[Fe(H_2O)_6]Br_3$

12. The state of hybridization of 'B' in BCl_3 is
(A) sp^3 (B) sp^2 (C) sp (D) sp^3d^2

13. The first emission line in the atomic spectrum of hydrogen in the Balmer series appears at
(A) $\frac{5R}{36}$ cm^{-1} (B) $\frac{3R}{4}$ cm^{-1} (C) $\frac{7R}{144}$ cm^{-1} (D) $\frac{9R}{400}$ cm^{-1}

14. The four quantum numbers of the valence electron of potassium are
(A) 4, 0, 1, $\frac{1}{2}$ (B) 4, 1, 0, $\frac{1}{2}$ (C) 4, 0, 0, $\frac{1}{2}$ (D) 4, 1, 1, $\frac{1}{2}$

15. Which of the following has magic number of protons and neutrons ?
(A) ${}_8O^{17}$ (B) ${}_3Al^{27}$ (C) ${}_9F^{17}$ (D) ${}_{20}Ca^{40}$

16. The activity of a radioactive nuclide X^{100} is 6.023 curie. If its disintegration constant is 3.7 x 10^4 sec^{-1}, the mass of X is
(A) 10^{-23}g (B) 10^{-15}g (C) 10^{-16}g (D) 10^{-14}g

17. According to VSEPR theory, the shape of the water molecule is
(A) Linear (B) Planar triangle
(C) Bent 'V' shape (D) Octahedral

18. The radii of F, F^-, O and O^{2-} are in the order
(A) $O^{2-} > F^- > O > F$ (B) $O^{2-} > F^- > F > O$
(C) $F^- > O^{2-} > F > O$ (D) $O^{2-} > O > F^- > F$

19. Which of the following has highest first ionization potential ?
(A) Oxygen (B) Carbon (C) Boron (D) Nitrogen

20. The standard reduction potentials at 298 K for the half cell reactions are given below. Find out the strongest reducing agent.
(A) $Zn^{2+}(aq) + 2e^- \rightarrow Zn(s) - 0.762$
(B) $Cr^{3+}(aq) + 3e^- \rightarrow Cr(s) - 0.74$
(C) $2H^+(aq) + 2e^- \rightarrow H_2(g) - 0.00$
(D) $Fe^{3+}(aq) + e- \rightarrow Fe^{2+}(aq) + 0.77$

21. According to Fajan's rule covalent bond is favoured by
(A) Small cation & large anion
(B) Small cation & small anion
(C) Large cation & large anion
(D) Large cation & small anion

22. The volume in litres of CO_2 liberated at STP when 10 grams of 90% pure lime - stone is heated completely is
(A) 2.016 (B) 20.16 (C) 2.24 (D) 22.4

23. A buffer solution is prepared by mixing 10 ml of 1.0 M acetic acid and 20 ml of 0.5 M sodium acetate and then diluted to 100ml with distilled water. If the pk_a of CH_3COOH is 4.76, what is the pH of the buffer solution prepared ?
(A) 2.84 (B) 4.76 (C) 4.34 (D) 5.21

24. The dissociation constant of a weak monobasic acid K is 1 x 10^{-5}. The pH of 0.1 M of that acid would be.
(A) 3 (B) Between 2-3 (C) 1 (D) 5

25. In the electrolytic extraction of aluminum, fused cryolite is added. It's role is
(A) To protect the anodes (B) To increase Al production
(C) To act as a reducing agent
(D) To make a conducting solution of Bauxite

26. ELECTRON is an alloy of
(A) Mg & Zn (B) Fe & Mg
(C) Ni & Zn (D) Al & Zn

27. The element which can displace other halogens from their compounds is
(A) Cl (B) Br (C) I (D) F

28. Gas 'A' is bubbled through slacked lime when a white precipitate is formed. On prolonged bubbling the precipitate dissolved. On heating the resultant solution, the white precipitate reappeared with the evolution of a gas 'B'. The gases 'A' and 'B' respectively are
(A) CO_2 & CO (B) CO & CO_2
(C) CO_2 & CO_2 (D) CO & CO

29. NITROLIM is a mixture of
(A) $Ca(CN)_2$ & graphite (B) CaC_2 & graphite
(C) $CaCN_2$ & graphite (D) $CaCN_2$ & N_2

30. Which of the following is used as a rocket fuel ?
(A) Nitrates $+O_3$ (B) Water gas $+O_3$
(C) Cyanogen $+C_2$ (D) Cyanogen $+O_3$

31. The homologue of ethyne is
(A) C_3H_8 (B) C_3H_4 (C) C_2H_4 (D) C_3H_6

32. In Wurtz reaction the reagent used is
(A) Na (B) Na/liq. NH_3
(C) Na/dry ether (D) Na/ dry alcohol

33. Benzene is obtained by fractional distillation of
(A) Light oil (B) Middle oil
(C) Anthracene oil (D) Heavy oil

34. Hydrolysis of trichloromethane with aqueous KOH gives
(A) Methanol (B) Chloral
(C) Acetylene (D) Potassium formate

35. The product obtained when acetaldehyde is treated with dilute NaOH is
(A) CH_3CH_2OH (B) CH_3COOH
(C) $CH_3-CH(OH)-CH_2CHO$ (D) $CH_3 - CH_3$

36. Which of the following does not participate in the carbylamine reaction ?
(A) Aniline (B) Chloroform (C) Ethanol (D) KOH

37. When H_2O_2 reacts with chlorine, the product obtained is
(A) ClO_2 (B) HOCl (C) O_2 (D) H_2

38. Reduction of nitrobenzene in the presence of Zn/ NH_4Cl gives
(A) Azobenzene (B) Aniline
(C) N-phenyldroxylamine (D) Hydrazobenzene

39. $KMnO_4$ oxidizes oxalic acid in acid medium. The number of CO_2 molecules produced as per the balanced equation is
(A) 10 (B) 8 (C) 6 (D) 3

40. The oxidation number of cobalt in $K_3[Co(NO_2)_6]$ is
(A) +6 (B) −3 (C) +3 (D) −6

41. The volume of 2.8g of carbon monoxide at 27°C and 0.821atm. pressure is (R=0.0821 lit. atm. $mol^{-1}k^{-1}$)
(A) 1.5 lit (B) 0.3 lit (C) 3 lit (D) 30 lit

42. A gaseous mixture of three gases A, B and C has a pressure of 10 atms. The total number of moles of all the gases is 10. If the partial pressures of A and B are 3.0 and 1.0 atm. respectively and if 'C' has a mol. wt. of 2.0 what is the weight of 'C' in gms present in the mixture ?
(A) 6 (B) 3 (C) 12 (D) 8

43. The weight of $AgNO_3$ (Mol.wt.170) present in 100 ml of its 0.25 M solution is
(A) 4.25 g (B) 42.5 g (C) 17 g (D) 1.7 g

44. For a cell, the cell reaction is $Mg(s) + Cu^{2+}(aq) \rightarrow Cu(s) + Mg^{2+}(aq)$. If the standard reduction potentials of Mg and Cu are −2.37 and +0.34 V respectively, the emf of the cell is
(A) 2.71V (B) −2.71V (C) 2.03V (D) −2.03V

45. For the reaction $C(s) + CO_2(g) \rightleftharpoons 2CO(g)$, the partial pressures of CO_2 and CO are 2.0 and 4.0 atm. respectively at equilibrium. What is the value of K_P for this reaction ?
(A) 0.5 (B) 4.0 (C) 8.0 (D) 32

46. After how many seconds will the concentration of the reactant in a first order reaction be halved if the rate constant is 1.155×10^{-3} sec^{-1} ?
(A) 600 (B) 100 (C) 60 (D) 10

47. If K_1 and K_2 are the equilibrium constant of the equilibria (i) and (ii) respectively, what is the relationship between the two constants ?
(i) $SO_2(g) + 1/2\, O_2(g) \rightleftharpoons SO_3(g)$, K_1
(ii) $2SO_3(g) \rightleftharpoons SO_2(g) + O_2$, K_2
(A) $K_1 = K_2$ (B) $K_1 = 1/K_2$
(C) $K_1 = K^2_1$ (D) $K^2_1 = 1/K_2$

48. The purest form of coal is
(A) Peat (B) Anthracite
(C) Lignite (D) Bituminous coal

49. For the given reaction, $N_2(g) + 3H_2(g) \rightarrow 2NH_3(g)$, $\Delta H = -92.3$ kJ/mole increase of temperature favours
(A) Forward reation (B) Backward reaction
(C) Both A & B (D) None of these

50. Heating an ore in the absence of air below its melting point is called
(A) Calcination (B) Smelting
(C) Roasting (D) Leaching

EAMCET - 1999 (Engineering)

1. Which of the following is an oxide mineral ?
(A) Zinc blende (B) Magnetite
(C) Dolamite (D) Carnallite

2. The units of electrochemical equivalent are :
(A) Gram coulomb (B) Gram $amp^{-1}sec^{-1}$
(B) Gram amp·sec^{-1} (D) Gram $amp^{-1}sec$

3. The number of acetylene molecules required to from one molecule of benzene is
(A) 1 (B) 2 (C) 3 (D) 4

4. At what temperature the density of heavy water will be maximum ?
(A) 4°C (B) 27°C (C) 0°C (D) 11.6°C

5. Which of the set of raw materials are used in the manufacture of Na_2CO_3 by Solvey process ?
(A) $Ca(OH)_2$, NH_3, CO_2 (B) $CaCl_2$, NH_3, CO_2
(C) NaOH, NH_3, CO_2 (D) NaCl, NH_3, CO_2

6. The half life of a first order reaction is 10 sec. What is its rate constant in sec^{-1} ?
(A) 0.0693 (B) 0.693 (C) 6.93 (D) 69.3

7. Absolute alcohol is prepared by the distillation of rectified spirit in the presence of
(A) C_2H_5OMg (B) $Ca(OH)_2$ (C) Na (D) $CaCO_3$

8. Which of the following pairs of ions contain the same

number of unpaired electrons ?

(A) Ni^{2+}, Co^{2+} (B) Mn^{2+}, Fe^{3+}

(C) Mn^{2+}, Ni^{2+} (D) Ti^{3+}, Co^{2+}

9. The number of C–C sigma bonds present in 1-butyne is

(A) 2 (B) 3 (C) 4 (D) 5

10. The following reaction is known to occur in the body

$CO_2 + H_2O \rightleftharpoons H_2CO_3 \rightleftharpoons H^+ + HCO_3^-$

If CO_2 escapes from the system

(A) H_2CO_3 concentration does not change

(B) The forward reaction will be promoted

(C) Hydrogen ion concentration increases

(D) Hydrogen ion concentration decreases

11. Hydrolysis of chloroform with aqueous potassium hydroxide gives

(A) $COCl_2$ (B) CH_3NC (C) HCOOK (D) CH_2ClOH

12. Chlorine is passed into dilute cold KOH solution. What are the oxidation numbers of chloring in the products

(A) –1, +5 (B) –1, +3 (C) +1, +7 (D) +1, –1

13. Chloroethane is reacted with alcoholic potassium hydroxide. The product formed is

(A) C_2H_4 (B) C_2H_4O (C) C_2H_6O (D) C_2H_6

14. Δx is given as, Δx = hydration energy - lattice energy. For a salt, Δx is found to be 500 kcal.mol^{-1}. Then on increasing the temperature, its solubility in water

(A) increases (B) decreases (C) remains same

(D) increases initially and then decreases

15. How many moles of potassim chlorate should be decomposed completely to obtain 67.2 litres of oxygen at STP ?

(A) 3 (B) 4 (C) 1 (D) 2

16. Tin stone is contaminated with Wolframite. Which one of the following methods can be employed to dress the tin stone ore ?

(A) levigation (C) flotation process (D) roasting

(B) electromagnetic separation

17. In which of the following reactions, heat liberated is known as standard heat of formation of CO_2 ?

(A) C(graphite) + O_2(g) $\rightarrow$ CO_2(g) + 94.05 k. cal

(B) 2CO(g) + O_2(g) $\rightarrow$ 2CO_2(g) + 135.5 k. cal

(C) C(diamond) + O_2(g) $\rightarrow$ CO_2(g) + 94.5 k. cal

(D) CH_4(g) + 2O_2(g) $\rightarrow$ CO_2(g) + 2H_2O(l) + 212.8 k. cal

18. The solvent used in the preparation of Grignard's reagent is

(A) dry ether (B) dry acetone

(C) dry alcohol (D) dry chloroform

19. O^{2-} and Si^{4+} are isoelectronic radius ions. If the ionic radius of O^{2-} is 1.4Å. the ionic radius of Si^{4+} will be

(A) 1.4Å (B) 0.41Å (C) 2.8Å (D) 1.5Å

20. Which of the following compounds undergo Aldol condensation ?

(A) CH_3CHO (B) $(CH_3)_3CHO$

(C) CCl_3CHO (D) HCHO

21. Acetic acid gives acetic anhydride when treated with

(A) Cl_2/P (B) Na (C) PCl_5 (D) P_2O_5

22. Which of the following reagents converts nitrobenzene to N-phenyl hydroxyl amine ?

(A) Sn/HCl (B) Zn/NaOH

(C) Zn/NH_4Cl (D) $LiAlH_4$

23. The electronic configuration of sodium is

(A) [Ne]$3s^2$ (B) [Ne]$3s^1$

(C) [Ar]$4s^1$ (D) [Ar]$4s^2$

24. The oxidation number of Mn in potassium permanganate is

(A) +6 (B) +7 (C) +5 (D) +8

25. Which of the following is not true ?

(A) Aluminium liberates hydrogen on treating with a base.

(B) Aluminium is used in the preparation of duralumin.

(C) Aluminium is extracted by the electrolysis of alumina in the presence of cryolite.

(D) Aluminium is a stong oxidising reagent.

26. The industrial preparation of nitric acid by Ostwald's process involves.

(A) Reduction of NH_3 (B) Oxidation of NH_3

(C) Hydrogenation of NH_3 (D) Hydrolysis of NH_3

27. In which of the following reactions, ozone acts as a reducing reagent ?

(A) PbS + 4O_3 $\rightarrow$ $PbSO_4$ + 4O_2

(B) BaO_2 + O_3 $\rightarrow$ BaO + 2O_2

(C) 2HCl + O_3 $\rightarrow$ Cl_2+ H_2O + O_2

(D) 2KI + O_3 + H_2O $\rightarrow$ 2KOH + I_2 + O_2

28. Aqueous solution of Carnallite gives positive test for

(A) Chloride ions only (B) Potassium ions only

(C) Potassium and chloride ions only

(D) Potassium, magnesium and chloride ions

29. Oxalic acid is heated with concentrated H_2SO_4. When the resultant gases are passed over red hot carbon, X is obtained. X is.

(A) CO (B) CH_4 (C) CO_2 (D) C_2H_4

30. Which of the following is a weak acid in its aqueous solution.

(A) H_2S (B) H_2Po (C) H_2Te (D) H_2Se

31. What is the oxidation state of 'Fe' in the product formed when acidified potassium ferro-cyanide $K_4[Fe(CN)_6]$ is treated with hydrogen peroxide ?

(A) +2 (B) +3 (C) +1 (D) +6

32. A burning strip of magnesium is introduced into a jar containing a gas. After some time the walls of the container is coated with carbon. The gas in the container is

(A) O_2 (B) N_2 (C) CO_2 (D) H_2O

33. Inorganic benzene contains .... elements

(A) C,H,Al (B) C,B,H (C) B,N,H (D) C,N,H

34. Which of the following is the anhydride of HNO_3 ?

(A) N_2O_4 (B) N_2O_5 (C) NO (D) N_2O_3

35. In Rutherford's alpha-ray scattering experiment, the

alpha particles are detected using a screen coated with
(A) Carbon black (B) Platinum black
(C) Zinc sulphide (D) Poly tetrafluoro ethene

36. The molecule having zero dipole moment is
(A) CH_3Cl (B) H_2O (C) CCl_4 (D) HCl

37. What is the numerical value of gas contant R, in J $mol^{-1}K^{-1}$?
(A) 0.0821 (B) 8.314 x 10^7 (C) 8.314 (D) 1.987

38. The enthyne is treated with 30% H_2SO_4 in the presence of 1% mercuric sulphate at 70°C, what is the product formed ?
(A) Ethene (B) ethanol
(C) ethanal (D) Ethylene glycol

39. Which of the following is used as a rocket fuel ?
(A) C_2H_2 (B) CH_4 (C) F_2 (D) N_2

40. Which of the following alkane can not be prepared by Wurtz reaction ?
(A) C_2H_6 (B) C_3H_8 (C) CH_4 (D) C_4H_{10}

41. The oxidation number of N in HN_3 is
(A) $+\frac{1}{3}$ (B) 0 (C) $-\frac{1}{3}$ (D) 1

42. n-Octane and n-nonane have
(A) same molecular weight
(B) similar chemical properties
(C) same molecular formula
(D) same boiling point

43. The activity of a radioactive element is 2 x 10^{10} disintegrations per second. After 10 minutes, the activity is reduced to 2 x 10^9 disintegrations per second. The disintegration constant of the radioactive element in min^{-1} is
(A) 2 (B) 2.303 (C) 10 (D) 0.2303

44. The pH of a 1 litre solution is 2. It is diluted with water till its pH become 4. How many litres of water is added ?
(A) 99 (B) 9 (C) 999 (D) 100

45. The unit of radioactivity is
(A) Fermi (B) Farad (C) Curie (D) Hertz

46. At 90°C, the concentration of H_3O^+ in pure water is 10^{-6} mol. lit^{-1}. What is the value of K_W at this temperature ?
(A) 10^{-4} (B) 10^{-8} (C) 10^{-14} (D) 10^{-12}

47. Among the first lines of Lyman, Balmer, Paschen and Brackett series in hydrogen atomic spectra, which has higher energy ?
(A) Paschen (B) Brackett (C) Lyman (D) Balmer

48. Which of the following covalent molecules is an exception to octet rule ?
(A) $BeCl_2$ (B) CO_2 (C) H_2O (D) CH_4

49. A mixture contains 16 grams of oxygen, 28 grams of nitrogen and 8 grams of methane. Total pressure of the mixture is 740 mm. What is the partial pressure of nitrogen in mm ?
(A) 185 (B) 370 (C) 555 (D) 740

50. A solid organic compound X on heating is converted directly to vapour phase which on cooling solidifies. The best method for purifying X is
(A) distillation
(B) distillation at reduced pressure
(C) sublimation (D) steam distillation

EAMCET - 1999 (Medical)

1. The two enzymes present in yeast that are responsible for the formation of ethyl alcohol from molasses in the fermentation process are
(A) invertase, zymase (B) invertase, diastase
(C) zymase, diastase (D) invertase, maltase

2. The amount of heat liberated in k cal, when one mole of aqueous NaOH reacts with one mole of aqueous HCl solution is
(A) 2.303 (B) 0.693 (C) 22.4 (D) 13.7

3. Which of the following reaction conditions are used for preparing diethyl ether from ethyl alcohol ?
(A) HI/200°C (B) Conc. H_2SO_4/140°C
(C) NaOH/70°C (D) $NaOC_2H_5$/27°C

4. The equilibrium constant for the reaction,
$H_2O(g) + CO(g) = H_2(g) + CO_2(g)$ is 64. If the rate constant for the forward reaction is 160, the rate constant for the backward reaction is
(A) 6.2 (B) 10.24 x 10^3 (C) 0.4 (D) 2.5

5. Aniline reacts with chloroform in the presence of alkali to give a product with very bad smell. This test is called
(A) Carbylamine test (B) silver mirror test
(C) Baeyer's test (D) Lucas test

6. An organic compound reacts (i) with metallic sodium to liberate hydrogen and (ii) with Na_2CO_3 solution to liberate CO_2. The compound is
(A) an alcohol (B) a carboxylic acid
(C) an ether (D) an ester

7. The compound formed when nitrobenzene is reduced with $LiAlH_4$ is
(A) $C_6H_5NH_2$ (B) C_6H_5NHOH
(C) $C_6H_5N = NC_6H_5$ (D) $C_6H_5NH\ NHC_6H_5$

8. At 27°C one mole of an ideal gas exerted a pressure of 0.821 atmospheres. What is its volume in litres ?
(R = 0.0821 lit. atm. $mol^{-1}K^{-1}$)
(A) 300 (B) 30 (C) 0.3 (D) 3

9. What is oxidation number of Cr in potassium dichromate ?
(A) +6 (B) +4 (C) +7 (D) +5

10. Which of the following ions has the same number of unpaired electrons as that of V^{3+} ion ?
(A) Ni^{2+} (B) Fe^{3+} (C) Cr^{3+} (D) Mn^{2+}

11. 18 C–H and 7 C–C sigma bonds are present in
(A) Cyclohexane (B) 3,3dimethyl pentane
(C) 2,2,3 trimethyl pentane (D) n-heptane

12. How many elecrons are present in the M-shell of the atom of an element with atomic number 24 ?
(A) 5 (B) 6 (C) 12 (D) 13

13. The pH of 1 litre NaOH solution is 12. What is the amount in grams of NaOH present in this solution ?
(A) 40 (B) 4 (C) 0.4 (D) 20

14. SiO_2 is soluble in
(A) HF (B) HCl (C) H_2SO_4 (D) HNO_3

15. In the nuclear reaction ${}^{27}_{13}Al + H^2_1 \xrightarrow{-\alpha} X$, X is
(A) ${}^{12}_{6}C$ (B) ${}^{30}_{14}Si$ (C) ${}^{25}_{12}Mg$ (D) ${}^{24}_{11}Na$

16. The conjugate acid N_3^- (azideion) is
(A) NH_2^- (B) N_2^- (C) NH_3 (D) HN_3

17. Which of the following form of sulphur can be prepared by pouring liquid sulphur into cold water ?
(A) monoclinic (B) rhombic (C) plastic (D) colloidal

18. Stainless steel vessels are made from steel containing
(A) Ni (B) Mn (C) W (D) Cr

19. If the dielectric constants of H_2O and D_2O are 82 and 80.5 respectively, the solubility of NaCl in D_2O compared to that in water is
(A) equal (B) slightly lower
(C) slightly higher (D) very high

20. Metals having ns^1 as the valence shell electronic configuration
(A) are highly electropositive
(B) have a first ionization potential more than 10 eV/atom
(C) act as strong oxidizing reagents
(D) are highly electronegative

21. A vessel contains equal volumes of SO_2 and CH_4. Through a small hole, the gases effused into vacuum. After 200 seconds, the total volume is reduced to half. What is the ratio of SO_2 and CH_4 remaining in the vessel ?
(A) 1:2 (B) 2:1 (C) 1:1 (D) 1:4

22. Solution of an organic compound with nitrogen, sulphur and chlorine obtained from Lassigne's test contains
(A) NaCN, Na_2S, NaCl (B) KCN, K_2S, KCl
(C) AgCl, NaCN, K_2S (D) NaNC, $FeSO_4$, KCl

23. A mixture of carbon monoxide, hydrogen and hydrocarbons is known as
(A) water gas (B) Carburetted water gas
(C) semi water gas (D) producer gas

24. Hydrolysis of SCl_4 forms
(A) SO_2 (B) H_2SO_6 (C) H_2SO_4 (D) H_2SO_3

25. The rms velocity of a gas at a particular temperature is found to be 12,240 cm. sec^{-1}. What is its most probable velocity in cm.sec^{-1} at the same temperature ?
(A) 10^3 (B) 10^4 (C) 11280 (D) 1128

26. Which of the following is used to reduce pure haematite to iron ?
(A) H_2 (B) CO (C) $LiAlH_4$ (D) H_2/Pd

27. Which of the following is used in the preparation of 'ammonal' ?
(A) Fe_2O_3 (B) Al powder
(C) $CaCN_2$ + C (graphite) (D) NaOH + CaO

28. What is formed when 3 parts of HCl is mixed with one part of HNO_3 ?
(A) NOCl (B) NH_4Cl (C) NO (D) NCl_3

29. Which of the following does not exhibit allotropy ?
(A) P (B) Bi (C) Sb (D) As

30. The pH of an aqueous solution of H_2O_2 is 6.0. Some chlorine is bubbled through this solution. The pH of the resultant solution will be
(A) 6 (B) 7 (C) less than 6 (D) more than 7

31. "Elecktron" is an alloy of
(A) Mg + Ni (B) Ni + Al (C) Mg + Al (D) Mg + Zn

32. Which of the following can form compounds with Xenon ?
(A) I_2 (B) F_2 (C) Br_2 (D) Cl

33. The test used for identifying carbon-carbon double bond in an alkene is
(A) Baeyers (B) Carbylamine
(C) Lassaigne (D) Silver mirror

34. Which of the following atomic orbitals is not directed along the axis ?
(A) P_X (B) $d_{x^2-y^2}$ (C) d_{xy} (D) d_{z^2}

35. The number of periods present in the long form of periodic table is
(A) 6 (B) 7 (C) 8 (D) 18

36. Which of the following is acetone oxime ?
(A) $(H_3C)(H)C{=}N{-}OH$
(B) $(H_3C\,CH_3)(H)C{=}N{-}OH$
(C) $(H)(H)C{=}N{-}OH$
(D) $(H_3C)(CH_3)C{=}N{-}OH$

37. 6 grams of urea is dissolved in 16.2 grams of water at 25°C. What is the vapour pressure in mm of the resultant solution ? The vapour pressure of water at 25°C is 24 mm.
(A) 2.4 (B) 21.6 (C) 23.6 (D) 18

38. Which of the following compounds liberates hydrogen gas on reacting with sodium metal ?
(A) C_2H_4 (B) C_2H_6 (C) C_6H_{14} (D) C_2H_2

39. Standard reduction potentials of Sn^{2+} / Sn and

Zn^{2+}/Zn electrodes are –0.14 and –0.74V respectively. What is the E° of the cell, Zn/Zn^{2+} // Sn^{2+}/Sn in volts ?
(A) –0.6 (B) –0.88 (C) 0.88 (D) 0.6

40. Benzene is a monocyclic compound . It is aromatic because it contains
(A) 6 carbons and 6 hydrogens (B) 3 pi bonds
(C) 6 p orbitals
(D) (4n+2) π electrons

41. What is the volume in litres of 0.1 M H_2SO_4 required to completely neutralize 1 litre 0.5 M NaOH ?
(A) 5 (B) 2.5 (C) 0.5 (D) 10

42. Which of the following compounds can exhibit functional isomerism ?
(A) C_2H_5Cl (B) C_5H_{12} (C) C_4H_8 (D) C_3H_8O

43. Radioactivity of an isotope falls to 1/8 of its original value in 48 days. Its half life period is
(A) 48 (B) 16 (C) 12 (D) 8

44. The number of sigma bonds in ethane formed by the overlap of sp^3 and s orbitals is
(A) 5 (B) 6 (C) 7 (D) 4

45. From the following data for the decomposition of N_2O_5 at 30°C, find out the rate constant (min^{-1})
Volume of O_2 after 10 minutes of the reaction = 90 ml
Volume of O_2 after completion of the reaction = 100 ml
(A) 2.303 (B) 0.2303 (C) 0.02303 (D) 23.03

46. In which of the following compounds the angle between two covalent bonds is the highest ?
(A) H_2O (B) H_2O_2 (C) CH_4 (D) NH_3

47. How many grams of phosphoric acid is required to completely neutralize 120 grams of sodium hydroxide ?
(A) 49 (B) 98 (C) 196 (D) 9.8

48. During the extraction of zinc, ZnO is obtained from zinc blende by
(A) roasting (B) cupellation
(C) poling (D) electrolytic reduction

49. Which of the following is most stable ?
(A) HF (B) HI (C) HCl (D) HBr

50. IUPAC name of $(C_2H_5)_4$ C is
(A) tetra ethyl methane (B) 2,2-dimethyl pentane
(C) 3,3-diethyl pentane (D) neopentane

EAMCET - 2000 (Engineering)

1. In which of the following reactions, hydrogen peroxide acts as a reducing agent ?
(A) $Ag_2O(moist) + H_2O_2 \rightarrow 2Ag + H_2O + O_2$
(B) $2KI + H_2O_2 + H_2SO_4 \rightarrow K_2SO_4 + I_2 + H_2O_2$
(C) $PbS + 4H_2O_2 \rightarrow PbSO_4 + 4H_2O$
(D) $Na_2SO_3 + H_2O_2 \rightarrow Na_2SO_4 + H_2O$

2. Brass is an alloy of
(A) Ag and Cu (B) Sn and Zn
(C) Cu and Sn (D) Cu and Zn

3. Which of the sixth group elements has the highest catenation power ?
(A) O (B) Se (C) S (D) Te

4. The half-life of radioactive nuclide is 100 years. What is its disintegration constant (in $year^{-1}$) ?
(A) 6.93 x 10^3 (B) 6.93 (C) 6.93 x 10^{-3} (D) 0.693

5. How many electrons are present in the M-shell of an atom of the element with atomic number Z = 24 ?
(A) 12 (B) 13 (C) 5 (D) 6

6. Which of the following reactions does not involve change in the oxidation state of the metal ?
(A) $VO^{2+} \rightarrow V_2O_3$ (B) $Na \rightarrow Na^+$
(C) $CrO_4^{2-} \rightarrow Cr_2O_7^{2-}$ (D) $Zn^{2+} \rightarrow Zn$

7. What is the volume in lit of carbondioxide liberated at STP, when 2.12 grams of sodium carbonate (mol. wt. = 106) is treated with excess dilute HCl ?
(A) 2.28 (B) 0.448 (C) 44.8 (D) 22.4

8. Which of the following is correct for an acid buffer ? (Salt = S; acid = A).
(A) $pK_a = pH + \log\frac{[S]}{[A]}$ (B) $pH = pK_a + \log\frac{[S]}{[A]}$
(C) $pKa = pH - \frac{A}{S}$ (D) $pH = pK_a + \log\frac{A}{S}$

9. X in the following reaction,
$2CH_3CHO \xrightarrow{\text{dil NaOH}} X$ is
(A) $CH_3 - CH(OH) - CHO$ (B) $CH_3 - C(=O) - CH_2 - CHO$
(C) $CH_3 - CH(OH) - CH_2 - CHO$
(D) $CH_3 - CH_2 - CH(OH) - CHO$

10. Which of the following converts acetic acid to acetyl chloride ?
(A) Cl_2/P (B) PCl_3 (C) NaCl (D) HCl

11. Nitrobenzene on reduction with Sn/HCl gives A which on reaction with acetyl chloride forms B. Which of the following is B ?
(A) $C_6H_5 - C(=O) - CH_3$ (B) $C_6H_5 - NH - C(=O) - CH_3$
(C) C_6H_5NHCl (D) $C_6H_5 - O - C(=O) - CH_3$

12. Which of the following converts acetone to acetone oxime ?
(A) $H_2N - NH_2$ (B) 2, 4 – DNP
(C) $C_6H_5N(CH_3)NH_2$ (D) NH_2OH

13. Which of the following is a secondary amine ?
(A) $C_6H_5 - NH_2$ (B) $CH_3 - NH_2$

(C) $(C_6H_5)_2 - N - CH_3$ (D) $C_6H_5 - NH - CH_3$

14. Chloroethane reacts with Y to form NaCl and Z. One mole of Z reacts with two moles of HI to form water and iodoethane. Which of the following is Y ?
(A) $C_2H_5OC_2H_5$ (B) C_2H_5ONa
(C) CH_3COOH (D) CH_3CHO

15. Which of the following is Lucas reagent ?
(A) Ammonical silver nitrate (B) Br_2/CCl_4
(C) $ZnCl_2$/Conc.HCl (D) Cold alkaline $KMnO_4$

16. Which of the following is a tertiary alcohol ?
(A) $CH_3 - CH(CH_3) - CH_2 - OH$
(B) $CH_3 - CH_2 - CH_2 - CH_2 - OH$
(C) $CH_3 - CH_2 - CH(CH_3) - CHO$
(D) $CH_3 - C(CH_3)_2 - OH$

17. The pH of an aqueous solution of a salt is 10. The salt is
(A) NaCl (B) NH_4Cl
(C) CH_3COONa (D) $(NH_4)_2SO_4$

18. Electronic configuration of group 13th elements is
(A) ns^1np^2 (B) ns^1np^3 (C) ns^2np^2 (D) ns^2np^1

19. Which of the following has the highest first ionization potential ?
(A) K (B) P (C) Al (D) Si

20. Which of the following pairs is correctly represented ?
(A) $BeCl_2$, Linear (B) NH_3, Linear
(C) CO_2, Tetrahedral (D) BF_3, Octahedral

21. What is the crystal structure of cesium chloride ?
(A) Body centered cubic (B) Face centered cubic
(C) Tetrahedral (D) Octahedral

22. Which of the following reacts with ethanol to form chloroform ?
(A) HCl (B) $CaOCl_2 / H_2O$ (C) $SOCl_2$ (D) PCl_5

23. A non-volatile solute A is dissolved in a volatile solvent B. The vapour pressure of the resultant solution is Ps. The vapour pressure of pure solvent is P°_B. If X is the mole fraction, which of the following relations is correct ?
(A) $P_S = P^\circ_B \times X_A$ (B) $P^\circ_B = P_S \times X_B$
(C) $P_S = P^\circ_B \times X_B$ (D) $P^\circ_B = P_S \times X_A$

24. The total pressure of a mixture of 6.4 grams of oxygen and 5.6 grams of nitrogen present in a 2 litre vessel is 1200 mm. What is partial pressure in mm of nitrogen ?
(A) 1200 (B) 600 (C) 900 (D) 200

25. The RMS velocity of an ideal gas at 300 K is 12240 cm. sec^{-1} What is its most probable velocity in cm.sec^{-1} at the same temperature ?
(A) 10000 (B) 11280 (C) 1000 (D) 12240

26. Calculate the heat of formation ΔH of CO in kcal, from the following data :
(i) C(graphite) + $O_2(g) \rightarrow CO_2(g)$ ΔH = –94 k cal
(ii) CO(g) + $1/2O_2(g) \rightarrow CO_2(g)$ Δ H = –68 k cal
(A) –13 (B) –26 (C) –162 (D) –81

27. What is the approximate quantity of electricity in coulombs required to deposit all the silver from 250 ml of 1 M $AgNO_3$ aqueous solution ? (At. wt. of Ag = 108)
(A) 48250 (B) 12062.5 (C) 96500 (D) 24125

28. A solid organic compound Y, on heating is converted to vapour state directly, which on cooling solidifies. Which of the following is the correct method of purifying Y ?
(A) Sublimation (B) Distillation
(C) Distillation at reduced pressure (D) Steam distillation

29. Chloroethane reacts with magnesium in dry ether to form X. When X is hydrolysed, a carbon compound Y and z are formed. Which of the following is Y ?
(A) C_2H_4 (B) C_2H_2 (C) C_2H_6 (D) C_6H_6

30. Ethene is reacted with Br_2/CCl_4 to form X. When X is reacted with alcoholic KOH, Y is formed X and Y are respectively
(A) $BrCH_2–CH_2Br$ and C_2H_2 (B) C_2H_5Br and C_2H_4
(C) C_2H_5Br and C_6H_6 (D) $C_2H_3Br_3$ and C_2H_4

31. Which of the following reacts with ammonical cuprous chloride ?
(A) C_2H_4 (B) C_6H_6 (C) CH_4 (D) C_2H_2

32. Which of the following hydrocarbons has least carbon-carbon bond length ?
(A) C_2H_6 (B) C_2H_4 (C) C_6H_6 (D) C_2H_2

33. What is the approximate calorific value (in Kcal m^{-3}) of water gas ?
(A) 500 (B) 2700 (C) 4000 (D) 1000

34. Which of the following is used in the preparation of soda water ?
(A) CO (B) CO_2 (C) SO_2 (D) HCl

35. What is the molecular formula of gypsum ?
(A) $CaSO_4H_2O$ (B) $CaSO_42H_2O$
(C) $2CaSO_4H_2O$ (D) $CaSO_43H_2O$

36. In the Castner's process of extraction of sodium, cathode is a/an
(A) Iron rod (B) Nickel rod
(C) Copper rod (D) Graphite rod

37. Which of the following is a sulphide ore ?
(A) Calamine (B) Cryolite
(B) Zinc blende (D) Haematite

38. Which of the following ions is colourless in its aqueous solution ?
(A) Ni^{2+} (B) Zn^{2+} (C) Ti^{3+} (D) Cu^{2+}

39. Which of the following is the strongest oxidizing agent ?
(A) F_2 (B) Br_2 (C) Cl_2 (D) I_2

40. Which of the following reacts with water to give ethane?

(A) CH_4 (B) C_2H_5MgBr
(C) C_2H_5OH (D) $C_2H_5OC_2H_5$

41. Which of the following reacts with BCl_3 to form diborane ?
(A) NH_3 (B) $LiAlH_4$ (C) Na–Hg (D) $K_2Cr_2O_7$

42. During the Serpeck's process, silica is eliminated as
(A) Si(solid) (B) SiO_2(vapour)
(C) SiO_2(soild) (D) Si (vapour)

43. What is the molecular formula of white phosphorous?
(A) P_2 (B) P_4 (C) P_5 (D) P_{16}

44. Which of the following is an acid ?
(A) Ca $(OH)_2$ (B) $P(OH)_3$ (C) NH_4OH (D) NaOH

45. What is the oxidation state of chlorine in hypochlorous acid ?
(A) +1 (B) +3 (C) +5 (D) +7

46. Which of the following has the lowest bond angle ?
(A) H_2S (B) NH_3 (C) H_2O (D) CH_4

47. Which of the following is an isotope of $_{32}Ge^{76}$?
(A) $_{33}As^{77}$ (B) $_{32}Ge^{77}$ (C) $_{34}Se^{77}$ (D) $_{35}Br^{80}$

48. What are the values of n[1] and n[2] respectively for H_β line in the Lyman series of hydrogen atomic spectrum?
(A) 3 and 5 (B) 2 and 3 (C) 1 and 3 (D) 2 and 4

49. Which catalyst is used in the manufacture of sulphuric acid by contact process ?
(A) Ni Powder (B) Platinised asbestos
(C) anhydrous Al_2O_3 (D) Al Powder

50. The equilibrium constant for the reaction
$H_2(g) + I_2(g) \rightleftharpoons 2HI(g)$ is 64 at certain temperature. The equilibrium concentrations of H_2 and HI are 2 mol. lit^{-1} and 16 mol lit^{-1} respectively. What is the equilibrium concentration in mol lit^{-1} of I_2 ?
(A) 16 (B) 4 (C) 8 (D) 2

EAMCET - 2000 (Medical)

1. What is the value of gas constant R In J mol^{-1} K^{-1} ?
(A) 8.314 (B) 0.0821 (C) 82.1 (D) 8.314×10^2

2. What is the volume in ml of 0.1 M Potassium permanganate solution required to completely oxidize 100ml of 0.5 M. ferrous sulphate solution in acid medium ?
(A) 20 (B) 200 (C) 50 (D) 100

3. What is the oxidation state of carbon in carbon dioxide ?
(A) +2 (B) +4 (C) +6 (D) +1

4. Which of the following nuclide is unstable ?
(A) $_7N^{14}$ (B) $_8O^{16}$ (C) $_5B^{10}$ (D) $_4Be^{10}$

5. The pk_a of a weak acid is 4.8. what is the ratio of [salt] / [acid], if a buffer of pH = 5.8 is to be prepared ?
(A) 1:1 (B) 10:1 (C) 2:1 (D) 1:10

6. The probability of finding an electron in an orbital is approximately ?
(A) 60% (B) 25% (C) 95% (D) 50%

7. Which of the following is a Lewis acid ?
(A) $HCOO^-$ (B) H_2SO_4 (C) SiF_4 (D) H_2S

8. What is the wavelength of H_β line of the Balmer series of hydrogen spectrum ? (R = Rydberg constant)
(A) 36/5R (B) 5R/36 (C) 3R/16 (D) 16/3R

9. The volume of 10 moles of an ideal gas is 1 litre at 27°C and 1 atm. Pressure. What is the volume in lit of 20 moles of the same gas at the same temperature and the pressure ?
(A) 2 (B) 4 (C) 1 (D) 8

10. Which of the following is used in the extraction of silver ?
(A) NaCN (B) Na_2SO_4 (C) NaCl (D) Na_2CO_3

11. Which of the following ions exhibits colour in aqueous solution ?
(A) Sc^{3+} (B) Ni^{2+} (C) Ti^{+4} (D) Zn^{+2}

12. What is the electrolyte used in the electrolytic method of preparation of fluorine?
(A) NaF + $CaCl_2$ (B) KHF_2 + LiF
(C) KF + $LiHF_2$ (D) $CaCl_2$ + KF

13. Lead impurity is removed from silver by ?
(A) Poling (B) Cupellation
(C) Distillation (D) Lavigation

14. Aniline on heating with X in the presence of KOH gives a product with very bad smell. Which of the following is X ?
(A) $CHCl_3$ (B) C_2H_5Cl (C) CH_3Cl (D) CCl_4

15. Chloral hydrate is dissolved in NaOH solution and distilled. What are the compounds obtained ?
(A) CH_3Cl,CH_3COONa
(B) $CHCl_3$, HCOONa, H_2O
(C) CH_2Cl_2, NaCl
(D) C_2H_5Cl, CH_3COONa

16. Bleaching action of chlorine requires the presence of
(A) O_2 (B) Moisture (C) Dry air (D) Sun light

17. Which of the following reaction conditions are used for the conversion of ethanol to ethylene ?
(A) Conc. H_2SO_4/70°C (B) Dil. H_2SO_4/140°C
(C) Dil. H_2SO_4/100°C (D) Conc.H_2SO_4/170°C

18. Which of the following has more acidic character ?
(A) H_2O (B) H_2Te (C) H_2S (D) H_2Se

19. Which of the following compounds is formed when ethanol reacts with acetic acid in the presence of concentrated H_2SO_4 ?
(A) $CH_3COOC_2H_5$ (B) $C_2H_5OC_2H_5$
(C) CH_3OCH_3 (D) CH_3CH_2CHO

20. In which of the following compounds, oxygen exhibits an oxidation state of +2 ?
(A) OF_2 (B) H_2SO_4 (C) H_2O (D) H_2O_2

21. Which of the following compounds is formed when acetaldehyde reacts with $LiAIH_4$?

(A) C_2H_5OH (B) CH_3COOH
(C) $C_2H_5OC_2H_5$ (D) CH_3COCH_3

22. Ethyl chloride reacts with X to form NaCl and Y. One mole of Y reacts with two moles of HI to form water and ethyl iodide. Which of the following is X ?
(A) C_2H_5ONa (B) $C_3H_5OC_2H_5$
(C) CH_2COOH (D) CH_3CHO

23. Which of the following is the correct order of ionic radii ?
(A) $Si^{4+} < Al^{3+} > Mg^{2+} > Na^+$
(B) $Na^+ < Mg^{2+} < Al^{3+} < Si^{4+}$
(C) $Al^{3+} < Si^{4+} > Na^+ > Mg^{2+}$
(D) $Na^+ > Mg^{2+} > Al^{3+} > Si^{4+}$

24. Which of the following is used to convert nitrobenzene to hydrazo benzene ?
(A) Zn/NH_4Cl (B) Zn/NaOH
(C) Sn/HCl (D) $LiAIH_4$

25. In which of the following compounds, the weight percentage of C and H are 80 and 20 respectively ?
(A) CH_4 (B) C_2H_2 (C) C_2H_6 (D) C_6H_6

26. What is the formula of diaspore ?
(A) $Al_2O_32H_2O$ (B) Al_2O_3
(C) $Al_2O_3H_2O$ (D) $AlPO_4$

27. Which of the following is calcium superphosphate ?
(A) $CaSO_4$ + CaO
(B) $Ca(H_2PO_4)2H_2O$ + $2(CaSO_2\ 2H_2O)$
(C) $Ca_3(PO_4)_2$ + $2CaSO_4$
(D) $Ca(H_2PO_4)_2H_2O$ + $CaSO_4$

28. What is the catalyst used for the manufacture of nitric acid by Ostwald's method ?
(A) Al_2O_3 (B) Fe (C) V_2O_5 (D) Platinum guage

29. Which of the following is an electron difficient compound ?
(A) B_2H_6 (B) KCl (C) NaCl (D) NaH

30. Element $_ZM^A$ emits one alpha particle followed by two beta particles. Which of the following is daughter element ?
(A) $z - 2M_4^{A-}$ (B) $z - 2M^A$
(C) zM^{A-4} (D) $z + 2M^{A-4}$

31. Which of the following aqueous solutions conducts electricity ?
(A) Urea (B) Glucose (C) Sucrose (D) NaCl

32. 500 ml of a solution contains 12.6 grams of oxalic acid (mol.wt.=126). 10 ml of this solution is diluted to 100ml in a flask. What is the molarity (in mol.lit^{-1}) of the resultant solution ?
(A) 0.2 (B) 0.1 (C) 0.02 (D) 0.01

33. Which of the following has the highest electronegativity ?
(A) Na (B) Cl (C) K (D) B

34. Which of the following has the least bond energy ?
(A) F_2 (B) H_2 (C) N_2 (D) O_2

35. Which of the following can be detected by silver mirror test ?
(A) CH_3COCH_3 (B) CH_3COOH
(C) C_2H_6 (D) CH_3CHO

36. Which of the following is used to convert nitrobenzene to azobenzene ?
(A) $LiAlH_4$ (B) Zn/NaOH
(C) Zn/NH_4Cl (D) Sn/HCl

37. Which of the following has the lowest boiling point ?
(A) HF (B) C_2H_5OH (C) CH_4 (D) H_2O

38. Acetic acid reacts with ethanol in the presence of H_2SO_4 to form X and water, which of the following is X ?
(A) $CH_3CH_2COOC_2H_5$ (B) CH_3COCH_3
(C) $CH_3COC_2H_5$ (D) $CH_3COOC_2H_5$

39. Methyl benzene can be prepared by reacting benzene with bromo methane in the presence of
(A) Anhydrous $AlCl_3$ (B) Br_2/CCl_4
(C) Ni/H_2m 70°C (D) Dil, H_2SO_4

40. 4 moles of HI is taken in a 1 litre closed vessel and heated till equilibrium is reached. At equilibrium the concentration of H_2 is 1 mol. lit^{-1}. Calculate the equilbrium constant for the reaction
(A) 4 (B) 0.5 (C) 2 (D) 0.25

41. The solution used for the preparation of hydrogen peroxide by electrolytic method is
(A) 0.2 N NaOH (B) 50% H_2SO_4
(C) 20 % NaOH (D) 5% NaCl

42. Heat of combustion of carbon is 96 Kcal. When some quantity of carbon is burnt in oxygen, 48 K cal of heat is liberated. What is the volume (in lit) of oxygen at STP reacted with this carbon ?
(A) 48 (B) 22.4 (C) 11.2 (D) 1

43. The rate of the reaction, $2A \rightarrow B$ is
(A) 1/2 (–d[A]/dt) (B) –d[A]/dt
(C) –d[B]/dt (D) +d[A]/dt

44. Benzene is used in the preparation of
(A) Polythene (B) Gammexane
(C) chloroform (D) Vinegar

45. In a reaction, sodium is heated with $BeCl_2$. In the reaction, sodium acts as
(A) Oxidizing agent (B) Reducing agent
(C) Hydrolysing agent (D) Precipitating agent

46. The molecular formula of plaster of paris is
(A) $2CaSO_4.H_2O$ (B) $2CaSO_42H_2O$
(C) $CaSO_42H_2O$ (D) $CaSO_43H_2O$

47. Which of the following is used in the manufacture of ethanol ?
(A) CH_4 (B) C_2H_4 (C) C_2H_6 (D) C_2H_2

48. What are the reaction conditions used for the formation of ethylene glycol from ethene ?

(A) Bromine water (B) Cold alkaline $KMnO_4$
(C) $SO_2/60°C$ (D) O_3/H_2O

49. If solid sodium hydroxide is reacted with carbon monoxide under pressure, the product obtained is
(A) CO_2 (B) CH_3ONa
(C) HCOONa (D) HCHO

50. What is the C–C bond length in angstroms in Diamond ?
(A) 2.0 (B) 5.2 (C) 1.54 (D) 3.35

EAMCET - 2001 (Engineering)

1. Equilibrium constant for the reaction $H_2O(g) + CO(g) \rightleftharpoons H_2(g) + CO_2(g)$ is 81. If the velocity constant of the forward reaction is 162 lit. mol^{-1} sec^{-1}, what is the velocity constant in lit. mol^{-1} sec^{-1} for the backward reaction ?
(A) 261 (B) 243 (C) 131.22 (D) 2

2. Consider the reaction, A → products. The reaction is completed in 100 minutes. The rate constant of this reaction at t_1 = 10 min is 10^{-2} min. What is the rate constant (in min^{-1}) at t_2 = 20 minutes ?
(A) 2 x 10^{-2} (B) 10^{-2} (C) 5 x 10^{-3} (D) 0.1

3. How many litres of oxygen at STP are required for complete combustion of 39 grams of liquid benzene (Atomic weights : C = 12; H = 1;O = 16)
(A) 84 (B) 22.4 (C) 42 (D) 11.2

4. 3 moles of ethanol reacts with one mole of phosphorus tribromide to form 3 moles of bromo-ethane and one mole of X. Which of the following is X ?
(A) HPO_3 (B) H_3PO_3 (C) H_3PO_4 (D) H_3PO_2

5. Absolute alcohol (100% alcohol) is prepared by distilling rectified spirit over ....
(A) Na (B) $CaCl_2$ (C) Mg (D) $Mg(OC_2H_5)_2$

6. Y in the following reactions is
$C_2H_5I + NaOC_2H_5 \rightarrow X + NaI$
$X + 2HI \rightarrow 2Y + H_2$
(A) C_2H_6 (B) C_2H_5I (C) C_2H_4 (D) $C_2H_5OC_2H_5$

7. Dry distillation of a mixture of calcium acetate and calcium formate forms ......
(A) Methanol (B) Ethanal (C) Ethanol (D) Acetone

8. When acetaldehyde is reacted with $LiAlH_4$, what is the product formed ?
(A) CH_3COOH (B) CH_3CH_2OH
(C) CH_3OH (D) HCOOH

9. Acetic acid is reacted with metallic sodium to form hydrogen and X. When X is heated with soda lime, Y and sodium carbonate are formed. Y is
(A) C_2H_6 (B) CH_4
(C) CH_3COONa (D) CH_3CONH_2

10. Consider the following reaction
$C_6H_5NO_2 \xrightarrow{Sn/HCl} X \xrightarrow{C_6H_5COCl} Y + HCl$. Y is
(A) Acetanilide (B) Benzanilide
(C) Azo-benzene (D) Hydrazo-benzene

11. The metal ion present in vitamin B_{12} is
(A) Fe^{2+} (B) Fe^{3+} (C) Co^{3+} (D) Co^{2+}

12. The energy of an electron present in Bohr's second orbit of hydrogen atom is
(A) –1312 J $atom^{-1}$ (B) –328 $kJmol^{-1}$
(C) –328 J mol^{-1} (D) –164 $kJmol^{-1}$

13. In the ground state, an element has 13 electrons in its "M-shell". The element is
(A) Copper (B) Chromium (C) Nickel (D) Iron

14. In a nucledie, one a m.u. of mass is dissipated into energy to bind its nucleous. The energy equivalent of this mass is
(A) 931.5 eV (B) 931.5 x 10^6 MeV
(C) 931.5 x 10^6 eV (D) 931.5 MV

15. X in the following reaction is
$^{26}_{12}Mg + ^{2}_{1}H \rightarrow ^{27}_{12}Mg + x$
(A) γ-ray (B) $^{1}_{0}n$ (C) $^{1}_{1}H$ (D) $^{2}_{1}D$

16. Let the electrongeativity, ionization energy and electron affinity be represented as EN, IP and EA respectively. Which of the following equation is correct according to Mulliken
(A) EN = IP X EA (B) EN = IP/EA
(C) EN = IP+EA/2 (D) EN = IP – EA

17. Which of the following reactions is an example of heterogeneous catalysis ?
(A) $2CO(g) + O_2(g) \xrightarrow{NO(g)} 2CO_2$
(B) $2SO_2(g) + O_2(g) \xrightarrow{NO(g)} 2SO_3$
(C) $2CO(g) + O_2(g) \xrightarrow{Pt(s)} 2CO_2$
(D) $CH_3CHO(g) \xrightarrow{I_2(g)} CH_4 + CO$

18. In which of the following reactions, H_2O_2 acts as a reducing reagent ?
(A) $Na_2SO_3(aq) + H_2O_2(aq) \rightarrow Na_2SO_4(s) + H_2O(l)$
(B) $PbO_2(s) + H_2O_2(aq) \rightarrow PbO(s) + H_2O(l) + O_2(g)$
(C) $2KI(aq) + H_2O_2(aq) \rightarrow 2KOH(aq) + I_2(s)$
(D) None of these

19. _____ process is used for the removal of hardness of water
(A) Calgon (B) Baeyer (C) Serpeck (D) Hoope

20. The electronic configuration of elements A, B and C are [He]$2s^1$, [Ne] $3s^1$ and [Ar] $4s^1$ respectively. Which of the following order is correct for the first ionization potentials in kJ mol^{-1} of A, B and C ?
(A) A > B > C (B) C > B > A
(C) B > C > A (D) C > A > B

21. In organic reactions, sodium in liquid ammonia is used as
(A) Reducing agent (B) Hydrolyzing agent
(C) Oxidizing agent (D) Precipitating agent

22. During the electrolytic reduction of alumina, the reaction at cathode is

(A) $2H_2O \rightarrow O_2 + 4H^+ + 4e^-$ (B) $3F^- \rightarrow 3F^- + 3e^-$
(C) $Al^{3+} + 3e^- \rightarrow Al$ (D) $2H^+ + 2e^- \rightarrow H_2$

23. What is the molecular formula of Borazole ?

(A) $B_3N_3H_6$ (B) $B_3N_3H_3$ (C) B_2H_6 (D) $B_6N_6H_6$

24. Which of the following order is correct for the bond energies of halogen molecules ?

(A) $I_2 > Cl_2 > Br_2$ (B) $Br_2 > Cl_2 > I_2$
(C) $I_2 > Br_2 > Cl_2$ (D) $Cl_2 > Br_2 > I_2$

25. Which of the following halogens liberates oxygen, when passed through a hot concentrated KOH solution ?

(A) Br_2 (B) F_2 (C) I_2 (D) Cl_2

26. The number of lone pair of electrons present on Xe in XeF_2 is

(A) 3 (B) 4 (C) 2 (D) 1

27. Which of the following ions exhibit highest magnetic moment ?

(A) Cu^{2+} (B) Ti^{3+} (C) Ni^{2+} (D) Mn^{2+}

28. Which of the following is mainly responsible for depletion of ozone layer ?

(A) Methane (B) Carbon dioxide
(C) Water (D) Chloro fluro carbons

29. The structural formula of 2-methly-2-butene is

(A) $CH_3 - CH = CH - CH_3$
(B) $CH_3 - CH(CH_3) - CH = CH_2$
(C) $CH_3 - CH_2 - C(CH_3) = CH_2$
(D) $CH_3 - CH = C(CH_3) - CH_3$

30. Which of the following pairs of compounds are functional isomers ?

(A) $CH_3CH_2CH_2OH, CH_3CH\ (OH)CH_3$
(B) $CH_3CH_2CH_2CH_2OH, (CH_3)_2CHCH_2OH$
(C) $CH_3CH_2CH_2OH, CH_3CH_2CH_2Cl$
(D) $CH_3CH_2CH_2OH, CH_3{-}O{-}CH_2CH_3$

31. Consider the following reactions

$C_2H_2 \xrightarrow[\text{Iron tube, 500°C}]{\text{Red hot}} A$

$A \xrightarrow[\text{70°C}]{\text{Conc. HNO}_3\text{ / Conc. H}_2\text{SO}_4} B$

$B \xrightarrow{LiAlH_4} C_6H_5.N = N.C_6H_5$. A and B are

(A) $A = C_2H_4;\ B = C_6H_6$
(B) $A = C_2H_6;\ B = C_6H_5NH_2$
(C) $A = C_2H_4;\ B = C_6H_5NH_2$
(D) $A = C_6H_6;\ B = C_6H_5NO_2$

32. Which of the following is used in the preparation of styrene ?

(A) CH_3CHO (B) P_2O_5 (C) CH_4 (D) C_6H_6

33. Ethanol, when reacted with PCl_5 gave A, $POCl_3$ and HCl. A reacts with silver nitrite to form B (major product) and AgCl. A and B are respectively

(A) C_2H_5Cl and $C_2H_5\ OC_2H_5$
(B) C_2H_6 and $C_2H_5\ OC_2H_5$
(C) C_2H_5Cl and $C_2H_5\ NO_2$ (D) C_2H_6 and $C_2H_5\ NO_2$

34. Chloroethane reacts with X to form diethyl ether. What is X ?

(A) C_2H_5ONa (B) $Na_2S_2O_3$ (C) NaOH (D) H_2SO_4

35. The catenation tendency of C, Si and Ge is in the order Ge < Si < C. The bond energies in KJ. mol^{-1} of C–C, Si–Si and Ge–Ge bonds, respectively are

(A) 167, 180, 348 (B) 180, 167, 348
(C) 348, 167, 180 (D) 348, 180, 167

36. Ionic radius (in Å) of As^{3+}, Sb^{3+} and Bi^{3+} follow the order

(A) $As^{3+} > Sb^{3+} > Bi^{3+}$ (B) $Sb^{3+} > Bi^{3+} > As^{3+}$
(C) $Bi^{3+} > As^{3+} > Sb^{3+}$ (D) $Bi^{3+} > SBb^{3+} > As^{3+}$

37. The element with atomic number 12 belongs to ___ group ___ period.

(A) I, third (B) III, third (C) II, third (D) II, second

38. Which of the following statements is true ?

(A) Hybridisation of the central atom in NH_3 and $CH_4$is sp^2
(B) $BeCl_2$ has "V" shape while SO_2 is linear
(C) SF_6 is octahedral and F–S–F bond angle is 90^0
(D) CO_2 has dipole moment

39. Bond dissociation energies of HF, HCl, HBr follow the order

(A) HCl > HBr > HF (B) HF > HBr > HCl
(C) HF > HCl > HBr (D) HBr > HCl > HF

40. What is the volume in litres of oxygen at STP required for complete combustion of 32 grams of CH_4? (molecular weight of CH_4 is 16).

(A) 22.4 (B) 179.2 (C) 44.8 (D) 89.6

41. What are the oxidation numbers of nitrogen in NH_4 NO_3 ?

(A) +3, –5 (B) –3, +5 (C) +3, –6 (D) +2, +2

42. The r.m.s. velocity of CO_2 at a temperature T (in Kelvin) is x $cm.sec^{-1}$. At what temperature (in Kelvin), the r.m.s. velocity of nitrous oxide would be 4x cm. sec^{-1} ?

(A) 16T (B) 2T (C) 4T (D) 32T

43. 'n' mole of an ideal gas at temperature T (in Kelvin) occupy "V" litres of volume, exerting a pressure of 'P' atmospheres. What is its concentration (in mol. lit') ? (Atomic weights of C, N and O are respectively 12, 14 and 16)

(A) P/RT (B) PT/R (C) RT/P (D) R/PT

44. 250 ml of a sodium carbonate solution contains 2.65 grams of Na_2CO_3. If 10 ml of this solution is diluted to one litre, what is the concentration of the resultant solution ? (molecular weight of Na_2CO_3 = 106)

(A) 0.1 M (B) 0.001 M (C) 0.01 M (D) 10^{-4}M

45. The pK_a of a weak acid is 4.8. What should be the ratio of (acid)/(salt), If a buffer of pH = 5.8 is required ?

(A) 0.1 (B) 10 (C) 1 (D) 2

46. Which of the following would produce a buffer solution when mixed in equal volumes ?

(A) 1 M CH_3COOH & 0.5 M NaOH
(B) 1 M CH_3COOH & 0.5 M HCl
(C) 1 M NH_4OH & 0.5 M NaOH
(D) 1 M NH_4Cl & 0.5 M HCl

47. One ampere of current is passed for 9650 second through molten $AlCl_3$. What is the weight in grams of Al deposited at cathode ? (Atomic weight of Al = 27)
(A) 0.09 (B) 90.0 (C) 0.9 (D) 9.0

48. Molten $CuCl_2$ is electrolysed using platinum electrodes. The reaction occurring at the anode is
(A) $2Cl^- \rightarrow Cl_2(g) + 2e^-$ (B) $Cl_2(g) + 2e^- \rightarrow 2Cl^-$
(C) $Cu^{2+} + 2e^- \rightarrow Cu(s)$ (D) $Cu(s) \rightarrow Cu^{2+} + 2e^-$

49. What are the products formed when ammonia is reacted with excess of chlorine ?
(A) N_2 and NCl_3 (B) N_2 and NCl_3
(C) N_2 and NH_4Cl (D) NCl_3 and HCl

50. What are the products formed when chlorine is passed through an aqueous hypo solution ?
(A) Na_2SO_3 + HCl +S (B) Na_2SO_3 + SO_3 +HCl
(C) Na_2SO_4 + HCl +S (D) Na_2SO_4 + HCl + SO_2

EAMCET - 2001 (Medical)

1. Which of the following is a tertiary alcohol ?
(A) $CH_3CH_2CH_2OH$ (B) $CH_3CH(OH)CH_3$
(C) $(CH_3)_2CHCH_2OH$ (D) $(CH_3)_3COH$

2. Williamson's synthesis is used for preparing
(A) Ethers (B) Aldehydes (C) Alcohols (D) Ketones

3. Ethanal is reacted with acidified $K_2Cr_2O_7$. What is the product formed ?
(A) C_2H_5OH (B) CH_3COOH
(C) C_2H_6 (D) CCl_3CHO

4. Which of the following compounds reacts with a saturated solution of sodium bisulphite to give colourless crystalline product ?
(A) CH_3CHO (B) $CHCl_3$ (C) C_2H_5OH (D) C_2H_6

5. The reagent used in the preparation of chloro-acetic acid from acetic acid is
(A) PCl_5 (B) Cl_2/redP (C) PCl_3 (D) $SOCl_2$

6. What is the product obtained when nitrobenzene is reacted with HCl in the presence of Sn .
(A) $C_6H_5N = NC_6H_5$ (B) $C_6H_5NH_2$
(C) $C_6H_5NH - NHC_6H_5$ (D) C_6H_5NHOH

7. The coordinaton number of Fe (II) in Oxyhaemoglobin is
(A) 6 (B) 4 (C) 8 (D) 10

8. Which of the following pairs of ions have the same electronic configuration ?
(A) Cr^{+3}, Fe^{+3} (B) Fe^{+3}, Mn^{+2}
(C) Fe^{+3}, CO^{+3} (D) Sc^{+3}, Cr^{+3}

9. The atomic number (Z) of an element is 25. If it is in its ground state, how many electrons are present in the "N" shell ?
(A) 13 (B) 2 (C) 15 (D) 3

10. Which of the following is an isobar of $^{14}_{6}C$?
(A) $^{13}_{6}C$ (B) $^{12}_{6}C$ (C) $^{14}_{7}N$ (D) $^{15}_{7}N$

11. What is the product formed when ozone reacts with mercury ?
(A) Hg_2O (B) HgO_2 (C) HgO (D) Hg_2O_2

12. What is the Oxidation state of chlorine in hypochlorous acid ?
(A) +7 (B) +5 (C) +3 (D) +1

13. In the presence of cobalt chloride $COCl_2$, bleaching powder decomposes to form
(A) $CaCO_3$ & O_3 (B) ClO_2 & CaO
(C) Cl_2O & CaO (D) $CaCl_2$ & O_2

14. What is the atomic number of the noble gas that reacts with fluorine to form several compounds ?
(A) 54 (B) 10 (C) 18 (D) 2

15. Which of the following is a diamagnetic ion ?
(A) Mn^{2+} (B) Sc^{3+} (C) Co^{2+} (D) Cu^{2+}

16. Which of the following regions of atmosphere contains Ozone ?
(A) Troposphere (B) Thermosphere
(C) Mesosphere (D) Stratosphere

17. Which of the following compounds is a chain isomer of 1-butanol ?
(A) 2-methyl-2-butanol (B) 2-methyl-1-butanol
(C) 3-methyl-2-butanol (D) 2-methyl-1-Propanol

18. _____test is used for detecting unsaturation in hydrocarbons.
(A) Silver mirror (B) Lassaigne's
(C) Carbylamine (D) Baeyer's

19. In the following reaction,

$$C_2H_2 \xrightarrow[HgSO_4/H_2SO_4,\ 60^\circ C]{H_2O} X \rightleftharpoons CH_3CHO$$, what is X ?

(A) CH_3CH_2OH (B) $CH_3 - O - CH_3$
(C) CH_3CH_2CHO (D) $H_2C = CHOH$

20. In the reaction, $C_6H_6 + R - Cl \xrightarrow{A} C_6H_5 - R + HCl$, A is
(A) NaOH (B) $FeCl_3$ (C) P_2O_5 (D) PCl_5

21. What is X in the following reaction ?
$C_2H_5Cl + X \rightarrow C_2H_5OH + KCl$
(A) $KHCO_3$ (B) Alcoholic KOH
(C) Aqueous KOH (D) K_2CO_3

22. Which of the following is a secondary alkyl halide ?
(A) $(CH_3)_3CCl$ (B) C_2H_5Cl
(C) $(CH_3)_2CHCl$ (D) $(CH_3)_2CHCH_2Cl$

23. Ethanol is heated to 170°C in the presence of concentrated H_2SO_4. The product formed is
(A) CH_2COOH (B) C_2H_6

(C) $C_2H_5OC_2H_5$ (D) C_2H_4

24. The radius of an atomic nucleus in generally expressed in units of
(A) Fermi (B) Tesla (C) Debye (D) Coulomb

25. In the reaction,
$CH_3COOH(aq) + {}^-OH(aq) \rightarrow CH_3COO^-(aq) + H_2O(l)$; $\Delta H = -55.2$ kJ.mol^{-1}.
What is the heat of dissociation (in kJ mol^{-1}) of acetic acid ?
(A) +2.1 (B) −57.3 (C) −35.1 (D) −176.1

26. Which of the following salts forms a "micelle" ?
(A) Sodium formate (B) Sodium acetate
(C) Sodium stearate (D) Sodium chloride

27. Which gas is liberated when alkaline formaldehyde solution is reacted with hydrogen peroxide in the presence of pyrogallol ?
(A) CH_4 (B) CO_2 (C) H_2 (D) O_2

28. 50 grams of calcium carbonate was completely burnt in air. What is the weight in grams of the residue ? (Atomic weights of Ca, C and O are 40, 12 and 16 respectively)
(A) 2.8 (B) 28 (C) 4.4 (D) 44

29. How many moles of acidified $FeSO_4$ can be completely oxidized by one mole of $KMnO_4$?
(A) 10 (B) 5 (C) 6 (D) 4

30. What are the conditions under which the relation between volume (V) and number of moles (n) of a gas are plotted? (P=pressure; T= temperature)
(A) Constant P & T (B) Constant T & V
(C) Constant P & V (D) Constant n & V

31. 7.5 grams of a gas occupy 5.6 litres of volume at STP. The gas is (Atomic weights of C, N and O are 12, 14 and 16 respectively)
(A) NO (B) N_2O (C) CO (D) CO_2

32. 250 ml of a sodium carbonate solution contains 2.65 grams of Na_2CO_3. 10ml of this solution is added to x ml of water to obtain 0.001 M Na_2CO_3 solution. What is the value of x in ml ?
(A) 9990 (B) 90 (C) 1000 (D) 990

33. The pH of a 0.1 M aqueous solution of a weak acid (HA) is 3. What is its degree of dissociation ?
(A) 1% (B) 10% (C) 50% (D) 25%

34. The pH range of methyl red indicator is
(A) 4.2 to 6.3 (B) 8.3 to 10.0
(C) 8.0 to 9.6 (D) 6.8 to 8.4

35. One Faraday of electricity is passed separately through one litre of one molar aqueous solutions of (i) $AgNO_3$ (ii) $SnCl_4$ and (iii) $CuSO_4$. The number of moles of Ag, Sn and Cu deposited at cathode are respectively
(A) 1.0, 0.25, 0.5 (B) 1.0, 0.5, 0.25
(C) 0.5, 1.0, 0.25 (D) 0.25, 0.5, 1.0

36. Aqueous copper sulphate solution is electrolysed using platinum electrodes. The electrode reaction occurring at cathode is
(A) $Cu^{2+}(aq) + 2e^- \rightarrow Cu(s)$
(B) $Cu(s) \rightarrow Cu^{2+}(aq) + 2e^-$
(C) $2H_2O(l) \rightarrow O_2(g) + 4H^+(aq) + 4e^-$
(D) $O_2(g) + 4H^+(aq) + 4e^- \rightarrow 2H_2O(l)$

37. 9.2 grams of $N_2O_4(g)$ is taken in a closed one litre vessel and heated till the following equilibrium is reached. $N_2O_4(g) \rightleftharpoons 2NO_2(g)$. At equilibrium 50% of $N_2O_4(g)$ is dissociated. What is the equilibrium constant in mol.lit^{-1} ? (Molecular weight of N_2O_4 is 92).
(A) 0.1 (B) 0.2 (C) 0.4 (D) 2

38. The rate of the reaction for A → products is 10 mol. lit^{-1} min^{-1} at time t_1=2 minutes. What will be the rate at time t_2 = 12 minute.
(A) more than 10 (B) less than 10 (C) 10 (D) 20

39. Sodium hexa-metaphosphate is known as.
(A) Calgon (B) Permutit (C) Natalite (D) Nitrolim

40. Consider the following reaction
$6NaOH + 3\,Cl_2 \rightarrow 5NaCl + A + 3H_2O$ What is the oxidation number of chlorine in "A"
(A) +5 (B) 1 (C) +3 (D) +1

41. Aqueous sodium hydroxide reacts with white phosphorous to form phosphine and
(A) Na_3PO_3 (B) P_2O_3 (C) NaH_2PO_2 (D) P_2O_5

42. Aluminium reacts with concentrated HCl and concentrated NaOH to liberate .... gases respectively.
(A) H_2 and O_2 (B) O_2 and H_2
(C) H_2 and H_2 (D) O_2 and O_2

43. Which of the following is used to remove silicon dioxide in the Serpeck's process of purification of bauxite ?
(A) CaO (B) Na_2CO_3 (C) Coke (D) Nickel

44. Which of the following allotropic forms of carbon is isomorphous with crystalline silicon ?
(A) Graphite (B) Coal (C) Coke (D) Diamond

45. Which of the following is used in the industrial preparation of nitric acid by Ostwald's process ?
(A) N_2O_5 (B) NH_3 (C) N_2 (D) NH_4NO_3

46. Which of the following is an oxyacid ?
(A) $Ba(OH)_2$ (B) $Mg(OH)_2$ (C) H_3PO_3 (D) HCl

47. The electronic configuration of four different elements is given below. Identify the 14th group element among these.
(A) [He] $2s^1$ (B) [Ne] $3s^2$
(C) [Ne] $3s^23p^2$ (D) [Ne] $3s^23p^5$

48. Which of the following elements shows different oxidation states ?
(A) Sodium (B) Fluorine (C) Chlorine (D) Potassium

49. In which of the following substances, hydrogen bonding is absent ?
(A) HF (B) H_2O (C) CCl_4 (D) Salicylaldehyde

50. Acetylene has .... bonds.
(A) 2σ, 2π (B) 2σ, 3π (C) 3σ, 2π (D) 3σ, 3π

EAMCET - 2002 (Engineering)

1. The energy of an electromagnetic radiation is 19.875×10^{-13} ergs. What is its wave number in cm^{-1} ? ($h = 6.625 \times 10^{-27}$ erg. sec; $C = 3 \times 10^{10}$ cm sec^{-1})
(A) 100 (B) 10,000 (C) 1,000 (D) 10^6

2. The products formed when heavy water is reacted with magnesium nitride, are
(A) NH_3, Mg (B) NH_3, Mg $(OD)_2$
(C) ND_2, Mg $(OH)_2$ (D) ND_3, Mg $(OD)_2$

3. The reagent used in the preparation of aspirin from salicylic acid is
(A) $SOCl_2$/ Pyridine (B) CH_3COOH/HCl
(C) $(CH_3CO)_2O$ conc.H_2SO_4 (D) $CH_3Cl/AlCl_3$

4. In the following reaction X, and Y respectively are

$$C_2H_5OH \xrightarrow{KmnO_4/H^+} X \xrightarrow[H_2SO_4\ \Delta]{Y} CH_3COOC_2H_5$$

(A) CH_3OH, C_2H_5OH (B) CH_3CHO, CH_3OH
(C) $CH_2 = CH_2$, CH_3COOH (D) CH_3COOH, C_2H_5OH

5. At 27°C, a closed vessel contains a mixture of equal weights of helium (mol. wt, = 4), methane (mol. wt = 16) and sulphur dioxide (mol. wt = 64). The pressure exerted by the mixture is 210 mm. If the partial pressures of helium, methane and sulphur dioxide are P_1, P_2 and P_3 respectively, which one of the following order is correct ?
(A) $P_3 > P_2 > P_1$ (B) $P_1 > P_2 > P_3$
(C) $P_1 > P_3 > P_2$ (D) $P_2 > P_3 > P_1$

6. The compound formed when gypsum is dissolved in aqueous ammonium sulphate solution is
(A) $CaSO_4NH_4ClH_2O$ (B) $CaCl_2.(NH_4)_2SO_4H_2O$
(C) $CaSO_4(NH_4)_2SO_4H_2O$ (D) $CaCl_2$. $NH_4Cl.2H_2O$

7. In the following reaction, A and B respectively are

$$A \xrightarrow{HBr} C_2H_5Br \xrightarrow{B} A$$

(A) C_2H_5OH and aq. KOH/Δ (B) C_2H_2 and PBr_3
(C) C_2H_4 and alcoholic KOH/Δ
(D) C_2H_5Cl and aq. KOH/Δ

8. 'X' litres of carbon monoxide is present at STP. It is completely oxidized to CO_2. The volume of CO_2 formed is 11.207 litres at S.T.P. What is the value of 'X' in litres ?
(A) 22.414 (B) 11.207 (C) 5.6035 (D) 44.828

9. The bond energies in kJ mol^{-1} of P–H, As–H and N–H are respectively
(A) 247, 138 & 389 (B) 247, 389 and 318
(C) 318, 389 & 247 (D) 318, 247 and 389

10. Which of the following represents the correct order of electro negativity ?
(A) P > O > N (B) N > P > O
(C) O > N > P (D) N > O > P

11. In the following reaction X and Y are respectively.

$$CH_3COOH + NH_3 \longrightarrow X \xrightarrow{\Delta} Y + H_2O$$

(A) CH_3CONH_2 ; CH_3COOH
(B) CH_3NH_2 ; CH_3CONH_2
(C) CH_3CONH_2 ; CH_4
(D) CH_3COONH_4 ; CH_3CONH_2

12. which of the following is the molecular formula for a tertiary amine ?
(A) C_2H_7N (B) C_3H_9N (C) CH_5N (D) None of these

13. 4 grams of an ideal gas occupies 5.6035 litres of volume at 546 K and 2 atm. pressure. What is its molecular weight ?
(A) 4 (B) 16 (C) 32 (D) 64

14. which of the following statements is correct with respect to basic character ?
(A) $PH_3 > P(CH_3)_3$ (B) $PH_3 = NH_3$
(C) $PH_3 > NH_3$ (D) $P(CH_3)_3 > PH_3$

15. The calculated magnetic moment (in Bohr magnetons) of Cu^{2+} ion is
(A) 2.6 (B) 3.4 (C) 1.73 (D) Zero

16. Which catalyst is used in the conversion of acetaldehyde to acetic acid ?
(A) Manganese acetate
(B) $LiAlH_4$ (C) H_2/Ni (D) Na/NH_3

17. 75 ml of 0.2 M HCl is mixed with 25 ml of 1 M HCl. To this solution, 300 ml of distilled water is added. What is the pH of the resultant solution ?
(A) 1 (B) 2 (C) 4 (D) 0.2

18. What are the products formed when ammonia reacts with excess of chlorine ?
(A) N_2 and CHl_3 (B) NCl_3 and HCl
(C) N_2 and NH_4Cl (D) N_2 and HCl

19. What is X in the following reaction ?

$$2CH_3COCH_3 \xrightarrow{Ba\ (OH)_2} X$$

(A) $CH_3 - C(OH)(CH_3) - CH_2 - C(=O) - CH_3$

(B) $CH_3 - C(OH)(CH_3) - CH_2 - CH_2 - CHO$

(C) $CH_3CH(CH_3)\ CH_2COCH_3$

(D) $CH_3 - CH(CH_3) - CH(OH) - C(=O) - CH_3$

20. Consider the following reaction,
$N_2(g) + 3H_2(g) \rightarrow 2NH_3(g)$
The rate of this reaction in terms of N_2 at T(K) is

$-\frac{d[N_2]}{dt} = 0.02$ mol. lit^{-1} sec^{-1}. What is the value of $-\frac{d[H_2]}{dt}$ (in units of mol.lit^{-1}sec^{-1}) at the same temperature?

(A) 0.02 (B) 50 (C) 0.06 (D) 0.04

21. The concentration of a 100 ml solution containing X grams of Na_2CO_3 (mol. wt.= 106) is Y M. The values of X and Y are respectively

(A) 2.12, 0.05 (B) 1.06, 0.2
(C) 1.06, 0.1 (D) 2.12, 0.1

22. One mole of fluorine is reacted with two moles of hot and concentrated KOH. The products formed are KF, H_2O and O_2. The molar ratio of KF, H_2O and O_2 respectively is

(A) 1 : 1 : 2 (B) 2 : 1 : 0.5 (C) 1 : 2 : 1 (D) 2 : 1 : 2

23. In chloroethane, the carbon bearing halogen is bonded to ... hydrogen (s). It is called... alkyl halide.

(A) Two, primary (B) Three, primary
(C) Two, secondary (D) One, tertiary

24. What is the reduction electrode potential (in volts.) of copper electrode when $[Cu^{2+}] = 0.01$ M in a solution at 25°C? (E° of Cu^{2+}/Cu electrode is +0.34 V)

(A) 0.3105 (B) 0.3695 (C) 0.3991 (D) 0.2809

25. Which of the following molecules contains both ionic and covalent bonds ?

(A) CH_2Cl_2 (B) K_2SO_4 (C) $BeCl_2$ (D) SO_2

26. 4 grams of a hydrocarbon on complete combustion gave 12.571 grams of CO_2 and 5.143 grams of water. What is the empirical formula of the hydrocarbon ?

(A) CH (B) CH_2 (C) CH_3 (D) C_2H_3

27. What is the reaction occurring at the anode in Down's process for the extraction of sodium ?

(A) $4OH \rightarrow 2H_2O + O_2 + 4e^-$
(B) $Na^+ + e^- \rightarrow Na$
(C) $2\,Cl^- \rightarrow Cl_2 + 2e^-$
(D) $NaOH \rightarrow Na^+ + OH^-$

28. What is the minimum quantity in grams of methyl iodide required for preparing one mole of ethane by Wurtz reaction ? (Atomic weight of iodine = 127)

(A) 326 (B) 284 (C) 142 (D) 568

29. Which of the following statements in not correct ?

(A) Physical adsorption decreases with increase in the temperature
(B) Physical adsorption is multilayered.
(C) Activation energy of physical adsorption is very high
(D) Enthalpy change of physical adsorption is about 20 kJ.mol^{-1}.

30. The products formed when diethyl ether is reacted with cold HI are :

(A) $C_2H_5I + C_2H_5OH$
(B) $2C_2H_5I + H_2O$
(C) $2C_2H_5OH$
(D) $C_2H_5 - O - O - C_2H_5 + H_2O$

31. Which of the following is a correct pair with respect to molecular formula of Xenon compound and hybridization state of Xenon in it ?

(A) XeF_4, sp^3 (B) XeF_2, sp
(C) XeF_2, sp^3d (D) XeF_4, sp^2

32. Which of the following statements is not correct ?

(A) Rydberg's constant and wave number have same unit.
(B) Lyman series of hydrogen spectrum occurs in the ultraviolet region
(C) The angular momentum of the electron in the ground state hydrogen atom is equal to h/2p
(D) The radius of first Bohr orbit of hydrogen atom is 2.116 x 10^{-8}cm

33. When bauxite powder is mixed with coke and reacted with nitrogen at 2075 K, carbon monoxide and X are formed. What is the gas formed when X is reacted with water ?

(A) N_2O (B) O_2 (C) NH_3 (D) N_2

34. What is the gas liberated when alkaline formaldehyde solution is treated with H_2O_2 ?

(A) CO_2 (B) O_2 (C) CH_4 (D) H_2

35. The chemicals used for preparing acetophenone are

(A) C_6H_6 (B) CH_3COCH_3
(C) CH_3COCl (D) Anhydrous $AlCl_3$

(A) A, B, C (B) B, C, D, (C) A, C, D (D) A, B, D

36. Which of the following statements is correct ?

(A) Bronsted - Lowry theory could not explain the acidic nature of BCl_3.
(B) The pH of 0.01 M NaOH solution is 2.
(C) The ionic product of water at 25°C is 10^{-10} mol^{-2}lit^{-2}
(D) The pH of a solution can be calculated using the equation pH = log [H$^+$]

37. "Fluorosis" disease is caused due to the reaction of ...... with excess of fluoride in the body.

(A) Fe (B) K (C) Ca (D) Mg

38. One mole of A(g) is heated to 200°C in a one litre closed flask, till the following equilibrium is reached. $A(g) \rightleftharpoons B(g)$ The rate of forward reaction at equilibrium is 0.02 mol. lit^{-1} min^{-1} what is the rate of the backward reaction at equilibrium ?

(A) 0.04 (B) 0.01 (C) 0.02 (D) 1

39. Iron sulphide is heated in air to form A, an oxide of sulphur. A is dissolved in water to give an acid. The basicity of this acid is

(A) 2 (B) 3 (C) 1 (D) Zero

40. The reaction conditions used for converting 1, 2-dibromo ethane to ethene are

(A) Zn, alcohol, Δ (B) KOH alcohol, Δ
(C) KOH water, Δ (D) Na, alcohol, Δ

41. Calculate the heat of combustion in kJ of methane from the following data

(i) C(graphite)+$2H_2$(g) $\rightarrow$ CH_4(g); ΔH = 74.8 kJ
(ii) C(graphite) + O_2(g) $\rightarrow$ CO_2(g); ΔH = –393.5 kJ

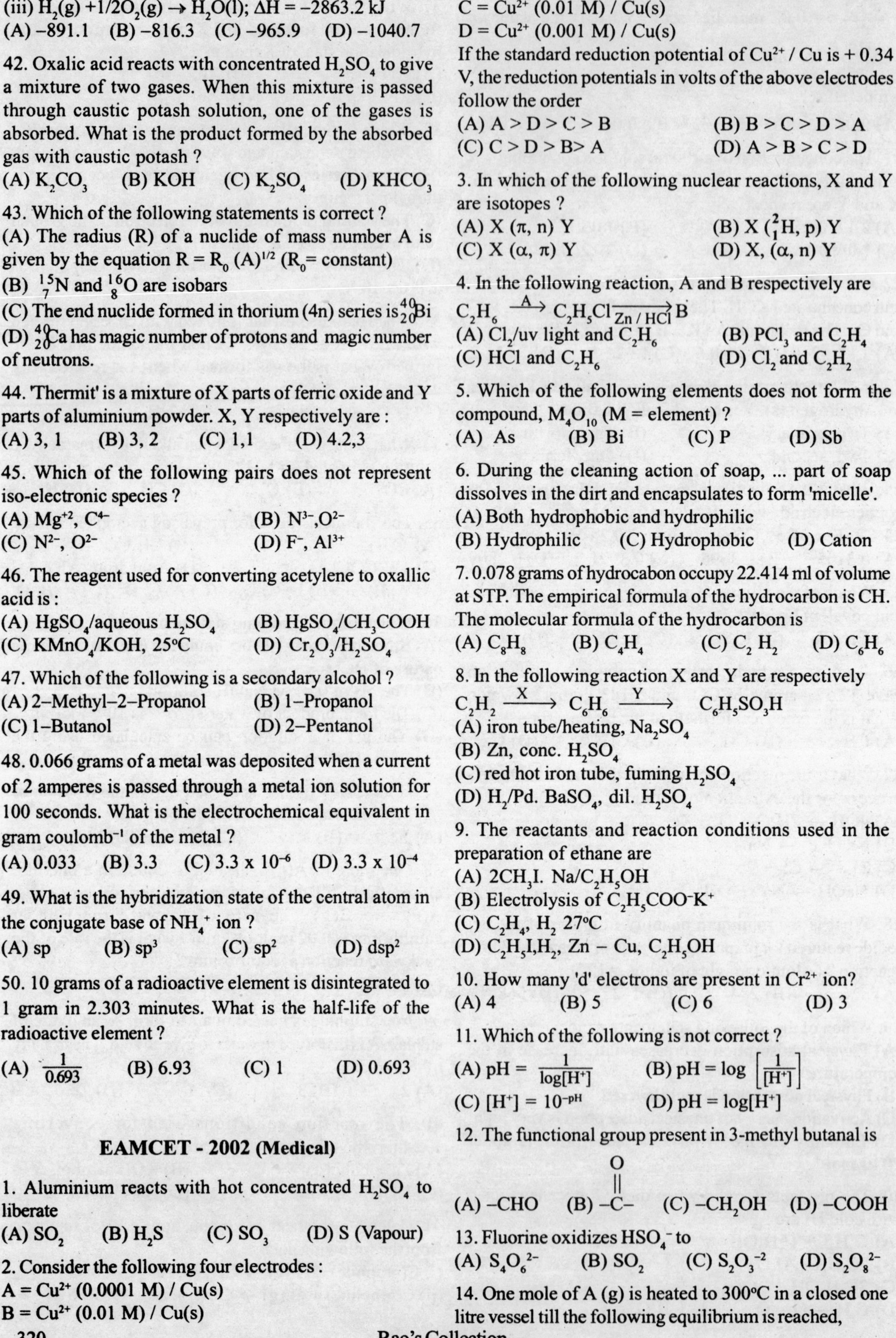

(iii) $H_2(g) + 1/2O_2(g) \rightarrow H_2O(l); \Delta H = -2863.2$ kJ
(A) –891.1 (B) –816.3 (C) –965.9 (D) –1040.7

42. Oxalic acid reacts with concentrated H_2SO_4 to give a mixture of two gases. When this mixture is passed through caustic potash solution, one of the gases is absorbed. What is the product formed by the absorbed gas with caustic potash ?
(A) K_2CO_3 (B) KOH (C) K_2SO_4 (D) $KHCO_3$

43. Which of the following statements is correct ?
(A) The radius (R) of a nuclide of mass number A is given by the equation $R = R_0 (A)^{1/2}$ (R_0= constant)
(B) ${}^{15}_{7}N$ and ${}^{16}_{8}O$ are isobars
(C) The end nuclide formed in thorium (4n) series is ${}^{40}_{20}Bi$
(D) ${}^{40}_{20}Ca$ has magic number of protons and magic number of neutrons.

44. 'Thermit' is a mixture of X parts of ferric oxide and Y parts of aluminium powder. X, Y respectively are :
(A) 3, 1 (B) 3, 2 (C) 1,1 (D) 4.2,3

45. Which of the following pairs does not represent iso-electronic species ?
(A) Mg^{+2}, C^{4-} (B) N^{3-},O^{2-}
(C) N^{2-}, O^{2-} (D) F^-, Al^{3+}

46. The reagent used for converting acetylene to oxallic acid is :
(A) $HgSO_4$/aqueous H_2SO_4 (B) $HgSO_4/CH_3COOH$
(C) $KMnO_4$/KOH, 25°C (D) Cr_2O_3/H_2SO_4

47. Which of the following is a secondary alcohol ?
(A) 2–Methyl–2–Propanol (B) 1–Propanol
(C) 1–Butanol (D) 2–Pentanol

48. 0.066 grams of a metal was deposited when a current of 2 amperes is passed through a metal ion solution for 100 seconds. What is the electrochemical equivalent in gram coulomb^{-1} of the metal ?
(A) 0.033 (B) 3.3 (C) 3.3 x 10^{-6} (D) 3.3 x 10^{-4}

49. What is the hybridization state of the central atom in the conjugate base of NH_4^+ ion ?
(A) sp (B) sp^3 (C) sp^2 (D) dsp^2

50. 10 grams of a radioactive element is disintegrated to 1 gram in 2.303 minutes. What is the half-life of the radioactive element ?
(A) $\frac{1}{0.693}$ (B) 6.93 (C) 1 (D) 0.693

EAMCET - 2002 (Medical)

1. Aluminium reacts with hot concentrated H_2SO_4 to liberate
(A) SO_2 (B) H_2S (C) SO_3 (D) S (Vapour)

2. Consider the following four electrodes :
A = Cu^{2+} (0.0001 M) / Cu(s)
B = Cu^{2+} (0.01 M) / Cu(s)
C = Cu^{2+} (0.01 M) / Cu(s)
D = Cu^{2+} (0.001 M) / Cu(s)
If the standard reduction potential of Cu^{2+} / Cu is + 0.34 V, the reduction potentials in volts of the above electrodes follow the order
(A) A > D > C > B (B) B > C > D > A
(C) C > D > B> A (D) A > B > C > D

3. In which of the following nuclear reactions, X and Y are isotopes ?
(A) X (π, n) Y (B) X (${}^{2}_{1}H$, p) Y
(C) X (α, π) Y (D) X, (α, n) Y

4. In the following reaction, A and B respectively are
$C_2H_6 \xrightarrow{A} C_2H_5Cl \xrightarrow[Zn/HCl]{} B$
(A) Cl_2/uv light and C_2H_6 (B) PCl_3 and C_2H_4
(C) HCl and C_2H_6 (D) Cl_2 and C_2H_2

5. Which of the following elements does not form the compound, M_4O_{10} (M = element) ?
(A) As (B) Bi (C) P (D) Sb

6. During the cleaning action of soap, ... part of soap dissolves in the dirt and encapsulates to form 'micelle'.
(A) Both hydrophobic and hydrophilic
(B) Hydrophilic (C) Hydrophobic (D) Cation

7. 0.078 grams of hydrocabon occupy 22.414 ml of volume at STP. The empirical formula of the hydrocarbon is CH. The molecular formula of the hydrocarbon is
(A) C_8H_8 (B) C_4H_4 (C) C_2H_2 (D) C_6H_6

8. In the following reaction X and Y are respectively
$C_2H_2 \xrightarrow{X} C_6H_6 \xrightarrow{Y} C_6H_5SO_3H$
(A) iron tube/heating, Na_2SO_4
(B) Zn, conc. H_2SO_4
(C) red hot iron tube, fuming H_2SO_4
(D) H_2/Pd. $BaSO_4$, dil. H_2SO_4

9. The reactants and reaction conditions used in the preparation of ethane are
(A) $2CH_3I$. Na/C_2H_5OH
(B) Electrolysis of $C_2H_5COO^-K^+$
(C) C_2H_4, H_2 27°C
(D) C_2H_5I,H_2, Zn – Cu, C_2H_5OH

10. How many 'd' electrons are present in Cr^{2+} ion?
(A) 4 (B) 5 (C) 6 (D) 3

11. Which of the following is not correct ?
(A) $pH = \frac{1}{\log[H^+]}$ (B) $pH = \log\left[\frac{1}{[H^+]}\right]$
(C) $[H^+] = 10^{-pH}$ (D) $pH = \log[H^+]$

12. The functional group present in 3-methyl butanal is
(A) –CHO (B) $-\overset{O}{\overset{||}{C}}-$ (C) $-CH_2OH$ (D) –COOH

13. Fluorine oxidizes HSO_4^- to
(A) $S_4O_6^{2-}$ (B) SO_2 (C) $S_2O_3^{-2}$ (D) $S_2O_8^{2-}$

14. One mole of A (g) is heated to 300°C in a closed one litre vessel till the following equilibrium is reached,

A(g) $\rightleftharpoons$ B(g). The equilibrium constant of this reaction at 300°C is 4. What is the concentration of B in mol. lit^{-1} at equilibrium?

(A) 0.2 (B) 0.6 (C) 0.8 (D) 0.1

15. Acetaldehyde is reduced with hydrogen in the presence of nickel. The compound formed is

(A) Ethanol (B) n-propanol

(C) Methanol (D) Isopropanol

16. Which of the following elements is a non-metal ?

(A) Sodium (B) Magnesium (C) Boron (D) Indium

17. Which of the following is correct for a first order reaction ? (K = rate constant; $t_{1/2}$ = half-life)

(A) $t_{1/2} = 0.693 \times k$ (B) $k.t_{1/2} = 1/0.693$

(C) $k.t_{1/2} = 0.693$ (D) $6.93 \times k \times t_{1/2} = 1$

18. In the following reaction, X and Y respectively are

$X \xrightarrow{aq.\ NaOH} CH_3COOH \xrightarrow{Y} (CH_3CO)_2O$ Solution

(A) CH_3CHO, PCl_5 (B) CH_3CN, P_2O_5

(C) CH_3CH_2OH, NaOAc (D) CH_3COCH_3; H_2SO_4

19. The number of unpaired electrons present in the first excited state of chlorine atom is

(A) 1 (B) 3 (C) 5 (D) 2

20. The half-life of a radioactive nuclide is 0.693 minutes. The time in minutes required for the disintegration of this nuclide from 10 grams to one gram is

(A) 6.93 (B) 2.303 (C) 0.693 (D) 1

21. In the following reaction, X and Y are respectively

$2C_2H_6 + 7O_2 \xrightarrow{Combustion} X + Y + heat$

(A) $2C_2H_5OH$, $6O_2$ (B) 4HCHO, $5H_2O$

(C) 4 CO, $10H_2O$ (D) $4CO_2$, $6H_2O$

22. When ammonium nitrate is gently heated, an oxide of nitrogen is formed. What is the oxidation state of nitrogen in the oxide formed ?

(A) +4 (B) +2 (C) +3 (D) +1

23. Two grams of sulphur is completely burnt in oxygen to from SO_2. In this reaction, what is the volume in litres of oxygen consumed at STP ? (At. wts of sulphur and oxygen are 32 and 16 respectively).

(A) $\frac{16}{22.414}$ (B) $\frac{22.414}{16}$

(C) $\frac{22.414}{32}$ (D) $\frac{32}{22.414}$

24. Sodium metal reacts with Al_2O_3 at high temperature to give a sodium compound X. X reacts with carbon dioxide in water to form Y. Y is...

(A) Na_2O_2 (B) Na_2 (C) Na_2CO_3 (D) $Na_2\ AlO_3$

25. Which of the following statements is correct ?

(A) '2s' orbital is spherical with two nodal planes.

(B) The de Broglie wavelength (λ) of a particle of mass 'm' and velocity 'V' is equal to mV/n

(C) The principal quantum number n indicates the shape of the orbital.

(D) The electronic configuration of phosphorous is [Ne] $3s^2, 3p_x^1, 3p_y^1, 3p_z^1$

26. The energy of an electromagnetic radiation is 3×10^{-12} ergs. What is its wavelength in nanometers ? (h = 6.625×10^{-27} erg. sec; c = 3×10^{10} cm. sec^{-1}).

(A) 400 (B) 228.3 (C) 3000 (D) 662.5

27. Drugs used for reducing fevers are called

(A) Anti-inflammatory (B) Analgesics

(C) Antipyretics (D) Antibiotics

28. Which of the following is a correct set for SiO_2 ?

(A) Linear, acidic (B) Linear, basic

(C) Tetrahedral, acidic (D) Angular, basic

29. What are the products formed when Li_2CO_3 undergoes decomposition ?

(A) $Li_2O + CO_2$ (B) $LiO_2 + CO$

(C) $Li_2O_2 + CO$ (D) $Li_2O + CO$

30. Aqueous NaCl solution is electrolyzed using platinum electrodes. What is the product formed at the cathode ?

(A) Na (B) H_2 (C) O2 (D) Cl_2

31. In Williamson synthesis, which one of the following reacts with sodium ethoxide to yield a symmetrical ether ?

(A) Ethyl iodide (B) Methyl iodide

(C) Ethanol (D) Acetaldehyde

32. An element M reacts with chlorine to form a compound X. The bond angle in X is 120° What is M ?

(A) Mg (B) N (C) Be (D) B

33. What is the number of sigma (σ) and pi (π) bonds present in sulphuric acid molecule ?

(A) $6\sigma, 2\pi$ (B) $6\sigma, 0\pi$ (C) $2\sigma, 4\pi$ (D) $2\sigma, 2\pi$

34. Which one of the following statements is not correct ?

(A) Cl^- is a Lewis acid

(B) The pH of 10^{-8} M HCl solution is less than 7

(C) The ionic product of water at 25°C is 10^{-14} $mol^2\ lit^{-2}$.

(D) Bronsted - Lowry theory could not explain the acidic character of $AlCl_3$.

35. The reagent used in the detection of fluoride in water is

(A) Zirconium - alizarin S (B) Phenolphthalein

(C) Bayer's reagent (D) Soda lime

36. Elements X and Y belong to the same group. Which of the following sets of atomic numbers represents X and Y ?

(A) 15, 30 (B) 20, 31 (C) 19,55 (D) 33,16

37. Equal volumes of 0.1 M $AgNO_3$ and 0.2 M NaCl solutions are mixed. The concentration of nitrate ions in the resultant mixture will be

(A) 0.05 M (B) 0.5M (C) 0.1M (D) 0.2 M

38. The reacting ion in the nitration of benzene is

(A) NO_2^- (B) NO_2^+ (C) NO_3^- (D) O_2^-

39. Heating chloroethane with ....yields a product containing an isocyanide group.

(A) AgCN (B) KCN (C) HCN (D) HNO_3

40. What are the products formed in the reaction of xenon hexafluoride with silicon dioxide ?
(A) $XeSiO_4 + HF$ (B) $XeF_2 + SiF_4$
(C) $XeOF_4 + SiF_4$ (D) $XeO_3 + SiF_2$

41. Which of the following is an example of exothermic reaction ?
(A) $C(graphid) + H_2O(g) \rightarrow CO(g) + H_2(g)\ \Delta H = +131.4\ kJ$
(B) $C(graphite) + 2S(s) + 91.9kJ \rightarrow CS_2(l)$
(C) $H_2(g) + Cl_2(g) \rightarrow 2HCl(g);\ \Delta H = -184.6\ kJ$
(D) $N_2(g) + O_2(g) \rightarrow 2NO(g);\ \Delta H = +180.8\ kJ$

42. When acetylene gas is passed through ....solution, a white precipitate is formed.
(A) Aqueous silver nitrate
(B) Ammonical cuprous chloride
(C) Ammonical silver nitrate
(D) Aqueous potassium permanganate

43. Which of the following is the correct set with reference to molecular formula, hybridization of central atom and shape of the molecule ?
(A) CO_2 ,sp^2, bent (B) H_2O,sp^2, bent
(C) $BeCl_2$, sp, linear (D) H_2O, sp^3, linear

44. Which of the following compounds undergoes hydrolysis during distillation to yield hydrogen peroxide ?
(A) $H_2S_2O_8$ (B) $H_2S_2O_6$ (C) HNO_3 (D) $H_4P_2O_7$

45. What is the density in g. lit^{-1} of CO_2 at 400 K and exerting a pressure of 0.0821 atm ?(R = 0.0821 lit. atm. $mol.^{-1}K^{-1}$)
(A) 2.5 (B) 44 (C) 0.01 (D) 0.11

46. Which of the following compounds converts methyl magnesium iodide to methane in one stop ?
(A) $C_2H_5OC_2H_5$ (B) C_2H_5Cl (C) C_2H_4 (D) C_2H_5OH

47. What is the oxidation state of Fe in the product formed when acidified potassium ferrocyanide is treated with H_2O_2 ?
(A) +2 (B) +6 (C) +1 (D) +3

48. If the r.m.s. velocity of a gas at 100 K is 10^4 cm. sec^{-1}, what is the temperature in°C at which the r.m.s. velocity will be 3 x 10^4 cm sec^{-1} ?
(A) 327 (B) 1217 (C) 900 (D) 627

49. What are X and Y respectively in the following reaction?

$$X \xrightarrow{PBr_3} C_2H_5Br \xrightarrow{AgOH / A} Y$$

(A) CH_3OH, C_2H_6
(B) C_2H_5OH, C_2H_5AgBr
(C) CH_3COOH, CH_3CH_2OH
(D) C_2H_5OH, C_2H_5OH

50. which one of the following is the correct order for the decomposition temperatures of $CaCO_3$ (X), $MgCO_3$ (Y) and $BaCO_3$ (Z) ?
(A) Z > X > Y (B) X > Y > Z
(C) Y > Z > X (D) X > Z > Y

EAMCET - 2003

1. Compound x is anhydride of sulphuric acid. The number of σ bonds and the number of π bonds respectively are
(A) 3,3 (B) 4,2 (C) 2,4 (D) 4,3

2. A mixture of sodium oxide and calcium oxide are disolved in water and saturated with excess carbon dioxide gas. The resulting solution is -----. It contains ----
(A) Basic, NaOH + $Ca(OH)_2$
(B) Neutral, Na_2CO_3 + $CaCO_3$
(C) Basic, Na_2CO_3 and $CaCO_3$
(D) Acidic NaOH and $CaCO_3$

3. Ferrous ion changes to X ion on reaction with acidified hydrogen peroxide. The number of d-electrons present in X and the magnetic moment in BM are respectively
(A) 6 and 6.93 (B) 5 and 5.92 (C) 5 and 4.9 (D) 4 and 5.92

4. The oxidation state of Xe in XeO_3 and the bond angle in it respectively are
(A) +6, 109° (B) +8, 103° (C) +6, 103° (D) +8, 120°

5. Which of the following is correct about the stability of the given ions ?
(A) $Pb^{2+} > Pb^{4+}$ (B) $Pb^{4+} > Pb^{2+}$
(C) $Si^{2+} > Si^{4+}$ (D) $Sn^{4+} > Sn^{2+}$

6. The product obtained at anode when 50% H_2SO_4 aqueous solution is electrolysed using platinum electrodes is
(A) H_2SO_3 (B) $H_2S_2O_8$ (C) O_2 (D) H_2

7. If the electron of a hydrogen atom is present in the first orbit, the total energy of the electron is
(A) $\frac{-e^2}{r}$ (B) $\frac{-e^2}{r^2}$ (C) $\frac{-e^2}{2r}$ (D) $\frac{-e^2}{2r^2}$

8. If the wave length of an electromagnetic radiation is 2000 Å. What is its energy in ergs ?
(A) 9.94 x 10^{-12} (B) 9.94 x 10^{-19}
(C) 4.97 x 10^{-12} (D) 4.97 x 10^{-19}

9. Which of the following equations represents Freundlich adsorption isotherm.
(A) $\frac{x}{m} = kp$ (B) $\frac{x}{m} = kp^n$
(C) $\log \frac{x}{m} = kp^n$ (D) $\log \frac{x}{m} = kn \log p$

10. Sulphur trioxide dissolves in heavy water to form a compound X. The hybridisation state of sulphur in X is
(A) sp^2 (B) sp^3 (C) sp (D) dsp^2

11. What are the products formed when an aqueous solution of magnesium bicarbonate is boiled ?
(A) $MgCO_3$, H_2O, CO_2 (B) $Mg(HCO_3)_2$ H_2O
(C) $Mg(OH)_2$, H_2O (D) $MgCO_3$, H_2O

12. Two oxides of nitrogen NO and NO_2 react together at 253k and form a compound of nitrogen X. X reacts with water to yield another compound of nitrogen Y. The shape of the anion of Y molecule is
(A) Tetrahedral (B) Triangular planar
(C) Square planar (D) Pyramidal

13. Consider the following reactions,

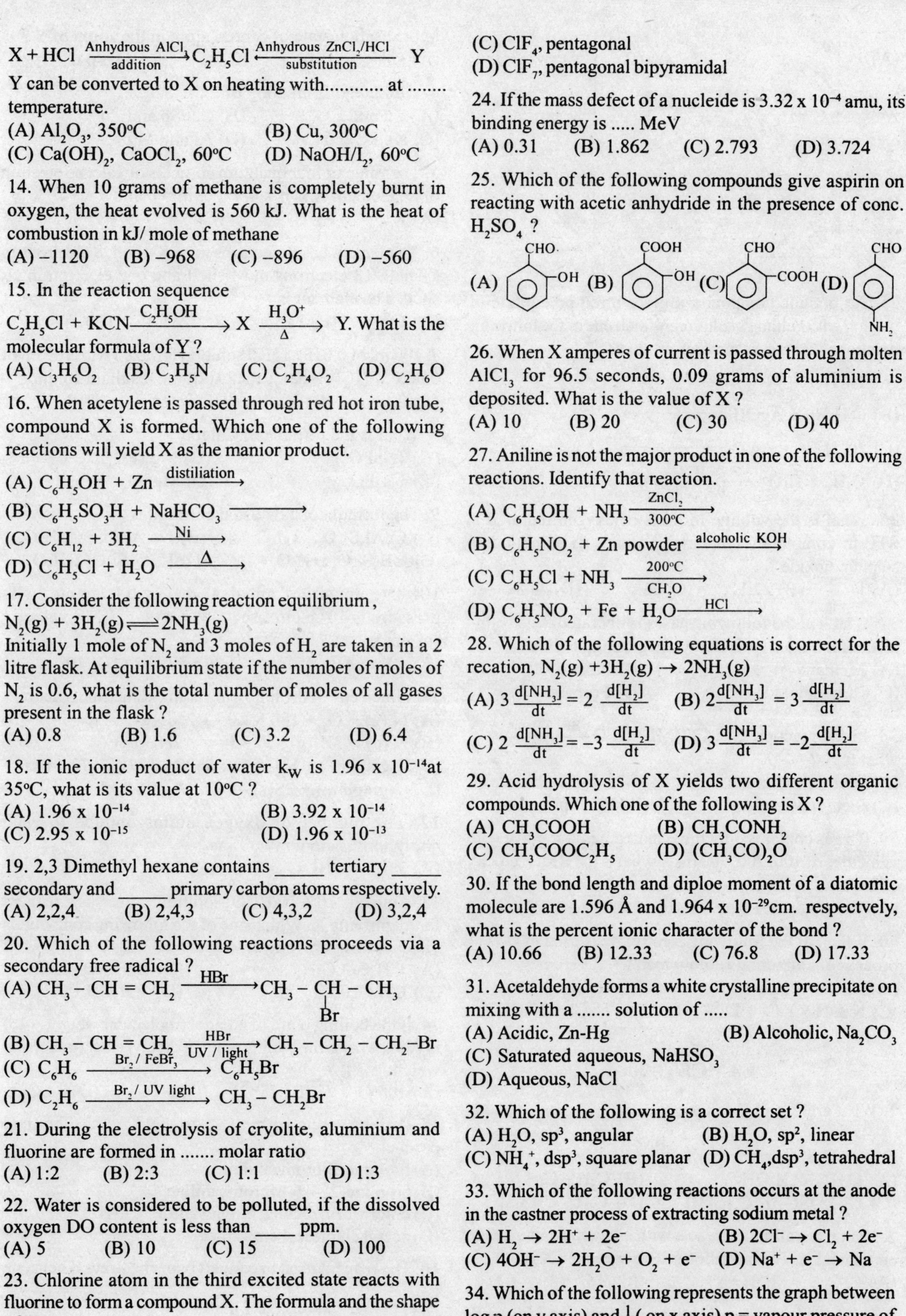

$X + HCl \xrightarrow[\text{addition}]{\text{Anhydrous } AlCl_3} C_2H_5Cl \xleftarrow[\text{substitution}]{\text{Anhydrous } ZnCl_2/HCl} Y$

Y can be converted to X on heating with............ at .... temperature.

(A) Al_2O_3, 350°C (B) Cu, 300°C
(C) $Ca(OH)_2$, $CaOCl_2$, 60°C (D) $NaOH/I_2$, 60°C

14. When 10 grams of methane is completely burnt in oxygen, the heat evolved is 560 kJ. What is the heat of combustion in kJ/ mole of methane

(A) –1120 (B) –968 (C) –896 (D) –560

15. In the reaction sequence

$C_2H_5Cl + KCN \xrightarrow{C_2H_5OH} X \xrightarrow[\Delta]{H_3O^+} Y$. What is the molecular formula of Y ?

(A) $C_3H_6O_2$ (B) C_3H_5N (C) $C_2H_4O_2$ (D) C_2H_6O

16. When acetylene is passed through red hot iron tube, compound X is formed. Which one of the following reactions will yield X as the manior product.

(A) $C_6H_5OH + Zn \xrightarrow{\text{distiliation}}$
(B) $C_6H_5SO_3H + NaHCO_3 \longrightarrow$
(C) $C_6H_{12} + 3H_2 \xrightarrow{Ni}$
(D) $C_6H_5Cl + H_2O \xrightarrow{\Delta}$

17. Consider the following reaction equilibrium,

$N_2(g) + 3H_2(g) \rightleftharpoons 2NH_3(g)$

Initially 1 mole of N_2 and 3 moles of H_2 are taken in a 2 litre flask. At equilibrium state if the number of moles of N_2 is 0.6, what is the total number of moles of all gases present in the flask ?

(A) 0.8 (B) 1.6 (C) 3.2 (D) 6.4

18. If the ionic product of water k_W is 1.96×10^{-14} at 35°C, what is its value at 10°C ?

(A) 1.96×10^{-14} (B) 3.92×10^{-14}
(C) 2.95×10^{-15} (D) 1.96×10^{-13}

19. 2,3 Dimethyl hexane contains _____ tertiary _____ secondary and _____ primary carbon atoms respectively.

(A) 2,2,4 (B) 2,4,3 (C) 4,3,2 (D) 3,2,4

20. Which of the following reactions proceeds via a secondary free radical ?

(A) $CH_3 - CH = CH_2 \xrightarrow{HBr} CH_3 - \underset{Br}{\underset{|}{C}H} - CH_3$
(B) $CH_3 - CH = CH_2 \xrightarrow[\text{UV / light}]{HBr} CH_3 - CH_2 - CH_2 - Br$
(C) $C_6H_6 \xrightarrow{Br_2 / FeBr_3} C_6H_5Br$
(D) $C_2H_6 \xrightarrow{Br_2 / \text{UV light}} CH_3 - CH_2Br$

21. During the electrolysis of cryolite, aluminium and fluorine are formed in .... molar ratio

(A) 1:2 (B) 2:3 (C) 1:1 (D) 1:3

22. Water is considered to be polluted, if the dissolved oxygen DO content is less than ____ ppm.

(A) 5 (B) 10 (C) 15 (D) 100

23. Chlorine atom in the third excited state reacts with fluorine to form a compound X. The formula and the shape of X are

(A) ClF_5, pentagonal (B) ClF_4, Tetrahedral
(C) ClF_4, pentagonal
(D) ClF_7, pentagonal bipyramidal

24. If the mass defect of a nucleide is 3.32×10^{-4} amu, its binding energy is ..... MeV

(A) 0.31 (B) 1.862 (C) 2.793 (D) 3.724

25. Which of the following compounds give aspirin on reacting with acetic anhydride in the presence of conc. H_2SO_4 ?

(A) CHO, OH (B) COOH, OH (C) CHO, COOH (D) CHO, NH_2

26. When X amperes of current is passed through molten $AlCl_3$ for 96.5 seconds, 0.09 grams of aluminium is deposited. What is the value of X ?

(A) 10 (B) 20 (C) 30 (D) 40

27. Aniline is not the major product in one of the following reactions. Identify that reaction.

(A) $C_6H_5OH + NH_3 \xrightarrow[300°C]{ZnCl_2}$
(B) $C_6H_5NO_2 + \text{Zn powder} \xrightarrow{\text{alcoholic KOH}}$
(C) $C_6H_5Cl + NH_3 \xrightarrow[Cu_2O]{200°C}$
(D) $C_6H_5NO_2 + Fe + H_2O \xrightarrow{HCl}$

28. Which of the following equations is correct for the recation, $N_2(g) + 3H_2(g) \rightarrow 2NH_3(g)$

(A) $3\frac{d[NH_3]}{dt} = 2\frac{d[H_2]}{dt}$ (B) $2\frac{d[NH_3]}{dt} = 3\frac{d[H_2]}{dt}$
(C) $2\frac{d[NH_3]}{dt} = -3\frac{d[H_2]}{dt}$ (D) $3\frac{d[NH_3]}{dt} = -2\frac{d[H_2]}{dt}$

29. Acid hydrolysis of X yields two different organic compounds. Which one of the following is X ?

(A) CH_3COOH (B) CH_3CONH_2
(C) $CH_3COOC_2H_5$ (D) $(CH_3CO)_2O$

30. If the bond length and diploe moment of a diatomic molecule are 1.596 Å and 1.964×10^{-29}cm. respectvely, what is the percent ionic character of the bond ?

(A) 10.66 (B) 12.33 (C) 76.8 (D) 17.33

31. Acetaldehyde forms a white crystalline precipitate on mixing with a .... solution of .....

(A) Acidic, Zn-Hg (B) Alcoholic, Na_2CO_3
(C) Saturated aqueous, $NaHSO_3$
(D) Aqueous, NaCl

32. Which of the following is a correct set ?

(A) H_2O, sp^3, angular (B) H_2O, sp^2, linear
(C) NH_4^+, dsp^3, square planar (D) CH_4, dsp^3, tetrahedral

33. Which of the following reactions occurs at the anode in the castner process of extracting sodium metal ?

(A) $H_2 \rightarrow 2H^+ + 2e^-$ (B) $2Cl^- \rightarrow Cl_2 + 2e^-$
(C) $4OH^- \rightarrow 2H_2O + O_2 + e^-$ (D) $Na^+ + e^- \rightarrow Na$

34. Which of the following represents the graph between log p (on y axis) and $\frac{1}{T}$(on x axis) p = vapour pressure of a liquid T=absolute temperature.

(A) (B)

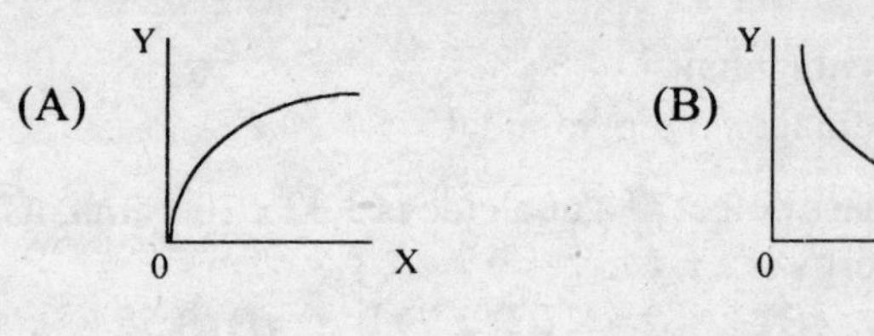

(C) (D)

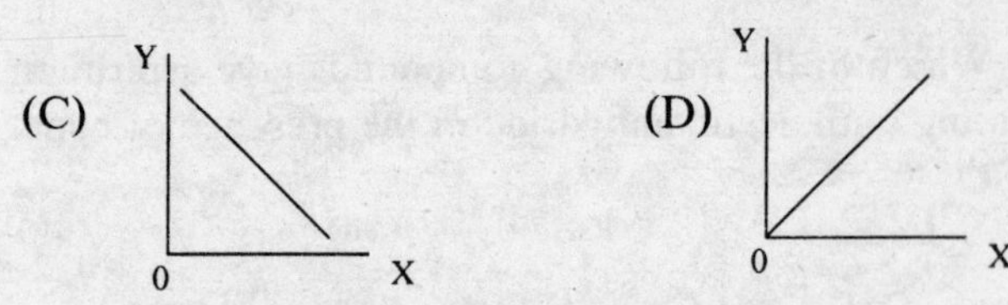

35. An organic compound x gives a red precipitate on heating with Fehling's solution. Which one of the following reactions yields x as a major product.

(A) $HCHO \xrightarrow[2.\ H_2O]{1.\ CH_3MgI}$

(B) $C_2H_5Br + AgOH \xrightarrow{\Delta}$

(C) $2C_2H_5Br + Ag_2O \xrightarrow{\Delta}$

(D) $C_2H_2 + H_2O \xrightarrow[1\%\ HgSO_4\ 60^{\circ}C]{40\%\ H_2SO_4}$

36. What is the volume in litres of oxygen required at STP to completely convert 1.5 moles of sulphur into sulphur dioxide ?

(A) 11.2 (B) 22.11 (C) 33.6 (D) 44.8

37. Which of the following gases is liberated when ethyl alcohol is heated with methyl magnesium iodide ?

(A) Methane (B) Ethane
(C) Carbondioxide (D) Propane

38. In the reaction $C_2H_5COC_2H_5 + CO \xrightarrow[150^{\circ}C]{BF_3,\ 50\ atm} X$ what is X ?

(A) Diethyl carbonate (B) Ethyl carbonate
(C) Diethyl peroxide (D) Ethyl propionate

39. If a gas contains only three molecules that move with velocities of 100, 200, 500 ms^{-1}, what is the RMS velocity of that gas in ms^{-1}.

(A) $100\sqrt{\frac{8}{3}}$ (B) $100\sqrt{30}$ (C) $100\sqrt{10}$ (D) $\frac{800}{3}$

40. Which of the following electrolyte is used in Downs process of extracting sodium metal ?

(A) $NaCl + CaCl_2 + kF$ (B) NaCl
(C) NaOH + KCl + kF (D) NaCl + NaOH

EAMCET - 2004

1. What are X and Y in the reaction,

$C_2H_4 + H_2SO_4 \xrightarrow{80^{\circ}C} X \xrightarrow{H_2O,\ \Delta} Y.$

(A) C_2H_6, C_2H_5OH (B) C_2H_4, C_2H_5SH
(C) $C_2H_5OSO_3H$, C_2H_5OH (D) C_2H_2, CH_3CHO

2. Xenon hexafluoride reacts with silica to from a xenon compound X. The oxidation state of xenon in X is.

(A) +2 (B) +4 (C) +6 (D) 0

3. Sulphur reacts with chlorine in 1:2 ratio and forms X. Hydrolysis of X gives a sulphur compound Y. What is the hybridisation state of central atom in the anion of Y ?

(A) sp^3 (B) sp (C) sp^2 (D) sp^3d

4. Gun metal is an alloy of ....

(A) Cu and Fe (B) Cu, Sn and Zn
(C) Ni, Fe and Cr (D) Al and Mg

5. The value of four quantum numbers of valence electron of an element X is n = 4, l = 0, m = 0 and s = $+\frac{1}{2}$. X is

(A) K (B) Ti (C) Na (D) Sc

6. An element has 2 electrons in its K-shell, 8 electrons in L-shell, 13 electrons in M-shell and one electron in N-shell. The element is

(A) Cr (B) Fe (C) V (D) Ti

7. 100 ml of 0.015 M HCl solution is mixed with 100 ml of 0.005 M HCl. What is the pH of the resultant solution ?

(A) 2.5 (B) 1.5 (C) 2 (D) 1

8. What is the formula of Calgon ?

(A) $(NaPO_3)_6$ (B) $Mg_3(PO_4)_2$
(C) Na_3PO_4 (D) $MgSO_4$

9. The formula of exhausted permutit is

(A) $CaAl_2Si_2O_8 . xH_2O$ (B) $Na_2Al_2Si_2O_8\ xH_2O$
(C) $CaB_2Si_2O_8\ xH_2O$ (D) $K_2Al_2Si_2O_8\ xH_2O$

10. The densityof an ideal gas is 0.03 $g.cm^{-3}$. Its pressure is $10^6\ g.cm^{-1}.sec^{-2}$. What is its RMS velocity (in $cm.sec^{-1}$) ?

(A) 10^3 (B) 8×10^4 (C) 10^8 (D) 10^4

11. Which one of the following is an exothermic reaction ?

(A) $N_2(g) + O_2 + 180.8\ kJ \rightarrow 2NO(g)$
(B) $N_2(g) + 3H_2(g) - 92\ kJ \rightarrow 2NH_3(g)$
(C) $C(graphite) + H_2O(g) \rightarrow CO(g) + H_2(g) - 131.4\ kJ$
(D) $C(graphite) + 2S(s) \rightarrow CS_2(l) - 91.9\ kJ$

12. The number of oxygen atoms bonded to each phosphorous atom in P_4O_{10} is

(A) 3 (B) 4 (C) 5 (D) 6

13. Aqueous 10% $NaHCO_3$ solution is used as a reagent for identifying X. Which one of the following compounds yields X on hydrolysis ?

(A) $CH_3CO_2C_2H_5$ (B) $C_2H_5–O–C_2H_5$
(C) CH_3CHO (D) CH_3CH_2OH

14. If the boiling point of ethanol (molecular weight = 46) is 78°C, what is the boiling point of diethyl ether (molecular weight = 74) ?

(A) 100°C (B) 78°C (C) 86°C (D) 34°C

15. Aniline reacts with .... to yield ....as the final product.

(A) bromine, 2-bromoaniline
(B) bromine, 2, 4, 6-tribromoaniline
(C) chloroform / KOH, ethylcyanide
(D) acetyl chloride, benzanilide

16. The reagent used to convert benzene to ethyl benzene is

(A) C_2H_5Cl, anhydrous $AlCl_3$
(B) C_2H_5Cl, aqueous $AlCl_3$

(C) C_2H_5OH, anhydrous $AlCl_3$ (D) C_2H_5Cl, $SOCl_2$

17. The product formed in the aldol condensation of acetaldehyde
(A) $CH_3CH_2CH(OH)CHO$
(B) $CH_3CH(OH)CH_2CHO$
(C) $CH_3CH(OH)COCH_3$ (D) $CH_3CH_2CH_2CHO$

18. Which one of the following is a lyophilic colloidal solution ?
(A) Smoke (B) Gold sol
(C) Starch aqueous solution (D) Cloud

19. What are X and Y respectively in the following reaction, $B_2H_6 + 2KOH + 2X \rightarrow 2Y + 6H_2$?
(A) K_2O, H_2O (B) H_2, H_3BO_3
(C) H_2O, KBO_2 (D) B_2O_3, H_3BO_3

20. Silica reacts with Mg to from a magnesium compound X. X reacts with dilute HCl and froms Y. Y is
(A) MgO (B) $MgCl_2$ (C) MgSiO (D) $SiCl_4$

21. One of the elements present in Carnalite shows flame colouration. The colour of the flame is
(A) Orange (B) Green (C) Yellow (D) Lilac

22. The electrolyte used in Castner's process of sodium extraction is
(A) Aqueous Na_2CO_3 (B) Aqueous NaOH
(C) NaCl + $CaCl_2$ (D) Fused anhydrous NaOH

23. Ammonical silver nitrate solution is called ....
(A) Fehling solution (B) Tollens reagent
(C) Benedict solution (D) Schiff's reagent

24. Which one of the following compounds decolourises cold alkaline potassium permanganate solution ?
(A) C_2H_6 (B) C_2H_5Cl (C) C_2H_4 (D) $C_2H_5OCH_3$

25. Wet ether is not used as a solvent in Wurtz reaction, because the water present in it ....
(A) Hydrolyses Rx to ROH (B) Reduces Rx to RH
(C) Destroy the Na metal (D) Reacts with R-R

26. $^{39}_{19}K$ and $^{40}_{20}Ca$ are
(A) Isotopes (B) Isobars
(C) Isotones (D) Isodiaphers

27. The first ionisation potentials is of four consecutive elements, present in the second period of the periodic table are 8.3, 11.3, 14.5 and 13.6 eV respectively. Which one of the following is the first ionisation potential (in eV) of nitrogen ?
(A) 13.6 (B) 11.2 (C) 8.3 (D) 14.5

28. In which of the following reactions, the concentration of reactant is equal to concentration of product at equilibrium (K = equilibrium constant)
(A) $A \rightleftharpoons B$; K = 0.01 (B) $R \rightleftharpoons P$; K=1
(C) $X \rightleftharpoons Y$; K = 10 (D) $L \rightleftharpoons J$; K = 0.025

29. 10 grams of $CaCO_3$ is completely decomposed to X and CaO. X is passed into an aqueous solution containing one mole of sodium carbonate. What is the number of moles of sodium bicarbonate formed ? (Mol. wts. : $CaCO_3$ = 100, Na_2CO_3 = 106; $NaHCO_3$ = 84)
(A) 0.2 (B) 0.1 (C) 0.01 (D) 10

30. 250 ml of a solution contains 6.3 grams of oxalic acid (mol.wt. =126). What is the volume (in litres) of water to be added to this solution to make it a 0.1 l solution ?
(A) 750 (B) 7.5 (C) 0.075 (D) 0.75

31. The standard reduction potentials of Ag, Cu, Co and Zn are 0.799, 0.337, -0.277 and -0.762 V respectively. Which of the following cells will have maximum cell e.m.f.?
(A) $Zn/Zn^{2+}(1M)//Cu^{2+}(1M)/Cu$
(B) $Zn/Zn^{2+}(1M)//Ag^{+}(1M)/Ag$
(C) $Cu/Cu^{2+}(1M)//Ag^{+}(1M)/Ag$
(D) $Zn/Zn^{2+}(1M)//Co^{2+}(1M)/Co$

32. Which of the following relation is correct for a first order reaction ? (k = rate constant; r = rate of reaction; c = conc. of reactant)
(A) $k = r \times c^2$ (B) $k = r \times c$
(B) $k = c / r$ (D) $k = r / c$

33. Which one of the following represents the reaction between fluorine and cold dilute NaOH ?
(A) $2F_2 + 4NaOH \rightarrow 4NaF + 2H_2O$
(B) $3F_2 + NaOH \rightarrow 5NaF + NaFO_3$
(C) $F_2 + 2NaOH \rightarrow NaF + NaOF + H_2O$
(D) $2F_2 + 2NaOH \rightarrow 2NaF + OF_2 + H_2O$

34. Which of the following is not correct ?
(A) One of the favourable conditions for the formation of cation is low ionisation potential.
(B) Coordination number of Ca in CsCl is 8
(C) Ionic bond is directional
(D) Ionic compounds have high melting and boiling points

35. Which of the following order is correct for the size of Fe^{3+}, Fe and Fe^{2+} ?
(A) $Fe < Fe^{2+} < Fe^{3+}$ (B) $Fe^{2+} < Fe^{3+} < Fe$
(C) $Fe < Fe^{3+} < Fe^{2+}$ (D) $Fe^{3+} < Fe^{2+} < Fe$

36. Match the following lists

List I	List II
(A) R-CHO	1. Aldehyde
(B) R-CO-R	2. Nitrile
(C) $R\text{-}CO_2H$	3. Ketone
(D) R-CN	4. Ester
	5. Carboxylic acid

The correct answer is

	(A)	(B)	(C)	(D)
(A)	1	3	5	2
(B)	3	1	5	2
(C)	1	2	3	4
(D)	4	1	5	3

37. Identify the correct sequence of increasing number of π-bonds in the structures of the following molecules
I. $H_2S_2O_5$ II. H_2SO_3 III. $H_2S_2O_6$
(A) I, II, III (B) II, III, I

(C) II, I, III (D) I, III, II

38. Study the following table

Compound (mol.wt.)	Weight of Compound (in g) taken
I. CO_2 (44)	4.4
II. NO_2 (46)	2.3
III. H_2O_2 (34)	6.8
IV. SO_2 (64)	1.6

Which two compounds have least weight of oxygen? (molecular weights of compunds are given in brackets)

(A) II and IV (B) I and III
(C) I and II (D) III and IV

39. Assertion (A) : 8 grams of mathane occupies 11.207 litres of volume at 273k and 1 atm pressure.

Reason (R) : One mole of any gas at STP occupies 22.414 litres of volume.

(A) A (B) B (C) C (D) D

40. What is the mass number of an atom with a nucleus of radius 5.6 Fermi ?

(A) 27 (B) 64 (C) 125 (D) 216

EAMCET - 2005 (Engineering)

1. In Fischer -Ringe's method of separation of noble gas mixture from air .... is used

(A) 90% $CaCO_3$ + 10% urea
(B) soda lime + potash solution
(C) coconut charcoal (C) 90% CaC_2 + 10 % $CaCl_2$

2.A complex compound of Co^{3+} with molecular formula $CoCl_x \cdot yNH_3$ gives a total of 3 ions when dissolved in water. How many Cl^- ions satisfy both primary and secondary valencies in this complex ?

(A) Zero (B) 4 (C) 3 (D) 1

3.The chemicals and the reaction conditions required for the preparation of ethane are

(A) CH_3CO_2Na, NaOH, CaO, Δ
(B) KOOC – CH = CH – COOK, electrolysis
(C) CH_3Cl, Na, H_2O (D) C_2H_5I, Zn–Cu, C_2H_5OH

4. Observe the following statements regarding purification of bauxite :

I. During Hall's process, silica is removed as Si (vapour)
II Bauxite ore contaminated with Fe_2O_3 is purified in Baeyer's process.
III. During Serpeck's process, AlN is formed.

The correct answer is :

(A) Only II and III are correct
(B) Only I and III are correct
(C) Only I and II are correct
(D) I, II and III are correct

5. Sodium is heated in air at 300°C to form X. X absorbs CO_2 and forms Na_2CO_3 and Y. Which of the following is Y?

(A) O_3 (B) H_2O_2 (C) O_2 (D) H_2

6. Identify D and B in the following reactions

$$A \xrightarrow{\text{aq. NaOH}} C_2H_5OH \xleftarrow{\text{AgOH}} B$$

(A) A = C_2H_5Cl, B = C_2H_5Cl
(B) A = C_2H_4 B = C_2H_5Cl
(C) A = C_2H_5Cl, B = C_2H_4
(D) A = C_2H_2, B = C_2H_6

7. Which one of the following reactions does not form gaseous product ?

(A) $Cl_2 + H_2O_2 \rightarrow$ (B) $PbS + H_2O_2 \rightarrow$
(C) Acidified $KMnO_4 + H_2O_2 \rightarrow$
(D) $PbO_2 + H_2O_2 \rightarrow$

8. Which of the following is an example for heterogeneous catalysis reaction ?

(A) Hydrolysis of liquid ester in the presence of aqueous mineral acid
(B) $2H_2O_2(l) \xrightarrow{Pt(s)} 2H_2O(l) + O_2(g)$
(C) Hydrolysis of aqueous sucrose solution in the pressence of aqueous mineral acid
(D) $2SO_2(g) + O_2(g) \xrightarrow{NO(g)} 2SO_3(g)$

9. In which of the following reactions, the product is an ether ?

(A) $C_2H_5Cl + C_2H_5ONa$
(B) $C_6H_6 + C_6H_5COCl$/anhydrous $AlCl_3$
(C) C_2H_5Cl + aq. KOH
(D) $C_6H_6 + CH_3COCl$/anhydrous $AlCl_3$

10. The atomic numbers of elements X, Y and Z are 19, 21 and 25 respectively. The number of electrons present in the "M" shells of these elements follow the order

(A) Y > Z > X (B) Z > Y > X
(C) X > Y > Z (D) Z > X > Y

11. Which of the following compounds is the reactant in Rosenmund's reduction ?

(A) CH_3COCl (B) CH_3CH_2Cl
(C) CH_3CHO (D) CH_3CO_2H

12. An electron is moving in Bohr's fourth orbit. Its de Broglie wavelength is λ. What is the circumference of the fourth orbit ?

(A) $4/\lambda$ (B) 4λ (C) 2λ (D) $2/\lambda$

13. The half lives of two radioactive nuclides A and B are 1 and 2 min. respectively. Equal weights of A and B are taken separately and allowed to disintegrate for 4 min. What will be the ratio of weight of A and B disintegrated ?

(A) 1 : 3 (B) 1 : 2 (C) 5 : 4 (D) 1 : 1

14. 3-Hydroxybutanal is formed when (X) reacts with Y in dilute (Z) solution. What are X, Y and Z ?

(A) CH_3CHO, CH_3CHO, NaOH
(B) $(CH_3)_2CO$, $(CH_3)_2CO$, HCl
(C) CH_3CHO, CH_3CHO NaCl
(D) CH_3CHO, $(CH_3)_2CO$, NaOH

15. Identify the correct order in which the covalent radius

of the follwong elements increases :

I Ti   II Ca   III Sc

(A) I, III, II   (B) II, I, III   (C) III, II, I   (D) I, II, III

16. Which of the following is a linear molecule ?

(A) CH_4 (B) SO_2 (C) H_2O (D) $BeCl_2$

17. Natalite is used as

(A) Preservative   (B) Insecticide

(C) Subatitute for petrol   (D) Anaesthetic

18. Which of the following is correct ?

(A) The correct order of ability to form ionic compounds among Na^+, Mg^{2+} and Al^{3+} is $Al^{3+} > Mg^{2+} < Na^+$

(B) According to VSEPR theory $SnCl_2$ is a linear molecule

(C) The rates of ionic reactions are very slow.

(D) The number of electrons present in the valence shell of S in SF_6 is 12

19. 'x' grams of calcim carbonate was completely burnt in air. The weight of the solid residue formed is 28g. What is the value of 'x' (in grams) ?

(A) 50   (B) 150   (C) 200   (D) 44

20. In the reaction,

$C_2H_5OH \xrightarrow[300°C]{Cu} X$ The molecular formula of X is :

(A) C_2H_6 (B) C_2H_4O (C) $C_4H_{10}O$ (D) C_4H_5O

21. A and B are ideal gases. The molecular weights of A and B are in the ratio of 1 : 4. The pressure of a gas mixture containing equal weights of A and B is P atm. What is the partial pressure (in atm) of B in the mixture ?

(A) $\frac{3P}{4}$ (B) $\frac{P}{2.5}$ (C) $\frac{P}{2}$ (D) $\frac{P}{5}$

22. The vapour pressure of water at 23°C is 19.8 mm, 0.1 mole of glucose is dissolved is 178.2 g of water. What is the vapour pressure (in mm) of the resultant solution ?

(A) 19.202   (B) 19.402   (C) 19.602   (D) 19.0

23. Which of the following is not correct ?

(A) The metal ion present in Vitamin B_{12} is Mg^{2+}

(B) Acetyl salicylic acid is known as aspirin

(C) The compound formed in the addition of oxygen to haemoglobin is called oxyhaemoglobin

(D) Chloropyhll is responsible for the synthesis of carbohydrates in plants

24. Assertion (A) : The pH of a buffer solution containing equal moles of acetic acid and sodium acetate is 4.8 (pK_a of acetic acid is 4.8)

Reson (R) : The ionic product of water at 25°C is 10^{-14} mol|. lit^{-2}.

The correct answer is

(A) (A) is not true but (R) is true

(B) (A) is true but (R) is not true

(C) Both (A) and (R) are true and (R) is not the correct explanation of (A)

(D) Both (A) and (R) are true and (R) is the correct explanation of (A)

25. What is the quantity of electricity (in coulombs) required to deposit all the silver from 250 ml of 1 M $AgNO_3$ solution ?

(A) 48250   (B) 4825.0   (C) 24125   (D) 2412.5

26. Which of the following is not an air pollutant ?

(A) CO   (B) NO   (C) N_2O (D) N_2

27. Which of the following is not correct ?

(A) Standard reduction potential of hydrogen electrode is zero volts.

(B) In the Nernst equation 'n' represents the number of electrons transferred in the electronde reaction

(C) The units of electrochemical equivalent are g. coulomb

(D) Aqueous solution of NaCl is an electrolyte

28. Observe the following reaction :

$2A + B \rightarrow C$

The rate of formation of C is 2.2 x 10^{-3} mol. lit^{-1}. What is the value of $-\frac{d[A]}{dt}$ (in mol. lit^{-1} mil^{-1}) ?

(A) 5.5 x 10^{-3} (B) 4.4 x 10^{-3}

(C) 1.1 x 10^{-3} (D) 2.2 x 10^{-3}

29. Which of the following compounds is soluble in benzene but almost insoluble in water ?

(A) $C_6H_5NO_2$ (B) CH_3CHO

(C) CH_3CO_2H (D) C_2H_5OH

30. At 550 K, the K_c for the following reaction is 10^4 mol^{-1} lit. $X(g) + Y(g) \rightleftharpoons Z(g)$

At equilibrium, it was observed that :

$$[X] = \frac{1}{2}[Y] = \frac{1}{2}[Z]$$

What is the value of [Z] (in mol lit^{-1}) at equilibrium ?

(A) 10^4 (B) 2 x 10^4 (C) 10^{-4} (D) 2 x 10^{-4}

31. Which of the following is not correct ?

(A) The heat of reaction at constant volume is denoted by ΔE

(B) The absolute value of enthalpy (H) can be determined experimentally

(C) Neutralisation process is always exothermic

(D) Dissolution of NH_4Cl in excess of water is an endothermic process

32. Which of the following is a pair of functional isomers?

(A) CH_3CO_2H, CH_3CHO

(B) $C_2H_5CO_2H$, $CH_3CO_2C_2H_5$

(C) $C_2H_5CO_2H$, $CH_3CO_2CH_3$

(D) CH_3COCH_3, CH_3CHO

33. The pH of a solutionof H_2O_2 is 6.0. Some chlorine gas is bubbled into this solution. Which of the following is correct ?

(A) Cl_2O is formed in the resultant solution

(B) The pH of resultant solution becomes less than 6.0 and oxygen gas is liberated

(C) Hydrogen gas is liberated from resultant solution

(D) The pH of resultant solution becomes 8.0

34. In which of the following reactions, MgO is not formed ?

(A) $Mg + B_2O_3$ ----→ (B) $Mg + NO$ ----→
(C) $Mg + dil\ HNO_3$ ----→ (D) $Mg + CO_2$ ----→

35. The compound prepared by a substitution reaction of benzene is :
(A) Hexabromo cyclohexane
(B) Cyclohexane (C) Glyoxal (D) Acetophenone

36. Which of the following is not correct ?
(A) The hybridisation of C in graphite is sp^2
(B) SiO_2 reacts with Na_2CO_3 and liberates CO
(C) The distance between the layers in graphite is 3.35×10^{-8}
(D) SiO_2 is used as acid flux

37. Which of the following is not correct ?
(A) Hydrolysis of NCl_3 gives NH_3 and HOCl
(B) A mixture of $Ca(H_2PO_4)$ and $CaSO_4.2H_2O$ is known as superphosphate of lime
(C) A mixture of $Ca(CN)_2$ and C is known as nitrolim
(D) Ammonia is used as refrigerant

38. Match the following lists :

List I	List II
(a) Benzene	1. Phosgene
(b) Ethylene	2. Silver mirror
(c) Acetaldehyde	3. Mustard gas
(d) Chloroform	4. (4n + 2) π electrons
	5. Carbylamine

The correct answer is :

	(a)	(b)	(c)	(d)
(A)	5	1	4	3
(B)	2	4	5	3
(C)	3	2	1	4
(D)	4	3	2	1

39. Which of the follwing is not correct ?
(A) Sodium thiosulphate reacts with iodine to form sodium sulphate
(B) Ozone is used to identify the presence of unsaturation in alkenes
(C) Sodium thiosulphate is soluble in water
(D) Iodine oxidizes sodium thiosulphate to sodium tetrathionate

40. Which of the following pairs of reactants does not form oxygen when they react with each other ?
(A) $CaOCl_2$, H_2SO_4
(B) Cl_2, NaOH solution (cold, dilute)
(C) Fe, H_2O (D) F_2, NaOH solution (hot, conc.)

EAMCET - 2005 (Medical)

1. Four grams of hydrocarbon (C_xH_y) on complete combustion gave 12 grams of CO_2. What is the empirical formula of the hydrocabon ? (C = 12: H = 1)
(A) C_3H_8 (B) CH (C) C_4H_9 (D) CH_3

2. Which of the following is obtained in a carbylamine reaction ?
(A) C_6H_5NC (B) C_6H_5CN (C) $COCl_2$ (D) $C_6H_5NH_2$

3. Which of the following molecules has zero dipolemoment ?
(A) H_2O (B) NH_3 (C) HCl (D) $BeCl_2$

4. Observe the following statements regarding isotones
I. ^{39}K and ^{40}Ca are isotones.
II. Nuclides having different atomic (Z) and mass numbers (A) but same number of neutrons (n) are called isotones.
III. ^{19}F and ^{23}Na are isotones.
(A) Only II and III are correct.
(B) Only I and III are correct.
(C) Only I and II are correct.
(D) I, II and III are correct.

5. $C_2H_2 + 2HCl + C_2H_4Cl_2$ is an example of .... reaction.
(A) Chlorination (B) Substitution
(C) Hydrogenation (D) Addition

6. Which of the following is not correct ?
(A) Nitric oxide in solid state exhibits diamagnetic property
(B) NH_3 is a weak reducing reagent compared to PH_3
(C) NH_3 is less stable than PH_3
(D) Hydrolysis of NCl_3 gives NH_3 and HOCl

7. S_2Cl_2 hydrolyses slowly to form HCl, SO_2 and X. Which of the following is X ?
(A) S (B) O_2 (C) H_2 (D) SO_3

8. SiO_2 is roasted with sodium carbonate. What is the gas liberated ?
(A) O_3 (B) CO_2 (C) O_2 (D) CO

9. Assertion (A) : The spin only magnetic moment of Sc^{3+} is 1.73 BM.
Reason :(R) The spin only magnetic moment (in BM) of an ion is equal to $\sqrt{n(n+2)}$.
Where n is the number of unpaired electrons present in the ion. The correct answer is :
(A) (A) is not true but (R) is true.
(B) (A) is true but (R) is not true.
(C) Both (A) and (R) true but (R) is not the correct explanation of (A).
(D) Both (A) and (R) true and (R) is the correct explanation of (A).

10. Which of the following is not correct ?
(A) Liquid helium is used as cryogenic liquid.
(B) Among noble gases, the occurance (percent by weight) of argon is highest in air.
(C) The hybridisation of Xe in XeF_4 is sp^3d^2.
(D) XeO_3, has four σ and four π bonds.

11. The compounds formed at anode in the eletrolysis of an aqueous solution of potassium acetate are :
(A) CH_4 and CO_2 (B) CH_4 and H_2
(C) C_2H_4 and CO_2 (D) C_2H_6 and CO_2

12. X reacts with aqueous NaOH solution to form Y and H_2. Aqueous solution of Y is heated to 323-333 K and on passing CO_2 into it, Na_2CO_3 and Z were formed. When Z is heated to 1200°C, Al_2O_3 is formed, X, Y and Z

respectively are:

(A) Al, $NaAlO_2$, $Al(OH)_3$ (B) Al, $Al(OH)_3$, $AlCl_3$
(C) Zn, Na_2ZnO_2, $Al(OH)_3$ (D) Al, $AlCl_3$, $NaAlO_2$

13. The value of .... is less for D_2O compared to that H_2O.
(A) dielectric constant at 20°C
(B) latent heat of vapourisation (J. g^{-1})
(C) boiling point (D) density (g. ml^{-1}) at 20°C

14. Which of the following is not correct regarding the electrolytic preparation of H_2O_2 ?
(A) sulphuric acid undergoes oxidation
(B) hydrogen is liberated at anode
(C) 50% H_2SO_4 is used (D) lead is used as cathode

15. At 600°C K_p for the following reaction is 1 atm.

$$X(g) \rightleftharpoons Y(g) + Z(g)$$

At equlibrium 50% of X(g) is dissociated. The total pressure of the equlibrium system is "P" atm. What is the partial pressure (in atm.) of X(g) at equilibrium ?
(A) 0.5 (B) 2 (C) 4 (D) 1

16. Identify X and Y in the following sequence of reaction

$$CH_3CHO \xrightarrow{HNO_3} X \xrightarrow[\Delta]{P_4O_{10}} Y$$

	X	Y
(A)	C_2H_5OH,	CH_2CO_2H
(B)	CH_3CO_2H	$CH_3CO_2CH_3$
(C)	CH_3CO_2H	$(CH_3CO)_2O$
(D)	C_2H_5OH	C_2H_4

17. Which of the following is correct ?
(A) Phenolpthalein is colourless in basic medium
(B) BF_3 is a Lewis base
(C) The conjugate base of H_2S is S^{2-}
(D) The pH of one litre solution containing 0.49 g of H_2SO_4 is 2.0

18. Match the following lists :

List I	List II
(a) Ethylene	1. Natalite
(b) Acetylene	2. Preservative
(c) Ethanol	3. Hawker's lamp
(d) Diethylether	4. Drug
	5. Polyethylene

answer is :

	(b)	(c)	(d)
(A)	5	1	4
(B)	5	3	2
(C)	5	1	2
(D)	3	2	1

19. Which of the following is correct ?
(A) Magnalium is an alloy of Mg and Zn
(B) Mg reacts with cold water and liberate hydrogen gas
(C) Sodium reduces CO_2 to carbon
(D) In the Castner's process of sodium extraction, NaCl is used as an electrolyte.

20. Hydrogen chloride and SO_2 are the side products in the reaction of ethanol with thionyl chloride. Which of the following is the main product in this reaction ?
(A) C_2H_5Cl (B) CH_3Cl
(C) C_2H_6 (D) $C_2H_5{-}O{-}C_2H_5$

21. The time taken for the completion of 90% of a first order reaction is "t" min. What is the time (in sec.) taken for the completion of 99% of the reaction ?
(A) 60 t (B) 120 t (C) t/30 (D) 2t

22. A compound X undergoes reduction with $LiAlH_4$ to yield X. When vapours of X are passed over freshly reduced copper at 300°C. X is formed. What is Y ?
(A) $CH_3{-}O{-}CH_3$ (B) CH_3CH_2OH
(C) CH_3CHO (D) CH_3COCH_3

23. The standard reduction potentials of Zn^{2+}/Zn and Cu^{2+}/Cu are –0.76 V and +0.34 V respectively. What is the cell e.m.f. (in V) of the following cell ? ($\frac{RT}{F}$=0.059)
Zn/Zn^{2+} (0.05 M) //Cu^{2+}(0.005 M)/Cu
(A) 1.041 (B) 1.1 (C) 1.0705 (D) 1.1295

24. Consider the following reaction :
$C_2H_5OX + C_2H_5I \xrightarrow{\Delta}$ X (Pleasant smelling liquid)
Which of the following reagents converts ethylbromide to X ?
(A) Dry silver powder (B) Ethyl chloride
(C) Dry silver oxide (D) Sodium

25. Which of the following is correct ?
(A) Hydrolysis of liquid ester in the presence of mineral acid is an example of heterogeneous catalysis reactions.
(B) Starch solution is a hydrosol
(C) Particle size of solute in true solutions is 10^{-3} m
(D) Catalyst unsergoes permanent chemical change

26. In an oxidation-reduction, MnO_4^- ion is converted to Mn^{2+}. What is the number of equivalents of $KMnO_4$ (mol. wt =156) present in 250 ml of 0.04 M $KMnO_4$ solution?
(A) 0.07 (B) 0.04 (C) 0.05 (D) 0.02

27. What is the ratio of kinetic energies of 3 g of hydrogen and 4 grams of oxygen at T (R) ?
(A) 24 : 1 (B) 1 : 6 (C) 6 : 1 (D) 12 : 1

28. In which part of atmosphere, Ozone layer is present?
(A) Thermosphere (B) Mososphere
(C) Troposphere (D) Stratosphere

29. Identify the correct order in which the ionic radius of the following ions increases?
(I) F^- (II) Na^+ (III) N^{3-}
(A) II, I, III (B) II, III, I (C) I, II, III (D) III, I, III

30. What is the lowest energy of the spectral line emitted by the hydrogen atom in the Lyman series ?
(h = Planck constant; c=Velocity of Light: R = Rydberg constant).
(A) $\frac{7hcR}{144}$ (B) $\frac{3hcR}{4}$ (C) $\frac{4hcR}{3}$ (D) $\frac{5hcR}{36}$

31. 1, 2-Dibromoethane reacts with alcoholic KOH to yield a product X. The hybridisation state of the carbons

present in X, respectively are :

(A) sp^3, sp^2 (B) sp^2, sp^2 (C) sp^2, sp^2 (D) sp, sp

32. Which of the following is used during the preparation of fluorine by Whytlaw Gray method ?

(A) NH_4F (B) molten KHF_2
(C) aqueous HF (D) aqueous KF

33. What are the reagent and reaction conditions used for converting ethylchloride to ethylnitrite (as the major product) ?

(A) $AgNO_2$, C_2H_5OH, H_2O, Δ
(B) KCN, H_2O, Δ (C) $NaNO_2$, HCl, 0°C
(D) KNO_2, C_2H_5OH, H_2O, Δ

34. Which of the following pairs liberate a gas when they react with each other ?

(A) Mg, CO_2 (B) Mg, B_2O_3
(C) Fused NaOH, C (D) SiO_2, Na

35. Which of the following reagents converts both acetaldehyde and acetone to alkanes ?

(A) Zn-Hg/conc. HCl (B) I_2/NaOH
(C) $LiAH_4$ (D) Ni/H_2

36. The heat of formation of CO(g) and CO_2(g) are $\Delta H = -110$ and $\Delta H = -393$ kJ. mol^{-1} respectively. What is the heat of reaction (ΔH) (in kJ mol^{-1}) for the following reaction ?

$$CO(g) + \frac{1}{2} O_2(g) \rightarrow CO_2(g)$$

(A) 504 (B) –283 (C) 142.5 (D) –504

37. What is the electrochemical equivalent (in g $coulomb^{-1}$) of silver ? (Ag = 108, F = Farady)

(A) 1/108F (B) F/108 (C) 108/F (D) 108 F

38. Which of the following is a favourable factor for cation formation ?

(A) Small atomic size (B) High electronegativity
(C) High electron affinity
(D) Low ionisation potential

39. Which of the following gives paracetamol on acetylation ?

(A) 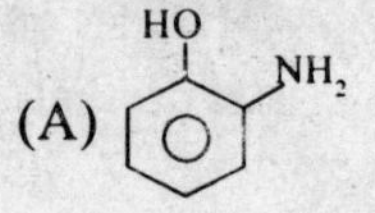

(B) 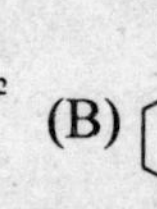

(C) 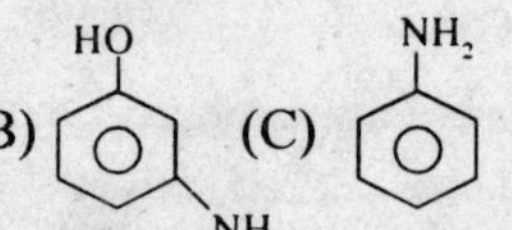

(D) 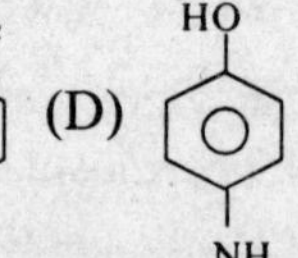

40. What is the wave length (in m) of a particle of mass 6.62×10^{-29} g moving with a velocity of 10^3 m. s^{-1} ?

(A) 6.62×10^{-4} (B) 6.62×10^{-3} (C) 10^{-5} (D) 10^5

EAMCET - 1990

1. (A)	11. (C)
2. (A)	12. (D)
3. (A)	13. (D)
4. (B)	14. (D)
5. (A)	15. (D)
6. (C)	16. (A)
7. (D)	17. (A)
8. (D)	18. (B)
9. (C)	19. (C)
10. (D)	20. (B)

Fill in the blanks

1. HCOOH.
2. CH_3COCl.
3. 0.1M.
4. zero .
5. 33 K. J.
6. $CuCl_2$.
7. 2.301 V .
8. 5 moles.
9. 3:7
10. $_6C^{14} + _1H^1$
11. 1/16
12. Aldehydes.
13. 44.
14. Copper
15. 1-Bromo 2-methyl proprane
16. Sc^{3+}.
17. IInd order.
18. Geometrical
19. S^{-2}.
20. 2
21. Paramethyl aniline.
22. 4.
23. Aluminium.
24. Nitrogen.
25. Negative.
26. V.
27. 12.
28. pH 6
29. 72
30. Sp^3d^2.

EAMCET - 1991

1. (B)	11. (B)
2. (A)	12. (B)
3. (A)	13. (A)
4. (C)	14. (A)
5. (D)	15. (D)
6. (D)	16. (A)
7. (B)	17. (B)
8. (D)	18. (A)
9. (C)	19. (B)
10. (D)	20. (B)

Fill in the blanks

1. same as
2. Chloroform
3. Annealing.
4. $C_2O_4^{2-}$.
5. Fluorine.
6. H_2SO_4.
7. Ethanol.
8. 70.
9. m = 0.
10. CS_2.
11. Hydrogen bond
12. Ethylene
13. 7Fm
14. 3-methyl heptane.
15. 5
16. CaO.
17. $_1^3T$
18. 0.00693 sec^{-1}
19. Hydrolysis.
20. 6.3 grams.
21. same
22. 3.
23. Aldehyde.
24. Eight.
25. Isotones.
26. Balmer
27. 0.5 moles.
28. Twice
29. SO_4^{2-}.
30. Lithium.

EAMCET - 1992

1. (D)	11. (A)
2. (D)	12. (C)
3. (D)	13. (D)
4. (A)	14. (A)
5. (B)	15. (B)
6. (A)	16. (D)
7. (A)	17. (A)
8. (A)	18. (A)
9. (C)	19. (B)
10. (A)	20. (D)

Fill up the blanks

1. I.P.
2. Fusion
3. sp^3
4. $C_6H_5 - NH_2$. Aniline
5. C_2H_4. Ethylene
6. Acetaldehyde
7. $_{56}Ba^{143}$
8. Sea water.
9. $CaSO_4.1/2H_2O$
10. $1s^22s^22p^63s^23p^64s^1$
11. Fluorine
12. Isotones
13. 12
14. +7.
15. spontaneous.
16. buffer solution.
17. phosgene.
18. N_2O.
19. Tollen's reagent.
20. N/8 molecules
21. 10°C
22. 1.67
23. 2methyl 1-butene or 2 methyl but-1-ene
24. Oxalic acid.
25. Greater
26. Lesser
27. even even
28. 55/63
29. 63
30. Octahedral .

EAMCET - 1993

1. (A)	11. (A)
2. (A)	12. (C)
3. (A)	13. (A)
4. (A)	14. (A)
5. (B)	15. (C)
6. (A)	16. (B)
7. (D)	17. (B)
8. (D)	18. (B)
9. (B)	19. (C)
10. (B)	20. (B)

Fill in the blanks

1. HF
2. Chloride ($CrCl_3$).
3. I_3^-
4. diethyl peroxide.
5. Poisonous
6. Alkali metal.
7. Lowers
8. Exothermic
9. Ethyl alcohal.
10. 3-methyl 2-butanol.
11. Direction
12. 1.0×10^{-14} $moles^2/lit.^2$.
13. +5.
14. Lowest
15. 2 units.
16. buffer
17. intermolecular hy
18. $CaSO_4$, 1/2 H_2O.
19. covalent.
20. $CaOCl_2$ (Calcium-oxi-dichloride)
21. CO
22. HBr.
23. Argon .
24. Neutral
25. Flourine.
26. $LiAlH_4$
27. d-orbitals
28. Unfilled (or) Vacant
29. f orbitals .
30. Bi-carbonate

EAMCET - 1994

1. (A)	11. (C)	21. (C)	31. (D)	41. (D)
2. (A)	12. (D)	22. (B)	32. (A)	42. (D)
3. (A)	13. (D)	23. (C)	33. (D)	43. (D)
4. (B)	14. (C)	24. (B)	34. (D)	44. (C)
5. (B)	15. (D)	25. (B)	35. (B)	45. (D)
6. (B)	16. (A)	26. (A)	36. (D)	46. (C)
7. (B)	17. (C)	27. (D)	37. (D)	47. (C)
8. (A)	18. (D)	28. (C)	38. (D)	48. (D)
9. (B)	19. (C)	29. (D)	39. (B)	49. (B)
10. (A)	20. (B)	30. (C)	40. (D)	50. (D)

EAMCET - 1995

1. (A)	11. (A)	21. (C)	31. (B)	41. (D)
2. (D)	12. (C)	22. (B)	32. (B)	42. (C)
3. (B)	13. (D)	23. (D)	33. (D)	43. (C)
4. (A)	14. (C)	24. (A)	34. (D)	44. (D)
5. (D)	15. (D)	25. (B)	35. (B)	45. (A)
6. (B)	16. (B)	26. (B)	36. (C)	46. (B)
7. (A)	17. (D)	27. (C)	37. (A)	47. (B)
8. (A)	18. (A)	28. (A)	38. (B)	48. (D)
9. (B)	19. (A)	29. (B)	39. (B)	49. (D)
10. (B)	20. (C)	30. (A)	40. (A)	50. (B)

EAMCET - 1996

1. (B) 11. (C) 21. (C) 31. (B) 41. (C)
2. (B) 12. (A) 22. (B) 32. (B) 42. (D)
3. (D) 13. (D) 23. (C) 33. (C) 43. (A)
4. (A) 14. (C) 24. (C) 34. (D) 44. (C)
5. (B) 15. (A) 25. (A) 35. (B) 45. (D)
6. (A) 16. (C) 26. (D) 36. (B) 46. (C)
7. (C) 17. (A) 27. (D) 37. (C) 47. (D)
8. (A) 18. (B) 28. (C) 38. (B) 48. (D)
9. (B) 19. (B) 29. (B) 39. (C) 49. (B)
10. (C) 20. (B) 30. (C) 40. (C) 50. (D)

EAMCET - 1997

1. (A) 11. (B) 21. (A) 31. (A)
2. (B) 12. (A) 22. (C) 32. (B)
3. (D) 13. (D) 23. (D) 33. (C)
4. (D) 14. (B) 24. (D) 34. (A)
5. (C) 15. (B) 25. (D) 35. (D)
6. (B) 16. (A) 26. (C)
7. (B) 17. (B) 27. (D)
8. (C) 18. (D) 28. (D)
9. (B) 19. (A) 29. (B)
10. (A) 20. (D) 30. (A)

EAMCET - 1998 (Engineering)

1. (A) 11. (B) 21. (B) 31. (C) 41. (B)
2. (D) 12. (C) 22. (C) 32. (C) 42. (C)
3. (B) 13. (D) 23. (C) 33. (C) 43. (D)
4. (D) 14. (A) 24. (A) 34. (A) 44. (C)
5. (A) 15. (C) 25. (D) 35. (D) 45. (B)
6. (A) 16. (C) 26. (B) 36. (B) 46. (D)
7. (C) 17. (D) 27. (A) 37. (A) 47. (D)
8. (A) 18. (B) 28. (D) 38. (B) 48. (D)
9. (C) 19. (C) 29. (C) 39. (B) 49. (B)
10. (D) 20. (A) 30. (A) 40. (B) 50. (C)

EAMCET - 1998 (Medical)

1. (C) 11. (C) 21. (A) 31. (B) 41. (C)
2. (C) 12. (B) 22. (A) 32. (C) 42. (C)
3. (D) 13. (A) 23. (B) 33. (A) 43. (A)
4. (D) 14. (C) 24. (A) 34. (D) 44. (A)
5. (B) 15. (D) 25. (D) 35. (C) 45. (C)
6. (B) 16. (B) 26. (A) 36. (C) 46. (A)
7. (A) 17. (C) 27. (D) 37. (C) 47. (D)
8. (D) 18. (A) 28. (C) 38. (C) 48. (B)
9. (B) 19. (D) 29. (C) 39. (A) 49. (B)
10. (A) 20. (A) 30. (D) 40. (C) 50. (A)

EAMCET - 1999 (Engineering)

1. (B) 11. (C) 21. (D) 31. (B) 41. (C)
2. (B) 12. (D) 22. (C) 32. (C) 42. (B)
3. (C) 13. (A) 23. (B) 33. (C) 43. (D)
4. (D) 14. (B) 24. (B) 34. (B) 44. (A)
5. (D) 15. (D) 25. (D) 35. (C) 45. (C)
6. (A) 16. (B) 26. (B) 36. (C) 46. (D)
7. (A) 17. (A) 27. (B) 37. (C) 47. (C)
8. (B) 18. (A) 28. (D) 38. (C) 48. (A)
9. (B) 19. (B) 29. (A) 39. (C) 49. (B)
10. (D) 20. (A) 30. (A) 40. (C) 50. (D)

EAMCET - 1999 (Medical)

1. (B) 11. (C) 21. (B) 31. (D) 41. (B)
2. (D) 12. (D) 22. (A) 32. (B) 42. (D)
3. (B) 13. (C) 23. (B) 33. (A) 43. (B)
4. (D) 14. (A) 24. (D) 34. (B) 44. (B)
5. (A) 15. (C) 25. (B) 35. (B) 45. (B)
6. (B) 16. (D) 26. (B) 36. (D) 46. (C)
7. (C) 17. (C) 27. (B) 37. (B) 47. (B)
8. (B) 18. (D) 28. (A) 38. (D) 48. (A)
9. (A) 19. (B) 29. (B) 39. (D) 49. (A)
10. (A) 20. (A) 30. (C) 40. (D) 50. (C)

EAMCET - 2000 (Engineering)

1. (A) 11. (B) 21. (A) 31. (D) 41. (B)
2. (D) 12. (D) 22. (B) 32. (D) 42. (B)
3. (C) 13. (D) 23. (C) 33. (B) 43. (B)
4. (C) 14. (B) 24. (B) 34. (B) 44. (B)
5. (B) 15. (C) 25. (A) 35. (B) 45. (A)
6. (C) 16. (D) 26. (B) 36. (A) 46. (A)
7. (B) 17. (C) 27. (D) 37. (B) 47. (B)
8. (B) 18. (D) 28. (A) 38. (B) 48. (C)
9. (C) 19. (B) 29. (C) 39. (A) 49. (B)
10. (B) 20. (A) 30. (A) 40. (B) 50. (D)

EAMCET - 2000 (Medical)

1. (A) 11. (B) 21. (A) 31. (D) 41. (B)
2. (D) 12. (B) 22. (A) 32. (C) 42. (C)
3. (B) 13. (B) 23. (D) 33. (B) 43. (A)
4. (D) 14. (A) 24. (B) 34. (A) 44. (B)
5. (B) 15. (B) 25. (C) 35. (D) 45. (B)
6. (C) 16. (B) 26. (C) 36. (A) 46. (A)
7. (C) 17. (D) 27. (B) 37. (C) 47. (B)
8. (D) 18. (B) 28. (D) 38. (D) 48. (B)
9. (A) 19. (A) 29. (A) 39. (A) 49. (C)
10. (A) 20. (A) 30. (C) 40. (D) 50. (C)

EAMCET - 2001 (Engineering)

1. (D) 11. (C) 21. (A) 31. (D) 41. (B)
2. (B) 12. (B) 22. (C) 32. (D) 42. (A)
3. (A) 13. (B) 23. (A) 33. (C) 43. (A)
4. (B) 14. (C) 24. (D) 34. (A) 44. (B)
5. (D) 15. (C) 25. (B) 35. (D) 45. (A)
6. (B) 16. (C) 26. (A) 36. (D) 46. (A)
7. (B) 17. (C) 27. (D) 37. (C) 47. (C)
8. (B) 18. (B) 28. (D) 38. (C) 48. (A)
9. (B) 19. (A) 29. (D) 39. (C) 49. (D)
10. (B) 20. (A) 30. (D) 40. (D) 50. (C)

EAMCET - 2001 (Medical)

1. (D) 11. (A) 21. (C) 31. (A) 41. (C)
2. (A) 12. (D) 22. (C) 32. (D) 42. (C)
3. (B) 13. (D) 23. (D) 33. (A) 43. (C)
4. (A) 14. (A) 24. (A) 34. (A) 44. (D)
5. (B) 15. (B) 25. (A) 35. (A) 45. (B)
6. (B) 16. (D) 26. (C) 36. (A) 46. (C)
7. (A) 17. (D) 27. (C) 37. (B) 47. (C)
8. (B) 18. (D) 28. (B) 38. (B) 48. (C)
9. (B) 19. (D) 29. (B) 39. (A) 49. (C)
10. (C) 20. (B) 30. (A) 40. (A) 50. (C)

EAMCET - 2002 (Engineering)

1. (B) 11. (D) 21. (C) 31. (C) 41. (A)
2. (D) 12. (B) 22. (B) 32. (D) 42. (A)
3. (C) 13. (B) 23. (A) 33. (C) 43. (D)
4. (D) 14. (D) 24. (D) 34. (D) 44. (A)
5. (B) 15. (C) 25. (B) 35. (C) 45. (C)
6. (C) 16. (A) 26. (B) 36. (A) 46. (C)
7. (C) 17. (A) 27. (C) 37. (C) 47. (D)
8. (B) 18. (B) 28. (B) 38. (C) 48. (D)
9. (D) 19. (A) 29. (C) 39. (A) 49. (B)
10. (C) 20. (C) 30. (A) 40. (A) 50. (D)

EAMCET - 2002 (Medical)

1. (A) 11. (A) 21. (D) 31. (A) 41. (C)
2. (B) 12. (A) 22. (D) 32. (D) 42. (C)
3. (B) 13. (D) 23. (B) 33. (A) 43. (C)
4. (A) 14. (C) 24. (C) 34. (A) 44. (A)
5. (B) 15. (A) 25. (D) 35. (A) 45. (D)
6. (C) 16. (C) 26. (D) 36. (C) 46. (D)
7. (D) 17. (C) 27. (C) 37. (A) 47. (D)
8. (C) 18. (B) 28. (C) 38. (B) 48. (D)
9. (D) 19. (B) 29. (A) 39. (A) 49. (D)
10. (A) 20. (B) 30. (B) 40. (C) 50. (A)

EAMCET - 2003

1. (A) 11. (A) 21. (B) 31. (C)
2. (C) 12. (B) 22. (A) 32. (A)
3. (B) 13. (A) 23. (D) 33. (C)
4. (C) 14. (C) 24. (A) 34. (C)
5. (A) 15. (A) 25. (B) 35. (D)
6. (B) 16. (A) 26. (A) 36. (C)
7. (C) 17. (C) 27. (B) 37. (A)
8. (A) 18. (C) 28. (C) 38. (D)
9. (B) 19. (A) 29. (C) 39. (C)
10. (B) 20. (B) 30. (C) 40. (A)

EAMCET - 2004

1. (C) 11. (B) 21. (D) 31. (B)
2. (C) 12. (B) 22. (D) 32. (D)
3. (A) 13. (A) 23. (B) 33. (D)
4. (B) 14. (D) 24. (C) 34. (C)
5. (A) 15. (B) 25. (C) 35. (D)
6. (A) 16. (A) 26. (C) 36. (A)
7. (C) 17. (B) 27. (D) 37. (B)
8. (A) 18. (C) 28. (B) 38. (A)
9. (A) 19. (C) 29. (A) 39. (A)
10. (D) 20. (B) 30. (D) 40. (B)

EAMCET - 2005 (Engineering)

1. (A) 11. (A) 21. (D) 31. (B)
2. (D) 12. (B) 22. (C) 32. (C)
3. (D) 13. (C) 23. (A) 33. (B)
4. (A) 14. (A) 24. (C) 34. (C)
5. (C) 15. (A) 25. (C) 35. (D)
6. (A) 16. (C) 26. (D) 36. (B)
7. (B) 17. (C) 27. (C) 37. (C)
8. (B) 18. (D) 28. (B) 38. (D)
9. (A) 19. (A) 29. (A) 39. (A)
10. (B) 20. (B) 30. (D) 40. (B)

EAMCET - 2005 (Medical)

1. (A) 11. (D) 21. (B) 31. (D)
2. (A) 12. (A) 22. (B) 32. (B)
3. (D) 13. (A) 23. (C) 33. (A)
4. (C) 14. (B) 24. (C) 34. (C)
5. (D) 15. (D) 25. (B) 35. (A)
6. (C) 16. (C) 26. (C) 36. (B)
7. (A) 17. (D) 27. (D) 37. (C)
8. (B) 18. (D) 28. (D) 38. (D)
9. (A) 19. (C) 29. (A) 39. (C)
10. (D) 20. (A) 30. (B) 40. (C)

IIT - 1990

1. Amongst the following, the most basic compound is
(A) Acetanilide (B) p-Nitroaniline
(C) Benzylamine (D) p-Cresol

2. Amongst the following elements (whose electronic configurations are given below), the one having the highest ionization energy is
(A) [Ne] $3s^2\ 3p^2$ (B) [Ar] $3d^{10}\ 4s^2\ 4p^3$
(C) [Ne] $3s^2\ 3p^1$ (D) [Ne] $3s^2\ 3p^3$

3. Aryl halides are less reactive towards nucleophilic substitution reaction as compared to alkyl halides due to
(A) resonance stabilization
(B) longer carbon-halogen bond
(C) the formation of less stable carbonium ion
(D) the inductive effect
(E) sp^2- hybridised carbon attached to the halogen.

4. The compounds used as refrigerant are
(A) NH_3 (B) CCl_4 (C) CF_4 (D) CF_2Cl_2 (E) CH_2F_2

5. The formation of cyanohydrin from a ketone is an example of
(A) Electrophilic addition
(B) Nucleophilic addition
(C) Nucleophilic substitution
(D) Electrophilic substitution.

6. Chlorination of toluene in the presence of light and heat followed by treatment with aqeous NaOH gives
(A) o-Cresol (B) p-Cresol
(C) 2, 4 Dihydroxytoluene (D) Benzoic acid

7. The density of neon will be highest at
(A) 273°C, 1 atm. (B) 273°C, 2 atm.
(C) STP (D) 0°C, 2 atm

8. Amongst the following hydroxides, the one which has the lowest value of K_{sp} at ordinary temperature (about 25°C) is
(A) $Mg(OH)_2$ (B) $Ca(OH)_2$
(C) $Ba(OH)_2$ (D) $Be(OH)_2$

9. The rate of diffusion of methane at a given temperature is twice that of a gas X. The molecular weight of X is
(A) 4.0 (B) 8.0 (C) 64.0 (D) 32.0

10. The oxidation number of phosphorus in $Ba(H_2PO_2)_2$ is
(A) +3 (B) +2 (C) +1 (D) –1

11. The enolic form of acetone contains
(A) 9 sigma bonds, 1 pi bond and 2 lone pairs
(B) 8 sigma bonds, 2 pi bonds and 2 lone pairs
(C) 10 sigma bonds, 1 pi bond and 1 lone pair
(D) 9 sigma bonds, 2 pi bonds and 1 lone pair

12. The freezing point of equimolal aqueous solution will be highest for
(A) $C_9H_5NH_3Cl$ (aniline hydrochloride)
(B) $Ca(NO_3)_2$ (C) $La(NO_3)_2$
(D) $C_6H_{12}O_6$ (glucose)

13. When zeolite, which is hydrated sodium aluminium silicate is treated with hard water the sodium ions are exchanged with
(A) H^+ ions (B) Ca^{2+} ions (C) SO_4^{2-} ions
(D) Mg^{2+} ions (E) OH^- ions.

14. The aqueous solution of following salts will be coloured in the case of
(A) $Zn(NO_3)_2$ (B) $LiNO_3$
(C) $Ca(NO_3)_2$ (D) $CrCl_3$

15. Of the following metals, the metals that can be obtained by electrolysis of the aqueous solution of their salts are
(A) Cu (B) Al (C) Ag (D) Mg

16. The shape of CH_3^+ is ....

17. The valence atomic orbitals on C in silver acetylide is .... hybridised.

18. The basicity of phosphorus acid(H_3PO_3) is ....

19. Phenol is acidic because of resonance stabilisation of its conjugate base namely ....

20. Fehling solution A consists of an aqueous solution of copper sulphate while Fehling solution B consists of an alkaline soultion of ....

IIT -1991

1. According to the kinetic theory of gases, for a diatomic molecule
(A) the root mean square velocity of the molecule is inversely proportional to the temperature.
(B) the pressure exerted by the gas is proportional to the mean velocity of the molecule.
(C) the pressure exerted by the gas is proportional to the root mean square velocity of the molecule.
(D) the mean translational kinetic energy of the molecule is proportional to the absolute temperature.

2. The difference between heats of reaction at constant pressure and constant volume for the reaction,
$2C_6H_6(l) + 15O_2(g) \rightarrow 12CO_2(g) + 6H_2O(l)$ at 25°C in kJ is
(A) –3.72 (B) +7.42 (C) –7.43 (D) +3.72

3. Among $Ni(CO)_4$, $[Ni(CN)_4]^{2-}$ and $NiCl_4^{2-}$
(A) $Ni(CO)_4$ and $NiCl_4^{2-}$ are diamagnetic and $[Ni(CN)_4]^{2-}$ is paramagnetic
(B) $Ni(CO)_4$ and $[Ni(CN)_4]^{2-}$ are diamagnetic and $NiCl_4^{2-}$ is paramagnetic
(C) $NiCl_4^{2-}$ and $[Ni(CN)_4]^{2-}$ are diamagnetic and $Ni(CO)_4$ is paramagnetic
(D) $Ni(CO)_4$ is diamagnetic and $NiCl_4^{2-}$ and $[Ni(CN)_4]^{2-}$ are paramagnetic.

4. The hybridization of carbon atoms in C–C single bond of $HC\equiv C-CH=CH_2$ is
(A) $sp-sp^2$ (B) sp^3-sp (C) sp^3-sp^3 (D) sp^2-sp^3

5. The oxidation states of the most electronegative

elements in the products of the reaction, BaO_2 with dil. H_2SO_4 are
(A) O and –1 (B) –1 and –2
(C) –2 and O (D)-2 and +1.

6. For the reaction, $PCl_5(g) \rightleftharpoons PCl_3(g) + Cl_2(g)$
The forward reaction at constant temperature is favoured by
(A) introducing an inert gas at constant pressure
(B) increasing the volume of the container
(C) introducing PCl_5 at constant volume
(D) introducing an inert gas at constant volume
(E) introducing chlorine gas at constant pressure

7. The linear structure is assumed by
(A) $SnCl_2$ (B) NCO^- (C) CS_2 (D) NO^{2+} (E) SO_2

8. The reagents NH_4Cl and aqueous NH_3 will precipitate
(A) Mg^{2+} (B) Zn^{2+} (C) Ca^{2+} (D) Al^{3+} (E) Bi^{3+}

9. There is no S–S bond in
(A) $S_2O_4^{2-}$ (B) $S_2O_5^{2-}$ (C) $S_2O_3^{2-}$ (D) $S_2O_7^{2-}$

10. The volume strength of 1.5 N H_2O_2 solution is
(A) 4.8 (B) 8.4 (C) 3.0 (D) 8.0

11. m-Chlorobenzaldehyde on reaction with conc. KOH at room temperature gives
(A) m-chlorobenzyl alcohol and m-hydroxybenzyl alcohol
(B) Potassium m-chlorobenzoate and m-ydroxybenzaldehyde
(C) m-hydroxybenzaldehyde and m-chlorobenzyl alcohol
(D) potassium m-chlorobenzoate and m-hydroxybenzyl alcohol

12. The reaction which proceeds in the forward direction is
(A) $SnCl_4 + Hg_2Cl_2 = SnCl_2 + 2HgCl_2$
(B) $Fe_2O_3 + 6HCl = 2FeCl_3 + 3H_2O$
(C) $NH_3 + H_2O + NaCl = NH_4Cl + NaOH$
(D) $2CuI + I_2 + 4K^+ = 2Cu^{2+} + 4KI$

13. 1-Chlorobutane on reaction with alcoholic potash gives
(A) 1-butene (B) 1-butanol
(C) 2-butene (D) 2-butanol

14. The interaction of elemental sulphur with Grignard reagent gives ....

15. The hydrolysis of alkyl substituted chlorosilanes gives ....

16. The mass of 1 x 10^{22} molecules of $CuSO_4.5H_2O$ is ....

IIT -1992

1. The type of hybrid orbitals used by the chlorine atom in ClO_2^- is
(A) sp (B) sp^2 (C) sp^3 (D) None of these.

2. Which of the following relates to photons both as wave motion and as a stream of particles ?
(A) Interference (B) $E = mc^2$
(C) Diffraction (D) $E = h\nu$

3. For an endothermic reaction where ΔH represents the enthalpy of the reaction in kJ/mol, the minimum value for the energy of activation will be
(A) more than ΔH (B) equal to ΔH
(C) less than ΔH (D) zero

4. The CN^- ion and N_2 are isoelectronic. But in contrast to CN^-, N_2 is chemically inert, because of
(A) unsymmetrical electron distribution
(B) low bond energy
(C) absence of bond polarity
(D) presence of more number of electrons in bonding orbitals.

5. Butanenitrile may be prepared by heating
(A) Propyl alcohol with KCN
(B) Butyl alcohol with KCN
(C) Butyl chloride with KCN
(D) Propyl chloride with KCN

6. For the redox reaction.
$MnO_4^- + C_2O_4^{2-} + H^+ \rightarrow Mn^{2+} + CO_2 + H_2O$
The correct coefficient for the balanced reaction:

	MnO_4^-	$C_2O_4^{2-}$	H^+
(A)	2	5	16
(B)	16	5	2
(C)	5	16	2
(D)	2	16	5

7. Hydrogenation of benzoyl chloride in the presence of Pd on $BaSO_4$ gives
(A) Benzoic acid (B) Phenol
(C) Benzyl alcohol (D) Benzaldehyde

8. The following equilibrium is established when hydrogen chloride is dissolved in acetic acid ,
$HCl + CH_3COOH \rightarrow Cl^- + CH_3COOH_2^+$
The set that characterises the conjugate acid-base Pairs is
(A) (HCl, CH_3COOH) and ($CH_3COOH_2^+$, Cl^-)
(B) (HCl, $CH_3COOH_2^+$) and CH_3COOH, Cl^-)
(C) ($CH_3COOH_2^+$, HCl) and (Cl^-, CH_3COOH)
(D) (HCl, Cl^-) and ($CH_3COOH_2^+$, CH_3COOH)

9. Which of the following solutions will have pH close to 1.0 ?
(A) 10 ml of (M/10) HCl + 90 ml of (M/10) NaOH
(B) 100 ml of (M/10) HCl +100 ml of (M/10) NaOH.
(C) 55 ml of (M/10) HCl + 45 ml of (M/10) NaOH
(D) 75 ml of (M/5) HCl + 25 ml of (M/5) NaOH.

10. The maximum possible number of hydrogen bonds a water molecule can form is
(A) 2 (B) 4 (C) 3 (D) 1

11. At constant volume, for a fixed number of moles of a gas, the pressure of the gas increases with rise of temperature due to
(A) increase in average molecular speed
(B) increased rate of collisions amongst molecules

(C) increase in molecular attraction
(D) decrease in mean free path.

12. Which of the following does not characterise X rays ?
(A) Deflected by electric and magnetic fields
(B) A radiation can ionise gases
(C) It causes ZnS to fluorescene
(D) Have wavelengths shorter than ultra-violet

13. The products of combustion of an aliphatic thiol (RSH) at 298 K are
(A) $CO_2(g)$, $H_2O(g)$ and $SO_2(g)$
(B) $CO_2(g)$, $H_2O(l)$ and $SO_2(g)$
(C) $CO_2(l)$, $H_2O(l)$ and $SO_2(g)$
(D) $CO_2(l)$, $H_2O(l)$ and $SO_2(g)$

14. The species that do not contain peroxide ions are
(A) PbO_2 (B) H_2O_2 (C) SrO_2 (D) BaO_2

15. Which of the following have identical bond order?
(A) CN^- (B) O_2^- (C) NO^+ (D) CN^+

16. Reaction of $R-\overset{O}{\overset{\|}{C}}-NH_2$ with a mixture of Br_2 and KOH gives $R-NH_2$ as the main product, the intermediates involved in this reaction are

(A) $R-\overset{O}{\overset{\|}{C}}-NHBr$ (B) RNHBr
(C) $R-N=C=O$ (D) $R-\overset{O}{\overset{\|}{C}}-NBr_2$

17. When nitrobenzene is treated with Br_2 in the presence of $FeBr_3$, the major product formed is m-bromonitrobenzene. Statements which are related to obtain the m-isomer are
(A) The electron density on meta carbon is more than that on ortho and para positions
(B) The intermediate carbonium ion formed after initial attack of Br^+ at the meta position is least destabilised.
(C) Loss of aromaticity when Br^+ attacks at the ortho and para positions and not at meta position.
(D) Easier loss of H^+ to regain aromaticity from the meta position than from ortho and para positions.

18. The molecules that will have dipole moment are
(A) 2,2-Dimethyl propane (B) Trans-2-pentene
(C) Cis-3-hexene (D) 2,2,3,3-Tetramethylbutane

19. Isomers which can be interconverted through rotation around a single bond are
(A) Enantiomers (B) Positional isomers.
(C) Conformers (D) Diastereomers

20. The statement that is not correct for the periodic classification of elements is
(A) The properties of elements are the periodic functions of their atomic numbers
(B) Non-metallic elements are lesser in number than metallic elements.
(C) The first ionisation energies along a period do not vary in a regular manner with increase in atomic number.
(D) For transition elements the d -subshells are filled with electrons monotonically with increase in atomic number.

21. Amongst the three isomers of nitrophenol, the one that is least soluble in water is ....

22. The structure of intermediate product formed by the oxidation of toluene with CrO_3 and acetic anhydride whose hydrolysis gives benzaldehyde is ....

23. Aliphatic ethers are purified by shaking with a solution of ferrous salt to remove .... which are formed on prolonged standing in contact with air.

24. In P_4O_{10}, the number of oxygen atoms bonded to each phosphorus atom is ....

25. The lead chamber process involves oxidationof SO_2 by atmospheric oxygen under the influence of .... as catalysts.

IIT -1993

1. The material used in solar cells contains
(A) Sn (B) Ti (C) Cs (D) Si

2. Examine the following two structures for the anilinium ion and choose the correct statement from the ones given below

I: $^+NH_3$ on benzene ring ⟶ II: NH_3 with + on ring

(A) I is not an acceptable canonical structure because carbonium ions are less stable than ammonium ions
(B) I is not acceptable canonical structure because it is not aromatic.
(C) II is not acceptable canonical structure because nitrogen has valence electrons.
(D) II is an acceptable canonical structure

3. Identify the intensive quantities from the following:
(A) Volume (B) Refractive index
(C) Enthalpy (D) Temperature

4. Pick out the isoelectronic structure from the following.
I : CH_3^+ II : H_3O^+ III : NH_3^+ IV : CH_3^-
(A) I and II (B) III and IV
(C) I and III (D) II,III and IV

5. The major role of fluorspar (CaF_2) which is added in small quantities in the electrolytic reduction of alumina dissolved in fused cryolite (Na_3AlF_6) is
(A) as a catalyst
(B) To make fused mixture very conducting
(C) to lower the temperature of the melt
(D) to decrease the rate of oxidation of carbon at the anode.

6. Which of the following alloys contain Cu and Zn?
(A) Gun metal (B) Type metal.
(C) Bronze (D) Brass

7. What is the decreasing order of strength of bases ?

OH^-, NH_2^-, $H—C \equiv C^-$ and $CH_3—CH_2^-$

(A) $CH_3—CH_2^- > NH_2^- > H—C\equiv C^- > OH^-$

(B) $H–C\equiv C^- > CH_3–CH_2^- > NH_2^- > OH^-$

(C) $OH^- > NH_2^- > H—C\equiv C^- > CH_3–CH_2^-$

(D) $NH_2^- > H–C\equiv C^- > OH^- > CH_3–CH_2^-$

8. Which of the following has the smallest heat of hydrogenation per mole ?

(A) cis–2–butene (B) 1, 3–butadiene

(C) 1–butene (D) trans–2–butene

9. Among the following which one has the highest paramagnetism ?

(A) $[Cr(H_2O)_6]^{3+}$ (B) $[Fe(H_2O)_6]^{2+}$

(C) $[Cu(H_2O)_6]^{2+}$ (D) $[Zn(H_2O)_6]^{2+}$

10. Equal weights of ethane and hydrogen are mixed in an empty container at 25°C. The fraction of total pressure exerted by hydrogen is

(A) 1:2 (B) 1:1 (C) 1:16 (D) 15:16

11. The light radiations with discrete quantities of energy are called ....

12. Wave functions of electrons in atoms and molecules are called ....

13. Ca^{2+} has smaller ionic radius than K^+ because it has ....

14. The $2p_x$, $2p_y$ and $2p_z$ orbitals of atoms have identical shapes but differ in their ....

15. A system is said to be .... if it can neither exchange matter nor energy with the surroundings.

16. The formula of deep red liquid formed on warming dichromate with KCl in concentrated sulphuric acid is ....

17. The heat content of the products is more than that of the reactants in an ....

18. Kolbe electrolysis of potassium succinate gives CO_2 and ....

19. The structure of enol form of $CH_3–CO–CH_2–CO–CH_3$ with intermolecular hydrogen bonding is

```
        O......H — O
        ||         |
CH3 — C — CH = C — CH3 .
```

20. Addition of water to acetylenic compounds is catalyzed by .... and ....

IIT -1994

1. The hydrolysis of trialkylchlorosilane, R_3SiCl, yields ....

2. Given that ΔT_f is the depression in freezing point of the solvent in a solution of a non-volatile solute of molaity, m, the quantity Lt $(\Delta T_f /m)$ is equal to .... $m \to O$

3. The kind of delocalization involving sigma bond orbitals is called ....

4. The outermost electronic configuration of Cr is ....

5. The two types of bonds present in B_2H_6 are covalent and ....

6. For a given reversible reaction, at a fixed temperature equilibrium constant K_P and K_C are related by ....

7. The type of magnetism exhibited by $(Mn(H_2O]^{2+}$ ion is ....

8. The bond dissociation energy needed to form the benzyl radical from toluene is .... than the formation of the methyl radical from methane.

9. One recently discovered allotrope of carbon (e.g.C_{60}) is commonly known as ....

10. For the reaction, $N_2(g) + 3H_2(g) \rightarrow 2NH_3(g)$, under certain conditions of temperature and partial pressure of the reactants, the rate of formation of NH_3 is 0.001kg hr^{-1}. The rate of conversion of H_2 under the same conditions is .... kg hr^{-1}.

11. The high melting point and insolubility in organic solvent of sulphanilic acid are due to its .... structure.

12. The compound $YBa_2Cu_3O_7$, which shows superconductivity, has copper in oxidation state .... Assume that the rare earth element, yttrium is in its usual +3 oxidation state.

13. Solubility of iodine in water is greatly increased by the addition of iodide ions because of the formation of ....

14. The IUPAC name of $[Co(NH_3)_6]Cl_3$ is ....

15. The IUPAC name of succinic acid is ....

IIT -1996

1. X ml of H_2 gas effuses through a hole in a container in 5 seconds. The time taken for the effusion of the same volume of the gas specified under identical conditions is

(A) 25 seconds : CO (B) 55 seconds : CO_2

(C) 10 seconds : He (D) 20 seconds : O_2

2. One mole of $N_2O_4(g)$ at 300 K is kept in a closed container under one atmosphere. It is heated to 600 K when 20% by mass of $N_2O_4(g)$ decomposes to $NO_2(g)$. The resultant pressure is

(A) 2.0 atm (B)1.0 atm (C) 1.2 atm (D) 2.4 atm

3. The molecular weight of benzoic acid in benzene as determined by depression in freezing point method corresponds to

(A) ionization of benzoic acid

(B) dimerization of benzoic acid

(C) trimerization of benzoic acid

(D) solution of benzoic acid

4. Which of the following halides is least stable and has doubtful existence ?

(A) SnI_4 (B) PbI_4 (C) Cl_4 (D) GeI_4

5. Which of the following oxides is neutral
(A) CO (B) SnO_2 (C) ZnO (D) SiO_2

6. Which of the following has maximum number of unpaired electrons ?
(A) V^{3+} (B) Fe^{2+} (C) Mg^{2+} (D) Ti^{3+}

7. The following acids have been arranged in the order of decreasing acid strength. Identify the correct order.
ClOH (I), BrOH (II), IOH (III)
(A) I > II > III (B) II > I > III
(C) III > II > I (D) I > III > II

8. $^{27}_{13}Al$ is a stable isotope. $^{29}_{13}Al$ is expected to disintegrate by
(A) α-emission (B) β-emission
(C) positron emission (D) proton emission

9. Among the following species identify the isostructural pairs
NF_3, NO_3^-, BF_3, H_3O^+, NH_3
(A) [NF_3, H_3O^+] and [NO_3^- , BF_3]
(B) [NF_3, NO_3^-] and [BF_3, H_3O^+]
(C) [NF_3, NH_3] and [NO_3^-, BF_3]
(D) [NF_3, H_3O^+] and [NH_3, BF_3]

10. The number and type of bonds between two carbon atoms in CaC_2 are ;
(A) one sigma and one π
(B) one sigma and two π
(C) one sigma and one and half π
(D) one σ bond

11. Which of the following statements is correct for $CsBr_3$?
(A) It contains Cs^+, and Br_3^- ions
(B) It is a covalent compound.
(C) It contains Cs^{3+} and Br^- ions
(D) It contains Cs^+, Br^- and Br_2 molecule.

12. Arrange the following compounds in order of increasing dipolemoment . Toluene (I), m-dichloro benzene (II), o-dichlorô benzene (III), p-dichloro benzene(IV)
(A) I < IV < II < III (B) IV < I < II < III
(C) IV < I < III < II (D) IV < II < I < III

13. KF combines with HF to form KHF_2. Thecompound contains the species.
(A) K^+, F^- and H^+ (B) K^+, F^- and HF
(C) K^+ and $[HF_2]^-$ (D) $[KHF]^+$ and F^-

14. Which compound is formed when excess of KCN is added to aqueous solution of copper sulphate ?
(A) $K[Cu(CN)_2]$ (B) $K_3[Cu(CN)_4]$
(C) $Cu(CN)_2$ (D) $K_2[Cu(CN)_4]$

15. Which compound does not dissolve in hot dilute HNO_3 ?
(A) HgS (B) PbS (C) CuS (D) CdS

16. Sodium thiosulphate is prepared by
(A) neutralising $H_2S_2O_3$ with NaOH
(B) reducing Na_2SO_4 solution with H_2S
(C) boiling Na_2SO_3 solution and S in alkaline medium
(D) boiling Na_2SO_3 solution with S in acidic medium.

17. A dilute aqueous solution of Na_2SO_4 is electrolysed using platinium electrodes. The products at the anode and cathode are
(A) O_2, H_2 (B) $S_2O_8^{2-}$, Na
(C) O_2, Na (D) $S_2O_8^{2-}$, H_2

18. The rate constant, the activation energy and the Arrhenius parameter of a chemical reaction at 25°C are $3.0 \times 10^{-4} s^{-1}$, 104.4 kJ mol^{-1} and $6.0 \times 10^{14} s^{-1}$ respectively. The value of rate constant as $T \to \infty$ is
(A) $2.0 \times 10^{18}\ s^{-1}$ (B) $6.0 \times 10^{14}\ s^{-1}$
(C) infinity (D) $3.6 \times 10^{30}\ s^{-1}$

19. The ratio between root mean square velocity of H_2 at 50 K and that of O_2 at 800 K is
(A) 1 (B) 1/4 (C) 4 (D) 2

20. The orbital angular momentum of electron in 2s orbital is
(A) $+\frac{1}{2}\frac{h}{2\pi}$ (B) zero (C) $\frac{h}{2\pi}$ (D) $\sqrt{2}\frac{h}{2\pi}$

21. In the following compounds,

OH / I (I) OH / CH_3 (II) OH / NO_2 (III) OH / NO_2 (IV)

the order of acidity is
(A) III > IV > I > II (B) I> IV> III> II
(C) II > I > III > IV (D) IV >III> I> II

22. In the Cannizzaro reaction given below,
$Ph-CHO \xrightarrow{OH^-} PH-CH_2OH + PhCO_2^-$, the slowest step is
(A) the abstraction of proton from the carboxylic acid.
(B) the attack of –OH at the carbonyl group
(C) the transfer of hydride to the carbonyl group
(D) the deprotonation of $Ph-CH_2OH$

23. The following compounds have been arranged in order of their increasing thermal stabilities. Identify the correct order
K_2CO_3(I); $MgCO_3$(II); $CaCO_3$(III); $BeCO_3$(IV)
(A) I < II < III < IV (B) IV< II< III < I
(C) IV < II < I < III (D) II < IV < III < I

24. Hydrolysis of one mole of peroxodisulphuric acid produces
(A) two moles of sulphuric acid
(B) two moles of peroxo monosulphuric acid
(C) one mole of sulphuric acid, one mole of peroxomonosulphuric acid and one mole of hydrogen
(D) one mole of sulphuric acid, one mole of peroxomono sulphuric acid and one mole of hydrogen peroxide

IIT -1997

1. During debromination of meso-dibromobutane, the major compound formed is
(A) cis-2-butene (B) trans-2-butene
(C) n-butane (D) 1-butene

2. In the following compounds,

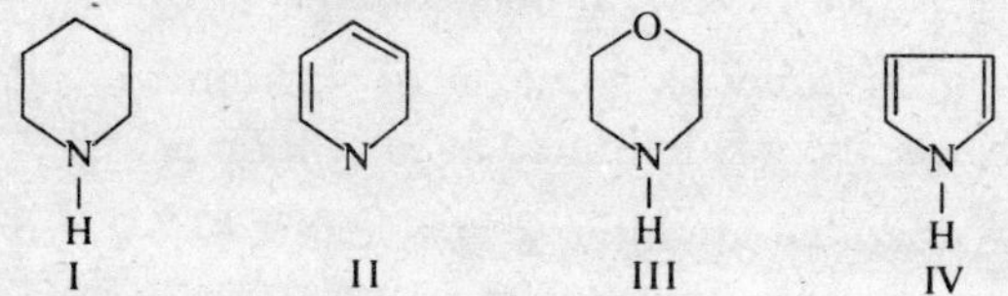

the order of basicity is
(A) IV > I > III > II (B) III > I > IV > II
(C) II > I > III > IV (D) I > III > II > IV

3. The number of possible enantiomeric pairs that can be produced during monochlorination of 2-methyl butane is
(A) 1 (B) 2 (C) 3 (D) 4

4. Among the following statements on the nitration of aromatic compounds, the false one is
(A) the rate of nitration of benzene is greater than that of hexadeuterobenzene
(B) the rate of nitration of benzene is almost the same as that of hexadeuterobenzene
(C) the rate of nitration of toluene is greater than that of benzene
(D) nitration is an electrophilic substitution reaction.

5. The intermediate during the addition HCl to propene in the presence of peroxide is
(A) $CH_3\dot{C}HCH_2Cl$ (B) $CH_3CH^+CH_3$
(C) $CH_3\dot{C}H_2CH_2$ (D) $CH_3CH_2\overset{+}{C}H_2$

6. The order of reactivity of the following alcohols is

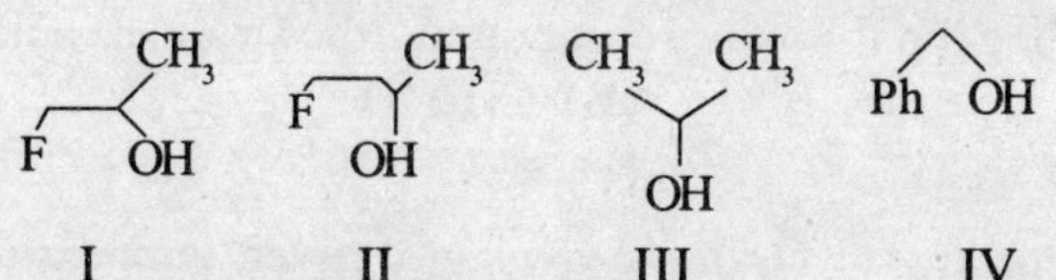

(A) I > II > III > IV (B) I > III > II > IV
(C) IV > III > II > I (D) IV > III > I > II

7. The correct statement about compounds A, B and C is

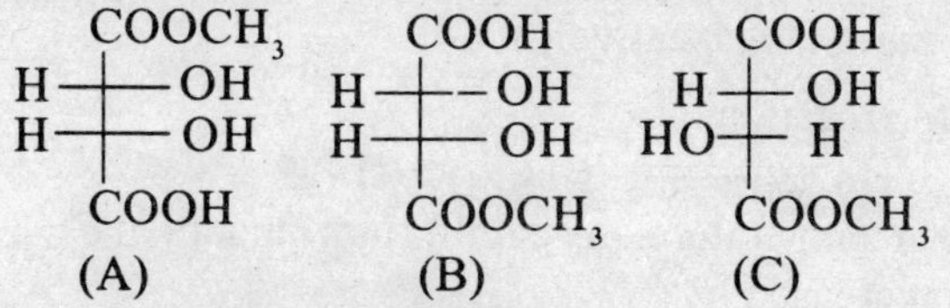

(A) A and B are identical
(B) A and B are disastermers
(C) A and C are enantiomers
(D) A and B are enantiomers

8. For a 'd'-electron, the orbital angular momentum is
(A) h (B) 2h (C) $\sqrt{6}$h (D) $\sqrt{2}$h

9. The standard reduction potential of Cu^{2+}/ Cu and Cu^{2+}/ Cu^+ are 0.337 and 0.153 V respectively. The standard electrode potential of Cu^+/ Cu half cell is
(A) 0.184 V (B) 0.827 V
(C) 0.521V (D) 0.490 V

10. The compressibility factor for an ideal gas is
(A) 1.5 (B) 1.0 (C) 2.0 (D) ∞

11. Which contains both polar and non-polar bonds?
(A) H_2O_2 (B) CH_4 (C) NH_4Cl (D) HCN

12. The critical temperature of water is higher than that of O_2 because the H_2O molecule has
(A) fewer electrons than O_2 (B) two covalent bonds
(C) V-shape (D) dipole moment

13. Which one of the following species is not a pseudohalide?
(A) CNO (B) RCOO (C) OCN (D) NNN

14. When cyclohexane is poured on water, flots, because
(A) cyclohexane is in 'crown' from
(B) cyclohexane is in 'boat' form
(C) cyclohexane is in 'chair' form
(D) cyclohexane is less dense than water.

15. Which of the following is an organometallic compound?
(A) Lithium methoxide (B) Lithium acetate
(C) Lithium dimethylamide (D) Methyl lithium

16. In the reaction of p-chlorotoluene with KNH_2 in liq. NH_3, the major product is
(A) o-toluidine (B) m-toluidine
(C) p-toluidine (D) p-chloroaniline

17. $(CH_3)_3$ CMgCl on reaction with D_2O produces
(A) $(CD_3)_3CD$ (B) $(CD_3)_3OD$
(C) $(CH_3)_3CD$ (D) $(CH_3)_3$ OD

18. Ammonium dichromate is used in some fireworks. The green coloured powder blown in air is
(A) CrO_3 (B) Cr_2O_3 (C) Cr (D) CrO (O_2)

19. Which one of the following compounds has sp^2 hybridization ?
(A) N_2O (B) CO (C) CO_2 (D) SO_2

20. The number of moles of $KMnO_4$ that will be needed to react with one mole of sulphite ion in acidic solution is
(A) $\frac{2}{5}$ (B) $\frac{3}{5}$ (C) $\frac{4}{5}$ (D) 1

21. Property of the alkaline earth metals that increases with their atomic number is
(A) solubility of their sulphates
(B) ionisation energy
(C) solubility of their hydroxides
(D) electronegativity

22. Nitrobenzene can be prepared from benzene by using a mixture of conc. HNO_3 and conc. H_2SO_4. In the nitrating mixture, HNO_3 acts as a
(A) base (B) acid
(C) reducing agent (D) catalyst

23. The first use of quantum theory to explain the structure of atom was made by
(A) Planck (B) Einstein (C) Heisenberg (D) Bohr

24. In the following groups,
–OAC –OMe –OSO_2Me –OSO_2CF_3
I II III IV
the order of leaving group ability is
(A) I > II > III > IV (B) IV > III > I > II
(C) III > II > I > IV (D) III > II > IV > I

25. Which of the following compounds is expected to be coloured ?
(A) MgF_2 (B) CuCl (C) Ag_2SO_4 (D) CuF_2

26. How many optically active stereoisomers are possible for butane-2,3-diol ?
(A) 1 (B) 2 (C) 3 (D) 4

27. Molar heat capacity of water in equilibrium with ice at constant pressure is
(A) zero (B) infinity (∞)
(C) 40.45 kJ K^{-1} mol^{-1} (D) 75.48 KJ^{-1} mol^{-1}

28. Standard molar enthalpy of formation of CO_2 is equal to
(A) zero
(B) the standard molar enthalpy of combustion of gaseous carbon
(C) the sum of standard molar enthalpies of formation of CO and O_2
(D) the standard molar enthalpy of combustion of carbon (graphite)

29. The incorrect statement among the following is :
(A) the first ionization potential of Na is less than the first ionization potential of Mg.
(B) the first ionization potential of Al is less than the first ionization potential of Mg.
(C) the second ionization potential of Mg is greater than the second ionization potential of Na
(D) the third ionization potential of Mg is greater than the third ionization potential of Al.

30. Among KO_2, AlO_2^-, BaO_2 and NO_2^+, unpaired electron is present in
(A) NO_2^+ and BaO_2 (B) KO_2 and AlO_2^-
(C) KO_2 only (D) BaO_2 only

31. In the sodium chloride structure, each Na^+ ion is surrounded by six Cl^- ions nearest neighbours and .... Na^+ ion nearest neighbours.

32. Among N_2O, SO_2, I_3^+ and I_3^-, the linear species are .... and ....

33. Among PCl_3, CH_3^+, NH_2^- and NF_3, ... is least reactive towards water.

34. The angle P–P–P in P_4 molecule is .... degree.

35. The absolute temperature of an ideal gas is .... to/ than the average kinetic energy of the gas molecules.

36. $(CH_3OH_2)^+$ is .... acidic than ....

37. On heating, $Rb(ICl_2)$ will decompose giving .... and ....

38. A solution of sodium in liquid ammonia at -33°C conducts electricity. On cooling, the conductivity of this solution ....

39. For a gaseous reaction, 2B→A, the equilibrium constant K_P is .... to/than K_C.

40. When an aqueous solution of sodium fluoride is electrolysed, the gas liberated at the anode is ....

41. In Arrhenius equation, k = A $exp^{-E/RT}$, A may be termed as the rate constant at ....

42. When Fe(s) is dissolved in aqueous hydrochloric acid in a closed vessel, the work done is ....

43. A liquid which is permanently supercooled is frequently called a ....

44. Enthalpy is an .... property.

45. In the reaction I + I_2 → I_3, the lewis acid is ....

46. Silver jewellery items tarnish slowly in the air due their reaction with ....

47. Compounds that formally contain Pb^{4+} are easily reduced to Pb^{2+}.The stability of lower oxidation state is due to ....

48. Glycerine contains one .... hydroxy group.

49. 1, 3-Butadiene with bromine in molar ratio generates predominantly ....

50. Vinylchloride on reaction with dimethyl copper gives ....

IIT -1998

1. Among the following compounds, the strongest acid is
(A) C_2H_6 (B) CH_3OH
(C) HC≡CH (D) C_6H_6

2. According to Grahm's law , at a given temperature, the rates of diffusion r_A/r_B of gases A and B is given by
(A) $(P_A/P_B)(M_A/M_B)^{1/2}$ (B) $(M_A/M_B)(P_A/P_B)^{1/2}$
(C) $(P_A/P_B)(M_B/M_A)^{1/2}$ (D) $(M_A/M_B)(P_B/P_A)^{1/2}$
(Where P and M are pressures and molecular weights of gases A and B respectively)

3. For the reaction,
$CO(g) + H_2O(g) \rightleftharpoons CO_2(g) + H_2(g)$,
at a given temperature, the equilibrium amount of $CO_2(g)$ can be increased by
(A) decreasing the volume of the container
(B) adding a suitable catalyst
(C) adding an inert gas
(D) increasing the amount of CO(g)

4. Which of the following will react with water ?
(A) $CHCl_3$ (B) Cl_3CCHO
(C) CCl_4 (D) $ClCH_2CH_2Cl$

5. The standard reduction potential values of three metallic cations, X, Y, Z are 0.52, –3.03 and –1.18V respectively. The order of reducing power of the corresponding metals is

(A) Z > Y > X (B) Z > X > Y
(C) Y > Z > X (D) X > Y > Z

6. Among the following compounds, which will react with acetone to give a product containing >C = N – ?

(A) $C_6H_5NH_2$ (B) $(CH_3)_3N$
(C) $C_6H_5NHC_6H_5$ (D) $C_6H_5NHNH_2$

7. The reaction of, $CH_3CH = CH-C_6H_4-OH$, with HBr gives

(A) $CH_3CHBrCH_2-C_6H_4-OH$

(B) $CH_3CH_2CHBr-C_6H_4-OH$

(C) $CH_3CHBrCH_2-C_6H_4-OH$

(D) $CH_3CH_2CHBr-C_6H_4-OH$

8. The energy of an electron in the first Bohr obrit of H atom is –13.6 eV. The possible energy value (s) of the excited state (s) for electrons in Bohr orbits of hydrogen is /are

(A) –3.4 eV (B) –4.2 eV
(C) –6.8 eV (D) +6.8.eV

9. In nitroprusside ion, the iron and NO exist as Fe^{II} and NO^+ rather than Fe^{III} and NO. These forms can be differentiated by

(A) Measuring the solid state magnetic moment
(B) estimating the concentration of iron
(C) measuring the concentration of CN^-
(D) thermally decomposing the compound.

10. Sodium nitrate decomposes above ~800°C to give,

(A) N_2 (B) O_2 (C) NO_2 (D) Na_2O

11. Decrease in atomic number is observed during

(A) alpha emission (B) beta emission
(C) positron emission (D) electron capture

12. Which of the following statements is (are) correct ?

(A) The electronic configuration of Cr is $[Ar]3d^5 4s^1$. (Atomic No. of Cr = 24)
(B) The magnetic quantum number may have negative value.
(C) In silver atom, 23 electrons have a spin of one type and 24 of theopposite type. (Atomic no. of Ag = 47)
(D) The oxidation state of nitrogen in HN_3 is –3.

13. Which of the following compounds will show geometrical isomerism ?

(A) 1-phenylpropene (B) 2-methyl-2-butene
(C) 2-butene (D) propene

14.The geometry and the type of hybrid orbital present about the central atom in BF_3 is

(A) linear, sp (B) trigonal planar, sp^2
(D) tetrahedral, sp^3 (C) pyramidal, sp^3

15. Benzenediazonium chloride on reaction with phenol in weakly basic medium gives

(A) chlorobenzene (B) benzene
(C) diphenyl ether (D) p-hydroxy azobenzene

16. Which of the following statement (s) is (are) correct when a mixture of NaCl and $K_2Cr_2O_7$ is gently warmed with conc.H_2SO_4 ?

(A) a deep red vapour is evolved.
(B) The vapour when passed into NaOH solution gives a yellow solution of Na_2CrO_4.
(C) Chlorine gas is evolved
(D) Chromyl chloride is formed

17. A new carbon-carbon bond formation is possible in

(A) Cannizzaro reaction
(B) Friedel Crafts alkylation
(C) Clemmensen reduction
(D) Reimer-Tiemann reaction

18. Among the following compounds, which will react with acetone to give a product containing > C = N – ?

(A) $C_6H_5NHC_6NH_5$ (B) $C_6H_5NHNH_2$
(C) $C_6H_5NH_2$ (D) $(CH_3)_3N$

19. For a first order reaction

(A) the degree of dissociation is equal to $(1-e^{-kt})$
(B) a plot of reciprocal concentration of the reactant vs. time gives a straight line
(C) the time taken for the completion of 75% reaction is thrice $t_{1/2}$ of the reaction
(D) the pre-exponential factor in Arrhenius equation has the dimention of time T^{-1}

20. Highly pure dilute solution of sodium in liquid ammonia

(A) shows blue colour
(B) exhibits electrical conductivity
(C) produces sodium amide
(D) produces hydrogen gas

21. Which of the following statement (s) is (are) correct with reference to the ferrous and ferric ions ?

(A) Fe^{3+} gives brown colour with potassium ferricyanide.
(B) Fe^{2+} gives blue precipitate with potassium ferricyanide
(C) Fe^{3+} gives red colour with potassimthiocyanate
(D) Fe^{2+} gives brown colour with ammonium thiocyanate

22. p–Chloroaniline and anilinium hydrochloride can be distinguished by

(A) Sandmeyer's reaction (B) $NaHCO_3$
(C) $AgNO_3$ (D) Carbylamine test

23. Which of the following undergo aldol condensation ?

(A) acetaldehyde (B) propanadehyde
(C) benzaldehyde (D) trideutero acetadehyde

24. The products of reaction of alcoholic silver nitrite with ethyl bromide are

(A) Ethane (B) Ethene (C) Nitroethane

(D) Ethyl alcohol (E) Ethyl nitrite

25. Tautomerism is exhibited by

(A) –CH = CH –OH (B) O= =O

(C) =O O (D) =O O

26. Which of the following statement (s) is (are) correct?

(A) The coordination number of each type of ion in CsCl is 8.

(B) A metal that crystallizes in bcc structure has a coordination number of 12.

(C) A unit cell of an ionic crystal shares some of its ions with other unit cells

(D) The length of the unit cell in NaCl is 552 pm. ($r_{Na^+} = 95$; $r_{Cl^-} = 181$ pm)

27. Which of the following statement (s) is (are) correct ?

(A) The pH of 1.0 x 10^{-8}M solution of HCl is 8.

(B) Autoprotolysis constant of water increases with temperature

(C) The conjugate base of $H_2PO_4^-$ is HPO_4^{2-}

(D) When a solution of a weak monoprotic acid is titrated against a strong base, at half neutralization point pH = (1/2)pK_a

28. Which of the following will react with water ?

(A) $CHCl_3$ (B) Cl_3CCHO

(C) CCl_4 (D) $ClCH_2CH_2Cl$

29. Benzyl chloride ($C_6H_5CH_2Cl$) can be prepared from toluene by chlorination with

(A) SO_2Cl_2 (B) $SOCl_2$ (C) Cl_2 (D) NaOCl

30. Addition of high proportions of manganese makes steel useful in making rails of railroads, because manganese

(A) can remove oxygen and sulphur

(B) gives hardness to steel

(C) helps the formation of oxides of iron

(D) can show highest oxidation state of +7

31. White phosphorus (P_4) has

(A) Six P-P single bonds

(B) Four P-P single bonds

(C) Four lone pairs of electrons

(D) PPP angle of 60°

IIT -1999

1. In the dichromate dianion

(A) All Cr — O bonds are equivalent

(B) 4 Cr — O bonds are equivalent

(C) 6 Cr — O bonds are equivalent

(D) all Cr — O bonds are non-equivalent

2. One mole of calcium phosphide on reaction with excess water gives

(A) two moles of phosphine

(B) one mole of phosphine

(C) two moles of phosphoric acid

(D) one mole of phosphorus pentoxide

3. The oxidation number of sulphur in S_8 , S_2F_2 ,H_2S respectively, are

(A) 0, + 1 and –2 (B) +2, + 1 and –2

(C) 0, + 1 and +2 (D) –2, + 1 and –2

4. On heating ammonium dichromate, the gas evolved is

(A) oxygen (B) ammonia

(C) nitrous oxide (D) nitrogen

5. In the commercial electrochemical process for aluminium extraction, the electrolyte used is

(A) a molten mixture of Al_2O_3 and Na_3AlF_6

(B) a molten mixture of AlO(OH) and $Al(OH)_3$

(C) $Al(OH)_3$ in NaOH solution

(D) an aqueous solution of $Al_2(SO_4)_3$

6. The geometry of H_2S and its dipole moment are

(A) angular and non-zero (B) angular and zero

(C) linear and non-zer (D) linear and zero

7. The geometry of H_2S and $Ni(PPh_3)_2Cl_2$ are

(A) both square planar

(B) tetrahedral and square planar respectively

(C) both tetrahedral

(D) square planar and tetrahedral, respectively

8. In compounds of type ECl_3, where E = B, P. As or Bi, the angles Cl –E – Cl for different E are in the Order

(A) B < P = As = Bi (B) B < P < As < Bi

(C) B > P = As = Bi (D) B > P >As > Bi

9. The electrons identified by quantum numbers n and *l*, (i) *l* =1 (ii) n = 4, *l*=0 (iii) n = 3, *l*=2 (iv) n =3, *l*=1 can be placed in order of increasing energy from the lowest to highest as

(A) (iv) < (ii) < (iii) < (i)

(B) (ii) < (iv) < (i) < (iii)

(C) (i) < (iii) < (ii) < (iv)

(D) (iii) < (i) < (iv) < (ii)

10. The number of neutrons accompanying the formation of $^{139}_{54}Xe$ and $^{94}_{38}Sr$ from the absorption of a slow neutron by $^{235}_{92}U$, followed by nuclear fission is,

(A) 0 (B) 1 (C) 2 (D) 3

11. The correct order of increasing C — O bond length of CO, CO_3^{2-}, CO_2 is

(A) CO_3^{2-} < CO_2< CO (B) CO_2 < CO_2 < CO

(C) CO < CO_3^{2-} < CO_2 (D) CO < CO_2 < CO_3^{2-}

12. A gas will approach ideal behaviour at

(A) low temperature and low pressure

(B) low temperature and high pressure

(C) high temperature and low pressure

(D) high temperature and high pressure

13. The normality of 0.3 M phosphorus acid (H_3PO_3) is,
(A) 0.1 (B) 0.9 (C) 0.3 (D) 0.6

14. The coordination number of a metal crystallising in a hexagonal close-packed structure is
(A) 8 (B) 6 (C) 12 (D) 4

15. A gas X at 1atm is bubbled through a solution containing a mixture of 1MY⁻ and 1MZ⁻ at 25°C. If the reduction potential is Z > Y > X , then,
(A) Y will oxidize both X and Z
(B) Y will oxidize X and not Z
(C) Y will oxidize Z and not X
(D) Y will reduce both X and Z

16. The pH of 0.1M solution of the following salts increases in the order
(A) $NaCl < NH_4Cl < NaCN < HCl$
(B) $HCl < NH_4Cl < NaCl < NaCN$
(C) $NaCN < NH_4Cl < NaCl < HCl$
(D) $HCl < NaCl < NaCN < NH_4Cl$

17. For the chemical reaction, $3X(g) + Y(g) \rightleftharpoons X_3Y(g)$, the amount of X_3Y at equilibrium is affected by
(A) temperature and pressure
(B) temperature only
(C) pressure only
(D) temperature, pressure and catalyst

18. In the compound, $CH_2 = CH - CH_2 - CH_2 - C \equiv CH$, the $C_2 - C_3$ bond is the type,
(A) $sp-sp^2$ (B) sp^3-sp^3 (C) $sp-sp$ (D) sp^2-sp^3

19. When propionic acid is treated with aqueous sodium bicarbonate, CO_2 is liberated. The 'C' of CO_2 comes from
(A) methyl group (B) carboxylic acid group
(C) methylene group (D) bicarbonate

20. The enol form of acetone,after treatment with D_2O, gives
(A) $CH_3 - \overset{OD}{\overset{|}{C}} - CH_2$ (B) $CD_3 - \overset{O}{\overset{||}{C}} - CD_3$
(C) $CH_2 = \overset{OH}{\overset{|}{C}} - CH_2D$ (D) $CD_2 - \overset{OD}{\overset{||}{C}} - CD_3$

21. A positive cabylamine test is given by
(A) N, N-dimethylaniline (B) 2,4-dimethylaniline
(C) N-methyl-o-methylaniline (D) p-methylbenzylamine

22. The optically active tartaric acid is named as D (+) tartaric acid because it has a positive
(A) optical rotation and is derived from D (+) glyceraldehyde
(B) optical rotation and is derived from D glucose
(C) pH in organic solvent
(D) optical rotation only when substituted by deuterium

23. A solution of (+) – 2-chloro-2-phenylethane in toluene recemises slowly in the presence of small amount of $SbCl_5$ due to the formation of
(A) carbanion (B) carbene
(C) free-radical (D) carbocation

24. The product (s) obtained via oxymercuration ($HgSO_4 + H_2SO_4$) of 1-butyne would be
(A) $CH_3 - CH_2 - \overset{O}{\overset{||}{C}} - CH_3$
(B) $CH_3 - CH_2 - CH_2 - CHO$
(C) $CH_3 - CH_2 - CHO + HCHO$
(D) $CH_3 - CH_2 - COOH + HCOOH$

25. The most unlikely representation of resonance structures of p-nitrophenoxide ion is
(A) 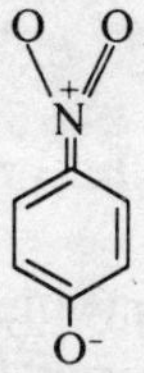 (B) 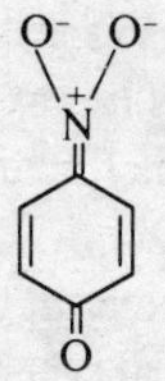 (C) 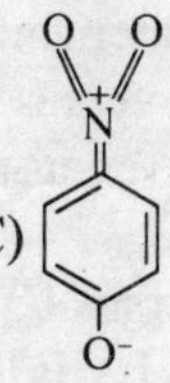 (D)

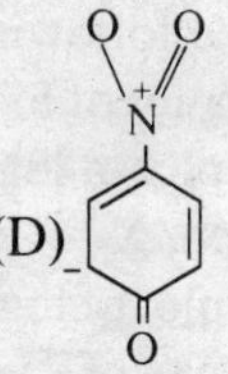

26. Ground state electronic configuration of nitrogen atom can be represented by
(A) 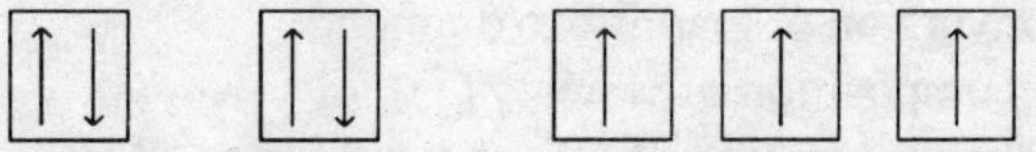 ↑↓ ↑↓ ↑ ↑ ↑
(B)  ↑↓ ↑↓ ↑ ↓ ↑
(C)  ↑↓ ↑↓ ↑ ↓ ↓
(D) 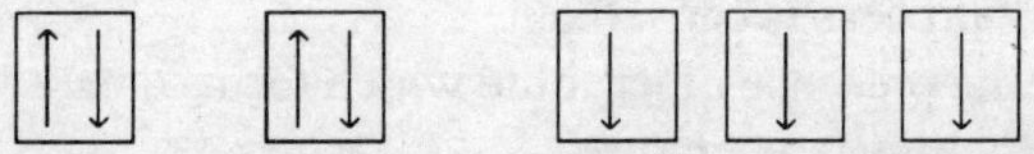 ↑↓ ↑↓ ↓ ↓ ↓

27. In the depression of freezing point experiment, it is found that
(A) only solute molecules solidify at the freezing point
(B) only solvent molecules solidify at the freezing point
(C) vapour pressure of the solution is less than that of pure solvent
(D) vapour pressure of the solution is more than that of pure solvent

28. The correct order of ionic radii is
(A) $K^+ > Cl^-$ (B) $P^{3+} > P^{5+}$
(C) $Ti^{4+} < Mn^{7+}$ (D) $^{35}Cl^- < ^{37}Cl^-$

29. Ammonia, on reaction with hypochlorite anion, can form
(A) N_2H_4 (B) HNO_2 (C) NO (D) NH_4Cl

30. A buffer solution can be prepared from a mixture of
(A) ammonia and sodium hydroxide in water
(B) sodium acetate and acetic acid in water
(C) sodium acetate and hydrochloric acid in water
(D) ammonia and ammonium chloride in water

31. An aromatic molecule will.
(A) have 4n π electrons (B) have (4n+2) π electrons
(C) be planar (D) be cyclic

32. Toluene, when treated with Br_2/Fe gives p-bromotoluene as the major product because the CH_3 group
(A) activates the ring by hyperconjugation

(B) is para directing
(C) is meta directing (D) deactivates the ring

33. The ether, $C_6H_5-O-CH_2-C_6H_5$ when treated with HI produces

(A) $C_6H_5-CH_2I$ (B) $C_6H_5-CH_2OH$
(C) C_6H_5-I (D) C_6H_5-OH

34. The following statement (s) is (are) correct
(A) A plot of log Kp versus I/T is linear
(B) A plot of log [X] versus time is linear for a first order reaction, $X \rightarrow P$
(C) A plot of log P versus I/T is linear at constant volume
(D) A plot of P versus I/V is linear at constant temperature

35. The following is (are) endothermic reaction (s)
(A) Combustion of methane
(B) Conversion of graphite to diamond
(C) Decomposition of water
(D) Dehydrogenation of ethane to ethylene

IIT -2000

1. Electrolytic reduction of alumina to aluminium by Hall-Heroult process is carried out
(A) in the presence of cryolite which forms a melt with lower melting temperature.
(B) in the presence of NaCl
(C) in the presence of fluorite
(D) In the presence of cryolite which forms a melt with higher melting temperature.

2. The number of P–O–P bonds in cyclic metaphosphoric acid is
(A) zero (B) two (C) three (D) four

3. The hybridisation of atomic orbitals of nitrogen in NO_2^+, NO_3^- and NH_4^+ are
(A) sp^2, sp and sp^3 respectively
(B) sp, sp^3 and sp_2 respectively
(C) sp, sp^2 and sp^3 respectively
(D) sp^2, sp^3 and sp respectively.

4. Which of the following has the most acidic hydrogen?
(A) 3-hexanone (B) 2, 4-hexanedione
(C) 2, 5-hexanedione (D) 2, 3-hexanedione

5. The correct order of acidic strength is
(A) $Ba_2O > MgO > Al_2O_3$ (B) $K_2O > CaO > MgO$
(C) $Cl_2O_7 > SO_2 > P_4O_{10}$ (D) $CO_2 > N_2O_5 > SO_3$

6. Ammonia can be dried by
(A) conc. H_2SO_4 (B) P_4O_{10}
(C) CaO (D) anhydrous $CaCl_2$

7. H_2O, H_2S, H_2Se and H_2Te, the one with the highest boiling point is
(A) H_2S because of hydrogen bonding
(B) H_2O because of hydrogen bonding
(C) H_2Te because of higher molecular weight
(D) H_2Se because of lower molecular weight

8. Which one of the following alkenes will react fastest with H_2 under catalytic hydrogenation condition ?
(A) RCH=CHR (R, R on top; H, H below)
(B) RCH=CHR (R, H on top; R, H below)
(C) $R_2C=CHR$ (R, R on top; R, H below)
(D) $R_2C=CR_2$

9. The order of reactivities of the following alkyl halides for a S_N2 reaction is
(A) RF > RCI > RBr >RI (B) RF > RBr > RCI >RI
(C) RCI > RBr > RF > RI (D) RI > RBr > RCI >RF

10. Molecular shapes of SF_4, CF_4 and XeF_4 are
(A) the same, with 2, 0 and 1 lone pairs of electrons respectively.
(B) the same,with 1, 1 and 1 lone pairs of electrons respectively
(C) different, with 0, 1 and 2 lone pairs of electrons respectively
(D) different,with 1, 0 and 2 lone pairs of electrons respectively.

11. The compressibility of a gas is less than unity at STP. Therefore
(A) $V_m = 22.4$ litres (B) $V_m = 44.8$ litres
(C) $V_m > 22.4$ litres (D) $V_m < 22.4$ litres

12. Among the following, the strongest base is
(A) $C_6H_5NH_2$ (B) p–NO_2–$C_6H_4NH_2$
(C) m–NO_2–$C_6H_4NH_2$ (D) $C_6H_5CH_2NH_2$

13. When two reactants, A & B are mixed to give products C & D, the reaction quotient, Q, at the initial stages of the reaction
(A) is zero (B) decreases with time
(C) is independent of time (D) increases with time

14. The rate constant for the reaction, $2N_2O_5 \rightarrow 4NO_2 + O_2$ 3.0×10^{-5} sec^{-1}. If the rate is 2.40×10^{-5} mol $litre^{-1}$ sec^{-1}, then the concentration of N_2O_5 (mol $litre^{-1}$) is
(A) 1.4 (B) 1.2 (C) 0.04 (D) 0.8

15. Propyne and propene can be distinguished by
(A) dil. $KMnO_4$ (B) $AgNO_3$ in ammonia
(C) conc. H_2SO_4 (D) Br_2 in CCl_4

16. At 100°C and 1 atm, if the density of liquid water is 1.0 g. cm^{-3} and that of water vapour is 0.0006 gm cm^{-3}, then the volume occupied by water in 1 litre of steam at that temperature is
(A) 6 cm^3 (B) 60 cm^3 (C) 0.6 cm^3 (D) 0.06 cm^3

17. The rms velocity of hydrogen is $\sqrt{7}$ times the rms velocity of nitrogen. If T is the temperature of the gas,
(A) $T(H_2) = T(N_2)$ (B) $T(H_2) > T(N_2)$
(C) $T(H_2) < T(N_2)$ (D) $T(H_2) - \sqrt{7}\, T(N_2)$

18. Which one of the following will most readily be dehydrated in acidic condition?

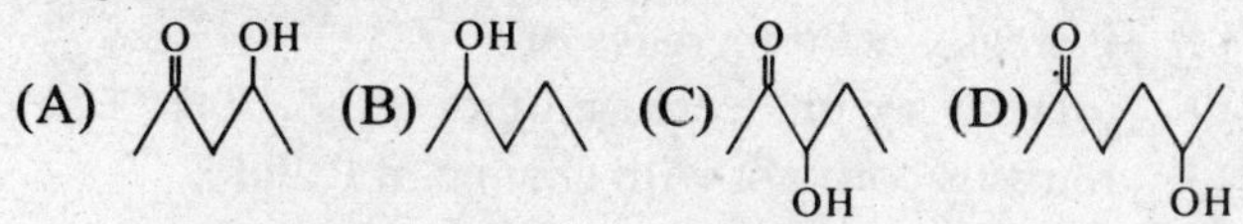

19. The ΔH_f^0 for $CO_2(g)$, $CO(g)$ and $H_2O(g)$ are -393.5, -110.5. and 241.8 kJ mol^{-1} respectively. The standard enthalpy change (in kJ) for the reaction $CO_2(g) + H_2(g) \rightarrow CO(g) + H_2O(g)$ is

(A) 524.1 (B) 41.2 (C) −262.5 (D) −41.2

20. Which of the following compounds will exhibit geometrical isomerism ?

(A) 1-phenyl-2-buutene (B) 3-phenyl-1-butene
(C) 2-phenyl-1-butene (D) 1,1-diphenyl-1-propene

21. For the reversible reaction, $N_2(g) + 3H_2(g) = 2NH_3(g)$ at 500°C, the value of K_p is 1.44x 10^{-5} when partial pressure is measured in atmospheres. The corresponding value of K_c, with concentration in mole $litre^{-1}$, is

(A) $1.44\times10^{-5} / (0.082 \times 773)^2$
(B) $1.44\times10^{-5} / (0.082 \times 500)^{-2}$
(C) $1.44\times10^{-5} / (8.314 \times 773)^{-2}$
(D) $1.44\times10^{-5} / (0.082 \times 773)^{-2}$

22. The electronic configuration of an element is $1s^2 2s^2 2p^6 3s^2 3p^6 3d^5 4s^1$. This represents the element's

(A) excited state (B) ground state
(C) cationic form (D) anionic form

23. The chemical processes in the production of steel from haematite ore involve

(A) reduction followed by oxidation
(B) oxidation followed by reduction.
(C) reduction (D) oxidation

24. Amongst the following, identify the species with an atom in +6 oxidation state

(A) MnO_4^- (B) $Cr(CN)_6^{3-}$ (C) NiF_6^{2-} (D) CrO_2Cl_2

25. Benzoyl chloride is prepared from benzoic acid by

(A) $SOCl_2$ (B) Cl_2, H_2O (C) Cl_2, hν (D) SO_2Cl_2

26. For the electrochemical cell, $M/M^+/X^-/X$, $E^o(M^+/M) = 0.44$ V and $E^o(X/X^-) = 0.33$ V. From this data one can deduce that

(A) $M + X \rightarrow M^+ + X^-$ is the spontaneous reaction
(B) $M^+ + X^- \rightarrow M + X$ is the spontaneous reaction
(C) $E_{cell} = 0.77$ V (D) $E_{cell} = -0.77$ V

27. The correct order of radii is

(A) N < Be < B (B) $F^- < O^{2-} < N^{3-}$
(C) Na < Li < K (D) $Fe^{3+} < Fe^2 < Fe^{4-}$

28. The number of nodal planes in P_x orbital is

(A) one (B) two (C) three (D) zero

29. The appropriate reagent for the following transformation is

O
CH_3
HO → HO
CH_2CH_3

(A) Zn (Hg), HCl (B) NH_2NH_2, OH^-
(C) H_2 / Ni (D) $NaBH_4$

30. Which of the following has the highest nucleophilicity?

(A) CH_3^- (B) NH_2^- (C) F^- (D) OH^-

31. A : The first ionization energy of Be is greater than that of B. R : 2p orbital is lower in energy than 2s

(A) A (B) B (C) C (D) D

32. A : 1-butene on reaction with HBr in the presence of a peroxide produces 1-bromobutane
R: It involves the formation of a primary radical

(A) A (B) B (C) C (D) D

33. A : The heat absorbed during the isothermal expansion of an ideal gas against vacuum is zero
R : The volume occupied by the molecules of an ideal gas is zero.

(A) A (B) B (C) C (D) D

34. A : The pressure of a fixed amount of an ideal gas is proportional to its temperature.
R : Frequency of collisions and their impact both increase in proportion to the square root of temperature

(A) A (B) B (C) C (D) D

35. A : Phenol is more reactive than benzene towards electrophilic substitution reaction.
R : In the case of phenol, the intermediate carbocation is more resonance stabilized.

(A) A (B) B (C) C (D) D

IIT- 2001

1. Satutrated solution of KNO_3 is used to make 'salt-bridge' because

(A) velocity of both K^+ and NO_3^- are nearly the same
(B) KNO_3 is highly soluble in water
(C) velocity of K^+ is greater than that of NO_3^-
(D) Velocity of NO_3^- is greater than that of K^+

2. The number of isomers for the compound with molecular formula, $C_2BrClFI$ is

(A) 3 (B) 4 (C) 5 (D) 6

3. The reaction of propene with HOCl proceeds via the addition of

(A) H^- in the first step (B) Cl^+ in the first step
(C) OH^- in the first step (D) Cl^- and OH^- in a single step

4. 1-Propanol and 2-Propanol can be best distinguished by

(A) oxidation by heating with copper followed by reaction with Fehling solution
(B) Oxidation with alkaline $KMnO_4$ followed by reaction with Fehling solution.
(C) oxidation with acidic dichromate followed by reaction with Fehling solution
(D) oxidation with concentrated H_2SO_4 followed with Fehling solution.

5. An S_N2 reaction at an asymmetric carbon of a compound always gives
(A) a mixture of distereomers
(B) an enantiomer of the substrate
(C) a product with opposite optical rotation
(D) a single stereoisomer

6. A mixture of benzaldehyde and formaldehyde on heating with aqueous NaOH solution gives
(A) benzyl alcohol and sodium formate
(B) sodium benzoate and methyl alcohol
(C) sodium benzoate and sodium formate
(D) benzyl alcohol and methyl alcohol

7. The correct order of basicities of the following compounds is

$CH_3-C(=NH)-NH_2$ (1) $CH_3-CH_2-NH_2$ (2) $(CH_3)_2NH$ (3) $CH_3-C(=O)-NH_2$ (4)

(A) 2 > 1 > 3 > 4 (B) 1 > 3 > 2 > 4
(C) 3 > 1 > 2 > 4 (D) 1 > 2 > 3 > 4

8. The compound that will react most readily with NaOH to form methanol is
(A) $(CH_3)_3S^+I^-$ (B) $(CH_3)_3CCl$
(C) $(CH_3)_4N^+I^-$ (D) CH_3OCH_3

9. In the presence of peroxide, hydrogen chloride and hydrogen iodide do not give anti-Markownikoff addition to alkenes because
(A) both are highly ionic
(B) one is oxidising and the other is reducing
(C) one of the steps is endothermic in both the cases
(D) all the steps are exothermic in both the cases

10. Me H Me / Me / H / H (structure)

Hydrogenation of the above compound in the presence of poisoned palladium catalyst gives
(A) an optically active compound
(B) an optically inactive compound
(C) a racemic mixture (D) a diastereomeric mixture

11. In a solid 'AB' having the NaCl structure, 'A' atoms occupy the corners of the cubic unit cell. If all the face-centred atoms along one of the axes are removed, then the resultant stoichiometry of the solid is
(A) A_4B_3 (B) A_3B_4 (C) AB_2 (D) A_2B

12. The wavelength associated with a golf ball weighing 200 g and moving at a speed of 5 m/h is of the order
(A) 10^{-10} m (B) 10^{-20} m (C) 10^{-30} m (D) 10^{-40} m

13. At constant temperature, the equilibrium constant (K_p) for the decomposition reaction, $N_2O_4 \rightleftharpoons 2NO_2$ is expressed by $K_p = (4x^2P)/(1-x^2)$, where P = pressure, x = extent of decomposition. Which one of the following statements is true ?
(A) K_p increases with increase of P
(B) K_p increases with increase of x
(C) K_p increases with decrease of x
(D) K_p remains constant with change in P and x.

14. If 'I' is the intensity of absorbed light and 'C' is the concentration of AB* for the photochemical process AB + hv → AB*, the rate of formation of AB is directly proportional to
(A) C (B) I (C) I^2 (D) C.I

15. The correct order of euivalent conductance at infinite dilution of LiCl, NaCl and KCl is
(A) NaCl > KCl > LiCl (C) LiCl > KCl > NaCl
(C) LiCl > NaCl > KCl (D) KCl > NaCl > LiCl

16. For a sparingly soluble salt A_pB_q, the relationship of its solubility product (L_S) with its solubility (S) is
(A) $L_S = S^{p+q} . p^p q^q$ (B) $L_S = S^{p+q} . p^q .q^p$
(C) $L_S = S^{pq} .p^p .q^q$ (D) $L_S = S^{pq} . (pq)^{p+q}$

17. In thermodynamics, a process is called reversible when
(A) surroundings and system change into each other
(B) there is no boundary between system and surroundings
(C) the surroundings are always in equilibrium with the system.
(D) the system changes into the surroundings spontaneously.

18. Which one of the following statement is false ?
(A) Change in the state is completely defined when the intial and final states are specified
(B) Work is a state function
(C) Temperature is a state function
(D) Work appears at the boundary of the system.

19. The quantum numbers +1/2 and −1/2 for the electron spin represent
(A) rotation of the electron in clockwise and anticlockwise direction respectively
(B) rotation of the electron in anticlockwise and clockwise direction respectively
(C) magnetic moment of the electron pointing up and down respectively
(D) two quantum mechanical spin states which have no classical analogue

20. The correct order of hybridization of the central atom in the following species, NH_3, $[PtCl_4]^{2-}$, PCl_5 and BCl_3 is
(A) dsp^2, dsp^3, sp^2 and sp^3 (B) sp^3, dsp^2, dsp^3, sp^2
(C) dsp^2, sp^2, sp^3, dsp^3 (D) dsp^2, sp^3, sp^2, dsp^3

21. The complex ion which has no 'd' electrons in the central metal atom is
(A) $[MnO_4]^-$ (B) $[Co(NH_3)_6]^{3-}$
(C) $[Fe(CN)_6]^{3-}$ (D) $[Cr(H_2O)_6]^{3+}$

22. In the standardization of $Na_2S_2O_3$ using $K_2Cr_2O_7$ by iodometry, the equivalent weight of $K_2Cr_2O_7$ is
(A) (molecular weight)/2 (B) (molecular weight)/6
(C) (molecular weight)/3

(D) Same as molecular weight

23. The set representing the correct order of first ionization potential is
(A) B > C > N (B) Ge > Si > C
(C) K > Na > Li (D) Be > Mg > Ca

24. An aqueous solution of 6.3 g oxalic acid dihydrate is made up to 250 ml. The volume of 0.1N NaOH required to completely neutralize 10 ml of this is
(A) 4 ml (B) 10 ml (C) 20 ml (D) 40 ml

25. The chemical composition of 'slag' formed during the smelting process in the extaction of copper is
(A) $Cu_2O + FeS$ (B) $FeSiO_3$
(C) $CuFeS_2$ (D) $Cu_2S + FeS$

26. The common features among the species CN^-, CO and NO^+ are
(A) bond order two and π-acceptors
(B) bond order three and isoelectronic
(C) bond order three and weak field ligands
(D) isoelectronic and weak field ligands

27. The number of S-S bonds in sulphur trioxide trimer (S_3O_9) is
(A) three (B) two (C) one (D) zero

28. The reaction, $3ClO^-(aq) \rightarrow ClO_3^-(aq) + 2Cl^-(aq)$, is an example of
(A) oxidation reaction (B) reduction reaction
(C) disproportionation reaction
(D) decomposition reaction

29. The set with correct order of acidity is
(A) $HClO < HClO_4 < HClO_3 < HClO_2$
(B) $HClO < HClO_2 < HClO_3 < HClO_4$
(C) $HClO_4 < HClO_3 < HClO_2 < HClO$
(D) $HClO_4 < HClO_2 < HClO_3 < HClO$

30. The root mean square velocity of an ideal gas at constant pressure varies with density (d) as
(A) d^2 (B) d (C) $\sqrt{d}$ (D) $1/\sqrt{d}$

31. A : Between $SiCl_4$ and CCl_4, only $SiCl_4$ reacts with water
R : $SiCl_4$ is ionic and CCl_4 is covalent.
(A) A (B) B (C) C (D) D

32. A : In any ionic solid [MX] with Schottky defects, the number of positive and negative ions are same
R : Equal number of cation and anion vacancies are present.
(A) A (B) B (C) C (D) D

33. A : Addition of bromine to trans-2-butene yields meso-2, 3-dibromobutane
R : Bromine addition to an alkene is an electrophilic addition
(A) A (B) B (C) C (D) D

34. A : In strongly acidic solutions, aniline becomes more reactive towards electrophilic reagents
R : The amino group being completely protonated in strongly acidic solution, the lone pair of electrons on the nitrogen is no longer available for resonance.
(A) A (B) B (C) C (D) D

35. A : Dimethyl sulphide is commonly used for the reduction of an ozonide of an alkene to get the carbonyl compounds.
R : It reduces the ozonide giving water soluble dimethyl sulphoxide and excess of it evaporates.
(A) A (B) B (C) C (D) D

IIT -2002

1. One mole of a non-ideal gas undergoes a change of state (2.0 atm, 3.0 L, 95K) → 4.0 atm, 5.0 L, 245K) with a change in internal energy, $\Delta U = 30.0$ L atm. The change in enthalpy (ΔH) of the process in L atm. is
(A) 40.0 (B) 42.3 (C) 44.0
(D) not defined, because pressure is not constant.

2. Consider the following equilibrium in a closed container
$$N_2O_4(g) \rightleftharpoons 2NO_2(g)$$
At a fixed temperature, the volume of the reaction container is halved. For this change, which of the following statements holds true regarding the equilibrium constant (K_P) and degree of dissociation (α) ?
(A) neither K_P nor α changes (B) both k_P and α change
(C) K_P changes, but α does not change
(D) K_P does not change, but α changes.

3. Consider the chemical reaction, $N_2(g) + 3H_2(g) \rightarrow 2NH_3(g)$. The rate of this reaction can be expressed in terms of time derivative of concentration of $N_2(g)$, $H_2(g)$ or $NH_3(g)$. Identify the correct relationship amongst the rate expressions
(A) Rate = $d[N_2]/dt = 1/3d\,[H_2]/dt = 1/2d\,[NH_3]/dt$
(B) Rate = $-d[N_2]/dt = -1/3d\,[H_2]/dt = 1/2d\,[NH_3]/dt$
(C) Rate = $-d[N_2]/dt = -3d\,[H_2]/dt = 2d\,[NH_3]/dt$
(D) Rate = $-d[N_2]/dt = -d\,[H_2]/dt = d\,[NH_3]/dt$

4. If the Nitrogen atom had electronic configuration $1s^7$. It would have energy lower than that of the normal ground state configuration $1s^2 2s^2 2p^3$, because the electrons would be closer to nucleus. Yet $1s^7$ is not observed because it vilates
(A) Pauli exclusion principle
(B) Heisenberg uncertainty principle (C) Hund's rule
(D) Bohr postulate of stationary orbits.

5. Rutherford's experiment, which established the nuclear model of the atom, used a beam of
(A) helium atoms, which impinged on a metal foil and got scattered
(B) β-particles which impinged on a metal foil and got absorbed.
(C) γ-rays which impinged on a metal foil and ejected electrons
(D) helium nuclei, which impinged on a metal foil and got scattered

6. When the temperature is increased, surface tension of water

(A) increases (B) decreases
(C) remains constant (D) shows irregular behaviour

7. Standard electrode potential data are useful for understanding the suitability of an oxidant in a redox titration. Some half cell reactions and their standard potentials are given below:

$MnO_4^-(aq.) + 8H^+(aq.) + 5e^- \rightarrow Mn^{2-}(aq.) + 4H_2O(l)$
$E^\circ = 1.51$ V

$Cr_2O_7^{2-}(aq.) + 14H^+ (aq) + 6e^- \rightarrow 2Cr^{3+}(aq.) + 7H_2O(l)$
$E^\circ = 1.38$ V

$Fe^{3+}(aq.) + e^- \rightarrow Fe^{2+}(aq.)$ $E^\circ = 0.77$ V

$Cl_2(g) + 2e^- \rightarrow 2Cl^-(aq.)$ $E^\circ = 1.40$ V

Identify the only incorrect statement regarding the quantitative estimation of aqueous $Fe(NO_3)_2$

(A) MnO_4^- can be used in aqueous H_2SO_4
(B) MnO_4^- can be used in aqueous HCl
(C) $Cr_2O_7^{2-}$ can be used in aqueous HCl
(D) $Cr_2O_7^{2-}$ can be used in aqueous H_2SO_4

8. Which of the following hydrocarbons has the lowest dipole moment?

(A) $H_3C(H)C=C(H)CH_3$ (cis) (B) $CH_3C \equiv CCH_3$
(C) $CH_3CH_2C \equiv CH$ (D) $CH_2 = CH-C \equiv CH$

9. A substance A_XB_Y crystallises in a face centered cubic (FCC) lattice in which atoms 'A' occupy each corner of the cube and atoms 'B' occopy the centers of each face of the cube. Identify the correct composition of the substance A_XB_Y

(A) AB_3 (B) A_4B_3
(C) A_3B (D) Composition can not be specified

10. Specify the coordination geometry around and hybridisation of N and B atoms in a 1:1 complex of BF_3 and NH_3

(A) N: pyramidal, sp^3; B:Planar, sp^2
(B) N: tetrahedral, sp^3; B:tetrahedral, sp^3
(C) N:pyramidal, sp^3; B:pyramidal, sp^3
(D) N:pyramidal, sp^3; B: tetrahedral, sp^3

11. Identify the correct order of boiling points of the following compounds

$CH_3CH_2CH_2CH_2OH$ (1) $CH_3CH_2CH_2CHO$ (2) $CH_3CH_2CH_2COOH$ (3)

(A) 1 > 2 > 3 (B) 3 > 1 > 2
(C) 1 > 3 > 2 (D) 3 > 2 > 1

12. Identify the reagent from the following list which can easily distinguish between 1-butyne and 2-butyne

(A) bromine,CCl_4 (B) H_2, Lindlar catalyst
(C) dilute H_2SO_4, $HgSO_4$
(D) ammoniacal Cu_2Cl_2 solution

13. Identify the correct order of acidic strengths of CO_2, CuO, CaO, H_2O

(A) $CaO < H_2O < CuO < CO_2$
(B) $CaO < CuO < H_2O < CO_2$
(C) $H_2O < CuO < CaO < CO_2$
(D) $H_2O < CO_2 < CaO < CuO$

14. Consider the following reaction,

$H_3C-CH(D)-CH(CH_3)-CH_3 + Br^\bullet \rightarrow$ 'X' + HBr

Identify the structure of the major product 'X'

(A) $H_3C-CH(D)-CH(CH_3)-\dot{C}H_2$ (B) $CH_3-CH(D)-\dot{C}(CH_3)-CH_3$
(C) $H_3C-\dot{C}(D)-CH(CH_3)-CH_3$ (D) $H_3C-\dot{C}H-CH(CH_3)-CH_3$

15. Identify the correct order of reactivity in electrophilic substitution reaction of the following compounds,

benzene (1), toluene (CH_3) (2), chlorobenzene (Cl) (3), nitrobenzene (NO_2) (4)

(A) 1 > 2 > 3 > 4 (B) 4 > 3 > 2 > 1
(C) 2 > 1 > 3 > 4 (D) 2 > 3 > 1 > 4

16. Polyphosphates are used as water softening agents because they

(A) form soluble complexes with cationic species
(B) form soluble complexes with anionic species
(C) precipitate anionic species
(D) precipitate cationic species

17. The nodal plane in the π-bond of ethene is located in

(A) the molecular plane
(B) a plane parallel to the molecular plane
(C) a plane perpendicular to the molecular plane which bisects the carbon-carbon σ-bond at right angle
(D) a plane perpendicular to the molecular plane which contains the carbon-carbon σ-bond

18. Identify the set of reagents / reaction conditions 'X' and 'Y' in the following set of transformations

$CH_3-CH_2-CH_2Br \xrightarrow{X}$ Product $\xrightarrow{Y} CH_3-CH(Br)-CH_3$

(A) X = dilute aqueous NaOH, 20°C; Y = HBr/acetic acid, 20°C
(B) X = concentrated alcoholic NaOH, 80°C; Y = HBr/acetic acid, 20°C
(C) X = dilute aqueous NaOH, 20°C; Y = $Br_2/CHCl_3$, 0°C
(D) X = concentrated alcoholic NaOH, 80°C; Y = $Br_2/CHCl_3$, 0°C

19. Compound 'A' (molecular formula C_3H_8O) is treated with acidified potassium dichromate to form a product 'B' (molecular formula: C_3H_6O). 'B' forms a shining silver mirror on warming with ammonical silver nitrate. 'B' when

treated with an aqeous solution of $H_2NCONHNH_2$. HCl and sodium acetate gives a product 'C'. Identify the structure of 'C'

(A) $CH_3CH_2CH = NNHCONH_2$

(B) $CH_3-\underset{\underset{CH_3}{|}}{C} = NNHCONH_2$

(C) $CH_3 -\underset{\underset{CH_3}{|}}{C} = NCONHNH_2$

(D) $CH_3CH_2CH = NCONHNH_2$

20. Identity the least stable ion amongst the following

(A) B^- (B) C^- (C) Li^- (D) Be^-

21. Which of the following molecular species has unpaired electron(s) ?

(A) N_2 (B) F_2 (C) O_2^- (D) O_2^{2-}

22. Which of the following acids has the smallest dissociation constant ?

(A) $CH_3CHFCOOH$ (B) FCH_2CH_2COOH

(C) $BrCH_2CH_2COOH$ (D) $CH_3CHBrCOOH$

23. Which of the following compounds exhibits stereoisomerism ?

(A) 2-methylbutene-1 (B) 3-methylbutyne-1

(C) 3-methylbutanoic acid (D) 2-methylbutanoic acid

24. Which of the following volume(V)- temperature (T) plots represents the behaviour of one mole of an ideal gas at one atmospheric pressure ?

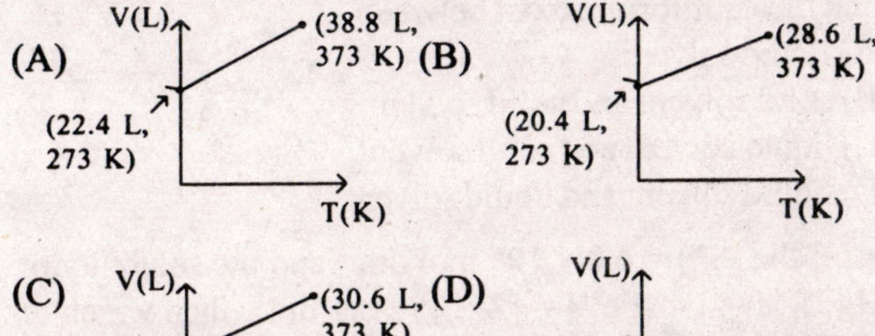

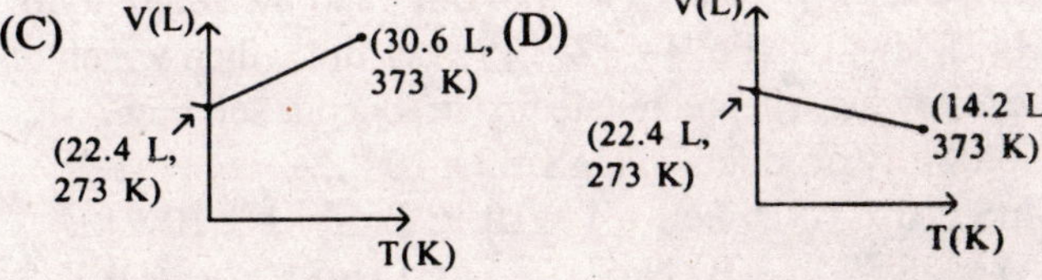

25. How many moles of electron weigh one kilogram ?

(A) 6.023×10^{23} (B) $\frac{1}{9.108} \times 10^{31}$

(C) $\frac{6.023}{9.108} \times 10^{54}$ (D) $\frac{1}{9.108 \times 6.023} \times 10^{8}$

26. A gas 'X' is passed through water to form a saturated solution. The aqueous solution on treatment with silver nitrate gives a white precipitate. The saturated aqueous solution also dissolves magnesium ribbon with evolution of a colourless gas 'Y'. Identify 'X' and 'Y'

(A) $X = CO_2, Y = Cl_2$ (B) $X = Cl_2, Y = CO_2$

(C) $X = Cl_2, Y = H_2$ (D) $X = H_2, Y = Cl_2$

27. Identify the correct order of solubility of Na_2S, CuS and Zns in aqueous medium

(A) CuS > ZnS > Na_2S (B) ZnS > Na_2S > CuS

(C) Na_2S > CuS > ZnS (D) Na_2S > ZnS > CuS

28. Anhydrous ferric chlloride is prepared by

(A) heating hydrated ferric chloride at a high temperature in a stream of air

(B) heating metallic iron in a stream of dry chlorine gas

(C) reaction of ferric oxide with hydrochloric acid

(D) reaction of metallic iron with hydrochloric acid.

29. An aqueous solution of a substance gives a white precipitate on treatment with dilute hydrochloric acid, which dissolves on heating. When hydrogen sulfide is passed through the hot acidic solution, a black precipitate is obtained. The substance is a

(A) Ag^+ salt (B) Pb^{2+} salt (C) Hg_2^{2+} salt (D) Cu^{2+} salt

30. Which of the following processes is used in the extractive metallurgy of magnesium ?

(A) fused salt electrolysis (B) self reduction

(C) aqueous solution electrolysis (D) thermite reduction

IIT -2003

1. F—⟨◯⟩—NO_2 $\xrightarrow[DMF\Delta]{(CH_3)_2NH}$ A

A $\xrightarrow[\text{(ii) } H_2 \text{ Catalytic Reduction}]{\text{(i) } NaNO_2 / HCl, 0 - 5°C}$ B; B is

(A) $(CH_3)_2$N—⟨◯⟩—NH_2 (B) CH_3HN—⟨◯⟩—NH_2

(C) H_2N —⟨◯⟩—NH_2 (D) F—⟨◯⟩— NH_2

2. (P / Q isomeric alkenes bearing $OCOCH_3$) $\xrightarrow[\text{Hydrolysis}]{\text{acidic}}$

Products formed by P and Q can be differentiated by

(A) NaOH (B) Cannizaro's reaction

(C) Hydrolysis (D) Fehings solution.

3. (Structure: COOH, OH, O_2N, OH, H)

When the compound made to react with 2 eq. of $NaNH_2$, the product formed will be

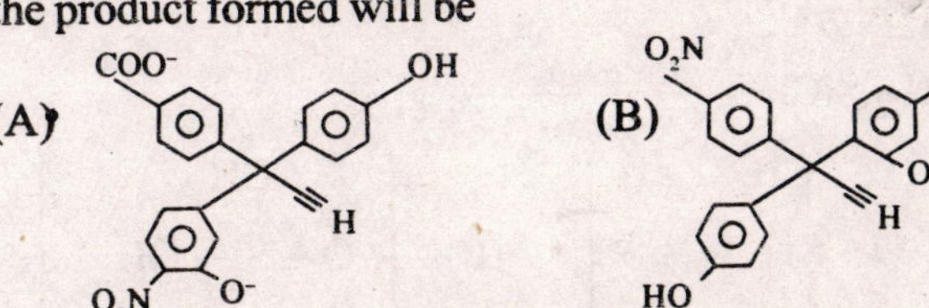

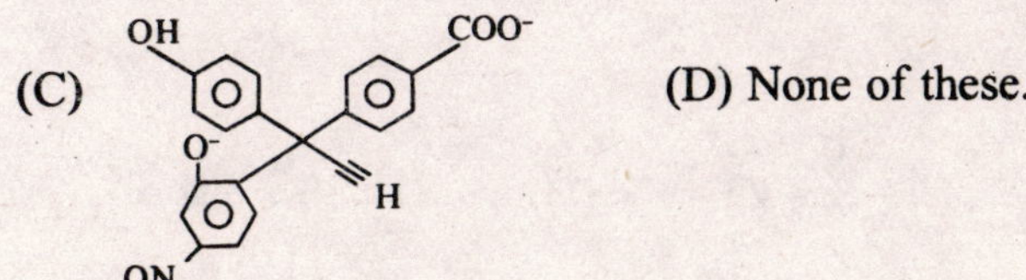

(D) None of these.

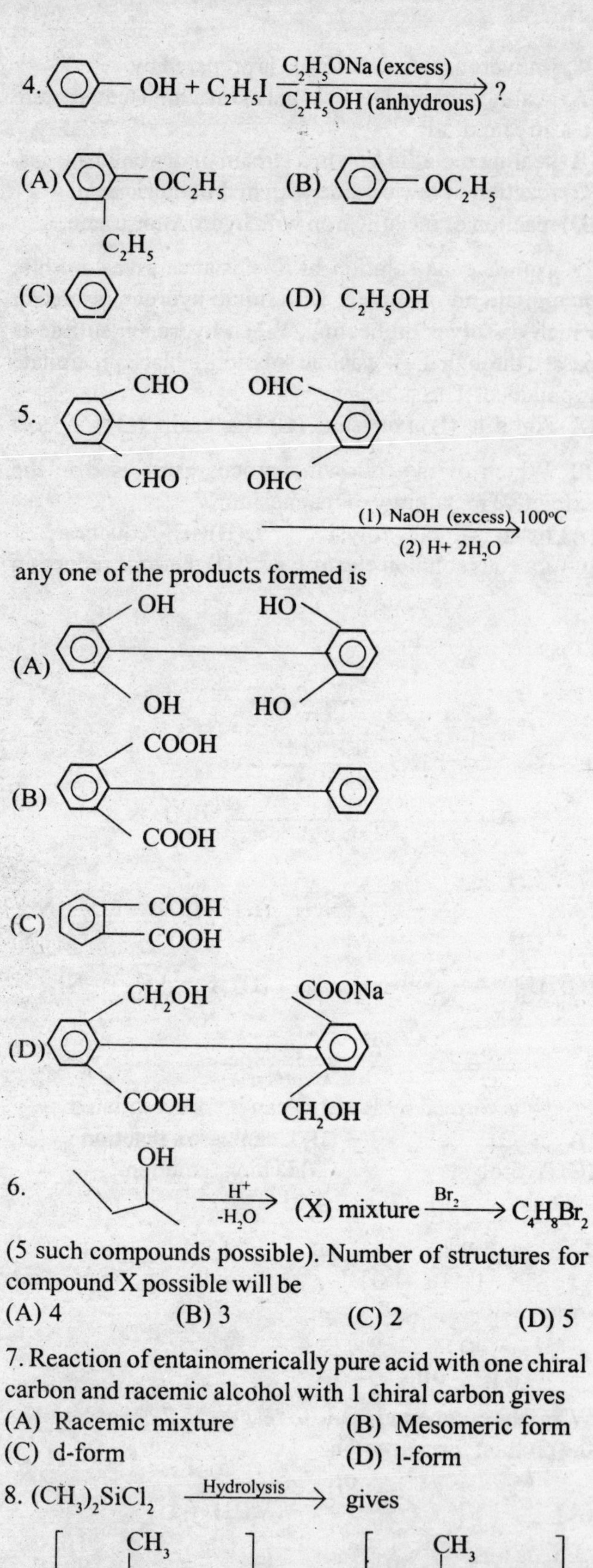

4. C_6H_5—OH + C_2H_5I $\xrightarrow[C_2H_5OH\ (anhydrous)]{C_2H_5ONa\ (excess)}$?

(A) $C_6H_4(OC_2H_5)(C_2H_5)$ (B) C_6H_5—OC_2H_5

(C) C_6H_6 (D) C_2H_5OH

5. CHO, CHO / OHC, OHC (biphenyl tetraaldehyde) $\xrightarrow[(2)\ H+\ 2H_2O]{(1)\ NaOH\ (excess)\ 100°C}$

any one of the products formed is

(A) OH, OH / HO, HO

(B) COOH, COOH

(C) COOH, COOH

(D) CH_2OH, COOH / COONa, CH_2OH

6. OH $\xrightarrow[-H_2O]{H^+}$ (X) mixture $\xrightarrow{Br_2}$ $C_4H_8Br_2$

(5 such compounds possible). Number of structures for compound X possible will be

(A) 4 (B) 3 (C) 2 (D) 5

7. Reaction of entainomerically pure acid with one chiral carbon and racemic alcohol with 1 chiral carbon gives

(A) Racemic mixture (B) Mesomeric form

(C) d-form (D) l-form

8. $(CH_3)_2SiCl_2$ $\xrightarrow{Hydrolysis}$ gives

(A) $[-O-Si(CH_3)_2-O-]_n$ (B) $[-O-Si(CH_3)_2-OH-]_n$

(C) $Cl-Si(O^-)(CH_3)-Cl-$ (D) $-O-Si(CH_3)(Cl)-CH_3-$

9. Left to right sp^2, sp^2, sp, sp hybridisation is present in

(A) $CH_2 == CH - C \equiv N$ (B) $CH \equiv C - C \equiv N$

(C) $CH_3 - CH_2 - C \equiv N$ (D) None of these

10. CH_3MgBr + Ethyl ester → which can be formed as product (excess)

(A) CHO—C(CHO)(CH_3)—CH_3 (B) HO—C(CH_3)(CH_3)—CH_3

(C) CHO—C(CH_3)(CH_3)—CH_3 (D) CHO—C(HO)(CHO)—CH_3

11. $C_6H_6 - C \equiv C - CH_3 \xrightarrow[H_2SO_4]{HgSO_4}$ A;A is

(A) C_2H_5—CO—CH_2—C_6H_5 (B) $C_6H_5CH_2CHO$

(C) $C_6H_5CHO+CH_3CHO$ (D) $C_6H_5 - CO - CH_3$

12. Which of the following equations gives the value of heat of formation (Δ H°) ?

(A) $H(g) + F(g) \rightarrow HF(l)$

(B) $\frac{1}{2}H_2(g) + \frac{1}{2}F_2(g) \rightarrow HF(g)$

(C) $H_2(g) + F_2(l) \rightarrow HF(g)$

(D) None of these

13. When the depression in freezing point is carried out, then the equilibrium exist between

(A) liquid solvents

(B) solid solvent and solid solvent

(C) liquid solvent and solid solvent

(D) liquid solvent and liquid solvent

14. If the $[S^{2-}] = 6.0 \times 10^{-2}$ mol dm^{-2} and the solubility of $[Hg^{2+}][Mn^{2+}] = [Fe^{2+}] = [Zn^{2+}] = 1x1\ dm^{-2}$, then which of the following will precipitate first in aqueous solution ? [Solubility products of HgS = 3×10^{-53}, MnS = 6×10^{-16}; FeS = 1×10^{-19}; ZnS = 1×10^{-22}]

(A) MgS (B) NaS (C) HgS (D) K_2S

15. Of NO_3^- , ClO_3^- , SO_3 , CO_3^{2-} , Isoelectronic and Isostructural species are

(A) ClO_3^- , NO_3^- (B) NO_3^-, CO_3^{2-}

(C) SO^3 , CO_3^- (D) ClO_3^- , CO_2^-

16. Of the following, maximum dipole moment is shown by

(A) CH_3Cl (B) $C_2H_2I_2$ (C) $C_2H_2Cl_2$ (D) CH_3I

17. Boric acid (H_3BO_3) is

(A) monobasic and weak lewis acid

(B) tribasic and weak lewis acid

(C) monobasic and strong lewis acid

(D) dibasic and weak lewis acid

18. For H_3PO_4 and H_3PO_3 , The correct choice is

(A) H_3PO_3 is reducing agent and diabasic

(B) H_3PO_3 is reducing agent and tribasic

(C) H_3PO_3 is reducing agent and monobasic

(D) None of these

19. Impure Gold $\xrightarrow[O_2]{KCN}$ (X)

(X) $\xrightarrow{Zn}$ (Y) + other product ; X and Y are

(A) $K[Au(CN)_2]$, K^+ , $[Zn(CN)_4]^{2-}$

(B) $[Au(CN)_2]^-$, $[Zn(CN)_4]^{2-}$

(C) $[Au(CN)_2]$, $[K(CN)_4]^-$ (D) None of these

20. In physical adsorption, the extent of absorbate increases with

(A) decrease in temperature

(B) increase in temperature

(C) decrease in pressure (D) All of these

21. Which is the correct statement for real gases ?

(A) Shape of the gas molecule is indefinite $\frac{PV}{nRT} = 1$

(B) Shape of the gas molecule is indefinite $\frac{PV}{nRT} > 1$

(C) Gases have definite size of molecules for real gases $\frac{PV}{nRT} > 1$

(D) None of these

22. (X) + dil. $H_2SO_4 \rightarrow$ (Y).
Y is an irritating smelling gas
(Y) + $K_2Cr_2O_7$ + dil. $H_2SO_4 \rightarrow$ Green transparent solution. X and Y are

(A) SO_3^{2-}, SO_2 (B) HCl , KNO_2

(C) KCl , SO_3 (D) None of these

23. If ^{23}Na is stable then ^{24}Na will undergo

(A) β-emission (B) α-emission

(C) γ-emission (D) α, β-emission

24. In which of the following, number of atoms are maximum ?

(A) 24 gm of Mg. (B) 20 gm of Ca

(C) 1 gm of H_2 (D) 24 gm of C

25. When MnO_2 is fused with KOH in presence of air, the compound formed and its colour is

(A) K_2MnO_4 and deep blue (B) K_2MnO_4 and deep red

(C) K_2MnO_4 and deep green (D) None of these

26. Find the rate constant (s^{-1}) of first order when the activity drops from 800 mol/dm³ to 50 mol/dm³ in 2 seconds.

(A) $1.40 \times 10^{-5}\ s^{-1}$ (B) $1.30 \times 10^{-4}\ s^{-1}$

(C) $1.30 \times 10^{-8}\ s^{-1}$ (D) $1.38\ s^{-1}$

27. A 2 litre solution (X) contain 0.02 mole of $[Co(NH_3)_5SO_4]Br$ and 0.02 mole $[Co(NH_3)_5Br]SO_4$. 1 litre of this solution (X) required
X + $AgNO_3$ (excess) → Y mole of AgBr ↓
X + $BaCl_2$ (excess) → Z mole of $BaSO_4$ ↓
The values of Y and Z are

(A) 0.02 , 0.03 (B) 0.00 , 0.01

(C) 0.01 , 0.01 (D) 0.01 , 0.02

28. In an electrolytic cell flow of e^- will be

(A) from anode to internal source

(B) from cathode to anode through internal source

(C) from cathode to anode (D) None of these

IIT - 2004

1. I^- reacts with MnO_4^- in alkaline medium. I^- change to

(A) I_2 (B) IO_4^- (C) IO_2^- (D) IO_3^-

2. The number of lone pairs in $XeOF_4$ is

(A) 0 (B) 1 (C) 2 (D) 3

3. The O–O linkage is present in

(A) $H_2S_2O_3$ (B) $H_2S_2O_8$ (C) $H_2S_4O_6$ (D) $H_2S_2O_6$

4. Which pair of the following compounds has elements in their highest oxidation state?

(A) $[Fe(CN)_6]^{3-}$ and $[Co(CN)_6]^{3-}$

(B) $[MnO_4]^-$ and CrO_2Cl_2

(C) $[MnO_4]^-$ and $[NiF_4]^{2-}$

(D) MnO_2 and TiO_2

5. Heating $(NH_4)_2Cr_2O_7$ yields a gas, which is also obtained by

(A) Mg_3N_2 in H_2O (B) heating NH_4NO_2

(C) heating NH_4NO_3 (D) Na_2O_2 in H_2O

6. Which of the following complexes has tetrahedral geometry ?

(A) $[Ni(CN)_4]^{2-}$ (B) $[Pd(CN)_4]^{2-}$

(C) $[NiCl_4]^{2-}$ (D) $[PdCl_4]^{2-}$

7. Spin magnetic moment in Bohr Magneton of $Hg[Co(SCN)_4]$ is

(A) $\sqrt{15}$ (B) $\sqrt{24}$ (C) $\sqrt{3}$ (D) $\sqrt{8}$

8. One of the conformations of n-butane is drawn in the given figure.

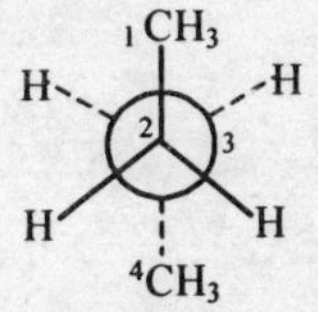

(Newman projection: $^1CH_3$ (top), H, H, H, H, $^4CH_3$ (bottom); carbons 2, 3)

Anticlockwise rotation of C_2 around C_2-C_3 bond by 120° will lead to

(A) gauche (B) staggered

(C) partially eclipsed (D) fully eclipsed

9. The number of chiral compounds obtained on monochlorination of 2-methyl butane is

(A) 2 (B) 4 (C) 6 (D) 8

10. 2-hexyne will give trans-2-hexene on heating with

(A) H_2, Pd/$BaSO_4$ (B) H_2, Pt

(C) $NaBH_4$ (D) Li, NH_3 in ethanol

11. 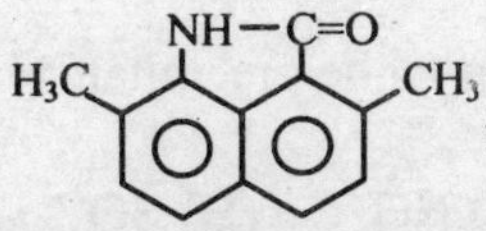 (structure: NH—C=O, H_3C, CH_3) is subjected to monobromination with Br_2/Fe. The major product is

(A) 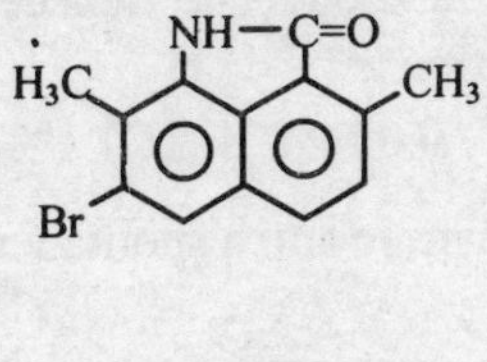 (structure: NH—C=O, H_3C, CH_3, Br) (B) 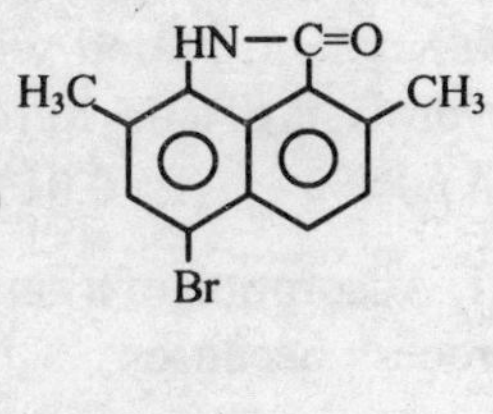 (structure: HN—C=O, H_3C, CH_3, Br)

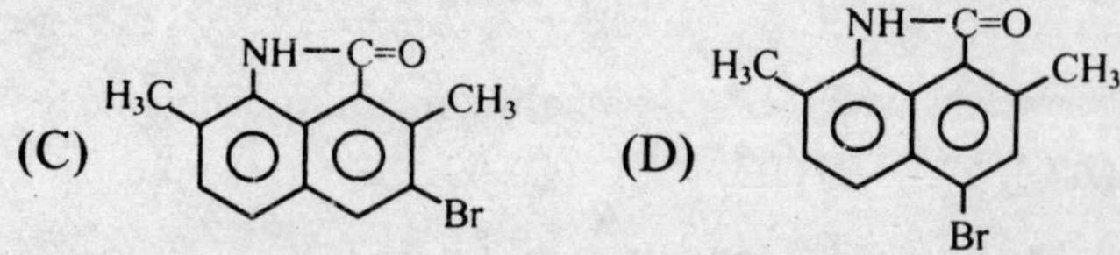

12.The order of reactivity of the following compounds with PhMgBr is

(I) PhCOPh (II) CH_3CHO (III) CH_3COCH_3

(A) (I) > (II) > (III) (B) (III) > (I) > (II)

(C) (II) > (I) > (III) (D) (II) > (III) > (I)

13. Which of the following product is formed on acid catalysed hydration of 2-phenyl propene?

(A) 1-phenyl-2-propanol (B) 1-phenyl-1-propanol

(C) 1-phenyl-3-propanol (D) 2-phenyl-2-propanol

14 Which of the following has same Bohr's radius as hydrogen atom in n = 1 shell ?

(A) n = 2, He^+ (B) n = 2 , Li^{2+}

(C) n = 3, Li^{2+} (D) n = 2, Be^{3+}

15. Which of the following statements is true about O_2 and O_2^+ on the basis of M.O.T ?

(A) is paramagnetic and its bond order is higher than O_2.

(B) is paramagnetic and its bond order is lower than O_2.

(C) is diamagnetic and its bond order is higher than O_2.

(D) is diamagnetic and its bond order is lower than O_2

16. An ideal monoatomic gas has molecular mass M and E is the average kinetic energy per mole, then which of the following relation holds true?

(A) $V_{rms} = \left(\frac{2E}{M}\right)^{\frac{1}{2}}$ (B) $V_{rms} = \left(\frac{M}{2E}\right)^{\frac{1}{2}}$

(C) $V_{rms} = \left(\frac{2}{3}\frac{M}{E}\right)^{\frac{1}{2}}$ (D) $V_{rms} = \left(\frac{3}{2}\frac{E}{M}\right)^{\frac{1}{2}}$

17. 0.004 M Na_2SO_4 is isotonic with 0.01 M glucose at equilibrium. What is the apparent degree of dissociation of Na_2SO_4?

(A) 25% (B) 50% (C) 75% (D) 85%

18.Pb and Sn metals are extracted from their ores by

(A) Self reduction and carbonreductionrespectively.

(B) Carbon reduction and self reduction respectively.

(C) Electrolytic and self reduction respectively.

(D) Self reduction and electrolytic reduction respectively.

19.Which of the following gives white precipitate with $MgCl_2$ on boiling ?

(A) CO_3^{2-} (B) HCO_3^- (C) NO_3^- (D) SO_4^{2-}

20.A weak acid HX has Ka = 1.0×10^{-5} and it forms a salt NaX when treated with caustic soda. The degree of hydrolysis of a 0.1 M NaX solution is

(A) 0.1% (B) 0.01% (C) 0.001% (D) 1%

21. Adsorption of a gas is exothermic and a spontaneous process because

(A) enthalpy increases (B) entropy increases

(C) entropy decreases (D) free energy increases

22. 2 moles of an ideal gas expand isothermally and reversibly at 300 K from V_1 = 1 litre to V_2 = 10 litres. The change in enthalpy in kJ mol^{-1} is

(A) 11.47 (B) -11.47 (C) 0 (D) 4.8

23. Enthalpy of vaporization of a liquid is 30 kJ mol^{-1} and the entropy change during the process is 75 J mol^{-1} K^{-1}. The boiling point of the liquid is

(A) 400 K (B) 500 K (C) 750 K (D) 3

24. In the first order reaction, X → Y ; the concentration of X changes from 0.1 M to 0.025 M in 40 min. The rate of reaction (mol/L/min) when concentration of X is 0.01 M, would be

(A) 3.47×10^{-5} (B) 3.47×10^{-4}

(C) 5.8×10^{-4} (D) 5.8×10^{-6}

25. The emf of the given cell,

Zn | Zn^{2+} (0.1 M) || Fe^{2+} (0.01 M) | Fe is 0.2905 V at 298 K. K_{eq} of this cell will be

(A) not available (B) $10^{0.32/0.059}$

(C) $10^{0.32/0.0295}$ (D) $e^{0.32/0.0295}$

26. Benzamide $\xrightarrow{POCl_3}$ (X).The product (X) is

(A) Aniline (B) Benzonitrile

(C) Benzylamine (D) Chlorobenzene

27.Acidic order of [structure: $\overset{+}{N}H_3$ (y) and COOH (x) on the same carbon of a chain ending in $\overset{+}{N}H_3$ (z)] is

(A) (z) > (y) > (x) (B) (y) > (x) > (z)

(C) (x) > (y) > (z) (D) (x) > (z) > (y)

28. Which pair of compounds gives Tollen's test ?

(A) Glucose & Fructose (B) Sucrose & Glucose

(C) Hexanal & Acetophenone (D)Fructose & Sucrose

IIT - 2005

1. The number of radial nodes on 3s and 2p orbitals respectively are:

(A) 2, 0 (B) 2, 1 (C) 2, 2 (D) 0, 2

2. Which of these has the maximum number of lone pairs of electron on the central atom?

(A) ClO (B) I (C) SF_4 (D) XeF_4

3. Ozone cannot oxidise

(A) KI (B) $KMnO_4$ (C) K_2MnO_4 (D) $FeSO_4$

4. Which allotrope of phosphorus is thermodynamically most stable?

(A) White (B) Red (C) Black (D) Yellow

5. C_6H_5-CHO + (X) $\xrightarrow[(2)\ 2H^+]{(1)\ CH_3COONa}$ C_6H_5-CH=CHCOOH

Identify the compound (X).

(A) CH_3COOH (B) $(CH_3CO)_2O$
(C) $BrCH_2COOH$ (D) CHOCOOH

6.Which type/s of isomerism is/are exhibited by the octahedral complex $Co(NH_3)_4Br_2Cl$?
(A) Geometrical and ionisation (B) Optical and ionisation
(C) Geometrical only (D) Optical only

7. The ratio of the rate of diffusion of He to that of methane at the same temperature and pressure would be
(A) 2 (B) 1 (C) 4 (D) 0.5

8. Which of the following ore contains copper as well as iron?
(A) Cuprite (B) Chalcocite
(C) Chalcopyrite (D) Malachite

9. 13.44 g $CuCl_2$ is dissolved in 1 kg of water. What is the elevation in boiling point? (Useful information: molecular weight of $CuCl_2$ = 134.4 g mol"1, K_b for water = 0.52 K kg mol"1).
(A) 0.05 K (B) 0.16 K (C) 0.10 K (D) 0.20 K

10. Which of the following gas is obtained as a product of the reaction when PbO_2 is heated with concentrated HNO_3?
(A) O_2 (B) N_2 (C) NO_2 (D) N_2O

11. Which compound has cubic close packed structure (FCC) with cation in alternate tetrahedral voids?
(A) NaCl (B) ZnS (C) Na_2O (D) CaF_2

12. 4"methylbenzene sulphonic acid
(A) ⌬-SO_2-OCO-CH_3 (B) ⌬-CO_2Na+ SO_2 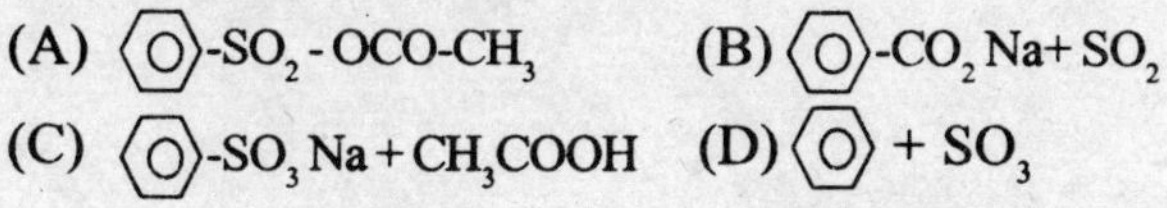
(C) ⌬-SO_3Na + CH_3COOH (D) ⌬ + SO_3

13 [1-bromo-2-chlorocyclobutane: □ with Cl and Br] + Na (2 equivalents) $\xrightarrow{\text{dry ether}}$? The product of the reaction is
(A) □-Cl (B) □-Br (C) □ (D) [bicyclobutane]

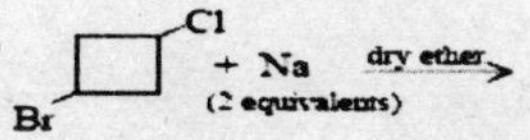

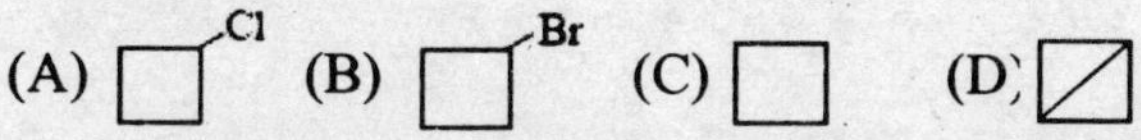

14. Which of the following is the best reagent to convert cyclohexanol to cyclohexene?
(A) Concentrated H_3PO_4 (B) Concentrated HCl
(C) Concentrated HBr
(D) Concentrated HCl + anhydrous $ZnCl_2$

15. The product of the reaction between

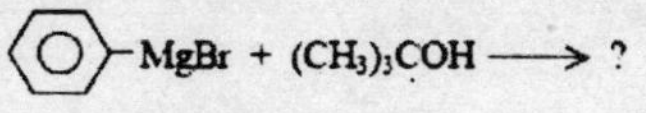

⌬-MgBr + $(CH_3)_3COH$ ⟶ ?

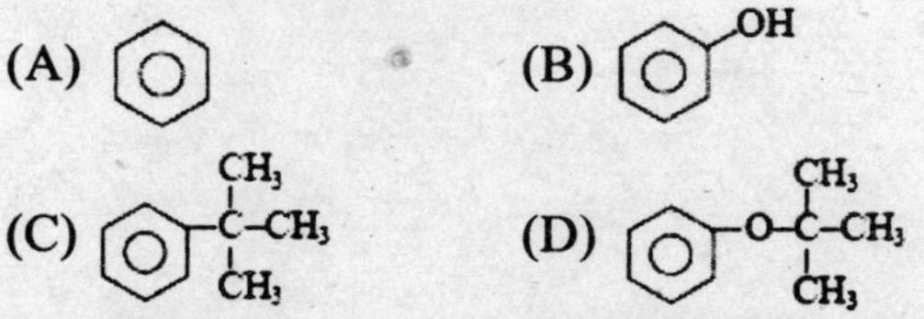

(A) ⌬ (B) ⌬-OH
(C) ⌬-C$(CH_3)_3$ (D) ⌬-O-C$(CH_3)_3$

16. The two forms of D"glucopyranose are
(A) epimers (B) anomers
(C) enantiomers (D) geometrical isomers

17. Which of the following is incorrect regarding the order of a chemical reaction?
(A) Order is determined experimentally.
(B) Order cannot be fractional.
(C) Order does not depend on sum of the stoichiometric coefficients.
(D) Order is equal to sum of the exponents of the concentration terms in the rate law expression of reaction.

18. Which of the following salt solutions (freshly prepared) will show same colour in aqueous solution?
(A) $VOCl_2$ and $CuCl_2$ (B) $VOCl_2$ and $FeCl_2$
(C) $FeCl_2$ and $MnCl_2$ (D) $FeCl_2$ and $CuCl_2$

19. Which of the following is true for the lyophilic sols?
(A) Irreversible sols.
(B) Coagulate readily on addition of an electrolyte.
(C) They are formed by inorganic compounds.
(D) Self stabilised sols.

20. Which of the following metal nitrate on reaction with KI gives a black precipitate and on further addition of KI gives an orange coloured solution?
(A) Hg^{2+} (B) Pb^{2+} (C) Cu^{2+} (D) Bi^{3+}

21. 0.1 mole of CH_3NH_2 (K_b = 5 × 10"4) and 0.08 mole of HCl are mixed and diluted to 1 l. What is the resultant $[H^+]$ at equilibrium?
(A) 8 × 10"11 (B) 1.25 × 10"4 (C) 1.6 × 10"11 (D) 5 × 10"2

22. 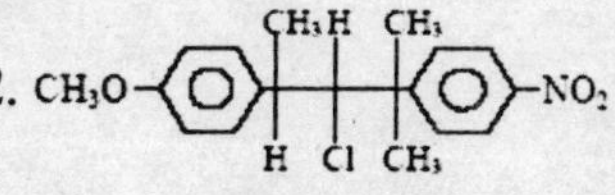 CH_3O-⌬-CH(CH_3)-CH(Cl)-C(CH_3)(CH_3)-⌬-NO_2 on hydrolysis in presence of aqueous acetone gives

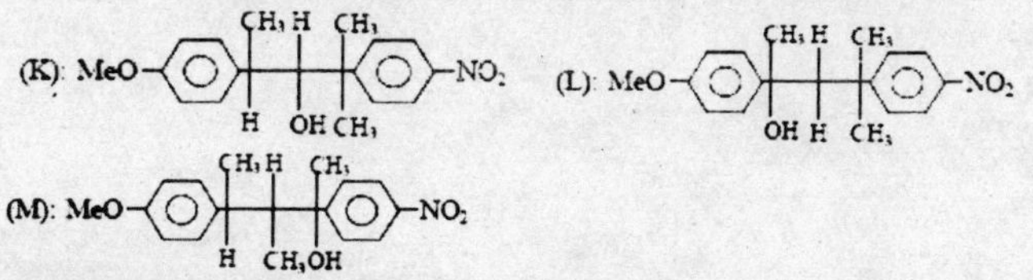

(K): MeO-⌬-CH(CH_3)-CH(OH)-C(CH_3)(CH_3)-⌬-NO_2
(L): MeO-⌬-C(CH_3)(OH)-CH_2-C(CH_3)(CH_3)-⌬-NO_2
(M): MeO-⌬-CH(CH_3)-CH(CH_3)-C(CH_3)(OH)-⌬-NO_2

(A) (K) and (L) (B) (K) only
(C) (K) and (M) (D) (M) only

23. In which of the following structures three oxygen atoms are shared by each SiO_4 tetrahedral unit?
(A) Sheet silicates (B) Pyrosilicates
(C) Single linear chain silicates
(D) Three"dimensional silicates

24. Which of these reagents will form n-propionic acid from CH_3-C(=O)-CH_2CH_3?
(A) NaOH + I_2, H^+ (B) Fehling's solution
(C) NaOH + NaI, H^+ (D) Tollen's reagent

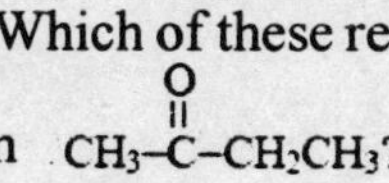

25. Which of the following is the least stable resonating structure of CH_2=CH-CH=CH-O-CH_3
(A) $\overset{\ominus}{C}H_2$-$\overset{\oplus}{C}H$-CH=CH-O-CH_3
(B) $\overset{\ominus}{C}H_2$-CH=CH-CH=$\overset{\oplus}{O}$-CH_3
(C) CH_2=CH-$\overset{\ominus}{C}H$-$\overset{\oplus}{C}H$-O-CH_3
(D) CH_2=CH-$\overset{\ominus}{C}H$-CH=$\overset{\oplus}{O}$-CH_3

26. 1 mol of a monoatomic gas at T K is expanded at constant external pressure of 1 atm adiabatically to double its volume from 1 L to 2 L. What is the final temperature?

(A) T (B) $\frac{T}{2^{\frac{5}{3}-1}}$ (C) $T-\frac{2}{0.0821\times 3}$ (D) $T+\frac{2}{0.0821\times 3}$

27. The reactions representing corrosion of iron are as follows:

$\frac{1}{2} O_2 + 2H^+ + 2e^- \rightarrow H_2O$; E° = 1.23 V

$Fe^{2+} + 2e^- \rightarrow Fe$; E° = 0.44 V

The ΔG° for the corrosion of iron is

(A) 0 (B) −76 kJ (C) −161 kJ (D) −322 kJ

28. When equimolar mixture of the two gases is mixed at "30°C, a blue liquid is formed. The blue liquid is

(A) N_2O (B) N_2O_3 (C) N_2O_5 (D) NO_2

IIT - 1990

1. (B)
2. (D)
3. (A) (E)
4. (A) (D)
5. (B)
6. (D)
7. (D)
8. (D)
9. (C)
10. (C)
11. (A)
12. (D)
13. (B) (D)
14. (C) (D)
15. (B) (D)
16. planar
17. sp
18. Two
19. phenoxide ion
20. Rochelle salt

IIT - 1991

1. (D)
2. (C)
3. (C)
4. (A)
5. (B)
6. (A) (B)
7. (B) (C) (D)
8. (D) (E)
9. (D)
10. (B)
11. (D)
12. (B)
13. (A)
14. R-S-Mg X
15. Silicones
16. 4.134

IIT - 1992

1. (C)
2. (D)
3. (A)
4. (C)
5. (D)
6. (A)
7. (D)
8. (D)
9. (D)
10. (B)
11. (A)
12. (A)
13. (B)
14. (A) (C)
15. (A) (C)
16. (A) (C)
17. (A) (D)
18. (B) (C)
19. (C)
20. (C)
21. o-nitrophenol
22. benzylidene
23. peroxides
24. four
25. oxidation of nitrogen

IIT - 1993

1. (D)
2. (C)
3. (B)
4. (B) (D)
5. (B)
6. (A) (D)
7. (A)
8. (D)
9. (B)
10. (D)
11. photons
12. orbitals
13. higher nuclear charge
14. orientation in space
15. isolated
16. CrO_2Cl_2
17. endothermic
18. ethene

O......H—O
||　　　|

19. $H_3C-C-CH=C-CH_3$
20. Hg^{2+}, H_2SO_4

IIT -1994

1. R -Si - O- SiR_3
2. K_C
3. hyper conjugation
4. $3d^5 4s^1$
5. three centre bond
6. $kp = k_C(RT)^{\Delta n}$
7. paramagnetism
8. less
9. fullerene
10. 1.82×10^{-4} Kg hr^{-1}
11. Zwitter ion
12. 7/3
13. I_3^-
14. hexaamminecobalt (III) chloride
15. butan-1, 4-dioic acid

IIT - 1996

1. (D)
2. (D)
3. (B)
4. (B)
5. (A)
6. (B)
7. (A)
8. (B)
9. (A)
10. (B)
11. (A)
12. (B)
13. (C)
14. (B)
15. (A)
16. (C)
17. (A)
18. (B)
19. (A)
20. (B)
21. (D)
22. (C)
23. (B)
24. (C)

IIT - 1997

1. (B)
2. (D)
3. (B)
4. (A)
5. (B)
6. (C)
7. (A)
8. (C)
9. (C)
10. (B)
11. (A)
12. (D)
13. (B)
14. (D)
15. (D)
16. (B)
17. (C)
18. (B)
19. (D)
20. (A)
21. (C)
22. (A)
23. (D)
24. (B)
25. (D)
26. (B)
27. (A)
28. (D)
29. (A)
30. (C)
31. 12
32. N_2O, I_3^-
33. NF_3
34. 60°
35. less
36. more, $CH_3 NH_2^+$
37. RbC*l*, IC*l*
38. decreases
39. less
40. oxygen
41. infinite temperature
42. O
43. glass
44. extensive
45. I_2
46. H_2S
47. inert pair effect
48. secondary
49. 1, 4-dibromobut-2-ene
50. propene

IIT - 1998	IIT - 1999	IIT - 2000	IIT - 2001	IIT - 2002	IIT - 2003
1. (B)	1. (C)	1. (A)	1. (A)	1. (C)	1. (A)
2. (C)	2. (A)	2. (C)	2. (D)	2. (D)	2. (D)
3. (D)	3. (A)	3. (C)	3. (B)	3. (B)	3. (A)
4. (B)	4. (D)	4. (B)	4. (A)	4. (A)	4. (B)
5. (C)	5. (A)	5. (C)	5. (C)	5. (D)	5. (D)
6. (D)	6. (A)	6. (C)	6. (A)	6. (B)	6. (B)
7. (B)	7. (C)	7. (B)	7. (B)	7. (B)	7. (A)
8. (A)	8. (D)	8. (A)	8. (C)	8. (B)	8. (A)
9. (A)	9. (A)	9. (D)	9. (C)	9. (A)	9. (A)
10. (B)	10. (D)	10. (D)	10. (B)	10. (B)	10. (B)
11. (A) (C) (D)	11. (D)	11. (D)	11. (B)	11. (B)	11. (A)
12. (A) (B) (C)	12. (C)	12. (D)	12. (C)	12. (D)	12. (B)
13. (A) (C)	13. (D)	13. (D)	13. (D)	13. (B)	13. (C)
14. (B)	14. (C)	14. (D)	14. (B)	14. (B)	14. (C)
15. (D)	15. (B)	15. (B)	15. (D)	15. (C)	15. (B)
16. (A) (B) (D)	16. (B)	16. (C)	16. (A)	16. (A)	16. (A)
17. (B) (D)	17. (A)	17. (C)	17. (C)	17. (A)	17. (A)
18. (B) (C)	18. (D)	18. (A)	18. (B)	18. (B)	18. (A)
19. (A) (D)	19. (D)	19. (B)	19. (D)	19. (A)	19. (B)
20. (A) (B)	20. (B)	20. (A)	20. (B)	20. (D)	20. (A)
21. (B) (C)	21. (B)	21. (D)	21. (A)	21. (C)	21. (C)
22. (C) (D)	22. (A)	22. (B)	22. (B)	22. (C)	22. (A)
23. (A) (B) (D)	23. (D)	23. (A)	23. (D)	23. (D)	23. (A)
24. (C) (E)	24. (A)	24. (D)	24. (D)	24. (C)	24. (D)
25. (A) (D)	25. (C)	25. (A)	25. (B)	25. (D)	25. (C)
26. (A) (C)	26. (A) (D)	26. (B)	26. (B)	26. (C)	26. (D)
27. (B) (C)	27. (B) (C)	27. (B)	27. (D)	27. (D)	27. (C)
28. (B) (D)	28. (B)	28. (A)	28. (C)	28. (B)	28. (B)
29. (B) (C)	29. (A) (D)	29. (B)	29. (B)	29. (B)	
30. (A) (B)	30. (B) (C) (D)	30. (A)	30. (D)	30. (A)	
31. (A) (C)	31. (B) (C) (D)	31. (C)	31. (C)		
	32. (A) (B)	32. (C)	32. (A)		
	33. (A) (D)	33. (D)	33. (B)		
	34. (A) (B) (D)	34. (B)	34. (D)		
	35. (B) (C) (D)	35. (A)	35. (A)		

IIT - 2004

1. (D)	5. (B)	9. (B)	13. (D)	17. (C)	21. (C)	25. (C)
2. (B)	6. (C)	10. (D)	14. (D)	18. (A)	22. (C)	26. (B)
3. (B)	7. (A)	11. (B)	15. (A)	19. (B)	23. (A)	27. (C)
4. (B)	8. (A)	12. (D)	16. (A)	20. (B)	24. (B)	28. (A)

IIT - 2005

1. (A)	5. (B)	9. (C)	13. (D)	17. (B)	21. (A)	25. (C)
2. (B)	6. (A)	10. (A)	14. (A)	18. (D)	22. (A)	26. (C)
3. (B)	7. (A)	11. (D)	15. (A)	19. (D)	23. (A)	27. (D)
4. (C)	8. (C)	12. (C)	16. (B)	20. (D)	24. (A)	28. (B)

JIPMER -1990

1. 10ml. of conc. H_2SO_4 (18 Molar) is diluted to 1 litre. The approximate strength of dilute acid is
(A) 18.0N (B) 0.09N (C) 0.18N (D) 0.36N

2. A compound which on reaction with aqueous nitrous acid at low temperature, produces nitroso amine is
(A) Diethyl amine (B) Triethyl amine
(C) Methyl amine (D) Ethyl amine

3. The radiations from naturally occurring radioactive substance, as seen after deflection by a magnetic field in one direction are
(A) Definitely α-rays (B) Definitely β-rays
(C) Both α and β- rays (D) Either α or β- rays

4. Which one of the following can be used in place of NH_4Cl for the identification of III group basic radicals
(A) $(NH_4)_2CO_3$ (B) NaCl
(C) NH_4NO_3 (D) $(NH_4)_2SO_4$

5. Ethylamine reacts with nitrous acid to form
(A) Methyl alcohol (B) Ethyl alcohol
(C) Ethane (D) Ethyl nitrate

6. In alkaline medium $KMnO_4$ reacts as
$2KMnO_4 + 2KOH \rightarrow 2K_2MnO_4 + H_2O + O$
Its equivalent weight will be
(A) 72.0 (B) 158.0 (C) 31.5 (D) 52.7

7. On hydrolysis of starch we obtain finally
(A) Glucose (B) Fructose
(C) Glucose and fructose (D) Sucrose

8. In benzene, the overlapping between carbon-carbon is of the type
(A) sp–sp (B) p–p (C) sp^3–sp^3 (D) sp^2–sp^2

9. Evaporation of a liquid is accompanied by :
(A) Increase in enthalpy (B) Decrease in enthalpy
(C) No change in free energy (D) Increase in entropy

10. The strongest electrophile of the following is
(A) NO^+ (B) NO_2^+ (C) H_3O^+ (D) RN_2

11. Which of the following is a polar molecule
(A) CO_2 (B) HCl (C) CCl_4 (D) BCl_2

12. Red phosphorus is chemically more denser and less reactive because
(A) It has polymeric structure (B) It is red in colour
(C) It is tetrahedral in structure
(D) It has octahedral structure

13. A diastereo isomer
(A) Is mirror image of an enantiomer
(B) Is optically inactive (C) Is optically active
(D) Does not contain a chiral centre.

14. Which of the following is not a salt derived from H_3PO_4
(A) NaH_2PO_3 (B) NaH_2PO_4
(C) Na_2HPO_4 (D) Na_3PO_4

15. The strongest Bronsted base of the following is
(A) ClO_2^- (B) ClO_4^- (C) ClO^- (D) ClO_3^-

16. Picric acid is
(A) Trinitro aniline (B) Trinitro toluene
(C) A volatile liquid (D) 2,4,6 trinitrophenol

17. 12 g. of Mg (At. wt. 24) will react completely with an acid to give
(A) One mole of O_2 (B) One half mole of H_2
(C) One mole of H_2 (D) None

18. Number of electrons in any shell is
(A) 2π (B) n (C) $2n^2$ (D) n^2

19. Iron is rendered passive by treatment with conc.
(A) HCl (B) HNO_3 (C) H_2SO_4 (D) H_3PO_4

20. In which of the following compounds does the ratio of anion to cation size have the lowest value
(A) CsI (B) LiI (C) CsF (D) LiF

21. How many gms of NaOH will be required to prepare 250 ml of 0.1 M solution
(A) 40gms (B) 10gms (C) 1gm (D) 4gms

22. Compound having lowest boiling point is
(A) HBr (B) HI (C) HF (D) HCl

23. Yellow phosphorus is kept in
(A) Water (B) Kerosene (C) Ether (D) Alcohol

24. Ce (Z = 58) is a member of
(A) s-block element (B) p- block elements
(C) d-block elements (D) f-block elements

25. Which of the following represents the correct set up of the four quantum numbers of 4s electron
(A) 4, 3, 2, +1/2 (B) 4, 2, 1, 0
(C) 4, 3, - 2, +1/2 (D) 4, 0, 0, -1/2

26. An enzyme which brings about the conversion of starch into maltose is
(A) Invertase (B) Diastase (C) Maltase (D) Zymase

27. The molecule having one unpaired electron is
(A) NO (B) CO (C) CN^- (D) O_2

28. The ion isoelectronic with CO is
(A) N_2^{2+} (B) O_2^{2-} (C) CN^- (D) O_2

29. Which of the following pairs of ions can not be separated by H_2S in dilute HCl ?
(A) Bi^{3+}, Sn^{4+} (B) Al^{3+}, Hg^{2+}
(C) Ni^{2+}, Cu^{2+} (D) Zn^{2+}, Cu^{2+}

30. Which of the following electron transitions in a hydrogen atom will require the largest amount of energy
(A) From n = 1 to n = 2 (B) From n = 2 to n = 3
(C) From n = ∞ to n = 1 (D) From n = 3 to n = 5

31. A gas has a volume of 27.3 ml at 0°C. What would be its volume at 10°C if the pressure remains unchanged
(A) 2.73 ml (B) 2.83 ml (C) 27.3 ml (D) 28.3 ml

32. $(CH_3)_3$ –C–NH_2 is a/an
(A) Primary amine (B) Secondary amine
(C) Tertiary amine (D) Ammonium salt

33. A molecule is divided into two halves which are mirror images of each other by s
(A) An axis of symmetry (B) A centre of symmetry

(C) A plane of symmetry (D) None

34. Dissociation constant of an acid is 1 x 10^{-10}. The pK_a of the acid is
(A) 4 (B) 5 (C) 10 (D) –10

35. The value of equilibrium constant for the reaction $H_2 + I_2 \rightleftharpoons 2NI$ is 48. What would be the value of equilibrium constant for the reverse reaction
(A) 0.04 (B) 0.08 (C) 0.01 (D) 0.02

36. Moderate electrical conductivity is shown by
(A) Silica (B) Graphite
(C) Diamond (D) Carborundum

37. Phenol may be distinguished from benzoic acid by reacting with
(A) Na_2CO_3 (B) RMgX (C) NaOH (D) $NaHCO_3$

38. In which of the following the central atom does not use sp^3 hybrid orbitals in its bonding
(A) NH_3^- (B) NF_3 (C) BeF_3^- (D) OH_3^+

39. Pure aniline is a
(A) Brown coloured liquid (B) Brown coloured solid
(C) Colourless liquid (D) Colourless solid

40. Which of the following substances reacts with both dilute and conc. H_2SO_4 to give the same product
(A) Cu (II) Oxide (B) $NaNO_3$
(C) Sucrose (D) Zinc

41. The ion that is isoelectronic with CO is
(A) O_2^- (B) N_2^+ (C) CN^- (D) None of these

42. The bond order of individual carbon bonds in benzene is
(A) One (B) Two
(C) Between one and two (D) One and two alternately

43. Which of the following hydrides has the lowest boiling point
(A) SbH_3 (B) AsH_3 (C) NH_3 (D) PH_3

44. If uranium (mass no. 238, atomic no. 92) emits an α-particle the product formed has mass number and atomic number respectively as
(A) 236 and 92 (B) 234 and 90
(C) 238 and 90 (D) 236 and 90

45. The element with highest second ionisation potential is
(A) O (B) F (C) C (D) N

46. The element with highest first ionisation potential is
(A) B (B) N (C) O (D) C

47. Which of the following gives acetone on ozonolysis ?
(A) 2-Butene (B) Propene
(C) 1-Butene (D) 2-Methyl-propene

48. Which of the following is not a petroleum product
(A) Paraffin wax (B) Kerosene oil
(C) Gasoline (D) Toluene

49. Which has the maximum electronegative character
(A) Fluorine (B) Astatine (C) Oxygen (D) Nitrogen

50. The nitration of nitrobenzne will give
(A) T.N.T. (B) 1,3 dinitrobenzene
(C) Picric acid (D) 1,4 dinitrobenzene

51. Colour is imparted to glass by mixing
(A) Synthetic dye (B) Metal oxide
(C) Oxide of non-metal (D) Colour

52. The least electro negative element is
(A) Sodium (B) Oxygen (C) Nitrogen (D) Sulphur

53. Which is soluble in acetone
(A) KI (B) NaI (C) LiI (D) NaCl

54. The total number of electrons in one molecule of CO_2 is :
(A) 22 (B) 44 (C) 66 (D) 88

55. If three electrons are lost by a metal ion M^{3+}, its final oxidation state would be
(A) 0 (B) +2 (C) +4 (D) +6

56. Which is not oxidised by MnO_2
(A) F^- (B) Cl^- (C) Br^- (D) I^-

57. The oxide that gives H_2O_2 on treatment with dilute acid is
(A) Na_2O_2 (B) TiO_2 (C) PbO_2 (D) MnO_2

58. The acid with highest pH is
(A) Formic acid (B) Acetic acid
(C) Monochloroacetic acid (D) Dichloroacetic acid

59. The linear structure is assumed by
(A) $SnCl_2$ (B) NCO^- (C) NO_2^+ (D) SO_2

60. Which of the following is not a nucleophile
(A) BF_3 (B) HSO_2^- (C) OH^- (D) NH_3

JIPMER - 1991

1. The number of moles of solute per kilogram of a solvent is called
(A) Normality (B) Mole fraction
(C) Molarity (D) Molality

2. With hot water, P_2O_5 gives
(A) H_2PO_3 (B) HPO_3 (C) H_3PO_3 (D) H_3PO_4

3. The principal quantum number of an atom is related to the
(A) Orbital angular momentum
(B) Size of the orbital (C) Spin angular momentum
(D) Orientation of the orbital in space

4. Which of the following exhibits optical activity
(A) Malonic acid (B) Maleic acid
(C) Glycollic acid (D) Malic acid

5. The linear molecule among the following is
(A) SO_2 (B) ClO_2 (C) CO_2 (D) NO_3

6. The compound that is not a Lewis acid is
(A) $AlCl_3$ (B) $SnCl_4$ (C) BF_3 (D) $BaCl_2$

7. Molecular formuia, C_3H_8O represents
(A) 2 Isomers (B) 3 Isomers

(C) 4 Isomers (D) 5 Isomers

8. Which has the minimum atomic radius
(A) K (B) F (C) N (D) Na

9. Ethyl alcohol is obtained when ethyl chloride is boiled with
(A) Alcoholic KOH (B) Aqueous KOH
(C) H_2O (D) H_2O_2

10. Mass of an individual atom of sulphur is
(A) 32 x 6.023 x 10^{23}g (B) 32 x 10^{-23}g.
(C) 32 g. (D) 5.31 x 10^{-23}g.

11. The compound with the highest boiling point is
(A) CH_3Br (B) CH_3I (C) CH_4 (D) CH_3OH

12. The compound which reacts faster with Lucas reagent is
(A) Butan-1-ol (B) Butan-2-ol
(C) 2-Methyl propan-1-ol (D) 2-Methyl propan-2-ol

13. The pH of 0.001 M solution of HCl is
(A) 3.0 (B) 10 (C) 1.0 (D) 2.0

14. The species that does not contain peroxide ions is
(A) SrO_2 (B) BaO_2 (C) PbO_2 (D) H_2O_2

15. If the pressure and temperature of 2 litres of CO_2 are doubled, the volume of CO_2 would become
(A) 8 litres (B) 2 litres (C) 5 litres (D) 4 litres

16. Ethyl alcohol is heated with conc. H_2SO_4. The product is
(A) $H_3C–CO–OC_2H_5$ (B) C_2H_6 (C) C_2H_4 (D) C_2H_2

17. The molecule that will have dipole moment is
(A) 2,2 dimethyl propane (B) 2,2,3,3 tetra methyl butane
(C) 3-hexene (D) Trans-2-pentene

18. A precipitate of calcium oxalate will not dissolve in ?
(A) Aqua regia (B) Acetic acid (C) HCl (D) HNO_3

19. An endothermic reaction is one in which
(A) Heat is absorbed (B) Heat is given out
(C) Heat is converted into electricity
(D) Heat is converted into mechanical work

20. Phenol does not react with
(A) $NaHCO_3$ (B) KOH (C) Na_2CO_3 (D) NaOH

21. Which has maximum percentage of chlorine
(A) Pyrene (B) PVC
(C) Gammexane (D) Ethylidene chloride

22. 30 ml of acid solution is neturalised by 15ml of 0.2 N base. The strength of the acid solution is
(A) 0.4N (B) 0.3N (C) 0.1N (D) 0.15N

23. Which of the following is most stable to heat
(A) HCl (B) HOCl (C) HBr (D) HI

24. Acidified $KMnO_4$ solution is decolourised by
(A) Bleaching powder (B) Mohr's salt
(C) Microcosmic salt (D) Laughing gas

25. An electrovalent compound is made up of
(A) Electrically charged atoms or groups of atoms
(B) Neutral atoms (C) Electrically charged molecules
(D) Neutral molecules

26. The weakest acid among the following is
(A) CH_3COOH (B) $ClCH_2COOH$
(C) Cl_2CHOOH (D) Cl_3CCOOH

27. When calcium acetate and calcium formate are heated together they form
(A) Acetone (B) Formic acid
(C) Acetic acid (D) Acetaldehyde

28. Nitrogen is chemically inert because
(A) Its electronic configuration is stable
(B) Its electronegativity is high
(C) Its molecular size is small
(D) Its bond dissociation energy is high

29. The cyanide ion, CN^- and N_2 are isoelectronic. But in contrast to CN^-, N_2 is chemically inert, because of
(A) Low bond energy (B) Absence of bond polarity
(C) Unsymmetrical electron distribution
(D) Presence of more number of electrons in bonding orbitals

30. Molten sodium chloride conducts electricity due to the presence of :
(A) Free molecules (B) Atoms of sodium and chlorine
(C) Free electrons (D) Free ions

31. An alloy of tin and copper is known as
(A) German silver (B) Brass
(C) Bronze (D) Solder

32. If n is the number of chiral atoms in a compound, the maximum number of stereo isomers that can exist is
(A) 2^{n+2} (B) 2^{n-1} (C) n^2 (D) 2^n

33. Which of the following is an acidic salt
(A) Na_2S (B) NaHS (C) Na_2SO_3 (D) Na_2SO_4

34. The volume of 0.1M H_2SO_4 that is needed to completely neutralize 30 ml of 0.2 M NaOH solution is
(A) 40 ml (B) 60ml (C) 30ml (D) 15 ml

35. Compound having zero dipole moment is
(A) H_2O (B) $BeCl_2$ (C) NH_3 (D) HCl

36. If two compounds have the same crystalline structures and analogous formulae, they are called :
(A) Isotopes (B) Allotropes (C) Isomers (D) Isomorphs

37. Which of the following can show positive and negative oxidation states
(A) F (B) Ne (C) Na (D) I

38. The reagent with which both acetaldehyde and acetone react easily is
(A) Fehling's solution (B) Grignard reagent
(C) Schiff's reagents (D) Tollen's reagent

39. The simplest formula of a compound containing 50% of element X (at. wt. 10) and 50% of element Y (at.wt = 20) is
(A) XY_3 (B) X_2Y_3 (C) XY (D) X_2Y

40. Which of the following forms molecular solid
(A) SO_2 (B) SiO_2 (C) SiC (D) Diamond

41. The hydrolysis of CH_3CHCl_2 gives ?
(A) CH_3CHO (B) CH_3COOH
(C) $CHCl_3$ (D) C_2H_5OH

42. When gases are heated from 20 to 40°C at constant pressure, the volumes
(A) Increase in the ratio of their molecular masses
(B) Increase but to different extent.
(C) Increase by the same magnitude (D) Become double

43. The volume of 0.1 M H_2SO_4 that is needed to completely neutralize 40ml of 0.2 M NaOH is
(A) 10ml (B) 20ml (C) 40ml (D) 80ml

44. Equal weights of hydrogen and helium are mixed in an empty container at 20°C. The fraction of total pressure exerted by hydrogen is
(A) 2/3 (B) 2/3 x 273/293 (C) 1/2 (D) 1/3

45. How many unpaired electrons are there in Ni^{2+}
(A) 0 (B) 2 (C) 4 (D) 8

46. Most chemically active form of phosphorus is
(A) Yellow (B) Violet (C) Black (D) Red

47. Which of the following has bond order 3
(A) C N (B) O^{-2} (C) N O^+ (D) C N^+

48. The p–p–p bond angle in white phosphorus is
(A) 60° (B) 109°28 (C) 120° (D) 90°

49. Any p-orbital can accommodate upto
(A) 4 electrons (B) 6 electrons
(C) Two electrons with parallel spins
(D) Two electrons with opposite spins

50. Which of the following basic radicals will not be precipitated by passing H_2S in concentrated acid solution
(A) As (B) Cd (C) Cu (D) Sb

51. Which of the molecule has pyramidal shape
(A) BF_3 (B) $BeCl_2$ (C) $SiCl_4$ (D) PCl_3

52. A gas can be liquefied by pressure alone when its temperature is
(A) Lower than its critical temperature
(B) Higher than its critical temperature
(C) Equal to its critical temperature (D) None

53. The process of covering iron sheets with a layer of zinc is called
(A) Galvanising (B) Zinc plating
(C) Tempering (D) Dressing

54. Which of the following dissolves in hot conc. NaOH solution
(A) Cu (B) Ag (C) Fe (D) Zn

55. Lunar caustic is
(A) AgCl (B) $AgNO_3$ (C) NaOH (D) KNO_3

56. As a result of osmosis, the volume of the solution
(A) Decreases (B) Increases
(C) Do not change (D) None

57. The compound that is most reactive towards electrophillic nitration is
(A) Toluene (B) Benzene
(C) Benzoic acid (D) Nitro benzene

58. Absolute alcohol contains
(A) 40% water (B) 20% water
(C) 5% water (D) No water

59. The bond angle in carbon tetrachloride is approx
(A) 120° (B) 180° (C) 90°C (D) 109°

60. Optical isomerism is shown by
(A) Butanol–1 (B) Butanol–2
(C) Isobutanol (D) Tert. butyl alcohol

JIPMER - 1992

1. 27 gm of Al will react completely with how many grams of oxygen
(A) 32g (B) 24g (C) 8g (D) 16g

2. Candles contain a mixture of
(A) Bees wax and stearic acid
(B) Paraffin wax and stearic acid
(C) Bees wax and paraffin wax
(D) Higher fatty acid

3. Ionic bonding is favoured by
(A) High charge on ion, small cation, large anion
(B) Low charge on ions, large cation, small anion
(C) Low charge on lons, large cation, small anion
(D) High charge on ions, large cation, small anion

4. Salicylic acid upon heating with soda lime forms
(A) Benzoic acid (B) Benzene
(C) Benzyl alcohol (D) Phenol

5. When chloroform is exposed to air and sunlight, it gives
(A) CH_3Cl (B) Mustard gas (C) CCl_4 (D) $COCl_2$

6. The process of separating colloidal particles from particles of molecular dimensions is
(A) Crystallisation (B) Filtration
(C) Dialysis (D) Sublimation

7. Isotopes of on element have
(A) Different mass numbers
(B) Different atomic numbers (C) Same mass numbers
(D) Different mass and atomic numbers

8. Butter is a colloid formed when
(A) Water is dispersed in fat
(B) Fat globules are dispersed in water
(C) Fat is dispersed in solid casein
(D) Suspension of casein in water

9. An electron that has the quantum number n = 3 and m = 2 ?
(A) Must have spin value +½ (B) Must have l=1
(C) Must have l= 0, 1 or 2 (D) Must have l=2

10. Phosphorus is manufactured from ?
(A) P_2O_5 (B) $Ca_3(PO_4)_2$ (C) Ca_3P_2 (D) $CaHPO_3$

11. Helium atom is twice heavier than a hydrogen

molecule. At 25°C the average K. E. of helium atom is :
(A) Four times that of hydrogen
(B) Twice that of hydrogen (C) Same as that of hydrogen
(D) Half that of hydrogen

12. Upon electrolyzing a dilute solution of H_2SO_4 between platinum electrodes, the gas evolved at the anode is
(A) O_2 (B) H_3 (C) SO_2 (D) SO_3

13. 2.76 gm of silver carbonate on being strongly heated yield a residue weighing
(A) 2.32 g (B) 2.64 g (C) 2.48 g (D) 2.16 g

14. Number of optical isomers possible for 2, 3 butandiol is
(A) 4 (B) 5 (C) 2 (D) 3

15. Be's 4th electron will have the four quantum numbers
(A) n = 1, l = 0, m = 0, s = ½
(B) n = 2, l = 0, m = 0, = –½
(C) n = 1, l = 0, m = 0, s = –½
(D) n = 2, l = 1 m = 0, s = ½

16. Iodoform on warming with silver powders gives ?
(A) Ethane (B) Methane
(C) Acetylene (D) Ethyl alcohol

17. A mixture of ethanol and phenol can be separated by ?
(A) Ion exchange (B) Sublimation
(C) Fitration (D) Fractional distillation

18. C–C bond length in benzene is
(A) 1.34 Å (B) 1.20 Å (C) 1,54 Å (D) 1.39 Å

19. Which of the following does not dissolve in aqua-regia ?
(A) Ir (B) Ag (C) Pt (D) Pd

20. The maximum number of isomers for an alkene with the molecular formula C_4H_8 is
(A) 4 (B) 5 (C) 2 (D) 3

21. If a molecule MX_2 has zero dipole moment, the sigma bonding orbitals used by M (Z < 21) are
(A) Pure p (B) sp hybrid
(C) sp^2 hybrid (D) sp^3 hybrid

22. The occurrence of different chemical compounds in the same crystalline form is known as
(A) Isomerism (B) Isomorphism
(C) Polymorphism (D) Enantiomorphism

23. Nitrogen dioxide can not be obtained by heating ?
(A) $Pb(NO_3)_2$ (B) $AgNO_3$ (C) KNO_3 (D) $Cu(NO_3)_2$

24. Which of following acid possesses oxidizing, reducing and complex forming properties
(A) HCl (B) HNO_2 (C) HNO_3 (D) H_2SO_4

25. The pH of a 10^{-3} M solution of HCl in water is
(A) 8 (B) –8
(C) Between 7 and 8 (D) Between 6 and 7

26. NH_3 can enter into complex formation due to
(A) Hydrogen bonding (B) Co-ordinate bond
(C) Covalent bond (D) Ionic bond

27. The chain isomers represented by formula C_6H_{14} are
(A) 3 (B) 6 (C) 5 (D) 4

28. The pH of a solution is 4. The H^+ ion concentration of the solution in gms/litre is
(A) 10^{-4} (B) 4 (C) 10^{-2} (D) 10^4

29. The carbon-magnesium linkage in Grignard reagent is
(A) A metallic bond (B) Co-ordinate bond
(C) Weakly polar bond (D) Highly polar covalent bond

30. The crystal of NaCl is
(A) Face centered cubic (B) Body centered cubic
(C) Hexagonal (D) Tetragonal

31. Phenolphthalein is not a good indicator for titrating
(A) NaOH against oxalic acid (B) $FeSO_4$ against $KMnO_4$
(C) NaOH against $KMnO_4$ (D) NaOH against H_2SO_4

32. Separation of two substances by fractional crystallisation depends upon their differences in
(A) Solubility (B) Crystalline shape
(C) Density (D) Volalitity

33. $C_4H_{10}O$ represents .... number of isomers
(A) 2 (B) 4 (C) 5 (D) 7

34. For preparing one litre of N/10 H_2SO_4 solution, the amount of H_2SO_4 needed is
(A) 100 gm (B) 4.9 gm (C) 98 gm (D) 10 gm

35. If the solubility product of lead iodide is 0.8×10^{-9} in water, its solubility in moles per litre is
(A) 2×10^{-3} (B) 1.26×10^{-3} (C) 8.94×10^{-5} (D) 4.47×10^{-5}

36. If 0.5 mole of $BaCl_2$ is mixed with 0.2 mole of Na_3PO_4, the maximum number of mole of $Ba_3(PO_4)_2$ that can be formed is
(A) 0.3 (B) 0.1 (C) 0.7 (D) 0.5

37. Walden inversion is due to
(A) S_N1 mechanism (B) Formation of carbanion
(C) S_N2 displacement (D) Formation of carbonium ion

38. For a spherical shaped orbital, the value of *l* is
(A) 1 (B) 2 (C) 3 (D) 0

39. Approximate atomic weight of an element is 26.89. If its equivalent is 8.9, the exact atomic weight of the element would be
(A) 25.89 (B) 8.9 (C) 26.7 (D) 17.8

40. Which of the following has the maximum ionization potential
(A) N (B) Ne (C) F (D) C

41. Which of the following compounds is covalent :
(A) CaO (B) H_2 (C) Na_2S (D) KCl

42. The conjugate acid of NH_2^- is
(A) NH_3 (B) NH_4^+ (C) NH_4OH (D) N_2H_4

43. Oxygen molecule shows
(A) Paramagnetism (B) Diamagnetism
(C) Ferromagnetism (D) None

44. The compound that gives acetaldehyde when heated with aqueous KOH is
(A) Chloroform (B) Ethylene chloride
(C) Carbon tetrachloride (D) Ethylidine chloride

45. The electronic configuration of alkali metals is
(A) ns^1 (B) $(n-1)d^0 ns^1$
(C) $(n-1)a^5ns^1$ (D) $(n-1)d^{10} ns^2$

46. The orbitals providing the most efficient overlap are
(A) sp^2-sp^2 (B) sp^3-sp^3 (C) sp–sp (D) s–s

47. Acidic hydrogen is present in
(A) Ethylene (B) Ethyne (C) Benzene (D) Ethane

48. H_3PO_4 dissociates in
(A) 3 stages (B) 4 stages (C) 1 stages (D) 2 stages

49. The valency shell of calcium contains
(A) 2 electrons (B) 4 electrons
(C) 6 electrons (D) 8 electrons

50. Which of the following is mono basic
(A) H_3PO_3 (B) H_2PO_2 (C) H_3PO_4 (D) $H_4P_2O_7$

51. Chloroform on nitration gives
(A) Tear gas (B) Water gas
(C) Producer gas (D) Laughing gas

52. Anti markownikoff's addition of HBr is not observed in
(A) Propene (B) Ethene (C) But-2-ene (D) Pent-2ene

53. Which of the following is expected to have the lowest boiling point
(A) n-pentane (B) n-Hexane
(C) Iso-pentane (D) Neo-pentane

54. Which of the following contains unpaired electrons
(A) O_2 (B) H_2O_2 (C) CO (D) N_2

55. Nitration of benzene is
(A) Nucleophilic substitution (B) Electrophilic addition
(C) Electrophilic substitution (D) Nucleophlic addition

56. Which of the following is generally bleached by bleaching power
(A) Straw (B) Ivory (C) Roll of cotton (D) None

57. Which of the following compounds will show negative test with phenyl hydrazine
(A) Acetone (B) Ethyl alcohol
(C) Benzophenone (D) Acetaldehyde

58. Which of the following is most acidic?
(A) O-chlorophenol (B) P-nitrophenol
(C) O-aminophenol (D) Chlorophenol

59. Ammonia reacts with excess of chlorine to form
(A) NOCl (B) $NOCl_2$ (C) NO_2Cl_2 (D) N_2OCl

60. The normality of 1 M solution of H_3PO_4 is
(A) 2 N (B) 3 N (C) 0.5 N (D) 1 N

JIPMER - 1993

1. In a photoelectric effect, the energy of the photon striking a metallic surface is 5.6×10^{-19} J. The kinetic energy of ejected electrons is 12.0×10^{-20} J. The work function is
(A) 4.4×10^{-19}J (B) 6.4×10^{-20}J
(C) 6.4×10^{-19}J (D) 6.8×10^{-19}J

2. Which of the following configurations corresponds to the element with maximum electron affinity ?
(A) [He] $2s^22p^5$ (B) [Ar] $4s^23s^{10} 4p^5$
(C) [Ne] $3s^23p^5$ (D) [Kr] $5s^24d^{10}5p^5$

3. The electronegativity difference between two atoms X and Y in a bond is 2.0, the percentage covalent character in the bond is about.
(A) 46% (B) 8% (C) 50% (D) 54%

4. Which of the following is an acidic oxide ?
(A) Mn_2O_7 (B) MnO (C) Mn_2O_3 (D) MnO_2

5. For the first order reaction the half life period is 20 min. What percentage of the reactant is left at the end of 100 minutes ?
(A)10.0% (B) 3.125% (C) 25% (D) 12.5%

6. A radioactive nucleus $^{218}_{84}Ra$ loses α and β particles and the daughter element has mass number 206 and atomic number. The number of α and β particles lost respectively are
(A) 5, 4 (B) 4, 3 (C) 5, 6 (D) 3, 4

7. For the reaction, 2 A(a) + B(g) ↔ 3C(g) + D(g.), two moles of each of A and B were taken in a flask. The following must be always true at equilibrium
(A) [A] = [B] (B) [A] > [B] (C) [B] = [D] (D) [A] < [B]

8. The commercially known compound gammexane is
(A) Freon (B) Chloral
(C) 2-phenyl ethanol (D) Benzene hexachloride

9. The IUPAC name of the compound ,
$(CH_3)_2CHCH = CHCH_2COOH$ is
(A) 4-methlyl-2-hexenoic acid
(B) 4-isopropyl-butenoic acid
(C) 5-methyl-3-hexenoic acid
(D) 5-carboxyl-2methyl pentene

10. Aniline undergoes condensation to form Schiff base on reacting with :
(A) Acetyl chloride (B) Ammoina
(C) Acetone (D) Benzaldehyde

11. Which of the following is most acidic?
(A) CH_3COOH (B) CH_2COOH with Cl on CH_2 (i.e. $ClCH_2COOH$)
(C) $ClCH_2CH_2COOH$ (D) CH_3CH_2COOH

12. In the presence of air & light, chloroform changes to:
(A) Phosgene (B) Acetylene
(C) Chloropicrin (D) Chloretone

13. The lanthanide contraction relates to
(A) nuclear masses (B) ionic radii
(C) valence electrons (D) the densities

14. The noble gas which does not form clathrate compounds is
(A) Argon (B) Neon (C) Xenon (D) Krypton

15. To an aqueous solution of I^- and Br^-, I_2 and Br_2 are added. The products of the reaction would be

($E^oBr_2/Br^- = 1.09$ V, $E^oI_2/I^- = 0.74$ V) ?
(A) I_2, Br^- (B) I^-, Br^- (C) Br_2, I^- (D) I_2; Br_2

16. Red lead is
(A) Pb_3O_4 (B) PbO (C) $PbCrO_4.PbO$ (D) PbO_2

17. Sulphuric acid dehydrates formic acid to give
(A) CO_2 (B) CO and CO_2 (C) C (D) CO

18. Isotonic solutions have the same
(A) boiling point (B) freezing point
(C) molar concentration (D) vapour pressure

19. Which of the following will react with Br_2 most readily?
(A) $CH_2 = CH_2$ (B) $CH_3CH = CHCH_3$
(C) $CH_3CH = CH_2$ (D) $(CH_3)_2C = C(CH_3)_2$

20. Which of the following compounds is oxidized to methyl ethyl ketone ?
(A) 1-Butanaol (B) 2-Butanol
(C) 2-Propanol (D) t-Butlyl alcohol

21. The least acidic among the following is
(A) Acetylene (B) Acetic acid (C) Phenol (D) Picric acid

22. An example of an azo dye is
(A) Indigo (B) Malachite green
(C) Alizarin (D) Orange-I

23. Propionamide when heated with a mixture of bromine and caustic alkali would give
(A) Propylamine (B) Propanol
(C) Propanal (D) Ethylamine

24. I dentify Z in the following reaction

$$CH_3CH = CH_2 \xrightarrow[\text{Peroxide}]{\text{HBr}} X \xrightarrow[\text{Ether}]{\text{Mg}} Y \xrightarrow[\text{(ii) } H_2O]{\text{(i) } CO_2} Z$$

(A) $(CH_3)_2CHCOOH$ (B) $CH_3CH_2COCH_3$
(C) $CH_3CH_2CH_2COOH$ (D) $CH_3CH_2COOCH_3$

25. Ozonolysis of an alkene A followed by decomposition with water and a reducing agent gave a mixture of two isomers of formula C_3H_6O. The structure of thealkene is
(A) $CH_3CH = CHCH_3$ (B) $(CH_3)_2C = CHCH_2CH(CH_3)_2$
(C) $CH_3CH_2CH = C(CH_3)_2$ (D) $(CH_3)_2)C = C(CH_3)_2$

26. The main factor for small B–F bonds in BF_3 is
(A) Three centered two electron bonds is BF_3
(B) Large electronegativity of fluorine
(C) $p\pi–p\pi$ back bonding (D) $p\pi–d\pi$ back bonding

27. A $1x10^{-6}$ M HCl solution is diluted to hundred times. The pH of the resulting solution is
(A) 7 (B) 6.96 (C) 8 (D) 6

28. The total number of isomers for the compound of the formula $C_4H_{10}O$ are
(A) 6 (B) 7 (C) 3 (D) 4

29. Acetic acid is reduced by $LiAlH_4$ to
(A) CH_3CH_2OH (B) CH_3CHO
(C) CH_3CH_3 (D) CH_3OH

30. Glycerol on warming with excess of hydrogen iodide gives
(A) Propylene (B) allyl iodide
(C) Isopropyl iodide (D) Nitro glycerine

31. Aspirin is
(A) Salicylic acid (B) Methyl salicylate
(C) Phenyl salicylate (D) Acetyl salicylic acid

32. When ethyl iodide is treated with sodium in the presence of dry ether, the product formed is
(A) Butylene (B) Butane (C) Ethylene (D) Acetylene

33. The enthalpy of vaporization for water is 41.2 x kJ mol^{-1}, ΔH for the reaction, $2H_2O(g) \rightarrow 2H_2O(l)$ is
(A) –82.4 kJ (B) –20.6 kJ (C) 82.4 kJ (D) 2.57 kJ

34. A buffer solution contains equal concentrations of X^- and HX. K_b for X^- is 10^{-10}. pH of the buffer is
(A) 10 (B) 14 (C) 4 (D) 7

35. For the cell reaction, $X(s) + 2Y^+ \rightarrow X^{2+} + 2Y$, K_c has been found to be 10^{12}. The E^0 cell is
(A) 0.708 V (B) 1.36 V (C) 0.354 V (D) 1.006 V

36. The coagulation power of an electrolyte for arsenic sulphide sol is maximum for
(A) $K_4[Fe(CN)_6]$ (B) HCl (C) $AlCl_3$ (D) Na_3PO_4

37. How many coulombs of electricity are required for the oxidation of 1 mol of H_2O to O_2?
(A) $1.93x10^5$ (B) $9.65x10^4$
(C) $3.86x10^5$ (D) $4.825x10^4$

38. The most basic among the following is
(A) La $(OH)_3$ (B) Lu $(OH)_3$
(C) $Ce(OH)_3$ (D) Al $(OH)_3$

39. The molar volume of methane at 819°C and 760 mm pressure is
(A) 89.6 (B) 5.6L (C) 22.4L (D) 44.8L

40. The difference between heats of reaction at constant pressure and constant volume for the reaction $2C_6H_6(l) + 15O_2(g) \rightarrow 12CO_2(g) + 6H_2O(l)$ at 27°C is :
(A) –7.48 kJ (B) 3.74 kJ (C) –3.74 kJ (D) 7.48 kJ

41. The movement of the colloidal particles under the influence of applied electric field is called
(A) Dialysis (B) Tyndall effect
(C) Osmosis (D) Electrophoresis

42. The density of acetone vapours (CH_3COCH_3) at 27°C and 0.82 atm is
(A) 2.6g/L (B) 1.93g/L (C) 3.86g/L (D) 1.25g/L

43. The complex $CoCl_3.4NH_3$ ionises to give
(A) 1 Cl^- ion (B) 2 Cl^- ion
(C) 3 Cl^- ion (D) no Cl^- ion

44. Which of the following molecule or molecular ion is paramagnetic
(A) B_2 (B) CN^- (C) O_2^- (D) N_2

45. Which of the following alkali metal ions has the lowest ionic mobility
(A) Li^+ (B) Cs^+ (C) K^+ (D) Na^+

46. Solubility product of lead bromide is $3.2x10^{-5}$ at a

certain temperature. Its solubility in moles per litre is
(A) $1.6x10^{-3}$ (B) $2.0x10^{-3}$ (C) $0.4x10^{-4}$ (D) $2.0x10^{-2}$

47. Deacon's process is used for the large scale preparation of
(A) O_3 (B) Cl_2 (C) N_2 (D) P_4

48. Of the different allotropic forms of phosphorus, the one which is most reactive is
(A) Violet phosphorus (B) Scarlet phosphorus
(C) Red phosphorus (D) White phosphorus

49. The halogen having some metallic character is
(A) F (B) I (C) Cl (D) Br

50. Which of the following ions has the smallest size
(A) S^{2-} (B) K^+ (C) Ca^{2+} (D) Cl^-

51. The K_C for an acid is $3.2x10^{-5}$. The degree of dissociation of the acid at concentration of 0.2 M is
(A) $6.0x10^{-2}$ (B) $1.6x10^{-2}$ (C) $4.0x10^{-4}$ (D) 0.04

52. The enthalpy of solution of anhydrous $CuSO_4$ and $CuSO_4.5H_2O$ are –15.89 and 2.80 kcal mol^{-1} respectively. The enthalpy of hydration of $CuSO_4$ to $CuSO_4.5H_2O$ is
(A) –18.69 kcal (B) 13.09 kcal
(C) –13.09 kcal (D) 18.69 kcal

53. When potassium dichromate is heated with conc. H_2SO_4 and a soluble metal chloride, the orange red vapours are produced. These are due to
(A) Cr_2O_3 (B) $CrOCl_2$ (C) CrO_4^{-2} (D) CrO_2Cl_2

54. When copper is heated with conc. H_2SO_4, it gives sulphur dioxide. In this reaction, sulphuric acid acts as
(A) An oxidizing agent (B) A reducing agent
(C) Dehydrating agent (D) Catalyst

55. An example of miscelle is
(A) Aqueous NaCl solution (B) Sodium stearate
(C) Gold sol (D) Milk

56. In the nitration of benzene, the reactive nitrating species is
(A) NO_2^- (B) NO_3^- (C) NO_2^+ (D) NO

57. In the reaction :

$$CH_3CH_2CH_2.COOH \xrightarrow{NaOH} X \xrightarrow[Heat]{Soda\ lime} Y \text{ is}$$

(A) C_4H_{10} (B) $CH_3CH_2CH_2CH_2OH$
(C) CH_3CH_2CHO (D) $CH_3CH_2CH_3$

58. Which of the following bases is found only in RNA and not in DNA?
(A) Adenine (B) Uracil (C) Thymine (D) Guanine

59. Which of the following has the lowest boiling point
(A) H_2O (B) H_2S (C) H_2Se (D) H_2Te

60. A 0.01 M solution of glucose in water freezes at –0.0186°C. A 0.01 M solution of NaCl in water will freeze at
(A) –0.0186°C (B) 0.0186°C
(C) –0.0372°C (D) 0°C

JIPMER - 1994

1. Which of the following metals shows allotropy
(A) K (B) Pb (C) Sn (D) Ca

2. Hyponitrous acid is
(A) HNO_3 (B) HNO_4 (C) HNO (D) HNO_2

3. Which of the following is a pseudohalide ?
(A) IF_5 (B) I_3^- (C) CN^- (D) ICl

4. The noble gas forming maximum number of compounds is
(A) Ne (B) Xe (C) Kr (D) Rn

5. Which of the following is used in photography ?
(A) Ag Br (B) $CuSO_4$ (C) Ag_2O (D) AgF

6. Crucible for the gravimetric estimation is made of
(A) Iron (B) Copper (C) Silica (D) Glass

7. Basicity of sulphuric acid is
(A) 1 (B) 2 (C) 3 (D) 4

8. What is the oxidation state of iron in sodium nitroprusside ?
(A) Zero (B) +1 (C) +2 (D) +3

9. Noble gas used in radiotherapy is
(A) Rn (B) Xe (C) Kr (D) Ar

10. Bromine is liberated when an aqueous KBr solution is treated with
(A) Cl_2 (B) I_2 (C) HI (D) SO_2

11. The hybrid orbitals involved in the formation of ethane are
(A) sp^2 (B) sp^3 (C) s (D) sp

12. The IUPAC name of $Cl_3C–CH_2–CHO$ is
(A) Chloral (B) Chloromethanal
(C) 3,3,3-trichloropropanal (D) 2,2,2-trichloropropanal

13. Fractional distillation is used in the processing of
(A) Petroleum (B) Coal tar
(C) Crude oil (D) All of these

14. How many conformations are there for ethane ?
(A) 5 (B) 6 (C) 2 (D) 3

15. An alkyl halide reacts with ammonia to give
(A) Amide (B) Cyanide (C) Amine (D) None of these

16. Which of the following compounds is known as freon ?
(A) CH_2F_2 (B) CF_4 (C) CCl_2F_2 (D) $CHCl_3$

17. The percentage of ethyl alcohol in rectified spirit is
(A) 100 (B) 59.6 (C) 95.6 (D) 50

18. Which of the following is used as an anaesthetic ?
(A) CH_3OCH_5 (B) $CH_3C_2H_5$
(C) $C_2H_5OC_2H_5$ (D) $C_3H_7OC_2H_5$

19. Which one of the following is used as a hypnotic ?
(A) Formaldehyde (B) Acetaldehyde
(C) Metaldehyde (D) Paraldehyde

20. A gas which burns with blue flame is
(A) CO (B) NO_2 (C) CO_2 (D) HCl

21. Heavy water is

(A) D_2O (B) $(H_2O)_n$ (C) H_2O (D) H_2O_2

22. Colour of solid iodine is

(A) Violet (B) Steel grey (C) Red (D) White

23. When the 3d orbital is complete the new electron will enter the

(A) 4p orbital (B) 4f orbital
(C) 4s orbital (D) 4d orbital

24. The Telluric helix was gives by

(A) De Chancourtois (B) Newlands
(C) L.Meyer (D) Mendeleev

25. Bond between Cl – Cl is a

(A) Co-ordinate (B) Ionic (C) Covalent (D) Metallic

26. Weight of $6.02x10^{23}$ molecules of CO_2 is

(A) 22g (B) 44g (C) 11g (D) 14g

27. Select a compound in which oxidation number of chlorine is +5

(A) HClO (B) $HClO_2$ (C) $HClO_3$ (D) $HClO_4$

28. Which of the following is not a solution ?

(A) Gold ring (B) Salt water (C) Air (D) Smoke

29. Pb_3O_4 is a

(A) Double oxide (B) Complex oxide
(C) Normal oxide (D) Mixed oxide

30. The hydrogen ion concentration in pure water is 10^{-7}g ions/litre. An aqueous solution will be acidic if its hydrogen ion concentration is .... g ions/litre is

(A) 10^{-6} (B) 10^{-7} (C) 10^{-8} (D) 10^{-9}

31. The compound that is not a Lewis acid is

(A) $BeCl_2$ (B) $SnCl_4$ (C) BF_3 (D) $AlCl_3$

32. In a cell $Zn/Zn^{2+} \| Cu^{2+}/Cu$ the positive electrode will be

(A) Cu (B) Cu^{2+} (C) Zn (D) Zn^{2+}

33. How many phases are present in a colloidal system ?

(A) 4 (B) 3 (C) 2 (D) 1

34. An example of a covalent crystalline solid is

(A) Al (B) Ar (C) NaF (D) Sillicon

35. A saline hydride of the following is

(A) CaH_2 (B) HF (C) H_2O (D) HI

36. Lunar caustic is

(A) $NaNO_2$ (B) $AgNO_3$ (C) $Cu(NO_3)_2$ (D) $Au(NO_3)_2$

37. Which of the following metals is found free in nature

(A) Na (B) Ca (C) Au (D) Ba

38. Rubies and sapphires contain

(A) Al_2O_3 (B) $Al_2(SO_4)_3$
(C) $Al_2(CO_3)_3$ (D) Al_2O_3 and aluminates

39. Amide may be converted into amines by the reaction named after.

(A) Claisen (B) Kolbe (C) Perkin (D) Hofmann

40. Colour obtained in biuret test for urea is

(A) green (B) blue (C) red (D) violet

41. Which test is used for detecting the presence of sugar in urine ?

(A) Biuret test (B) Lassaigne's test
(C) Dumas teast (D) Molisch test

42. The reactions $C_6H_6 + CH_3Cl \xrightarrow{AlCl_3} C_6H_5CH_3 + HCl$ is called

(A) Wurtz Fittig reaction (B) Kolbe's reaction
(C) Friedel-Crafts reaction (D) None of these

43. Which of the following has a smell of bitter almonds ?

(A) Phenol (B) Benzoic acid
(C) Aniline (D) Benzaldehyde

44. Which of the following is a molecular disease ?

(A) Cancer (B) Allergy
(C) Sickle cell anemia (D) German measles

45. Which of the following contains lead ?

(A) Soda glass (B) Potash glass
(C) Flint glass (D) Bottle glass

46. The oxidation state of oxygen atom in potassium superoxide is

(A) 0 (B) –½ (C) –1 (D) –2

47. The kinetic energy of 1 mole of O_2 gas at O°C is

(A) 150 Cal (B) 273.2 Cal
(C) 543.8 Cal (D) 814.3 Cal

48. Which of the following species has trigonal planar shape ?

(A) $BeCl_2$ (B) : CH_2 (C) CH_3^+ (D) BF_4^+

49. The ratio of C_p/C_v for inert gases is

(A) 1.33 (B) 1.66 (C) 2.13 (D) 1.99

50. The catalyst often used in the hydrogenation of oils is

(A) Mo (B) Pd (C) Fe (D) Ni

JIPMER - 1995

1. Propionic acid is subjected to reduction with hydroiodic acid in the presence of a little of P. The product formed is

(A) Propane (B) Butane (C) Ethane (D) None of these

2. Which of the following structures of carboxylic acid accounts for the acidic nature ?

(A) $R-\overset{+}{C}\langle^{O^-}_{OH}$ (B) $R-\overset{+}{C}\langle^{OH}_{OH}$

(C) $R-C\langle^{\,=O}_{OH}$ (D) None of these.

3. Glycine may be classified as all of the following except

(A) An acid (B) A base
(C) A Zwitter ion (D) An optically active acid

4. Which of the following is expected to be optically active ?

(A) $CH_3CH = CHCH_3$ (B) $(CH_3)_4C$
(C) $C_2H_5CH(CH_3)C_3H_7$ (D) (C_2H_2) $CHCH_3$

5. The second ionization potential of an element M is the

energy required to
(A) Remove one mole of electron from one mole of gaseous anion
(B) Remove one mole of electron from one mole of monovalent gaseous cations of the element
(C) Remove one mole of electron from one mole of gaseouc cation of the element.
(D) Remove 2 moles of electron from one mole of gaseous atoms.

6. Which of the following is mainly responsible for the viscous nature of glycerol ?
(A) van der Waall forces (B) Covalent bonds
(C) Hydrogen bonds (D) Ionic forces

7. The type of hybridization that can explain the bonding in C_2H_4 is
(A) sp^2 (B) dsp^2 (C) sp^3d (D) sp^3

8. The interparticle forces beween linear chains in Nylon-6,6 are
(A) H-bonds (B) Covalent bonds
(C) Ionic bonds (D) Co-ordinate bonds

9. The monomer unit of polyethene is
(A) $CF_2 = CF_2$ (B) $CH_2 = CHCN$
(C) $CH_2 = CHCl$ (D) $CH_2 = CH_2$

10. Five valence electrons of P are labelled as

[AB] (3s) [X | Y | Z] (3p)

If the spin quantum of B and Z is +1/2, the group of electrons with three of the quantum numbers same are
(A) AB (B) AB, XYZ.
(C) XYZ, AZ. (D) AB,XYZ, BY

11. A random motion in a zig-zag manner of colloidal particles is called
(A) Coagulation (B) Osmosis
(C) Electrophoresis (D) Brownian movement

12. The deficiency of vitamin D causes
(A) Rickets (B) Xerosis
(C) Night-blindness (D) Loss of appetite

13. The product formed by heating glycerol at 380 K is
(A) Oxalic acid (B) Acrolein
(C) Formic acid (D) Alkyl alcohol

14. Aniline is usually purified by
(A) By the addition of oxalic acid
(B) Chromatographic technique (C) Steam distillation
(D) Fractional crystallization

15. The reagent used for the protection of amino group during the nitration of aniline is
(A) Acetic acid (B) $SOCl_2$/pyridine
(C) PCl_5 (D) Acetic anhydride

16. The oxidation product of 2-propanol, with hot conc. HNO_3 is
(A) Propanal (B) Ethanoic acid
(C) Propanone (D) None of these

17. An organic halide shaken with aqueous NaOH followed by the addition of dil. HNO_3 and silver nitrate solution gives white ppt. The compound can be
(A) $C_6H_5CH_2Cl$ (B) $C_6H_4(CH_3)Br$
(C) C_6H_5Cl (D) None of these

18. The coagulating power of an electrolyte for arsenic sulphide decreases in the order of
(A) $Na^+ > Al^{3+} > Ba^{2+}$ (B) $SO_4^{2-} > PO_4^{3-} > Cl^-$
(C) $PO_4^{-3} > SO_4^{2-} > Cl^-$ (D) $Al^{3+} > Ba^{2+} > Na^+$

19. The IUPAC name of the compound $CH_3CH(OH)CH_2\underset{CH_3}{\underset{|}{C}}-(OH)\ CH_3$ is
(A) 1,3,3-trimethyl propan1,3diol
(B) 1,1dimethyl butan-1,3-diol
(C) 2-methyl pentan-2,4-diol (D) None of these

20. Which of the following statements is not characteristic of free radical chain reaction ?
(A) It is usually sensitive to change in solvent polarity
(B) It gives major product derived from most stable free radical
(C) It proceeds in three main steps like initiation, propagation and termination
(D) It may be initiated by U.V. light.

21. Which of the following cannot be taken to exemplify a chiral structure ?
(A) A screw (B) A shoe
(C) A screw driver (D) None of these

22. Which of the following compounds represents acrylonitrile ?
(A) Prop-2ene nitrile (B) Vinyl cyanide
(C) Cyno ethane (D) All of these

23. A solution of weak acid is diluted by adding an equal volume of water. Which of the following will not change ?
(A) The value of $[H_3O^+]$ (B) Strength of the acid
(C) pH of the solution (D) None of these

24. Which of the following solutions can act as a buffer ?
(A) 0.1 molar aq. $NaCl^-$
(B) 0.2 molar aq. CH_3COOH + 0.1 molar NaOH
(C) 0.1 molar aq. ammonium acetate
(D) None of these

25. Which of the following is incorrect for an ideal solution ?
(A) ΔS mixing = 0 (B) ΔV mix = 0
(C) ΔH mix = 0 (D) It obeys Raoult's law

26. Which of the following statements about galvanic cell is not correct ?
(A) Reduction occurs at cathode (B) Anode is positive
(C) Oxidation occurs at the electrode with lower reduction potential. (D) Cathode is positive.

27. What is E° for electrode represented by Pt, O_2 (1atm.)/ $2H^+$(Im) ?
(A) Zero (B) 0.118 V (C) 0.018 V (D) Unpredictable

28. If bond energies of $N \equiv N$, H–H and N–H bonds are 945, 437 and 389 kJ respectively, then ΔH for the following gaseous reaction $N_2 + 3H_2 \rightarrow 2NH_3$ is
(A) –156kJ (B) –78kJ (C) –1478kJ (D) +1478 kJ

29. A broad spectrum antibiotic of the following is

(A) Paracetamol (B) Aspirin
(C) Chloroamphenicol (D) Penicillin

30. Which of the following species is planar ?
(A) NH_3 (B) CO_3^{-2} (C) PCl_3 (D) None of these

31. Which of the following gaseous atoms has the highest value of IE ?
(A) Mg (B) P (C) Si (D) Al

32. A characteristic feature of transition elements is that
(A) They show inert pair effect
(B) They do not form complex ions
(C) They show variable oxidation states
(D) They all exhibit diamagnetism

33. An electronic transition from Is orbital of an atom causes
(A) Emission of energy (B) Absorption of energy
(C) Both emission and absorption of energy
(D) Unpredictable behaviour

34. The reaction
$CH_3COCl \xrightarrow[Pd/BaSO_4]{2H} CH_3CHO + HCl$ is called
(A) Reimer and Tiemann's reaction
(B) Rosenmund's reaction
(C) Cannizzaro's reaction
(D) Reformatsky's reaction

35. A strong beam of light passed through a colloidal solution is
(A) Scattered (B) Reflected
(C) Refracted (D) Absorbed completely

36. Which of the following represents surfactant molecule?
(A) H_2O (B) $C_{17}H_{36}$
(C) $C_{17}H_{25}COO^-Na^+$ (D) None of these

37. $X \xrightarrow{\alpha} Y \xrightarrow{\beta} Z \xrightarrow{\beta} W$ In the above sequence of reaction, the elements which are isotopes of each other are
(A) X and Z (B) Y and Z
(C) X and W (D) None of these

38. Which of the following ions has the highest magnetic moment ?
(A) Mn^{2+} (B) Zn^{2+} (C) Sc^{3+} (D) Ti^{3+}

39. Which of the following compounds does not exist ?
(A) $BiCl_5$ (B) $SbCl_5$ (C) $AsCl_5$ (D) $SbCl_3$

40. Of the following substances the one which does not cantain oxygen is
(A) Epsom salt (B) Bauxite
(C) Cryolite (D) Dolomite

41. All elements in third period have
(A) 3 shells
(B) 3 complete sub-shells (C) an atomic number 3
(D) 3 electrons less than the octet

42. Which of the following complexes will give white precipitate with $BaCl_2(aq)$?
(A) $[Cr(NH_3)_5SO_4]Cl$ (B) $[CO(NH_2)_4SO_4]NO_2$
(C) $[Cr(NH_3)_5Cl]SO_4$ (D) Both (A) and (C)

43. The co-ordination and oxidation number of X in the compound respectively $[X(SO_4)(NH_3)_5]$ Cl will be
(A) 10 and 3 (B) 6 and 4 (C) 2 and 6 (D) 6 and 3

44. Which of the following is called sulphuric anhydride ?
(A) $H_2S_2O_3$ (B) SO_2 (C) $H_2S_2O_7$ (D) SO_3

45. When ammonia is bubbled through copper sulphate solution, and a deep blue solution results. The deep blue colour is due to formation of
(A) $Cu(OH)_2$ (B) Cu^{2+} ions
(C) $[Cu(NH_3)_4]^{2+}$ (D) None of these

46. The chief source of iodine, in which it is present as sodium iodate is
(A) Carnallite (B) Sea weeds (C) Caliche
(D) Iodine never exists as sodium iodate

47. A bottle of dry ammonia and a bottle of dry hydrogen chloride connected through a long tube are opened simultaneously. At both ends the white fumes of ammonium chloride are formed. The first ring will be
(A) At the center of the tube (B) Near the ammonia bottle
(C) Near the hydrogen chloride bottle
(D) Throughout the length of the tube.

48. Consider the reactions :
$C(s) + 2H_2(g) \rightarrow CH_4(g), \Delta H = -x$ kcal
$C(g) + 4H(g) \rightarrow CH_4(g), \Delta H = -x_1$ kcal
$CH_4(g) \rightarrow CH_3(g) + H(g), \Delta H = -y_1$ kcal
The bond energy of C–H bond is
(A) y kcal mol^{-1} (B) x/4 kcal mol^{-1}
(C) x_1 kcal mol^{-1} (D) $x_1/4$ kcalmol^{-1}

49. The energy of a photon is
(A) Directly proportional to its velocity
(B) Directly proportional to frequency
(C) Inversely proportional to the wave number
(D) Directly proportional to its wavelength

50. The order of the reaction occurring by following mechanism should be
(i) $A_2 \rightarrow A + A$ (fast)
(ii) $A + B_2 \leftrightarrow AB + B$ (slow)
(iii) $A + B \rightarrow$ (fast)
(A) $1\frac{1}{2}$ (B) 2 (C) $3\frac{1}{2}$ (D) None of these

JIPMER - 1996

1. Which of the following cannot be hydrolysed ?
(A) $SiCl_4$ (B) CCl_4 (C) $GeCl_4$ (D) $SnCl_4$

2. When formaldehyde is heated with ammonia, the compound formed is
(A) methylamine (B) amino formaldehyde
(C) hexamethylenetetramine (D) Formaline

3. Which one of following is not found in alkenes ?
(A) Chain isomerism (B) Geometirical isomerism
(C) Tautomerism (D) Position isomerism

4. The β– decay of $^{24}_{11}Na$ will produce an isotope of
(A) Mg (B) Na (C) NO (D) any of these

5. Acid hydrolysis of fixed oil gives
(A) Glycerol + detergent (B) Glycerol + soap
(C)Glycerol + fatty acid (D) Glycerol + aldehydes

6. Which of the following is the weakest base
(A) NH_3 (B) $C_2H_5NH_2$
(C) $C_6H_5NH_2$ (D) unpredictable.

7. Which of the following alcohols cannot be dehydrated to give alkene
(A) Tert-butyl alcohol (B) Neopentyl alcohol
(C) Isopropyl alcohol (D) All of these can be dehydrated

8. Isopropyl alcohol on mild oxidation gives
(A) Acetone (B) Propanoic acid
(C) Propene (D) Propane.

9. A paste of bleaching powder on heating with ethanol gives
(A) Carbon tetra chloride (B) None of these
(C) Chloroform (D) Ethylene

10. A mixture of equal parts of enantiomers is called
(A) Configuration (B) Resolution
(C) Recemic mixture (D) Meso mixture

11. Ethyl amine reacts with HNO_2 to form
(A) Ethanol (B) Methanal (C) Methanol (D) Ethanal

12. Which of the following acids will be able to give silver mirror test
(A) Acetic acid (B) Butyric acid (C) Formic acid
(D) Carboxylic acids do not give this test

13. Which of the following quantities are equal at equilibrium under the particular set of conditions
(A) ΔH and ΔG (B) ΔH and ΔS
(C) ΔG and ΔS (D) none of the above

14. A chemical reaction will not occur at all if the conditions are chosen so that
(A) Enthalpy change and entropy change are both positive
(B) Enthalpy change is negative and entropy change is positive
(C) Enthalpy change and entropy change are both negative
(D) Enthalpy change is positive and entropy change is negative

15. Weak electrolytes are only partly dissociated in solutions. The extent to which they are disassociated in a given solution is known as the degree of dissociation;. Which of the following statements for the degree of dissociation in aqueous solutions is true ?
(A) It is directly proportional to the initial concentration of the electrolyte
(B) It is independent of the initial concentration of the electrolyte
(C) It is inversely proportional to the initial concentration of the electrolyte
(D) It depends on the equilibrium concentration of the electrolyte

16. Which of the substances, Na, Hg, S, Pt and graphite, can be used as electrodes in electrolytic cells having aqueous solutions ?
(A) Hg, Pt and graphite (B) Na, S and Hg
(C) Pt and graphite only (D) Na and S only

17. Which of the following ions in aqueous medium has orange colour.
(A) MnO_4^- (B) MnO_4^{2-} (C) Cr^{+3} (D) $Cr_2O_7^{2-}$

18. The most stable oxidation state of Fe is
(A) +2 (B) + 3 (C) –2 (D) –3

19. The most abundant transition element is
(A) W (B) Mn (C) Cr (D) Fe

20. Which of the following is used for galvanizing iron
(A) Cr (B) Zn (C) Ni (D) Sn

21. Which of the following carbonate will not decompose on heating
(A) Na_2CO_3 (B) $CdCO_3$ (C) $CaCO_3$ (D) $BaCO_3$

22. Which of the following does not have S–S bond
(A) $S_2O_3^{2-}$ (B) $S_1O_5^{-2}$ (C) $S_2O_4^{2-}$ (D) $S_2O_7^{2-}$

23. In which of the following states of matter the average distance between the molecules lies between 10^{-5} cm to 10^{-7} cm ?
(A) Solid (B) Liquid (C) Gas (D) none of these

24. The volume of the resulting solution obtained by mixing 25 mL of acetone and 25 mL of $CHCl_3$ will be
(A) 50 mL (B) <50 mL (C) >50 mL (D) unpredictable

25. Out of the following hydrogen halides, which one has the highest boiling point ?
(A) HCl (B) HF (C) HI (D) HBr

26. Which one of the following is not an explosive ?
(A) nitroglycerin (B) o-aminotoluene
(C) dynamite (D) trinitrotoluene

27. Lucas reagent is a mixture of
(A) conc. HCl and hydrated $ZnCl_2$
(B) conc. HNO_3 and hydrous $ZnCl_2$
(C) conc. HCl and anhydrous $ZnCl_2$
(D) conc. HNO_2 and anhydrous $ZnCl_2$

28. For a hypothetical reaction aA+bB → product, the rate law is $r = k[A]^x[B]^y$, then
(A) (a +b) > (x + y)
(B) (a +b) = (x + y) (C) (a +b) < (x + y)
(D) (A + b) may not be equal to (x + y)

29. The order of a reaction with respect to one of the reacting components Y is zero. It implies that
(A) The rate of a reaction does not vary with temperature.
(B) The reaction rate is independent of the conc. of Y
(C) The reaction must proceed at a constant rate.
(D) The rate of formation of activated complex is zero.

30. For a hypothetical reaction A → L, the rate expression is $r = \frac{-d[A]}{dt}$ The negative sign represent.
(A) That the rate is negative
(B) Decrease in concentration of reactant.
(C) Attractive forces between the reactant.

(D) None is correct.

31. For an endothermic reaction carried out in a closed vessel :
(A) $\Delta H > 0$ (B) $\Delta H < 0$ (C) $\Delta H = \Delta E$ (D) $\Delta H = 0$

32. Be exhibits the diagonal relationship with
(A) Si (B) Al (C) Mg (D) Li

33. Which of the following hydroxides is a strongest base ?
(A) Li OH (B) Na OH (C) KOH (D) $Ca(OH)_2$

34. Which of the following triatomic molecules has the largest angle around the central atom
(A) H_2S (B) H_2Se (C) H_2O (D) SO_2

35. The hybrid state of sulphur is SO_4^{-2} is
(A) sp^3 (B) sp^2 (C) sp (D) none

36. An element X has five electrons in p-orbitals of valence shell and also belongs to fifth period. Its molecular formula should be
(A) X (B) X_2 (C) X_3 (D) Unpredictable

37. The most electronegative element of 1st period is :
(A) Li (B) He (C) H (D) Unpredictable

38. A 3d-electron specified by s = +1/2 can have a magnetic quantum number
(A) +3 (B) +2 (C) 4+ (D) –3

39. The K.E. of electron in He^+ is maximum in
(A) 2nd orbit (B) 3rd orbit
(C) 1st orbit (D) In orbit with $n = \infty$

40. The existence of line spectrum of hydrogen provides an evidence that
(A) The electrons do not lose energy while revolving in stationary states
(B) The energy of electron is quantised
(C) There is absorption of energy
(D) There is an emission of energy when hydrogen is heated

41. The basic principle involved in hydrogen bomb is
(A) Nuclear fission (B) Nuclear fusion
(C) Nuclear disintegration (D) None of these

42. For the transformation ${}_7N^{14} + ? \longrightarrow {}_6C^{14} + {}_1H^1$. The bombarding particle is
(A) Deuteron (B) Electron (C) Proton (D) Neutron

43. Emulsifier is an agent which
(A) Homogenises the emulsion
(B) Stabilises the emulsion
(C) Accelerates the dispersion (D) Breaks up the emulsion

44. The hybrid state of positively charged carbon in Vinyl ($CH_2 = CH^+$) cation is
(A) sp (B) unpredictable (C) sp^3 (D) sp^2

45. Which of the following molecules has shortest C–C single bond length
(A) Propane (B) Propyne (C) Propene
(D) All will have same bond length.

46. Two electrodes A^{2+}/A and B^+/B have reduction potentials (E^o) 0.62 V and –1.25 V respectively. The cell voltage will be
(A) 0.63 V (B) –0.63V
(C) 1.87V (D) None of the above

47. A solution of HCl is 5×10^{-8} M. Its pH will be
(A) 7 (B) 8–log 5 (C) 8 (D) between 6 and 7

48. The conjugate acid of CH_3NH_2 is
(A) CH_3OH (B) NH_2 (C) $CH_3NH_3^+$ (D) CH_3NH^-

49. Identify the product B in the following reaction

$$CH_3CHO \xrightarrow{CH_3.MgI} A \xrightarrow[\text{Hydrolysis}]{} B$$

(A) $(CH_3)_2CHOH$ (B) $(CH_3)_3COH$
(C) CH_3OH (D) CH_3CH_2OH

50. The conductivity of an aqueous solution of a strong electrolyte
(A) decreases with increase in dilution
(B) remains constant at all concentration
(C) bears no relationship with concentration
(D) increases with dilution

51. Out of molality (m), molarity (M), formality (F), normality (N) and mole fraction (x), the two independent of temperature are
(A) m, N (B) F, N (C) N, m (D) m, x

52. The movement of colloidal particles with reference to dispersion medium when subjected to an electrical potential is called
(A) Electro osmosis (B) electrophoresis
(C) Tyndall effect (D) Brownian movement

53. Which of the following has the highest protective power on lyophobic colloids
(A) gelatin (B) sodium oleate
(C) gum arabic (D) starch

54. Which of the following compounds would exhibit co-dination isomerism
(A) $[Cr(en)_2]NO_2$ (B) $[Cr(H_2O)_6]Cl_3$
(C) $[Cr(NH_3)_6][Co(CN)_6]$ (D) None of the above

55. At constant volume for a fixed number of moles of a gas, the pressure of the gas increses with the rise in temperature due to
(A) increase in mean free path
(B) increae in average molecular speed
(C) increase in molecular attraction
(D) increae in rate of collisions amongst molecules

56. (i) $Na_2S + 2HCl \longrightarrow 2NaCl + H_2S$; $\Delta H = -X_1$ kJ
(ii) $H_2(g) + S(s) \longrightarrow H_2S(g)$; $\Delta H = -X_2$kJ
(iii) $Mg(HS)_2 + 2H_2O \longrightarrow Mg(OH)_2 + 2H_2S$; $\Delta H = -X_3$kJ
The value of ΔH_f of hydrogen sulphide is
(A) $-X_3$ kJ mol^{-1} (B) $-X_1$kJ mol^{-1}
(C) $-X_2$ kJ mol^{-1} (D) none of these

57. In alkenes, the negative part of the addendum adds on to the carbon atom joined to the least number of hydrogen atoms. This statement is related to
(A) Peroxide effect (B) Baeyer's strain theory
(C) Thiele's theory (D) Markownikoff's Rule

58. Which one is the electron deficient compound ?
(A) BCl_3 (B) ClF_3 (C) NH_3 (D) PCl_3

59. The blue complex formed on addition of conc. NH_4OH solution to a Cu^{2+} salt solution has the structure
(A) $[Cu(NH_3)_2]^{2+}$ (B) $[Cu(NH_3)_4]^{2+}$
(C) $[Cu(NH_4)_4]^{2+}$ (D) $[Cu(NH_4)_2]^{2+}$

60. The equilibrium constant in a reversible chemical reaction at a given temperature
(A) depends on the initial concentrations of the reactants
(B) in not characteristic of the reaction
(C) does not depend on the initial concentration of reactants
(D) depends on the concentration of one of the products at equilibrium.

JIPMER - 1997

1. The shape of the hydrides of Group XV elements is
(A) Tetrahedral (B) Trigonal planar
(C) Square planar (D) Pyramidal

2. Cassiterite is concentrated by
(A) Gravity separation (B) Roasting
(C) Electromagnetic separation (D) All the above

3. AgCl is soluble in
(A) Sodium cyanide (B) Ethanol (C) Water (D) Ether

4. The most volatile of the following compounds is
(A) H_2O (B) CH_3CH_2OH
(C) $CH_3–O–CH_3$ (D) $C_2H_5–O–C_2H_5$

5. The chemical name for grain alcohol is
(A) Methanol (B) Ethanol
(C) Amyl alcohol (D) Butyl alcohol

6. Which of the following is not a water soluble vitamin ?
(A) Ascorbic acid (C) (B) Calciferol (D)
(C) Thiamine (B_1) (D) Riboflavin (B_2)

7. Acetic and formic acids do not differ in their action on
(A) Ammonical $AgNO_3$ (B) $NaHCO_3$
(C) Phosphorus pentachloride (D) conc. H_2SO_4

8. Acetaldehyde on treatment with HCN forms a cyanohydrin. This on hydrolysis gives
(A) α–hydroxy propanoic acid (B) Acetic acid
(C) Acetone (D) Ethanol

9. A carbonyl compound undergoes the following reactions
(i) With saturated $NaHSO_3$, it gives a white precipitate.
(ii) It does not give pink colour with Schiff's reagent
(iii) On reduction it gives a secondary alcohol
(iv) With iodine and sodium carbonate, it gives iodoform.
The compound is.
(A) $C_2H_5COCH_3$ (B) $C_3H_7COC_3H_7$
(C) $C_2H_5COC_2H_5$ (D) $C_2H_5COC_3H_7$

10. Benzyl alcohol is formed form benzaldehyde by
(A) Friedel-Crafts reaction (B) Cannizzaro's reaction
(C) Kolbe's reaction (D) Reimer-Tiemann reaction

11. Dynamite is a mixture of nitroglycerine
(A) And cellulose acetate
(B) Saw dust and NH_4NO_3 (C) And cellulose nitrate
(D) Cellulose nitrate and Vaseline

12. When ethene is passed into cold dilute alkaline $KMnO_4$, the product is
(A) Glycol (B) Methanol
(C) Acetylene (D) Two molecules of CO_2

13. If the rate of a chemical reaction between gases is independent of pressure the order of the reaction is.
(A) 2 (B) 3 (C) 0 (D) 1

14. If at 250°C the time for half-change for the thermal dissociation of N_2O is 5.7 hours, the specific rate constant is given by
(A) $\frac{2}{5.7}$ hour^{-1} (B) $\frac{ln2}{5.7}$ hour^{-1}
(C) $ln\ \frac{1}{5.7}$ hour^{-1} (D) 0.5 hour^{-1}

15. The units for the equilibrium constant K_P for the reaction. $N_2(g) + 3H_2(g) \leftrightarrow 2NH_3(g)$ are
(A) atm^{-1} (B) atm^{-1} (C) atm (D) atm^2

16. $CH_3.CH = C = CH_2$ has
(A) Both sp and sp^2 hybridised atoms
(B) Only sp^2 hybridised carbon atoms
(C) Only sp hybridisered carbon atoms
(D) sp, sp^2 and sp^3 hybridised atoms

17. Energy of the atomic orbitals in a particular energy level is in the order.
(A) $s < p < d < f$ (B) $s > p > d > f$
(C) $p < d < f < s$ (D) $f > d > s > P$

18. Which of the following statements is wrong ?
(A) In resonance, the resonating structures exist as an equilibrium mixture
(B) The dextro and laevo forms of an optically active substance are not super-posable
(C) In tautomerism, the tautomers exist as an equilibrium mixture
(D) A branched chain isomer has a lower boiling point than the corresponding straight chain compound

19. H_2 produced from dil. H_2SO_4 by zinc is (i) collected over water and (ii) over Hg ,then
(A) The volume of gas collected was the same.
(B) The amount of hydrogen collected over mercury was greater.
(C) The amount of hydrogen collected over mercury was the same as the amount of the gas collected over water.
(D) The amount of hydrogen collected over water was greater

20. The volume occupied by 1 mole of a gas at 0.5 atm and 335 K is
(A) 55*l* (B) 45*l* (C) 50*l* (D) 60*l*

21. When the same amount of Al is treated with an excess

of HCl and NaOH respectively the volumes of hydrogen evolved are in the ratio of
(A) 1 : 2 (B) 2 : 1 (C) 1 : 1 (D) 2 : 3

22. The inversion temperature for a van der Waal's gas is given by
(A) $Ti = \frac{2a}{Rb}$ (B) $Ti = \frac{a}{Rb}$
(C) $Ti = \frac{a}{2Rb}$ (D) Ti = 0.5 T Boyle

23. Azodyes are
(A) Mordant dyes (B) Basic dyes
(C) Acid dyes (D) Ingrain dyes

24. The compound with the smell of peppermint is
(A) Ethyl acetate (B) Ethyl benzoate
(C) Methyl acetate (D) Methyl salicylate

25. Which of the following substances possesses an odour of bitter almonds ?
(A) Aniline (B) Acetone
(C) Benzaldehyde (D) Benzene sulphonic acid

26. Which of the following methods cannot be used for the preparation of a ketone ?
(A) Hydrolysis of a cyanide
(B) Oxidation of secondary alcohols
(C) Heating of a volatile acid with MnO_2 at 100°C
(D) Dehydrogenation of secondary alcohols

27. Picric acid is
(A) m-nitrophenol (B) Aminophenol
(C) p-nitrophenol (D) 2,4,6-trinitrophenol

28. Phenol is a stronger acid than
(A) Chloroacetic acid (B) 2-nitrophenol
(C) Ethanol (D) Acetic acid

29. When glycerol is heated with excess of oxalic acid at 110°C, it produces
(A) Acetic acid (B) Acetaldehyde
(C) Formic acid (D) Allyl alcohol

30. Conversion of n-hexane to benzene in the presence of chromic oxide on Al_2O_3 support is called
(A) Hydrogenation (B)Dehydrogenation
(C) Isomerisation (D) Aromatisation

31. Heavy water is obtained by
(A) Distillation of water
(B) Removing Ca and Mg salts from water
(C) Dissolving heavy salts in water
(D) Prolonged electrolysis of water

32.Which of the following tests is used to identify H_2O_2 ?
(A) Tailing of mercury (B) Perchromic acid
(C) Dichromate paper (D) Lead acetate paper

33. Which of the following molecules / ions cannot exist
(A) H_2 (B) H_2^- (C) H_2^{+2} (D) H_2^+

34. Iron carbonate ore known as
(A) Haematite (B) Siderite (C) Magnesite (D) Iron pyrites

35. Lead pipes are not used for water pipe lines due to
(A) Plumbosolvency (B) Toxicity
(C) Action of CO_2 (D) Precipitation of $Pb(OH)_2$

36. Cement is
(A) Calcium silicates + aluminium silicates + gypsum
(B) Calcium silicates + magnesium silicates
(C) Calcium silicates + calcium aluminates
(D) Calcium silicates + magnesium silicates + gypsum

37. Which of the following isoelectronic species has the smallest size ?
(A) Ca^{2+} (B) Ti^{4+} (C) S^{2-} (D) Ar

38. Which of the following is not true for the statement "An element with high electronegativity is generally"
(A) Having a high electron affinity
(B) Having a high ionization potential
(C) A non metal (D) A metal

39. The value of IE_1, IE_2, IE_3, IE_4, and IE_5 of an element are 7.1, 14.3, 34.5, 46.8 and 162.2 eV respectively. The element is likely to be.
(A) Na (B) Si (C) F (D) Ca

40. The element with the highest first ionization potential is.
(A) Nitrogen (B) Oxygen (C) Boron (D) Carbon

41. To which radioactive series does ^{207}Pb belong ?
(A) ^{232}Th (B) ^{237}Np (C) ^{227}Ac (D) ^{238}U

42. The continuum is an atomic spectrum is due to
(A) Interaction of anion with an atom
(B) Wave property of an electron
(C) Instability of the excited state
(D)Ionization of the atom

43. After the emission of a ----, the daughter element formed is placed one group to the left of the parent element
(A) β-ray (B) α-ray (C) Positron (D) γ-ray

44. The unrelated member in the following group is
(A) Neutron (B) Electron (C) Position (D) Cyclotron

45. For a basic buffer solution the pH is 9.3. The pH can be increased by
(A) Increasing the concentration of the base
(B) Decreasing the concentration of the base
(C) Increasing the concentration of the salt
(D) Decreasing the K_b value of the base

46. The pH of an aqueous solution is 10.8. It
(A) Turns yellow with phenolphthalein
(B) Has no action on phenolphthalein
(C) Turns pink with phenolphthalein
(D) Turns from red to yellow with phenolphthalein

47. The tyndall effect associated with colloidal particles is due to
(A) The electrical charge on them
(B) Scattering of light
(C) Absorption of light (D) Reflection of light

48. The equivalent weight of Mohr's salt is
(A) 278 (B) 392 (C) 56 (D) 158

49. Which of the following is observed when some solvent is added to a dilute solution of a non-volatile solute ?
(A) Increase of vapour pressure
(B) Depression of freezing point

(C) Elevation of boiling point
(D) Increase of osmotic pressure

50. The most favourable conditions for the alcoholic fermentation of sugar are
(A) Low concentration of sugar solution, high temperature and absence of air.
(B) Low concentration of sugar solution, low temperature and plenty of air supply.
(C) High concentration of sugar solution, low temperature and plenty of air.
(D) High concentration of sugar solution, low temperature and absence of air.

51. Which of the following is not a nucleophile ?
(A) CH_3OH (B) SO_3 (C) NH_3 (D) H_2O

52. In the hydrogenation reactions given below the hydridisation state of the carbon atom changes from

$C_2H_2 \xrightarrow{H_2/Ni} C_2H_4 \xrightarrow{H_2/Ni} C_6H_6$

(A) $sp - sp^2 - sp^3$ (B) $sp^3 - sp^2 - sp$
(C) $sp^2 - sp - sp^3$ (D) $sp^3 - sp - sp^2$

53. Which of the following is not a periodic property ?
(A) Density (B) Atomic volume
(C) Mass number (D) Ionization energy

54. The alloy of copper which does not contain zinc is
(A) Brass (B) Bronze
(C) Gun metal (D) German sliver

55. Which of the following gases is colourless and odourless ?
(A) HCl (B) NH_3 (C) CO (D) H_2S

56. The superior quality laboratory glass apparatus are made of
(A) Soda-lime glass (B) Hard glass
(C) Flint glass (D) Pyrex glass

57. Ortho and Para hydrogen
(A) Have different atomic number
(B) Differ in the spin of their nuclei
(C) Have different masses
(D) Do not differ in the spin of their electrons

58. When warm oxalic acid is titrated against $KMnO_4$ in acid medium, the reaction which is slow initially becomes brisk. This is due to
(A) Promoter (B) Catalysis
(C) Auto-catalysis (D) None of the above

59. Which of the following statements regarding the kinetic theory of matter is incorrect ?
(A) The velocity of the particles is proportional to the pressure of the particles.
(B) The collisions between the particles are elastic
(C) The velocity of the particles is proportional to the square root of the temperature.
(D) The collisions between the particles and the walls of the vessels are elastic.

60. In the dissociation of HI(g), if 20% of HI is dissociated, the equilibrium constant will be.
(A) 0.0156 (B) 64 (C) 0.0625 (D) 16

JIPMER - 1998

1. Red phosphorus is chemically much less reactive than white phosphorus because
(A) It does not contain tetrahedral P_4 molecules
(B) It does not contain p-p bonds
(C) It has a polymeric structure
(D) It has a linear structure

2. Which of the following compounds is sparingly soluble in ammonia ?
(A) AgCl (B) $CuCl_2$ (C) AgI (D) AgBr

3. The element which forms ions in the dimeric state is
(A) Mercury (B) Silver (C) Copper (D) Nickel

4. The carbonate which does not leave a residue on heating is
(A) Na_2CO_3 (B) Ag_2CO_3 (C) $CuCO_3$ (D) $(NH_4)_2CO_3$

5. When 2-butylanthraquinol dissolved in a mixture of benzene and cyclohexanol is oxidised, we obtain
(A) Ethanol (B) Hydrogen peroxide
(C) Anthracene (D) None of these

6. Hydrogen has a tendency to gain one electron to acquire helium configuration. In this respect it resembles
(A) Halogens (B) Actinides
(C) Transition elements (D) Alkali metals

7. pK_a of a weak acid is defined as
(A) $\log_{10}K_a$ (B) $\frac{1}{\log_{10}K_a}$ (C) $\log_{10}\frac{1}{K_a}$ (D) $-\log_{10}\frac{1}{K_a}$

8. The radiations from a naturally occurring radio element, as seen after deflection in a magnetic field in one direction, are
(A) α rays only (B) β rays only
(C) Both α and β rays (D) Either α or β rays

9. Neutrons are obtained by
(A) Bombardment of Be with α-particles
(B) radioactive disintegration of uranium
(C) Bombardment of Ra with β-particles
(D) None of the above

10. Which of the following sub-atomic particles is not present in an atom ?
(A) Electron (B) Positron (C) Neutron (D) Proton

11. The kind of radioactive decay that does not lead to the formation of a daughter nucleus which is an isobar of the parent nucleus ?
(A) α-ray (B) β-ray (C) Positron (D) Electron capture

12. p-orbitals in a given shell can accommodate upto
(A) Four electrons (B) Two electrons with parallel spins
(C) Six electrons (D) Two electrons with opposite spins

13. In the isotopes of an element
(A) The mass number remains the same
(B) Both atomic number and mass number are conserved

(C) The atomic number remains the same

(D) Neither the atomic number nor mass number is conserved.

14. The correct set of four quantum numbers for the valence (outermost) electron of rubidium (z = 37) is

(A) 5, 1, 1, +1/2 (B) 6, 0, 0, +1/2

(C) 5, 0, 0, +1/2 (D) 5, 1, 0, +1/2

15. Which of the following requires minimum ionization energy ?

(A) As (B) Br (C) Ge (D) Se

16. When a molten ionic hydride is electrolysed

(A) Hydrogen is liberated at the anode

(B) There is no reaction

(C) Hydrogen is liberated at the cathode

(D) H^- ions produced migrate to the cathode

17. The oxidation number of nickel in $K_4[Ni(CN)_4]$ is

(A) –2 (B) –1 (C) +3 (D) 0

18. The smallest among the following ions is

(A) Ba^{+2} (B) Al^{3+} (C) Na^+ (D) Mg^{+2}

19. A vessel is filled with a mixture of oxygen and nitrogen. At what ratio of the partial pressures, the masses of the gases will be identical ?

(A) $pO_2 = pN_2$ (B) $pO_2 = 0.5\ pN_2$

(C) $pO_2 = 0.875\ pN_2$ (D) $pO_2 = 1.14\ pN_2$

20. 1 litre of a gas is at a pressure of 10^{-6} mm of mercury at 25°C. How many molecules are present in the vessel ?

(A) 3.2×10^6 (B) 3.2×10^{10}

(C) 3.2×10^{13} (D) 3.2×10^4

21. According to kinetic theory of gases

(A) There are no inter molecular attractions

(B) The velocity of the molecules decreases with each collision

(C) There are inter molecular attractions

(D) Molecules have considerable volume.

22. Calculate the heat of formation of $PCl_5(s)$ from the following data :

$2P(s) + 3Cl_2(g) \rightarrow 2PCl_3(l);\ \Delta H = -151.8kcal$

$PCl_3(l) + Cl_2(g) \rightarrow PCl_5(s);\ \Delta H = -32.8kcal$

(A) –108.7 kcal (B) +108.7 kcal

(B) –184.6 kcal (D) + 184.6 kcal

23. Lithopone is

(A) $ZnS + BaSO_4$ (B) $BaS + ZnSO_4$

(C) $ZnS + CaSO_4$ (D) $ZnSO_4 + CaS$

24. Which of the following silver compounds finds maximum use in photography ?

(A) AgI (B) $AgNO_3$ (C) AgCl (D) AgBr

25. Which of the following substances when dissolved in water will give a solution that does not conduct electricity ?

(A) Hydrogen chloride (B) Potassium Hydroxide

(C) Sodium acetate (D) Urea

26. Monomers required for Buna-s-rubber are

(A) Butadiene and isobutylene (B) Chloroprene

(C) Butadiene and styrene (D) Styrene

27. Which of the following does not undergo charring when warmed with con.H_2SO_4 ?

(A) Glucose (B) Sucrose

(C) Tartaric acid (D) Oxalic acid

28. In the presence of a dilute base, two molecules of an aldehyde condense to form

(A) An acid (B) A β-hydroxy aldehyde

(C) A β-hydroxy ketone (D) An alcohol

29. Sodium phenoxide when treated with CO_2 under pressure gives

(A) [benzene ring with –COOH and –OH on adjacent positions] (B) [benzene ring with –OH and –COOH in meta positions]

(C) HOOC–[benzene ring]–OH (D) None of these

30. Which of the following is false regarding diethyl ether ?

(A) It is soluble in water

(B) It does not possess a dipole moment

(C) It is soluble in conc.H_2SO_4 (D) It is inflammable

31. Chlorine reacts with ethanol to give

(A) Ethyl chloride (B) Chloroform

(C) Acetaldehyde (D) Chloral

32. Which of the following methods is not used for the purification of liquids ?

(A) Chromatography (B) Steam distillation

(C) Distillation (D) Sublimation

33. The monomer units of a polypeptide are

(A) Dienes (B) Diamines (C) Aldehydes (D) Aminoacids

34. The weakest base among the following is.

(A) CH_3NH_2 (B) $(CH_3)_2NH$ (C) $(CH_3)_3N$ (D) $C_6N_5NH_2$

35. The addition of an inert gas at constant volume to an equilibrium mixture will favour

(A) The forward reaction (B) The reverse reaction

(C) Neither reactions (D) All the above

36. If at 3000 K the equilibrium pressures of CO_2, CO and O_2 are 0.6, 0.4 and 0.2 atmospheres respectively. K_P for the reaction, $2CO_2 \leftrightarrow 2CO + O_2$ is

(A) 0.133 (B) 0.177 (C) 0.074 (D) 0.0533

37. The alkaline hydrolysis of an ester represented by $RCOOR' + OH^- \longrightarrow RCOO^- + R'OH$ belongs to

(A) First order, bimolecular reaction

(B) Second order but not bimolecular reaction

(C) Bimolecular but zero order reaction

(D) Bimolecular and second order reaction

38. The compound which reacts fastest with Lucas reagent is

(A) Butan-1-ol (B) Butan-2-ol

(C) 2-methyl propan-1-ol (D) 2-methylpropan-2-ol

39. When ethyne is passed through dilute H_2SO_4 containing $HgSO_4$, it forms
(A) Ethanal (B) Ethene (C) Ethanol (D) Ethanoic acid

40. Which of the following statements regarding a chemical equilibrium is wrong ?
(A) An equilibrium can be shifted by altering the temperature or pressure.
(B) An equilibrium is dynamic.
(C) The same state of an equilibrium is reached whether one starts with the reactants or the products.
(D) The forward reaction is favoured by the addition of a catalyst.

41. For the reaction A + B $\longrightarrow$ Products, doubling the concentration of A the rate of the reaction is doubled, but on doubling the concentration of B rate remains unaltered. The over all order of the reaction is
(A) 1 (B) 0 (C) 2 (D) 3

42. The osmotic pressures of a dilute solution of a non-volatile solute is
(A) Inversely proportional to its temperature on the Kelvin scale.
(B) Directly proportional to its temperature on the centigrade scale.
(C) Directly proportional to its temperature on the Kelvin scale.
(D) Inversely proportional to its temperature on the centigrade scale.

43. The compound which does not possess a peroxide linkage is
(A) Na_2O_2 (B) CrO_5 (C) H_2SO_5 (D) PbO_2

44. The oxide which is not reduced by hydrogen is
(A) Fe_2O_3 (B) P_4O_{10} (C) Ag_2O (D) K_2O

45. 20 ml of a N solution of $KMnO_4$ just reacts with 20 ml of solution of oxalic acid. The weight of oxalic acid crystals in 1litre of the solution is
(A) 31.5g (B) 126 g (C) 63 g (D) 6.3 g

46. Colloidal solutions of gold prepared by different methods have different colours owing to
(A) The difference in the size of the colloidal particles.
(B) The fact that gold exhibits variable valency of +1 and +3.
(C) Different concentrations of gold
(D) Presence of different types of foreign particles depending upon the method of preparation of the colloid.

47. The statement, "If 0.003 moles of a gas are dissolved in 900g of water under a pressure of 1 atmosphere, 0.006 moles will be dissolved under a pressure of 2 atmospheres", illustrates
(A) Dalton's law of partial pressure
(B) Graham's law (C) Raoult's law (D) Henry's law

48. The weight of sodium carbonate required to prepare 500 ml of a N/2 solution is
(A) 53 g (B) 6.125g (C) 13.25g (D) 26.5 g

49. The fundamental basis of the modern Periodic table is that elements are
(A) Arranged in the order of increasing atomic weights.
(B) Grouped according to chemical properties
(C) Arranged in the order of increasing number of neutrons in the atomic nucleus
(D) Arranged in the order of increasing number of protons in the nucleus.

50. Which is the strongest of the following acids ?
(A) $HClO_4$ (B) H_2SO_4 (C) HCl (D) HNO_3

51. pH of a solution is 5. Its hydroxyl ion concentration is
(A) 10^{-5} (B) 10^{-9} (C) 5 (D) 10

52. In the viscose process the solvent mixture used for cellulose is
(A) Ether and alcohol
(B) Copper sulphate and ammonia
(C) Sodium hydroxide and carbon disulphide
(D) Acetic acid and acetic anhydride

53. The reaction of benzoyl chloride with phenol in alkaline medium gives phenyl benzoate. It is
(A) Schotten-Baumann reaction
(B) Friedel-Crafts reaction
(C) Perkin's reaction (D) Kolbe's reaction

54. The structural formula of benzyl phenyl ketone is
(A) $C_6H_5COC_6H_5$ (B) $C_6H_5CO.COC_6H_5$
(C) $C_6H_5CH_2COC_6H_5$ (D) $C_6H_5CH_2COCH_2C_6H_5$

55. Benzaldehyde undergoes benzoin condensation in the presence of.
(A) Potassium cyanide (B) Conc.H_2SO_4
(C) Sodium metal (D) Anhydrous $ZnCl_2$

56. In the presence of HCl, two molecules of acetone condense to form.
(A) Mesityl oxide (B) Phorone
(C) 4-methyl-2-pentanone (D) 4-methyl-2- pentanoic acid

57. The reaction in which phenol condenses with chloroform in the presence of alkali is known as
(A) Reimer-Tiemann reaction
(B) Houben-Hoesch reaction
(C) Etard reaction (D) Cannizzaro's reaction

58. Which of the following pairs of reagents will not give rise to an ether ?
(A) $C_2H_5Br + (CH_3)_3CONa$ (B) $C_2H_5Br + C_2H_5ONa$
(C) $C_6H_5Br + C_3H_7ONa$ (D) $C_2H_5Br + C_6H_5ONa$

59. The compound which forms a thermosetting polymer with methanal is
(A) Benzene (B) Phenyl amine
(C) Benzaldehyde (D) Phenol

60. The type of glass used in making lenses and prisms is
(A) A flint glass (B) Jena glass
(C) Pyrex glass (D) Quartz glass

JIPMER - 1999

1. Propene can be converted to 1-propanol. Indicate which set of reagents among the following is ideal for effecting the above conversion of propene to 1-propanol
(A) Osmium tetraoxide (B) C_3H_6 and alkaline H_2O_2
(C) Alkaline $KMnO_4$ (D) Ozone and Zn dust

2. Which of the following reactants would form 2,2-dibromo-propane on interaction ?
(A) $CH_3CH = CHBr + HBR$ (B) $CH = CH + 2HBr$
(C) $CH_3CH = CH_2 + HBr$ (D) $CH_3C \equiv CH + HBr$

3. Phosphorous has the oxidation state of +3 in
(A) Ortho-phosphorous acid
(B) meta-phosphoric acid (C) Ortho-phosphoric acid
(D) Hypo-phosphorous acid

4. Which of the following does not give nitrogen dioxide on heating ?
(A) $Cu(NO_3)_2$ (B) $AgNO_3$ (C) KNO_3 (D) $Pb(NO_3)_2$

5. In which of the following neutralisation reactions, the heat of neutralisation will be the highest ?
(A) HCl and NaOH (B) NH_4OH and HCl
(C) NaOH and CH_3COOH (D) CH_3COOH and NH_4OH

6. Hypophosphorous acid (H_3PO_2) is
(A) Monobasic (B) Neutral (C) Dibasic (D) Tribasic

7. Oxygen is obtained by heating
(A) Ammonium dichromate (B) Sodium peroxide
(C) Potassium permanganate (D) Bleaching power

8. Bromine water cannot be decolouised by
(A) $H_4P_2O_5$ (B) $H_4P_2O_7$ (C) $Na_2S_2O_3$ (D) NaOH

9. Ozone can be easily detected by the use of
(A) Ag (B) Hg (C) AgCl (D) H_2O_2

10. Chloroform when treated with acetone gives
(A) Chloroprene (B) Chloretone
(C) Carbylamine (D) Acetylene

11. Carbon tetrachloride upon treatment with Fe and dilute HCl gives
(A) Methane (B) Methylenechloride
(C) Methylchloride (D) Chloroform

12. An organic compound A reacts with sodium to evolve hydrogen and with oxidising agents to yield a ketone. The formula of A could be
(A) CH_3-CH_2OH (B) $CH_3-CHOH-CH_3$
(C) $C(CH_3)_3CHO$ (D) $OH-CH_2-CH_2-OH$

13. The IUPAC name of the compound ,

$$CH_3 - \underset{\displaystyle CH_3}{\underset{|}{CH}} - CH = CH - CH_2OH$$ is

(A) 4-Methyl pent-2-en-1-ol
(B) 2-Methyl-4-pentenol
(C) 4-Methyl-2-pentenol
(D) 2-Methyl-4-en-1-ol

14. The electronic configuration of sulphide ion is
(A) $1s^22s^22p^63s^23p^4$ (B) $1s^22s^22p^63s^23p^3$
(C) $1s^22s^22p^63s^23p^6$ (D) $1s^22s^22p^63s^23p^5$

15. With dsp^2 or sp^2d hybridisation, the geometry of the molecule will be.
(A) Trigonal planar (B) Octahedral
(C) Tetrahedral (D) Square planar

16. Of the following hydrides which one has the lowest boiling point ?
(A) HBr (B) HI (C) HF (D) HCl

17. The non-linear molecule amongst the following is
(A) C_2H_2 (B) OF_2 (C) CO_2 (D) Cl_2

18. The correct order of electron affinity of F, Cl, Br, and I is.
(A) F > Cl > Br > I (B) I > Br > Cl > F
(C) Cl > F > Br > I (D) Br > Cl > F > I

19. In P^{3-}, S^{2-}, Cl^- ions the increasing order of size is
(A) Cl^-, S^{2-}, P^{3-} (B) P^{3-}, S^{2-}, Cl^-
(C) S^{2-}, Cl^{-1}, P^3 (D) P^{3-}, S^{2-}, S^{2-}

20. Experimental evidence for the existence of the atomic nucleus comes from.
(A) Millikan's oil-drop experiment
(B) Alpha ray scattering by a thin metal foil
(C) Atomic emission spectroscopy
(D) The magnetic bending of cathode rays

21. The only oxide formed by an element X has the formula X_2O_3, while element Y forms a hydride YH_2. On the basis of this information the most likely formula for the stable compound formed between X and Y is.
(A) XY (B) X_2Y (C) X_3Y_2 (D) X_2Y_3

22. The temperature at which a real gas obeys the ideal gas laws over a wide range of pressure is called.
(A) Critical temperature (B) Inversion temperature
(C) Reduced temperature (D) Boyle's temperature

23. The radius of the third Bohr orbit is 4.77 Å. The radius of the second orbit will be.
(A) 2.12 Å (B) 1.59 Å (C) 1.16 Å (D) 3.18 Å

24. In sodium fusion test, the nitrogen of an organic compound is converted into.
(A) Sodium cyanide (B) Sodium nitrate
(C) Sodamide (D) Sodium nitrite

25. Natural rubber is a polymer of
(A) Ethylene (B) Acrylic acid
(C) 1,4 Buta diene (D) Isoprene

26. Which of the following compounds will react with sodium ?
(A) CH_3Cl (B) C_2H_2 (C) C_2H_6 (D) C_2H_4

27. Which chloro-derivative of benzene among the following would undergo hydrolysis most readily with aqueous sodium hydroxide to furnish the corresponding hydroxy derivative ?

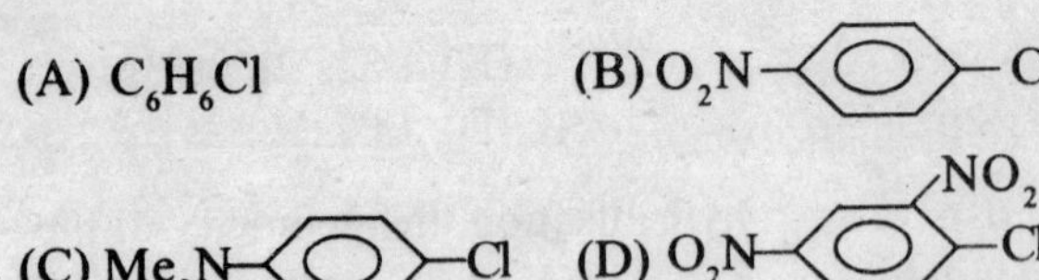

(C) $Me_2N-C_6H_4-Cl$ (D) $O_2N-C_6H_2(NO_2)_2-Cl$

28.The compound of formula NH_2-CH_2-COOH may behave

(A) Only as an acid (B) Only as base
(C) As both acid and base (D) As neither acid nor base

29. Which is the correct structure of the organic product formed when acetic acid is treated with thionylchloride ?

(A) $CH_3-\overset{O}{\overset{||}{C}}-O-\underset{\underset{O}{||}}{S}-Cl$ (B) $CH_3-\overset{O}{\overset{||}{C}}-Cl$

(C) $CH_3-\overset{O}{\overset{||}{C}}-OCl$ (D) $ClCH_2-\overset{O}{\overset{||}{C}}-O-SOCl$

30. Which of the following would lose water when exposed to atomophere ?

(A) Caustic soda (B) Concentrated H_2SO_4
(C) Anhydrous sodium carbonate
(D) A saturated solution of washing soda

31. The Lowry-Bronsted theory defines the acid as a molecule or an ion which

(A) Donates a proton
(B) Releases hydrogen in its reaction with metals
(C) Contains a replaceable hydrogen
(D) Accepts a proton

32. How much the hydrogen ion concentration in a solution be changed for the pH of the solution to increase by I ?

(A) It must be increased by 1 mole/litre
(B) It must be diminished to 1/10 th of its original value
(C) It must be increased 10 times
(D) It must be diminished by 1 mole/litre

33. If 15g of a solute in 100g of water makes the solution freeze at $-1°C$, then 30g of the same solute in 100g of water will make the solution freeze at

(A) 0°C (B) 2°C (C) $-2°C$ (D) $-0.5°C$

34. Joule-Thomson Expansion of an ideal gas produces.

(A) No change in temperature (B) Heating (C) Cooling
(D) Heating above the inversion temperature

35. Which of the following statements is wrong ? The enthalphy of a reaction depends upon

(A) The physical state of the reactants and products
(B) The temperature of the reaction
(C) The number of moles of reactants
(D) The path by which the reactants are converted into products

36. In CH_3CH_2OH the bond that undergoes heterolytic cleavage most readily is

(A) C – H (B) O – H (C) C – C (D) C – O

37. A compound X of the formula C_3H_8O yields a compound C_3H_6O on oxidation. To which of the following class of compounds could X belong ?

(A) Aldehyde (B) Secondary alcohol
(C) Alkene (D) Tertiary alcohol

38. Bromine is liberated when aqueous solution of potassium bromide is treated with

(A) I_2 (B) Cl_2 (C) SO_2 (D) dil. H_2SO_4

39. Which one of the following oxides of chlorine is obtained by passing dry chlorine over silver perchlorate at 190°C ?

(A) ClO_3 (B) ClO_4 (C) Cl_2O (D) ClO_2

40. Sodium acetate reacts with acetyl chloride to form

(A) Acetone (B) Acetic anhydride
(C) Sodium formate (D) Acetic acid

41. Upon hydration with $HgSO_4$ and dilute H_2SO_4, propyne gives mainly

(A) Propanal (B) Propane
(C) Propyl alcohol (D) Propanone

42. When sodium acetate is added to acetic acid it

(A) Decreases ionization of acetic acid
(B) Increases ionization of sodium acetate
(C) Increases ionization of acetic acid
(D) Is independent of concentration

43. Cu^{2+} and Ag^+ are present in the same solution. To precipitate one of the two ions and leave the other in solution, we should add

(A) $HNO_3(aq)$ (B) $H_2S(aq)$
(C) $HCl(aq)$ (D) $NH_4NO_3(aq)$

44. The volume of the reaction vessel containing an equilibrium mixture of the reaction $SO_2Cl_2\ (g) \longleftrightarrow SO_2(g) + Cl_2(g)$, is increased. When the equilibrium is re-established the amount of

(A) $SO_2Cl_2(g)$ will have increased
(B) $Cl_2(g)$ will have remained unchanged
(C) $SO_2(g)$ will have decreased
(D) $Cl_2(g)$ will have increased.

45. Monocarboxlic acid can be obtained by hydrolysing

(A) Aldehydes (B) Alkylcyanides
(C) Amides in presence of alkali (D) Ethers

46. The relationship between acetylene and benzene is comparable to the relationship between propyne and

(A) Dimethyl benzene (B) Neoprene
(C) Mesitylene (D) Propyl benzene

47. If α is the degree of dissociation of PCl_5 at equilibrium for the reaction $PCl_5 \longleftrightarrow PCl_3 + Cl_2$, the total number of moles at equilibrium is

(A) $1 + \alpha$ (B) 3 (C) 1 (D) $1 - \alpha$

48. For the dissociation of $N_2O_4(g) \longleftrightarrow 2NO_2$, K_P is

(A) Equal to K_C (B) Less than K_C
(C) Greater than K_C (D) None of these

49. An isotone of ${}^{30}_{14}Si$ is

(A) ${}^{31}_{15}P$ (B) P^{32}_{15} (C) S^{31}_{16} (D) ${}^{34}_{17}S$

50. A subshell with $l = 2$ can take up a maximum of -------- electrons
(A) 2 (B) 6 (C) 10 (D) 14

51. The ion with the smallest radius is
(A) Si^{4+} (B) Al^{3+} (C) Mg^{2+} (D) P^{3-}

52. Which of the following ions has the highest polarizing power ?
(A) Na^+ (B) Ca^{2+} (C) Mg^{2+} (D) Al^{3+}

53. An aldehyde can be obtained from an acid chloride by
(A) Rosenmund's reduction (B) Wolf-Kischner reduction
(C) Clemmensen's reduction
(D) Meerwin-Pondorff-Verley reduction

54. Cresol when distilled with ------ is converted into toluene
(A) Al_2O_3 (B) Zinc dust
(C) $LiAlH_4$ (D) Acetic anhydride

55. Diethyl ether when treated with Cl_2 in the dark gives mostly
(A) C_2H_5Cl (B) $CH_3CHClOCH_2CH_3$
(C) $CH_2Cl.CH_2O.CH_2CH_2Cl$ (D) $CH_3CHClOCHClCH_3$

56. A compound x when boiled with aqueous Na_2CO_3 gives glycol. The compound x is
(A) ethylene (B) ethylene oxide
(C) ethylene bromide (D) cthyl hydrogen sulphate

57. The adsorption of hydrogen by platinum is known as
(A) Hydrogenation (B) Reduction
(C) Occlusion (D) Dehydrogenation

58. The approximate composition of ordinary glass is
(A) Na_2O CaO $6SiO_2$ (B) Na_2O CaO SiO_2
(C) K_2O CaO $6SiO_2$ (D) Na_2O CaO $\frac{1}{2}SiO_2$

59. Distillation under reduced pressure is used to purify liquids which
(A) decompose below their boiling point
(B) are explosives (C) are highly volatile
(D) have high boiling point

60. The reaction of alkanes with Br_2 in the presence of ultra violet light is
(A) An elimination reaction (B) An addition reaction
(C) A substitution reaction (D) A rearrangement reaction

JIPMER - 2000

1. The most dangerous method of preparing hydrogen would be by the action of HCl on
(A) Fe (B) Zn (C) Al (D) K

2. The ionic carbide of the following is
(A) SiC (B) CaC_2 (C) ZnC (D) TiC

3. Pb pipes are not used for drinking water supply because
(A) A layer of lead dioxide is deposited over pipes
(B) Lead reacts with air to form litharge
(C) Lead reacts with water containing air to form $Pb(OH)_2$
(D) Lead forms basic lead carbonate

4. PbO_2 is
(A) Basic (B) Acidic (C) Neutral (D) Amphoteric

5. N_2 combines with a metal to from
(A) Nitrite (B) Nitrosyl chloride (C) Nitride (D) Nitrate

6. Sulphuric acid reacts with PCl_5 to yield
(A) Sulphur monochloride (B) Phosphoric acid
(C) Sulphuryl chloride (D) Thionyl chloride

7. Which of the following is not oxidized by MnO_2
(A) Cl^- (B) I^- (C) Br^- (D) F^-

8. Which of the following halogens has basic properties
(A) Bromine (B) Fluorine (C) Chlorine (D) Iodine

9. The metal that can not displace hydrogen from an acid is
(A) Hg (B) Zn (C) Ca (D) Al

10. Carbanions initiate
(A) Substitution reactions (B) Addition reactions
(C) Both (D) None

11. In the case of homologous series of alkanes, which one of the following statements is not correct
(A) The members of the series have similar chemical properties
(B) The members of the series have the general formula C_nH_{2n+2}, where n is an integer
(C) The members of the series are isomers of each other
(D) The difference between any two successive members of the series correcponds to 14 unit of relative atomic mass

12. Butyne-1 and butyne -2 can be distinguished by
(A) Ammonical Cu_2Cl_2 (B) Alkaline $KMnO_4$
(C) Benedict reagent (D) Br_2 water

13. Compound formed when ethyl amine is heated with chloroform in the presence of KOH is
(A) Ethyl isocyanide (B) Ethyl chloride
(C) Ethanol (D) Diethy ether

14. A mixture of camphor and benzoic acid can be easily separated by
(A) Fractional crystallization (B) Chemical method
(C) Sublimation (D) Extraction with solvent

15. The vant Hoff factor calculated from association data is always . . . than calculated from dissociation data
(A) Same (B) More or less (C) Less (D) More

16. The activation energy for a reaction is 9.0 k cal/mol. The increase in the rate constant when its temperature is increased from 298 K to 308 K is
(A) 63% (B) 50% (C) 100% (D)10%

17. According to Le Chatelier's principle adding heat to a solid and a liquid in an equilibrium with endothermic nature will cause the
(A) Temperature to rise (B) Temperature to fall
(C) Amount of solid to decrease
(D) Amount of liquid to decrease

18. Equilibrium concentration of HI, I_2 and H_2 is 0.7, 0.1 and 0.1 M respectively. The equilibrium constant for the

reaction, $I_2 + H_2 \leftrightarrow 2HI$ is
(A) 0.49 (B) 0.36 (C) 36 (D) 49

19. Although CO is neutral, it shows acidic nature on at high pressure and temperature with
(A) LiOH (B) NaOH (C) $Ca(OH)_2$ (D) $Mg(OH)_2$

20. Which of the following may be added to one litre of water to act as a buffer
(A) One mole of $HC_2H_3O_2$ and 0.5 mole of NaOH
(B) One mole of NH_4Cl and one mole of HCl
(C) One mole of NH_4OH and one mole of NaOH
(D) One mole of $HC_2H_3O_2$ and one mole of HCl

21. The pH of a 0.01M solution of acetic acid having degree of dissociation 12.5% is
(A) 3.723 (B) 4.509 (C) 5.623 (D) 2.903

22. If the pH of a solution of an alkali metal hydroxide is 13.6, the concentration of hydroxide is
(A) Between 0.1M and 1M (B) More than 1M
(C) Less than 0.001M (D) Between 0.01M and 1M

23. Heat of neutralization of the acid-base reaction is 57.32 kJ for
(A) HNO_3 + LiOH (B) HCOOH + KOH
(C) HCl + NH_4OH (D) CH_3COOH + NaOH

24. The unit cell cube length for LiCl (just like NaCl structure) is 5.14 Å. Assuming anion-anion contact, the ionic radius for chloride ion is
(A) 2.8 Å (B) 1.815 Å (C) 4.815 Å (D) 3.8 Å

25. Joule-Thomson expansion is
(A) Iosbaric (B) Iso enthalpic
(C) Isothermal (D) Iso entropic

26. What is the pH for a neutral solution at the normal temperature of the human body
(A) 7.2 (B) 14.0 (C) 6.8 (D) 6.0

27. The decomposition of H_2O_2 can be slowed down by the addition of a small amount of phosphoric acid which acts as
(A) Promoter (B) Inhibitor (C) Detainer (D) Stopper

28. V vs T curves at constant pressure P_1 and P_2 for an ideal gas are shown in figure. Which relation is correct

V
P_1
P_2
T

(A) $P_1=P_2$ (B) $P_1>P_2$ (C) $P_1<P_2$ (D) All

29. The de-Broglie wavelength associated with a material particle is
(A) Directly proportional to momentum
(B) Inversely proportional to its energy
(C) Directly proportional to its energy
(D) Inversely proportional to momentum

30. Which of the following sets has the same number of unpaired electrons in their ground state
(A) Cl^-, Fe^{3+}, Cr^{3+} (B) Na^+, Mg^{2+}, Al
(C) Na, P, Cl (D) N, P, V

31. A radioactive substance has a constant activity of 2000 disintegration/minute. The material is separated into two fractions, one of which has an initial activity of 1000 disintegrations per second while the other fraction decays with $t_{1/2}$= 24 hours. The total activity in both samples after 48 hours of separation is
(A) 1000 (B) 1500 (C) 2000 (D) 1250

32. Which species does not exist
(A) $(SiCl_6)^{2-}$ (B) $(CCl_6)^{-2}$ (C) $(GeCl^6)^{2-}$ (D) $(SnCl_6)^{2-}$

33. Which of the following is a molecule with residual bonding capacity
(A) CH_4 (B) N_2 (C) $BeCl_2$ (D) NaCl

34. The velency shell of calcium contains
(A) 8 electrons (B) 6 electrons
(C) 4 electrons (D) 2 electrons

35. Uranium decays into a stable isotope of
(A) Np (B) Pb (C) Rn (D) Ra

36. How many grams are present in one mole of Ag
(A) 108.6 (B) None (C) 10.29 (D) 107.9

37. When H_2S reacts with halogens, the halogens
(A) Form sulphur halides (B) Are oxidised
(C) Are reduced (D) None of these

38. The oxidation potential of a hydrogen electrode at pH = 10 and P_{H_2}=1 is
(A) 0.059V (B) +0.59V (C) 0.00V (D) 0.51V

39. The ratio of energy of photons of 3000 Å and 6000 Å wavelength radiations is
(A) 1/2 (B) 1/4 (C) 4 (D) 2

40. As a result of osmosis the volume of solution
(A) Increases (B) Decreases
(C) remains constant (D) Increases or decreases

41. Isopropyl alcohol on passing over heated copper at 300°C gives
(A) Acetone (B) Acetaldehyde
(C) Propylene (D) None

42. The organic compounds A and B react with sodium metal and release H_2 gas. A and B react with each other to give ethyl acetate. A and B are
(A) CH_3COOH and CH_3OH
(B) HCOOH and C_2H_5OH
(C) CH_3COOH and HCOOH
(D) CH_3COOH and C_2H_5OH

43. Which of the following reacts with $COCl_2$ to give phenyl isocyanate
(A) Chlorobenzene (B)Nitrobenzene
(C) Aminophenol (D)Aniline

44. Chlorobenzene is prepared commercially by
(A) Raschig process (B) Wurtz Fittig reaction
(C) Friedel-Craft's reaction (D) Grignard reaction

45. Xylenes on oxidation with acidic $KMnO_4$ gives
(A) Terphthalic acid (B) Phthalic acid
(C) Isophthalic acid (D) All

46. Which of the following is least basic

(A) BCl_3 (B) BF_3 (C) BI_3 (D) BBr_3

47. Which solution has the highest pH value

(A) 1 M KOH (B) 1 M H_2SO_4

(C) Chlorine water (D) Water containing carbon dioxide

48. The volume of a gas is 100 ml at 100°C. If the pressure is kept constant, at what temperature will the gas have a volume of 200 ml

(A) 50°C (B) 446°C (C) 473°C (D) 200°C

49. An element X with the electronic configuration $1s^2 2s^2 2p^6 3s^2$ would be expected to from the chloride with the formula

(A) XCl (B) X_2Cl (C) XCl_3 (D) XCl_2

50. For the equilibrium $N_2 + 3H_2 \rightleftharpoons 2NH_3$, K_c at 1000 K is 2.37×10^{-3}. If at equilibrium $[N_2] = 2M$, $[H_2] = 3M$, the concentration of NH_3 is

(A) 0.00358 M (B) 0.0358 M (C) 0.358 M (D) 3.38

51. On heating NaCl + $K_2Cr_2O_7$ + conc.H_2SO_4 , the gas that forms is

(A) O_2 (B) Cl_2 (C) $CrOCl_2$ (D) CrO_2Cl_2

52. The distinguishing test for triple bond contaning acidic hydrogen is

(A) $Ag(NH_3)_2^+$ (B) Br_2 in CCl_4

(C) Alkaline $KMnO_4$ (D) None of these

53. 1,3-butadiene has

(A) sp and sp^2 hybridised C-atoms

(B) sp, sp^2 and sp^3 hybridised C-atoms

(C) Only sp^2 hydridised C-atoms

(D) Only sp hybridised C-atoms

54. Vinyl chloride reacts with HCl to form

(A) Tetrachloro ethylene

(B) 1,1-dichloro ethane (C) 1,2-dichloro ethane

(D) Mixture of 1,2 and 1,1-dichloro ethane.

55. HBr reacts fastest with

(A) 2- methyl propan-1-ol (B) Propan 2-ol

(C) Propan-1-ol (D) 2-methylpropan-2-ol

56. When HCHO is treated with C_6H_5CHO in presence of NaOH, the products expected are

(A) CH_3OH and C_6H_5COONa

(B) CH_3OH and HCOO Na

(C) HCOONa and $C_6H_5CH_2OH$

(D) $C_6H_5CH_2OH$ and C_6H_5COONa

57. $CH_3CH_2\,COOH \xrightarrow[Fe]{Cl_2} X \xrightarrow[KOH]{alc.} Y$.

The compound Y is

(A) $CH_3CHClCOOH$ (B) $CH_2 = CHCOOH$

(C) CH_3CH_2CN (D) CH_3CH_2OH

58. What is the end product in the following sequence of reactions,

$C_2H_5NH_2 \xrightarrow{HNO_2} A \xrightarrow{PCl_5} B \xrightarrow{NH_3} C$

(A) Methyl amine (B) Acetamide

(C) Ethyl cyanide (D) Ethylamine

59. The change in optical rotation of freshly prepared solutions of sugar with time is known as

(A) Maturation (B) Specific rotation

(C) Inversion (D) None of these

60. When benzene reacts with an alkyl chloride (RCl) in presence of anhydrous $AlCl_3$

(A) $C_6H_5AlCl_3$ is produced (B) C_6H_5Cl is produced

(C) C_6H_5R is produced (D) Nothing happens

JIPMER - 2001

1. When a catalyst is added to an equilibrium, the

(A) Rate of forward reaction is increased and that of backward reaction is decreased

(B) Equilibrium concentrations are unchanged

(C) Value of equilibrium constant is decreased

(D) Equilibrium concentrations are increased.

2. If the rate of a first order reaction is 0.6932×10^{-2} mol L^{-1} min^{-1} and the initial concentration of the reactants is 1 M, $T_{1/2}$ is equal to

(A) 6.932 minute (B) 100 minute

(C) 0.6932×10^{-3} minute (D) 0.6932×10^{-2} minute

3. A reversible reaction $H_2 + Cl_2 \rightleftharpoons 2HCl$ is carried out in one litre flask. If the same reaction is carried out in two litre flask, the equilibrium constant will be

(A) Halved (B) Same (C) Decreased (D) Doubled

4. For the reaction

$2NO_2(g) \rightleftharpoons 2NO(g) + O_2(g)$, $K_C = 1.8 \times 10^{-6}$ at 185°C,

At 185°C, the K_C for $NO(g) + 1/2\, O_2(g) \rightleftharpoons 2NO_2(g)$ is

(A) 1.95×10^{-3} (B) 1.95×10^{3}

(C) 7.5×10^{2} (D) 0.9×10^{6}

5. The compound HCl behaves as . . . in the reaction,

$HCl + HF \rightarrow H_2^+Cl + F^-$

(A) Weak base (B) Weak acid

(C) Strong base (D) Strong acid

6. A solution of sodium borate has a pH of approximately

(A) =7 (B) Between 4 to 5 (C) < 7 (D) > 7

7. The number of moles of hydroxide (OH^-) ion in 0.3 litre of 0.005 M solution of $Ba(OH)_2$ is

(A) 0.0050 (B) 0.0030 (C) 0.0015 (D) 0.0075

8. Self-ionisation of liquid ammonia occurs as, $2NH_3 \rightleftarrows NH^+_4 + NH^-_2$; $K = 10^{-10}$. In this solvent, the species that acts as an acid is

(A) NH^+_4 (B) NH_3

(C) Any species that will form NH^+_4 (D) Any of these

9. In an exothermic reaction, the reactants

(A) Are at a higher temperature than the products

(B) Have less energy than the products

(C) Have more energy than the products

(D) Have the same energy as the products

10. For the given reaction,

$C_6H_6(l) + \frac{15}{2}O_2(g) \rightarrow 3H_2O(l) + 6CO_2(g)$; $\Delta H = -3264.6$kJ heat evolved when 39g of C_6H_6 are burnt is

(A) 2448.45 kJ (B) 6528.2 kJ
(C) 1632.3 kJ (D) 816.15 kJ

11. For an exothermic reaction, the equilibrium constant
(A) Decreases with increase of P
(B) Increases with increase of temperature
(C) Increases with increase of P
(D) Decreases with increase of temperature

12. A and B are two identical vessels. The vessel A contains 15g of ethane at 298 K and 1 atm. The vessel B contains 75 g gas X_2 at the same temperature and pressure. The vapour density of X_2 is
(A) 150 (B) 75 (C) 300 (D) 35.5

13. Two vessels having equal volume contain molecular hydrogen at one atmosphere and helium at two atmosphere pressure respectively. If both samples are at the same temperature, the mean velocity of hydrogen molecule is
(A) Half that of helium (B) Twice that of helium
(C) Equal to that of helium (D) $\sqrt{2}$ times that of helium

14. Which property is different for neutral atoms of the two isotopes of the same element
(A) Number of protons (B) Atomic number
(C) Number of neutrons (D) None

15. Radioactivity of neptunium stops when it is converted to
(A) Th (B) Pb (C) Bi (D) Rn

16. In which of the following, the angle between the two covalent bonds is the greatest
(A) H_2O (B) NH_3 (C) CH_4 (D) CO_2

17. Which combination is best explained by the co-ordinate covalent bond
(A) $H^+ + H_2O$ (B) Cl + Cl (C) $Mg + \frac{1}{2}O_2$ (D) $H_2 + I_2$

18. Which of the following pairs of atoms or ions will have the same electronic configuration
(A) F^+ and Ne (B) Li^+ and He^-
(C) Cl^- and Ar (D) Na and K

19. The equivalent weight of iron in Fe_2O_3 would be
(A) 121.1 (B) 127.5 (C) 112 (D) 26.66

20. Vapour density of a metal chloride is 66. Its oxide contains 53% metal. The atomic weight of the metal is
(A) 2.706 (B) 27.06 (C) 54 (D) 21

21. The reaction $H_2S + H_2O_2 \rightarrow 2H_2O + S$ shows
(A) Oxidizing action of H_2O_2
(B) Reducing action of H_2O_2
(C) Alkaline nature of H_2O_2 (D) Acidic nature of H_2O_2

22. Which of the following solutions will show highest resistance during the passage of current
(A) 0.05 N NaCl (B) 2 N NaCl
(C) 0.1 N NaCl (D) 1 N NaCl

23. The voltage of a cell whose half cells are given below is $Mg^{2+} + 2e^- \rightarrow Mg\ (s);\ E^o = -2.37$ V
$Cu^{2+} + 2e^- \rightarrow Cu\ (s);\ E^o = +0.34$ V
(A) 1.36 (B) +2.71 (C) -2.03 (D) +2.71

24. Which of the following salts shows maximum osmotic pressure in its 1 m solution
(A) Na_2SO_4 (B) $AgNO_3$ (C) $MgCl_2$ (D) $(NH_4)_3PO_4$

25. In the following sequence of reactions,

$$CH_3CH_2CH_2Br \xrightarrow{\text{KOH (alc.)}} (A) \xrightarrow{\text{HBr}} (B) \xrightarrow{\text{KOH (aq.)}} (C)$$

The product (C) is
(A) Propyne (B) Propene
(C) Propan–2–ol (D) Propan–1–ol

26. Diethyl ether on heating with conc. HI gives two moles of
(A) Methyl iodide (B) Ethyl iodide
(C) Iodoform (D) Ethanol

27. Aldehydes are isomeric with
(A) Ethers (B) Ketones (C) Alcohols (D) None

28. Acetaldehyde undergoes self condensation in the presence of aluminium ethoxide to give ethyl acetate. This reaction is called
(A) Aldol condensation (B) Cannizzaro's reaction
(C) Tischenko's reaction (D) Perkin reaction

29. Identify the product Z in the series,

$$CH_3CN \xrightarrow{Na\ /\ C_2H_5OH} X \xrightarrow{HNO_2} Y \xrightarrow{[O]} Z$$

(A) CH_3CH_2 NHOH (B) CH_3COOH
(C) $CH_3CH_2CONH_2$ (D) CH_3CHO

30. Aniline on heating with benzaldehyde produces
(A) Unsaturated Acid (B) Azoxy benzene
(C) Schiff's base (D) Benzoin

31. A mixture of benzene and aniline can be separated by
(A) Hot water (B) Dil HCl
(C) Dil NaOH (D) Alcohol

32. When alcohol reacts with an acid, it produces
(A) Secondary alcohol (B) Alkane
(C) Higher acid (D) Ester and water

33. K_C for $H_2(g) + I_2(g) \rightleftharpoons 2HI(g)$ is 64. If the volume of the container is reduced to one-half of its original volume, the value of the equilibrium constant will be
(A) 32 (B) 16 (C) +28 (D) 64

34. The compound which forms a dative bond with ammonia is
(A) CCl_4 (B) BCl_3 (C) $MgCl_2$ (D) NaCl

35. Which of the following is not a nucleophile
(A) HSO^-_3 (B) BF_3 (C) Ammonia (D) OH^-

36. The dissociation constant for HCN is 5×10^{-10}. The pH of the solution prepared by mixing 1.5 mole of HCN and 0.15 moles of KCN in water and making up the total volume to 0.5 dm^3 is
(A) 8.302 (B) 10.302 (C) 7.302 (D) 9.302

37. The oxidation of toluene to bezaldehyde by chromyl chloride is called
(A) Reimer-Tiemann reaction (B) Etard reaction
(C) Cannizzaro's reaction (D) Wurtz reaction

38. To obtain chromium from chromic oxide Cr_2O_3, the method used is
(A) Electrolytic reduction (B) Carbon reduction
(C) Alumino-thermic process
(D) Carbon monoxide reduction.

39. Deuterium resembles hydrogen in chemical properties but reacts
(A) More vigorously than hydrogen
(B) Faster than hydrogen
(C) Slower than hydrogen (D) Just as hydrogen

40. An ore of potassium is
(A) Carnallite (B) Cryolite (C) Bauxite (D) Dolomite

41. Diborane reacts with water to form
(A) $H_3BO_3 + H_2$ (B) H_2 (C) HBO_2 (D) H_3BO_3

42. Litharge is
(A) $Pb(CH_3COO)_2$ (B) Pb_3O_4 (C) PbO_2 (D) PbO

43. Tin, a silvery white metal exists in
(A) Two allotropic forms (B) Five allotropic forms
(C) Three allotropic forms (D) Four allotropic forms

44. Concentrated nitric acid reacts with iodine to give
(A) HOI (B) HI (C) $HOIO_3$ (D) $HOIO_2$

45. H_2SO_4 acts as dehydrating agent in its reaction with
(A) $H_2C_2O_4$ (B) $Ba(OH)_2$ (C) KOH (D) Zn

46. Bromine is liberated when an aqueous solution of KBr is treated with
(A) Dilute H_2SO_4 (B) I_2 (C) Cl_2 (D) SO_2

47. Which of the following burns to form an oxide which is a gas at room temperature
(A) Sodium (B) Hydrogen (C) S (D) P

48. The nitrate of which metal leaves metallic globules on strong heating
(A) $Pb(NO_3)_2$ (B) $NaNO_3$ (C) $AgNO_3$ (D) $Cu(NO_3)_2$

49. Red lead is an example of a/an . . . oxide
(A) Basic (B) Super (C) Mixed (D) Amphoteric

50. Conjugated double bond is present in
(A) 1, 3-butadiene (B) Butene
(C) Isobutene (D) Propene

51. A hydrocarbon X adds on one mole of hydrogen to give another hydrocarbon and decolourises bromine water. X reacts with $KMnO_4$ in presence of acid to give two moles of the same carboxylic acid. The structure of X is
(A) $CH_2 = CH–CH_2CH_2CH_3$
(B) $CH_3CH_2CH_2–CH = CHCH_3$
(C) $CH_3CH_2\ CH = CHCH_2CH_3$
(D) $CH_3CH = CHCH_2CH_3$

52. If Ksp for $HgSO_4$ is 6.4×10^{-5}, the solubility of its salt is
(A) 8×10^{-6} (B) 8×10^{-3}
(C) 6.4×10^{-5} (D) 6.4×10^{-3}

53. Which of the following transition elements shows the highest oxidation state ?
(A) Cr (B) V (C) Fe (D) Mn

54. The structure of PF_5 molecule is
(A) trigonal bipyramidal (B) pentagonal bipyranidal
(C) square planar (D) tetrahedral

55. The molarity of a solution having 18g of glucose (mole wt. =180) dissolved in 500g of water is
(A) 0.5 M (B) 0.1 M (C) 2.2 M (D) 0.2 M

56. When acetone reacts with bleaching powder, the compound formed is
(A) CH_2COCH (B) CH_4CHO
(C) $CHCl_3$ (D) CCl_4

57. Which of the following is one of the best methods to test the presence of water in a liquid
(A) use of anhydrous copper sulphate
(B) use of litmus paper (C) taste (D) smell

58. The number of unpaired electrons in ferrous ion is
(A) 5 (B) 4 (C) 3 (D) 2

59. Gammexane is the name of
(A) benzene hexa chloride (B) chlorobenzene
(C) benzoyl chloride (D) DDT

60. Number of water molecules in Mohr's salt is
(A) 6 (B) 5 (C) 8 (D) 7

JIPMER - 2002

1. Arrange S, P, As in order of increasing ionization energy.
(A) As < S < P (B) As < P < S
(C) S < P < As (D) P < S < As

2. π– bond is formed
(A) By transfer of electrons
(B) By sidewise overlapping of half filled p-orbitals.
(C) By the overlapping of atomic orbitals on the axis of nuclei.
(D) By overlapping of s-orbitals with p-orbitlas.

3. Which of the following arrangements of molecules is correct on the basis of their dipole moments ?
(A) $BF_3 > NF_3 > NH_3$ (B) $NF_3 > BF_3 > NH_3$
(C) $NH_3 > BF_3 > NF_3$ (D) $NH_3 > NF_3 > BF_3$

4. Enthalphy of neutralization of NH_4OH and HCl is numerically.
(A) 57.1 kJ mol^{-1} (B) < 57.1 KJ mol^{-1}
(C) > 57.1 kJ Mol^{-1} (D) Zero

5. Which one of the following is an acid salt ?
(A) $NaHSO_3$ (B) Na_2SO_4 (C) Na_2S (D) Na_2SO_3

6. The ionic product of water will increase if
(A) Pressure is decreased (B) H^+ ions are added
(C) OH^- ions are added (D) Temperature is increased

7. Which of the following statements is wrong for gases ?
(A) Volume of a gas is equal to volume of container

confining the gas.
(B) Confined gas exerts uniform pressure on the walls of its container in all directions
(C) Gases do not have a definite shape and volume.
(D) Mass of a gas cannot be determined by weighing a container in which it is enclosed.

8. For a reaction $2N_2O_5 \longrightarrow 4NO_2 + O_2$ rate and rate constant are 1.02 x 10^{-4} mol $L^{-1}S^{-1}$ and 3.4 x 10^{-5} sec^{-1} respectively. The conc. of N_2O_5 at that time will be.
(A) 1.732 mol L^{-1} (B) 3 mol L^{-1}
(C) 1.02 x 10^{-4} mol L^{-1} (D) 3.2 x 105 mol L^{-1}

9. Azeotropic mixture of HCl and water has
(A) 84% HCl (B) 22.2% HCl
(C) 63% HCl (D) 20.2% HCl

10. Which of the following is not a reducing agent ?
(A) CO_2 (B) NO_2 (C) SO_2 (D) H_2O_2

11. The standard reduction potential of Li^+/Li, Ba^{2+}/Ba, Na^+ / Na and Mg^{2+} / Mg are −3.05, −2.73, −2.71 and −2.37 volts respectively. Which one of the following is the strongest oxidizing agent ?
(A) Na^+ (B) Li^+ (C) Ba^{+2} (D) Mg^{2+}

12. Milk is a colloid of the type
(A) A solid dispersed in liquid
(B) A gas dispersed in liquid
(C) Liquid dispersed in liquid
(D) Some sugar dispersed in water

13. Green vitriol is
(A) $CuSO_4.5H_2O$ (B) $CaSO_4.7H_2O$
(C) $FeSO_4.7H_2O$ (D) $FeSO_4.5H_2O$

14. Which of the following has the smallest size
(A) Al^{3+} (B) Si^{4+} (C) Mg^{2+} (D) Na^+

15. X-rays are produced due to
(A) Bombardment of α–particles on solids
(B) Bombardment of γ-rays on solids
(C) Bombardment of electrons on solids
(D) Bombardment of neutrons on solids

16. In 1 gram of a metal oxide, metal precipitated is 0.68gram. what is the equivalent weight of the metal
(A) 17 (B) 34 (C) 68 (D) 52

17. What happens when an α–particle is emitted
(A) Mass number decreases by 4 units and atomic number decreases by one unit.
(B) Mass number decreases by 12 units and atomic number decreases by 4 units.
(C) Only mass number decreases.
(D) Only atomic number decreases.

18. If an element in M^{3+} form has electronic configuration [Ar] $3d^1$, the element is
(A) Ti (B) Ti (C) Ca (D) Sc

19. The structure of NH_3 is
(A) Tetrahedral (B) Octahedral
(C) Square planer (D) Pyramidal

20. Rate of diffusion of a gas X is twice the gas Y. If the molecular weight of X is 8, what is the molecular weight of Y.
(A) 8 (B) 16 (C) 32 (D) None of these

21. When ethylene glycol reacts with periodic acid, the compound formed is called
(A) Formaldehyde (B) Acetaldehyde
(C) Formic acid (D) Acetic acid

22. Fatty acids are
(A) Long chain alkanonic acids
(B) Aromatic carboxylic acids
(C) Unsaturated dicarboxylic acids
(D) Aromatic dicarboxylic acids

23. M is the molecular mass of $KMnO_4$. The equivalent weight of $KMnO_4$ when it is converted into K_2MnO_4 is.
(A) M (B) M/3 (C) M/15 (D) M/7

24. Which law directly explains the law of conservationof mass ?
(A) Daltons law (B) Avogadors law
(C) Beizlius law (D) None of the above

25. If the uncertainty in the momentum of an electron is $1.0x10^{-5}kg.ms^{-1}$, the uncertainty in its position is (h = 6.62 x $10^{-34}kgm^2s^{-1}$)
(A) 1.05 x 10^{-28}m (B) 1.05 x 10^{-26}m
(C) 5.27 x 10^{-30}m (D) 5.25 x 10^{-28}m

26. What is the packing fraction of $^{56}_{26}Fe$?
(A) −14.167 (B) −73.90 (C) +14.167 (D) +73.90

27. IUPAC of name of $CH_3-\underset{|}{\overset{OH}{C}}H-CH_2-\overset{CH_3}{\overset{|}{C}}HCHO$ is
(A) 4-Hydroxy - 2- methylpentanal
(B) 4-Hydroxy - 4 methylpentanal
(C) 2-Methyl pent - 4 - *ol* - 1 al (D) None of these

28. Which of the following molecules contains asymmetric carbon atom ?
(A) $CH_3CHClCOOH$ (B) CH_3-CH_2-COOH
(C) $CH_3CH_2CH_2COOH$ (D) $Cl_2CHCOOH$

29. The most stable carbonium ion among the following is
(A) $C_6H_5\overset{+}{C}HC_6H_5$ (B) $C_6H_5\overset{+}{C}H_2$
(C) $CH_3\overset{+}{C}H_2$ (D) $C_6H_5CH_2\overset{+}{C}H_2$

30. In which of the following resonance is possible ?
(A) $CH_3-CO-CH_3$
(B) $CH_2 = CH-CH_2-CHO$ (C) $CH_2 = CH-CHO$
(D) $CH_2 = CH-CH_2-CH = CH_2$

31. Most reactive hydrocabon of the following is
(A) Ethane (B) Methane (C) Ethene (D) Ethyne

32. $CH_3C_6H_5 \xrightarrow[H_2SO_4]{K_2Cr_2O_7}$ Y; Y is
(A) Benzaldehyde (B) Toluene
(C) Benzoic acid (D) Ethyl benzene

33. Reaction of ethyne with HCN in the presence of $Ba(CN)_2$ is a
(A) Nucleophilic addition reaction

(B) Free radical addition reaction
(C) Electrophilic addition reaction
(D) Electrophilic substitution reaction

34. Empiricial formula of a hydrocarbon containing 80% carbon and 20% hydrogen is
(A) CH (B) CH_2 (C) CH_3 (D) CH_4

35. $CH = CH \xrightarrow[Pressure]{Ni(CN)_2} X$; X is
(A) Benzene (B) Ethane
(C) Cyclo octatetraene (D) Cyclohexane

36. Chloroform reacts with NaOH to give
(A) CH_3 COONa (B) Sodium Oxalate
(C) CH_3 OH (D) HCOONa

37. Which of the following will not be soluble in sodium carbonate solution ?
(A) O_2N, OH, COOH, NO_2 (B) COOH (C) OH, NO_2 (D) SO_3OH

38. $CH \equiv CH \xrightarrow[H_2SO_4]{H_2O/Hg^{2+}} X \xrightarrow{LiAlH_4} Y \xrightarrow{P_4/Br_2} Z$; Z is
(A) Ethylene bromide (B) Ethanol
(C) Ethyl bromide (D)Ethylidene bromide

39. How many isomeric forms of C_7H_9N contain a benzene ring ?
(A) 6 (B) 7 (C) 4 (D) 5

40. Magnetic nature of the compounds of alkaline earth metals is following
(A) Diamagnetic (B) Paramagnetic
(C) Ferromagnetic (D) Antiferromagnetic

41. Which of the following has the largest size ?
(A) Al^{2+} (B) Al^{3+} (C) Al (D) Al^{+}

42. Producer gas is
(A) $CO + N_2$ (B) $CH_4 + H_2 + CO + CO_2$
(C) $CO + H_2 + N_2$ (D) $CH_4 + CO + H_2$

43. The ion that can be precipitated by HCl as well as H_2S is
(A) Pb^{2+} (B) Fe^{3+} (C) Zn^{2+} (D) Cu^{2+}

44. Which of the following is isoelectronic and has the same structure as that of N_2O ?
(A) NO_2 (B) CO_2 (C) N_3H (D) H_2O

45. Which of the following sets is in the increasing order of the ionic character.
(A) $PbCl_4 < PbCl_2 < CaCl_2 < NaCl$
(B) $PbCl_2 < PbCl_4 < CaCl_2 < NaCl$
(C) $PbCl_2 < PbCl_4 < NaCl < CaCl_2$
(D) $PbCl_4 < PbCl_2 < NaCl_2 < CaCl_2$

46. Which of the following noble gas is least polarisable ?
(A) He (B) Xe (C) Ar (D) Ne

47. The elements of group 14 have 4 valance electrons. Out of these, the element having maximum tendency to exhibit +2 oxidation state is
(A) Sn (B) C (C) Ge (D) Pb

48. When lead nitrate is heated it gives
(A) N_2O_5 (B) N_2O (C) NO_2 (D) NO

49. Soild PCl_5 exits as
(A) PCl_5 (B) PCl_4^+ (C) PCl_6^- (D) PCl^+_4 and PCl^-_6

50. Which complex has the square planar structure ?
(A) $Ni(CO)_4$ (B) $[NiCl_4]^{2-}$
(C) $[Ni(H_2O)_6]^{2+}$ (D) $[Cu(NH_3)_4]^{2+}$

51. Which of the following hydride is capable of showing CONFORMATIONS?
(A) $NH_2–NH_2$ (B) B_2H_6 (C) CH_4 (D) None of these

52. A 20 volume H_2O_2 solution has a strength of about
(A) 6% (B) 3% (C) 30% (D) 10%

53. The principal constituent of pyrex glass is
(A) Cl (B) Pb (C) B (D) Zn

54. Semi water gas is
(A) $CO + H_2 + N_2$ (B) $CO + H_2$
(C) $CO + N_2$ (D) None

55. Which of the following is not a drying and dehydrating agent ?
(A) P_2O_5 (B) Silica gel
(C) Hydrated $CaCl_2$ (D) Conc. H_2SO_4

56. Which of the following is a chelating ligand
(A) H_2O: (B) OH^- (C) DMG (D) Cl^-

57. The highest oxidation state shown by Mn is
(A) +4 (B) +7 (C) +3 (D) +2

58. The substance which is added to ore to remove impurities during smelting is known as
(A) Catalyst (B) Gangue (C) Flux (D) Slag

59. The formula of plaster of paris is
(A) $CaSO_4 + H_2O$ (B) $CaSO_4$
(C) $CaSO_4 + 2H_2$ (D) $2CaSO_4 + H_2O$

60. The chemical formula for lunar caustic is
(A) $AgNO_3$ (B) AgCl (C) Ag_2S (D) Ag_2SO_4

JIPMER - 2003

1.Toluene on reaction with N-bromo-succinamidegives
(A) o-bromomethylbenzene (B) p-bromomethylbenzene
(C) m-bromomethylbenzene (D) Phenyl bromomethane

2.The boiling points of alcohols are much higher than hydrocarbons of comparable molecular masses due to
(A) Dipole-dipole interactions
(B) Intramolecular H-bonding
(C) Intermolecular H-bonding
(D) van der Waal's forces of attraction

3. An optically active compound reacts with hydroxylamine to form an oxime and also gives a positive haloform test. What is the structure of the compound ?
(A) $(CH_3)_2CHCOCH_2CH_3$

(B) $CH_3CH_2CH_2COCH_2CH_3$
(C) $CH_3CH_2CH(CH_3)COCH_3$
(D) $(CH_3)_2CHCH_2COCH_3$

4.Ethylalcohol can be obtained in one step from acetic acid by reduction with
(A) Nickel and hydrogen (B) Zinc and acetic acid
(C) $LiAlH_4$ (D) HI and red Phosphorous

5. Hydrolysis of benzonitrile gives
(A) Benzoic acid (B) Aniline
(C) Benzylamine (D) Benzene

6. Oxidation number of cobalt in $K[Co(CO)_4]$ is
(A) -1 (B) +1 (C) -3 (D) +3

7. For measuring the conductivity of an electrolyte,it's solution should be prepared in
(A) Tap water (B) Conductivity water
(C) Pplywater (D) Distilled water

8. A reaction involving two different reactants
(A) Can never be a unimolecular reaction
(B) Can never be a bimolecular reaction
(C) Can never be a first order reaction
(D) Can never be a second order reaction

9. During α-decay N/P ratio
(A) Increases (B) Decreases
(C) May increase or decrease (D) Remain same

10. Milk is a colloid in which
(A) A liquid is dispersed in liquid
(B) A gas is dispersed in liquid
(C) A solid is dispersed in liquid
(D) Some sugar is dispersed in water

11. High purity copper is obtained by
(A) Hydrogen reduction (B) Carbon reduction
(C) Thermite process (D) Electrolytic reduction

12. Addition of ethanol to 3,3-dimethylbut-1-ene in presence of mercuric triflouracetate followed by reduction with $NaBH_4$ gives
(A) 3,3-dimethylbutan-2-ol (B) 2,3-dimethylbutan-3-ol
(C) 2-ethoxy-2,3-dimethylbutane
(D) 3-ethoxy-2,2-dimethylbutane

13. R-$COCH_2OH$ will show
(A) Functional isomerism (B) Ring-chain isomerism
(C) Keto-enol tautomerism (D) None of these

14. A bottle of dry NH_3 and a bottle of dry HCl are connected through a long tube are opened simultaneously at both ends. The white NH_4Cl ring first formed will be
(A) Near the ammonia bottle
(B) Throughout the length of the tube
(C) Near the HCl bottle (D) At the centre of the tube

15. Which gas may be collected over water
(A) N_2 (B) HCl (C) SO_2 (D) NH_3

16. One requires energy En to remove a nucleon and an energy Ee to remove an electron from the orbit of an atom, then
(A) $E_n \geq E_e$ (B) $E_n > E_e$ (C) $E_n = E_e$ (D) $E_n < E_e$

17. Which of the following is an addition polymer ?
(A) Polythene (B) Teflon
(C) PVC (D) All the above

18.The enzyme which converts glucose and fructose into ethyl alcohol is
(A) Diastase (B) Zymase (C) Maltase (D) Invertase

19.The vitamine which is water soluble and antioxidant is
(A) Vitamin E (B) Vitamin C
(C) Vitamin D (D) Vitamin B_1

20. Structurally biodegradable detergent should contain
(A) Normal alkyl chain (B) Branched alkyl chain
C) Cyclohexyl side chain (D) Phenyl side chain

21. If NA is Avogadro's number then number of valence electrons in 4.2 g of nitride ion is
(A) 2.4 N_A (B) 3.2 N_A (C) 4.2 N_A (D) 1.6 N_A

22.Pressure remaining the same, the volume of a given mass of an ideal gas increases for every degree centrigrade rise in temperature by definite fraction of its volume at
(A) its critical temperature (B) Absolute zero
(C) Its Boyle temperature (D) O°C

23. $_{92}U^{238}$ is an isotope of uranium. The number of electrons it can contain is
(A) 126 (B) 136 (C) 146 (D) 156

24. The member of Actinium series (4n+3) is
(A) Th^{232} (B) Po^{212} (C) Ac^{225} (D) Rn^{219}

25. Which mode of expressing concentration is independent of temperature ?
(A) Molarity (B) Molality (C) Normality (D) Formality

26. Conversion of KMnO4 to MnSO4 is a process of
(A) Dehydration (B) Oxidation
(C) Both oxidation and reduction (D) Reduction

27. The reduction electrode potential E, of 0.1 M solution of M= ions (E^o_{RP} = -2.36) is
(A) -2.41 (B) +2.41 (C) -4.82 (D) None of these

28. The weight of NaOH required to neutralise 100 ml 0.1 N HCl is
(A) 0.04 g (B) 0.4 g (C) 4.0 g (D) 2.0 g

29. If a thin slice of sugar beat is placed in concentrated solution of NaCl, then
(A) Sugar beat will loose water from its cells
(B) Sugar beat will absorb water from solution
(C) Sugar beat will neither absorb nor loose water
(D) Sugar beat will dissolve in solution

30.The arrangement of the first two layers, one above the other in hcp and ccp arrangements is
(A) Exactly same in both cases
(B) Different from each other
(C) Partly same and partly different
(D) Nothing definite

31. The spectrum produced by white light is
(A) Absorption spectrum (B) Emission spectrum

(C) Continuous spectrum
(D) Both emission and continuous spectrum

32. Molten sodium chloride conducts electricity due to the presence of
(A) Free sodium and chloride atoms (B) Free molecules
(C) Free sodium and chloride ions (D) Free electrons

33. A solution of $CaCl_2$ is 0.5 mol/litre, then the moles of chloride ions in 500 ml will be
(A) 0.25 (B) 0.5 (C) 0.75 (D) 1.0

34.The bond dissociation energies of gaseous H_2, Cl_2, and HCl are 104, 58 and 103 respectively. The enthalpy of formation of HCl gas would be
(A) -11 kcal (B) -22 kcal (C) -44 kcal (D) -88 kcal

35. A cylinder with a movable piston contains liquid water with water vapour at 25°C. Which one of the following operations results in a decrease in the equilibrium vapour pressure ?
(A) Removing a small amount of liquid water
(B) Removing a small amount of vapour
(C) Dissolving salt in water
(D) Moving the piston downward a short distance

37. The most stable conformation of n-butane is
(A) Staggered-antii (B) Eclipsed
(C) Skew-boat (D) Gauche

38. The first Nobel prize in chemistry was awarded to
(A) Madam Curie (B) Baeyer
(C) Hoffmann (D) Van't Hoff

39. The typical reaction of olefinic bond is
(A) Nucleophilic addition reactions
(B) Nucleophilic substitution reactions
(C) Electrophilic addition reactions
(D) Electrophilic substitution reactions

40. High octane gasoline contains
(A) Large proportions of straight chain hydrocarbons
(B) Large proportions of branched chain hydrocarbons
(C) Large proportions of isooctane only
(D) Large proportions of n-octane

41. A hydrocarbon reacts with hypochlorous acid to give 1-chloro-2-hydroxyethane. The hydrocarbon is
(A)Methane (B) Ethane (C) Ethene (D) Ethyne

42. Quartz is a/an
(A) Three dimentional silicate (B) Sheet silicate
(C) Ortho silicate (D) None of these

43.Among the following statements the incorrect one is
(A) Calamine and siderite are carbonates
(B) Malachite and azurite are the ores of copper
(C) Argentite and cuprite are oxides
(D) Zinc blende and iron pyrites are sulphides

44. Which of the following has exceptionally high melting point ?
(A) NaOH (B) NaCl (C) KCl (D) MgO

45. Alloy is an example of
(A) Aerosol (B) Solid sol (C) Gel (D) Emulsion

46. $[Ti(H_2O)_6]^{+3}$ is purple in colour because it is complimentary colour of
(A) Green (B) Red (C) Blue (D) Greenish-yellow

47. The coordination and oxidation number of X in the compound $[X(SO_4)(NH_3)_5]$ respectively are
(A) 6 , 3 (B) 6 ,4 (C) 2 ,6 (D) 10 ,3

48. Which of the following reagents would be able to remove sulphate ion from the solution containing both sulphate and chloride ions ?
(A) $Ba(OH)_2$ (B) NaOH (C) KOH (D) $BaSO_4$

49. Which of the following properties is not true regarding organic compounds ?
(A) They have high melting and boiling points
(B) They are generally insoluble in water
(C) They are generally covalent compounds
(D) They usually show isomerism

50. Sublimation can not be used for the purification of
(A) Benzoic acid (B) Camphor
(C) Naphthalene (D) Urea

51. The isotope of hydrogen which is radioactive is
(A) Tritium (B) Parahydrogen
(C) Deuterium (D) Nascent hydrogen

52. The ionisation energy of nitrogen is larger than that of oxygen because of
(A) The size of the nitrogen atom being smaller
(B) Greater penetration effect
(C) Greater attraction of electrons by the nucleus
(D) The half filled p-orbitals possess extra stability in N

53. Fires, that result from combustion of alkali metals can be extinguished by
(A) Kerosene (B) Water (C) CCl_4 (D) Sand

54. Which of the following ions has the largest radius ?
(A) Be^{2+} (B) Mg^{2+} (C) Ca^{2+} (D) Sr^{2+}

55.In which of the following elements +1 oxidation state is more stable than +3 ?
(A) Al (B) B (C) Tl (D) Ga

56. The shape of tetrahalides of group 14 elements is
(A) Square planar (B) Tetrahedral
(C) Trigonal bipyramidal (D) Octahedral

57. Which of the following elements is most metallic ?
(A) P (B) Bi (C) As (D) Sb

58. The number of lone pairs of electrons present on each S atom in S_8 molecule are
(A) 1 (B) 2 (C) 4 (D) None of these

59. The strongest reducing agent among the following is
(A) Chlorine (B) Cl^- (C) Bromine (D) Br^-

60. The noble gas which was discovered first in the Sun and then on the Earth is
(A) Helium (B) Neon (C) Argon (D) Xenon

JIPMER - 2004

1. Magnesium has polarizing power closer to that of
(A) Lithium (B) Sodium (C) Caesium (D) Potassium

2. The hydration energy of Mg^{2+} ions is larger than that of :
(A) Al^{3+} (B) Na^{+} (C) Be^{2+} (D) Mg^{3+}

3. In which case +1 oxidation state is stable than +3 ?
(A) Ga (B) Al (C) Tl (D) B

4. An ionic compound is :
(A) CCl_4 (B) $SnCl_2$ (C) $SiCl_4$ (D) $GeCl_4$

5. Which is not an oxyacid of chlorine :
(A) HClO (B) $HClO_3$ (C) $HClO_5$ (D) $HClO_2$

6. Which one is not an acid salt :
(A) NaH_2PO_4 (B) NaH_2PO_2
(C) NaH_2PO_3 (D) None

7. SO_2 reduces
(A) Mg (B) H_2S (C) $KMnO_4$ (D) All

8. If the energy difference between the states n = 2 and n =3 is E eV. in hydrogen atom. The ionisation potential of H atom is :
(A) 3.2 E (B) 5.6 E (C) 7.2 E (D) 13.2 E

9. The bond angle in H_2O is approximately :
(A) 80ºC (B) 105ºC (C) 120ºC (D) 180ºC

10. Compound which form dative bond with NH_3 :
(A) NaCl (B) BCl_3 (C) CH_4 (D) CCl_4

11. The bond angle in ammonia is :
(A) 109º28' (B) 90º (C) 107º (D) 105º

12. In ethane, ethylene and acetylene, the C – H bond energy is :
(A) Greatest in ethane (B) Greatest in ethylene
(C) Greatest in acetylen (D) Same in all

13. If 75% quantity of a radioactive isotope disintegrates in 2 hours, its half life would be :
(A) 1 hour (B) 45 minutes (C) 30 minutes (D) 15 minutes

14. A sample of rock from moon contains equal number of atoms of uranium and lead ($t_{1/2}$ for U = 4.5 x 10^{-3} years). The age of the rock would be :
(A) 4.5 x 10^{-3} years (B) 9.0 x 10^{2} years
(C) 13.5 x 10^{3} years (D) 2.25 xx 10^{2} years

15. A flask filled with CCl_4 was weighed at a temperature and pressure. The flask was then filled with oxygen at the same temperture and pressure. The mass of CCl_4 vapour would be about :
(A) The same as that of the oxygen
(B) One-fifth as heavy as oxygen
(C) Five times as heavy as oxygen
(D) Twice as heagy as oxygen

16. The rate of diffusion of SO_2 and O_2 are in the ratio (S=32, O=16)
(A) 1 : 1/√2 (B) 1 : 32 (C) 1 : 4 (D) 1 : 2

17. When sodium extract is prepared, generally the substance ignites :
(A) Organic compound (B) O_2 (C) H_2 (D) Na

18. If a compound contains two oxygen atoms, four carbon atoms and number of hydrogen atoms is double of carbon atoms, the vapour density of it is :
(A) 88 (B) 44 (C) 132 (D) 72

19. The compound that is most rective towards electrophilic nitration is :
(A) Toluene (B) Benzene
(C) Benzoic acid (D) Nitrobenzen

20. An optically active compound is :
(A) CH_3CH_2COOH (B) $CH_3CHOHCOOH$
(C) $HOOCCH_2COOH$ (D) $C_2H_5COCOOH$

21. Which of the following ions is most satble:
(A) $CH_3CH\overset{\oplus}{}CH_3$ (B) $CH_3CH_2\overset{\oplus}{C}H_2$

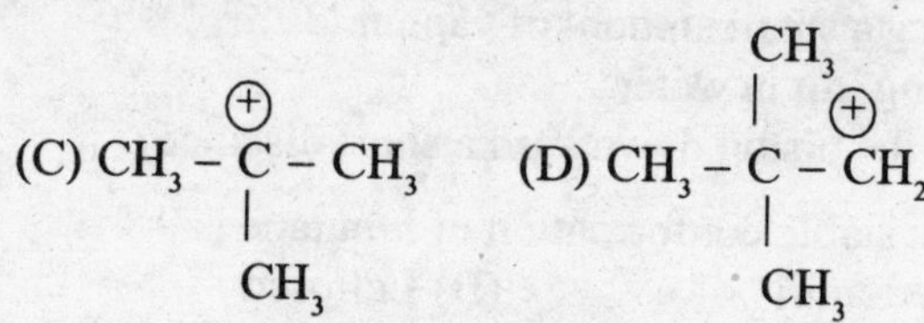

22. The most reactive halogen in the halogenation of alkenes is :
(A) Cl_2 (B) I_2 (C) Br_2 (D) All are eqaual

23. The compound 1, 2 - butadiene has :
(A) sp, sp^2 and sp^3 hybridized carbon atom
(B) Both sp and sp^2 hybridized carbon atoms
(C) Only sp hybridized carbon atom
(D) Only sp^2 hybridized carbon atoms

24. To produce 160 g of oxygen, the number of moles of water required to be electrolysed is :
(A) 2.5 (B) 10 (C) 5 (D) 20

25. How many mole of Fe^{2+} ions are formed, when excess of iron is treated with 50 ml of 4.0 M HCl under inert atmosphere ? Assume no change in volume :
(A) 0.4 (B) 0.1 (C) 0.2 (D) 0.8

26. Which has maximum freezing point :
(A) 1 molar NaCl solution (B) 1 molar KCl solution
(C) 1 molar of $CaCl_2$ solution (D) 1 molar of urea solution

27. K_{sp} = 1.2 x 10^{-5} of M_2SO_4 (M^+ is monovalent metal ion) at 298 K. The maximum concentration of M^+ ions that could be attained in a saturated solution of this solid at 298 K is :
(A) 3.46 x 10^{-3} M (B) 7.0 x 10^{-3} M
(C) 2.88 x 10^{-2} M (D) 14.4 x 10^{-3} M

28. The polyprotic acid is :
(A) HCl (B) $HClO_4$ (C) H_3PO_4 (D) HNO_3

29. The extent of ionisation increases :
(A) with increase in con. of solute
(B) On adding water (C) On decreasing temperature

(D) On stirring solution

30. pH of a solution of the mixture of 0.1 N HCl and 0.1 N CH_3COOH, is :
(A) 1 (B) 2 (C) 1.7 (D) None

31. Which statement is incorrect ? The d-block elements
(A) Have high melting points, boiling points and tensile strength
(B) Have atomic radii larger than s- and p- block elements
(C) Exhibit catalytic properties
(D) Have variable oxidation states

32. Mercury is transported in metal containers made of :
(A) Iron (B) Silver (C) Aluminium (D) Lead

33. Inert gases do not react with other elements because
(A) They are mono atomic
(B) They are found in abundance
(C) They have small size of atom
(D) They have comletely paired up and stable electron shells

34. The metal that reacts with both acids and alkalis with the liberation of hydrogen is :
(A) Cu (B) Al (C) Zn (D) Pb

35. The number of ions formed when cuprammonium sulphate dissolves in water is :
(A) Zero (B) Two (C) One (D) Four

36. Solder is an alloy of :
(A) Zn an Cu (B) Cu and Pb
(C) Pb and Sn (D) Fe and Zn

37. The correct statement is :
(A) Galena is the ore of mercury
(B) Dolomite is the ore of zinc
(C) Pyrolusite is the ore of iron
(D) Cassiterite is the ore of tin

38. In fermentation by zymase, alcohol and CO_2 are obtained form the following sugar :
(A) Invert sugar (B) Glucose
(C) Fructose (D) All of these

39. Dehydration of methanol with conc. H_2SO_4 at 410 K gives :
(A) Ethane (B) Dimethyl ether (C) HCHO (D) All

40. The products of combustion of an aliphatic thiol (RSH) at 298 K are:
(A) $CO_2(g)$, $H_2O(l)$ and $SO_2(g)$
(B) $CO_2(g)$, $H_2O(g)$ and $SO_2(g)$
(C) $CO_2(l)$, $H_2O(l)$ and $SO_2(g)$
(D) $CO_2(g)$, $H_2O(l)$ and $SO_2(l)$

41. Which will form two oximes with NH_2OH :
(A) CH_3. CH_2 $COCH_2CH_3$
(B) $\begin{matrix} H \\ H \end{matrix} > C = O$
(C) CH_3. $COCH_3$ (D) CH_3. CH_2. CO. CH_3

42. Decarboxylation of maloinic acid gives :
(A) HCHO (B) COOH – COOH
(C) CH_4 (D) CH_3COOH

43. Diazonium salt decomoses as,
$C_6H_5N_2^+ Cl^- \rightarrow C_6H_5Cl + N_2$. At 0°C, the evolution of N_2 becomes two times faster when the initial concentration of the salt is doubled. Therfore it is :
(A) Independent of the initial concentration of the salt
(B) A zero order reaction
(C) A firs order reaction (D) A second order reaction

44. The rection of aniline with acetyl chloride in presence of NaOH gives :
(A) Aniline hydrochloride (B) A red dye
(C) Acetanilide (D) p-chloroaniline

45. $H_2(g) + Cl_2(g) = 2HCl\ (g)$; ΔH 298 K = –22.06 kcal. For this reaction, ΔE is equal to
(A) – 22.06 – 2 x 298 x 4 kcal
(B) – 22.06 kcal
(C) – 22.06 + 2 x 10^{-3} x 298 x 2 k cal
(D) – 22.06 – 2 x 298 x 4 kcal

46. The reaction is impossible if :
(A) ΔH is –ve and ΔS is +ve
(B) ΔH is +ve and ΔS is +ve
(C) ΔH is –ve and ΔH is –ve
(D) ΔH is +ve and ΔS is –ve

47. Rate of a reaction :
(A) Incerases with increase in temperature
(B) Decreaes with increase in temperature
(C) Does not depend on temperature
(D) Does not depend on concentration

48. For the reaction, A + B $\rightarrow$ Products, on doubling the initial concentration of B keeping that of A fixed the rate is doubled; and on doubling the initial concentrations of A and B the rate is increaesd by a factor 4. The rate law is given by :
(A) k $[A]^2$ $[B]^2$ (B) k [A] [B]
(C) k $[B]^2$ [A] (D) k $[A]^2$ [B]

49. If the half cell reation, A + e $\rightarrow$ A^-, has a large negative standard reduction potential, if follows that :
(A) A is readily reduced (B) A is readily oxidised
(C) A^- is readily reduced (D) A^- is readily oxidised

50. The highest oxidation number of X is in :
(A) HXO (B) HXO_2 (C) HXO_3 (D) HXO_4

51. For the redox reaction,
$MnO_4^- + C_2O_4^{2-} + H^+ \rightarrow Mn^{2+} + CO_2 + H_2O$
the correct coefficient of the reactants for the balanced reaction are :

	MnO_4^-	$C_2O_4^{2-}$	H^+
(A)	5	16	2
(B)	2	16	5
(C)	16	5	2
(D)	2	5	16

52. The standard EMF for the cell reaction Zn + Cu^{2+} = Cu + Zn^{2+} is 1.10 volt at 25°C. The EMF of the cell

reaction, when 0.1 M Cu^{2+} and 0.1 M Zn^{2+} solutions are used, at 25°C is :

(A) 0.110V (B) 1.10V (C) –1.10V (D) –0.110V

53. In a reaction between acid and iodine in which zinc iodide is formed, which is oxidezed:

(A) Zinc ion (B) Iondie ions (C) Zinc atom (D) Iodine

54. Charge on 1 g ion of N^{3-} is :

(A) F (B) 3F (C) 4F (D) 2F

55. The eq. wt. of K_2CrO_4 as an oxidizing agent in acid medium is :

(A) Mol. wt./2 (B) (2 x mol.wt.)/3
(C) (Mol. wt.)/3 (D) (Mol. wt.)/6

56. Which process occurs in the electrolysis of aqueous solution of nickel chloride at nickel anode :

(A) $Nk^{2+} + 2e \rightarrow Ni$ (B) $2H^+ + 2e \rightarrow H_2$
(C) $2Cl^- \rightarrow Cl_2 + 2e$ (D) $Ni \rightarrow Ni^{2+} + 2e$

57. Oxidation number of sulphur in Na_2SO_4 is

(A) +4 (B) +2 (C) -2 (D) +6

58. $HgCl_2$ and $SnCl_2$ cannot coexist in a solution due to:

(A) common ion effect (B) Solubility product
(C) Redox change (D) All

59. Gold number of a lyophilic sol refers that :

(A) Lower is its value, the greater is the protecting power
(B) Larger is its value, the greater is the protecting power
(C) Larger is its value, the greater is the peptizing power
(D) Lower is its value, the greater is the peptizing power

60. Which one is a phospholipid ?

(A) Kelphalin (B) Lecithin (C) Cephalin (D) All

JIPMER -1990

1. (D) 31. (D)
2. (A) 32. (A)
3. (D) 33. (C)
4. (C) 34. (C)
5. (B) 35. (D)
6. (B) 36. (B)
7. (A) 37. (D)
8. (D) 38. (C)
9. (A) 39. (C)
10. (B) 40. (A)
11. (B) 41. (D)
12. (A) 42. (C)
13. (C) 43. (D)
14. (A) 44. (B)
15. (C) 45. (A)
16. (D) 46. (B)
17. (B) 47. (D)
18. (C) 48. (D)
19. (B) 49. (A)
20. (C) 50. (B)
21. (C) 51. (B)
22. (D) 52. (A)
23. (A) 53. (C)
24. (D) 54. (A)
25. (D) 55. (D)
26. (B) 56. (A)
27. (A) 57. (A)
28. (C) 58. (B)
29. (A) 59. (C)
30. (A) 60. (A)

JIPMER -1991

1. (D) 31. (C)
2. (D) 32. (D)
3. (B) 33. (B)
4. (D) 34. (C)
5. (C) 35. (B)
6. (D) 36. (D)
7. (B) 37. (D)
8. (B) 38. (B)
9. (B) 39. (A)
10. (D) 40. (D)
11. (D) 41. (A)
12. (D) 42. (B)
13. (A) 43. (C)
14. (A) 44. (A)
15. (B) 45. (B)
16. (C) 46. (A)
17. (C) 47. (C)
18. (B) 48. (A)
19. (A) 49. (D)
20. (A) 50. (B)
21. (A) 51. (D)
22. (C) 52. (A)
23. (B) 53. (A)
24. (B) 54. (D)
25. (A) 55. (B)
26. (A) 56. (B)
27. (D) 57. (A)
28. (D) 58. (D)
29. (B) 59. (D)
30. (D) 60. (B)

JIPMER -1992

1. (B) 31. (B)
2. (B) 32. (A)
3. (C) 33. (D)
4. (D) 34. (B)
5. (D) 35. (B)
6. (C) 36. (B)
7. (A) 37. (C)
8. (B) 38. (D)
9. (D) 39. (C)
10. (B) 40. (B)
11. (C) 41. (B)
12. (A) 42. (A)
13. (D) 43. (A)
14. (D) 44. (D)
15. (B) 45. (A)
16. (C) 46. (C)
17. (D) 47. (B)
18. (D) 48. (D)
19. (A) 49. (B)
20. (D) 50. (B)
21. (C) 51. (A)
22. (B) 52. (C)
23. (C) 53. (D)
24. (B) 54. (A)
25. (D) 55. (C)
26. (B) 56. (C)
27. (C) 57. (B)
28. (A) 58. (B)
29. (D) 59. (A)
30. (B) 60. (B)

JIPMER -1993

1. (A) 31. (D)
2. (C) 32. (B)
3. (D) 33. (C)
4. (A) 34. (C)
5. (B) 35. (C)
6. (D) 36. (C)
7. (B) 37. (A)
8. (D) 38. (A)
9. (C) 39. (A)
10. (D) 40. (A)
11. (B) 41. (D)
12. (A) 42. (B)
13. (B) 43. (A)
14. (B) 44. (C)
15. (A) 45. (A)
16. (A) 46. (D)
17. (C) 47. (B)
18. (C) 48. (D)
19. (D) 49. (D)
20. (B) 50. (C)
21. (A) 51. (D)
22. (D) 52. (A)
23. (D) 53. (D)
24. (C) 54. (A)
25. (C) 55. (B)
26. (C) 56. (C)
27. (B) 57. (D)
28. (B) 58. (B)
29. (A) 59. (B)
30. (C) 60. (C)

JIPMER -1994

1. (C) 31. (A)
2. (C) 32. (C)
3. (C) 33. (C)
4. (B) 34. (D)
5. (A) 35. (A)
6. (C) 36. (B)
7. (B) 37. (C)
8. (C) 38. (D)
9. (A) 39. (D)
10. (A) 40. (D)
11. (B) 41. (D)
12. (C) 42. (C)
13. (D) 43. (D)
14. (D) 44. (A)
15. (C) 45. (C)
16. (C) 46. (C)
17. (C) 47. (D)
18. (C) 48. (C)
19. (D) 49. (B)
20. (A) 50. (D)
21. (A)
22. (A)
23. (A)
24. (A)
25. (C)
26. (B)
27. (C)
28. (D)
29. (D)
30. (A)

JIPMER -1995

1. (A) 21. (C) 41. (A)
2. (A) 22. (B) 42. (C)
3. (D) 23. (A) 43. (D)
4. (C) 24. (B) 44. (C)
5. (B) 25. (A) 45. (C)
6. (C) 26. (B) 46. (C)
7. (A) 27. (A) 47. (C)
8. (A) 28. (B) 48. (A)
9. (D) 29. (C) 49. (B)
10. (A) 30. (B) 50. (A)
11. (D) 31. (D)
12. (A) 32. (C)
13. (B) 33. (B)
14. (C) 34. (B)
15. (D) 35. (A)
16. (B) 36. (C)
17. (A) 37. (C)
18. (D) 38. (A)
19. (C) 39. (A)
20. (A) 40. (C)

JIPMER -1996

1. (B) 21. (A) 41. (B)
2. (C) 22. (D) 42. (D)
3. (C) 23. (A) 43. (B)
4. (A) 24. (B) 44. (A)
5. (C) 25. (B) 45. (B)
6. (C) 26. (B) 46. (C)
7. (B) 27. (C) 47. (D)
8. (A) 28. (D) 48. (C)
9. (C) 29. (B) 49. (A)
10. (C) 30. (B) 50. (A)
11. (A) 31. (A) 51. (D)
12. (C) 32. (B) 52. (B)
13. (D) 33. (C) 53. (A)
14. (D) 34. (D) 54. (C)
15. (A) 35. (A) 55. (B)
16. (A) 36. (B) 56. (C)
17. (D) 37. (C) 57. (D)
18. (B) 38. (B) 58. (A)
19. (D) 39. (C) 59. (B)
20. (B) 40. (B) 60. (C)

JIPMER -1997

1. (D) 21. (C) 41. (C)
2. (D) 22. (A) 42. (D)
3. (A) 23. (D) 43. (C)
4. (C) 24. (B) 44. (D)
5. (B) 25. (C) 45. (C)
6. (B) 26. (A) 46. (C)
7. (B) 27. (D) 47. (B)
8. (A) 28. (C) 48. (B)
9. (A) 29. (C) 49. (A)
10. (B) 30. (D) 50. (B)
11. (B) 31. (D) 51. (B)
12. (A) 32. (B) 52. (A)
13. (C) 33. (C) 53. (C)
14. (B) 34. (B) 54. (C)
15. (A) 35. (A) 55. (C)
16. (D) 36. (A) 56. (D)
17. (A) 37. (B) 57. (B)
18. (A) 38. (D) 58. (C)
19. (B) 39. (B) 59. (A)
20. (A) 40. (A) 60. (A)

JIPMER -1998

Q	Ans	Q	Ans
1.	(C)	31.	(D)
2.	(C)	32.	(D)
3.	(A)	33.	(D)
4.	(D)	34.	(D)
5.	(B)	35.	(C)
6.	(A)	36.	(C)
7.	(C)	37.	(D)
8.	(D)	38.	(D)
9.	(A)	39.	(A)
10.	(B)	40.	(D)
11.	(A)	41.	(A)
12.	(C)	42.	(A)
13.	(C)	43.	(D)
14.	(C)	44.	(D)
15.	(C)	45.	(C)
16.	(A)	46.	(A)
17.	(D)	47.	(D)
18.	(B)	48.	(C)
19.	(C)	49.	(D)
20.	(C)	50.	(A)
21.	(A)	51.	(B)
22.	(A)	52.	(C)
23.	(A)	53.	(A)
24.	(D)	54.	(C)
25.	(D)	55.	(A)
26.	(C)	56.	(A)
27.	(D)	57.	(A)
28.	(B)	58.	(C)
29.	(A)	59.	(D)
30.	(A)	60.	(A)

JIPMER -1999

Q	Ans	Q	Ans
1.	(B)	31.	(A)
2.	(D)	32.	(B)
3.	(A)	33.	(C)
4.	(C)	34.	(A)
5.	(A)	35.	(D)
6.	(A)	36.	(B)
7.	(C)	37.	(B)
8.	(B)	38.	(B)
9.	(B)	39.	(D)
10.	(B)	40.	(B)
11.	(C)	41.	(D)
12.	(B)	42.	(A)
13.	(A)	43.	(C)
14.	(C)	44.	(D)
15.	(D)	45.	(B)
16.	(D)	46.	(C)
17.	(B)	47.	(A)
18.	(C)	48.	(C)
19.	(A)	49.	(A)
20.	(B)	50.	(C)
21.	(D)	51.	(A)
22.	(D)	52.	(D)
23.	(A)	53.	(A)
24.	(A)	54.	(B)
25.	(D)	55.	(D)
26.	(B)	56.	(C)
27.	(C)	57.	(C)
28.	(C)	58.	(A)
29.	(B)	59.	(A)
30.	(D)	60.	(C)

JIPMER -2000

Q	Ans	Q	Ans
1.	(D)	31.	(C)
2.	(B)	32.	(B)
3.	(C)	33.	(C)
4.	(D)	34.	(D)
5.	(C)	35.	(B)
6.	(C)	36.	(D)
7.	(D)	37.	(C)
8.	(D)	38.	(B)
9.	(A)	39.	(D)
10.	(C)	40.	(D)
11.	(C)	41.	(A)
12.	(A)	42.	(D)
13.	(A)	43.	(D)
14.	(B)	44.	(A)
15.	(C)	45.	(D)
16.	(A)	46.	(C)
17.	(C)	47.	(A)
18.	(D)	48.	(C)
19.	(B)	49.	(D)
20.	(A)	50.	(C)
21.	(D)	51.	(D)
22.	(A)	52.	(A)
23.	(A)	53.	(C)
24.	(B)	54.	(B)
25.	(B)	55.	(D)
26.	(C)	56.	(C)
27.	(B)	57.	(B)
28.	(C)	58.	(D)
29.	(D)	59.	(A)
30.	(D)	60.	(C)

JIPMER -2001

Q	Ans	Q	Ans
1.	(B)	31.	(B)
2.	(B)	32.	(D)
3.	(B)	33.	(D)
4.	(C)	34.	(B)
5.	(A)	35.	(B)
6.	(D)	36.	(A)
7.	(B)	37.	(B)
8.	(A)	38.	(C)
9.	(C)	39.	(C)
10.	(C)	40.	(A)
11.	(D)	41.	(A)
12.	(B)	42.	(D)
13.	(D)	43.	(C)
14.	(C)	44.	(D)
15.	(C)	45.	(A)
16.	(D)	46.	(C)
17.	(A)	47.	(C)
18.	(C)	48.	(C)
19.	(D)	49.	(C)
20.	(B)	50.	(A)
21.	(A)	51.	(C)
22.	(B)	52.	(B)
23.	(B)	53.	(D)
24.	(D)	54.	(A)
25.	(C)	55.	(D)
26.	(B)	56.	(C)
27.	(B)	57.	(A)
28.	(C)	58.	(B)
29.	(B)	59.	(A)
30.	(C)	60.	(A)

JIPMER -2002

Q	Ans	Q	Ans
1.	(A)	31.	(C)
2.	(B)	32.	(C)
3.	(D)	33.	(A)
4.	(B)	34.	(C)
5.	(A)	35.	(C)
6.	(D)	36.	(D)
7.	(D)	37.	(C)
8.	(B)	38.	(C)
9.	(D)	39.	(D)
10.	(A)	40.	(A)
11.	(D)	41.	(C)
12.	(C)	42.	(A)
13.	(C)	43.	(A)
14.	(B)	44.	(B)
15.	(C)	45.	(A)
16.	(A)	46.	(A)
17.	(A)	47.	(D)
18.	(A)	48.	(C)
19.	(D)	49.	(D)
20.	(C)	50.	(D)
21.	(A)	51.	(A)
22.	(A)	52.	(A)
23.	(A)	53.	(C)
24.	(A)	54.	(A)
25.	(C)	55.	(C)
26.	(A)	56.	(C)
27.	(A)	57.	(B)
28.	(A)	58.	(C)
29.	(A)	59.	(D)
30.	(C)	60.	(A)

JIPMER - 2003

Q	Ans	Q	Ans	Q	Ans
1.	(D)	21.	(A)	41.	(C)
2.	(C)	22.	(D)	42.	(A)
3.	(C)	23.	(C)	43.	(C)
4.	(C)	24.	(D)	44.	(D)
5.	(A)	25.	(B)	45.	(B)
6.	(A)	26.	(D)	46.	(D)
7.	(B)	27.	(A)	47.	(A)
8.	(A)	28.	(B)	48.	(A)
9.	(A)	29.	(A)	49.	(A)
10.	(A)	30.	(D)	50.	(D)
11.	(D)	31.	(D)	51.	(A)
12.	(D)	32.	(C)	52.	(D)
13.	(B)	33.	(B)	53.	(C)
14.	(C)	34.	(B)	54.	(D)
15.	(A)	35.	(C)	55.	(C)
16.	(B)	36.	(D)	56.	(B)
17.	(D)	37.	(A)	57.	(B)
18.	(B)	38.	(D)	58.	(B)
19.	(B)	39.	(C)	59.	(D)
20.	(D)	40.	(B)	60.	(A)

JIPMER - 2004

Q	Ans	Q	Ans	Q	Ans
1.	(A)	21.	(C)	41.	(D)
2.	(B)	22.	(B)	42.	(C)
3.	(C)	23.	(A)	43.	(C)
4.	(B)	24.	(C)	44.	(C)
5.	(C)	25.	(B)	45.	(B)
6.	(B)	26.	(D)	46.	(D)
7.	(C)	27.	(C)	47.	(A)
8.	(C)	28.	(C)	48.	(B)
9.	(B)	29.	(B)	49.	(C)
10.	(B)	30.	(A)	50.	(D)
11.	(D)	31.	(B)	51.	(D)
12.	(C)	32.	(A)	52.	(B)
13.	(A)	33.	(D)	53.	(C)
14.	(A)	34.	(C)	54.	(B)
15.	(C)	35.	(B)	55.	(C)
16.	(A)	36.	(C)	56.	(D)
17.	(C)	37.	(D)	57.	(D)
18.	(B)	38.	(B)	58.	(C)
19.	(A)	39.	(B)	59.	(D)
20.	(B)	40.	(A)	60.	(D)

1995

1. Elements of group III A has characteristic electronic configuration as
(A) (n-l) d^2ns^2 (B) (n-l)d^2, ns^2, np^1
(C) ns^2 np^1 (D) ns^2 np^2

2. Correct order of decreasing electron affinity of halogens (F, Cl, Br, 1) is
(A) I, Br, F, Cl (B) I, Br, Cl, F
(C) F, Cl, Br, 1 (D) Cl, F, Br, 1

3. Which of the following will not show allotropic forms?
(A) Silicon (B) Tin (C) Carbon (D) None of these

4. Lithopone is a mixture of
(A) $CaS + ZnSO_4$ (B) $CaSO_4 + ZnS$
(C) $BaS + ZnSO_4$ (D) $BaSO_4 + ZnS$

5. Gold exhibits the valency as
(A) +1, +2 and +3 (B) +1 and +3
(C) +2 and +3 (D) +1 and +2

6. Borax-bead test depends on the formation of
(A) Metal borate (B) Pyroborate
(C) Boron oxide (D) Boron metal

7. The temperature at which a real gas obeys the ideal gas laws over a wide range of pressure is
(A) reduced temperature (B) inversion temperature
(C) Boyle temperaute (D) Critical temperature

8. Which of the following is used for taking x-rays spectra of digestive track ?
(A) $BaSO_4$ (B) $MgSO_4$ (C) $SrSO_4$ (D) $CaSO_4$

9. A trend common to both group (ii) and (vi) in the periodic table as atomic number increases is
(A) maximum valence increases
(B) reactivity increases (C) atomic radius increases
(D) oxidising power increases

10. Which is preferred for use in gas analysis to absorb carbon monoxide ?
(A) Charcoal (B) Silica gel (C) Pyrogallol
(D) Ammoniacal cuprous chloride

11. Which one of the following will form crimson red colour on the flame
(A) $BaCl_2$ (B) $CaCl_2$ $6H_2O$
(C) $SrCl_2$ (D) none of these

12. The elements in the periodic table which readily form cations are ?
(A) inner transition metals (B) alkali metals
(C) noble gases (D) transition metals

13. Ionisation potential is the energy required for the loss of valence electron from the valence shell of the neutral gaseous atom. What charge is left on the particle after ionisation
(A) +ve charge (B) Negative charge
(C) Neutral (D) No charge

14. Which one of the following is an electron deficient compound ?
(A) NCl_3 (B) PCl_3 (C) BCl_3 (D) NH_3

15. The smallest atom among the follwoing is
(A) Ba (B) Cu (C) Cs (D) O

16. Elements in the same vertical group of the periodic table have genrally the same
(A) number of electron in the outer shell of their atoms
(B) number of isotopes (C) electronic configuration
(D) atomic number

17. Iodine can displace chlorine under suitable conditions from
(A) potassium chlorate (B) Bleaching powder
(C) Potassium chloride (D) None of these

18. Which one the following does not exist at all ?
(A) $SbCl_2$ (B) AsH_3 (C) PH_4I (D) As_2O_3

19. The formula of a phosphate is MH_2PO_4. The formula of its oxide is
(A) M_2O_3 (B) MO_2 (C) MO (D) M_2O

20. Nitrogen forms several oxides. This illustrates the law of
(A) Reciprocal proportions
(B) Multiple proportions
(C) Both the laws of multiple and reciprocal proportions

21. Among the following groups of elements the one whose elements can have positive as well as negative oxidation states are
(A) Na, Mg, Al (B) H, Cl, Br (C) H,F,O (D) He,Li,Be

22. By fixation of nitrogen is meant
(A) Conversion of nitrogen into important nitrogenous compounds
(B) Producing nitrogen from calcium nitride
(C) Manufacturing nitrogen from air
(D) Liquefaction of nitrogen

23. .... is formed when iodide salt is treated with Con.H_2SO_4
(A) IO (B) IO_3^- (C) HI (D) I_2

24. Most reactive allotropic form of phosphorus is
(A) black phosphorus (B) violet phosphorus
(C) red phosphorus (D) white phosphorus

25. Which of the following hydride of nitrogen is acidic in nature ?
(A) N_2H_2 (B) N_2H_4 (C) N_3H (D) NH_3

26. Ozone is made from oxygen by
(A) conversion under slight pressure
(B) Silent electric discharge

(C) Oxidation at high temperature
(D) Oxidising using catalyst

27. $_6C^{14}$ in the upper layer of the atmosphere is produced by the action of neutrons on
(A) O_8^{16} (B) C_6^{12} (C) N_7^{14} (D) O_8^{17}

28. Permissible values of m (magnetic quantum number) for 1 = 2, are
(A) 1, 2 (B) ±2, ±1.0 (C) 2, 1, 0 (D) −2, −1, 0

29. The number of electrons in an atom of an element is equal to its
(A) electron affinity (B) equlvalent weight
(C) atomic number (D) atomic weight

30. Heisenberg's uncertainty principle states that
(A) moving charged particles resemble electromagnetic waves in their behaviour
(B) simultaneous position and momentum of a particle cannot be found precisely
(C) Neither the position nor the momentum can be precisely determined
(D) Moving bodies exhibit both particle and wave character

31. The compound with zero dipole moment is
(A) Methylene chloride (B) Trichlororoethane
(C) Chloromethane (D) Tetrachloromethane

32. The carbon carbon bond in cthene are formed by
(A) sp^2–sp^2 and s overlapping
(B) sp^2–sp^2 and sp^2 overlapping
(C) sp^2–sp^2 and sp overlapping
(D) sp^2–sp^2 and p–p overlapping

33. Which of the following molecules has a non polar bond?
(A) H_2O (B) AgCl (C) HF (D) F_2

34. Boiling point of a compound does not depend on the
(A) size of the molecule
(B) solubility of the compound in water
(C) Polarity of its molecule
(D) Hydrogen bonding in the compound

35. The projectile experiencing minimum repulsive forces on approaching a particular nucleus is
(A) positrons (B) protons
(C) neutrons (D) alpha particle

36. The elements of 2nd short period are called
(A) Inert elements (B) Transition elements
(C) Typical elements (D) None of the above

37. The statement "if 0.02 mole of a gas is dissolved in 100g. of water under 1 atm. pressure, 0.04 mole will be dissolved if the pressure is 2 atmospheres" illustrates
(A) Henry's law (B) Dalton's law of partial pressures
(C) Raoult's law (D) None of these

38. In the equation of state of an ideal gas PV = nRT, the value of the universal gas constant would depend only on
(A) the units in which P and V are expressed
(B) the temperature of the gas
(C) the pressure of the gas
(D) the nature of the gas

39. The relative density of a gas is its density
(A) relative to helium (B) relative to oxygen
(C) relative to nitrogen (D) relative to hydrogen

40. The mean free path of a gas molecule is the distance
(A) between two opposite walls of a container
(B) which a molecule travels in one second
(C) through which a molecule moves between two consecutive collisions
(D) None of these

41. According to Boyle's law
(A) $\frac{P_1}{P_2} = \frac{T_1}{T_2}$ at constant volume
(B) $\frac{V_1}{V_2} = \frac{T_1}{T_2}$ at constant pressure
(C) $\frac{P_1}{P_2} = \frac{V_2}{V_1}$ at constant temperature
(D) $\frac{P_1}{P_2} = \frac{V_1}{V_2}$ at constant temperature

42. Which one of the following occupies the least volume at S.T.P. ?
(A) 1 gram mole of carbondioxide
(B) 1 gram equivalent of chlorine
(C) 1 gram equivalent of oxygen
(D) 1 gram equivalent of hydrogen

43. The weight of iron (at wt. 56) displaced by the quantity of current which displaced 11200 C.C. of hydrogen at N.T.P. will be
(A) 5.6 g (B) 28 g (C) 56 g (D) none of these

44. The amount of water formed when 2 g of hydrogen burns in 24g of oxygen
(A) 10 g (B) 26 g (C) 2.6 g (D) None

45. Conc. sulphuric acid is 90% by mass with density 1.8 g per ml. In order to prepare 2 litres 0.2 m sulphuric acid the volume of the concentrated acid required is
(A) 49 ml (B) 39.2 ml (C) 24.2 ml (D) 19.6 ml

46. A maxima or minima obtained in the temperature composition curve of a mixture of two liquids indicates
(A) A eutectic formation (B) an azeotropic mixture
(C) that the liquids are partially miscible at the maximum or minimum
(D) that the liquids are immiscible with one another

47. 500 ml of 0.5 M $AgNO_3$ solution is added to sodium

chloride solution which contains 58 g of NaCl, the number of moles of silver chloride precipitated is
(A) 1.5 mole (B) 1 mole (C) 0.5 mole (D) 0.25 mole

48. Two solutions A and B are separated by a semipermeable membrane. Solution A has a greater vapour pressure than solution B. Water will be flowing
(A) from B to A (B) from A to B
(C) equal amount of water flows from A and B
(D) none of these

49. The mole is the amount of substance which contains as many elementary particles as there are atoms in ..... kg of carbon[12].
(A) 12 kg (B) 1.2 kg (C) 0.12 kg (D) 0.012 kg

50. At higher altitudes boiling point of water is lowered. This is because
(A) Atm. pressure is high (B) The temperature is low
(C) Atm. pressure is low (D) none of these

51. The volume of a decinormal solution of an acid which contains 1 gram equivalent of the acid
(A) 15 litres (B) 10 litres (C) 5 litres (d) 1 litre

52. The I.U.P.A.C. name for crotonaldehyde is
(A) 2al-propene-1 (B) But-2-ene-1 al
(B) Propene - al (D) Butane 2-ene-1 al

53. A certain compound contains 0.08% by weight of sulphur. Assuming that there is one atom of sulphur per molecule of the compound, the approximate molecular weight of the compound would be
(A) 256 (B) 10000 (C) 20000 (D) 40000

54. The Beilstein test is used to detect the presence of ...... in the organic compound
(A) Halogen (B) Nitrogen
(C) Sulphur (D) all of these

55. Optical activity of a compound is due to the presence of
(A) Molecular symmetry
(B) plane of symmetry in the molecule
(C) asymmetric centre in the molecule
(D) none of these

56. An alkyl halide may be converted into alcohol by ?
(A) Dehydration (B) Substitutin
(C) Elimination (D) addition

57. General formula of alkynes is
(A) CnH_{2n-2} (B) CnH_{2n+2}
(C) CnH_{2n} (D) none of these

58. The addition of HCl to pentene-1 will give as major product
(A) 2-chloro 2 methyl butane (B) 2-chloropentane
(C) 1-chloropentane (D) none of these

59. When isobutyl magnesium iodide is treated with water, the product produced is ?
(A) isobutyl alcohol (B) isobutene
(C) isobutane (D) n-butane

60. The product of reaction between ethyl magnesium bromide and methanal on hydrolysis is
(A) 2-propanone (B) propanol-2
(C) Ethanol (D) propanol-1

61. Tbe correct order of reactivity of alkyl halide is
(A) $C_2H_5Cl > C_2H_5l > C_2H_5Br$
(B) $C_2H_5l > C_2H_5Cl > C_2H_5Br$
(C) $C_2H_5l > C_2H_5Br > C_2H_5Cl$
(D) $C_2H_5Cl > C_2H_5Br > C_2H_5l$

62. Which of the following can give iodoform reaction ?
(a) CH_3OH (B) $CH_3.CH_2.CH_2OH$
(C) $CH_3.CH_2.CO.CH_2.CH_3$ (D) $CH_3.CO\text{-}CH_2CH_3$

63. If two halogen atoms are attached to same carbon atom of an alkane, it is termed as
(A) polymethylenedilhalide (B) alkyli dene halide
(C) alkylene halide (D) none of these

64. Which of the following is present in alcoholic beverages ?
(A) Methyl alcohol (B) Isopropyl alcohol
(C) amyl alcohol (D) Ethyl alcohol

65. Sodium metal reacts with
(A) Ethanal (B) Propene
(C) Ethanol (D) ethyl ether

66. Absolute alcohol from rectified spirit can be obtained
(A) by putting rectified spirit in contact with quick lime followed a distillation over quick lime
(B) by fractional distillation (C) by steam distillation
(D) cannot be produced

67. Dehydrogenation of alcohol can be brought about by heating alcohol with
(A) Zn-Cu (B) Al_2O_3
(C) AlCO KOH (D) Cu at 575 K

68. Glycerol with excess of Hl yeilds
(A) 2-iodopropane (B) allyl iodide
(C) 1,2,3 tri-iodopropane (D) propene

69. $C_6H_5NH_2 \longrightarrow A \longrightarrow C$. The product C is
(A) $C_6H_5CH_2CN$ (B) C_6H_5CN
(C) C_6H_5COOH (D) $C_6H_5CH_2NH_2$

70. In an ideal solution of two components A and B, the intermolecular forces between A and A, B and B and A and B will be
(A) different (B) the same (C) an average
(D) one will be greater than the other

71. The value of pK_W at 298 k is
(A) –14 (B) 7 (C) 14 (D) 110^{-14}

72. For the reaction $2HI \rightleftharpoons H_2 + I_2$ the equilibrium constant K, at 290°C is 0.22. Then the equlibrium constant for $H_2 + I_2 \rightleftharpoons 2HI$ will be
(A) 0.25 (B) 0.0022 (C) 45.45 (D) none of these

73. Which one of the following is Lewi's acid ?
(A) OH^- (B) H_2O (C) NH_3 (D) $AlCl_3$

74. A solution of sodium borate is likely to have pH
(A) between 4 and 5 (B) 7
(C) less than 7 (D) more than 7

75. According to the law of mass action, the rate of a chemical reaction is directly proprotional to
(A) Nature of the medium (B) Equlibrium constant
(C) Molar concentration of the reactant
(D) Nature of the reactant

76. The pH of a solution is 5.0. To this solution sufficient acid is added to decrease the pH to 2. The increase in hydrogen ion concentration is
(A) 3 times (B) 5 times (C) 100 times (D) 1000 times

77. In which of the following the change in entropy is positive
(A) $H_2(g) + O_2(g) = H_2O(l)$
(B) $MgCO_3(S) = MgO(S) + CO_2(g)$
(C) $N_2(g) + 2H_2(g) = N_2H_4(g)$
(D) $H_2(g) + O_2(g) = H_2O(g)$

78. The enthalpy per mole of a mono atomic gas is
(A) 5/2 RT (B) RT (C) 3/2 RT (D) none of these

79. The heat of combustion of rhombic sulphur and monoclinic sulphur are 70.960 Kcal and 71.030 Kcal respectively. The heat of transition of $S_{rhombic} \longrightarrow S_{monoclinic}$ is
(A) + 0.7 Kcal (B) −0.7 Kcal
(C) 71.030 Kcal (D) 70.960 Kcal

80. A thermodynamic quantity is
(A) A quantity which is used in measuring thermal change
(B) A quantity whose value depends only on the state of the system
(C) one which obeys all the laws of thermo dynamics
(D) one which is used in thermo chemistry

81. The equivalent weight of stannous chloride (Mol wt = 190) in the following reaction
$SnCl_2 + 2HgCl_2 \rightleftharpoons Hg_2Cl_2 + SnCl_4$ is
(A) 95 (B) 190 (C) 47.5 (D) none of these

82. The process in which chemical change accompanies the passage of current is known as
(A) Non-electrolytic conduction
(B) electrolytic conduction (C) conduction
(D) metallic conduction

83. The equivalent weight of metal is 4. If the vapour density of its volatile chloride is 59.5, then its valency will be
(A) 4 (B) 3 (C) 2 (D) 1

84. For a redox reaction to preceed spotaneously in a given direction, the e.m.f. should
(A) have either +ve sign (B) be zero
(C) have −ve sign (D) have +ve sign

85. Which of the following statement is false ?
(A) Oxidation number during reduction decreases
(B) Oxidation and reduction always go side by side
(C) Oxidation agent is itself reduced
(D) All reactions are oxidation and reduction reactions

86. Oxidation number of S_8 is
(A) +4 (B) 0 (C) 8 (D) −2

87. Reduction involves
(A) Decrease in the valency of negative part
(B) Increase in the valency of negative part
(C) Gain of electrons (D) Loss of electrons

88. The process of protecting iron by coating with zinc is called
(A) Rusting (B) Corrosion (C) smelting (D) galvanizing

89. Which one of the following is most commonly used as a primary standard oxidising agent
(A) $K_2Cr_2O_7$ (B) HNO_3 (C) $KMnO_4$ (D) Iodine

90. Colloids are stable because of
(A) Brownian movement (B) Tyndall effect
(C) Like charges carried by particles
(D) both a and c

91. Solution of brown sugar can be decolourised by passing through which one of the following
(A) Animal black (B) Graphite
(C) Coke (D) Carbon black

92. The catalyst used in contact process of manufacture of sulphuric acid is
(A) oxides of nitrogen (B) spongy iron
(C) platinised asbestos or vanadium pentoxide
(D) none of these

93. $Fe(OH)_3$ precipitate in presence of water containing some ferric chloride becomes colloidal on shaking. This is an example of
(A) cataphoresis (B) peptization
(C) electro-osmosis (D) coagulation

94. Adsorption of hydrogen by palladium is known as
(A) oxidation (B) reduction (C) occlusion (D) hydrogen

95. Rinmann's green is the green ash obtained in the case of
(A) Barium salts (B) calcium salts

(C) zinc salts (D) aluminum salts

96. The essential constituent of amalgam is
(A) silver (B) mercury
(C) an alkali metal (D) an alkaline earth metal

97. On heating copper nitrate strongly, which one of the following is the final product
(A) copper nitrite (B) cuprous oxide
(C) cupric oxide (D) copper

98. Tritium is obtained by
(A) Action of NaOH on Al
(B) Action of H_2SO_4 on zinc
(C) Passing steam over heated coke
(D) Nuclear reaction

99. The tendency towards complex formation is maximum in
(A) f-block elements (B) d-block elements
(C) p-block elements (D) s-block elements

100. To test whether a given clear liquid is water, of the following methods, the best one could be
(A) Moisten the litmus paper with liquid and see for the colour change
(B) Taste the liquid (C) smell the liquid
(D) Add anhydrous copper sulphate and look for the colour change

101. In the metallurgy of copper, from the matrix, blister copper is obtained by
(A) Electrolytic refining (B) Poling
(C) Bessemeristaion (D) Roasting

102. The chief ore of 'Aluminium' is
(A) Bauxite (B) kaolin (C) Felspar (D) Cryolite

103. What holds good when aluminium and a zinc salt react with excess of sodium hydroxide
(A) The metal dissolves with the evolution of hydrogen
(B) white precipitate is formed
(C) hydroxide of these metals are formed and sodium metal is set free
(D) No change takes place

104. Bauxite containig impurities of silica is concentrated by
(A) Serpeck's process (B) gravity process
(C) Hall's process (D) Baeyer's process

105. Desiccator frequently contains either anhydrous $CaCl_2$ or
(A) NaOH (B) H_2SO_4(conc) (C) HCI (D) any of these

106. Sodium carbonate cannot be used in place of ammonium carbonate for identification of group V radicals. This is because
(A) sodium ion will react with acid radical
(B) magnesium will be precipitated
(C) sodium ions will interfere in the detection of magnesium
(D) Concentration of CO_3^{2-} is very low

107. A solution contains the following ions Al^{3+}, Fe^{3+}, Cr^{3+}, Mn^{2+}, Zn^{2+}. The addition of NH_4Cl and NH_4OH will precipitate
(A) Mn^{2+}, Fe^{3+} (B) Cr^{2+}, Fe^{3+}
(C) Fe^{3+}, Al^{3+}, Cr^{3+} (D) All of these

108. Sometimes yellow turbidity appears while passing H_2S gas even in the absence of group II radicals. This is because
(A) of the oxidation of H_2S by some acid radical
(B) group III radicals are precipitated
(C) Sulphur is present in the mixture
(D) group IV radicals are precipitated

109. In the chromyl chloride test, the reddish brown vapours are of
(A) Cr_2O_3 (B) CrO_2Cl_2 (C) CrO_3 (D) $CrCl_3$

110. When phenol is treated with excess of bromine water it gives ?
(A) 2, 4, dibromophenol
(B) 2,4,6 tri-bromo phenol
(C) m-bromophenol
(D) mixture of ortho and parabromophenol

111. Which of the following reagents cannot distinguish between aldehyde and ketone ?
(A) Schiff's reagent (B) $NaHSO_3$ solution
(C) Tollen's reagent (D) Benedict's solution

112. Which of the following is produced when acetaldehyde reacts with a dilute alkali ?
(A) $CH_3CH(OH)CH_2CHO$ (B) CH_3CH_2OH
(C) CH_3COONa (D) none of these

113. Fehling's solution-2 is
(A) sodium potassium tartarate
(B) sodium potassium oxalate
(C) sodium oxalate (D) sodium potassium citrate

114. Cyanohydrin of which of these compounds yields lactic acid, on hydrolysis
(A) Acetone (B) propanal
(C) Formaldehyde (D) Acetaldehyde

115. Propanoic acid on warming with chlorine in presence of red phosphorus yields
(A) CH_3CH_2COCl (B) CH_3CH_2Cl
(C) $CH_2Cl.CH_2COOH$ (D) $CH_3CHCl.COOH$

116. Acid syntnesised from its elements is
(A) Oxalic acid (B) Aceticacid
(C) Formic acid (D) None of these

117. A compound which on reaction with aqueous nitrous acid at low temperature produces an oily nitrosoamine is
(A) Dimethyl amine (B) Tri-methylamme

(C) methylamine (D) Ethylamine

118. The best source of vitamin 'A' is
(A) Wheat (B) oranges (C) Beans (D) Carrots

1996

1. Maximum number of electrons that can be accommodated in 'N' shell of the atom is
(A) 32 (B) 8 (C) 18 (D) 9

2. Total number of orbital association with orbit number n is
(A) n^2-2 (B) n^2+2 (C) n^2 (D) $2n^2$

3. Atoms with the same number of neutrons are known as
(A) isobars (B) isotopes
(C) isotones (D) none of these

4. If 3 compounds H_2S, SO_2 and H_2O are taken as an example, the law illustrated is
(A) Law of reciprocal proportions
(B) Law of multiple proportions
(C) Law of definite proportions
(D) All of these

5. Meson was discovered by
(A) Chadwick (B) Powell (C) Yukawa (D) Crompton

6. How many moles of electrons are there in a mole of water
(A) 10 mole (B) 8 moles (C) 6 moles (D) 1 moles

7. The positron is nearly as heavy as
(A) electrons (B) deuterons
(C) α particles (D) protons

8. The shape of $CHCl_3$ molecule is
(A) Trigonal blpyramid (B) Tetrahedral
(C) pyramidal (D) Linear

9. A cyclotron cannot accelerate
(A) Positrons (B) Neutrons
(C) Protons (D) Deuterons

10. One curie of activity is equivalent to
(A) 3.7×10^4 disintegrations per second
(B) 10^6 disintegrations per second
(C) 3.7×10^7 disintegrations per second
(D) 3.7×10^{10} disintegrations per second

11. Isobars are produced as a result of the emission of
(A) Beta particle (B) X-rays
(C) Alpha particle (D) Gammarays

12. The nuclear reaction given below is of the type

$${}^{27}_{13}Al + {}^{4}_{2}He \longrightarrow {}^{30}_{14}Sl + {}^{1}_{1}P$$

(A) Transmutation (B) Chemical
(C) Fission (D) Fusion

13. Real gases approach ideal behaviour at
(A) High temperature and high pressure
(B) Critical temperature
(C) low temperature and low pressure
(D) High temperature and low pressure

14. The volume of air expelled from a vessel contaning 400 ml at 7°C when heated to 27°c at the same pressure wilt be
(A) 400ml (B) 40 ml (C) 20ml (D) 28.57 ml

15. Certain volume of gas exerts on its walls some pressure at a particular temperature. If has been found that by reducing the volume of the gas half of its original value, the pressure becomes twice of the initial value at constant temperature. This happens because
(A) More number of molecules strike the surface per second
(B) gas molecules attract one another
(C) weight of the gas increases with pressure
(D) velocity of the gas molecules decreases

16. The total number of molecules contained in 22.4 litres of hydrogen at N.T.P is
(A) 6.023×10^{23} (B) 2
(C) $2 \times 6.023 \times 10^{23}$ (D) 3.0165×10^{23}

17. The van der Waal's equation explains the behaviour of
(A) vapours (B) ideal gas
(C) non-real gas (D) real gas

18. Which will be heavier ?
(A) one gram mole of ammonia
(B) one gram atom of sulphur
(C) one gram mole of CO_2
(D) one gram atom of oxygen

19. Volume of carbon dioxide at S.T.P evolved by the complete decomposition of 1 g. of calcium carbonate
(A) 11200 ml (B) 22400 ml (C) 2240 ml (D) 224 ml

20. The concentration of acetic acid ($Ka = 1.8 \times 10^{-5}$) required to give 3.5×10^{-4} moles /litre H^+ is
(A) 194 moles/litre (B) 1.94×10^{-1} mole/litre
(C) 6.8×10^{-3} mole/litre (D) 6.8 moles/1 litre

21. The degree of dissociation of sodium chloride, if the observed and theoretical value of molecular weight is found to be 31.80 and 58.50 respectievely would be
(A) 100% (B) 90% (C) 83.96% (D) 83.90%

22. How many grams of Na_2SO_3 are needed to prepare five litres of an 8% solution (by mass), density = 1.075 g/ml ?
(A) 400 g (B) 40 g (C) 43.0g (D) 430 g

23. Hydrocyanic acid ($K_a = 4.8 \times 10^{-10}$) is dissociated to an extent of............. in 1.0 M solution
(A) 0.022% (B) 0.0022% (C) .0048% (D) none of these

24. Two solutions have different osmotic pressures. The solution of higher osmotic pressure is called
(A) Hypotonic solution (B) Hypertonic solution
(C) Isotonic solution (D) None of these

25. Aqueous solution of glucose was prepared by dissolving 18g of glucose in 90 g of water. The relative lowering of vapour pressure is equal to
(A) 20 (B) 196 (C) 1.1 (D) 0.0196

26. The van't Hoff factor for a dilute solution of $K_3[Fe(CN)_6]$ is likely to be
(A) 1/4 (B) 5 (C) 10 (D) 4

27. On the Pauling's electronegativity scale, the element next to F is
(A) Ne (B) Cl (C) O (D) He

28. Purest form of silica is
(A) Flint (B) sand stone (C) Quartz (D) Keisulghur

29. Crooks galss used for making lenses contains...... which cuts off ultraviolet rays harmful to eyes
(A) TlO_2 (B) V_2O_5 (C) CeO_2 (D) BaO

30. In the periodic table the element with atomic number 30 is expected to be present in
(A) IB (B) IIB (C) IIA (D) IA

31. The lightest metal among the following is
(A) Mg (B) Li (C) Na (D) Ca

32. In order to prepare 'Sorel's Cement' magnesium oxide is moistened with saturated solution of
(A) Magnesium peroxide (B) Magnesium hydroxide
(C) Magnesium chloride (D) Magnesium sulphate

33. Man dies in an atmosphere of carbon monoxide because
(A) It combines with the haemoglobin of the blood making it incapable of absorbing oxygen
(B) It dries up the blood
(C) It combines with the oxygen present in the body to form CO_2
(D) It reduces the organic matter of the tissues

34. Alkali metals are usually kept under
(A) Petrol (B) Carbon tetra chloride
(C) Kerosene (D) Absolute alcohol

35. In which of the following groups are there three elements belonging to same period
(A) Cr, In, Pb (B) Pt and I
(C) Li, Na, K (D) Ca, Cr, Zn

36. In the electrolytic refining of impure aluminium, the electrolyte is
(A) fused mixture of fluorides of sodium, barium and aluminium saturated with alumina
(B) fused aluminium fluoride alumina
(C) fused aluminium nitrate
(D) Fused mixture of aluminium chloride and aluminium fluoride

37. Which of the following does not give metalic oxide on heating ?
(A) LI_2CO_3 (B) K_2CO_3 (C) $MgCO_3$ (D) $ZnCO_3$

38. Which one of the following substances is not used as a water softener for permanent hard water
(A) washing soda is (B) sodium hexa metaphosphate
(C) sodium hydroxide (D) calcium hydroxide

39. Protium atom contains no
(A) neutron (B) powell (C) electron (D) proton

40. Ionisation energy of hydrogen is
(A) +1312 KJ/mole (B) -1312 KJ/mole
(C) +2624 KJ/mole (D) none

41. Which of the following ions have higher polarising power ?
(A) Al^{3+} (B) Mg^{2+} (C) Ca^{2+} (D) Na^+

42. The specific heat of an element is 0.1. Its approximate atomic weight will be
(A) 4 (B) 40 (C) 0.4 (D) none of these

43. Oxidation state of C in CH_2Cl_2 is
(A) 0 (B) +2 (C) +4 (D) –4

44. H_2Br and Hl can reduce sulphuric acid. HCl can reduce $KMnO_4$ and HF can reduce
(A) $KMnO_4$ (B) $K_2Cr_2O_7$ (C) H_2SO_4 (D) none of these

45. No oxygen is liberated when O_3 is treated with
(A) Hg (B) $SnCl_2$ + HCl (C) HCl (D) Kl

46. Dilute solution of HF cannot be concentrated beyond 36% by distilling only, because
(A) HF is less acidic (B) HF is non-volatile
(C) HF forms constant boiling mixture
(D) HF is highly soluble in water

47. How many lone pairs are associated with xenon in xenon difluoride ?
(A) One (B) two (C) three (D) four

48. The lightest element with all the p electrons unpaired is
(A) O (B) F (C) C (D) N

49. Which one of the following is called a stranger gas
(A) Xenon (B) Krypton (C) Helium (D) Neon

50. The most abundant element in the earth's crust is
(A) aluminium (B) Silicon (C) oxygen (D) iron

51. Which of the following halide is explosive ?
(A) $BICl_3$ (B) $SbCl_3$ (C) NCl_3 (D) PCl_3

52. When $C_2Cr_2O_7$ is converted into K_2CrO_4, the change in oxidation number is

(A) 4 (B) 6 (C) 0 (D) 3

53. The transition elements have
(A) completely filled d-orbitals and partly filled higher s orbitals
(B) completely filled d-orbitals and completely filled s orbitals
(C) partly filled d-orbitals and completely filled higher s orbitals
(D) all of these

54. Dichromate ions in alkaline medium exist as
(A) CrO_3 (B) CrO_4^{2-} (C) Cr^{3+} (D) none of these

55. How many unpaired electrons are there in the electronic structure for Cu
(a) 5 (b) 3 (c) 1 (D) none

56. Chlorophyll contains 2% by weight of magnesium. The number of magnesium atoms in 2g. of chlorophyll is
(A) 10^{24} (B) 6.023 x 10^{23} (C) 10^{21} (D) 10^{23}

57. 0.092 g. of a volatile substance on vapourisation displaced 44.8 ml of dry air at S.T.P molecular weight of the substance will be
(A) 92 (B) 64 (C) 46 (D) none of these

58. In the Dumas method for determining the nitrogen content of an organic compound, the nitrogen content is determined inthe form of
(A) ammonium sulphate (B) gascous nitrogen
(C) sodium cynide (D) gaseous ammonia

59. Number of atoms of carbon present in 1.8 g of glucose is
(A) 66.023 x 10^{23} (B) 6.023 x 10^{23}
(C) 6x6.023x10^{23} (D) 6

60. The I.U.P.A.C. name for the compound

$$CH_3 - \underset{OH}{CH} - CH_2 - CH_2 - \underset{Br}{CH} - CHO$$ is

(A) 2-hydroxy-5-bromo-6-hexanal
(B) 2-bromo-5-hydroxy-1-hexanal
(C) 1-bromo-4-hydroxy pentanal
(D) none of above

61. Simple eno's have a tendency to be converted into the more stable keto form by migrating H of-OH to the adjacent doubley bonded carbon atom. This phenomenon is example of
(A) isomerism (B) resonance
(C) tantomerism (D) metamerism

62. Which one of the following is an optically active compound ?
(A) $CH_3COCOOH$ (B) HOOC CH_2COOH
(C) $CH_3CH(OH)COOH$ (D) CH_3CH_2COOH

63. Which one of the following acids is the.strongest
(A) propanoic acid (B) benzoic acid
(C) Formic acid (D) acetic acid

64. Nitrating mixture is a mixture of
(A) KNO_3 + Con HCl (B) KNO_3 +Con.H_2SO_4
(C) Conc. HNO_3 and Conc HCl
(D) Conc. HNO_3 and Conc.H_2SO_4

65. Propanone and propanal are
(A) chain isomers (B) position isomers
(C) functional isomers (D) geometric isomers

66. Which of the hybrid orbitals are present in the compound $CH_2 = C = CH_2$?
(A) sp^2 and sp^3 (B) sp and sp^3
(C) sp and sp^2 (D) sp only

67. Amongst $FeCl_3$, Cl^+, Cl^- and $Cl^•$, the free radical is
(A) $Cl^•$ (B) Cl^+ (C) Cl^- (D) $FeCl_3$

68. Which of the following reactions does not take place by electrophillic substitution
(A) $C_6H_6 + HNO_3 \rightarrow C_6H_6NO_2 + H_2O$
(B) $C_6H_6 + H_2SO_4 \rightarrow C_6H_5SO_3H + H_2O$
(C) $C_6H_6 + Br_2 \rightarrow C_6H_5Br + HBr$
(D) $C_6H_6 + 3Cl_2 \rightarrow C_6H_6Cl_6$

69. A neutral divalent carbon intermediate poduced by the removal of two attached atoms is known as
(A) Carbenes (B) Frec radical
(C) Carbonium ion (D) Carbanions

70. Chlorination of benzene nucleus takes place in presence of Lewi's acid $FeCl_3$. The attacking species is
(A) $FeCl_4^-$ or Cl^+ (B) Cl^- (C) Cl^+ (D) $Cl^•$

71. Which fraction of coal tar is rich in arenes
(A) Heavy oil (B) Green oil
(C) Light oil (D) Carbolic oil

72. When toluene is treated with methyl chloride in presence of $AlCl_3$, we get
(A) m-xylene (B) 1, 3, 5 rimethyl benzene
(C) o and p-xylene (D) Diphenyl methane

73. The six carbon atoms of benzene are of
(A) three type (B) two types
(C) different types (D) same type

74. Which of the following finds use in the manufacture of candle ?
(A) paraffin wax (B) Lanolin wax
(C) Bee wax (D) All of these

75. Acidic hydrogen is present in
(A) Ethyne (B) Ethane (C) Ethene (D) All these

76. Ethylene dichloride and ethylidene dichloride are isomeric compounds. Identify the statement which is not applicable. Both of them
(A) respond to Beilstein test

(B) react with aqueous KOH to give same product
(C) react with alcoholic KOH to produce same product
(D) are dihalides

77. Phenol and carboxylic acid can be distinguished by the reagent
(A) $NaHCO_3$ (B) Na (C) both (D) none

78. Ethyl ether when reacted with chlorine in dark yields
(A) CH_3 CHClO CHCl CH_3
(B) $ClCH_2$ CH_2O CH_2 CH_2Cl
(C) $CH_3CH_2OCH_2$ CH_2Cl (D) CH_3CHCl O CH_2 CH_3

79. Aldol condensation between which of the following compounds followed by dehydration gives mesityl oxide
(A) two molecules of acetaldehyde
(B) acetaldehyde and formaldehyde
(C) Acetone and acetaldehyde
(D) two molecules of acetone

80. $CH_3.COOH \xrightarrow{Ca(OH)_2} A \xrightarrow{Ca(HCOO)_2} B \xrightarrow{NH_2OH} C$. In the above series of reactions, compound C is
(A) $CH_3CH_2CH = NOH$ (B) $H_2C = NOH$
(C) $CH_3CH = NOH$ (D) $> C = NOH$

81. The most probable structural formula for the compound whose formula is C_3H_6O and which gives silver mirror teat is
(A) CH_3COCH_3 (B) CH_3CH_2CHO
(C) $CH_2 = CH\ CH_2OH$ (D) none of these

82. When calcium acetate and calcium formate together is subjected to dry distillation, the product is
(A) acetone (B) formaldehyde(C) acetaldehyde (D) none

83. Which of the following substances possese the odour of bitter almonds
(A) benzaldehyde (B) benzene sulphonamide
(C) aniline (D) acetone

84. Benzaldehyde can be obtained by the hydrolysis of
(A) C_6H_5Cl (B) $C_6H_5CCl_3$
(C) $C_6H_5CHCl_2$ (D) $C_6H_5CH_2Cl$

85. C_6H_5-CH_2-$CH_2COOH \xrightarrow{Cl_2/P} A \xrightarrow{AlC\ KOH} B$ compound B is
(A) C_6H_5-CH_2-CH_2-COCl (B) C_6H_5-CH(OH) - CH_2.COOH
(C) C_6H_5-CH=CH.COOH (D) C_6H_5-CH_2-CH_2-COOH

86. When CH_3CONH_2 is heated with PCl_5, the product is
(A) CH_3COCCl_3 (B) CH_3COCl
(C) CH_3CN (D) $CH_3CCl_2NH_2$

87. When ammonium acetate is strongly heated the product formed is
(A) acetonitrile (B) formamide (C) acetamide (D) none

88. Which of the following acids, is hydroxy acid
(A) Lactic acid (B) Adipicacid
(C) Maleic acid (D) succinic acid

89. Acetamide is treated separately with the following reagents, which one would give methylamine
(A) $Br_2 + KOH$ (B) Conc. H_2SO_4 (C) PCl_5 (D) P_2O_5

90. An acid can be converted to alcohol by reduction with
(A) $LiAlH_4$ (B) H_2 in presence of catalyst
(C) nascent hydrogen (D) none of these

91. Which method of purification is represented by the equation $Ti + 2I_2 \longrightarrow Ti\ I_4 \longrightarrow Ti_{(Pure)} + 2I_2$
(A) Van-Arkel method (B) zone refining
(C) cupellation (D) poling

92. A naturally occurring substance from which metal can be extracted profitably is
(A) ore (B) flux (C) mineral (D) gangue

93. Radium was obtained from the mineral
(A) Rutile (B) Chalcopyrite
(C) Limestone (D) Pitchblende

94. H_2SO_4 cannot be used in place of HCl for preparing original solution in the analysis of cations because
(A) It results in the precipitation of some of cations
(B) It is strong dehydrating agent
(C) It is strong oxidising agent
(D) none of these

95. Ionic reaction takes place in
(A) solution (B) gaseous state
(C) solid state (D) liquid state

96. The sotability of $BaSO_4$ (molecular weight 233) at 20°c is 2.33×10^{-4} g/100 cc. The solubility ionised is
(A) 1×1^{-10} (B) 5.43×10^{-8} (C) 1×10^{-12} (D) 5.42×10^{-6}

97. The solubility coefficient of copper sulphate at 30°C is 25g per 100g. of H_2O. 18% solution of the salt will be
(A) supersaturated (B) supercooled
(C) saturated (D) unsaturated

98. Decomposition of PCl_5 at 323 K is represented by $PCl_5(g) \rightleftharpoons PCl_3(g) + Cl_2(g)$. If the degree of dissociation of PCl_5 is α and the total pressure developed is P, then the partial pressure of PCl_3 at the given temperature is
(A) $p\alpha$ (B) $(\alpha/1 + \alpha)p$
(C) $(1-\alpha) / (1+\alpha)p$ (D) none of these

99. Which of the following can act as both Lowry Bronsted acid and base
(A) BF_3 (B) H_3O^+ (C) HSO_3^- (D) CO_3^{2-}

100. An example of a buffer solution is
(A) NaOH and KOH (B) CH_3COOH and CH_3COONa
(C) KOH and KCl (D) K_2SO_4 and H_2SO_4

101. Which of the following procedures leads to a change

in the value of equilibrium constant of a chemical reaction
(A) Replacement of a catalyst
(B) A change in the concentration of the reactants and products
(C) A change in the pressure
(D) A change in the temperature

102. Ionic product of water at 90°c is 10^{-12}. A solution with hydrogen ion concentration of 10^{-4} at that temperature is
(A) Basic (B) Highly basic (C) Acidic (D) Neutral

103. Which one of the following systems is an example of a closed system
(A) some amount of hot water enclosed in a closed container which is not insulated
(B) Hot water contained in an open vessel
(C) some amount of water in equilibrium with its vapour in a closed and insulated vessel
(D) none of these

104. The heat of a reaction at constant volume is equal to
(a) $E_R - E_P$ (B) $H_R - H_P$ (C) $H_P - H_R$ (D) $E_P - E_R$

105. If the concentration units are reduced by 'K' time, then the value of rate constant of first order will
(A) increases by K times (B) decreases by factor of K
(C) not alter (D) decreases by factor of K

106. The value of E for the half cell in which the reaction $Zn^{2+}((.01M) + 2e - Zn_{(3)}$ takes place at 25°c is ? (Given $Zn^{2+} + 2e \rightarrow Zn\ E_0 = 0.76V$
(A) –0.74V (B) –0.79V (C) –0.82V (D) –0.79V

107. Half time of 1st order reaction is t 1/2. The half life period will
(A) Be inversely proportional to square of concentration
(B) Be directly proportional to square of conentration
(C) Increase with increase in temperature
(D) Decrease with increase in temperature

108. Cell constant has the dimension
(A) cm^2 (B) cm mol $séc^{-1}$ (C) cm^{-1} (D) cm

109. The number of moles of sodium and chlorine formed at the appropriate electrodes when 1 Faraday of electricity is passed through molten sodium chloride is
(A) 1 mole of sodium and 1/2 mole of chlorine
(B) 1/2 mole of sodium and 1/2 mole of chlorine
(C) 1 mole of sodium and 2 moles of chlorine
(D) 1 mole of sodium and 1 mole of chlorine

110. The series of spectral lines which appear in the visible region of hydrogen spectrum is called
(A) Lyman series (B) Pfund series
(C) Balmer series (D) Brackett series

111. The poison for platinum catalyst is
(A) carbon (B) Arsenic (C) phsophorus (D) sulphur

112. Methylamine with nitrous acid yields
(A) dimethyl ether (B) methyl nitriteand dimethylether
(C) methyl alcohol (D) methyl nitrite

113. Exact order of basic strength of RNH_2, R_2NH, R_3N and $NH_3(R = CH_3)$ is
(A) $NH_3 > R_3N > R_2NH > RNH_2$
(B) $R_3N > R_2NH > RNH_2 > NH_3$
(C) $RNH_2 > R_2NH > R_3N > NH_3$
(D) $R_2NH > RNH_2 > R_3N > NH_3$

114. Nitrobenzene on electrolytic reduction in weak acidic medium yeilds
(A) P-hydroxyaniline (B) Nitrosobenzene
(C) N-phenyl hydroxylamine (D) Anilie

115. Compounds of the type $R\text{-}CH<^{OR}_{OR}$ are known as
(A) hemiacetals (B) acetals
(C) ethers (D) acid anhydrides

116. Iodine containing hormone is
(A) Insulin (B) Estrene (C) Adrenaline (D) Cortisole

117. Vitamin 'A' deficiency leads to disease known as
(A) Scurvy (B) T.B
(C) Night blindness (D) Beri-Beri

118. Ethylene oxide on hydrolysis yields
(A) Ethanal (B) Ethyl alcohol
(C) ethylene glycol (D) none of these

119. The compound required for the formation of a thermosetting polymer with methanal is
(A) Benzaldehyde (B) phenol
(C) Benzene (D) Phenylamine

120. The bond angle in cyclopropane is
(A) 90 (B) 108 (C) 60 (D) none of these

1997

1. If the Pauli's exculsion principle, were not valid the electronic configuration of nitrogen would have been
(A) $1s^7$ (B) $1s^2, 2s^2, 2p_x^1, 2p_y^2$
(C) $1s^2, 2s^2, 2p_x^2, 2p_y^1$ (D) $1s^2, 2s^2, 2p_x^1, 2p_y^1$

2. How does the energy of an electron of an electron atom depends on the orbital quantum number for a constant value of principal quantum number
(A) It remains constant (B) It cannot be predicted
(C) It increases with the increase in the value
(D) It diminishes with the value of 1

3. The p, sub shell, d subshell etc. orbitals which are close in energy, in the absence of magnetic field are equal in energy. Such orbitals of energy are known as
(A) generate orbitals (B) p-orbitals

(C) degenerate orbitals (D) none of these

4. The azimuthal quantum number l = 3, m can have
(A) 5 values (B) 7 values
(C) only one value (D) 3 values

5. The energy level that would allow the hydrogen atom to absorb a photon but not to emit a photon
(A) 2p (B) 3s (C) 2s (D) 1s

6. When atoms are excited, while in a magnetic field, some spectra lines are observed to be spilt very slightly. This splitting is called
(A) The levelling effect (B) The Stark effect
(C) The Zeeman effcct (D) None of these

7. Number of f orbitals associated with n = 5 is
(A) 5 (B) 7 (C) 9 (D) 14

8. Electromagnetic radiation having the maximum wave-length is
(A) Radiowave (B) X-ray (C) Infra-red (D) Gamma ray

9. (N+l) value for 4f is
(A) 7 (B) 6 (C) 5 (D) 4

10. In which molecule BF, or NH_3 is the value of dipole moment greater ?
(A) In NH_3 (B) In BF_3
(C) Dipole moment is equal (D) none of these

11. Carhon tetrachloride has no net dipole moment because of
(A) similar sizes of carbon and chlorine atoms
(B) similar electronegativites of the atoms
(C) its planar structure
(D) its symmetrical tetrahedral structure

12. There is deviation of the normal tetrahedral bond angle 109°28' in H_2O (105°), though the central atom is sp^3 hybridised. This is explained as given below.
(A) Because of the difference in the electronegativity of the atoms
(B) Because of the difference in the size of the atoms
(C) Because of the repulsion due to the nonbonded electrons
(D) Because of repulsion due to the bonded pairs only

13. In which of the following is the angle greatest between the two covalent bonds ?
(A) NH_3 (B) H_2O (C) CO_2 (D) CH_4

14. One 's' orbital and one 'p' orbital hybridise. Then one gets
(A) two hybrid orbitals which are mutually perpendicular
(B) three orbitals in the same plane
(C) two orbitals at 90° (D) two hybrid orbitals at 180°

15. The bond energy increases in the order
(A) $s < p < sp < sp^3 < sp^2$ (B) $s < p < sp^2 < sp^3 < sp$
(C) $s < p < sp < sp^2 < sp^3$ (D) $sp^3 < sp^2 < sp < p < s$

16. The types of bonds present in $CuSO_4 . 5H_2O$ are
(A) electro-valent and covalent
(B) electrovalent coavalent and co-ordinate covalent
(C) electrovalent only (D) covalent only

17. The number of π bonds in nitrogen molecule is
(A) one (B) two (C) three (D) none

18. Which of the following statements is wrong ?
(A) sp^3d^2 hybrid orbitals are directed towards the corners of octahedron
(B) sp^3d^2 hybrid orbitals are all directed at 90° to each other
(C) sp^2 hybrid orbitals arc formed from the mixing of two 'p' and one 's' atomic orbitals
(D) Hybridization is mixing of atomic orbitals prior to their combining into molecular orbitals

19. Which one of the following has a pyramidal shape
(A) CO_3^{2-} (B) NO_3^- (C) PCl_3 (D) SO_3

20. The conditions most favourable for electo-valency are
(A) high charge on ions, large cation and small anion
(B) low charge on ions, small cation and large anion
(C) Low charge on ions, large cation and small anion

21. Loss of a β-particle from an atom brings about the following change
(A) decrease of one nucleon (B) decrease of atomic number
(C) increase of one proton (D) increase of one neutron

22. The man made radio active disintegration series is
(A) 4n + 2 series (B) 4n+1 series
(C) 4n+3 series (D) none

23. A nuclear reaction must be balanced in terms of
(A) Number of electrons (B) Only energy
(C) Only mass (D) Mass and energy

24. At 27°c a gas was compressed to half of its volume. To what temperature must be now heated, so that it occupies its original volume ?
(A) 327 K (B) 600°C (C) 327°C (D) 54°C

25. The number of moles of impure potassium chlorate of 75% purity required to produce 24 g. of oxygen is
(A) 2.5 moles (B) 1 mole (C) 0.667 mole (D) 2 moles

26. Sum of mole fractions of the two componants of a solution is always
(A) less than one (B) not fixed
(C) more than one (D) exactly one

27. Which of the following 0.1 M aqueous solution will have the lowest freezing point ?
(A) urea (B) glucose
(C) potassium sulphate (D) sodium chloride

28. At 298 K the solubility of sodium chloride is 36.0 g in 100 g of water. Then the concentration of saturated solution in water in percent is
(A) 26.47% (B) 56.25% (C) 36% (D) none of these

29. The solubility of a substance at a given temperature is the maximum amount of the substance that can be dissolved by
(A) 1 kg of solvent (B) 10 g of solvent
(C) 1 gram of solvent (D) 100 g.of solvent

30. Which gas is evolved during fermentation
(A) CO_2 (B) H_2 (C) O_2 (D) CO

31. Oxides of the elements of group III are generally
(A) Neutral oxide (B) Amphoteric oxide
(C) Acidic oxide (D) Basic oxide

32. The chloride ion is isoelectronic with potassium. The size of chloride ion is..............than/to potassium ion
(A) equal (B) smaller or greater (C) greater (D) smaller

33. Bleaching action of hydrogen peroxide is due to its
(A) Acidic nature (B) Thermal instability
(C) Reducing property (D) Oxidising property

34. When hydrogen peroxide is reacted with acidified $K_2Cr_2O_7$ a blue coloured compund is produced. It is called
(A) Perchromic acid (B) Chromium trioxide
(C) Chromium sulphate (D) Potassium chromate

35. Most favourable condition, for formation of anion is
(A) Lower value of electronegativity
(B) Higher value of electron affinity
(C) Lower value of electron affinity
(D) Lower value of ionisation potential

36. Molecular formula of Glauber's salt is
(A) $FeSO_4.7H_2O$ (B) $Na_2SO_4.10\ H_2O$
(C) $CuSO_4.\ 5H_2O$ (D) $ZnSO_4.\ 7H_2O$

37. Which one of the following statements regarding H_2O_2 is false ?
(A) It is more stable in a basic solution
(B) It behaves as a reducing agent towards acidified $KMnO_4$
(C) It is a strong oxidising agent as well as reducing agent in acidic and also in basic medium
(D) It is decomposed in the presence of MnO_2

38. Free hydrogen is found in
(A) water (B) water ges (C) acids (D) marsh gas

39. A sample of hydrogen peroxide is labelled as 10 volume. Its percentage strength is nearly
(A) 20% (B) 90% (C) 10% (D) 3%

40. The oxide that gives hydrogen peroxide on treatment with a dilute acid is
(A) MnO_2 (B) TiO_2 (C) PbO_2 (D) Na_2O_2

41. Elements whose atoms have 3 outermost main energy levels incomplete are known as
(A) normal elements (B) inner transition elements
(C) p-block elements (D) s-block elements

42. The correct electronic configuration of chromium atom (At. No:24) in the ground state is
(A) $[Ar]3d^6s^0$ (B) $[Ar]4d^5\ 4s^1$
(C) $[Ar]3d^54s^1$ (D) $[Ar]3d^44s^2$

43. Which of the following is false with respect to hydrogen atom ?
(A) It has no neutron
(B) It exists in two isotopic forms only
(C) It's electronic configuration is $1s^1$
(D) It is the lightest atom

44. The ionisation energy of nitrogen is larger than that of oxygen because
(A) there is greater attraction of electrons towards the nucleus
(B) half -filled p-orbitals of nitrogen have extra stability
(C) the size of nitrogen atom is smaller.
(D) none of these

45. The tendency of an atom in a molecule to attract electrons towards itself is called.
(A) electronegativity
(B) ability of the atom to get oxidised
(C) ionisation potential (D) electron affinity

46. Which one of the following electronic structures indicates metallic character ?
(A) 2,8,8 (B) 2,8,7 (C) 2,8,2 (D) 2,8,4

47. Oxygen has an oxidation state of +2 in
(A) F_2O (B) SO_2 (C) H_2O_2 (D) H_2O

48. Among the following gases, which one is least polarisable ?
(A) Argon (B) Krypton (C) Helium (D) Neon

49. Oxygen can be obtained by heating
(A) Fe_2O_3 (B) Fe_3O_4 (C) BaO_2 (D) Na_2O_2

50. Cp/Cv for noble gases is
(A) 1.80 (B) 1.33 (C) 1.66 (D) 1.43

51. Which one of the following is the most acidic ?
(A) $CH_2ClCOOH$ (B) CCl_3COOH
(C) CH_3COOH (D) CH_3CH_2OH

52. Relative acidity of the following is in the order
(A) ROH > RCOOH > H_2CO_3 > C_6H_5OH > H_2O
(B) RCOOH > C_6H_5OH > ROH > H_2CO_3 > H_2O
(C) RCOOH > H_2CO_3 > C_6H_5OH > H_2O > ROH
(D) RCOOH > ROH > H_2CO_3 > C_6H_5OH > H_2O

53. House hold gaseous fuel (L.P.G) mainly contains

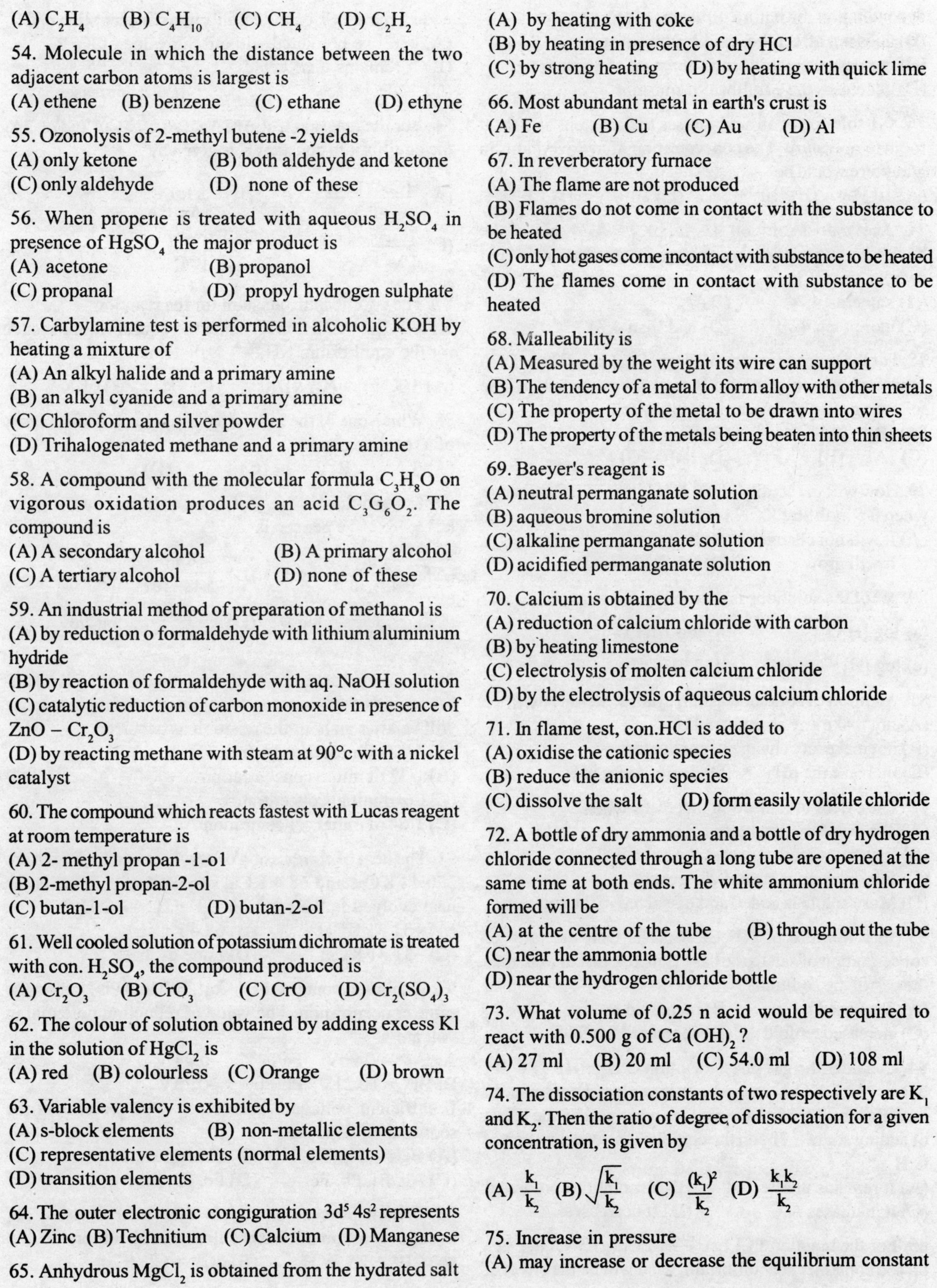

(A) C_2H_4 (B) C_4H_{10} (C) CH_4 (D) C_2H_2

54. Molecule in which the distance between the two adjacent carbon atoms is largest is
(A) ethene (B) benzene (C) ethane (D) ethyne

55. Ozonolysis of 2-methyl butene -2 yields
(A) only ketone (B) both aldehyde and ketone
(C) only aldehyde (D) none of these

56. When propene is treated with aqueous H_2SO_4 in presence of $HgSO_4$ the major product is
(A) acetone (B) propanol
(C) propanal (D) propyl hydrogen sulphate

57. Carbylamine test is performed in alcoholic KOH by heating a mixture of
(A) An alkyl halide and a primary amine
(B) an alkyl cyanide and a primary amine
(C) Chloroform and silver powder
(D) Trihalogenated methane and a primary amine

58. A compound with the molecular formula C_3H_8O on vigorous oxidation produces an acid $C_3G_6O_2$. The compound is
(A) A secondary alcohol (B) A primary alcohol
(C) A tertiary alcohol (D) none of these

59. An industrial method of preparation of methanol is
(A) by reduction o formaldehyde with lithium aluminium hydride
(B) by reaction of formaldehyde with aq. NaOH solution
(C) catalytic reduction of carbon monoxide in presence of $ZnO - Cr_2O_3$
(D) by reacting methane with steam at 90°c with a nickel catalyst

60. The compound which reacts fastest with Lucas reagent at room temperature is
(A) 2- methyl propan -1-o1
(B) 2-methyl propan-2-ol
(C) butan-1-ol (D) butan-2-ol

61. Well cooled solution of potassium dichromate is treated with con. H_2SO_4, the compound produced is
(A) Cr_2O_3 (B) CrO_3 (C) CrO (D) $Cr_2(SO_4)_3$

62. The colour of solution obtained by adding excess Kl in the solution of $HgCl_2$ is
(A) red (B) colourless (C) Orange (D) brown

63. Variable valency is exhibited by
(A) s-block elements (B) non-metallic elements
(C) representative elements (normal elements)
(D) transition elements

64. The outer electronic congiguration $3d^5 4s^2$ represents
(A) Zinc (B) Technitium (C) Calcium (D) Manganese

65. Anhydrous $MgCl_2$ is obtained from the hydrated salt
(A) by heating with coke
(B) by heating in presence of dry HCl
(C) by strong heating (D) by heating with quick lime

66. Most abundant metal in earth's crust is
(A) Fe (B) Cu (C) Au (D) Al

67. In reverberatory furnace
(A) The flame are not produced
(B) Flames do not come in contact with the substance to be heated
(C) only hot gases come incontact with substance to be heated
(D) The flames come in contact with substance to be heated

68. Malleability is
(A) Measured by the weight its wire can support
(B) The tendency of a metal to form alloy with other metals
(C) The property of the metal to be drawn into wires
(D) The property of the metals being beaten into thin sheets

69. Baeyer's reagent is
(A) neutral permanganate solution
(B) aqueous bromine solution
(C) alkaline permanganate solution
(D) acidified permanganate solution

70. Calcium is obtained by the
(A) reduction of calcium chloride with carbon
(B) by heating limestone
(C) electrolysis of molten calcium chloride
(D) by the electrolysis of aqueous calcium chloride

71. In flame test, con.HCl is added to
(A) oxidise the cationic species
(B) reduce the anionic species
(C) dissolve the salt (D) form easily volatile chloride

72. A bottle of dry ammonia and a bottle of dry hydrogen chloride connected through a long tube are opened at the same time at both ends. The white ammonium chloride formed will be
(A) at the centre of the tube (B) through out the tube
(C) near the ammonia bottle
(D) near the hydrogen chloride bottle

73. What volume of 0.25 n acid would be required to react with 0.500 g of Ca $(OH)_2$?
(A) 27 ml (B) 20 ml (C) 54.0 ml (D) 108 ml

74. The dissociation constants of two respectively are K_1 and K_2. Then the ratio of degree of dissociation for a given concentration, is given by
(A) $\frac{k_1}{k_2}$ (B) $\sqrt{\frac{k_1}{k_2}}$ (C) $\frac{(k_1)^2}{k_2}$ (D) $\frac{k_1k_2}{k_2}$

75. Increase in pressure
(A) may increase or decrease the equilibrium constant

depending on the nature of the reaction
(B) does not affect the equilibrium constant
(C) increase the equilibrium constant of a reaction
(D) decreases the equilibrium constant

76. 0.1 solution of an acid ionises to the extent of 10% at room temperature. The concentration of hydroxyl ions in g/ion/litre would be
(A) 10^{-11} (B) 10^{-10} (C) 10^{-2} (D) 10^{-12}

77. Acetic acid solution has its $pK_a = 4.74$. pH of the buffer having equimolar quantities of acetic acid and sodium acetate is
(A) sameras 4.74 (B) 7
(C) more than 4.74 (D) less than 4.74

78. For the gaseous reaction $2A(g)+B(g) \rightleftharpoons 3C(g)+D(g)$, two moles each of A and B were taken container. The following must always be at equilibrium
(A) [B] > [A] (B) [B]< [A]
(C) [A] = [B] (D) [B] > [C]

79. How will the acidity of 0.2 N solution of HCN change when 0.5 mol/litre KCN is added
(A) It will not change (B) It depends on the time
(C) It will grow (D) it will diminish

80. pH of the solutions is defined by the relation
(a) $\log [H_3O^+]$ (b) $\log 1/[H_3O^+]$
(c) log [H] (d) $1/[H_3O^+]$

81. Addition of solid potassium cyanide to water will
(A) not make any change mpH+
(B) not make any change in conductivity
(C) increase the pH (D) decrease the pH

82. The extent of ionisation increases when
(A) Temperature. decreased
(B) Water is added in excess
(C) The solution is stirred vigorously
(D) More solute is added and dissolved in the solution

83. In a mixture of NH_4Cl and NH_4OH, the ratio of concentration of salt to the base is increased ten fold, then the p^{OH} of the solution
(A) decreased by one (B) increased by one
(C) increased tenfold (D) decreased ten fold

84. Consider the gas phase reaction $C_2H_4 + H_2 \quad C_2H_6$, $\Delta H = -32.7$ KCal. carried out in a closed container at equilibrium. At a fixed temp ture what will be the effect of adding some C_2H_5, on the equilibrium concentration of C_2H_4
(A) It remains unaffected (B) It can not be predicted
(C) It increases (D) It decreases

85. For the reaction $PCl_5(g) \rightleftharpoons PCl_3(g) + Cl_2(g)$ in a closed container at equilibrium, what will be the effect of adding more PCl_5 on the equilibrium concentration of $Cl_{2(g)}$
(A) can't be predicted without the value of Kp ?
(B) It remains unaffected
(C) It decreases (D) It increases

86. For the reaction $Cu(s)+2Ag^+(aq) \rightleftharpoons Cu^{2+}(aq) + 2Ag$ the equilibrium constant K is given by
(A) $\frac{[Cu^{2+}][Ag]^2}{[Cu][Ag^+]^2}$ (B) $\frac{[Ag]^2}{[Ag^+]^2}$
(C) $\frac{[Cu^{2+}]}{[Ag^+]^2}$ (D) $\frac{[Cu^{2+}]}{[Cu]}$

87. The equilibrium constant for the reaction $N_2 + 3H_2 \rightleftharpoons 2NH_3$ is K. Then the equilibrium constant for the equilibrium $NH_3 \rightleftharpoons N_2 + H_2$ is
(A) $1/K^2$ (B) $\sqrt{1/K}$ (C) $\sqrt{K}$ (D) 1/K

88. Which one of the following is incorrect. Molecularity of a reaction can be
(A) 3 (B) 2 (C) 1 (D) 0

89. The two-third life $(t_{2/3})$ of a first order reaction in which k = 5.48x1014 per sec is
(A) $\frac{2.303}{5.48 \times 10^{-14}} \log 3$ (B) $\frac{2.303}{5.48 \times 10^{-14}} \log 2$
(C) $\frac{2.303}{5.48 \times 10^{-14}} \log 1/3$ (D) $\frac{2.303}{5.48 \times 10^{-14}} \log 2/3$

90. In a first order reaction the substance recting has half-life period of ten minutes. What fraction of the substance will be after an hour the reaction as occurred
(A) 1/12 of initial concentration
(B) 1/32 of initial concentration
(C) 1/6 of initial concentration
(D) 1/64 of initial concentration

91. The heat of combustion of ethylene and hydrogen are 370.44 KCal and 68.4 KCal respectievely. What is the heat evolved in the reaction $C_2H_4+H_2 \rightarrow C_2H_6$?
(A) +31.31 KCal (B) 68.4 KCal
(C) –31.31 KCal (D) none of these

92. A solution contains Fe^{2+}, Ag^+, Bi^{3+} and Pb^{2+} ions in the same concentration. The value of reduction potential in volt are
$Ag/Ag^+ = 0.80V$ $Fe/Fe^{2+} = -0.44$ V
$Bl/Bl^{3+} = +0.21V$ $Pb/Pb^{2+} = -0.13V$
If sufficient voltage is applied to depc metal then the sequence of deposite
(A) Bi, Pb, Ag, Fe (B) Pb, Ag, Bi, Fe
(C) Ag, Bi, Pb, Fe (D) Fe, Pb, Bi, Ag

93. Which process occurs in the electrolysis of an, aqueous solution of Nickel chloride at Nickle anode
(A) $2H^+ + 2e \rightarrow H_2$ (B) $2Cl^- = Cl_2 + 2e$

(C) $Ni^{2+} + 2e = Nl$ (D) $Ni \rightarrow Ni^{2+} + 2e$

94. When a piece of copper wire is immersed in a solution of silver nitrate, the colour of the solution becomes blue. This is due to the
(A) fomation of a complex (B) oxidation of Ag
(C) oxidation of Cu (D) reduction of Cu

95. The standard cell potential for the cell, Zn/Zn^{2+} (1M)//Cri^{2+} (1M); Cu, given $E^o_{Cu} = 0.34V$ and $E^o_m = -0.76$ V is
(A) –1.10V (B) –0.42V (C) 1.10V (D) none of these

96. The units of equivalent conductance is
(A) ohm^{-1} cm^2 gm eq^{-1} (B) ohm cm^2 gm Eq^{-1}
(C) ohm cm gm eq^{-1} (D) ohm^{-1} cm^{-1} gm Eq^{-1}

97. Conductivities at inifinte dilution of NH_4Cl, NaOH and NaCl are 129.8, 217.4 and 108.9 mhos respectively. If the equivalent conductance of N/100 solution of NH_4OH is 9.33 mhos, then degree of dissociation of NH_4OH at this temperature is
(A) 3.9% (B) 0.1% (C) 39.0% (D) –39%

98. In the cell Cu / Cu^{2+} // Ag^+ / Ag, the cell reaction is
(A) $Cu + 2Ag^+ \rightarrow Cu^{2+} + 2Ag$
(B) $Cu + 2Ag \rightarrow Cu^{2+} + 2Ag^+$
(C) $Cu^{2+} + 2Ag^+ \rightarrow Cu + 2Ag$ (D) none of these

99. If a direct current deposits 19.5 g of potassium (At.wt.39) in one minute, the number of grams of alumimnm (at.wt.27) deposited by the same current during the same time interval would be
(A) 9 (B) 4.5 (C) 13.5 (D) 27

100. In the reaction between sodium and chlorine to form sodium chloride
(A) chloride ion is reduced (B) chlorine atom is reduced
(C) sodium atom is reduced (D) sodium ion is reduced

101. In the cell Zn/Zn^{2+} : Pb^{2+}//Pb if the E, $Zn^{2+}/Zn = -0.76$ V and E, pb^{2+}, pb = –13 V, then cell e.m.f. is
(A) +0.74 V (B) –0.63V (C) –0.89V (D) 0.89 V

102. The sp.heat of an element is 0.1 and its equivalent weight is 31.6. The correct atomic weight of the element will be
(A) 126.4 (B) 31.5 (C) 63.2 (D) none of these

103. 0.5 g of zinc liberates 0.484 g of copper from its salt solution. If the equivalent weight of zinc is 32.5, the equivalent weight of Cu is
(A) 54.5 (B) 48.5 (C) 31.46 (D) 71

104. Indicate in which of the following process the nitrogen is oxidised
(A) $NO_3^- \rightarrow NH_4^+$ (B) $NO_3^- \rightarrow NO$
(C) $NH_4^+ \rightarrow N_2$ (D) none

105. Which of the following is a redox reaction
(A) $2CuI_2 \rightarrow Cu_2I_2 + I_2$
(B) $NaCl + AgNO_3 \rightarrow AgCl + NaNO_3$
(C) $Cr_2(SO_4)_3 + 6KOH \rightarrow K_2SO_4 + 2K_2S + 3H_2O$
(D) $NH_4Cl + NaOH \rightarrow NH_3 + NaCl + H_2O$

106. When the same amount of zinc is treated with excess of sulphuric acid and excess of sodium hydroxide, separately the ratio of volumes of hydrogen evolved is
(A) 2:1 (B) 3:4 (C) 1:1 (D) 1:2

107. A mixture contains 2 moles of N_2 and 6 moles of H_2 gases. If the pressure of the mixture is 4 atm, the partial pressure of nitrogen in the mixture will be
(A) 6 atm (B) 4 atm (C) 2 atm (D) 1 atm

108. $HCHO \xrightarrow{reduction} A \xrightarrow{P + I_2} B \xrightarrow{KCN} C \xrightarrow{H_3O^+} D$. "D"is
(A) Acetamide (B) Acetic acid
(B) ethylamine (D) none of these

109. "Tri-oxane" is obtained, when
(A) Acetaldehyde is treated with H_2SO_4
(B) Acetaldehyde is treated with dry HCl gas
(C) Aqueous solution of formaldehyde is evaporated to dryness
(D) Aqueous solution of formaldehyde mixed with dil H_2SO_4, is allowed to stand

110. Ketones on reaction with $NH_2 - CO.NH.NH_2$ form well defined crystalline compounds called
(A) Schiff's base (B) semi carbazones
(C) Hydrazones (D) oximes

111. Which of the following are examples of aldol condensation ?

(A) $2HCHO \xrightarrow[NaOH]{alkali} C_6H_5COONa + C_6H_5CH_2OH$

(B) $2C_6H_5CHO \xrightarrow[NaOH]{alkali} C_6H_5COONa + C_6H_5-CH_2OH$

(C) $2CH_3CHO \xrightarrow{dil\ NaOH} CH_3CHOH\ CH,CHO$

(D) $2CH_3CO.CH_3 \xrightarrow{dilNaOH} CH_3C(OH) + (CH_3)CH_2COCH_3$

112. Benzene diazonium chloride reacts with hypophosphorous acid to give
(A) phenyl phosphite (B) phenyl phosphate
(C) benzene (D) phenol

113. Primary, secondary and tertiary amines may be separated by using
(A) Diethyl oxalate (B) Acetyl chloride
(C) Benzene sulphonyl chloride (D) Sulphonyl chloride

114. The molecule (ion) having one unpaired electron is
(A) CN^{-1} (B) O_2 (C) NO (D) CO

115. In the crystals of Zn, bonds are
(A) Metallic (B) Co-ordinate (C) Covalent (D) Ionic

116. Atomic hydrogen produces formaldehyde when it reacts with

(A) acetylene (B) sulphur
(C) carbon monoxide (D) Oxygen

117. Sliver nitrate reacts with RBr to give a
(A) mixture of RNO_2 and R – O – N = O
(B) RNO_2 (C) RONO (D) none of these

118. I.U.P.A.C name of the compound having the formula

$$H_3C - \underset{CH_3}{\overset{CH_3}{\underset{|}{\overset{|}{C}}}} - Cl$$

(A) 2-methyl-2-chloropropane
(B) 2-methyl-2-propyl chloride
(C) ter-butyl chloride (D) iso-butyl chloride

119. Propionic acid when treated with bromine in presence of phosphorus forms
(A) $CH_3CHBrCOOH$ (B) $CH_3CHBrCOBr$
(C) CH_2BrCH_2COOH (D) CH_3CH_2COBr

120. The reagent with which both acetaldehyde and acetone react easily is
(A) Schiff's reagent (B) Tollen's reagent
(C) Fehling's solution (D) Grignard reagent

1998

1. In Carius method 0.369g.of an organic compound gave 0.345 g. of silver bromide. The percentage of bromine in the organic compound is
(A) 50.5% (B) 23.13% (C) 39.78% (D) none of these

2. The I.U.P.A.C name of the following compound

$$H_3C - \underset{CH_3}{\overset{CH_3}{\underset{|}{\overset{|}{C}}}} - CH = CH_2 \text{ is}$$

(A)3,3,3,trimethyl-l-propane (B) 2,2, dimethyl-3-butene
(C) 3,3-dimethyl-1-butene (D) l,l,l-trimethyl-2-propane

3. Kjeldahl's method is used for the estimation of
(A) Halogen (B) phosphorus (C) Nitrogen (D) sulphur

4. Though NH_4Cl contains nitrogen it does not show the Lassaigne's test for nitrogen, because
(A) It is an ionic compound
(B) It is an inorganic compound
(C) This is an exception
(D) The conditions are not favourable

5. H, CH_3, C_2H_3, C_3H_7 show positive inductive effect. Correct order of increasing inductive effect is
(A) H > CH_3 > C_2H_5 > C_3H_7
(B) CH_3 > C_2H_5 > C_3H_7 > H
(C) H > CH_3 > C_3H_7 > C_2H_5
(D) all show inductive effect to the same extent

6. An electron donating group orients the new eletrophile to
(A) para position (B) ortho and paraposition
(C) meta position (D) ortho position

7. Which of the following has the highest boiling point
(A) CH_3COOH (B) CH_3OCH_3
(C) CH_3CN (D) C_2H_5OH

8. Unlike olefines, benzene (though unsaturated) does not give addition reactions easily. This is because
(A) There are no π electrons in olefins
(B) There is no sp^2 hybridisation in olefines
(C) π electrons of benzne are delocaised
(D) none of these

9. The main source of many of the organic compounds is
(A) Coal tar (B) petroleum (C) both (D) none

10. Freon is
(A) CH_2Cl_2 (B) $CHCl_2F$ (C) $CHClF_2$ (D) CCl_2F_2

11. When tin is treated with conc. HNO_3
(A) It is converted into meta-stannic acid
(B) It is rendered passive
(C) It is converted into stannic nitrate
(D) It is converted into stannous nitrate

12. The most electronegative element has the following electronic configuration in the outermost shell
(A) $2s^22p^5$ (B) ns^2np^6 (C) ns^2np^3 (D) ns^2np^4

13. Which form of coal has the maximum percentage of carbon
(A) Anthracite (B) Bituminous (C) Peat (D) Lignite

14. Which one of the following is correct
(A) Boron and Aluminium show allotropy
(B) B_2O_3 alone is acidic while Al_2O_3 is amphoteric
(C) Boron halides are all monomeric while those of Al is dimeric
(D) Boron is the only element of the family which exhibits allotropy

15. Which of the following ions has the greatest coagulating power when added to negative As_2S_3 sol
(A) PO_4^{3-} (B) Al^{3+} (C) Mg^{2+} (D) Na^+

16. Carbon is in the lowest state in
(A) CO_2 (B) CF_4 (C) CCl_4 (D) CH_4

17. Slaking is the process of adding water to
(A) $CaSO_4$ (B) $CaCl_2$ (C) $CaCO_3$ (D) CaO

18. The ore of "boron" is
(A) Borax (B) Bauxite
(C) colemanite (D) both A and C

19. Sodium meta bisuiphite is

(A) $Na_2S_2O_7$ (B) Na_2SO_3
(C) $Na_2S_2O_35H_2O$ (D) $Na_2S_2O_5$

20. In the manufacture of lead pencil .... is mixed with clay
(A) Lamp black (B) Bone black (C) graphite (D) Lead

21. Silicon is chiefly found in
(A) sand (B) Lime stone (C) coal (D) Sea water

22. Among the hydrides of group VI of the periodic table water has the highest boiling point. This is because
(A) water is a polar molecule (B) the O-H bond is weak
(C) there is hydrogen bonding (D) there is hybridisation

23. In the manufacture of HNO_3 from NH_3 by the Ostwald's process ammonia is first oxidised to
(A) nitrogen dioxide (B) nitrogen pentoxide
(C) nitric oxide (D) nitrous oxide

24. The acid character of oxyacid of chlorine is in the order
(A) $HClO > HClO_2 > HClO_3 > HClO_4$
(B) $HClO > HClO_4 > HClO_2 > HClO_3$
(C) $HClO_4 > HClO_3 > HClO_2 > HClO$
(D) $HClO_3 > HClO_4 > HClO_2 > HClO$

25. Bleaching powder is an example of
(A) mixed slat (B) complex salt
(C) double salt (D) none of these

26. Artificial rain is prouced by seeding cloud with
(A) Silver nitrate (B) Copper sulphate
(C) Potassium iodide (D) Silver iodide

27. When SO_2 is passed through an aqueous soluion of iodine, the solution becomes colourless.This is because of
(A) the formation of HI (B) formation of iodic acid
(C) the bleaching action of SO_2
(D) formation of sulphur and sulphuric acid

28. When potassium nitrate is heated, the product produced is
(A) Nitrogen dioxide together with oxygen
(B) Nitrogen dioxide (C) Nitrous oxide (D) Oxygeh

29. The oxidation state of nitrogen in NH_4NO_3 is
(A) +3 and +5 respectively
(B) +5 in both cases (C) -3 in both radical
(D) −3 in NH_4^+ and +5 in NO_3^-

30. When a mixture of ammonium chloride and sodium nitrite is heated, the gas evolved is
(A) NH_3 (B) Nitrous oxide (C) Nitrogen (D) Oxygen

31. The acidic hydride of nitrogen is
(A) N_3H (B) H_2H_4 (C) NH_3 (D) none

32. HI is a powerful reducing agent. It reduces H_2SO_4 to
(A) SO_2, S and H_2S (B) H_2S (C) S (D) SO_2

33. An acid anhydride among the following is
(A) CO_2 (B) PbO_2 (C) N_2O_5 (D) CO

34. If for 'g' sub-shell l = 4 then it will show degeneracy of
(A) 11 (B) 9 (C) 7 (D) 5

35. The total number of elections with l = 2 will be
(A) 6 (B) 10 (C) 3 (D) zero

36. The angular momentum of electrons in the hydrogen atom that can be possible is
(A) 2h (B) h x π (C) h/2π (D) h /4π

37. The total energy of the atomic electron is
(A) May be less or more than zero depending upon the nature of the orbital involved
(B) less than zero (C) more than zero
(D) none of the above

38. The pair of molecules forming the strongest intermolecular hydrogen bonding is
(A) HCOOH and CH_3COOH
(B) H_2O_2 and H_2O (C) SIH_4 and SIF_4
(D) CH_3COCH_3 and $CHCl_3$

39. Among the following, the species which has zero bond order
(A) He_2^+ (B) He_2 (C) H_2^+ (D) H_2

40. The substance which contains both ionic and covalent bonds is
(A) N_2O (B) NaCN (C) Cl_2 (D) CH_4

41. The strength of the sigma bounds formed by s-s and p-p overlap are in the order of
(A) s-p > s-s > p-p (B) s-s > p-p > s-p
(C) s-s > s-p > p-p (D) p-p > s-p > s-s

42. Polarisations is the distortion of the shape of an anion by adjacently placed cation.Which of the following statement is correct ?
(A) A small anion is likely to bring about a large degree of polarisation
(B) A small anion is likely to undergo a high degree of polarisation
(C) Maximum polarisation is brought about by a cation of high charge
(D) Minimum polarisation is brought about by a cation of low radius

43. Plutonium Pu^{239}_{94} undergoes a fission reaction by absorbing a thermal neutron and producing 3 neutrons. One product is Kr^{90}_{36}. The other product is
(A) Ce^{147}_{58} (B) Ce^{150}_{58} (C) Ce^{140}_{58} (D) none of these

44. γ-rays are not deflected by electric and magnetic fields. This is because
(A) They are charged particles to the same extent.

(B) They have very high ionising power
(C) The electric and magnetic fields are not strong
(D) They are not charged

45. Which of the following is non fissionable
(A) U^{239} (B) U^{238} (C) U^{235} (D) none of these

46. The time taken for the disintegration of 75% of a radio active material is 32 minutes. Its half life is
(A) 24 minutes (B) 16 minutes
(C) 8 minutes (D) none of these

47. The volumes of gases involved in chemical reactions are in the ratio of whole numbers. This is known as
(A) Charle's law (B) Gay-lussac's law
(C) Avogadro's law (D) Boyle's law

48. The kinetic theory of gas predicts that the total kinetic energy of the gas depends upon
(A) volume of the gas
(B) P.V and temperature of the gas
(C) pressure of the gas (D) temperature of the gas

49. The rate of diffusion of a gas is half the rate of diffusion of helium. The molecular weight of the gas is
(A) 20 (B) 4 (C) 12 (D) 16

50. r.m.s velocity is always
(A) sometimes more and sometimes less
(B) equal to most probable velocity
(C) more than the average velocity
(D) less than the average velocity

51. At what temperature the R.M.S velocity of chlorine gas will be equal to that of SO_2 at S.T.P
(A) 273°c (B) 302.9°c (C) 29.9°c (D) 0°c

52. If the molecules of a gas were point masses i.e. the molecules occupied no volume, the mean free path would be
(A) 10^{16}cm (B) 10^4 cm (C) 10^8 cm (D) infinite

53. Which of the following is an example of a solid-solid system ?
(A) synthetic gems (B) cake
(C) pumice stone (D) smoke

54. A molal solution is one that contains one mole of a solute in
(A) 1000 gm of the solvent (B) one litre of the solvent
(C) one litre of the solution (D) 22.4 litres of the solution

55. Ideal solution is formed when its components
(A) have both zero heat mixing and zero volume change
(B) can-be converted into ideal gases
(C) have only zero heat of mixing
(D) have only zero volume change

56. The azeotropic mixture of water (b.p 100°c) and HCl (b. p. 85°c) boils at 108.5°c. When this mixture is distilled it is possible to obtain
(A) pure water as well as pure HCl
(B) neither HCl nor H_2O in their pure states
(C) pure hydrogen chloride (D) pure water

57. The most stable oxidation state of iron is
(A) +5 (B) +3 (C) +2 (D) 0

58. Purple colour is imparted to glass by adding
(A) $AuCl_3$ (B) Cobalt oxide (C) MnO_2 (D) Cu_2O

59. Platinum, palladium, Indium, etc are called noble metals, because
(A) They are shining, lustrous and pleasing to look at
(B) They are found in native state
(C) Alfred Nobel discovered them
(D) They are inert towards many common reagents

60. Most stable oxidation state of gold is
(A) +5 (B) +3 (C) +2 (D) +1

61. Which one of the following is not coloured
(A) NI^{2+} (B) Fe^{3+} (C) CU^{2+} (D) Cu^+

62. Hydrogen gas will not reduce
(A) heated stannic oxide (B) heated aluminium oxide
(C) heated cupric oxide (D) heated ferric oxide

63. Extraction of silver from commercial lead is possible by
(A) Haber's process (B) Clarke's process
(C) Mond's process (D) Parke's process

64. In the metallurgy of.iron, lime stone is added to the blast furnace. The calcium ions are used up forming
(A) metallic calcium (B) calcium carbonate
(C) slag (D) gangue

65. In the manufacture of steel by the open hearth process if the cast iron contains manganese impurity but no phosphorus, the furnace is lined with
(A) lime (B) silica (C) limestone (D) magnesia

66. When $BaCl_2$ and Na_2SO_3 are mixed in presence of bromine water, the precipitate formed is
(A) BaS_2O_3 (B) BaS (C) $BaSO_4$ (D) $BaSO_3$

67. Which of the following combines with Fe(ll) ions to form a brown complex
(A) N_2O_3 (B) NO_2 (C) N_2O (D) NO

68. When barium chloride is added to sodium sulphite solution, a white precipitate is formed
(A) It may or may not be soluble in hydro chloric acid
(B) It is insoluble in hydrochloric acid
(C) It is soluble in hydrochloric acid
(D) none of the above

69. Which one can be used in place of NH_4Cl for the idenitification of group III radicals
(A) $(NH_4)_2CO_3$ (B) NaCl

(C) NH_4NO_3 (D) $(NH_4)_2C_2O_4$

70. When dilute HCl is added to a solution of a thiosulphate, turbidity is observed. This is due to
(A) formation of chloride (B) formation of sulphate
(C) evolution of SO_2 (D) formation of sulphur

71. $CH_3CH_2I \xrightarrow{KCN} A \xrightarrow[\text{partial}]{H^+, H_2O} B \xrightarrow{Br_2/KOH} C$. The product C is
(A) $C_3H_7NH_2$ (B) C_2H_5COOH
(C) $C_2H_5NH_2$ (D) $C_2H_2CONH_2$

72. Vegetable oils are
(A) Unsaturated fatty acids
(B) glycerides of saturated fatty acids
(C) Essential oils obtained from plants
(D) glycerides of unsaturated fatty acids

73. The following reaction is known as
$C_2H_5ONa + C_2H_5Br \rightarrow C_2H_5 - O - C_2H_5 + NaBr$
(A) Williamson's synthesis (B) Gabriel synthesis
(C) Kolbe's synthesis (D) Wurtz synthesis

74. Which compound is obtained when mixture of calcium acetate and calcium formate is subjected to dry distillation
(A) acetone (B) Acetic acid
(C) acetaldehyde (D) formaldehyde

75. Which one of the following reduces Tollen's reagent
(A) cold citric acid (B) oxalic acid
(C) Acetic acid (D) Tartaric acid

76. Salol, used as an intestinal antiseptic is
(A) methyl salicylate (B) methyl salicylic acid
(C) phenyl salicylate (D) acetyl salicyclic acid

77. Which one of the following is a better acetylating agent
(A) Acetyl chloride (B) Benzoyl chloride
(C) Acetic anhydride (D) Acetamide

78. Propanamide yields propanoic acid with the evolution of N_2 by reacting with
(A) NaOH + Br_2 (B) $LiAlH_4$ (C) HNO_2 (D) NaOH

79. When CH_3CONH_2 is heated with PCl_5, the product is
(A) CCl_3CONH_2 (B) CH_3COCl
(C) CH_3CN (D) $CH_3CCl_2NH_2$

80. The reagent commonly used for converting toulene into benzene is
(A) Chlorine water (B) Conc.H_2SO_4
(C) Alkaline $KMnO_4$ (D) Fenton's reagent

81. The building block of a crystal is known as
(A) lattice points (B) lattice plane
(C) unit cell (D) none of these

82. In molecular crystals the force which holds them together is
(A) strong coulombic force of attraction
(B) van der Waal's force (C) London forces
(D) none of these

83. For the reaction
$Cu(s) + 2Ag^+(aq) \rightleftharpoons Cu^{2+} + 2Ag(g)$ The equilibrium constant KCl is given by

(A) $\frac{[Cu^{2+}][Ag]^2}{[Cu][Ag^+]^2}$ (B) $\frac{[Ag]^2}{[Ag^+]^2}$ (C) $\frac{[Cu^{2+}]}{[Ag^+]^2}$ (D) $\frac{[Cu^{2+}]}{[Cu]}$

84. How must the H^+ ion concentration in a solution be changed for pH of solution to decrease by unity
(A) It must be diminished by one tenth of its orignial volume
(B) It must be increased by 10 times
(C) It must be decreased by 10 times
(D) It must be increased by 1 mole/litre

85. If equal volumes of two solutions of pH 2 and 3 are mixed, then the pH of the mixtrue will be
(A) 5 (B) 10 (C) 2.2596 (D) 2.5

86. The conjugate base of the acid 'H_2S' is
(A) S (B) S^{2-} (C) HS^- (D) none of these

87. The dissociation constants of a weak acid and a weak base constituting the salt are equal. The pH of the solution ofthe salt is
(A) greater than seven (B) less than seven
(C) seven (D) zero

88. pH of 10^{-8} solution of HCl in water is
(A) between 7 and 8 (B) between 6 and 7
(C) 8 (D) – 8

89. The precipitate of CaF_2 ($Ksp = 1.7 x 10^{-10}$) is obtained when equal volumes of the followin are mixed.
(A) 10^{-5} M Ca^{2+} +10^{-3} MF^{-1} (B) 10^{-3} Ca^{2+} + 10^{-5} MF^-
(C) 10^{-14} M Ca^{2+} + 10^{-4} MF^{-1}(D) 10^{-2} M ca $^{2+}$ +10^{-3}MF

90. Which of the following will favour the formation of hydrogen iodide
$H_2(g) + I_2(g) \rightleftharpoons 2HI(g)$; $\Delta H_f = 25.9$ KJ/mol
(A) pressure is decreased (B) temperature is increased
(C) volume is increased (D) pressure is incresed

91. For a certain reaction, the value of equilibrium constant is found to decrease at higher temperatures. This shows that
(A) the reaction is exothermic
(B) the reaction is endothermic
(C) the equilibrium concentration are not altered
(D) none of these

92. A solution of NaCl in contact with atmosphere has a pH of about
(A) 7 (B) 14 (C) 3.5 (D) 5

93. Hydrolysis constant of a salt of a weak acid and weak base is inversely proportional to

(A) Dissociation constant of weak base
(B) Dissociation constant ot both weak acid and weak base
(C) Dissociation constant of weak acid
(D) ionic product of water

94. Acid chloride reacts with alcohol to form ester. This is carried out in presence of a base like pyridine.This is to
(A) remove the HCI formed during the reaction
(B) slow down the reaction
(C) remove the excess alcohol (D) none of these

95. Purification of sodium chloride by passage of hydrogen chloride through brine is based on
(A) Le-Chateller's principle (B) Displacement law
(C) common ion effect (D) distribution effect

96. The equilibrium constant for the reaction $2SO_2 + O_2 \rightleftharpoons 2SO_3$ at 1000 k is 3.5 atm. What would be the partial pressure of oxygen gas if equal volumes of SO_2 and SO_3 are to be formed at equilibrium
(A) 1.75 atm (B) 0.29 atm (C) 3.5 atm (D) 1.87 atm

97. Ionic product of water at 298 K
(A) remains constant (B) can not be predicted
(C) Increases with temperature
(D) Decreases with temperature

98. Calcium carbonate dissociates in the following manner $CaCO_3(s) \rightleftharpoons CaO(s) + CO_2(g)$ $\Delta H = 110$ KJ in a closed vessel. The pressure of CO_2
(A) decreases if temperature is raised
(B) increases if an inert gas is pumped keeping the temperature constant
(C) increases if temperature is raised
(D) Increases on adding a catalyst

99. From the following data, the heat of formation of $C_2H_5OH(l)$ is ....Kcal.
$C_2H_5OH(l) + 3O_2(g) \rightarrow 2CO_2 + 3H_2O$
$\Delta H = -326.7$Kcal
$3H_2 + 3/2O_2 \rightarrow 3H_2O$ $\Delta H = -205.1$ Kcal
$2CO_2 \rightarrow 2C(gr) + 2O(g) + 2O_2(g)$ $\Delta H = 188.4$ Kcal
(A) –343.7 Kcal (B) – 66.5 Kcal
(C) 719.9 Kcal (D) None of these

100. Joule is a unit of
(A) potential difference (B) charge (C) current (D) energy

101. Activation energy and threshold energy are related as
(A) activation energy = Threshold energy + Energy actually possesed by particles
(B) Threshold energy = Activation energy + energy barrier
(C) Activation energy = Threshold energy– Energy actually possessed by paricles
(D) none of these

102. The heat of formation of $H_2O(l)$ is 68.0 k.Cal. The heat of formation $H_2O(g)$ is likely to be
(a) –68.0 k Cal (b) –69.4 k Cal
(c) 80.0 k Cal (D) –58.3 k Cal

103. Which of the following is true concerning first order reaction
(A) The rate is proportional to temperature
(B) The half life is independent of the initial concentration
(C) the rate falls to zero after two half-lives
(D) The rate is independent of initial concentration

104. The pH value of the following half cell is........ PtH_2/H_2SO_4. The electrode potential is –0.3V
(A) 6 (B) 4 (C) 3 (D) 5.08

105. One mole of N_2H_4 loses 10 moles of electrons to form a new compound y. Assuming that all nitrogen appear in the new compound, what is the oxidation state of nitrogen in y. There is no change in the oxidation state of hydrogen
(A) +5 (B) –1 (C) –3 (D) +3

106. Rusting of iron in moist air involves
(A) neither loss nor gain of electron
(B) hydration of iron (C) loss of electron by iron
(D) gain of electrons by iron

107. Greater the standard reduction potential of an element
(A) weaker it is as an oxidising agent
(B) neither -weak nor strong as an oxidising agent
(C) stronger it is as an oxidising agent
(D) there is no such tendency

108. Which one is wrong. The oxidation number of an atom in a species can be
(A) cannot be fractional (B) positive
(C) negative (D) zero

109. The E.M.F of a Daniel Cell, Zn/Zn^{2+} // Cu^{2+} / Cu can be increases by
(A) Increasing the surface area of the electrodes
(B) decreasing the surface area of the electrode
(C) increasing the concentration of zinc sulphate solution
(D) Increasing the concentration of copper sulphate solution

110. The variation of electrode potential with concentration of metallic ion in solution is given by
(A) Nernst equation (B) Kohlrausch relationship
(C) Arrhemus equation (D) van derWaal's equation

111. Higher the charge on the precipitating ion, the greater is the precipitating power. This rule is known as
(A) Hess's law (B) Raoult's law
(C) Kohlrausch's law (D) Hardy-Schulz rule

112. The cleaning action of soap is because of
(A) the alkanity of the soap (B) ionisation in water
(C) the solubility in water

(D) the formation of an emulsion of dirt and grease with water in presence of soap

113. Which of the following is not a colloid
(A) bread (B) blood (C) juice (D) milk

114. Which of the following is not correct
(A) Temperature affects enzyme activity
(B) pH of the solution does not affect enzyme activity
(C) Enzymes are proteins produced by living organisms
(D) They catalyse specific reactions

115. Glycine can be obtained from formaldehyde by
(A) Willainson's synthesis (B) The diazo reaction
(C) Streker synthesis (D) phthalimide synthesis

116. Which one of the following will induce sleep and unconsciousness if taken in large quantities
(A) psychedelic drugs (B) Hypnotics
(C) Antipyretics (D) Narcotics

117. When H_2S is passed through ferric chloride solution
(A) Ferric chloride is reduced to ferrous chloride and a precipitate of sulphur is formed
(B) a precipitate of ferrous sulphide is formed
(C) a precipitate of ferric sulphie is formed
(D) none of the above

118. A metal ion M^{3+} after loss of three electrons in a reaction, will have an oxidation number equal to
(A) +3 (B) +4 (C) +2 (D) 0

119. When the pressure gradient and temperatare are held constant the rates of diffusion of CO and N_2 will be in the ratio
(A) 2:1 (B) 1:4 (C) 1:1 (D) 1:2

120. How many σ and π bonds are there in the molecule of tetracyano-ethylene$(CN)_2C = C(CN)_2$
(A) Nine σ and sevens π (B) Five σ and eight π
(C) Nine σ and nine π (D) Five σ and nine π

1999

1. Which of the following is an efflorescent material ?
(A) copper nitrate (B) potassium sulphate
(C) washing soda (D) calcium chloride

2. Hydration energy of Mg^{2+} ion is larger than that of
(A) Na^+ (B) Cr^{3+} (C) Al^{3+} (D) Be^{2+}

3. Alkali metal has the lowest
(A) Electron affinity (B) Electronegativity
(C) lonization energy (D) all the above

4. Which of the following has got incompletely filled f-subshell ?
(A) Lawrencium (B) Tantalum
(C) Gadolinium (D) Lutetium

5. A mixture of chromic oxide and Ferric oxide is used as a catalyst in the manufacture of
(A) Hydrogen (B) H_2O_2 (C) ammonia (D) SO_2

6. Which of the followine oxides will react with both Hydrochloric acid and sodium hyroxide ?
(A) ZnO (B) N_2O_3 (C) Na_2O (D) CaO

7. Which is the product formed when sulphur dioxide is passed through a solution of washing soda ?
(A) $NaHSO_3$ (B) $NaHSO_4$ (C) Na_2SO_3 (D) Na_2SO_4

8. The element with the highest first ionization energy among the folowing is
(A) Nitrogen (B) Carbon (C) Boron (D) Beryllium

9. What volume of Oxygen at STP will be obtained by the decomposition of one litre 100 volume H_2O_2 ?
(A) 1 litre (B) 10 litre (C) 100 litre (D) 200 litre

10. The ion that is isoelectronic with carbon monoxide molecule is
(A) O (B) N_2^+ (C) CN^- (D) O_2^-

11. Types of bonds preseat in N_2O_5 are
(A) ionic and coordinate (B) only convalent
(C) ionic and covalent (D)covalent and coordinate

12. A dibasic acid of phosphorus
(A) metaphosphoric acid (B) pyrophosphoric acid
(C) phosphorus acid (D) phosphoric acid

13. Which one of the following noble gases is least polarizable ?
(A) Ar (B) Kr (C) He (D) Ne

14. The formula of hypochlorous acid is
(A) $HClO_3$ (B) $HClO_4$ (C) HClO (D) $HClO_2$

15. Oxygen has got positive oxidation state in
(A) $H_2S_2O_2$ (B) $HBrO_2$ (C) OF_2 (D) I_2O_5

16. Oxidising action decreases in the order
(A) Cl > Br > l > F (B) l > F > Cl >Br
(C) F > Cl > Br > l (D) l > Br > Cl > F

17. Which of the following can have positive and negative oxidation states in its compounds ?
(A) Iodine (B) Aluminium (C) Sodium (D) Fluorine

18. The Bohr model of atom could successfully explain
(A) the spectra of atoms containing more thae one electron
(B) the spectra of hydrogen and hydrogen like atoms.
(C) the spectra of all atoms
(D) the spectra of hydrogen atoms

19. Solar spectrum is an example for
(A) continuous spectrum (B) band spectrum
(C) absorption spectrum (D) atomic spectrum

20. Total number of electrons in one molecule of

carbondioxide is
(A) 24 (B) 66 (C) 44 (D) 22

21. The value of n = 4. Then that electronic shell in the atom contain——-orbitals
(A) 4 (B) 32 (C) 16 (d) 8

22. Which series lies in the visible region
(A) Brackett (B) Paschen (C) Lyman (D) Balmer

23. Which one of the following has Hie highest boiling point ?
(A) CH_3CH_2OH (B) CH_3OCH_3 (C) H_2O (D) HBr

24. Which one of the following has the highest dipole moment ?
(A) Ethylalcohol (B) Hydrogen chloride
(C) Hydrogen fluoride (D) water

25.Bond order of the species O_2, O_2^+, O_2^{2+} and O_2^{2-} increases in the order
(A) $O_2^{2+} < O_2^+ < O_2 < O_2^{2-}$
(B) $O_2^+ < O_2^{2+} < O_2 < O_2^{2-}$
(C) $O^{2-} < O_2 < O_2^+ < O_2^{2+}$
(D) $O_2 < O_2^+ < O_2^{2+} < O_2^{2-}$

26. The energy equivalent of 2.0 mg mass defect is
(A) $1.5 x 10^{20}$ ergs (B) $1.8 x 10^{18}$ ergs
(C) $1.8 x 10^{4}$ ergs. (D) $9 x 10^{-19}$ ergs

27. The starting element in the actinium series is
(A) U^{238}_{92} (B) Pa^{231}_{91} (C) U^{235}_{92} (D) AC^{227}_{89}

28. The average velocity of an ideal gas molecule at 27°C is 0.3 ms^{-1}. The average velocity at 927°C will be
(A) 3.0 $m.s^{-1}$ (B) 0.9 $m.s^{-1}$ (C) 0.3 $m.s^{-1}$ (D) 0.6 ms^{-1}

29. A gas mixture contains equal masses of hydrogen and helium at a pressure of 3 atmospheres. What is the partial pressure of hydrogen ?
(A) 2.25 atm (B) 2 atm (C) 1.5 atm (D) 1 atm

30. Freezing point of a solvent was lowered by 3.72°c when 60g urea (NH_2-CO-NH_2) was dissolved in 1000g of solvent, cryoscopic constant of the solvent will be
(A) 37.2° (B) 7.44° (C) 3.72° (D) 0.372°

31. One molar solution of which of the following substances will show maximum osmotic pressure ?
(A) $MgCl_2$ (B) $(NH_4)_3 PO_4$ (C) $AgNO_3$ (D) NH_4Cl

32. 600 gms of a compound on analysis gave C = 24g. H = 4g and O = 32g. what is its empirical formula ?
(A) CH_2O (B) CH_2O_2 (C) $C_2H_2O_2$ (D) $C_2H_4O_2$

33. In which of the following chromatographic methods, the mobile and stationary phases are liquids ?
(A) paper chromatography (B) gas liquid chromatography
(C) column chromtography
(D) thin layer chromatrography

34. Insulin contains about 0.08% by weight of sulphur. What would be the approximate molecular mass of insulin if one sulphur atom is present in a molecule of insulin ?
(A) 2560 (B) 25,600 (C) 40,000 (D) 20,000

35. The active species formed from nitrating mixture is
(A) NO_2^+ (B) NO_3^+ (C) NO_2 (D) NO^+

36. The most stable ion among the following is
(A) $(CH_3)_3C^+$ (B) $C_6H_5CH_2^+$
(C) $CH_3CH_2^+$ (D) $(CH_3)_2CH^+$

37. Greatest acidic character is exhibited by
(A) sulphonic acid (B) carboxylic acid
(C) phenol (D) alcohol

38. Which of the following shows geometrical isomerism?
(A) $CH_2 = C Cl_2$ (B) CHCl = CHCl
(C) $CH_3CH_2CH = CH_2$ (D) $(CH_3)_2C = CH_2$

39. Which among the following will undergo nucleophilic substitution reaction ?
(A) Haloalkanes (B) Alcohol
(C) Aldehydes (D) Arenes

40. Only one monochloro derivative is possible for
(A) neopentane (B) quinol
(C) para xylene (D) all of these

41. Monocarboxylic acids can be functional group isomers of
(A) esters (B) dihydric, alcohols
(C) aldehydes (D) none of these

42. Metamerism can be exhibited by
(A) secondary amines (B) ketones
(C) ethers (D) All of these

43. The chief constituent of Gobar gas is
(A) Hydrogen sulphidel (B) Carbon monoxide
(C) Propane (D) Methane

44. Clemmensen's reduction can be used to prepare
(A) alcohol (B) alkane (C) aldehyde (D) carboxylicacid

45. Markownikoff's rule is applicable for
(A) additions of halogen to unsymmetrical alkenes
(B) addition of hydrogen halide to unsymmetrical alkenes
(C) dehydrohalogenation of alkylhalides
(D) none of the above

46. One of the reagents used for the preparation of DDT is
(A) hydrochloric acid (B) chlorobenzene
(C) ethanol (D) dichlorobenzene

47. Which of the following statements is correct ?
(A) C_2H_5Br on treatment with sodium ethoxide forms $C_2H_5OC_2H_5$
(B) C_2H_5Br reacts with AgCN to form ethyl cyanide
(C) C_2H_5Br reacts with alcoholic KOH to form C_2H_5OH

(D) C_2H_5Br on treatment with metallic sodium gives ethane

48. A compound containing elements X and Y crystillises in the cubic pattern with 'X' atoms at the comers and 'Y' atoms at the face centres. What is the simplest formulae of the compound ?
(A) XY_2 (B) X_3Y_4 (C) X_2Y (D) XY_3

49. The metal which finds exclusive application as semic conductor is
(A) Ga (B) As (C) Au (D) Ge

50. The defect which arises when equal number of cations and anions miss from the crystal lattice
(A) Electronic defect (B) Schottky defect
(C) stoichiometric defect (D) Frenkel defect

51. pH of a 0.10 M Nitric acid is
(A) 0.1 (B) -1 (C) 2 (D) - log $[1x10^{-1}]$

52. The heat evolved when 500 cc of 0.1 M sulphuric acid is mixed with 600 cc of 0.05 M sodium hydroxide solution
(A) 1.71 KJ (B) 1.903 KJ (C) 5.71 KJ (D) 2.85

53. In the reaction $CaCO_3(g) \rightleftharpoons CaO(g) + CO_2(g)$ the equilibrium constant depends upon the concentrations of
(A) $CaCO_3$, CaO and CO_2 (B) CO_2
(C) CaO and CO_2 (D) $CaCO_3$

54. One mole of HI was heated to 400°C so as to attain equilibrium. The equilibrium mixture contained 0.2 mole iodine. What will be the equilibrium constant ?
(A) 0.0625 (B) 0.1111 (C) 16 (D) 9

55. K for the reaction
$2NO(g) + Cl_2(g) \rightleftharpoons 2NOCl(g)$ was found to be $1.9x10^3$ atm^{-1} at 25°c. What is the value of K_c for this reaction ?
(A) $1.9x10^3$ l mol^{-1} (B) $4.6x10^4$ $mol.l^{-1}$
(C) $1.9x10^3$ $mol.l^{-1}$ (D) $4.6x10^4$ l. mol^{-1}

56. K_a value for 1M acetic acid is $1.8x10^{-5}$. What percentage of acetic acid is dissociated in this solution ?
(A) 8.4% (B) 0.42% (C) 4.2% (D) 42%

57. What will be the heat of formation of methane if the heat of combusion of carbon is '–x' kJ. Heat of formation of water is '–y' kJ and heat of combustion of methane is 'z' kJ ?
(A) (–x–2y–z) kJ (B) (z–x–2y) kJ
(C) (–x–y +z) kJ (D) (–z–x–2y)kJ

58. For a reaction to be spontaneous at all temperatures
(A) ΔG-ve, ΔH-ve & ΔS-ve
(B) ΔG -ve, ΔH -ve & ΔS +ve
(C) ΔG -ve, ΔH +ve & ΔS +ve
(D) ΔG +ve, ΔH -ve & ΔS +ve

59. Calculate the work done when a gas expands from 40 litres to 60 litres at a constant pressure of 15 atmospheres
(A) -600 lit. atm (B) 600 lit. atm
(C) 300 lit. atm (D) -240 lit. atm

60. An endothermic compoud among the following
(A) HI (B) SO_3 (C) CS_2 (D) CO_2

61. ΔG is zero in a reaction. It is
(A) in a state of equilibrium (B) spontaneous
(C) exothermic (D) reversible

62. A compound that can act both as oxidising agent and reducing agent is
(A) H_2S (B) H_2O_2 (C) $KMnO_4$ (D) $H_2C_2O_4$

63. Which of the following reactions is an example for redox reactions ?
(A) $Ca(HCO_3) + Ca(OH)_2 \rightarrow 2CaCO_3 + 2H_2O$
(B) $10HNO_3 + I_2 \rightarrow 2HIO_3 + 10NO_2 + 4H_2O$
(C) $AgNO_3 + NaCl \rightarrow AgCl + NaNO_3$
(D) $Na_2CO_3 + SiO_2 \rightarrow Na_2SiO_3 + CO_2$

64. Oxidation number of sulphur in sulphur monochloride is
(A) zero (B) one (C) two (D) three

65. Iron can displace copper from its salt solution. This means that
(A) activity co-efficient of copper is greater than that of iron
(B) copper and iron are metals
(C) Iron has greater reduction electrode potential than copper
(D) Copper has greater reduction electrode potential than iron

66. Which of the following compounds is coloured ?
(A) $VOSO_4$ (B) $CoCl_2$ (C) $TiCl_3$ (D) all are coloured

67. Which of the following shows highest magnetic moment ?
(A) Co^{3+} (B) Fe^{3+} (C) V^{3+} (D) Cr^{3+}

68. A chemical used for etching copper and silver is
(A) $FeCl_3$ (B) $ZnSO_4$ (C) $AgNO_3$ (D) $CuSO_4$

69. Which of the following has maximum number of unpaired electrons ?
(A) Cobalt (B) Iron (C) Chromium (D) Manganese

70. Calomel is a compound having chemical formula
(A) $HgSO_4$ (B) Hg_2Cl_2 (C) $HgCl_2$ (D) HgI_2

71. An alloy which does not contain copper is
(A) Brass (B) Bell metal (C) Solder (D) Bronze

72. Which of the following metal ions will give a rose-red precipitate with dimethyl glyoxime ?
(A) Ni^{2+} (B) Co^{2+} (C) Mg^{2+} (D) Ca^{2+}

73. What is the molar concentration of a solution obtained by adding 50 ml water in 20 ml 2.5M HNO_3

(A) 0.5M (B) 1.25M (C) 1M (D) 0.71M

74. Diethylether on heating with concentrated hydroiodic acid gives
(A) methyliodide (B) ethyl iodide
(C) ethanol (D) iodoform

75. A mixture of naphthalene and benzoic acid can be separated using
(A) ether (B) benzene (C) alcohol (D) water

76. Formula of aspirin
(A) [benzene ring with $OCOCH_3$ and $COOH$ in ortho positions] (B) [benzene ring with OH and COC_6H_5 in ortho positions]
(C) [benzene ring with OH and $NHCOCH_3$ in para positions] (D) [benzene ring with OC_2H_5 and $NHCOCH_3$ in para positions]

77. Which of the following statements is not true for the compound $CH_3CH = CHCOOH$?
(A) It exhibits geometrical isomerism
(B) It will not give effervescence with sodium bicarbonate
(C) It will decolourise bromine water
(D) It will undergo ozonolysis

78. Which of the following yields a positively charged sol ?
(A) starch (B) Gold (C) $Fe(OH)_3$ (D) As_2S_3

79. Colloidal system made up of two liquids is called
(A) gel (B) foam (C) sol (D) emulsion

80. Action of nitrous acid on ethyl amine could give
(A) C_2H_5OH (B) $C_2H_5NO_2$ (C) C_2H_6 (D) NH_3

81. Which of the following is a polymer containing nitrogen ?
(A) Nylon (B) Terylene
(C) Polyvinyl chloride (D) Bakelite

82. Which of the following is a heterocyclic aminoacid ?
(A) phenylalanine (B) tryptophan (C) glycine (D) alanine

83. If the concentration of CrO_4^{2-} ions in a saturated solution of silver chromate is 2 x 10^{-4} M, what is the solubility product of Silver Chromate ?
(A) $16x10^{-12}$ (B) $32x10^{-12}$ (C) $4x10^{-4}$ (D) $8x10^{-12}$

2000

1. Which of the following molecules can act as an oxidizing as well as a reducing agent ? (Atomic Weights C = 12, H = 1, O = 16)
(A) H_2O_2 (B) F_2 (C) H_2SO_4 (D) H_2S (E) SO_3

2. The three-dimensional lattice of zeolites consist of
(A) $[AlSi_3O_8]^-$ (B) only SiO_2
(C) $[Si_2O_5]_n^{2n-}$ (D) $[SiO_3]_n^{2n-}$ (E) $[Si_2O_7]_6$

3. Producer gas is a mixture of
(A) $CO + N_2$ (B) $CO_2 + H_2$
(C) $CO + H_2 + O_2$ (D) $CO + H_2$ (E) $H_2 + CO_2$

4. The element with electronic configuiationof its atom $1s^2 2s^2 2p^6 3s^2 3p^6 3d^{10} 4s^1$ is
(A) Ni (B) Zn (C) Cu (D) Fe (E) Co

5. Which one of the following is triple superphosphate ?
(A) $CaHPO_4 . 2H_2O$ (B) $Ca(H_2PO_4)_2$
(C) $Ca_3(PO_4)_2$ (D) $Ca(H_2PO_4)_2 + CaSO_4$
(E) $[Ca_3(PO_4)_2 . CaF_2]$

6. Which one of the following elements is most reactive ?
(A) Xe (B) Kr (C) Ar (D) Ne (E) He

7. The lowest boiling point of helium is due to its
(A) weak van der Waals forces between atom
(B) small size (C) inertness (D) gaseous nature
(E) high polarisability

8. According to Bohr's theory the energy required for the transition of H atom from $n_1 = 6$ to $n_2 = 8$ state is
(A) equal to the energy required for the transitioni from $n_1 = 7$ to $n_2 = 9$ state
(B) equal to the energy required for the tranisiton from $n_1 = 5$ to $n_2 = 7$ state
(C) larger than in A (D) less than in A (E) less than D

9. The shape of gaseous $SnCl_2$ is
(A) T-shaped (B) distorted tetrahedral
(C) tetrahedral (D) linear (E) angular

10. Which of the following molecules will have unequal bond lengths ?
(A) SiF_4 (B) SF_6 (C) PF_5 (D) BF_3 (E) NF_3

11. Which of the following substances has the least ionic character ?
(A) $CdCl_2$ (B) $MgCl_2$
(C) $HgCl_2$ (D) $FeCl_2$ (E) $ZnCl_2$

12. The dimensions of pressure is
(A) $ML^{-2}T^{-2}$ (B) ML^{-2}
(C) MLT^{-2} (D) $ML^{-1}T^{-2}$ (E) ML^2T^{-2}

13. The dimensions of viscosity coefficient are
(A) MLT (B) MLT^{-2}
(C) $ML^{-1}T^{-1}$ (D) MLT^{-1} (E) $ML^{-1}T$

14. Complete combustion of 0.858 g. of compound X gives 2.63 g. of CO_2 and 1.28 g of H_2O. The lowest molecular weight which X can have, is
(A) 22 g (B) 172 g (C) 129 g (D) 86 g (E) 43 g

15. Geometrical isomerism is possible in case of
(A) propene (B) 2-butene (C) pentene
(D) 2-butyne (E) 1-butene

16. Which of the following will exhibit highest boiling point?
(A) $CH_3CH_2CH_2CH(CH_3)OH$
(B) $CH_3CH_2C(CH_3)_2OH$ (C) $CH_3C(CH_3)_2CH_2OH$
(D) $CH_3CH_2OCH_2CH_3$ (E) $CH_3CH_2CH_2CH_2CH_2OH$

17. How many isomers can C_5H_{12} have ?
(A) 1 (B) 5 (C) 3 (D) 4 (E) 2

18. Which amino acid is achiral ?
(A) valine (B) proline (C) alanine
(D) Histidine (E) none of the above

19. Reduction of carbonyl compounds with hydrazine in presence of strong base is called
(A) Beckmann rearrangement
(B) Meerwein-Pondorf reduction
(C) Wolf-kishner reduction
(D) Clemmensen's reduction (E) Cannizaro's reaction

20. An alkane C_7H_{16} is produced by the reaction of lithium di (3-pentyl) cuprate with ethyl bromide. The structural formula of the product is
(A) 3-methyl hexane (B) 2 - methyl hexane (C) heptane
(D) 3-ethyl pentane (E) 2-ethyl pentane

21. When propyne is treated with dilute sulphuric acid in presence of mercury (II) sulphate, the major product is
(A) 2-propanol (B) propanal (C) propanol
(D) propene (D) acetone

22. An alkene C_4H_8, was treated with ozone and then with zinc and water to give acetone and formaldehyde. The alkene is
(A) 2-methy-2-butene (B) 2-methyl propene
(C) 2-methyl butene (D) Propanone (E) 1-butene

23. The product formed in the following reaction is $CH_3CH(CH_3)CH = CH_2 + HBr \rightarrow$ product
(A) $(CH_3)_2C(Br)CH_2CH_3$
(B) $CH_3CH(CH_3)CH(Br)CH_2CH_3$
(C) $CH_3CH(CH_3)CH(Br)CH_2(Br)$
(D) $(CH_3)_2CHCH(Br)CH_3$ (E) $(CH_3)_2CHCH_2CH_2Br$

24. n-Butyl benzene on oxidation will give
(A) benzyl alcohol (B) benzaldehyde
(C) 4-phenyl butanoic acid (D) benzoic acid
(E) butanoic acid

25. Transition metals usually exhibit highest oxidation states in their
(A) iodides (B) hydrides
(C) chlorides (D) fluorides (E) bromides

26. The number of isomers exhibited by $[Cr(NH_3)_3Cl_3]$ is
(A) 6 (B) 5 (C) 4 (D) 3 (E) 2

27. For the square planar complex [M(a) (b) (c) (d)] where M= central metal a, b, c and d are monodentate ligands) the number of possible geometrical isomers are
(A) 6 (B) 4 (C) 3 (D) 2 (E)

28. Of the following species, one which can form transition metal organometallics is
(A) RNC (B) CO_3^{2-} (C) CN (D) NO (D) CO_2

29. Among lanthanide ions, the most stable +2 oxidation state is exhibited by
(A) Gd (B) Nb (C) Sm (D) Eu (E) Ce

30. Nickel is purfied by thermal decomposition of its
(A) carbonyl (B) iodide
(C) hydride (D) chloride (E) azide

31. The most stable +2 oxidation state is exhibited by
(A) Pb (B) Si (C) Os (D) Fe (E) Sn

32. The product of following reacton is

(pent-1-ene skeletal structure) $\xrightarrow{BH_3/THF;\ 2H_2O_2,\ OH^-}$

(A) pentane (B) 1,2 - pentan-di-ol (C) pent-2-one
(D) pentanol (E) 2-pentanol

33. Glycerol is
(A) 2,3-dihydroxy propane
(B) 1,2,3,4-tetrahydroxy butane (C) glyceraldyde
(D) 1,3-dihydroxy propane (E) 2,3-dihydroxy propanol

34. Propanal on reaction with dilute sodium hydroxide forms
(A) $CH_3CH_2CH_2CH(OH)CH_2CHO$
(B) $CH_3CH_2CH(OH)CH(CH_3)CHO$
(C) CH_3CH_2COOH (D) $CH_3CH_2CH_2CH_2CH_2CHO$
(E) $CH_3CH_2CH(OH)CH_2CH_2CHO$

35. Which one of the following will be most basic
(A) p-Nitroaniline (B) p-Methylaniline
(C) Benzylamine (D) Aniline (b) p-Methoxyaniline

36. Gabriel synthesis is used for the synthesis of
(A) aldehydes (B) acids (C) tertiary amines
(D) primary amines (E) secondary amines

37. What kind of bonding is responsible for the secondary structure of proteins ?
(A) ionic bonding (B) van der Waal forces
(C) amino acid sequence (D) covalent bonding
(E) hydrogen bonding

38. Which of the following compound is found most abundantly in nature ?
(A) Starch (B) Cellulose (C) Maltose
(D) Fructose (E) Glucose

39. Sanger's reagent is used for the idenitification of
(A) side chain of the amino acids
(B) molecular weight of the peptide chain
(C) number of amino acids in peptide chain
(D) N- terminal of peptide chain
(E) C-terminal of a peptide chain

40. What. structural feature distinguishes proline from other natural α–aminoacids ?
(A) it is a dicarboxylic acid
(B) it has a secondary amine group
(C) it contains two amino groups
(D) it is optically inacitve (E) it contains aromatic group

41. Soft drink and baby feeding bottles are generaly made up of
(A) polyamide (B) polystyrene
(C).polyester (D) polyurethane (E) polyurea

42. The beta and alpha glucose have different specific rotations. When either is dissolved in water, their rotation changes until the same fixed value results. This is called
(A) anomerization (B) mutarotation (C) inversion
(D) epimerization (E) recemization

43. Streptomycin, a well known antibiotic, is a derivative of
(A) purines (B) terpenes (C) alkaloids
(D) peptides (E) carbohydrates

44. Which of the following is the most stable form of cylohexane ?
(A) half chair (B) chair
(C) boat (D) planar (E) twist boat

45. Galvanization of iron denotes coating with
(A) Zn (B) Pb (C) Cd (D) Sn (E) Al

2001

1. The increasing order of atomic radius for the elements Na, Rb, K and Mg is
(A) Mg < Na < K < Rb (B) Rb < K < Mg < Rb
(C) Na < Mg < K < Rb (D) K < Na < Mg < Rb
(E) Na < K < Mg < Rb

2. In the modern long form of the periodic table, elements are arranged in the increasing order of
(A) Number of neutrons (B) Metallic character
(C) Mass number (D) Atomic number
(E) Atomic mass

3. Limestone is not used in which of the following manufacturing proceses ?
(A) Portland cement
(B) Solvay process of sodium carbonate
(C) Iron from haematite
(D) Ordinary (soda-lime) glass
(E) Phosphorous from phosphorite

4. Colemanite is
(A) $Co(BO_2)_2$ (B) H_3BO_4
(C) $NaBO_2$ (D) $Ca_2B_6O_{11}.\ 5H_2O$
(E) $Na_2B_4O_7.\ 10H_2O$

5. Which one of the following sets gives the correct arrangement of the compounds involved, based on their bond strength ?
(A) HF > HBr > HCl > HI (B) HF < HCl > HBr > HI
(C) HCl > HF > HBr > HI (D) HF > HCl > HBr > HI
(E) HI > HBr > HCl > HF

6. The reaction in which hydrogen peroxide acts as a reducing agent is
(A) $2FeSO_4 + H_2SO_4 + H_2O_2 \rightarrow Fe(SO_4)_3 + 2H_2O$
(B) $H_2SO_4 + H_2O_2 \rightarrow 2Ag + H_2O_2 + O_2$
(C) $Ag_2O + H_2O_2 \rightarrow 2Ag + H_2O + O_2$
(D) $PbS + 4H_2O_2 \rightarrow PbSO_4 + 4H_2O$
(E) $2KI + H_2O_2 \rightarrow 2KOH + I_2$

7. The basicity of phosphorous acid is
(A) 5 (B) 4 (C) 3 (D) 2 (E) 1

8. Match list I with list II and select the answer using the code given

Code	List I	Code	List II
a	XeF_4	1	distorted octahedral
b	XeF_6	2	tetrahedral
c	XeO_3	3	square planar
d	XeO_4	4	trigonal pyramidal

(A) a-2, b-3, c-4, d-1 (B) a-3, b-1, c-4, d-2
(C) a-1, b-4, c-2, d-3 (D) a-2, b-3, c-1, d-4
(E) a-4, b-1, c-3, d-2

9. The van der Waals equation reduces itself to the ideal gas equation at
(A) High pressure alone
(B) Hig hpressure and high temperature
(C) Low pressure and high temperature
(D) Low pressure and low temperature
(E) High pressure and low temperature

10. The de Broglie wavelength associated with a ball of mass 200g. and moving at a speed of 5 metres/hour, is of the order of ($h = 6.625 \times 10^{-34}$ Js)
(A) 10^{-35}m (B) 10^{-30}m (C) 10^{-35}m
(D) 10^{-20}m (E) 10^{-15}m

11. The orbital diagram in which the Aufbau principle is violated is

	2s	2p		
(A)	↑↓	↑	↑	
(B)	↑↓	↑	↑	↑
(C)	↑↓	↑↓	↑	
(D)	↑↓	↑↓	↑	↑
(E)	↑↓	↑↓	↑↓	↑

12. The formation of a single mono-substitued product by benzene is attributed to its resonance structure. If the C-C bond distance in ethane is 154 pm and that in ethene 134 pm, the C-C bond distance in benzene would be
(A) longer than 134 pm but shorter than 154 pm
(B) longer than 154 pm
(C) 154 pm (C) 134 pm (D) shorter that 134 pm.

13. The hybrid orbitals of carbon atoms numbered, 1 and. 2 (superscript) in the compound $H_2C' = C^2 = CH_2$ are
(A) Sp^2 and Sp^3 respectively
(B) Sp^2 and Sp respectively (C) Sp^2 only
(D) Sp only (E) Sp and Sp^3 respectively

14. The correct order of bond angles in the molecules, H_2O, NH_3, CH_4 and CO_2 is
(A) $H_2O < NH_3 > CO_2 > CH_4$
(B) $H_2O > NH_3 < CH_4 > CO_2$
(C) $CO_2 > CH_4 > NH_3 > H_2O$
(D) $H_2O > NH_3 > CH_4 > CO_2$
(E) $H_2O < NH_3 < CO_2 < CH_4$

15. A wooden articraft and freshly cut tree, show the acitvity of 7.8 and 15.6 disintegrations min^{-1} g of carbon of ^{14}C =5760 years respectively. The age of the articraft is
(A) 5.760 years (B) 5760 x (15.6-7.8) years
(C) 5760 x 78/15.6 years (D) 5760 x 3 x 15.6/27.8 years
(E) 5760 years

16. The members W, X, Y and Z of naturally occurring radioactive series are related as,
$$w \xrightarrow{-\alpha} x \xrightarrow{-\beta} y \xrightarrow{-\beta} z$$
The isotopes among the above are
(A) Y and Z (B) X and Z (C) W and Z
(D) W and Y (E) W and X

17. Reaction of calcium with water is represented by the equation $Ca + 2H_2O \rightarrow Ca(OH)_2 + H_2$. What volume of hydrogen, measured at STP. would be liberated when 8 grams of calcium completely react with water (atomic mass of Ca is 40) ?
(A) $0.4 cm^3$ (B) $0.20 cm^3$ (C) $4480 cm^3$
(D) $2240 cm^3$ (E) $1120 cm^3$

18. In a flask of volume, V litre. 0.2 mol of oxygen, 0.4 mol of nitrogen, 0.1 mol of ammonia and 0.3.mol of helium are enclosed at 27°C. If the total pressure exerted by these non-reacting gases is one atmosphere, the partial pressure exerted by nitrogen is
(A) 0.2 atmosphere (B) 0.3 atmosphere
(C) 0.4 atmosphere (D) 1 atmosphere
(D) 0.1 atmosphere

19. A 1% aqueous solution (mass by volume) of a certain substance is isotonic with a 3% solution of glucose (molar mass 18θ) at a given temperature. The molar mass of the substance is
(A) 540 (B) 360 (C) 180 (D) 120 (e) 60

20. The aqueous solution that has the lowest vapour pressure at a given temperature is
(A) 0.1 molal acetic acid (B) 0.1 molal glucose
(C) 0.1 molal sodium chloride
(D) 0.1 molal barium chloride
(E) 0.1 molal sodium phosphate

21. Assuming complete dissociation, the vant Hoff factor for a dilute aqueous solution of sodium chloride is
(A) 2.0 (B) 1.5 (C) 1.0 (D) 0.5 (E) zero

22. IUPAC name of $H_3C–O–C_2H_5$ is
(A) Methyl ethyl ether (B) Ethoxy methane
(C) Ethyl methyl oxide (D) Methoxy ethane
(E) Ethyl methyl ether

23. In the Kjeldahl's method of estimating nitrogen in an organic compond, nitrogen is first converted to
(A) Ammonium nitrate (B) Nitrogen dioxide
(C) Ammonium sulphate (D) Nitric acid
(E) Nitrogen gas

24. In a hydrocarbon the mass ratio of hydrogen to carbon is 1:3. The empirical formula of the hydrocarbon is
(A) CH_4 (B) CH_3 (C) C_2H (D) CH_2 (E) CH

25. Electrophile involved in the formation of nitrobenzene from benzene is
(A) NO (B) $HNO_3 + H_2SO_4$
(C) HNO_3 (D) NO^+_2 (E) NO^-_2

26. A molecule is said to be chiral if it
(A) Cannot be superimposed on its mirror image
(B) Exhibits cis-trans isomerism
(C) Is highly symmetrical
(D) Has a centre of symmetry
(E) Has a plane of symmetry

27. 1-chloropropane and 2-chloropropane are
(A) Geometrical isomers (B) Optical isomers
(C) Functional isomers (D) Chain isomers
(E) Position isomers

28. $CaC_2 + 2H_2O \rightarrow X + Ca(OH)_2$ $X + H_2O \xrightarrow{H_2SO_4 + HgSO_4, 60°C} y$ identify X and Y in the above reaction sequence.
(A) CH_4 and $H_2C = O$ (B) C_2H_2 and CH_3CHO
(C) C_2H_4 and CH_3COOH
(D) C_2H_4 and HCOOH (E) C_2H_2 and CH_3 COOH

29. The negative part of the addendum (molecule to be added) adds on to the carbon atom (of the double/triple bond) containng the least number of hydrogen atoms is
(A) Markownikoff's rule (B) Friedel-Craft's role
(C) Kekule's rule (D) Peroxide effect
(E) Anti-Markownikoff's rule

30. A fuel has the same knocking property as mixture of 70% iso octane (2,2,4- trimethyl pentane) and 30% n- heptane by volume. The octane number of the fuel is
(A) 30 (B) 40 (C) 50 (D) 60 (E) 70

31. Identify the product Y in the reaction sequence :

$CH_3CHO + CH_3Mg \xrightarrow{\text{dry ether}} X \xrightarrow{HOH} Y$

(A) CH_3CH_2CHO (B) $(CH_3)_3COH$
(C) $(CH_3)_2CHOH$ (D) CH_3CH_2OH (D) CH_3OH

32. The common oxidation state shown by lanthanides in their compounds is
(A) +7 (B) +6 (C) +5 (D) +3 (D) +1

33. The primary valency of iron in $K_4(Fe(CN)_6)$ is
(A) 6 (B) 4 (C) 3 (D) 2 (E) 1

34. When a mixture of solid NaCl and solid $K_2Cr_2O_7$ is heated with concentrated sulphuric acid, the orange red vapours obtained are due to
(A) Chromyl chloride (B) Chromic, sulphate
(C) Chromic oxide (D) Chromic chloride
(E) Chromous chloride

35. The element that does not form a monoxide is
(A) Germanium (B) Lead (C) Tin
(D) Carbon (E) Silicon

36. A compound of a metal ion M^{x+} (Z = 24) has a spin magnetic moment of $\sqrt{15}$ Bohr Magnetons. The number of unpaired electrons in the compound are
(A) 6 (B) 3 (C) 5 (D) 4 (E) 2

37. Calomel is
(A) $HgCl_2$ (B) Hg_2Cl_2 (C) K_2HgI_4 (C) HgO (D) HgI_2

38. Purified bauxite is converted to metallic aluminium by
(A) Hydrometallugry (B) Electrolytic reduction
(C) Electrolytic oxidation (D) Reduction with zinc
(E) Reduction with carbon

39. The alloy best suited for making metre scales is
(A) Chromium tungsten steel
(B) Tungsten steel (C) Alnico
(D) Invar (E) Stainless steel

40. Electromagnetic separation is used in the concentration of
(A) Zinc blende (B) Cinnabar (C) Cassiterite
(D) Bauxite (E) Copper pyrites

41. The wrong statement about glycerol is
(A) It is a tertiary alcohol
(B) It is used in the manufacture of explosives
(C) Acidified $KMnO_4$ converts it to oxalic acid
(D) It is generally obtained by the hydrolysis of oils and fats
(E) It is trihydric alcohol

42. Picric acid is
(A) 2,4,6-trinitrophenol (B) 1,3,5-trinitrobenzene
(C) para-nitrophenol (D) 2,4,6-tribromophenol
(D) 2,4,6-trinitrotoluene

43. Saponification of ethylbenzoate with caustic soda as alkali, gives
(A) Benzoic acid and sodium ethoxide
(B) Phenol and ethanoic acid
(C) Sodiumbenzoxide and ethanoic acid
(D) Benzyl alcohol and ethanoic acid
(E) sodium benzoate and ethanol

44. According to the second law of thermodynamics, a process (reaction) is spontaneous, if during the process
(A) ΔH system > 0 (B) ΔS universe = ΔS system
(C) ΔS universe < 0 (D) ΔS universe > 0
(E) ΔS universe = 0

45. The decomposition of a substance follows first order kinetics. If its concentration is reduced 50 to 1/8th of its initial value. in 24 minutes, the rate constant of the decomposition process is
(A) log (1/24) min^{-1} (B) 0.692/24 min^{-1}
(C) 1/24 log (1/8) min^{-1} (D) 2.303/24 log (8) min^{-1}
(E) 2.303/24 log (8/1) min^{-1}

46. For a first order reaction, the plot of log k agains 1/T is straight line. The slope of the line equal to
(A) –Ea/R (B) –2.303/Ea.R (C) –Ea/2.303
(D) Ea/2.303R (E) –Ea/2.303

47. The standard hydrogen etectrode potential is zero because
(A) It has been. measured accurately
(B) Hydrogen is reluctant to donate eleetrons
(C) It has been defined that way
(D) There is no potential difference between the electode and the solution
(E) Hydrogen ions acquire electrons from a platinum electrode

48. BF_3 is an acid accoiding to
(A) Lewis concept (B) Ostwald (C) Henderson
(D) Arrhenius concept (E) Bronsted-Lowry concept

49. The difference between ΔH and ΔE for the combustion of methane at 27°C will be (in Joule/ mol)
(A) 8.314 x 300 x (–2) (B) 8.314 x 300 x (1)
(C) 8.314 x 27 x (–2) (D) 8.314 x 27 x (–3)
(E) 8.314 x 300 x (–3)

50. The half-cell (electrode) reactions, with their appropriate standard reduction potentials are: (i) Pb^{2+} + 2e⇔ Pb (E° = 0.13 V); (ii) Ag^+ + e ⇔Ag(E° = +0.80 V) . In the light of the data given, which one of the following reactions will take place ?
(A) 2Ag + Pb^{2}+ → $2Ag^+$ + Pb
(B) $2Ag^+$ + Pb → Pb^{2+} + 2Ag

(C) $2H^+ + 2Ag \rightarrow Ag^+ + H_2$
(D) $Pb^{2+} + H_2 \rightarrow 2H^+ + Pb$ (E) $Pb^{2+} + 2Ag \rightarrow 2Ag^+ + Pb$

51. Substances which behave as normal electrolytes, in solutions at low concentrations and exhibit collodial properties at higher concentrations, are called
(A) macromolecular colloids (B) associated colloids
(C) gels (D) lyophilic colloids (E) emulsions

52. Specific conductance of 0.1 m sodium chioide solution is 1.06x 10^{-2} ohm^{-1} cm^{-1}. Its molar conductance in ohm^{-1} cm^{-1} mol^{-1} is
(A) 53 (B) 5.3x10^2 (C) 1.060x10^{-2}
(D) 1.06x10^3 (E) 1.06x10^4

53. The heat of neutralisation is maximum when
(A) sodium hydroxide is neutralised by hydrochloric acid
(B) sodium hydroxide is neutralized by formic acid
(C) ammonium hydroxide is neutralized by hydrochloric acid
(D) ammonium hydroxide is neutralized by acetic acid
(E) sodium hydroxide is neutralized by acetic acid

54. The hydrogen ion concentration in mol/dm^3 in a 0.2 M solution of a weak acid, HA (k_a = 2x10^5), is close to
(A) 2x10^{-2} (B) 2x10^{-1} (C) 2x10^{-5}
(D) 2x10^{-4} (E) 2x10^{-3}

55. Terylene is obtained by the condensation polymerization of
(A) Styrene and 1,3-butadiene
(B) Phenol and formaldehyde
(C) Urea and formaldehyde
(D) Ethylene glycol and terephthalic acid
(E) Hexamethylene diamine and adipic acid

56. Isoelectric point of an amino acid is. the pH at which
(A) The amino acid molecules would not migrate. either to the anode or to the cathode, and exist as neutral dipolar ions
(B) The amino acid molecules can migrate either to the anode or to the cathode
(C) The amino acid molecules move alternately to the anode and the cathode
(D) The amino acid molecules migrate to the anode
(E) The amino acid molecules migrate to the cathode

57. Which one of the following bases is not found in RNA?
(A) Cytosine (B) Uracil (C) Thymine
(D) Adenine (E) Guanine

58. Reduction of H_3C-NC with hydrogen in presence of Ni/Pt as catalyst, gives
(A) $CH_3CH_2NH_2$ (B) $(CH_3)_3N$
(C) $CH_3CH_2CH_2NH_2$ (D) CH_3NH_2 (E) CH_3NHCH_3

59. The aqueous solution of which one of the following mixtures can give a buffer solution ?
(A) HNO_3 + KOH (B) HNO_3 +KNO_3
(C) KOH + KCl (D) HCl + NaCl
(E) CH_3COOH + CH_3COONa

60. Given that, the solubility product, K_{sp}, of AgCl is 1.8x10^{-10}, the concentration of Cl ions that must just be exceeded before AgCl will precipitate from a solution containing 4 x 10^{-3} M Ag ions is
(A) 1.8x10^{-8} M (B) lx10^{-8} M (C) 2x10^{-8} M
(D) 4.5x10^{-8} M (E) 4x10^{-8} M

61. The substance that causes temporary hardness in water is
(A) $Mg(HCO_3)_2$ (B) NaHCO
(C) $CaCO_3$ (D) $MgCO_3$ (D) $CaSO_4$

62. The range of radius ratio (cationic to anionic) for an octahedral arrangement of ions in an ionic solid is
(A) 0.255 – 0.414 (B) 0.414 – 0.732
(C) 0.732-1.000 (D) 0 – 0.155 (E) 0.155 – 0.255

63. The metal that is present in vitamin B_{12} is
(A) Co (B) Fe (C) Zn (D) Ca (E) Mg

64. One of the oxidants used with liquid propellants is
(A) Phosphorous pentoxide (B) Nitrogen tetroxide (N_2O_4)
(C) Sulphuric acid (D) Nitro cellulose
(E) Ammonium perchlorate

2002

1. Eka aluminium and Ekasilicon are known as
(A) Neutron and Magnesium
(B) Proton and Silicon (C) Iron and Sulphur
(D) aluminium and Silicon (E) Gallium and Germanium

2. Davy and Faraday proved that
(A) the strength of graphite is minimum compared to platinum
(B) graphite is very hard
(C) bond length of graphite is equal to bo length of carbon containing compound
(D) diamond is a form of carbon
(E) the bond lengths of carbon containing compounds are always equal = 2

3. BF_3 is used as a catalyst in several industrial processes due to its
(A) strong Lewis acid nature
(B) weak Lewis acid character (C) reducing nature
(C) strong reducmg agent (E) weak reducing agent

4. The first element of a group in the periodic table of elements often shows the resemblance to the second element of the neighbouring group on the right, which is
(A) electronegativity (B) ionization energy
(C) parallel relationship (D) diagonal relationship

(E) none of the above

5. As the alkaline earth metals (except Be) tend to lose their valence electrons readily they act as
(A) strong oxidising agent (B) strong reducing agent
(C) bases (D) weak oxidizing agent
(E) weak reducing agent

6. Noble gases are group of elements which exhibit very
(A) much diamagnetic properties
(B) much paramagnetic properties
(C) minimum electronegativity
(D) low chemical activity (E) high chemical activity

7. The triatomic species of elemental oxygen is known as
(A) Diazone (B) Ozone (C) Triozone
(D) Azone (E) polyzone

8. When chlorine water is exposed to sunlight the colour change that occur is
(A) light blue to colourless
(B) colourless to greenish yellow
(C) greenish yellow to colourless
(D) colourless to brown (E) brown to colourless

9. The three important oxidation states of phosphorus are
(A) -3, +4 and -4 (B) -3, +3 and +4
(C) -3,+3 and +2 (D) -3, +3 and +5 (E) -3, +3 and -5

10. The quantum numbers n = 2, $l = 1$ represent
(A) 3d orbital (B) 3s orbital
(C) 1s orbital (D) 2s orbital (E) 2p orbital

11. Accurate determination of atomic masses is done with the instrument called as
(A) Colorimeter (B) Calorimeter
(C) Atomci absorption (D) Mass spectrometer
(E) Spectrophotometer

12. The phenomenon of radioactivity arises from the
(A) stable nuclei (B) decay of unstable nuclei
(C) binary fission (D) nuclear fission (E) nuclear fusion

13. All the nuclei from the initial element to the final element constitute a series which is called
(A) disintegration series (B) nuclear series
(C) γ-series (D) β- series (E) β-γ series

14. The atmospheric pressure is sum of the
(A) vapour pressure of chemical and vapour pressure of volatiles
(B) pressure created on to atmospheric molecules
(C) vapour pressure of organic substances and vapour pressure of water
(D) pressure on the biomolecules
(E) vapour pressure of atmospheric constituents

15. The process used for the extraction of sodium is called
(A) Thermite process (B) Haber's process
(C) Down's process (D) Serpeck's process
(E) Baeyer's process

16. Use of hot air balloons in sports and for meterological observations is an application of
(A) Charle's Law (B) Brown's Law
(C) Kelvin's Law (D) Newton's Law
(E) Boyle's Law

17. "Equal volumes of all gases at the same temperature and pressure contain equal number of particles" This statement is a direct consequence of
(A) Partial law of pressure
(B) Ideal gas equation (C) Charle's law
(D) Partial law of volume (E) Perfect gas law

18. When the temperature is raised, the viscosity of liquid decreases, this is because
(A) increased attraction between molecules
(B) decreased covalent and hydrogen bond forces
(C) increase in temperature increases the average kinetic energy of molecules which overcomes the attraction force between them
(D) decreaesed volume of the solution

19. The interionic attraction depends on interation of
(A) solute-solute (B) solvent-solvent
(C) the charges (D) atomic particles (E) none

20. The vicosity of the solvent depends on
(A) solute-olvent interaci (B) density of the liquid
(C) solvent-solvent interaction
(D) isothemic nature (E) solute-solvent interaction

21. When the concentration is expressed as the number of moles of a solute per litre of solution it is known as
(A) mole fraction (B) mass percentage
(C) molarity (D) normality (E) molarity

22. When a solute is present in trace quantities the following expression is used
(A) microgram percent (B) nanopani percent
(C) parts per million (D) gram per million
(E) milligram percent

23. Chromatomanhy is valuable method for the separation, isolation, purification and identification of the constituents of a mixture and it is based on general principle of
(A) interphase separation
(B) phase operation (C) phase contact
(D) phase rule (E) phase distribution

24. If the light waves pass through a Nicol prism, then all the oscillations occur only in one plane, such beam of light is called as
(A) polarized light (B) optical light
(C) plane polarized light
(D) non-polarized light (E) rotational light

25. Chirality of a carbon compound is because of its

(A) divalent nature of carbon
(B) trivalent nature of carbon
(C) pentavalent nature of carbon
(D) tetrahedral nature of carbon
(E) monovalent nature of carbon

26. Electrophilic substitution of ($C_6H_5NH_2$) with bromine gives
(A) 4-bromo aniline (B) 3 - bromo aniline
(C) 2 - bromo aniline (D) 1,4,6-tribromo aniline
(E) 2,4,6 - tribromo aniline

27. Write the products of the addition reaction of

$C = C + XY \rightarrow$

(A) $>C(X) - C(Y)<$ (B) $X - C = C - Y$

(C) $>C = C(X) - Y$ (D) $Y - >C - C< - X$

(E) 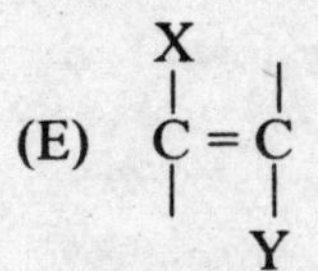

28. Benzene is a monocyclic arene whereas napthalene is
(A) optical (B) non-optical
(C) bicyclic (D) tricyclic (E) tetracyclic

29. What happens when naphthalene balls are put inside kerosene ?
(A) dissolves easily (B) does not dissolve (C) precipitates
(D) dissolves upon heating (E) no reaction

30. The trivial names of the propanone and 3 - methyl-2-butanone are
(A) Diisopropyl ketone and acetone
(B) Acetone and diisopropyl ketone
(C) Aldone and butane
(D) Acetone and methyl isopropyl ketone
(E) None of the above

31. Addition of Grignard reagents to an aldehyde or ketone gives an
(a) anomer (b) ether (c) ester (d) dimer (e) alcohol

32. Fehling's solution is composed of
(A) Potassium sodium tartarate
(B) Glycol and KOH (C) Tartarate and sodium hydroxide
(D) Copper ions in acid medium
(E) Alkaline copper (ii) ions complexed with sodium potassium taratarate

33. Benzenehexa chloride has a trade name called
(A) Calomel (B) Gammexane
(C) Treon (D) Preon (E) None of th above

34. The major binding force of diamond, silicon and quartz is
(A) van der Waals force
(B) Non-covalent bond force
(C) Covalent bond force
(D) Electrical attraction (E) Electrostatic force

35. The acid that results when a base accepts a proton is called
(A) conjugate protonated base
(B) Lewis base (C) Conjugate acid of the base
(D) Conjugate base of the acid
(E) None of the-above

36. At dynamic equilibrium two changes occur at the same rate so that the mass on both sides of the equilibrium undergoes no change, only when
(A) $\Delta G = 0$ (B) $\Delta G = -2$ (C) $\Delta G = 2$
(D) $\Delta G = +1$ (E) $\Delta G = -1$

37. Write the equilibrium constant K. for
$CH_3COOH + H_2O = H_3O^+ + CH_3COO^-$

(A) $K = \frac{[H_3O^+][H_2O]}{[CH_3COO^-][CH_3COOH]}$

(B) $K = \frac{[H_3O^+][CH_3COO^-]}{[H_2O][CH_3COOH]}$

(C) $K = \frac{[H_3O^+]+[CH_3COOH]}{[H_2O][CH_3COO^-]}$

(D) $K = \frac{[H_2O]+[CH_3COO^-]}{[H_2O][CH_3COOH]}$

(E) none of these

38. $E_2 = E_1 + q + w$ or $\Delta E = q + w$ is the mathematical statement of
(A) adiabatic change (B) non-isothermal change
(C) second law of thermodynamics
(D) internal energy (E) first law of thermodynamics

39. If a mole of H_2 molecule is treated to high temperature the following reaction takes place
(A) $2H_2(g) + 436J = 2H_2$ (B) $H_2 + H_2 = H^+ + H^+$
(C) $H_2(g) + 436KJ = H(g) + H(g)$
(D) $H2(g) + 820KJ = 2H_2(g)$
(E) None of the above

40. If the system does not exchange heat with the surroundings the process is
(A) partial thermal (B) Isothermal
(C) Isomeic (D) Non-adiabatic (E) adiabatic

41. Mechanical work is specially important in systems that contain
(A) solid-solid (B) amalgam (C) gases
(D) solid-liquid (E) liquid - liquid

42. Corrosion is basically a
(A) interaction
(B) union between light metal and heavy metal
(C) glycosidic interaction

(D) altered reaction in presence of H_2O
(E) electrochemical phenomenon

43. If steel is heated to a temperature well below red heat and is then cooled slowly, the process is called
(A) annealing (B) tempering and hardening
(C) tempering (D) hardening (E) softening

44. Bell metal is an alloy of
(A) Cu and Zn (B) Fe, Cr, Ni (C) Cu and Ni
(D) Fe and Ni (E) Cu and Sn

45. The colour of $CoCl_3$, $5NH_3$, H_2O is
(A) green (B) violet (C) pink
(D) orange yellow (E) orange

46. Many transition metals and their compounds exhibit
(A) paramagnetic behaviour (B) diamagnetic behavior
(C) perimagnetic behaviour (D) van der Waals behaviour
(E) all of the above

47. $ArH + N_2 + H_3PO_3 + HX$ are the products of the following reaction
(A) $ArNX + H_3PO_4 + NADH^+$
(B) $ArN + H_3PO_2 + 2H_2$
(C) $ArN_2X + H_3PO_2 + HOH$ (D) $ArN + H_2PO_2 + H_2$
(E) $ArNX + H_2 + H_3PO_2$

48. Corundum is an ore of
(A) Sodium (B) Iron
(C) copper (D) Aluminium (E) Boron

49. The metal used for making radiation shield is
(A) Zinc (B) Lead (C) Aluminium
(D) Silicon (E) Iron

50. Colligative properties of solutions are those properties which depend on the
(A) nature of particles only
(B) number of particles only
(C) volume of the solution (D) shapes of particles
(E) nature of the solvent

51. The term PLATFORMMING is
(A) flat sheet of platinum (B) platinum coating
(C) platinum manufacturing
(D) platinum used as a catalyst
(E) platinum used in painting

52. Diazomum salts + Cu_2Cl_2 + HCl → (chlorobenzene: benzene ring with Cl) the reaction is known as
(A) Perkinseaction (B) Substitution reaction
(C) Grignard reaction (D) Chlorination
(E) Sandmeyer's reaction

53. The unshared pair of electrons on a cyanide ion can act as
(A) catalytic centre (B) cationic centre
(C) nucleophilic centre (D) isocyanide centre
(E) amido centre

54. Which of the following has an ester linkage ?
(A) PVC (B) rubber
(C) nylon (D) bakelite (E) terylene

55. Oils are converted into fats by
(A) Hydrogenation (B) Dehydrogenation
(C) Hydrogenolysis (D) Hydration
(E) Decarboxylation

56. Tranquilizers are substances used for the treatment of
(A) mental diseases (B) physical disorders
(C) blood infection (D) cancer (E) AIDS

57. The use of chemicals for treatment of diseases are called as
(A) Angiotherapy (B) Physiotherapy
(C) Chemotherapy (D) Homoeotherapy
(E) Isotherm therapy

58. The chemical substances used to bring down body temperature in high fever are known as
(A) antipyretics (B) analgesic
(C) tranquilizers (D) disinfectants (E) antiseptic

59. The IUPAC name of n-butyl chloride is
(A) ter-Butylchloride (B) 2-Methylbutane
(C) 2-Ethylbutane (D) 1-Chlorobutane
(E) n-Chlorobutane

60. The most common minerals of phosphorus are
(A) borax and fluorapatite
(B) hydroxy apatite and colemanite
(C) hydroxy apatite and fluorapatite
(D) hydroxy apatite and kemite
(E) colemanite and fluorapatite

61. Water or dilute mineral acids on metal can give
(A) Dihydrogen (B) Trihydrogen (C) D_2
(D) Monohydrogen (E) Tritium

2003

1. Water softening by Clarke's process uses
(A) calcium bicarbonate
(B) sodium bicarbonate (C) potash alum
(D) calcium hydroxide (D) calcium bicarbonate

2. Inorganic graphite is
(A) $Fe(CO)_5$ (B) P_4S_3 (C) $B_3N_3H_6$ (D) B_3N_3 (E) SiC

3. Which one of the following materials is used in high temperature thermometers ?
(A) sodium amalgam
(B) alloy of sodium and potassium
(C) high carbon steel

(D) crystalline sodium chloride (E) molten aluminium

4. Development of a photographic plate involves the use of
(A) a neutralizing agent (B) a precipitating agent
(C) a reducing agent (D) an oxidizing agent
(E) a complexing agent

5. In any period the valency of an element with respect to oxygen
(A) increases one by one from IA to IVA and then decreases from VA to-VIIA one by one
(B) decreases one by one from IA to IVA and then increases from VA to VIIA one by one
(C) increases one by one from IA to VIIA
(D) decreases one by one from lAtoVIlA
(E) changes randomly

6. The normality of "10 volume" Hydrogen peroxide is
(A) 17.8 (B) 0.88 (C) 1.78 (D) 0.176 (E) 3.52

7. The substance used in smoke screen is
(A) calcium phosphide (B) calcium fluoride
(C) calcium phosphate (D) sodium chloride
(E) zinc phosphate

8. Ozone is formed in th upper atmosphere from oxygen by the action of
(A) free radicals (B) nitrogen oxides
(C) thermal radiation from sunlight
(D) ultraviolet rays (E) cosmic rays

9. Nitrogen show different oxidation states in the range
(A) –3 to +3 (B) +3 to +5
(C) 0 to +5 (D) –3 to +5 (E) –5 to +3

10. H_3PO_4 phosphorous acid is
(A) a monoproticacid (B) not acidic
(C) not soluble in water (D) a diprotic acid
(E) a triprotic acid

11. An example of a neutral oxide is
(A) CaO (B) ZnO (C) SO_2 (D) NO (E) CO_2

12. Two electrons in the same atomic orbital can be disinguished by their
(A) magnetic quantum number
(B) spin quantum number
(C) orbital quantum number
(D) principal quantum number
(E) azimuthal quantum number

13. lf the ionization energy of hydrogen is 2.1810^{-19} J, the first ionization energy of He^+ ion is
(A) 8.72×10^{-19}J (B) 0.545×10^{-19}J
(C) 1.09×10^{-19}J (D) 2.18×10^{-19}J (E) 4.36×10^{-19}J

14. The emission spectrum of hydrogen is found to satisfy the expression for the energy change ΔE (in Joules) such that $\Delta E = 2.18\times10^{-18}\ (1/n_1^2 - 1/n_2^2)$ J where n_1 = 1,2,3,... and n_2 = 2,3,4 .... The spectral lines correspond to Paschen series if
(A) n_1 = 3 and n_2 = 4,5,6 (B) n_1 = 1 and n_2 = 3,4,5
(C) n_1 = 2 and n_2 = 3,4,5 (D) n_1 = 1 and n_2 = 2,3,4
(E) n_1 = 1 and n_2 = infinity

15. From elementary molecular orbital theory we can deduce the electronic configuration of the singly positive nitrogen molecular ion N_2^+ as
(A) $\sigma_{1s}^2\ \sigma_{1s}^{*2}\ \sigma_{2s}^2\ \sigma_{2s}^{*2}\ \sigma_{2p}^4\ \sigma_{2p}^1$
(B) $\sigma_{1s}^2\ \sigma_{1s}^{*2}\ \sigma_{2s}^2\ \sigma_{2s}^{*2}\ \sigma_{2p}^1\ \sigma_{2p}^3$
(C) $\sigma_{1s}^2\ \sigma_{1s}^{*2}\ \sigma_{2s}^2\ \sigma_{2s}^{*2}\ \sigma_{2p}^2\ \sigma_{2p}^4$
(D) $\sigma_{1s}^2\ \sigma_{1s}^{*2}\ \sigma_{2s}^2\ \sigma_{2s}^{*2}\ \sigma_{2s}^{*2}\ \sigma_{2p}^1\ \sigma_{2p}^2$
(E) $\sigma_{1s}^2\ \sigma_{1s}^{*2}\ \sigma_{2s}^2\ \sigma_{2s}^{*2}\ \sigma_{2p}^2\ \sigma_{2p}^2\ \sigma_{2p}^2\ \sigma_{2p}^{*1}$

16. Which of the following bonds require the largest amount of bond energy to dissociate the atoms concerned ?
(A) N = N bond in N_2 (B) O = O bond in O_2
(C) C – C bond in ethane
(D) H–H bond in H_2 (E) C – H bond in CH_4

17. Which one of the following has the smallest bond angle?
(A) H_2O (B) CH_4 (C) SO_2 (D) NH_3 (E) BeF_2

18. Given that a radioactive species decays according to the exponential law $N = N_0 e\lambda$.. The half life of the species is
(A) λ/ln2 (B) N_0 (C) ln2/λ (D) λ (E) N_0

19. The process, ${}_1H^2 + {}_1H^3 \rightarrow {}_2He^4 + {}_0n^1$ represents the type of reaction known as
(A) autocatalytic (B) fission (C) thermal
(D) fusion (E) chemical

20. Which of the following graph best represents the variation of pressure (P) with volume (V) of a gas at a fixed temperture ?

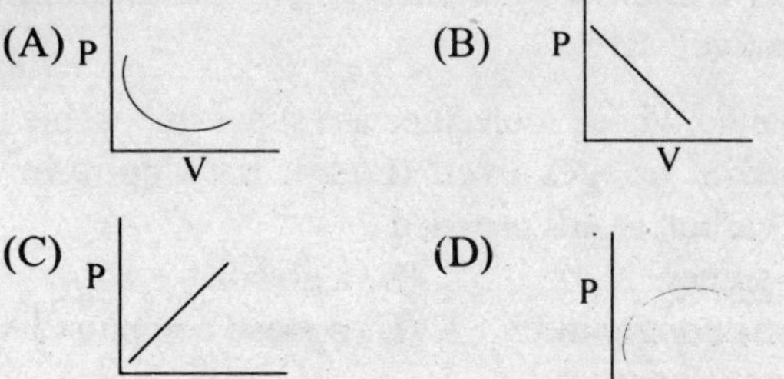

(E) none of the above

21. Under identical conditions of temperature and pressure the ratio of the rates of effusion of O_2 and CO_2 gases given by

(A) $\dfrac{\text{rate of effusion of oxygen}}{\text{rate of effusion of carbon dioxide}} = 0.87$

(B) $\dfrac{\text{rate of effusion of oxygen}}{\text{rate of effusion of carbon dioxide}} = 1.17$

(C) $\frac{\text{rate of effusion of oxygen}}{\text{rate of effusion of carbon dioxide}} = 8.7$

(D) $\frac{\text{rate of effusion of oxygen}}{\text{rate of effusion of carbon dioxide}} = 0.117$

(E) $\frac{\text{rate of effusion of oxygen}}{\text{rate of effusion of carbon dioxide}} = 11.7$

22. The expression for van der Waals equation for the gas law is
(A) $nRT = (P + n^2a/v^2)(v - nb)$
(B) $nRT = (P + a/v^2)(v+b)$ (C) $nRT = (P - a/v^2)(v+b)$
(D) $nRT = (P + a/v^2)(v-b)$ (E) $nRT = (P - a/v^2)(v-b)$

23. What is the effect of the addition of sugar on the boiling and freezing points of water ?
(A) Boiling point decreases and fieezing point increases
(B) Boiling point increases and freezing point decreases
(C) Boiling point increases and the freezing point is not affected
(D) Both boiling point and freezing point increase
(E) Both boiling point and freezing point decrease

24. Name the alkene with the molecular formula $C_{10}H_{20}$
(A) Decane (B) Heptene (C) Nonene
(D) Dodecene (E) Undecene

25. The empircal formula of a compound is CH_2O and its molecular weight is 120. The molecular formula of the compound is
(A) C_4H_2O (B) CH_2O (C) $C_4H_8O_4$
(D) $C_3H_6O_3$ (E) $C_2H_4O_2$

26. In Victor Mayor's method 0.2 gm of an organic substance displaced 56 ml of air at STP. The molecular weight of the compound is
(A) 80 (B) 28 (C) 280 (D) 56 (E) 112

27. In electrophilic substitution reactions nitrobenzene is
(A) meta-directing (B) otho-directing
(C) para-directing
(D) not reactive and does not undergo any substitution
(E) non-selective

28. A compound whose molecules are super imposable on their mirror images even though they contain asymmetric carbon atoms is called
(A) a threo isomer (B) a glycol
(C) an eutectic compound (D) a meso compound
(E) an erythro isomer

29. Which of the followilig statement is wrong ?
(A) n-Butane and 2-Methyl propane are chain isomers
(B) Acetone and propinaldehyde are functional isomers
(C) Acetyl acetone exhibits tautomerism
(D) Diethyiketone and methyl propyl ketone are position isomers
(E) 2-Chloropentane and l-Chloropentane are position isomers

30. Of the following the compound possessing optical isomerism is
(A) $CH_2OHCHOHCH_2OH$
(B) CCl_2BrF (C) $CClF_2$ (D) $CH_3CHOHC_2H_5$
(E) CH_3CH_2OH

31. Hydrolysis of the ozanide of 1-butene gives
(A) Propionaldehyde and Formaldehyde
(B) Acetaldehyde only
(C) Acetaldehyde and oxalic acid
(D) Ethylene only
(E) Acetaldehyde and Formaldehyde

32. The chemical added to leaded petrol to prevent the deposition of lead in the combustion chamber is
(A) Tetra ethyl lead (B) Mercaptan (C) n-Heptene
(D) Isooctane (E) Ethylenedibromide

33. What is the major product of the following reaction
$CH_3C \equiv C - CH_2 - CH_3 \xrightarrow{\text{1 mole of } Cl_2}$?

(A) $CH_3-CCl_2-CCl_2-CH_2CH_3$

(B) $(Cl)(CH_3)C=C(CH_2CH_3)(Cl)$

(C) $CH_3-CCl_2-CH_2CH_2CH_3$

(D) $CH_3CH_2-CCl_2-CH_2CH_3$

(E) $(Cl)(CH_3)C=C(Cl)(CH_2CH_3)$

34. Allyl chloride on dehydro chlorination gives
(A) Allyl alcohol (B) Acetone (C) Acetyl chloride
(D) Propadiene (E) Propylene

35. A crystalline solid
(A) undergoes deformation of its geometry easily
(B) has an-irregular 3- dimensional arrangement
(C) softens slowly

(D) changes abruptly from solid to liquid
(E) has no definite melting point

36. One mole of a compound AB reacts with one mole of a compound CD according to the equation AB + CD⇌AD +CB. When equilibrium had been established it was found that 3/4 mole each of reactants AB and CD had been converted to AD and CB There is no change in volume. The equilibrium constant for the reaction is
(A) 16 (B) 9 (C) 16/9 (D) 1/9 (E) 9/16

37. Which of the following statement regarding catalyst in an equilibrium process is correct ?
(A) It decreases the rate of backward process
(B) It decreases the activation energy of direct process and increases the activation energy of backward process
(C) It increases the rate of forward reaction only
(D) It hastens the easy attainment of equilibriurn
(E) It increases the activation energy of both the forward and backward process

38. In thermodynamics which one of the following is not an intensive property
(A) volume (B) temperature (C) surface tension
(D) pressure (E) density

39. The free energy change for the following reactions are given below,
$C_2H_2(g) + 5/2O_2(g) \rightarrow 2CO_2(g) + H_2O(1)\ \Delta G = -1234$ KJ
$C(s) + O_2(g) \rightarrow CO_2(g)\ \Delta G^\circ = -394$ KJ
$H_2(g) + O_2(g) \rightarrow H_2O(l)\ \Delta G^\circ = -237$ kJ
What is the standard free energy change for the reaction
$H_2(g) + 2C(s) \rightarrow C_2H_2(g)$?
(A) 603 kJ (B) 209 kJ (C) +2259 kJ
(D) -2259 kJ (E) -209 kJ

40. The time taken for the completion of 3/4 of a first order reaction is
(A) (2.303/k) log 1/4 (B) (2.303/0.75) log k
(C) (2.303/k) log 4/3 (D) (2.0303/k) log 3/4
(E) (2.303/k) log 4

41. The standard reduction potential for two reactions are given below A gCl(s) + $e^- \rightarrow Ag(s) + Cl^-$ (aq) E° = 0.22 V
$Ag^+(aq) + e^- \rightarrow Ag(s) + E^\circ = 0.80V$. The solubility product of AgCl under standard conditions of temperature is given by,
(A) 3.2×10^{-8} (B) 1.5×10^{-10}
(C) 1.6×10^{-5} (D) 1.5×10^{-8} (E) 3.2×10^{-10}

42. 75% of a first order reaction was completed in 32 minutes. When was 50% of the reaction completed ?
(A) 8 minutes (B) 14 minutes (C) 32 minutes
(D) 16 minutes (E) 24 minutes

43. Unpaired electrons in Ni^{++}, (Z = 28) is
(A) 10 (B) 8 (C) 4 (D) 2 (E) 0

44. Lanthanide contraction is due to increase in
(A) atomic radius (B) size of 4f orbitals
(C) effective nuclear charge (D) atomic number
(E) shielding by 4f electrons

45. All the following statements about the transitional elements are true except that
(A) most of the transitional elements show only one valence state
(B) all the transitional elements have partly filled d orbitals
(C) most of the transitional elements show pronounced catalytic acitivity
(D) in aqueous solution many of their simple ions are coloured
(E) all of the transitional elements are predominantly metallic

46. In Nessler's reagent for the detection of ammonia the active species is
(a) Hg_2O (b) Hg^{2+} (c) Hg_2I_2 (d) $Hg_2I^{2-}_4$ (e) HgI_2^{2-}

47. A 0.001 molal solution of $[Pt(NH_3)_4Cl_4]$ in water had a freezing point depression of 0.0054°C. If K_f for water is 1.80, the correct formulation for the molecule is
(A) $[Pt(NH_3)_4Cl_2]Cl_3$ (B) $[Pt(NH_3)Cl]Cl_3$
(C) $[Pt(NH_3)_4Cl_3]Cl$ (D) $[Pt(NH_3)_4Cl_4]$
(e) none of the above

48. In the reaction, $2KMnO_4 + HCl \rightarrow 2MnCl_2 + 2KCl + 8H_2O$, the reduction product is
(A) KCl (B) HCl (C) Cl_2 (D) $MnCl_2$ (E) H_2O

49. Carnalite on electrolysis gives
(A) Al and Cl_2 (B) Mg and Cl_2 (C) K and CO_2
(D) Ca and Cl_2 (E) Na and CO_2

50. Ferrosilicon is used in the steel industry as
(A) a reducing agent (B) a cutting tool
(C) alloying agent (D) a flux
(E) scavenger of hydrogen

51. 5 ml of N-HCl, 20 ml of $N/2H_2SO_4$ and 30 ml of N/3 HNO_3 are mixed together and volume made to one litre. The normality of the resulting solution
(A) N/5 (B) N/10 (C) N/20 (D) N/40 (E) N/25

52. An electric current is passed through silver voltameter connected to a water voltameter. The cathode of the silver voltameter weighed 0.108 g more at the end of the electrolysis. The volume of oxygen evolved at STP is
(A) 5.6 cm^3 (B) 11.2 cm^3 (C) 22:4 cm^3
(D) 56 cm^3 (E) 550 cm^3

53. The standard reduction potentials for two half cell reactions are given below
$Cd^{2+}(aq) + 2e^- \rightarrow Cd(s)\ E^\circ = -0.40$ V
$Ag^+(aq) + e^- \rightarrow Ag(s)\ E^\circ = 0.80V$

The standard free energy change for the reaction $2Ag^+(aq) + Cd(s) \rightarrow 2Ag(s) + Cd^{2+}(aq)$ is
(A) 0 (B) 231.6 kJ (C) -231.6 kJ
(D) -115.8 kJ (E) 115.8 kJ

54. Which one of the following material conducts electricity ?
(A) barium sulphate (B) molten sulphur
(C) fused potassium chloride
(D) diamond (E) crystalline sodium chloride

55. Alkyl halides react with metallic sodium in dry ether producing
(A) alkenes with triple the number of carbon atoms
(B) alkenes with same number of carbon atom
(C) alkenes with double the number of carbon atoms
(D) alkanes with same number of carbon atoms
(E) alkanes with double the number of carbon atoms

56. The volumes of gases H_2, CH_4, CO_2 and NH_3 adsorbed by 1 gm of activated charcol at 298K are in the order;
(A) $CO_2 > NH_3 > H_2 > CH_4$ (B) $NH_3 > CO_2 > CH_4 > H_2$
(C) $CO_2 > NH_3 > CH_4 = H_2$ (D) $H_2 > CH_4 > CO_2 > NH_3$
(E) $CH_4 > CO_2 > NH_3 > H_2$

57. Identify the reactant X and the product Y.
$CH_3 - CO - CH_3 + X \rightarrow (CH_3)_3C - OH$
(A) X = MgCl (OH); Y = $(CH_3)_2CHOH$
(B) X = C_2H_5MgCl Y = $(CH_3)_3$ C- OH
(C) X = CH_3 MgCl; Y = $(CH_3)_3$ C- OH
(D) X = CH_3 MgCl; Y = $C_2H_5COOCH_3$
(E) X = $MgCl_2$ Y = CH_3 CH = CH_2

58. The ester among the following is
(A) Sodium acetate (B) Calcium formate
(C) Methyl butanoate (D) Calcium lactate
(E) Ammonium acetate

59. Benzaldehyde condenses with N,N-dimethylaniline in presence of anhydrous $ZnCl_2$ to give
(A) Malachite green (B) Buffer yellow (C) Anil
(D) Michler's ketone (E) Azo dye

60. Benzoic acid, when heated with soda lime yields
(A) beznyl alcohol (B) cyclohexane
(C) Toluene (D) benzene (E) benzaldehyde

61. Which of the following is currently used as a tyre cord ?
(A) Polypropylene (B) Bakelite
(C) Nylon-6 (D) Terelene (E) Polyethylene

62. The pH value of the solution in which a particular amino acid does not migrate under the influence of an electric field is called the
(A) neutralization point (B) effusion
(C) isoelectric point (D) eutectic point
(E) yielding point

63. Accepting the definition that an acid is a proton donor, the acid in the following reaction
$NH_3 + H_2O \rightarrow NH_4^+ \; OH^-$ is
(A) H_2O (B) OH^- (C) NH_4^+ (D) H (E) NH_3

64. When aniline is heated with chloroform and alcoholic KOH it gives
(A) C_6H_5NC (B) $(C_6H_5)_4N^+OH^-$
(C) $(C_6H_5)_2NCl$ (D) $(C_6H_5)_3N$
(E) $(C_6H_5)_2NH$

65. Identify the product, Z in the following reaction scheme
$$C_6H_5NH_2 \xrightarrow{(Ac_2)_2O} X \xrightarrow{Br_2/CCl_4} Y \xrightarrow{HOH} Z$$
(A) o-Bromo acetanilide
(B) o-Bromo acetophenone
(C) p-Bromo acetanilide
(D) p-Bromo acetophenone (E) p-Bromoaniline

2004

1. In a given shell, the order of screening effect is
(A) d > f < s > p (B) f > p > s > d
(C) s > p > d > f (D) f > d > p > s (E) p < d < s <f

2. Which one of the following compounds gives methane on treatment with water ?
(A) VC (B) SiC (C) B_4C (D) Al_4C_3 (D) CaC_2

3. The hardest substance amongs the following is
(A) Graphite (B) Be_2C
(C) B_4C_3 (D) SiC (E) Titanium

4. The H-O-O bond angle in H_2O_2 is
(A) 104.5° (B) 106° (C) 97° (D) 107.28° (E) 109.28°

5. H_2O_2 acts as an oxidizing agent in
(A) Alkaline mediium
(B) Alkaline and neutral medium
(C) Acidic and alkaline medium
(D) Neutral medium (E) Acidic medium

6. Which of the following does not exist in free form ?
(A) BBr_3 (B) BH_3 (C) BF_3 (D) BCl_3
(E) None of the above

7. A mixture of NO_2^- and N_2O_4 has a vapour density of 38.3 at 300 K. What is the number of moles of NO_2 in 100g of the mixture ?
(A) 3.4 (B) 3.86 (C) 0.437 (D) 0.043 (E) 4.4

8. A sample of phosphorous trichloride (PCl_3) contains 1:4 moles of the substance. How many atoms are there in the sample ?
(A) 3.372×10^{24} (B) 2.409×10^{24}
(C) 4 (D) 5.6 (E) 8.431×10^{23}

9. Which of the following electronic configuration is not

possible according to Hund's rule ?

(A) $1s^2\ 2s^2\ 2p_x2$ (B) $1s^2\ 2s^2\ 2p_x2\ 2p_y^1\ 2p_z^1$
(C) $ls^2 2s^2$ (D) $ls^2\ 2s^1$
(E) $1s^2\ 2s^2\ 2p_x^1\ 2p_y^1\ 2p_z^1$

10. Time taken for an electron to complete one revolution in the Bohr orbit of hydrogen atom is

(A) $4\pi^2mr^2/nh$ (B) $nh/4\pi^2mr^2$ (C) $2\pi mr/n^2h^2$
(D) $h/2\pi mr$ (E) none of these

11. The atomic number of an element is derived from

(A) Number of neutrons (B) Number of isotopes
(C) Number of nucleons (D) Number of electrons
(E) Number of protons

12. The paramagnetic property of the oxygen molecule is due to the presence of unpaired electrons present in

(A) $(\pi^*_{2Pz})^1$ and $(\pi_{2Pz})^1$
(B) $(\pi^*_{2Py})^1$ and $(\pi_{2Py})^1$
(C) $(\pi^*_{2Py})^1$ and $(\pi^*_{2Pz})^1$
(D) $(\sigma_{2PX})^1$ and $(\sigma_{2Py})^1$
(E) $(\sigma_{2PX})^1$ and $(\pi_{2PX})^1$

13. Among the compounds, BF_3, NCl_3, H_2S, SF_4 and $BeCl_2$ identify the ones in which the central atom has the same type of hybridization

(A) BF_3, NCl_3 and H_2S (B) SF_4 and $BeCl_2$
(C) NCl_3 and H_2S (D) BF_3 and NCl_3
(E) H_2S and $BeCl_2$

14. Among the species : CO_2, CH_3COO^-, CO, CO_3^{2-} HCHO, which has the weakest carbon-oxygen bond ?

(A) CO_2 (B) CH_3COO^- (C) CO (D) CO_3^{2-} (E) HCHO

15. The boiling point of methanoi is greater than that of methyl thiol, because

(A) There is no hydrogen bonding in methanol and intermolecular hydrogen bonding in methyl thiol
(B) There is intramolecular hydrogen bonding in methanol and no hydrogen bonding in methyl thiol
(C) There is no hydrogen bonding in methanol and intramolecular hydrogen bonding in methyl thiol
(D) There is intramolecular hydrogen bonding in methanol and intermolecular hydrogen bonding in methyl thiol
(E) There is intermolecular hydrogen bonding in methanol and no hydrogen bonding in methyl thiol

16. If a molecule MX_3 has zero dipole moment, the sigma bonding orbital used by M are

(A) sp^3d^2 hybrid (B) sp^3-hybrid (C) sp^2-hybrid
(D) sp^3d-hybrid (E) sp-hybrid

17. The number of α and β-particles emitted respectively during the transformation of $Th^{232}_{90} \rightarrow Pb^{208}_{82}$

(A) 4.6 (B) 6.4 (C) 6.8 (D) 3.6 (E) 6.3

18. To which of the following mixtures Dalton's law of partial pressures is not applicable ?

(A) CO and CO_2 (B) CO_2 and N_2
(C) CH_4 and C_2H_6 (D) HCl and NH_3
(E) O_2 and N_2

19. A gas mixture contains 50% helium and 50% methane by volume. What is the percent by 'weight of methane in the mixture ?

(A) 75% (B) 80.03%
(C) 19.97 % (D) 20.05% (E) 50%

20. N_2 is found in a litre flask under 100 kP_a pressure and O_2, is found in another 3 litre flask under 320 kP_a pressure. If the two flasks are connected, the resultant pressure is

(A) 310 kP_a (B) 210 kP_a
(C) 420 kP_a (D) 365 kP_a (E) 265 kP_a

21. Which of the following colligative properties can provide molar mass of 21. proteins (or polymers or colloids) with, greater precision ?

(A) Depression in freezing point
(B) Osmotic pressure (C) Rast's method
(D) Relative lowering of vapour pressure
(E) Elevation of boiling point

22. 116 mg of a compound on vaporisation in a Victor Meyer's apparatus displaces 44.8 ml of air measured at S.T.P. The molecular weight of the compound is

(A) 46.4 (B) 44.8 (C) 58 (D) 232 (E) 116

23. An orgnic compound contains 49.3% carbon, 6.84% hydrogen and its vapour density is 73. Molecular formula of the compound is

(A) $C_6H_{10}O_4$ (B) $C_4H_{10}O_2$
(C) C_6H_9O (D) $C_3H_{10}O_2$ (E) $C_3H_8O_2$

24. A compound has an empirical formula C_2H_4O. An independent analysis gave a value of 132.16 for its molecula mass. What is the correct molecular formula ?

(A) $C_4H_8O_5$ (B) $C_6H_{12}O_3$ (C) C_7O_3
(D) $C_{10}H_{12}$ (E) $C_4H_4O_5$

25. Photochemical chlorination of alkane is initiated by a process of

(A) Peroxidation (B) Homolysis
(C) Pyrolysis (D) Substitution (E) Craking

26. Orbital interaction between the sigma bonds of a substituent group and a neighbouring pi orbital is known as

(A) Electric quadrupole interactions
(B) Dipole-dipole interactions (C) Steric effect.
(D) Inductive effect (E) Hyperconjugation

27. Which of the following compounds will exhibit cis-trans isomerism ?

(A) Butanol (B) Butanone (C) 2-Butanol
(D) 2-Butyne (E) 2-Butene

28. How many enantiomer pairs are obtained by monochlorination of 2, 3-dimethylbutane ?

(A) One (B) Three (C) Two (D) Four (E) Nil

29. Among the following compounds $C_3H_7NH_2$, NH_3, CH_3NH_2, $C_2H_5NH_2$ and $C_6H_5NH_2$, the least basic compound is
(A) CH_3NH_2 (B) $C_6H_5NH_2$ (C) $C_6H_5NH_2$
(D) $C_3H_7NH_2$ (E) NH_3

30. What happens when a mixture of acetylene and hydrogen is passed over heated Lindlar's catalyst ?
(A) Ethylene is formed
(B) Acetylene and ethane are formed
(C) Ethylene and water are formed
(D) Ethane and water are formed
(E) None of the above

31. 1,2-Dibromoethane when heated with alcoholic potash gives
(A) Ethylene (B) Methane
(C) Ethane (D) Acetylene (E) None of the above

32. An alkyl bromide produces a single alkene when it reacts with sodium ethoxide and ethanol. This alkene undergoes hydrogenation and produces 2-methylbutane. What is the identity of the alkyl bromide ?
(A) 1-bromo-2-methylbutane
(B) 2-bromo-2-methylbutane
(C) 2-bromopentane (E) 1-bromobutane
(D) 1-Bromo -2, 2-dimethylpropane

33. An organic compound A(C_4H_9Cl) on reaction with Na/diethyl ether gives a hydrocarbon, which on monochlorination gives only one chloro derivative. A is
(A) Isobutyl chloride (B) s-butyl chloride
(C) t, butyl chloride (D) n-butyl chloride
(E) Non of these

34. The number of tetrahedral voids in the unit cell of a face centered cubic lattice of similar atoms is
(A) 12 (B) 10 (C) 8 (D) 6 (E) 4

35. 40 mg of pure sodium hydroxide is dissolved in 10 litres of distilled water. The pH of the solution is
(A) 11 (B) 12 (C) 8 (D) 9 (E) 10

36. The principal buffer present in human blood
(A) $H_2CO_3 + HCO_3^-$
(B) $Na_2HPO_4 + Na_3PO_4$
(C) $CH_3COOH + CH_3COONa$
(D) $H_3PO_4 + NaH_2PO_4$
(E) $NaH_2PO_4 + NaHPO_4$

37. Ammonium carbonate when heated to 200°C gives amixture of NH_3 and CO_2 vapour with a density of 13.0. What is the degree of dissociation of ammonium carbonate ?
(A) 5/2 (B) 1 (C) 2 (D) 1/2 (E) 3/2

38. Consider the following reversible reaction at equilibrium,

$2H_2O(g) \rightarrow 2H_2(g) + O_2(g)$; $\Delta H = 24.7$ kJ

Which one of the following changes in conditions will lead to maximum decomposition of $H_2O(g)$?
(A) Increasing temperature and decreasing pressure
(B) Increasing temperature at constant pressure
(C) Increasing pressure at constant temperature
(D) Increasing both temperature and pressure
(E) Decreasing temperature and increasing pressure

39. Values of ΔH and ΔS for five differenct reactiosn are given below

Reaction	ΔH (kJ mol^{-1})	ΔS (JK1 mol^{-1})
I	+98.0	+14.8
II	−55.5	−84.6
III	+28.3	−17.0
IV	−40.5	+24.6
V	+34.7	0.0

On the basis of these values predict which one of these will be spontaneous at all temperatures ?
(A) Reaction IV (B) Reaction V
(C) Reaction I (D) Reaction II (E) Reaction III

40. The difference between heats of reaction at constant pressure and constant volume for the reaction
$2C_6H_6(l) + 15O_2(g) \rightarrow 12CO_2(g) + 6H_2O(l)$ at 25°C in kJ mol^{-1} is
(A) −7.43 (B) +3.72 (C) −3.72 (D) +7.43 (E) −3.27

41. A cylinder of gas supplied by Bharat Petroleum is assumed to.contain 14 kg of butane. If a normal family requires 20,000 kJ of enerey per day for cooking, butane gas in the cylinder last for.... days
(ΔH_c of $C_4H_{10} = -2658$ kJ per mole)
(A) 32 days (B) 40 days (C) 50 days
(D) 20 days (E) 15 days

42. Guigret's green is known as
(A) FeO, Cr_2O_3 (B) $FeCO_3.Cr_2O_3$
(C) Cu_2O_3 (D) $FeO_3.2H_2O$ (E) $Cr_2O_2.2H_2O$

43. Vanadium (III) oxide is a strong
(A) Reducing agent (B) Oxidizing agent
(C) Drying agent (D) Precipitating agent
(E) Wetting agent

44. The lanthanide contraction relates to
(A) Valence electrons (B) Oxidation states
(C) lonization energy (D) Atomic radil
(E) Atomic as well as M^{3+} radii

45. When one mol of each of the following complex salts is treated with excess of $AgNO_3$ which of them gives maximum amount of AgCl ?
(A) $[Co(NH_3)_4Cl_2]Cl$ (B) $Na_2\ [PtCl_6]$
(C) $[Co(NH_3)_2Cl_3]$ (D) $[Co(NH_3)_6]Cl_3$
(E) $[Co(NH_3)_5Cl]Cl_2$

46. Which of the following species is expected to show

the highest magnetic moment ?
(Atomic numbers: Cr = 24; Mn = 25; Co = 27; Ni = 28; Cu = 29)
(A) Ni^{2+} (B) Co^{2+} (C) Cu^{2+} (D) Mn^{2+} (E) Cr^{2+}

47. Which process of purification is represented by the following scheme ?
$Ti + 2I_2 \rightarrow TiI_4 \rightarrow Ti + I_2$
(A) Electrolytic refining (B) Poling (C) Zone refining
(D) Cupellation (E) Van-Arkel process

48. Which of the following is not a mineral of iron ?
(A) Magnetite (B) Siderite (C) Smithsonite
(D) Limonite (E) Haematite

49. The weight of pure NaOH required to prepare 250 cm^3 of 0.1 N solutions is
(A) 10 g (B) 5 g (C) 2 g (D) 1 g (E) 4 g

50. In the brown ring test for the nitrate ion, the brown colour is due to
(A) $[Fe(H_2O)_6]^+$ (B) $[Fe(H_2O)_2(NO)_2]^{3+}$
(C) $[Fe(H_2O)_4(NO)_2]^+$ (D) $[Fe(H_2O)_5NO]^{2+}$
(E) $[Fe(H_2O)_5NO]^{3+}$

51. Which of the following sulphides when heated strongly in air gives the corresponding metal ?
(A) Cu_2S (B) Fe_2S_3 (C) HgS (D) CuS (E) FeS

52. When a standard solution of NaOH is left in air for a few hours,
(A) Strength will decrease (B) Strength wil increase
(C) The concentration of Na^+ ions will decrease
(D) A precipitate will form (E) All are wrong

53. The volumes of 4N HCl and ION HCl required to make 1 litre of 6N HCl are
(A) 0.80 litre of 4N HCl and 0.20 litre of 10N HCl
(B) 0.50 litre of 4N HCl and 0.50 litre of 10N HCl
(C) 0.75 litre of 10N HCl and 0.25 litre of 4N HCl
(D) 0.25 litre of 4 N HCl and 0.75 litre of 10N HCl
(E) 0.67 litre of 4N HCl and 0.33 litre of 10N HCl

54. Ethylene glycol reacts with excess of PCl_5 to give
(A) 1, 1,1-trichloroethane
(B) 1,1,2,2-tetrachloroethane
(C) 2, 2-dichloroethane (D) 1, 1-dichloroethane
(E) 1, 2-dichloroethane

55. The reaction of an ester RCOOR' with an alcohol R" OH in the presence of an acid
(A) R"COOR (B) RCOOR" (C) R'COOR"
(D) RCOOH (E) R'COOH

56. Which of the following undergoes Cannizzaro's reaction ?
(A) $C_6H_5CH_2Cl$ (B) C_6H_5OH
(C) C_6H_5CHO (D) CH_3CHO
(E) CH_3COCH_3

57. Aldehydes and Ketones can be reduced to corresponding hydrocarbons by
(A) Refluxing with soda amalgam and water
(B) Refluxing with Zinc amalgam and Conc. HCl
(C) Passing the vapour untier heated PbO_2
(D) Refluxing with water
(E) Refluxing with strong acid

58. RCOOH on treatment with PCl_5 and KCN, is subjected to hydrolysis followed by Clemmensen's reduction, the product obtained is
(A) RCOCN (B) RCN (C) R-OH
(D) RCH_2-COCl (E) RCH_2-COOH

59. The end product of the reaction
Ethyl amine $\xrightarrow{HNO_2}$ A $\xrightarrow{PCl_5}$ B $\xrightarrow{KCN}$ C is
(A) Propane nitrite (B) Triethyl amine
(C) Methyl amine (D) Ethyl amine
(E) Diethyl amine

60. Which of the following does not reduce Tollen's reagent ?
(A) HCOOH (B) $C_6H_5NO_2$ (C) CH_3CHO
(D) C_6H_5NHOH (E) None

61. Teflon is a polymer of
(A) Difluorodichloro ethane
(B) Tetrafluoro propene (C) Tetrafluoro ethane
(D) Trifluoro ethene (E) Difluro ethene

62. Which of the following is not correct regarding terylene ?
(A) It is also called dacron (B) Thermosetting plastic
(C) Step-growth polymer (D) Synthetic fibre
(D) Condensation polymer

63. An alteration in the base sequence of nucleic acid molecule is called
(A) Flocculation (B) Dislocation
(C) Duplication (D) Mutation (E) Replication

64. Carbohydrates are stored in human body as the polysaccharide
(A) Galactose (B) Fructose
(C) Starch (D) Glucose (E) Glycogen

65. Which of the hydrogen halides forms salts like KHX_2 (Where X is a halogen atom) ?
(A) HBr (B) HI (C) HCl (D) HF (E) All the above

Kerala - 2005

1. The value of PV/T for one mole of an ideal gas is nearly equal to
(A) 2 J $mol^{-1}K^{-1}$ (B) 8.3 eal $mol^{-1}K^{-1}$
(c) 4.2 J $mol^{-1}K^{-1}$ (D) 2 cal $mol^{-1}K^{-1}$

2. The temperature of equal masses of three different

liquids A, B and C are 12°C, 19°C and 28°C respectively. The temperature when A and B are mixed is 16°C and when B and C are mixed is 23°C. The temperature when A and C are mixed is
(A) 18.2°C (B) 22°C (C) 20.2°C
(D) 24.2°C (E) 20.8°C

3. The time period of the seconds hand of a watch is
(A) 1 hour (B) 1s (C) 12 hours
(D) 1 minute (E) 0.1 hour

4. Identify the product in the following sequence 3, 4, 5-tribromoaniline $\xrightarrow[(2)\ H_3PO_2]{(1)\ \text{diazotization}}$
(A) 3,4,5- tribromobenzene (B) 1,2,3- tribromobenzene
(C) 2,4, 6- tribromobenzene
(D) 3,4,5- tribromonitrobenzene (E) 3,4,5- tribromophenol

5. Among the amines a) $C_6H_5NH_2$ b) CH_3NH_2 c) $(CH_3)_2NH$ d) $(CH_3)_3N$ the order of basicity is
(A) a<d<b<c (B) d<c<b<a (C) a>b>c>d
(D) b<c<d<a (E) d<c<b<a

6. In biological systems, the RNA molecules direct the synthesis of specific, proteins which are characteristic of each kind of organism. This process is known as -
(A) Transcription (B) Mutation
(C) Replication (D) Translation (E) Floculation

7. In Antarctica ozone depletion is due to the formation of following compound
(A) acrolein (B) peroxy acetyl nitrate (C) SO_2 and SO_3
(D) chlorine nitrate (E) formaldehyde

8. 100 g $CaCO_3$ is treated with 1 litre of 1-N HCl. What would be the weight of CO_2 liberated after the completion of the reaction?
(A) 5.5 g (B) 11 g (C) 22g (D) 33 g (E) 44g

9. The relationship between the energy E_1 of the radiation with a wave length 8000Å and the energy E_2 of the radiation with a wavelength 16000Å is
(A) $E_1 = 6E_2$ (B) $E_1 = 2E_2$ (C) $E_1 = 4E_2$
(D) $E_1 = {}^1/_2E_2$ (E) $E_1 = E_2$

10. If the molecule of HCl were totally polar, the expected value of dipole moment is 6.12 D, (debye), but the experimental value of dipole moment was 1.03 D. Calculate the percentage ionic character
(A) 17 (B) 83 (C) 50 (D) Zero (E) 90

11. Which one of the following molecules has the smallest bond angle?
(A) NH_3 (B) PH_3 (C) H_2O (D) H_2Se (E) H_2S

12. If the absolute temperature of a gas is doubled and the pressure is reduced to one half, the volume of the gas will
(A) remain unchanged (B) be doubled
(C) increase fourfold (D) be halved
(E) be reduced to one-fourth

13. When electric current is passed through an ionic hydride in molten state
(A) hydrogen is liberated at anode
(B) hydrogen is liberated at cathode
(C) no change takes place
(D) hydride ion migrates towards cathode
(E) hydride ion remains in solution

14. The order of first ionization energies of the elements Li, Be, B, Na is
(A) Li>Be>B>Na (B) Be> B> Li> Na
(C) Na> Li> Be (D) Be> Li> B> Na
(E) B> Be> Li> Na

15. The froth-floatation process is based upon
(A) The difference in the specific gravity of ore and gangue particles
(B) The magnetic properties of gangue and ore
(C) Preferential wetting of gangue particles by oil
(D) Preferential wetting of ore particles by oil
(E) The solubility of ore particles in suitable reagent

16. Which one of the following statements is true for all the alkali metals ?
(A) Their nitrates decompose on heating to give NO_2 and O_2
(B) Their carbonates decompose on heating of give CO_2 and the metal oxide
(C) They react with oxygen to give mainly the oxide M_2O
(D) They react with halogens to give the halides MX
(E) They react with nitrogen to give nitrides

17. The order of acidic strength of boron trihalides is
(A) $BF_3<BCl_3<BCr_3<BI_3$ (B) $BI_3<BBr_3<BCl_3<BF_3$
(C) $BCl_3<BBr_3<BI_3<BF_3$ (D) $BBr_3<BCl_3<BCF_3<BI_3$
(E) $BF_3<BI_3<BCI_3<BBr_3$

18. The carbide which reacts with water to form propyne is
(A) CaC_2 (B) SiC (C) Mg_2C_3 (D) Al_4C_3 (E) Be_2C

19. Effective magnetic moment of Sc^{3+} ion is
(A) 1.72 (B) 0 (C) 5.92 (D) 2.83 (E) 3.87

20. Potassium permanganate acts as an oxidant in alkaline and acidic media. The final products formed from $KMnO_4$ in the two conditions are respectively
(A) MnO_2- and Mn^{3+} (B) Mn^{3+} and Mn^{2+}
(C) Mn^{2+} and Mn^{2+} (D) MnO_2 and Mn^{2+}
(E) Mn^{+2}, MnO_2

21. Calculate the mass loss in the following

$${}^2_1H + {}^3_1H \longrightarrow {}^4_2He + {}^1_0n$$

(A) 0.018 amu (B) 0.18 amu (C) 0.0018 amu
(D) 1.8 amu (E) 18 amu

22. Match list I and list II and select the correct answer using the code given below the lists

List I	List II
Nuclear reactor components	Substance used
1. Moderator	a. Uranium
2. Control rods	b. Graphite
3. Fuel rods	c. Boron
4. Coolant	d. Lead
	e. Sodium

codes :
(A) 1-b, 2-a, 3-c, 4-e (B) 1-b, 2-c, 3-a, 4-e
(C) 1-c, 2-b, 3-a, 4-e (D) 1-c, 2-d, 3-a, 4-b
(E) 1-d, 2-c, 3-b, 4-a

23. ΔH and ΔS for a reaction are +30.558 kJ mol^{-1} and 0.066 k Jk^{-1} at 1 atm pressure. The temperature at which free energy change is equal to zero and the nature of the reaction below this temperature are
(A) 483 K, spontaneous (B) 443 K, non-spontaneous
(C) 443 K, spontaneous (D) 463 K, non-spontaneous
(E) 463 K, spontaneous

24. What would be the heat released when an aqueous solution containing 0.5 mole of HNO_3 is mixed with 0.3 of OH^- (Enthalpy of neutralization is 57.1 kJ)?
(A) 28.5 kJ (B) 17.1 kJ
(C) 45.7 kJ (D) 1.7 kJ (E) 2.85 kJ

25. $A(g) + 3B(g) \rightleftharpoons 4C(g)$. Initial concentration of A is equal to that of B. The equilibrium concentrations of A and C are equal kc =
(A) 0.08 (B) 0.8 (C) 8 (D) 80 (E) 1/8

26. Two moles of PCl_5 is heated in a closed vessel of 2 litre capacity. When the equilibrium is attained 40% of it has been found to be dissociated. What is the K_c in mol/ dm^3 ?
(A) 0.532 (B) 0.266 (C) 0.133 (D) 0.174 (E) 0.25

27. Dry air is passed through a solution containing 10g of a solute in 90g of water and then through pure water. The loss in weights of solution is 2.5 and that of pure solvent is 0.05 g Calculate the molecular weight of the solute?
(A) 50 (B) 180 (C) 102 (D) 25 (E) 51

28. The van't Hoff factor for $BaCl_2$ at 0.01 M concentration is 1.98. The percentage of dissociation fo $BaCl_2$ at this concentration is
(A) 49 (B) 69 (C) 89 (D) 98 (E) 100

29. The standard electrode potentials of Ag^+/Ag is + 0.80 V and Cu^+/Cu is + 0.34 V. These electrodes are connected through a salt bride and if
(A) Copper electrode acts as a cathode then E° cell is +0.46 V
(B) Silver electrode acts as anode then E° cell is - 0.34 V
(C) Copper electrode acts as anode then E° cell is + 0.46V
(D) Silver electrode acts as a cathode then E° cell is - 0.34 V
(E) Silver electrode acts as anode and E° cell is + 1.14V

30. For the reaction $2Na_2O_5(g) \longrightarrow 4NO_2(g) + O_2(g)$ if the concentration of NO_2 increases by 5.2×10^{-3} M in 100 seconds then the rate of the reactions is
(A) 1.3×10^{-5} Ms^{-1} (B) $5 \times 10^{-4} Ms^{-1}$ (C) $7.6 \times 10^{-4} Ms^{-1}$
(D) $2 \times 10^{-4} Ms^{-1}$ (E) $2.5 \times 10^{-4} Ms^{-1}$

31. A first order reaction is 10% complete in 20 min. The time taken for 19% completion is
(A) 30 min (B) 40 min (C) 50 min
(D) 38 min (E) 45 min

32. Lyophilic sols are more stable than lyophobic sols because the particles
(A) are positively charged (B) are negatively charged
(C) are solvated (D) repel each other
(E) are heavy

33. Potassium stearate is obtained by the saponification of an oil or fat. It has the formula $CH_3 - (CH_2)_{16} - COO^- K^+$. The molecule has a lyophobic end (CH_3) and a lyophilic end $COO^- K^+$. Potassium stearate is an example for
(A) lyophobic colloid (B) lyophilic colloid
(C) multimolecular colloid (D) macromolecular colloid
(E) associated colloid or micelle

34. (a) $K_4[Fe(CN)_6]$ (b) $K_3[Cr(CN)_6]$ (c) $K_3[Co(CN)_6]$ (d) $K_2[Ni(CN)_4]$. Select the complexes, which are diamagnetic
(A) (a), (b) and (c) (B) (b), (c) and (d)
(C) (a), (c) and (d) (D) (a), (b) and (d)
(E) (b) and (d)

35. Which is not true of the coordination compound $[Co(en)_2Cl_2]Cl$?
(A) exhibits geometrical isomerism
(B) exhibits optical isomerism
(C) exhibits ionization isomerism
(D) is an octahedral complex (E) is a cationic complex

36. The IUPAC name of the compound is

$$HOOC-CH_2-\underset{\displaystyle COOH}{\underset{|}{CH}}-CH_2-CH_2-COOH$$

(A) 2(carboxymethyl)-pentane-1, 5-dioic acid
(B) 3-carboxyhexane-1, 6-dioic acid
(C) butane-1, 2, 4-tricarboxylic acid
(D) 4-carboxyhexane-1, 6-dioic acid
(E) 1, 2-dicarboxy-pentanoic acid

37. How much of sulphur is present in an organic compound, if 0.53 g of the compound gave 1.158 g of

$BaSO_4$ on analysis?
(A) 10% (B) 15% (C) 20% (D) 25% (E) 30%

38. An alkene having the molecular formula C_9H_{18} on ozonolysis gives 2, 2-dimethyl-propanal and 2-butanone. The alkene is
(A) 2,2, 4-trimethyl-3-hexene
(B) 2,2,6-trimethyl-3-hexane (C) 2,3,4-trimethyl-2-hexane
(D) 2,2,4-trimethyl-2-hexene (E) 2,2-dimethyl-3 heptene

39. Observe the following reactions and predict the nature of A and B

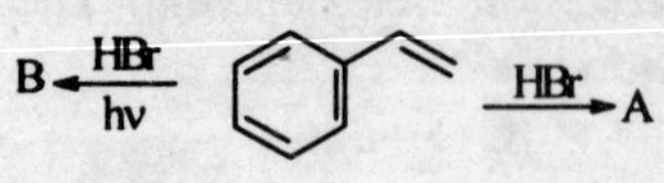

B $\xleftarrow[hv]{HBr}$ (styrene) $\xrightarrow{HBr}$ A

(A) A and B both are (1-bromoethyl)benzene [Br]
(B) A and B both are (2-bromoethyl)benzene [Br]
(C) A is (1-bromoethyl)benzene [Br] and B is (2-bromoethyl)benzene [Br]
(D) A is (2-bromoethyl)benzene [Br] and B is (1-bromoethyl)benzene [Br]
(E) A is 1-bromo-4-(2-bromoethyl)benzene [Br, Br] and B is (2-bromoethyl)benzene [Br]

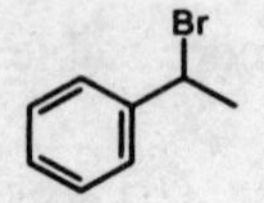

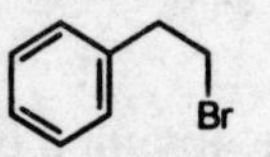

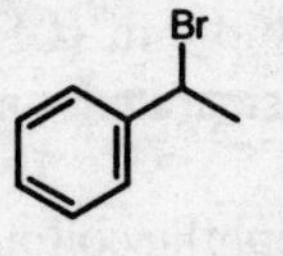

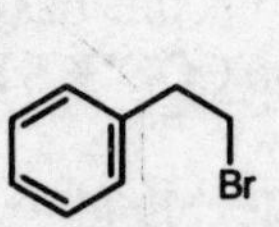

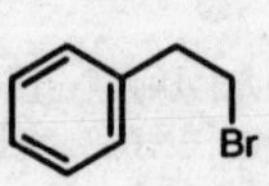

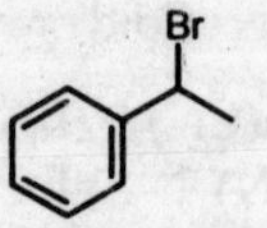

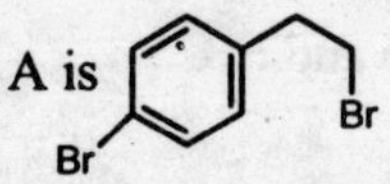

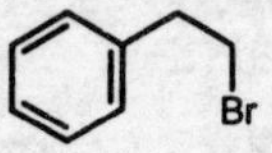

40. The name of the compound is (structure)
(A) (2Z, 4Z)- 2, 4-hexadiene
(B) (2Z, 4E)-2,4-hexadiene (C) (2E, 4Z)-2,4-hexadiene
(D) (4E, 4E)-2,4-hexadiene (E) (2E, 4E)-2,4-hexadiene

41. When 32.25g of ethyl chloride is subjected to dehydrohalogenation reaction, the yield of the alkene formed is 50%. The mass of the product formed is (atomic mass of chlorine is 35.5)
(A) 14 g (B) 28g (C) 64.5 g (D) 56 g (E) 7 g

42. Chlorination of toluene in presence of light and heat followed by treatment with aqueous NaOH give.
(A) o-cresol (B) p-cresol
(C) mixture of o-cresol and p-cresol
(D) benzyl alcohol (E) 1,3,5, - trihydroxy toluene

43. Nitration of aniline in strongly acidic medium result in the formation of m- nitroaniline also. This is because
(A) Amino group is meta orienting during electrophilic substitution reaction
(B) Nitro group goes always to the meta position irrespective of the substituents
(C) Nitration of aniline is a nucleophilic substitution reaction in strongly acidic medium
(D) In strongly acidic conditions aniline is present as anilinium ion
(E) Strong acids generate nitrite anion which can attack only the meta position

44. How many sigma and pi bonds are present in toluene?
(A) $3\pi + 8\sigma$ (B) $3\pi + 6\sigma$ (C) $3\pi + 15\sigma$
(D) $6\pi + 3\sigma$ (E) $6\pi + 6\sigma$

45. Which of the following Fischer's projection formula is identical to D- glyceraldehydes?

(A) top: CH_2OH; left: HO; right: CHO; bottom: H
(B) top: CH_2OH; left: H; right: OH; bottom: CHO
(C) top: CHO; left: HO; right: CH_2OH; bottom: H
(D) top: CHO; left: H; right: CH_2OH; bottom: OH
(E) top: H; left: HO; right: CH_2OH; bottom: CHO

46. CH_3-CHO+HCN → A.
Compound A on hydrolysis gives
(A) CH_3-CH_2-COOH (B) CH_3-CH_2-CH_2-NH_2
(C) CH_3-CO-COOH (D) CH_3-CO- CH=NOH
(E) CH_3—CH(OH)—COOH

47. Which of the following does not undergo Cannizzaro's reaction ?
(A) Benzaldehyde (B) 2-Methylpropanal
(C) p-methoxybenzaldehyde
(D) 2,2-dimethylpropanal (E) Formaladehyde

KERALA - 1995

1. (C)	31. (D)	61. (C)	91. (D)
2. (D)	32. (D)	62. (D)	92. (C)
3. (A)	33. (D)	63. (B)	93. (B)
4. (D)	34. (B)	64. (D)	94. (C)
5. (B)	35. (C)	65. (C)	95. (C)
6. (A)	36. (D)	66. (A)	96. (B)
7. (C)	37. (A)	67. (D)	97. (C)
8. (A)	38. (A)	68. (A)	98. (D)
9. (C)	39. (D)	69. (A)	99. (B)
10. (D)	40. (C)	70. (B)	100. (D)
11. (C)	41. (C)	71. (C)	101. (C)
12. (B)	42. (C)	72. (C)	102. (A)
13. (A)	43. (B)	73. (D)	103. (A)
14. (C)	44. (B)	74. (D)	104. (A)
15. (D)	45. (C)	75. (C)	105. (B)
16. (A)	46. (B)	76. (D)	106. (B)
17. (A)	47. (D)	77. (B)	107. (C)
18. (A)	48. (B)	78. (A)	108. (A)
19. (D)	49. (D)	79. (A)	109. (B)
20. (B)	50. (C)	80. (B)	110. (B)
21. (B)	51. (C)	81. (A)	111. (B)
22. (A)	52. (A)	82. (B)	112. (A)
23. (D)	53. (D)	83. (B)	113. (A)
24. (D)	54. (A)	84. (D)	114. (D)
25. (C)	55. (C)	85. (D)	115. (D)
26. (B)	56. (B)	86. (B)	116. (A)
27. (C)	57. (A)	87. (C)	117. (A)
28. (B)	58. (B)	88. (D)	118. (D)
29. (C)	59. (C)	89. (A)	
30. (B)	60. (D)	90. (D)	

KERALA - 1996

1. (A)	31. (B)	61. (C)	91. (A)
2. (C)	32. (C)	62. (C)	92. (A)
3. (C)	33. (A)	63. (C)	93. (D)
4. (A)	34. (C)	64. (D)	94. (A)
5. (C)	35. (D)	65. (C)	95. (A)
6. (A)	36. (A)	66. (C)	96. (A)
7. (A)	37. (B)	67. (A)	97. (D)
8. (B)	38. (D)	68. (D)	98. (B)
9. (B)	39. (A)	69. (A)	99. (C)
10. (D)	40. (A)	70. (A)	100. (B)
11. (A)	41. (A)	71. (C)	101. (D)
12. (A)	42. (B)	72. (C)	102. (C)
13. (D)	43. (A)	73. (D)	103. (A)
14. (D)	44. (D)	74. (A)	104. (B)
15. (A)	45. (B)	75. (A)	105. (C)
16. (A)	46. (C)	76. (B)	106. (C)
17. (D)	47. (C)	77. (A)	107. (D)
18. (A)	48. (D)	78. (A)	108. (C)
19. (D)	49. (A)	79. (D)	109. (B)
20. (D)	50. (A)	80. (C)	110. (C)
21. (C)	51. (C)	81. (B)	111. (B)
22. (D)	52. (C)	82. (C)	112. (B)
23. (A)	53. (C)	83. (A)	113. (D)
24. (B)	54. (B)	84. (C)	114. (D)
25. (D)	55. (C)	85. (C)	115. (B)
26. (D)	56. (C)	86. (C)	116. (A)
27. (C)	57. (C)	87. (C)	117. (C)
28. (C)	58. (B)	88. (A)	118. (C)
29. (C)	59. (C)	89. (A)	119. (B)
30. (B)	60. (B)	90. (A)	120. (C)

KERALA - 1997

1. (A)	19. (C)	37. (A)	55. (B)	73. (C)	91. (C)	109. (D)
2. (C)	20. (A)	38. (B)	56. (A)	74. (B)	92. (C)	110. (B)
3. (C)	21. (C)	39. (D)	57. (D)	75. (B)	93. (D)	111. (C)
4. (B)	22. (B)	40. (D)	58. (B)	76. (D)	94. (C)	112. (C)
5. (D)	23. (D)	41. (B)	59. (C)	77. (A)	95. (C)	113. (C)
6. (C)	24. (C)	42. (C)	60. (B)	78. (A)	96. (A)	114. (C)
7. (B)	25. (C)	43. (B)	61. (B)	79. (D)	97. (A)	115. (A)
8. (A)	26. (D)	44. (B)	62. (B)	80. (B)	98. (A)	116. (C)
9. (A)	27. (C)	45. (A)	63. (D)	81. (C)	99. (B)	117. (A)
10. (A)	28. (A)	46. (C)	64. (D)	82. (B)	100. (B)	118. (A)
11. (D)	29. (D)	47. (A)	65. (B)	83. (B)	101. (A)	119. (A)
12. (D)	30. (A)	48. (C)	66. (D)	84. (C)	102. (C)	120. (D)
13. (C)	31. (D)	49. (C)	67. (D)	85. (B)	103. (C)	
14. (D)	32. (C)	50. (C)	68. (D)	86. (A)	104. (C)	
15. (C)	33. (D)	51. (B)	69. (C)	87. (B)	105. (A)	
16. (B)	34. (A)	52. (C)	70. (C)	88. (D)	106. (C)	
17. (B)	35. (B)	53. (B)	71. (D)	89. (A)	107. (D)	
18. (B)	36. (B)	54. (C)	72. (D)	90. (D)	108. (B)	

KERALA - 1998

Q.	Ans.	Q.	Ans.	Q.	Ans.	Q.	Ans.
1.	(C)	31.	(A)	61.	(D)	91.	(A)
2.	(C)	32.	(A)	62.	(B)	92.	(A)
3.	(C)	33.	(A)	63.	(D)	93.	(B)
4.	(B)	34.	(B)	64.	(C)	94.	(A)
5.	(A)	35.	(B)	65.	(B)	95.	(C)
6.	(B)	36.	(C)	66.	(C)	96.	(B)
7.	(A)	37.	(B)	67.	(D)	97.	(C)
8.	(C)	38.	(A)	68.	(C)	98.	(C)
9.	(C)	39.	(B)	69.	(C)	99.	(B)
10.	(D)	40.	(B)	70.	(D)	100.	(D)
11.	(A)	41.	(D)	71.	(C)	101.	(C)
12.	(A)	42.	(C)	72.	(D)	102.	(D)
13.	(A)	43.	(A)	73.	(A)	103.	(A)
14.	(A)	44.	(D)	74.	(C)	104.	(D)
15.	(B)	45.	(B)	75.	(D)	105.	(D)
16.	(D)	46.	(B)	76.	(C)	106.	(C)
17.	(D)	47.	(B)	77.	(C)	107.	(C)
18.	(C)	48.	(D)	78.	(C)	108.	(A)
19.	(D)	49.	(D)	79.	(C)	109.	(D)
20.	(C)	50.	(C)	80.	(C)	110.	(A)
21.	(A)	51.	(C)	81.	(B)	111.	(D)
22.	(C)	52.	(D)	82.	(C)	112.	(D)
23.	(C)	53.	(A)	83.	(C)	113.	(A)
24.	(C)	54.	(A)	84.	(B)	114.	(B)
25.	(A)	55.	(A)	85.	(C)	115.	(A)
26.	(D)	56.	(B)	86.	(C)	116.	(B)
27.	(A)	57.	(B)	87.	(C)	117.	(A)
28.	(D)	58.	(C)	88.	(B)	118.	(A)
29.	(D)	59.	(D)	89.	(D)	119.	(C)
30.	(C)	60.	(B)	90.	(B)	120.	(C)

KERALA - 1999

Q.	Ans.	Q.	Ans.	Q.	Ans.	Q.	Ans.
1.	(C)	22.	(D)	43.	(D)	64.	(B)
2.	(A)	23.	(C)	44.	(B)	65.	(D)
3.	(B)	24.	(C)	45.	(B)	66.	(D)
4.	(C)	25.	(C)	46.	(B)	67.	(B)
5.	(A)	26.	(B)	47.	(A)	68.	(A)
6.	(A)	27.	(C)	48.	(D)	69.	(C)
7.	(C)	28.	(D)	49.	(D)	70.	(B)
8.	(A)	29.	(B)	50.	(B)	71.	(C)
9.	(C)	30.	(C)	51.	(D)	72.	(A)
10.	(A)	31.	(B)	52.	(A)	73.	(D)
11.	(D)	32.	(A)	53.	(B)	74.	(B)
12.	(C)	33.	(A)	54.	(B)	75.	(D)
13.	(C)	34.	(C)	55.	(D)	76.	(A)
14.	(C)	35.	(A)	56.	(B)	77.	(B)
15.	(C)	36.	(A)	57.	(B)	78.	(C)
16.	(C)	37.	(A)	58.	(B)	79.	(D)
17.	(A)	38.	(B)	59.	(C)	80.	(A)
18.	(B)	39.	(A)	60.	(C)	81.	(A)
19.	(A)	40.	(D)	61.	(A)	82.	(B)
20.	(D)	41.	(A)	62.	(B)	83.	(B)
21.	(C)	42.	(D)	63.	(B)		

KERALA - 2000

Q.	Ans.	Q.	Ans.	Q.	Ans.
1.	(A)	16.	(E)	31.	(A)
2.	(A)	17.	(C)	32.	(E)
3.	(A)	18.	(E)	33.	(E)
4.	(C)	19.	(C)	34.	(E)
5.	(B)	20.	(D)	35.	(C)
6.	(B)	21.	(D)	36.	(D)
7.	(A)	22.	(B)	37.	(E)
8.	(D)	23.	(A)	38.	(D)
9.	(E)	24.	(D)	39.	(D)
10.	(C)	25.	(D)	40.	(B)
11.	(C)	26.	(E)	41.	(D)
12.	(D)	27.	(C)	42.	(B)
13.	(C)	28.	(A)	43.	(E)
14.	(D)	29.	(D)	44.	(B)
15.	(B)	30.	(A)	45.	(A)

KERALA - 2001

Q.	Ans.	Q.	Ans.	Q.	Ans.	Q.	Ans.
1.	(B)	17.	(C)	33.	(D)	49.	(A)
2.	(D)	18.	(C)	34.	(A)	50.	(B)
3.	(E)	19.	(E)	35.	(E)	51.	(B)
4.	(D)	20.	(E)	36.	(B)	52.	(C)
5.	(D)	21.	(D)	37.	(C)	53.	(C)
6.	(C)	22.	(D)	38.	(B)	54.	(E)
7.	(D)	23.	(C)	39.	(D)	55.	(D)
8.	(B)	24.	(A)	40.	(C)	56.	(A)
9.	(C)	25.	(D)	41.	(A)	57.	(C)
10.	(B)	26.	(A)	42.	(A)	58.	(E)
11.	(C)	27.	(E)	43.	(E)	59.	(E)
12.	(A)	28.	(B)	44.	(D)	60.	(D)
13.	(B)	29.	(A)	45.	(E)	61.	(A)
14.	(C)	30.	(E)	46.	(C)	62.	(B)
15.	(E)	31.	(C)	47.	(C)	63.	(A)
16.	(C)	32.	(D)	48.	(A)	64.	(B)

KERALA - 2002

Q	Ans	Q	Ans	Q	Ans	Q	Ans
1.	(E)	17.	(B)	33.	(B)	49.	(D)
2.	(D)	18.	(C)	34.	(A)	50.	(B)
3.	(A)	19.	(C)	35.	(C)	51.	(D)
4.	(D)	20.	(B)	36.	(A)	52.	(E)
5.	(B)	21.	(E)	37.	(B)	53.	(C)
6.	(D)	22.	(D)	38.	(E)	54.	(E)
7.	(B)	23.	(A)	39.	(C)	55.	(A)
8.	(C)	24.	(C)	40.	(E)	56.	(A)
9.	(D)	25.	(D)	41.	(C)	57.	(C)
10.	(E)	26.	(E)	42.	(E)	58.	(A)
11.	(D)	27.	(A)	43.	(C)	59.	(D)
12.	(B)	28.	(C)	44.	(E)	60.	(B)
13.	(A)	29.	(A)	45.	(C)	61.	(A)
14.	(E)	30.	(D)	46.	(A)		
15.	(C)	31.	(E)	47.	(C)		
16.	(D)	32.	(E)	48.	(D)		

KERALA - 2003

Q	Ans	Q	Ans	Q	Ans	Q	Ans
1.	(D)	18.	(C)	35.	(D)	52.	(A)
2.	(D)	19.	(D)	36.	(B)	53.	(C)
3.	(B)	20.	(A)	37.	(A)	54.	(C)
4.	(C)	21.	(B)	38.	(A)	55.	(E)
5.	(A)	22.	(A)	39.	(B)	56.	(B)
6.	(C)	23.	(B)	40.	(E)	57.	(C)
7.	(A)	24.	(A)	41.	(B)	58.	(C)
8.	(D)	25.	(C)	42.	(D)	59.	(A)
9.	(D)	26.	(A)	43.	(D)	60.	(D)
10.	(D)	27.	(C)	44.	(C)	61.	(C)
11.	(D)	28.	(D)	45.	(A)	62.	(C)
12.	(B)	29.	(D)	46.	(D)	63.	(D)
13.	(A)	30.	(D)	47.	(A)	64.	(A)
14.	(A)	31.	(A)	48.	(D)	65.	(E)
15.	(A)	32.	(E)	49.	(B)		
16.	(A)	33.	(B)	50.	(C)		
17.	(A)	34.	(D)	51.	(D)		

KERALA - 2004

Q	Ans	Q	Ans	Q	Ans	Q	Ans
1.	(C)	18.	(D)	35.	(E)	52.	(A)
2.	(D)	19.	(B)	36.	(A)	53.	(E)
3.	(C)	20.	(E)	37.	(B)	54.	(E)
4.	(C)	21.	(B)	38.	(A)	55.	(B)
5.	(C)	22.	(C)	39.	(A)	56.	(C)
6.	(B)	23.	(C)	40.	(A)	57.	(B)
7.	(C)	24.	(B)	41.	(A)	58.	(E)
8.	(A)	25.	(B)	42.	(E)	59.	(A)
9.	(A)	26.	(E)	43.	(A)	60.	(B)
10.	(A)	27.	(E)	44.	(E)	61.	(C)
11.	(E)	28.	(C)	45.	(D)	62.	(B)
12.	(C)	29.	(B)	46.	(D)	63.	(D)
13.	(C)	30.	(A)	47.	(E)	64.	(E)
14.	(B)	31.	(D)	48.	(C)	65.	(D)
15.	(E)	32.	(A)	49.	(C)		
16.	(C)	33.	(C)	50.	(D)		
17.	(B)	34.	(C)	51.	(B)		

KERALA - 2005

Q	Ans	Q	Ans	Q	Ans
1.	(D)	17.	(A)	33.	(E)
2.	(C)	18.	(C)	34.	(C)
3.	(D)	19.	(B)	35.	(C)
4.	(B)	20.	(D)	36.	(B)
5.	(A)	21.	(A)	37.	(E)
6.	(D)	22.	(B)	38.	(A)
7.	(D)	23.	(D)	39.	(C)
8.	(C)	24.	(B)	40.	(A)
9.	(B)	25.	(C)	41.	(E)
10.	(A)	26.	(B)	42.	(D)
11.	(D)	27.	(C)	43.	(D)
12.	(E)	28.	(A)	44.	(C)
13.	(A)	29.	(C)	45.	(B)
14.	(B)	30.	(A)	46.	(E)
15.	(D)	31.	(B)	47.	(B)
16.	(C)	32.	(C)		

PMT - 1993

1. The electrons would go to lower energy levels first and then to higher energy levels according to which of the following
(A) Hund's rule of maximum multiplicity
(B) Aufbau principle
(C) Pauli's exclusion principle
(D) Heisenberg's uncertainty principle

2. The electronic configuration of chromium is
(A) [Ne] $3s^2$ $3p^6$ $3d^4$ $4s^2$ (B) [Ne] $3s^2$ $3p^6$ $3d^5$ $4s^1$
(C) [Ne] $3s^2$ $3p^6$ $4s^2$ $4p^4$ (D) [Ne] $3s^2$ $3p^6$ $3d^1$ $4s^2$ $4p^3$

3. The shape of 'p' orbital is
(A) elliptical (B) spherical
(C) dumb-bell (D) none of these

4. Among the bonds formed by a chlorine atom with atoms of hydrogen, chlorine, sodium and carbon, the strongest bond is formed between
(A) Na --- Cl (B) C - Cl
(C) H - Cl (D) Cl -Cl

5. Of the following electronegativity of beryllium is approximately equal to that of
(A) Aluminium (B) Boron
(C) Magnesium (D) Sodium

6. Element with electronic configuration $1s^2$ $2s^2$ $2p^6$ $3s^2$ belongs to a/an
(A) Metal (B) Non-metal
(C) Metalloid (D) Inert-gas

7. Which of the following mixtures forms an acid buffer ?
(A) $NH_4OH + NH_4Cl$
(B) $H_2CO_3 + (NH_4)_2CO_3$
(C) NaOH + HCL
(D) $CH_3COOH + CH_3COONa$

8. When NH_4Cl is added to NH_4OH solution, the dissociation of ammonium hydroxide is reduced. It is due to
(A) Common ion effect (B) Hydrolysis
(C) Oxidation (D) Reduction

9. The high boiling point of ethanol (78.2°C) compared to dimethyl ether (-23.6°C), though both having the same molecular formula C_2H_6O, is due to
(A) Hydrogen bonding
(B) Ionic bonding
(C) Co-ordinate Co-valent bonding
(D) Resonance

10. When acetaldehyde reacts with PCl_5, the resulting compound is
(A) Ethyl chloride (B) Ethylene chloride
(C) Ethylidene chloride (D) Trichloro acetaldehyde

11. Which of the following noble gases is not found in atmosphere
(A) Ne (B) Ar (C) Rn (D) Kr

12. Helium is added to the Oxygen supply used by deep sea divers because
(A) it is less soluble in blood than Nitrogen at high pressure
(B) it is lighter than Nitrogen
(C) it is readily miscible with Oxygen
(D) it is less poisonous than Nitrogen

13. If NaOH is added to an aqueous solution of Zinc ions, a white precipitate appears and on adding excess NaOH the precipitate dissolves. In this solution Zinc exists in the
(A) Cationic Part (B) Anionic Part
(C) Both in Cationic and Anionic Parts
(D) There is no Zinc in the solution

14. Nessler's reagent is
(A) K_2HgI_4 (B) K_2HgI_4 + KOH
(C) K_2HgI_2 + KOH (D) K_2HgI_4 + Hg

15. 0.16 gm. of a dibasic acid required 25 ml. decinormal NaOH solution for complete neutralisation. The molecular weight of the acid is
(A) 128 (B) 256 (C) 32 (D) 64

16. A mixture of benzene and toluene form
(A) an ideal solution (B) non ideal solution
(C) suspension (D) emulsion

17. The colligative properties of a solution depends on
(A) nature of the solvent used
(B) number of solute particles present
(C) chemical nature of the solute particles present in it .
(D) nature of both the solvent and solute

18. The experimental molecular weight of an electrolyte will always be less than its calculated value because the value of Van't Haff factor, 'i' is
(A) less than 1 (B) greater than 1
(C) equivalent than 1 (D) Zero

19. Tetragonal crystal system has the following unit cell dimensions
(A) $a = b = c$
$\alpha = \beta = \gamma = 90°$
(B) $a = b \neq c$
$\alpha = \beta = \gamma = 90°$
(C) $a \neq b \neq c$
$\alpha = \beta = \gamma = 90°$
(D) $a = b \neq c$
$\alpha = \beta = 90°; \gamma = 120°$

20. A colloidal system in which gas bubbles are dispersed in a liquid is known as
(A) Aerosol (B) Emulsion
(C) Foam (D) Sol

21. The fibre obtained by the condensation of hexamethylene diamine and adipic acid is
(A) Dacron (B) Nylon 6, 6
(C) Rayon (D) Teflon

22. Which one among the following is a thermosetting plastic ?
(A) PVC (B) PVA
(C) Bakelite (D) Perspex

23. Which one of the following is the reagent used to identify glucose ?
(A) Ammonical silver nitrate
(B) Sodium ethtoxide
(C) Neutral ferric chloride
(D) Chloroform and alcoholic KOH

24. Which of the following statements is true ? The entropy of the universe ?
(A) increases and tends towards maximmum value.
(B) decreases and tends to be zero
(C) remains constant.
(D) decreases and increases with a periodic rate.

25. The overall heat change in a chemical reaction at constant volume is carried out
(A) in one step at constant temperature and pressure
(B) in several steps at constant temperature and pressure.
(C) in one step at constant temperature only.
(D) in one or several steps at constant pressure.

26. The minimum energy required for molecules to enter into the reaction is called
(A) potential energy (B) kinetic energy
(C) nuclear energy (D) activation energy

27. The inversion of Cane Sugar into Glucose and Fructose, according to the equation

$$\underset{\text{Cane Sugar}}{C_{12}H_{22}O_{11}} + H_2O \rightarrow \underset{\text{Glucose}}{C_6H_{12}O_6} + \underset{\text{Fructose}}{C_6H_{12}O_6}$$

is an example of
(A) First order reaction (B) Second order reaction
(C) Third order reaction (D) Zero order reaction

28. An isolated system is that system in which
(A) there is no exchange of mass or energy with the surroundings
(B) there is no exchange of energy with the surrounding
(C) there is exchange of mass and energy with the surroundings
(D) there is exchange of mass with the surroundings

29. For an ideal gas, the value of $\left(\frac{\partial E}{\partial V}\right)_T$ is
(A) Positive (B) Zero
(C) Negative (D) Interchangeable

30. When ethanol is refluxed with conc. H_2SO_4 at 150 - 170° C, the resulting compound is
(A) Ethylene (B) Ethyl hydrogen sulphate
(C) Diethylether (D) Acetaldehyde

31. When treated with ammonical cuprous chloride, which one among the following forms copper derivative ?
(A) C_2H_2 (B) C_6H_6 (C) C_2H_6 (D) C_2H_4

32. Out of the following fractions of petroleum, the one having the lowest boiling point is
(A) Kerosene oil (B) Diesel oil
(B) Gasoline (D) Heavy oil

33. Which one among the following is assigned an octane number of Zero ?
(A) Iso-Octane (B) n-Heptane
(C) Iso-heptane (D) 2-Methyl octane

34. Which one among the following represents an amide ?
(A) 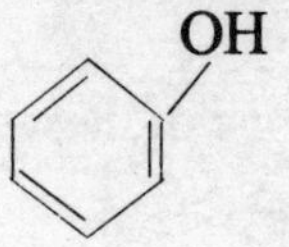 OH (B) 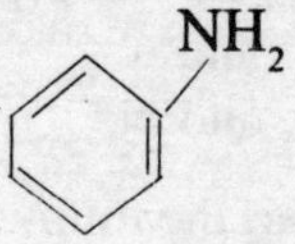 NH_2
(C) 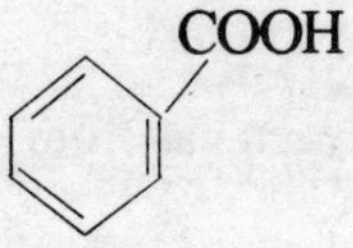 COOH (D) 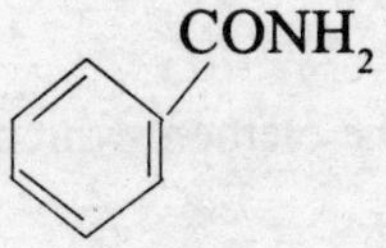 $CONH_2$

35. Enzymes are
(A) Vitamins (B) Proteins
(C) Carbohydrates (D) Lipids

36. Sucrose on hydrolysis gives
(A) One molecule each of Glucose and Fructose
(B) Two molecules of Glucose
(C) Two molecules of Fructose
(D) One molecule each of Glucose and Mannose

37. Which one of the following proteins transports oxygen in the blood stream ?
(A) Myoglobin (B) Insulin
(C) Albumin (D) Haemoglobin

38. Hydrogen-oxygen fuel cells are used in space craft to suppluy
(A) Power for heat and light (B) Power for pressure
(C) Oxygen (D) Water

39. Prevention of corrosion of Iron metal by Zinc coating is called
(A) Electrolysis (B) Photo-electrolysis
(C) Galvanisation (D) Cathodic protection

40. The most electro-positive amongst the alkaline earth metals is
(A) Berelium (B) Magnesium
(C) Calcium (D) Barium

41. Which element of the following has the largest first ionization energy
(A) Nitrogen (B) Oxygen (C) Boron (D) Carbon

42. $[Co(NH_3)_4Cl_2]NO_2$ and $[Co(NH_3)_4Cl.NO_2]Cl$ are .... isomers
(A) Geometrical (B) Optical
(C) Linkage (D) Ionization

43. IUPAC name of complex $K_3[Al\,(C_2O_4)_3]$ is
(A) Potassiumalumino oxalate
(B) Potassium trioxalatoaluminate (III)
(C) Poatassiumaluminium (III) oxalate
(D) Potassiumtrioxalatoaluminate (VI)

44. Which of the following ions has the least magnetic moment ?
(A) Co^{3+} (B) Fe^{2+} (C) Cu^{2+} (D) Ni^{2+}

45. Which of the following molecules is trigonal bipyramidal ?
(A) BF_3 (B) CH_4 (C) PCl_5 (D) SF_6

46. Calcium carbide reacts with nitrogen to form
(A) $CaCN_2$ and C (B) $CaCN_2$
(C) $CaCN_2$ and N_2 (D) none of these

47. Which of the following is not an ore of lead
(A) Calamine (B) Cerrusite
(C) Galena (D) Anglesite

48. The element which liberates oxygen gas from water is
(A) P (B) Na (C) F (D) I

49. Na_2CO_3 can be manufactured by Solvey's Process but K_2CO_3 can not be prepared because
(A) $KHCO_3$ is more soluble than $NaHCO_3$
(B) $KHCO_3$ is less soluble than $NaHCO_3$.
(C) K_2CO_3 is more soluble
(D) K_2CO_3 is less soluble

50. If M is the molecular weight of $KMnO_4$, its equivalent weight when it is converted into K_2MnO_4 will be
(A) M (B) M/3 (C) M/5 (D) M/7

51. On the addition of a solution containing CrO_4^{2-} ions to the solution of Ba^{2+}, Sr^{2+} and Ca^{2+} ions, the precipitate obtained first will be of
(A) $CaCrO_4$ (B) $SrCrO_4$
(C) $BaCrO_4$ (D) A mixture of all the three

52. When Acetone is heated with Hydroxyl amine, the compound formed is
(A) Cyanohydrin (B) Oxime
(C) Semicarbazone (D) Hydrazone

53. When methyl cyanide is hydrolysed in presence of alkali, the product is
(A) $CO_2 + H_2O$ (B) Acetic acid
(C) Acetamide (D) Methane

54. Ammonia molecule is formed due to the following hybridized orbitals
(A) dsp^2 (B) sp^3 (C) sp^3d (D) d^2sp

55. The high boiling point of water is due to the presence of
(A) dative bond (B) covalent bond
(C) hydrogen bond (D) van der Waals forces

56. Among the following outermost configurations of transition metals, which shows the highest oxidation state is
(A) $3d^5\ 4s^2$ (B) $3d^6\ 4s^2$ (C) $3d^3\ 4s^2$ (D) $3d^5\ 4s^1$

57. When 100 ml N-NaOH solution and 10 ml. of 10N-sulphuric acid solution are mixed together, the resulting solution will be
(A) alkaline (B) weakly acidic
(C) strongly acidic (D) neutral

58. Which of the following is a pseudo halogen
(A) IF_7 (B) $(CN)_2$ (C) ICl_2 (D) I_3^-

59. Deacon's Process is used for the manufacture of
(A) Bleaching Powder (B) Sulphuric Acid
(C) Nitric Acid (D) Chlorine

60. HBr and Hg reduce sulphuric acid, HCl can reduce $KMnO_4$ and HF can reduce
(A) $KMnO_4$ (B) $K_2Cr_2O_7$ (C) H_2SO_4 (D) none of these

61. When ammonium dichromate is heated, the gas formed is
(A) N_2 (B) O_2 (C) H_2 (D) NH_3

62. NO_2 gas can not be obtained by heating
(A) KNO_3 (B) $Pb(NO_3)_2$ (C) $Cu(NO_3)$ (D) $AgNO_3$

63. Which of the following salts is insoluble in water at room temperature but soluble in boiling water ?
(A) $CaCl_2$ (B) $BaCl_2$ (C) $SrCl_2$ (D) $PbCl_2$

64. When chloroform reacts with ethylamine in presence of alcoholic KOH, the compound formed is
(A) Ethyl cyanide (B) Ethyl isocyanide
(C) Formic acid (D) Phosgene

65. The product of the following reaction is

$$C_6H_6 + Cl_2 \xrightarrow{\text{sun light}} ?$$

(A) C_6H_5Cl (B) O - $C_6H_4\ Cl_2$
(C) $C_6H_6Cl_6$ (D) P - $C_6H_4\ Cl_2$

66. Which of the following statements is correct ? A catalyst is a substance which
(A) does not influence the reaction rate
(B) increases the rate of forward reaction
(C) increases both rate of forward reaction and reverse reaction equally
(D) alters the equilibrium constant of a reaction

67. According to Langmuir adsorption isotherm, the amount of gas adsorbed at very high pressures
(A) reaches a constant limiting value
(B) goes on increasing with pressure
(C) goes on decreasing with pressure
(D) increases first and decreases later with pressure

68. An example of halide ores is
(A) Galena (B) Bauxite (C) Cinnabar (D) Cryolite

69. After partial roasting, the sulphide of copper is reduced by
(A) Self-reduction (B) Cynide Process
(C) Reduction by carbon (D) Electrolysis

70. Carbolic acid is
(A) Phenol (B) Methanol
(C) Formic acid (D) Acetic acid

71. Addition of bromine water to an aqueous solution of phenol gives
(A) o- Bromophenol (B) p-Bromophenol
(C) m-Bromophenol (D) 2, 4, 6, - tribromophenol

72. The compound formed by the hydrolysis of calcium carbide is
(A) Acetylene (B) Ethane

(C) Ethene (D) Ethylene glycol

73. For an ionic crystal of the general formula AX and coordination number 6, the value of radius ratio will be

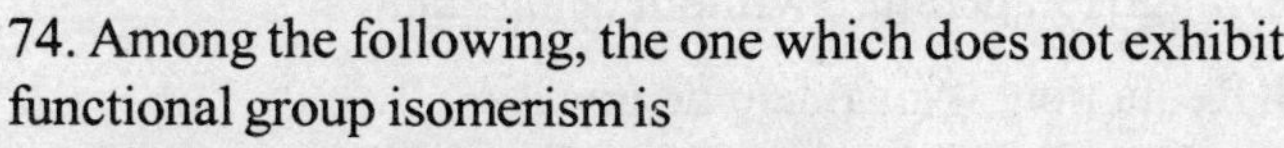

(A) in between 0.41 and 0.22
(B) greater than 0.73
(C) in between 0.73 and 0.41
(D) less than 0.22

74. Among the following, the one which does not exhibit functional group isomerism is
(A) C_4H_8 (B) $C_4H_{10}O$ (C) C_2H_6O (D) C_3H_8O

75. The hybrid orbitals used by carbon in benzene belong to
(A) sp^3 (B) sp^2 (C) sp (D) dsp^3

76. The reaction rate at a given temperature becomes slower when
(A) the free energy of activation is higher
(B) the free energy of activation is lower
(C) entropy changes
(D) initial concentration of the reactants remains constant

77. When one ampere of current flows for 1 sec. through a conductor, this quantity of electricity is known as
(A) E. M. F. (B) Ohm
(C) Faraday (D) Coulomb

78. Which of the following statements is correct ? Galvanic cell cenverts
(A) chemical energy into electrical energy
(B) electrical energy into chemical energy
(C) metal from its elemental state to the combined state
(D) electrolyte into individual ions

79. Space lattice of CaF_2 is
(A) face centered cubic (B) body centered cubic
(C) simple cubic (D) hexagonal close packing

80. The existence of a substance in more than one solid modifications is known as
(A) Polymorphism (B) Isomorphism
(C) Allotropy (D) Enantiomorphism

81. Stable nuclides are those whose n/p ratio is
(A) n/p > 1 (B) n/p < 1
(C) n/p = 1 (D) n/p =2

82. A radium $_{88}Ra^{224}$ isotope, on emission of an α-particle gives rise to a new element whose mass number and atomic number will be
(A) 220 and 86 (B) 225 and 87
(C) 228 and 88 (D) 224 and 86

83. The basic principle of the hydrogen bomb formation is
(A) spallation (B) nuclear fusion
(C) nuclear fission (D) disintegration

84. Hoffmann's hypobromite reaction affords a method of
(A) Preparing a tertiary amine
(B) Preparing a mixture of amines
(C) Stepping down a series
(D) Stepping up a series

85. Which of the following is an explosive ?
(A) Nitrobenzene (B) Nitrophenol
(C) Nitromethane (D) Trinitrobenzene

86. Polymerisation in which two or more chemically different monomers take part is called
(A) Addition polymerisation
(B) Co-polymerisation
(C) Chain polymerisation
(D) Homo polymerisation

87. Natural rubber is basically a polymer of
(A) Neoprene (B) Isoprene
(C) Chloroprene (D) Butadiene

88. Which of the following is the smallest cation ?
(A) Ca^{2+} (B) Al^{3+} (C) Na^{+} (D) Mg^{2+}

89. The radio-isotope of carbon that is used for knowing the appropriate time of death of a planet is
(A) C_6^{12} (B) C_6^{13} (C) C_6^{14} (D) C_6^{16}

90. In which of the following equilibria, the value of K_p is less than K_c?
(A) $H_2 + I_2 \rightleftharpoons 2HI$
(B) $N_2 + 3H_2 \rightleftharpoons 2NH_3$
(C) $N_2 + O_2 \rightleftharpoons 2NO$.
(D) $CO + H_2O \rightleftharpoons CO_2 + H_2$

91. The number of gram molercules of a substance present in unit volume is termed as
(A) activity (B) normal solution
(C) molar concentration (D) active mass

92. According to Le Chatelier's Principle, an increase in temperature on the reaction $N_2 + O_2 \rightleftharpoons 2NO + 43200$ cal. will
(A) increase the yield of NO
(B) decrease the yield of NO
(C) not effect the yield of NO
(D) not help the reaction to proceed in forward direction

93. The pH of a 0.02 M solution of hydrochloric acid is
(A) 2.0 (B) 1.7 (C) 0.3 (D) 2.2

94. Which one among the following is Williamson's synthesis ?

(A) $(CH_3)_2C=O \xrightarrow[\text{conc. HCl}]{\text{Zn - Hg}} CH_3\text{-}CH_2\text{-}CH_3$

(B) $CH_3 - CHO \xrightarrow[-H_2O]{\text{dil. NaOH}} CH_3 - CH = CH - CHO$

(C) $C_2H_5I + C_2H_5ONa \rightarrow C_2H_5 - O - C_2H_5 + NaI$

(D) $HCHO \xrightarrow{NaOH} HCOONa + CH_3OH$

95. The product of the reaction between ammonia and formaldehyde is
(A) Urotropine (B) Formamide
(C) Paraformaldehyde (D) Methanol

96. Which one of the following is most metallic ?
(A) Sb (B) Bi (C) P (D) As

97. An element having electronic configuration [Ar] $3d^2$ $4s^2$ belongs to
(A) s-block elements (B) p-block elements
(C) d-block elements (D) f-block elements

98. A compound has the empirical formuls CH_2O. Its vapour density is 30. Its molecular formula is
(A) $C_2H_4O_2$ (B) C_2H_6O
(C) $C_2H_6O_2$ (D) C_2H_4O

99. The IUPAC name of the Compound

$$\begin{array}{l} CH_3 \\ | \\ CH_3 - CH - CH - CH_3 \quad \text{is} \\ \qquad\qquad\ \ | \\ \qquad\quad\ C_2H_5 \end{array}$$

(A) 2-methyl-3-ethyl butane
(B) 3,4-dimethyl pentane
(C) 2-Ethyl-3-methyl butane
(D) 2,3-dimethyl pentane

100. Which of the following is correct for 2p orbitals ?
(A) n = 1, l = 2 (B) n = 1, l = 0
(C) n = 2, l = 1 (D) n = 2, l = 0

PMT - 1994

1. If the amount of radioactive substance is increased three times, the number of atoms disintegrated per unit time would
(A) remain one third (B) not change
(C) be double (D) be triple

2. The brown ring complex compound is formulated as $[Fe(H_2O_5)_5NO]SO_4$, the oxidation state of iron is
(A) + 1 (B) - 1 (C) + 2 (D) + 3

3. Colligative properties of a solution depend upon
(A) nature of solute
(B) nature of solvent
(C) the relative number of solute and solvent particles
(D) none of these

4. If 5.85 gms of NaCl are dissolved in 90 gms of water, the mole fraction of NaCl is
(A) 0.2 (B) 0.196 (C) 0.1 (D) 0.01

5. 120 gms of urea are present in 5 litre solution, the active mass of urea is
(A) 0.4 (B) 0.08 (C) 0.2 (D) 0.06

6. Starch is the polymer of
(A) Glucose (B) Fructose
(C) Both (D) None of them

7. In the formation of SO_3 according to the following equilibrium, $2SO_2 + O_2 \rightleftharpoons 2SO_3$ + heat which condition favour the forwared reaction ?
(A) low pressure
(B) low concentraon of oxygen
(C) low temp and use of a catalyst
(D) high temperature without using catalyst

8. Reduction of an aldehyde gives a
(A) Secondary Alcohol (B) Tertiary Alcohol
(C) Primary Alcohol (D) Mono-carboxylic Acid

9. Aromatic properties of benzene are proved by
(A) Aromatic Sextet Theroy
(B) Resonance Theroy
(C) Molecular Orbital Theory
(D) All of the above

10. The reaction between an alkyl halide and sodium metal in dry ether is called
(A) Clemmensen Reduction (B) Kolbe's Reaction
(C) Wurtz's Reaction (D) Cannizaro's Reaction

11. Marsh gas primarily contains
(A) H_2S (B) CO (C) C_2H_2 (D) CH_4

12. General formula of alkenes is
(A) C_nH_{2n+2} (B) C_nH_{2n}
(C) F_nH_{2n-2} (D) $C_{2n+2}H_n$

13. The tetrahedral nature of carbon was first given by
(A) Kekule (B) Vont-Hoff and Lebel
(C) Lewis (D) Wohler

14. Natural rubber is a
(A) Poly ester (B) Poly amide
(C) Poly isoprene (D) Poly saccharide

15. The structural unit, $-C \equiv C-$ is present in
(A) Ethyne (B) Glycerine (C) Ethene (D) Butene

16. Dehydration of glycerol gives
(A) Propene (B) Acrolein (C) Propane (D) Benzene

17. An organic compound contains C = 40%, O = 53.34% and H = 6.66%. The empirical formula of compound is
(A) CH_2O (B) CHO (C) CH_4O_2 (D) C_2H_2O

18. The total number of electrons in one molecule of CO_2 is
(A) 22 (B) 44 (C) 66 (D) 88

19. The atomic number of an element is always equal to
(A) Weight of the nucleus
(B) Atomic weight divided by 2
(C) Number of neutrons in the nucleus
(D) Electrical charge of the nucleus

20. The bleaching action of the bleaching powder is due to the liberation of
(A) chlorine (B) molelcular oxygen
(C) nascent oxygen (D) calcium carbonate

21. Which of the following is isoelectronic with carbon

atom

(A) O^{2-} (B) N^+ (C) Na^+ (D) Al^{3+}

22. The screeing effect of inner electrons of the nucleus causes
(A) a decrease in the ionisation potential
(B) an increase in the ionisation potential
(C) no effect on the ionisation potential
(D) an increase in the attraction of the nucleus to the electrons

23. An atom of an element has electronic configuration, 2, 8, 1. Which of the following statements is correct ?
(A) the element is of non-metallic nature
(B) the element's valency is 7
(C) the element exists as a diatomic moleucle
(D) the element forms a basic oxide

24. When SO_2 is passed through an aqueous H_2S, sulphur is precipitated. In this reaction SO_2 acts as
(A) an oxidising agent (B) a reducing agent
(C) an acid (D) a complexing agent

25. When $E^o_{Ag+/Ag} = 0.8$ volt and $E^o_{Zn^{2+}/Zn} = -0.76$ volt, which of the following is correct ?
(A) Ag^+ can be reduced by H_2
(B) Ag can oxidise H^2 into H^+
(C) Zn^{2+} can be reduced by H_2
(D) Ag can reduce Zn^{2+} ion

26. When CH_3COONa is added to aqueous CH_3COOH solution
(A) pH value remains unchanged
(B) pH value becomes zero
(C) pH value decreases
(D) pH value increases

27. The solubility product of AgCl is 1.44×10^{-4} at 100°C. The solubility of silver chloride in boiling water may be
(A) 0.72×10^{-4} M (B) 1.20×10^{-2} M
(C) 0.72×10^{-2} M (D) 1.20×10^{-4} M

28. In the following reaction,
$3Br_2 + 6CO_3^{2-} + 3H_2O \rightarrow 5Br^- + BrO_3^- + 6HCO_3^-$
(A) Bromine is oxidised and carbonate is reduced.
(B) Bromine is reduced and carbonate is oxidised.
(C) Bromine is neither reduced nor oxidised.
(D) Bromine is reduced and oxidised.

29. When electricity is passed through a solution of aluminium chloride, 13.5 gms of Al are deposited. The number of Faraday must be :
(A) 1.0 (B) 0.5 (C) 2.0 (D) 1.5

30. Which will not affect the degree of ionisation ?
(A) Temperature (B) Concentration
(C) Type of solvent (D) Current

31. The p^{OH} of beer is 10.0. The hydrogen ion concentration will be
(A) 10^{-2} (B) 10^{-4} (C) 10^{-8} (D) 10^{-10}

32. Heat of neutralisation of NH_4OH and HCl is
(A) equal to 13.7 K. Cal.
(B) more than 13.7 K.Cal.
(C) less than 13.7 K. Cal.
(D) more than one is correct

33. In the combustion of 2.0 gm. of methane 25 K.cal. heat is liberated, heat of combustion of methane would be
(A) 100 K.cal. (B) 200 K.cal.
(C) 300 K.cal. (D) 400 K.cal.

34. Which of the following relations is correct ?
(A) $\Delta H = \Delta G - T\Delta S$
(B) $\Delta H = \Delta G + T\Delta S$
(C) $\Delta H = -\Delta G - T\Delta S$
(D) $\Delta H = -\Delta G + T\Delta S$

35. Which one of the following is a good conductor of elctricity ?
(A) diamond (B) graphite
(D) silicon (C) amorphous carbon

36. The colour of $K_2Cr_2O_7$ changes from orange red to lemon yellow on treatment with aqueous KOH because of
(A) the reduction of Cr^{VI} to Cr^{III}
(B) the formation of chromium hydroxide
(C) the conversion of dichromate to chromate
(D) the oxidation of potassium hydroxide to potassium

37. The electronic configuration of Mn^{2+} is
(A) $[Ar]3d^4\ 4s^1$ (B) $[Ar]3d^5\ 4s^0$
(C) $[Ar]4d^5\ 5s^0$ (D) $[Ar]3d^3\ 4s^2$

38. The pair of co-ordination compounds $[Co(NH_3)_5Br]SO_4$ and $[Co(NH_3)_5SO_4]$ Br is an example for
(A) Structural isomerism
(B) Ionisation isomerism
(C) Geometrical isomerism
(D) Co-ordination isomerism

39. The correct nomenclature for $Fe_4[Fe(CN_6)]_3$ is
(A) Ferroso ferric cyanide
(B) Ferric-Ferrous hexa cyanate
(C) Iron (III) hexa cyanoferrate (II)
(D) Hexa cyanoferrate (III - II)

40. Which of the following will give a white precipitate with $AgNO_3$ in aqueous medium ?
(A) $[Co(NH_3)_5Cl](NO_2)_2$ (B) $[Pt(NH_3)_2Cl_2]$
(C) $[Pt(en)Cl_2]$ (D) $[Pt(NH_3)_4]Cl_2$

41. The d-block element consists mostly of
(A) monovalent metals
(B) all non-metals
(C) elements which generally form stoichiometric metal oxide
(D) many metals with catalytic properties

42. The electroplating of chromium is undertaken because
(A) electroysis of chromium is easier
(B) chromium can form alloys with other metals

(C) chromium gives protective and decorative coating to the base metal
(D) of the high reactivity of metallic chromium

43. The passivity of iron in concentrated nitric acid is due to
(A) a thin oxide layer coating on the metal.
(B) ferric nitrate coating on the metal
(C) ammonium nitrate coating on the metal
(D) a hydride coating on the metal

44. The action of steam on heated iron is represented as
(A) $3Fe + 4H_2O \rightarrow Fe_3O_4 + 4H_2$
(B) $2Fe + 3H_2O \rightarrow Fe_2O_3 + 3H_2$
(C) $Fe + H_2O \rightarrow FeO + H_2$
(D) $2Fe + H_2O + O_2 \rightarrow Fe_2O_3 + H_2$

45. Which of the following solutions will not hydrolyse ?
(A) KNO_3 (B) KCN
(C) Pot. Succinnate (D) Pot. Carbonate

46. Butter is
(A) Water dispersed in fat
(B) Water dispersed in oil
(C) Fat dispersed in milk.
(D) Fat dispersed in water

47. The unit for specific reaction rate of second order reaction is
(A) Sec^{-1} (B) mole $litre^{-1}$ sec^{-1}
(C) $litre^2$ $mole^{-2}$ sec^{-1} (D) litre $mole^{-1}$ sec^{-1}

48. Half life period of second order reaction is
(A) Proportional to initial concentration of reactants
(B) Independent of initial concentration of reactants
(C) Inversely proportional to initial concntration of reactants
(D) Inversely proportional to square of initial concentration of reactants

49. Oxidation of oxalic acid by aciditied $KMnO_4$ is an example of auto catalysis, it is due to presence of
(A) Mn^{+2} (B) K^+ (C) SO_4^{-2} (D) MnO_4^-

50. Which of the following makes the lyophilic sol. unstable ?
(A) Dialysis
(B) Addition of electrolyte
(C) Addition of alcohol
(D) Addition of alcohol and electrolyte both

51. Which of the following is a buffer solution ?
(A) $CH_3COOH + CH_3COONa$
(B) KOH + HCl
(C) $HNO_3 + NaNO_3$
(D) HCl + NaCl

52. Of the following the one having a square planar structure is
(A) NH_4^+ (B) BF_4^- (C) XeF_4 (D) CCl_4

53. In the phenomenon of osmosis, the membrane allows passage to
(A) solute molecules only
(B) solvent molecules only
(C) both solute and solvent molecules
(D) none of these

54. Pick out from the following complex compounds, a poor electrolytic conductor solution
(A) K_2 $[PtCl_6]$ (B) $[Co(NH_3)_3(NO_2)_3]$
(C) $K_4[Fe(CN)_6]$ (D) $[Cu(NH_3)_4]SO_4$

55. Which of the following is insoluble in water and acid both ?
(A) $CaCO_3$ (B) $Zn(NO_3)_2$
(C) $BaSO_4$ (D) $ZnCO_3$

56. On addition of aqueous NaOH to a salt solution, a white gelatinous precipitate is formed, which dissolves in excess alkali. The salt solution contains
(A) Barium ions (B) Iron ions
(C) Chromium ions (D) Aluminium ions

57. A mixture containing Cu^{2+} and Ni^{2+} can be separated for identification by
(A) passing H_2S in acid medium
(B) passing H_2S in alkaline medium
(C) passing H_2S in neutral medium
(D) passing H_2S in dry mixture

58. The indicator used in the titration of acetic acid with sodium hydroxide for quantitative estimation is
(A) Methyl red
(B) Methyl orange
(C) Phenolphthalein
(D) A mixture of methyl red and methyl orange

59. What weight of Sodium Hydroxide is required to neutralise 100 ml of 0.1 N - HCl ?
(A) 4.0 gm (B) 0.04 gm (C) 0.4 gm (D) 2.0 gm

60. The I. U. P. A. C. name of

$$CH_3 - \underset{\displaystyle CH_3}{\underset{|}{CH}} - CH_3$$ is

(A) Butane (B) Isopropyl methane
(C) 2-methyl propane (D) Di methyl ethane

61. n-propyl alcohol and iso propyl alcohol are examples of
(A) Tautomerism (B) Geometrical isomerism
(C) Position isomerism (D) Chain isomerism

62. Optical Isomerism is shown by
(A) Butanol -1 (B) Butanol -2
(C) Butene - 1 (D) Butene - 2

63. Calcium Carbide reacts with water to produce
(A) Methane (B) Ethane (C) Ethylene (D) Acetylene

64. Heating together of sodium ethoxide and ethyl iodide will give
(A) Acetaldehyde (B) Acetic Acid
(C) Ether (D) Ethyl Alcohol

65. The relation between Kp and Kc is correctly shown

as
(A) $Kc = Kp (RT)^{\Delta n}$ (B) $Kp = Kc (RT)^{-\Delta n}$
(C) $Kp = Kc (RT)^{\Delta n}$ (D) $Kc = Kp (RT)^{-\Delta n}$

66. Reverberatory furnance is employed in the metallurgical process mainly for
(A) reduction of oxide ores
(B) smelting of sulphide ores
(C) conversion of chloride to sulphate
(D) getting magnetic materials

67. The process of ore dressing is carried out to
(A) convert the ore to oxide
(B) remove the siliceous materials
(C) add flux to the mineral
(D) remove the poisonous impurities

68. In the manufacture of metallic sodium by the fused salt electrolysis (Down's process), a small amount of calcium chloride is added to
(A) improve the electrical conduction
(B) increase the temperature of electrolysis
(C) bring bown the melt temperature
(D) stablilize the metallic sodium

69. An important ore for iron is
(A) galena (B) haematite
(C) malachite (D) pitch blende

70. In the cyanide process for the extraction of silver, sodium cyanide is used to
(A) convert silver into a soluble silver complex
(B) reduce silver
(C) pricipitate silver
(D) oxidise silver

71. Which one of the following is the anhydride of nitric acid ?
(A) N_2O_3 (B) N_2O_5 (C) NO (D) NO_2

72. In the following reaction
$P_4 + 3NaOH + 3H_2O \rightarrow PH_3 + 3NaH_2PO_2$
(A) Phosphorus is oxidised
(B) Phosphorus is oxidised and reduced
(C) Phosphorus is reduced
(D) Sodium is oxidised

73. The reactivity of the alkali metal sodium with water, is made use of
(A) in drying of ammonia solution
(B) as a general drying agent
(C) in drying of alcohols
(D) in drying of benzene

74. Fluorine forms chemical compounds with
(A) He (B) Ne (C) Ar (D) Xe

75. Which of the following is produced when acetylene is passed in dilute H_2SO_4 in the presnece of $HgSO_4$?
(A) C_2H_5OH (B) CH_3COCH_3
(C) CH_3CHO (D) Mercury Carbide

76. Baeyer's Reagent is
(A) Alkaline $KMnO_4$ (B) Ammonical $AgNO_3$
(C) Ammonical $CuSO_4$ (D) Acidic $CaSO_4$

77. The possibility of finding an electron in an orbital was conceived by
(A) Heisenberg (B) Schrodinger
(C) Rutherford (D) Bohr

78. In which molecule, are all atoms coplanar ?
(A) CH_4 (B) BF_3 (C) PF_3 (D) NH_3

79. Which of the following does not have the hydrogen bond ?
(A) Phenol (B) Liquid NH_3 (C) Water (D) Liquid HCl

80. Ionic bonds are usually formed by combination of elements with
(A) high ionisation potential and high electron affinity
(B) high ionisation potential and low electron affinity
(C) low ionisation potential and high electron affinity
(D) low ionisation potential and low electron affinity

81. The angle between two covalent bonds is maximum in
(A) CH_4 (B) H_2O (C) CO_2 (D) SO_2

82. If a molecule MX_3 has zero dipole moment, the sigma bonding orbitals used by M (At. No. 21) are
(A) pure p (B) sp hybrid
(C) sp^2 hybrid (D) sp^3 hybrid

83. CO_2 is iso-structural with
(A) SO_2 (B) $HgCl_2$ (C) $SnCl_2$ (D) all above

84. Crystals can be classified into .... basic crystal babits
(A) 3 (B) 7 (C) 4 (D) 14

85. Hexagonal close packed arrangement of ions is decscribed as
(A) ABC ABA (B) ABC ABC
(C) ABABAB (D) ABBAB

86. Strongest Bronsted base among the following anions is
(A) ClO^- (B) ClO_2^- (C) ClO_3^- (D) ClO_4^-

87. pH of the solution is 4. The hydroxide ion concentration of the solution would be
(A) 10^{-2} (B) 10^{-12} (C) 10^{-4} (D) 10^{-10}

88. If the half life of a radioactive element is 30 minutes, one sixteenth of original quantity of element will remain unchanged after
(A) One hour (B) 16 hours (C) 4 hours (D) 2 hours

89. Which of the following is added to fuel as an anti-knock agent ?
(A) Lead-Bromide (B) Zinc Bromide
(C) Tetra ethyl lead and ethyl bromide
(D) Lead oxide

90. The reaction between a primary amine, chloroform and alcoholic KOH is called
(A) Wurtz Reaction (B) Frankland Reaction
(C) Cannizaro's Reaction (D) Carbylamine Reaction

91. Phosgene is a common name of
(A) Carbonyl chloride
(B) Carbon dioxide and phosgene
(C) Phosphoryl chloride
(D) Carbon tetrachloride

92. C_2H_5OH can be differentiated from CH_3OH by
(A) Reaction with HCl (B) Reaction with NH_3
(C) By iodoform test (D) By solubility in water

93. The reaction between an alcohol and an acid is called
(A) Esterification (B) Saponification
(C) Hydrolysis (D) Hydrogenation

94. The acid present in vinegar is
(A) HCl (B) HNO_3 (C) CH_3COOH (D) H_2SO_4

95. Benzene hexachloride is used as
(A) Dye (B) Antimalarial drug
(C) Antibiotic (D) Insecticide

96. Mark the correct statement
(A) Methyl amine is acidic
(B) Methyl amine is less basic than NH_3
(C) Methyl amine is a stronger base than NH_3
(D) Methyl amine forms salts with bases

97. Hydrolysis of acetamide gives
(A) Acetic acid (B) Acetaldehyde
(C) Methylamine (D) Formic acid

98. Glucose gives a silver mirror with Tollen's reagent. This shows the presence of
(A) Alcoholic group (B) Acid group
(C) Aldehyde group (D) Ketonic group

99. RNA does not contain
(A) Uracil (B) Ribose (C) Thymine (D) Phosphate

100. Hydrolysis of proteins in the presence of enzymes produces
(A) Amino acids (B) Aromatic acids
(C) Hydroxy acids (D) Dicarboxylic acids

PMT - 1995

1. Which of the following compounds gives red precipitate with $AgNO_3$?
(A) NaBr (B) $NaNO_3$ (C) KI (D) K_2CrO_4

2. When zinc reacts with very dilute nitric acid it produces
(A) NO (B) NH_4NO_3 (C) NO_2 (D) H_2

3. White lead is
(A) $PbCO_3$ (B) $PbCO_3.\ PbO$
(C) $Pb(OH)_2.\ 2PbCO_3$ (D) $PbSO_4.\ PbO$

4. To which block is related an element having electronic configuration $1s^2\ 2s^2\ 2p^6\ 3s^2\ 3p^6\ 3d^{10}\ 4s^1$ in the periodic table ?
(A) d-block (B) f-block (C) s-block (D) p-block

5. Alkali metals in each period have
(A) smallest size
(B) lowest ionization potential
(C) highest ionization potential
(D) highest electronegativity

6. The electronic configuration of an element is $1s^2\ 2s^2\ 2p^6\ 3s^2\ 3p^6\ 3d^{10}\ 4s^2\ 4p^3$.
Its properties would be similar to which of the following elements ?
(A) Nitrogen (B) Chlorine
(C) Boron (D) Oxygen

7. The C-Cl bond in chlorobenzene as compared with C-Cl bond in methyl chloride is
(A) longer and weaker (B) shorter and weaker
(C) shorter and stronger (D) longer and stronger

8. For Phenol, which of the following statements is correct ?
(A) It is insoluble in water
(B) It has lower melting point compared to aromatic hydrocarbons of comparable molecular weight
(C) It has higher boiling point than toluene
(D) It does not show acidic property

9. Which of the following types of isomerism is shown by Pentanone ?
(A) Chain isomerism (B) Position isomerism
(C) Functional isomerism (D) All of these

10. In Williamson's synthesis, ethoxy ethane is prepared by
(A) ethyl alcohol with sulphuric acid
(B) passing ethanol over heated alumina
(C) sodium ethoxide with ethyl bromide
(D) ethyl iodide and dry silver oxide

11. In aldehydes and ketones, carbon of carbonyl group is
(A) sp^3 hybridised (B) sp^2 hybridised
(C) sp hybridised (D) unhybridised

12. The I. U. P. A. C. name of

$$CH_3-\underset{\displaystyle CH}{\overset{\displaystyle CH_3\ |}{}}\!-\overset{\displaystyle O\ \|}{C}-CH_2-CH_2OH$$ is

(A) 1-hydroxy 4-methyl 3-pentanone
(B) 2-methyl 5-hydroxy 3-pentanone
(C) 4-methyl 3-oxo 1-pentanol
(D) hexanol-1, one-3

13. Which statement is not correct ?
(A) Ethyl amine has higher boiling point than propane
(B) Methyl amine is more basic than ammonia
(C) Amines form hydrogen bond
(D) Dimethyl amine is less basic than methyl amine

14. Which of the following gives primary amine on reduction ?
(A) $CH_3-CH_2-\overset{O\ \|}{N}\rightarrow O$
(B) $CH_3-CH_2-O-N=O$
(C) $CH_3\ CH_2\ NO_3$

(D) None of these

15. The number of unpaired electrons in Ni^{++} is
(A) 2 (B) 4 (C) 8 (D) zero

16. Half life of redioactive element is 100 years. The time in which it disintegrates to 50% of its mass, will be
(A) 50 yrs (B) 200 yrs (C) 100 yrs (D) 25 yrs

17. Which one of the following has both electovalent and covalent bonds ?
(A) NaCl (B) NaOH (C) H_2O (D) HCl gas

18. For which of the following reactions Kp = Kc ?
(A) $H_2 + I_2 \rightleftharpoons 2HI$ (B) $2SO_2 + O_2 \rightleftharpoons 2SO_3$
(C) $PCl_3 + Cl_2 \rightleftharpoons PCl_5$ (D) $N_2 + 3H_2 \rightleftharpoons 2NH_3$

19. Low pressure is favourable for which of the following reactions ?
(A) $PCl_5 \rightleftharpoons PCl_3 + Cl_2$ (B) $N_2 + 3H_2 \rightleftharpoons 2NH_3$
(C) $2SO_2 + O_2 \rightleftharpoons 2SO_3$ (D) $N_2 + O_2 \rightleftharpoons 2NO$

20. When water is added to quick lime, the reaction is
(A) exothermic (B) endothermic
(C) explosive (D) none of these

21. The element with atomic number 55 belong to which block of the periodic table ?
(A) d-block (B) f-block
(C) s-block (D) p-block

22.which of the following has largest size ?
(A) Al (B) Al^+ (C) Al^{2+} (D) Al^{3+}

23. With the increase in atomic weights, melting points of the alkali metals
(A) increase (B) decrease
(C) remain constant (D) do not show definite trend

24. Which element from V group gives most basic compound with hydrogen?
(A) Nitrogen (B) Bismuth
(C)Arsenic (D) Phosphorus

25.Which is the weakest acid out of HF,HCl,HBr and HI ?
(A) HBr (B) HI (C) HF (D) HCl

26. Which element out of He, Ar, Kr, and Xe forms least number of compounds ?
(A) He (B) Ar (C) Kr (D) Xe

27. Which of the alkaline earth meatals is strongest reducing agent ?
(A) Ba (B) Mg (C) Ca (D) Sr

28. Choose the correct answer for transition elements
(A) Transition elements have low melting points
(B) Transition elements do not show catalytic activity
(C) Transition elements exhibit variable oxidation states
(D) Transition elements exhibit inert pair effect

29. Which of the ions having following electronic configuration would have maximum magnetic moment ?
(A) $1s^2 2s^2 2p^6 3s^2 3p^6 3d^3$ (B) $1s^2 2s^2 2p^6 3s^2 3p^6 3d^5$
(C) $1s^2 2s^2 2p^6 3s^2 3p^6 3d^7$ (D) $1s^2 2s^2 2p^6 3s^2 3p^6 3d^9$

30. Which of the following elements does not belong to first transition series ?
(A) Ag (B) Cu (C) Fe (D) V

31. MnO_4^- on reduction in acidic medium forms
(A) MnO_4^{-2} (B) Mn (C) MnO_2 (D) Mn^{++}

32. Conversion of glucose into ethylalcohol is made by
(A) Acid (B) Enzyme
(C) Hydroxylamine (D) Phenyl hydrazine

33. Oxidation number of P in $Mg_2P_2O_7$ is
(A) + 3 (B) + 2 (C) + 5 (D) - 3

34. In the reaction,
$3 Br_2 + 6 CO_3^{-2} + 3H_2O = 5 Br^- + BrO_3^- + 6HCO_3^-$
(A) Bromine is oxidised, carbonate is reduced
(B) Both bromine and carbonate are reduced
(C) Bromine is neither rcduced nor oxidised
(D) Bromine is reduced as well as oxidised

35. In the cell Zn / Zn^{2+} // Cu^{2+} / Cu the negative electrode is
(A) Cu (B) Cu^{2+} (C) Zn (D) Zn^{2+}

36. CO_2 is isostructural with which of the following ?
(A) $HgCl_2$ (B) H_2O (C) $SnCl_2$ (D) NO_2

37. The number of unpaired electrons in Fe^{+++} (Z= 26) is
(A) 3 (B) 4 (C) 5 (D) 6

38. Heat of formation in the reaction,
$H_2 + Cl_2 \rightarrow 2HCl + 44$ K. cal. is
(A) 44 K. cal (B) 44000 K. cal.
(C) 22 K. cal. (D) 11 K. cal.

39. 96500 coulombs of electric current liberates from $CuSO_4$ solution
(A) 63.5 gm Cu (B) 31.75 gm Cu
(C) 96500 gm Cu (D) 100 gm Cu

40. Electrode potentials of five elements A, B, C, D and E are respectively, -1.36, -0.32,. 0, -1.26 and -0.42. The reactivity order of these elements are in the order of
(A) A D E B and C (B) C B E D and A
(C) B D E A and C (D) C A E D and B

41. pH of 0.001 M Na OH is
(A) 10^{11} (B) 11 (C) 10^{-3} (D) 3

42. The size of the colloidal particles is in between
(A) $10^{-7} - 10^{-9}$ cm (B) $10^{-9} - 10^{-11}$ cm
(C) $10^{-5} - 10^{-7}$ cm (D) $10^{-2} - 10^{-3}$ cm

43. Which of the following is most acidic ?
(A) Formic acid (B) Chloro acetic acid
(C) Propionic acid (D) None of these

44. Which of the following is an example of condensation polymers ?
(A) Polythene (B) P. V. C. (C) Orlon (D) Terylene

45. Natural rubber is formed by polymerisaion of which of the following ?
(A) Chloroprene (B) Styrene (C) Phenol (D) Isoprene

46. Which of the following is the thermosetting plastic ?
(A) Nylon (B) Bakelite (C) Polythene (D) Polystyrene

47. Which of the following is responsible for maintaining blood sugar level in human body ?
(A) Riboflavin (B) Insulin (C) Fats (D) Hormone

48. Proteins are composed of
(A) vitamins (B) mineral salts
(C) α – amino acids (D) carbohydrates

49. The base adenine occurs in
(A) DNA only (B) RNA only
(C) DNA and RNA both (D) Protein

50. Which one of the following molecules possesses dipole moment ?
(A) Carbon tetrachloride (B) Water
(C) Methane (D) Nitrogen

51. The number of sigma (σ) bonds in 1-butene is
(A) 8 (B) 10 (C) 11 (D) 12

52. A sample of gasoline contains 81% isooctane and 19% n-heptane. It's octane number will be
(A) 100 (B) 62 (C) 19 (D) 81

53. The Natural Petroleum contains
(A) Saturated hydrocarbons
(B) Cyclic saturated hydrocarbons
(C) Compounds of sulphur
(D) All of these

54.The reaction of benzene with methyl chloride in presence of anhydrous aluminium chloride is called as
(A) Friedel Craft's reaction
(B) Riemer-Tieman reaction
(C) Gattermann Koch reaction
(D) None of these

55. The preparation of ethane by electrolysis of aqueous solution of potassium acetate is called as
(A) Wurtz reaction (B) Sabatier-Senderen's reaction
(C) Kolbe's synthesis (D) Grignard reaction

56. Formation of 2-butene as major product by dehydration of 2-butanol is according to
(A) Markownikov rule (B) saytzeff rule
(C) Peroxide effect (D) Anti Markownikov rule

57. Carbon-halogen bond is strongest among the following
(A) CH_3F (B) CH_3I (C) CH_3Cl (D) CH_3Br

58. The stsbility of lyophillic colloids is due to which of the following ?
(A) Charge on their particles
(B) Large size of their particles
(C) Smaller size of their particles
(D) A layer of medium of dispersion on their particles

59. Milk is
(A) Fat dispersed in water
(B) Water dispersed in fat
(C) Water dispersed n oil
(D) Fat dispersed in fat

60. In an exothermic reaction ΔH is
(A) Positive (B) Negative
(C) Zero (D) Both positive and negative

61. Which of the following is the correct equation ?
(A) $\Delta G = T\Delta S - \Delta H$ (B) $\Delta G = \Delta H - S\Delta T$
(C) $\Delta G = \Delta H + T\Delta S$ (D) $\Delta G = \Delta H - T\Delta S$

62. Heat of neutralization of an acid with a base is 13.7 K. cal. when
(A) both acid and base are weak
(B) acid is weak and base is strong
(C) acid is strong and base is weak
(D) both acid and base are strong

63. The unit of rate constant for first order reaction is
(A) s^{-1} (B) mole lit^{-1} s^{-1}
(C) mole s^{-1} (D) lit $mole^{-1}$ s^{-1}

64. The law of mass action was enunciated by
(A) Guldberg and Waage (B) Bodenstein
(C) Berthelot (D) Graham

65. If the equilibrium constant of the reaction, $2HI \rightleftharpoons H_2 + I_2$ is 0.25, the equilibrium contant for the reaction, $H_2 + I_2 \rightleftharpoons 2HI$ would be
(A) 1.0 (B) 2.0 (C) 3.0 (D) 4.0

66. Conjugate base of HPO_4^{2-} is
(A) H_3PO_4 (B) H_3PO_3 (C) PO_4^{3-} (D) $H_2PO_4^-$

67. Which of the following is the strongest acid ?
(A) CH_3COOH (B) $ClCH_2COOH$
(C) $Cl_2CHCOOH$ (D) Cl_3CCOOH

68. The solubility of $PbCl_2$ is
(A) $\sqrt{Ksp}$ (B) $\sqrt[3]{Ksp}$ (C) $\sqrt[3]{Ksp/4}$ (D) $\sqrt{8Ksp}$

69. The ppt. of CaF_2 ($Ksp = 1.7 \times 10^{-10}$) is obtained when equal volume of which of the following are mixed ?
(A) 10^{-1} M Ca^{2+} + 10^{-4} M F^- (B) 10^{-2} M Ca^{2+} + 10^{-3} M F^-
(C) 10^{-2} M Ca^{2+} + 10^{-5} M F^- (D) 10^{-3} M Ca^{2+} + 10^{-5} M F^-

70. Which of the following reacts with Phenol to give Salicylaldehyde after hydrolysis ?
(A) Dichloro methane (B) Trichloro methane
(C) Methyl chloride (D) None of these

71. Which of the following is most soluble in water ?
(A) Normal butyl alcohol (B) Iso butyl alcohol
(C) Tertiary butyl alcohol (D) Secondary butyl alcohol

72. Ethyl alcohol exhibits acidic chracter on reacting with
(A) Acetic acid (B) Sodium metal
(C) Hydrogen iodide (D) Acidic potassium dichromate

73. Phenol is obtained by heating aqueous solution of
(A) aniline (B) benzene diazonium chloride
(C) benzoic acid (D) none of these

74. Which type of isomers are
$[Co(NH_3)_5Br]SO_4$ and $[Co(NH_3)_5SO_4]Br$?
(A) Hydrate isomers (B) Ionisation isomers
(C) Ligand isomers (D) Coordination isomers

75. Correct I.U.P.A.C. nomenclature of $K_4[Fe(CN)_6]$ is
(A) Potassium Ferro Cyanide
(B) Tetra potassium Ferrous Cyanide
(C) Potassium Ferri Cyanide
(D) Potassiumhexacyanoferrate (II)

76. Which of the following ions forms most stable complex compound ?
(A) Fe^{++} (B) Mn^{++} (C) Cu^{++} (D) Ni^{++}

77. Which of the following compounds turns black on addition of ammonium hydroxide ?
(A) $PbCl_2$ (B) Hg_2Cl_2 (C) $HgCl_2$ (D) AgCl

78. On adding excess of ammonium hydroxide to a copper sulphate solution
(A) Blue precipitate of copper hydroxide is obtained
(B) Black precipitate of copper oxide is obtained
(C) A deep blue solution is obtained
(D) No change is observed

79. On passing H_2S gas in II group, some-times the solution turns milky. It indicates the presence of
(A) Acidic salt (B) Oxidising agent
(C) Thiosulphate (D) Reducing agent

80. Equivalent weight of crystalline oxallic acid is :
(A) 53 (B) 45 (C) 90 (D) 63

81. The sulphate of a metal has the formula $M_2(SO_4)_3$. The formula for its phosphate will be
(A) $M(HPO_4)_2$ (B) $M_3(PO_4)_2$
(C) $M_2(PO_4)_3$ (D) MPO_4

82. The octet rule is not valid for which of the following molecules ?
(A) CO_2 (B) H_2O (C) CO (D) O_2

83. How many σ and π bonds are there in the molecule of tetracyano ethylene ?

$$\begin{matrix} N \equiv C \\ \\ N \equiv C \end{matrix} \Big\rangle C = C \Big\langle \begin{matrix} C \equiv N \\ \\ C \equiv N \end{matrix}$$

(A) nine σ , nine π (B) five σ , nine π
(C) nine σ , seven π (D) five σ , eight π

84. Chlorine atom differs from chloride ion in the number of which of the following ?
(A) Protons (B) Neutrons
(C) Electrons (D) Both protons and electrons

85. For n = 3 energy level, the number of possible orbitals (all kinds) are
(A) 1 (B) 3 (C) 4 (D) 9

86. When the 3d orbital is complete, the new electrons will enter in
(A) 4s orbital (B) 4d orbital
(C) 4p orbital (D) 4f orbital

87. The radius of an atom is of the order of
(A) 10^{-10} cm (B) 10^{-13} cm
(C) 10^{-15} cm (D) 10^{-8} cm

88. On which of the following principles hydrogen bomb is based ?
(A) Nuclear fission (B) Nuclear fusion
(C) Artificial radioactivity (D) Natural radioactivity

89. Which one of the following is an isotone of $^{76}_{32}Ge$?
(A) $^{77}_{32}Ge$ (B) $^{78}_{33}As$ (C) $^{77}_{34}Se$ (D) $^{78}_{34}Se$

90. Bragg's law is given by the equation
(A) $n\lambda = 2\theta \sin\theta$ (B) $n\lambda = 2d \sin\theta$
(C) $2n\lambda = d \sin\theta$ (D) $n\frac{\theta}{2} = \frac{d}{2} \sin\theta$

91. The impurities associated with minerals used in metallurgy are called collectively
(A) Gangue (B) Ore (C) Slag (D) Flux

92. Main process for the manufacture of sodium carbonate is
(A) Castner Process (B) Solvay Process
(C) Down's Process (D) Nelson Process

93. Aluminium reacts with caustic soda to form
(A) aluminium hydroxide
(B) aluminium oxide
(C) sodium meta aluminate
(D) sodium tetra aluminate

94. Rusting of iron needs
(A) dry air
(B) air and water
(C) distilled water and carbon dioxide
(D) oxygen and carbon dioxide

95. Electronic configuration of Cu^{29} is
(A) $1s^2 2s^2 2p^6 3s^2 3p^5 3d^{10} 4s^2$
(B) $1s^2 2s^2 2p^6 3s^1 3p^6 3d^{10} 4s^1$
(C) $1s^2 2s^2 2p^6 3s^2 3p^6 3d^{10} 4s^1$
(D) $1s^2 2s^2 2p^6 3s^2 3p^5 3d^{11} 4s^1$

96. Which of the following molecules has a linear structure ?
(A) H_2O (B) CO_2 (C) NH_3 (D) CH_4

97. 4.0 gram of caustic soda is dissolved in 100 cc solution. The normality of solution is
(A) 1.0 (B) 0.1 (C) 0.5 (D) 4.0

98. The structure of chloromethane is
(A) Tetrahedral (B) Trigonal
(C) Linear (D) Hexagonal

99. I.U.P.A.C. name of $CH_2 = CH - CH_2Cl$ is
(A) Allyl chloride (B) 1- chloro 3-propene
(C) Vinyl chloride (D) 3 - chloro 1-propene

100. On analysis a saturated hydrocarbon is found to contain 83.70 percent carbon and 16.30% hydrogen. The empirical formula will be
(At. wt. of C=12, At. wt. of H=1)
(A) C_3H_7 (B) C_6H_{12}
(C) C_3H_6 (D) C_3H_8

PMT - 1996

1. The order of the magnitude of first ionisation potentials of Be, B, N and O is

(A) Be > B > N > O (B) B > Be > O > N
(C) N > O > Be > B (D) N > Be > O > B

2. The order of magnitude of ionic radii of ions Na^+, Mg^{2+}, Al^{3+} and Si^{4+} is

(A) $Na^+ < Mg^{2+} < Al^{3+} < Si^{4+}$
(B) $Mg^{2+} > Na^+ > Al^{3+} > Si^{4+}$
(C) $Al^{3+} > Na^+ > Si^{4+} > Mg^{2+}$
(D) $Na^+ > Mg^{2+} > Al^{3+} > Si^{4+}$

3. The order of the magnitude of ionic radii of ions N^{3-}, O^{2-} and F^- is

(A) $N^{3-} > O^{2-} > F^-$ (B) $N^{3-} < O^{2-} < F^-$
(C) $N^{3-} > O^{2-} > F^-$ (D) $N^{3-} < O^{2-} < F^-$

4. Elements of the same group in periodic table have the same

(A) Electronic configuration
(B) Atomic weight
(C) Atomic number
(D) Number of electrons in the outer most shell

5 In the reaction sequence,

$$CaC_2 \xrightarrow{H_2O} A \xrightarrow[Hg^{++}]{\text{dil } H_2SO_4} B \xrightarrow[Ni]{H_2} C$$

the product C is,

(A) CH_3OH (B) CH_3CHO
(C) C_2H_5OH (D) C_2H_4

6. Which of the following substances is used as an anti-knock compound ?

(A) Tetra ethyl lead (B) Lead tetrachloride
(C) Lead acetate (D) Ethyl acetate

7. Lucas reagent is

(A) conc. HNO_3 + anhydrous $ZnCl_2$
(B) conc. HNO_3 + anhydous $MgCl_2$
(C) conc. HCl + anhydrous $ZnCl_2$
(D) dil HCl + hydrated $ZnCl_2$

8. Reaction of an alkyl halide with sodium alkoxide yields ?

(A) an alcohol (B) a ketone
(C) an ether (D) a carboxylic acid

9. Which one of the following does NOT give iodoform test ?

(A) Methyl alcohol (B) Ethyl alcohol
(C) Acetaldehyde (D) Acetone

10. The products of the reaction between $C_6H_5NH_2$, $CHCl_3$ and KOH are

(A) C_6H_5NC + KCl
(B) $C_6H_5OH + NH_4Cl + H_2O$
(C) $C_6H_5Cl + NH_4Cl$ + KCl (D) C_6H_5CN + KCl

11. The percentage of ethyl alcohol present in rectified spirit is

(A) 95.6 (B) 100.0 (C) 75.0 (D) 85.5

12. A buffer solution contains 0.1 M of acetic acid and 0.1 M of sodium acetate. What will be its pH, if pKa of acetic acid is 4.75 ?

(A) 5.00 (B) 5.25 (C) 4.00 (D) 4.75

13. Which of the following cannot be hydrolysed ?

(A) A salt of weak acid and strong base
(B) A salt of strong acid and weak base
(C) A salt of weak acid and weak base
(D) A salt of strong acid and strong base

14. Which of the following statements is true ?

(A) The conjugate base of a weak acid is a strong base
(B) The conjugate base of a weak acid is a weak base
(C) The conjugate base of a strong acid is a strong base
(D) The base and its conjugate acid react to form a neutral solution

15. The heat change for the following reaction, C (s) + 2S (s) $\rightarrow CS_2$ (*l*) is known as

(A) heat of vaporization (B) heat of solution
(C) heat of fusion (D) heat of formation

16. When a solid changes into liquid, entropy

(A) becomes zero (B) becomes minimum
(C) increases (D) remains the same

17. Which of the following expressions is correct ?

(A) H = E - PV (B) P = E + HV
(C) E = H + PV (D) H = E + PV

18. The term $-\left(\frac{dc}{dt}\right)$ in a rate equation refers to the

(A) concentration of the reactant
(B) decrease in concentration of the reactant with time
(C) increase in concentration of the reactant with time
(D) velocity constant of the reaction

19. If the rate expression for a chemical reaction is given by Rate = k $[A]^m$ $[B]^n$

(A) the order of the reaction is m + n
(B) the order of the reaction is m
(C) the order of the reaction is n
(D) the order of the reaction is m - n

20. The electronic configuration in the outermost shell of the halogen aom is

(A) s^2p^6 (B) s^2p^4 (C) s^2p^5 (D) s^2p^3

21. Element of atomic number 23 is placed in periodic table in

(A) s-block (B) p-block (C) d-block (D) f-block

22. Alkali metals are

(A) Li, Na, Be, Mg, Cs (B) Li, Na, K, Rb, Cs
(C) Na, K, Mg, Ca, Rb (D) K, Rb, Cs, Ba, Sr

23. The atomic number of an element is 11. Its oxide will be

(A) Acidic (B) Basic
(C) Acidic and Basic both (D) Neutral

24. The commercial production of sodium carbonate is done by

(A) Lead Chamber process (B) Haber's process
(C) Solvay's process (D) Castner'sprocess

25. The formula of washing soda is
(A) $Na_2CO_3.10H_2O$ (B) Na_2CO_3
(C) Na_2CO_3. $7H_2O$ (D) Na_2CO_3. H_2O

26. The metal which reacts with water at room temperature is
(A) Copper (B) Iron
(C) Magnesium (D) Sodium

27. 'Alkaline earth meatals are
(A) Li, Be, K, Mg, Ca (B) Be, Mg, Ca, Sr, Ba
(C) Be, K, Mg, Ca, Sr (D) Be, Mg, Ca, K, Rb

28. The composition formula of gypsum is
(A) $CaSO_4$ (B) $CaSO_4$. $2H_2O$
(C) $(CaSO_4)_2$. H_2O (D) $2CaSO_4$. H_2O

29. The formula of a phosphate of a metal, M is MPO_4. The formula of its nitrate will be
(A) MNO_3 (B) $M(NO_3)_2$
(C) $M(NO_3)_3$ (D) $M_2(NO_3)_3$

30. How many molecules are there in the unit cell of sodium chloride ?
(A) 2 (B) 4 (C) 6 (D) 8

31. What type of point defect is found in crystal of sodium chloride ?
(A) Linear defect (B) Impurity defect
(C) Frenkel difect (D) Schottky defect

32. There exists on γ-rays
(A) positive charge (B) negative charge
(C) no charge
(D) sometimes positive charge, some times negative charge

33. Which element is the end product of each natural radioactive series ?
(A) Pb (B) C (C) Sn (D) Bi

34. For artificial transmutation of nuclei, the most effective one is
(A) Proton (B) Deutron
(C) Helium nuclei (D) Neutron

35. Which of the following does not take place by α-decay ?
(A) ${}^{238}_{92}U \rightarrow {}^{234}_{90}Th$ (B) ${}^{232}_{90}Th \rightarrow {}^{228}_{88}Ra$
(C) ${}^{226}_{88}Ra \rightarrow {}^{222}_{86}Rn$ (D) ${}^{213}_{83}Bi \rightarrow {}^{213}_{84}Po$

36. A chemical reaction is in equilibrium when
(A) the rate of forward reaction is equal to the rate of backward reaction
(B) the formation of products is minimised
(C) the reactants are completely transformed into products
(D) equal amount of reactants and products are present

37. According to Le Chatelier's principle which of the following factors influence a chemical system ?
(A) concentration only
(B) Pressure only
(C) Temperature only
(D) Concentration, Pressure and Temperature

38. For the following gaseous reaction, $H_2 + I_2 \rightarrow 2HI$. the equilibrium constant.
(A) Kp > Kc (B) Kp < Kc (C) Kp = Kc (D) $Kp = \frac{1}{Kc}$

39. Sodium chloride is purified by passing hydrogen chloride gas in an impure solution of sodium chloride. It is based on
(A) buffer action (B) common ion effect
(C) association of salt (D) hydrolysis of salt

40. The formula of a phosphate of a metal, M, is MNH_4PO_4. The formula of its sulphate will be :
(A) M_2SO_4 (B) $M_2(SO_4)_3$
(C) MSO_4 (D) $M(SO_4)_2$

41. The formula of an oxyacid of phosphorus is H_3PO_4. It is a
(A) dibasic acid (B) monobasic acid
(C) tribasic acid (D) tetrabasic acid

42. Unit of first order rate constant is
(A) mole litre second (B) $mole^{-1}$ $litre^{-1}$ $second^{-1}$
(C) mole $litre^{-1}$ $second^{-1}$ (D) $second^{-1}$, $minute^{-1}$ etc.

43. The oxidation number of chromium in $K_2Cr_2O_7$ is
(A) +2 (B) -2 (C) -6 (D) +6

44. The enzyme, which converts glucose and fructose into ethyl alcohol is
(A) diastase (B) invertase (C) zymase (D) maltase

45. Enzymes are
(A) Oils (B) Fatty acids (C) Proteins (D) Minerals

46. Protein is a polymer of
(A) Glucose (B) Terephthalic acid
(C) Amino acids (D) Glycol

47. When formaldehyde is heated with ammonia, the compound formed is
(A) formaldehyde ammonia (B) methyl amine
(C) hexamethylene tetramine (D) formalin

48. The empirical formula of a compound is CH_2. One molucule of this compound has a mass of 42 g. Its molucular formula is
(A) C_3H_8 (B) C_3H_6 (C) CH_2 (D) C_2H_2

49. The I.U.P.A.C. name of CH_3CHO is :
(A) Ethane (B) Ethanol
(C) Ethanal (D) Acetaldehyde

50. 1, 2-Butadiene contains
(A) sp and sp^2 hybridised carbon atoms
(B) only sp hybridised carbon atoms
(C) only sp^2 hybridised carbon atoms
(D) sp, sp^2 and sp^3 hybridised carbon atoms

51. Dehydrohalogenation of an alkyl halide is
(A) an addition reaction (B) a substitution reaction
(C) an elimination reaction (D) an oxidation reaction

52. I. U. P. A. C. name of the compound

$CH_3 - CH = \underset{\displaystyle CH_2 - CH_3}{\overset{|}{C}} - CH_3$ is

(A) 2-ethyl-2-butene (B) 3-ethyl-2-butene
(C) 3-methyl-3-pentene (D) 3-methyl-2-pentene

53. Benzene is prepared in laboratory from which one of the following compounds ?
(A) C_6H_5COONa (B) $C_6H_5SO_3H$
(C) $C_6H_5N_2Cl$ (D) C_6H_5OH

54. A gas decolourises alkaline $KMnO_4$ but does not give precipitate with ammonical silver nitrate, the gas is
(A) Methane (B) Ethane
(C) Ethene (D) Acetylene

55. When ethyl alcohol is heated with conc. H_2SO_4, the product obtained is
(A) $H_3C - \underset{\displaystyle O}{\overset{}{\underset{||}{C}}} - OC_2H_5$ (B) C_2H_6
(C) C_2H_4 (D) C_2H_2

56. The reaction

$CH_2 = CH_2 + H_2 \xrightarrow[250^\circ - 300^\circ C]{Ni} CH_3 - CH_3$ is called

(A) Wurtz reaction
(B) Carbylamine reaction
(C) Kolbe's reaction
(D) Sabatier Senderens reaction

57. When 96500 coulomb of electricity is passed through a copper sulphate solution, the amount of copper deposited will be
(A) 0.25 mol (B) 0.50 mol
(C) 1.00 mol (D) 2.00 mol

58. The increase in equivalent conductance of an electrolyte solution with dilution is due to the increase in
(A) molecular attraction
(B) degree of association of the electrolyte
(C) ionic attraction
(D) degree of ionisation of the electrolyte

59. Iron displaces copper from its salt solution, because
(A) atomic number of iron is less than that of copper
(B) the standard reduction potential of iron is less than that of copper
(C) the standard reduction potential of iron is more than that of copper
(D) the iron salt is more soluble in water than the copper salt

60. In chemical adsorption, how many layers are adsorbed ?
(A) Zero (B) One (C) Two (D) Multi

61. What is the role of a catalyst in a catalysed reaction ?
(A) Lowers the activation energy
(B) Increases the activation energy
(C) Affects the free energy change
(D) Affects the enthalpy change

62. For coagulating As_2S_3 colloidal sol., which of the following will have the lowest coagulation value ?
(A) NaCl (B) KCl (C) $BaCl_2$ (D) $AlCl_3$

63. Am important oxide ore of iron is
(A) Haematite (B) Siderite
(C) Limestone (D) Pyrites

64. The molecular formula of sulphur is :
(A) S (B) S_2 (C) S_4 (D) S_8

65. The atomic number of an element is 22. The highest oxidation state exhibited by it in its compounds is :
(A) 1 (B) 2 (C) 3 (D) 4

66. The complex compound $[Co(NH_3)_3\ NO_2\ ClCN]$ is named as
(A) Cyanonitrochlorotriamminecobalt (III)
(B) Triamminechlorocyanonitrocobalt (III)
(C) Nitrochlorocyanotriamminecobalt (III)
(D) Triamminenitrochlorocyanocobalt (III)

67. Pb^{++}, Cu^{++} and Ni^{++} ions are present in a given acidic solution. On passing hydrogen sulphide gas through this solution, the avilable precipitate will contain
(A) PbS and NiS (B) PbS and CuS
(C) CuS and ZnS (D) CuS and NiS

68. Al^{3+}, Fe^{3+}, Zn^{2+} and Ni^{2+} ions are present in an acidic solution. Excess of NH_4Cl is added following by addition of NH_4OH solution. The avilable precipitate will contain
(A) $Zn(OH)_2$ and $Ni(OH)_2$
(B) $Al(OH)_3$ and $Fe(OH)_3$
(C) $Zn(OH)_2$ and $Al(OH)_3$
(D) $Ni(OH)_2$ and $Fe(OH)_3$

69. 0.53 gm. of Na_2CO_3 has been dissolved in 100 ml. of sodium carbonate solution. The normality of the solution will be
(A) $\frac{N}{5}$ (B) $\frac{N}{2}$ (C) $\frac{N}{10}$ (D) N

70. The amount of crystalline oxalic acid required to prepare 250 ml. N/10 oxalic acid solution will be
(A) 1.575 gm (B) 3.150 gm
(C) 6.300 gm (D) 0.787 gm

71. Cryolite is an ore of
(A) zinc (B) aluminium (C) iorn (D) silver

72. The molucular formula of bleaching powder is
(A) $CaCl_2$ (B) Ca_2OCl_2
(C) $CaOCl_2$ (D) $CaCl_2.\ CaOCl$

73. The strongest oxidising agent amongst F_2, Cl_2, Br_2 and I_2 is
(A) F_2 (B) I_2 (C) Cl (D) Br_2

74. 64 gm of an organic compound contains 24 gm of carbon, 8 gm of hydrogen and the rest oxygen. The empirical formula of the compound is
(A) CH_2O (B) C_2H_4O
(C) CH_4O (D) $C_2H_8O_2$

75. The metal which is extracted by Parke's process is
(A) Iron (B) Lead (C) Copper (D) Silver

76. Sodium metal is extracted by
(A) Heating sodium oxide with carbon
(B) Heating sodium oxide with hydrogen
(C) Electrolysis of aqueous solution of sodium chloride
(D) Electrolysis of fused sodium chloride

77. Brass is an alloy of
(A) Gold and Copper (B) Silver and Zinc
(C) Copper and Zinc (D) Copper and Aluminium

78. Which of the following expressions gives the de Broglie relationship ?
(A) $h = \frac{\lambda}{mv}$ (B) $\lambda = \frac{h}{mv}$ (C) $\lambda = \frac{m}{hv}$ (D) $\lambda = \frac{v}{mh}$

79. A covalent bond between two atoms is formed by which of the following ?
(A) Electron nuclear attraction
(B) Electron sharing (C) Electron transfer
(D) Electrostatic attraction

80. Hybridisation involves
(A) removal of an electron pair
(B) addition of an electron pair
(C) mixing up of atomic orbitals
(D) separation of orbitals

81. On the basis of Molecular Orbital theory which molecule does not exist ?
(A) H_2 (B) He_2^+ (C) He_2 (D) Li_2

82. Which of the following compounds in liquid state does not have hydrogen bonding ?
(A) NH_3 (B) C_6H_6 (C) H_2O (D) HF

83. The molality of a solution is
(A) number of moles of solute per 1000 ml of the solvent
(B) number of moles of solute per 1000 gm of the solvent
(C) number of moles of solute per 1000 ml of the solution
(D) number of gram equivalents of solute per 1000 ml of the solution

84. An ideal solution is that which
(A) has no connection with Raoult's law
(B) obeys Raoult's law
(C) shows positive deviation from Raoult's law
(D) shows negative deviation from Raoult's law

85. Which of the following is a colligative property ?
(A) Boiling point (B) Freezing point
(C) Melting point (D) Osmotic pressure

86. Which of the following reagents can distinguish between an aldehyde and a ketone ?
(A) Fehling's solution (B) H_2SO_4 solution
(C) $NaHSO_4$ (D) NH_3

87. Dry distillation of the mixture of calcium formate and calcium acetate gives
(A) Formaldehyde (B) Fomic acid
(C) Acetone (D) Acetaldehyde

88. Which of the following will react with acetamide to give methyl amine ?
(A) PCl_5 (B) NaOH + Br_2
(C) Soda lime (D) H_2SO_4

89. The general formula $(RCO)_2O$ represents
(A) a ketone (B) an ether
(C) an acid anhydride (D) an ester

90. Prolonged heating of ammonium cyanate gives
(A) nitrogen (B) carbon - dioxide
(C) biuret (D) ammonium carbonate

91. The chemical name of the insecticide Gammexane is
(A) D. D. T. (B) Hexachloro benzene
(C) Chloral (D) Benzene hexachloride

92. Primary amine and secondary amine are distinguished by
(A) Br_2 / KOH (B) $HClO_4$ (C) HNO_2 (D) NH_3

93. On heating glucose with Fehling's solution, a precipitate is obtained whose colour is
(A) Black (B) White (C) Yellow (D) Red

94. When aniline reacts with $NaNO_2$ and dil. HCl at 0° - 5^oC, the product formed is
(A) Nitroaniline (B) Benzene diazonium chloride
(C) Benzene (D) Tri-nitroaniline

95. Which of the following describes the shape of orbital ?
(A) Magnetic quantum number
(B) Principal quantum number
(C) Azimuthal quantum number
(D) Spin quantum number

96. The number of orbitals in a 2p subshell is
(A) 1 (B) 2 (C) 3 (D) 6

97. How many unpaired electrons are present in an atom of electronic configuration $1s^2\ 2s^2\ 2p^4$?
(A) 1 (B) 2 (C) 3 (D) 4

98. Which of the following will produce the maximum depression in freezing point of its aqueous solution ?
(A) 0.1 M glucose
(B) 0.1 M sodium chloride
(C) 0.1 M barium chloride
(D) 0.1 M magnesium sulphate

99. How many space lattices are obtainable from different crystal systems ?
(A) 7 (B) 14 (C) 32 (D) 230

100. What type of lattice is found in potassium chloride crystal ?
(A) Simple cubic (B) Simple tetragonal
(C) Face centred cubic (D) Body centred cubic

PMT - 1997

1. If the impurity in a metal has a greater affinity for oxygen and is more easily oxidised than the metal, then the purification of metal may be carried out by
(A) Electrolytic refining (B) cupellation
(C) Poling (D) Zone refining

2. Aluminium has a great affinity for oxygen and its oxidation is an exothermic process, This fact is made use of in
(A) Preparing thin foils of aluminium
(B) Thermite welding
(C) Making utensils
(D) Preparing duralumin alloy

3. Point out the wrong statement. In a given period of the periodic table, the s-block element has, in general, a lower value of
(A) Ionisation energy (B) Electronegativity
(C) Atomic radius (D) Electron affinity

4. The electron affinities of halogens are F=322, Cl=349, Br = 324 and I = 295 KJ mol^{-1}respectively. The higher value for Cl as compared to that of F is due to
(A) Smaller electronegativity of F
(B) Weaker electron-electron repulsion in Cl
(C) Higher atomic radius of F
(D) More vacant P-subshell in Cl

5. Which of the following is the correct order of decreasing basic nature of the oxides
(A) Al_2O_3, MgO, Cl_2O_7, SO_3
(B) MgO, Al_2O_3, SO_3, Cl_2O_7
(C) Cl_2O_7, SO_3, Al_2O_3, MgO
(D) SO_3, Cl_2O_7, MgO, Al_2O_3

6. In the Presence of iodine catalyst chlorine reacts with acetic acid to form

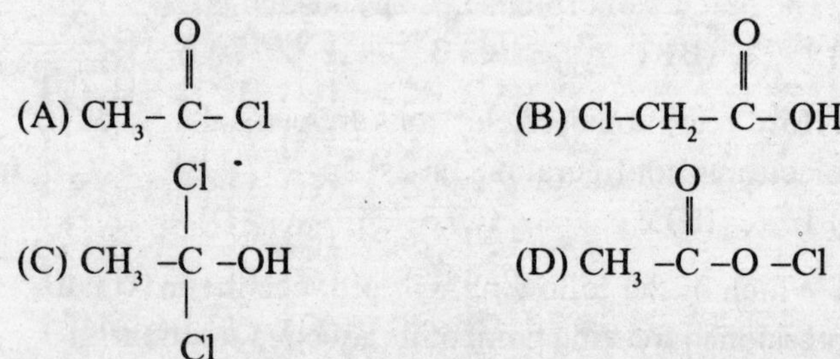

7. Which is the strongest acid (pka value is given)
(A) HCOOH (3.77) (B) C_6H_5COOH (4.22)
(C) CH_3COOH (4.71) (D) CH_3CH_2COOH (4.88)

8. Which of the following is most stable ?
(A) $CH_3\ \overset{+}{C}H_2$ (B) $(CH_3)_2\ \overset{+}{C}H$
(C) $(CH_3)_3\ C^+$ (D) $(CH_3)_2$ CH –CH_2 – $^+CH_2$

9. Lucas reagent is
(A) Conc. HNO_3 + anhydrous $ZnCl_2$
(B) Conc. HNO_3 + anhydrous $MgCl_2$
(C) Conc. HCl + anhydrous $ZnCl_2$
(D) Dil. HCl + hydrous $ZnCl_2$

10. Point out the correct systematic (IUPAC) name of $K_4[Fe\ (CN)_6]$
(A) Potassiumferrocyanide
(B) Potassiumhexacyanoferrate (II)
(C) Tetrapotassiumhexacyanoferrate (III)
(D) Tetrapotassiumferrous hexacyanide

11. Diaamminedichloroplatinum(II) complex has the formula
(A) [Pt. $Cl_2(NH_3)_2$] (B) [Pt.R.$(NH_2)_2$] Cl_2
(C) Pt.[$Cl_2(NH_3)_2$] (D) Pt.[R.$(NH_2)_2$] Cl_2

12. The 3d-elements show variable oxidation states. The maximum oxidation state shown by the element Mn
(A) +4 (B) +5 (C) +6 (D) +7

13. The tendency of 3d-metal ions to form stable complexes is due to their
(A) High charge /size ratio and vacant d-orbitals
(B) Variable oxidation state
(C) Strong electronegative nature
(D) Very low ionization energies

14. The 3d-metal ions are paramagnetic in nature because
(A) They are reducing agents
(B) They form coloured salts
(C) They have one or more paired s-electrons
(D) They have one or more unpaired d-electrons

15. Potassium permanganate acts as an oxidant in neutral, alkaline as well as acidic media. The final products obtained from it in the three conditions are, respectively
(A) MnO^{2-}, MnO_2, Mn^{2+} (B) MnO_4^{2-}, Mn^{3+}, Mn^{2+}
(C) MnO_2, MnO_4^{2-}, Mn^{3+} (D) MnO, MnO_2 Mn^{2+}

16. Out of the following compounds, the one which forms hydrogen bond is
(A) H_2O (B) $CHCl_3$ (C) CH_4 (D) NaCl

17. Out of the following hybrid orbitals, the one which forms the bond at an angle of 120° is
(A) d^2sp^3 (B) sp^3 (C) sp^2 (D) sp

18. As the p character increases the bond angle in hybrid orbitals formed by s and p atomic orbitals
(A) Decreases (B) Increases
(C) Doubles (D) Remains unchanged

19. Out of the following electronic configurations, the one of a transition element is
(A) $1s^22s^2p^63s^2p^64s^2$ (B) $1s^22s^2p^63s^2p^6d^{10}4s^2p^1$
(C) $1s^12s^2p^63s^2p^6d^{10}4s^2p^6$ (D) $1s^22s^2p^63s^2p^6d^34s^2$

20. As compared to covalent compounds, the electrovalent compounds generally have
(A) Low melting point and high boiling point
(B) High melting point and high boiling point
(C) Low melting point and low boiling point
(D) High melting point and low boiling point

21. One mole of solute A is dissolved in a given volume of a solvent. The association of the solute takes place

according to $nA \rightleftharpoons (A)_n$. The Vant Hoff's factor is expressed as

(A) $i = 1 - x$ (B) $i = 1 + \frac{x}{n}$

(C) $i = \frac{1 - x + \frac{x}{n}}{1}$ (D) $i = 1$

22. Freezing point of water is 0°C. At what temperature will one molal solution of NaCl freeze if NaCl is taken to be completely dissociated (mole depression constant of water = 1.86)

(A) 0°C (B) 3.72°C (C) - 3.72°C (D) - 1.86°C

23. The number of moles of solute in its solution is 20 and total number of moles are 80. The mole fraction of solute is

(A) 1 (B) 0.75 (C) 2.5 (D) 0.25

24. Structure of NaCl crystal is

(A) Tetragonal (B) Cubic

(C) Orthorhombic (D) Monoclinic

25. Bravais lattices are of

(A) 8 types (B) 12 types (C) 14 types (D) 9 types

26. The pair of complex compounds $[Cr(H_2O)_6] Cl_3$ and $[Cr (H_2O)_5 Cl] Cl_2H_2O$ are an example of

(A) Linkage isomerism (B) Ionisation isomerism

(C) Coordination isomerism (D) Hydrate isomerism

27. The geometry of the complex $[Cu (NH_3)_4^{2+}]$ is

(A) Triangular bipyramid (B) Octahedral

(C) Square planar (D) Tetrahedral

28. An anoin solution gives a white ppt. with $AgNO_3$ solution. The ppt. dissolves in dilute ammonia due to the formation of

(A) $Ag NO_3$ (B) NH_4NO_3

(C) $[Ag(NH_3)_2] Cl$ (D) $[Ag (NH_3)_2] Br$

29. pH of water is 7. When a substance Y is dissolved in water, the pH becomes 13. The substance Y is a salt of

(A) Strong acid and weak base

(B) Strong acid and strong base

(C) Weak acid and weak base

(D) Weak acid and strong base

30. Which of the following relations between free energy, enthalpy and entropy is correct

(A) $\Delta H = \Delta G + T\Delta S$ (B) $\Delta H = \Delta G - T\Delta S$

(C) $\Delta H = \Delta G + \frac{\Delta S}{T}$ (D) $\Delta H = T\Delta G + \Delta S$

31. Heat of reaction at constant pressure is given by

(A) $E_P - E_R$ (B) $E_R - E_P$ (C) $H_P - H_R$ (D) $H_R - H_P$

32. If enthalpy of B be greaterthan that of A, the reaction $A \rightarrow B$ will be

(A) Exothermic (B) Endothermic

(C) Instantaneous (D) Spontaneous

33. In an isochoric process, the increase in internal energy is

(A) Equal to the work done

(B) Equal to the heat absorbed

(C) Equal to he heat evolved

(D) Equal to the sum of the heat absorbed and work done

34. The positive value of ΔS indicates that

(A) The system becomes less disordered

(B) The system becomes more disordered

(C) The system is in equlibrium position

(D) The system tends to reach at equilibrium position

35. The rate at wihch substances react depends on their

(A) Atomic weight (B) Molecular weight

(C) equivalent weight (D) Active mass

36. For the reaction $N_2 + 3H_2 \rightleftharpoons 2NH_3 + Heat$

(A) $K_P = K_C$ (B) $K_P = K_C RT$

(C) $K_P = K_C (RT)^{-2}$ (D) $K_P = K_C (RT)^{-1}$

37. Which of the properties remains unchanged on descending a group in the periodic table

(A) Valence electrons (B) Metallic character

(C) Atomic size (D) Density

38. The atomic radii of alkali metals (M) lie in the order Li <Na< K < Rb but the radii of M^+ ions in aqueous solution lie in the reverse order $Li^+ > Na^+ > K^+ > Rb^+$. What is the reason for this reverse order (on going from Li to Rb)

(A) Gradual increase in ionisation energy

(B) Increasing weakness of the metallic bond

(C) Increasing electropositive character

(D) Decreasing degree of hydration

39. Which of the following can be represented by the configuration $[Kr] 5s^2$?

(A) Ba (B) Ra (C) Ca (D) Sr

40. Point out the incorrect statement regarding Be (group-II)

(A) It forms an ionic carbide

(B) Its carbonate decomposes on heating

(C) Its halides are covalent

(D) It is easily attacked by water

41. Beryllium differs from the rest of the members of its family (Groups- II) in many ways. The reason for this is its

(A) Large size and lower ionisation energy

(B) Small size and higer electronegativity

(C) Small size and lower electronegativity

(D) Large size and largest ionic radius

42. The number of electrons in the outermost shell of the 3d-transition elements generally remains

(A) $(n - 1) d^n$ (B) nd^n (C) ns^2 (D) $(n -1) s^2$

43. When temperature is raised, the rate of the reaction increases because of

(A) Lowering of activation energy

(B) Increase in the number of collisions

(C) Decrease in the number of activated molecules

(D) Decrease in the number of collisions

44. The half-life period of a first order reaction is 100 sec. The rate constant of the reaction is
(A) 6.93×10^{-3} sec^{-1} (B) 6.93×10^{-4} sec^{-1}
(C) 0.693 sec^{-1} (D) 69.3 sec^{-1}

45. Which of the following tendencies remains unchanged on going down in the nitrogen family ?
(A) Highest oxidation state (B) Non-metallic character
(C) Stability of hydrides (D) Physical state

46. Point out in which of following properties, oxygen differs from the rest of the members of its family (Group - VIA)
(A) Polymorphism
(B) Formation of hydrides
(C) High value of ionisation energies
(D) Oxidation states (2,4,6)

47. Which of the following properties increases on going down from F to I in group VII-A of the pereiodic table
(A) Electronegativity (B) Volatile nature
(C) Ionic radius (D) Oxidising power

48. Ammonical silver nitrate solution reacts with C_2H_2 to from
(A) Silver mirror (B) Silver oxide
(C) Silver formate (D) Silver acetylide

49. The reaction

$C_{10}H_{22} \xrightarrow{900\,K} C_4H_8 + C_6H_{14}$ is called

(A) Pyrolysis (B) Dissociation
(C) Cracking (D) Fractionation

50. Acetylene reacts with dilute H_2SO_4 and Hg^{2+} to give the stable product
(A) Formaldehyde (B) Acetic acid
(C) Vinyl alcohol (D) Acetaldehyde

51. For a given alkyl group the densites of the halides follow the order :
(A) RI < RBr < RCl (B) RI < RCl < RBr
(C) RBr < RI < RCI (D) RCI < RBr < RI

52. n-propyl chloride reacts with alcoholic KOH to give
(A) Propene (B) Propyl alcohol
(C) Propyne (D) 2-methyl propene

53. Acetaldehyde reacts with Cl_2 (in excess) to give
(A) Chloral (B) Chloroform
(C) Acetic acid (D) Trichloroacetic acid

54. 2p orbitals have
(A) n = 1, l = 2 (B) n = 1, l = 0
(C) n = 2, l = 1 (D) n = 2, l = 0

55. Aufbau principle is not satisfied by
(A) Cr and Mg (B) Cu and Na
(C) Cr and Cl (D) Cu and Ag

56. Amongst the following electrodes, the one with zero electrode potential is
(A) Calomel electrode
(B) Standard hydrogen electrode
(C) Glass electrode (D) Gas electrode

57. Which of the following is correct expression for electrode potential of a cell

(A) $E = E^o \ \frac{RT}{nF} \ln \frac{[product]}{[reactant]}$

(B) $E = E^o + \frac{RT}{F} \ln \frac{[product]}{[reactant]}$

(C) $E = E^o \ \frac{RT}{nF} \ln \frac{[reactant]}{[product]}$

(D) $E = - \frac{RT}{F} \ln \frac{[product]}{[reactant]}$

58. Calculate standard free energy change for the reaction,

$$\frac{1}{2}Cu(s) + \frac{1}{2}Cl_2 \rightleftharpoons \frac{1}{2}Cu^{2+} + Cl^-$$

taking place at 25°C in a cell whose standard e.m.f. is 1.02 volts
(A) 96500 J (B) -49215 J (C) - 98430 J (D) 9843 J

59. When the temperature is lowered and pressure is raised, the adsorption of a gas on a solid
(A) Decreases (B) Increases
(C) Remains unaffected
(D) Decreases first and then increases

60. On adding a few drops of dil. HCl to freshly precipitated ferric hydroxide, a red coloured colloid is obtained. This process is called
(A) Peptization (B) Dialysis
(C) Dissolution (D) Protective action

61. Platinised asbestos helps in the formation of SO_3 from SO_2 and O_2. But, if even a small amout of As_2O_3 is present, the platinised asbestos does not help in the formation of SO_3. As_2O_3 acts here as
(A) A positive catalyst (B) A negative catalyst
(C) An autocatlalyst (D) A poison

62. Which among the following metals is most abundant in the earth's crust ?
(A) Na (B) Mg (C) Al (D) Fe

63. An ore like zinc blende is concentrated by
(A) Froth floatation (B) Magnetic separation
(C) Leaching (D) Washing with water

64. Which of the following is obtained as slag in the metallurgy of copper (from copper pyrites) ?
(A) $CaSiO_3$ (B) $Ca_3(PO_4)_2$
(C) $FeSiO_3$ (D) $MnSiO_3$

65. Neutrons are found in atoms of all elements except in
(A) Argon (B) Hydrogen (C) Chlorine (D) Oxygen

66. The atomic weight of an element is 39. The number of neutrons in its nucleus is one more than the number of protons. The number of protons, neutrons and electrons respectively in its atom would be
(A) 19, 20, 19 (B) 19, 19, 20
(C) 20, 19, 19 (D) 20, 19, 20

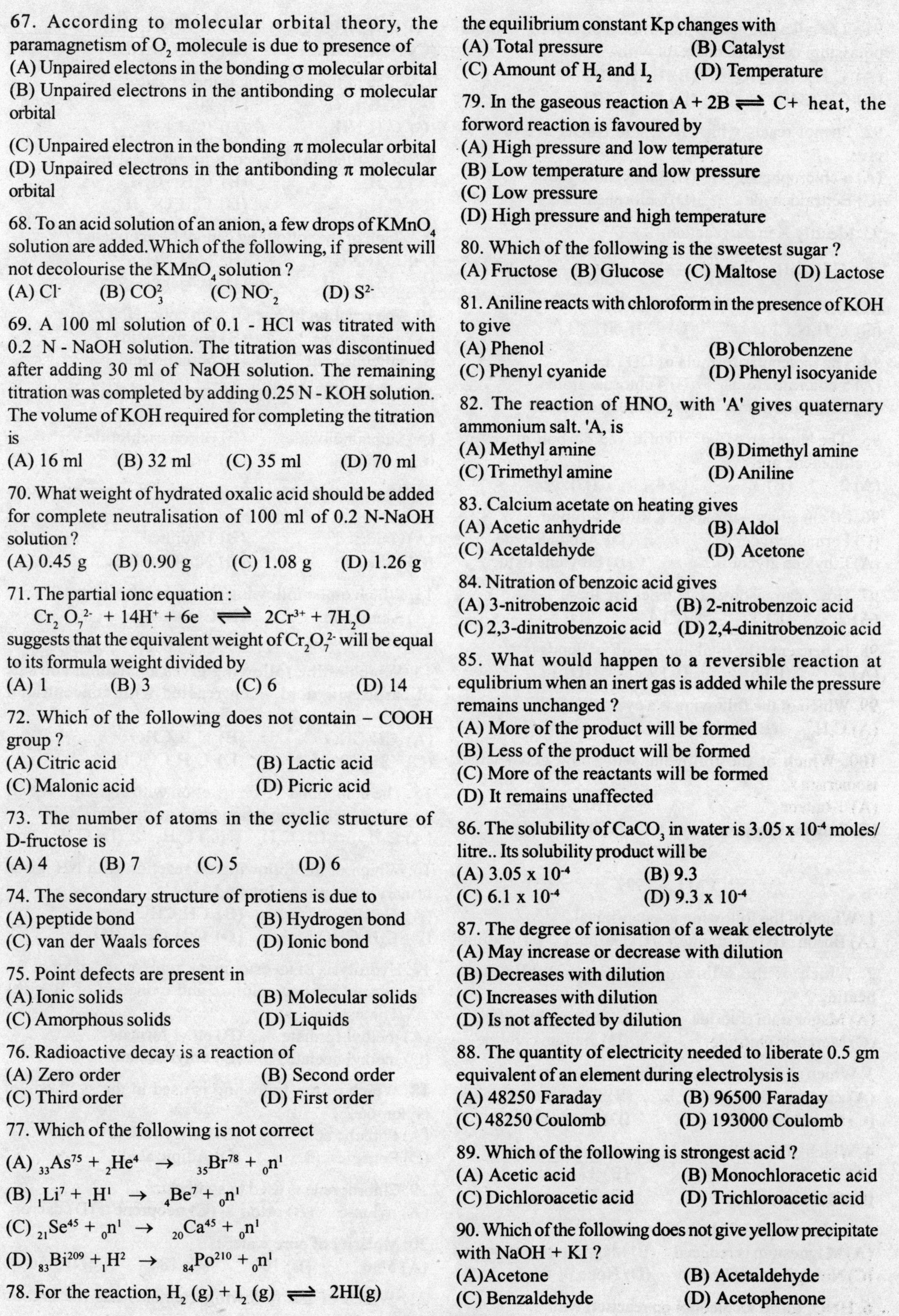

67. According to molecular orbital theory, the paramagnetism of O_2 molecule is due to presence of
(A) Unpaired electons in the bonding σ molecular orbital
(B) Unpaired electrons in the antibonding σ molecular orbital
(C) Unpaired electron in the bonding π molecular orbital
(D) Unpaired electrons in the antibonding π molecular orbital

68. To an acid solution of an anion, a few drops of $KMnO_4$ solution are added. Which of the following, if present will not decolourise the $KMnO_4$ solution ?
(A) Cl^- (B) CO_3^{2-} (C) NO_2^- (D) S^{2-}

69. A 100 ml solution of 0.1 - HCl was titrated with 0.2 N - NaOH solution. The titration was discontinued after adding 30 ml of NaOH solution. The remaining titration was completed by adding 0.25 N - KOH solution. The volume of KOH required for completing the titration is
(A) 16 ml (B) 32 ml (C) 35 ml (D) 70 ml

70. What weight of hydrated oxalic acid should be added for complete neutralisation of 100 ml of 0.2 N-NaOH solution ?
(A) 0.45 g (B) 0.90 g (C) 1.08 g (D) 1.26 g

71. The partial ionc equation :
$Cr_2O_7^{2-} + 14H^+ + 6e \rightleftharpoons 2Cr^{3+} + 7H_2O$
suggests that the equivalent weight of $Cr_2O_7^{2-}$ will be equal to its formula weight divided by
(A) 1 (B) 3 (C) 6 (D) 14

72. Which of the following does not contain – COOH group ?
(A) Citric acid (B) Lactic acid
(C) Malonic acid (D) Picric acid

73. The number of atoms in the cyclic structure of D-fructose is
(A) 4 (B) 7 (C) 5 (D) 6

74. The secondary structure of protiens is due to
(A) peptide bond (B) Hydrogen bond
(C) van der Waals forces (D) Ionic bond

75. Point defects are present in
(A) Ionic solids (B) Molecular solids
(C) Amorphous solids (D) Liquids

76. Radioactive decay is a reaction of
(A) Zero order (B) Second order
(C) Third order (D) First order

77. Which of the following is not correct
(A) ${}_{33}As^{75} + {}_{2}He^{4} \rightarrow {}_{35}Br^{78} + {}_{0}n^{1}$
(B) ${}_{1}Li^{7} + {}_{1}H^{1} \rightarrow {}_{4}Be^{7} + {}_{0}n^{1}$
(C) ${}_{21}Se^{45} + {}_{0}n^{1} \rightarrow {}_{20}Ca^{45} + {}_{0}n^{1}$
(D) ${}_{83}Bi^{209} + {}_{1}H^{2} \rightarrow {}_{84}Po^{210} + {}_{0}n^{1}$

78. For the reaction, $H_2(g) + I_2(g) \rightleftharpoons 2HI(g)$ the equilibrium constant Kp changes with
(A) Total pressure (B) Catalyst
(C) Amount of H_2 and I_2 (D) Temperature

79. In the gaseous reaction $A + 2B \rightleftharpoons C +$ heat, the forword reaction is favoured by
(A) High pressure and low temperature
(B) Low temperature and low pressure
(C) Low pressure
(D) High pressure and high temperature

80. Which of the following is the sweetest sugar ?
(A) Fructose (B) Glucose (C) Maltose (D) Lactose

81. Aniline reacts with chloroform in the presence of KOH to give
(A) Phenol (B) Chlorobenzene
(C) Phenyl cyanide (D) Phenyl isocyanide

82. The reaction of HNO_2 with 'A' gives quaternary ammonium salt. 'A, is
(A) Methyl amine (B) Dimethyl amine
(C) Trimethyl amine (D) Aniline

83. Calcium acetate on heating gives
(A) Acetic anhydride (B) Aldol
(C) Acetaldehyde (D) Acetone

84. Nitration of benzoic acid gives
(A) 3-nitrobenzoic acid (B) 2-nitrobenzoic acid
(C) 2,3-dinitrobenzoic acid (D) 2,4-dinitrobenzoic acid

85. What would happen to a reversible reaction at equlibrium when an inert gas is added while the pressure remains unchanged ?
(A) More of the product will be formed
(B) Less of the product will be formed
(C) More of the reactants will be formed
(D) It remains unaffected

86. The solubility of $CaCO_3$ in water is 3.05×10^{-4} moles/litre.. Its solubility product will be
(A) 3.05×10^{-4} (B) 9.3
(C) 6.1×10^{-4} (D) 9.3×10^{-4}

87. The degree of ionisation of a weak electrolyte
(A) May increase or decrease with dilution
(B) Decreases with dilution
(C) Increases with dilution
(D) Is not affected by dilution

88. The quantity of electricity needed to liberate 0.5 gm equivalent of an element during electrolysis is
(A) 48250 Faraday (B) 96500 Faraday
(C) 48250 Coulomb (D) 193000 Coulomb

89. Which of the following is strongest acid ?
(A) Acetic acid (B) Monochloracetic acid
(C) Dichloroacetic acid (D) Trichloroacetic acid

90. Which of the following does not give yellow precipitate with NaOH + KI ?
(A) Acetone (B) Acetaldehyde
(C) Benzaldehyde (D) Acetophenone

91. The alkaline $CuSO_4$ solution containing sodium potassium tarta does not react with
(A) $C_6H_5CH_2CHO$ (B) C_6H_5CHO
(C) CH_3CHO (D) C_2H_5CHO

92. Phenol reacts with $CHCl_3$ and NaOH at 340 K to give
(A) o-chlorophenol (B) Salicylaldehyde
(C) Benzaldehyde (D) Chlorobenzene

93. Identify X in the reaction,

$$C_6H_5NH_2 \xrightarrow[280\,K]{NaNO_2/HCl} Y \xrightarrow[300\,K]{H_2O/H^+} X$$

(A) C_6H_5Cl (B) C_6H_5OH
(C) $C_6H_5N_2Cl$ (D) C_6H_5NHOH

94. The molecular formula of DDT has
(A) 5 chlorine atoms (B) 4 chlorine atoms
(C) 3 chlorine atoms (D) 2 chlorine atoms

95. The number of sp^3 hybridized carbon atoms in cyclohexene are
(A) 2 (B) 3 (C) 4 (D) 6

96. Ethene gives with acidic $KMnO_4$ solution
(C) Formaldehyde (D) Acetaldehyde
(A) Ethylene glycol (B) Ethylene oxide

97. How many isomeric butanes are these ?
(A) 2 (B) 3 (C) 4 (D) 5

98. In benzene, the total number of σ bonds is
(A) 3 (B) 6 (C) 9 (D) 12

99. Which of the following is a cycloalkane ?
(A) C_6H_{10} (B) C_6H_8 (C) C_6H_{14} (D) C_6H_{12}

100. Which of the following will show geometrical isomerism :
(A) 1-butene (B) 2-butene
(C) 2-methyl propene (D) Propene

PMT - 1999

1. Which of the following is a non-metal ?
(A) Boron (B) Aluminium (C) Gallium (D) Indium

2. Which of the following substances sublimes on heating ?
(A) Magnesium chloride (B) Silver chloride
(C) Mercuric chloride (D) Sodium chloride

3. Which of the following sulphides is yellow ?
(A) Zinc sulphide (B) Cadmium sulphide
(C) Nickel sulphide (D) Lead sulphide

4. Which of the following is not a transition metal ?
(A) Lead (B) Tungsten
(C) Chromium (D) Titanium

5. In the reaction $3Mg + N_2 \rightarrow Mg_3N_2$
(A) Magnesium is reduced (B) Magnesium is oxidized
(C) Nitrogen is oxidized (D) None of these

6. HNO_2 forms an alcohol on reaction with
(A) $C_6H_5NHCH_3$ (B) $C_6H_5NH_2$
(C) $(CH_3)_2NH$ (D) $CH_3CH_2NH_2$

7. Which of the following is most basic ?
(A) $(CH_3)_3N$ (B) NH_3
(C) $C_6H_5NH_2$ (D) $(CH_3)_2NH$

8. The distillation of phenol with zinc dust gives
(A) C_6H_6 (B) $C_6H_5-C_6H_5$
(C) C_6H_{12} (D) $C_6H_5OC_6H_5$

9. Which of the following cannot be acetylated ?
(A) $C_2H_5NH_2$ (B) $C_6H_5NH_2$
(C) $(CH_3)_2NH$ (D) $(CH_3)_3N$

10. The reaction of phenol with conc. HNO_3. gives
(A) picric acid (B) benzoic acid
(C) salicylic acid (D) phthalic acid

11. Which of the following compounds has zero dipole moment ?
(A) Sulphur dioxide (B) Boron trichloride
(C) Ammonia (D) Water

12. When sodium chloride is dissolved in water, the sodium ion is
(A) Oxidized (B) Hydrated
(C) Rduced (D) None of these

13. Which of the following is an s-block element ?
(A) Aluminium (B) Chromium
(C) Niobium (D) Potassium

14. Which of the following gives an alcohol and salt of carboxylic acid when reacted with concentrated NaOH ?
(A) CH_3CHO (B) C_6H_5CHO
(C) CH_3COCH_3 (D) $C_6H_5COCH_3$

15. The compound 'X' on reaction with $HgSO_4 + H_2SO_4$ gives Y which on oxidation gives acetic acid. 'X' is
(A) C_3H_4 (B) C_4H_6 (C) C_2H_2 (D) C_2H_4

16. Which of the following on reaction with NH_3 gives urinary antiseptic compound ?
(A) HCHO (B) CH_3CHO
(C) C_6H_5CHO (D) $C_6H_5CH_2CHO$

17. Hydrolysis of an ester gives acid 'A' and alcohol 'B'. 'A' reduces Fehling solution and oxidation of 'B' gives 'A'. The ester is
(A) methyl formate (B) ethyl formate
(C) methyl acetate (D) ethyl acetate

18. Which of the following is used in the synthesis of Nylon-66 ?
(A) Phthalic acid (B) Salicylic acid.
(C) Benzoicacid (D) Adipic acid

19. Chloroprene is used to synthesize
(A) nylon-6 (B) orlon (C) neoprene (D) dacron

20. Molarity of pure water is
(A) 55.6 (B) 100 (C) 18 (D) 5.56

21. Which of the following elements has the highest

electronegativity ?
(A) Chlorine (B) Fluorine (C) Oxygen (D) Nitrogen

22. The correct electroninc configuration of Ti (Z=22) atom is
(A) $1s^2\ 2s^2\ 2p^6\ 3s^2\ 3p^6\ 3d^4$
(B) $1s^2\ 2s^2\ 2p^6\ 3s^2\ 3p^6\ 4s^2\ 3d^2$
(C) $1s^2\ 2s^2\ 2p^6\ 3s^2\ 3p^6\ 4s^2\ 4p^2$
(D) $1s^2\ 2s^2\ 2p^6\ 3s^2\ 3p^6\ 4s^1\ 3d^3$

23. de-Broglie equation is
(A) $n\lambda = 2d\sin\theta$ (B) $E = hv$ (C) $E = mc^2$ (D) $\lambda = \frac{h}{mv}$

24. The hybrid orbitals used by central atoms in $BeCl_2$, BCl_3 and CCl_4 molecules are respectively
(A) sp^2, sp^3 and sp (B) sp, sp^2 and sp^3
(C) sp^3, sp and sp^2 (D) sp^2, sp and sp^3

25. Amongst ClF_3, BF_3 and NH_3 molecules the one with non-planar geomtry is
(A) NH_3 (B) BF_3 (C) ClF_3 (D) None of these

26. 200 ml of a solution contains 5.85 g dissolved sodium chloride. The concentration of the solution will be
(A) 1 molar (B) 2 molar (C) 0.5 molar (D) 0.25 molar

27. In the following nuclear reactions,
$_7N^{14} + {_2He^4} \rightarrow {_8O^{17}} + X_1$ and $_{13}Al^{27} + {_1D^2} \rightarrow {_{14}Si^{28}} + X_2$
X_1 and X_2 are respectively
(A) $_1H^1$ and $_0n^1$ (B) $_0n^1$ and $_1H^1$
(C) $_2He^4$ and $_0n^1$ (D) $_0n^1$ and $_2He^4$

28. $_{95}Am^{241}$ and $_{90}Th^{234}$ belong respectively to
(A) 4n + 1 and 4n + 3 readioactive disintegration series
(B) 4n and 4n + 1 radioactive disintegration series
(C) 4n + 1 and 4n + 2 radioactive disintegration series
(D) 4n + 1 and 4n readioactive disintegration series

29. The half-life period of a radioactive substance is 8 years. After 16 years, the mass of the substance will reduce from starting 16.0 g to
(A) 8.0 g (B) 6.0 g (C) 4.0 g (D) 2.0 g

30. In the sequence of following nuclear reactions,

$$_{92}X^{238} \xrightarrow{-\alpha} Y \xrightarrow{-\beta} Z \xrightarrow{-\beta} L \xrightarrow{-n\alpha} {_{84}M^{218}}$$

the value of n will be
(A) 3 (B) 4 (C) 5 (D) 6

31. The number of ionic, covalent and coordinate bonds in NH_4Cl are respectively
(A) 1, 3 and 1 (B) 1, 3 and 2
(C) 1, 2 and 3 (D) 1, 1 and 3

32. 20 ml of HCl solution requires 19.85 ml of 0.01 M NaOH solution for complete neutralization. The molarity of HCl solution is
(A) 0.0099 (B) 0.099 (C) 0.99 (D) 9.9

33. How many grams of NaOH will be required to neutralize 12.2 grams of benzoic acid ?
(A) 40 gms (B) 4 gms (C) 16 gms (D) 12.2 gms

34. Element showing the phenomenon of allotropy is
(A) Aluminium (B) Tin (C) Lead (D) Copper

35. In the chemical reaction, $Cl_2 + H_2S \rightarrow 2HCl + S$, the oxidation number of sulphur changes from
(A) 0 to 2 (B) 2 to 0 (C) - 2 to 0 (D) -2 to -1

36. In transition elements, the orbitals partially filled with electrons are
(A) s-orbitals (B) p-orbitals
(C) p-orbitals (D) f-orbitals

37. How many ml of 1 M H_2SO_4 is required to neutralize 10 ml of 1 M NaOH solution ?
(A) 5.0 ml (B) 10.0 ml (C) 20.0 ml (D) 2.5 ml

38. The last member of inert gas elements is
(A) Helium (B) Neon (C) Argon (D) Radon

39. The metal that does not give the borax bead test is
(A) Lead (B) Manganese
(C) Chromium (D) Nickel

40. Which of the following is known as Lunar Caustic in fused state ?
(A) Silver nitrate (B) Silver Sulphate
(C) Silver Chloride (D) Sodium Sulphate

41. The common ion effect is shown by a solution containing
(A) Barium chloride and Barium nitrate
(B) Sodium chloride and Hydrochloric acid
(C) Sodium cyanide and Silver cyanide
(D) Ammonium chloride and Ammonium hydroxide

42. Excess of NaOH solution reacts with zinc to form
(A) Zinc hydroxide (B) Zinc oxide
(C) Sodium zincate (D) Disodium zinc

43. 50 ml water is added to a 50 ml solution of $Ba(OH)_2$ of strength 0.01 M. The pH value of the resulting solution will be
(A) 8 (B) 10 (C) 12 (D) 6

44. Amongst the following solutions, the buffer solution is
(A) NH_4OH + HCl solution
(B) NaOH + HCl solution
(C) $NH_4Cl + NH_4OH$ solution
(D) NH_4Cl + NaOH solution

45. Amongst the following, the one having characteristics of Lewis acid is
(A) ClF_3 (B) BF_3 (C) NCl_3 (D) BrF_3

46. At 20°C temperature, the concentration of Ag^+ ions in a saturated solution of Ag_2CrO_4 is 1.5×10^{-4} mole/litre. The solubility product of Ag_2CrO_4 at 20°C temperature will be
(A) 16.87×10^{-14} (B) 16.87×10^{-13}
(C) 16.87×10^{-12} (D) 16.87×10^{-15}

47. A chemical reaction cannot occur at all if its
(A) ΔH value is positive and ΔS value is negative
(B) ΔH value is negative and ΔS value is positive
(C) ΔH value and ΔS value are negative but $\Delta H > T\Delta S$
(D) ΔH value and ΔS value are positive but $\Delta H < T\Delta S$

48. The values of bond dissociation energies of gaseous

H_2, Cl_2 and HCl are respectively 104, 58 and 103 Kcals. The enthalpy of formation of gaseous HCl will be
(A) 22 kcals/mole (B) 44 kcals/mole
(C) - 22 kcals/mole (D) - 44 kcals/mole

49. Amongst the following chemical reactions, the unimolecular reaction is
(A) $H_2 + Cl_2 \rightarrow 2HCl$ (B) $PCl_3 + Cl_2 \rightarrow PCl_5$
(C) $N_2O_5 \rightarrow N_2O_4 + \frac{1}{2}O_2$ (D) $2HI \rightarrow H_2 + I_2$

50. The unit of rate cnstant of a second order chemical reaction is
(A) $mole^{-1}$ $litre^{-1}$ $second^{-1}$ (B) mole $litre^{-1}$ $second^{-1}$
(C) mole litre second (D) $mole^{-1}$ litre $second^{-1}$

51. The unit of rate constant of a first order chemical reaction is
(A) mole $litre^{-1}$ (B) $second^{-1}$
(C) $mole^{-1}$ $second^{-1}$ (D) $mole^{-1}$ $second^{-2}$

52. Which of the following compounds exhibits optical isomerism ?
(A) $CH_3CH_2CH_2OH$ (B) $CH_3CHOHCH_3$
(C) CH_3CH_2COOH (D) $CH_3CHOHCOOH$

53. The number of possible isomers for compound $C_2H_3Cl_2Br$ is
(A) 2 (B) 3 (C) 3 (D) 5

54. The number of π bonds in hex-1-ene-3-yne is
(A) 1 (B) 2 (C) 3 (D) 4

55. The IUPAC name of croton aldehyde is
(A) prop-2-ene-1-al (B) propenal
(C) but-2-ene-1-al (D) butenal

56. In the modern periodic table the place of the element with atomic number 31 is in
(A) s-block (B) d-block (C) p-block (D) f-block

57. A white solid 'A' on heating gives off a gas which truns lime water milky. The residue is yellow when hot but turns white on cooling. This solid 'A' is
(A) Lead sulphate (B) Lead carbonate
(C) Zinc sulphate (D) Zinc carbonate

58. Permanent magnet is made from
(A) Cast iron (B) Steel
(C) Wrought iron (D) All of these

59. The activity of carbon-14 in a piece of an ancient wood is only 12.5%. If the half-life period of carbon-14 is 5760 years, the age of the piece of wood will be (log 2 = 0.3010)
(A) 17.281 x 10^2 years (B) 172.81 x 10^2 years
(C) 1.7281 x 10^2 years (D) 1728.1 x 10^2 years

60. The end product of (4n +1) radioactive disintegration series is
(A) $_{83}Bi^{209}$ (B) $_{84}Po^{210}$
(C) $_{82}Pb^{208}$ (D) $_{82}Pb^{207}$

61. For the following chemical reaction $2X + Y \rightleftharpoons Z$, the expression of equilibrium constant will be
(A) $K_c = \frac{[X]^2\ [Y]}{[Z]}$ (B) $K_c = \frac{[X]\ [Y]^2}{[Z]}$
(C) $K_c = \frac{[Z]}{[X]^2\ [Y]}$ (D) $K_c = \frac{[Z]}{[X]\ [Y]^2}$

62. Natural rubber is a polymer of
(A) isoprene (B) ethylene glycol
(C) starch (D) acrylonitrile

63. Carbohydrates are stored in human body as
(A) glucose (B) glycogen (C) starch (D) fructose

64. The deficiency of vitamin B_1 causes
(A) beri-beri (B) scurvy (C) rickets (D) anaemia

65. Which of the following is not present in nucleic acids ?
(A) Uracil (B) 2-Aminopyridine
(C) Thymine (D) Adenine

66. The constituents of an atomic nucleus are
(A) neutron and proton
(B) proton and electron
(C) proton, neutron and electron
(D) neutron and electron

67. The mass number of an anion, X^{3-}, is 14. If there are ten electrons in the anion, the number of neutrons in the nucleus of atom, X_2 of the element will be
(A) 10 (B) 14 (C) 7 (D) 5

68. The fusion of sodium with aniline gives mainly
(A) NaCN (B) NaN_3 (C) NaSCN (D) $NaNO_2$

69. Isopropyl alcohol is obtained by reacting which of the following alkenes with conc. H_2SO_4 and H_2O ?
(A) Ethene (B) Propylene
(C) 2-Methyl propene (D) Isoprene

70. The compound C_3H_4 has a triple-bond, which is indicated by its reaction with
(A) bromine water (B) Bayer,s reagent
(C) Fehling solution (D) ammonical silver nitrate

71. Anti-Markownikoff addition of HBr is not observed for
(A) but-2-ene (B) pent-2-ene
(C) propene (D) but-1-ene

72. Gasoline is obtained from crude petrolium oil by its
(A) fractional distillation (B) vacuum distillation
(C) steam distillation (D) pyrolysis

73. The reaction of CH_3MgI with acetone and hydrolysis of the resulting pruduct gives
(A) $CH_3CH_2CH_2OH$ (B) $CH_3CH_2CH_2CH_2OH$
(C) $(CH_3)_2CHOH$ (D) $(CH_3)_3COH$

74. The atomic mass of an element is 12.00719 amu. If there are 6 neutrons in the nucleus of the atom of the element, the binding energy per nucleon of the nucleus

will be

(e^-=0.00055 amu; p = 1.00814 amu; n = 1.00893 amu)

(A) 764 MeV (B) 0.764 MeV

(C) 7.64 MeV (D) 76.4 MeV

75. The expression for Bohr's radius of an atom is

(A) $r = \frac{n^2h^2}{4\pi^2me^4z^2}$ (B) $r = \frac{n^2h^2}{4\pi^2 m e^2 z}$

(C) $r = \frac{n^2h^2}{4\pi^2 m e^2 z^2}$ (D) $r = \frac{n^2h^2}{4\pi^2 m^2 e^2 z^2}$

76. The energy of an electron revolving in n^{th} Bohr's orbit of an atom is given by the expression

(A) $En = -\frac{2\pi^2 m^4 e^2 z^2}{n^2h^2}$ (B) $En = -\frac{2\pi^2 m e^2 z^2}{n^2h^2}$

(C) $En = -\frac{2\pi^2 m e^4 z^2}{n^2h^2}$ (D) $En = -\frac{2\pi m^2 e^2 z^4}{n^2h^2}$

77. For Na (Z=11) atom, the four quantum number of last electron are

(A) $n = 3; l = 1; m = 1; s = +\frac{1}{2}$

(B) $n = 3; l = 0; m = 0; s = +\frac{1}{2}$

(C) $n = 3; l = 0; m = 1; s = +\frac{1}{2}$

(D) $n = 3; l = 1; m = 1; s = -\frac{1}{2}$

78. Which of the following does not give alkane ?

(A) Electrolysis of concentrated sodium acetate solution

(B) Reaction of CH_3I with Na in ether

(C) Reaction of sodium acetate with soda-lime

(D) Reaction of ethylchloride with alcoholic KOH

79. LPG is a mixture of

(A) $C_6H_{12} + C_6H_6$ (B) $C_4H_{10} + C_3H_8$

(C) $C_2H_4 + C_2H_2$ (D) $C_2H_4 + CH_4$

80. Which of the following reacts first with Lucas reagent ?

(A) C_3H_7OH (B) $(CH_3)_2CHOH$

(C) $(CH_3)_3CHO$ (D) C_6H_5OH

81. Which of the following gives a positive iodoform test ?

(A) CH_3OH (B) $(CH_3)_2CHOH$

(C) $(CH_3)_3CHO$ (D) C_3H_7OH

82. A compound X on reaction with chloroform and NaOH gives a compound with a very unpleasant odour. X is

(A) $C_6H_5CH_2NHCH_3$ (B) $C_6H_5NHCH_3$

(C) $C_6H_5CONH_2$ (D) $C_6H_5NH_2$

83. The expression for rate constant of a first order chemical reaction is

(A) $k = \frac{x}{t}$ (B) $k = \frac{2.303}{t}\log_{10}\left(\frac{a}{a - x}\right)$

(C) $k = \frac{1}{t}\frac{x}{a(a - x)}$ (D) $k = \frac{1}{2t}\left[\frac{1}{(a - x)^2} - \frac{1}{a^2}\right]$

84. In acidic medium, equivalent weight of potassium permanganate is equal to

(A) molecular weight of $KMnO_4$

(B) $\frac{1}{5}$ x molecular weight of $KMnO_4$

(C) $\frac{1}{3}$ x molecular weight of $KMnO_4$

(D) $\frac{1}{2}$ x molecular weight of $KMnO_4$

85. The process of separation of a colloid from crystalloid through semi-permeable membrane is known as

(A) dialysis (B) coagulation

(C) ultrafiltration (D) titration

86. Amongst the following chemical reactions, the one representing homogeneous catalysis is

(A) $N_2(g) + 3H_2(g) \xrightarrow{Fe} 2NH_3(g)$

(B) $2SO_2(g) + O_2(g) \xrightarrow{2NO} 2SO_3 + 2NO(g)$

(C) $CO(g) + 3H_2(g) \xrightarrow{Ni} CH_4(g) + H_2O$

(D) $2SO_2(g) + O_2(g) \xrightarrow{V_2O_5} 2SO_3(g)$

87. chloride of which of the following elements is paramagnetic ?

(A) Zinc (B) Cobalt (C) Silver (D) Mercury

88. Which of the following elements is paramagnetic ?

(A) Nitrogen (B) Hydrogen (C) Oxygen (D) Chlorine

89. What is the oxidation number of chromium in potassium dichromate ?

(A) +2 (B) +3 (C) + 6 (D) –4

90. The radium and uranium atoms, in a sample of uranium mineral are in the ratio of 1 2.8 x 10^6, If half-life period of radium is 1620 years, the half-life period of uranium will be

(A) 45.3 x 10^9 years (B) 45.3 x 10^{10} years

(C) 4.53 x 10^9 years (D) 4.53 x 10^{10} years

91. For maximum production of SO_3 in the following chemical reaction, $2SO_2 + O_2 \rightleftharpoons 2SO_3; \Delta H = -Q$ kcals the favourable conditions will be

(A) high temperature and low pressure

(B) low temperature and low pressure

(C) high temperature and high pressure

(D) low temperature and high pressure

92. For maximum production of NH_3 in the following chemical reaction, $N_2 + 3H_2 \rightleftharpoons 2NH_3 + Q$ kcals the favourable conditions will be

(A) low temperature and low pressure

(B) high temperature and low pressure

(C) low temerature and high pressure

(D) high temperature and high pressure

93. Addition of ammoinum chloride in the solution of ammonium hydroxide results in

(A) increase of OH^- ion concentration in solution

(B) decrease of OH^- ion concentration in solution

(C) increase of NH_4^+ and OH^- ion concentration in solution

(D) increase of dissociation of NH_4OH

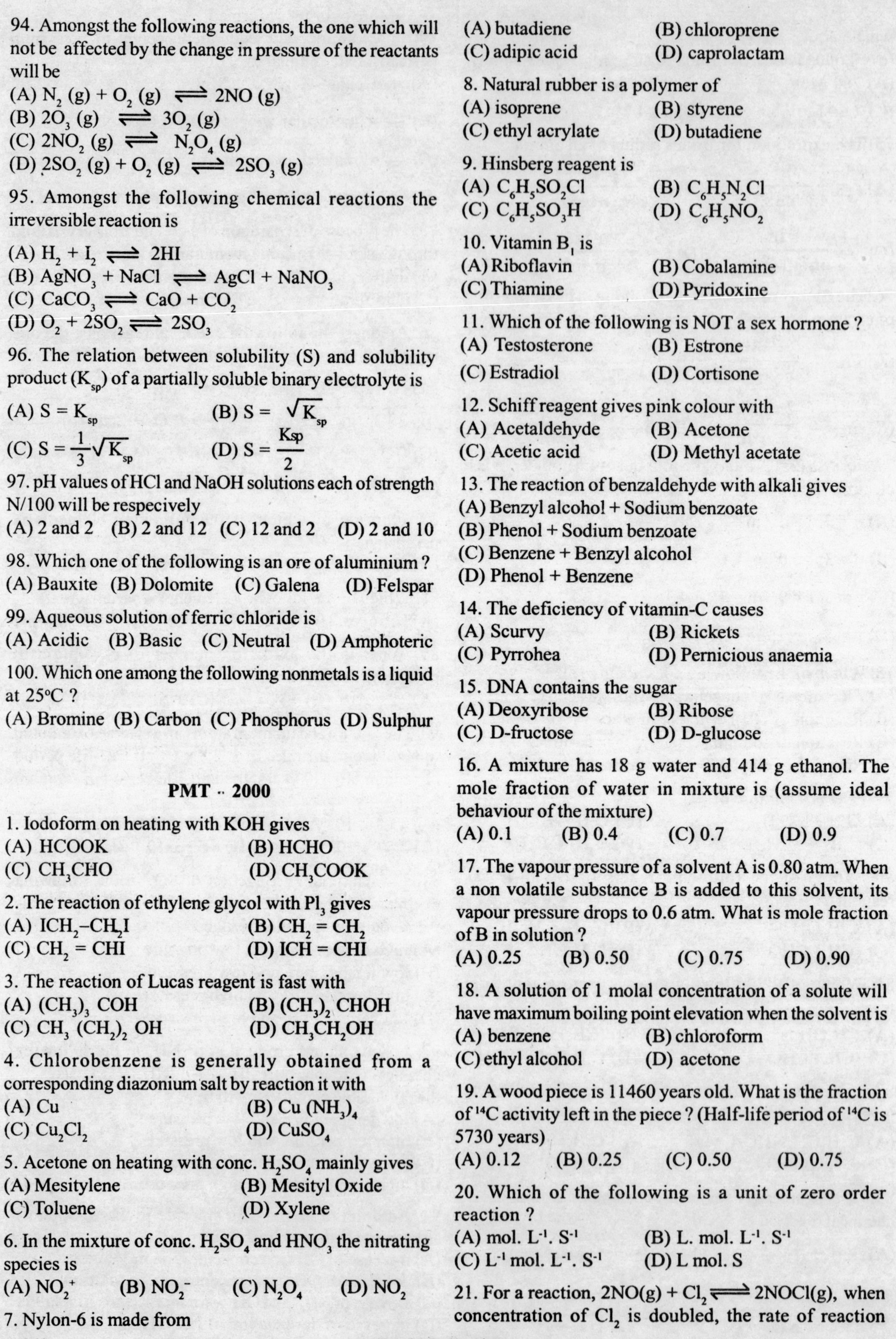

94. Amongst the following reactions, the one which will not be affected by the change in pressure of the reactants will be
(A) $N_2 (g) + O_2 (g) \rightleftharpoons 2NO (g)$
(B) $2O_3 (g) \rightleftharpoons 3O_2 (g)$
(C) $2NO_2 (g) \rightleftharpoons N_2O_4 (g)$
(D) $2SO_2 (g) + O_2 (g) \rightleftharpoons 2SO_3 (g)$

95. Amongst the following chemical reactions the irreversible reaction is
(A) $H_2 + I_2 \rightleftharpoons 2HI$
(B) $AgNO_3 + NaCl \rightleftharpoons AgCl + NaNO_3$
(C) $CaCO_3 \rightleftharpoons CaO + CO_2$
(D) $O_2 + 2SO_2 \rightleftharpoons 2SO_3$

96. The relation between solubility (S) and solubility product (K_{sp}) of a partially soluble binary electrolyte is
(A) $S = K_{sp}$ (B) $S = \sqrt{K_{sp}}$
(C) $S = \frac{1}{3}\sqrt{K_{sp}}$ (D) $S = \frac{Ksp}{2}$

97. pH values of HCl and NaOH solutions each of strength N/100 will be respecively
(A) 2 and 2 (B) 2 and 12 (C) 12 and 2 (D) 2 and 10

98. Which one of the following is an ore of aluminium ?
(A) Bauxite (B) Dolomite (C) Galena (D) Felspar

99. Aqueous solution of ferric chloride is
(A) Acidic (B) Basic (C) Neutral (D) Amphoteric

100. Which one among the following nonmetals is a liquid at 25°C ?
(A) Bromine (B) Carbon (C) Phosphorus (D) Sulphur

PMT - 2000

1. Iodoform on heating with KOH gives
(A) HCOOK (B) HCHO
(C) CH_3CHO (D) CH_3COOK

2. The reaction of ethylene glycol with PI_3 gives
(A) ICH_2-CH_2I (B) $CH_2 = CH_2$
(C) $CH_2 = CHI$ (D) $ICH = CHI$

3. The reaction of Lucas reagent is fast with
(A) $(CH_3)_3 COH$ (B) $(CH_3)_2 CHOH$
(C) $CH_3 (CH_2)_2 OH$ (D) CH_3CH_2OH

4. Chlorobenzene is generally obtained from a corresponding diazonium salt by reaction it with
(A) Cu (B) $Cu (NH_3)_4$
(C) Cu_2Cl_2 (D) $CuSO_4$

5. Acetone on heating with conc. H_2SO_4 mainly gives
(A) Mesitylene (B) Mesityl Oxide
(C) Toluene (D) Xylene

6. In the mixture of conc. H_2SO_4 and HNO_3 the nitrating species is
(A) NO_2^+ (B) NO_2^- (C) N_2O_4 (D) NO_2

7. Nylon-6 is made from
(A) butadiene (B) chloroprene
(C) adipic acid (D) caprolactam

8. Natural rubber is a polymer of
(A) isoprene (B) styrene
(C) ethyl acrylate (D) butadiene

9. Hinsberg reagent is
(A) $C_6H_5SO_2Cl$ (B) $C_6H_5N_2Cl$
(C) $C_6H_5SO_3H$ (D) $C_6H_5NO_2$

10. Vitamin B_1 is
(A) Riboflavin (B) Cobalamine
(C) Thiamine (D) Pyridoxine

11. Which of the following is NOT a sex hormone ?
(A) Testosterone (B) Estrone
(C) Estradiol (D) Cortisone

12. Schiff reagent gives pink colour with
(A) Acetaldehyde (B) Acetone
(C) Acetic acid (D) Methyl acetate

13. The reaction of benzaldehyde with alkali gives
(A) Benzyl alcohol + Sodium benzoate
(B) Phenol + Sodium benzoate
(C) Benzene + Benzyl alcohol
(D) Phenol + Benzene

14. The deficiency of vitamin-C causes
(A) Scurvy (B) Rickets
(C) Pyrrohea (D) Pernicious anaemia

15. DNA contains the sugar
(A) Deoxyribose (B) Ribose
(C) D-fructose (D) D-glucose

16. A mixture has 18 g water and 414 g ethanol. The mole fraction of water in mixture is (assume ideal behaviour of the mixture)
(A) 0.1 (B) 0.4 (C) 0.7 (D) 0.9

17. The vapour pressure of a solvent A is 0.80 atm. When a non volatile substance B is added to this solvent, its vapour pressure drops to 0.6 atm. What is mole fraction of B in solution ?
(A) 0.25 (B) 0.50 (C) 0.75 (D) 0.90

18. A solution of 1 molal concentration of a solute will have maximum boiling point elevation when the solvent is
(A) benzene (B) chloroform
(C) ethyl alcohol (D) acetone

19. A wood piece is 11460 years old. What is the fraction of ^{14}C activity left in the piece ? (Half-life period of ^{14}C is 5730 years)
(A) 0.12 (B) 0.25 (C) 0.50 (D) 0.75

20. Which of the following is a unit of zero order reaction ?
(A) $mol. L^{-1}. S^{-1}$ (B) $L. mol. L^{-1}. S^{-1}$
(C) $L^{-1} mol. L^{-1}. S^{-1}$ (D) L mol. S

21. For a reaction, $2NO(g) + Cl_2 \rightleftharpoons 2NOCl(g)$, when concentration of Cl_2 is doubled, the rate of reaction

becomes two-times of the original. When the concentration of NO is doubled the rate becomes four times. What is the order of the reaction ?

(A) 1 (B) 2 (C) 3 (D) 4

22. The first ionization energy for Li is 5.4 eV and electron affinity of Cl is 3.61 eV. The ΔH (in kJ/mol) for the reaction Li (g) + Cl (g) $\rightarrow$ $Li^+ + Cl^-$ is (if resulting ions do not combine with each other)

($1 eV = 1.6 \times 10^{-19} J$)

(A) 170 (B) 270 (C) 70 (D) 100

23. In which of the following pairs the two molecules have identical bond order ?

(A) N_2, O_2^{2+} (B) N_2, O_2^-

(C) N_2^-, O_2 (D) O_2^{2-}, N_2

24. The dipolc moment of HBr is 1.6×10^{-30} c.m. and interatomic spacing is 1. Å. The % ionic character of HBr is

(A) 7 (B) 10 (C) 15 (D) 27

25. The number of unit cells in 58.5 g of NaCl is nearly

(A) 6×10^{20} (B) 3×10^{22}

(C) 1.5×10^{23} (D) 0.5×10^{24}

26. In the nuclear reaction ${}^{9}_{4}Be$ (p, α) X, the X is

(A) ${}^{4}_{2}He$ (B) ${}^{6}_{3}Li$ (C) ${}^{7}_{3}Li$ (D) ${}^{8}_{4}Be$

27. Which of the following does NOT contain number of neutrons equal to that of ${}^{40}_{18}Ar$?

(A) ${}^{41}_{19}K$ (B) ${}^{43}_{21}Sc$ (C) ${}^{40}_{21}Sc$ (D) ${}^{42}_{20}Ca$

28. The half life of ${}^{90}_{38}Sr$ is 20 years. If its sample having initial activity of 8000 dis/min is taken, what would be its activity after 80 years ?

(A) 500 dis/min (B) 800 dis/min

(C) 1000 dis/min (D) 1600 dis/min

29. Which of the following does NOT show positive deviation from Raoults law ?

(A) Benzene - Chloroform (B) Benzene - Acentone

(C) Benzene - Ethanol

(D) Benzene - Carbon tetrachloride

30. Which of the following equilibria will shift to right side on increasing the temperature ?

(A) $CO (g) + H_2O (g) \rightleftharpoons CO_2 (g) + H_2O (g)$

(B) $2SO_2 (g) + O_2 (g) \rightleftharpoons 3SO_3 (g)$

(C) $H_2O (g) \rightleftharpoons H_2 (g) + 1/2 (O_2) (g)$

(D) $4HCl (g) + O_2 (g) \rightleftharpoons 2H_2O (g) + 2Cl_2 (g)$

31. When 3 mole of A and 1 mole of B are mixed in 1 litre vessel the following reaction takes place, $A (g) + B (g) \rightleftharpoons 2C (g)$. The equilibrium constant for the reaction is

(A) 0.12 (B) 0.25 (C) 0.50 (D) 0.75

32. The equilibrium $SO_2Cl_2 (g) \rightleftharpoons SO_2 (g) + Cl_2 (g)$ is attained at 25°C in a closed container and an inert gas helium is introduced, which of the following statements is correct ?

(A) More chlorine is formed

(B) Concentration of SO_2 is reduced

(C) More SO_2Cl_2 is formed

(D) Concentration of SO_2Cl_2. SO_2 and Cl_2 does not change

33. The number of octahedral sites per sphere in a fcc structure is

(A) 1 (B) 2 (C) 4 (D) 8

34. The packing factor for a body centred cube is

(A) 0.42 (B) 0.53 (C) 0.68 (D) 0.82

35. Which of the following has Frenkel defect ?

(A) Sodium chloride (B) Graphite

(C) Silver bromide (D) Diamond

36. In polysaccharides the linkage connecting monosaccharide units is called

(A) glycoside linkage (B) nucleoside linkage

(C) glycogen linkage (D) peptide linkage

37. The number of possible alkynes with molecular formula C_5H_8 is

(A) 2 (B) 3 (C) 4 (D) 5

38. The IUPAC name of acraldehyde is

(A) but-2-ene-1-al (B) propenol

(C) prop-2-ene-1-al (D) propenyle aldehyde

39. Which of the following will exhibit optical isomerism ?

(A) CH_3CH_2COOH (B) $CH_3CHOHCOOH$

(C) $CH_3CHOHCH_3$ (D) $CH_3CH_2CH_2OH$

40. Toluene has

(A) 6σ and 3π bond (B) 9σ and 3π bond

(C) 9σ and 6π bond (D) 15σ and 3π bond

41. An organic compound has a triple bond and no double bonds. It can be tested by

(A) bromine water (B) Bayer's reagent

(C) Fehling solution (D) Ammonical silver nitrate

42. Which of the folliwng has highest octane number ?

(A) n-Hexane (B) n- Heptane

(C) n-Pentane (D) 2, 2, 4-Trimethyl pentane

43. Which of the following is an antiknocking compound ?

(A) $Pb (C_2H_5)_4$ (B) $Sn (C_2H_5)_4$

(C) $Zn (C_2 H_5)_2$ (D) $Hg (CH_3)_2$

44. Which of the alkenes on ozonolysis yields acetone ?

(A) $CH_3 - CH = CH - CH_3$

(B) $(CH_3)_2 - C = C - (CH_3)_2$

(C) $CH_2 = CH - CH_2 - CH_3$

(D) $CH_3 - CH = CH_2$

45. In chlorination of benzene, the reactive species is

(A) Cl_2^- (B) Cl_2 (C) Cl^+ (D) Cl^-

46. The number of moles of proton which can be easily given by butyne-1 (1 mole) is

(A) 1 (B) 2 (C) 3 (D) 6

47. The addition of HBr is easiest with

(A) $CH_2 = CHCl$ (B) $ClCH = CHCl$

(C) $CH_3 - CH = CH_2$ (D) $(CH_3)_2 C = CH_2$

48. The pH of 10^{-8} M HCl is
(A) < 7 (B) 7 (C) 8 (D) >7

49. For an endothermic reaction, where ΔH represents the enthalpy of reaction in kJ/mol, the minimum value for the energy of activation will be
(A) less than ΔH (B) zero
(C) more than ΔH (D) equal to ΔH

50. Which of the following does NOT result in an increase in the entropy ?
(A) conversion of ice to water
(B) crystallization of sucrose from a solution
(C) rusting of iron
(D) vaporization of camphor

51. As lead storage battery is charged
(A) lead dioxide dissolves
(B) sulphuric acid is regenerated
(C) lead electrode becomes coated with lead sulphate
(D) the concentration of sulphuric acid decreases

52. Blood may be purified by
(A) dialysis (B) electro-osmosis
(C) coagulation (D) filtration

53. A 1M solution of glucose reaches dissociation equilibrium according to equation given below. 6 HCHO $\rightleftharpoons$ $C_6H_{12}O_6$. What is the concentration of HCHO at equilibrium if equilibrium constant is 6×10^{22}
(A) 1.6×10^{-8} M (B) 6.2×10^{5} M
(C) 3.2×10^{-4} M (D) 1.6×10^{-4} M

54. What is $[H^+]$ of a solution that is 0.01 M in HCN and 0.02 M in NaCN ? (Ka for HCN = 6.2×10^{-10})
(A) 3.1×10^{10} (B) 6.2×10^{5}
(C) 6.2×10^{-10} (D) 3.1×10^{-10}

55. Which of the following is most effective in causing the coagulation of ferric hydroxide sol ?
(A) KCl (B) KNO_3 (C) K_2SO_4 (D) $K_3Fe(CN)_6$

56. The e.m.f. of cell Ni / Ni^{2+} (1.0 M) // Au^{3+} (1.0 M)/ Au (E^0 = -0.25 V for Ni^{2+}/Ni; 1.5 V for Au^{3+}/Au) is
(A) 1.25 V (B) -1.25 V (C) 1.75 V (D) 2.00 V

57. How many grams of CaC_2O_4 will dissolve in distilled water to make one litre of saturated solution ? (Ksp for $CaC_2O_4 = 2.5 \times 10^{-9}$ mole^2 litre^{-2})
(A) 0.0064 g (B) 0.1028 g
(C) 0.1280 g (D) 0.2056

58. Which of the following represents the electronic confguration of the most electropostive element ?
(A) [He] $2S^1$ (B) [He] $2S^2$
(C) [Xe] $6S^1$ (D) [Xe] $2S^2$

59. Mark the oxide which is amphoteric in character
(A) SnO_2 (B) CaO (C) CO_2 (D) SiO_2

60. In which of the following, the inert pair effect is most prominent ?
(A) C (B) Si (C) Ge (D) Pb

61. The alkali metal that reacts with nitrogen directly to form nitride is
(A) K (B) Rb (C) Li (D) Na

62. Zinc and Mercury do not show variable valency like d-block elements because
(A) they are soft
(B) their d-shells are complete
(C) they have only two electrons in the outer most subshell
(D) their d-shells are incomplete

63. Which of the following is the smallest cation ?
(A) Ca^{2+} (B) Al^{3+} (C) Na^+ (D) Mg^{2+}

64. The halogen that is most easily reduced is
(A) F_2 (B) Cl_2 (C) Br_2 (D) I_2

65. The valency of Cr in the complex $[Cr(H_2O)_4Cl_2]^+$ is
(A) 1 (B) 3 (C) 5 (D) 6

66. Which of the following is expected to be a paramagnetic complex ?
(A) $[Ni(H_2O)_6]^{2+}$ (B) $[Ni(CO)_4]$
(C) $[Zn(NH_3)_4]^{2+}$ (D) $[CO(NH_3)_6]$

67. Which of the following does not give formaldehyde on heating or distillation ?
(A) Formalin (B) Trioxan
(C) Para-aldehyde (D) Para-formaldehyde

68. A cyanohydrin of a compound X on hydrolysis gives lactic acid. X is
(A) $(CH_3)_2CO$ (B) $C_6H_5CH_2CHO$
(C) HCHO (D) CH_3CHO

69. Which of the following reacts with $NaNO_2 + HCl$ to give alcohol/phenol ?
(A) $C_6H_5CH_2NHCH_3$ (B) $(CH_3)_2NH$
(C) $CH_3CH_2NH_2$ (D) $C_6H_5NH_2$

70. Which of the following is strongest Lewis base ?
(A) CH_3^- (B) NH_2^- (C) OH^- (D) F^-

71. For the reaction, $N_2 + 3H_2 \rightarrow 2NH_3$ if $\frac{\Delta[NH_3]}{\Delta t} = 2 \times 10^{-4}$ mol. L^{-1} S^{-1}, the value of $\frac{-\Delta[H_2]}{\Delta t}$ would be
(A) 1×10^{-4} mol L^{-1} S^{-1} (B) 3×10^{-4} mol L^{-1} S^{-1}
(C) 4×10^{-4} mol L^{-1} S^{-1} (D) 6×10^{-4} mol L^{-1} S^{-1}

72. What is the minimum concentration of SO_4^{2-} required to precipitate $BaSO_4$ in a solution contataining 1.0×10^{-4} mol. Ba^{2+} ? Ksp for $BaSO_4 = 4 \times 10^{-11}$.
(A) 4×10^{-10} M (B) 2×10^{-7} M
(C) 4×10^{-6} M (D) 2×10^{-3} M

73. What is the potential of a cell containing two hydrogen electrodes, the negative one in contact with 10^{-8} MH$^+$ and the positive one C in contact with 0.025 M H^+ ?
(A) 0.18 V (B) 0.28 V (C) 0.38 V (D) 0.48 V

74. The number of unpaired electrons in a Nickel atom

(ground state) are (Atomic No. of Ni 28)
(A) 2 (B) 3 (C) 5 (D) 7

75. What is the maximum number of electrons which can be accommodated in an atom in which the highest principal quantum number value is 4 ?
(A) 10 (B) 18 (C) 36 (D) 54

76. In Balmer series of hydrogen atom spectrum which electronic transition causes third line ?
(A) Fourth Bohr orbit to second one
(B) Fourth Bohr orbit to first one
(C) fifth Bohr orbit to second one
(D) Fifth Bohr orbit to first one

77. Phenolphthalein is most suitable indicator for the titration of
(A) CH_3COOH and NH_4OH
(B) CH_3COOH and NaOH
(C) HCl and NH_4OH
(D) H_2CO_3 and NH_4OH

78. Ring test for nitrate is performed by adding freshly prepared ferrous sulphate solution and conc. H_2SO_4. The brown ring in this test is due to
(A) $FeSO_4 . NO_2$ (B) $FeSO_4 . NO$
(C) $FeSO_4$ (D) $Fe_2 (SO_4)_3$

79. Heating pyrites to remove sulphur is called
(A) Smelting (B) Calcination
(C) Liquation (D) Roasting

80. In the metallurgy of which of the following, cupellation process is used ?
(A) Iron (B) Aluminium (C) Copper (D) Silver

81. Cryolite is used in the electrolytic extraction of aluminium.
(A) to obtain more aluminium (B) to decompose bauxite
(C) to protect anodes (D) as a reducing agent

82. The important ore of lead is
(A) Chalcopyrite (B) Haematite
(C) Galena (D) Bauxite

83. The iron obtained from blast furnace is
(A) pig iron (B) wrought iron (C) soft iron (D) cast iron

84. The ionic radii of N^{3-}, O^{2-}, F^- and Na^+ are in the order
(A) $N^{3-} > O^{2-} > F^- > Na^+$ (B) $N^{3-} > Na^+ > O^{2-} > F^-$
(C) $Na^+ > O^{2-} > N^{3-} > F^-$ (D) $O^{2-} > F^- > Na^+ > N^{3-}$

85. Going from Fluorine to chlorine, bromine and iodine the electronegativity
(A) increases (B) decreases
(C) first decreases then increases (D) changes randomly

86. The first ionisation potential on Na, Mg, Al and Si are in the order
(A) Na < Mg > Al < Si (B) Na > Mg > Al < Si
(C) Na > Mg > Al > Si (D) Na < Mg < Al < Si

87. In the Nitrogen family the H-M-H bond angle in the hydrides (MH_3) gradually becomes closer to 90° on going from N to Sb. This shows that
(A) the basic strength of hydrides increases.
(B) the bond energies of M-H bonds increase.
(C) almost pure orbitals are used for M-H bonding.
(D) the bond pairs of electrons come nearer to the central atom.

88. Energy of electron of hydrogen atom in second Bohr orbit is
(A) -5.44×10^{-19} J (B) -5.44×10^{-19} KJ
(C) -5.44×10^{-19} cal (D) -5.44×10^{-19} eV

89. The number of possible resonance structures for CO_3^{2-} is
(A) 2 (B) 3 (C) 6 (D) 9

90. Which of the following molecules is linear ?
(A) NO_2^- (B) SCl_2 (C) SO_2 (D) NO_2^+

91. IUPAC name of $Na_3[Co(NO_2)_6]$ is
(A) Sodiumhexanitritocobalt (III)
(B) Sodiumcobaltinitrite
(C) Sodiumhexanitritocobaltate (III)
(D) Sodiumhexanitritocobaltate (II)

92. Which of the following is not precipitated as sulphide by passing H_2S in presence of conc. HCl ?
(A) copper (B) arsenic (C) cadmium (D) lead

93. Which of the following sulphates is insoluble in water ?
(A) $CuSO_4$ (B) $CdSO_4$ (C) $PbSO_4$ (D) $Bi_2(SO_4)_3$

94. A deep brown gas is formed by mixing two colourless gases which are
(A) NO_2 and O_2 (B) N_2O and NO
(C) NO and O_2 (D) NH_3 and HCl

95. A group of atoms can function as a ligand only when
(A) it is a negatively charged ion
(B) it is a small molecule
(C) it has an unshared electron pair
(D) it is a positively charged ion

96. According to the postulates of Werner for coordinate compounds
(A) Primary valency is ionizable
(B) Secondary valency is ionizable
(C) Primary and secondary valencies are non-ionizable
(D) only primary valency is nonionizable

97. Elements which generally exhibit multiple oxidation states and whose ions are usually coloured are
(A) metalloids (B) non-metals
(C) transition elements (D) gases

98. The transition netals mostly are
(A) diamagnetic (B) paramagnetic
(C) neither diamagnetic nor paramagnetic
(D) both diamagnetic and paramagnetic

99. The correct statement in respect of d-block elements is
(A) they form coloured ions and complex salts
(B) they are all metals

(C) they show variable valency
(D) all above statements are correct

100. Reduction of $KMnO_4$ by oxalic acid in acidic medium, the oxidation number of Mn changes. What is the magnitude of this change ?
(A) from 7 to 2 (B) from 6 to 2
(C) from 5 to 2 (D) from 7 to 4

PMT - 2001

1. I. U. P. C. name of the following compound is

$$CH_3-\underset{CH_3}{\overset{CH_3}{\underset{|}{\overset{|}{C}}}}-CH=CH_2$$

(A) 3, 3, 3 trimethyl - 1 propene
(B) 1, 1, 1 trimethyl - 2 propene
(C) 3, 3 dimethyl 1 - butene
(D) 2, 2 dimethyl - 3 butene

2. Which of the following compounds shows tautomerism ?
(A) HCHO (B) CH_3 CHO
(C) CH_3 $CONH_2$ (D) HCOOH

3. The empirical formula of an organic compound is CH_2. One mole of this compound has a mass of 42 gm. Its molecular formula is
(A) C_2H_2 (B) C_3H_8 (C) CH_2 (D) C_3H_6

4. The number of tertiary carbon atoms in the compound $(CH_3)_2$ $CHCH_2CH$ $(CH_3)_2$ is
(A) 2 (B) 3 (C) 1 (D) 4

5. Carbon - Carnon distance in benzene is nearly
(A) 1.54 Å (B) 1.40 Å (C) 1.33 Å (D) 1.25 Å

6. The compound which has one isopropyl group is
(A) 2, 2, 3 - trimethyl pentane
(B) 2 - methyl pentane
(C) 2, 2, 3, 3 - tetramethyl pentane
(D) 3, 3, - dimethyl pentane

7. When methyl bromide is heated with Zn it gives
(A) CH_4 (B) C_2 H_6 (C) C_2 H_4 (D) CH_3OH

8. The AS_2S_3 sol is charged due to adsorption of
(A) H^+ (B) AS^{3+} (C) S^{2-} (D) OH^-

9. H_2S does not precipitate .... ion in the presence of 1 M acetic acid
(A) Sb (III) (B) Sn (II) (C) Bi (III) (D) Zn (II)

10. The protection of steel by chrome plating is due to
(A) Cathodic protection
(B) Anodic protection
(C) covering of steel surface
(D) Formation of alloy with iron

11. The outer electronic configuration of chromium is
(A) $4s^1$ $3d^5$ (B) $4s^2$ $3d^4$ (C) $4s^0$ $3d^6$ (D) $4s^2$ $3d^5$

12. Which of the following transition metals shows the highest oxidation state ?
(A) Sc (B) Ti (C) Mn (D) Zn

13. In potassium manganate, the oxidation state of manganese is
(A) +5 (B) +6 (C) +7 (D) +8

14. Which of the following is an organo-metallic compound ?
(A) Lithium acetate (B) Lithium carbide
(C) Lithium ethoxide (D) Ethyl lithium

15. In India, on the occasion of marriage, the fireworks are used to give coloured flames. The salt of which one of the following metals is used to obtain green flame for this purose?
(A) Na (B) K (C) Ba (D) Ca

16. A gas decolourises alkaline $KMnO_4$ but does not give precipitate with ammonical $AgNO_3$. The gas is
(A) C_2 H_6 (B) C_2 H_2 (C) C_2 H_4 (D) CH_4

17. When phenol is distilled with Zn dust it gives
(A) Benzaldehyde (B) Benzoic acid
(C) Toluene (D) Benzene

18. Liquid hydrocarbon is converted into a mixture of gaseous hydrocarbons by
(A) Cracking (B) Hydrolysis (C) Oxidation
(D) Distillation under reduced pressure.

19. Which of the following compounds can exhibit linkage isomerism ?
(A) [Co $(en)_2$ NO_2 Cl] Br
(B) [Co $(en)_3$] Cl_3
(C) [Co $(NH_3)_6$] [Cr $(CN)_6$]
(D) [Co $(NH_3)_5$ Cl] Br_2

20. Which of the following metals is extracted by the electrometallurgical method ?
(A) Cu (B) Fe (C) Na (D) Ag

21. A solution of sodium metal in liquid ammonia is a strong reducing agent due to the presence of
(A) Sodium atoms (B) Sodium hydride
(C) Sodium amide (D) Solvated electrons

22. Aluminium is refined by
(A) Serpeck's process (B) Baeyer's process
(C) Hall's process (D) Hoope's process

23. Reduction of acetyl chloride with H_2 in presence of Pd gives
(A) CH_3 $COCH_3$ (B) C_2 H_5 OH
(C) CH_3 COOH (D) CH_3 CHO

24. Amongst the following, the most basic compound is
(A) Acetanilide (B) P-nitroaniline
(C) Benzyl amine (D) Aniline

25. The insecticide gammexane is
(A) DDT (B) Benzene hexachloride

(C) Hexachloroethane (D) Chloral

26. Natural rubber is a polymer of
(A) Polyethylene (B) PVC
(C) Isoprene (D) Insulin

27. Which of the following is not a polymer ?
(A) Neoprene (B) Orlon (C) ICl (D) Teflon

28. Nylon is a
(A) Polyamide (B) Polyester
(C) Polyethylene (D) Polypropylene

29. The enzyme pepsin hydrolyses
(A) Polysaccharides to monosaccharides
(B) Fats to fatty acids
(C) Glucose to ethyl alcohol
(D) Proteins to amino acids

30. Rutherford's α- particle scattering experiment proved that atom has
(A) Electrons (B) Neutron
(C) Nucleus (D) Orbitals

31. In 1 mole of NaCl, the protons present are
(A) 6 mole (B) 11 mole (C) 17 mole (D)28 mole

32. What causes nitrogen to be chemically inert ?
(A) Absence of bond polarity
(B) Short internuclear distance
(C) High bond energy
(D) Multiple bond formation in the molecule

33. Heating of pyrite ore in air for oxidation of sulphur is called
(A) Roasting (B) Calcination
(C) Smelting (D) Concentration

34. For which of the following ores, froth flotation method is used for concentration
(A) Horn silver (B) Bauxite
(C) Cinnabar (D) Haematite

35. Which of the following compounds is explosive in nature ?
(A) Phosphorous trichloride (B) Nitrogen trichloride
(C) Hyponitrousacid (D) Nitrosyl chloride

36. Which of the following does not exhibit the periodicity in properities of the elements ?
(A) Ionisation energy (B) n / p ratio
(C) Electronegativity (D) Atomic radius

37. Which of the following elements exhibits the most basic properties ?
(A) F (B) Cl (C) Br (D) I

38. The correct sequence of decrease in size of the following species is
(A) $I > I^+ > I^-$ (B) $I > I^- > I^+$
(C) $I^+ > I^- > I$ (D) $I^- > I > I^+$

39. The decreasing order of the ionisation potential in the following elements is
(A) Ne > Cl > P > S > Al > Mg
(B) Ne > Cl > S > P > Mg > Al
(C) Ne > Cl > P > S > Mg > Al
(D) Ne > Cl > S > P > Al > Mg

40. To obtain bromine in the manufacture of bromine from sea water, the mother liquor containing bromide is treated with
(A) CO_2 (B) Cl_2 (C) I_2 (D) H_2O

41. The helical structure of proteins is stabilized by
(A) Peptide bonds (B) Dipeptide bond
(C) Hydrogen bond (D) van der Waal's forces

42. Mutation of DNA occurs due to changes in the sequence of the following
(A) Bases (B) Ribose units
(C) Phosphate (D) Sugar units

43. The optically inactive amino acid is
(A) Lysine (B) Glycine
(C) Arginine (D) Alanine

44. The oxidation number of cobalt in $K[Co(CO)_4]$ is
(A) +1 (B) -1 (C) +3 (D) -3

45. The I. U. P. C. name of $K_3[Ir(C_2O_4)_3]$ is
(A) Potassiumtris (oxalato)iridium (III)
(B) Potassiumtris (oxalato)iridate (III)
(C) Potassiumtrioxalatoiridium (III)
(D) Potassiumtrioxalato iridate (III)

46. The pH of 10^{-7} M NaOH is
(A) 7.01 (B) Between 7 and 8
(C) Between 9 and 10 (D) Greater than 10

47. For the functioning of enzymes, which of the following statements is not correct ?
(A) They are substrate specific
(B) An optimum temperature is needed
(C) An optimum pH is needed
(D) They always increase activation energy

48. The oxidation state of nitrogen is highest in
(A) N_3H (B) NH_2OH (C) N_2H_4 (D) NH_3

49. The oxidation state of oxygen is zero in
(A) SO_2 (B) H_2O_2 (C) CO (D) O_3

50. For conversion C (graphite) $\rightarrow$C (diamond), the ΔS is
(A) Zero (B) Positive (C) Negative (D) Unknown

51. In which of the following $\Delta E = \Delta H$?
(A) $N_2O_4 (g) \rightleftharpoons 2NO_2 (g)$
(B) $2SO_2 (g) + O_2 (g) \rightleftharpoons 2SO_3 (g)$
(C) $H_2 (g) + I_2 (g) \rightleftharpoons 2HI (g)$
(D) $H_2 (g) + \frac{1}{2} O_2 (g) \rightleftharpoons H_2O (l)$

52. The bond dissociation energies of O H_2, Cl_2 and HCl are 104, 58 and 103 K cal respectively. The enthalpy of formation of HCl is
(A) +44 K cal (B) +22 K cal
(C) -22 K cal (D) -44 K cal

53. For melting of 3 moles of water at 0°C the ΔG^o is

(A) Zero (B) +ve (C) -ve (D) Unpredictable

54. 2.5 F of electricity are passed through a $CuSO_4$ solution. The number of gm equivalent of Cu deposited on anode is

(A) Zero (B) 1.25 (C) 2.5 (D) 5.0

55. For a first order reaction velocity constant, K = 10. Two third life for it would be

(A) 1100 s (B) 2200 s (C) 3300 s (D) 4400 s

56. The reaction $2NO_2 (g) + O_2 (g) \rightleftharpoons 2NO_2 (g)$ is of first order. If the volume of the reaction vessel is redcued to 1/3, the rate of reaction would be

(A) 3 times (B) 6 times
(C) 1/3 times (D) 2/3 times

57. For which one of the following, the octane number is zero ?

(A) n-heptane (B) n-hexane
(C) Cyclohexane (D) Isooctane

58. The starting material for the preparation of styrene is

(A) Ethane (B) Ethene
(C) Ethyne (D) Vinyl chloride

59. The compound which on reaction with aqueous HNO_2 at low temperature produces an oily nitrosoamine is

(A) Methyl amine (B) Ethyl amine
(C) Diethyl amine (D) Aniline

60. Benzene reacts with chlorine to form benzene hexachloride in the presence of

(A) $AlCl_3$ (B) Ni (C) Sunlight (D) Zn

61. Colloids can be purified by

(A) Dialysis (B) Electrolysis
(C) Distillation (D) Chromatography

62. Identify 'Z' in the following series,

$C_2H_5I \xrightarrow{\text{alc. KOH}} X \xrightarrow{Br_2} Y \xrightarrow{KCN} Z$

(A) $CH_3 - CH_2CN$ (B) $CN - CH_2 - CH_2 CN$
(C) $Br\ CH_2 - CH_2\ CN$ (D) $Br\ CH = CHCN$

63. The reaction,

$C_2H_5\ ONa + Br\ C_2H_5 \rightarrow C_2H_5 - O - C_2H_5 + NaBr$ is called

(A) Frankland reaction (B) Wurtz reaction
(C) Williamson's synthesis (D) Cannizaro reaction

64. When chloroform reacts with aniline and alcoholic KOH, it gives

(A) Phenyl cyanate (B) Phenyl isocyanate
(C) Phenyl cyanide (D) Phenyl isocyanide

65. Primary, secondary and tertiary alcohols may be distinguished by

(A) Fehling solution test (B) Hoffmann test
(C) Beilstein test (D) Victor- Meyer test

66. The carbon dating is based on

(A) ${}^{15}_{6}C$ (B) ${}^{14}_{6}C$ (C) ${}^{13}_{6}C$ (D) ${}^{11}_{6}C$

67. Which of the following is not deflected by magnetic field ?

(A) Deuteron (B) Positron
(C) Proton (D) Photon

68. For transmutation of the ${}^{232}_{90}Th$ to ${}^{208}_{82}Pb$ the number of α and β particles emitted are respectively

(A) 6,4 (B) 7,3 (C) 4,6 (D) 5,5

69. Which of the following can be used to convert ${}^{14}_{7}N$ into ${}^{17}_{8}O$?

(A) Dcuteron (B) Proton
(C) α – particle (D) Neutron

70. Which of the following equilibriums is not shifted by increase in the pressure ?

(A) $H_2 (g) + I_2 (g) \rightleftharpoons 2HI (g)$
(B) $N_2 (g) + 3H_2 (g) \rightleftharpoons 2NH_3 (g)$
(C) $2CO (g) + O_2 (g) \rightleftharpoons 2CO_2 (g)$
(D) $2C (s) + O_2 (g) \rightleftharpoons 2CO (g)$

71. The pH of water decreases with

(A) increase in temperature (B) increase in pressure
(C) decrease in temperature (D) decrease in pressure

72. $SO_2 (g) + \frac{1}{2} O_2(g) \rightleftharpoons SO_3 (g)$, equilibrium constant = K_1

$2SO_3 (g) \rightleftharpoons 2SO_2 (g) + O_2 (g)$ equilibrium constant = K_2

Which of the following relation is correct ?

(A) $K_2 = K_1^2$ (B) $K_2 = K_1^{-2}$
(C) $K_2 = K^1$ (D) $K_2 = K^{-1}$

73. If for $H_2 (g) + \frac{1}{2} S_2 (g) \rightleftharpoons H_2S (g)$ and $H_2 (g) + Br_2 (g) \rightleftharpoons 2HBr (g)$, the equilibrium costants are K_1 and K_2 respectively. the reaction $Br_2 (g) + H_2S (g) \rightleftharpoons 2HBr (g) + \frac{1}{2} S_2 (s)$ would have equilibrium constant

(A) $K_1 \times K_2$ (B) K_1 / K_2 (C) K_2 / K_1 (D) K^2 / K_1

74. At room temperature HCl is a gas while HF is a low boiling liquid. This is because

(A) H-F bond is covalent (B) H-F bond is ionic
(C) H-F has metallic bond
(D) H-F has hydrogen bond

75. Which of the following is the life saving mixture for an asthma patient ?

(A) Mixture of xenon and nitrogen
(B) Mixture of argon and oxygen
(C) Mixture of helium and oxygen
(D) Mixture of neon and oxygen

76. Which of the following inert gases forms a large number of compounds ?

(A) Kr (B) Xe (C) Ne (D) Ar

77. An aqueous solution of an inorganic salt on treatment with HCl gives a white precipitate. This solution contains

(A) Hg_2^{+2} (B) Hg^{2+} (C) Zn^{2+} (D) Cd^{2+}

78. Ferric ion forms a prussian blue coloured precipitate due to formation of

(A) $K_4\ [Fe(CN)_6]$ (B) $Fe_4[Fe(CN)_6]_3$

(C) $Fe(CNS)_3$ (D) $K_3[Fe(CN)_6]$

79. On performing a borax-bead test with a given inorganic mixture for qualitative analysis, the colour of the bead was found to be emerald green both in oxidising and reducing flame. It indicates the possibility of the presence of

(A) Cr^{+3} (B) Cu^{+2} (C) Co^{+2} (D) Ni^{+2}

80. The % of silver in German silver is

(A) 100 % silver (B) 50 % silver
(C) 25 % silver (D) 0 % silver

81. The aqueous solution of which one of the following is basic ?

(A) HOCl (B) $NaHSO_4$
(C) NH_4NO_3 (D) NaOCl

82. Diethyl ether on heating with conc. HI gives two moles of

(A) Iodoform (B) Methyl iodide
(C) Ethyl iodide (D) Ethanol

83. Schiff's reagent is

(A) P-rosaniline hydrochloride solution decolourised with sulphurous acid
(B) Magenta solution decolourised with Cl_2
(C) Ammonical silver nitrate solution
(D) Alkaline $KMnO_4$ solution

84. Which of the following is radioactive ?

(A) Para - hydrogen (B) Protium
(C) Deuterium (D) Tritium

85. The volume of 0.1 M H_2SO_4 that is needed to completely neutralise 30 ml of 0.2 M NaOH solution is

(a) 15 ml (B) 30 ml (C) 40 ml (D) 60 ml

86. The equivalent weight of $KMnO_4$ in alkaline medium will be

(A) 79.00 (B) 158.00 (C) 31.60 (D) 52.66

87. Which one of the following is an incorrect statement ?

(A) The ionisation potential of beryllium is greater than that of boron
(B) The electronegativity of fluorine is greater than that of chlorine
(C) The ionisation potential of nitrogen is greater than that of oxygen
(D) The electron affinity of fluorine is greather than that of chlorine

88. The radius of first Bohr's orbit for hydrogen is 0.53 Å.The radius of Bohr's third orbit would be

(A) 0.79 Å (B) 1.59 Å (C) 3.18 Å (D) 4.77 Å

89. The third line in Balmer series corresponds to an electronic transition between which Bohr's orbits in hydrogen ?

(A) $5 \rightarrow 3$ (B) $5 \rightarrow 2$ (C) $4 \rightarrow 3$ (D) $4 \rightarrow 2$

90. Which of the following has maximum number of unpaired electrons (atomic no. of Fe 26) ?

(A) Fe (B) Fe (ii)
(C) Fe (iii) (D) Fe (iv)

91. In $[Co(NH_3)_6]Cl_3$ the number of covalent bonds is

(A) 3 (B) 6 (C) 9 (D) 18

92. Which of the following is not tetrahedral ?

(A) SCl_4 (B) SO_4^{2-} (C) $Ni(Co)_4$ (D) $NiCl_4^{2-}$

93. The bond order is not three for ?

(A) N_2^+ (B) O_2^{2+} (C) N_2 (D) NO^+

94. The molarity of a solution made by mixing 50 ml of conc. H_2SO_4 (36N) with 50 ml of warter is

(A) 36 M (B) 18 M (C) 9 M (D) 6 M

95. 171 g of cane sugar ($C_{12}H_{22}O_{11}$) is dissolved in 1 litre water. The molarity of the solution is

(A) 2.0 M (B) 1.0 M
(C) 0.5 M (D) 0.25 M

96. The vapour pressure of a pure liquid 'A' is 0.80 atm. When a non-volatile solute 'B' is dissolved in 'A' its vapour pressure become 0.60 atm. The mole fraction of 'B' in the solution is

(A) 0.125 (B) 0.25 (C) 0.50 (D) 0.75

97. In CsCl structure, the coordination number of Cs is

(A) Equal to that of Cl, that is 6
(B) Equal to that of Cl, that is 8
(C) Not equal to that of Cl, that is 6
(D) Not equal to that of Cl, that is 8

98. The number of octahedral sites per sphere arranged in a face centred cubic structure, is

(A) 1 (B) 2 (C) 3 (D) 4

99. 1 mole of crystalline NaCl will have .... number of unit cells

(A) 1.506×10^{23} (B) 3.012×10^{23}
(C) 4.518×10^{23} (D) 6.023×10^{23}

100. Which of the following does not show negative deviation from Rault's law ?

(A) Acetone - chloroform (B) Acctone - Benzene
(C) Chloroform - Ether (D) Chloroform - Benzene

PMT - 2002

1. Which of the following is Bragg's equation ?
(A) $2n\lambda = d. \sin\theta$ (B) $n.\theta/2=d/2. \sin\theta$
(C) $n\lambda = 2\theta. \sin\theta$ (D) $n\lambda = 2d. \sin\theta$

2. Due to Frenkel defect the density of ionic solids
(A) Decreases (B) Increases
(C) Does not change (D) Changes

3. Body centred cube lattice has a coordination number
(A) 8 (B) 12 (C) 6 (D) 4

4. The ages of most ancient formations are estimated by
(A) Potassium-argon method
(B) Corbon-14 dating method
(C) Radium-silicone method (D) Uranium-lead method

5. Hydrogen bomb is based on the principle of
(A) Nuclear fission (B) Nuclear fusion
(C) Nuclear explosion (D) None of the above

6. Penetrating power of an α–particle is
(A) Less than β-rays (B) More than γ-rays
(C) More than γ-rays (D) None of these

7. When the concentration of 'A' in the reaction $A+B \rightleftharpoons AB$ is doubled, the rate of reaction will be
(A) Doubled (B) Decreased by half
(C) Unchanged (D) Increased by four times

8. The rate of reaction of which of the following is not affected by pressure
(A) $PCl_3 + Cl_2 \rightleftharpoons PCl_5$ (B) $N_2 + 3H_2 \rightleftharpoons 2NH_3$
(C) $N_2 + O_2 \rightleftharpoons 2NO$ (D) $2SO_2 + O_2 \rightleftharpoons 2SO_3$

9. A solute is soluble in two immiscible liquids which are present in the mixture. The concentration of the solute in upper layer mixture will be
(A) More than in the lower layer
(B) In fixed ratio with that in the lower layer
(C) Same as in the lower layer
(D) Less than in the lower layer

10. Dehydrogenation of $CH_3-\underset{\substack{|\\OH}}{CH}-CH_3$ gives
(A) Acetone (B) Acetaldehyde
(C) Acetic acid (D) Acetylene

11. In the sequence of the following reactions,
$CH_3OH \xrightarrow{HI} CH_3I \xrightarrow{KCN}$
$CH_3CN \xrightarrow{Reduction} X \xrightarrow{HNO_2} Y$
X and Y are respectively
(A) $CH_3CH_2NH_2$ and CH_3CH_2OH
(B) $CH_3CH_2NH_2$ and CH_3COOH
(C) CH_3CH_2OH and CH_3CHO
(D) CH_3OCH_3 and CH_3CHO

12. Tollen's reagent is
(A) Alkaline mercuric chloride solution
(B) Ammonical silver nitrate solution
(C) Ammonium citrate solution
(D) Alkaline potassium permanganate solution

13. In the formation of methane molecule, carbon makes use of
(A) sp hydridised orbitals (B) sp^2 hybridised orbitals
(C) sp^3 hybridised orbitals (D) Unhybridised orbitals

14. In the sequence of the following reactions,
$CH_3CH_2OH \xrightarrow[\text{Chromic Acid}]{[O]} X \xrightarrow[\text{Chromic Acid}]{[O]} CH_3COOH$
X is
(A) CH_3COCH_3 (B) CH_3CHO
(C) CH_3OCH_3 (D) CH_3CH_2COOH

15. In the following reaction,
$C_2H_5OC_2H_5 + 4H \xrightarrow{\text{Red P + HI}} 2X + H_2O$ X is
(A) Butane (B) Propane (C) Ethane (D) Ethene

16. The reaction,
$2CH_3CHO \xrightarrow{Al(OC_2H_5)_3} CH_3COOC_2H_5$ is known as
(A) Tishenko reaction (B) Cannizzaro's reaction
(C) Knoevenagel reaction (D) Wurtz's reaction

17. According to Werner's theory of coordinate compounds
(A) Primary valency is ionisable
(B) Secondary valency is ionisable
(C) Primary and secondary valencies are non ionisable
(D) Only primary valency is non-ionisable

18. IUPAC name of $K_3[Al(C_2O_4)_3]$ is
(A) Potassiumaluminium(III)oxalate
(B) Potassiumaluminooxalate
(C) Potassiumtrixalatoaluminate(III)
(D) Potassiumtrixalatoaluminate(VI)

19. Which one of the following will not show geometrical isomerism ?
(A) $[Cr(NH_3]_4Cl_2)Cl$ (B) $[Co(en)_2Cl_2]Cl$
(C) $[Co(NH_3)_5NO_2]Cl_2$ (D) $[Pt(NH_3)_2Cl_2]$

20. According to Lewis the ligands are
(A) Acidic in nature (B) Basic in nature
(C) Neither acidic nor basic
(D) Some are acidic and others are basic

21. The most stable complex among the following is
(A) $K_3[Al(C_2O_4)_3]$ (B) $[Pt(en)_2]Cl_2$
(C) $Ag(NH_3)_2Cl$ (D) $K_2[Ni(EDTA)]$

22. Nitrate is confirmed by ring test. The brown colour of the ring is due to the formation of
(A) Ferrous nitrate (B) Ferrous nitrite
(C) $FeSO_4NO$ (D) $FeSO_4NO_2$

23. Amongst the following acids, the strongest acid is
(A) CH_3COCH_3 (B) CH_3OCH_3
(C) $ClCH_2COOH$ (D) Cl_3CCOOH

24. Hydrolysis of acetyl chloride gives
(A) C_2H_5OH (B) CH_3COOH
(C) CH_3COCH_3 (D) CH_3OCH_3

25. Reduction of methyl isocyanide gives
(A) Ethylamine (B) Methylamine
(C) Dimethylamine (D) Trimethylamine

26. Mark the correct statement
(A) I group basic radicals precipitate as chlorides
(B) IV group basic radicals precipitate as sulphides
(C) V group basic radicals precipitate as carbonates
(D) All the above statements are correct

27. What volume of water is to be added to 50 ml of 4N NaOH solution to obtain a 1 N solution ?
(A) 100 ml (B) 150 ml (C) 200 ml (D) 250 ml

28. 50 ml of 10 N H_2SO_4, 25 ml 12 N HCl and 40 ml 5 N HNO_3 were mixed together and the volume of the mixture was made upto 1000 ml by adding water. The normality of this obtained solution will be
(A) 1 N (B) 2 N (C) 3 N (D) 4 N

29. Proteins contain mainly
(A) C, H and O (B) C, H, O and N
(C) C, H and N (D) N, H and O

30. Proteines are hydrolysed by enzymes to give
(A) Dihydroxy acids (B) Aminoacids
(C) Amines (D) Carboxylic acids

31. In the reaction, $CH_3NH_2 + 3CHCl_3 + 3\ KOH \longrightarrow X + 3KCl + 3H_2O$. X is
(A) CH_3NC (B) CH_3CH_2NC
(C) CH_3COOH (D) CH_3CH_2COOH

32. Polythene is an addition polymer of
(A) Ethene (B) Acetylene (C) Methane (D) Ethane

33. The nucleus of tritium contains
(A) 1 proton + 1 neutron (B) 1 proton + 3 neutrons
(C) 1 proton + 0 neutron (D) 1 proton + 2 neutrons

34. Number of unpaired electrons in the electronic configuration $1s^2, 2s^2, sp^4$ is
(A) 2 (B) 3 (C) 4 (D) 6

35. The modern atomic weight scale is based on
(A) C^{12} (B) O^{16} (C) H^1 (D) C^{13}

36. Bohr model of atom is contradicted by
(A) Pauli's exclusion principle
(B) Planck quantum theory
(C) Heisenberg uncertainty principle
(D) All the above

37. The electronic configuration of nitrogen has three unpaired electrons according to
(A) Hund's rule (B) Aufbau principle
(C) Pauli's exclusion principle (D) Heisenberg's principle

38. The spectrum of He is expected to be similar to that of
(A) He^+ (B) Li^+ (C) H (D) Na

39. P in PCl_5 is hybridized as
(A) sp^3 (B) dsp^2 (C) ds^3p (D) sp^3d

40. Which of the following has zero dipole moment ?
(A) ClF (B) PCl_3 (C) SiF_4 (D) $CFCl_3$

41. Among the following the weakest base is
(A) CH_3O^- (B) Cl^- (C) H^- (D) CH_3^-

42. Acidic buffer solution is produced on mixing the following aqueous solutions of
(A) $NH_4Cl + NH_4OH$ (B) $CH_3COONa + CH_3COOH$
(C) NaCl + NaOH (D) $Na_2SO_4 + H_2SO_4$

43. A solution of $MgCl_2$ in water has pH
(A) < 7 (B) > 7 (C) 7 (D) 14.2

44. When H_2S gas is passed through the HCl containing aqueous solutions of $CuCl_2$, $HgCl_2$, $BiCl_2$ and $CoCl_2$, it does not precipitate out
(A) CuS (B) HgS (C) Bi_2S_3 (D) CoS

45. An adiabatic expansion of an ideal gas, always has
(A) Increases in temperature (B) $\Delta H = O$
(C) Q = O (D) W = O

46. Which of the following expressions represents the first law of thermodynamics ?
(A) $Q = \Delta E - W$ (B) $\Delta E = Q - W$
(C) $\Delta H = Q + W$ (D) $\Delta E = \Delta H + P\Delta V$

47. The heat of neutralization will be highest for
(A) NH_4OH and HCl (B) NH_4OH and HNO_3
(C) KOH and CH_3COOH (D) KOH and HCl

48. Gibbs free energy G, enthalpy H and entropy S are interrelated as
(A) G – TS = H (B) G = S = H
(C) G = H + TS (D) G = H – TS

49. Half life of a radioactive substance which disintegrates by 75% in 60 minutes, will be
(A) 120 min (B) 30 min (C) 45 min (D) 20 min

50. A reaction having equal activation energies for forward and reverse reaction has
(A) $\Delta H = O$ (B) $\Delta S = O$
(C) Zero order (D) None of the above

51. A solution of H_2SO_4 whose 9.8 gm is dissolved in 2 litres of water has molarity
(A) 0.1 M (B) 0.05 M (C) 0.01 M (D) 0.2 M

52. Isotonic solutions have the same
(A) Density (B) Molar concentration
(C) Normality (D) Optical rotation

53. Colligative properties of a solution depend upon
(A) Nature of solvent (B) Nature of solute
(C) Relative numbers of solute and solvent particles in solution
(D) None of the above

54. Collision theory is applicable to
(A) first order reactions (B) Zero order reactions
(C) Bimolecular reactions
(D) Intra-molecular reactions

55. The reaction $2N_2O_5 \rightleftharpoons 2N_2O_4 + O_2$ is
(A) Bimolecular and second order
(B) Unimolecular and first order

(C) Bimolecular and first order
(D) Bimolecular and zero order

56. At anode in the electrolysis of fused NaCl
(A) Na is oxidized (B) Cl^- is oxidized
(C) Cl is reduced (D) Na is reduced

57. General formulae of alkenes and alkyl radicals are respectively
(A) C_nH_{2n} and C_nH_{2n+1} (B) C_nH_{2n} and C_nH_{2n+2}
(C) C_nH_{2n-1} and C_nH_{2n} (D) C_nH_{2n+1} & C_nH_{2n+2}

58. Ethane is formed by the reaction of methyl iodide and sodium metal in dry ether solution. This reaction is known as
(A) Clemensen reduction (B) Kolbe's reaction
(C) Wurtz reaction (D) Cannizzaro's reaction

59. Which of the following compounds cannot be prepared by Wurtz reaction ?
(A) C_3H_8 (B) C_4H_{10} (C) CH_4 (D) C_2H_6

60. The reaction,
$$CH_3.CH = CH_2 \xrightarrow[H^+]{CO + H_2O} CH_3 - \underset{\displaystyle COOH}{\underset{|}{CH}} - CH_3$$
is known as
(A) Wurtz reaction (B) Koch reaction
(C) Clemensen reduction (D) Kolbe's reaction

61.In the reaction
$$CH_3CH = CH_2 + H_2O + [O] \xrightarrow[Acid]{KMnO_4}$$
$$CH_3 - \underset{OH}{\underset{|}{CH}} - \underset{OH}{\underset{|}{CH_2}} \xrightarrow{[O]} X + HCOOH, \text{ X is}$$
(A) CH_3CH_2COOH (B) CH_3COOH
(C) CH_3CH_2CHO (D) CH_3CH_2OH

62. Successive alkanes differ by
(A) $>CH_2$ (B) $\equiv CH$ (C) $-CH_3$ (D) C_2H_4

63. The process in which higher hydrocarbons are broken down into lower hydrocarbons by controlled pyrolysis, is called
(A) Hydrolysis (B) Cracking
C) Oxidation (D) Reduction

64. The reaction,
$$C_6H_5OH + CHCl_3 + 3NaOH \longrightarrow o\text{-}HOC_6H_4CHO + 3NaCl + 2H_2O$$
is called
(A) Carbylamine reaction (B) Cannizzaro's reaction
(C) Reimer Tiemann reaction (D) Wurtz reaction

65. At cathode, the electrolysis of aqueous Na_2SO_4 gives
(A) SO_3 (B) SO_2 (C) Na (D) H_2

66. In galvanic cell, the salt bridge is used to
(A) Reduce the electric resistance in the cell
(B) Separate cathode from anode
(C) Complete the circuit
(D) Carry salts for the chemical reaction

67. The reference electrode is made by using
(A) $ZnCl_2$ (B) $CuSO_4$ (C) $HgCl_2$ (D) Hg_2Cl_2

68. Hydrolysis of ethyl acetate is catalysed by aqueous
(A) Na_2SO_4 (B) K_2SO_4 (C) H_2SO_4 (D) $BaSO_4$

69. Surface tension of lyophilic sols is
(A) Lower than that of H_2O
(B) More than that of H_2O
(C) Equal to that of H_2O (D) None of the above

70. Milk is a colloid in which a
(A) Gas is dispersed in a liquid
(B) Liquid is dispersed in a liquid
(C) Solid is dispersed in a liquid
(D) Sugar is dispersed in a liquid

71. Adsorbed acetic acid on activated charcoal is
(A) Adsorber (B) Absorber
(C) Adsorbent (D) Adsorbate

72. An important ore of copper is
(A) Bauxite (B) Galena (C) Siderite (D) Chalcopyrite

73. Cryolite is used in the electrolytic extraction of aluminum
(A) To obtain more aluminium (B) To reduce bauxite
(C) To decompose bauxite (D) To protect anode

74. Heating pyrite ores to remove sulphur is known as
(A) Liquation (B) Smelting (C) Calcination (D) Roasting

75. Cupellation process is used in the metallurgy of
(A) Copper (B) Silver (C) Iron (D) None of these

76. The reaction of NaCl and $K_2Cr_2O_7$ mixture on heating with conc. H_2SO_4 gives
(A) CrO_2Cl_2 (B) $CrOCl_2$ (C) CrO_2Cl (D) $(CrOCl)_2$

77. Which of the following sets is an example of non-typical transition elements ?
(A) Li K, Na (B) Be, Al, Pb
(C) Zn, Cd, Hg (D) Ba, Ca, Se

78. An organic compound containing carbon, hydrogen and oxygen contains 52.20% carbon and 13.04% hydrogen. Vapour density of the compound is 23. Its molecular formula will be
(A) C_2H_6O (B) C_3H_8O (C) C_4H_8O (D) $C_5H_{10}O$

79. The empirical formula of a compound is CH_2. One mole of the compound has a mass of 42 gm. Its molecular formula is
(A) C_3H_6 (B) C_3H_8 (C) CH_2 (D) C_2H_2

80. Compounds $CH_3CH_2OCH_2CH_3$ and $CH_3OCH_2CH_2CH_3$ are
(A) Tautomers (B) Metamers
(C) Functional isomers (D) Optical isomers

81. General formula of alkanes is
(A) C_nH_{2n+1} (B) C_nH_{2n+2} (C) C_nH_{2n-1} (D) C_nH_{2n}

82. Which of the following is a correct statement ?

(A) A mineral cannot be an ore
(B) An ore cannot be a mineral
(C) All ores are minerals (D) All minerals are ores

83. Parke's process is used
(A) To extract silver from argentiferrous lead
(B) To extract silver by forming amalgam
(C) To extract silver using NaCN
(D) To extract copper using $CuFeS_2$

84. Isomers of propanoic acid are
(A) CH_3COOCH_3 and C_3H_7OH
(B) $HCOOC_2H_5$ and CH_3COOCH_3
(C) $HCOOC_2H_5$ and $C_3H_7COOCH_3$
(D) C_3H_7OH and CH_3COCH_3

85. The name of $H_3C - \underset{\displaystyle CH_3}{\underset{|}{CH}} - \underset{\displaystyle OH}{\underset{|}{CH}} - CH_3$ according to IUPAC nomenclature system is
(A) Butanol (B) 2-Methyl butanol-3
(C) 3-Methyl butanol-2 (D) Pentanol

86. The name of $Cl\ H_2C - \underset{\displaystyle Br}{\underset{|}{C}} = \underset{\displaystyle Br}{\underset{|}{C}} - CH_2Cl$ according to IUPAC nomenclature system is
(A) 2, 3-dibromo-1, 4-dichloro-2-butene
(B) 1, 4-dibromo-2, 3-dichlorobutene-2
(C) Dichlorodibromobutene (D) Dichlorodibromobutane

87. The functional isomer of ethyl alcohol is
(A) CH_3OCH_3 (B) CH_3COCH_3
(C) CH_3COOH (D) CH_3CH_2CHO

88. The electronegativity of an element in a period of periodic table from left to right
(A) Increases (B) Remains constant
(C) Decreases (D) Changes irregularly

89. Electron affinity depends on
(A) Atomic size (B) Nuclear charge
(C) Atomic number
(D) Atomic size and nuclear charge

90. Which of the following orders is correct for the first ionization potential of B, C and N ?
(A) $N > C < B$ (B) $N < C < B$
(C) $B > C > N$ (D) $N > C > B$

91. The ionic radii of Li^+, Na^+, K^+ are in the following order
(A) $K^+ > Na^+ > Li^+$ (B) $K^+ > Na^+ < Li^+$
(C) $K^+ < Na^+ < Li^+$ (D) $Li^+ > Na^+ < K^+$

92. Which of the following elements is most metallic ?
(A) P (B) As (C) Sb (D) Bi

93. In view of their low ionization energies the alkali metals are
(A) Weak oxidizing agents
(B) Strong reducing agents
(C) Strong oxidizing agents (D) Weak reducing agents

94. The electron affinity of noble gas elements is
(A) Low (B) Nearly zero (C) High (D) Very high

95. The thermal stability of alkaline earth metal carbonates $MgCO_3$, $CaCO_3$, $BaCO_3$ and $SrCO_3$ decrease as
(A) $BaCO_3 > SrCO_3 > CaCO_3 > MgCO_3$
(B) $CaCO_3 > SrCO_3 > MgCO_3 > BaCO_3$
(C) $BaCO_3 > SrCO_3 > MgCO_3 > CaCO_3$
(D) $MgCO_3 > CaCO_3 > SrCO_3 > BaCO_3$

96. Of the following compounds, the most acidic one is
(A) AS_2O_3 (B) P_2O_5 (C) Sb_2O_3 (D) Bi_2O_3

97. Which of the following electronic configurations is that of a transition element ?
(A) $1s^22s^22p^63s^23p^63d^{10}4s^24p^3$
(B) $1s^22s^22p^63s^23p^63d^24s^2$
(C) $1s^22s^22p^63s^23p^63d^{10}4s^24p^6$
(D) $1s^22s^22p^63s^23p^64s^2$

98. A transition element normally
(A) Forms coloured compounds (B) Has variable valance
(C) Has incomplete d-orbitals
(D) Has all the above characteristics

99. The oxidation number of chromium in potassium dichromate is
(A) +6 (B) +3 (C) +5 (D) +7

100. In alkaline condition $KMnO_4$ reacts as $2KMnO_4 + 2KOH \rightarrow 2K_2MnO_4 + H_2O + O$ the equivalent weight of $KMnO_4$ would be (Atomic mass of K = 39, Mn = 55, O = 16)
(A) 158.0 (B) 79.0 (C) 52.7 (D) 31.6

PMT - 2003

1. The most widely used and safe sugar substitute now a days is
(A) Saccharin (B) Fructose
(C) Cyclodextrin (D) Aspartame

2. Amino acid, isoleucine has the structural formula
(A) $NH_2 - \underset{\displaystyle CH(CH_3)_2}{\underset{|}{CH}} - COOH$ (B) $NH_2 - \underset{\displaystyle CH_3}{\underset{|}{CH}} - COOH$
(C) $NH_2 - \underset{\displaystyle CH_3-CH-C_2H_5}{\underset{|}{CH}} - COOH$ (D) $NH_2 - \underset{\displaystyle CH(C_2H_5)_2}{\underset{|}{CH}} - COOH$

3. Which of the following reactions is catalysed by the enzyme maltase ?
(A) Starch - Maltose (B) Maltose - Glucose
(C) Lactose - Maltose (D) Maltose - Glucose + Fructose

4. Bromination of propene by N-bromosuccinimidy yields
(A) $CH_3 - \underset{\displaystyle Br}{\underset{|}{C}} = CH_2$ (B) $BrCH_2 - CH = CH_2$
(C) $CH_3 - \underset{\displaystyle Br}{\underset{|}{CH}} - CH_2Br$ (D) $BrCH_2 - CH_2 = CH_2Br$

5. The main product of deoxygenation of the carboxylated phenol obtained by Reimer Tiemann method would be

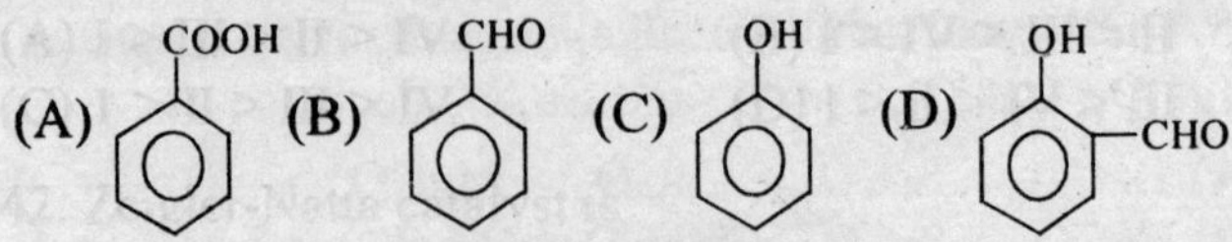

6. On Liebermann nitroso test, phenol gives

(A) Yellow colour (B) Red-brown colour
(C) Red colour (D) Blue colour

7. Phenol on reaction with bromine water will give

(A) OH, Br (B) OH, Br

(C) OH, Br + OH, Br (D) OH, Br, Br, Br

8. Ethyl alcohol when heated at 165-170°C in presence of excess of conc. H_2SO_4 gives.

(A) $C_2H_5HSO_4$ (B) $CH_2 = CH_2$
(C) $C_2H_5OC_2H_5$ (D) $(C_2H_5)_2SO_4$

9. Picric acid has the molecular formula

(A) SO_3H, O_2N, NO_2, NO_2 (B) COOH, O_2N, NO_2, NO_2

(C) OH, O_2N, NO_2, NO_2 (D) $O = P \langle ^{OH}_{OH}$, O_2N, NO_2, NO_2

10. Acetone and chloroform react in presence of a strong base to yield

(A) $\frac{CH_3}{CH_3}\rangle C \langle \frac{OH}{COOH}$ (B) $CH_3 - CO - CCl_3$

(C) $CCl_3 - \underset{\underset{O}{\|}}{C} - CCl_3$ (D) $\frac{CH_3}{CH_3}\rangle C \langle \frac{CCl_3}{OH}$

11. In the reaction,
$2Fe^{3+}(aq) + Sn^{2+}(aq) \rightarrow 2Fe^{2+}(aq) + A$. A belongs to

(A) $Sn^{4+}(aq)$ (B) $Sn^{2+}(aq)$ (C) $Sn^{3+}(aq)$ (D) Sn^0

12. A chemical reaction is catalysed by 'X', 'X'

(A) Decreases the enthalpy of the reaction
(B) Increases the activation energy
(C) Does not effect the equilibrium constant of the reaction
(D) Decreases the velocity constant of the reaction

13. Brownian movement is due to

(A) Attraction force between colloidal particles and the molecules of dispersion medium
(B) Convection currents in lliquid
(C) Impact of the molecules of dispersion medium on colloidal particles
(D) Changes in heat energy in liquid state.

14. Calamine is a mineral which contains

(A) $ZnSiO_3$ (B) $ZnCO_3$ (C) ZnS (D) ZnO

15. Which of the following methods is used for the concentration of copper pyrite

(A) Froth flotation method
(B) Gravitational method
(C) Electrical magnetic method (D) All of these

16. Iron obtained from blast furnace is known as

(A) Wrought iron (B) Pig iron
(C) Cast iron (D) Steel

17. The correct order of second ionization potential of C, N, O and F is

(A) O > N > F > C (B) C > N > O > F
(C) O > F > N > C (D) F > O > N > C

18. Manganese belongs to

(A) 'p' block (B) 's' block
(C) 'd' block (D) 'f' block

19. In Cl^- ion the number of electrons is

(A) 17 (B) 18 (C) 19 (D) 40

20. The electronic configuration, $1s^2 2s^2 2p^6 3s^2 3p^5 4s^1$ shows

(A) Lowers state of chlorine atom
(B) Excited state of argon
(C) Excited state of chlorine atom
(D) Lowest state of potassium

21. The number of uncoupled electrons in cobalt atom (at no. 27) is

(A) 2 (B) 3 (C) 4 (D) 8

22. If the velocity of hydrogen molecule is 5×10^4 cm. sec the de-Broglie wave length of it will be

(A) 2A (B) 4A (C) 5A (D) 100 A

23. Which of the following molecules has dipole moment zero ?

(A) Water (B) Beryllium difluoride
(C) Sulphur dioxide (D) Ammonia

24. BCl_3 is planar molecule because in this molecule boron (B) is

(A) sp hybridised (B) sp^2 hybridised
(C) sp^3 hybridised (D) None of these

25. 2.0 molal solution is obtained by dissolving 0.5 mole of solute in

(A) 250 g. of solvent (B) 500 ml of solvent
(C) 250 ml of solution (D) 250 ml of solvent

26. Which one is the correct order of size of the iodine species

(A) $I > I^- > I^+$ (B) $I > I^+ > I^-$
(C) $I^+ > I^- > I$ (D) $I^- > I > I^+$

27. Which element has the highest electro negativity ?
(A) Na (B) Ne (C) F (D) H

28. With respect to chlorine, hydrogen will be
(A) Electropositive (B) Electronegative
(C) Neutral (D) None of the above

29. Which of the following hydroxides is the strongest base ?
(A) $Mg(OH)_2$ (B) $Be(OH)_2$
(C) $Ca(OH)_2$ (D) $Ba(OH)_2$

30. 500 ml of solution contains 4.9 g of H_3PO_4 in water. Calculate the normality of the solution.
(A) 3.0 (B) 1.0 (C) 0.3 (D) 0.4

31. Chloroform when dissolved in ethyl alcohol, a solution is obtained which shows
(A) Ideal properties
(B) Non experimental verification of Henery law
(C) Negative deviation from Raoult's law
(D) Positive deviation from Raoult's law

32. Assuming the 90% dissociation of a decimolar solution of each of the following salt, which has the highest value of osmotic pressure ?
(A) Aluminum sulphate
(B) Potassium Sulphate (C) Barium Chloride
(D) The mixture of equal volumes of (B) + (C)

33. Which of the following is not a colligative property ?
(A) Lowering of vapour pressure
(B) Depression of freezing point
(C) Osmotic pressure (D) Elevation of boiling point

34. Number of σ and π bonds in butenyne would be respectively
(A) 7σ and 5π (B) 7σ and 3π
(C) 4σ and 5π (D) 4σ and 3π

35. Which of the following compounds does not show geometrical isomerism ?
(A) $(CH_3)(CH_3)C=C(Cl)(Br)$ (B) $(H)(D)C=C(Cl)(Br)$
(C) $(CH_3)(H)C=C(Cl)(Br)$ (D) $(Cl)(Br)C=C(Cl)(Br)$

36. Metamers of ethyl propionate are
(A) CH_3COOCH_3 and $C_3H_4COOC_3H_7$
(B) C_4H_9COOH and C_3H_7 $COOCH_3$
(C) C_4H_9COOH and $HCOOC_4H_9$
(D) $CH_3COOC_3H_7$ and $C_3H_7COOCH_3$

37. The correct IUPAC name for $CH_2=CH-CH_2-C\equiv CH$ is -
(A) Pent–1–en-1–yne (B) Pent–1–en-4–yne
(C) Pent–1, 4–enyne (D) Pent–4, 1–enyne

38. The chief constituent of Marsh gas is
(A) Butane (B) Propane (C) Methane (D) Isobutene

39. Lewisite, the poisonous gas is obtained by reacting the following
(A) $CH\equiv CH$ and $AsCl_3$ (B) $CH_2=CH_2$ and $AsCl_3$
(C) $CH_2=CH_2$ and NOCl (D) $CH\equiv CH$ and NOCl

40. Major product obtained by passing the vapours of benzene over V_2O_5 at 450°C would be
(A) CHCOOH ‖ CHCOOH (B) maleic anhydride (CHC(=O), ‖, CHC(=O), bridged by O)
(C) p-benzoquinone (D) phenol (OH)

41. Which of the following compounds on ozonolysis followed by hydrolysis will provide CH_3COCH_3 and CH_3CHO only ?
(A) $CH_3-CH=C(CH_3)_2$
(B) $CH_3-CH=CH-CH_3$
(C) $CH_3-C(CH_3)=CH-CH_2-CH=CH-CH_3$
(D) $CH_3-C(CH_3)=C(CH_3)-CH_3$

42. Best reagent to distinguish between propene and propyne would be
(A) Solution of bromine in CCl_4 (B) Bromine water
(C) Alkaline solution of $KMnO_4$
(D) Ammonical solution of $AgNO_3$

43. In gaseous reaction,
$C_2H_4 + H_2 \rightleftharpoons C_2H_6$, $\Delta H = -32.7$ kcal. the concentration of C_2H_4 can be reduced by
(A) Increasing the pressure
(B) Reducing the temperature
(C) Increasing the H concentration (D) All of these

44. In the chemical reaction,
$SnCl_2 + 2Cl^- \rightarrow SnCl_4 + 2e^-$, which of the following is acting as Lewis acid ?
(A) $SnCl_4$ (B) Cl^- (C) $SnCl_2$ (D) None of these

45. Which of the following is conjugated acid of water in the reaction $H_2SO_4 + H_2O \rightleftharpoons H_3O^+ + HSO_4^-$
(A) SO_4^{2-} (B) HSO_4^- (C) H_2O (D) H_3O^+

46. Which aqueous solution has the pH value less than 7
(A) NaOH (B) $FeCl_3$ (C) NaCN (D) KNO_3

47. 50ml of 2N CH_3COOH ($K_a = 10^{-5}$) is mixed with 10 ml of 1 N sodium acetate. pH obtained is about
(A) 7 (B) 6 (C) 5 (D) 8

48. Phenolphthalein is not a right indicator for which of the following titrations between
(A) Potassium carbonate and hydrochloric acid
(B) Potassium hydroxide and sulphuric acid
(C) Sodium hydroxide and oxalic acid
(D) None of these

49. The K_{sp} values of CuS, Ag_2S and HgS are 10^{-31}, 10^{-42} and 10^{-54} respectively. What is the right sequence of solubility of them in water ?

(A) HgS > Ag_2S > CuS (B) HgS > CuS > Ag_2S
(C) Ag_2S > CuS > HgS (D) Ag_2S > HgS > CuS

50. For NaCl, the K_{sp} = 36 $mol^2/litre^2$, the molar concentration of it will be

(A) $\frac{1}{36}$ M (B) $\frac{1}{16}$ M (C) 36M (D) 6M

51. According to Hess's law the thermal effect of a reaction depends upon

(A) Intermediate state of the reaction
(B) Final condition of the reacting substances
(C) Initial condition of reactions
(D) Initial and final conditions of the reacting sub-stances.

52. Which of the following equations is the correct one ?

(A) $\Delta G = \Delta H - T\Delta S$ (B) G = H – S T
(C) G = T S – H (D) G = H + T S

53. For a spontaneous reaction ΔG (free energy) is

(A) Zero (B) Negative (C) Infinite (D) Positive

54. $2CH_3-\underset{\underset{O}{||}}{C}-OC_2H_5 \xrightarrow{C_2H_5ONa} CH_3-\underset{\underset{O}{||}}{C}-CH_2-\underset{\underset{O}{||}}{C}-OC_2H_5 + C_2H_5OH$

the above reaction is

(A) Claisen schmidt condensation
(B) Perkin reaction
(C) Claisen condensation
(D) Dieckmann reaction

55. Formaldehyde reacts with ammonia to form Urotropine which has the formula

(A) $(CH_2)_4N_6$ (B) $(CH_2)_3N_4$
(C) $(CH_2)_6N_4$ (D) $(CH_2)_4N_3$

56. $CH_3\underset{\underset{O}{||}}{C}-OCH_3 + C_2H_5OH \xrightarrow{H^+ \text{ or } OH^-} CH_3\underset{\underset{O}{||}}{C}-O-C_2H_5 + CH_3OH$, the reaction is

(A) Claisen schmidt reaction (B) Schmidt reaction
(C) Esterification (D) Trans-esterification

57. Which of the following chemical transformations is the Arndt Eistert reaction used ?

(A) $R-CH_2OH \rightarrow R-OH$
(B) $R-CH_2-COOH \rightarrow R-COOH$
(C) $R-OH \rightarrow R-CH_2-OH$
(D) $R-COOH \rightarrow R-CH_2-COOH$

58. The components of the Ziegler-Natta catalyst used for obtaining poly propylene from propylene are

(A) $Ti(C_2H_5)_3 + AlCl_3$ (B) $TiCl_4 + Al(C_2H_5)_3$
(C) $TiCl_3 + Al(C_2H_5)_3$ (D) $Ti(C_2H_5) + AlCl_3$

59. Hinsberg's reagent is

(A) C_6H_5-COCl (B) $C_6H_5-SO_2Cl$
(C) $CH_3-C_6H_4-COCl$ (D) $CH_3-C_6H_4-SO_2Cl$

60. Monomer of the Teflon polymer is

(A) $CF_2 = CF_2$ (B) $CF_2 = CCl_2$
(C) $CF_2 = CFCl$ (D) $CF_2 = CH_2$

61. The oxidation number of "Fe" in $K_4[Fe(CN)_6]$ is.

(A) 0 (B) 2 (C) 3 (D) 5

62. $K_3[Al(C_2O_4)_3]$ is called

(A) Pot. trioxalatealuminate
(B) Pot. aluminooxalate (C) Potalumino (III) oxalate
(D) Pot. trioxalatoaluminate (III)

63. Lithium tertrahydrido aluminate is correctly represented as

(A) $Al_2[LiH_4]_3$ (B) $Al[LiH_4]$
(C) $Li[AlH_4]$ (D) $Li[AlH_4]_2$

64. The wt. of solute (W) required to prepare a standard solution of solutes of known normality in a definite volume is given by (N- normality , E- equivalent weight of solute, V- volume of solution)

(A) $W = \frac{NEV}{1000}$ (B) $W = \frac{N.E}{1000}$
(C) W = 0.01 NEV (D) $W = \frac{0.01NE}{V}$

65. In which of the following groups, group reagent H_2S gas is common in the analysis of basic radicals

(A) III and IV group (B) II and IV group
(C) II and III group (D) I and II group

66. The correct IUPAC name for

(benzene ring with CHO at top, NO_2 ortho, OCH_3 para)

is

(A) 4-mehtoxy 2- nitro benzaldehyde
(B) 4-methoxy-6 nitro-benzaldehyde
(C) 4-formyl 1-3 nitro-anisole
(D) 2 formyl-5-methoxy - nitrobenzene

67. A sample of petrol has an octane number of 40, it means that its knocking power matches with the following mixture

(A) 40% petrol + 60% iso octane
(B) 40% n-heptane + 60% iso octane
(C) 60% n-heptane + 40% iso octane
(D) 60% petrol + 40% iso octanc

68. The specific rotation of equilibrium mixture of α-D glucose and β-D glucose is

(A) +112° (B) +19° (C) +52° (D) +66.5°

69. What would be the molecular formula of the alkane, if 5.6 litres of which was found to weigh 11g. at normal

temperature and pressure ?

(A) C_3H_8 (B) C_2H_6 (C) C_4H_{10} (D) C_5H_{12}

70. 36g of water and 828g. of ethyl alcohol form a mixture of ideal nature. What is the mole fraction of water in this mixture ?

(A) 0.9 (B) 0.7 (C) 0.4 (D) 0.1

71. For the strength "S" in gram per litre for a substance in a solution is related to its normality (N) and the eq. wt. (E) by

(A) $S = \frac{N}{E}$ (B) $S = \frac{E}{N}$ (C) S = N, E (D) None

72. The molecular weight of a tribasic acid is M. The equivalent weight of the acid will be

(A) $\frac{M}{2}$ (B) 2M (C) $\frac{M}{3}$ (D) $\sqrt{\frac{M}{3}}$

73. Copper metal is crystallized in face centered cubic cell unit. Calculate the atomic radius of copper atom if the axial distance is 3.615Å and there are 4 atoms of copper in a unit cell

(A) 1.80 A (B) 1.05 A (C) 1.28 A (D) 1.65 A

74. A parental nucleus 'X' after consecutive expulsion of two β particles, one after the other gives $^{14}_{7}N$. What is the number of neutrons in X.

(A) 7 (B) 9 (C) 14 (D) 15

75. In the form of energy, one atomic mass unit (amu) is equal to

(A) 1000 J (B) 1.592 x 10 erg
(C) 93.1.1 meV (D) 10 erg

76. In a radioactive transformation

$R \xrightarrow{-\alpha} X \xrightarrow{-\beta} Y \xrightarrow{-\beta} Z$

Which is the relation between the nucleus of R and Z ?

(A) Isotopic (B) Isobaric
(C) Isoneutronic (D) Isomeric

77. Which of the following particle is used in the transformation of $^{27}_{13}Al$ into $^{31}_{15}P$

(A) Proton (B) Neutron
(C) Alpha particle (D) Deutron

78. 87.5% dissociation of the nuclei of a radioactive substance is completed in 3 hours. What will be its half life period ?

(A) 2 Hours (B) 1 Hour
(C) 30 Minutes (D) 90 Minutes

79. If the equilibrium constant of the reaction, $2AB \leftrightarrows A_2 + B_2$ is 49, what is the value of equilibrium constant for the reaction $AB \leftrightarrows \frac{1}{2}A_2 + \frac{1}{2}B_2$?

(A) 49 (B) 7 (C) 0.01 (D) 2401

80. For the chemical reaction,

$PCl_3(g) + Cl_2(g) \leftrightarrows PCl_5(g)$, the value of K_c at 452°C is 26 $mol^{-1}/ litre^{-1}$. The value of K_p at this temperature is

(A) 0.506 Atm. pressure (B) 5.06 Atm. pressure
(C) 0.0506 Atm. pressure (D) None of these

81. Diagonal relationship occurs between

(A) B and Al (B) B and C (C) B and Ge (D) B and Si

82. Which of the following has the largest size?

(A) Al^{+1} (B) Al^{+3} (C) Al (D) Al^{+2}

83. Which of the following does not show the variable oxidation state ?

(A) Cu (B) Fe (C) Zn (D) Mn

84. The general electronic configuration of transitional elements is

(A) $(n-1)d^{1-10} ns^1$ (B) $(n-1)d^{1-5}$
(C) $(n-1)d^{1-10} ns^{1 \text{ or } 2}$ (D) $(n-1)d^{1-10} ns^{0.1.2}$

85. Transition elements are related to

(A) 'P' block (B) 's' block
(C) 'd' block (D) None of the above

86. Which transition metal is used for the reduction of steam to hydrogen ?

(A) Se (B) Fe (C) Mg (D) Pt

87. Zn belongs to

(A) 11th group (B) 12th group
(C) 17th group (D) 18th group

88. What change occurs in atomic volumes on moving from left to right in series of transition elements ?

(A) Volumes increase (B) Volumes decrease
(C) Volumes remain same (D) None

89. Which of the following is the formula of saltpeter ?

(A) KNO_3 (B) NaCl (C) $NaNO_3$ (D) $NaCO_3$

90. The formula of Felspar is

(A) $K_2O\ Al_2O_3\ 6SiO_2$ (B) $MgAl_2O_4$
(C) Na_3AlF_6 (D) $K_2O_3 . Al_2O_3\ 6SiO_2$

91. For the neutralization of 1 mol of H_2SO_4 with 2 mol of NaOH in dilute solution, the heat evolved is

(A) 57.0 KJ (B) Unpredictable
(C) More than 57.0 KJ (D) Less than 57.0 KJ

92. Which of the following expressions is correct for the first order reactions (Co refers to initial concentration and $t_{1/2}$ refers to half life time

(A) $t_{1/2} \propto Co^0$ (B) $t_{1/2} \propto Co^{-1}$
(C) $t_{1/2} \propto Co$ (D) $t_{1/2} \propto Co^{1/2}$

93. For the reaction, A + B → product, it is found that if the concentration A is doubled, velocity increases four times but when concentrations of B is doubled velocity of reaction is not effected. The velocity equation is

(A) Velocity = $K/4 [A]^2$ (B) Velocity = K [A] [B]
(C) Velocity = $K [A]^2 [B]^0$ (D) Velocity = $K [A]^2 [B]^2$

94. If in the acidic solution of copper sulphate and alkaline solution of copper cyanide respectively, same electric current is flown for a definite time, which of the following statements is correct ?

(A) Amount of copper deposited in both the solution is the same.

(B) Amount of copper deposited in alkaline copper cyanide

solution is less.
(C) Amount of copper deposited in acidic copper sulphate solution is less. (D) None of these

95. The following four colourless solutions of different salts are taken separately in four beakers. A copper metal piece is dropped in each of the solution. which of the four solutions is changed into blue colour after some time ?
(A) Zinc nitrate solution (B) Barium nitrate solution
(C) Silver nitrate solution (D) Sodium nitrate solution

96. What is the emf of the cell, whose half cells are given below ?
$Mg^{2+} + 2e^- \leftrightarrows Mg(s)$ $E^o = -2.37$ V
$Cu^{2+} + 2e^- \leftrightarrows Cu(s)$ $E^o = +0.34$ V
(A) -2.30 V (B)1.336 V (C) 2.71 V (D) 2.03 V

97. Which of the following reactions is a redox reaction ?
(A) $2AgNO_3 + BaCl_2 = 2AgCl + Ba(NO_3)_2$
(B) $BaO_2 + H_2SO_4 = BaSO_4 + H_2O_2$
(C) $P_2O_5 + 2H_2O = H_4P_2O_7$
(D) $Cu + 2AgNO_3 = 2Ag + Cu(NO_3)_2$

98. Formula of Chilisaltpeter ore of sodium metal is
(A) Na_2CO_3 (B) $NaNO_3$ (C) KNO_3 (D) NaCl

99. In complex compounds, ligand behaves as
(A) Neither acceptor nor donar (B) Electron pair donar
(C) Electron pair acceptor (D) None of these

100. The oxidation of the 'Fe' atom in $[Fe(CO)_5]$ is
(A) –1 (B) +4 (C) –3 (D) None of these

PMT - 2004

1. The triad of nuclei which is isotonic is
(A) ${}_6C^{14}$, ${}_7N^{15}$, ${}_9F^{17}$ (B) ${}_6C^{14}$, ${}_7N^{14}$, ${}_9F^{17}$
(C) ${}_6C^{12}$, ${}_7N^{14}$, ${}_9F^{19}$ (D) ${}_6C^{14}$, ${}_7N^{14}$, ${}_9F^{19}$

2. Alkali metals act as
(A) Good reducing agent
(B) Good dehydrating agent
(C) Good oxidising agent
(D) None of these

3. Which of the followng halides is most acidic
(A) PCl_3 (B) $BiCl_3$ (C) $SbCl_3$ (D) CCl_4

4. A hydride of nitrogen which is acidic is
(A) NH_3 (B) N_2H_2 (C) N_3H (D) N_2H_4

5. The equation $\lambda = h/mv$ was deduced by
(A) Planck (B) Newton
(C) Heisenberg (D) de-Broglie

6. The radius of electron in the first excited state of hydrogen atom is
(A) a_0 (B) $4a_0$ (C) $2a_0$ (D) $8a_0$

7. The de-Broglie equation applies
(A) to elecrons only (B) to neutrons only
(C) to protons only
(D) All the material objects in motion

8. Which of the following contains both ionic and covalent bond
(A) H_2O (B) C_6H_5Cl (C) NaOH (D) CO_2

9. Which of the following has a linear structure ?
(A) CCl_4 (B) C_2H_2 (C) SO_2 (D) C_2H_4

10. Among the following compounds the one that is planer and has central atom with sp^2-hybridization is
(A) H_2CO_3 (B) BF_3 (C) SiF_4 (D) $HClO_2$

11. The molecule with is pyramid shape is
(A) PCl_3 (B) CO_3^{2-} (C) SO_3 (D) NO_3^-

12. The maximum number of water molecules that one water molecule can hold through hydrogen bonding is :
(A) 2 (B) 6 (C) 4 (D) 8

13. Formation of acetylene from ethylene is an example of
(A) Elimination reaction
(B) Substitution reaction
(C) Addition reaction
(D) Condensation reaction

14. Which of the following compounds exhibit stereoisomerism ?
(A) 2-methyl-butane 1 (B) 3-methyl-butanoic acid
(C) 3-methyl-butyne 1 (D) 2-methyl butanoic acid

15. Anti-Markownikoff's addition is not observed in
(A) Propene (B) Butene-1
(C) Butene-2 (D) Pentene-1

16. The reaction of an aromatic halogen compound with an alkyl halides in presence of sodium and ether is called
(A) Wurtz reaction (B) Sandmeyer's reaction
(C) Wurtz-Fittig reaction (D) Kolbe reaction

17. Which one is organometallic compound ?
(A) Lithium methoxide (B) Lithium dimethyl amide
(C) Lithium acetate (D) Methyl lithium

18. Which one belongs to 3d-transition series ?
(A) Copper (B) Gold (C) Cobalt (D) Silver

19. Manganese exhibits maximum oxidation state in
(A) K_2MnO_4 (B) MnO_2
(C) $KMnO_4$ (D) Mn_3O_4

20. β-particles is emitted is radioactivity by
(A) Conversion of proton to neutron
(B) Conversion of neutron to proton
(C) Outermost orbit
(D) β-particles not emitted

21. For a dilute solution Raoult's law states that
(A) The lowering of vapour pressure is equal to the mole fraction of solute
(B) The relative lowering of vapour pressure is equal to mole fraction of solute
(C) The relative lowering of vapour pressure is proportional to amount of solute in solution
(D) The vapour pressure of the solution is equal to mole fraction of the solvent

22. The co-ordination number of a metal crystallizing in a hexagonal close packed structure is
(A) 12 (B) 8 (C) 4 (D) 6

23. With anhydrous zinc chloride, ethylene glycol gives
(A) Formaldehyde (B) Acetylene
(C) Acetaldehyde (D) Acetone

24. Strongest acid is
(A) $HC \equiv CH$ (B) C_2H_6
(C) C_6H_6 (D) CH_3OH

25. Carbolic acid is
(A) Phenol (B) Ethanol
(C) Methanol (D) Benzyl alcohol

26. Iron is manufactured from the ore
(A) Cryolite (B) Haematite
(C) Bauxite (D) Chalcopyrite

27. Which of the element is alloyed with copper to from brass ?
(A) Lead (B) Zinc
(C) Bismuth (D) Antimony

28. In the reversible reaction two substances are in equilibrium. If the concentration of each one is doubled, the equilibrium constant will be
(A) Reduced to its half of its original value
(B) Reduced to its one-fourth of its original value
(C) Doubled (D) Constant

29. The pH of an aqueous solution containing $[H^+]$ concentration $=3.0 \times 10^{-3}$ M. The pH of the solution is
(A) 2.471 (B) 3.0
(C) 2.523 (D) None of these

30. The concentration of KI and KCl in certain solution containing both is 0.001 M each. If 20 ml of this solution is added to 20 ml. of a saturated solution of AgI in water ? What will happen ?
(A) AgCl will be precipitated
(B) AgI will be precipitated
(C) Both AgCl and AgI will be precipitated
(D) There will be no precipitate

31. A precipitate is formed when
(A) A solution becomes saturated
(B) The ionic product is nearly equal to the solubility product
(C) The ionic product is less than solubility product
(D) The ionic product exceeds the solubility product.

32. For the reaction
$$PCl_3(g) + Cl_2(g) \rightleftharpoons PCl_5(g)$$
the position of equilibrium can be shifted to the right by
(A) Increasing the temperature
(B) Doubling the volume
(C) Addition of Cl_2 at constant volume
(D) Addition of equimolar quantities of PCl_3 and PCl_5

33. The enthalpy change (ΔH) for the process
$$N_2H_4(g) \longrightarrow 2N(g) + 4H(g)$$
is 1724 KJ mol^{-1}. If the bond energy of N-H bond in ammonia is 391 KJ mol^{-1}. What is the bond energy of N-N bond is N_2H_4?
(A) 160 KJ mol^{-1} (B) 391 KJ mol^{-1}
(C) 1173 KJ mol^{-1} (D) 320 KJ mol^{-1}

34. Which of the following would be expected to have the largest entropy per mole ?
(A) $SO_2Cl_2(s)$ (B) $SO_2Cl_2(g)$
(C) $SO_2Cl_2(l)$ (D) $SO_2(g)$

35. The velocity constant of a reaction at 290 K was found to be 3.2×10^{-3}. At 300 K it will be
(A) 1.28×10^{-2} (B) 6.4×10^{-3}
(C) 9.6×10^{-3} (D) 3.2×10^{-4}

36. For the first order reaction A $\rightarrow$ products, the half-life time is 100 sec. The rate constant of reaction is
(A) 6.9×10^{-2} sec^{-1} (B) 6.93×10^{-3} sec^{-1}
(C) 6.93×10^{-4} sec^{-1} (D) 6.93×10^{-1} sec^{-1}

37. Which one of the following gives positive Iodoform and positive Tollen's test ?
(A) Formeldehyde (B) Acetone
(C) Acetaldehyde (D) Propanaldehyde

38. The oxidation of benzyl chloride with lead nitrate gives
(A) Benzyl alcohol (B) Benzoic acid
(C) Benzaldehyde (D) p-chlorobenzaldehyde

39. Dacron is an example of
(A) Polyamide (B) Polyester
(C) Polyurethane (D) Polypropylene

40. Nucleic acid is a polymer of
(A) Nucleosides (B) α-amino acids
(C) Nucleotides (D) Glucose

41. Sodium alkyl benzene sulphonate is used as
(A) Soap (B) Fertilizers
(C) Pesticides (D) Detergents

42. When a lead storage battery is discharged
(A) SO_2 is evolved (B) Lead sulphate is consumed
(C) Lead is formed (D) Sulphuric acid is consumed

43. The amount of electricity that can deposit 0.8gm of silver from $AgNO_3$ solution is
(A) 1 Ampere (B) 1 Faraday
(C) 1 Coulomb (D) None of these

44. If the ΔG of a cell reaction
$$AgCl + e^- \longrightarrow Ag + Cl^-$$
is -21.20 KJ; the standard e.m.f., of cell is
(A) 0.229 V (B) 0.220V
(C) - 0.220 V (D) -0.110 V

45. The equilvalent weight of KIO_3 in the reaction
$$2Cr(OH)_3 + 4OH^- + KIO_3 \rightarrow 2CrO_4^{2-} + 5H_2O + KI$$ is
(A) Mole wt. (B) $\frac{\text{Mole wt.}}{5}$
(C) $\frac{\text{Mole wt.}}{2}$ (D) $\frac{\text{Mole wt.}}{3}$

46. Gold number gives
(A) the amount of gold present in the colloid
(B) the amount of gold required to break the colloid
(C) the amount of gold required to protect the colloide
(D) None of the above

47. Milk is an example of
(A) Pure solution (B) Gel
(C) Emulsion (D) Suspension

48. Which one has sp^2-hybridisaton
(A) CO_2 (B) N_2O (C) SO_2 (D) CO

49. Acetonitrile is
(A) C_2H_5CN (B) CH_3CN
(C) CH_3COCN (D) $C_6H_5CH_2CN$

50. Potassium ferrocynide is a
(A) Normal salt (B) Mixed salt
(C) Double salt (D) Complex salt

PMT - 1993

1. (B) 11. (C) 21. (B) 31. (A) 41. (A) 51. (C) 61. (A) 71. (D) 81. (C) 91. (C)
2. (B) 12. (A) 22. (C) 32. (B) 42. (D) 52. (B) 62. (A) 72. (A) 82. (A) 92. (B)
3. (C) 13. (B) 23. (A) 33. (B) 43. (B) 53. (C) 63. (D) 73. (C) 83. (B) 93. (B)
4. (A) 14. (B) 24. (A) 34. (D) 44. (C) 54. (B) 64. (B) 74. (A) 84. (C) 94. (C)
5. (A) 15. (A) 25. (D) 35. (B) 45. (C) 55. (C) 65. (C) 75. (B) 85. (D) 95. (A)
6. (A) 16. (A) 26. (D) 36. (A) 46. (A) 56. (A) 66. (C) 76. (A) 86. (B) 96. (B)
7. (D) 17. (B) 27. (A) 37. (D) 47. (A) 57. (D) 67. (A) 77. (D) 87. (B) 97. (C)
8. (A) 18. (B) 28. (A) 38. (B) 48. (C) 58. (B) 68. (D) 78. (A) 88. (B) 98. (A)
9. (A) 19. (B) 29. (B) 39. (C) 49. (A) 59. (D) 69. (A) 79. (C) 89. (C) 99. (D)
10. (C) 20. (C) 30. (A) 40. (D) 50. (A) 60. (D) 70. (A) 80. (C) 90. (B) 100.(C)

PMT - 1994

1. (D) 11. (D) 21. (B) 31. (B) 41. (D) 51. (A) 61. (C) 71. (B) 81. (C) 91. (A)
2. (C) 12. (B) 22. (A) 32. (C) 42. (C) 52. (C) 62. (B) 72. (B) 82. (C) 92. (C)
3. (C) 13. (B) 23. (D) 33. (B) 43. (A) 53. (B) 63. (D) 73. (B) 83. (B) 93. (A)
4. (B) 14. (C) 24. (A) 34. (B) 44. (A) 54. (B) 64. (C) 74. (D) 84. (B) 94. (C)
5. (A) 15. (A) 25. (A) 35. (B) 45. (A) 55. (C) 65. (C) 75. (C) 85. (C) 95. (D)
6. (A) 16. (B) 26. (D) 36. (C) 46. (A) 56. (D) 66. (B) 76. (A) 86. (A) 96. (C)
7. (C) 17. (A) 27. (B) 37. (B) 47. (D) 57. (A) 67. (B) 77. (C) 87. (D) 97. (A)
8. (C) 18. (A) 28. (D) 38. (B) 48. (C) 58. (C) 68. (C) 78. (B) 88. (D) 98. (C)
9. (D) 19. (D) 29. (D) 39. (C) 49. (A) 59. (C) 69. (B) 79. (D) 89. (C) 99. (C)
10. (C) 20. (A) 30. (D) 40. (D) 50. (A) 60. (C) 70. (A) 80. (C) 90. (D) 100.(A)

PMT - 1995

1. (D) 11. (B) 21. (C) 31. (D) 41. (B) 51. (C) 61. (D) 71. (C) 81. (D) 91. (A)
2. (B) 12. (A) 22. (A) 32. (B) 42. (C) 52. (D) 62. (D) 72. (B) 82. (C) 92. (B)
3. (C) 13. (D) 23. (B) 33. (C) 43. (B) 53. (D) 63. (A) 73. (B) 83. (A) 93. (C)
4. (A) 14. (A) 24. (A) 34. (D) 44. (D) 54. (A) 64. (A) 74. (B) 84. (C) 94. (B)
5. (B) 15. (A) 25. (C) 35. (C) 45. (D) 55. (C) 65. (D) 75. (D) 85. (D) 95. (C)
6. (A) 16. (C) 26. (A) 36. (A) 46. (B) 56. (B) 66. (C) 76. (C) 86. (C) 96. (B)
7. (C) 17. (B) 27. (A) 37. (C) 47. (B) 57. (A) 67. (D) 77. (B) 87. (D) 97. (A)
8. (C) 18. (A) 28. (C) 38. (B) 48. (C) 58. (D) 68. (C) 78. (C) 88. (B) 98. (A)
9. (D) 19. (A) 29. (B) 39. (B) 49. (C) 59. (A) 69. (B) 79. (B) 89. (D) 99. (D)
10. (C) 20. (A) 30. (A) 40. (A) 50. (B) 60. (B) 70. (B) 80. (D) 90. (B) 100.(A)

PMT - 1996

1. (C) 11. (A) 21. (C) 31. (D) 41. (A) 51. (C) 61. (A) 71. (B) 81. (C) 91. (D)
2. (D) 12. (D) 22. (B) 32. (C) 42. (D) 52. (D) 62. (D) 72. (C) 82. (B) 92. (C)
3. (A) 13. (D) 23. (B) 33. (A) 43. (D) 53. (A) 63. (A) 73. (A) 83. (B) 93. (D)
4. (D) 14. (A) 24. (C) 34. (D) 44. (C) 54. (C) 64. (D) 74. (C) 84. (B) 94. (B)
5. (C) 15. (D) 25. (A) 35. (D) 45. (C) 55. (C) 65. (D) 75. (D) 85. (D) 95. (C)
6. (A) 16. (C) 26. (D) 36. (A) 46. (C) 56. (D) 66. (B) 76. (D) 86. (A) 96. (C)
7. (C) 17. (D) 27. (B) 37. (D) 47. (C) 57. (B) 67. (B) 77. (C) 87. (D) 97. (B)
8. (C) 18. (B) 28. (B) 38. (C) 48. (B) 58. (D) 68. (D) 78. (B) 88. (B) 98. (C)
9. (A) 19. (A) 29. (C) 39. (B) 49. (C) 59. (B) 69. (C) 79. (B) 89. (C) 99. (B)
10. (A) 20. (C) 30. (B) 40. (C) 50. (D) 60. (B) 70. (A) 80. (C) 90. (C) 100.(C)

PMT - 1997

1. (B) 11. (A) 21. (C) 31. (C) 41. (B) 51. (D) 61. (D) 71. (C) 81. (D) 91. (B)
2. (B) 12. (D) 22. (C) 32. (B) 42. (D) 52. (A) 62. (C) 72. (D) 82. (C) 92. (B)
3. (A) 13. (A) 23. (D) 33. (B) 43. (B) 53. (A) 63. (A) 73. (D) 83. (D) 93. (B)
4. (B) 14. (D) 24. (B) 34. (B) 44. (A) 54. (C) 64. (C) 74. (B) 84. (A) 94. (A)
5. (B) 15. (D) 25. (C) 35. (D) 45. (A) 55. (D) 65. (B) 75. (A) 85. (D) 95. (C)
6. (B) 16. (A) 26. (D) 36. (C) 46. (D) 56. (B) 66. (A) 76. (D) 86. (D) 96. (C)
7. (A) 17. (C) 27. (D) 37. (A) 47. (C) 57. (A) 67. (D) 77. (C) 87. (C) 97. (A)
8. (C) 18. (A) 28. (C) 38. (D) 48. (D) 58. (C) 68. (B) 78. (D) 88. (C) 98. (B)
9. (C) 19. (D) 29. (D) 39. (D) 49. (C) 59. (B) 69. (A) 79. (A) 89. (D) 99. (D)
10. (B) 20. (B) 30. (A) 40. (D) 50. (D) 60. (A) 70. (D) 80. (A) 90. (C) 100.(B)

PMT - 1999

1. (A)	11. (B)	21. (B)	31. (A)	41. (D)	51. (B)	61. (D)	71. (A)	81. (B)	91. (D)
2. (C)	12. (B)	22. (B)	32. (A)	42. (C)	52. (D)	62. (A)	72. (A)	82. (D)	92. (C)
3. (B)	13. (D)	23. (D)	33. (B)	43. (C)	53. (B)	63. (B)	73. (D)	83. (B)	93. (B)
4. (A)	14. (B)	24. (B)	34. (B)	44. (C)	54. (C)	64. (A)	74. (C)	84. (B)	94. (A)
5. (B)	15. (C)	25. (A)	35. (C)	45. (B)	55. (C)	65. (B)	75. (B)	85. (A)	95. (B)
6. (D)	16. (A)	26. (C)	36. (C)	46. (C)	56. (C)	66. (A)	76. (C)	86. (B)	96. (B)
7. (D)	17. (A)	27. (A)	37. (A)	47. (A)	57. (D)	67. (C)	77. (B)	87. (B)	97. (B)
8. (A)	18. (D)	28. (C)	38. (D)	48. (C)	58. (B)	68. (A)	78. (D)	88. (C)	98. (A)
9. (D)	19. (C)	29. (C)	39. (A)	49. (C)	59. (B)	69. (B)	79. (B)	89. (C)	99. (A)
10. (A)	20. (A)	30. (B)	40. (A)	50. (D)	60. (A)	70. (D)	80. (C)	90. (C)	100.(A)

PMT - 2000

1. (A)	11. (D)	21. (C)	31. (A)	41. (D)	51. (B)	61. (C)	71. (B)	81. (B)	91. (C)
2. (B)	12. (A)	22. (A)	32. (A)	42. (D)	52. (A)	62. (B)	72. (C)	82. (C)	92. (C)
3. (A)	13. (A)	23. (A)	33. (D)	43. (A)	53. (D)	63. (B)	73. (A)	83. (A)	93. (C)
4. (C)	14. (A)	24. (B)	34. (C)	44. (B)	54. (D)	64. (A)	74. (A)	84. (A)	94. (C)
5. (A)	15. (A)	25. (C)	35. (C)	45. (C)	55. (D)	65. (B)	75. (C)	85. (B)	95. (C)
6. (A)	16. (A)	26. (B)	36. (A)	46. (A)	56. (C)	66. (A)	76. (C)	86. (A)	96. (A)
7. (D)	17. (A)	27. (C)	37. (B)	47. (D)	57. (A)	67. (C)	77. (B)	87. (C)	97. (C)
8. (A)	18. (B)	28. (A)	38. (C)	48. (A)	58. (C)	68. (D)	78. (B)	88. (A)	98. (B)
9. (A)	19. (B)	29. (A)	39. (B)	49. (C)	59. (A)	69. (C)	79. (D)	89. (B)	99. (D)
10. (C)	20. (A)	30. (C)	40. (D)	50. (B)	60. (D)	70. (C)	80. (D)	90. (D)	100.(A)

PMT - 2001

1. (C)	11. (A)	21. (D)	31. (D)	41. (C)	51. (C)	61. (A)	71. (A)	81. (D)	91. (D)
2. (C)	12. (C)	22. (D)	32. (C)	42. (A)	52. (C)	62. (B)	72. (B)	82. (C)	92. (A)
3. (D)	13. (B)	23. (B)	33. (A)	43. (B)	53. (A)	63. (C)	73. (C)	83. (A)	93. (A)
4. (A)	14. (D)	24. (C)	34. (C)	44. (B)	54. (A)	64. (D)	74. (D)	84. (D)	94. (C)
5. (B)	15. (C)	25. (B)	35. (B)	45. (D)	55. (A)	65. (D)	75. (C)	85. (B)	95. (C)
6. (B)	16. (C)	26. (C)	36. (B)	46. (B)	56. (A)	66. (B)	76. (B)	86. (B)	96. (C)
7. (B)	17. (D)	27. (C)	37. (D)	47. (D)	57. (A)	67. (D)	77. (A)	87. (D)	97. (B)
8. (C)	18. (A)	28. (A)	38. (D)	48. (A)	58. (D)	68. (C)	78. (B)	88. (D)	98. (A)
9. (D)	19. (A)	29. (D)	39. (C)	49. (D)	59. (C)	69. (C)	79. (A)	89. (B)	99. (A)
10. (D)	20. (C)	30. (C)	40. (B)	50. (B)	60. (C)	70. (A)	80. (D)	90. (C)	100. (B)

PMT - 2002

1. (D)	11. (A)	21. (D)	31. (A)	41. (B)	51. (B)	61. (B)	71. (D)	81. (B)	91. (A)
2. (C)	12. (B)	22. (C)	32. (A)	42. (B)	52. (B)	62. (A)	72. (D)	82. (C)	92. (D)
3. (A)	13. (C)	23. (D)	33. (D)	43. (C)	53. (C)	63. (B)	73. (C)	83. (A)	93. (B)
4. (D)	14. (B)	24. (B)	34. (A)	44. (D)	54. (C)	64. (C)	74. (D)	84. (B)	94. (B)
5. (B)	15. (C)	25. (C)	35. (A)	45. (C)	55. (C)	65. (D)	75. (B)	85. (C)	95. (A)
6. (A)	16. (A)	26. (D)	36. (C)	46. (A)	56. (B)	66. (C)	76. (A)	86. (A)	96. (B)
7. (A)	17. (A)	27. (B)	37. (A)	47. (D)	57. (A)	67. (D)	77. (C)	87. (A)	97. (B)
8. (C)	18. (C)	28. (A)	38. (B)	48. (D)	58. (C)	68. (C)	78. (A)	88. (A)	98. (D)
9. (B)	19. (C)	29. (B)	39. (D)	49. (B)	59. (C)	69. (A)	79. (A)	89. (D)	99. (A)
10. (A)	20. (B)	30. (B)	40. (C)	50. (A)	60. (B)	70. (B)	80. (B)	90. (D)	100.(A)

PMT - 2003

1. (D)	11. (A)	21. (B)	31. (D)	41. (A)	51. (D)	61. (B)	71. (C)	81. (D)	91. (A)
2. (C)	12. (C)	22. (B)	32. (A)	42. (D)	52. (A)	62. (D)	72. (C)	82. (C)	92. (A)
3. (B)	13. (C)	23. (B)	33. (A)	43. (D)	53. (B)	63. (C)	73. (C)	83. (C)	93. (C)
4. (B)	14. (B)	24. (B)	34. (B)	44. (C)	54. (C)	64. (A)	74. (B)	84. (C)	94. (B)
5. (B)	15. (D)	25. (A)	35. (A)	45. (D)	55. (C)	65. (B)	75. (C)	85. (C)	95. (C)
6. (C)	16. (B)	26. (D)	36. (D)	46. (B)	56. (D)	66. (A)	76. (A)	86. (B)	96. (C)
7. (D)	17. (C)	27. (C)	37. (B)	47. (B)	57. (D)	67. (C)	77. (C)	87. (B)	97. (D)
8. (B)	18. (C)	28. (A)	38. (C)	48. (D)	58. (B)	68. (C)	78. (B)	88. (D)	98. (B)
9. (C)	19. (B)	29. (D)	39. (A)	49. (C)	59. (B)	69. (A)	79. (B)	89. (A)	99. (B)
10. (D)	20. (B)	30. (C)	40. (B)	50. (D)	60. (A)	70. (D)	80. (D)	90. (A)	100.(D)

PMT - 2004

1. (A)	11. (A)	21. (B)	31. (D)	41. (D)
2. (A)	12. (C)	22. (A)	32. (C)	42. (D)
3. (A)	13. (C)	23. (B)	33. (A)	43. (B)
4. (C)	14. (D)	24. (D)	34. (D)	44. (B)
5. (D)	15. (C)	25. (A)	35. (B)	45. (D)
6. (B)	16. (C)	26. (B)	36. (B)	46. (D)
7. (D)	17. (D)	27. (B)	37. (C)	47. (C)
8. (C)	18. (A)	28. (D)	38. (C)	48. (C)
9. (B)	19. (C)	29. (C)	39. (B)	49. (B)
10. (A)	20. (B)	30. (B)	40. (C)	50. (D)

PET - 1993

1. Which of the following sets of elements has the strongest tendency to form anions ?
(A) As, Se, Br (B) Sb, Te, I (C) N, O, F (D) P, S, Cl

2. Electron affinity is the
(A) Energy required to take out an electron from an isolated gaseous atom
(B) Energy absorbed when an electron is added to an isolated atom in the gaseous state
(C) Energy released when an electron is added to an isolated atom in the gaseous state
(D) Power of an atom to attract an electron to itself

3. The heat of neutralization of a strong acid by a strong base is nearly equal to
(A) +57.32 J (B) −57.32 kJ (C) +13.7 J (D) −13.7 kJ

4. In exothermic reactions
(A) ΔE is zero (B) ΔH is negative
(C) ΔS is zero (D) ΔG is zero

5. Decrease of free energy of a reacting system indicates to a/an
(A) Exothermic reaction (B) Equilibrium reaction
(C) Spontaneous reaction (D) Slow reaction

6. For the reaction $H_2(g) + Br_2(g) \rightarrow 2HBr(g)$, the experimental data suggest, rate = $k[H_2][Br_2]^{1/2}$ The molecularity and order of the reaction are respectively
(A) $2, \frac{3}{2}$ (B) $\frac{3}{2}\ \frac{3}{2}$ (C) 1, 1 (D) $1, \frac{1}{2}$

7. The rate of the reaction,
$CCl_3CHO + NO \rightarrow CHCl_3 + NO + CO$ is given by equation, rate = k $[CCl_3CHO]$ [NO]. If concentrations are expressed in moles/litre the units of k are
(A) Litre $mole^{-1}$ sec^{-1} (B) $Second^{-1}$
(C) $Litre^2$ $mole^{-2}$ sec^{-1} (D) Mole $litre^{-1}$ sec^{-1}

8. The name of compound having the formula, $ClCH_2CH_2COOH$, is
(A) 3-chloropropanoic acid (B) 2-chloropropanoic acid
(C) 2-chloroethanoic acid (D) Chlorosuccinic acid

9. The structural formula of isobutyl chloride is
(A) $CH_3CH_2CH_2CH_2Cl$ (B) $(CH_3)_2CH_2CH_2Cl$
(C) $CH_3CH_2CHClCH_3$ (D) $(CH_3)_3$ C–Cl

10. An isomer of ethanol is
(A) Ethanal (B) Methanol
(C) Diethly ether (D) Dimethly ether

11. The shape of ethylene molecule is
(A) Tetrahedral (B) Pyramidal (C) Planar (D) Linear

12. Halogenation of alkanes is an example of
(A) Electrophilic substitution
(B) Nucleophilic substitution
(C) Free-radical substitution (D) Oxidation

13. Carbon black, which is used in making printers ink, is obtained by decomposition of
(A) Acetylene (B) Benzene
(C) Carbon tetrachloride (D) Methane

14. The end product C, of the following reaction sequence is

$$CaO + C \xrightarrow{heat} A \xrightarrow{H_2O} B \xrightarrow[HgSO_4]{dill\ H_2SO_4} C$$

(A) Acetaldehyde (B) Ethanol
(C) Ethylene (D) Ethyl hydrogen sulphate

15. For reaction, $2A + B \rightarrow$ Products, the active mass of B is kept constant and that of A is doubled. The rate of reaction will then
(A) Decrease 2 times (B) Decrease 4 times
(C) Increase 2 times (D) Increase 4 times

16. In a reaction, $2A + B \rightarrow A_2B$, the reactant A will disappear at
(A) Twice the rate that B will decrease
(B) The same rate that A_2B will from
(C) Half the rate that B will decease
(D) The same rate that B will decrease

17. The addition of tetraethyl lead to petrol
(A) Lowers its octane number
(B) Raises its octane number
(C) May raise or lower the octane number
(D) Has no ettect on octane number

18. One Faraday of electricity will liberate one gram atom of the metal from a solution of
(A) $AuCl_3$ (B) $BaCl_2$ (C) $CuSO_4$ (D) NaCl

19. EMF of the cell Ni / Ni^{2+} (1.0 M) // Au^{3+} (1.0 M) / Au (where E° for Ni^{2+}/Ni is −0.25 V; E° for Au^{3+}/Au is 1.50 V) is
(A) +1.75 V (B) +4.0 V (C) +1.25 V (D) −1.75 V

20. When lead storage battery is charged
(A) Lead dioxide dissolves
(B) Sulphuric acid is regenerated
(C) The lead electrode becomes coated with lead sulphate
(D) The amount of sulpuric acid decreases

21. Which of the following statements is not correct ?
(A) Physical adsorption is reversible
(B) Physical adsorption is due to van der Waal's forces
(C) Chemical adsorption decreases at high temperature and low pressure
(D) Adsorption energy for a chemical adsorption is generally greater than that of physical adsorption

22. The main structural feature of proteins is
(A) Ester linkage (B) Ether linkage
(C) Peptide linkage (D) Ketone group

23. On complete hydrolysis of starch, we finally get
(A) Glucose and fructose (B) Sucrose
(C) Glucose (D) Fructose

24. Enzymes are
(A) Living organisms (B) Dead organisms
(C) Complex nitrogenous substances produced in living cells

(D) None of these

25. The most important food reserves of animals and plants are
(A) Carbohydrates (B) Proteins
(C) Vitamins (D) Fats

26. Which of the following pairs is not correctly matched ?
(A) Effective atomic number of Pt in $[PtCl_6]^{2-}$ – 84
(B) Absorption peak for $[Cr^{III}(NH_3)_6]^{3+}$ is at $21680 cm^{-1}$
(C) Crystal field stabilization energy of d^2 in weak ligand field = (–) 0.8 Δ_o
(D) Example of weak ligand field for d^5 configuration – M $[Mn^{II}F_6]^{4-}$

27. Which of the following is not true for metal carbonyls ?
(A) Metal carbonyls are single bonded species.
(B) $d_\pi - p_\pi$ overlap is observed in metal carbonyls
(C) The oxidation state of the metal in the carbonyls is zero
(D) The secondary carbonyls are obtained from photo-decomposition

28. In which of the following compounds,Manganese achieves its highest oxidation state ?
(A) $KMnO_4$ (B) K_2MnO_4 (C) MnO_3 (D) Mn_3O_4

29. Which of the following is not true for transition metals ?
(A) They crystallize with body-centered cubic hexagonal close-packed structures only
(B) They are malleable and ductile
(C) They have high boiling and melting points
(D) They show variable oxidation states, although not always

30. Which of the following is not true for ligand-metal complex ?
(A) Larger the permanent dipole moment of ligand, the more stable is the bond
(B) Larger the ligand, the more stable is the metal-ligand bond
(C) Highly charged ligand forms stronger bond
(D) Greater the ionization potential of central metal, the stronger is the bond

31. Which one of the isomeric-pairs shows ionization isomerism ?
(A) $[Co(NH_3)_6] [Cr(CN)_6]$ and $[Cr(NH_3)_6] [Co(CN)_6]$
(B) $[Cr(H_2O)_6] Cl_3$ and $[Cr(H_2O)_5Cl]Cl_2.H_2O$
(C) $[Pt(NH_3)_2Cl_2]$ and $[(Pt(NH_3)_5Br] [Pt Cl_4]$
(D) $[Co(NH_3)_5Br] SO_4$ and $[Co(NH)_3)_5SO_4] Br$

32. The substance used in the thermite process of reducing metal ores is
(A) Aluminium (B) Thorium
(C) Heated platinum (D) Carbon

33. Which of the following do not respond to the borax bead test ?
(A) Cobalt salts (B) Copper salts
(C) Aluminium salts (D) Nickel salts

34. The brown ring test is performed for the qualitative detection of
(A) Bromides (B) Iodides
(C) Nitrates (D) Phosphates

35. AgCl precipitate dissolves in ammonia due to the formation of
(A) $[Ag(NH_3)_2]OH$ (B) $[Ag(NH_3)_2]Cl$
(C) $[Ag(NH_4)_2]OH$ (D) $[Ag(NH_4)_2]Cl$

36. Which of the following statements is wrong ?
(A) pH + pOH = 14 at room temperature
(B) The commonest primary standards for standardization of acids are Na_2CO_3 or borax
(C) $[H^+] [OH^-] = 10^{-14}$ at 22°C
(D) Alkalies are generally standardized against acetic acid or borax

37. In the metallurgy of zinc, the zinc dust obtained from roasting and reduction of zinc sulphide contains some ZnO. How is this removed ?
(A) Absorbance of UV light and re-emission of white light is employed
(B) Shock-cooling by contact with a shower of molten lead is done
(C) X-ray method is used (D) Smelting is employed

38. In the Down's process of obtaining sodium, $CaCl_2$ is added.
(A) To lower the operating temperature
(B) To act as a reducing agent
(C) To get high yield of sodium
(D) To improve the quality of sodium obtained

39. In the periodic table, the metallic character of elements.
(A) Decreases from left to right across a period and on descending a group
(B) Decreases from left to right across a period and increases on descending a group
(C) Increases from left to right across a period and on descending a group
(D) Increases from left to right across a period and decreases on descending a group

40. Among the following, a natural polymer is
(A) Cellulose (B) P.V.C.
(C) Teflon (D) Polyethylene

41. Which of the following can be used as monomer in a polymerization reaction ?
(A) CH_3CH_2Cl (B) CH_3CH_2OH (C) C_6H_6 (D) C_3H_6

42. The reaction of benzene with chlorine in the presence of iron gives.
(A) Benzyl chloride (B) Benzoyl chloride
(C) Benzene hexachloride (D) Chlorobenzene

43. When phenol is treated with excess of bromine water, it gives
(A) Bromophenol (B) 2,4,6-Tribromophenol
(C) m-Bromophenol (D) O-and-p- Bromophenol

44. Nitro group in nitrobenzene is
(A) Ortho-directing (B) Meta-directing
(C) Para-directing (D) Ortho-and para-directing

45. A metallic element has a cubic lattice. Each edge of the unit cell is 2 Å. The density of the metal is $2.5 gcm^{-3}$. The number of unit cells in 200 g of the metal are
(A) 1×10^{25} (B) 1×10^{24} (C) 1×10^{22} (D) 1×10^{20}

46. $^{231}_{89}Ac$ gives $^{207}_{82}Pb$ after emission of some α and β particles. The number of such α and β particles are respectively
(A) 5, 6 (B) 6, 5 (C) 7, 5 (D) 5, 7

47. The decay constant of a radioactive element is 3×10^{-6} min^{-1}. Its half life is
(A) 2.31×10^5 min. (B) 2.31×10^6 min.
(C) 2.31×10^{-6} min. (D) 2.31×10^{-7} min.

48. The $^{14}_{6}C$ in upper atmosphere is generated by the nuclear reaction
(A) $^{14}_{7}N + ^{1}_{1}H \longrightarrow ^{14}_{6}C + ^{0}_{+1}e + ^{1}_{1}H$
(B) $^{14}_{7}N \longrightarrow ^{14}_{6}C + ^{0}_{+1}e$
(C) $^{14}_{7}N + ^{1}_{0}n \longrightarrow ^{14}_{6}C + ^{1}_{1}H$
(D) $^{14}_{7}N + ^{3}_{1}H \longrightarrow ^{14}_{6}C + ^{4}_{2}He$

49. An equilibrium mixture for the reaction,
$$2H_2S(g) \rightleftharpoons 2H_2(g) + S_2(g)$$
had 0.5 mole H_2S, 0.1 mole of H_2 0.4 mole S_2 in a one litre vessel. The equilibrium constant of this reaction is given by
(A) 0.016 mole $litre^{-1}$ (B) 0.160 mole $litre^{-1}$
(C) 0.004 mole $litre^{-1}$ (D) 0.080 mole $litre^{-1}$

50. What conditions will favour the exothermic ammonia synthesis reaction, $N_2(g) + 3H_2(g) \rightleftharpoons 2NH_3(g)$
(A) High temperature, high pressure
(B) High temperature, low pressure
(C) Low temperature, high pressure
(D) Low temperature, low pressure

51. Which of the following is most effective in coagulating a ferric hydroxide sol ?
(A) KCl (B) KNO_3 (C) K_2SO_4 (D) $K_3[Fe(CN)_6]$

52. The froth-floatation process is used for the concentration of
(A) Phosphate ores (B) Sulphide ores
(C) Halide ores (D) Oxide ores

53. In the manufacture of iron from haematite, limestone is added to act as
(A) An oxidizing agent (B) A reducing agent
(C) Flux (D) Slag

54. The interatomic distances in H_2 and Cl_2 molecules are 74 and 198 pm respectively. The bond length of HCl is
(A) 272 pm (B) 136 pm (C) 124 pm (D) 248 pm

55. Which of the following liquid-pairs shows a positive deviation from Raoult's law ?
(A) Acetone-chloroform (B) Benzene-methanol
(C) Water-hydrochloric acid (D) Water-nitric acid

56. The molarity of a solution prepared by dissolving 7.1 g of Na_2SO_4 in 100 ml water is
(A) 0.5 M (B) 0.05 M (C) 2.0 M (D) 1.0 M

57. Acetic acid dissolved in benzene shows a molecular mass of
(A) 30 (B) 60 (C) 120 (D) 240

58. On dissolving 1 mole of each of the following acids in 1 litre water, the acid which does not give a solution of strength 1 N is
(A) Nitric acid (B) Phosphoric acid
(C) Hydrochloric acid (D) Perchloric acid

59. The reaction of an aldehyde with hydroxylamine gives a product which is called
(A) Semicarbazone (B) Oxime
(C) Aminohydroxide (D) Hydrazone

60. When acetaldehyde is heated with Fehling solution, it gives a red precipitate of
(A) Cu (B) CuO (C) Cu_2O (D) $Cu(OH)_2$

61. Calcium acetate on heating gives
(A) Acetic anhydride (B) Acetone
(C) Acetaldehyde (D) Methane

62. The two gaseous equilibria $SO_2(g)+\frac{1}{2}O_2(g) \rightleftharpoons SO_3(g)$ and $2SO_3(g) \rightleftharpoons SO_2(g)+O_2(g)$ have equlibrium constants K_1 and K_2 respectively at 298°K. Which of the following relationships between K_1 and K_2 is correct ?
(A) $K_1 = K_2$ (B) $K_2 = K_1^2$ (C) $K_2 = 1/K_1^2$ (D) $K_2 = 1/K_1$

63. Which of the following solutions will have pH close to 1.0 ?
(A) 100 ml of $\frac{M}{10}$ HCl + 100 ml of $\frac{M}{10}$ NaOH
(B) 55 ml of $\frac{M}{10}$ HCl + 45ml of $\frac{M}{10}$ NaOH
(C) 10 ml of $\frac{M}{10}$ HCl + 90 ml of $\frac{M}{10}$ NaOH
(D) 75 ml of $\frac{M}{5}$ HCl + 25 ml of $\frac{M}{5}$ NaOH

64. Colloidal solutions are purified by
(A) Coagulation (B) Dialysis
(C) Electro-osmosis (D) Filtration

65. How many grams of CaC_2O_4 will dissolve in distilled water to make one litre of saturated solution? Solubility product of CaC_2O_4 is 2.5×10^{-9} $mole^2$ $litre^{-2}$ and its molecular weight is 128.
(A) 0.0064 g. (B) 0.0128 g. (C) 0.0032 g. (D) 0.0640 g.

66. Acetamide is treated separately with the following reagents. Which one of these would give methyl amine ?
(A) Conc. H_2SO_4 (B) Soda lime
(C) PCl_5 (D) NaOH + Br_2

67. Alkaline hydrolysis of an ester is called
(A) Polymerisation (B) Saponification
(C) Neutralisation (D) Esterification

68. A nauseating smell in the carbylamine test for primary

amines is due to the formation of
(A) Isocyanide (B) Chloroform
(C) Cyanide (D) D.D.T.

69. Ethylamine reacts with nitrous acid to give off
(A) Ethane (B) Nitrogen
(C) Ammonia (D) Nitrous oxide

70. The addition of HCl will not suppress the ionization of
(A) Acetic acid (B) Benzoic acid
(C) H_2S (D) Sulphuric acid

71. In the crystal of CsCl the nearest neighbours of each Cs ion are
(A) Six Cs ions (B) Eight Cs ions
(C) Six chloride ions (D) Eight chloride ions

72. The number of atoms in a face centred cubic unit cell of a monoatomic substance is
(A) 2 (B) 4 (C) 6 (D) 8

73. A match box exhibits
(A) Cubic geometry (B) Monoclinic geometry
(C) Orthorhombic geometry (D) Tetragonal geometry

74. Which of the following statements is not correct for a noble gas ?
(A) Half-life of Rn is only 3-8 days
(B) Ar is used in electric bulb
(C) Kr is obtained during radioactive disintegration
(D) He is used in producing very low temperatures

75. The highest magnetic moments is shown by the transition metal ion with the outer electronic configuration
(A) $3d^2$ (B) $3d^5$ (C) $3d^7$ (D) $3d^9$

76. Which one of the following transition metals exhibits the highest oxidation state ?
(A) Pd (B) Os (C) Cr (D) Mn

77. For an endothermic reaction where ΔH represents the enthalpy of the reaction in KJ/mol, the minimum value for the energy of activation will be
(A) Less than ΔH (B) Zero
(C) More than ΔH (D) Equal to ΔH

78. When ice melts into water, entropy
(A) Becomes zero (B) Decreases
(C) Increases (D) Remains the same

79. The ionization energy of nitrogen is more than that of oxygen because
(A) Nitrogen contains less number of electrons
(B) Nitrogen is less electronegative
(C) Nitrogen has half-filled p-orbitals
(D) Nitrogen is left to the oxygen in the same period of the periodic table

80. Which of the following is the biggest ion ?
(A) Mg^{+2} (B) Na^+ (C) Al^{+3} (D) Ba^{+2}

81. Aluminium is diagonally related to (in periodic table)
(A) Li (B) C (C) B (D) Be

82. Which of the following does not illustrate the anomalous properties of lithium ?
(A) The melting point and boiling point of Li are comparatively high
(B) Li is much softer than the other Group 1 metals
(C) Li forms a nitride Li_3N, unlike Group 1 metals
(D) The ion of Li and its compounds are more heavily hydrated than those of the rest of the group

83. The right order of the solubility of sulphates of alkaline earth metals in water is
(A) Be > Mg > Ca > Sr > Ba
(B) Be > Ca > Mg > Ba > Sr
(C) Mg > Be > Ba > Ca > Sr
(D) Mg > Ca > Ba > Be > Sr

84. Which of the following fluorides does not exist ?
(A) AsF_5 (B) SbF_5 (C) NF_5 (D) PF_5

85. The highest catenation ability is shown by
(A) Oxygen (B) Sulphur (C) Selenium (D) tellurium

86. What is the value of carbonate hardness of water sample if 100 ml. of it took 5.00 ml. of 0.09 N HCl solution? (Na_2CO_3 = 106)
(A) 4.50 mg – eq / litre (B) 447.00 mg – eq/litre
(C) 0.042 mg – eq/litre (D) 1.80 mg – eq/litre

87. The electronic configuration (outermost) of Mn^{2+} ion (atomic no. of Mn 25) in its ground state is
(A) $3d^3\ 4s^2$ (B) $3d^2\ 4s^2\ 4p^2$ (C) $3d^5\ 4s^0$ (D) $3d^4\ 4s^1$

88. If electron; hydrogen, helium and neon nuclei are all moving with the velocity of light, wavelengths associated with these particles are in the order
(A) Electron > hydrogen > helium > neon
(B) Electron > helium > hydrogen > neon
(C) Electron < helium < hydrogen < neon
(D) Neon < hydrogen < helium < electron

89. The maximum number of hydrogen bonds formed by a water molecule in ice is
(A) 4 (B) 3 (C) 2 (D) 1

90. The number of anti-bonding electrons in O_2^{2-} is
(A) 4 (B) 6 (C) 8 (D) 10

91. The hybridization of carbons of C–C single bond of $HC \equiv C–CH = CH_2$ is
(A) $sp^3 - sp^3$ (B) $sp^2 - sp^2$ (C) $sp^3 - sp$ (D) $sp - sp^2$

92. The molecule which has the largest dipole moment is
(A) CHI_3 (B) CCl_4 (C) CH_4 (D) $CHCl_3$

93. When chloroform is exposed to light and air it gives
(A) Chlorine gas (B) Carbon tetrachloride
(C) Methyl chloride (D) Phosgene gas

94. Hydrogen bonding is maximum in
(A) Ethanol (B) Diethyl ether
(C) Ethyl acetate (D) Triethyl amine

95. Ca has atomic no. 20 and atomic weight 40 which of the following statements is not correct about Ca atom ?
(A) The number of electrons is same as the number of neutrons
(B) The number of nucleons is double of number of

electrons

(C) The number of protons is half of the number of neutrons

(D) The number of nucleons is double of the atomic number

96. The four quantum numbers of the outermost electron of K (atomic no. 19) are

(A) $n = 2, l = 0, m = 0, s = +\frac{1}{2}$

(B) $n = 4, l = 0, m = 0, s = \frac{1}{2}$

(C) $n = 3, l = 1, m = 1, s = +\frac{1}{2}$

(D) $n = 4, l = 2, m = -1, s = +\frac{1}{2}$

97. The most acidic compound among the following is

(A) CH_3COOH (B) $CH_3CH_2CH_2OH$

(C) CH_3CH_2OH (D) C_6H_5OH

98. Which of the following pairs is not correctly matched ?

(A) A halogen which is liquid at room temperature-bromine

(B) The most electronegative element-fluorine

(C) The most reactive halogen-fluorine

(D) The strongest oxidizing halogen-iodine

99. An organic compound contains 40% carbon and 6.66% hydrogen. The empirical formula of the compound is

(A) CHO (B) CHO_2 (C) CH_2 (D) CH_2O

100. Insulin contains 3.4% sulphur. The minimum molecular weight of insulin is

(A) 350 (B) 470 (C) 560 (D) 940

PET - 1994

1. What is the emf of the cell, whose half cells are given below, $Mg^{2+} + 2e^- \rightleftharpoons Mg(s)$ E = −2.37 V

$Cu^{2+} + 2e^- \rightleftharpoons Cu(s)$ E = + 0.34 V

(A) 2.71V (B) 2.03V (C) - 2.30V (D) 1.336V

2. If the solubility products of AgCl and AgBr are 1.2×10^{-10} and 3.5×10^{-13} respectively, then the relation between the solubilities (denoted by the symbol 'S') of these salts can correctly be represented as

(A) S of AgBr is equal to that of AgCl

(B) S of AgBr is less than that of AgCl

(C) S of AgBr is greater then that of AgCl

(D) S of AgBr is 106 greater than that of AgCl

3. When an electron revolves in stationary orbit, then

(A) it absorbs energy (B) It gains kinetic energy

(C) It emits radiation (D) Its energy remains constant

4. A sample of petrol contains 22% n-heptane and 78% isooctane. Its octane number is

(A) 56 (B) 44 (C) 22 (D) 78

5. Which of the following compounds will give a positive haloform test and a positive Fehling test ?

(A) Acetone (B) Acetic acid

(C) Ethanol (D) Acetadehyde

6. The energy evolved is highest for which of the following reactions

(A) $F + e^- \rightarrow F^-$ (B) $Cl + e^- \rightarrow Cl^-$

(C) $S + 2e^- \rightarrow S^{2-}$ (D) $O + 2e^- \rightarrow O^{2-}$

7. Which of the following pentoses will be optically active

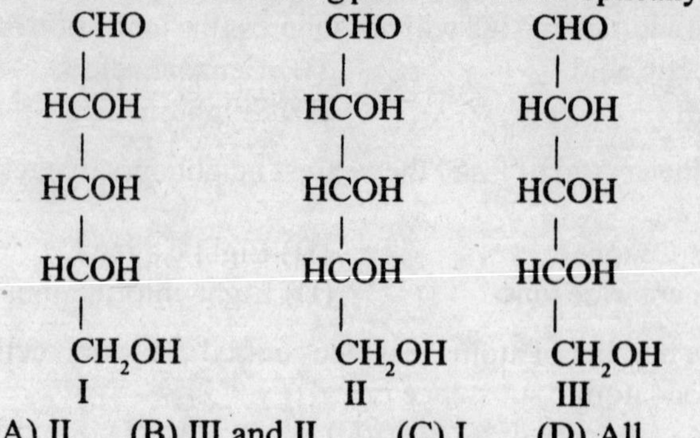

(A) II (B) III and II (C) I (D) All

8. For α-amino acids having the structure,

$R-CH(NH_2)-CO_2H$

Which of the following statements are true ?

(a) Water solubility is maximum at a pH when concentrations of anions and cations are equal

(b) They give ninhydrin test

(c) On reacting with nitrous acid give off N_2

(A) a and b (B) a (C) b and c (D) All

9. The linkage present in proteins and peptides is

(A) $-C(=O)-O-C(=O)-$ (B) $-C(=O)-O-$

(C) $-NH-$ (D) $-C(=O)-NH-$

10. The number of optical isomers of an organic compound having 'n' asymmetric carbon atoms will be

(A) 2^{n+1} (B) n^2 (C) 2^n (D) 2^{n-1}

11. The monomer or natural polymer rubber is

(A) Isoprene (B) Ethylene

(C) Vinyl chloride (D) Acrolein

12. Which of the following polymers will not catch fire

(A) $(CF_2 - CF_2)_n$ (B) $(CH_2 - CH_2)_n$

(C) $(CH(Cl) - CH(Cl))_n$ (D) $(CH_2 - CH(Cl))_n$

13. In the metallurgical extraction of zinc from ZnO, the reducing agent used is

(A) Carbon dioxide (B) Nitric oxide

(C) Carbon monoxide (D) Sulphur dioxide

14. Cupellation is used for the purification of

(A) Na (B) Ag (C) Pb (D) Al

15. Chemical reduction is not suitable for converting

(A) Bauxite into aluminium (B) Cuprite into copper

(C) Haematite into iron (D) Zinc oxide into zinc

16. A molal solution is one that contains one mole of solute in
(A) 1000 grams of the solvent
(B) One litre of the solution
(C) One litre of the solvent (D) 22.4 litres of the solution

17. A cup is plated with silver by passing 965 Coulombs of electricity. The amount of silver deposited is
(A) 1.0787 gms (B) 1.002 gms
(C) 9.65 gms (D) 107.87 gms

18. In which of the following gaseous equilibria, K_P is less than K_C ?
(A) $2SO_2 + O_2 \rightleftharpoons SO_3$ (B) $2HI \rightleftharpoons H_2 + I_2$
(C) $N_2O_4 \rightleftharpoons 2\ NO_2$ (D) $N_2 + O_2 \rightleftharpoons 2NO$

19. When a radioactive element emits a β-particle the daughter element formed will have its
(A) Mass number one unit less
(B) Atomic number one unit more
(C) Atomic number one unit less
(D) Mass number one unit more

20. The total number of electrons present in all the p-orbitals of bromine are
(A) Five (B) Eighteen
(C) Seventeen (D) Thirty-five

21. The reaction $2ZnS + 3O_2 \rightarrow 2ZnO + 2SO_2$ in the metallurgical process of zinc is called
(A) Calcination (B) Cupellation
(C) Smelting (D) Roasting

22. In which of the following, ionic radius of chromium would be smallest ?
(A) $CrCl_3$ (B) CrF_3 (C) $K_2Cr_2O_4$ (D) CrO_2

23. Which of the following has highest first ionization energy ?
(A) Sulphur (B) Oxygen (C) Nitrogen (D) Phosphorus

24. The electronic configuration, $1s^2, 2s^22p^6, 3s^23p^6, 3d^6$ corresponds to
(A) Co (B) Ge (C) Mn^{2+} (D) Fe^{2+}

25. Which of the following mixtures in aqueous solution acts as a buffer ?
(A) $H_2SO_4 + K_2SO_4$ (B) $NH_4OH + NH_4Cl$
(C) $CH_3COOH + NaCl$ (D) $HNO_3 + KNO_3$

26. The decomposition temperatures is maximum for
(A) $BaCO_3$ (B) $SrCO_3$
(C) $CaCO_3$ (D) $MgCO_3$

27. On Pauling scale which of the following does not have electronegativity ≥ 3.0
(A) Chlorine (B) Bromine (C) Oxygen (D) Nitrogen

28. Which of the following exhibits highest solubility in water?
(A) AsH_3 (B) SbH_3 (C) NH_3 (D) PH_3

29. In which of the following pairs, the difference between the covalent radii of the two metals is maximum ?
(A) K, Ca (B) Mn, Fe (C) Co, Ni (D) Cr, Mn

30. Which of the following dissociates to give H^+ most easily ?
(A) H_2Te (B) H_2Se (C) H_2O (D) H_2S

31. Which of the following oxides is most basic ?
(A) Na_2O (B) Al_2O_3 (C) SO_3 (D) SO_2

32. Which of the following oxidises H_2O to oxygen ?
(A) Bromine (B) Iodine (C) Chloride (D) Fluorine

33. Which of the following has density greater than water?
(A) Li (B) Na (C) K (D) Rb

34. Which of the following is not obtained by direct reaction of constituent elements ?
(A) XeO_3 (B) XeF_6 (C) XeF_2 (D) XeF_4

35. Which of the following is most soluble in water ?
(A) $MgSO_4$ (B) $CaSO_4$ (C) $SrSO_4$ (D) $BaSO_4$

36. Which of the following has highest boiling point ?
(A) AsH_3 (B) SbH_3 (C) NH_3 (D) PH_3

37. Which of the following statements is not correct about the electronic configuration of gaseous chromium atom ?
(A) It has 5 electrons in 3d and one electron in 4s orbitals
(B) The principal quantum numbers of its valence electrons are 3 and 4
(C) It has 6 electrons in 3d orbital
(D) Its valence electrons have quantum no. '*l*', 0 and 2

38. Which of the following statements is correct about equivalent weight of $KMnO_4$?
(A) It is one third of its molecular weight in alkaline medium
(B) It is one fifth of its molecular weight in alkaline medium
(C) It is equal to its molecular weight in acidic medium
(D) It is one third of its molecular weight in acidic medium

39. The reaction of $K_2Cr_2O_7$ with NaCl and conc. H_2SO_4 gives
(A) $CrO_2\ Cl_2$ (B) Cr_2O_3 (C) $CrCl_3$ (D) $CrOCl_2$

40. Which of the following compounds is colourless ?
(A) $Cu_2\ (CH_3COO)_4.\ H_2O$
(B) $CuSO_4.\ 5H_2O$ (C) Cu_2Cl_2
(D) $[Cu\ (NH_3)_4]\ SO_4.\ 4H_2O$

41. A current of 2.5 amperes was passed through a copper sulphate solution for 6 minutes 26 seconds. The amount of copper deposited is
(Atomic weight of copper = 63.5
1Faraday = 96500 Coulmbs)
(A) 6.35 gms (B) 0.3175 gms
(C) 0.635 gms (D) 3.175 gms

42. Which of the following properties is different for neutral atoms of the isotopes of the same elements ?
(A) Mass (B) Atomic number
(C) Number of electrons (D) General chemical reactions

43. Which of the following is a Lewis acid
(A) BF_3 (B) Cl^- (C) NH_3 (D) H_2O

44. A 500 ml solution of 2 M sodium hydroxide is added to a 500 ml solution of 2 M hydrochloric acid and the mixture is well stirred. The rise in temperature T_1 is

recorded. The experiment is then repeated using 250 ml of each solution and the rise in temperature T_2 is recorded. Then which of the following statements is correct

(A) $T_1 = T_2$ (B) $T_1 = 2T_2$
(C) $T_1 = 4T_2$ (D) $T_2 = 2T_1$

45. The hydrogen ion concentration of a 0.006 M benzoie acid solution is ($K_a = 6 \times 10^{-5}$)

(A) 0.6×10^{-4} (B) 6×10^{-4}
(C) 6×10^{-5} (D) 3.6×10^{-4}

46. The number of isomers possible for square planar complex, $K_2[PdClBr_2(SCN)]$ is

(A) 2 (B) 3 (C) 4 (D) 6

47. I.U.P.A.C. name of $[Co(NH_3)_3(H_2O)Cl]Cl_2$ is

(A) Chlorodiamminediaquocobalt(II)chloride
(B) Diamminediaquachlorocobalt(II)chloride
(C) Diaquachlorodiamminecobalt(III)chloride
(D) Triamminediaquochlorocobalt. (III)chloride

48. Which of the following exhibits highest molar conductivity ?

(A) $[Co(NH_3)_6]Cl_3$ (B) $[Co(NH_3)_5Cl]Cl_2$
(C) $[Co(NH_3)_4Cl_2]Cl$ (D) $[Co(NH_3)_3Cl_3]$

49. In tris(ethylene diamine)cobalt(III)chloride the coordination number of cobalt is

(A) 3 (B) 4 (C) 6 (D) 7

50. In the reaction $Ar - OH + Rx \xrightarrow{alkali} A$; A is

(A) An aldehyde (B) An aryl chloride
(C) An ether (D) A ketone

51. The synthetic polymer 'terylene' is a

(A) Polyether (B) Polythene (C) Polyester (D) Polyamide

52. An aromatic compound is insoluble in water, it dissolves in aq. NaOH but does not dissolve in aq. $NaHCO_3$. It gives deep red colouration with $FeCl_3$. The compound is

(A) An alcohol (B) A phenol
(C) An aldehyde (D) A ketone

53. Which of the following conditions is favourable for the production of ammonia by Haber process ?

(A) Continuous removal of ammonia
(B) High concentration of reactants
(C) Low temperature and high pressure
(D) All the above

54. Which of the following is soluble in yellow ammonium sulphide

(A) SnS (B) PbS (C) CuS (D) CdS

55. To neutralize 25 ml of 0.25 M Na_2CO_3 solution, how much volume of 0.5 M HCl is required ?

(A) 12.5 ml (B) 25 ml (C) 37.5 ml (D) 50 ml

56. Sodium borate on reaction with H_2SO_4 (conc.) and C_2H_5OH gives a compound 'A' which burns with green flame. The compound 'A' is

(A) $H_2B_4O_7$ (B) $(C_2H_5)_2B_4O_7$
(C) H_3BO_4 (D) $(C_2H_5)_3BO_3$

57. The correct IUPAC name for the compound

$$CH_3-CH_2-CH_2-CH_2-CH_2-\underset{|}{\overset{CH_3\ |}{C}}H-\underset{CH_2-CH_2-CH_3}{\overset{CH_3}{C}}-CH_2-CH_3$$

(A) 3, 4-dimethyl-3-n-propyl nonane
(B) 6, 7-dimethyl-7-ethyl decane
(C) 6, 7-dmethyl-7-n-propylnonane
(D) 4, 5-dimethyl-4-ethyldecane

58. In the reaction,

$$\underset{H}{\overset{Br}{}}\!\!>\underset{1}{C}=\underset{2}{C}<\!\!\underset{H}{\overset{Br}{}} \xrightarrow[\text{Catalyst}]{H_2} Br\underset{3}{C}H_2\underset{4}{C}H_2Br$$

The hybridization states of carbon atoms 1, 2, 3 and 4 are

(A) 1 and 2 sp^2; 3 and 4 sp^3
(B) 1 and 2 sp^2; 3 and 4 sp
(C) 1, 2, 3 and 4 sp (D) 1, 2 sp^3; 3, 4 sp^2

59. The reagent which is used to distinguish between propene and propyne is

(A) Ammoniacal $AgNO_3$ (B) Ozone
(C) Bromine (D) Alkaline $KMnO_4$

60. Which of the following reactions would be the best for the formation of 2-bromobutane ?

(1) $CH_3CH = CH\ CH_2CH_3 \xrightarrow{HBr}$

(2) $CH_3CH_2CH_2 = CH_2 \xrightarrow{HBr}$

(3) $CH_3CHCH = CHCH_3 \xrightarrow{Br_2}$

(4) $CH_3CH_2CH = CH_2 \xrightarrow[\text{Peroxide}]{HBr}$

(A) 1 (B) 2 (C) 3 (D) 4

61. The reaction

Phenol (OH) $\xrightarrow[\text{aq. NaOH } 70^{\circ}C]{CHCl_3}$ o-hydroxybenzaldehyde (OH, –CHO)

is called

(A) Wittig reaction (B) Williamson's synthesis
(C) Wurtz-Fittig reaction
(D) Reimer-Tiemann Reaction

62. Diethylether and methyl-n-propyl ether are

(A) Metamers (B) Chain isomers
(C) Position isomers (D) Functional isomers

63. In the reaction, $2A + B \rightarrow A_2B$, if the concentration of A is doubled and of B is halved, then the rate of the reaction will

(A) Increase by four times (B) Decrease by two times
(C) Increase by two times (D) Remain the same

64. Radius of the isoelectronic species

(A) Increases with the increase of nuclear charge
(B) Decreases with the increase of nuclear charge

(C) Is the same for all
(D) First increases and then decreases

65. An element occurring in the BCC structure has 12.08×10^{23} unit cells. The total number of atoms of the element in these cells will be
(A) 6.04×10^{23} (B) 12.08×10^{23}
(C) 24.16×10^{23} (D) 36.18×10^{23}

66. Bauxite ore is concentrated by
(A) Chemical separation (B) Hydraulic separation
(C) Froth floation
(D) Electromagnetic separation

67. Which one of the following reactions would produce secondary alcohol ?
(A) $C_6H_5-\overset{O}{\overset{\|}{C}}-CH_3 \xrightarrow[2.\ H^+]{1.\ CH_3MgBr}$
(B) $C_6H_5-\overset{O}{\overset{\|}{C}}-CH_3 \xrightarrow[2.\ H^+]{1.\ LiAlH_4}$
(C) $CH_3-CHO \xrightarrow[2.\ H^+]{1.\ LiAlH_4}$
(D) $C_6H_5-\overset{O}{\overset{\|}{C}}-CH_3 \xrightarrow[2.\ Br_2]{1.\ ^-OH}$

68. In a compound CH_3COCl, which type of orbitals have been used by the circled carbon in bond formation
(A) p (B) Sp (C) Sp^2 (D) Sp^3

69. The order of increasing reactivity towards HCl of the following compounds
1. $CH_2 = CH_2$ 2. $(CH_3)_2\ CH = CH_2$
3. $CH_3CH = CHCH_3$ will be
(A) $1 < 2 < 3$ (B) $1 < 3 < 2$
(C) $3 < 2 < 1$ (D) $2 < 1 < 3$

70. In order to complete the reaction 1-pentyne to (a) 4-octyne (b) cis-4-octene, a and b will be
(a) (b)
1. $NaNH_2$: CH_3CH_2Br : H_2(one mole) Pd or Ni
2. $NaNH_2$: $CH_3CH_2CH_2Br$: H_2 : (two moles) Pd or Ni
3. $NaNH_2$: $CH_3CH_2CH_2Br$: H_2 (one mole) Pd or Ni
4. $NaNH_2$: $CH_3CH_2CH_2Br$: $BH_3.H_2O_2.OH^-$
(A) 1 (B) 2 (C) 3 (D) 4

71. Among the following, which property is commonly exhibited by a covalent compound ?
(A) High solubility in water
(B) High electrical conductance
(C) Low boiling point (D) High melting point

72. In which of the following compounds sP^2 hybrid orbitals are used ?
(A) NH_3 (B) BeH_2 (C) BCl_3 (D) CH_4

73. The bond angle in water molecule is
(A) 120° (B) 109°, 28' (C) 90° (D) 104° 27'

74. On the basis of the electrochemical theory of aqueous corrosion, the reaction occurring at the cathode is
(A) $O_2\ (g) + 4H^+(aq) + 4e^- \rightarrow 2H_2O_{(l)}$
(B) $Fe(s) \rightarrow Fe^{2+}(aq) + 2e^-$
(C) $Fe^{2+}(aq) \rightarrow Fe^{3+}(aq) + e^-$
(D) $H_2\ (g) + 2OH^-\ (aq) \rightarrow 2H_2O_{(l)} + 2e^-$

75. In which of the following compounds intramolecular hydrogen bond is present ?
(A) Salicylaldehyde (B) Hydrogen sulphide
(C) Ethyl alcohol (D) Water

76. Lucas test is used to distinguish between
(A) 1°, 2° and 3° alcohols (B) 1°, 2° and 3° amines
(C) Aldehydes and ketones (D) Alkenes and alkynes

77. For an exothermic reaction
(A) H of the products is equal to H of the reactants
(B) H of the products is less than H of the reactants
(C) H of the products is more than H of the reactants
(D) ΔH is always positive

78. The units of rate constant of a first order reaction are
(A) Mol lit^{-1} sec^{-1} (B) Sec^{-1}
(C) Mol^{-1} sec (D) Litre sec^{-1} mol^{-1}

79. In the adsorption of acetic acid on activated charcoal, the acetic acid is an
(A) Adsorbent (B) Adsorbate
(C) Adsorber (D) Absorber

80. In a reversible reaction, a catalyst
(A) Increases the rate of the forward and backward reactions equally
(B) Increases the rate of the forward reaction
(C) Increases the rate of the backward reaction
(D) None of above

81. The ability of an ion to bring about coagulation of a given colloid depends upon
(A) Its size
(B) The magnitude of its charge only
(C) The sign of its charge
(D) Both magnitude and sign of its charges

82. The enthalpy change for the transition of liquid water to steam, $\Delta H_{vap} = 37.3$ kJ mol^{-1} at 373° K. The entropy change for the process is
(A) 111.9 J mol^{-1} K^{-1} (B) 37.3J mol^{-1} K^{-1}
(C) 100 J mol^{-1} K^{-1} (D) 74.6 J mol^{-1} K^{-1}

83. In which of the following processes, shape-selective catalysis is occurring ?
(A) Polymerisation of ethylene
(B) Synthesis of ammonia
(C) Conversion of alcohol to gasoline
(D) Synthesis of methanol from CO and H_2

84. Which of the following solutions will have the highest osmotic pressure ?
(A) 0.1 M Sucrose (B) 0.1 M Sodium chloride
(C) 0.1 M Barium chloride (D) 0.1 M Urea

85. The four quantum numbers for the electron in the outermost orbital of potassium are

(A) $n = 4, l = 1, m = 0, S = +\frac{1}{2}$

(B) $n = 4, l = 0, m = 0, S = +\frac{1}{2}$

(C) $n = 4, l = 0, m = 0, S = -\frac{1}{2}$

(D) $n = 4, l = 0, m = +1, S = -\frac{1}{2}$

86. The half-life of radium (226) is 1620 years. The time taken to convert 10 grams of radium to 1.25 grams is
(A) 3240 years (B) 4860 years
(C) 810 years (D) 1620 years

87. The rate constant (K') of one reaction is double of the rate constant (K") of another reaction. Then the relationship between the corresponding activation energies of the two reactions (Ea' and Ea") will be
(A) Ea' > Ea" (B) Ea' = Ea"
(C) Ea' < Ea" (D) Ea' = 4 Ea"

88. The P-P orbital overlapping is present in the following molecule
(A) Hydrogen (B) Hydrogen bromide
(C) Hydrogen chlorine (D) Chlorine

89. Frenkel defect is caused due to
(A) An extra negative ion occupying an interstitial position in the lattice
(B) An ion missing from the normal lattice site creating a vacancy
(C) An extra positive ion occupying an interstitial position in the lattice
(D) The shift of a positive ion from its normal lattice site to an interstitial site

90. The minimum energy required for a molecule to enter into a reaction is called
(A) Potential energy (B) Kinetic energy
(C) Nuclear energy (D) Activation energy

91. The vapour pressure of a solution is
(A) Directly proportional to the mole fraction of the solvent
(B) Inversely proportional to the mole fraction of the solvent
(C) Directly proportional to the mole fraction of the solute
(D) Inversely proportional to the mole fraction of the solute

92. In the presence of a dilute base C_6H_5CHO and CH_3CHO react together to give a product. The product is
(A) $C_6H_5CH_2OH$ (B) $C_6H_5CH = CHCHO$
(C) $C_6H_5CH_3$ (D) $C_6H_5CH_2CH_2OH$

93. Oxalic acid on being heated upto 90° with conc H_2SO_4 forms
(A) $HCOOH + CO_2$ (B) $CO_2 + H_2O$
(C) $CO_2 + CO + H_2O$ (D) $HCOOH + CO$

94. Acetamide on being heated with P_2O_5 gives
(A) Methyl Cyanide (B) Methyl amine
(C) Acetone (D) Acetic acid

95. The pH of a solution is 2. If its pH is to be raised to 4, then the $[H^+]$ of the original solution has to be
(A) Doubled (B) Halved
(C) Increased hundred times
(D) Decreased hundred times

96. In the reaction $CH_3NH_2 + X + KOH \rightarrow CH_3NC$ (highly offensive odour). X is
(A) CH_3Cl (B) CCl_4 (C) CH_2Cl_2 (D) $CHCl_3$

97. Which of the following is Friedel Crafts reaction ?
(A) $C_6H_6 + FeCl_3 + Cl_2 \rightarrow C_6H_5Cl$
(B) $C_6H_5CHO + CH_3CHO + KOH \rightarrow C_6H_5CH = CH\text{-}CHO$
(C) $C_6H_6 + CH_3COCl + AlCl_3 \rightarrow C_6H_5 - \overset{\overset{\displaystyle O}{\|}}{C} - CH_3$
(D) $C_6H_5OH + CHCl_3 + KOH \rightarrow$ $C_6H_4(OH)(CHO)$ (o-hydroxybenzaldehyde)

98. In the reaction,
$ArNH_2 \xrightarrow[\text{Cold}]{NaNO_2 + HCl} A \xrightarrow{CuBr} B,$
the compound B is
(A) Ar - Br (B) Ar - NO_2
(C) Ar - Cl (D) Ar - Ar

99. In the reaction,
$C_2H_5MgBr + H_2C\text{—}CH_2$ (epoxide, bridged by O) $\xrightarrow{H_2O}$ A, A is
(A) $C_2H_5CH_2\ CHO$ (B) $C_2H_5CH_2CH_2OH$
(C) $C_2H_5CH_2OH$ (D) C_2H_5CHO

100. The enzyme which converts glucose into ethanol is
(A) Diastase (B) Maltase
(C) Invertase (D) Zymase

PET - 1995

1. A reaction rate constant is given by $k = 1.2 \times 10^{14} e^{-25000/RT} \sec^{-1}$. It means
(A) log k versus log 1/T will give a straight line with slope as - 25000
(B) log k versus 1/T will give a straight line
(C) log k versus log T will give a straight line with slope as - 25000
(D) log k versus T will give a straight line with slope as 25000

2. The decomposition of N_2O_5 is a first order reaction represented by
$N_2O_5 \rightarrow N_2O_4 + \frac{1}{2} O_2$
After 15 minutes, the volume of O_2 produced is 9 ml and at the end of the reaction 35 ml. The rate constant is equal to
(A) $\frac{1}{15} \ln \frac{35}{44}$ (B) $\frac{1}{15} \ln \frac{44}{26}$
(C) $\frac{1}{15} \ln \frac{44}{35}$ (D) $\frac{1}{15} \ln \frac{35}{26}$

3. Electrolysis is a process in which the cations, anions of the electrolyte are
(A) charged (B) discharged
(C) hydrated (D) hydrolysed

4. Which of the following is aspirin
(A) Methyl salicylate (B) Ethyl salicylate
(C) Acetyl salicylic acid (D) Ortho-hydoxy benzoic acid

5. When acetamide is reacted with Br_2 and caustic soda, we get
(A) Methyl amine (B) Ethane
(C) Acetic acid (D) Bromo acetic acid

6. Name of the reaction,
$C_{10}H_{22} \xrightarrow{900\ K} C_4H_8 + C_6H_{14}$ is
(A) Alkylation (B) Cracking
(C) Dehydrogenation (D) Fractionation

7. Ethylene reacts with alkaline $KMnO_4$ to form -
(A) Ethyl alcohol (B) Ethylene glycol
(C) Oxalic acid (D) HCHO

8. $C_6H_6 + CH_3Cl \xrightarrow[AlCl_3]{Anhydrous} C_6H_5CH_3 + HCl$
is an example of
(A) Friedel Craft's reaction (B) Kolbe's synthesis
(C) Wurtz reaction (D) Grignard reaction

9. Which of the following compound is used as a refrigerant ?
(A) CF_4 (B) CCl_2F_2 (C) Acetone (D) CCl_4

10. $C_2H_5Cl + KCN \rightarrow X \xrightarrow{Hydrolysis} Y$.
X and Y are -
(A) C_2H_6 and C_2H_5CN (B) C_2H_5CN and C_2H_6
(C) C_2H_5CN and $C_2H_5CH_2NH_2$
(D) C_2H_5CN and C_2H_5COOH

11. The product of the reaction,
CH_3CH_2OH + Cu (Reduced) $\xrightarrow{300^oC}$ is
(A) CH_3CHO (B) CH_3COOH (C) C_3H_6 (D) CH_3COCH_3

12. Calcium acetate on dry distillation gives
(A) Formaldehyde (B) Acetaldehyde
(C) Acetone (D) Acetic anhydride

13. The insecticide Gammexane is
(A) Parathion (B) Chloral
(C) D. D. T. (D) Benzene hexachloride

14. The reaction of HCOOH with Conc. H_2SO_4 gives
(A) CO_2 (B) CO (C) Oxalic acid (D) Acetic acid

15. Lucas reagent is
(A) Conc. HCl + Anhydrous $ZnCl_2$
(B) Dil. HCl + $ZnCl_2$ (C) H_2SO_4 + $ZnCl_2$
(D) Conc. HCl + Zn

16. Iodoform is formed on warming I_2 and NaOH with
(A) HCOOH (B) C_6H_5
(C) C_2H_5OH (D) CH_3OH

17. An isomer of ethanol is
(A) Methanol (B) Dimethyl ether
(C) Diethyl ether (D) Ethylene glycol

18. How many isomers of alkane, C_6H_{14} has
(A) 4 (B) 5 (C) 6 (D) 7

19. Which carbohydrates has highest abundance in human blood
(A) d-fructose (B) d-glucose (C) Sucrose (D) Lactose

20. Fat is a/an
(A) Ester (B) Hydrocarbon (C) Acid (D) Alcohol

21. Vitamin A is present in
(A) Cord liver oil (B) Carrot
(C) Milk (D) In all the above

22. Ascorbic acid is a
(A) Vitamin (B) Enzyme (C) Protein (D) Carbohydrate

23. Which one is not present in R.N.A.
(A) Ribose (B) Phosphate (C) Uracil (D) Thymine

24. Natural rubber is a polymer derived from
(A) Isoprene (B) Propylene (C) Butadiene (D) Ethylene

25. Both methane and ethane may be obtained in one step reaction from
(A) C_2H_4 (B) CH_3I (C) CH_3OH (D) C_2H_5OH

26. The relation between Δ G and E for a cell is, $\Delta G = -nFE$, the cell reaction will be spontaneous, if
(A) E is negative (B) E is positive
(C) G is negative (D) G is positive

27. Faraday has the dimensions of
(A) Coulombs (B) Coulomb equivalent
(C) Coulomb per equivalent
(D) Coulomb per degree Kelvin

28. Zinc displaces copper from the solution of its salt because
(A) Gibbs free energy of zinc is less than that of copper
(B) Atomic number of zinc is more than that of copper
(C) Zinc salt is more soluble in water than the copper salt
(D) Zinc is placed higher than copper in electrochemical series

29. For an adiabatic process
(where q, w, p and ΔV denote heat, work, pressure and change in volume respectively) the relationship is
(A) q = 0 (B) q = – w
(C) q = + w (D) p Δ V = 0

30. Decrease of free energy of a reacting system points to a/an
(A) Spontaneous reaction (B) Isothermal reaction
(C) Exothermic reaction (D) Equilibrium reaction

31. When a solid is converted into liquid, entropy
(A) Becomes zero (B) Remains the same
(C) Decreases (D) Increases

32. The laughing gas is
(A) Nitrogen peroxide (B) Nitrogen pentaoxide
(C) Nitric oxide (D) Nitrous oxide

33. In the periodic table, the elements are arranged in the order of
(A) Increasing volume (B) Increasing atomic number
(C) Increasing mass (D) Increasing density

34. The electron affinity is the largest for

(A) F (B) Cl (C) Br (D) I

35. The highest oxidation state is shown by that transition element wich has outermost configuration as
(A) d^5s^2 (B) d^8s^2 (C) d^3s^2 (D) d^5s^1

36. Coordination number of Cu in complex $[Cu(H_2O)_4]^{++}$ is-
(A) 4 (B) 3 (C) 2 (D) 1

37. Complex $[Pt(NH_3)_6]\,Cl_4$ furnishes
(A) 5 ions (B) 4 ions (C) 3 ions (D) 2 ions

38. The strongest ligand in the following is
(A) HO^- (B) F^- (C) CN^- (D) Br^-

39. The coordination number of Pt in $[Pt(NH_3)_4\,Cl_2]^{++}$ ion is
(A) 2 (B) 4 (C) 6 (D) 8

40. If the magnetic moment of $[Ag(CN_2]^{-1}$ is zero, then the number of unpaired electrons will be
(A) 1 (B) 2 (C) 3 (D) Zero

41. $[Co(NH_3)_4Cl_2]NO_2$ and $[Co(NH_3)_4\,Cl\,NO_2]Cl$ are .... isomers
(A) Linkage (B) Ionisation
(C) Geometrical (D) Optical

42. Carbon and carbon monoxide burn in oxygen to form carbon dioxide according to the following reactions
$C + O_2 \rightarrow CO_2$, $\Delta H = -\ 394$ kJ mol^{-1}
$2CO + O_2 \rightarrow 2CO_2$, $\Delta H = -569$ kJ mol^{-1}
The heat of formation of 1 mole of carbon monoxide is thus
(A) – 175.0 kJ mol^{-1} (B) – 87.5 kJ mol^{-1}
(C) – 219.0 kJ mol^{-1} (D) –109.5 kJ mol^{-1}

43. The average life period of a radioactive element is the reciprocal of its
(A) Half-life period (B) Disintegration constant
(C) Number of atoms present at any time
(D) Number of neutrons

44. Uranium ultimately decays into a stable isotope of
(A) Lead (B) Neptunium (C) Radium (D) Carbon

45. An isotope $^{X}_{Y}A$ undergoes a series of m alpha and n beta disintegration to form a stable isotope $^{X-32}_{Y-10}B$. The value of m and n are respectively
(A) 6 and 8 (B) 8 and 10 (C) 5 and 8 (D) 8 and 6

46. A substance used as a moderator in a nuclear reactors is
(A) Cadmium (B) Uranium - 235
(C) Lead (D) Heavy water

47. The atomic radii in periodic table among elements from right to left
(A) Decreases (B) Increases (C) Remain constant
(D) First decreases and then increases

48. Element having atomic number 17 is placed in
(A) I-group (B) V-group
(C) VIII-group (D) VII-group

49. Lithium shows diagonal relationship with
(A) Be (B) Ce (C) Ne (D) Mg

50. Which of the following forms coloured ion ?
(A) Metalloids (B) Non-metals
(C) p-Block elements (D) Transition elements

51. Blister copper is
(A) 1% impure copper (B) Copper alloy
(C) Pure copper (D) Copper ore

52. The purest form of iron is
(A) Cast iron (B) Pig iron (C) wrought iron (D) Steel

53. Prussian blue is
(A) $Fe_4[Fe(CN)_6]_3$ (B) $Fe_2[Fe(CN)_6]$
(C) $Fe_3[Fe(CN)_6]$ (D) $Fe[Fe(CN)_6]_2$

54. Emulsifier
(A) Stabilizes the emulsion
(B) Accelerates the dispersion
(C) Homogenesis the emulsion
(D) Aids the flocculation of emulsion

55. The catalyst remains unchanged at the end of the reaction as regards
(A) Chemical composition, quantity & physical state
(B) Quantity and physical state
(C) Chemical composition and physical state
(D) Quantity and chemical composition

56. In a reaction involving hydrolysis of an organic chloride in presence of large excess of water,
$RCl + H_2O \rightarrow ROH + HCl$
(A) Molecularity is 2, order of reaction is also 2
(B) Molecularity is 2, order of reaction is 1
(C) Molecularity is 1 order of reaction is 2
(D) Molecularity is 1, order of reaction is also 1

57. An increase in the rate of a reaction for a rise in temperature is due to
(A) The increase in the number of collisions
(B) The decrease in the number of activated molecules
(C) The shortening of the mean free path
(D) The lowering of the activation energy

58. Consider the two gaseous equilbria involving SO_2 and the corresponding equilibrium constants at 290°K
$SO_2(g) + \frac{1}{2}O_2(g) \rightleftharpoons SO_3(g);\ K_1$
$2SO_3(g) \rightleftharpoons 2SO_2(g) + O_2(g);\ K_2$
The value of the equilibrium constants are related by
(A) $K_2 = K_1$ (B) $K_2 = K_1^2$
(C) $K_2 = 1/K_1^2$ (D) $K_2 = 1/K_1$

59. Given that the dissociation constant for H_2O is $K_W = 1 \times 10^{-14}$ mole^2 litre^{-2}, what is the pH of a 0.001 molar KOH solution ?
(A) 10^{-11} (B) 3 (C) 14 (D) 11

60. Which of the following can act both as Bronsted acid and a base
(A) H_3O^+ (B) SO_4^{2-} (C) Cl^- (D) HCO_3^-

61. If the solubility product of $BaSO_4$ is 1.5×10^{-9} in water, its solubility in moles per litre is
(A) 1.5×10^{-9} (B) 3.9×10^{-5}
(C) 7.5×10^{-5} (D) 1.5×10^{-5}

62. In the reaction,
$2SO_2 + O_2 \xrightarrow[As_2O_3]{pt} 2SO_3$,
As_2O_3 acts as a/an
(A) Auto catalyst (B) Poison
(C) Promotor (D) Positive Catalyst

63. On dissolving common salt in water, the
(A) Boiling point of water remains same
(B) Nothing can be predicted
(C) Boiling point of water decreases
(D) Boiling point of water increases

64. The liquid pair, benzene-toluene shows
(A) Positive deviation from Raoult's Law
(B) Irregular deviation from Raoult's Law
(C) Negative deviation from Raoult's Law
(D) Practically no deviation from Raoult's Law

65. Hoope's process is used in the refining of
(A) Al (B) Zn (C) Ag (D) Cu

66. Red lead is
(A) PbO_2 (B) Pb_4O_3 (C) Pb_3O_4 (D) PbO

67. Solder metal is
(A) Pb + Sb (B) Pb + Sn (C) Cu + Sn (D) Cu + Zn

68. In the manufacture of iron from haematite the function of lime stone is that of a
(A) Slag (B) Gangue (C) Reducing agent (D) Flux

69. In solid argon the atoms are held together
(A) By ionic bonds (B) By hydrogen bonds
(C) By van der Waals forces (D) By hydrophobic bonds

70. Oxygen is soluble in
(A) Acid (B) Alkaline pyrogallol
(C) Alcohol (D) Acidic pyrogallol

71. Monazite is a source of
(A) Ne (B) Ra (C) Kr (D) Th

72. A match box exhibits
(A) Monoclinic geometry (B) Cubic geometry
(C) Orthorhombic geometry (D) Tetragonal geometry

73. The following is NOT a function of an impurity present in a crystal
(A) Contributing to scattering
(B) Establishing thermal equilibrium
(C) Having tendency to diffuse
(D) Introducing new electronic energy levels

74. Tollen's reagent is
(A) Ammonical cuprous chloride
(B) Ammonical cuprous oxide
(C) Ammonical silver bromide
(D) Ammonical silver nitrate

75. Electronic configuration of chromium element (Atomic Number 24) is
(A) $1s^2 2s^2 2p^6 3s^2 3p^6 3d^6$ (B) $1s^2 2s^2 2p^6 3s^2 3p^6 4s^2 p^4$
(C) $1s^2 2s^2 2p^6 3s^2 3p^6 4s^2 3d^4$ (D) $1s^2 2s^2 2p^6 3s^2 3p^6 4s^1 3d^5$

76. For an electron imposible combination of quantum numbers is
(A) $n = 3, 1 = 2, m_l = -2, m_S = +\frac{1}{2}$
(B) $n = 3, 1 = 2, m_l = -3, m_S = +\frac{1}{2}$
(C) $n = 4, 1 = 0, m_l = 0, m_S = -\frac{1}{2}$
(D) $n = 5, 1 = 3, m_l = 0, m_S = \frac{1}{2}$

77. Which of the following equilibria is not affected by a pressure change ?
(A) $2NO_2(g) \rightleftharpoons N_2O_4(g)$
(B) $CO_2(g) + H_2O(1) \rightleftharpoons CO_2$ solution
(C) $H_2(g) + I_2(g) \rightleftharpoons 2HI(g)$
(D) $2O_3(g) \rightleftharpoons 3O_2(g)$

78. In a chemical reaction, equilibrium is established when
(A) Velocity of opposing reaction is the same as that of the forward one
(B) Order of the reaction is infinite
(C) Concentrations of the reactants and products are equal
(D) Reaction ceases to generate heat

79. In a NaCl crystal each Na^+ is surrounded by how many nearest neighbouring equally spaced Cl^- ?
(A) 2 (B) 4 (C) 6 (D) 8

80. Which is NOT a property of solids ?
(A) The diffusion of solids is very slow
(B) Solids have definite volume
(C) Solids are always crystalline in nature
(D) Solids have high density and low compressibility

81. The molarity of solution prepared by adding 7.1 grams of Na_2SO_4 (formula weight 142 amu) to water to make 100 ml volume is
(A) 2.0 M (B) 1.0 M (C) 0.5 M (D) 0.05 M

82. When the azimuthal quantum number has a value of 1 = 0, the shape of the orbital is
(A) Dumb-bell (B) Unsymmetrical
(C) Rectangular (D) Spherical

83. The formation of a chemical bond is accompanied by
(A) Decrease in energy (B) Increase in energy
(C) Neither increase nor decrease in energy
(D) None of the above

84. Which of the following molecules has a linear structure ?
(A) CO_2 (B) H_2O (C) SO_2 (D) H_2O_2

85. Hydrogen bonding is formed in compounds containing hydrogen and
(A) Metal atoms with d-orbitals occupied
(B) Metalloids
(C) Highly electronegative atoms
(D) Highly electropositive atoms

86. Covalent molecules are usually held in a crystal structure by
(A) Dipole-dipole attraction (B) electrostatic attraction

(C) Hydrogen bonds (D) van der Waal's attraction

87. The PCl_5 molecule is a result of the hybridization of
(A) spd^3 (B) sp^2d^3 (C) sp^2d^2 (D) sp^3d

88. A compound has 50% carbon, 50% oxygen and approximate molecular weight of 290. Its molecular formula is
(A) CO (B) C_4O_3 (C) $C_{12}O_9$ (D) C_3O_3

89. Identify B in the reaction,

Acetamide $\xrightarrow[\Delta]{P_2O_5}$ A $\xrightarrow[Sn / HCl]{4H}$ B

(A) CH_3NH_2 (B) $CH_3CH_2NH_2$
(C) CH_3CN (D) CH_3COONH_4

90. In the reaction,

$R-CH_2-CCl_2-R \xrightarrow{Reagent} R-C \equiv C-R$

the reagent is
(A) Na amalgam (B) HCl and H_2O
(C) Alcoholic KOH (D) Aqueous KOH

91. The I.U.P.A.C. name for
$CH_3\ CH(CH_3)CH\ =CH\ -CH_2-COOH$ is
(A) 5- methyl 3-Hexenoic acid
(B) 5- carboxyl 2-methyl pentene-4
(C) 4- iosopropyle butene 3-oic acid-1
(D) None of these

92. Which of the following hybridisation has maximum s-charaters ?
(A) Sp^3 (B) Sp^2 (C) Sp (D) None

93. The hybridization of carbon atom in C - C single bond of $HC \equiv C-CH=CH_2$ is
(A) Sp - Sp^2 (B) Sp^3 - Sp
(C) Sp^3 - Sp^3 (D) Sp^2 - Sp^3

94. Which of the following reagents distinguishes between aldehyde and Ketone ?
(A) Fehling solution (B) H_2SO_4
(C) $NaHSO_3$ (D) NH_3

95. Which of the following oxides is white but becomes yellow on heating ?
(A) FeO (B) ZnO (C) AgO (D) Ag_2O

96. Which of the following ions gives coloured solution ?
(A) Cu^+ (B) Zn^{++} (C) Ag^+ (D) Fe^{++}

97. In Borax bead test deep blue coloured bead is obtained due to
(A) Ni (B) Co (C) Cr (D) Cu

98. On heating $Mn(OH)_2$ with PbO_2 and conc. HNO_3, pink colour is obtained due to the formation of
(A) $KMnO_4$ (B) $HMnO_4$ (C) $Pb(MnO_4)_2$ (D) $PbMnO_4$

99. 15 ml of N/10 NaOH solution completely neutralizes 12 ml of H_2SO_4 solution. The normality of H_2SO_4 solution will be
(A) N/8 (B) N (C) N/5 (D) N/10

100. In the titration of CH_3COOH with NaOH the indicator used is
(A) Methyl orange (B) Methyl red
(C) Phenolphthalein (D) Litmus

PET - 1996

1. The order of stability of carbonium ions is
(A) Iso-propyl > tert-butyl > ethyl > methyl
(B) Tert-butyl > ethyl > isopropyl > methyl
(C) Methyl > ethyl > iso-propyl > tertbutyl
(D) Tert-butyl > iso-propyl > ethyl > methyl

2. In the nitration of benzene with conc. $H_2SO_4 + HNO_3$, the attack on ring is made by
(A) NO_2^- (B) NO_3^- (C) NO_2^+ (D) NO_2

3. The final product obtained by distilling ethyl alcohol with excess of chlorine and $Ca(OH)_2$ is
(A) CH_3CHO (B) CCl_3CHO
(C) $CHCl_3$ (D) $(CH_3)_2O$

4. For converting aniline into chlorobenzene, which of the following reagents is not used ?
(A) Cl_2 (B) HCl (C) HNO_2 (D) CuCl

5. Sodium phenoxide reacts with CO_2 at 400 K and 4-7 atm pressure to give
(A) Catechol (B) Benzoic acid
(C) Sodium salicylate (D) Salicylaldehyde

6. Which of the following does not give yellow precipitate with I_2 and NaOH ?
(A) CH_3COCH_3 (B) HCHO
(C) C_2H_5OH (D) CH_3CHO

7. Acetylene reacts with HCl and $HgCl_2$ to give
(A) Vinyl chloride (B) Dichloroethane
(C) Ethyl chloride (D) Ethylidene chloride

8. The thermal decomposition of a compound is first order. If a sample of the compound decomposes 50% in 120 minutes, in what time will it undergo 90% decomposition ?
(A) Nearly 240 minutes (B) Nearly 480 minutes
(C) Nearly 450 minutes (D) Nearly 400 minutes

9. The order of a reaction with rate equals to $K\,C_A^{3/2}\,C_B^{-1/2}$ is
(A) 2 (B) 1 (C) - 1/2 (D) 3/2

10. The unit of cell constant is
(A) $ohm^{-1}cm^{-1}$ (B) ohm cm (C) cm (D) cm^{-1}

11. The chemical reaction taking place at the anode of a cell is
(A) Oxidation (B) Hydrolysis
(C) Ionization (D) Reduction

12. In a dry cell the reaction which takes place at the graphite anode is
(A) $Zn^{2+} + 2e^- \rightarrow Zn(s)$ (B) $Zn(s) \rightarrow Zn^{2+} + 2e^-$
(C) $Mn^{2+} + 2e- \rightarrow Mn\ (s)$
(D) $Mn(s) \rightarrow Mn^+ + e^- + 1.5\ V$

13. Milk is an example of
(A) Sol (B) Gel (C) Emulsion (D) Suspension

14. Fog is a colloidal system of
(A) Gas in gas (B) Solid in gas
(C) A gas in liquid (D) Liquid in gas

15. The order of reactivity of halogen acids for reaction with C_2H_5OH is
(A) HCl > HBr > HI (B) HI > HBr > HCl
(C) HBr > HI > HCl (D) HBr > HCl > HI

16. The reaction of C_2H_5OH with H_2SO_4 does not give
(A) Acetylene (B) Ethyl hydrogen supphate
(C) Ethylene (D) Diethyl ether

17. The pH of 10^{-5} M aqueous solution of NaOH is
(A) 5 (B) 7 (C) 9 (D) 11

18. In a spontaneous process, the entropy of the system and its surroundings
(A) Increases (B) Remains constant
(C) Equals zero (D) Decreases

19. In physical adsorption, the gas molecules are held on solid surface by
(A) Chemical forces (B) Electrostatic forces
(C) Gravitational forces (D) van der Waal's forces

20. The natural materials from which an element can be extracted economically are called
(A) Ores (B) Minerals
(C) Gangue (D) None of the above

21. When a metal is to be extracted from its ore, if the gangue associated with the ore is silica, then
(A) An acidic flux is needed
(B) a basic flux is needed
(C) Both acidic and basic flux are needed
(D) Neither of them is needed

22. An example of an oxide ore is
(A) Bauxite (B) Malachite (C) Zinc blende (D) Feldspar

23. In the Hoopes process for refining of Aluminium, the fused materials form three different layers and they remain separated during electrolysis also. This is because
(A) The 3 layers have different densities
(B) The 3 layers are maintained at different temperatures
(C) The upper layer is kept attracted by the cathode and the lower layer is kept attracted by the anode
(D) There is special arrangement in the cell to keep the layers separate

24. Which of the following is the odd one out ?
(A) Potassium ferricyanide
(B) Potassium ferrocyanide
(C) Ferrous ammonium sulphate
(D) Tetrammine copper (II) sulphate

25. Generally, a group of atoms can function as a ligand if
(A) They are positively charged ions
(B) They are free radicals
(C) They are either neutral molecules or negatively charged ions
(D) None of the above

26. Which of the following is not an organometallic compound ?
(A) Ethyl magnesium bromide (B) Tetraethyl lead
(C) Sodium ethoxide (D) Trimethyl aluminium

27. Of the ions Zn^{2+}, Ni^{2+} and Cr^{3+} [At. Nos – Zn = 30, Ni = 28, Cr = 24]
(A) Only Zn^{2+} is colourless and Ni^{2+} and Cr^{3+} are coloured
(B) All three are colourless (C) All three are coloured
(D) Only Ni^{2+} is coloured and Zn^{2+} and Cr^{3+} are colouress

28. Paraldehyde is trimer of
(A) Benzaldehyde (B) Propionaldehyde
(B) Formaldehyde (D) Acetaldehyde

29. Which of the following on reaction with conc. NaOH gives an alcohol ?
(A) Methanal (B) Ethanal (C) Propanal (D) Butanal

30. Lactic acid on heating with conc. H_2SO_4 gives
(A) Acrylic acid (B) Propionic acid
(C) Acetic acid (D) Formic acid

31. Which of the following is most acidic ?
(A) Acetic acid (B) Chloroacetic acid
(C) Dichloroacetic acid (D) Trichloroacetic acid

32. Acetal is obtained by reacting in the presence of dry HCl an alcohol with
(A) Aldehyde (B) Ketone
(C) Ether (D) Carboxylic acid

33. Acetone is easily oxidized with
(A) Tollen's reagent (B) Fehling solution
(C) Acidic dichromate solution (D) Benedict's solution

34. Which of the following is not a constituent of RNA ?
(A) Adenine (B) Pyridine (C) d-Ribose (D) Phosphate

35. The valence shell electronic configuration of alkali metals is
(A) ns^2np^1 (B) ns^1
(C) (n-1) p^6ns^2 (D) (n - 1) d^2ns^2

36. The ratio between radii of He^+ ion and H atom is
(A) 1/2 (B) 1 (C) 3/2 (D) 2

37. Common oxidation state of Scandium, a transition element, is/are [At. No. - Sc = 21]
(A) + 4 (B) +1 (C) +2 & +3 (D) +4 & +1

38. Which of the following is not correct about transition metals ?
(A) They can form ionic or covalent compounds
(B) Their melting and boiling points are high
(C) Their compounds are generally coloured
(D) They do not exhibit variable valency.

39. Which of the following noble gases does not have an octet of electrons in its outermost shell ?
(A) Argon (B) Helium (C) Neon (D) Radon

40. Acidified potassium dichromate is treated with hydrogen sulphide. In the reaction the oxidation number of chromium
(A) Increases from +3 to +6
(B) Decreases from +6 to +3
(C) Remains unchanged (D) Decreases from +6 to +2

41. When $KMnO_4$ reacts with acidified $FeSO_4$
(A) Only $FeSO_4$ is oxidized (B) Only $KMnO_4$ is oxidized
(C) $FeSO_4$ is oxidized and $KMnO_4$ is reduced
(D) None of the above

42. The ion that can be precipitated by HCl as well as H_2S is
(A) Zn^{2+} (B) Cu^{2+} (C) Pb^{2+} (D) Fe^{3+}

43. 100 cm^3 of 0.1 N HCl solution is mixed with 100 cm^3 of 0.2 N NaOH solution the resulting solution is
(A) 0.1 N and the solution is acidic
(B) 0.1 N and the solution is basic
(C) 0.05 N and the solution is basic
(D) 0.05 N and the solution is acidic

44. Which of the following is converted into an alcohol on treatment with HNO_2 ?
(A) Methyl amine (B) Aniline
(C) Diethyl amine (D) Triethyl amine

45. The compound used in the manufacture of terylene is
(A) Ethylene (B) Vinyl chloride
(C) Ethylene glycol (D) Adipic acid

46. Which of the following gives RNC, when reacted with $CHCl_3$ and KOH ?
(A) R_3N (B) $R_4N^+Cl^-$ (C) RNH_2 (D) R_2NH

47. Maltose is made of
(A) Glucose and fructose molecules
(B) Two molecules of sucrose
(C) Two molecules of glucose
(D) Two molecules of fructose

48. Which of the following is not used as an explosive ?
(A) Trinitrotoluene (B) Trinitrobenzene
(C) Picric acid (D) Nitrobenzene

49. Secondary structure of proteins is due to
(A) Peptide bond (B) Hydrogen bond
(C) Covalent bond (D) Coordinate bond

50. For preparing 0.1 N solution of a compound from its impure sample of which the percentage purity is known, the weight of the substance required will be
(A) Same as theoretical weight
(B) More than the theoretical weight
(C) Less than the theoretical weight
(D) None of the above

51. A salt on treatment with dil. HCl gives a pungent smelling gas and a yellow precipitate. The salt gives green flame when tested. The salt solution gives a yellow precipitate with potassium chromate. The salt is
(A) $NiSO_4$ (B) BaS_2O_3 (C) PbS_2O_3 (D) $CuSO_4$

52. The co-ordination number of a metal in co-ordination compounds is
(A) Same as primary valency
(B) Sum of primary and secondary valencies
(C) Same as secondary valency
(D) None of the above

53. Which of the following will have geometrical isomers ?
(A) 2-Methylpropene (B) 2-Butene
(C) 1-Butene (C) Propene

54. Which statement is true for cyclohexane?
(A) It has two possible isomers
(B) It has three conformations
(C) Boat conformation is most stable
(D) Chair and boat conformations differ in energy by 44KJ/mol

55. The average distance of an electron in an atom from its nucleus is of the order of
(A) 10^{-10} m (B) 10^{-15}m (C) 10^6 m (D) 10^{-6}m

56. The maximum probability of finding an electron in the d_{xy} orbital is
(A) Along the x-axis (B) Along the y-axis
(C) Along the x-axis and y axes
(D) At an angle of 45° from the x and y axes

57. The maximum number of electrons accommodated in 5f orbitals are
(A) 5 (B) 10 (C) 14 (D) 18

58. All electrons on the 4p sub shell must be characterized by the quantum number (s)
(A) N = 4, m = 0, s = ± 1/2 (B) l = 1
(C) l = O, s = ± 1/2 (D) s = ± 1/2

59. Which is the correct electron dot structure of N_2O molecule ?
(A) $:N=N=\underset{\cdot\cdot}{\ddot{O}}$ (B) $:N\equiv N^+-\underset{\cdot\cdot}{\ddot{O}}:^-$
(C) $\underset{\cdot\cdot}{\ddot{N}}=\ddot{N}=\underset{\cdot\cdot}{\ddot{O}}$ (D) $:N=N=\underset{\cdot\cdot}{\dot{O}}\vdots$

60. If the solution boils at a temperature T_1 and the solvent at a temperature T_2 the elevation of boiling point is given by
(A) $T_1 + T_2$ (B) $T_1 - T_2$ (C) $T_2 - T_1$ (D) $T_1 \div T_2$

61. Of the following the ion with the smallest ionic radius is
(A) Ti^{3+} (B) Ti^{4+} (C) K^+ (D) Ca^{2+}

62. Due to Frenkel defect, the density of ionic solids
(A) Increases (B) Decreases
(C) Does not change (D) Changes

63. The coordination number of Na^+ in cubic closest packing (CCP) structure of NaCl is
(A) 4 (B) 6 (C) 8 (D) 12

64. Which of the following statements is false for isopentane?
(A) It has three CH_3 groups
(B) It has one CH_2 group (C) It has one CH group
(D) It has a carbon which is not bonded to hydrogen

65. The number of σ bonds in o-xylene is
(A) 6 (B) 9 (C) 12 (D) 18

66. A hydrocarbon has C, 85.72% and remaining H. The hydrocarbon is
(A) C_2H_2 (B) CH_4 (C) C_2H_4 (D) C_2H_6

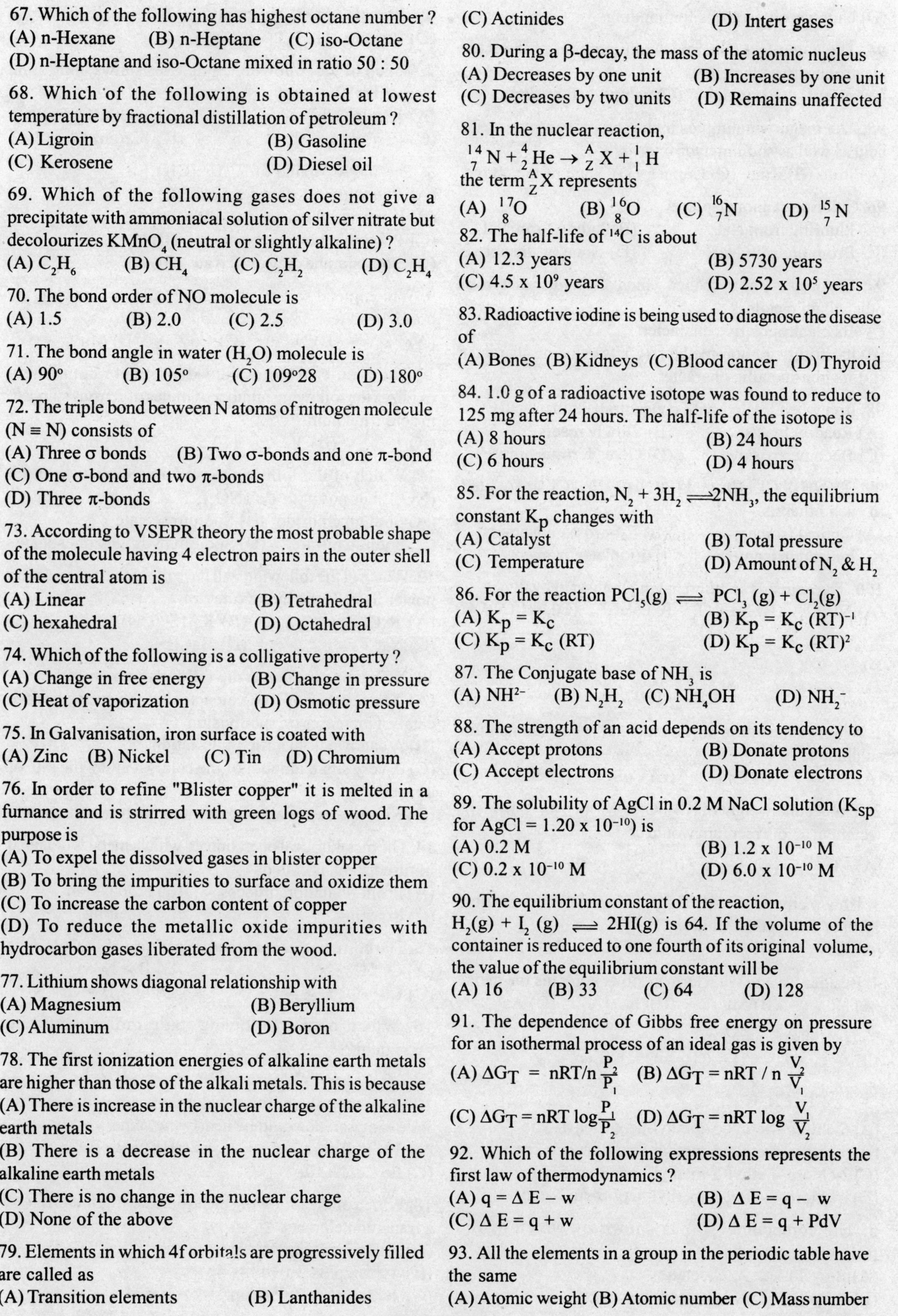

67. Which of the following has highest octane number ?
(A) n-Hexane (B) n-Heptane (C) iso-Octane
(D) n-Heptane and iso-Octane mixed in ratio 50 : 50

68. Which of the following is obtained at lowest temperature by fractional distillation of petroleum ?
(A) Ligroin (B) Gasoline
(C) Kerosene (D) Diesel oil

69. Which of the following gases does not give a precipitate with ammoniacal solution of silver nitrate but decolourizes $KMnO_4$ (neutral or slightly alkaline) ?
(A) C_2H_6 (B) CH_4 (C) C_2H_2 (D) C_2H_4

70. The bond order of NO molecule is
(A) 1.5 (B) 2.0 (C) 2.5 (D) 3.0

71. The bond angle in water (H_2O) molecule is
(A) 90° (B) 105° (C) 109°28 (D) 180°

72. The triple bond between N atoms of nitrogen molecule ($N \equiv N$) consists of
(A) Three σ bonds (B) Two σ-bonds and one π-bond
(C) One σ-bond and two π-bonds
(D) Three π-bonds

73. According to VSEPR theory the most probable shape of the molecule having 4 electron pairs in the outer shell of the central atom is
(A) Linear (B) Tetrahedral
(C) hexahedral (D) Octahedral

74. Which of the following is a colligative property ?
(A) Change in free energy (B) Change in pressure
(C) Heat of vaporization (D) Osmotic pressure

75. In Galvanisation, iron surface is coated with
(A) Zinc (B) Nickel (C) Tin (D) Chromium

76. In order to refine "Blister copper" it is melted in a furnance and is strirred with green logs of wood. The purpose is
(A) To expel the dissolved gases in blister copper
(B) To bring the impurities to surface and oxidize them
(C) To increase the carbon content of copper
(D) To reduce the metallic oxide impurities with hydrocarbon gases liberated from the wood.

77. Lithium shows diagonal relationship with
(A) Magnesium (B) Beryllium
(C) Aluminum (D) Boron

78. The first ionization energies of alkaline earth metals are higher than those of the alkali metals. This is because
(A) There is increase in the nuclear charge of the alkaline earth metals
(B) There is a decrease in the nuclear charge of the alkaline earth metals
(C) There is no change in the nuclear charge
(D) None of the above

79. Elements in which 4f orbitals are progressively filled are called as
(A) Transition elements (B) Lanthanides
(C) Actinides (D) Intert gases

80. During a β-decay, the mass of the atomic nucleus
(A) Decreases by one unit (B) Increases by one unit
(C) Decreases by two units (D) Remains unaffected

81. In the nuclear reaction,
$^{14}_{7}N + ^{4}_{2}He \rightarrow ^{A}_{Z}X + ^{1}_{1}H$
the term $^{A}_{Z}X$ represents
(A) $^{17}_{8}O$ (B) $^{16}_{8}O$ (C) $^{16}_{7}N$ (D) $^{15}_{7}N$

82. The half-life of ^{14}C is about
(A) 12.3 years (B) 5730 years
(C) 4.5×10^9 years (D) 2.52×10^5 years

83. Radioactive iodine is being used to diagnose the disease of
(A) Bones (B) Kidneys (C) Blood cancer (D) Thyroid

84. 1.0 g of a radioactive isotope was found to reduce to 125 mg after 24 hours. The half-life of the isotope is
(A) 8 hours (B) 24 hours
(C) 6 hours (D) 4 hours

85. For the reaction, $N_2 + 3H_2 \rightleftharpoons 2NH_3$, the equilibrium constant K_p changes with
(A) Catalyst (B) Total pressure
(C) Temperature (D) Amount of N_2 & H_2

86. For the reaction $PCl_5(g) \rightleftharpoons PCl_3(g) + Cl_2(g)$
(A) $K_p = K_c$ (B) $K_p = K_c(RT)^{-1}$
(C) $K_p = K_c(RT)$ (D) $K_p = K_c(RT)^2$

87. The Conjugate base of NH_3 is
(A) NH^{2-} (B) N_2H_2 (C) NH_4OH (D) NH_2^-

88. The strength of an acid depends on its tendency to
(A) Accept protons (B) Donate protons
(C) Accept electrons (D) Donate electrons

89. The solubility of AgCl in 0.2 M NaCl solution (K_{sp} for AgCl = 1.20×10^{-10}) is
(A) 0.2 M (B) 1.2×10^{-10} M
(C) 0.2×10^{-10} M (D) 6.0×10^{-10} M

90. The equilibrium constant of the reaction, $H_2(g) + I_2(g) \rightleftharpoons 2HI(g)$ is 64. If the volume of the container is reduced to one fourth of its original volume, the value of the equilibrium constant will be
(A) 16 (B) 33 (C) 64 (D) 128

91. The dependence of Gibbs free energy on pressure for an isothermal process of an ideal gas is given by
(A) $\Delta G_T = nRT/n \frac{P_2}{P_1}$ (B) $\Delta G_T = nRT / n \frac{V_2}{V_1}$
(C) $\Delta G_T = nRT \log \frac{P_1}{P_2}$ (D) $\Delta G_T = nRT \log \frac{V_1}{V_2}$

92. Which of the following expressions represents the first law of thermodynamics ?
(A) $q = \Delta E - w$ (B) $\Delta E = q - w$
(C) $\Delta E = q + w$ (D) $\Delta E = q + PdV$

93. All the elements in a group in the periodic table have the same
(A) Atomic weight (B) Atomic number (C) Mass number

(D) Number of electrons for bonding

94. The outermost electronic configuration of the most electronegative element is
(A) ns^2np^5 (B) ns^2np^6 (C) ns^2np^3 (D) ns^2np^4

95. The metal which gives hydrogen on treatment with acid as well as sodium hydroxide is
(A) Iron (B) Zinc (C) Copper (D) None of the above

96. Chlorine cannot displace
(A) Fluorine from NaF (B) Iodine form NaI
(C) Bromine from NaBr (D) None of the above

97. The property of hydrogen which distinguishes it from other alkali metals is
(A) Its electropositive character
(B) Its affinity for non-metals (C) Its reducing character
(D) Its non-metallic character

98. Iron when treated with concentrated nitric acid
(A) Readily reacts (B) Slowly reacts
(C) Becomes passive (D) Gives ferrous nitrate

99. $[Pt(NH_3)_4 Cl_2] Br_2$ and $[Pt(NH_3)_4 Br_2] Cl_2$ are related to each other as
(A) Optical isomers (B) Co-ordinate isomers
(C) Ionization isomers (D) Linkage isomers

100. The aqueous solution having highest pH value is
(A) NH_4Cl (B) Na_2CO_3 (C) NaCl (D) CH_3COOH

PET - 1997

1. Which of the following is soluble in yellow ammonium sulphide?
(A) SnS (B) PbS (C) CuS (D) CdS

2. If molecular weight of $KMnO_4$ is 'M' then its equivalent weight in acidic medium would be
(A) M (B) $\frac{M}{2}$ (C) $\frac{M}{5}$ (D) $\frac{M}{4}$

3. How many ml of 1M H_2SO_4 is required to neutralise 10ml of 1M NaOH solution?
(A) 2.5 (B) 5.0 (C) 10.0 (D) 20.0

4. Besides Zn and Cu, German silver contains the metal
(A) Ni (B) Mg (C) Sn (D) Ag

5. The IUPAC name of

$$\underset{}{CH_3} - \underset{|\atop Cl}{C} = \underset{|\atop CH_3}{C} - \underset{|\atop C_2H_5}{CH} - CH_2 - C \equiv CH \text{ is}$$

(A) 6-chloro-4-ethyl-5-methyl-hept-5-en-1-yne
(B) 6-chloro-4-ethyl-5-methyl-hept-1-yne-5-ene
(C) 2-chloro-4-ethyl-3-methyl-hept-2-en-6-yne
(D) 2-chloro-4-ethyl-3-methyl-hept-6-yn-2-ene

6. The compound [structure] is known by which of the following names?
(A) Bicyclo - [2, 2, 2] octane
(B) Bicyclo - [2, 2, 1] octane
(C) Bicyclo - [1, 2, 1] octane
(D) Bicyclo - [1, 1, 1] octane

7. Which of the following complexes shows ionization isomerism?
(A) $[Cr(en)_3Cl_3]$ (B) $[Co(NH_3)_5Br]SO_4$
(C) $[Cr(NH_3)_6Cl_3]$ (D) $[Cr(en)_2Cl_2]$

8. The IUPAC name of $K_3 [Fe(CN)_6]$ is
(A) Potassiumferrohexacyanate (II)
(B) Potassiumferrocyanide (II)
(C) Potassiumhexaferrocyanate (III)
(D) Potassiumhexacyanoferrate (III)

9. Which metal salt gives a violet coloured bead in the borax bead test?
(A) Fe^{2+} (B) Ni^{2+} (C) Co^{2+} (D) Mn^{2+}

10. One mole of potassium dichromate completely oxidizes the following number of moles of ferrous sulphate in acidic medium
(A) 1 (B) 3 (C) 5 (D) 6

11. Which of the following gives a precipitate with Pb $(NO_3)_2$ but not with Ba $(NO_3)_2$?
(A) Sodium chloride (B) Sodium acetate
(C) Sodium nitrate (D) Sodium hydrogen phosphate

12. Which of the following will have the highest coagulation power for a Ferric hydroxide Sol?
(A) K_2CrO_4 (B) $K_3 [Fe(CN)_6]$
(C) NaCl (D) $BaCl_2$

13. Which one is false in the following statements?
(A) The number of free valencies on the surface of the catalyst increases on subdivision
(B) A catalyst is specific in its action
(C) A very small amount of the catalyst alters the rate of a reaction
(D) Ni is used as catalyst in the manufacture of ammonia

14. The metallurgical operation, in which, metal is obtained in fused state, is called :
(A) Froth floatation process (B) Calcination
(C) Roasting (D) Smelting

15. The correct order of electron affinity of B, C, N, O is
(A) $O > C > N > B$ (B) $B > N > C > O$
(C) $O > C > B > N$ (D) $O > B > C > N$

16. Which of the following ionic radius would be maximum?
(A) O^{2-} (B) Mg^{2+} (C) C^{4-} (D) N^{3-}

17. The atomic radius of elements of which of the following series would be nearly the same
(A) Na K Rb Cs (B) Li Be B C
(C) Fe Co, Ni Cu (D) F Cl Br I

18. Which of the following electronic configurations is of a transition element?
(A) $1s^22s^22p^63s^23p^64s^2$
(B) $1s^22s^22p^23s^23p^63d^{10}4s^24p^1$
(C) $1s^22s^22p^63p^63d^{10}4s^24p^6$

(D) $1s^2 2s^2 2p^6 3s^2 3p^6 3d^2 4s^2$

19. Which of the following metals is obtained by electrolytic reduction process ?
(A) Ag (B) Al (C) Fe (D) Cu

20. Who developed the long from of periodic table ?
(A) Lothar Meyer (B) Niels Bohr
(C) Mendeleev (D) Moseley

21. The correct sequence of elements in decreasing order of first ionization energy is
(A) Al > Mg > Na (B) Mg > Al > Na
(C) Na > Mg > Al (D) Mg > Na > Al

22. Which of the following statements is correct ?
(A) X^- ion is larger in size than X atom
(B) X^+ ion is larger in size than X atom
(C) X^+ ion is larger in size than X^- ion
(D) X^+ and X^- ions are equal in size

23. The molecule having bond order 3, is
(A) O^2 (B) He_2^+ (C) H_2 (D) N_2

24. Point out incorrect statement about resonance
(A) In resonance structures, there should not be the same number of electron pairs
(B) Resonance structures should differ only in the location of electrons around the constituent atoms
(C) Resonance structures should have equal energy
(D) In resonance structures, the constituent atoms should be in the same position

25. Which one is false in the following statements ?
(A) Each carbon in benzene is in sp^2 hybridization
(B) Each carbon in ethylene is in sp^2 hybridization
(C) Each carbon in acetylene is in sp^3 hybridization
(D) Each carbon in ethane is in sp^3 hybridization

26. The hydrogen bond is strongest in
(A) O – H ..... S (B) S – H ..... O
(C) F – H ..... F (D) F – H ..... O

27. The geometry of ClO_3^-, according to Valence Shell Electron Pair Repulsion (VSEPR) theory will be
(A) Planar Triangle (B) Pyramidal
(C) tetrahedral (D) Square planar

28. The ground state electronic configuration of Chromium is
(A) $[Ar]3d^5\ 4s^1$ (B) $[Ar]3d^4\ 4s^2$
(C) $[Ar]3d^6\ 4s^0$ (D) $[Kr]4d^5\ 4s^1$

29. An electron has principal quantum number 3. The number of its (i) sub-shells and (ii) orbitals would be respectively
(A) 3 and 5 (B) 3 and 7
(C) 3 and 9 (D) 2 and 5

30. The uncertainty principal and the concept of wave nature of matter was proposed by .... and .... respectively.
(A) Heisenberg, Planck
(B) Planck, Heisenberg
(C) Heisenberg, de Broglie
(D) De Broglie, Heisenberg

31. Which of the following statements is incorrect ?
(A) Mass defect is related with binding energy
(B) 'Meson' was discovered by Yukawa
(C) The size of the nucleus is of the order of 10^{-12} – 10^{-13} cm
(D) Magnetic quantum number is a measure of 'orbital' angular momentum of the electron

32. Empirical formula of a hydrocarbon containing 80% carbon and 20% hydrogen is
(A) CH_3 (B) CH_4 (C) CH (D) CH_2

33. The total number of possible isomeric trimethyl benzene is
(A) 2 (B) 3 (C) 4 (D) 6

34. Propene when heated with chlorine at about 500° forms
(A) $CH_2Cl - CH = CH_2$ (B) $CH_3 - CHCl - CH_2Cl$
(C) $CH_2Cl - CHCl - CH_2Cl$ (D) All the above

35. In the reaction,
$$CH_3-C \equiv C-CH_3 \xrightarrow[\text{(ii) } H_2O/Zn]{\text{(i) X}} CH_3-\underset{\underset{O}{\|}}{C}-\underset{\underset{O}{\|}}{C}-CH_3$$
X is
(A) HNO_3 (B) O_2 (C) O_3 (D) $KMnO_4$

36. Chloroprene is used in making
(A) Synthetic rubber (B) Plastic
(C) Petrol (D) All of the above

37. Gammexane is
(A) Chloral (B) Hexachloroethane
(C) D.D.T. (D) Benzene hexachloride

38. Identify Z in the following series :
$$C_2H_5I \xrightarrow{KOH} X \xrightarrow{Br_2} Y \xrightarrow{KCN} Z$$
(A) CH_3CH_2CN (B) $CN-CH_2-CH_2CN$
(C) $Br-CH_2-CH_2CN$ (D) $Br-CH = CHCN$

39. An element (X) forms compounds of the formula XCl_3, X_2O_5 and Ca_3X_2 but does not form XCl_5 which of the following is the element X ?
(A) B (B) Al (C) N (D) P

40. Which of the following oxides of nitrogen reacts with ferrous sulphate to form a dark brown compound ?
(A) NO_2 (B) N_2O_3 (C) N_2O (D) NO

41. Which of the following halogens does not form oxyacid ?
(A) Fluorine (B) Chlorine
(C) Bromine (D) Iodine

42. The strongest of the following acids is
(A) HF (B) HI (C) HBr (D) HCl

43. Nessler's reagent is
(A) $KHgI_4$ (B) $K_2HgI_4 + NH_4OH$
(C) $K_2HgI_4 + KOH$ (D) $KHgI_4 + NH_4OH$

44. Helium is used with oxygen in the apparatus of divers because it

(A) is lighter than nitrogen
(B) Is not soluble in blood at high pressure
(C) Is easily available
(D) Is less reactive than nitrogen

45. Gradual addition of electronic shells in the noble gases causes a decrease in their
(A) Ionization energy (B) Atomic radius
(C) Boiling point (D) Density

46. Very dilute nitric acid reacts with zinc to form zinc nitrate and
(A) Ammonium nitrate (B) NO_2
(C) NO (D) N_2O

47. By passing H_2S gas in acidified $KMnO_4$ solution, we get
(A) K_2SO_3 (B) MnO_2 (C) K_2S (D) S

48. Unlike other halogens, Fluorine does not show higher oxidation states, because
(A) It is highly electronegative
(B) It has no d-orbitals
(C) Its atomic radius is very small
(D) The F^- ion is stable and isoelectronic with neon

49. Which of the following hydrides has the lowest boiling point ?
(A) H_2Se (B) H_2Te (C) H_2O (D) H_2S

50. In which of the following complexes, oxidation state of the metal is zero ?
(A) $[Pt(NH_3)_2Cl_2]$ (B) $[Cr(CO)_6]$
(C) $[Cr(NH_3)_3Cl_3]$ (D) $[Cr(en)_2Cl_2]$

51. Ethyl bromide reacts with lead-sodium alloy to form
(A) Tetraethyl lead (B) Tetraethyl bromide
(C) (A) and (B) both (D) None of the above

52. If chloroform is left open in air in the presence of sun-rays it gives
(A) Mustard gas (B) Lewisite
(C) Carbon tetrachloride (D) Carbonyl chloride

53. Which of the following compounds gives a ketone with Grignard reagent ?
(A) Formaldehyde (B) Ethyl alcohol
(C) Methyl cyanide (D) Methyl iodide

54. Which of the following substances is used as an anti-knock compound ?
(A) Tetraethyl lead (B) Lead tetrachloride
(C) Lead acetate (D) Ethyl acetate

55. Which of the following compounds gives a positive iodoform test ?
(A) Pentanal (B) 1-Phenyl ethanol
(C) 2-Phenyl ethanol (D) 3- Pentanol

56. For preparing a buffer solution of pH 6 by mixing sodium acetate and acetic acid, the ratio of the concentration of salt and acid should be ($K_a = 10^{-5}$)
(A) 1 : 10 (B) 10 : 1 (C) 100 : 1 (D) 1 : 100

57. For which one of the following reactions, Kp = Kc ?
(A) $N_2 + 3H_2 \rightleftharpoons 2NH_3$ (B) $N_2 + O_2 \rightleftharpoons 2NO$
(C) $PCl_5 \rightleftharpoons PCl_3 + Cl_2$ (D) $2SO_3 \rightleftharpoons 2SO_2 + O_2$

58. The equilibrium constant for the reversible reaction, $N_2 + 3H_2 \rightleftharpoons 2NH_3$ is K and for the reaction $\frac{1}{2}N_2 + \frac{3}{2}H_2 \rightleftharpoons NH_3$ the equilibrium constant is K', K and K' will be related as
(A) K = K' (B) $K' = \sqrt{K}$
(C) $K = \sqrt{K'}$ (D) K x K' = 1

59. At 20°C the Ag^+ ion concentration in a saturated solution of Ag_2CrO_4 is 1.5 x 10^{-4} mole/litre at 20°C. The solubility product of Ag_2CrO_4 would be
(A) 3.3750 x 10^{-12} (B) 1.6875 x 10^{-10}
(C) 1.6875 x 10^{-12} (D) 1.6875 x 10^{-11}

60. For the reaction, : $CO(g) + \frac{1}{2} O_2(g) \rightarrow CO_2(g)$, which one of the statements is correct at constant T and P ?
(A) $\Delta H > \Delta E$
(B) $\Delta H = \Delta E$
(C) $\Delta H < \Delta E$
(D) ΔH is independent of physical state of the reagents

61. Of the following anions, the strongest Bronsted base is
(A) ClO_3^- (B) ClO_4^- (C) ClO^- (D) ClO_2

62. Degree of dissociation of 0.1N CH_3COOH is (dissociation const. = 1 x 10^{-5})
(A) 10^{-5} (B) 10^{-4} (C) 10^{-3} (D) 10^{-2}

63. The bond dissociation energies of gaseous H_2, Cl_2 and HCl are 140, 58 and 103 kcal respectively. The enthalpy of formation of HCl gas would be
(A) –44 kcal (B) –88 kcal
(C) –22 kcal (D) –11 kcal

64. Mark the correct statement
(A) For a chemical reaction to be feasible ΔG should be positive
(B) The total energy of an isolated system is constant
(C) For a chemical reaction to be feasible, ΔG should be zero
(D) Entropy is a measure of order in a system

65. Which one of the following is not an ideal solution ?
(A) $C_2H_5Br + C_2H_5I$
(B) $C_6H_5Cl + C_6H_5Br$
(C) $C_6H_6 + C_6H_5CH_3$
(D) $C_2H_5I + C_2H_5OH$

66. The end product of (4n + 2) radioactive disintegration series is
(A) $^{208}_{82}Pb$ (B) $^{206}_{82}Pb$ (C) $^{207}_{82}Pb$ (D) $^{210}_{83}Bi$

67. "The relative lowering of the vapour pressures is equal to the mole fraction of the solute". The law is called
(A) Henry's law (B) Raoult's law
(C) Ostwald's law (D) Arrhenius's law

68. The order of osmotic pressures of equimolar solutions of $BaCl_2$ NaCl and glucose will be

(A) Glucose > $BaCl_2$ > NaCl
(B) $BaCl_2$ > NaCl > Glucose
(C) NaCl > $BaCl_2$ > Glucose
(D) Glucose > NaCl > $BaCl_2$

69. Molal depression constant for water is 1.86°C. The freezing point of a 0.05 molal solution of a non-electrolyte in water is
(A) −1.86°C (B) −0.93°C
(C) −0.093°C (D) 0.93°C

70. The correct statement in the following is
(A) In ionic compounds having Frenkel defect, the ratio r^+/r^- is high
(B) The ionic crystal of AgBr has Schottky defect
(C) The unit cell having crystal parameters $a = b \neq c$, $\alpha = \beta = 90°$, $\gamma = 120°$ is hexagonal
(D) The coordination number of Na^+ ion in NaCl is 4

71. The compound which reacts fastest with Lucas reagent at room temperature is
(A) Butane -1-ol (B) 2-Methyl Propan-2-ol
(C) 2-Methyl Propan-1-ol (D) Butan -2-ol

72. Methyl ter-butyl ether on heating with HI of one molar concentration gives
(A) $CH_3I + (CH_3)_3COH$ (B) $CH_3OH+(CH_3)_3Cl$
(C) $CH_3I + (CH_3)_3Cl$ (D) None of the above

73. The product of the following reaction,

$CH_3—C(CH_3)_2—CH=CH_2$ $\xrightarrow[\text{(ii) NaBH}_4]{\text{(i) Hg (OAC)}_2\text{, H}_2\text{O}}$

is

(A) $CH_3—C(CH_3)_2—CH(OH)—CH_3$

(B) $CH_3—C(CH_3)_2—CH_2—CH_2OH$

(C) 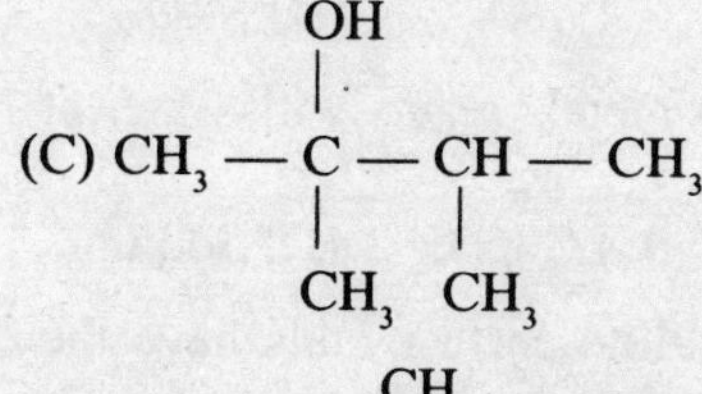

(D) 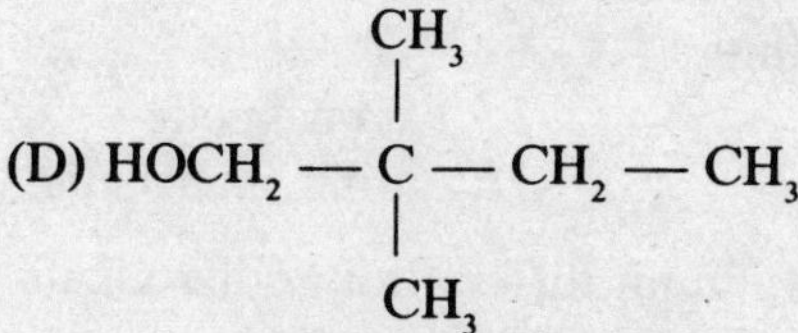

74. Reaction of acetaldehyde with HCN followed by hydrolysis gives a compound which shows
(A) Optical isomerism (B) Geometrical isomerism
(C) Metamerism (D) Tautomerism

75. A compound A has a molecular formula C_2Cl_3OH. It reduces Fehling solution and on oxidation gives a monocarboxylic acid B. A is obtained by action of chlorine on ethyl alcohol. A is
(A) CH_3Cl (B) Chloroacetic acid
(C) Chloral (D) $CHCl_3$

76. Formaldehyde when treated with KOH gives methanol and potassium formate. The reaction is known as
(A) Perkin reaction (B) Claisen reaction
(C) Cannizzaro reaction (D) Knoevenagel reaction

77. The number of NaCl molecules in unit cell of its crystal is
(A) 2 (B) 4 (C) 6 (D) 8

78. What is 'X' in the following nuclear reaction ?
${}_7N^{14} + {}_1H^1 \rightarrow {}_8O^{15}$ + 'X'
(A) ${}_0n^1$ (B) ${}_{-1}e^0$ (C) ${}_{+1}e^0$ (D) γ

79. The radioactive isotope which is used to estimate the age of ancient geological formations is
(A) Fe^{59} (B) C^{14} (C) O^8 (D) Pb^{212}

80. The half-life period of a radioactive element is 140 days. In 560 days a sample of this element would be reduced to .... of its initial mass.
(A) 1/2 (B) 1/4 (C) 1/8 (D) 1/16

81. On the basis of Le Chatelier's principle, predict which of the following conditions would be unfavourable for the formation of SO_3 ? Given that, $2SO_2 + O_2 \rightleftharpoons 2SO_3$, $\Delta H = -42$ kcal
(A) Low temperature (B) High pressure
(C) High temperature (D) High concentration of SO_2

82. The product D of the reaction,

$CH_3Cl \xrightarrow{KCN} (A) \xrightarrow{H_2O} (B) \xrightarrow{NH_3} (C) \xrightarrow{\Delta} (D)$ is

(A) $CH_3CH_2NH_2$ (B) CH_3CN
(C) $HCONH_2$ (D) CH_3CONH_2

83. When phenol is heated with phthalic anhydride in the presence of concentrated H_2SO_4, we get
(A) Benzene (B) Benzoic acid
(C) Phthalic acid (D) Phenolphthalein

84. The number of possible structural isomers of $C_4H_{10}O$ is
(A) 5 (B) 6 (C) 7 (D) 8

85. Which of the following compounds will exhibit geometrical isomerism ?
(A) 2-Butene (B) 2-Butyne
(C) 2-Butanol (D) Butanal

86. If the radius of first Bohr-orbit be a°, then the radius of third orbit would be
(A) 3 x a° (B) 6 x a° (C) 9 x a° (D) 1/9 x a°

87. The following reaction

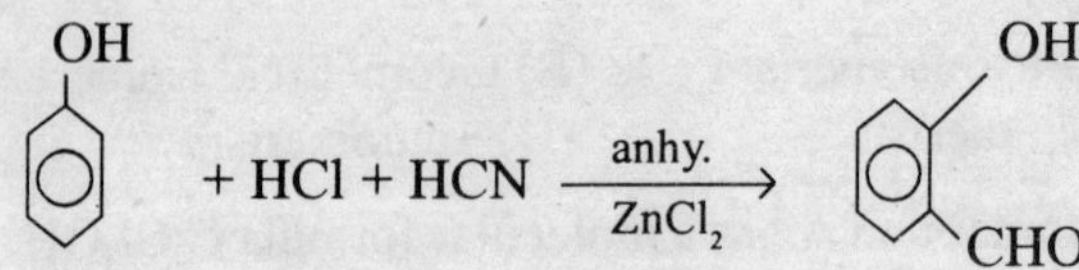

is known as

(A) Perkin reaction (B) Gattermann reaction
(C) Kolbe reaction (D) Gattermann-Koch reaction

88. $(Cl-C_6H_4)_2C(H)-CCl_3$

The above structural formula refers to

(A) DDT (B) RNA (C) BHC (D) DNA

89. The shape of methane molecule is

(A) Tetrahedral (B) Octahedral
(C) Linear (D) Triangular

90. The main structural feature of protein is

(A) Ester linkage (B) Ether linkage
(C) Peptide linkage (D) All of the above

91. Bond length between carbon in ethylene molecule is

(A) 1.19 Å (B) 2.4 Å (C) 1.54 Å (D) 1.35 Å

92. Reimer-Tiemann reaction involves a

(A) Carbanion intermediate
(B) Carbonium ion intermediate
(C) Carbene intermediate
(D) Free radical intermediate

93. A reaction is not feasible, if

(A) ΔH is positive and ΔS is also positive
(B) ΔH is negative and ΔS is also negative
(C) ΔH is negative and ΔS is positive
(D) ΔH is positive and ΔS is negative

94. A first order reaction is 75% complete after 32 minutes. When was 50% of the reaction completed ?

(A) 16 minutes (B) 8 minutes
(C) 4 minutes (D) 32 minutes

95. The first order rate constant for the decomposition of N_2O_5 is 6.2×10^{-4} $second^{-1}$. The half life period for this decomposition is

(A) 1117.7 (B) 111.7 (C) 223.4 (D) 160.9

96. The Arrhenius equation expressing the effect of temperature on the rate constant is

(A) $k = e^{-E_a/RT}$ (B) $k = \frac{E_a}{RT}$
(C) $k = \log_e \frac{E_a}{RT}$ (D) $k = Ae^{-E_a/RT}$

97. Which one of the following statements is correct ?

(A) The oxidation number of oxygen in KO_2 is zero
(B) The specific conductance of an electrolyte solution decreases with increase in dilution
(C) Sn^{2+} oxidizes Fe^{3+}
(D) Zn / $ZnSO_4$ is a reference electrode

98. What would be the weight of deposited silver on passing 965 coulombs of electricity in a solution of silver nitrate ?

(A) 1.08 gm (B) 2.16 gm
(C) 0.54 gm (D) 0.27 gm

99. Mark the false statement

(A) Nernst equation for single electrode $E = E^o - RT/nF \ln a[M^{n+}]$
(B) The efficiency of a hydrogen oxygen fuel cell is 23%
(C) A salt bridge is used to eliminate liquid junction potential
(D) The Gibbs free energy change, ΔG is related with electromotive force E, as $\Delta G = -nFE$

100. Point out the false statement

(A) Brownian movement and Tyndall effect is shown by colloidal systems.
(B) Gold number is a measure of the protective power of a lyophilic colloid.
(C) The colloidal solution of a liquid in liquid is called is gel.
(D) Hardy-Schulze rule is related with coagulation.

PET - 1998

1. In the following, the element with the highest electropositivity is

(A) Barium (B) Chromium
(C) Copper (D) Caesium

2. The stability of the following alkali metal chlorides follows the order

(A) NaCl > KCl > LiCl > CsCl
(B) LiCl > KCl > NaCl > CsCl
(C) CsCl > KCl > NaCl > LiCl
(D) KCl > CsCl > NaCl > CiCl

3. Which of the following ligands forms a chelate ?

(A) Acetate (B) Oxalate
(C) Cyanide (D) Ammonia

4. For which one of the following properties of halogens, the sequence, F > Cl > Br > I holds good ?

(A) Electronaffinity (B) Electronegativity
(C) Atomic radius (D) Boiling point

5. The electronic configuration of a metal M is $1s^2 2s^2 2p^6 3s^1$. The formula of its oxide will be

(A) M_2O_3 (B) MO_2 (C) MO (D) M_2O

6. Which of the following metals reacts with water at room temperature ?

(A) Nickel (B) Silver (C) Copper (D) Sodium

7. The compounds of alkaline earth metals have the following magnetic nature

(A) Diamagnetic (B) Paramagnetic
(C) Ferromagnetic (D) Antiferromagnetic

8. In certain matters Lithium differs from other alkali metals, the main reason for this is

(A) Small size of Li atom and Li^+ ion
(B) Extremely high electropositivity of Li

(C) Greater hardness of Li
(D) Hydration of Li^+ ion

9. For the first order reaction with the constant k, which expression gives the half-life period ?
(Initial concentration = a)
(A) $\frac{\ln^2}{k}$ (B) $\frac{1}{k_a}$ (C) $\frac{0.693}{k_a}$ (D) $\frac{3}{2k_a^2}$

10. The oxidation number of chromium in $K_2Cr_2O_7$ is
(A) –6 (B) + 6 (C) + 2 (D) – 2

11. In which of the following Tyndall effect is not observed ?
(A) Suspensions (B) Emulsions
(C) Sugar solution (D) Gold sol

12. Which of the following is a lyophilic colloid ?
(A) Fog (B) Blood (C) Milk (D) Gum

13. Which of the following kinds of catalysis can be explained by the adsorption theory ?
(A) Homogeneous catalysis (B) Acid-base catalysis
(C) Heterogeneous catalysis (D) Enzyme catalysis

14. k_p / k_c for the reaction $CO(g) + \frac{1}{2}O_2(g) \rightleftharpoons CO_2(g)$ is
(A) 1 (B) RT (C) $1/\sqrt{RT}$ (D) $(RT)^{1/2}$

15. The metal extracted by electrolysis of its fused salt is
(A) Sodium (B) Copper (C) Iron (D) Lead

16. The rate constant is given by the equation, $k = pze^{-E/RT}$. Which factor should register a decrease for the reaction to proceed more rapidly ?
(A) T (B) Z (C) E (D) P

17. For a reaction of the type, A + B → products, it is observed that doubling the concentration of A causes the reaction rate to be four times as great, but doubling the amount of B does not affect the rate. The rate equation is
(A) Rate = k [A] [B] (B) Rate = $k/4.[A]^2$
(C) Rate = $k [A]^2 [B]$ (D) Rate = $k [A]^2 [B]^2$

18. In which cell, the free energy of a chemical reaction is directly converted into electricity ?
(A) Fuel cell (B) Lead storage battery
(C) Leclanche cell (D) Concentration cell

19. The number of coulombs required for the deposition of 107.870 g of silver is
(A) 96,500 (B) 45, 250
(C) 1,93,000 (D) 10,000

20. Heterolytic fission of an organic covalent bond gives
(A) Free radicals (B) Both cation and anion
(C) Only cation (D) Only anion

21. The compound most reactive in reaction towards electrophilic reagent is
(A) Nitrobenzene (B) Acetophenone
(C) Toluene (D) Benzene

22. Which of the following expressions gives the de Broglie relationship ?
(A) $\lambda = \frac{h}{mv} = p$ (B) $\lambda = \frac{h}{mv}$
(C) $\lambda = \frac{h}{mp}$ (D) $\lambda m = \frac{v}{p}$

23. What is the electronic configuration of Cu^{2+} (z = 29) ?
(A) [Ar] $4s^1 3d^8$ (B) [Ar] $4s^2 3d^{10} 4p^1$
(C) [Ar] $4s^1 3d^{10}$ (D) [Ar] $3d^9$

24. Which of the following orbitals has a dumb-bell shape ?
(A) s (B) p (C) d (D) f

25. The alcohol which does not give a stable compound on dehydration is
(A) Ethyl alcohol (B) Methyl alcohol
(C) n-Propyl alcohol (D) n-Butyl alcohol

26. Which compound is not a lipid ?
(A) Cerebroside (B) Cephalin
(C) Lecithin (D) Lysine

27. The maximum number of electrons in an atom with l = 2 and n = 3 is
(A) 2 (B) 6 (C) 12 (D) 10

28. Which of the following statements regarding covalent bond is not true ?
(A) The strength of the bond depends upon the extent of overlapping
(B) The electrons are shared between atoms
(C) The bond is non-directional
(D) The bond formed may or may not be polar

29. On going from right to left in a period in the periodic table the electronegativity of the elements
(A) Increases
(B) Decreases
(C) Remains unchanged
(D) Decreases first then increases

30. In the periodic table, the element with atomic number 16 is placed in the group
(A) Third (B) Fourth (C) Fifth (D) Sixth

31. Metals belonging to the same group in the periodic table are
(A) Magnesium and barium
(B) Magnesium and sodium
(C) Magnesium and copper
(D) Magnesium and potassium

32. In the following, the element with the highest electronegativity is
(A) Carbon (B) Magnesium
(C) Oxygen (D) Sulphur

33. The electronic configuration in the outermost sheil of halogens is
(A) s^2p^3 (B) s^2p^6 (C) s^2p^4 (D) s^2p^5

34. The Ionic radii N^{3-}, O^{2-} F^- and Na^+ follow the order
(A) $N^{3-} > O^{2-} > F^- > Na^+$

(B) $N^{3-} > Na^+ > O^{2-} > F^-$
(C) $Na^+ > O^{2-} > N^{3-} > F^-$
(D) $O^{2-} > F^- > Na^+ > N^{3-}$

35. Elements with outer electronic configuration ns^2np^6 are
(A) Alkaline earth metals (B) Transition elements
(C) Chalcogens (D) Noble gases

36. Directed bonds in water form an angle of
(A) 90° (B) 120° (C) 105° (D) 60°

37. Hydrogen bonding is not present in
(A) Hydrogen (B) Hydrogen Fluride
(C) Glycerine (D) Water

38. sp^3 hybridization leads to which shape of the molecule ?
(A) Tetrahedron (B) Octahedron
(C) Linear (D) Plane triangle

39. The important oxide ore of iron is
(A) Siderite (B) Haematite
(C) Pyrites (D) Bauxite

40. The metal extracted by the cyanide process is
(A) Silver (B) Copper
(C) Iron (D) Sodium

41. Malachite is an ore of
(A) Iron (B) Zinc (C) Copper (D) Mercury

42. German silver is an alloy of
(A) Silver, copper and gold
(B) Copper, zinc and nickel
(C) Copper, zinc and tin
(D) Iron, chromium and nickel

43. Which of the following compounds gives dye test ?
(A) Diphenylamine (B) Ethylamine
(C) Aniline (D) Methylamine

44. What happens when 2-hydroxy benzoic acid is distilled with zinc dust ? It gives
(A) Benzaldehyde (B) A Polymeric compound
(C) Phenol (D) Benzoic acid

45. What amount of bromine will be required to convert 2g of phenol into 2,4,6 tribromophenol
(A) 4.00 (B) 6.00 (C) 10.22 (D) 20.44

46. Amongst the following compounds which is most reactive towards nitration ?
(A) Benzene (B) Nitrobenzene
(C) Toluene (D) Chlorobenzene

47. The reaction in which sodium cyanide is used is
(A) Perkin reaction (B) Reimer-Tiemann reaction
(C) Benzoin condensation
(D) Rosenmund reaction

48. The number of molecules in 16gm of methane is
(A) 3.0×10^{23} (B) 6.02×10^{23}
(C) $\frac{16}{6.02} \times 10^{23}$ (D) $\frac{16}{3.0} \times 10^{23}$

49. An unknown compound A has the molecular formula C_4H_6. When 'A' is treated with an excess of Br_2, a new substance 'B' with formula $C_4H_6Br_4$ is formed. 'A' forms a white precipitate with ammoniacal silver nitrate solution. A may be
(A) Butene-1 (B) Butene-2
(C) Butyne-1 (D) Butyne-2

50. Reaction of nitrous acid with aliphatic primary amine in the cold gives
(A) A diazonium salt (B) An alcohol
(C) a nitrile (D) a dye

51. Teflon is a polymer of
(A) Acetylene (B) Tetrafluoroethene
(C) Vinyl chloride (D) Ethylene

52. Which of the following will have the highest boiling point at 1 atm pressure ?
(A) 0.1 M NaCl (B) 0.1 M Sucrose
(C) 0.1 M $BaCl_2$ (D) 0.1 M Glucose

53. Diamond is an example of
(A) Covalent solid (B) Electrovalent solid
(C) Solid with hydrogen bonding (D) Glass

54. Which element is the end product of every natural radioactive series ?
(A) C (B) Bi (C) Pb (D) Sn

55. In radioactive decay, which one of the following moves the fastest ?
(A) α-particle (B) β-particle
(C) γ-rays (D) Positron

56. Which of the following notations shows the product incorrectly ?
(A) $^{242}_{96}Cm(\alpha,2n)\,^{243}_{97}Bk$ (B) $^{10}_{5}B(\alpha,n)\,^{13}_{7}N$
(C) $^{14}_{7}N(n,p)\,^{14}_{6}C$ (D) $^{28}_{14}Si(d,n)\,^{29}_{15}P$

57. In terms of energy, 1 a.m.u. is equal to
(A) 100 J (B) 931.1 MeV
(C) 931.1 kcal (D) 10^7 erg

58. Methanal and phenol react in the presence of base to give
(A) Dacron (B) Nylon - 66
(C) Bakelite (D) Polyethylene

59. Lucas reagent is
(A) Tin + Hydrochloric acid
(B) Zinc + Hydrochloric acid
(C) Zinc + Caustic soda
(D) Tin + Caustic soda

60. The compound formed as a result of potassium permanganate oxidation of ethylbenzene is
(A) Benzoic acid (B) Benzyl alcohol
(C) Benzophenone (D) Acetophenone

61. What will be the molality of a solution having 18 g of glucose (mol. wt. = 180) dissolved in 500 g of water ?
(A) 1m (B) 0.5 m (C) 0.2 m (D) 2 m

62. An aqueous solution freezes at $-0.186^{o}C$ ($k_f = 1.86^{o}$; $k_b = 0.512^{o}$) What is the elevation in boiling point ?
(A) 0.512/1.86 (B) 0.0512 (C) 0.186 (D) 0.512

63. Which of the following is a colligative property ?
(A) Surface Tension (B) Osmotic Pressure
(C) Viscosity (D) Refractive Index

64. How many kinds of space lattices are possible in a crystal ?
(A) 23 (B) 7 (C) 230 (D) 14

65. Potassium crystallizes with a
(A) Simple cubic lattice
(B) Face-centred cubic lattice
(C) Body- centred cubic lattice
(D) Orthorhombic lattice

66. In the Nitrogen family the H-M-H bond angle in the hydrides MH_3 gradually becomes closer to 90° on going from N to Sb. This shows that gradually
(A) The basic strength of hydrides increases
(B) Almost pure p-orbitals are used for M-H bonding
(C) The bond energies of M-H bond increase
(D) The bond pairs of electrons become nearer to the central atom

67. In acidic medium potassium dichromate acts as an oxidant according to the equation
$Cr_2O_7^{2-} + 14 H^+ + 6e^- \rightarrow 2Cr^{3+} + 7H_2O$
What is the equivalent of $K_2Cr_2O_7$ (mol.wt.=M)
(A) M (B) M/2 (C) M/3 (D) M/6

68. Which type of isomerism is shown by the complex compounds $[Co(NH_3)_5Br]SO_4$ and $[Co(NH_3)_5SO_4]Br$?
(A) Co-ordination (B) Optical
(C) Ionic (D) Linkage

69. 50ml 10N – H_2SO_4, 25ml 12 N – HCl and 40 ml 5N– HNO_3 were mixed together and the volume of the mixture was made 1000 ml by adding water. The normality of the resultant solution will be
(A) 1N (B) 2N (C) 3N (D) 4N

70. H_2S gas was passed through an acidic solution containing Pb^{2+}, Zn^{2+}, Cu^{2+} and Ni^{2+} ions. The precipitate will consist of
(A) ZnS and PbS (B) PbS and NiS
(C) NiS and CuS (D) CuS and PbS

71. The oxidation state of nickel in Ni $(CO)_4$ will be
(A) 1 (B) Zero (C) 2 (D) 4

72. The molecular formula of sulphur is
(A) S_6 (B) S_8 (C) S_2 (D) S_4

73. All the elements of oxygen family are
(A) Non-metals (B) Metalloids
(C) Radioactive (D) Polymorphic

74. Iodine is liberated when KI solution is added to an aqueous solution of
(A) $ZnSO_4$ (B) $CuSO_4$
(C) $(NH_4)_2SO_4$ (D) Na_2SO_4

75. In acidic medium, one mole of MnO_4^- accepts how many moles of electrons in a redox process ?
(A) 1 (B) 3 (C) 5 (D) 6

76. The IUPAC name of the compound having the formula, $Cl_3C. CH_2CHO$ is
(A) 3,3,3-trichloropropanal
(B) 1,1,1-trichloropropanal
(C) 2,2,2-trichloropropanal (D) Chloral

77. When ethanal is treated with Fehling's solution. It give a precipitate of
(A) Cu (B) CuO (C) Cu_2O (D) $Cu_2O + Cu_2O_3$

78. For the homogeneous gas reaction,
$4NH_3 + 5O_2 \rightleftharpoons 4NO + 6H_2O$, the equilibrium constant k_C has the dimension of
(A) conc.$^{+10}$ (B) conc.$^{+1}$
(C) conc.$^{-1}$ (D) It is dimensionless

79. Formation of SO_3 takes place according to the reaction, $2SO_2 + O_2 \rightleftharpoons 2SO_3$; $\Delta H = -45.2$ kcal
Which of the following factors favours the formation of SO_3 ?
(A) Increase in temperature (B) Increase in pressure
(C) Removal of oxygen (D) Increase in volume

80. In the equilibrium reaction, 2HI (g) $\rightleftharpoons$ $H_2 + I_2$, which of the following expressions is true ?
(A) $k_P = k_C$ (B) $k_C = 2k_P$
(C) $k_P > k_C$ (D) $k_C = k_P (RT)^2$

81. What will be the pH of a 10^{-8} M HCl solution ?
(A) 6.98 (B) 14.0 (C) 8.0 (D) 7.0

82. Which of the following can act both as Bronsted acid and a Bronsted base ?
(A) Na_2CO_3 (B) OH^- (C) HCO_3^- (D) NH_3

83. What will be the sum of pH and pOH in an aqueous solution ?
(A) 7 (B) pk_W (C) Zero (D) 1

84. Given that the dissociation constant for H_2O is $k_W = 1 \times 10^{-14}$ mol.^{2} L^{-2}, what is the pH of a 0.001 M KOH solution ?
(A) 11 (B) 3 (C) 14 (D) 10^{-11}

85. Carbolic acid is
(A) Phenyl acetate (B) Salol
(C) Phenol (D) Phenyl benzoate

86. Which of the following reactions is a method for the conversion of a ketone into a hydrocarbon ?
(A) Cannizzaro reaction
(B) Aldol condensation
(C) Reimer-Tiemann reaction
(D) Wolff - Kishner reduction

87. Which reaction is suitable for the preparation of α-chloroacetic acid ?
(A) Hell-Volhard Zelinsky reaction
(B) Nef reaction
(C) Stephen's reactions

(D) Perkin reaction

88. Which compound is known as oil of winter green ?
(A) Phenyl acetate (B) Methyl salicylate
(C) Phenyl benzoate (D) Phenyl salicylate

89. The compound on reaction with cold nitrous acid giving oily nitrosoamine is
(A) Trimethylamine (B) Triethylamine
(C) Methylamine (D) Dimethylamine

90. In presence of acid, hydrolysis of methyl cyanide gives
(A) Acetic acid (B) Methylamine
(C) Methyl acid (D) Formic acid

91. The relation between enthalpy (H), pressure (P) volume (V) and internal energy (E) is given by
(A) E = H + PV (B) H = E + PV
(C) H = E – PV (D) H = E + P + V

92. In an adiabatic process which of the following is true ?
(A) q = + w (B) q = 0
(C) Δ E = q (D) P Δ V = 0

93. What is the free energy ΔG, when 1.0 mole of water at 100°C and 1 atm pressure is converted into steam at 100°C and 1 atm pressure ?
(A) 540 Cal (B) –9800 Cal
(C) 9800 Cal (D) 0 Cal

94. The standard Gibbs free energy change. ΔG^o is related to equilibrium constant, k_p, as
(A) $k_p = -RT\ ln\ \Delta G^o$ (B) $k_p = \left(\frac{e}{Rt}\right) \Delta G^o$
(C) $k_p = \frac{\Delta G^o}{RT}$ (D) $k_p = e\ \frac{\Delta G^o}{RT}$

95. Which of the following contains isoprene units ?
(A) Polyethylene (B) Dacron
(C) Natural rubber (D) Nylon - 66

96. The acid, showing salt-like character in aqueous solution is
(A) Acetic acid (B) Benzoic acid
(C) Formic acid (D) α –Amino acetic acid

97. Which of the following reactions is most suitable for the preparation of n-propyl benzene ?
(A) Friedel-Crafts reaction (B) Wurtz reaction
(C) Wurtz-Fittig faction (D) Grignard reaction

98. The reagent used for the dehydration of an alcohol is
(A) Aluminium oxide (B) Sodium chloride
(C) Phosphorus pentachloride (D) Calcium chloride

99. Which substance is not present in nucleic acid
(A) Cytosine (B) Adenine
(C) Thymine (D) Guanidine

100. Which of the following is not an amino acid ?
(A) Glycine (B) Alanine
(C) Histidine (D) Benzididine

PET - 1999

1. What is NOT true about polymers ?
(A) Polymers scatter light
(B) Polymers do not carry any charge
(C) Polymers have high viscosity
(D) Polymers have low molecular weight

2. The basis of the mode of their formation, the polymers can be classified
(A) As addition polymers only
(B) As condensation polymers only
(C) As copolymers
(D) Both as addition and condensation polymers

3. The commonest disaccharide has the molecular formula
(A) $C_{10}H_{20}O_{10}$ (B) $C_{18}H_{32}O_{11}$
(C) $C_{12}H_{22}O_{11}$ (D) $C_{10}H_{18}O_9$

4. Which carbohydrate is an essential constituent of plant cells ?
(A) Starch (B) Cellulose
(C) Sucrose (D) Vitamins

5. Which of the following halogens is solid at room temperature?
(A) Bromine (B) Fluorine
(C) Chlorine (D) Iodine

6. Which of the following properties is not of transition elements?
(A) Colour (B) Paramagnetism
(C) Fixed valency (D) None of the above

7. Which of the following chemicals contains chlorine ?
(A) Fischer salt (B) Epsom salt
(C) Fremy's salt (D) Spirit of salt

8. Which of the following is manufactured from the molecular nitrogen by bacteria?
(A) NO_3 (B) NO_2
(C) Amino acids (D) Ammonia

9. The bond-order in O_2^+ is
(A) 1.5 (B) 2 (C) 3 (D) 2.5

10. Which statement is not correct?
(A) Potassium permanganate is a powerful oxidising substance
(B) Potassium permanganate is a weaker oxidising substance than potassium dichromate
(C) Potassium permanganate isa stronger oxidising substance than potassium dichromate
(D) Potassium dichromate oxidises a secondary alcohol into a ketone

11. The enzyme pepsin hydrolyses
(A) Proteins to amino acids
(B) Fats to fatty acids
(C) Glucose to ethyl alcohol
(D) Polysaccharides to monosaccharides

12. Which of the following gives maximum energy in metabolic processes ?

(A) Lipids (B) Carbohydrates
(C) Protein (D) Vitamins

13. Enzymes in the living systems
(A) Provide energy
(B) Provide immunity
(C) Transport oxygen
(D) Catalyse biological processes

14. Which of the following will be octahedral?
(A) PCl_5 (B) BO_3^{3-} (C) SF_6 (D) BF_4^-

15. The molar solution of sulphuric acid is equal to
(A) N solution (B) 2N solution
(C) N/2 solution (D) 3N solution

16. In which of the following, tendency towards formation of coloured ions is maximum?
(A) s-block elements (B) d-blcok elements
(C) p-block elements (D) f-block elements

17. The equivalent weight of potassium permanganate for acid solution is
(A) 52.16 (B) 79 (C) 158 (D) 31.6

18. If 100 ml of 1N-sulphuric acid were mixed with 100 ml of 1M-sodium hydroxide, the solution will be
(A) Acidic (B) Basic
(C) Neutral (D) Slightly acidic

19. The example of dsp^2 hybridisation is
(A) $Zn(NH_3)_4^{2+}$ (B) FeF_6^{3-}
(C) $Fe(CN)_6^{3-}$ (D) $Ni(CN)_4^{2-}$

20. An α-particle is identical with
(A) Proton (B) Neutron
(C) Helium nucleus (D) Electron

21. Heaviest particle of the following is
(A) Meson (B) Neutron
(C) Proton (D) Electron

22. The ratio of specific charge of a proton and an α-particle is
(A) 2 1 (B) 1 2 (C) 1 4 (D) 1 1

23. The symbol X in the following equation, $_{11}Na^{23} + {_1}H^1 \rightarrow {_{12}}Mg^{23} + X$, is
(A) A neutron (B) A deutron
(C) A positron (D) An α-particle

24. The half-life of $_{92}U^{238}$ is 4.5 x 10^9 years. After how many years, the amount of $_{92}U^{238}$ will be reduced to half of its present amount ?
(A) 4.5 x 10^9 years (B) 4.5 x $10^{4.5}$ years
(C) 9.0 x 10^9 years (D) 13.5 x 10^9 years

25. The rate constant of a first order reaction whose half-life is 480 seconds, is
(A) 2.88 x 10^{-3} sec^{-1} (B) 1.44 x 10^{-3} sec^{-1}
(C) 1.44 sec^{-1} (D) 0.72 x 10^{-3} sec^{-1}

26. The conversion of A → B, follows second order kinetics. Doubling the concentration of A will increase the rate of formation of B by a factor
(A) 1/4 (B) 2 (C) 1/2 (D) 4

27. The given reaction, $2FeCl_3 + SnCl_2 \rightarrow 2FeCl_2 + SnCl_4$ is an example of
(A) First order reaction (B) Second order reaction
(C) Third order reaction (D) None of these

28. If reaction between A and B to give C shows first order kinetics in A and second order in B, the rate equation can be written as
(A) Rate = k [A] $[B]^{1/2}$ (B) Rate = k $[A]^{1/2}$ [B]
(C) Rate = k [A] $[B]^2$ (D) Rate = k $[A]^2$ [B]

29. The enzyme which can catalyse the conversion of glucose to ethanol is
(A) Zymase (B) Invertase
(C) Maltase (D) Diastase

30. For the following homogeneous reaction,
$A + B \longrightarrow C$
the unit of rate constant is
(A) Sec^{-1} (B) Sec^{-1} mole
(C) Sec^{-1} $mole^{-1}$ (D) Sec

31. The temperature coefficient of most of the reactions lies between
(A) 1 and 4 (B) 2 and 4 (C) 1 and 3 (D) 2 and 3

32. If formaldehyde and KOH are heated then we get
(A) Acetylene (B) Methane
(C) Methyl alcohol (D) Ethyl formate

33. Which of the following reagents reacts differently with HCHO, CH_3CHO and CH_3COCH_3
(A) NH_2OH (B) NH_3
(C) HCN (D) NH_2NH_2

34. Identify the product C in the series,

$$CH_3CN \xrightarrow{Na/\ C_2H_5OH} A \xrightarrow{HNO_2} B \xrightarrow{\text{Tollen's reagent}} C$$

(A) CH_3COOH (B) CH_3CH_2NHOH
(C) CH_3CONH_2 (D) CH_3CHO

35. Lower carboxylic acids are soluble in water due to
(A) Low molecular weight (B) Hydrogen bonding
(C) Dissociation into ions (D) Easy hydrolysis

36. Ethyl acetate reacts with CH_3MgBr to form
(A) Secondary alcohol (B) Tertiary alcohol
(C) Primary alcohol and acid (D) Acid

37. Which of the following is not a synthetic polymer ?
(A) Nylon (B) Cellophane
(C) Polyethylene (D) PVC

38. The amine which does not react with acetyl chloride is
(A) CH_3NH_2 (B) $(CH_3)_2NH$
(C) $(CH_3)_3N$ (D) None of these

39. Ethyl bromide on treatment with alcoholic KOH gives
(A) Acetic acid (B) Ethane
(C) Ethylene (D) Ethanol

40. $AgNO_3$ does not give precipitate with chloroform because
(A) $CHCi_3$ is insoluble in water
(B) $CHCl_3$ does not ionize in water
(C) $CHCl_3$ is an organic compound
(D) $AgNO_3$ is insoluble in $CHCl_3$

41. Benzene reacts with chlorine to form benzene hexachloride in presence of
(A) Nickel (B) $AlCl_3$
(C) Bright Sunlight (D) Zinc

42. Which of the following alkyl halides is used as a methylating agent ?
(A) C_2H_5Cl (B) C_2H_5Br
(C) C_2H_5I (D) CH_3I

43. Gammexane is
(A) D.D.T. (B) Benzene hexachloride (B.H.C.)
(C) Chloral (D) Hexachlorobenzene

44. Prevention of corrosion of iron by Zn coating is called
(A) Galvanization (B) Cathodic protection
(C) Electrolysis (D) Photoelectrolysis

45. Hydrogen-Oxygen fuel cells are used in space craft to supply
(A) Power for heat and light (B) Power for pressure
(C) Oxygen (D) Water

46. The reaction is spontaneous if the cell potential is
(A) Positive (B) Negative
(C) Zero (D) Infinite

47. At constant T and P which one of the following statements is correct for the reactions,

$CO(g) + \frac{1}{2} O_2(g) \rightarrow CO_2(g)$

(A) $\Delta H = \Delta E$ (B) $\Delta H < \Delta E$ (C) $\Delta H > \Delta E$
(D) ΔH is independent of the physical state of the reactant of that compounds

48. Heat of neutralisation of strong acid against strong base is constant and equals to
(A) 13.7 kcal (B) 57 kJ
(C) 5.7×10^{-4} J (D) All of the above

49. Which of the following substances consists of only one element?
(A) Diamond (B) Glass
(C) Marble (D) Sand

50. Which of the following substances can be used for drying gases?
(A) Calcium carbonate (B) Sodium carbonate
(C) Sodium bicarbonate (D) Calcium oxide

51. Which of the following metals is sometimes found native in nature?
(A) Aluminium (B) Copper
(C) Iron (D) Magnesium

52. Evaporation of water is
(A) A process where no heat changes occur
(B) A process accompanied by chemical reaction
(C) An exothermic change
(D) An endothermic change

53. Which of the following metals is preserved in water?
(A) Phosphorus (B) Sodium
(C) Zinc (D) Iron

54. An organic compound with C = 40% and H = 6.7% will have the empirical formula
(A) CH_2 (B) CH_2O (C) $C_3H_6O_3$ (D) $C_2H_4O_2$

55. The IUPAC name of the compound, is

$CH_3 - \underset{\displaystyle CH_3}{\underset{|}{CH}} - CH_2 - CH_2 - Cl$

(A) 1-chloro-3-methylbutane
(B) 2-methyl-4-chlorobutane
(C) 2-methyl-1-chlorobutane
(D) 1-chlorobutane

56. Optical active isomers but not mirror images are called
(A) Enantiomers (B) Mesomers
(C) Tautomers (D) Diastereoisomers

57. An organic compound will show optical isomerism if
(A) Two groups attached to C atom are different
(B) Four groups attached to C atom are different
(C) Three groups attached to C atom are different
(D) All the groups attached to C atom are same

58. The arrangement of hydrogen atoms around carbon in the methane molecule is
(A) Triangular (B) Square planar
(C) Tetrahedral (D) Hexagonal

59. The process which is catalysed by one of the products is called
(A) Acid-base catalysis (B) Auto-catalysis
(C) Negative catalysis (D) None of the above

60. Which of the following requires a catalyst ?
(A) $S + O_2 \rightarrow SO_2$ (B) $2SO_2 + O_2 \rightarrow 2SO_3$
(C) $C + O_2 \rightarrow CO_2$ (D) All of the above

61. In the Ostwald's process for the manufacture of HNO_3 the catalyst used is
(A) Ni (B) Pt (C) Mo (D) Fe

62. Which is correct statement about Proton ?
(A) Proton is nucleus of deuterium
(B) Proton is an alpha particle
(C) Proton is ionized hydrogen molecule
(D) Proton is ionized hydrogen

63. In a given atom, no two electrons can have the same values of all the quantum numbers. This is called
(A) Hunds Rule of maximum multiplicity
(B) Aufbau Principle
(C) Heisenberg's uncertainty Principle
(D) Pauli's Exclusion Principle

64. Magnetic quantum number specifies
(A) Orientation of orbitals (B) Nuclear stability
(C) Size of orbitals (D) Shape of orbitals

65. An alkyne has the general formula
(A) C_nH_{2n} (B) C_nH_{2n+2}
(C) C_nH_{2n-2} (D) C_nH_{2n+1}

66. In its reaction with silver nitrate acetylene shows
(A) Oxidising property (B) Reducing property
(C) Basic Property (D) Acidic Property

67. A gas decolourises alkaline $KMnO_4$ solution but does not give precipitate with $AgNO_3$, it is
(A) C_2H_2 (B) C_2H_6 (C) CH_4 (D) C_2H_4

68. Heating a mixture of sodium benzoate and soda lime gives
(A) Benzene (B) Methane
(C) Sodium benzoate (D) Calcium benzoate

69. Octane number is zero for
(A) n-heptane (B) iso-octane
(C) n-hexane (D) iso-heptane

70. For preparing an alkane, a concentrated aqueous solution of sodium or potassium salt of a saturated carboxylic acid is subjected to
(A) Hydrolysis (B) Oxidation
(C) Hydrogenation (D) Electrolysis

71. An arsenious sulphide sol carries a negative charge. The maximum precipitating power for this sol is possessed by
(A) Na_3PO_4 (B) $AlCl_3$ (C) K_2SO_4 (D) $CaCl_2$

72. Tyndall effect would be observed in a
(A) Solution (B) Solvent
(C) Colloidal Solution (D) Precipitate

73. Adsorption is multi layer in the case of
(A) Physical adsorption (B) Chemisorption
(C) Both (D) None of both

74. How many Cl^- ions are there around Na^+ ions in NaCl crystal ?
(A) 3 (B) 4 (C) 6 (D) 8

75. The existence of a substance in more than one solid modifications is known as
(A) Isomorphism (B) Polymorphism
(C) Amorphism (D) Allotropy

76. Which has the smallest size?
(A) Al^{3+} (B) P^{5+} (C) Na^+ (D) Mg^{2+}

77. Which of the following metals exhibits more than one oxidation state?
(A) Na (B) Mg (C) Fe (D) Al

78. Which one of the following ions is colourless?
(A) Ni^{2+} (B) Fe^{3+} (C) Cu^+ (D) Co^{2+}

79. Which of the following is paramagnetic ?
(A) CO (B) NO^+ (C) O_2 (D) CN^-

80. Acidified solution of chromic acid on treatment with hydrogen peroxide yields
(A) $CrO_3 + H_2O + O_2$ (B) $Cr_2O_3+H_2O+O_2$
(C) $CrO_5 + H_2O$ (D) $H_2Cr_2O_7+H_2O+O_2$

81. Which ion is detected by Nessler's reagent ?
(A) NH_4^+ (B) MnO_4^- (C) PO_4^{3-} (D) CrO_4^{2-}

82. The correct formula of permanganic acid is
(A) $HMnO_4$ (B) $HMnO_5$
(C) H_2MnO_4 (D) H_2MnO_3

83. Who among the following first prepared a stable compound of noble gas?
(A) Rutherford (B) Rayleigh
(C) Ramsay (D) Neil Bartlett

84. Lead sulphate is soluble
(A) In conc. Nitric acid
(B) In conc. Hydrochloric acid
(C) In a solution of ammonium acetate
(D) In water

85. What will be the molarity of a solution containing 5 g of sodium hydroxide in 250 ml solution?
(A) 0.5 (B) 1.0 (C) 2.0 (D) 0.1

86. The energy of an eleetron in nth orbit of hydrogen atom is
(A) $\frac{13.6}{n^4}$ eV (B) $\frac{13.6}{n^3}$ eV
(C) $\frac{13.6}{n^2}$ eV (D) $\frac{13.6}{n}$ eV

87. If wavelength of photon is 2.2×10^{-11} m, $h=6.6\times10^{-34}$ Js, then momentum of photon is
(A) 3×10^{-23} Kg ms^{-1} (B) 3.33×10^{22} Kg ms^{-1}
(C) 1.452×10^{-44} Kg ms^{-1} (D) 6.89×10^{43} Kg ms^{-1}

88. The pH of a 10^{-8} molar solution of HCl in water is
(A) 8 (B) -8
(C) Between 7 and 8 (D) Between 6 and 7

89. Which of the following salts when dissolved in water hydrolyse?
(A) KCl (B) Na_2SO_4
(C) NaCl (D) NH_4Cl

90. What is the correct representation for the solubility product of SnS_2?
(A) $[Sn^{2+}][S^{2-}]^2$ (B) $[Sn^{4+}][S^{2-}]^2$
(C) $[Sn^{2+}]2S^{2-}]$ (D) $[Sn^{4+}]^2S^{2-}]^2$

91. Which of the following noble gases is least polarizable?
(A) He (B) Ne (C) Ar (D) Xe

92. Epsom salt is
(A) $MgSO_4$ $7H_2O$ (B) $CaSO_4$ H_2O
(C) $MgSO_4$ $2H_2O$ (D) $BaSO_4$ $2H_2O$

93. Among the alkaline earth metals, the element forming predominantly covalent compound is
(A) Barium (B) Strontium
(C) Calcium (D) Berrylium

94. In which of the following, Oxygen is collected ?
(A) Mercury (B) Kerosene oil
(C) Ethanol (D) Water

95. In a double bond connecting two atoms there is a sharing of
(A) 1 electrons (B) 2 electrons
(C) 4 electrons (D) All electrons

96. The maximum number of hydrogen bonds in which a water molecule can participate is
(A) 1 (B) 2 (C) 3 (D) 4

97. A molal solution is one that contains 1 mole of a solute in
(A) 1000 g of the solvent (B) One litre of the solvent
(C) One litre of the solution (D) 22.4 litre of the solution

98. Which of the following is not a colligative property?
(A) Osmotic pressure (B) Elevation in boiling point
(C) Vapour pressure (D) Depression in freezing point

99. Solutions with same osmotic pressure are called
(A) Isotonic (B) Normal
(C) Hypertonic (D) Hypotonic

100. A monoprotic acid in a 0.1 M solution ionizes to 0.001%. Its ionization constant is
(A) 1.0×10^{-3} (B) 1.0×10^{-6}
(C) 1.0×10^{-8} (D) 1.0×10^{-11}

PET - 2000

1. If 2.0 g of a radioactive substance has half-life of 7 days, the half-life of 1 g sample is
(A) 7 days (B) 14 days (C) 28 days (D) 35 days

2. In which of the following, oxygen has an oxidation state of + 2 ?
(A) H_2O (B) OF_2 (C) H_2O_2 (D) CO_2

3. Which molecule has dipole moment ?
(A) BF_3 (B) HF (C) CCl_4 (D) C_6H_6

4. Gold number is related with
(A) Colloids (B) Radioactivity
(C) Gas equation (D) Kinetic energy

5. The equation $\Delta x. \Delta p \geq h/4\pi$ shows
(A) De Broglie relation
(B) Heisenberg's uncertainty principle
(C) Aufbau principle (D) Hund's rule

6. Which will be the proper alternative in place of A in the following equation ?
$2Fe^{3+} (aq) + Sn^{2+} (aq) \rightarrow 2 Fe^{2+} (aq) + A$
(A) Sn^{2+} (B) Sn^{0} (C) Sn^{4+} (D) Sn^{3+}

7. By which law, energy produced in nuclear reaction is given ?
(A) Graham's law (B) Charles's law
(C) Gay lussac's law (D) Einstein's law

8. An electron has kinetic energy 2.8×10^{-23}J. de Broglie wavelength will be nearly ($m_e = 9.1 \times 10^{-31}$ kg)
(A) 9.28×10^{-4}m (B) 9.28×10^{-7}m
(C) 9.28×10^{-8}m (D) 9.28×10^{-10}m

9. $E^o = \frac{RT}{nF} ln K_{eq}$. The above equation is called
(A) Gibb's equation (B) Gibbs - Helmholtz equation
(C) Nernst equation (D) van der Waal's equation

10. Inversion of sucrose is
(A) Zero order reaction (B) First order reaction
(C) Second order reaction (D) Third order reaction

11. For the reaction, $2HI \rightleftharpoons H_2 + I_2$, the rate of the reaction is proportional to $[HI]^2$, this means that the reaction is
(A) Unimolecular (B) Bimolecular
(C) Of first order (D) Of second order

12. Chloroform on reduction with Zinc and HCl forms
(A) Methylene chloride (B) Phosgene
(C) Formic acid (D) Chloropicrine

13. From which of the following, tertiary butyl alcohol is obtained by the action of methyl magnesium iodide ?
(A) HCHO (B) CH_3CHO
(C) CH_3COCH_3 (D) CO_2

14. The compound that forms on heating formaldehyde with ammonia is
(A) Methyl amine (B) Amino formaldehyde
(C) Hexa-methylene tetramine(D) Formalin

15. An organic compound, on reaction with saturated aqueous solution of $NaHCO_3$ gives effervescence, the compound would be ?
(A) Acetic Acid (B) Ethyl Alcohol
(C) Alkane (D) Alkene

16. Ethyl amine on heating with CS_2 in presence of Hg_2Cl_2 forms
(A) $(C_2H_5)_2$ CS (B) C_2H_5 $(CS)_2$
(C) C_2H_5NCS (D) $(C_2H_5)_2S$

17. The haemoglobin contains 0.33% of iron by weight. Its molecular weight is 67200. The number of iron atoms in each molecule is [At. wt. of Fe = 56]
(A) 2 (B) 3 (C) 4 (D) 5

18. The best source of vitamin - A is
(A) Oranges (B) Beans (C) Carrots (D) Wheat

19. The natural dye in the following is
(A) Alizarin (B) Methyl orange
(C) Azodyes (D) Rhodamine

20. Example of colloidal solution is
(A) Milk (B) Aqueous solution of urea
(C) Alkaline solution of benzoic acid
(D) None of the above

21. Which of the following acts both as oxidizing and reducing agent ?
(A) $KMnO_4$ (B) H_2O_2 (C) HNO_3 (D) $K_2Cr_2O_7$

22. pH of a 10^{-3} M solution of hydrochloric acid will be
(A) 1.3 (B) 2.0 (C) 3.0 (D) 4.5

23. The oxidation number of oxygen in ozone is

(A) −3 (B) −2 (C) 0 (D) + 3

24. Which of the following is most basic ?
(A) CH_3NH_2 (B) $(C_2H_5)_3$ N
(C) $(CH_3)_3$ N (D) $(CH_3)_2$ NH

25. Oxidation of toluene with chromyl chloride gives
(A) Benzoic acid (B) Acetophenone
(C) Benzyl alcohol (D) Benzaldehyde

26. The number of primary amines of formula, $C_4H_{11}N$ is
(A) 1 (B) 2 (C) 3 (D) 4

27. The bond order of individual carbon-carbon bond in benzene is
(A) One (B) Two
(C) Between one and Two (D) One and two alternately

28. In nitration of Benzene with conc. HNO_3 and conc. H_2SO_4 the species which takes active part is
(A) Nitrite ion (B) Nitrate ion
(C) Nitronium ion (D) Nitrogen peroxide

29. To which of the following, the mass of an element deposited at an electrode is directly proportional ?
(A) Atomic weight (B) Equivalent weight
(C) Molecular weight (D) Atomic number

30. What will be the de Broglie wavelength of an electron moving with a velocity of 1.20×10^5 ms^{-1}?
(A) 6.626×10^{-9} (B) 6.018×10^{-7}
(C) 6.068×10^{-9} (D) 3.133×10^{-37}

31. What will be the half-life period of a nucleus at the end of 4.2 days, N = 0.798 No ?
(A) 15 days (B) 10 days
(C) 12.83 days (D) 20 days

32. Ferrous oxide has a cubic structure and each edge of the unit cell is 5.0 Å. Assuming density of the oxide as 4.0 g-cm^{-3}, the number of Fe^{2+} and O^{2-} ions present in each unit cell will be
(A) Four Fe^{2+} and two O^{2-}
(B) Four Fe^{2+} and four O^{2-}
(C) Two Fe^{2+} and four O^{2-}
(D) Three Fe^{2+} and three O^{2-}

33. From the solution of which of the following, one Faraday of electricity will liberate one gram atom of metal ?
(A) $CuSO_4$ (B) $AlCl_3$ (C) NaCl (D) $BaCl_2$

34. If the concentration is expressed in moles per litre, the unit of the rate constant for a first order reaction is
(A) mol L^{-1} S^{-1} (B) mol L^{-1} (C) S^{-1} (D) mol^{-1}

35. Brownian movement is
(A) Zig-zag motion of the colloidal particles
(B) Migration of colloidal particles under the influence of electric field
(C) Scattering of light by colloidal particles
(D) None of the above

36. What is the product obtained in the reaction of $HgCl_2$ and Hg $(CN)_2$?
(A) $(CN)_2$ (B) Hg (CN)Cl (C) Hg [Hg $(CN)_2Cl_2$]
(D) Addition compound, $HgCl_2$. Hg $(CN)_2$

37. XeF_6 on hydrolysis gives
(A) XeO_3 (B) XeO (C) XeO_2 (D) Xe

38. Which of the following should be most stable ?
(A) H (B) H^- (C) H_2^+ (D) H^+

39. Oxidation number of carbon in CH_3–Cl is
(A) −3 (B) −2 (C) −1 (D) 0

40. 0.1914 g of an organic acid is dissolved in about 20 ml of water. 25 ml of 0.12 N NaOH is required for the complete neutralization of the acid solution. The equivalent weight of the acid is
(A) 65 (B) 64 (C) 63.80 (D) 62.50

41. Which of the following is least ionic ?
(A) $BaCl_2$ (B) $C_6H_5N^+$ H_3Cl^-
(B) C_2H_5Cl (D) KCl

42. Which of the following is correct order of the size of iodine species ?
(A) $I^+ > I^- > I$ (B) $I^- > I > I^+$
(C) $I > I^- > I^+$ (D) $I > I^+ > I^-$

43. Molecular formula of Glauber's salt is
(A) $MgSO_4$ $7H_2O$ (B) $CuSO_4.5H_2O$
(C) $FeSO_4.7H_2O$ (D) $Na_2SO_4.10H_2O$

44. In which of the following, dissociation of NH_4OH, will be minimum ?
(A) NaOH (B) H_2O (C) NH_4Cl (D) NaCl

45. Sodium sulphate dissolves in water with evolution of heat. Consider a saturated solution of sodium sulphate. If the temperature is raised then according to Le Chatelier principle
(A) The solution will become supersaturated
(B) More solid will dissolve
(C) Some solid will precipitate out form the solution
(D) Solution concentration will remain unchanged

46. If an iron wire is placed in a solution of copper sulphate
(A) Copper will precipitate out
(B) Iron will dissolve and precipitate out thereafter
(C) Iron and copper both will go into solution
(D) No reaction will take place

47. The metal that is extracted from sea water is
(A) Mg (B) Ca (C) Fe (D) Ni

48. Which of the following oxides reacts with HCl and NaOH ?
(A) N_2O_5 (B) CO_2 (C) CaO (D) ZnO

49. The most basic element is
(A) Fluorine (B) Iodine (C) Chlorine (D) Bromine

50. The heat evolved in the combustion of benzene is given by the equation,
$C_6H_6(l) + 7\frac{1}{2}\ O_2(g) \rightarrow 3H_2O(l) + 6CO_2\ (g)$

$\Delta H = 781.0$ kcal mol^{-1}. Which of the following quantities of heat energy will be evolved when 39g of benzene is burnt in an open container ?
(A) 1562.0 kcal mol^{-1} (B) 586.75 kcal mol^{-1}
(C) 122.25 kcal mol^{-1} (D) 390.5 kcal mol^{-1}

51. Energy required to dissociate 4g of gaseous hydrogen into free gaseous atoms is 208 kcal at 25°C. The bond energy of H–H will be
(A) 104 kcal (B) 10.4 kcal
(C) 1040 kcal (D) 52 kcal

52. Which of the following is correct for the first law of thermodynamics ?
(A) $\Delta E = q + W$ (B) $\Delta E = q + P$
(C) $W = \Delta E - q$ (D) $\Delta E = q - W$

53. Who had proposed the following equation, $\Delta G = \Delta H - T \Delta S$?
(A) Boltzmann (B) Faraday
(C) Gibbs-Helmholtz (D) Thomson

54. Which of the following units represents the largest amount of energy ?
(A) Calorie (B) Joule (C) Erg (D) Electron volt

55. In the equilibrium, $N_2 + 3H_2 \rightleftharpoons 2NH_3 + 22$ kcal, the formation of ammonia is favoured by
(A) Increasing the pressure
(B) Increasing the temperature
(C) Decreasing the pressure
(D) Adding ammonia

56. For $N_2 + 3H_2 \rightleftharpoons 2NH_3$ + heat
(A) $K_P = K_C$ (B) $K_P = K_C RT$
(C) $K_P = K_C (RT)^{-2}$ (D) $K_P = K_C (RT)^{-1}$

57. Which is the best reducing agent ?
(A) Br^- (B) I^- (C) F^- (D) Cl^-

58. Which of the following sulphides is only completely precipitated when the acidic solution is made dilute ?
(A) HgS (B) PbS (C) CdS (D) CuS

59. Which of the following substances consists of only one element ?
(A) Marble (B) Sand (C) Diamond (D) Glass

60. Of the four oxyacids of chlorine the strongest oxidising agent in dilute aqueous solution is
(A) $HClO_2$ (B) HOCl (C) $HClO_4$ (D) $HClO_3$

61. An organic compound contains 49.3% carbon, 6.84% Hydrogen and its vapour density is 73. Molecular formula of the compound is
(A) $C_3H_5O_2$ (B) $C_6H_{10}O_4$
(C) $C_3H_{10}O_2$ (D) $C_4H_{10}O_2$

62. The compound $C_4H_{10}O$ can show
(A) Metamerism (B) Functional isomerism
(C) Position isomerism (D) All types

63. The IUPAC name of, $CH_3OC_2H_5$ is
(A) Methyl ethyl ether (B) Ethyl-methyl ether
(C) Methoxy ethane (D) Ethoxy methane

64. The number of sigma and pi bonds in 1-butene 3-yne are
(A) 8 sigma and 2 pi (B) 6 sigma and 4 pi
(C) 5 sigma and 5 pi (D) 7 sigma and 3 pi

65. Which of the following is most stable ?
(A) Propane (B) Butane (C) Methane (D) Ethane

66. Ammonium hydroxide is a weak base because
(A) It has low vapour pressure
(B) It is only slightly ionized
(C) It is not a hydroxide of any metal
(D) It has low density

67. What will happen if carbon tetrachloride is treated with silver nitrate ?
(A) A white precipitate of silver chloride will be formed
(B) NO_2 will be evolved
(C) Carbon tetrachloride will dissolve silver nitrate
(D) Nothing will happen

68. The conjugate acid of NH_2^- is
(A) NH_3 (B) N_2H_4 (C) NH_3 (D) NH_4OH

69. Which one is an example of an octahedral complex ?
(A) FeF_6^{3-} (B) $Zn(NH_3)_4^{2+}$
(C) $Ni(CN)_4^{2-}$ (D) $Cu(NH_3)_4^{2+}$

70. From which of the following, the hydration energy of Mg^{2+} is larger ?
(A) Na^+ (B) Al^{3+} (C) Be^{2+} (D) Cr^{3+}

71. The compound which gives only acetaldehyde on ozonolysis is
(A) Butene-1 (B) Butene-2
(C) Ethylene (D) Propylene

72. Which of the following is used to distinguish ethylene and acetylene ?
(A) Alkaline $KMnO_4$ (B) Bromine water
(C) Ammoniacal cuprous chloride (D) Conc. H_2SO_4

73. When chloroform is left open in light and air, it forms
(A) Carbon tetrachloride (B) Phosgene
(C) Methylene chloride (D) Formic Acid

74. Carbons in the compound 2-butene-4-yne are
(A) sp hybridized (B) sp^2 hybridized
(C) sp and sp^2 hybridized (D) sp, sp^2 and sp^3 hybridized

75. Distinction is made between ethyl alcohol and methyl alcohol by
(A) Chloroform test (B) Victor Meyer test
(C) Rate of Esterification (D) Iodoform test

76. Fenton reagent is
(A) $FeSO_4 + H_2O_2$ (B) Zn + HCl
(C) Sn + HCl (D) None of the above

77. Chloramphenicol is used in the treatment of which of the following ?
(A) Typhoid (B) Pneumonia
(C) Headache and fever (D) Bronchitis

78. SLV - 3 uses .... propellants

(A) Solid-Liquid (B) Biliquid
(C) Solid (D) Liquid

79. The most important energy carrier in all the living cells is
(A) ADP (B) UDP (C) AMP (D) ATP

80. Which is used for the formation of nylon 66 ?
(A) Sulphurous acid (B) Phthalic acid
(C) Sulphur hexafluoride (D) Adipic acid

81. The petrol having octane number 80 has
(A) 20% normal heptane + 80% iso-octane
(B) 80% normal heptane + 20% iso-octane
(C) 20% normal heptane +80% normal octane
(D) 80% normal heptane +20% normal octane

82. In the reaction,
$CH_3CH_2COOH \xrightarrow{Cl_2\,/\,red\,P} A \xrightarrow{Alc.\ KOH} B$
The compound B is
(A) CH_3CH_2OH (B) CH_3CH_2COCl
(C) $CH_2 = CHCOOH$ (D) $CH_3–CHClCOOH$

83. Which of the following combines with Fe^{2+} ions to form a brown complex ?
(A) N_2O_3 (B) N_2O_5 (C) N_2O (D) NO

84. In the equation
$4M+8CN^- + 2H_2O + O_2 \rightarrow 4\,[M(CN_2)]^- + 4OH^-$
the metal M is
(A) Copper (B) Iron (C) Gold (D) Zinc

85. Incorrect statement regarding rusting is
(A) Oxygen gas is reduced to oxide ions
(B) Metallic iron is oxidized to Fe^{3+} ions
(C) Metallic iron is reduced to Fe^{2+} ions
(D) Yellowish - brown product is formed

86. The second most abundant element in the universe is
(A) Silicon (B) Oxygen
(C) Nitrogen (D) Helium

87. Which statement is not true ?
(A) Ni $(CO)_4$ is diamagnetic
(B) BI_3 is a stronger Lewis acid then BF_3
(C) Graphite conducts electricity whereas diamond does not
(D) CCl_4 is hydrolysed whereas BCl_3 is inert

88. If a 4 1 mixture of N_2 and O_2 is taken and 1% of Argon is added to it, the vapour pressure of mixed gases will
(A) Increase (B) Decrease
(C) Remain constant (D) First increase and then decrease

89. The solution of sodium carbonate has pH
(A) Equal to 7 (B) Equal to Zero
(C) Greater than 7 (D) Less than 7

90. If the pH of a solution is 2, its normality will be
(A) 2 N (B) 0.5 N (C) 0.01 N (D) 0.02N

91. In which of the following molecules, is the covalent bond is most polar ?
(A) HCl (B) H_2 (C) HI (D) HBr

92. The solution of sugar in water contains free/hydrated
(A) Atoms (B) Molecules
(C) Ions (D) Atoms and molecules

93. Among the following compounds which one is planar in shape ?
(A) Benzene (B) Isobutane
(C) Methane (D) Acetylene

94. Which of the following is not used as a lewis acid ?
(A) KCl (B) BF_3 (C) $SnCl_4$ (D) $FeCl_3$

95. Malachite is an ore of which metal ?
(A) Copper (B) Zinc (C) Iron (D) Silver

96. Which of the following metals will not form an amalgam ?
(A) Gold (B) Silver (C) Zinc (D) Iron

97. Which of the following is obtained when nitrogen reacts with calcium carbide ?
(A) Calcium Carbonate (B) Calcium cyanamide
(C) Calcium acetate (D) Calcium cyanate

98. The number of antibonding electron pairs in O_2^{2-} molecular ion on the basis of molecular orbital theory is
(A) 4 (B) 3 (C) 2 (D) 5

99. Conjugate base of NH_4^+ is
(A) NH_2^- (B) H_3O^+ (C) NH_3 (D) H_2O

100. The correct order of the increasing ionic character is
(A) $BeCl_2 < BaCl_2 < MgCl_2 < CaCl_2$
(B) $BeCl_2 < CaCl_2 < MgCl_2 < BeCl_2$
(C) $BeCl_2 < MgCl_2 < CaCl_2 < BaCl_2$
(D) $BeCl_2 < MgCl_2 < BaCl_2 < CaCl_2$

PET- 2001

1. The heat of neutralization of HCl and NaOH is
(A) +57.3 kJ (B) Zero
(C) –57.3 kJ (D) None of these

2. The heat of neutralization of HCl by NaOH under certain condition is –55.9 kJ and that of HCN by NaOH is –12.1 kJ The heat of ionization of HCN is
(A) -68.0 kj mol^{-1} (B) -43.8 kj mol^{-1}
(C) 68.0 kj mol^{-1} (D) 43.8 kj mol^{-1}

3. Hess's law is an application of
(A) Ist law of thermodynamics
(B) IInd law of thermodynamics
(C) Entropy change (D) Free energy change

4. Free energy change for a reversible process is
(A) More than zero (B) Less than zero
(C) Equal to zero (D) None of these

5. Spontaneity of a chemical reaction is decided by the negative change in
(A) Entropy (B) Free energy

(C) Internal energy (D) Enthalpy

6. For a first order reaction, the unit of rate constant is
(A) mol L^{-1} S^{-1} (B) mol^{-1} L^{-1} (C) s^{-1} (D) mol $L^{-1}s$

7.The order of a reaction with rate equal to $kC_A^{+3/2}$ $C_B^{-1/2}$ will be
(A) 2 (B) 1 (C) – 1/2 (D) 3/2

8. For the first order reaction, A $\rightarrow$ Product, that half-life time is 100 seconds. The rate constant of the reaction is
(A) 6.93 x $10^{-2}s^{-1}$ (B) 6.93 x $10^{-4}s^{-1}$
(C) 6.93 x $10^{-3}s$ (D) 6.93s

9. The solubility of Ag in NaI solution is less than that in pure water because
(A) Solubility product of AgI is less than that of NaI
(B) The temperature of the solution decreases
(C) AgI forms complex with NaI
(D) Of common Ion effect

10. The solubility product of a sparingly soluble salt AB at room temperature is 1.21 x $10^{-6}m^2$. Its molar solubility is
(A) 1.21 x 10^{-6}M (B) 1.1 x 10^{-4} M
(C) 1.1 x 10^{-3} M (D) 1.21 x 10^{-3} M

11. An aqueous solution of CH_3COONa will be
(A) Acidic (B) Alkaline
(C) Neutral (D) None of these

12. The solubility of A_2X_3 is y mol dm^{-3}. Its solubility product is
(A) 6 y^4 (B) 64 y^4 (C) 36 y^5 (D) 108 y^5

13. Which of the following sets of quantum numbers is not possible ?
(A) n = 3 l = +2 m = 0 $s = +\frac{1}{2}$
(B) n = 3 l = +0 m = 0 $s = -\frac{1}{2}$
(C) n = 3 l = +0 m = –1 $s = +\frac{1}{2}$
(D) n = 3 l = +1 m = 0 $s = -\frac{1}{2}$

14. H-bonding is most extensive in
(A) Ethyl Chloride (B) Triethyl amine
(C) Ethanol (D) Diethyl ether

15. According to VSEPR theory, the most probable shape of the molecule having 4 electron pairs around the central atom is
(A) Octahedral (B) Linear
(C) Hexahedral (D) Tetrahedral

16. Which of the following is a non-colligative property ?
(A) Elevation of boiling point
(B) Depression freezing point
(C) Osmotic pressure (D) Optical activity

17. The law stating that "the relative lowering of vapour pressure is equal to the mole fraction of solute" in the solution is known as
(A) Faraday's law (B) Raoult's law
(C) Van't Hoff's law (D) None of these

18. For a reaction, 2A + B $\rightarrow$ Products, doubling the initial concentration of both the reactants increases the rate by a factor of 8 and doubling the concentration of B alone doubles the rate. The rate law for the reaction is
(A) r = K [A] [B] (C) r = K $[A]^2$ $[B]^2$
(C) r = K [A] $[B]^2$ (D) r = K $[A]^2$ [B]

19. The oxidation number of C in CO_2 is
(A) –2 (B) +2 (C) –4 (D) +4

20. One faraday of electricity is equal to
(A) 96500 Coulomb (B) 10^6 Coulomb
(C) 3.7 x 10^6 Coulomb (D) 6.23 x 10^{23} Coulomb

21. Which of the following compounds will not undergo decomposition on passing electricity through aqueous solution ?
(A) Sodium Bromide (B) Sodium Acetate
(C) Sugar (D) Sodium Chloride

22. Which is an emulsion among the following ?
(A) Butter (B) Air (C) Milk (D) Wood

23. The pH of an aqueous solution containing $[H]^+$ = 3 x 10^{-3} M is
(A) 2.471 (B) 2.523 (C) 3.0 (D) –3

24. The process of extracting the metal form its ore is called
(A) Leaching (B) Metallurgy
(C) Refining (D) Concentration

25. The number of close neighbour in a body centred cubic lattice of identical sphere is
(A) 8 (B) 6 (C) 4 (D) 2

26. The number of equidistant oppositely charged ions in a sodium chloride crystal is
(A) 8 (B) 6 (C) 4 (D) 2

27. The atomic orbitals are progressively filled in the order of increasing energy. This principle is called as
(A) Hund's rule (B) Aufbau principle
(C) Exclusion principle (D) Broglie rule

28. The correct configuration of copper with atomic number 29 is
(A) [Ar] $3d^{10}4s^1$ (B) [Ar] $3d^94s^2$
(C) [Ar] $4s^1$ (D) [Ar] 4s2

29. 0.126 g of an acid requires 20 ml of 0.1 N NaOH for complete neutralization. The equivalent weight of the acid is
(A) 45 (B) 53 (C) 40 (D) 63

30. The ratio of amounts of H_2S needed to precipitate all the metal ions from 100 ml of 1 M $AgNO_3$ and 100 ml of 1 M $CuSO_4$ is
(A) 1 2 (B) 2 1 (C) Zero (D) Infinity

31. Copper pyrites ore is concentrated by
(A) Electro-magnetic method (B) Gravity method
(C) Froth floatation method (D) All the above methods

32. The purpose of smelting an ore is to
(A) Obtain an alloy (B) Separate volatile impurities

(C) Reduce it (D) Oxidise it

33. Aluminium is diagonally related to
(A) Li (B) Be (C) C (D) B

34. Which of the following species contain non directional bond ?
(A) NH_4 (B) BeF_2 (C) NH_3 (D) CsCl

35. The formation of adduct between NH_3 and BF_3 involves the formation of
(A) Ionic bond (B) Covalent bond
(C) Coordinate bond (D) Hydrogen bond

36. Which one gives positive silver mirror test ?
(A) Sucrose (B) Glucose
(C) Starch (D) Fructose

37. Which compound is aspirin ?
(A) Acetamide (B) Salicyl amide
(C) Acetyl Salicylic acid (D) Salicylic acid

38. The "middle oil" fraction of coal tar distillation contains
(A) Naphthalene (B) Xylene
(C) Benzene (D) Anthracene

39. Which of the following statements about proteins is not true ?
(A) Amino acid residues join together to make a protein molecule
(B) Proteins are polymers with formula $(C_6H_{10}O_5)n$
(C) Eggs are rich in protein
(D) Pulses are good sources of protein

40. Ethylene may be obtained by dehydration of which of the following with concentrated H_2SO_4 at 160-170°C
(A) $CH_3CH_2CH_2OH$ (B) $(CH_3)_2CHCH_2OH$
(C) C_2H_5OH (D) CH_3 OH

41. Alkyl halides react with Mg in dry ether to form
(A) Magnesium halide
(B) Grignard's reagent
(C) Alkene (D) Alkyne

42. Reduction of an aldehyde gives
(A) Mono-carboxylic acid (B) Primary alcohol
(C) Secondary alcohol (D) Tertiary alcohol

43. Which among the following is the strongest acid ?
(A) Mono-chloroacid (B) Butyric acid
(C) Acetic acid (D) Propionic acid

44. Natural rubber is a polymer of
(A) Ethane (B) Propane
(C) Styrene (D) Isoprene

45. Which polymer is used in making rayon?
(A) Polyester (B) Nylon
(C) Cellulose (D) Starch

46. For the reaction $PCl_3 (g) + Cl_2 (g) \rightleftharpoons PCl_5 (g)$ The value of K_c at 250° C is 26. The value of K_p at this temperature will be
(A) 0.61 (B) 0.57 (C) 0.83 (D) 0.46

47. The arrangement ABC, ABC, ABC .... is referred as
(A) Octahedral close packing
(B) Hexagonal close packing
(C) Tetragonal close packing (D) Cubic close packing

48. A radioactive nucleide 'X' decays at the rate of 1.00×10^5 disintegration $s^{-1}g^{-1}$. Radium decays at the rate of 8.70×10^{10} disintegration $s^{-1}g^{-1}$. The activity of 'X' in millicuries (m ci g^{-1}) is
(A) 0.027 (B) 0.270 (C) 0.00270 (D) 0.000270

49. If half-life of a certain radioactive nucleus is 1000s, the disintegration constant is
(A) $6.93 \times 10^{-4} s^{-1}$ (B) $6.93 \times 10^3 s$
(C) $6.93 \times 10^2 s^{-1}$ (D) 6.93×10^{-4} s

50. In the Thorium series $_{90}Th^{232}$ loses a total of 6 α–particles and 4 β–particles in ten stages. The final isotope produced in the series is
(A) $_{82}Pb^{209}$ (B) $_{83}Bi^{209}$ (C) $_{82}Pb^{208}$ (D) $_{82}Pb^{206}$

51. Radioactive carbon dating was discovered by
(A) J. Willard Gibbs (B) W. Nernst
(C) W.F. Libby (D) G.N. Lewis

52. In a chemical reaction equilibrium is established when
(A) Velocity of opposing reaction is the same as that of forward reaction
(B) Opposing reaction ceases
(C) Concentration of reactants and Products are equal
(D) Reaction ceases to generate heat

53. According to Le Chatelier's principle adding heat to a solid and liquid in equilibrium will cause the
(A) Amount of solid to decrease
(B) Amount of liquid to decrease
(C) Temperature to rise (D) Temperature to fall

54. Alkaline earth metals belong to the
(A) s- block in periodic table
(B) p- block in periodic table
(C) d-block in periodic table
(D) f-block in periodic table

55. An element of atomic number 29 belongs to the .... block in periodic table
(A) s - block (B) p - block
(C) d - block (D) f- block

56. Which of the following has the largest radius ?
(A) O^{2-} (B) F^- (C) Mg^{2+} (D) Na^+

57. The ionization potential is lowest for the
(A) Halogens (B) Inert gases
(C) Alkaline earth metals (D) Alkali metals

58. Which of the following elements has the maximum electron affinity ?
(A) F (B) Cl (C) Br (D) I

59. Variable valency is shown by
(A) Metallic elements (B) Normal elements
(C) Transition elements (D) Non-metal elements

60. The lightest metal is

(A) Ca (B) Na (C) Li (D) Mg

61. The strongest reducing agents of the following is
(A) K (B) Al (C) Mg (D) Br

62. Among the following, outermost configurations of transition metals, which shows the highest oxidation state ?
(A) $3d^5 4s^2$ (B) $3d^6 4s^2$ (C) $3d^3 4s^2$ (D) $3d^5 4s^1$

63. Manganese achieves its maximum oxidation state in its compounds
(A) MnO_2 (B) Mn_3O_4 (C) $KMnO_4$ (D) K_2MnO_4

64. In the following the plaster of paris is
(A) $CaSO_4.H_2O$ (B) $CaSO_4.2H_2O$
(C) $CaSO_4.\frac{1}{2}H_2O$ (D) $CaSO_4.\ 1\frac{1}{2}\ H_2O$

65. The dissociation constant for water is 1 x 10^{-14} mol^2 $litre^{-2}$. What is the pH of a 0.01 M KOH solution ?
(A) 10^{-11} (B) 10^{-3} (C) 3 (D) 11

66. Which buffer solution out of the following will have pH > 7 ?
(A) $CH_3COOH + CH_3COONa$ (B) HCOOH + HCOOK
(C) CH_3COONH_4 (D) $NH_4OH + NH_4Cl$

67. Hydrolysis of acetamide gives
(A) Methyl amine (B) Formic acid
(C) Acetic acid (D) Acetadehyde

68. An example of a double salt is
(A) $K_4\ [Fe(CN)]_6$ (B) Potash Alum
(C) Bleaching powder (D) Hypo

69. Which of the following elements does not form amalgam ?
(A) Zinc (B) Cobalt (C) Fe (D) Silver

70. An aqueous solution of ammonia consists of
(A) H^+ (B) OH^- (C) NH_4^+ (D) NH_4^+ and OH^-

71. Aqua regia is a mixture of
(A) $3HCl + 1HNO_3$ (B) $H_3PO_4 + H_2SO_4$
(C) $3HNO_3 + 1HCl$ (D) $HCl + CH_3COOH$

72. Leakage of which gas was responsible for the Bhopal tragedy in 1984 ?
(A) $CH_3 - N = C = O$ (B) $CH_3 - C - N = S$
(C) $CHCl_3$ (D) C_6H_5COCl

73. Ethylene reacts with bromine to form
(A) $Br - CH_2 - CH_3$ (B) $CH_3 - C - Br_3$
(C) $Br - CH_2 - CH_2Br$ (D) $CHBr_3$

74. The correct I.U.P.A.C. name of

$$H - \overset{\overset{\displaystyle O}{\|}}{C} - CHO$$ is

(A) Formyl methanal (B) 1,2 -ethanedione
(C) Formyl methanoate (D) 1,2 -ehanedial

75. Which of the following is a thermoplastic polymer?
(A) Polythene (B) Vulcanized rubber
(C) Bakelite (D) Melamine-formaldehyde

76. What is the maximum of open chain structures possible for C_4H_8 ?
(A) 2 (B) 3 (C) 4 (D) 1

77. Which of the following alkanes contains primary, secondary, tertiary and quarternary carbon atoms together ?
(A) $(CH_3)_3CH$ (B) $(C_2H_5)_3$ CH
(C) $(CH_3)_3CCH_2CH(CH_3)_2$ (D) $(CH_3)_4C$

78. The disaccharide present in milk is
(A) Maltose (B) Lactose
(C) Sucrose (D) Cellobiose

79. Which of the following is a trihydric alcohol ?
(A) Ethylene glycol (B) Trimethylene glycol
(C) Pinacol (D) Glycerol

80. The proton and neutron are collectively called as
(A) Meson (B) Nucleons
(C) Deutron (D) Positron

81. Which of the following is not a polymer ?
(A) DDT (B) Starch (C) Silk (D) DNA

82. The shape of methane molecule is
(A) Tetrahedral (B) Triangular
(C) Planar (D) Octahedral

83. Which catalyst is used in Friedel-Crafts synthesis ?
(A) Finely divided Fe (B) Platinized Fe
(C) Anhydrous $AlCl_3$ (D) Alcoholic KOH

84. In the following the strongest base is
(A) AsH_3 (B) SbH_3 (C) NH_3 (D) PH_3

85. $FeSO_4$ forms brown ring with
(A) NO (B) N_2O_3 (C) NO_3 (D) NO_2

86. Precipitate of AgCl dissolves in NH_4OH due to formation of a complex
(A) $[Ag(NH_4)_2]Cl$ (B) $[Ag(NH_4)_3]Cl$
(C) $[Ag(NH_3)_2]Cl$ (D) $[Ag(NH_3)_2OH]$

87. $[CO(NH_3)_4Cl_2]\ NO_2$ and $[CO(NH_3)_4Cl\ NO_2]Cl$ are isomers of which type ?
(A) Geometrical (B) Optical
(C) Linkage (D) Ionization

88. How many carbon atoms in the molecule, $HCOO - (CHOH)_2 - COOH$ are asymmetric ?
(A) 1 (B) 2 (C) 3 (D) None of these

89. The reaction between an alkyl halide and sodium metal in dry ether is called
(A) Clemmensen reduction (B) Kolbe's reaction
(C) Wurtz reaction (D) Cannizaro's reaction

90. Which of the following is an antiknock compound ?
(A) Tetraethyl lead (B) Lead tetrachloride
(C) Lead Acetate (D) Ethyl Acetate

91. Which compound is not inflammable ?
(A) CH_4 (B) C_6H_6 (C) CCl_4 (D) C_2H_5OH

92. Action of HNO_2 on ethylamine gives
(A) Ethane (B) Ethanol

(C) Ammonia (D) Nitromethane

93. Which one of the following is not isomeric with diethyl ether?
(A) n-propyl methyl ether (B) 1-butanol
(C) 2 methyl -2proranol (D) Butanone

94. The reaction of caustic soda with oils and fats is known as
(A) Neutralization (B) Saponification
(C) Oxidation (D) Esterification

95. Which compound is obtained on passing ethanol vapour on heated Al_2O_3?
(A) Acetaldehyde (B) Ethane
(C) Ethyl ether (D) Acetone

96. Which of the following statements is false for alkali metals ?
(A) Li is the strongest reducing agent
(B) Na is amphoteric in nature
(C) Li^+ is exceptionally small
(D) All alkali metals give blue solution in liquid ammonia

97. Which of the following ions has largest heat of hydration ?
(A) Ba^{2+} (B) K^+ (C) Li^+ (D) Be^{2+}

98. Oxidation state +1 for phosphorus is found in
(A) H_3PO_3 (B) H_3PO_4 (C) H_3PO_2 (D) $H_4P_2O_7$

99. The number of ions per mole of a complex $[CoCl_2.5NH_3]$ in aqueous solution will be
(A) Nine (B) Four (C) Three (D) Two

100. Which one has the highest percentage of nitrogen ?
(A) CAN (B) Calcium nitrate
(C) Urea (D) Ammonium nitrate

PET - 2002

1. Which one has the highest boiling point ?
(A) Diethyl ether (B) Chloroform
(C) Acetone (D) Ethyl alcohol

2. In benzene, all the six C — C bonds have the same length because of
(A) Isomerism (B) Inductive effect
(C) Tautomerism (D) sp^2 hybridisation

3. O-nitrophenol is more volatile than p-nitrophenol due to
(A) Intramolecular H-bonding
(B) Intermolecular H-bonding
(C) Resonance (D) Inductive effect

4. Which of the following species possesses the highest bond order
(A) O_2 (B) O_2^+ (C) O_2^- (D) O_2^{-2}

5. The solubility of a gas in water depends on
(A) Nature of the gas (B) Temperature
(C) Pressure of the gas (D) All of the above

6. The condition for an ideal solution is
(A) ΔH mixing = 0 (B) ΔV mixing $\neq 0$
(C) $P_A \neq P_{A^\circ} \times A$ (D) ΔH mixing > 1

7. The boiling point of a solution of 0.11 gm of a substance in 15 gm of ether was found to be 0.1°C higher than that of the pure ether. The molecular weight of the substance will be (K_b = 2.16)
(A) 148 (B) 158 (C) 168 (D) 178

8. Acetic acid dissolved in benzene shows a molecular weight of
(A) 60 (B) 120 (C) 180 (D) 240

9. Which one of the following pairs is not correctly matched ?
(A) Rutherford - Proton (B) J.J. Thomsom - Electron
(C) J. Chadwick - Neutron (D) Bohr - Isotope

10. Which of the following is not isoelectronic ?
(A) O^{2-} (B) Cl^- (C) Na^+ (D) Mg^{2+}

11. Wurtz reaction involves the interaction of alkyl halides in dry ether with
(A) Copper (B) Platinum (C) Sodium (D) Zinc

12. Which process converts olefins into parafins ?
(A) Halogenation (B) Dehydration
(C) Hydrogenation (D) Hydrolysis

13. The H — C — H bond angle in CH_4 is
(A) 109°28 (B) 107°28 (C) 90° (D) 180°

14. Which one has the lowest octane number ?
(A) n-hexane (B) n-hexadecane
(C) Iso-octane (D) n-heptane

15. Which gas is used for cutting and welding the metals ?
(A) Acetylene (B) Methane
(C) Carbon dioxide (D) Neon

16. The correct order of increasing energy of atomic orbitals is
(A) 4f < 5p < 5d < 6s (B) 5p < 5d < 4f < 6d
(C) 5p < 4f < 6s < 5d (D) 5p < 6s < 4f < 5d

17. The value of the energy for the first excited state of hydrogen atom will be
(A) –13.6 eV (B) –3.40 eV (C) –1.51 eV (D) –0.85 eV

18. The number of dative bonds in sulphuric acid molecules is
(A) 0 (B) 1 (C) 2 (D) 4

19. The correct sequence of decrease in the bond angle of the following hydrides is
(A) $SbH_3 > AsH_3 > PH_3 > NH_3$
(B) $NH_3 > PH_3 > AsH_3 > SbH_3$
(C) $NH_3 > AsH_3 > PH_3 > SbH_3$
(D) $PH_3 > NH_3 >$ AsH3 $> SbH_3$

20. Paramagnetism is exhibited by molecules
(A) Not attracted in a magnetic filed
(B) Containing only paired electrons
(C) Carrying a positive charge
(D) Containing unpaired electrons

21. Which of the following metals is present in chlorophyll
(A) Chromium (B) Cobalt (C) Magnesium (D) Iron

22. During extraction of silver, which of the following is formed
(A) $Na_4[Ag(CN)_2]$ (B) Na $[Ag(CN)_2]$
(C) $Na_2[Ag(CN)_2]$ (D) None of these

23. The frequency of yellow light having wavelength 600 nm is
(A) 5.0 x 10^{14} Hz (B) 2.5 x 10^7 Hz
(C) 5.0 x 10^7 Hz (D) 2.5 x 10^{14} Hz

24. The oxidation state of Fe in $K_4[Fe(CN)_6]$ is
(A) +2 (B) +6 (C) + 3 (D) +4

25. Which of the following reactions is not exothermic
(A) $C(s) + O_2(g) \rightarrow CO_2(g)$
(B) $C(s) + 2S(s) \rightarrow CS_2\ (g)$
(C) $CH_4(g) + 2O_2(g) \rightarrow CO_2(g) + 2H_2O(l)$
(D) $CO(g) + \frac{1}{2} O_2(g) \rightarrow CO_2(g)$

26. The time of completion of 90% of a first order reaction is approximately
(A) 1.1 times that of half-life
(B) 2.2 times that of half-life
(C) 3.3 times that of half-life
(D) 4.4 times that of half-life

27. The rate law for the reaction ,
$2N_2O_5 \rightarrow 4\ NO_2 + O_2$ is
(A) $r = K\ [N_2O_5]$ (B) $r = K[N_2O_5]^2$
(C) $r = K\ [N_2O_5]^0$ (D) $r = K\ [NO_2]^4[O_2]$

28. The half-life period for a first order reaction is 693 seconds. The rate constant for this reaction would be
(A) 0.1 sec^{-1} (B) 0.01 sec^{-1}
(C) 0.001 sec^{-1} (D) 0.0001 sec^{-1}

29. Isoprene is a valuable substance for making
(A) Propene (B) Liquid fuel
(C) Synthetic rubber (D) Petrol

30. Mustard gas is obtained by
(A) Treating sulphur chloride with ethylene
(B) The action of dilute acids on mustard seeds
(C) Treating ethylene with mustard oil
(D) None of these

31. Which one of the following salts gives an acidic solution in water
(A) CH_3COONa (B) NH_4Cl
(C) NaCl (D) CH_3COONH_4

32. The solubility product of $BaSO_4$ is 1.3 x 10^{-9}. The solubility of this salt in pure water will be
(A) 1.69 x 10^{-9} mol $litre^{-1}$ (B) 1.69 x 10^{-18} mol $litre^{-1}$
(C) 3.6 x 10^{-5} mol $litre^{-1}$ (D) 3.6 x 10^{-9} mol $litre^{-1}$

33. The favourable conditions for a spontaneous reaction are
(A) $T\Delta S = \Delta H$, $\Delta H = -ve$, $\Delta S = -ve$
(B) $T\Delta S > \Delta H$, $\Delta H = +ve$, $\Delta S = +ve$
(C) $T\Delta S > \Delta H$, $\Delta H = +ve$, $\Delta S = -ve$
(D) $T\Delta S = \Delta H$, $\Delta H = +ve$, $\Delta S = +ve$

34. Heat exchanged in a chemical reaction at constant temperature and constant pressure is termed as
(A) Internal energy (B) Enthalpy
(C) Entropy (D) Free energy

35. The entropy change involved in the conversion of 1 mole of liquid water at 373K to vapour at the same temperature will be (ΔH_{vap} = 2,257 kJ/gm)
(A) 0.129 kJ (B) 0.120 kJ
(C) 0.119 kJ (D) 0.109 kJ

36. The diazonium salts are the reaction products in presence of excess of mineral acid with nitrous acid and
(A) Primary aliphatic amine
(B) Secondary aliphatic amine
(C) Primary aromatic amine
(D) Tertiary aliphatic amine

37. $\begin{matrix} H_3C \\ H \end{matrix} > C = C < \begin{matrix} CH_3 \\ H \end{matrix}$ and $\begin{matrix} H_3C \\ H \end{matrix} > C = C < \begin{matrix} H \\ CH_3 \end{matrix}$
exhibit which isomerism ?
(A) Position isomerism (B) Geometrical isomerism
(C) Optical isomerism (D) Functional isomerism

38. Deficiency of which vitamin causes rickets
(A) Vitamin C (B) Vitamin B
(C) Vitamin A (D) Vitamin K

39. Fehling solution for the detection of aldehyde group consists of $CuSO_4$ solution and
(A) Alkaline KOH solution
(B) Sodium potassium tartrate
(C) Alcoholic KOH solution
(D) Ammonium hydroxide solution

40. $C_6H_6 + Cl_2 \rightarrow C_6H_6Cl_6$ This reaction takes place when
(A) Chlorine (aq) is shaken with benzene
(B) the mixture is exposed to sunlight
(C) Chlorine is bubbled through benzene
(D) the mixture is kept in dark

41. An organic compound has been found to possess the empirical formula CH_2O and molecular weight 90. Its molecular formula is
(A) $C_3H_6O_3$ (B) $C_2H_4O_2$ (C) $C_6H_{12}O_6$ (D) CH_2O

42. The compound 'A' when treated with HNO_3 (in presence of H_2SO_4) gives compound 'B', which is then reduced with Sn and HCl to aniline ? The compound 'A' is
(A) Ethane (B) Acetamide (C) Toulene (D) Benzene

43. Which of the following radio isotopes is used in the treatment of blood cancer ?
(A) P^{32} (B) Co^{60} (C) I^{131} (D) Na^{24}

44. A glass bulb is filled with NO_2 gas and immersed in an icebath at 0°C, which becomes colourless after some time. This colourless gas will be
(A) NO_2 (B) N_2O (C) N_2O_4 (D) N_2O_5

45. The favourable conditions for Haber's process are
(A) High temperature and low pressure

(B) Low temperature and low pressure
(C) Low temperature and high pressure
(D) High temperature and high pressure

46. The correct relationship between K_P and K_C is
(A) $K_P = K_C (RT)$ (B) $K_P = K_C (\Delta n)$
(C) $K_P = K_C (RT)^{\Delta n}$ (D) $K_P = K_C (RT)^{1/\Delta n}$

47. Which of the following is not a Lewis acid ?
(A) CO (B) $SiCl_4$ (C) SO_3 (D) Zn^{2+}

48. Which of the following is the buffer solution of strong acidic nature ?
(A) $HCOOH + HCOO^-$ (B) $CH_3COOH + CH_3COO^-$
(C) $H_2C_2O_4 + C_2O_4^{2-}$ (D) $H_3BO_3 + BO_3^{3-}$

49. The compound 'A' treated with methyl alcohol and drops of H_2SO_4 and heating with water gave smell of wintergreen. The compound 'A' is
(A) Succinic acid (B) Salicylic acid
(C) Tartaric acid (D) Oxalic acid

50. The compound 'A' when treated with ceric-ammonium nitrate solution gives yellow ppt. The compound 'A' is
(A) Acid (B) Alkane (C) Alcohol (D) Aldehyde

51. The interionic distance for cesium chloride crystal will be :
(A) a (B) $\frac{a}{2}$ (C) $\frac{\sqrt{3}a}{2}$ (D) $\frac{2a}{\sqrt{3}}$

52. Which one of the following crystals does not exhibit Frenkel defect ?
(A) AgBr (B) AgCl (C) KBr (D) ZnS

53. Which one of the following metal oxides is anti-ferromagnetic in nature
(A) VO_2 (B) CrO_2 (C) MnO_2 (D) TiO_2

54. The nuclear reaction,
$^{63}_{29}Cu + ^{4}_{2}He \rightarrow ^{37}_{17}Cl + 14\,^{1}_{1}H + 16\,^{1}_{0}n$
is referred to as
(A) Spallation reaction (B) Fusion reaction
(C) Fission reaction (D) Chain reaction

55. In smelting of iron, which of the following reactions takes place in Blast furnace at 400°C - 600°C ?
(A) $FeO + SiO_2 \rightarrow Fe\,SiO_3$
(B) $CaO + SiO_2 \rightarrow CaSiO_3$
(C) $2FeS + 3O_2 \rightarrow 2Fe + SO_2$
(D) $Fe_2O_3 + 3CO \rightarrow 2Fe + 3CO_2$

56. Which of the following is insoluble in water ?
(A) Na_2CO_3 (B) $CaCO_3$
(C) $ZnCO_3$ (D) $Al_2(CO_3)_3$

57. 20 ml of a solution of a weak monobasic acid neutralizes 22.18 ml of a solution of NaOH and 20 ml of N/10 HCl neutralises 21.5 ml of the same NaOH solution. The normality of the acid is nearly
(A) 10 N (B) 1 N (C) 0.10 N (D) 100 N

58. Which of the following reactions occurs at the cathode during the charging of a lead storage battery
(A) $Pb^{2+} + 2e^- \rightarrow Pb$ (B) $Pb^{2+} + SO_4^{2-} \rightarrow PbSO_4$
(C) $Pb \rightarrow Pb^{2+} + 2e^-$
(D) $PbSO_4 + 2H_2O \rightarrow PbO_2 + 4H^+ + SO_4^{2-} + 2e^-$

59. ^{226}Ra disintegrates at such a rate that after 3160 years only one-fourth of its original amount remains. The half-life of ^{226}Ra will be
(A) 790 years (B) 3160 years
(C) 1580 years (D) 6230 years

60. The proper rays for radiocarbon dating are
(A) UV rays (B) IR rays
(C) Cosmic ray (D) X-rays

61. Two elements whose electronegativities are 1.2 and 3.0, the bond formed between them would be
(A) Coordinate (B) Metallic (C) Ionic (D) Covalent

62. Which of the following is correct ?
(A) A mineral cannot be an ore
(B) An ore cannot be a mineral
(C) All minerals are ores
(D) All ores are minerals

63. The common method of extraction of metals from oxide ores is
(A) Reduction with Carbon
(B) Reduction with Hydrogen
(C) Reduction with Aluminium (D) Electrolytic method

64. The neutron was discovered by
(A) J.J. Thomson (B) G.T. Seaborg
(C) E. Rutherford (D) James Chadwick

65. Colourless solutions of the following four salts are placed separately in four different test tubes and a strip of Copper is dipped in each one of the these. Which solution will turn blue
(A) $Zn(NO_3)_2$ (B) $ZnSO_4$ (C) KNO_3 (D) $AgNO_3$

66. In order to prepare one litre normal solution of $KMnO_4$, how many gm of $KMnO_4$ are required if the solution is to be used in acid medium for oxidation ?
(A) 158 gm (B) 31.60 gm (C) 62.0 gm (D) 790 gm

67. Conc. H_2SO_4 has a density of 1.98 gm/ml and is 98% H_2SO_4 by weight. Its normality is
(A) 2 N (B) 19.8 N (C) 39.6 (D) 98 N

68. Which of the following groups shows the highest boiling point ?
(A) H_2Se (B) H_2Te (C) H_2O (D) H_2S

69. The chemical composition of 'Cryolite' mineral is
(A) Al_2O_3 (B) $Al_2O_3.12H_2O$
(C) $KAlSi_3O_8$ (D) Na_3AlF_6

70. Which of the following halogens exhibits only one oxidation state in its compounds ?
(A) Bromine (B) Chlorine (C) Fluorine (D) Iodine

71. Bleaching action of SO_2 is due to its
(A) Oxidising property (B) Acidic property
(C) Reducing property (D) Basic property

72. Sea weeds are an important source of
(A) Iron (B) Chlorine (C) Iodine (D) Bromine

73. The coloured discharge tube for advertisements mainly contains
(A) Xenon (Xe) (B) Argon (Ar)
(C) Helium (He) (D) Neon (Ne)

74. Which of the following has the least ionization potential ?
(A) Lithium (Li) (B) Helium (He)
(C) Nitrogen (N) (D) Zinc (Zn)

75. Which gives green colour in fireworks ?
(A) Barium (B) Calcium (C) Sodium (D) Potassium

76. If NaOH is added to an aqueous solution of zinc ions, a white precipitate appears and on adding excess of NaOH, the precipitate dissolves. In this solution zinc exists in the
(A) Cationic part (B) Anionic part
(C) Both in cationic and anionic parts
(D) There is no zinc in the solution

77. Which of the following lead oxides is 'Sindhur' ?
(A) Pb_2O_3 (B) Pb_3O_4 (C) PbO (D) PbO_2

78. The chemical formula of Pyruvic acid is
(A) $HOCH_2COOH$ (B) $CH_3COCOOH$
(C) CH_3COH_2CCOOH (D) CH_3COCH_3

79. The conversion of methylamine into methyl isocyanide with alc. KOH and $CHCl_3$ is termed as
(A) Friedel-Crafts reaction (B) Carbylamine reaction
(C) Wurtz reaction (D) Kolbe's reaction

80. Acetamide when treated with bromine and alkali yields
(A) Methylamine (B) Methyl bromide
(C) Acetone (D) Acetaldehyde

81. Lewisite is
(A) $Cl.CH = CH\ As\langle^{Cl}_{Cl}$ (B) $CH_2 = CH.As\langle^{Cl}_{Cl}$

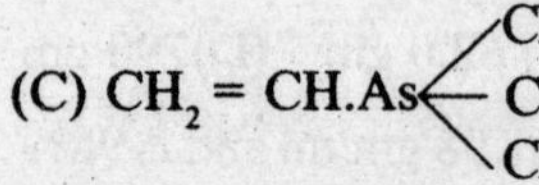

(C) $CH_2 = CH.As \leftarrow$ Cl, Cl, Cl (D) $AsCl_3$

82. Why is chloroform stored in dark coloured bottles ?
(A) To prevent it from oxidation to form phosgene
(B) To prevent its reaction with glass
(C) To prevent evaporation
(D) To prevent from moisture

83. Denatured spirit is mainly used as
(A) Drug in medicine (B) Good fuel
(C) Solvent in preparing varnishes
(D) Material in preparing beverages

84. General formula C_nH_{2n+2} represents
(A) Alkanes (B) Alkenes (C) Aikynes (D) Cycloalkanes

85. Liquid Ammonia is used for refrigeration because
(A) It has a high dipole moment
(B) It has a high heat of vapourisation
(C) It is basic (D) It is a stable compound

86. When ammonia is passed over heated copper oxide, the metallic copper is obtained. The reaction shows that ammonia is
(A) A reducing agent (B) A nitrating agent
(C) A dehydrating agent (D) An oxidizing agent

87. Oleum is chemically represented as
(A) $H_2S_2O_7$ (B) $H_2S_2O_8$ (C) H_2SO_3 (D) H_2SO_5

88. Bakelite is a product formed from
(A) Reaction of polypropylene with acid
(B) Reaction of formaldehyde with phenol
(C) Reaction of polyethylene with phenol
(D) It is a natural product

89. Iodine value is related to
(A) Fats and oils (B) Alcohols
(C) Esters (D) Hydrocarbons

90. How is [benzene ring with CH_3 at top, O_2N and NO_2 on either side, NO_2 at bottom] widely used ?
(A) Insecticide (B) Drug (C) Explosive (D) Dye

91. D.D.T. stands for
(A) Diphenyl dimethyl trichloroethane
(B) Dichloro diphenyl trichloroethane
(C) Dihydro dichloro trichloroethane
(D) Diamino diphenyl toluene

92. A graph plotted between log K Vs. 1/T for calculating activation energy is shown by

(A) 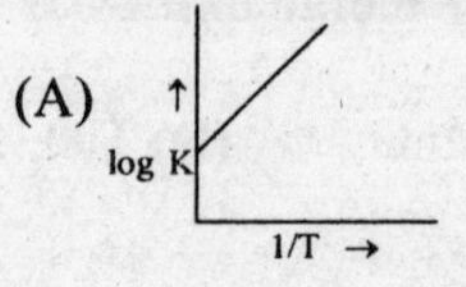

(B) 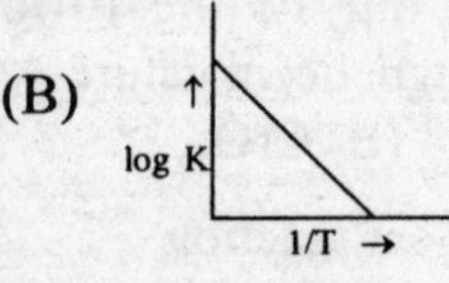

(C) 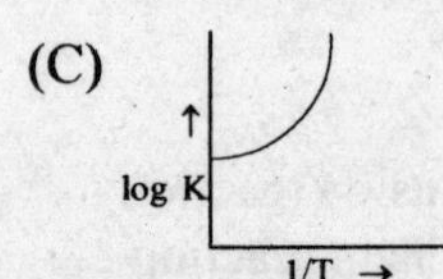

(D) 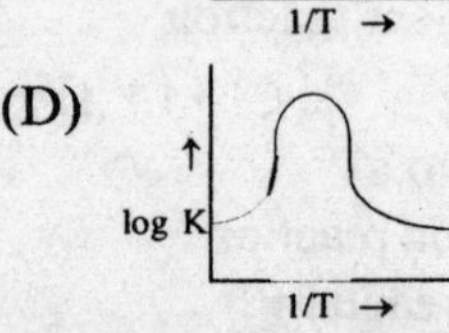

93. The relationship between standard reduction potential of cell and equilibrium constant is shown by
(A) $E^\circ_{cell} = \frac{n}{0.059} \log K_C$ (B) $E^\circ_{cell} = \frac{0.059}{n} \log K_C$
(C) $E^\circ_{cell} = 0.059 \log K_C$ (D) $E^\circ_{cell} = \frac{\log K_C}{n}$

94. The oxidation number of S in $H_2S_2O_8$ is
(A) +2 (B) +4 (C) +6 (D) +7

95. The standard cell potential of $Zn / Zn^{2+}(aq) // Cu^{2+}(aq) / Cu$ cell is 1.10 V. The maximum work obtained by this cell will be
(A) 106.15 kilo joule (B) –212.30 kilo joule
(C) –318.45 kilo joule (D) –424.60 kilo joule

96. Which of the following is an incorrect statement for physisorption ?
(A) It requires activation energy
(B) It is a reversible process
(C) It requires less heat of adsorption

(D) It takes place at low temperature

97. Which of the following substances is not used for preparing lyophilic sols ?
(A) Gelatin (B) Metal sulphide (C) Starch (D) Gum

98. The catalyst used in the manufacture of methanol from water gas is
(A) V_2O_5 (B) Ni + Mo (C) ZnO + Cr_2O_3 (D) Pt + W

99. Atomic orbitals of carbon in diamond are
(A) sp^3 hybridized (B) sp^2 hybridized
(C) Sp hybridized (D) Not hybridized

100. An aqueous solution carbonate is alkaline because sodium carbonate is a salt of
(A) Weak acid and strong base
(B) Weak acid and weak base
(C) Strong acid and weak base
(D) Strong acid and strong base

PET - 2003

1. The reaction,

$$CH_3COOH + Cl_2 \rightarrow ClCH_2COOH + HCl$$

is called
(A) Stephen reaction (B) Rosenmund reaction
(C) Hell - Volhard - Zelinsky raction
(D) Hunsdiecker reaction

2. Benzaldehyde on reaction with acetophenone in the presence of sodium hydroxide solution gives
(A) $C_6H_5CH = CHC_2H_5$
(B) $C_6H_5 - CH = CH - CO - C_6H_5$
(C) $C_2H_5COCH_2C_2H_5$
(D) $C_6H_5CH(OH)COC_2H_5$

3. Acetonitrile is
(A) CH_3COCN (B) C_2H_5CN
(C) CH_3CN (D) $C_2H_5CH_2CN$

4. Reaction

$$R_2C = O \xrightarrow{H_2NNH_2} R_2CH_2 + N_2$$

is called
(A) Tischenko reaction (B) Sandmeyer reaction
(C) Reformatsky reaction
(D) Wolff-Kishner reaction

5. Which of the follwing statements is false in the context of enzyme?
(A) Temperature affect their functions
(B) pH affects their functions
(C) They always increase activation energy
(D) Their reactions are specific

6. The example of a protein is
(A) Cephalin (B) Lecithin (C) Insulin (D) Narvone

7. Oils and fats together are called
(A) Polymer (B) Lipids (C) Enzymes (D) Soaps

8. Sodium potassium salts of fatty acids are called
(A) Terpenes (B) Proteins (C) Sugars (D) Soaps

9. Urea can be identified by
(A) Biuret test (B) Benedict test
(C) Lucas test (D) Mullicken test

10. Which of the following polymers contains nitrogen?
(A) PVC (B) Terylene (C) Nylon (D) Teflon

11. The element whose ionizatioin potential second is less than its ionization potential first
(A) Be (B) Li (C) He (D) None of the above

12. Increasing order of density is
(A) K<Li<Na<Rb<Cs (B) Cs<Rb<K<Na
(C) Li<Na<K,Rb<Cs (D) Li<k<Na<Rb<Cs

13. Which of the following has lowest fusion temperature ?
(A) Naphthalene (B) NaCl (C) Diamond (D) Cr

14. Which of the follwoing elements is alloyed with copper to form brass ?
(A) Zinc (B) Antimony (C) Lead (D) Bismuth

15. Which of the following transition metal ions shows highest magnetic moment having the following outer electronic configuration ?
(A) $3d^8$ (B) $3d^2$ (C) $3d^5$ (D) $3d^-$

16. Most powerful oxidizing property of manganese is shown by which of the follwoing oxidatiion state ?
(A) Mn (+4) (B) Mn (+7)
(C) Mn (+2) (D) Mn (+3)

17. Which of the following is most stable hydride ?
(A) NH_3(b) (B) AsH_3 (C) PH_3 (D) BiH_3

18. For which of the following, the bond angle will be minimum ?
(A) H_2S (B) H_2O (C) H_2Se (D) H_2Te

19. The ion formed with the reaction of nitrous acid and hydrochloric acid is
(A) Nitrosonium ion (B) Nitronium ion
(C) Nitrate ion (D) Nitracidium ion

20. Co-ordination number of Fe in the complexes $[Fe(CN)_6]^{4-}$, $[Fe(CN)_6]^{3-}$ and $[FeCl_4]^-$ would be respectively.
(A) 6, 3, 4 (B) 6, 6, 4 (C) 2, 3, 3 (D) 6, 4, 6

21. The effective atomic number of cobalt in the complex $[Co(NH_3)_6]^{3+}$ is
(A) 24 (B) 26 (C) 33 (D) 36

22. The correct name of $[Pt(NH_3)_4Cl_2]$ $[PtCl_4]$ is
(A). Tetraaminedichloroplatinium(iv)tetrachloro platinate(ii)
(B) Dichlorotetraamineplatinium(IV)tetrachloro platirate(II)
(C) Tetrachloro platinium (ii) tetra amine dichloro platinate (iv)
(D) Tetra chloro platinium (ii) dichloro tetra amine platinate (iv)

23. Which compound has the highest boiling point?
(A) Diethylether (B) Ethanol
(C) Methanol (D) Acetone

24. Which has a pyramidal structure ?
(A) Ethanol (B) Water
(C) Acetylene (D) Trimethylamine

25. The vapour pressure of liquid A is 0.80 atm. On mixing a non volatile B to A, its vapour pressure becomes 0.6 atm. The mole fraction of B in the solution is
(A) 0.75 (B) 0.50 (C) 0.125 (D) 0.25

26. Which of the follwoing mixtures does not show positive deviation from Raoult's law ?
(A) Acetone - ethanol (B) Benzene - acetone
(C) Acetone - chloroform (D) Water - ethanol

27. Equimolar solutions of $BaCl_2$, NaCl and sucrose have osmotic pressure in the order :
(A) NaCl > $BaCl_2$ > sucorose
(B) $BaCl_2$ > NaCl > Sucrose
(C) Sucrose > NaCl > $BaCl_2$
(D) $BaCl_2$ > Sucrose > NaCl

28. In solid CsCl each Cl^- is closely packed with how many Cs^+ ?
(A) 8 (B) 6 (C) 10 (D) 2

29. In the transformation of ${}^{232}_{90}Th$ to ${}^{208}_{82}Pb$. The particles emitted are
(A) 4α, 3β (B) 6α, 4β (C) 2α, 4β (D) 4α, 6β

30. Which of the following can convert (transmutation) ${}^{14}_{7}N$ in to ${}^{17}_{8}O$?
(A) α- particle (B) Neutron
(C) Proton (D) deutron

31. Radioactive isotope of hydrogen is
(A) deuterium (B) Ortho hydrogen
(C) Tritium (D) Para hydrogen

32. A tenfold increase in pressure on the rection, $N_2(g) + 3H_2(g) \rightarrow 2NH_3(g)$ at equitabrium makes K_p
(A) Unchanged (B) five times
(C) two times (D) ten times

33. For equilibrium, $2NO(g) \rightarrow N_2(g) + O_2(g)$, which of the following is not correct ?
(A) Equilibrium constant does not change with pressure at constant temperature.
(B) Equilibrium constant does not change with volume at constant temperature.
(C) At high pressure dissociation of NO is favoured.
(D) At high temperature dissociation of NO is favoured.

34. Which statement is not true regarding alkanes ?
(A) Many alkanes are soluble in water
(B) At room temprature some alkanes are liquids some solids and some gases
(C) All alkanes have a lower density than water
(D) All alkanes burn

35. Which of the following alkenes gives only acetic acid on oxidation with potassium permangnate solution ?
(A) Propene (B) 2-butene
(C) 1-butene (D) Ethylene

36. Carbolic acid is
(A) Methanol (B) Benzyl alcohol
(C) Phenol (D) Ethanol

37. Reaction,

$$C_6H_5OH + CH_2 = O \xrightarrow{\text{acid or base}} o\text{-}HOC_6H_4CH_2OH + p\text{-}HOC_6H_4CH_2OH$$

is called
(A) Claisen condensation (B) Mannich reaction
(C) Benzoin condensation
(D) Lederer Manasse reaction

38. Which isomer of cyclohexane hexachloride is a powerful insecticide?
(A) σ (B) δ (C) γ (D) β

39. Which compound requires chloral in its synthesis ?
(A) Michler's Ketone (B) D.D.T.
(C) Aldrin (D) Gammexane

40. When phenol is heated with phthalic anhydride in concentrated sulphuric acid and the hot reaction mixture is poured into a dilute solution of sodium hydroxide the product formed is
(A) Indigo (B) Alizarin
(C) Fluorescein (D) Phenolphthalein

41. Which halide will be least reactive with the hydrolysis point of view?
(A) Allyl chloride (B) t- Butyl chloride
(B) Methyl chloride (D) Vinyl chloride

42. Acetamide on heating with P_2O_5 gives
(A) Methyl cyanide (B) Methanol
(C) Ethylamine (D) Methylamine

43. Which test is performed to identify the secondary amine ?
(A) Lucas test (B) Ethylamine
(C) Molisch test (D) Liebermann nitroso test

44. Glyoxal is
(A) OHC - CHO (B) OHC - CH_2OH
(C) $HOCH_2$ - CH_2OH (D) HOOCH - COOH

45. For the raection, C (graphite) → C (diamond), ΔS is
(A) Zero (B) Positive (C) Negative (D) None

46. For complete neutralization of HCl with NaOH, the heat of neutralization is
(A) +57.32 KJ mol (B) −13.70 KJ mol
(C) −57.32 KJ mol (D) +13.70 KJ mol

47. For which of the following, ΔE = ΔH
(A) $H_2(g) + \frac{1}{2} O_2(g) \rightarrow H_2O(l)$
(B) $2SO_2(g) + O_2(g) \rightarrow 2SO_3(g)$
(C) $H_2(g) + I_2(g) \rightarrow 2HI(g)$

(D) $N_2O_4(g) \rightarrow 2NO_2(g)$

48. A gaseous reaction has rate = [A] [B]. If volume of reaction vessel is reduced to $\frac{1}{4}$, the rate becomes in comparision to original

(A) $\frac{1}{8}$ (B) $\frac{1}{16}$ (C) 8 times (D) 16 times

49. In a reaction reactant 'A' decomposes 10% in 1 hour, 20% in 2 hour and 30% in 3 hour, the unit of rate constant of this reaction is

(A) S^{-1} (B) Litre mol^{-1} S^{-1}
(C) mol $litre^{-1}$ S^{-1} (D) $Litre^2$ $mole^{-2}$ S^{-1}

50. The rate of a reaction is doubled for every 10°C rise in temperature. If temperature is increased from 10°C to 100°C. The increase in the rate of reation is

(A) 512 times (B) 1024 times
(C) 256 times (D) 128 times

51. Which of the following is not used to make salt bridge ?

(A) KCl (B) KNO_3 (C) NH_4NO_3 (D) CH_3COOK

52. The electricity needed to liberate one gram equivalent of Cu^{2+} is

(A) 96500 amperes (B) 1 Faraday
(C) 96500 Faraday (D) 1 Coulomb

53. When lead storage battery is charged

(A) Lead is deposited on lead electrode
(B) $PbSO_4$ is deposited on lead electrode
(C) PbO_2 is dissolved
(D) H_2SO_4 is regenerated

54. For the redox reaction,
$MnO_4^- + C_2O_4^{-2} + H^+ \rightarrow Mn^{+2} + H_2O$, the correct coefficients of the reactants for the balanced reaction are

	MnO_4^-	$C_2O_4^{-2}$	H^+
(A)	2	5	16
(B)	16	5	2
(C)	2	16	5
(D)	5	16	2

55. The possible number of Optical isomers in $[Co(en)_2Cl_2]^{+2}$ are

(A) 2 (B) 3 (C) 4 (D) 6

56. π-bonding is not involved in

(A) Ferrocene (B) Dibenzene chromium
(C) Zeise salt (D) Grignard reagent

57. In which of the following, the addition of $AgNO_3$ does not result precipitation of silver chloride AgCl ?

(A) $[Co(NH_3)_3Cl_3]$ (B) $[Co(NH_3)_5Cl]Cl_2$
(C) $[Co(NH_3)_4Cl_2]Cl$ (D) $[Cr(NH_3)_6]Cl_3$

58. A solution of a salt in dilute sulphuric acid produces deep blue colour with starch iodine solution. It shows the presence of which of the following

(A) Br^- (B) I^- (C) NO_3^- (D) NO_2^-

59. To a 25 ml of H_2O_2 solution, excess of acidified solution of KI was added. The liberated I_2 require 20ml of 0.3 M hypo solution of H_2O_2 will be

(A) 1.44ml (B) 2.42ml (C) 1.34ml (D) 1.52ml

60. Which of the following acids adds to propene in the presence of peroxide to give anti-Markownikoff 's product ?

(A) HBr (B) HI (C) HCl (D) HF

61. The radius of first Bohr's orbit for hydrogen is 0.53 Å. The size of third Bohr's orbit is

(A) 4.77 Å (B) 3.18 Å (C) 1.59 Å (D) None

62. The frequency corresponding to transition n = 2 to n = 1 in hydrogen atom is

(A) 17.66×10^{10} HZ (B) 40.57×10^{24} HZ
(C) 31.57×10^{14} HZ (D) 24.66×10^{14} HZ

63. The number of urrpaired electrons is maximum in (Atomic number : Ti. 22, V.23, Cr. 24, Fe.26)

(A) Fe (B) Cr (C) Ti (D)V

64. A 200g golf ball is moving with a speed of 5m per hour. The associated wave length is (h = 6.625×10^{-34} J sec)

(A) 10^{-40} m (B)10^{-50} m (C) 10^{-30} m (D) 10^{-10} m

65. Orbital angular momentum for a d-electron is :

(A) $\frac{12h}{2\pi}$ (B) $\frac{\sqrt{6}h}{2\pi}$ (C) $\frac{6h}{2\pi}$ (D) $\frac{\sqrt{12}h}{2\pi}$

66. General configuration of ultimate and penultimate shell is (n-1) s^2 (n-1) p^6 (n-1) d^x ns^2. If n = 4 & x = 5 then number of protons in the nucleus will be

(A)> 25 (B) < 24 (C) 25 (D) None.

67. The number of electrons and protons in the third alkaline earth metal will be

	e	p		e	p		e	p		e	p
(A)	20	20	(B)	20	18	(C)	18	18	(D)	20	19

68. in which of the following pairs, molecules have bond order three and are isoelectronic ?

(A) CN^-,CO (B) CO, O_2^+
(C) CN, O_2^+ (D) NO^+, CO

69. Which of the follwing has dsp^2 hydridization ?

(A) NH_4^+ (B) $[PtCl_4]^{2-}$ (C) C_2H_2 (D) SCl_4

70. Hydrogen bond is strongest in

(A) H - F --------H (B) H ------------ O
(C) O - H ------- O (D) H ------------ N

71. In ClF_3, the hybridized state of Cl is

(A) sp^3d (B) sp^3 (C) sp^2 (D) sp^3d^2

72. In which of the following, metallic bond is storngest ?

(A) Fe (B) Cr (C) Sc (D) Mn

73. The method for the preparation of synthetic petroleum is

(A) Cannizzaro reaction (B) Skraup synthesis
(C) Fisher-Tropsch method (D) Curtius reaction

74. Which of the following compounds will be optically active ?

(A) Chloroacetic acid (B) Meso-tartaric acid
(C) Lactic acid (D) Oxalic acid

75. For which category of the compunds, the general formula C_nH_{2n-2} is ?
(A) Alkenes (B) Alkanes
(C) Alkynes (D) None of the above

76. Which of the following is not an emulsion?
(A) Ice Cream (B) Cloud (C) Butter (D) Milk

77. Which of the follwing is not an 'ore' of magnesium ?
(A) Dolomite (B) Carnallite
(C) Gypsum (D) Magnesite

78. A process used for the concentration of ore is
(A) Electrolysis (B) Froth floatation
(C) Roasting (D) None

79. Which metal is used as a reducing agent in smelting ?
(A) C (B) Zn (C) Al (D) none of these

80. In electro-refining, the impure metal is made
(A) Cathode (B) Anode
(C) Both of the above (D) None of these

81. Parke's process is used in the extraction of
(A) Silver (B) Zinc (C) Iron (D) Sodium

82. Which of the following pairs of atomic numbers represent s-block element
(A) 7,15 (B) 6,12 (C) 9,17 (D) 3,12

83. The element whose valancy is 7 would be
(A) V (B) Mn (C) C (D) Cr

84. Propyne on reaction with water in the presence of $HgSO_4$ and H_2SO_4 gives
(A) Acetic acid (B) Acetone
(C) Acetaldehyde (D) Methyl acetate

85. Which hydrocarbon will be most stable ?
(A) Methane (B) Ethane (C) Propane (D) Butane

86. Which can not be synthesized by the Wurtz method ?
(A) Methane (B) Butane (C) Ethane (D) Propane

87. Two moles of NH_3 when introduced into a previously evacuated vessel (one liter),partially dissociate into N_2 and H_2. The equilibrium constant is
(A) 27/64 $mol^2\ l^{-2}$ (B) 3/4 $mol^2\ l^{-2}$
(C) 27/16 $mol^2\ l^{-2}$ (D) 27/1 $mol^2\ l^{-2}$

88. The pH of 0.1 M NaOH is
(A) 11 (B) 13 (C) 10 (D) 12

89. How many grams of CaC_2O_4 (molecular weight =128) on dissolving in distilled water will give a saturated solution ? [$K_{sp}(CaC_2O_4) = 2.5 \times 10^{-9}$ mol^2 $liter^{-2}$]
(A) 1.2800g (B) 0.0128g
(C) 0.1280g (D) 0.0064g

90. A buffer solution contains equal concentration of X^- and HX. The k for HX is 10^{-8}. The pH of the buffer solution is
(A) 3 (B) 8 (C)10 (D)14

91. Which of the following makes basic solution ?
(A) NaOCl (B) HOCl (C) $NaHSO_4$ (D) None

92. Which of the following solutions will have pH close to 1.0 ?
(A) 55ml of $\frac{M}{10}$ HCl + 45ml of $\frac{M}{10}$ NaOH
(B) 100ml of $\frac{M}{10}$ HCl +100ml of $\frac{M}{10}$ NaOH
(C) 10ml of $\frac{M}{10}$ HCl + 90ml of $\frac{M}{10}$ NaOH
(D) 75ml of $\frac{M}{10}$ HCl + 25ml of $\frac{M}{10}$ NaOH

93. Which of the follwoing statements is wrong ?
(A) pH + pOH = 14 at room temperature
(B) The commonest primary standards for standardization of acids are Na_2CO_3 or borax
(C) $[H^+][OH^-] = 10^{-14}$ at 22°C
(D) Alkalies are generally standardized against acetic acid or borax.

94. To what pH must a solution containing a precipitate of $Cr(OH)_3$ be adjusted so that all of precipitate dissolves when Cr^{3+}= 0.1mol/L, $k_{sp} = 6 \times 10^{-31}$
(A) Upto 4.0 (B) Upto 4.1
(C) Upto 4.3 (D) Upto 4.2

95. The dissociation energy of methane is 360 KCal mol^{-1} and that of ethane is 620 KCal mol^{-1} the C–C bond energy (KCal mol^{-1}) is
(A) 210 (B) 130 (C) 180 (D) 80

96. E^o_{RP} for Fe^{+2}/Fe and Sn^{+2}/Sn are 0.44 and - 0.14 volts respectively. The standard e.m.f. for cell, $Fe^{+2} + Sn \rightarrow Sn^{+2} + Fe$ is
(A) +0.30 V (B) +0.58 V
(C) –0.58 V (D) –0.30 V

97. In XeO_3 and XeF_6, the oxidation state of Xe is
(A) +4 (B) +6 (C) +2 (D) +3

98. A group of acidic oxide is
(A) Na_2O, Al_2O_3 (B) ZnO, Al_2O_3
(C) Li_2O, MgO (D) CrO_3, Mn_2O_7

99. Which of the following can not form micelle
(A) Sodium lauryl sulphate (B) Sodium stearate
(C) Sodium alkyl benzene sulphonate
(D) Sodium benzoate

100. Which of the follwing does not give sol ?
(A) Electrical disintegration (B) Peptization
(C) Solvent exchange (D) Electrophoresis

PET - 2004

1. The orbital angular momentum of an electron in 2s-orbital is
(A) $\frac{1}{2}\frac{h}{2\pi}$ (B) $\frac{h}{2\pi}$ (C) $\sqrt{2}\frac{h}{2\pi}$ (D) Zero

2. The nucleus of an element contain 9 protons. Its valency would be
(A) 1 (B) 3 (C) 2 (D) 5

3. Carnalite is

(A) KCl (B) $LiAl(SiO_3)_2$
(C) $KCl.MgCl_2.6H_2O$ (D) $MgCl_2.6H_2O$

4. Chile Saltpetre is
(A) $NaNO_3$ (B) KNO_3
(C) Na_2SO_4 (D) Na_2S_2O

5. The colour of hydrogen is
(A) Red (B) Yellow (C) Green (D) Colourless

6. The electronic configuration of outer-most shell of the most electronegative elements
(A) ns^2np^3 (B) ns^2np^4
(C) ns^2np^5 (D) ns^2np^6

7. The correct set os quantum number for the unpaired electron of Cl atom is
(A) n = 2 $l = 1$ m = 0
(B) n = 2 $l = 1$ m = 1
(C) n = 3 $l = 1$ m = 1
(D) n = 3 $l = 0$ m = 0

8. Splitting of spectral lines under the influence of magnetic field is called
(A) Zeeman effect (B) Stark effect
(C) Photoelectric effect (D) None

9. The bond length in the species O_2, O_2^+ and O_2^- are in the order of
(A) $O_2^+ > O_2 > O_2^-$ (B) $O_2^+ > O_2^- > O_2$
(C) $O_2 > O_2^+ > O_2^-$ (D) $O_2^- > O_2 > O_2^+$

10. sp^3d^2 hybrid orbitals are
(A) Linear bipyramidal (B) Pentagonal
(C) Trigonal bipyramidal (D) Octahedral

11. The total number of electrons that takes part in forming bonds in N_2 is
(A) 2 (B) 4 (C) 6 (D) 10

12. Maximum number of H-bonds formed by a water molecule is
(A) 2 (B) 8 (C) 1 (D) 4

13. The number of α– and β–particles emitted when a radioactive element ${}_{90}F^{232}$ changes into ${}_{86}G^{220}$ will be
(A) 5 and 4 (B) 2 and 3
(C) 3 and 2 (D) 4 and 1

14. The compressibility factor of a gas is less than 1 at S.T.P. Its molar volume V_m will be
(A) $V_m > 22.42$ (B) $V_m < 22.42$
(C) $V_m = 22.42$ (D) None

15. The compressibility factor for an ideal gas is
(A) 1.5 (B) 1.0 (C) 2.0 (D) ∞

16. Which of the following will be most easily attacked by an electrophile.
(A) chlorobenzene (Cl on benzene ring) (B) phenol (OH on benzene ring) (C) toluene (CH_3 on benzene ring) (D) benzene

17. Addition of HCl to vinyl chloride gives 1, 1-dichloroethane because of
(A) Mesomeric effect of Cl
(B) Inductive effect of Cl
(C) Restricted rotation around double bond
(D) None of these

18. Hydrocarbon C_6H_6 decolourise Br_2 water and gives ppt. with ammonical $AgNO_3$ Hydrocarbon can be
(A) 1, 3, 5 Cyclohexatriene (B) 1, 5 Hexadiyne
(C) 2, 4 Hexadiyne (D) None

19. In reaction

$$C_6H_5CH_3 \xrightarrow{\text{Oxidation}} A \xrightarrow{\text{NaOH}} B \xrightarrow{\text{Sodalime}} C$$

Then C is
(A) C_6H_6 (B) C_6H_5OH
(C) C_6H_5COONa (D) C_6H_5ONa

20. The IUPAC name of Gamaxene is
(A) Benzene hexachloride
(B) Hexachlorobenzene
(C) 1, 2, 3, 4, 5, 6 hexachlorobenzene
(D) 1, 2, 3, 4, 5, 6 hexachlorocyclohexane

21. At normal temperature iodoform is
(A) Thick viscous liquid (B) Gas
(C) Volatile liquid (D) Solid

22. The compound added to prevent chloroform to form phosgene gas is
(A) C_2H_5OH (B) CH_3COOH
(C) CH_3COCH_3 (D) CH_3OH

23. The formula of Freon-n is
(A) CH_2Cl_2 (B) CCl_2F_2
(C) CH_2F_2 (D) $CClF_4$

24. A solution of $Al_2(SO_4)_3$ { d = 1.253gm/ml} contain 22% salt by weight. The molarity, normality and molality of the solution is
(A) 0.805M, 4.48N, 0.825M
(B) 0.825M, 48.3N 0.805M
(C) 4.83M, 4.83N, 4.83M
(D) None

25. Osmotic pressure of urea solution at 10°C is 500 mm. Osmotic pressure of the solution becomes 105.3 mm. When it is diluted and temperature raised to 25°C. The extent of dilution is
(A) 6 times (B) 5 times
(C) 7 times (D) 4 times

26. 50 ml of 1 M oxalic acid is shaken with 0.5gm of wood charcoal. The final concentration of the solution after adsorption is 0.5 M. Amount of oxalic acid adsorbed per gm of charcoal is
(A) 3.45 gm (B) 3.15 gm
(C) 6.30 gm (D) None

27. The edge of unit cell of fcc Xe crystal is 620pm. The

radius of Xe atom is
(A) 219.25 pm (B) 235.16 pm
(C) 189.37 pm (D) 209.87 pm

28. Which is not the buffer solution ?
(A) $CH_3COOH + CH_3COONa$
(B) $NH_4Cl + NH_4OH$
(C) CH_3COONH_4 (D) Borax + Boric acid

29. In the reaction

$$A_2(g) + 4B_2(g) \rightleftharpoons 2AB_4(g)$$

$\Delta H > 0$ the formation of AB_4 will be favoured at
(A) Low temperature, high pressure
(B) High temperature, low pressure
(C) Low temperature, low pressure
(D) High temperature, high pressure

30. For which of the reaction $\Delta H = \Delta E$?
(A) $N_2 + 3H_2 \rightleftharpoons 2NH_3$
(B) $N_2 + O_2 \rightarrow 2NO$
(C) $2Na + Cl_2 \rightarrow 2NaCl$
(D) $PCl_5 \rightarrow PCl_3 + Cl_2$

31. Which of the following is most acidic ?
(A) Picric acid (B) p-nitrophenol
(C) m-nitrophenol (D) o,p' dinitrophenol

32. Silver mirror test can be used to distinguish between
(A) Ketone and acid (B) Phenol and acid
(C) Aldehyde and acid (D) Alcohol and phenol

33. Three molecules of acetone in the presence of dry HCl from
(A) Mesitylene (B) Phorone
(C) Glyoxal (D) Mesityl oxide

34. Which reaction is used for the preparation of α-bromoacetic acid ?
(A) Kolbe's Reaction
(B) Reimer Tiemann Reaction
(C) Hell Volhard Zelinsky Reaction
(D) Perkin's Reaction

35. Which is correct for adiabatic process ?
(A) $P\Delta V = 0$ (B) $q = -W$ (C) $q = 0$ (D) $\Delta E = q$

36. The unit of velocity constant in case of zero order reaction is
(A) Concentration x $Time^{-1}$
(B) $Concentration^{-1}$ x $Time^{-1}$
(C) Concentration x $Time^{2}$
(D) $Concentration^{-1}$ x Time

37. In the reaction $I_2 + 2S_2O_3^{2-} \rightarrow 2I^- + S_4O_6^{2-}$ equivalent weight of iodine will be equal to
(A) 1/2 of molecular weight
(B) Molecular weight
(C) 1/4 of molecular weight (D) None

38. In the reaction $P + NaOH \rightarrow PH_3 + NaH_2PO_2$
(A) P is oxidised only (B) P is reduced only
(C) P is oxidized as well as reduced
(D) Na is reduced

39. On addition of one ml of 10% NaCl solution to 10 ml gold sol in the presence of 0.25gm of starch. the coagulation is just prevent. Starch has gold number
(A) 0.025 (B) 0.25 (C) 2.5 (D) None

40. The nature of anode rays depends upon
(A) Nature of electrode
(B) Nature of residual gas
(C) Nature of discharge tube
(D) All the above

41. How many unpaired electrons are there in Ni^{2-}
(A) 2 (B) 4 (C) 5 (D) 0

42. Cinnabar is an ore of
(A) Hg (B) Cu (C) Zn (D) Pb

43. Which of the following salt is insoluble in water ?
(A) $CuSO_4$ (B) $CdSO_4$
(C) $PbSO_4$ (D) $Bi_2(SO_4)_3$

44. Oil of vitriol is
(A) H_2SO_4 (B) H_2SO_3
(C) $H_2S_2O_7$ (D) $H_2S_2O_8$

45. Allyl cyanide contains------- σ– and π-bonds
(A) 9σ, 3π (B) 9σ, 9π
(C) 3σ, 4π (D) 5σ, 7π

46. Polymer used in bullet proof glass is
(A) Lexane (B) PMMA (C) Nomex (D) Kevlar

47. Which of the following is used for preparation of nylon ?
(A) Butadiene (B) Adipic acid
(C) Ethylene (D) Isoprene

48. The Purine base in RNA is
(A) Cytocine (B) Guanine
(C) Uracil (D) Thymine

49. The main structural feature of proteins is
(A) Ether linkage (B) Ester linkage
(C) Peptide linkage (D) None of these

50. The bond in $K_4[Fe(CN)_6]$ are
(A) All ionic (B) All covalent
(C) Ionic and covalent
(D) Ionic, covalent and co-ordinate covalent

PET - 1993

1. (C) 11. (C) 21. (C) 31. (D) 41. (D) 51. (D) 61. (B) 71. (D) 81. (D) 91. (D)
2. (C) 12. (C) 22. (C) 32. (A) 42. (D) 52. (B) 62. (C) 72. (B) 82. (B) 92. (D)
3. (B) 13. (D) 23. (C) 33. (C) 43. (B) 53. (C) 63. (D) 73. (C) 83. (A) 93. (D)
4. (B) 14. (A) 24. (C) 34. (C) 44. (B) 54. (B) 64. (B) 74. (D) 84. (C) 94. (A)
5. (C) 15. (D) 25. (A) 35. (B) 45. (A) 55. (B) 65. (A) 75. (B) 85. (B) 95. (C)
6. (A) 16. (A) 26. (A) 36. (D) 46. (B) 56. (A) 66. (D) 76. (B) 86. (A) 96. (B)
7. (A) 17. (B) 27. (A) 37. (D) 47. (A) 57. (C) 67. (B) 77. (D) 87. (C) 97. (A)
8. (A) 18. (D) 28. (A) 38. (A) 48. (C) 58. (B) 68. (A) 78. (C) 88. (A) 98. (D)
9. (B) 19. (A) 29. (A) 39. (B) 49. (A) 59. (B) 69. (B) 79. (C) 89. (A) 99. (D)
10. (D) 20. (D) 30. (B) 40. (A) 50. (C) 60. (C) 70. (D) 80. (D) 90. (C) 100.(D)

PET - 1994

1. (A) 11. (A) 21. (D) 31. (A) 41. (B) 51. (C) 61. (D) 71. (C) 81. (D) 91. (A)
2. (B) 12. (A) 22. (C) 32. (D) 42. (A) 52. (B) 62. (A) 72. (C) 82. (C) 92. (B)
3. (D) 13. (C) 23. (C) 33. (D) 43. (A) 53. (C) 63. (C) 73. (D) 83. (C) 93. (C)
4. (D) 14. (B) 24. (D) 34. (A) 44. (A) 54. (D) 64. (B) 74. (A) 84. (C) 94. (A)
5. (D) 15. (A) 25. (B) 35. (A) 45. (B) 55. (B) 65. (C) 75. (A) 85. (B) 95. (D)
6. (B) 16. (A) 26. (A) 36. (B) 46. (A) 56. (D) 66. (A) 76. (A) 86. (B) 96. (D)
7. (D) 17. (A) 27. (B) 37. (C) 47. (D) 57. (D) 67. (B) 77. (B) 87. (C) 97. (C)
8. (D) 18. (A) 28. (C) 38. (A) 48. (A) 58. (A) 68. (C) 78. (A) 88. (D) 98. (A)
9. (D) 19. (B) 29. (A) 39. (A) 49. (C) 59. (A) 69. (C) 79. (B) 89. (D) 99. (B)
10. (C) 20. (C) 30. (A) 40. (C) 50. (C) 60. (A) 70. (C) 80. (A) 90. (D) 100.(D)

PET - 1995

1. (B) 11. (A) 21. (D) 31. (D) 41. (B) 51. (A) 61. (B) 71. (D) 81. (C) 91. (A)
2. (D) 12. (C) 22. (A) 32. (D) 42. (D) 52. (C) 62. (B) 72. (C) 82. (D) 92. (C)
3. (B) 13. (D) 23. (D) 33. (B) 43. (B) 53. (A) 63. (D) 73. (A) 83. (A) 93. (A)
4. (C) 14. (B) 24. (A) 34. (B) 44. (A) 54. (A) 64. (D) 74. (D) 84. (A) 94. (A)
5. (A) 15. (A) 25. (B) 35. (A) 45. (D) 55. (D) 65. (A) 75. (D) 85. (C) 95. (B)
6. (B) 16. (C) 26. (C) 36. (A) 46. (D) 56. (B) 66. (C) 76. (B) 86. (D) 96. (D)
7. (B) 17. (B) 27. (C) 37. (A) 47. (B) 57. (A) 67. (B) 77. (C) 87. (D) 97. (B)
8. (A) 18. (B) 28. (D) 38. (C) 48. (C) 58. (C) 68. (D) 78. (A) 88. (C) 98. (B)
9. (B) 19. (B) 29. (A) 39. (C) 49. (D) 59. (D) 69. (C) 79. (C) 89. (B) 99. (A)
10. (D) 20. (A) 30. (A) 40. (D) 50. (D) 60. (D) 70. (B) 80. (C) 90. (C) 100.(C)

PET - 1996

1. (D) 11. (A) 21. (B) 31. (D) 41. (C) 51. (C) 61. (B) 71. (B) 81. (A) 91. (A)
2. (C) 12. (B) 22. (A) 32. (A) 42. (C) 52. (C) 62. (C) 72. (C) 82. (B) 92. (B)
3. (C) 13. (C) 23. (C) 33. (C) 43. (C) 53. (B) 63. (B) 73. (B) 83. (D) 93. (D)
4. (A) 14. (C) 24. (C) 34. (B) 44. (A) 54. (D) 64. (D) 74. (D) 84. (A) 94. (A)
5. (C) 15. (B) 25. (C) 35. (B) 45. (C) 55. (A) 65. (D) 75. (A) 85. (C) 95. (B)
6. (B) 16. (A) 26. (C) 36. (A) 46. (C) 56. (D) 66. (C) 76. (D) 86. (C) 96. (A)
7. (A) 17. (C) 27. (A) 37. (C) 47. (C) 57. (C) 67. (C) 77. (A) 87. (D) 97. (D)
8. (D) 18. (A) 28. (D) 38. (D) 48. (D) 58. (B) 68. (B) 78. (A) 88. (B) 98. (C)
9. (B) 19. (D) 29. (A) 39. (B) 49. (B) 59. (B) 69. (D) 79. (B) 89. (D) 99. (C)
10. (D) 20. (A) 30. (D) 40. (B) 50. (B) 60. (B) 70. (C) 80. (D) 90. (C) 100.(B)

PET - 1997

1. (A) 11. (D) 21. (B) 31. (D) 41. (A) 51. (A) 61. (C) 71. (B) 81. (C) 91. (D)
2. (C) 12. (B) 22. (A) 32. (A) 42. (B) 52. (D) 62. (D) 72. (A) 82. (D) 92. (C)
3. (B) 13. (D) 23. (D) 33. (B) 43. (C) 53. (C) 63. (C) 73. (A) 83. (D) 93. (D)
4. (A) 14. (D) 24. (A) 34. (A) 44. (B) 54. (A) 64. (B) 74. (A) 84. (C) 94. (A)
5. (C) 15. (A) 25. (C) 35. (C) 45. (A) 55. (B) 65. (D) 75. (C) 85. (A) 95. (A)
6. (A) 16. (C) 26. (C) 36. (A) 46. (A) 56. (B) 66. (B) 76. (C) 86. (C) 96. (D)
7. (B) 17. (C) 27. (B) 37. (D) 47. (D) 57. (B) 67. (B) 77. (B) 87. (B) 97. (B)
8. (D) 18. (D) 28. (A) 38. (B) 48. (B) 58. (B) 68. (B) 78. (D) 88. (A) 98. (A)
9. (D) 19. (B) 29. (C) 39. (C) 49. (D) 59. (C) 69. (C) 79. (B) 89. (A) 99. (B)
10. (D) 20. (B) 30. (C) 40. (D) 50. (B) 60. (C) 70. (C) 80. (D) 90. (C) 100.(C)

PET - 1998

1. (D) 11. (C) 21. (C) 31. (A) 41. (C) 51. (B) 61. (C) 71. (B) 81. (A) 91. (B)
2. (C) 12. (D) 22. (B) 32. (C) 42. (B) 52. (C) 62. (B) 72. (B) 82. (C) 92. (B)
3. (B) 13. (C) 23. (D) 33. (D) 43. (C) 53. (A) 63. (B) 73. (D) 83. (B) 93. (D)
4. (B) 14. (C) 24. (B) 34. (A) 44. (D) 54. (C) 64. (D) 74. (B) 84. (A) 94. (D)
5. (D) 15. (A) 25. (B) 35. (D) 45. (C) 55. (C) 65. (C) 75. (C) 85. (C) 95. (C)
6. (D) 16. (C) 26. (D) 36. (C) 46. (C) 56. (A) 66. (B) 76. (A) 86. (D) 96. (D)
7. (A) 17. (B) 27. (D) 37. (A) 47. (C) 57. (B) 67. (D) 77. (C) 87. (A) 97. (C)
8. (A) 18. (A) 28. (C) 38. (A) 48. (B) 58. (C) 68. (C) 78. (B) 88. (B) 98. (A)
9. (A) 19. (A) 29. (B) 39. (B) 49. (C) 59. (B) 69. (A) 79. (B) 89. (D) 99. (D)
10. (B) 20. (B) 30. (D) 40. (A) 50. (B) 60. (A) 70. (D) 80. (A) 90. (A) 100.(D)

PET - 1999

1. (A) 11. (A) 21. (B) 31. (D) 41. (C) 51. (C) 61. (B) 71. (B) 81. (A) 91. (A)
2. (D) 12. (A) 22. (B) 32. (C) 42. (D) 52. (D) 62. (D) 72. (C) 82. (A) 92. (A)
3. (C) 13. (D) 23. (A) 33. (B) 43. (B) 53. (A) 63. (D) 73. (A) 83. (D) 93. (D)
4. (B) 14. (C) 24. (A) 34. (A) 44. (A) 54. (B) 64. (A) 74. (C) 84. (B) 94. (D)
5. (D) 15. (B) 25. (B) 35. (B) 45. (A) 55. (A) 65. (C) 75. (D) 85. (A) 95. (C)
6. (C) 16. (B) 26. (D) 36. (B) 46. (A) 56. (D) 66. (D) 76. (B) 86. (C) 96. (D)
7. (D) 17. (D) 27. (C) 37. (B) 47. (B) 57. (B) 67. (D) 77. (C) 87. (A) 97. (A)
8. (A) 18. (C) 28. (C) 38. (C) 48. (D) 58. (C) 68. (A) 78. (C) 88. (D) 98. (C)
9. (D) 19. (D) 29. (A) 39. (C) 49. (A) 59. (B) 69. (A) 79. (C) 89. (D) 99. (A)
10. (B) 20. (C) 30. (C) 40. (B) 50. (D) 60. (B) 70. (D) 80. (B) 90. (B) 100.(D)

PET - 2000

1. (A) 11. (D) 21. (B) 31. (C) 41. (B) 51. (A) 61. (B) 71. (B) 81. (A) 91. (A)
2. (B) 12. (A) 22. (C) 32. (B) 42. (B) 52. (D) 62. (D) 72. (C) 82. (C) 92. (B)
3. (B) 13. (C) 23. (C) 33. (C) 43. (D) 53. (C) 63. (C) 73. (B) 83. (D) 93. (A)
4. (A) 14. (C) 24. (D) 34. (C) 44. (C) 54. (A) 64. (D) 74. (D) 84. (C) 94. (A)
5. (B) 15. (A) 25. (D) 35. (A) 45. (C) 55. (A) 65. (B) 75. (D) 85. (C) 95. (A)
6. (C) 16. (C) 26. (D) 36. (A) 46. (A) 56. (C) 66. (B) 76. (A) 86. (A) 96. (D)
7. (D) 17. (C) 27. (C) 37. (A) 47. (A) 57. (B) 67. (D) 77. (A) 87. (D) 97. (B)
8. (C) 18. (C) 28. (C) 38. (B) 48. (D) 58. (C) 68. (C) 78. (A) 88. (C) 98. (A)
9. (C) 19. (A) 29. (B) 39. (B) 49. (B) 59. (C) 69. (A) 79. (D) 89. (C) 99. (C)
10. (B) 20. (A) 30. (A) 40. (C) 50. (D) 60. (B) 70. (A) 80. (D) 90. (C) 100.(C)

PET - 2001

1. (C) 11. (B) 21. (C) 31. (C) 41. (B) 51. (C) 61. (A) 71. (A) 81. (D) 91. (C)
2. (B) 12. (D) 22. (C) 32. (C) 42. (B) 52. (A) 62. (A) 72. (A) 82. (A) 92. (B)
3. (A) 13. (C) 23. (B) 33. (B) 43. (A) 53. (A) 63. (C) 73. (C) 83. (C) 93. (D)
4. (C) 14. (C) 24. (B) 34. (D) 44. (D) 54. (A) 64. (C) 74. (D) 84. (C) 94. (B)
5. (B) 15. (D) 25. (A) 35. (C) 45. (C) 55. (C) 65. (D) 75. (A) 85. (A) 95. (C)
6. (C) 16. (D) 26. (B) 36. (B) 46. (A) 56. (A) 66. (D) 76. (B) 86. (C) 96. (B)
7. (B) 17. (B) 27. (B) 37. (C) 47. (D) 57. (D) 67. (C) 77. (C) 87. (D) 97. (D)
8. (C) 18. (D) 28. (A) 38. (A) 48. (D) 58. (B) 68. (B) 78. (B) 88. (B) 98. (C)
9. (D) 19. (D) 29. (D) 39. (B) 49. (A) 59. (C) 69. (C) 79. (D) 89. (C) 99. (D)
10. (C) 20. (A) 30. (A) 40. (C) 50. (C) 60. (C) 70. (D) 80. (B) 90. (A) 100.(C)

PET - 2002

1. (D) 11. (C) 21. (C) 31. (B) 41. (A) 51. (C) 61. (C) 71. (C) 81. (A) 91. (B)
2. (D) 12. (C) 22. (B) 32. (C) 42. (D) 52. (C) 62. (D) 72. (C) 82. (A) 92. (B)
3. (A) 13. (A) 23. (A) 33. (B) 43. (B) 53. (C) 63. (A) 73. (D) 83. (C) 93. (B)
4. (B) 14. (D) 24. (A) 34. (B) 44. (C) 54. (A) 64. (D) 74. (A) 84. (A) 94. (C)
5. (D) 15. (A) 25. (B) 35. (D) 45. (C) 55. (D) 65. (D) 75. (A) 85. (B) 95. (B)
6. (A) 16. (D) 26. (C) 36. (C) 46. (C) 56. (B) 66. (B) 76. (B) 86. (A) 96. (A)
7. (B) 17. (B) 27. (A) 37. (B) 47. (A) 57. (C) 67. (C) 77. (B) 87. (A) 97. (B)
8. (B) 18. (C) 28. (C) 38. (A) 48. (A) 58. (D) 68. (C) 78. (B) 88. (B) 98. (C)
9. (D) 19. (B) 29. (C) 39. (B) 49. (B) 59. (C) 69. (D) 79. (B) 89. (A) 99. (A)
10. (B) 20. (D) 30. (A) 40. (B) 50. (C) 60. (C) 70. (C) 80. (A) 90. (C) 100.(A)

PET - 2003

1. (C)	11. (D)	21. (D)	31. (C)	41. (D)	51. (D)	61. (A)	71. (A)	81. (A)	91. (A)
2. (B)	12. (D)	22. (B)	32. (A)	42. (A)	52. (B)	62. (D)	72. (B)	82. (D)	92. (C)
3. (C)	13. (A)	23. (C)	33. (C)	43. (D)	53. (D)	63. (B)	73. (C)	83. (B)	93. (D)
4. (D)	14. (A)	24. (D)	34. (A)	44. (A)	54. (A)	64. (C)	74. (C)	84. (B)	94. (D)
5. (C)	15. (C)	25. (D)	35. (B)	45. (C)	55. (A)	65. (B)	75. (C)	85. (A)	95. (D)
6. (C)	16. (B)	26. (C)	36. (C)	46. (C)	56. (D)	66. (C)	76. (B)	86. (A)	96. (D)
7. (B)	17. (B)	27. (B)	37. (D)	47. (C)	57. (A)	67. (A)	77. (C)	87. (C)	97. (B)
8. (D)	18. (D)	28. (A)	38. (C)	48. (D)	58. (D)	68. (A)	78. (B)	88. (B)	98. (D)
9. (A)	19. (C)	29. (B)	39. (B)	49. (A)	59. (C)	69. (B)	79. (A)	89. (D)	99. (D)
10. (C)	20. (B)	30. (A)	40. (D)	50. (A)	60. (A)	70. (A)	80. (B)	90. (B)	100. (D)

PET - 2004

1. (D)	11. (C)	21. (D)	31. (A)	41. (A)
2. (A)	12. (D)	22. (A)	32. (C)	42. (A)
3. (C)	13. (C)	23. (B)	33. (B)	43. (C)
4. (B)	14. (B)	24. (A)	34. (C)	44. (A)
5. (D)	15. (B)	25. (B)	35. (C)	45. (A)
6. (C)	16. (B)	26. (C)	36. (A)	46. (B)
7. (C)	17. (D)	27. (A)	37. (A)	47. (B)
8. (A)	18. (D)	28. (A)	38. (D)	48. (B)
9. (A)	19. (A)	29. (A)	39. (D)	49. (C)
10. (D)	20. (C)	30. (B)	40. (B)	50. (D)

TNPCEE -1995

1. The electronic configuration of the element with atomic number 29 is
(A) [Ar] $4s^4 3d^8$ (B) [Ar] $4s^1 3d^{10}$
(C) [Ar] $4s^2 3d^9$ (D) [Kr] $4s^2 3d^9$

2. Water gas is a mixture of
(A) CH_4 and H_2 (B) CH_4 and H_2O
(C) CO and H_2 (D) CO_2 and H_2

3. In the aluminothermic process aluminium plays the role of
(A) a dehydrating agent (B) a flux
(C) a reducing agent (D) an oxidisingagent

4. The products obtined when $KClO_3$ is heated with iodine in nitric acid solution are
(A) $KI + O_3$ (B) $KCl + ClO_2$
(C) $KI + O_2$ (D) $KIO_3 + Cl_2$

5. Iodine dissolves in KI due to the formation of
(A) IO_3^- (B) I_3^- (C) I^- (D) I_3^{3-}

6. The bond between phosphorous and oxygen atom in phosphorous acid is
(A) metallic bond (B) ionic bond
(C) covalent bond (D) co-ordinate bond

7. The composition of superphosphate of lime is
(A) $Ca_3(PO_4)_2 + CaSO_4$
(B) $Ca(H_2PO_4)_2\ H_2O + CaSO_4\ 2H_2O$
(C) $Ca_2(PO_4)_2 + CaSO_4\ 2H_2O$
(D) $Ca(H_2PO_4)_2 + Na_3PO_4$

8. $HBO_3 \xrightarrow[\Delta]{433}$?
(A) $Na_2B_4O_7$ (B) B_2O_3 (C) H_3BO_3 (D) $N_2B_4O_2$

9. The molecular formula of Colemenite is
(A) $Ca_2B_6O_{11}.\ 5H_2O$ (B) $Na_2B_4O_7.\ 10H_2O$
(C) $Ca_2B_6O_{11}$ (D) $Ca_2B_6O_{11}.\ 4H_2O$

10. In the manufacture of glass some broken pieces of glass are added to the raw material. The reason is that it
(A) reduces the coefficient of expansion
(B) melts easily and helps on melting the rest ot the change
(C) solidifies and helps to solidify other substances
(D) purifies the mixture

11. The products obtained on boiling a mixture of white phosphorus and 40% sodium hydroxide solution in an inert atmosphere of carbon dioxide are
(A) P_2O_5 and NaH_2PO_2 (B) P_2O_5 and Na_3PO_4
(C) PH_3 and NaH_2PO_2 (D) PH_3 and Na_3PO_4

12. The elements which occupy the peaks of the atomic volume-vs-atomic number curve are
(A) Na, K, Rb (B) Ne, Ar, Kr
(C) Cl, Br, I (D) Fr, Co, Ni

13. Ionisation energy of Nitrogen is greater than that of oxygen because of
(A) half-filled 2p orbitals of O atom
(B) completely filled 2p orbitals of O atom
(C) half-filled 2p orbitals of N atom
(D) completely filled 2p orbitals of N atom

14. Identify the optically active compound
(A) tertiary butyl alcohol (B) aniline
(C) glyceraldehyde (D) glycine

15. Which of the follwing pairs are chain isomers ?
(A) 1- Butanol and ethoxyethane
(B) 1- Butanol and methoxypropane
(C) 1- Butanol and 2-methyl-1-propanol
(D) 1-Butanol and 2-butanol

16. The number of asymmetric carbon atoms in glucose is
(A) 1 (B) 3 (C) 4 (D) 6

17. The conversion of vegetable oils to solids is effected by
(A) esterification (B) hydrolysis
(C) oxidation (D) hydrogenation

18. Ehanol reacts with moist bleaching powder to yield
(A) CH_3COCH_3 (B) $CHCl_3$ (C) CCl_3 (D) CH_3Cl

19. Lucas reagent is a mixture of
(A) anhydrous zinc chloride and conc. HCl
(B) anhydrous chloride of aluminium and zinc
(C) zinc chloride and dil. HCl
(D) anhydrous aluminium chloride and conc. HCl

20. Anisole is another name for
(A) diphenyl ether (B) ethyl phenyl ether
(C) methyl phenyl ether (D) diethyl ether

21. Compound A of molecular formula C_2H_2O is a sweet smelling liquid A. On reaction with acidified $K_2Cr_2O_7$ gives compound B of molecular formula C_2H_2O. B reduces Tolulens reagent. Then A and B are respectively
(A) diphenyl carbinol and behzophenone
(B) benzyl alcohol and benzaldehyde
(C) methyl phenyl carbinol and acetophenone
(D) benzaldehyde and benzoic acid

22. Phorone is a compound obtained by the reaction of
(A) acetone and gaseous hydrogen chloride
(B) acetaldehyde and gaseous hydrogen chloride
(C) acetone and conc. H_2SO_4
(D) acetone and iodine

23. Cannizzaro reaction is undergone by
(A) aliphatic and aromatic aldehydes without α-hydrogen atom
(B) aliphatic and aromatic aldehydes with α-hydrogen atom
(C) aromatic aldehydes only (D) aliphatic aldehydes only

24. When phthalic acid is heated with soda lime it forms
(A) phthalic anhydride (B) benzene
(C) sodium benzoate (D) calcium phthalate

25. Which among the following is a strong acid but can be easily oxidised ?
(A) $CH_3CH_2CH_2COOH$ (B) CH_3CH_2COOH
(C) CH_3COOH (D) HCOOH

26. For a given value of the azimuthal quantum number l, the total number of allowed values of the magnetic quantum number m_l is
(A) $2l - 1$ (B) $2(2l + 1)$ (C) $2l + 1$ (D) $2l$

27. The ionisation potential of the hydrogen atom is 13.6eV. The energy of the atom in the n = 2 state is
(A) 3.4 eV (B) – 3.4 eV (C) –13.6 eV (D) 13.6 eV

28. The molarity of 9.8 gram of sulphuric acid in 2 dm^3 solution is
(A) 0.5 M (B) 0.005 M (C) 0.05 M (D) 0.1 M

29. Which of the following mixtures can serve as a buffer solution ?
(A) H_3PO_4 and Na_3PO_4 (B) CH_3COOH and CH_3COONa
(C) NaOH and NaCl (D) HCl and NaCl

30. The conjugate acid of the base OH is
(A) H_2 (B) H^+ (C) H_3O^+ (D) H_2O

31. The pH of a solution is 5.00. Sufficient acid is added to it so that its final pH is 3.00. The H^+ concentration now becomes
(A) 100 times (B) 10 times (C) 1000 times (D) 2 times

32. When one mole of sulphuric acid is completely neutralised by sodium hydroxide, the enthalpy of neutralisation will be
(A) –28.7 kJ (B) –172.2 kJ (C) –114.8 kJ (D) –57.4 kJ

33. For the reaction

$C_6H_5COOH(s) + \frac{1}{2}7O_2(g) \rightarrow 7CO_2(g) + 3H_2O(l)$ the relationship between ΔH and ΔE is

(A) $\Delta H = \Delta E - \frac{1}{2}RT$ (B) $\Delta H = \Delta E + \lceil 2RT$

(C) $\Delta H = \Delta E + \frac{1}{2}RT$ (D) $\Delta H = \Delta E - \lceil 2RT$

34. The reaction used in time bomb is the oxidation of glycerol by
(A) concentrated HNO_3 (B) fentons reagent
(C) bismuth and sodium nitrate (D) solid $KMnO_4$

35. The IUPAC name of potassium ferrocyanide is
(A) potassium hexacyano argentite (III)
(B) ferro hexacyano potassium (II)
(C) potassium hexacyano ferrate (II)
(D) potassium hexacyano ferrate (III)

36. Tin reacts with dil. NHO_3 to give
(A) stannous nitrate and ammonium nitrate
(B) meta stannic acid and nitrogen dioxide
(C) nitrous oxide (D) nitric oxide

37. Solder is an alloy of
(A) Sn, Zn, Cu (B) Pb. Sn (C) Cu, Zn (D) Fe, Mn, C

38. Brown ring test is done to confirm the presence of
(A) nitrate (B) carbonate (C) chloride (D) sulphate

39. 30 cm^3 of 0.1 N HCl was required to neutralise 20 m^3 of NaOH. The normality of NaOH is
(A) 1.5 N (B) 0.005 N (C) 0.1 N (D) 0.15 N

40. When a small amount of a salt is treated with dil HCl and warmed a colourless gas emanates with the smell of a rotten egg and it turns most lead acetate later black. This indicates
(A) absence of sulphide and carbonate
(B) presence of sulphide
(C) presence of carbonate (D) absence of chloride

41. The rate constant of a first order reaction is of 0.0693 sec. The half life period of the reaction is
(A) 0.01 sec (B) 0.1 sec (C) 10 sec (D) 100 sec

42. The order and the molecularity of iodination of acidified acetone are respectively.
(A) 1 and 2 (B) 0 and 2 (C) 2 and 2 (D) 2 and 0

43. 1 m solution of HCl conducts more electricity than 1 m solution of acetic acid. This is because
(A) HCl is a mineral acid and acetic acid is an organic acid
(B) HCl furnishes more ions than acetic acid
(C) HCl furnishes less ions than acetic acid
(D) HCl is highly soluble in water

44. The oxidation number of P in Na_3PO_4 is
(A) –3 (B) –5 (C) + 3 (D) +5

45. The migration of colloidal particles under the influence of an electric field is known as
(A) Electrophoresis (B) Brownian movement
(C) Electro-osmosis (D) Tyndall effect

46. The catalyst used in the hydrogenation of oils and fats is
(A) Pt (B) Ni (C) Fe (D) V_2O_5

47. The role of the catalyst in a reaction is to
(A) accelerate the forward and backward reactions equally
(B) accelerate the forward reaction only
(C) increases the yield of the products
(D) increases activation energy

48. $CH_3COCl \xrightarrow[+H]{?} CH_3CHO + HCl$
(A) $Pd/BaSO_4$ (B) Al_2Cl_6 (C) $LiAlH_4$ (D) Ni

49. A carboxylic acid on treatment with lime water gives a salt A. On dry distillation A gives a compound of molecular formula C_3H_6O which shows the iodoform test. The carboxylic acid is
(A) HOOC-COOH (B) CH_3COOH
(C) C_6H_5COOH (D) HCOOH

50. An organic compound that liberates nitrogen gas with nitrous acid and produces a poisonous gas with chloroform in presence of alkali is a
(A) quaternary ammonium salt (B) primary amine
(C) tertiary amine (D) secondary amine

51. $C_6H_5N_2Cl \xrightarrow{Sn/HCl} ?$
(A) $C_6H_5NHNH_2$ (B) C_6H_5OH (C) $C_6H_5NH_2$ (D) C_6H_5Cl

52. When nitrobenzene is reduced with zinc and aqueous sodium hydroxide we get
(A) nitrosobenzene (B) phenyl hydroxylamine
(C) hydrazobenzene (D) aniline

53. Ethylene dichloride is one of the raw materials in the manufacture of
(A) thiokal (B) dacron (C) neoprene (D) bakelite

54. The raw materials used in the manufacture of nylon are
(A) cellulose and basic copper sulphate
(B) cellulose and acetic acid
(C) terephthalic acid and ethylene glycol
(D) adipic acid and hexamethylene diamine

55. The hormone essential for the regulation of carbohydrates in the body is
(A) oxytocin (B) insulin (C) adrenalin (D) thyroxin

56. An example for a vat dye is
(A) magents (B) triphenyl methane dye
(C) indigo (D) alizarin

57. The commonly used colouring agent in Rose milk is
(A) rhodamine -B (B) rosaniline
(C) orange-2 (D) methyl orange

TNPCEE -1996

1. Which of the following has the highest electron affinity value ?
(A) Halogens (B) Actinides
(C) Lanthanides (D) Inert gases

2. Amatol is a mixture of
(A) T.N.T and Ammonium nitrate
(B) T.N.T. and Sodium nitrate
(C) Picric acid and Ammonium nitrate
(D) Picric acid and Sodium nitrate

3. The metal which gives brown colour in the oxidising flame in borax bead test is
(A) Cobalt (B) Chromium (C) Nickel (D) Iron

4. Jena glass is mixture of
(A) Sodium and Barium borosilicates
(B) Zinc and Aluminium borosilicates
(C) Sodium and Aluminium borasilicates
(D) Zinc and Barium borosilicates

5. In boric acid, the orbitals of boron undergo
(A) p-p hybridisation (B) sp hybridisation
(C) sp^3 hybridisation (D) sp^2 hybridisation

6. Fire clay contains
(A) Al_2O_3. SiO_2 (B) ZnO. SiO_2
(C) B_2O_3. SiO_2 (D) Na_2O. SiO_2

7. The molecular formula of Borax is
(A) $N_2B_4O_7$ (B) $Na_2B_4O_7$ $10H_2O$
(C) $Ca_2B_6O_{11}$ $5H_2O$ (D) H_2BO_3

8. Potassium chlorate on heating gives
(A) H_2 (B) Cl_2 (C) O_2 (D) ClO_2

9. The composition of triple superphosphate is
(A) $Ca(H_2PO_4)_2$ (B) $Ca_3(PO_4)_2 + Na_3PO_4$
(C) $Ca_3(PO_4)_2 + CaSO_4$ (D) $Ca_3(PO_4)_2$

10. In the carbon-reduction process lime prevents the formation of
(A) $PbSiO_3$ and CO_2 (B) $PbSiO_3$ and $PbSO_4$
(C) $CaSiO_3$ and $PbiSiO_3$ (D) $CaSiO_3$ and $PbSO_4$

11. Electronic configuration of silver (Ag47) is
(A) [Xe] $6s^2$ $4f^{14}$ $5d^9$ (B) [Kr] $5s^1$ $4d^{10}$
(C) [Xe] $6s^1$ $4f^{14}$ $5d^{10}$ (D) [Kr] $5s^1$ $4d^9$

12. In Holme's signals, the gas liberated along with phosphine is
(A) propane (B) acetylene (C) methane (D) ethylene

13. For the principal quantum number 3 and the azimuthual quantum number 2, how many values does the magnetic quantum number have ?
(A) One (B) Three (C) Five (D) Seven

14. What is the molarity of the solution containing 6.3 g of nitric acid in 0.5 dm^3 solution ?
(A) 0.4 m (B) 0.5 m (C) 0.2 m (D) 1 m

15. Which one of the following compounds is not opically active ?
(A) Bromochloroiodomethane (B) 2-butanol
(C) Glyceraldehyde (D) Glycine

16. The alcohol used in the manufacture of dynamite is
(A) Glycerol (B) Glycol
(C) Ethyl alcohol (D) Methyl alcohol

17. Isopropyl alcohol on oxidation gives
(A) n-propyl alcohol (B) Formaldehyde
(C) Acetaldehyde (D) Acetone

18. Which of the following is used to detect OH group in organic compounds ?
(A) Benzyl chloride (B) Benzoyl chloride
(C) Formyl chloride (D) Acetyl chloride

19. Freezing mixture of ethylene glycol with water is sold under the name
(A) Prestone (B) Lysol (C) Spirit (D) Westron

20. Ethylene glycol is an example of a

(A) Aromatic alchol (B) Trihydric alchol
(C) Dihydric alchol (D) Monohydric alchol

21. Phenol gives benzene on distillation with
(A) Sodium (B) Zinc (C) Palladium (D) Platinum

22. Phenol $\xrightarrow[\Delta]{?}$ Quinone
(A) K_2CO_3 + conc H_2SO_4 (B) $KMnO_4$ + dil H_2SO_4
(C) CrO_2Cl_2 (D) K_2CrO_4 + conc H_2SO_4

23. Ethers are
(A) strongly basic (B) strongly acidic
(C) weakly basic (D) weakly acidic

24. Addition product of acetaldehy and HCN, on hydrolysis gives
(A) Fumaric acid (B) Maleic acid
(C) Lactic acid (D) Malonic acid

25. Acetone when treated with conc. H_2SO_4 gives
(A) Propylene (B) Mesitylene
(C) Ethylene (D) Acetylene

26. Formaldehyde reacts with Iodine and Sodium hydroxide to give
(A) Sodium formate (B) Sodium acetate
(C) Methyl iodide (D) Iodoform

27. Anthranilic acid is
(A) o-nitrobenzoic acid (B) p-aminobenzoic acid
(C) m-aminobenzoic acid (D) o-aminobenzoic acid

28. Methyl salicylate is used as
(A) antiseptic (B) sprain and ache reliever
(C) lubricant (D) sedative

29. The product obtained in Claisen ester condensation reaction is
(A) Ethyl acetate (B) Ethyl acetoacetate
(C) Benzylidene acetone (D) Cinnamaldehyde

30. A is a compound with formula C_2H_5ON. This undergoes Hoffmann's degradation giving B which on reaction with nitrous acid gives C. Compound B gives carbylamine reaction and C produes red colour in Victor Meyer's Test. A and C can be
(A) Methylamine and Ethanol
(B) Acetamide and Methanol
(C) Methanol and Ethanol
(D) Benzamide and Benzyl alcohol

31. Which one of the following is known as the secondary valency of the central metal atom in a co-ordination compound ?
(A) The oxidation state of the metal
(B) The co-ordination number
(C) The valency that is satisfied by negative ions
(D) Group number of the metal

32. A compound used in the preparation of hair dye is
(A) $KClO_3$ (B) $Na_2S_2O_3$ (C) K_2CrO_4 (D) $AgNO_3$

33. Gun metal is an alloy of
(A) Sn, Zn, Cu (B) Pb, Mn, Cu
(C) Sn, Cu, Pb (D) Sn, Zn, Ca

34. The ores of which metals are concern treated by froth floatation process ?
(A) Chromium, nickel and gold
(B) Manganese, lead and nickel
(C) Nickel, chromium and manganese
(D) Lead, nickel and silver

35. Alloy used in production of rock crushing machinery is
(A) Chrome-Vanadium steel (B) Manganese steel
(C) Platinite (D) Invar

36. $20cm^3$ of sodium carbonate is neutralised by $40cm^3$ of 0.1 N HCl.The amount of sodium carbonate in $1dm^3$ solution is
(A) 5.3 g (B) 10.6 g (C) 106 g (D) 53 g

37. p-hydroxy azo benzene is obtained when benzene diazonium chloride is coupled with
(A) Phenol (B) Methylamine
(C) Aniline (D) Nitrobenzene

38. In Hinsberg's method, the mixture of amines is treated with
(A) NItrous acid (B) Methyl chloride
(C) Diethyl oxalate (D) Benzene sulphonic chloride

39. Schiff's bases are obtained when aniline reacts with
(A) aromatic alcohol (B) aromatic aldehyde
(C) aromatic acid (D) aromatic amide

40. When benzophenone is heated with potassium hydroxide it gives
(A) Toluene (B) Cresol (C) Benzene (D) Phenol

41. When nitromethane is reduced in neutral medium (Zn dust, NH_4Cl) it gives
(A) NH_2 - NH_2 (B) CH_3NO
(C) CH_3NHOH (D) CH_3NH_2

42. Which of the following is an example for thermosetting plastic ?
(A) Bakelite (B) Polyvinyl chloride
(C) Polystyrene (D) Polythene

43. Which of the following is uded as a coagulating agent for rubber latex ?
(A) Lactic acid (B) Acetic acid
(C) Benzoic acid (D) Formic acid

44. Enzyme involved in clotting of blood is
(A) Maltase (B) Diastase
(C) Thrombin (D) Zymase

45. The hormone which is secreted by the posterior pituitary gland is
(A) Progesterone (B) Oxytocin
(C) Estrogen (D) Insulin

46. Kesari dal used as an adulterant in other pulses causes
(A) Cancer (B) Crippling paralysis
(C) Blurred vision (D) Diarrhoea

47. To which class of dyes does Alizarin belong to ?
(A) Mordant dye (B) Acid dye
(C) Baisc dye (D) Vat dye

48. In the reaction $N_2 + 3H_2 \rightleftharpoons 2NH_3$
(A) increase in pressure favours backward reaction
(B) pressure has no effect on reaction
(C) decrease in pressure favours forward reaction
(D) increase in pressure favours forward reaction

49. At 730 K temperature K_p of the reaction $H_2 + I_2 \rightleftharpoons 2HI$ is 49. The valued of K_c is
(A) 49 (B) .0245 (C) 98 (D) 24.5

50. Calculate the pH of a buffer solution containing 0.4 M acetic acid and 0.04 M sodium acetate, given that pK_a of acetic acid at 296° K is 4.74.
(A) 5.74 (B) 1.74 (C) 3.74 (D) 2.74

51.Identify the Lewi's acid in the following
(A) BF_3 (B) HCOOH
(C) H_2SO_4 (D) C_2H_5OH

52. What is the pH value of 0.001M NaOH solution ?
(A) 0 (B) 2 (C) 12 (D) 14

53. In the reaction $N_2 + 3H_2 \rightleftharpoons 2NH_3$, $\Delta H = -92$kJ, the heat of formation of NH_3 is
(A) +46 kJ (B) +92 kJ (C) –92 kJ (D) –46 kJ

54. Enthalpy of sublimation of iodine from the equations is
$H_2(g) + I_2(g) \rightarrow 2HI(g), \Delta H = -92$ kJ
$H_2(g) + I_2(s) \rightarrow 2HI(g), \Delta H = -92$ kJ
(A) +42.7 kJ mol^{-1} (B) –42.7 kJ mol^{-1}
(C) +61.1 kJ mol^{-1} (D) –61.1 kJ mol^{-1}

55. A first order reaction is 40% complete at the end of 20 minutes. In how many minutes will the reaction be 80% complete ? (The rate constant K = 4.25 x 10^{-4} sec^{-1})
(A) 5419 log 1.25 (B) 5419 log 5
(C) 90.31 log 1.25 (D) 90.31 log 5

56. The oxidation number of sulphur in $H_2S_2O_7$ is
(A) 0 (B) +6 (C) +2 (D) +4

57. What is the electrochemical equivalent of copper, if 0.1978 g of copper is deposited by a current of 0.2 amp in 50 min ?
(A) 2.3296 x 10^{-4} g (B) 3.296 x 10^{-4} g
(C) 4.6592 x 10^{-4} g (D) 2.3296 x 10^{-3} g

58. To which of the following types of colloidal systems does opal belong ?
(A) Solid in liquid (B) Liquid in gas
(C) Liquid in liquid (D) Liquid in solid

59. The catalyst employed in the hydration of carbon dioxide is
(A) Diastase (B) Carbonic anhydrase
(C) Zymase (D) Invertase

60. Which one of the following is a poison of platinum catalyst ?
(A) HgO (B) V_2O_3 (C) As_2O_3 (D) ZnO

TNPCEE -1997

1. The valence shell electronic configuration of copper is
(A) [Ar] $3d^{10}$ $4s^1$ (B) [Ar] $3d^{10}$ $4s^0$
(C) [Ar] $3d^9$ $4s^2$ (D) [Ar] $3d^9$ $4s^1$

2. Water gas is composed of
(A) CO_2, H_2CO (B) CO, H_2, CH_4, CO_2
(C) CO, H_2, N_2, CO_2 (D) CO, H_2, CH_4

3. 6Hg + dil. $8HNO_3 \rightarrow 3Hg_2(NO_3)_2$ + ?
(A) $2NO_2 + 4H_2$ (B) $2NO_2 + 2H_2O + 2H_2$
(C) $2NO \uparrow + 4H_2O$ (D) $2NO_2 + 2H_2O$

4. Phosphine burns in air of oxygen readily to give
(A) phosphorous pentoxide (B) metaphosphoric acid
(C) red phosphorus and hydrogen
(D) phosphorous trichloride

5. The oxidation state of phosphorus in orthophosphoric acid is
(A) +2 (B) +3 (C) +4 (D) +5

6. Among the hydrogen halides the H-F bond and the H-I bond are
(A) of equal strength and the weakest
(B) of equal strength and the strongest respectively
(C) the weakest and the strongest respectively
(D) the strongest and the weakest respectively

7. If n is the number of ions from each molecule of the electrolyte the relation between Vant Hoff's factor i and the degree of dissociation α is given by
(A) $\alpha = 1 - i\,(n - 1)$ (B) $i = 1 - \alpha\,(n - 1)$
(C) $\alpha = 1 + i\,(n - 1)$ (D) $i = 1 + \alpha\,(n - 1)$

8. At room temperature the vapour pressure of a pure solvent is 0.6 atm and the vapour pressure of the solution is 0.4 atm. Then the mole fraction of the solvent is
(A) 0.25 (B) 0.33 (C) 0.67 (D) 0.75

9. If Kb and Kf are respectively the ebulloscopic and cryoscopic constants of a solvent, then the ratio of the elevation in the boiling point of the solution to the depression in the freezing point of the solution is equal to
(A) Kb/(Kf – Kb) (B) Kf/(Kf – Kb)
(C) Kb / Kf (D) Kf / Kb

10. The IUPAC name of $[Cr(H_2O)_5Cl]Cl_2$ is
(A) pentaaquochlorochromium (II) chloride
(B) pentachloroaquochromium (II) chloride

(C) pentaaquochlorochromium (III) chloride
(D) pentachloroaquochromium (III) chloride

11. The IUPAC name of $CH_3-N(CH_3)-CH_2CH_2CH_3$

(A) 1-(N, N-dimethyl amino) propane
(B) 3-(N, N-dimethyl amino) propane
(C) N, N-dimethyl propyl amine
(D) dimethyl propyl amine

12. In 2,2 dimethyl propane, bromide is attached to a
(A) hydrogen atom (B) primary carbon atom
(C) secondary carbon atom (D) tertiary carbon atom

13. The product formed when phenyl chloride reacts with methyl chloride in presence of anhydrous $AlCl_3$ is
(A) benzyl chloride (B) m-chlorotoluene
(C) mixture of o-and p-chlorotoluene (D) toluene

14. The optical activity is externally compensated in
(A) mesotartaric acid (B) d-tartaric acid
(C) recemic mixture of tartaric acid (D) glycine

15. Which of the following on ozonolysis following by reduction can give mixture of ketone and aldehyde ?
(A) n-butylene (B) isobutylene
(C) 3-methyl-1-butene (D) 2-butene

16. The compound used in the treatment of whooping cough is
(A) phenyl benzoate (B) methyl benzoate
(C) ethyl benzoate (D) benzyl benzoate

17. The concentration of HI, I_2 and H_2 at equilibrium are 0.7, 0.1 and 0.1 mole/litre. The equilibrium constants K_c and K_p for the reaction $H_2 + I_2 \rightleftharpoons 2HI$ have the respective values
(A) 49 and 49 (B) 7 and 7
(C) 49 and 1/49 (D) 7 and 1/7

18. In the reaction
$A + 2B \rightleftharpoons C + D$ + heat. $\Delta H = -ve$, if the temperature is increased
(A) both the reverse and forward reaction will be favoured
(B) the reverse reaction will be favoured
(C) the forward reaction will be favoured
(D) there will be no effect on the reaction

19. Which one of the following has the highest pK_a value among them ?
(A) ehanol (B) benzoic acid
(C) acetic acid (D) phenol

20. From the acid base reaction represented as, $BF_3 + NH_3 \rightleftharpoons BF_3^- + NH_3^+$ it follows that BF_3 and NH_3 are
(A) Lewis base and Lewis base respectively
(B) Lewis base and Lewis acid respectively
(C) Arrhenius acid and Arrhenius base respectively
(D) Lowry-Bronsted bases

21. An electrolyte that suppresses the dissociation of CH_3COOH in the reaction
$CH_3COOH + H_2O \rightleftharpoons CH_3COO^- + H_3O^+$ is
(A) sodium acetate (B) sodium nitrate
(C) sodium hydroxide (D) sodium chloride

22. The pH of 0.005 M solution of H_2SO_4 is
(A) one (B) two (C) three (D) four

23. Given that the enthalpy of hydration of $CuSO_4$ to $CuSO_4.5H_2O$ is –66.5 kJ mol^{-1} and the enthalpy of solution of anhydrous copper sulphate in water is –78.2 kJ mol^{-1} what is the enthalpy of solution of $CuSO_4.5H_2O$?
(A) +144.7 kJ mol^{-1} (B) +11.7 kJ mol^{-1}
(C) –11.7 kJ mol^{-1} (D) –144.7 kJ mol^{-1}

24. The enthalpy of solution when one mole of SO_3 is dissolved in one mole of H_2O to get H_2SO_4 is –131 kJ. If the standard ethalphy of formation of H_2SO_4(l) and H_2O(l) are –812 kJ and –286kJ, respectively, what is the standard enthalpy of formation of SO_3(g) ?
(A) +971 kJ (B) –971 kJ (C) +395 kJ (D) –395 kJ

25. For which of the following reaction, ΔH and ΔD are the same?
(A) Formation of CO_2 (B) Combustion of CH_4
(C) Vapourisation of H_2O
(D) Neutralisation of HCl by NaOH

26. Glycerol on heating with $KHSO_4$ gives
(A) formic aid (B) acrolein
(C) allyl alcohol (D) propylene

27. An alcohol of the formula $C_4H_{10}O$ on oxidation gives a compound C_4H_8O which answers the iodoform test. Then the alcohol is
(A) sec-butyl alcohol (B) tert-butyl alcohol
(C) iso butyl alcohol (D) n-butyl alcohol

28. Which one of the following is used as a coolant in aircraft ?
(A) ethylene glycol (B) ethyl alcohol
(C) benzyl alcohol (D) glycerol

29. Ethylene glycol can be prepared by
(A) oxidation of ethylene with water at 200 C under pressure
(B) hydrolysis of 1, 2-dibromoethane with aqueous sodium carbonate solution
(C) oxidation of ethylene chlorohydrin with alkaline solution of $KMnO_4$
(D) action of Baeyer's reagent on ethylene diamine

30. C_2H_6O (A) is a sweet smeling liquid, readily forming chloroform with Cl_2. It is oxidised by acid $K_2Cr_2O_7$ to give B which undergoes aldol condensation. What could be A and B respectively ?
(A) Dimethyl ether and acetone

(B) Dimethyl ether and acetaldehyde
(C) Ethyl alcohol and acetaldehyde
(D) Ethyl alcohol and acetone

31. Which one of the following are example of aromatic ethers ?
(A) CH_3–O–C_6H_5 and CH_3–O–C_2H_5
(B) CH_3–O–CH_3 and CH_3–O–C_2H_5
(C) C_6H_5–O–C_6H_5 and C_6H_5–O–CH_3
(D) C_6H_5–O–C_6H_5 and C_2H_5–O–C_2H_5

32. Compound A with molecular formula $C_4H_{10}O$ does not give hydrogen with sodium, but gives C_2H_5I only with excess HI. Then the compound A is
(A) CH_3–CH_2–CH_2–CH_2–OH (B) CH_3–CH_2–O–CH_2–CH_3
(C) CH_3–CH_2–CH_2–O–CH_3 (D) CH_3–O–CH–$(CH_3)_2$

33. The reagent which can be used to differentiate acetone and acetophenone is
(A) $NaHSO_3$ (B) HCN
(C) Zn/Hg with HCl (D) I_2 and NaOH

34. Mesitylene is obtained by the action of acetone with
(A) Cl_2 (B) $K_2Cr_2O_7$
(C) dry HCl (D) con H_2SO_4

35. The compound which gives addition product with ammonia is
(A) acetone (B) formaldehyde
(C) acetyl chloride (D) acetaldehyde

36. Which of the following statements is correct ?
(A) Both benzaldehyde and acetaldehyde form Schiff's base with primary amines
(B) Benzaldehyde reduces with Fehling's solution while acetaldehyde does not
(C) Neither Benzaldehyde nor acetaldehyde reacts with caustic soda
(D) Benzaldehyde does not undergo polymerisation while acetaldehyde does

37. As we move from La to Lu in the perodic table, the atomic radius and the ionic radius of trivalent lanthanides
(A) decrease steadily (B) increase steadily
(C) increase and decrease respectively
(D) decrease and increase respectively

38. The important ores of silver are
(A) carnierite, pentlandite & pyrargyrite
(B) pyrargyrite, chloragyrite and argentite
(C) cinnabar and cuprite
(D) calamine and argentite

39. In the ash test of a simple salt analysis the pink ash indicates the presence of
(A) Al^{3+} (B) Zn^{2+} (C) Mg^{2+} (D) Pb^{2+}

40. The suitable indicator for the titration of weak acid against strong base is
(A) methyl red (B) phenolphthalein
(C) bromothymol blue (D) methyl orange

41. If k is the rate constant, the half-life period of a first order reaction is equal to
(A) 2.303 log k (B) 0.6932 k
(C) 0.6932/k (D) k/0.6932

42. The frequency factor becomes equal to rate constant of a reaction when the energy of activation is
(A) one (B) zero
(C) lesser than one (D) greater than one

43. If 70% of A and 60% of B remained at the end of the reaction 2A + B $\rightarrow$ C + D, the order of the reaction is
(A) one (B) two (C) three (D) four

44. The migration of colloidal particles under the influence of electric field is known as
(A) electrophoresis (B) dialysis
(C) electro osmosis (D) electro dialysis

45. Gas-in-solid colloidal form is known as
(A) gel (B) solid foam (C) aerosol (D) foam

46. In the reaction
$2H_2O_2 \xrightarrow{\text{glycerol}} 2H_2O + O_2\uparrow$, glycerol is a
(A) enzyme catalyst (B) auto catalyst
(C) negative catalyst (D) positive catalyst

47. When 9649.5 coulombs of electricity is passed through a solution of $AgNO_3$ (atomic weigth of Ag = 107.87), the silver deposited is
(A) 1.0787 g (B) 107.87 g (C) 10.787 g (D) 107.87g/9649.5

48. Which one of the following is prepared by Kolf method
(A) Picric acid (B) Cinnamic acid
(C) Salicylic acid (D) Benzoic acid

49. Compound A of molecular formula $C_3H_6O_3$ is optically active and gives effervescence with $NaHCO_3$. It gets reduced by HI to B of molecular formula $C_3H_6O_2$. B when warmed with alcohol and con.H_2SO_4 gives a fruity odour. A and B can be respectively
(A) glycolic acid and acetic acid
(B) lactic acid and pyruvic acid
(C) lactic acid and propionic acid
(D) glycolic acid and propionic acid

50. Which one of the following compounds does not reduce Tollen's reagent ?
(A) CH_3CHO (B) CH_3COOH (C) HCHO (D) HCOOH

51. The formation of benzamide from amine and benzoyl chloride in the presence of NaOH, is called
(A) Diazotisation reaction
(B) Schortten-Baumann reaction
(C) Condensation reaction (D) Carbylamine reaction

52. The reactants of Mustard oil reaction are
(A) methyl amine and hydrogen chloride
(B) methyl amine and chloroform
(C) methyl amine and carbon dioxide

(D) methyl amine and carbon disulphide

53. When nitromethane is boiled with con.HCl it is hydrolysed to form
(A) formic acid and hydroxylamine
(B) methyl hydroxylamine and hydrogen
(C) formic acid and ammonia
(D) formic acid and nitrogen

54. Structure of phenyl nitromethane is

(A) 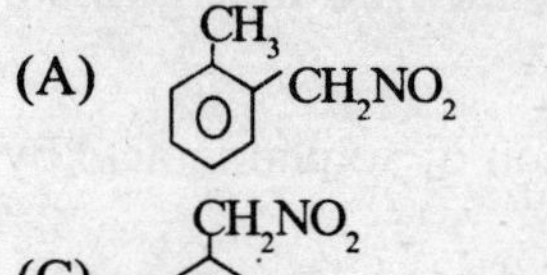 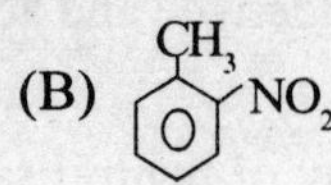 (B) CH_3, NO_2

(C)  (D) NO_2

55. The initiator used in the preparation of low density polyethylene is
(A) $(C_2H_5)_3Al + TiCl_3$ (B) O_2
(C) $(C_2H_5)_3Al$ (D) $TiCl_4$

56. (a) Glucose is oxidised by bromine water
(b) Fructose is oxidised by bromine water
(c) Glucose rotates plane polarised light in clockwise direction
(d) Fructose rotates plane polarised light in anti clockwise direction
Which of the above statements are true ?
(A) (a), (c) and (d) only (B) (b), (c) and (d) only
(C) (a) and (c) only (D) (c) and (d) only

57. Which of the following is a peptide linkage ?
(A) $-COO-NH_2$ (B) $-CO-NH_2$
(C) $-COO-NH-$ (D) $-CO-NH-$

58. Which of the following is used as a dye ?
(A) p-hydroxy azobenzene (B) benzamide
(C) aniline (D) benzene diazonium chloride

59. Zinc displaces copper from $CuSO_4$ solution because zinc is
(A) less dense than copper
(B) more electropositive than copper
(C) electropositive while copper is electropositive
(D) less electropositive than copper

60. The structure of $[Cu^{II}(NH_3)_4]SO_4$ is
(A) triangular bipyramid (B) squareplanar
(C) octahedral (D) tetrahedral

TNPCEE -1998

1. The electronic configuration and valency of mercury are
(A) $[Xe]\ 4f^{15}\ 5d^9\ 6s^2$; 1, 2 (B) $[Xe]\ 4f^{12}\ 5d^{11}\ 6s^3$; 1, 3
(C) $[Xe]\ 4f^{14}\ 5d^{10}\ 6s^2$; 1, 2 (D) $[Xe]\ 4f^{14}\ 5d^9\ 6s^3$; 3

2. A mixture used in Holme's signal is of
(A) $Ca_3(PO_4)_2$ and $CaCO_3$ (B) Ca_3P_2 and $CaCO_3$
(C) $Ca_3(PO_4)_2$ and CaC_2 (D) Ca_3P_2 and CaC_2

3. With which of the following the ionisation potential of elements will decrease ?
(a) Increase in the nuclear charge
(b) Increase in the distance between nucleus and valance electron
(c) Decrease in the screening effect
(A) (a) and (c) only (B) (b) only
(C) (b) and (c) only (D) (a) and (b) only

4. On heating at 593 K, H_3PO_4 gives
(A) phosphorus pentoxide (B) metaphosphoric acid
(C) pyrophosphoric acid (D) phosphorus acid

5. The freezing point of a solution containing some non volatile solute in 30g of water is 272.07 K. Ferrzing point of water is 0°C. K_f of water is 1.86 kgK mol^{-1}. Molecular weight of the solute is 200 g mol^{-1}. What is the weight of the solute ?
(A) 2g (B) 3g (C) 4g (D) 5g

6. The vapour pressure of a pure solvent at 27°C is 0.75 atm. What is the vapour pressure of the solution containing non-volatile solute at the same temperature if the mole fraction of the solute is 0.70 ?
(A) 0.225 atm (B) 0.560 atm
(C) 0.440 atm (D) 0.785 atm

7. Which of the following is/are true relating to the determination of osmotic pressure by Berkeley-Hartley method ?
(a) Osmotic pressure is not recorded directly and the method is long
(b) Concentration of the solution changes during the measurement of osmotic pressure
(c) Strain on the semipermeable membrane is very little
(A) (a) and (b) only (B) (c) only
(C) (b) and (c) only (D) (a) and (c) only

8. In the Haber's process of synthesis of ammonia, as and when ammonia is formed it is removed from the system because this act
(A) raises the temperature (B) lowers the pressure
(C) favours the backward reaction
(D) favours the forward reaction

9. Two examples for basic buffer solutions are
(A) NH_4OH and NaCl (B) NH_4OH and $(NH_4)_2CO_3$
(C) NH_4OH and NH_4Cl (D) NaOH and NH_4Cl

10. A Henderson equation in usual notations is
(A) $pK_b = pOH - \log \frac{Base}{Salt}$ (B) $pH = pK_b + \log \frac{Salt}{Base}$
(C) $pOH = pK_b + \log \frac{Salt}{Base}$ (D) $pH = pK_a - \log \frac{Acid}{Salt}$

11. What is the degree of dissociation of 0.02 M acetic

acid solution given that the dissociation constant for acetic acid is

(A) 0.003 (B) 0.0672 (C) 0.03 (D) 1.342 x 10^{-2}

12. In which of the following equilibrium reactions $K_p = K_c$?

(A) $N_2(g) + 3H_2(g) \rightleftharpoons 3NH_3(g)$

(B) $2SO_2(g) + O_2(g) \rightleftharpoons 3SO_3(g)$

(C) $H_2(g) + I_2(g) \rightleftharpoons 2HI(g)$

(D) $PCl_5(g) \rightleftharpoons PCl_3(g) + Cl_2(g)$

13. A solution of sodium acetate

(A) is neutral to litmus

(B) turns methyl orange into red

(C) turns red litmus into blue

(D) turns blue litmus into red

14. (a) 3-pentaone (b) acetophenone
(c) propanone (d) 2-pentanone

With the above set of IUPAC names which set of common names match ?

(A) (a) diethyl ketone (b) acetone
(c) ethyl methyl ketone (d) dimethyl ketone

(B) (a) dimethyl ketone (b) methyl phenyl ketone
(c) diethyl ketone (d) diphenyl ketone

(C) (a) diphenyl ketone (b) ethyl methyl ketone
(c) acetone (d) methyl-n-propyl ketone

(D) (a) diethyl ketone (b) methyl phenyl ketone
(c) dimethyl ketone (d) methyl-n-propyl ketone

15. The rate of S_N2 reaction depends on the concentration of

(A) nucleophile only (B) substrate only

(C) intermediate only (D) both substrate and nucleophile

16. $CH_3-CH_2-CH_2-O-CH_3$ and $CH_3-CH_2-O-CH_2-CH_3$ are example for

(A) functional isomerism (B) metamerism

(C) chain isomerism (D) position isomerism

17. Which of the following is not correct ?

(A) Aliphatic nitro compounds do not undergo electrophile substitution

(B) Aliphatic nitro compounds do not show tautomerism

(C) Aromatic nitro compounds do not show tautomerism

(D) Aromatic nitro compounds do not react with nitrous acid

18. A compound with asymmetric carbon atom is

(A) $CH_2(NH_2)COOH$ (B) $CH_3CHOHCH_3$

(C) $CH_3CHOHCOOH$ (D) CH_3CH_2COOH

19. 2-Butene on ozonolysis followed by hydrolysis gives

(A) Propionaldehyde (B) Butyraldehyde

(C) Acetaldehyde (D) Formaldehyde

20. The compound which can add with Grignard reagent and give ester on hydrolysis is

(A) silver acetate (B) ethyl chloroformate

(C) acetyl chloride (D) ethyl formate

21. Which of the following is correct ?

(A) Formaldehyde undergoes chlorination while acetaldehyde does not undergo chlorination

(B) With Grignard reagent formaldehyde forms primary alcohol while acetaldehyde forms secondary alcohol

(C) Formaldehyde and not acetaldehyde, undergoes iodoform test

(D) Both formaldehyde and acetaldehyde undergo aldol condensation

22. The enthalpy of neutralisation of sulphuric acid by sodium hydroxide is

(A) +57.32 kJ (B) –28.66 kJ (C) –39.5 kJ (D) –57.32 kJ

23. If $C(gr) + O_2(g) \rightarrow CO_2(g)$, $\Delta H = -394$ kJ &
$CO(g) + O_2(g) \rightarrow CO_2(g)$, $\Delta H = -284$ kJ,
then the heat of formation of CO is

(A) +110 kJ (B) –110 kJ (C) + 678 kJ (D) – 678 kJ

24. The enthalpy of formation of CO_2 and H_2O are –394kJ and –286 kJ. The enthalpy of combustion of methane is –890 kJ. The enthalpy of formation of methane is

(A) –152 kJ (B) +760 kJ (C) +76 kJ (D) –76 kJ

25. According to Arrhenius equation relating rate constant k, temperature T and the collision frequency A of a reaction, when the energy of activation if doubled, (k/A)

(A) is squared (B) is doubled

(C) is halved (D) becomes square root of the original

26. The order of oxidation reaction of potassium iodide by potassium persulphate is

(A) 0 (B) 1 (C) 2 (D) 3

27. What is the value of the rate constant of the first order reaction if the time required for completion of 90% of this reaction is 46.06 minutes ?

(A) 6.932 x 10^{-2} min^{-1} (B) 1 x 10^{-1} min^{-1}

(C) 5 x 10^{-2} min^{-1} (D) 2 x 10^{-2} min^{-1}

28. The alcohol which is used in ointments as antiseptic and local anaesthetic is

(A) ethanol (B) glycol (C) phenol (D) benzyl alcohol

29. When excess of ethyl alcohol is treated with concentrated sulphuric acid

(A) at 314 K ethylene is formed

(B) at 413 K ethylene is formed

(C) at 443 K diethylether is formed

(D) at 413 K diethylether is formed

30. Glycerol is oxidised to form the product

(A) glyceraldehyde and dihydroxyacetone with bromine water as oxidising agent

(B) oxalic acid with bismuth nitrate as oxidising agent

(C) mesoxalic acid with conc. HNO_3 as oxidising agent

(D) glyceric acid and mesoxalic acid with dilute HNO_3 as oxidising agent

31. In Victor Meyer's test the red colour is given by
(A) tertiary butyl alcohol (B) secondary butyl alcohol
(C) isobutyl alcohol (D) isopropyl alcohol

32. 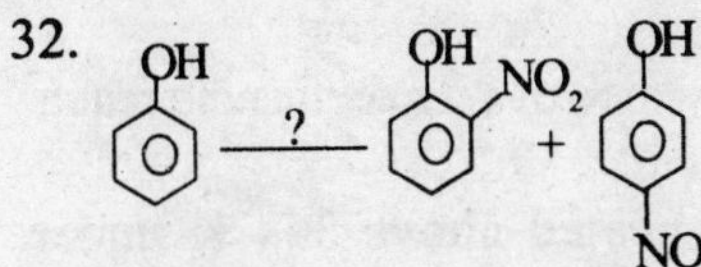

(A) dil. HNO_3 (B) conc. H_2SO_4
(C) HONO (D) conc. HNO_3

33. High boiling point of Phenol is due to
(A) intra molecular hydrogen bonding
(B) inter molecular hydrogen bonding
(C) covalent bond (D) ionic bond

34. Which one of the following reacts with phenol to give anisole ?
(A) Nitro methane (B) Amino methane
(C) Benzene diazonium chloride (D) Diazomethane

35. Name the reaction that converts benzaldehyde into cinnamic acid by condensation
(A) Fridel-Crafts reaction
(B) Schottern-baumann reaction
(C) Knoevenagel reaction (D) Coupling reaction

36. Which one of the following is used as an urianry antiseptic ?
(A) hexamethylene tetramine (B) formalin
(C) para formaldehyde (D) urethene

37. A colourless aromatic liquid (A) of molecular formula C_8H_8O, when reduced with sodium and alcohol gives a neutral compound (B) of molecular formula $C_8H_{10}O$. What are A and B, respectively ?
(A) Diphenyl carbinol and acetophenone
(B) Phenyl methyl carbinol & benzaldehyde
(C) Acetophenone & phenyl methyl carbinol
(D) Benzaldehyde and benzoic acid

38. Iodine value is a measure of which of the following in a fat or an oil ?
(A) Number of carboxylic groups
(B) Degree of saponification
(C) Number of saturated bonds
(D) Degree of unsaturations

39. The carboxylic acid present in rancid butter is
(A) propionic acid (B) butyric acid
(C) acetic acid (D) formic acid

40. The reagent which can convert benzyl chloride into benzoic acid is
(A) alkaline $KMnO_4$ (B) aqueous $Pb(NO_3)_2$
(C) alcoholic KOH (D) aqueous NaOH

41. Lactic acid on heating with dil. H_2SO_4 at 403 K gives
(A) pyruvic acid + water (B) lactide + water
(C) acetic acid + formic acid
(D) acetaldehyde + formic acid

42. Which one of the following has melting point 1725 K, boiling point 3003 K and specific gravity 8.9 ?
(A) Manganese (B) Nickel (C) Chromium (D) Copper

43. The IUPAC name Potassiumhexacyanoferrate(II) represents the coordination compound of formula
(A) $K_3[Fe^{II}(CN)_6]$ (B) $K_4[Fe^{II}(CN)_4]$
(C) $K_4[Fe^{II}(CN)_6]$ (D) $K_3[Fe^{II}(CN)_4]$

44. $2Cr + 6\underset{\text{hot conc.}}{H_2SO_4} \rightarrow 6H_2O + ?$

(A) $Cr_2(SO_4)_3 + 3SO_2\uparrow$ (B) $CrO_3 + SO_2\uparrow$
(C) $CrSO_4 + H_2\uparrow$ (D) $Cr_2O_3 + SO_3\uparrow$

45. A coordination compound cannot contain
(A) a simple cation and a comlex anion
(B) a complex cation & a complex anion
(C) a complex cation and a simple anion
(D) a simple cation and a simple anion

46. For which three lanthanides the +3 ions have the 4f subshell empty, half filled and competely filled, respectively
(A) Lu, La, Gd (B) La, Gd, Lu
(C) Gd, La, Lu (D) La, Lu, Gd

47. Monel metal is an alloy of zinc consisting of
(A) 67% Zn, 30% Cu and the rest Fe and Mn
(B) 30% Zn, 50% Cu and the rest Ni
(C) 90% Zn, 4% Al and the rest Cu and Mn
(D) 25% Zn, 25% Ni and the rest Cu

48. In the confirmatory tests for basic radicals, if no precipitate is obtained by adding potassium chromate solution to a few drops of the given solution, then what is confirmed is
(A) Calcium (B) Magnesium (C) Copper (D) Aluminium

49. With increase in temperature the conductivity of metals and the conductivity of electrolytes
(A) decrease and increase respectively (B) both decrease
(C) both increase (D) increase and decrease respectively

50. $As_2O_3 + 3H_2S \rightarrow \underset{\text{(colloid)}}{As_2S_3} + 3H_2O$

This is an example for preparation of colloid by
(A) peptisation (B) reduction
(C) oxidation (D) double decomposition

51. Examples of colloids in the state of liquid in solid dispersion medium are
(A) white of egg and ink (B) cheese and curd
(C) milk and whipped cream (D) curd and milk

52. Which of the following is/are ture ?
(a) Intermediate Compound Formation catalysis
(b) Adsorption theory explains the heterogeneous catalysis
(c) Platinum is used as a catalyst in the heterogeneous catalysis

(d) In the Haber process for manufacture of NH_3, molybdenum acts as a catalyst and iron as a promoter
(A) (a) and (b) only (B) (c) only
(C) (d) only (D) (c) and (d) only

53. When benzene diazonium chloride is reduced with tin and hydrochloric acid, the product is
(A) azobenzene (B) phenyl hydroxylamine
(C) aniline (D) phenyl hydrazine

54. The reaction between a primary amine, chloroform and alcoholic KOH is known as
(A) Carbylamine reaction (B) Wurtz reaction
(C) Hoffmann's reaction (D) Sandmayer reaction

55. Aminobenzene forms Schiff's base with
(A) acetic anhydride (B) benzaldehyde
(C) benzyl chloride (D) benzyl alcohol

56. Compared to the boiling points of alkanes and alkyl halides the boiling points of nitroalkanes are
(A) higher due to greater polar character of nitro group
(B) lower due to lower polar character of nitro group
(C) lower due to greater polar character of nitro group
(D) higher due to lower polar character of nitro group

57. Electrolytic reduction of nitrobenzene in conc. sulphuric acid solution produced
(A) m-nitrobenzene sulphonic acid
(B) aniline with azobenzene as an intermediate
(C) phenyl hydroxylamine rearranges to p-aminophenol
(D) phenyl hydroxylamine, with nitrosobenzene as an intermediate

58. The organic compound used in the coagulation of rubber latex is
(A) $C_6H_5CH_2OH$ (B) HCOOH
(C) CH_3COOH (D) CH_3–CHOH–COOH

59. Which of the following is not true regarding fructose ?
(A) It reduces Tollen's reagent and Fehling's solution
(B) It has five –OH groups
(C) It shows all reactions of hydroxyl group but not carbonyl group
(D) Its molecular formula is $C_6H_{12}O_6$

60. Cyanocobalamine is the chemical name of
(A) Vitamin B_{12} (B) Vitamin B_6
(C) Vitamin B_1 (D) Vitamin B_2

TNPCEE -1999

1. In the 15th group elements as we move form N to Bi, trivalency gets more pronounced because
(A) s-orbital contain two electrons and p-orbital contains 3 electrons.
(B) of half filled configuration of p-orbital
(C) the two electrons in the s-orbital become inert
(D) all the inner shells are completely filled

2. Which of the following statement(s) is/are wrong ?
(a) With conc. H_2SO_4, formic acid does not react but acetic acid reacts
(b) With Chlorine in the presence of P/I_2, acetic acid reacts but not formic acid
(c) When formic acid is heated above 344 K under pressure, it gives $CO_2 + H_2$; but acetic acid is stable to heat
(A) (a) only (B) (a) and (c) only
(C) (b) only (D) (a) and (b) only

3. What is the colourless liquid compound of specific gravity 1.57 and boiling point 349K, which reacts with water forming phosphorus acid ?
(A) Phosphorus trichloride (B) Ortho phosphoric acid
(C) Phosphorus pentachloride (D) Phosphine

4. $P + 5HNO_3 \rightarrow H_3PO_4 + 5NO_2 \uparrow + H_2O$
(A) CO_2 (B) N_2 (C) I_2 (D) Cl_2

5. An queous solution containing 6 g of a non-volatile solute in 60 g of water boils at 373.51 K at atmospheric pressure. If the boiling point of pure water at the same pressure is 373 K, what is the molecular weight of the solute ?
(For water K_f= 1.86 kg K mol^{-1}, and K_b = 0.51 kg K mol^{-1})
(A) 100 g mol^{-1} (B) 200 g mol^{-1}
(C) 300 g mol^{-1} (D) 50 g mol^{-1}

6. The depression of freezing point of a 0.5 molal solution of non-volatile solute in water is
(A) 3.72 K (B) 0.093 K (C) 0.93 K (D) 9.3 K

7. P and P° are the vapour pressures of a solution and the pure solvent, respectively; n_1 and n_2 are the number of moles of the solvent and solute respectively. Then

(A) $P^o = \frac{n_1}{n_1+n_2} P$ (B) $P = \frac{n_1}{n_1+n_2} P^o$

(C) $P = \frac{n_2}{n_1+n_2} P^o$ (D) $\frac{P^o - P}{P^o} = \frac{n_1}{n_1+n_2}$

8. Let K_C be the equilibrium constant for the reaction $PCl_5(g) \rightleftharpoons PCl_3(g) + Cl_2(g)$. If the pressure is doubled without changing the temperature,
(A) K_C is unaffected (B) K_C is doubled
(C) the forward reaction is favoured
(D) the reverse reaction is favoured

9. $2NO(g) + O_2(g) \longrightarrow 2NO_2(g)$. For this reaction, the number of molecules reacting in the elementary step is
(A) zero (B) one (C) two (D) three

10. 1 mole of H_2 and 1 mole of I_2 are heated in a closed vessel till equilibrium is attained. If the mixture contains 0.4 mole of HI, then for the formation of HI the value of K_C is

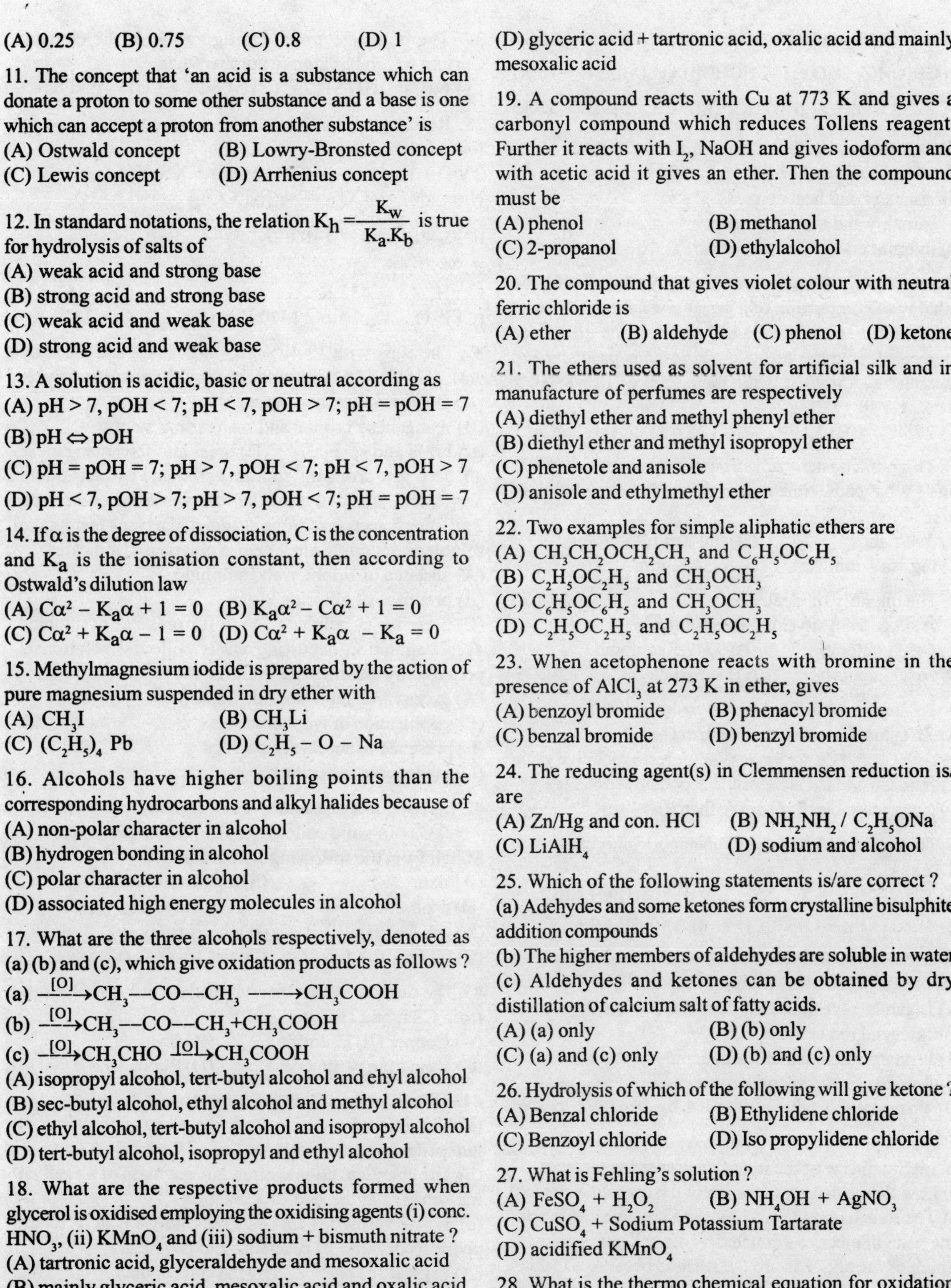

(A) 0.25 (B) 0.75 (C) 0.8 (D) 1

11. The concept that 'an acid is a substance which can donate a proton to some other substance and a base is one which can accept a proton from another substance' is
(A) Ostwald concept (B) Lowry-Bronsted concept
(C) Lewis concept (D) Arrhenius concept

12. In standard notations, the relation $K_h = \frac{K_w}{K_a.K_b}$ is true for hydrolysis of salts of
(A) weak acid and strong base
(B) strong acid and strong base
(C) weak acid and weak base
(D) strong acid and weak base

13. A solution is acidic, basic or neutral according as
(A) pH > 7, pOH < 7; pH < 7, pOH > 7; pH = pOH = 7
(B) pH ⇔ pOH
(C) pH = pOH = 7; pH > 7, pOH < 7; pH < 7, pOH > 7
(D) pH < 7, pOH > 7; pH > 7, pOH < 7; pH = pOH = 7

14. If α is the degree of dissociation, C is the concentration and K_a is the ionisation constant, then according to Ostwald's dilution law
(A) $C\alpha^2 - K_a\alpha + 1 = 0$ (B) $K_a\alpha^2 - C\alpha^2 + 1 = 0$
(C) $C\alpha^2 + K_a\alpha - 1 = 0$ (D) $C\alpha^2 + K_a\alpha - K_a = 0$

15. Methylmagnesium iodide is prepared by the action of pure magnesium suspended in dry ether with
(A) CH_3I (B) CH_3Li
(C) $(C_2H_5)_4$ Pb (D) $C_2H_5 - O - Na$

16. Alcohols have higher boiling points than the corresponding hydrocarbons and alkyl halides because of
(A) non-polar character in alcohol
(B) hydrogen bonding in alcohol
(C) polar character in alcohol
(D) associated high energy molecules in alcohol

17. What are the three alcohols respectively, denoted as (a) (b) and (c), which give oxidation products as follows ?
(a) $\xrightarrow{[O]} CH_3—CO—CH_3 \longrightarrow CH_3COOH$
(b) $\xrightarrow{[O]} CH_3—CO—CH_3 + CH_3COOH$
(c) $\xrightarrow{[O]} CH_3CHO \xrightarrow{[O]} CH_3COOH$
(A) isopropyl alcohol, tert-butyl alcohol and ehyl alcohol
(B) sec-butyl alcohol, ethyl alcohol and methyl alcohol
(C) ethyl alcohol, tert-butyl alcohol and isopropyl alcohol
(D) tert-butyl alcohol, isopropyl and ethyl alcohol

18. What are the respective products formed when glycerol is oxidised employing the oxidising agents (i) conc. HNO_3, (ii) $KMnO_4$ and (iii) sodium + bismuth nitrate ?
(A) tartronic acid, glyceraldehyde and mesoxalic acid
(B) mainly glyceric acid, mesoxalic acid and oxalic acid
(C) oxalic acid, mainly glyceric acid and glyceraldehyde + dihydroxyacetone
(D) glyceric acid + tartronic acid, oxalic acid and mainly mesoxalic acid

19. A compound reacts with Cu at 773 K and gives a carbonyl compound which reduces Tollens reagent. Further it reacts with I_2, NaOH and gives iodoform and with acetic acid it gives an ether. Then the compound must be
(A) phenol (B) methanol
(C) 2-propanol (D) ethylalcohol

20. The compound that gives violet colour with neutral ferric chloride is
(A) ether (B) aldehyde (C) phenol (D) ketone

21. The ethers used as solvent for artificial silk and in manufacture of perfumes are respectively
(A) diethyl ether and methyl phenyl ether
(B) diethyl ether and methyl isopropyl ether
(C) phenetole and anisole
(D) anisole and ethylmethyl ether

22. Two examples for simple aliphatic ethers are
(A) $CH_3CH_2OCH_2CH_3$ and $C_6H_5OC_6H_5$
(B) $C_2H_5OC_2H_5$ and CH_3OCH_3
(C) $C_6H_5OC_6H_5$ and CH_3OCH_3
(D) $C_2H_5OC_2H_5$ and $C_2H_5OC_2H_5$

23. When acetophenone reacts with bromine in the presence of $AlCl_3$ at 273 K in ether, gives
(A) benzoyl bromide (B) phenacyl bromide
(C) benzal bromide (D) benzyl bromide

24. The reducing agent(s) in Clemmensen reduction is/are
(A) Zn/Hg and con. HCl (B) NH_2NH_2 / C_2H_5ONa
(C) $LiAlH_4$ (D) sodium and alcohol

25. Which of the following statements is/are correct ?
(a) Adehydes and some ketones form crystalline bisulphite addition compounds
(b) The higher members of aldehydes are soluble in water
(c) Aldehydes and ketones can be obtained by dry distillation of calcium salt of fatty acids.
(A) (a) only (B) (b) only
(C) (a) and (c) only (D) (b) and (c) only

26. Hydrolysis of which of the following will give ketone ?
(A) Benzal chloride (B) Ethylidene chloride
(C) Benzoyl chloride (D) Iso propylidene chloride

27. What is Fehling's solution ?
(A) $FeSO_4 + H_2O_2$ (B) $NH_4OH + AgNO_3$
(C) $CuSO_4$ + Sodium Potassium Tartarate
(D) acidified $KMnO_4$

28. What is the thermo chemical equation for oxidation of methanol to formaldehyde ?
(A) $CH_3OH(l) + \alpha O_2(g) \rightarrow HCHO(l) + H_2O(l)$, ΔH = -163kJ

(B) $2CH_3OH(l) + O_2(g) \rightarrow 2HCHO(l) + 2H_2O(l)$, $\Delta H = -326kJ$

(C) $CH_3OH(l) + \alpha O_2(g) \rightarrow HCHO(l) + H_2O(l)$, $\Delta H = +163kJ$

(D) $2CH_3OH(l) + O_2(g) \rightarrow 2HCHO(l) + 2H_2O(l)$, $\Delta H = +326kJ$

29. Examples for extensive properties of systems are
(A) temperature and enthalpy
(B) viscosity and heat capacity
(C) enthalpy and internal energy
(D) internal energy an temperature

30. At a certain temperature, the differene between the enthalpy of combustion of a gas at constant volume and at constant pressure is 8.314 kJ mol^{-1}. If the difference between the number of moles of gaseous products and reactants is -3.5, what is the temperature? (R =8.314 x 10^{-3} kJ/K)
(A) 285°C (B) 12.7°C (C) 25.4°C (D) 25°C

31. The rate constant of a first order reaction is 4.606 min^{-1}. What is the time required for 75% completion of the reaction ?
(A) 3.495 min^{-1} (B) log 2 min^{-1}
(C) log 16/9 min^{-1} (D) log $2/\sqrt{3} min^{-1}$

32. The quantity of electricity required to deposit 10.79 gm of silver from silver nitrate solution is
(A) 6495 coulomb (B) 9649.5 coulomb
(C) 964.95 coulomb (D) 1 Faraday

33. Zn displaces Cu from $CuSO_4$ solution because
(A) Zn is more electropositive than Cu
(B) both Cu and Zn are having same electro positive value
(C) both Cu and Zn are less electropositive than Hg
(D) Cu is more electropositive than Zn

34. The chemical formula of Pentaaquochlorochromium (III) chloride is
(A) $Cr[(H_2O)_5Cl]Cl_2$ (B) $[Cr(H_2O)_5Cl]Cl_3$
(C) $[Cr(H_2O)_3(H_2O)_2Cl]Cl$ (D) $[Cr(H_2O)_5Cl]Cl_2$

35. Which of the following is not conforming to Werner's theory of co-ordination compounds ?
(A) Ligands have unshared pair of electrons donated to central metal ion or atom
(B) Primary valencies are directional in nature
(C) Ligands satisfy secondary valencies
(D) Promary valency is satisfied by negative ions of metals

36. (a) Small size of the ion of the transition element
(b) Low charge density of the ion of the transition element
(c) The ions having vacant d-orbitals which can accept lone pairs of electrons donated by other ligands
Which of the above is/are the reasons because of which the transition elements form many complex compounds ?
(A) (a) and (c) only (B) (b) only
(C) (b) and (c) only (D) (c) only

37. The outer electronic configuration of the element Curium which has atomic number 96 is
(A) $6d^2 7s^2$ (B) $5f^{10} 7s^2$ (C) $5f^7 6d^1 7s^2$ (D) $4f^1 5d^1 6s^2$

38. Refining of Nickel by Mond's process is done as follows
$2NiO + H_2 + CO \longrightarrow 2Ni(crude) + X + Y$
$Ni(crude) + 4CO \longrightarrow [Ni(CO)_4]$
$[Ni(CO)_4] \xrightarrow{Z} Ni(pure) + 4CO$
X, Y, Z in the above are respectively
(A) H_2, CO_3, 453 K (B) H_2O, CO_2, 453 K
(C) H_2O, CO_2, below 1370 K (D) H_2O, CO, 1370 K

39. The alloy with 20-40% Zn, 60-80% Cu and the alloy with 5%C, 65-75% Fe, and the rest Mn are respectively
(A) German silver and brass
(B) manganese bronze and manganese steel
(C) brass and spiegal (D) brass and ferromanganese

40. To a few drops of 'sodium carbonate extract' dilute acetic acid is added until the effervescence ceases and 2 ml of lead acetate solution is added. If a black precipitate soluble in hot dilute nitric acid is obtained, it indicates
(A) absence of sulphate and sulphide
(B) presence of chloride
(C) presence of sulphate (D) presence of sulphide

41. Coagulation occurring while colloidal solution is prepared by chemical methods is due to
(A) excess amount of soluble impurities
(B) stabilisation of lyophobic sols
(C) presence of soluble impurities
(D) inadequate impurities

42. (a) Lyophobic colloid (b) Condensation method
(c) Gas-in-solid colloid (d) Liquid-in-solid colloid
Match from the following with the above :
(i) Foam (ii) Sulphur in water
(iii) peptisation (iv) Hydrolysis (v) Curd
(A) (v), (iv), (i), (ii) (B) (ii), (iv), (i), (v)
(C) (i), (iv), (v), (ii) (D) (ii), (iii), (v), (i)

43. The catalyst used in the manufacture of Vinyl Chloride from C_2H_2 and HCl is
(A) Copper (II) Chloride (B) Iron
(C) Vanadium pentoxide (D) Platinum

44. Which of the following is/are true ?
(a) The rate of reaction in homogeneous reactions is independent of the concentraton of the catalyst.
(b) The chemical composition but not mass of a catalyst may change at the end of a reaction.
(c) In heterogeneous catalysis the catalyst is in different phase from those of reactants and products
(A) (a) and (c) only (B) (b) and (c) only
(C) (c) only (D) (a) only

45. Salicylaldehyde is formed by the reaction of $CHCl_3$

and NaOH with

(A) aniline (B) phenol

(C) benzaldehyde (D) toluene

46. When lactic acid is heated with dilute sulphuric acid at 403 K, it gives

(A) acetaldehyde and formic acid (B) propylene

(C) ethyl alcohol and carbon dioxide (D) lactyl sulphate

47. $RCONH_2 + Br_2 + (a) \rightarrow RNH_2 + 2KBr + (b) + 2H_2O$

If the above describes a Hoffmann's reacton, (a) and (b) are respectively

(A) $2K_2CO_3$ and 2KOH (B) K_2CO_3 and KOH

(C) 4KOH and K_2CO_3 (D) KOH and $2K_2CO_3$

48. The product formed when oxalic acid is reduced with Zn/H_2SO_4 is

(A) glycollic acid (B) mesoxalic acid

(C) glycol (D) glyoxalic acid

49. The compound used as a preservative for tomato ketchup is

(A) benzyl benzoate (B) methyl benzoate

(C) sodium benzoate (D) benzyl acetate

50. Which among the following is the most basic ?

(A) aniline (B) ethyl amine

(C) benzyl amine (D) ammonia

51. Which of the following is not a true statement about aniline ?

(A) On oxidation with potassium dichromate and sulphuric acid aniline gives carbylamine

(B) Aniline reacts with strong acids to form crystalline salts

(C) Aniline condenses with aromatic aldehyde to form Schiff's base

(D) Aniline's boiling point is higher than that of benzene

52. benzene diazonium chloride when boiled with water produces

(A) m-cresol (B) salicylic acid

(C) benzyl alcohol (D) carbolic acid

53. The amine used as an inhibitor of corrosion of iron is

(A) allyl amine (B) phenyl amine

(C) diamyl amine (D) ethyl amine

54. The reduction of nitrobenzene with zinc dust and ammonium chloride gives

(A) phenyl hydroxylamine (B) aniline

(C) azobenzene (D) hydrazobenzene

55. Poly ethylene terephthalate is prepared by the action of ethylene glycol with

(A) 1, 3 benzene dicarboxylic acid (B) methyl salicylate

(C) 1, 4 benzene dicarboxylic acid

(D) 1, 2 benzene dicarboxylic acid

56. Which of the following statements is wrong ?

(A) Suplha drugs are effective against gram positive and gram negative bacteria

(B) Malachite green is an example for nitrodyes

(C) Compared to synthetic fibres, natural fibres have high affinity for sulphur and vat dyes

(D) Thermoplastics are linear polymers; thermosetting plastics are cross-linked

57. Which of the following is not a function of carbohydrates in living organisms ?

(A) Producing energy necessary for functioning of living body and doing work

(B) Being present in plants and serving as food for grazing animals

(C) Storing chemical energy in the form of glycogen in spleen

(D) Supporting plant structure

58. Nerve disorders, rickets, night blindness and beriberi are disorders caused by deficiencies respectively in vitamins

(A) B_{12}, E, A and D (B) B_6, B_2, C and D

(C) D, B_6, K and B (D) B_6, D, A and B

59. The bonds usually present in explosives are

(A) N–Cl, O–Cl, N–O, N–N (B) N–S, N–K, N–P, N–R

(C) N–Cl, N–O, O–Cl, N–S (D) N–N, N–S, N–O, N–Cl

60. With the IUPAC names (a) 1-bromo-2,2dimethyl propane, (b) 2-iodo propane and (c) 2-chloro-2-methyl propane correctly match the common names (i) iso-propyl iodide, (ii) tert-butyl chloride and (iii) neopentyl bromide

(A) (a) - (i), (b) -(iii), (c) -(ii)

(B) (a) - (ii), (b) -(i), (c) -(iii)

(C) (a) - (iii), (b) -(i), (c) -(ii)

(D) (a) - (iii), (b) -(ii), (c) -(i)

TNPCEE -2000

1. The element which does not contain 5f electron is

(A) Thorium (B) Curium (C) Uranium (D) Plutonium

2. The expected electronic configuration of Cu should be

(A) $[Ar]3d^8 4s^2$ (B) $[Kr]5s^2 5p^3$

(C) $[Ar] 3d^9 4s^2$ (D) $[Ar]3d^{10} 4s^1$

3. The trend in the anionic radii among halide ions is

(A) $Br^- > F^- > I^- > Cl^-$ (B) $I^- > Br^- > Cl^- > F^-$

(C) $Cl^- > F^- > Br^- > I^-$ (D) $F^- > Cl^- > Br^- > I^-$

4. The number of co-ordinate bond present in $H_4P_2O_7$ is

(A) 5 (B) 4 (C) 3 (D) 2

5. Which of the following hydrides is most stable ?

(A) AsH_3 (B) PH_3 (C) NH_3 (D) BiH_3

6. The degree of dissociation of a weak acid HA in 0.1 M solution is 0.1%. Its dissociation constant ($K\alpha$) is

(A) 1×10^{-14} (B) 1×10^{-7} (C) 1×10^{-10} (D) 1×10^{-3}

7. When two solutions of different concentrations are separated by a semi permeable membrane, the solvent

molecules diffuse from the dilute solution side to the concentrated solution side. This occurs due to the
(A) lower mole fraction of the solvent
(B) higher mole fraction of the solute
(C) higher mole fraction of the solvent
(D) large size holes in semi permeable membrane.

8. Which of the following solutions will not have abnormal colligative property ?
(A) HCl (B) $CuSO_4$ (C) Sucrose (D) NaCl

9. The relative lowering of vapour pressure of a dilute solution at 27°C is 0.44. The mole fraction of solvent is
(A) 0.56 (B) 0.75 (C) 0.66 (D) 0.44

10. The observed and theoretical osmotic pressure of 1% $Ca(NO_3)_2$ is 8.1 and 3.0 respectively. The degree of dissociation of $Ca(NO_3)_2$ is
(A) 0.21 (B) 0.425 (C) 0.75 (D) 0.85

11. Abnormal colligative properties of solutes (e.g. : Benzoic acid) are observed wherever the solute molecules associate themselves. This association is due to the
(A) van der Waal's forces of attraction
(B) formation of intra molecular hydrogen bonding
(C) formation of intermolecular hydrogen bonding
(D) dipoles interaction

12. Identify the optically active compound
(A) glyceraldehyde (B) isopropyl chloride
(C) chloro iodo methane sulphonic acid
(D) benzoic acid

13. Dehydrohalogenation of tertiary butyl bromide with alcoholic alkali is an example for
(A) E_1 Reaction (B) S_N1 Reaction
(C) S_N2 Reaction (D) E_2 Reaction

14. How many ether isomers are possible for the formula $C_4H_{10}O$?
(A) One (B) Two (C) Three (D) Four

15. The organic mixture used as a substitute for petrol is
(A) alcohol + KOH (B) alcohol + pyridine
(C) alcohol + diethyl ether (D) alcohol + anisole

16. Which compound has the least boiling point ?
(A) alkane (B) aldehyde (C) alkyl halide (D) alcohol

17. Which of the following gives highest calorific value when the same quantity is completely combusted ?
(A) Naphthalene (B) p-Xylene (C) Toluene (D) Benzene

18. Chloropicrin is
(A) CCl_3NO_2 (B) CCl_3NH_2 (C) CCl_2N_2Cl (D) CCl_3NOH

19. The compound that cannot be prepared by using alkyl-magnesium-halide is
(A) $CH_3-CH_2-CH_2-COOH$ (B) H–COOH
(C) CH_3-CH_2-COOH (D) CH_3-COOH

20. The addition of third group elements to fourth group elements causes
(A) Reduction in conductivity of IV gp-elements
(B) n as well as p type conductivity
(C) p-Type conductivity (D) n-Type conductivity

21. The amount of caustic soda in grams required to make one litre solution with pH of 12 is
(A) 0.20 (B) 0.28 (C) 0.40 (D) 0.56

22. In which of the following K_p is less than K_c ?
(A) $N_2O_4(g) \rightleftharpoons 2NO_2(g)$
(B) $2SO_2(g) + O_2(g) \rightleftharpoons 2SO_3(g)$
(C) $PCl_5(g) \rightleftharpoons PCl_3(g) + Cl_2(g)$
(D) $H_2(g) + I_2(g) \rightleftharpoons 2HI(g)$

23. The reaction $H_2(g) + O_2(g)$ is a
(A) heterogeneous reaction (B) chain reaction
(C) reversible reaction (D) irreversible reaction

24. The equilibrium constant K_C for the reaction $2HI(g) \rightleftharpoons H_2(g) + I_2(g)$ at 250°C is 4. What would be the K_C for the reaction $H_2(g) + I_2(g) \rightleftharpoons 2HI(g)$ at 250°C
(A) 2 (B) 16 (C) 0.25 (D) 4

25. During combustion the cylinder of a gasoline engine the piston is pushed and ultimately the car moves. This is due to
(A) the hot gases expand in the cylinder
(B) combustible nature of petrol
(C) the increase in temperature of the cylinder
(D) the energy released from petrol

26. The enthalpy of combustion of ethylene at constant volume is -1406 kJ mol^{-1} at 27°C. What is the enthalpy of combustion of ethylene at constant pressure at the same temperature ?
(A) +.703 kJ (B) −1411 kJ (C) +1406 kJ (D) +1411 kJ

27. For a first order reaction, the time required for 99.9% completion is about how many times its half life period ?
(A) 1/10 (B) 1000 (C) 10 (D) 100

28. Decomposition of HI on gold is .... reaction
(A) second order (B) first order
(C) zero order (D) pseudo first order

29. If half life period of a first order reaction is 20 minutes then the time taken for 1 kg of the substance reduced into 250 gm is
(A) 20 min. (B) 30 min. (C) 40 min. (D) 60 min.

30. Which of the following compounds exhibits non-linear curve when its equivalent conductance is plotted against square root of concentration ?
(A) potassium chloride (B) sodium chloride
(C) sodium sulphate (D) ammonium hydroxide

31. The unit of electrochemical equivalent is
(A) Kg $Coulomb^{-1}$ (B) Coulomb

(C) Coulomb Kg^{-1} (D) Kg coulomb

32. Which is Luca's reagent ?
(A) Conc. HNO_3 and anhydrous $ZnCl_2$
(B) Conc. HNO_3 and hydrated $ZnCl_2$
(C) Conc. HCl and anhydrous $ZnCl_2$
(D) Conc. HCl and hydrated $ZnCl_2$

33. Primary alcohols can be prepared by treating Grignard reagent with
(A) ethylformate (B) acetone
(C) formaldehyde (D) acetaldehyde

34. The addition of ethylene glycol to water in vehicle radiator
(A) increases the freezing point and decreases the boiling point of water.
(B) increases the boiling point and decreases the boiling point of water
(C) decreases the freezing point of water
(D) increases the boiling point of water

35. Which one of the following acids does not contain –COOH group ?
(A) tartaric acid (B) citric acid
(C) picric acid (D) pyruvic acid

36. The boiling point of phenol is high because of
(A) intermolecular hydrogen bonding (B) resonance
(C) covalent characteristics (D) aromatic property

37. Acetone on saturation with hydrogen chloride gives
(A) chloroform (B) mesityl oxide
(C) chloro ethane (D) methyl chloride

38. Hexa methylene tetramine is used as an
(A) analgesic (B) antipyretic
(C) antibiotic (D) antiseptic

39. The compound which does not reduce Fehling solution is
(A) C_6H_5CHO (B) CH_3CHO (C) HCHO (D) HCOOH

40. When acetaldehyde reacts with Fehing's solution, the colour change is
(A) blue to colourless (B) blue to green
(C) red to blue (D) blue to red

41. Which one of the following acids contains both an aldehyde and a carboxyl group ?
(A) lactic acid (B) oxalic acid
(C) acetic acid (D) formic acid

42. The acid present is beef is
(A) palmitic acid (B) stearic acid
(C) butyric acid (D) acetic acid

43. High altitude airfcraft components are made from
(A) Neoprene (B) Thiockol
(C) Natural rubber (D) Buna N rubber

44. Which one of the following reactions proceeds with molybdenum supported iron catalyst ?
(A) $CH{\equiv}CH + HCl \rightarrow CH_2{=}CH{-}Cl$
(B) $CO{+}H_2 \rightarrow CH_3OH$
(C) $N_2{+}3H_2 \rightleftharpoons 2NH_3$ (D) $SO_2 + O_2 \rightleftharpoons SO_3$

45. The percentage composition of "constantine" is
(A) 25% Ni + 50% Cu + 25% Zn (B) 40% Ni + 60% Cu
(C) 25% Ni + 75% Cu (D) 46% Ni + 54% Fe

46. Which of the following ligands can satisfy both the primary and secondary vaency of a metal atom ?
(A) NO_2 (B) NO (C) CO (D) NH_3

47. The metal that absorbs large volume of hydrogen on the surface is
(A) Nickel (B) Chromium (C) Copper (D) Zinc

48. The aqueous solution of potassium ferrocyanide furnishes
(A) Fe^{2+}, $[K(CN)_2]^-$ ions (B) K^+, $[Fe(CN)_6]^{4+}$ ions
(C) K^+, $[Fe(CN)_6]^{4-}$ ions (D) K^+, Fe^{2+} and CN^- ions

49. Which of the following reacions is correct ?
(A) $Fe^{2+} + K_3[Fe(CN)_6] \rightarrow$ Blue colouration
(B) $Cu^{2+} + NH_4OH \rightarrow$ Chocolate brown precipitate
(C) $Fe^{2+} + NH_4CNS \rightarrow$ Blood red colouration
(D) $Fe^{2+} + K_4[Fe(CN)_6] \rightarrow$ Blue colouration

50. Silver dissolves in a solution of an alkali cyanide (KCN) in the presence of oxygen to form
(A) $[Ag(CN)_3]^-$ (B) $[Ag(CN)_4]^{3-}$
(C) $[Ag(CN)_2]^-$ (D) $[Ag(CN)_2]^{2-}$

51. The final concentration of a solution made by mixing 50 N of 250 ml oxalic acid with 500 ml distilled water in a 1000 ml capacity standard measuring flask is
(A) 0.1375 M (B) 0.2750 M (C) 0.1375 N (D) 0.2750 N

52. Comparing with methyl amine aniline is
(A) acid (B) neutral
(C) weaker base (D) stronger base

53. What is the product formed when phenyl hydroxylamine is rearranged ?
(A) nitroso phenol (B) para amino phenol
(C) meta amino phenol (D) ortho amino phenol

54. Which of the following will form yellow oily liquid of nitrosoamine with HNO_2 ?
(A) N-N dimethyl amino ethane
(B) N-methyl amino ethane
(C) tert-butylamine (D) sec. butylamine

55. Which of the following can form salt with NaOH ?
(A) CH_3NH_2 (B) CH_3NO_2 (C) $C_6H_5NH_2$ (D) $C_6H_5NO_2$

56. Zwitter ion has
(A) keto group (B) both amino and carboxylic group
(C) carboxylic group (D) amino group

57. Conversion of – NO_2 group into – NH_2 groupis

effected by

(A) Zn / NH_4OH (B) Zn / CH_3OH
(C) Sn / HCl (D) Sn / NaOH

58. For the maturation of RBC the essential vitamin required is
(A) vitamin - K (B) ergocalciferol (vitamin D)
(C) cyano cobalamine (D) pyridoxin

59. Vanaspathi is prepared from oils by the reaction called
(A) dehydrogenation (B) hydrogenation
(C) hydration (D) hydrolysis

60. Which of the following colloids do not stay homogeneous indefinitely ?
(A) rubber latex (B) misty fog
(C) soap solution (D) starch solution

TNPCEE -2001

1. The atomic radius increases as we move down a group because
(A) atomic number increases (B) atomic mass increases
(C) additive electrons are accommodated in new electron level
(D) effective nuclear charge increases

2. When orthophosphoric acid heated to 523K the product formed is
(A) phosphorus pentoxide (B) phosphorus acid
(C) metaphosphoric acid (D) pyrophosphoric acid

3. Among the 15th group elements, as we move from nitrogen to bismuth, pentavalency becomes less pronounced and trivalency becomes more pronounced due to
(A) large ionisation energy (B) high electronegativity
(C) inert pair effect (D) non metallic character

4. The number of asymmetric carbon atoms in fructose and the number of optical isomers possible are
(A) five and two (B) four and twelve
(C) three and eight (D) two and six

5. Among the following the true property about $(CH_3)_3C^+$ is
(A) does not undergo hydrolysis (B) not planar
(C) 'C' is sp^2 hybridised (D) electrophile can attack 'C'

6. The number of ether metamers represented by molecular formula $C_4H_{10}O$ is
(A) 1 (B) 2 (C) 3 (D) 4

7. Ethyl acetate can be obtained when methyl magnesium iodide reacts with
(A) carbon dioxide (B) acetyl chloride
(C) ethyl chloroformate (D) ethyl formate

8. The vapour pressure of a solution is equal to that of atmospheric pressure at the
(A) melting point (B) boiling point
(C) elevation of boiling point (D) freezing point

9. The vapour pressuer of a solution (P) and the vapour pressure of the pure solvent (P^0) are related to each other as
(A) $P^0 = PX_2$ (B) $P = P^0X_1$ (C) $P^0 = PX_1$ (D) $P = P^0X_2$

10. The depression of freezing point is directly proportional to
(A) molality of the solvent (B) molality of the solution
(C) molarity of the solution (D) mole fraction of the solution

11. The addition of arsenic to germanium makes the latter a
(A) extrinsic semiconductor (B) mixed conductor
(C) intrinsic semiconductor (D) metallic conductor

12. $H_2 + I_2 \rightleftharpoons 2HI$. Which one of the following is not affecting the rate of this reaction ?
(A) catalyst (B) concentration change
(C) temperature change (D) pressuer change

13. The enthalpy of neutralisation of weak acid by strong base is
(A) > −57.32 kJ (B) +57.32 kJ (C) − 57.32 kJ
(D) = −57.32 kJ + enthalpy of ionisation of weak acid

14. pOH for a weak base and its salt by Henderson equation is =

(A) $pK_b + \log \frac{Base}{Salt}$ (B) $pK_a + \log \frac{Acid}{Salt}$

(C) $pK_b + \log \frac{Salt}{Base}$ (D) $pK_a + \log \frac{Salt}{Acid}$

15. To suppress the dissociation of acetic acid the compound added to it is
(A) sodium nitrate (B) sodium carbonate
(C) sodium acetate (D) sodium oxalate

16. In the Haber process iron is used to
(A) shift the equilibrium to the right
(B) attain the equilibrium more quickly
(C) increase the dissociation of ammonia
(D) increase yield of ammonia

17. The heat of formation of sulphur dioxide is −297kJ. The heat of decomposition of sulphur dioxide is
(A) −594 kJ (B) 594 kJ (C) 297 kJ (D) −297 kJ

18. The rate of the reaction intermadiates can be determined by the study of
(A) solvent effects (B) concentration of the reactants
(C) temperature effects (D) catalyst effects

19. The reaction 2A+B→ products, follows the rate law, rate = $k[A]^2$. The orders with respect to A and B are respectively
(A) 2.0 (B) 0.2 (C) 1.1 (D) 0.1

20. For a first order reaction, a plot of log(a-x) against time is a straight line with a negative slope equal to

(A) -2.303/K (B) 2.303/K (C) -2.303K (D) K/-2.303

21. Which of the following statements regarding d-block elements are wrong ?

(1) The frist element of the first transition series is chromium

(2) The colour of anhydrous $CuSO_4$ is blue

(3) "Spitting of Silver" can be prevented by covering the surface of molten silver with charcoal

(4) Iodine liberated in a reaction can be estimated by titration against a standard thiosulphate solution

(A) (1), (2) and (3) (B) (3) and (4) only
(C) (2) and (3) only (D) (1) and (2) only

22. Which of the following has square planar structure ?

(A) $Ca_2[Fe^{II}(CN)_6]$ (B) $[Cu^{II}(NH_3)_4]SO_4$
(C) $K_3[Fe^{III}(CN)_6]$ (D) $K_4[Fe^{II}(CN)_6]$

23. What is the primary valency of iron in the complex $[Fe(H_2O)_6]SO_4$?

(A) 1 (B) 2 (C) 4 (D) 6

24. Which of the following elements does not show variable oxidation states ?

(A) Titanium (B) Zinc (C) Iron (D) Copper

25. The reason for the stability of Gd^{+3} ion is

(A) 4f sub shell -empty (B) 4f sub shell -half filled
(C) possess the general electronic configuration of noble gases
(D) 4f sub shell -completely filled

26. The alloy used as a levelling agent in the production of steel is

(A) spiegel (B) ferromanganese
(C) nichrome (D) stainless steel

27. The electrolyte used in the electrolytic refining of chromium is

(A) Chromium chloride (B) Chromic acid
(C) Chromium sulphate (D) Chromyl chloride

28. 3.92g of ferrous ammonium sulphate crystals are dissolved in 100 ml of water. 20 ml of this solution requires 18 ml of potassium permanganate during titration for complete oxidation. The weight of $KMnO_4$ present in litre of solution is

(A) 3.476g (B) 1.238g (C) 12.38g (D) 34.76g

29. Accoding to the Quinonoid theory, the colour change of an acid-base indicator arises as a result of

(A) common ion effect (B) pH variation
(C) ionisation (D) structural change

30. Ethylene glycol is oxidised by acidified $KMnO_4$ to give

(A) glyoxalic acid (B) formic acid
(C) glycolic acid (D) oxalic acid

31. The chemical that undergoes self oxidation and self reduction in the same reaction is

(A) acetic acid (B) formaldehyde
(C) acetone (D) benzyl alcohol

32. When ethyl alcohol is heated with con. H_2SO_4 at 443K, ethylene is formed by

(A) intramolecular dehydration
(B) intermolecular dehydration
(C) intermolecular hydration
(D) intramolecular hydration

33. Glycerol when treated with sodium and bismuth nitrate, mainly gives

(A) tartronic acid (B) oxalic acid
(C) meso oxalic acid (D) glyceric acid

34. The compound which reacts fastest with Lucas reagent at room temperature is

(A) 2-methyl-2-propanol (B) 2-methyl-1-propanol
(C) 2-butanol (D) 1-butanol

35. The siliceous material used to absorb nitroglycerine is

(A) quartz (B) crystobalite
(C) kieselghur (D) tridimite

36. The name of the reaction for the preparation of p-hydroxyazobenzene from phenol is

(A) Schotten-Baumann reaction (B) Kolbe reactin
(C) Coupling reaction (D) diazotisation

37. An organic compound with molecular formula $C_4H_{10}O$ does not react with sodium. With excess of HI, it gives rise to only one type of alkyl iodide. What is that compound ?

(A) 2-methoxyl propane (B) 2^0-butyl alcohol
(C) n-butyl alcohol (D) ethoxy ethane

38. The reagent used for the preparation of higher ethers from halogenated ethers is

(A) Grignard reagent (B) dry silver oxide
(C) sodium alkoxide (D) conc. H_2SO_4

39. Which of the following statements regarding chemical properties of acetophenone are wrong ?

(1) It is reduced to phenyl methyl carbinol by sodium and ethanol

(2) It is oxidised to benzoic acid with acidified $KMnO_4$

(3) It does not undergo electrophilic substitution like nitration at meta position.

(4) It does not undergo iodoform reaction with iodine and alkali

(A) (1) and (2) (B) (3) and (4)
(C) (2) and (4) (D) (1) and (3)

40. The aldehyde that is formed during photosynthesis in plants is

(A) phenyl methanal (B) propenal
(C) acetaldehyde (D) methanal

41. Benzophenone can be converted into benzene using

(A) acidified dichromate (B) sodium amalgam in water
(C) anhydrous $AlCl_3$ (D) fused alkali

42. Benzaldehyde with chlorine in the absence of catalyst gives
(A) benzoyl chloride
(B) ortho and para chloro benzaldehyde
(C) o-chloro benzaldehyde (D) m-chloro benzaldehyde

43. The formation of cyanohydrin from a ketone is an example of
(A) electrophilic addition (B) electrophilic substitution
(C) nucleophilic addition (D) nucleophilic substitution

44. In the mechanism of Hoffmann reaction which intermediate rearranges to alkyl isocyanide ?
(A) amide (B) nitroso (C) nitrene (D) bromamide

45. Oxalic acid when reduced with Zinc and H_2SO_4 gives
(A) glycol (B) glycollic acid
(C) glyoxal (D) glyoxalic acid

46. A compound used in the preparation of fluorescein is
(A) terephthalic acid (B) phthalic anhydride
(C) phthalic acid (D) phthalimide

47. Carboxylic acids are more acidic than phenol and alcohol because of
(A) resonance stabilisation of their conjugate base
(B) highly acidic hydrogen (C) formation of dimers
(D) intermolecular hydrogen bonding

48. Lactic acid on oxidation by alkaline potassium permanganate gives
(A) propionic acid (B) cinnamic acid
(C) pyruvic acid (D) tartaric acid

49. The equivalent conductivity of 0.1 M weak acid is 100 times less than that at infinite dilution. The degree of dissociation is
(A) 0.001 (B) 0.01 (C) 10 (D) 100

50. Colloidal gold is prepared by
(A) hydrolysis (B) Bredig's arc method
(C) peptisation (D) mechanical dispersion

51. The migration of colloidal solute particles in a colloidal solution, when an electric current is applied to the solution is known as
(A) electrodialysis (B) electrophoresis
(C) electro-osmosis (D) Brownian movement

52. The catalyst/promoter used in the manufacture of CH_3OH from CO and H_2 is
(A) Platinum (B) ZnO/Cr_2O_3 (C) MO (D) Fe

53. $C(gr) + O_2(g) \rightarrow CO_2(g)$ The value of ΔH for this reaction is
(A) –394 kJ (B) –493 kJ (C) –286 kJ (D) +51 kJ

54. Which one of the following structures corresponds to Schiff's base ?
(A) $C_6H_5N{=}CHC_6H_5$ (B) $C_6H_5NHCOC_6H_5$
(C) $C_6H_5\text{-}N = NH\text{-}C_6H_5$ (D) $C_6H_5\text{-}NH\text{-}NH\text{-}C_6H_5$

55. Benzylamine reacts with nitrous acid to form
(A) phenol (B) benzyl alcohol
(C) benzene (D) azobenzene

56. Benzene diazonium chloride is reduced to benzene by
(A) phosphine (B) hypophosphoric acid
(C) hypophosphorus acid (D) phosphorous acid

57. The final product obtained when nitrobenzene is electrolytically reduced in conc. H_2SO_4 solution is
(A) nitrosobenzene (B) p-amino phenol
(C) aniline (D) phenyl hydroxylamine

58. Primary nitro compounds react with nitrous acid to form nitrolic acids which dissolve in NaOH giving
(A) red solution (B) colourless solution
(C) blue solution (D) yellow solution

59. The number of amino acids found in proteins that a human body can synthesize is
(A) 5 (B) 10 (C) 14 (D) 20

60. Chrysoidene belongs to which type of dye ?
(A) azo (B) nitro (C) nitroso (D) anthraquinone

TNPCEE -2002

1. The black compound formed during the reaction between sodium thiosulphate and silver nitrate is
(A) Silver sulphite (Ag_2SO_3)
(B) Silver sulphate (Ag_2SO_4)
(C) Silver sulphide (Ag_2S)
(D) Silver thiosulphate ($Ag_2S_2O_3$)

2. Which plant essentially requires the fertilizer, potassium nitrate ?
(A) tomato (B) coconut (C) rice (D) tobacco

3. Which one of the following is an explosive and readily hydrolysed ?
(A) N_2O_3 (B) NH_3 (C) NCl_3 (D) NF_3

4. The basicity of phosphorus acid, ortho phosphoric acid and meta phosphoric acid are respectively
(A) 1, 2 and 3 (B) 3, 2 and 1
(C) 2, 3 and 1 (D) 2, 1 and 3

5. Which of the following is correctly matched ?
(A) Ionic radius of F^- – 0.095 nm
(B) C – Cl bond length – 0.176 nm
(C) Ionic radius of Na+ – 0.136 nm
(D) C – C bond length – 0.077 nm

6. The half life of a radioactive element is 90 minutes. What percentage of radioactive element will be present after 6 hours of decomposition ?

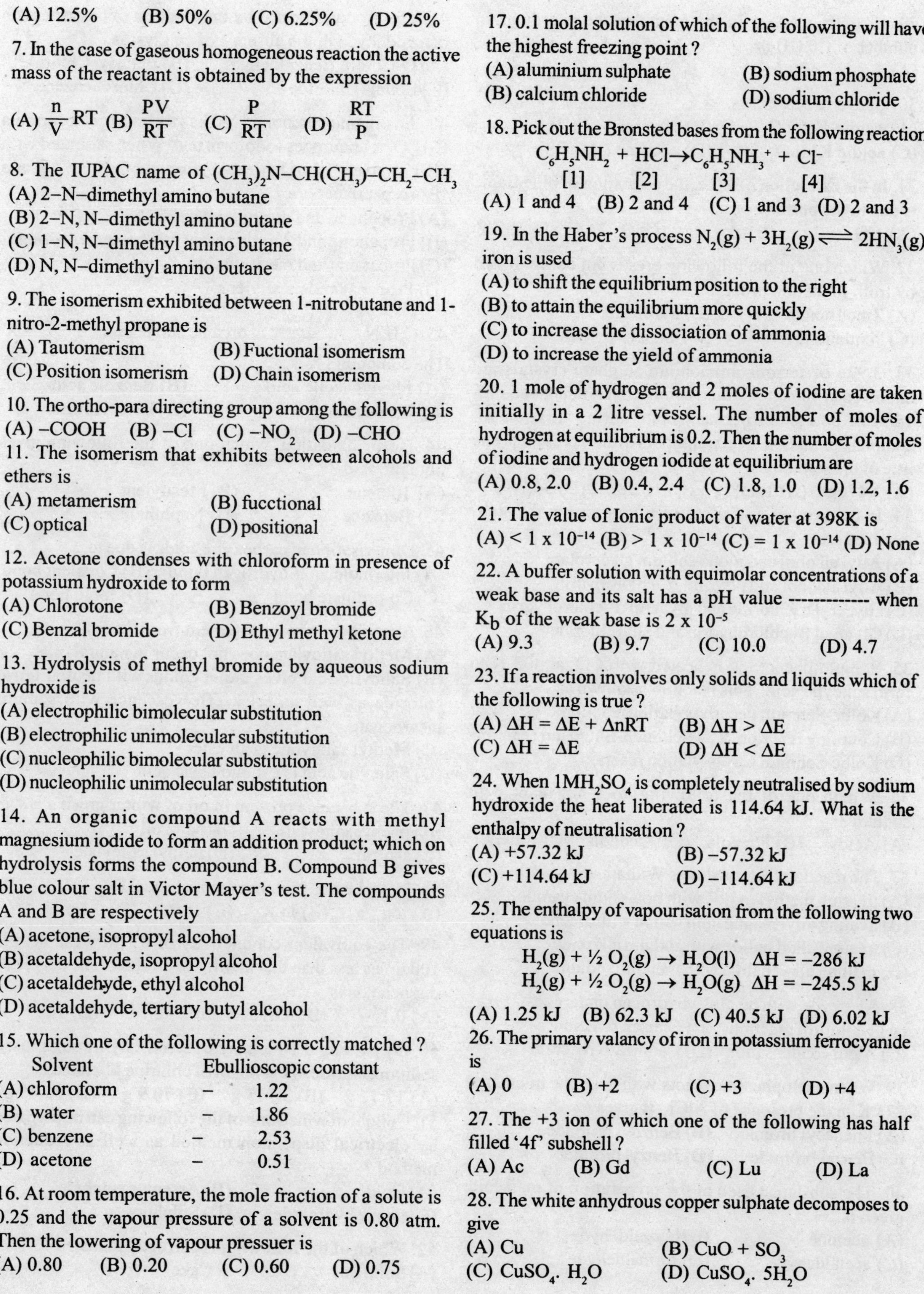

(A) 12.5% (B) 50% (C) 6.25% (D) 25%

7. In the case of gaseous homogeneous reaction the active mass of the reactant is obtained by the expression

(A) $\frac{n}{V}RT$ (B) $\frac{PV}{RT}$ (C) $\frac{P}{RT}$ (D) $\frac{RT}{P}$

8. The IUPAC name of $(CH_3)_2N–CH(CH_3)–CH_2–CH_3$
(A) 2–N–dimethyl amino butane
(B) 2–N, N–dimethyl amino butane
(C) 1–N, N–dimethyl amino butane
(D) N, N–dimethyl amino butane

9. The isomerism exhibited between 1-nitrobutane and 1-nitro-2-methyl propane is
(A) Tautomerism (B) Fuctional isomerism
(C) Position isomerism (D) Chain isomerism

10. The ortho-para directing group among the following is
(A) –COOH (B) –Cl (C) $–NO_2$ (D) –CHO

11. The isomerism that exhibits between alcohols and ethers is
(A) metamerism (B) fucctional
(C) optical (D) positional

12. Acetone condenses with chloroform in presence of potassium hydroxide to form
(A) Chlorotone (B) Benzoyl bromide
(C) Benzal bromide (D) Ethyl methyl ketone

13. Hydrolysis of methyl bromide by aqueous sodium hydroxide is
(A) electrophilic bimolecular substitution
(B) electrophilic unimolecular substitution
(C) nucleophilic bimolecular substitution
(D) nucleophilic unimolecular substitution

14. An organic compound A reacts with methyl magnesium iodide to form an addition product; which on hydrolysis forms the compound B. Compound B gives blue colour salt in Victor Mayer's test. The compounds A and B are respectively
(A) acetone, isopropyl alcohol
(B) acetaldehyde, isopropyl alcohol
(C) acetaldehyde, ethyl alcohol
(D) acetaldehyde, tertiary butyl alcohol

15. Which one of the following is correctly matched ?

	Solvent		Ebullioscopic constant
(A)	chloroform	–	1.22
(B)	water	–	1.86
(C)	benzene	–	2.53
(D)	acetone	–	0.51

16. At room temperature, the mole fraction of a solute is 0.25 and the vapour pressure of a solvent is 0.80 atm. Then the lowering of vapour pressuer is
(A) 0.80 (B) 0.20 (C) 0.60 (D) 0.75

17. 0.1 molal solution of which of the following will have the highest freezing point ?
(A) aluminium sulphate (B) sodium phosphate
(B) calcium chloride (D) sodium chloride

18. Pick out the Bronsted bases from the following reaction

$$\underset{[1]}{C_6H_5NH_2} + \underset{[2]}{HCl} \rightarrow \underset{[3]}{C_6H_5NH_3^+} + \underset{[4]}{Cl^-}$$

(A) 1 and 4 (B) 2 and 4 (C) 1 and 3 (D) 2 and 3

19. In the Haber's process $N_2(g) + 3H_2(g) \rightleftharpoons 2HN_3(g)$ iron is used
(A) to shift the equilibrium position to the right
(B) to attain the equilibrium more quickly
(C) to increase the dissociation of ammonia
(D) to increase the yield of ammonia

20. 1 mole of hydrogen and 2 moles of iodine are taken initially in a 2 litre vessel. The number of moles of hydrogen at equilibrium is 0.2. Then the number of moles of iodine and hydrogen iodide at equilibrium are
(A) 0.8, 2.0 (B) 0.4, 2.4 (C) 1.8, 1.0 (D) 1.2, 1.6

21. The value of Ionic product of water at 398K is
(A) $< 1 \times 10^{-14}$ (B) $> 1 \times 10^{-14}$ (C) $= 1 \times 10^{-14}$ (D) None

22. A buffer solution with equimolar concentrations of a weak base and its salt has a pH value ------------ when K_b of the weak base is 2×10^{-5}
(A) 9.3 (B) 9.7 (C) 10.0 (D) 4.7

23. If a reaction involves only solids and liquids which of the following is true ?
(A) $\Delta H = \Delta E + \Delta nRT$ (B) $\Delta H > \Delta E$
(C) $\Delta H = \Delta E$ (D) $\Delta H < \Delta E$

24. When $1MH_2SO_4$ is completely neutralised by sodium hydroxide the heat liberated is 114.64 kJ. What is the enthalpy of neutralisation ?
(A) +57.32 kJ (B) –57.32 kJ
(C) +114.64 kJ (D) –114.64 kJ

25. The enthalpy of vapourisation from the following two equations is

$H_2(g) + ½ O_2(g) \rightarrow H_2O(l)$ $\Delta H = -286$ kJ
$H_2(g) + ½ O_2(g) \rightarrow H_2O(g)$ $\Delta H = -245.5$ kJ

(A) 1.25 kJ (B) 62.3 kJ (C) 40.5 kJ (D) 6.02 kJ

26. The primary valancy of iron in potassium ferrocyanide is
(A) 0 (B) +2 (C) +3 (D) +4

27. The +3 ion of which one of the following has half filled '4f' subshell ?
(A) Ac (B) Gd (C) Lu (D) La

28. The white anhydrous copper sulphate decomposes to give
(A) Cu (B) $CuO + SO_3$
(C) $CuSO_4 . H_2O$ (D) $CuSO_4 . 5H_2O$

29. $[Cu(NH_3)_4]SO_4$: In this complex the coordination number of Cu (II) is
(A) 3 (B) 6 (C) 4 (D) 2

30. Baeyer's reagent is
(A) acidic $K_2Cr_2O_7$ (B) alkaline $K_2Cr_2O_7$
(C) acidic $KMnO_4$ (D) alkaline $KMnO_4$

31. In the extraction of Zinc, the formation of blue flame is due to the burning of
(A) CO (B) Zn (C) C (D) ZnO

32. Which one of the following ores is not concentrated by froth floatation process ?
(A) Zinc Blende (B) Pyrolusite
(C) Pentlandite (D) Copper pyrites

33. 3.92g of ferrous ammonium sulphate crystals are dissolved in 100 ml of water. 20ml of this solution requires 18ml of potassium permanganate during titration for complete oxidation. The weight of $KMnO_4$ present is one litre of the solution is
(A) 3.476g (B) 1.238 g (C) 12.38g (D) 34.76 g

34. In the synthesis of glycerol from propene the steps involved are
(A) Allyl alcohol and mono sodium glycerolate
(B) Allyl alcohol and glycerol α–chlorohydrin
(C) Glyceryl trichloride and glycerol α–chlorohydrin
(D) Glycerol β–chlorohydrin and allyl chloride

35. Sodium phenoxide is heated with CO_2 at 403 K to form salicylic acid. This reaction is known as
(A) Kolbe-Schmidt decarboxylation reaction
(B) Coupling reaction (C) Schotten-Baumann reaction
(D) Kolbe-Schmidt carboxylation reaction

36. Which one of the following does not react with sodium ?
(A) Acids (B) Phenols (C) Alcohols (D) Ethers

37. The reaction that illustrates Williamson's synthesis is
(A) treating methyl iodide with potassium cyanide
(B) treating allyl magnesium halide with alcohol
(C) treating alkyl halide with sodium alkoxide
(D) treating alkyl halide with metallic sodium

38. Aldehyde with no alpha hydrogen undergoes
(A) Haloform reaction (B) Claisen reaction
(C) Aldol condensation (D) Cannizaro reaction

39. When acetophenone reacts with bromine in ether at 273°K in the presence of $AlCl_3$, it gives
(A) Phenacyl bromide (B) Benzoyl bromide
(C) Benzal bromide (D) Benzyl bromide

40. The substance used in the preparation of malachite green is
(A) acetone (B) benzaldehyde
(C) acetaldehyde (D) formaldehyde

41. The product formed by the reaction of chlorine with benzaldehyde in the absence of catalyst is
(A) O-Chloro Benzaldehyde (B) Benzoyl Chloride
(C) Benzyl Chloride (D) Chlorobenzene

42. An organic compound 'A' has the molecular formula C_3H_6O. It undergoes Iodoform test. When saturated with HCl, it gives 'B' of molecular formula $C_9H_{14}O$. 'A' and 'B' respectively are
(A) Propanone and mesitylene oxide
(B) Propanone and 2, 6-dimethyl 2, 5 heptadiene-4-one
(C) Propanone and mesityloxide
(D) Propanal and mesitylene

43. $C_6H_5N_2Cl \xrightarrow[\text{Cu Powder}]{\text{aq. KCN}}$ organic compound $\xrightarrow[\text{hydrolysis}]{\text{dil HCl}}$ X.
The compound X is
(A) Phenyl acetic acid (B) Benzoic acid
(C) Benzo nitrile (D) Benzo nitrite

44. By air oxidation, which one of the following gives phthalic acid ?
(A) Toluene (B) Mesitylene
(C) Benzene (D) Naphthalene

45. Dimerisation in carboxylic acids is due to
(A) Intermolecular hydrogen bond (B) Covalent bond
(C) Co-ordinate bond (D) Ionic bond

46. Identify the wrong statement from the following
(A) Methyl salicylate does not occur in natural oils
(B) Salicylic acid gives violet colour with neutral ferric chloride as well as brisk effervescence with sodium bicarbonate
(C) Methyl salicylate is an ester
(D) Salicylic acid is a mono basic acid

47. The substance present in oil of winter green's is
(A) phenyl salicylate (B) palm oil
(C) aspirin (D) nethyl salicylate

48. In Arrhenius plot, intercept is equal to
(A) $\log_{10}$ a (B) ln K (C) ln A (D) – Ea / R

49. The equivalent conductivity of 0.1 M weak acid is 100 times less than that at infinite dilution. The degree of dissociation is
(A) 0.1 (B) 0.01 (C) 10 (D) 100

50. By passing 0.1 Faraday of electricity through fused sodium chloride, the amount of chlorine liberated is
(A) 17.77 g (B) 3.545 g (C) 70.9 g (D) 35.45 g

51. Colloid of which one of the following can be prepared by electrical dispersion method as well as reduction method ?
(A) Gold (B) Arsenius sulphide
(C) Ferric hydroxide (D) Sulphur

52. Which of the following is correctly matched ?
(A) Solid Sol – Cake

(B) Aerosol – Smoke
(C) Foam – Mist
(D) Emulsion – Curd

53. The catalyst used in the manufacture of nitric acid in Ostwald's process is
(A) Fe (B) Cu_2Cl_2 (C) V_2O_5 (D) Pt

54. Consider the following statements about the basic nature of amines
(1) aliphatic amines are stronger than ammonia
(2) aliphatic amines are weaker than ammonia
(3) aromatic amines are much less basic than aliphatic amines
(4) aromatic amines are much greater basic than aliphatic amines
Which of the following is correct ?
(A) (3) and (4) are true (B) (1) and (4) are not true
(C) (1) and (3) are true (D) (1) and (4) are true

55. Which reagent reacts with nitro methane to form methyl hydroxylamine ?
(A) Sn / HCl (B) Zn / NaOH
(C) Zn / NH_4Cl (D) Zn / HCl

56. Nitrobenzene on electrolytic reduction gives
(A) nitroso benzene (B) nitroaniline
(C) p-aminophenol (D) aniline

57. High altitude aircraft components are made from
(A) GR. A. rubber (B) Neoprene
(C) P.V.C. (D) Thiokol

58. In fructose the possible optical isomers are
(A) 4 (B) 8 (C) 12 (D) 16

59. α–Tocopherol is found in
(A) tomatoes (B) peanuts
(C) lemon (D) vegetable oils

60. The chemical used in oinments as a local anaesthetic and antiseptic is
(A) CH_3OH (B) CH_3CH_2OH
(C) C_6H_5OH (D) $C_6H_5CH_2OH$

TNPCEE -2003

1. If Mg^{+2} and O^{-2} are isoelectronic, then which one of the following statements is true?
(A) Mg^{+2} and O^{-2} have same number of protons
(B) Mg^{+2} and O^{-2} have same ionic radii
(C) Mg^{+2} ion is larger than O^{-2} ion
(D) O^{-2} ion is large than Mg^{2+} ion

2. Which one the following statements is true ?
(A) Dry chlorine does not combine with Chromium
(B) Dilute sulphuric acid does not react with the Chromium
(C) Dilute nitric acid does not react with the Chromium
(D) Dilute hydrochloric acid does not react with the chromium

3. The basicity of phosphorus acid is
(A) Zero (B) One (C) Two (D) Three

4. Phosphorus pentoxide (P_2O_5) is purified by which of the following methods ?
(A) Crystallisation from ethanol (B) Sublimation at 633 K
(C) Distillation at 733 K (D) Crystallisation from water

5. The IUPAC name of CH_2=CH-CHO is
(A) Ethylene methanal (B) Vinyl methanal
(C) Propenal (D) Acrolein

6. The IUPAC name of Cumene is
(A) 2-phenyl propylene (B) 2 phenyl propane
(C) benzyl propane (D) phenyl propane

7. The nitrogen content of urea is
(A) 46.6% (B) 32.6% (C) 64.6% (D) 23.3%

8. Terylene belongs to which of the following groups ?
(A) Polyester (B) Polyamine
(C) Polyamide (D) Polyacid

9. The type of isomerism in CH_3-NO_2 and CH_3-O-N=O is
(A) Metamerism (B) Position isomerism
(C) Chain isomerism (D) Functional isomerism

10. How many forms of optically active isomers are possible for tartaric acid ?
(A) one (B) two (C) three (D) four

11. Carbocation is planar because it is
(A) s-s overlap (B) sp^2 hybridised
(C) sp^3 hybridised (D) sp hybridised

12. The functional isomer of n-butyl alcohol is
(A) Ethyl methyl ketone (B) 2-methyl-1-propanol
(C) Ethoxy ethane (D) 2-methyl-2-propanol

13. Ethylamine reacts with $CHCl_3$ and alcoholic KOH to give
(A) Acetic acid (B) Ethyl alcohol
(C) Ethyl isocyanide (D) Ethyl cyanide

14. The selective passage of small ions and molecules through a membrane is called
(A) reverse osmosis (B) electrodialysis
(C) osmosis (D) dialysis

15. The Osmotic pressure of a solution is
(A) Equal to nRT (B) Equal to CT
(C) Directly proportional to CRT (D) Equal to CRT

16. The observed and theoretical osmotic pressure values of 1% aqueous $CaCl_2$ solution are $2.7x10^5$ Pa nd $1x10^5$ Pa respectively. The degree of dissociation of $CaCl_2$ is
(A) 0.65 (B) 0.75 (C) 0.85 (D) 0.95

17. The molal boilling point elevation constant of chloroform is
(A) 3.26 (B) 3.36 (C) 3.62 (D) 3.63

18. The value of equilibrium constant Kp for the reaction $H_2(g) + I_2(g) \rightleftharpoons 2HI(g)$ is 64 at 373K. The value of equilibrium constant when the volume of the container is doubled at the same temperature is
(A) 16 (B) 32 (C) 64 (D) 128

19. The active mass of 8.5 gram of ammonia placed in a 2 litre container is
(A) 0.25 (B) 0.50 (C) 0.75 (D) 0.85

20. The enthalpy of neutralisation of strong acid by a strong base is constant and equal to
(A) –27.32 J $equi^{-1}$ (B) 57.32 kJ $equi^{-1}$
(C) –57.32 kJ $equi^{-1}$ (D) –57.32 J $equi^{-1}$

21. In which of the following there is no sharp change in pH at the end point when titrated against each other ?
(A) $H_2C_2O_4$ Vs Na_2CO_3 (B) HCl Vs NH_4OH
(C) $H_2C_2O_4$ Vs NaOH (D) H_2SO_4 Vs Na_2CO_3

22. The value of ionic product K_W for water at 298K is
(A) $1x10^{-7}$ mol^2 (B) $14x10^{-4}$ mol^3
(C) $1x10^{-14}$ $mol^2 dim^{-6}$ (D) $7x10^{-1}$ dim^{-6}

23. The change in the strength of the acid when the medium changes is called
(A) Ionisation effect (B) levelling effect
(C) Ionic effect (D) Screening effect

24. Which one of the following is a fusion reaction ?
(A) $CuSO_4(s)$+ aq $\rightarrow$ $CuSO_4(aq)$, $\Delta H = -78.2$ kJ
(B) $H_2O(l) \rightarrow H_2O(g)$, $\Delta H = 40.5$ kJ
(C) $H_2O(s) \rightarrow H_2O(l)$, $\Delta H = 6.02$ kJ
(D) $I_2(s) \rightarrow I_2(g)$, $\Delta H = 62.3$ kJ

25. The enthalpies of formation of methanol. formaldehyde and water are –239kJ, –116kJ and –286kJ, respectively. The enthalpy of oxidation of methanol to formaldehyde and water is
(A) –136 kJ (B) –163 kJ (C) –165 kJ (D) –176 kJ

26. For which of the following reactions ΔH and ΔE values are equal ?
(A) Combustion of hydrogen (B) Combustion of methane
(C) Formation of carbondioxide (D) Formation of ethylene

27. The rate of a chemical reaction $-\frac{dx}{dt}$ at different time intervals are as follows

Time(mts):	0	10	20	30
Rate :	$1.8x10^{-2}$	$1.8x10^{-2}$	$1.8x10^{-2}$	$1.8x10^{-2}$

The above reaction is
(A) Zero order reaction (B) First order reaction
(C) Second order reaction (D) Third order reaction

28. Hydrolysis of sucrose in the presence of mineral acids is an example for
(A) First order reaction (B) Pseudo-first order reaction
(C) Unimolecular reaction (D) Bimolecular reaction

29. The rate constant of a first order reaction is 4.606 min^{-1}. What is the time required for 75% completion of the reaction ?
(A) log 16/9 min (B) log 2/√3 min
(C) log 2 min (D) 3.495 min

30. The coordination complex in which the central metal ion has +3 oxidation state among the following is
(A) $Cu_2[(Fe(CN)_6]$ (B) $[Co(NH_3)_6]Cl_3$
(C) $[Pt(NH_3)_2Cl_2]$ (D) $[Ni(NH_3)_6]SO_4$

31. The charge on the complex ion is equal to
(A) charges on the metal ion
(B) charges on the ligand minus metal ion
(C) charges on the ligand
(D) sum of the charges on the metal ion and the ligands

32. Which of the following statements are true ?
(a) Lanthanides do not conduct heat and electricity
(b) The lanthanide element with no 4f electron is lanthanum
(c) The common oxidation state of uranium is +6
(d) While moving from Ac to Ir, the ionic radii decreases
(A) (a) and (d) (B) (a) and (c)
(C) (a), (b) and (c) (D) (b), (c) and (d)

33.Which is called "Philosopher's Wool" ?
(A) Ag_2O (B) CuO (C) ZnO (D) Cr_2O_3

34. Which of the following melts and forms a red liquid on heating ?
(A) $AgNO_3$ (B) AgI (C) AgBr (D) AgCl

35. 1.96g of crystalline ferrous ammonium sulphate is present in 2 litres of a solution. The volume of 0.0025N potassium permanganate solution required for the oxidation of 20ml of the above solution is
(A) 15ml (B) 20ml (C) 25ml (D) 30ml

36. Methyl magnesium iodide reacts with ethyl chloroformate and then hydrolysed to form
(A) Acetone (B) Acetic acid
(C) Methyl acetate (D) Ethyl acetate

37. Catalytic dehydrogenation of tertiary butyl alcohol gives
(A) propene (B) isobutylene
(C) dimethyl ketone (D) acetaldehyde

38. The one that does not undergo oxidation under normal conditions is
(A) benzyl alcohol (B) primary alcohols
(C) secondary alcohols (D) tertiary alcohols

39. Isopropyl alcohol undergoes
(A) Condensation with hydroxylamine (B) Schiff's test
(C) Iodoform reaction (D) Bisulphite addition

40. Phenetole is

(A) phenolic ether (B) aromatic aldehyde
(C) aromatic alcohol (D) anisole

41. The structure of Catechol is

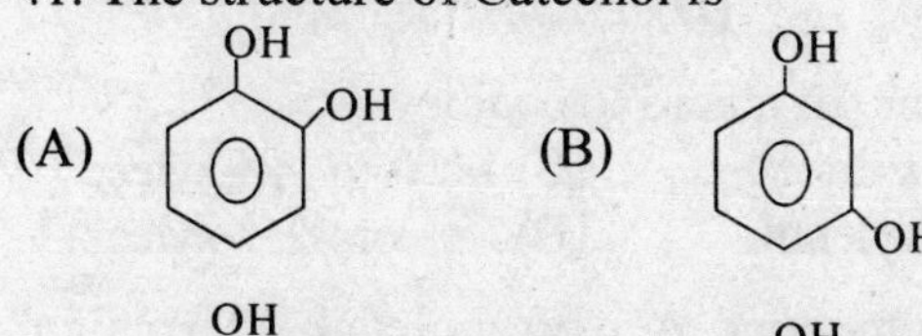

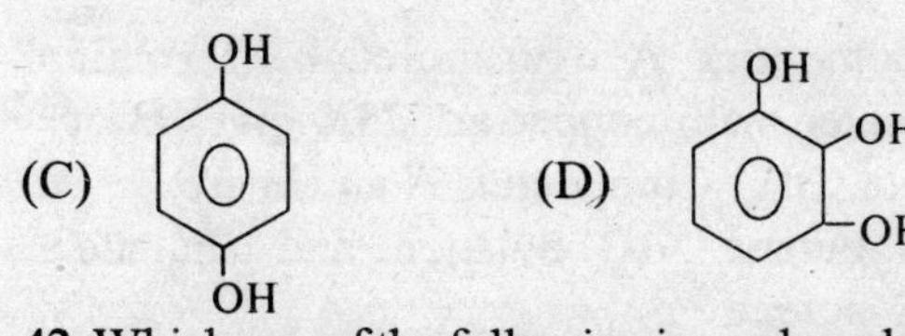

42. Which one of the following is used as a high explosive ?
(A) Oxalic acid (B) Picric acid
(C) Nitric acid (D) Benzoic acid

43. Which of the following statements is wrong about ethers ?
(A) Lighter than Water (B) Forms hydrogen bonding
(C) Highly volatile
(D) Possess lower boiling point than alcohol

44. The reaction between aromatic aldehyde and aliphatic aldehyde is called as
(A) Kolbe reaction (B) Perkin reaction
(C) Claisen reaction (D) Knoevenagel reaction

45. Match the following
(a) Sulphonal (i) Benzaldehyde
(b) Malachite green (ii) Acetaldehyde
(c) Perfumery (iii) Acetone
(d) Paraldehyde (iv) Benzophenone
(A) (a)-(i), (b)-(iii), (c)-(ii), (d)-(iv)
(B) (a)-(iii), (b)-(i), (c)-(ii), (d)-(iv)
(C) (a)-(i), (b)-(iii), (c)-(iv), (d)-(ii)
(D) (a)-(iii), (b)-(i), (c)-(iv), (d)-(ii)

46. Both acetaldehyde and acetone will react with
(A) Schiff's reagent (B) Grignard reagent
(C) Tollen's reagent (D) Fehling's solution

47. Which of the following acids cannot be prepared by the hydrolysis of nitriles ?
(A) Lactic acid (B) Benzoic acid
(C) Acetic acid (D) Formic acid

48. The formula for chemical compound in the oil of winter green is

(A) OCH_3, COOH (B) OH, $COOCH_3$

(C) COOH, $COOCH_3$ (D) OH, $\overset{}{C}OCH_3$ (=O)

49. Which of the following is used in the preservation of pickles ?
(A) Methyl oxalate (B) Methyl salicylate
(C) Sodium benzoate (D) Sodium acetate

50. 0.5 ampere current is passed for one hour through the aqueous solutions of $CuSO_4$, $ZnSO_4$, $AgNO_3$ and H_2SO_4. The substance deposited in more amount at the cathode is [Atomic weight: Cu = 63.54; Ag = 107.9; Zn = 65.38; H = 1.008]
(A) Cu (B) Zn (C) Ag (D) H

51. Which of the following statements is false regarding the reaction between $KMnO_4$ and Con. HCl ?
(A) Cl^- ion is oxidised (B) Cl^- ion has lost one electron
(C) Mn has gained 5 electrons (D) Mn has lost 5 electrons

52. The migration of colloidal particles under the influence of electric field is known as
(A) Electro-dialysis (B) Electro-osmosis
(C) Dialysis (D) Eectrophoresis

53. The name of the catalysis in the following reaction is
$2SO_2(g) + O_2(g) \xrightarrow{NO(g)} 2SO_3(g)$
(A) Heterogeneous catalysis (B) Homogeneous catalysis
(C) Negative catalysis (D) Positive catalysis

54. Which of the following statements are correct ?
(a) Nitroalkanes cannot be reduced in acid medium.
(b) Nitroalkanes cannot be halogenated in the α-position
(c) Nitroalkanes undergo condensation reaction with aldehydes and ketones
(d) When reduced in a neutral medium nitroalkanes yield alkyl hydroxylamine.
(A) (a), (b) and (d) (B) (b), (c) and (d)
(C) (c) and (d) only (D) (a) and (b) only

55. The resonance structures of benzene diazonium chloride can be given as

(A) $\left[C_6H_5-\overset{+}{N}\equiv\ddot{N}\leftrightarrow C_6H_5\,\ddot{N}=\overset{+}{N}:\right]Cl^-$

(B) $\left[C_6H_5-\overset{+}{N}\equiv\ddot{N}\leftrightarrow C_6H_5\,\ddot{N}=\overset{+}{N}\right]Cl^-$

(C) $\left[C_6H_5-\ddot{N}\equiv\overset{+}{N}\leftrightarrow C_6H_5\,\ddot{N}=\overset{+}{N}\right]Cl^-$

(D) $\left[C_6H_5-\ddot{N}\equiv\overset{+}{N}\leftrightarrow C_6H_5\,\ddot{N}=\overset{+}{N}:\right]Cl^-$

56. Which of the following statement is not true regarding aniline ?
(A) it reacts readily with carbonyl chloride
(B) it is miscible with water
(C) when heated with reactive metals metallic derivatives

are formed
(D) it is less basic than ethylamine

57. Carbylamine $\xleftarrow[\text{alc.KOH}]{CHCl_3}$ C_7H_9N $\xrightarrow{NaNO_2/HCl}$ alcohol

Carbylamine $\xleftarrow[\text{alc.KOH}]{CHCl_3}$ C_7H_9N $\xrightarrow{NaNO_2/HCl}$ diazonium chloride
A

A and B are respectively
(A) benzylamine and touiden B(B) toluidene and benzylamine
(C) benzylamine and aniline (D) aniline and benzylamine

58. Aliphatic amines are more basic than aromatic amines due to
(A) Alkyl group increasing the electron density on the nitrogen
(B) Lewis acidity of aliphatic amines
(C) Aromatic group increasing the electron density on the nitrogen
(D) Higher molecular weight of aromatic rings

59. Read the following statements and identify the correct answer
(a) Deficiency of vitamin A leads to night blindness
(b) The chemical name of vitamin C is ascorbic acid
(c) Coagulation of blood is the function of vitamin E
(d) Vitamin D is rich in tomatoes
(A) (a), (b) & (c) are true (B) (b), (c) & (d) are false
(C) Only (b) & (c) are true (D) Only (c) & (d) are false

60. The vitamin responsible for the formation of bones is
(A) Vitamin D (B) Vitamin B
(C) Vitamin K (D) Vitamin B_{12}

TNPCEE -2004

1. Arrange Ca, Cs and F elements in decreasing order of electronegativity
(A) F > Ca > Cs (B) Cs > Ca > F
(C) Ca > F > Cs (D) F > Cs > Ca

2. Ionisation energy of Au is greater than Ag because
(A) Au is an inert metal (B) Au has incomplete 4f orbital
(C) Atomic number of Au is 79
(D) In Au, 4f electrons have lower screening effect.

3. Which of the following does not belong to periodic properties ?
(A) atomic radius (B) ionisation energy
(C) electronegativity (D) activation energy

4. Which of the following is used as a source of hydrogen gas ?
(A) natural gas (B) water gas
(C) producer gas (D) gobar gas

5. Which of the following is used in making lithopone ?
(A) $ZnCO_3$ (B) $Na_2S_2O_3$ (C) $CuSO_4$ (D) $ZnSO_4$

6. Outer electronic configuration of four elements A, B, C and D is given below. Which among them can form multiple bonds ?
(A) A....$[He]2s^22p^3$ (B) B.....$[Ne]3s^23p^3$
(C) C....$[Ar]4s^24p^3$ (D) D.....$[Kr]5s^25p^3$

7. The IUPAC name of chloroxelenol is
(A) 3-chloro-4, 5-xylenol (B) 4-chloro-2, 5-xylenol
(C) 4-chloro-3, 5-xylenol (D) 3-chloro-2, 5-xylenol

8. An organic compound A with molecular formula $C_4H_{10}O$, when heated with copper at 573K gives B of molecular formula C_4H_8. Compounds A and B are
(A) butanol and 2-butene (B) btuanone and 1-butene
(C) butanal and 1-butene
(D) 2-methyl-2-propanol and 2-methylpropene

9. Which of the following has high probability of undergoings S_N1 mechanism ?
(A) benzyl alcohol (B) isopropyl alcohol
(C) n-butyl alcohol (D) n-propyl alcohol

10. In racemic mixture, optical activity disappears due to
(A) the presence of a plane of symmetry
(B) external compensation (C) internal compensation
(D) the absence of asymmetric carbon atom

11. $C_6H_6 + Cl_2 \xrightarrow{FeCl_3} A \xrightarrow{Fe, Cl_2} B + C$ compounds B and C are
(A) p-dichloro benzene + benzyl chloride
(B) p-dichloro benzene + m-dichloro benzene
(C) o-dichloro benzene + p-dichloro benzere
(D) o-dichloro benzene + m-dichloro benzene

12. CH_3MgI reacts with compound A to form ethyl acetate. Compoound A is
(A) ethyl formate (B) acetyl Chloride
(C) ethyl chloroformate (D) formyl chloride

13. Benzene diazonium chloride is obtained using aniline. sodium nitrite and dil.HCl at
(A) room temperature (B) 100°C
(C) 50°C (D) ice cold condition

14. Chloretone is formed when chloroform reacts with
(A) C_6H_5CHO (B) CH_3COCH_3
(C) CH_3CHO (D) HCHO

15. Which of the following represents phenacyl bromide ?

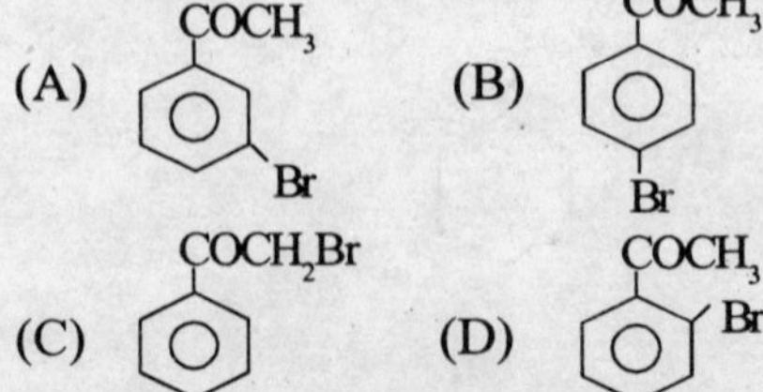

16. The molar concentration of 85g of ammonia dissolved in 1 lit. of water. kept in $2dm^3$ container is

(A) 1 mol. dm^{-3} (B) 2 mol dm^{-3}
(C) 2.5 mol. dm^{-3} (D) 5 mol dm^{-3}

17. What is the osmotic pressure (atm) of a solution containing 6.84g of sucrose per litre at 400K ? The molecular weight of sucrose and the value of gas constant (R) are 342 and 0.082 lit. atm. $mole^{-1} K^{-1}$ respectively.
(A) 0.328 (B) 0.0328 (C) 0.656 (D) 0.164

18. The freezing point of one molal aqueous solution will be the highest for
(A) glucose (B) sodium phosphate
(C) calcium chloride (D) potassium chloride

19. What will be the pH of the buffer solution containing 0.2 mole of acetic acid and 0.1 mole of sodium acetate per litre ? The dissociation constant of acetic acid is 1.8×10^{-5}.
(A) 3.274 (B) 5.674 (C) 4.474 (D) 6.874

20. At 50°C (323 K) ionic product of water k_w
(A) will be equal to 7
(B) will be less than 1×10^{-14} $mole^2/dm^6$
(C) will be greater than 1×10^{-14} $mole^2/dm^6$
(D) will be equal to 1×10^{-14} $mole^2/dm^6$

21. $PCl_5 \rightleftharpoons PCl_3 + Cl_2$. In the above equilibrium, forward reaction can be favoured by
(A) increase in pressure at equlibrium (B) addition of PCl_5
(C) addition of suitable catalyst (D) addition of PCl_3

22. Le-Chatelier-Braun principle applies
(A) to all equilibrium reactions (B) to all liquid reactions
(C) to all chemical reactions (D) only to gaseous reactions

23.

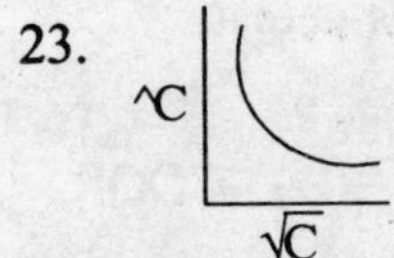

The above plot is applicable for
(A) CH_3COOH (B) H_2SO_4 (C) KCl (D) NaOH

24. Which one of the following is wrongly metched ?
(A) 0.1 M HCl – pH = 1
(B) 0.1 M HNO_3 – pH = 1
(C) 0.1 M CH_3COOH – pH = 1
(D) 0.05 M H_2SO_4 – pH = 1

25. 20 gms of NaOH should be dissolved in how many ml of water to get a NaOH solution with a pH of 14 ?
(A) 250 ml (B) 400 ml (C) 500 ml (D) 1000 ml

26. The change in the value of enthalpy of formation of ammonia is
(A) + 91 kJ (B) –91kJ (C) + 45.5 kJ (D) – 45.5 kJ

27. The enthalpy of combustion of butane is –2890 kJ at STP. The amount of heat liberated when 1.12×10^{-2} m^3 of butene is completely burnt at STP is
(A) –1445 kJ (B) 1445 kJ (C) –2890 kJ (D) 2890 kJ

28. In an exothermic reversible reaction, the increase in temperature
(A) does not affect the rate of reaction
(B) increases the rate of forward and the backward reaction to the same extent
(C) increases the rate of backward reaction
(D) increases the rate of forward rection

29. A graph of log(a-x) in the Y axis and time in X axis is drawn, for a first order reaction. The graph pattern is

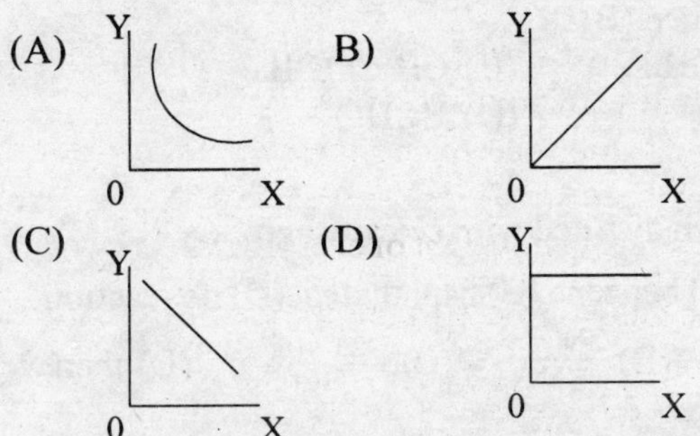

30. What will be the rate constant of the first order reaction if its half life period is 15 minutes ?
(A) 0.0246 min^{-1} (B) 0.0642 min^{-1}
(C) 0.834 min^{-1} (D) 0.0462 min^{-1}

31. Which of the follwoing metallic ion has five unpaired electrons in its 3d orbital and exhibits pale green colour ?
(A) cobalt (B) chromium (C) manganese (D) copper

32. What is the primary valency of iron in the complex $[Fe(H_2O)_6]SO_4$?
(A) 1 (B) 2 (C) 4 (D) 6

33. CuO and NiO are present together as a mixture. From this nickel oxide can be recovered by using
(A) Water (B) Dilute H_2SO_4
(C) BaO_2 and Mg powde (D) Water gas

34. In the distillation method of purification of zinc, the residual liquid contains metals
(1) iron (2) lead (3) cadmium (4) arsenic
The correct answer is
(A) (1) & (3) (B) (2) & (3) (C) (2) & (4) (D) (1) & (2)

35. In paints, CeO_2 is used as
(A) Reducing agent (B) Oxidising agent
(C) Mordant (D) Opacifier

36. Which of the following ore is not concentrated by froth floatation process ?
(A) zinc blende (B) copper glance
(C) pentlandite (D) pyrolusite

37. Ethylene is formed from ethyl alcohol when heated with conc. H_2SO_4 at 443 K by

(A) Intramolecular dehydration
(B) Intermolecular dehydration
(C) Intermolecular hydration (D) Intramolecular hydration

38. The types of alcoholic groups present in glycerol are
(A) two secondary and one tertiary groups
(B) three primary groups (C) two primary groups
(D) two primary and one secondary groups

39. When glycerol is heated, the product is
(A) allyl alcohol (B) acetaldehyde
(C) acrolein (D) acetone

40. (A) and (B) on controlled oxidation form (C) and (D). (B) is homologue and comes after (A). Molecular mass of (C) is 30. Then (B) is
(A) CH_3CH_2OH (B) CH_3COOH
(C) HCOOH (D) HCHO

41. (OH-benzene) when distilled with zinc dust gives
(A) toluene (B) benzene (C) naphthalene(D) no reaction

42. (A) $\xrightarrow{\text{Zn dust}}$ (B) $\xrightarrow[\text{AlCl}_3]{\text{anhydrous}}$ (C) $\xrightarrow[\text{HCl}]{\text{Zn / Hg}}$ (D) then, (A) and (D) are
(A) Benzene, Toluene (B) Phenol, biphenyl
(C) Phenol, ethyl benzene (D) Benzene, Phenol

43. Which of the following occurs when ethoxy ethane is treated with chlorine in the dark ?
(A) substitution takes place only at β-carbon atom
(B) substitution takes place only at α-carbon
(C) formation of perchlorodiethyl ether
(D) no reaction

44. Which one of the following is used in the manufacture of Cordite ?
(A) $C_6H_5-CO-CH_3$ (B) $C_6H_5-CO-C_6H_5$
(C) CH_3-O-CH_3 (D) $CH_3-CO-CH_3$

45. An organic amide which does not undergo Hoffmann's degradation reaction is
(A) benzamide (B) methanamide
(C) acetamide (D) propionamide

46. Which of the following about oxalic acid is true ?
(A) It is stable to heat (B) It is poisonous
(C) It gives only one type of salt with sodium hydroxide
(D) It is a weaker acid than acetic acid

47. Bee's wax is
(A) $CH_3(CH_2)_{16}COOC_{34}H_{53}$
(B) $CH_3(CH_2)_{14}COOC_{31}H_{63}$
(C) $CH_3(CH_2)_{16}\ COOC_{25}H_{53}$
(D) $CH_3(CH_2)_{14}\ COOC_{16}H_{33}$

48. Mesotartaric acid is
(A) a racemic mixture (B) optically inactive (C) optically active
(D) having a plane of symmetry and optically inactive

49. Phthalic acid is obtained by the oxidation of naphthalene in the presence of catalyst
(A) Pd (B) P_2O_5 (C) Pt (D) V_2O_5

50. Unit of equivalent conductance ^c is
(A) S m² gm equivalent (B) S $metre^{-1}$
(C) S m^2 $mole^{-1}$ (D) S m^2gm/equivalent

51. Collodial solution of eye lotion contains
(A) milk of magnesia (B) aluminium sulphate
(C) silver and gelatin (D) titanium oxide

52. Which of the following colloidal substance is used as germicide ?
(A) colloidal ferric hydroxide (B) colloidal silver
(C) colloidal sulphur (D) milk of magnesia

53. All lower amines have characteristic and distinctive
(A) fish like odour (B) foul smell of rancid butter
(C) rotten egg smell (D) fruity odour

54. A secondary amine is a compound in which
(A) two hydrogen atoms of NH_3 are replaced by two alkyl groups
(B) the $-NH_2$ group is attached at the secondary carbon atom
(C) the $-NH_2$ group is attached at the second carbon atom
(D) two $-NH_2$ groups are present in it

55. Compounds used as intermediate in the production of explosive and detergents are
(A) diazonium salts (B) nitro alkanes
(C) nitro arenes (D) amines

56. Glucose reacts with acetic anhydride and forms
(A) tri acetate (B) di acetate
(C) hexa acetate (D) penta acetate

57. Which of the folowing is Peptide linkage ?
(A) –CO– (B) –CO–NH– (C) $-CONH_2$–(D) –COOR

58. Amino acid which is not optically active is
(A) $CH_2(\overset{+}{N}H_3)COO^-$ (B) $CH_3CH(\overset{+}{N}H_3)COO^-$
(C) $(CH_3)_2CHCH_2CH(\overset{+}{N}H_3)COO^-$ (D) $(CH_3)_2CHCH(\overset{+}{N}H_3)COO^-$

59. Compound used to prepare malachite green is
(A) HCHO (B) C_6H_5CHO (C) CH_3CHO (D) CH_3COCH_3

60. When acetaldehyde reacts with ammonia th final product is
(A) Aldehyde amine (B) Urotropine
(C) Aldehyde ammonia (D) Aldimine

TNPCEE -1995

1. (B)	21. (B)	41. (C)
2. (C)	22. (A)	42. (B)
3. (C)	23. (A)	43. (B)
4. (D)	24. (B)	44. (D)
5. (B)	25. (D)	45. (A)
6. (D)	26. (C)	46. (B)
7. (B)	27. (B)	47. (A)
8. (D)	28. (C)	48. (A)
9. (A)	29. (B)	49. (B)
10. (B)	30. (D)	50. (B)
11. (C)	31. (A)	51. (A)
12. (A)	32. (C)	52. (C)
13. (C)	33. (A)	53. (A)
14. (C)	34. (D)	54. (D)
15. (C)	35. (C)	55. (B)
16. (C)	36. (A)	56. (C)
17. (D)	37. (B)	57. (A)
18. (B)	38. (A)	
19. (A)	39. (D)	
20. (C)	40. (B)	

TNPCEE -1996

1. (A)	21. (B)	41. (C)
2. (A)	22. (C)	42. (A)
3. (C)	23. (B)	43. (B)
4. (D)	24. (C)	44. (C)
5. (D)	25. (D)	45. (B)
6. (A)	26. (A)	46. (B)
7. (B)	27. (D)	47. (A)
8. (B)	28. (B)	48. (D)
9. (A)	29. (B)	49. (A)
10. (B)	30. (B)	50. (C)
11. (B)	31. (B)	51. (A)
12. (B)	32. (D)	52. (C)
13. (C)	33. (A)	53. (B)
14. (D)	34. (D)	54. (B)
15. (D)	35. (B)	55. (B)
16. (A)	36. (B)	56. (B)
17. (D)	37. (A)	57. (B)
18. (D)	38. (D)	58. (C)
19. (A)	39. (B)	59. (B)
20. (C)	40. (C)	60. (C)

TNPCEE -1997

1. (A)	21. (A)	41. (C)
2. (C)	22. (B)	42. (B)
3. (C)	23. (C)	43. (A)
4. (A)	24. (D)	44. (A)
5. (D)	25. (A)	45. (B)
6. (D)	26. (B)	46. (C)
7. (D)	27. (A)	47. (C)
8. (C)	28. (A)	48. (B)
9. (C)	29. (B)	49. (C)
10. (C)	30. (C)	50. (B)
11. (A)	31. (C)	51. (B)
12. (B)	32. (B)	52. (D)
13. (C)	33. (A)	53. (A)
14. (C)	34. (D)	54. (B)
15. (B)	35. (D)	55. (B)
16. (D)	36. (D)	56. (A)
17. (A)	37. (A)	57. (D)
18. (B)	38. (B)	58. (A)
19. (A)	39. (C)	59. (B)
20. (A)	40. (B)	60. (B)

TNPCEE -1998

1. (C)	21. (B)	41. (D)
2. (D)	22. (D)	42. (B)
3. (B)	23. (B)	43. (C)
4. (C)	24. (D)	44. (A)
5. (B)	25. (A)	45. (D)
6. (A)	26. (C)	46. (B)
7. (B)	27. (A)	47. (A)
8. (D)	28. (D)	48. (A)
9. (B)	29. (D)	49. (A)
10. (D)	30. (A)	50. (D)
11. (C)	31. (C)	51. (B)
12. (C)	32. (A)	52. (D)
13. (C)	33. (B)	53. (D)
14. (C)	34. (D)	54. (A)
15. (D)	35. (C)	55. (B)
16. (B)	36. (A)	56. (A)
17. (B)	37. (C)	57. (C)
18. (C)	38. (D)	58. (C)
19. (C)	39. (B)	59. (C)
20. (B)	40. (A)	60. (A)

TNPCEE -1999

1. (C)	21. (A)	41. (A)
2. (B)	22. (B)	42. (B)
3. (A)	23. (B)	43. (A)
4. (C)	24. (B)	44. (C)
5. (A)	25. (C)	45. (B)
6. (C)	26. (D)	46. (A)
7. (B)	27. (C)	47. (C)
8. (D)	28. (A)	48. (A)
9. (C)	29. (C)	49. (C)
10. (A)	30. (A)	50. (B)
11. (B)	31. (B)	51. (A)
12. (C)	32. (B)	52. (D)
13. (D)	33. (A)	53. (C)
14. (D)	34. (D)	54. (A)
15. (A)	35. (B)	55. (C)
16. (B)	36. (A)	56. (B)
17. (A)	37. (C)	57. (C)
18. (D)	38. (B)	58. (D)
19. (D)	39. (C)	59. (A)
20. (C)	40. (D)	60. (C)

TNPCEE -2000

1. (A)	21. (C)	41. (D)
2. (C)	22. (B)	42. (B)
3. (B)	23. (D)	43. (D)
4. (D)	24. (C)	44. (C)
5. (C)	25. (D)	45. (B)
6. (B)	26. (B)	46. (A)
7. (C)	27. (C)	47. (A)
8. (C)	28. (C)	48. (C)
9. (A)	29. (C)	49. (A)
10. (B)	30. (D)	50. (C)
11. (C)	31. (A)	51. (C)
12. (C)	32. (C)	52. (C)
13. (A)	33. (C)	53. (B)
14. (C)	34. (C)	54. (B)
15. (C)	35. (C)	55. (B)
16. (A)	36. (A)	56. (B)
17. (A)	37. (B)	57. (C)
18. (A)	38. (D)	58. (C)
19. (B)	39. (D)	59. (B)
20. (C)	40. (D)	60. (A)

TNPCEE -2001

1. (C)	21. (D)	41. (D)
2. (D)	22. (B)	42. (A)
3. (C)	23. (B)	43. (C)
4. (C)	24. (B)	44. (D)
5. (C)	25. (B)	45. (B)
6. (B)	26. (A)	46. (B)
7. (C)	27. (A)	47. (A)
8. (B)	28. (D)	48. (C)
9. (B)	29. (D)	49. (D)
10. (B,C)	30. (B)	50. (B)
11. (A)	31. (D)	51. (B)
12. (D)	32. (A)	52. (B)
13. (D)	33. (C)	53. (A)
14. (C)	34. (A)	54. (A)
15. (C)	35. (C)	55. (B)
16. (B)	36. (C)	56. (C)
17. (C)	37. (A)	57. (B)
18. (A)	38. (A)	58. (A)
19. (A)	39. (B)	59. (B)
20. (D)	40. (D)	60. (A)

TNPCEE -2002

1. (C)	21. (B)	41. (B)
2. (D)	22. (A)	42. (B)
3. (C)	23. (C)	43. (B)
4. (C)	24. (B)	44. (D)
5. (B)	25. (C)	45. (A)
6. (C)	26. (B)	46. (A)
7. (C)	27. (B)	47. (D)
8. (B)	28. (B)	48. (C)
9. (D)	29. (C)	49. (B)
10. (B)	30. (D)	50. (B)
11. (B)	31. (A)	51. (A)
12. (A)	32. (B)	52. (B)
13. (C)	33. (A)	53. (D)
14. (B)	34. (B)	54. (C)
15. (C)	35. (D)	55. (C)
16. (B)	36. (D)	56. (C)
17. (D)	37. (C)	57. (A)
18. (A)	38. (D)	58. (B)
19. (B)	39. (A)	59. (D)
20. (D)	40. (B)	60. (D)

TNPCEE -2003

1. (D)	21. (A)	41. (A)
2. (C)	22. (C)	42. (B)
3. (C)	23. (B)	43. (B)
4. (B)	24. (C)	44. (C)
5. (C)	25. (B)	45. (D)
6. (B)	26. (C)	46. (B)
7. (A)	27. (A)	47. (D)
8. (A)	28. (B)	48. (D)
9. (D)	29. (C)	49. (C)
10. (B)	30. (B)	50. (C)
11. (B)	31. (D)	51. (D)
12. (C)	32. (D)	52. (D)
13. (C)	33. (C)	53. (B)
14. (D)	34. (C)	54. (C)
15. (D)	35. (B)	55. (A)
16. (C)	36. (D)	56. (B)
17. (D)	37. (B)	57. (A)
18. (C)	38. (D)	58. (A)
19. (A)	39. (C)	59. (D)
20. (C)	40. (A)	60. (A)

TNPCEE -2004

1. (A)	21. (B)	41. (B)
2. (D)	22. (A)	42. (C)
3. (D)	23. (A)	43. (B)
4. (B)	24. (C)	44. (D)
5. (D)	25. (C)	45. (B)
6. (A)	26. (D)	46. (B)
7. (C)	27. (B)	47. (B)
8. (D)	28. (C)	48. (D)
9. (B)	29. (C)	49. (D)
10. (B)	30. (D)	50. (D)
11. (C)	31. (C)	51. (D)
12. (C)	32. (B)	52. (C)
13. (D)	33. (C)	53. (A)
14. (B)	34. (D)	54. (A)
15. (C)	35. (D)	55. (B)
16. (D)	36. (D)	56. (D)
17. (C)	37. (A)	57. (B)
18. (A)	38. (D)	58. (A)
19. (C)	39. (C)	59. (B)
20. (C)	40. (A)	60. (D)